Foundations of Colorectal Cancer

Edited by

Alejandro Pazos Sierra
Full Professor, Faculty of Computer Science, University of A Coruna; Coordinator, Galician Research Network of Colorectal Cancer (REGICC), A Coruna, Galicia, Spain

Academic Press is an imprint of Elsevier
125 London Wall, London EC2Y 5AS, United Kingdom
525 B Street, Suite 1650, San Diego, CA 92101, United States
50 Hampshire Street, 5th Floor, Cambridge, MA 02139, United States
The Boulevard, Langford Lane, Kidlington, Oxford OX5 1GB, United Kingdom

Copyright © 2022 Elsevier Inc. All rights reserved.

No part of this publication may be reproduced or transmitted in any form or by any means, electronic or mechanical, including photocopying, recording, or any information storage and retrieval system, without permission in writing from the publisher. Details on how to seek permission, further information about the Publisher's permissions policies and our arrangements with organizations such as the Copyright Clearance Center and the Copyright Licensing Agency, can be found at our website: www.elsevier.com/permissions.

This book and the individual contributions contained in it are protected under copyright by the Publisher (other than as may be noted herein).

Notices

Knowledge and best practice in this field are constantly changing. As new research and experience broaden our understanding, changes in research methods, professional practices, or medical treatment may become necessary.

Practitioners and researchers must always rely on their own experience and knowledge in evaluating and using any information, methods, compounds, or experiments described herein. In using such information or methods they should be mindful of their own safety and the safety of others, including parties for whom they have a professional responsibility.

To the fullest extent of the law, neither the Publisher nor the authors, contributors, or editors, assume any liability for any injury and/or damage to persons or property as a matter of products liability, negligence or otherwise, or from any use or operation of any methods, products, instructions, or ideas contained in the material herein.

Library of Congress Cataloging-in-Publication Data
A catalog record for this book is available from the Library of Congress

British Library Cataloguing-in-Publication Data
A catalogue record for this book is available from the British Library

ISBN: 978-0-323-90055-3

For information on all Academic Press publications
visit our website at https://www.elsevier.com/books-and-journals

Publisher: Stacy Masucci
Acquisitions Editor: Rafael E. Teixeira
Editorial Project Manager: Sam W. Young
Production Project Manager: Stalin Viswanathan
Cover Designer: Miles Hitchen

Typeset by STRAIVE, India

Contents

Contributors xix
Editor's note xxv
Prologue xxvii
Acknowledgments xxix

Section A
Epidemiological studies in CRC

1. Incidence and mortality of CRC 1

*María Teresa Seoane Pillado,
Sonia Pértega Díaz, Vanesa Balboa Barreiro,
and Cristina González Martín*

Incidence 3
Mortality 3
Discussion 15
References 15

2. Prognosis and follow-up of CRC patients: Role of diagnostic and therapeutic delay 17

*Sonia Pértega Díaz, Vanesa Balboa Barreiro,
María Teresa Seoane Pillado,
Mohammed Alhayek-Aí,
Paloma González Santamaría, and
Cristina González Martín*

Survival and prognosis 17
Follow-up strategies 18
Quality of life and long-term sequelae 19
Impact of diagnostic and therapeutic delay on stage and prognosis 20
References 21

Section B
Clinical manifestations and disease detection

3. Primary prevention of CRC 27

*José Luis Ulla Rocha,
Raquel Sardina Ferreiro,
Rosa Fungueiriño Suarez, and
Juan Turnes Vázquez*

Introduction 27
Epidemiology and CRC 27
Environment and CRC 28
Diet 29
Foods recommended to be reduced 29
   Processed and red meat 29
Food and drink recommended to be increased 30
   Dairy products 30
   Dietary fiber and whole grain cereals 30
   Fish 30
   White and lean meat 31
   Coffee 31
   Garlic 31
Lifestyle behaviors 31
   Unhealthy practices 31
   Healthy practices 32
Therapeutic strategies linked to lowering CRC risk 32
   Salicylic acetyl acid (SAA) 32
   Nonsteroidal antiinflammatory drugs (NSAIDs) 33
   Calcium and magnesium supplementation 33

Vitamin D supplementation	33
B-group vitamins	34
Folic acid and folate	34
Antioxidants	34
Hormone replacement therapy	34
Statins	34
Bisphosphonates	35
Metformin	35
Chondroitin sulfate and glucosamine	35
Angiotensin II inhibitors	35
Dietary patterns and CRC	35
Relationship between dietary index and CRC	35
Summary: Evidence and recommendations	36
References	37

4. Early onset of CRC — 41

Andrés Dacal Rivas, Eva Martí Marqués, and Leopoldo López Rosés

Introduction	41
Exposure to environmental factors	41
Microbiota alteration	42
Hereditary factors	42
Molecular differences	42
Clinical features	43
Conclusions	44
References	44

5. Population-based universal screening for CRC: Secondary prevention — 45

Loretta De Chiara, María Gallardo-Gómez, Vicent Hernández, and Oscar J. Cordero

Secondary prevention and screening concept	45
Conditions for population screening: WHO general principles	45
Colorectal cancer screening	45
Colorectal cancer screening tests	47
Colonoscopy	47
Flexible sigmoidoscopy	47
CT scan colonography	48
Fecal occult blood test (FOBT)	48
Other noninvasive tests: Stool or blood tests	49
Requirements for the implementation of a screening program	49
Evaluation of screening programs for their implementation: Efficacy, effectiveness, and efficiency	50
Problems related to a colorectal cancer screening program	51
Situation of screening in Spain	51
Implementation of CRC screening in Spain	52
Current situation of CRC screening in Spain: Cancer screening program network, 2019	52
Current situation of CRC screening in Galicia: *Xunta de Galicia*, 2019	53
Strengths and areas for improvement in Spanish screening programs	53
References	53

6. The role of primary care in early referral of CRC patients — 57

Francisco Javier Maestro Saavedra, Carla Blanco Vázquez, Alain García de Castro, and Begoña Bravo Bueno

The role of primary care in the early release of patients with CRC	58
Prevention and promotion of health	58
Early diagnosis	59
Symptomatology in colorectal cancer	59
Tests for early detection of colorectal cancer	59
Stool occult blood test	60
Sigmoidoscopy	60
Colonoscopy	60
Indications	60
Preparation for tests	61
Technique of performing a colonoscopy	61
Evaluation of results	61
Monitoring of index or precursor injuries (polyps)	63
Surveillance in special situations	63
Criteria for referral to specialized care	65
CRC of family/hereditary characteristics	65
References	67

7. Rapid diagnostic pathways for patients with suspected CRC — 69

Alejandro Ledo Rodríguez and Ismael Said-Criado

Clinical pathways	69
Rapid diagnosis pathways for colorectal cancer	70
Introduction	70
Implementation	70
Published research	71
Meeting referral criteria	71

Cancer detection rates	71
Percentages of diagnosed cancer cases	73
Are RDPs sensitive and specific?	73
Effect on care timing	73
Effect on tumor stage and survival	74
Possibilities for increasing effectiveness of CRC rapid pathways	74
Conclusions	75
References	76

Section C
Diagnosis and staging

Section C.I
Current diagnosis methods in colorectan cancer

8. Colon capsule endoscopy — 83

Ramiro Manuel Macenlle García

Introduction	83
Technical characteristics of the second-generation colon capsule	83
Preparation of the patient	83
Comparative studies between colon capsule endoscopy and colonoscopy	84
Comparative studies between CCE and colonography by computerized axial tomography	86
Current indications and contraindications of the colon capsule endoscopy: Future perspectives	86
Conclusions	86
References	87

9. Endoscopic ultrasound in CRC — 89

*José Luis Ulla Rocha,
Raquel Sardina Ferreiro, and
Juan Turnes Vázquez*

Outline of technique	89
EUS for rectal cancer staging	89
Degree of invasion of intestinal wall (uT)	89
Affectation of local adenopathies (uN)	91
Significance of the circumferential resection margin	91
EUS for assessment of neoadjuvant therapy response	92
Interventional EUS	92

EUS-FNA initial diagnosis and recurrence	92
Drainage of postsurgical fluid collections	93
Extrahepatic bile duct drainage	93
Conclusion	94
References	94

10. CT-Colonography (CTC): Technical requirements, indications and current status — 95

*Concepción Crespo García,
María Jose Martinez-Sapiña Llanas, and
Susana A. Otero Muinelo*

Introduction	95
Technical requirements	95
Preparation of the colon	95
Distension of the colon	95
Image acquisition	96
Software and tools	97
Indications for virtual colonoscopy	98
Clinically significant findings in CTC	98
Polyps	98
Cancers	100
Submucosal lesions	102
Chronic diverticular disease	103
Other	104
Conclusions	104
References	106

11. Rectal pathology: Findings at CT-Colonography — 109

*María Jose Martinez-Sapiña Llanas,
Susana A. Otero Muinelo, and
Concepción Crespo García*

Introduction	109
Technical considerations	109
Rectal pathology	110
Polyps	111
Villous tumors	111
Malignant neoplasms	112
Submucosal lesions	112
Vascular lesions	114
Hypertrophied anal papilla	115
Inflammatory conditions	115
Postoperative changes	115
Pseudolesions and artifacts	116
Other lesions	117
Conclusion	117
References	118

12. Applications of [18F] FDG PET and PET/CT in colorectal carcinoma — 119

Sofía Rodríguez Martínez de Llano, Fernando Zelaya Reinquet, Shirly Margarita Nieves Maldonado, Sara Seijas Marcos, and Paulino Pais Silva

Introduction	119
Staging	120
Restaging	122
Recurrent disease detection	122
Preoperative evaluation of potentially resectable disease	123
Others: Incidental diagnosis of colorectal cancer. Assessment of treatment response	124
Incidental diagnosis of colorectal cancer	124
Assessment of treatment response	124
New perspectives	124
References	127

13. Magnetic resonance imaging (MRI) in staging and restaging after neoadjuvant therapy — 131

Cristina Méndez Díaz, Esther Rodríguez García, and Rafaela Soler Fernández

Introduction	131
Total mesorectal excision	131
Key anatomical references	132
Mesorectal fascia and anterior peritoneal reflection	133
Upper rectum	134
Middle rectum	134
Lower rectum	134
Lymph nodes	134
Local staging	134
T-staging	134
T1 and T2 tumors	135
T3 tumors	136
T4 tumors	137
N-staging	137
Restaging after neoadjuvant treatment	138
T-restaging	139
N-restaging	142
The structured radiology report	143
Location	143
Morphology	143
T-staging	144
Extramural extension of the tumor	144
Distance to mesorectal fascia	144
Adenopathies and mesorectal tumor deposits	144
Extramesorectal adenopathies	144
Restaging after neoadjuvant treatment	144
Locoregional recurrence	145
References	146

14. Histopathological diagnosis of CRC — 149

Álvaro Gómez Castro and Héctor Lázare Iglesias

Introduction	149
Histopathological diagnosis of CRC in biopsies and endoscopic polypectomies	149
Anatomopathological diagnosis of CRC in surgical specimens	151
Histopathological types of colorectal neoplasms	153
Biomarkers in CRC	154
References	155

15. Colonoscopy: Technique and quality factors — 157

Ramón Vázquez Dourado, Leticia García Diéguez, and Javier Castro Alvariño

Historical introduction	157
Colorectal cancer	157
Colonoscopy technique	157
Endoscopic description	158
Magnification techniques	160
Classification of colorectal polyps	161
Serrated pathway in CRC	162
Colonoscopy in the diagnosis of CRC in medium-risk populations	163
Colonoscopy in the diagnosis of CRC in high-risk populations	164
Colonoscopy quality indicators	164
Endoscopic iconography	165
References	168

16. Endoscopic diagnosis of preneoplastic and neoplastic lesions — 171

Beatriz Romero-Mosquera, Alfonso Martínez-Turnes, and Vicent Hernández

Introduction	171
Identification of lesions	171
Quality of colonoscopy	171
High-definition colonoscopy	171
Conventional chromoendoscopy	171
Virtual chromoendoscopy	172
Use of ancillary equipment and new colonoscopes	172
Characterization of lesions	173

Morphological evaluation	173
Evaluation of the glandular pattern	174
WASP classification	177
Role of colonoscopy in the diagnostic of advanced colonic lesions	177
References	178

17. Oligometastatic disease — 181

José Luis Ulla Rocha and Paloma Sosa Fajardo

18. Hereditary nonpolyposis CRC — 183

M. Lidia Vazquez-Tunas

Introduction	183
Genetic basis	183
Molecular identification of Lynch syndrome	184
Microsatellite instability analysis	184
Immunohistochemistry for reparative proteins in tumor tissue	185
Hypermethylation of *MLH1*	185
Detection of germline mutations in genomic DNA	185
Clinical features	185
Diagnosis of Lynch syndrome (HNPCC) and criteria for referral to a genetic counseling unit (GCU)	187
Molecular study strategy	188
Predictive models	189
Clinical controls and surveillance	189
Follow-up of healthy individuals in families with HNPCC	189
Follow-up in patients with colorectal cancer and HNPCC	190
General recommendations	190
Chemoprevention	191
Systemic treatment	191
Other entities	191
Colorectal cancer type X	191
Lynch-like syndrome	191
Conclusions	191
References	192

19. Hereditary polyposis CRC — 195

José Manuel Mera Calviño and Enrique González de la Ballina González

Adenomatous polyposis syndromes: Familial adenomatous polyposis and MUTYH-associated polyposis	195
Genetics	195
Clinical symptoms	195
Diagnostics	196
Monitoring of healthy individuals with classic PAF	197
Follow-up of healthy individuals with AFAF	198
Follow-up of healthy individuals with MAP	198
Treatment of FAP	198
Treatment of AFAP	199
Treatment of FAP	199
Chemoprevention	199
Other inherited adenomatous polyposis syndromes	199
Gardner syndrome	199
Turcot syndrome (glioma-polyposis)	199
Hamartomatous polyposis syndrome	200
Peutz–Jeghers syndrome (PJS)	200
Juvenile polyposis syndrome	200
Cowden syndrome	200
Bannayan–Ruvalcaba–Riley syndrome	201
Rare hereditary hamartomatous polyposis syndromes	201
Serrated polyposis syndrome (SPS)	201
References	202

Section C.II
New tools of diagnosis in CRC

20. Diagnostic, prognostic, predictive and therapeutic molecular biomarkers in CRC: Understanding the present and foreseeing the future — 207

Alberto Veiga, Francisco Queipo, Germán Bou, Alfonso Cepeda-Emiliani, and Ángel Concha

Introduction	207
What are biomarkers? Concept and classification	208
Specific to a distinct type of tumor originating from a specific organ	208
Generic "pan-cancer" biomarkers	209
Cell surface membrane antigens	209
Molecular alterations of oncogenic genes (oncogenes)	209
Other biomarkers	210
Determination of biomarkers in CRC	210
Blood and stool biomarkers for CRC screening	211
Circulating tumor cells (CTCs)	212
Exosomes	212
Cancer-educated platelets (CEPs)	212
Messenger ribonucleic acid (mRNA)	213
Small noncoding ribonucleic acids (miRNAs)	213
Proteins	213

Free tumor deoxyribonucleic acid
(ctDNA) 213
RAS: K-RAS and N-RAS, the current key for
precision medicine in CRC 215
B-RAF mutation in colorectal cancer 215
EGFR/HER FAMILY/TP53-APC/
β-CATENINS/C-MET as prognostic
indicators 216
HER-2 216
P53 216
APC/β-CATENINS 216
C-MET 217
MSI, its role in CRC 217
miRNAs, small players with large
potential 218
Aberrant DNA methylation in CRC 218
Microbial markers in CRC prognosis and
therapeutics 219
New insights in CRC biomarkers, recent
advances, and future challenges 220
"Undruggable" RAS 220
The role of mass sequencing
techniques 220
The continuing value of the
anatomopathological report 220
Molecular staging of CRC 220
Advantages of liquid biopsy 221
Study of NTRK rearrangements 221
TRIM67/P53 axis 221
Immunotherapy in colon cancer 221
Tumor microenvironment in CRC 222
Concluding remarks 223
References 223

21. Bioinformatic tools for research in CRC 231

Virginia Mato-Abad, Alejandro Pazos, Cristian R. Munteanu, Jose Liñares-Blanco, Sara Alvarez-Gonzalez, José M. Vázquez-Naya, Nieves Pedreira, Jorge Amigo, and Carlos Fernandez-Lozano

Introduction 231
Methodologies for obtaining primary
data in cancer 232
High-throughput genotyping (HTG) 232
Next-generation sequencing (NGS) 233
Bioinformatics tools for cancer research:
Applications in colorectal cancer 233
Databases 233
Complex networks 239
Biomedical tools for assessing the risk of
suffering from colorectal cancer 240

Deep learning for medical imaging in
colorectal cancer 242
Conclusions 245
References 246

22. Omics-based biomarkers for CRC 249

María Gallardo-Gómez, Paula Álvarez-Chaver, Alberto Cepeda, Patricia Regal, Alexandre Lamas, and Loretta De Chiara

Introduction 249
Biological samples for the discovery of
biomarkers and their clinical
application 249
Genomics approaches for colorectal
cancer biomarker discovery and
validation 250
De novo identification of genetic
biomarkers 251
Approaches for targeted validation of
genomic biomarkers 251
Proposed genetic biomarkers for colorectal
cancer 252
Transcriptomics approaches for colorectal
cancer biomarker discovery and
validation 252
Identification of transcriptomic
biomarkers 252
Proposed mRNAs as colorectal cancer
biomarkers 253
Proposed miRNAs as colorectal cancer
biomarkers 253
Epigenomics approaches for colorectal
cancer biomarkers discovery and
validation 254
De novo identification of differentially
methylated patterns 255
Analysis of targeted methylation
biomarkers 255
Whole genome methylation profiling 255
Proposed DNA methylation biomarkers for
CRC 255
Proteomics approaches for colorectal
cancer biomarkers discovery and
validation 256
Identification of proteomic
biomarkers 256
Targeted validation of protein
biomarkers 257
Proposed protein biomarkers for CRC 258
Other omics technologies for the discovery
of biomarkers 258

Phases for the development of clinical
    biomarkers   259
  Biomarker discovery: Exploratory
      analysis   259
  Development of the validation assay:
      Technical/analytical validation   259
  Validation of the clinical utility   260
  Clinical translation   261
References   261

Section D
Treatment

Section D.I
Endoscopic treatment

23. Multidisciplinary committee for a comprehensive approach to CRC patients 269

Ramón Vázquez Dourado, Pedro Carpintero Briones, and Javier Castro Alvariño

Introduction   269
Clinical oncology committees   269
Factors influencing the performance of
    committee activity   271
Effect of the multidisciplinary approach on
    colon and rectal cancer   272
References   272

24. Endoscopic treatment of preneoplastic or early lesions 275

Leticia García Diéguez, Ramón Vázquez Dourado, and Javier Castro Alvariño

Introduction   275
Endoscopic polypectomy   275
Main advanced techniques for endoscopic
    treatment of colorectal tumors   278
  Endoscopic mucosal resection (EMR)   278
  Endoscopic submucosal dissection
      (ESD)   280
  Other techniques   280
References   281

25. Endoscopic surveillance 283

Nereida Fernández Fernández, Antonio Rodríguez-D'Jesús, and Arantza Germade Martínez

Surveillance of preneoplastic colonic
    lesions   283
  Initial colonoscopy   283
  Characteristics of basal polyps   283
  Risk groups   286
  When to end follow-up   287
Nonhereditary polyposis: Nonhereditary
    attenuated polyposis and serrated
    polyposis   288
Nonhereditary attenuated polyposis   288
  Definition   288
  Management   288
  Recommendations for family screening   289
Serrated polyposis   289
  Definition   289
  Endoscopic types and characteristics   289
  Epidemiology and risk factors   290
  Endoscopic monitoring   290
  Family screenings   291
Follow-up of pT1 neoplastic lesions after
    endoscopic treatment   291
  Histology   291
  Resection margin after polypectomy   291
  Submucosal infiltration   291
  Tumor budding   292
  Lymphovascular infiltration   292
  Interaction between factors   292
  Follow-up   292
References   292

26. Advanced endoscopy in colorectal cancer: Colorectal prostheses 295

María Teresa Vázquez Rey, Benito González Conde, Ignacio Couto Worner, and Pedro A. Alonso Aguirre

Introduction   295
Indications and contraindications   295
  Indications   295
  Contraindications   296
Endoscopic technique   296
Results and complications   298
References   299

Section D.II
Surgical treatment

27. Emergency surgery for CRC — 301

*Alejandra García Novoa,
Alba Gómez Dovigo,
Tatiana María Civeira Taboada, and
José Francisco Noguera Aguilar*

Diagnosis in the emergency department	303
Diagnostic tests	303
Treatment of complicated colorectal carcinoma	304
Type of surgery	304
References	305

28. Innovation and new technologies in colorectal cancer UNIVEC device development experience — 307

*Alba Gómez Dovigo,
Alejandra García Novoa,
Javier Aguirrezabalaga González,
José Francisco Noguera Aguilar, and
Alberto Centeno Cortés*

Introduction	307
Minimally invasive surgery	307
Instruments, innovative devices, and robotic surgery	308
Highly skilled laparoscopic instruments	308
Innovation in imaging systems	309
Robotic surgery	310
Robotic camera controller	311
Flexible single incision surgery	311
References	313

29. Colon cancer surgery — 317

*FernandoFernández López,
Jesús P. Paredes Cotoré, and
Manuel Bustamante Montalvo*

Surgical treatment of nonmetastatic colon cancer	317
General surgical technique	317
Surgical technique according to specific locations and special situations	318
Treatment of colon cancer presenting as a surgical emergency	318
Surgical options for urgent CRC	319
Recommended resections depending on the location of the tumor	319
Palliative procedures	319
Laparoscopic approach to colon cancer	320
Contraindications for laparoscopic surgery	320
References	320

30. Rectal cancer surgery — 323

*Jesús P. Paredes Cotoré,
FernandoFernández López, and
Manuel Bustamante Montalvo*

Surgical treatment of rectal cancer	323
Preoperative assessment and staging	323
Oncological and technical principles	324
Surgical techniques for rectal cancer	325
Anterior resection	325
Intersphincteric resection	326
Transanal mesorectal resection (TME, "down-to-up" approach)	326
Dysfunctionalizing stoma	326
Abdomino-perineal amputation	327
Hartmann-type resection	327
Local resection	327
Laparoscopic surgery for rectal cancer	328
Robotic surgery	328
Nonsurgical treatment	328
References	329

31. Liver metastases from CRC: A treatment paradigm — 331

*Manuel Bustamante Montalvo,
Francisco Javier González Rodríguez, and
Sergio Manuel Estévez Fernández*

Introduction	331
Treatment of metastatic disease	331
Patient-related factors	332
Tumor-related factors	332
Anatomical factors	333
Abdominal computed tomography (CT)	333
Magnetic resonance imaging (MRI)	333
Positron emission tomography	333
Colonoscopy	334
Three-dimensional (3D) planning	334
Surgical technique	334
Neoadjuvant chemotherapy	335
Preoperative portal embolization and associating liver partition and portal vein ligation for staged hepatectomy (ALPPS)	336
Local treatment for metastatic disease	337
Regional chemotherapy via the hepatic artery	337
Radiotherapy	337
Tumor ablation	337

Treatment of recurrence	340
Chemotherapy after resection	341
Follow-up after resection	341
Conclusions	341
References	341

32. Pulmonary metastasectomy for CRC 343

Rodrigo A.S. Sardenberg and Diego Gonzalez-Rivas

Introduction	343
Results	343
Discussion	344
Patient selection	345
Pathophysiology	346
Symptoms	346
Preoperative evaluation	347
Prognostic factors	349
Surgical technique	351
New perspectives	351
Conclusions	352
References	352

Section D.III
Pharmacological and radiotherapeutic treatment

33. Molecularly targeted therapy in metastatic CRC 357

Juan Ruiz-Bañobre, Elena Brozos-Vázquez, Francisca Vázquez-Rivera, Yolanda Vidal-Insua, Rafael López-López, and Sonia Candamio-Folgar

Introduction	357
Anti-EGFR antibodies	357
Inmmune checkpoint inhibitors	358
BRAF tyrosine kinase inhibitors	359
HER-2 blockade	359
Tyrosine kinase inhibitors	360
KRAS inhibitors	361
Conclusion	361
Acknowledgments	361
References	361

34. A roadmap for medical treatment of metastatic CRC 365

Gala Martínez-Bernal, Julia Martínez-Pérez, and Manuel Valladares-Ayerbes

Background	365
Roadmap for treatment	365
First-line chemotherapy	366
Selection of bevacizumab or anti-EGFR in combination with chemotherapy	368
Triplet combinations of 5-fluorouracilo, oxaliplatin, and irinotecan	370
Duration and intensity of first-line treatment	370
Second-line treatment strategies	371
Treatment-refractory metastatic CRC	371
BRAF V600E-variant metastatic CRC	372
Immunotherapy	373
Chemotherapy and surgery of metastases	374
Management of colorectal cancer and peritoneal metastases	374
References	375

35. Adjuvant chemotherapy for colorectal cancer 381

Marta Covela Rúa, Silvia Varela Ferreiro, and Begoña Campos Balea

Introduction	381
Adjuvant treatment of colon cancer	382
Stage II	383
Stage III	383
Duration of treatment	385
Resected stage IV	386
Elderly patients	386
Conclusions	386
Neoadjuvant/adjuvant treatment of rectal cancer	386
Neoadjuvant treatment	387
Induction and consolidation chemotherapies associated with neoadjuvant treatment	388
Adjuvant treatment	388
Elderly patients	389
Conclusions	389
References	390

36. Oral administration of cytostatic drugs in the treatment of CRC 391

Carmen Álvarez Lorenzo, Martina Lema Oreiro, and Ángel Concheiro Nine

Treatment of colorectal cancer (CRC) via the oral route	391
Systemic treatment	393
Physicochemical properties of antitumor drugs	393
Absorption	394

Distribution and elimination	396
PEGylation	396
Nanostructures	397
Drug dosage forms for colonic delivery	398
Time-dependent release	399
pH-dependent release	399
Enzyme-activated release	399
References	400

37. Neoadjuvant, adjuvant, and intraoperative radiotherapy for rectal cancer — 403

Ana María Carballo Castro, Paula Peleteiro Higuero, Begoña Taboada Valladares, Patricia Calvo Crespo, Jesús Paredes Cotoré, Roberto García Figueiras, and Antonio Gómez Caamaño

Introduction	403
Adjuvant treatment	403
Neoadjuvant treatment	404
Neoadjuvant radiotherapy	404
Neoadjuvant radiotherapy vs. adjuvant radiochemotherapy	405
Neoadjuvant radiochemotherapy	405
Interval between neoadjuvant treatment and surgery	405
Neoadjuvant radiochemotherapy vs. adjuvant radiochemotherapy	406
Neoadjuvant radiochemotherapy vs. neoadjuvant radiotherapy	407
Short-course neoadjuvant radiotherapy vs. long-course neoadjuvant radiochemotherapy	407
Watch and wait	409
Total neoadjuvant treatment	409
Neoadjuvant chemotherapy vs. neoadjuvant radiochemotherapy	409
Meta-analyses and systematic reviews of supplementary treatment of rectal cancer	410
New agents in combination with radiotherapy	411
Capecitabine	411
Oxaliplatin	412
Irinotecan	413
Biological agents	413
Intraoperative radiotherapy (IORT)	413
References	415
Further reading	418

38. Radiotherapy (stereotactic body radiotherapy) for oligometastatic disease — 421

Paula Peleteiro Higuero, Patricia Calvo Crespo, and Ana María Carballo Castro

Role of SBRT in lung metastases from CRC	422
Role of SBRT in lymph node metastases from CRC	422
ROLE of SBRT in liver metastases from CRC	423
Conclusions	426
References	426

39. Palliative radiotherapy in CRC — 429

Patricia Calvo-Crespo, Begoña Taboada-Valladares, and Antonio Gómez-Caamaño

Introduction	429
Retrospective studies	429
Prospective studies	430
Systematic reviews	431
Conclusions	432
References	432

40. Immunology and immunotherapy in CRC — 435

Oscar J. Cordero, Rubén Varela-Calviño, Begoña Graña-Suárez, and Alba García-López

Immunity, inflammation, intestinal microbiota, and colorectal cancer	435
Inflammatory bowel disease and dysbiosis	436
Dysbiosis. Oral microbiota	437
Other risk factors and their relationship with inflammation	438
NSAIDs, inflammation, and CRC	440
Suppression mechanisms of the immune system in the tumor microenvironment	441
Immunotherapy in CRC	443
Molecular subtypes consensus in CRC (meaning of MMR and MSI)	443
Key immunotherapeutic assays in CRC	443
Other immunotherapeutic approaches tested in humans	444
Other monoclonal antibodies	444

Cell therapies	445	Enterocutaneous fistula	472
Vaccinations	445	Generalized peritonitis	472
References	446	Sepsis and septic shock	473
		Empirical antibiotic treatment for an IAI	474

Section D.IV
Anesthetic treatment and postoperative management

41. Multimodal rehabilitation: Pre- and intraoperative optimization in CRC surgery — 457

Manuel Núñez Deben, Miguel Pereira Loureiro, Vanesa Vilanova Vázquez, and Gerardo Baños Rodríguez

Introduction	457
Preoperative	457
Factors that cannot be influenced preoperatively	458
Factors that can be influenced preoperatively	458
Intraoperative	460
Perioperative management of nausea and vomiting	460
Drugs used to achieve rapid recovery	461
Opioid free analgesia	462
Perioperative fluid therapy	462
Baseline fluid and electrolyte needs	462
Changes in the surgical patient	463
Calculation of perioperative fluid loss and replenishment	463
Types of fluids	464
Fluid management schemes	465
Restrictive vs liberal therapy	465
Hemodynamic monitoring	466
Advanced monitoring	467
Static variables	467
Dynamic fluid response variables	467
Emergency surgery	468
References	468

42. Postoperative control: Complications and management in critical care units — 471

Susana López Piñeiro

Introduction	471
Suture dehiscence with anastomotic leak	471
Intraabdominal abscess	471

Enterocutaneous fistula	472
Generalized peritonitis	472
Sepsis and septic shock	473
Empirical antibiotic treatment for an IAI	474
Microbiology of an intraabdominal infection depending on its origin	474
Risk factors for poor outcomes	474
Empirical antibiotic treatment	475
Empirical antifungal treatment	476
Duration of antibiotic treatment	476
Abdominal compartment syndrome	476
References	478

43. Pain units: Symptom control — 479

Pilar Díaz Parada and M. del Carmen Corujeira Rivera

Acute postoperative pain	479
Strategies for APP management	480
Most frequent causes of inadequate analgesia	481
Analgesic techniques for CRC surgery	481
Chronic pain in colorectal cancer	484
Frequent types of pain in colorectal cancer	485
Pain secondary to chemotherapy	485
Pain secondary to radiotherapy	486
Treatment of baseline pain	486
Analgesic coadjuvants	487
Treatment of breakthrough pain	488
Interventional techniques	488
Radiotherapy in the treatment of pain from bone metastases	490
Complementary treatments	490
Conclusions	490
References	490

Section E
Microbiota, molecular and biological mechanisms of CRC

44. The role of intestinal microbiota in the colorectal carcinogenesis — 495

Alejandra Cardelle-Cobas, Beatriz I. Vázquez, José Luis Ulla Rocha, Carlos N. Franco, Margarita Poza, Nieves Martínez Lago, and Luis M. Antón Aparicio

Introduction	495
Human gut microbiota	495
Diet and human gut microbiota	497

Advanced technologies for the human gut microbiota analysis	498
Human gut microbiota and colorectal cancer	499
Bacteria	500
Fungi	504
Virus	506
Plasmids	508
References	508

45. Genetic susceptibility to CRC — 513

Ceres Fernández-Rozadilla, Anael López-Novo, Ángel Carracedo, and Clara Ruiz-Ponte

Introduction	513
High penetrance rare variants: Hereditary colorectal cancer predisposition syndromes	513
Other hereditary cancer syndromes	514
Candidate CRC susceptibility genes and NGS	514
Low-penetrance variants and association studies	515
Candidate gene studies	515
Genome-wide association studies (GWAS)	516
One step further: strategies for causal variant identification and fine-mapping	516
Multiomic integration	517
Polygenic risk scores	517
References	517

46. Signaling pathways in CRC — 519

Víctor Sacristán Santos, Nieves Martínez Lago, Carla Pazos García, Alejandro Pazos García, and Luis M. Antón Aparicio

Introduction	519
Signaling pathways	521
TGF-β and its role in colon cancer	521
Smads and its role in colon cancer	522
EGFR and its role in colon cancer	523
MAP kinase (MAPK) pathways and their role in colon cancer	524
PI3K-AKT-mTOR and its role in colon cancer	525
PI3K-AKT-mTOR and its role in colon cancer	526
References	526

47. CRC: A Darwinian model of cellular immunoselection — 529

Mónica Bernal, Natalia Aptsiauri, María Otero, Ángel Concha, Federico Garrido, and Francisco Ruíz-Cabello

Cancer immunoedition: From immunosurveillance to tumor escape	529
Tumor escape strategies	531
Defects in the presentation of tumor antigens: Alterations in HLA class I molecules	532
Colorectal carcinomas with microsatellite instability: Role of the B2M gene	533
Mutations in the B2m gene: An immunoselection model in MSI-H carcinogenesis	534
Clinical implication of B2m gene mutations in CRCs with MSI-H	535
The prognostic value of tumor lymphocyte infiltration: The importance of the "Immunoscore"	536
References	537

48. Epithelial-mesenchymal transition and CRC — 543

Angélica Figueroa

Introduction	543
Epithelial plasticity	543
Mechanisms responsible for E-cadherin inactivation during tumor progression	545
Mutations in the CDH1 gene	545
Epigenetic silencing of the CDH1 gene by methylation of the promoter	546
Transcriptional repressors of the E-cadherin promoter	546
Posttranscriptional regulators of EMT	547
Posttranslational regulators of E-cadherin	547
Conclusions	549
References	549

49. Colorectal carcinoma: From molecular pathology to clinical practice — 551

Catuxa Celeiro Muñoz, María Sánchez Ares, and José Ramón Antúnez López

Introduction	551

Colorectal cancer: Molecular-genetic context	551
Chromosomal instability	552
Microsatellite instability	552
CpG island methylator phenotype (CIMP)	553
Microsatellite instability and Lynch syndrome	553
Methods for determination of microsatellite instability	553
Characterization of Lynch syndrome	555
Emerging concepts in Lynch syndrome	556
Differential diagnosis of Lynch syndrome versus sporadic colorectal carcinoma with MSI	556
Clinical application of molecular diagnostics: Biomarker analysis	558
Identification of microsatellite instability in colorectal carcinoma	558
Predicting response to anti-EGFR therapy in metastatic CRC	559
Emerging biomarkers in colorectal carcinoma	559
Role of liquid biopsy in CRC	559
Final considerations: the pathologist's point of view	559
References	560

Section F
Biobanks

50. The role of biobanks in the study of colorectal carcinoma   565

Vanesa Val Varela, Orlando Fernández Lago, Paula Vieiro Balo, Joaquín González-Carreró, Lydia Fraga Fontoira, and Máximo Fraga Rodríguez

Biobanks and biomedical research	565
Organization and operation of the biobank	566
Biobank orientation	567
Integration of diagnostic anatomical pathology files in biobanks	567
The role of biobanks in cancer research	568
Tumor banking and high-throughput techniques	568
Colorectal cancer and biobanks	569
Conclusions	570
References	570

Section G
Approach to the sequelae and consequences of CRC

51. Psychological approach and emotional management in CRC   575

Lucía Álvarez-Santullano, Lucía Barcia, Alba Burundarena, Marcos Calvo, Ainhoa Carrasco, Rosalía Fernández, Miriam Rojas, and Rosa Trillo

Keys and tools in the relationship between the health care team and the patient	575
Searching for health information on the Internet	576
Psychological aspects in the different phases of the disease	576
Emotional aspects in the prediagnostic phase	576
Diagnostic phase	577
Treatment phase	577
Physical sequelae of the treatments	578
Ostomy	578
Impact on quality of life	578
Emotional impact	579
Impact on body image	579
Impact on sexuality	580
Factors to take into account for improving this situation	581
Psychological aspects of the SARS-CoV-2 pandemic (COVID-19)	582
References	582

52. Nutritional status in patients with CRC: Assessment and recommendations   585

María Teresa García Rodríguez

Introduction	585
Related complications	585
Causes of malnutrition	586
Assessment of nutritional status in colorectal cancer patients	586
Dietary recommendations and nutritional support	589
Recommendations for symptoms that hinder food intake	590
References	591

53. Fecal incontinence and CRC — 593

José Luis Ulla Rocha, Pablo Parada Vázquez, Raquel Sardina Ferreiro, and Juan Turnes Vázquez

Incontinence in relation to colorectal cancer	593
Surgery type	593
Radiotherapy damage	593
Assessment	594
Clinical history and physical examination	594
Colonoscopy	594
Endoanal ultrasonography	594
Anorectal manometry	594
Magnetic nuclear resonance	594
Defecography	595
Treatment and rehabilitation	595
Dietary regime and other general measures	595
Specific medications for incontinence	596
Specific treatments for radiation proctitis	596
Biofeedback	596
Pelvic floor rehabilitation	596
Transanal irrigation	596
Sacral nerve stimulation	596
Posterior tibial nerve stimulation	596
Corrective surgery to the sphincters	597
Conclusion	597
References	597

54. Ostomy care — 599

Alba María Arceo Vilas, Antonio Jurjo Sieira, and Silvia Louzao Méndez

Introduction	599
Gastrointestinal ostomies	599
Ileostomy	599
Colostomies	601
Main complications of gastrointestinal ostomies	601
Immediate or early complications	601
Late complications	602
Skin complications	604
Preoperative consultation	605
Stoma care and hygiene	605
Device types	605
Skin protectors	606
Devices for controlling evacuations	607
Stoma marking	607
Postoperative follow-up	607
Ostomy consultation at discharge	607
References	608

55. Sexual dysfunction among patients with ostomies — 609

María Teresa García Rodríguez, Adriana Barreiro Trillo, and Sonia Pértega Díaz

Introduction	609
Sexual impairment in patients with ostomies	609
Assessment of sexual dysfunction	610
General questionnaires	610
Specific questionnaires	612
Recommendations to cope with the fear of sexual intimacy	613
Health education: Help to adapt to change	613
References	614

Section H
Ethical and legal aspects in CRC

56. Ethical and legal aspects in CRC: Research and clinical assistance — 619

Natalia Cal Purriños, Isaac Martínez Bendayán, and Aliuska Duardo Sánchez

Introduction: The relationship between ethics and law	619
Ethical principles and research ethics committees	619
Regulations governing biomedical research	620
The new regulatory framework for personal data protection	624
New technologies and research with health data	626
References	627

Index — 629

Contributors

Numbers in parentheses indicate the pages on which the author's contributions begin.

José Francisco Noguera Aguilar (303,307), General Surgery Service of CHUAC; Coloproctology Unit—CHUAC, Head of the General Surgery Department of the University Hospital of A Coruña, A Coruña, Spain

Mohammed Alhayek-Aí (17), Ophthalmology Department, Hospital de la Santa Creu i Sant Pau, Barcelona, Spain

Pedro A. Alonso Aguirre (295), Department of Gastroenterology, University Hospital Complex of A Coruña (CHUAC), SERGAS, A Coruña, Spain

Paula Álvarez-Chaver (249), Proteomics Unit, Service of Structural Determination, Proteomics and Genomics, Center for Scientific and Technological Research Support (CACTI), University of Vigo, Vigo, Spain

Sara Alvarez-Gonzalez (231), Department of Computer Science and Information Technologies, Faculty of Computer Science, CITIC-Research Center of Information and Communication Technologies, Universidade da Coruna, Biomedical Research Institute of A Coruña (INIBIC), A Coruña, Spain

Lucía Álvarez-Santullano (575), Spanish Association Against Cancer, Madrid, Spain

Javier Castro Alvariño (157, 269, 275), Gastroenterology Department, Ferrol University Hospital Complex, A Coruña, Spain

Jorge Amigo (231), Galician Public Foundation of Genomic Medicine, A Coruña, Spain

Luis M. Antón Aparicio (495,519), Medical Oncology Service, A Coruña University Hospital (CHUAC), A Coruña, Spain

Natalia Aptsiauri (529), Department of Biochemistry and Molecular Biology III/Immunology, University of Granada; Institute of Biosanitary Research of Granada (IBS), Granada, Spain

María Sánchez Ares (551), Health Research Institute of Santiago de Compostela Foundation, Santiago de Compostela, Spain

Begoña Campos Balea (381), Medical Oncology Service, Lucus Augusti Hospital, Lugo, Spain

Paula Vieiro Balo (565), Santiago de Compostela University Hospital Complex Biobank, Santiago, Spain

Lucía Barcia (575), Spanish Association Against Cancer, Madrid, Spain

Germán Bou (207), Department of Microbiology, University Hospital Complex of A Coruña (CHUAC). Microbiology Research Group, Biomedical Research Institute of A Coruña (INIBIC), A Coruña, Spain

Vanesa Balboa Barreiro (3,17), Research Support Unit, Institute for Biomedical Research of A Coruña (INIBIC), University Hospital Complex of A Coruña (CHUAC), SERGAS, University of A Coruña, A Coruña, Spain

Isaac Martínez Bendayán (619), Research Ethics Committee A Coruña, Ferrol; Cardiology Department, University Hospital of A Coruña, A Coruña and Cee Health Area, SERGAS, A Coruña; Structural and Congenital Heart Disease Research Group, Institute for Biomedical Research A Coruña—INIBIC, A Coruña, Spain

Mónica Bernal (529), UGC of Laboratories, Hospital Universitario Virgen de las Nieves; Department of Biochemistry and Molecular Biology III/Immunology, University of Granada, Granada, Spain

Pedro Carpintero Briones (269), Gastroenterology Department, Ferrol University Hospital Complex, Ferrol, Spain

Elena Brozos-Vázquez (357), Medical Oncology Department, University Clinical Hospital of Santiago de Compostela, University of Santiago de Compostela (USC); Translational Medical Oncology Group (Oncomet), Health Research Institute of Santiago (IDIS), University Clinical Hospital of Santiago de Compostela, University of Santiago de Compostela (USC), CIBERONC, Santiago de Compostela, Spain

Begoña Bravo Bueno (57), Resident Physician in Family and Community Medicine, Elviña-Mesoiro Health Center, Teaching Unit of the Xerencia Xestión Integrada de A Coruña, Galician Health Service, Santiago de Compostela, Spain

Alba Burundarena (575), Spanish Association Against Cancer, Madrid, Spain

Antonio Gómez Caamaño (403), Department of Radiation Oncology, Santiago de Compostela University Hospital Complex, Santiago de Compostela, Spain

José Manuel Mera Calviño (195), Gastoenterology Service. Pontevedra University Hospital Complex, Pontevedra, Spain

Marcos Calvo (575), Spanish Association Against Cancer, Madrid, Spain

Sonia Candamio-Folgar (357), Medical Oncology Department, University Clinical Hospital of Santiago de Compostela, University of Santiago de Compostela (USC); Translational Medical Oncology Group (Oncomet), Health Research Institute of Santiago (IDIS), University Clinical Hospital of Santiago de Compostela, University of Santiago de Compostela (USC), CIBERONC, Santiago de Compostela, Spain

Alejandra Cardelle-Cobas (495), Department of Analytical Chemistry, Nutrition and Bromatology, University of Santiago de Compostela, Lugo, Spain

Ángel Carracedo (513), Genomic Medicine Group, Santiago de Compostela; Health Research Institute of Santiago de Compostela, A Coruña; Santiago de Compostela University, Santiago de Compostela; Galician Public Foundation of Genomic Medicine, Santiago de Compostela; Center for Biomedical Research Network on Rare Diseases, Barcelona, Spain

Ainhoa Carrasco (575), Spanish Association Against Cancer, Madrid, Spain

Álvaro Gómez Castro (149), Head of the Pathology Department, Pontevedra Hospital Complex, Pontevedra, Spain

Ana María Carballo Castro (403,421), Department of Radiation Oncology, Complexo Hospitalario Universitario de Santiago de Compostela, A Coruña, Spain

Alberto Cepeda (249), Department of Analytical Chemistry, Nutrition and Bromatology, University of Santiago de Compostela, Lugo, Spain

Alfonso Cepeda-Emiliani (207), Department of Morphological Sciences, School of Medicine and Dentistry, University of Santiago de Compostela, A Coruña, Spain

Ángel Concha (207,529), Pathology Anatomy and Pathology Department, A Coruña University Hospital Complex; Biobank A Coruña, Institute of Biosanitary Research of A Coruña (INIBIC); Department of Anatomical Pathology, University Hospital Complex of A Coruña (CHUAC). Biobank of A Coruña, Biomedical Research Institute of A Coruña (INIBIC), A Coruña, Spain

Benito González Conde (295), Department of Gastroenterology, University Hospital Complex of A Coruña (CHUAC), SERGAS, A Coruña, Spain

Oscar J. Cordero (45,435), University of Santiago de Compostela, Department of Biochemistry and Molecular Biology, Santiago de Compostela, Spain

Alberto Centeno Cortés (307), CTF-XXIAC, A Coruña, Spain

Jesús Paredes Cotoré (317,323,403), Coloproctology Unit of the General Surgery Service, Associate Professor in Health Sciences, Department of Surgery, Faculty of Medicine; Department of Surgery, Complexo Hospitalario Universitario de Santiago de Compostela, A Coruña, Spain

Patricia Calvo Crespo (403,421,429), Department of Radiation Oncology, Complexo Hospitalario Universitario de Santiago de Compostela, A Coruña, Spain

Alain García de Castro (57), Resident Physician in Family and Community Medicine, Elviña-Mesoiro Health Center, Teaching Unit of the Xerencia Xestión Integrada de A Coruña, Galician Health Service, Santiago de Compostela, Spain

Loretta De Chiara (45,249), Department of Biochemistry, Genetics and Immunology; Singular Research Center of Galicia (CINBIO), University of Vigo, Vigo, Spain

Enrique González de la Ballina González (195), Gastoenterology Service. Pontevedra University Hospital Complex, Pontevedra, Spain

Sofía Rodríguez Martínez de Llano (119), Department of Nuclear Medicine, Oncology Centre of Galicia, A Coruña, Spain

Manuel Núñez Deben (457), Department of Anesthesiology and Reanimation, Vigo University Hospital Complex, Vigo, Spain

M. del Carmen Corujeira Rivera (479), Anestesiology Service. Vigo University Hospital Complex, Vigo, Spain

Cristina Méndez Díaz (131), Department of Radiology, A Coruña University Hospital Complex, A Coruña, Spain

Sonia Pértega Díaz (3,17,609), Research Support Unit, Institute for Biomedical Research of A Coruña (INIBIC), University Hospital Complex of A Coruña (CHUAC), SERGAS, University of A Coruña, A Coruña, Spain

Leticia García Diéguez (157,275), Gastroenterology Department, Ferrol University Hospital Complex, A Coruña, Spain

Antonio Rodríguez D'Jesús (283), Department of Gastroenterology, Xerencia de Xestión Integrada de Vigo SERGAS, Vigo; Research Group on Digestive Diseases, Galicia Sur Health Research Institute (IIS Galicia Sur), SERGAS-UVIGO, Pontevedra, Spain

Ramón Vázquez Dourado (157,269,275), Gastroenterology Department, Ferrol University Hospital Complex, A Coruña, Spain

Alba Gómez Dovigo (303,307), General Surgery Service, Quironsalud A Coruña Hospital, A Coruña, Spain

Paloma Sosa Fajardo (181), Radiation Oncology Unit, Universitary Hospital Complex Santiago de Compostela, Santiago de Compostela, Spain

Nereida Fernández Fernández (283), Department of Gastroenterology, Xerencia de Xestión Integrada de Vigo, SERGAS, Vigo; Research Group on Digestive Diseases, Galicia Sur Health Research Institute (IIS Galicia Sur), SERGAS-UVIGO, Pontevedra, Spain

Rafaela Soler Fernández (131), Department of Radiology, A Coruña University Hospital Complex, A Coruña, Spain

Rosalía Fernández (51), Spanish Association Against Cancer, Madrid, Spain

Sergio Manuel Estévez Fernández (31), Álvaro Cunqueiro Hospital of Vigo, Vigo, Spain

Carlos Fernandez-Lozano (231), Department of Computer Science and Information Technologies, Faculty of Computer Science, CITIC-Research Center of Information and Communication Technologies, Universidade da Coruna, Biomedical Research Institute of A Coruña (INIBIC), A Coruña, Spain

Ceres Fernández-Rozadilla (513), Genomic Medicine Group, Santiago de Compostela University (USC); Health Research Institute of Santiago de Compostela, Spain

Raquel Sardina Ferreiro (27,89,593), Department of Internal Medicine, Arquitecto Marcide Hospital, Ferrol, Spain

Silvia Varela Ferreiro (381), Medical Oncology Service, Lucus Augusti Hospital, Lugo, Spain

Roberto García Figueiras (403), Department of Radiology, Complexo Hospitalario Universitario de Santiago de Compostela, A Coruña, Spain

Angélica Figueroa (543), Epithelial Plasticity and Metastasis Group, A Coruña Biomedical Research Institute, A Coruña University Hospital Complex, University of A Coruña, A Coruña, Spain

Lydia Fraga Fontoira (565), Santiago de Compostela University Hospital Complex Biobank, Santiago, Spain

Carlos M.N Franco (495), Department of Analytical Chemistry, Nutrition and Bromatology, University of Santiago de Compostela, Lugo, Spain

María Gallardo-Gómez (45,249), Department of Biochemistry, Genetics and Immunology; Singular Research Center of Galicia (CINBIO), University of Vigo, Vigo, Spain

Concepción Crespo García (95,109), Department of Radiology, A Coruña University Hospital, A Coruña, Spain

Esther Rodríguez García (131), Department of Radiology, A Coruña University Hospital Complex, A Coruña, Spain

Alba García-López (435), Galician Health System (Sergas). Service of Pharmacy, Universitary Hospital Complex of Santiago de Compostela (CHUS), Santiago, Spain

Federico Garrido (529), UGC of Laboratories, Hospital Universitario Virgen de las Nieves; Department of Biochemistry and Molecular Biology III/Immunology, University of Granada; Institute of Biosanitary Research of Granada (IBS), Granada, Spain

Javier Aguirrezabalaga González (307), HBP Unit—CHUAC, A Coruña, Spain

Joaquín González-Carreró (565), Institute of Health Research in Southern Galicia Biobank, Galicia, Spain

Diego Gonzalez-Rivas (343), Uniportal VATS Training Program, Shanghai Pulmonary Hospital, Shanghai, People's Republic of China; Ernst Von Bergmann Hospital, Berlin, Germany; Department of Thoracic Surgery and Lung Transplantation, A Coruña University Hospital; Minimally Invasive Thoracic Surgery Unit (UCTMI), A Coruña, Spain

Begoña Graña-Suárez (435), Galician Health System (Sergas), Service of Oncology, Universitary Hospital Complex of A Coruña (CHUAC), Santiago, Spain

Vicent Hernández (45,171), Digestive System Service, South Galicia Health Research Institute, Álvaro Cunqueiro Hospital, EOXI of Vigo; Department of Gastroenterology, Xerencia Xestion Integrada de Vigo, SERGAS; Research Group in Digestive Diseases, Galicia Sur Health Research Institute (IIS Galicia Sur), SERGAS-UVIGO, Vigo, Spain

Paula Peleteiro Higuero (403,421), Department of Radiation Oncology, Complexo Hospitalario Universitario de Santiago de Compostela, A Coruña, Spain

Héctor Lázare Iglesias (149), Specialist in Pathological Anatomy and Pathology Santiago de Compostela University Hospital Complex, Santiago de Compostela, Spain

Orlando Fernández Lago (565), Santiago de Compostela University Hospital Complex Biobank, Santiago, Spain

Alexandre Lamas (249), Department of Analytical Chemistry, Nutrition and Bromatology, University of Santiago de Compostela, Lugo, Spain

Jose Liñares-Blanco (231), Department of Computer Science and Information Technologies, Faculty of Computer Science, CITIC-Research Center of Information and Communication Technologies, Universidade da Coruna, Biomedical Research Institute of A Coruña (INIBIC), A Coruña, Spain

María Jose Martinez-Sapiña Llanas (95,109), Department of Radiology, A Coruña University Hospital, A Coruña, Spain

Fernando Fernández López (317,323), Coloproctology Unit of the General Surgery Service, Associate Professor in Health Sciences, Department of Surgery, Faculty of Medicine, University of Santiago de Compostela, A Coruña, Spain

José Ramón Antúnez López (551), Pathological Anatomy Department, University Clinical Hospital of Santiago de Compostela, Santiago de Compostela, Spain

Rafael López-López (357), Medical Oncology Department, University Clinical Hospital of Santiago de Compostela, University of Santiago de Compostela (USC); Translational Medical Oncology Group (Oncomet), Health Research Institute of Santiago (IDIS), University Clinical Hospital of Santiago de Compostela, University of Santiago de Compostela (USC), CIBERONC, Santiago de Compostela, Spain

Anael López-Novo (513), Genomic Medicine Group, Santiago de Compostela University (USC); Health Research Institute of Santiago de Compostela (IDIS), Santiago de Compostela, Spain

Carmen Álvarez Lorenzo (391), Department of Pharmacology, Pharmacy and Pharmaceutical Technology, Faculty of Pharmacy, University of Santiago de Compostela, Λ Coruña, Spain

Miguel Pereira Loureiro (457), Department of Anesthesiology and Reanimation, Vigo University Hospital Complex, Vigo, Spain

Ramiro Manuel Macenlle García (83), Digestive System Service Ourense University Hospital Complex, Ourense, Spain

Shirly Margarita Nieves Maldonado (119), Department of Nuclear Medicine, Oncology Centre of Galicia, A Coruña, Spain

Sara Seijas Marcos (119), Department of Nuclear Medicine, Oncology Centre of Galicia, A Coruña, Spain

Eva Martí Marqués (41), Lucus Augusti University Hospital, Lugo, Spain

Cristina González Martín (3,17), Department of Health Sciences, Faculty of Nursing and Podiatry, Universidade da Coruña, Ferrol, Spain

Arantza Germade Martínez (283), Department of Gastroenterology, Xerencia de Xestión Integrada de Vigo, SERGAS, Vigo; Research Group on Digestive Diseases, Galicia Sur Health Research Institute (IIS Galicia Sur), SERGAS-UVIGO, Pontevedra, Spain

Nieves Martínez Lago (495,519), Medical Oncology Service, A Coruña University Hospital (CHUAC), A Coruña, Spain

Gala Martínez-Bernal (365), Medical Oncology Unit, Virgen del Rocío University Hospital, Sevilla, Spain

Julia Martínez-Pérez (365), Medical Oncology Unit, Virgen del Rocío University Hospital, Sevilla, Spain

Virginia Mato-Abad (231), Department of Computer Science and Information Technologies, Faculty of Computer Science, CITIC-Research Center of Information and Communication Technologies, Universidade da Coruna, Biomedical Research Institute of A Coruña (INIBIC), A Coruña, Spain

Silvia Louzao Méndez (599), University Hospital Complex of A Coruña, A Coruña, Spain

Manuel Bustamante Montalvo (317,323,331), Head of the Department of General and Digestive Surgery, University Hospital Complex of Santiago de Compostela, Santiago de Compostela, Spain

Beatriz Romero Mosquera (171), Department of Gastroenterology, Xerencia Xestion Integrada de Vigo SERGAS; Research Group in Digestive Diseases, Galicia Sur Health Research Institute (IIS Galicia Sur), SERGAS-UVIGO, Vigo, Spain

Catuxa Celeiro Muñoz (551), Department of Pathological Anatomy, University Clinical Hospital of Santiago, Santiago de Compostela, Spain

Cristian R. Munteanu (231), Department of Computer Science and Information Technologies, Faculty of Computer Science, CITIC-Research Center of Information and Communication Technologies, Universidade da Coruna, Biomedical Research Institute of A Coruña (INIBIC), A Coruña, Spain

Ángel Concheiro Nine (391), Department of Pharmacology, Pharmacy and Pharmaceutical Technology, Faculty of Pharmacy, University of Santiago de Compostela, A Coruña, Spain

Alejandra García Novoa (303,307), General Surgery Service, Quironsalud A Coruña Hospital, A Coruña, Spain

Martina Lema Oreiro (391), Pharmaceutical Services Management Service, S.X. de Farmacia, General Directorate of Health Care, Consellería de Sanidade, SERGAS, A Coruña, Spain

Francisco Ruíz-Cabello Osuna (529), UGC of Laboratories, Hospital Universitario Virgen de las Nieves; Department of Biochemistry and Molecular Biology

III/Immunology, University of Granada; Institute of Biosanitary Research of Granada (IBS), Granada, Spain

María Otero (529), Pathological Anatomy Service Santiago de Compostela University Hospital Complex; Institute of Biosanitary Research of Santiago de Compostela (IDIS), Santiago de Compostela, Spain

Susana A. Otero Muinelo (95,109), Department of Radiology, A Coruña University Hospital, A Coruña, Spain

Pilar Díaz Parada (479), Anestesiology Service, Pontevedra University Hospital Complex, Pontevedra, Spain

Jesús P. Paredes Cotoré (317), Coloproctology Unit of the General Surgery Service Associate Professor in Health Sciences, Department of Surgery, Faculty of Medicine, University of Santiago de Compostela, A Coruña, Spain

Alejandro Pazos (231), Department of Computer Science and Information Technologies, Faculty of Computer Science, CITIC-Research Center of Information and Communication Technologies, Universidade da Coruna, Biomedical Research Institute of A Coruña (INIBIC), A Coruña, Spain

Alejandro Pazos García (519), Medical University of Bialystock (MUB), Białystok, Poland

Carla Pazos García (519), New Vision University (NVU), Tbilisi, Georgia

Nieves Pedreira (231), Department of Computer Science and Information Technologies, Faculty of Computer Science, CITIC-Research Center of Information and Communication Technologies, Universidade da Coruna, Biomedical Research Institute of A Coruña (INIBIC), A Coruña, Spain

María Teresa Seoane Pillado (3,17), Department of Health Sciences, Faculty of Nursing and Podiatry, Universidade da Coruña, Ferrol, Spain

Susana López Piñeiro (471), Anaesthesia and Resuscitation Department, Hospital Complex of Pontevedra, Pontevedra, Spain

Margarita Poza (495), Microbiology Research Group, University Hospital Complex (CHUAC)—Institute of Biomedical Research (INIBIC), University of A Coruña (UDC), A Coruña, Spain

Natalia Cal Purriños (619), Novoa Santos Foundation; Institute for Biomedical Research A Coruña—INIBIC, A Coruña; Research Ethics Committee A Coruña, Ferrol, Spain

Francisco Queipo (207), Department of Anatomical Pathology, University Hospital Complex of A Coruña (CHUAC), A Coruña, Spain

Patricia Regal (249), Department of Analytical Chemistry, Nutrition and Bromatology, University of Santiago de Compostela, Lugo, Spain

Fernando Zelaya Reinquet (119), Department of Nuclear Medicine, Oncology Centre of Galicia, A Coruña, Spain

María Teresa Vázquez Rey (295), Department of Gastroenterology, University Hospital Complex of A Coruña (CHUAC), SERGAS, A Coruña, Spain

Andrés Dacal Rivas (41), Lucus Augusti University Hospital, Lugo, Spain

José Luis Ulla Rocha (27,89,181,495,593), Gastroenterology Department, Digestive Disease Unit, Pontevedra University Hospital Complex, Pontevedra, Spain

Alejandro Ledo Rodríguez (69), Gastroenterology Department, Integrated Health Area of Pontevedra-O Salnés, Pontevedra, Spain

Francisco Javier González Rodríguez (331), Santiago de Compostela University Hospital Complex, Santiago de Compostela, Spain

Gerardo Baños Rodríguez (457), Department of Anesthesiology and Reanimation, Vigo University Hospital Complex, Vigo, Spain

María Teresa García Rodríguez (585,609), Research in Nursing and Health Care, Institute of Biomedical Research of A Coruña (INIBIC), A Coruña University Hospital Complex (CHUAC), SERGAS, A Coruña, Spain

Máximo Fraga Rodríguez (565), Santiago de Compostela University Hospital Complex Biobank, Santiago, Spain

Miriam Rojas (575), Spanish Association Against Cancer, Madrid, Spain

Leopoldo López Rosés (41), Lucus Augusti University Hospital, Lugo, Spain

Marta Covela Rúa (381), Medical Oncology Service, Lucus Augusti Hospital, Lugo, Spain

Juan Ruiz-Bañobre (357), Medical Oncology Department, University Clinical Hospital of Santiago de Compostela, University of Santiago de Compostela (USC); Translational Medical Oncology Group (Oncomet), Health Research Institute of Santiago (IDIS), University Clinical Hospital of Santiago de Compostela, University of Santiago de Compostela (USC), CIBERONC, Santiago de Compostela, Spain

Clara Ruiz-Ponte (513), Genomic Medicine Group, Santiago de Compostela; Health Research Institute of Santiago de Compostela (IDIS); Galician Public Foundation of Genomic Medicine, Santiago de Compostela; Center for Biomedical Research Network on Rare Diseases (CIBERER), Barcelona, Spain

Francisco Javier Maestro Saavedra (57), Specialist in Family and Community Medicine, Elviña-Mesoiro Health Center, Xerencia Xestión Integrada de A Coruña, Galician Health Service, Santiago de Compostela, Spain

Víctor Sacristán Santos (519), Medical Oncology Service, A Coruña University Hospital (CHUAC), A Coruña, Spain

Ismael Said-Criado (69), Emergency Department, Integrated Health Area of Vigo, Health Research Institute Galicia Sur, Vigo, Spain

Aliuska Duardo Sánchez (619), Department of Public Law, Faculty of Law UPV/EHU, G.I Chair in Law and the Human Genome, Leioa, Spain

Paloma González Santamaría (17), Meicende Health Centre, SERGAS, A Coruña, Spain

Rodrigo A.S. Sardenberg (343), Thoracic Surgeon Hospital Alemão Oswaldo Cruz and United Health Group; Center for Advanced Research at University of Great Lakes São José do Rio Preto, São Paulo, Brazil

Antonio Jurjo Sieira (599), Casa do Mar of A Coruña Health Center, A Coruña, Spain

Paulino Pais Silva (119), Department of Nuclear Medicine, Oncology Centre of Galicia, A Coruña, Spain

Rosa Fungueiriño Suarez (27), Preventive Medicine Service, Pontevedra University Hospital Complex, Pontevedra, Spain

Tatiana María Civeira Taboada (303), Coloproctology Unit—CHUAC, Head of the General Surgery Department of the University Hospital of A Coruña, A Coruña, Spain

Adriana Barreiro Trillo (609), University Hospital Complex of A Coruña (CHUAC), A Coruña, Spain

Rosa Trillo (575), Spanish Association Against Cancer, Madrid, Spain

M. Lidia Vázquez-Tunas (183), Medical Oncology Services, Álvaro Cunqueiro Hospital, Vigo University Hospital Complex, Pondevedra, Spain

Alfonso Martínez Turnes (171), Department of Gastroenterology, Vigo Integrated Management Services, SERGAS; Research Group in Digestive Diseases, Galicia Sur Health Research Institute (IIS Galicia Sur), SERGAS-UVIGO, Vigo, Spain

Begoña Taboada Valladares (403,429), Department of Radiation Oncology, Complexo Hospitalario Universitario de Santiago de Compostela, A Coruña, Spain

Manuel Valladares-Ayerbes (365), Medical Oncology Unit, Virgen del Rocío University Hospital, Sevilla, Spain

Vanesa Val Varela (656), Institute of Health Research in Southern Galicia Biobank, Galicia, Spain

Rubén Varela-Calviño (435), University of Santiago de Compostela, Department of Biochemistry and Molecular Biology, Santiago de Compostela, Spain

Beatriz I. Vázquez (44), Department of Analytical Chemistry, Nutrition and Bromatology, University of Santiago de Compostela, Lugo, Spain

Carla Blanco Vázquez (57), Resident Physician in Family and Community Medicine, Elviña-Mesoiro Health Center, Teaching Unit of the Xerencia Xestión Integrada de A Coruña, Galician Health Service, Santiago de Compostela, Spain

Juan Turnes Vázquez (27,89,593), Gastroenterology Department, Digestive Disease Unit, Pontevedra University Hospital Complex, Pontevedra, Spain

Pablo Parada Vázquez (593), Gastroenterology Department, Pontevedra University Hospital Complex, Pontevedra, Spain

Vanesa Vilanova Vázquez (457), Department of Anesthesiology and Reanimation, Vigo University Hospital Complex, Vigo, Spain

José M. Vázquez-Naya (231), Department of Computer Science and Information Technologies, Faculty of Computer Science, CITIC-Research Center of Information and Communication Technologies, Universidade da Coruna, Biomedical Research Institute of A Coruña (INIBIC), A Coruña, Spain

Francisca Vázquez-Rivera (357), Medical Oncology Department, University Clinical Hospital of Santiago de Compostela, University of Santiago de Compostela (USC); Translational Medical Oncology Group (Oncomet), Health Research Institute of Santiago (IDIS), University Clinical Hospital of Santiago de Compostela, University of Santiago de Compostela (USC), CIBERONC, Santiago de Compostela, Spain

Alberto Veiga (207), Department of Anatomical Pathology, University Hospital Complex of Ferrol (CHUF), A Coruña, Spain

Yolanda Vidal-Ínsua (357), Medical Oncology Department, University Clinical Hospital of Santiago de Compostela, University of Santiago de Compostela (USC); Translational Medical Oncology Group (Oncomet), Health Research Institute of Santiago (IDIS), University Clinical Hospital of Santiago de Compostela, University of Santiago de Compostela (USC), CIBERONC, Santiago de Compostela, Spain

Alba María Arceo Vilas (599), University Hospital Complex of A Coruña, A Coruña, Spain

Ignacio Couto Worner (295), Department of Gastroenterology, University Hospital Complex of A Coruña (CHUAC), SERGAS, A Coruña, Spain

Editor's note

Future health care for people will be characterized by three events: (1) placing the patient at the center of the health care process instead of the medical staff or the disease itself; (2) the change of dimension in the etiology of pathological processes, from systemic to organic, then to tissue, later to cellular and, lastly, to molecular (it has gone from a "macro" scale to a "micro" scale, and even "nanometric"), without ruling out the possibility of reaching an atomic or subatomic level; (3) an exponential growth of far-reaching scientific and technological advances, with particular emphasis on the new knowledge generated passing "from the muses to the theater," from research laboratories to application in the health care sector. The complete sequencing of the human genome has been taken as a turning point, having been considered, unsuccessfully in many cases, as a roadmap with which to address the knowledge and explanation of all types of diseases.[1]

These last two events, the etiopathogenic scale and the knowledge generated by scientific and technological advances, will be greatly influenced by the ongoing progress made in what is known as "omics" (genomics, proteomics, metabolomics, exposomics, etc.) in "evidence-based medicine," in the robotization of central services and operating rooms, in "nanomedicine," and in the multiple diagnostic imaging techniques (3D, 4D, microarrays, etc.). We are talking about the arrival of "Precision Medicine" and "Personalized Medicine" on the health care scenario.[2, 3] Although in the literature it is very common for these terms to be used interchangeably, here "personalized medicine" is that in which the processes of diagnosis, therapy, control, and follow-up are tailored to each person, depending on their "omic" characteristics (genomics, proteomics, metabolomics, etc.) and their interaction with their own environment. "Precision medicine," on the other hand, refers to when we carry out health care activities in the least iatrogenic and intrusive way, with the fewest possible adverse effects, even resorting to nano-scale techniques. For example, administering nanomolecules that specifically act on specific cells, organelles, or proteins; using micro or nano surgical incisions to avoid infections or to restore mutations of DNA molecules, etc. Although the term 'personalized medicine' was already in use, it became known internationally in 2015 when former US President Barack Obama made it one of the main objectives of his mandate, in order to create a new model for researching pathological processes based on the patient and the ecosystem in which they operate.[4]

In medicine, the use of Information and Communication Technologies (ICT) has greatly helped in improving efficiency in health care and will continue to do so in the future, with every day bringing more differentiated and earlier interventions. It will also make the therapeutic acts as personalized and precise as possible.

Thanks to ICT, genomics, robotics, and artificial Intelligence have become particularly relevant because all the earlier mentioned will make it necessary to intelligently manage huge amounts of data, information, and knowledge (hereinafter DIK), in a context that is now known as "big data," which is nothing more than the huge amount of DIK that is generated thanks to scientific and technological advances, which must be available 24/7 in a safe and efficient manner. We refer to Data as the value that a variable can take; for example, body temperature $41°C$. Information would be when the Data takes on a meaning; for example, body temperature of $41°C$ means fever. And we will understand by knowledge when the data and information acquire a useful characteristic; for example, a body temperature of $41°C$ means fever and measures must be taken to reduce this temperature (administer antipyretic drugs, etc.) because if it increases or is maintained for a long time it can cause damage in other areas of the body's economy, such as the neurological system.

When dealing with the management of complex diseases such as colorectal cancer, it is essential to do so from a holistic, multidisciplinary, and translational perspective. To address this challenge, the appearance of different highly necessary and complementary tools has been crucial: bioinformatics, ultrasequencing, robotization, surgical advances, new diagnostic imaging techniques, advances in pharmacogenomics, etc.

Different intelligent computer systems capable of analyzing DIK in an integrated way are currently being developed, in order to offer solutions for all stages of the patient care process, taking into account that it will be necessary to integrate DIK from other heterogeneous sources, such as the exposome, nutrition, social environment, Internet of things (IoT) devices, etc.

In recent years, the health care paradigm has started to change, and preventions, diagnoses, treatments, controls, and follow-ups are being carried out thanks to the "omic data" and their integration with other types of data (medical history, nutrition, environmental, etc.) that also affect the pathological process of each individual. It is not only the symptomatologic data that need to be tended to. The use of techniques such as liquid biopsy for the identification of different molecular biomarkers, as well as genetic tests in different types of cancer for knowing which patients will respond to a certain treatment, are advances that are in tune with what we can expect to achieve in the future, but that, without well-established plans by the heads of the competent authorities, will probably be in vain.

There are many examples of new use cases for Personalized and Precision Medicine. One of the most promising is the development of biomarker kits for the early diagnosis of different diseases, such as colorectal cancer. Kits of this type are already available and offer the possibility of detecting the absence of genetic mutations, such as KRAS, BRAF, NRAS, and PIK3CA, which will provoke a certain choice of adjuvant therapy for each case. The objective is to have the possibility of using genetic or epigenetic biomarkers for establishing a diagnosis, prognosis, and possible response to treatment in patients.[5] But new questions are also arising from the ethical point of view. An important aspect is the excessively early diagnosis of certain pathologies, such as cancer. Using biomarkers, it is possible to establish that a person is going to develop cancer years in advance of its clinical expression, but no type of therapeutic solution can be offered since it is not possible to locate the cancer due to the excessively initial stage and it not being sufficiently expressed. This situation can cause future patients a state of unnecessary anxiety that can very negatively affect their quality of life and can even accelerate and/or worsen the development of the disease in question. As Oduncu said: "Everything the doctor tells the patient should be true! But not everything that is true must be told to the patient."[6]

References

1. Lander ES, Linton LM, Birren B, et al. Initial sequencing and analysis of the human genome. *Nature* 2001;**409**(6822):860–921. https://doi.org/10.1038/35057062.
2. ABC-Salud. *La medicina de precisión, la próxima revolución en biomedicina*; 2018. Published online https://www.abc.es/salud/enfermedades/abci-medicina-precision-proxima-revolucion-biomedicina-201807261905_noticia.html.
3. Bernardo A. *¿Qué es la medicina de precisión?*; 2015. Published online https://blogthinkbig.com/la-medicina-precision.
4. Obama B. *Precision medicine initiative*; 2015. Published online https://obamawhitehouse.archives.gov/precision-medicine.
5. Vacante M, Borzì AM, Basile F, Biondi A. Biomarkers in colorectal cancer: current clinical utility and future perspectives. *World J Clin Cases* 2018;**6**(15):869–81. https://doi.org/10.12998/wjcc.v6.i15.869.
6. Oduncu P. The role of ethical principles in cancer medicine and palliative care. In: *Presented at the: NVU conference*; 2019.

Prologue

The health care crisis caused by the outbreak of COVID-19 has helped to underline the central role played by science and scientific research in all aspects of our society, particularly with regard to health care.

It can be said that this pandemic has put scientific and technical progress in the spotlight. Government administrations cannot and must not limit themselves to supporting research and science intermittently.

At the Xunta de Galicia [Regional government of Galicia] we are totally aware of the importance of resolutely and continuously supporting our investigators' projects through several initiatives, whose main objective is to accompany them and back their research throughout their professional careers.

The support we have given to the Red Gallega de Investigación en Cáncer Colorrectal (REGICC) [Galician Network of Research in Colorectal Cancer] over the past 15 years is an example of our commitment. This network comprises over 100 clinicians specializing in different areas, investigators in several fields, and patient associations who work together in an attempt to improve the quality of health care for patients suffering from this disease.

Foundations of Colorectal Cancer is the result of this cooperation; it takes a multidisciplinary approach to the disease, with 56 chapters written by experts in each subject, more than 150 in total, most of whom belong to or work with the REGICC.

It is unquestionably a tremendously complex project that can be used both in clinical practice and teaching and training activities and is also informative for patients, caregivers, relatives, or the general public.

I would like to thank the editor and authors for their enormous effort, also down to our science and investigators who contribute to social progress and improving our quality of life.

Román Rodríguez González
Head of Department of Culture, Education and Universities, Xunta de Galicia, Santiago de Compostela, Spain

Acknowledgments

I would like to thank the authors of the chapters for their selfless collaboration. Special thanks to the coordinators of the different sections, headed by Dr. José Luis Ulla Rocha. They have gone through this endeavor completely altruistically *knowing* that much of their work will soon sink into the anonymity of oblivion. The true recognition will be their contribution to improving understanding of this very serious disease, the quality of care that each colorectal cancer patient should receive, and the increase in the sustainability of public health services that their contribution implies.

Moreover, I would like to thank the several organizations that have made this possible through the financing of R+D+i activities of the RNASA-IMEDIR group that I lead. So thanks to the "Collaborative Project in Genomic Data Integration (CICLOGEN)" PI17/01826 funded by the Carlos III Health Institute from the Spanish National plan for Scientific and Technical Research and Innovation 2013–2016 and the European Regional Development Funds (FEDER)—"A way to build Europe" and the General Directorate of Culture, Education and University Management of Xunta de Galicia, Spain (Ref. ED431D 2017/16), the "Galician Network for Colorectal Cancer Research, Spain" (Ref. ED431D 2017/23), and Competitive Reference Groups, Spain (Ref. ED431C 2018/49). CITIC (UDC's Research Center accredited by Galician University System) funded by "Consellería de Cultura, Educación e Universidades from Xunta de Galicia, Spain," supported through ERDF Funds, Spain, ERDF Operational Programme Galicia 2014–2020, and by "Secretaría Xeral de Universidades, Spain" (Grant ED431G 2019/01). Also, thanks to the Spanish Ministry of Economy and Competitiveness via funding of the unique installation BIOCAI (UNLC08-1E-002, UNLC13-13-3503), the European Regional Development Funds (FEDER) and INIBIC (A Coruña Biomedical Research Institute). Finally, thanks to SEGAS (Galician Health Service) for the collaboration of the healthcare personnel of this institution in the preparation of the different chapters.

Section A

Epidemiological studies in CRC

Chapter 1

Incidence and mortality of CRC

María Teresa Seoane Pillado[a], Sonia Pértega Díaz[b], Vanesa Balboa Barreiro[b], and Cristina González Martín[a]
[a]*Department of Health Sciences, Faculty of Nursing and Podiatry, Universidade da Coruña, Ferrol, Spain,* [b]*Research Support Unit, Institute for Biomedical Research of A Coruña (INIBIC), University Hospital Complex of A Coruña (CHUAC), SERGAS, University of A Coruña, A Coruña, Spain*

Incidence

Colorectal cancer is now the third most common tumor in men, after lung and prostate cancer, and the second most common in women, after breast cancer. According to data from the GLOBOCAN 2018 project[1] (http://globocan.iarc.fr/), which updates previously published estimates of cancer incidence and mortality from 2012, 1,849,518 new cases of colorectal cancer were diagnosed worldwide in 2018: 1,026,215 cases in males (10.8% of all incident cases of cancer) and 823,303 in females (9.5% of the total).[1] This results in a standardized incidence rate (using the world standard population) of 19.7 cases per 100,000 population (23.6 per 100,000 in males and 16.3 per 100,000 in females) (Table 1.1).[1,2]

Almost 60% of colorectal cancer cases occur in developed countries. Thus the highest incidence rates are found in Australia/New Zealand (41.7 per 100,000 in males and 32.1 per 100,000 in females) and in Southern European countries (40.4 and 24.1 per 100,000, respectively), while the lowest rates are found in Africa (except in the south) and South-Central Asia, with adjusted incidence rates below 7.0 per 100,000 in males and 4.0 per 100,000 in females (Fig. 1.1).[1,2]

As we have already mentioned, in 2018, the highest incidence in Europe is recorded in Southern European countries, followed by the Nordic countries (37.5 and 27.3 per 100,000, in men and women, respectively) and Central and Eastern Europe (37.5 and 23.2 per 100,000, respectively). We can also see that the incidence in all regions of Europe is above the global average (23.6 and 16.3 per 100,000, respectively) (Fig. 1.1).[1,2]

In Europe, colorectal cancer remains one of the most frequent tumors, with 499,667 incident cases in 2018, representing 11.8% of all new cancer cases diagnosed that year. These figures make it the third most frequent tumor in men (271,600 incident cases, 12.1% of the total), behind prostate and lung cancer, and the second most frequent in women (228,067 incident cases, 11.5% of the total), behind breast cancer. The standardized incidence rate of colorectal cancer (using the standard European population) for 2018 in Europe would be 30.0 per 100,000 (37.5 per 100,000 in males and 24.2 per 100,000 in females) (Table 1.2).[1,2]

This same study reflects a great variability in the incidence of this tumor between different European countries. Thus, in men, the highest incidence rates are recorded in Hungary (70.6 per 100,000), Slovakia (60.7 per 100,000), and Slovenia (58.9 per 100,000). In women, they would correspond to Norway (39.3 per 100,000), Hungary (36.8 per 100,000), and Denmark (36.6 per 100,000). The lowest incidence rates, on the other hand, would correspond to Albania (9.6 per 100,000 in males and 7.2 per 100,000 in females), Montenegro (22.7 per 100,000 in males and 14.9 per 100,000 in females), and Austria (26.3 per 100,000 in males and 16.1 per 100,000 in females)[1,2] (Fig. 1.2).

Data from the GLOBOCAN project show that, in Spain, 37,172 new cases of colorectal cancer were recorded in 2018 (13.7% of the total of 270,363 malignant tumors): 22,744 in men (14.5% of the total of 155,971 tumors) and 14,428 in women (12.6% of the total of 114,392 tumors). It is the cancer with the highest incidence in both sexes, and the second highest in men, behind prostate cancer, and also in women, behind breast cancer (Table 1.3, Fig. 1.3).[1,2]

Mortality

Colorectal cancer is also one of the tumors that causes the highest mortality. Thus data from the GLOBOCAN project (http://globocan.iarc.fr) estimate that in 2018, there were around 880,792 deaths from colorectal cancer worldwide.[1,2] These deaths accounted for 9.22% of deaths caused by malignant tumors in the same year, making it the second most common cause of cancer death globally, after lung cancer. In line with the incidence data, mortality data are higher in men than in women. Thus colorectal cancer mortality in 2018 was 484,224 deaths in men (9.22% of deaths from malignant

TABLE 1.1 Incidence of cancer worldwide in 2018, according to tumor location.

Cancer	Total cases	Total %	AT total	Men cases	Men %	AT men	Women cases	Women %	AT women
Lip, oral cavity	354,864	1.96	4.00	246,420	2.61	5.8	108,444	1.26	2.3
Salivary glands	52,799	0.29	0.59	29,256	0.31	0.69	23,543	0.27	0.51
Oropharynx	92,887	0.51	1.10	74,472	0.79	1.8	18,415	0.21	0.4
Nasopharynx	129,079	0.71	1.50	93,416	0.99	2.2	35,663	0.41	0.82
Hypopharynx	80,608	0.45	0.91	67,496	0.71	1.6	13,112	0.15	0.29
Esophagus	572,034	3.16	6.30	399,699	4.23	9.3	172,335	2.00	3.5
Stomach	1,033,701	5.72	11.10	683,754	7.23	15.7	349,947	4.08	7
Colorectum	1,849,518	10.23	19.70	1,026,215	10.85	23.6	823,303	9.55	16.3
Liver	841,080	4.65	9.30	596,574	6.31	13.9	244,506	2.84	4.9
Gallbladder	219,420	1.21	2.30	97,396	1.03	2.2	122,024	1.42	2.4
Pancreas	458,918	2.54	4.80	243,033	2.57	5.5	215,885	2.50	4
Larynx	177,422	0.98	2.00	154,977	1.64	3.6	22,445	0.26	0.48
Lung	2,093,876	11.58	22.50	1,368,524	14.47	31.5	725,352	8.41	14.6
Melanoma of skin	287,723	1.59	3.10	150,698	1.59	3.5	137,025	1.59	2.9
Mesothelioma	30,443	0.17	0.31	21,662	0.23	0.48	8781	0.10	0.18
Kaposi sarcoma	41,799	0.23	0.50	28,248	0.30	0.68	13,551	0.16	0.33
Breast	2,088,849	11.55	46.30	–	–	–	2,088,849	24.23	46.3
Vulva	44,235	0.24	0.88	–	–	–	44,235	0.51	0.88
Vagina	17,600	0.10	0.37	–	–	–	17,600	0.20	0.37
Cervix uteri	569,847	3.15	13.10	–	–	–	569,847	6.61	13.1
Corpus uteri	382,069	2.11	8.40	–	–	–	382,069	4.43	8.4
Ovary	295,414	1.63	6.60	–	–	–	295,414	3.43	6.6
Penis	34,475	0.19	0.80	34,475	0.36	0.8	–	–	–
Prostate	1,276,106	7.06	29.30	1,276,106	13.49	29.3	–	–	–
Testis	71,105	0.39	1.70	71,105	0.75	1.7	–	–	–
Kidney	403,262	2.23	4.50	254,507	2.69	6	148,755	1.73	3.1
Bladder	549,393	3.04	5.70	424,082	4.48	9.6	125,311	1.45	2.4
Brain, central nervous system	296,851	1.64	3.50	162,534	1.72	3.9	134,317	1.56	3.1
Thyroid	567,233	3.14	6.70	130,889	1.38	3.1	436,344	5.06	10.2
Hodgkin lymphoma	79,990	0.44	0.97	46,559	0.49	1.1	33,431	0.39	0.8
Non-Hodgkin lymphoma	509,590	2.82	5.70	284,713	3.01	6.7	224,877	2.61	4.7
Multiple myeloma	159,985	0.88	1.70	89,897	0.95	2.1	70,088	0.81	1.4
Leukemia	437,033	2.42	5.20	249,454	2.64	6.1	187,579	2.18	4.3
All cancers	18,078,957	100.00	197.90	9,456,418	100.00	218.6	8,622,539	100.00	182.6

%: Percentage of total incident cases of malignant tumors; AT: age-adjusted incidence rate (world population) (per 100,000).
Source: GLOBOCAN 2018, International Agency for Research on Cancer-World Health Organization. Estimated cancer incidence, mortality, prevalence and disability-adjusted life years (DALYs) worldwide in 2018. Global cancer observatory. Accessed 2 December 2019. https://gco.iarc.fr/.

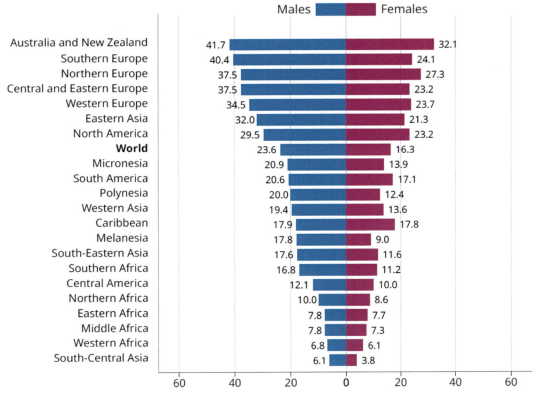

FIG. 1.1 Age-adjusted incidence rates of colorectal cancer worldwide in 2018. *(Source: GLOBOCAN 2018, International Agency for Research on Cancer-World Health Organization. Estimated cancer incidence, mortality, prevalence and disability-adjusted life years (DALYs) worldwide in 2018. Global cancer observatory. Accessed 2 December 2019. https://gco.iarc.fr/.)*

TABLE 1.2 Incidence of cancer in Europe in 2018, according to tumor location.

Cancer	Total cases	Total %	AT total	Men cases	Men %	AT men	Women cases	Women %	AT women
Lip, oral cavity	61,885	1.46	4.40	42,181	1.88	6.8	19,704	0.99	2.30
Salivary glands	9487	0.22	0.66	5480	0.24	0.82	4007	0.20	0.53
Oropharynx	27,974	0.66	2.20	21,414	0.95	3.6	6560	0.33	0.93
Nasopharynx	5019	0.12	0.44	3683	0.16	0.68	1336	0.07	0.22
Hypopharynx	17	0.00	1.30	15,137	0.67	2.5	1840	0.09	0.25
Esophagus	52,964	1.25	3.40	40,667	1.81	6	12,297	0.62	1.30
Stomach	133,133	3.15	8.10	81,611	3.63	11.4	51,522	2.60	5.40
Colorectum	499,667	11.81	30.00	271,600	12.08	37.5	228,067	11.51	24.20
Liver	82,466	1.95	5.10	55,825	2.48	8	26,641	1.34	2.70
Gallbladder	33,716	0.80	1.80	15,229	0.68	2	18,487	0.93	1.70
Pancreas	132,559	3.13	7.70	67,206	2.99	9.3	65,353	3.30	6.30
Larynx	39,875	0.94	2.90	35,245	1.57	5.5	4630	0.23	0.62

Continued

TABLE 1.2 Incidence of cancer in Europe in 2018, according to tumor location—cont'd

Cancer	Total cases	Total %	AT total	Men cases	Men %	AT men	Women cases	Women %	AT women
Lung	470,039	11.11	29.80	311,843	13.87	44.3	158,196	7.98	18.30
Melanoma of skin	144,209	3.41	11.20	71,168	3.17	11.5	73,041	3.68	11.30
Mesothelioma	13,197	0.31	0.72	10,060	0.45	1.2	3137	0.16	0.33
Kaposi sarcoma	2733	0.06	0.19	2102	0.09	0.34	631	0.03	0.06
Breast	522,513	12.35	74.40	–	–	–	522,513	26.36	74.40
Vulva	16,395	0.39	1.70	–	–	–	16,395	0.83	1.70
Vagina	2796	0.07	0.32	–	–	–	2796	0.14	0.32
Cervix uteri	61,072	1.44	11.20	–	–	–	61,072	3.08	11.20
Corpus uteri	121,578	2.87	15.80	–	–	–	121,578	6.13	15.80
Ovary	67,771	1.60	9.50	–	–	–	67,771	3.42	9.50
Penis	6324	0.15	0.89	6324	0.28	0.89	–	–	–
Prostate	449,761	10.63	62.10	449,761	20.01	62.1	–	–	–
Testis	23,987	0.57	6.20	23,987	1.07	6.2	–	–	–
Kidney	136,515	3.23	9.60	84,928	3.78	13.2	51,587	2.60	6.50
Bladder	197,105	4.66	11.30	153,849	6.85	20.2	43,256	2.18	4.30
Brain, central nervous system	64,639	1.53	5.60	35,276	1.57	6.6	29,363	1.48	4.70
Thyroid	78,418	1.85	7.50	18,007	0.80	3.4	60,411	3.05	11.40
Hodgkin lymphoma	19,193	0.45	2.40	10,460	0.47	2.6	8733	0.44	2.10
Non-Hodgkin lymphoma	115,118	2.72	8.10	62,387	2.78	9.8	52,731	2.66	6.60
Multiple myeloma	48,297	1.14	2.90	26,336	1.17	3.6	21,961	1.11	2.30
Leukemia	94,780	2.24	7.40	53,259	2.37	9	41,521	2.09	6.00
All cancers	4,229,662	100.00	281.50	2,247,518	100.00	323.7	1,982,144	100.00	253.30

%: Percentage of total incident cases of malignant tumors; AT: age-adjusted incidence rate (world population) (per 100,000).
Source: GLOBOCAN 2018, International Agency for Research on Cancer-World Health Organization. Estimated cancer incidence, mortality, prevalence and disability-adjusted life years (DALYs) worldwide in 2018. Global cancer observatory. Accessed 2 December 2019. https://gco.iarc.fr/.

tumors) compared to 396,568 deaths in women (9.51% of all cancer deaths), with age-adjusted mortality rates (using the world standard population) of 10.8 per 100,000 and 7.2 per 100,000, respectively (Table 1.4).[1,2]

Although mortality data are geographically more homogeneous than incidence data, there are also some differences. Thus more developed countries have higher mortality rates than less developed countries. The highest mortality rates for both sexes are in Central and Eastern Europe (20.5 per 100,000 for males, 11.9 per 100,000 for females), while the lowest are in South Asia (4.6 and 2.6, respectively) (Fig. 1.4).[1,2]

In Europe, according to data from the GLOBOCAN project, a total of 242,483 deaths from colorectal cancer were recorded in 2018 (129,706 in men and 112,777 in women), representing 12.5% of deaths from malignant tumors recorded that year. Thus, on this continent, colorectal cancer is the second most frequent cause of death from cancer. In men, it remains the second most common cause of cancer death (after lung cancer), and the third most common cause of cancer death in women (after breast and lung cancer). The age-adjusted mortality rate (using the world standard population) was 12.6 per 100,000 (16.2 per 100,000 in males and 9.8 per 100,000 in females) (Table 1.5).[1,2]

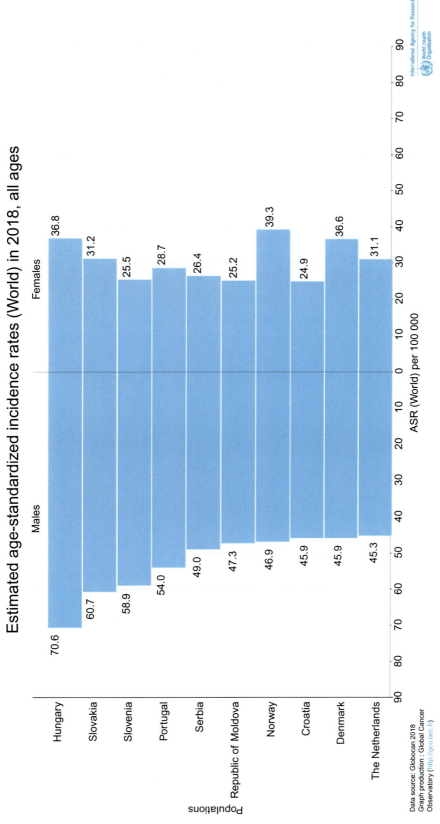

FIG. 1.2 Age-adjusted incidence rates of colorectal cancer in Europe in 2018. (Source: GLOBOCAN 2018, International Agency for Research on Cancer-World Health Organization. *Estimated cancer incidence, mortality, prevalence and disability-adjusted life years (DALYs) worldwide in 2018. Global cancer observatory. Accessed 2 December 2019. https://gco.iarc.fr/.*)

TABLE 1.3 Incidence of cancer in Spain in 2018, according to tumor location.

Cancer	Total cases	Total %	AT total	Men cases	Men %	AT men	Women cases	Women %	AT women
Lip, oral cavity	4526	1.67	4.2	2897	1.86	6.1	1629	1.42	2.5
Salivary glands	635	0.23	0.62	383	0.25	0.8	252	0.22	0.46
Oropharynx	1165	0.43	1.4	972	0.62	2.4	193	0.17	0.43
Nasopharynx	375	0.14	0.54	272	0.17	0.78	103	0.09	0.31
Hypopharynx	854	0.32	1	772	0.49	1.9	82	0.07	0.2
Esophagus	2311	0.85	2.3	1909	1.22	4.2	402	0.35	0.68
Stomach	7684	2.84	6.6	4761	3.05	9.2	2923	2.56	4.3
Colorectum	37,172	13.75	33.4	22,744	14.58	45.2	14,428	12.61	23.3
Liver	6630	2.45	6.5	4976	3.19	10.9	1654	1.45	2.4
Gallbladder	2740	1.01	2	1512	0.97	2.6	1228	1.07	1.5
Pancreas	7765	2.87	6.6	3988	2.56	7.9	3777	3.30	5.4
Larynx	2689	0.99	2.9	2416	1.55	5.4	273	0.24	0.63
Lung	27,351	10.12	27	20,437	13.10	42.1	6914	6.04	14
Melanoma of skin	5319	1.97	6.4	2228	1.43	5.5	3091	2.70	7.2
Mesothelioma	535	0.20	0.47	412	0.26	0.77	123	0.11	0.22
Kaposi sarcoma	300	0.11	0.34	241	0.15	0.61	59	0.05	0.08
Breast	32,825	12.14	75.4				32,825	28.70	75.4
Vulva	1155	0.43	2.1				1155	1.01	2.1
Vagina	150	0.06	0.24				150	0.13	0.24
Cervix uteri	1942	0.72	5.2				1942	1.70	5.2
Corpus uteri	6784	2.51	13.7				6784	5.93	13.7
Ovary	3427	1.27	7.5				3427	3.00	7.5
Penis	543	0.20	1	543	0.35	1		0.00	
Prostate	31,728	11.74	73.1	31,728	20.34	73.1		0.00	
Testis	1110	0.41	5.1	1110	0.71	5.1		0.00	
Kidney	8075	2.99	8.7	5773	3.70	13.2	2302	2.01	4.5
Bladder	18,268	6.76	15.6	14,793	9.48	27.5	3475	3.04	5.6
Brain, central nervous system	4281	1.58	5.6	2308	1.48	6.5	1973	1.72	4.7
Thyroid	4801	1.78	7.1	1037	0.66	3	3764	3.29	11.2
Hodgkin lymphoma	954	0.35	2	488	0.31	2.1	466	0.41	2
Non-Hodgkin lymphoma	7811	2.89	8.7	4060	2.60	9.8	3751	3.28	7.6
Multiple myeloma	3261	1.21	2.8	1855	1.19	3.5	1406	1.23	2.2
Leukemia	5839	2.16	6.9	3364	2.16	8.4	2475	2.16	5.5
All cancers	270,363	100.00	272.3	155,971	100.00	328.6	114,392	100.00	227.1

%: Percentage of total incident cases of malignant tumors; AT: age-adjusted incidence rate (world population) (per 100,000).
Source: GLOBOCAN 2018, International Agency for Research on Cancer-World Health Organization. Estimated cancer incidence, mortality, prevalence and disability-adjusted life years (DALYs) worldwide in 2018. Global cancer observatory. Accessed 2 December 2019. https://gco.iarc.fr/.

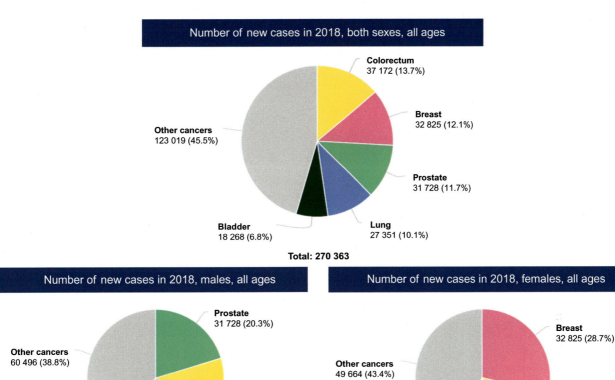

FIG. 1.3 Cases and incidence rates of colorectal cancer in Spain in 2018. *(Source: GLOBOCAN 2018, International Agency for Research on Cancer-World Health Organization. Estimated cancer incidence, mortality, prevalence and disability-adjusted life years (DALYs) worldwide in 2018. Global cancer observatory. Accessed 2 December 2019. https://gco.iarc.fr/.)*

TABLE 1.4 Cancer mortality worldwide in the year 2018, according to tumor location.

Cancer	Total cases	Total %	AT total	Men cases	Men %	AT men	Women cases	Women %	AT women
Lip, oral cavity	177,384	1.86	2.00	119,693	2.22	2.80	57,691	1.38	1.20
Salivary glands	22,176	0.23	0.24	13,440	0.25	0.31	8736	0.21	0.18
Oropharynx	51,005	0.53	0.57	42,116	0.78	0.99	8889	0.21	0.19
Nasopharynx	72,987	0.76	0.84	54,280	1.01	1.30	18,707	0.45	0.41
Hypopharynx	34,984	0.37	0.39	29,415	0.55	0.69	5569	0.13	0.12
Esophagus	508,585	5.32	5.50	357,190	6.63	8.30	151,395	3.63	3.00
Stomach	782,685	8.19	8.20	513,555	9.54	11.70	269,130	6.45	5.20
Colorectum	880,792	9.22	8.90	484,224	8.99	10.80	396,568	9.51	7.20
Liver	781,631	8.18	8.50	548,375	10.18	12.70	233,256	5.59	4.60
Gallbladder	165,087	1.73	1.70	70,168	1.30	1.60	94,919	2.28	1.80

Continued

TABLE 1.4 Cancer mortality worldwide in the year 2018, according to tumor location—cont'd

Cancer	Total cases	Total %	AT total	Men cases	Men %	AT men	Women cases	Women %	AT women
Pancreas	432,242	4.52	4.40	226,910	4.21	5.10	205,332	4.92	3.80
Larynx	94,771	0.99	1.00	81,806	1.52	1.90	12,965	0.31	0.27
Lung	1,761,007	18.43	18.60	1,184,947	22.00	27.10	576,060	13.82	11.20
Melanoma of skin	60,712	0.64	0.63	34,831	0.65	0.78	25,881	0.62	0.50
Mesothelioma	25,576	0.27	0.26	18,332	0.34	0.40	7244	0.17	0.14
Kaposi sarcoma	19,902	0.21	0.24	13,117	0.24	0.32	6785	0.16	0.16
Breast	626,679	6.56	13.00	–	–	–	626,679	15.03	13.00
Vulva	15,222	0.16	0.27	–	–	–	15,222	0.37	0.27
Vagina	8062	0.08	0.16	–	–	–	8062	0.19	0.16
Cervix uteri	311,365	3.26	6.90	–	–	–	311,365	7.47	6.90
Corpus uteri	89,929	0.94	1.80	–	–	–	89,929	2.16	1.80
Ovary	184,799	1.93	3.90	–	–	–	184,799	4.43	3.90
Penis	15,138	0.16	0.35	15,138	0.28	0.35	–	–	–
Prostate	358,989	3.76	7.60	358,989	6.67	7.60	–	–	–
Testis	9507	0.10	0.23	9507	0.18	0.23	–	–	–
Kidney	175,098	1.83	1.80	113,822	2.11	2.60	61,276	1.47	1.10
Bladder	199,922	2.09	1.90	148,270	2.75	3.20	51,652	1.24	0.87
Brain, central nervous system	241,037	2.52	2.80	135,843	2.52	3.20	105,194	2.52	2.30
Thyroid	41,071	0.43	0.42	15,557	0.29	0.35	25,514	0.61	0.49
Hodgkin lymphoma	26,167	0.27	0.30	15,770	0.29	0.37	10,397	0.25	0.22
Non-Hodgkin lymphoma	248,724	2.60	2.60	145,969	2.71	3.30	102,755	2.46	2.00
Multiple myeloma	106,105	1.11	1.10	58,825	1.09	1.30	47,280	1.13	0.89
Leukemia	309,006	3.23	3.50	179,518	3.33	4.20	129,488	3.11	2.80
All cancers	9,555,027	100.00	101.10	5,385,640	100.00	122.70	4,169,387	100.00	83.10

%: Percentage of total incident cases of malignant tumors; AT: age-adjusted incidence rate (world population) (per 100,000).
Source: GLOBOCAN 2018, International Agency for Research on Cancer-World Health Organization. Estimated cancer incidence, mortality, prevalence and disability-adjusted life years (DALYs) worldwide in 2018. Global cancer observatory. Accessed 2 December 2019. https://gco.iarc.fr/.

In Spain, GLOBOCAN data put the number of deaths from colorectal cancer in 2018 at 16,683: 10,038 in men (15.6% of all cancer deaths) and 6645 in women (14.9% of the total). The overall mortality rate, standardized by world population, was 12.0 per 100,000 (18.1 per 100,000 in men and 9.5 per 100,000 in women). These figures place Spain above the European average for colorectal cancer mortality in men and below the average for mortality in women (Table 1.6).[1,2]

Regarding the situation in Galicia, according to data published by the Dirección Xeral de Saúde Pública—Conselleria de Sanidade—Xunta de Galicia in the SIMCA (Sistema de Información sobre Mortalidade por Cancro en Galicia, http://saudepublica.melisa.gal/),[3] we can observe that for the year 2017, the world population standardized rate for colon cancer was 8.5 per 100,000 (11.7 in men and 5.8 in women) and 3.6 per 100,000 (5.6 in men and 1.9 in women) for rectal cancer.

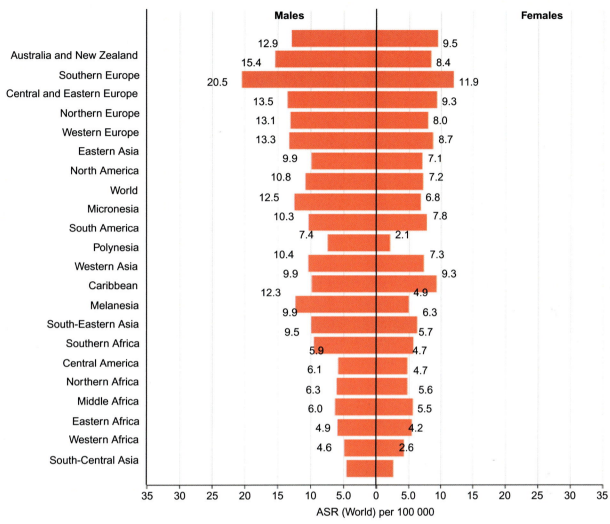

FIG. 1.4 Adjusted mortality rates for colorectal cancer worldwide in 2018. *(Source: GLOBOCAN 2018, International Agency for Research on Cancer-World Health Organization. Estimated cancer incidence, mortality, prevalence and disability-adjusted life years (DALYs) worldwide in 2018. Global cancer observatory. Accessed 2 December 2019. https://gco.iarc.fr/.)*

TABLE 1.5 Cancer mortality in Europe in 2018, according to tumor location.

Cancer	Total cases	Total %	AT total	Men cases	Men %	AT men	Women cases	Women %	AT women
Lip, oral cavity	24,063	1.24	1.70	17,620	1.62	2.80	6443	0.75	0.67
Salivary glands	4035	0.21	0.23	2606	0.24	0.35	1429	0.17	0.13
Oropharynx	12,991	0.67	0.96	10,708	0.99	1.70	2283	0.27	0.28
Nasopharynx	2568	0.13	0.19	1883	0.17	0.31	685	0.08	0.09
Hypopharynx	9502	0.49	0.70	8568	0.79	1.40	934	0.11	0.12
Esophagus	45,061	2.32	2.80	34,932	3.22	5.00	10,129	1.18	0.94
Stomach	102,167	5.26	5.90	61,880	5.70	8.40	40,287	4.70	3.90
Colorectum	242,483	12.48	12.60	129,706	11.95	16.20	112,777	13.15	9.80
Liver	77,375	3.98	4.40	50,365	4.64	6.80	27,010	3.15	2.40

Continued

TABLE 1.5 Cancer mortality in Europe in 2018, according to tumor location—cont'd

Cancer	Total cases	Total %	AT total	Men cases	Men %	AT men	Women cases	Women %	AT women
Gallbladder	24,862	1.28	1.30	10,428	0.96	1.30	14,434	1.68	1.30
Pancreas	128,045	6.59	7.20	65,016	5.99	8.80	63,029	7.35	5.70
Larynx	19,577	1.01	1.30	17,629	1.62	2.60	1948	0.23	0.22
Lung	387,913	19.96	23.50	267,316	24.62	36.80	120,597	14.06	13.00
Melanoma of skin	27,147	1.40	1.70	15,243	1.40	2.20	11,904	1.39	1.30
Mesothelioma	11,953	0.62	0.61	9235	0.85	1.10	2718	0.32	0.26
Kaposi sarcoma	447	0.02	0.02	295	0.03	0.04	152	0.02	0.01
Breast	137,707	7.09	14.90	–	–	–	137,707	16.05	14.90
Vulva	6258	0.32	0.51	–	–	–	6258	0.73	0.51
Vagina	1298	0.07	0.12	–	–	–	1298	0.15	0.12
Cervix uteri	25,829	1.33	3.80	–	–	–	25,829	3.01	3.80
Corpus uteri	29,628	1.52	2.90	–	–	–	29,638	3.45	2.90
Ovary	44,576	2.29	5.10	–	–	–	44,576	5.20	5.10
Penis	1823	0.09	0.25	1823	0.17	0.25	–	–	–
Prostate	107,315	5.52	11.30	107,315	9.89	11.30	–	–	–
Testis	1646	0.08	0.36	1646	0.15	0.36	–	–	–
Kidney	54,709	2.82	3.10	35,092	3.23	4.70	19,617	2.29	1.80
Bladder	64,966	3.34	3.00	49,309	4.54	5.60	15,657	1.83	1.20
Brain, central nervous system	53,027	2.73	4.00	29,150	2.69	4.90	23,877	2.78	3.20
Thyroid	6988	0.36	0.38	2784	0.26	0.38	4204	0.49	0.38
Hodgkin lymphoma	4307	0.22	0.33	2422	0.22	0.42	1885	0.22	0.26
Non-Hodgkin lymphoma	48,096	2.47	2.60	26,304	2.42	3.40	21,792	2.54	2.00
Multiple myeloma	30,860	1.59	1.60	16,114	1.48	2.00	14,746	1.72	1.30
Leukemia	61,476	3.16	3.60	34,031	3.13	4.60	27,445	3.20	2.80
All cancers	1,943,478	100.00	111.30	1,085,592	100.00	144.00	857,886	100.00	86.70

%: Percentage of total incident cases of malignant tumors; AT: age-adjusted incidence rate (world population) (per 100,000).
Source: GLOBOCAN 2018, International Agency for Research on Cancer-World Health Organization. Estimated cancer incidence, mortality, prevalence and disability-adjusted life years (DALYs) worldwide in 2018. Global cancer observatory. Accessed 2 December 2019. https://gco.iarc.fr/.

TABLE 1.6 Cancer mortality in Spain in 2018, according to tumor location.

Cancer	Total cases	Total %	AT total	Men cases	Men %	AT men	Women cases	Women %	AT women
Lip, oral cavity	1211	1.07	1.10	771	1.12	1.60	440	0.99	0.60
Salivary glands	260	0.23	0.18	178	0.26	0.29	82	0.18	0.10
Oropharynx	601	0.53	0.67	514	0.75	1.20	87	0.19	0.17
Nasopharynx	183	0.16	0.19	126	0.18	0.29	57	0.13	0.10

TABLE 1.6 Cancer mortality in Spain in 2018, according to tumor location—cont'd

Cancer	Total cases	Total %	AT total	Men cases	Men %	AT men	Women cases	Women %	AT women
Hypopharynx	322	0.28	0.36	298	0.43	0.69	24	0.05	0.05
Esophagus	2026	1.78	1.90	1696	2.46	3.50	330	0.74	0.49
Stomach	5609	4.94	4.30	3413	4.95	6.10	2196	4.92	2.80
Colorectum	16,683	14.69	12.00	10,038	14.56	16.80	6645	14.88	8.00
Liver	5569	4.90	4.70	3872	5.62	7.60	1697	3.80	2.00
Gallbladder	1476	1.30	1.00	740	1.07	1.20	736	1.65	0.83
Pancreas	7279	6.41	5.90	3708	5.38	7.10	3571	8.00	4.80
Larynx	1273	1.12	1.10	1176	1.71	2.20	97	0.22	0.18
Lung	22,896	20.16	21.20	17,559	25.48	34.40	5337	11.95	10.00
Melanoma of skin	1171	1.03	1.10	636	0.92	1.30	535	1.20	0.88
Mesothelioma	479	0.42	0.39	370	0.54	0.65	109	0.24	0.18
Kaposi sarcoma	39	0.03	0.03	28	0.04	0.06	11	0.02	0.01
Breast	6421	5.65	10.60	–	–	–	6421	14.38	10.60
Vulva	356	0.31	0.36	–	–	–	356	0.80	0.36
Vagina	63	0.06	0.07	–	–	–	63	0.14	0.07
Cervix uteri	825	0.73	1.70	–	–	–	825	1.85	1.70
Corpus uteri	1660	1.46	2.20	–	–	–	1660	3.72	2.20
Ovary	2123	1.87	3.70	–	–	–	2123	4.75	3.70
Penis	143	0.13	0.26	143	0.21	0.26	–	–	–
Prostate	5793	5.10	7.40	5793	8.41	7.40	–	–	–
Testis	50	0.04	0.17	50	0.07	0.17	–	–	–
Kidney	2861	2.52	2.40	2001	2.90	3.70	860	1.93	1.20
Bladder	5680	5.00	3.50	4576	6.64	6.60	1104	2.47	1.10
Brain, central nervous system	3211	2.83	s3.50	1760	2.55	4.20	1451	3.25	2.90
Thyroid	370	0.33	0.30	155	0.22	0.30	215	0.48	0.29
Hodgkin lymphoma	203	0.18	0.21	108	0.16	0.26	95	0.21	0.17
Non-Hodgkin lymphoma	3044	2.68	2.40	1649	2.39	3.00	1395	3.12	1.90
Multiple myeloma	2114	1.86	1.40	1128	1.64	1.80	986	2.21	1.20
Leukemia	3884	3.42	3.20	2199	3.19	4.10	1685	3.77	2.40
All cancers	113,584	100.00	92.30	68,919	100.00	124.80	44,665	100.00	65.90

%: Percentage of total incident cases of malignant tumors; AT: age-adjusted incidence rate (world population) (per 100,000).
Source: GLOBOCAN 2018, International Agency for Research on Cancer-World Health Organization. Estimated cancer incidence, mortality, prevalence and disability-adjusted life years (DALYs) worldwide in 2018. Global cancer observatory. Accessed 2 December 2019. https://gco.iarc.fr/.

FIG. 1.5 Evolution of the global population-adjusted rate of colon cancer in Galicia 1980–2017. *(Source: SIMCA, Department of Health-Xunta de Galicia. Sistema de Información sobre Mortalidade por Cancro en Galicia. Sistema de Información sobre Mortalidade por Cancro en Galicia. Accessed 2 December 2019, 2019. http://saudepublica.melisa.gal/.)*

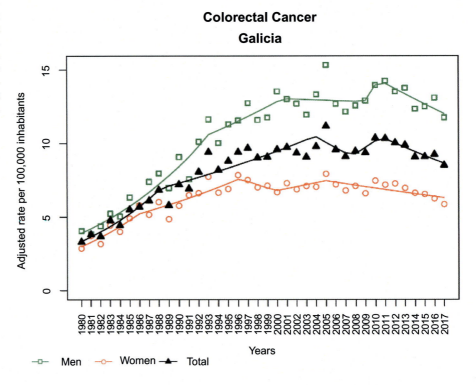

FIG. 1.6 Evolution of the global population-adjusted rate of rectal cancer in Galicia 1980–2017. *(Source: SIMCA, Department of Health-Xunta de Galicia. Sistema de Información sobre Mortalidade por Cancro en Galicia. Sistema de Información sobre Mortalidade por Cancro en Galicia. Accessed 2 December 2019, 2019. http://saudepublica.melisa.gal/.)*

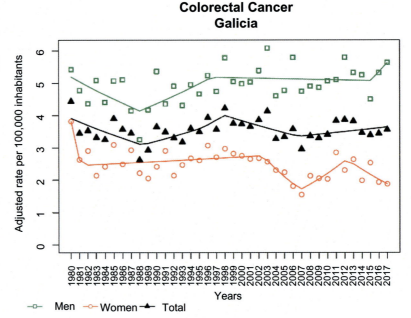

If we analyze the evolution in Galicia since 1980, we can observe the current downward trend in mortality from colon cancer in both men and women (Fig. 1.5), while there is an increase in men and a decrease in women for rectal cancer (Fig. 1.6).

Finally, regarding future projections, a study published by Araghi M et al. in 2019[4] predicts the trend in mortality from colon and rectal cancer until 2035 by adjusting age–period–cohort models using mortality data from the World Health Organization (WHO) for the period 1989–2016. This study concludes that mortality rates from colon cancer will decrease in 32 of the 42 countries studied, most of which are located in Asia, Europe, North America, and Oceania. In Spain, the decrease in colon cancer mortality rate is estimated at −12.4%. In Latin American and Caribbean countries, no such downward trend in colon cancer mortality rates is observed. Rectal cancer rates follow a pattern similar to colon cancer, although significant increases are seen in Costa Rica, Australia, the United States, Ireland, and Canada. In Spain, the rate is expected to increase by 10.0%.[4]

On average, between 2013 and 2035, mortality rates for colon and rectal cancer are predicted to decrease only slightly, by 8.4% and 5.1%, respectively, from 7.1 per 100,000 in 2013 to 6.5 per 100,000 in 2035 for colon cancer, and from 3.9 to 3.7 per 100,000 for rectal cancer.[4]

Discussion

The GLOBOCAN database, compiled by IARC (International Agency for Research on Cancer), is updated regularly, providing estimates of national cancer incidence and mortality. The incidence data used by GLOBOCAN 2018 are collected from population-based cancer registries, which are usually national, although in some cases, such as in developing countries, they may cover smaller areas such as urban areas. Mortality statistics are collected and published by WHO. Virtually all countries in Europe and the Americas have death registration systems, although this is not the case in African and Asian countries, where such systems do not exist in a large number of countries.

We estimate that there were 18.1 million new cases of cancer and 9.6 million cancer deaths globally in 2018. There is approximately a 20% risk of getting cancer before the age of 75 and a 10% chance of dying from cancer.[1,2]

Colorectal cancer is the third most common cancer worldwide and the fourth most common cause of cancer death. Having concrete and up-to-date data and being able to make predictions gives us the ability to plan future health actions and raise awareness of the need to control the action of this type of disease.

Several studies have shown that colorectal cancer mortality rates have been declining steadily for at least two decades in many high-income countries in North America, Oceania, and Northern and Western Europe. On the other hand, an increase in mortality rates is observed in less developed countries (countries in Asia, Africa, and Latin America), where rates have been historically low.

Epidemiological analysis of colorectal cancer incidence and mortality data can provide valuable information for the organization of health services in an increasingly aging population with a longer life expectancy. Preventive measures that can reduce the impact of colorectal cancer as a health problem will include the promotion of physical exercise, an adequate body mass index, a healthy diet, and reduced alcohol and tobacco consumption.[5–8] Countries with more resources may also benefit from the implementation of screening programs to detect precancerous pathology and tumors at earlier stages of the disease.[9]

Predicting future developments in colorectal cancer leads us to assume that although a global decline in colorectal cancer mortality rates is observed, the number of deaths is expected to increase due to population growth and aging. Reductions in mortality rates will likely be due to improved early detection services and specialized care accessibility.

References

1. International Agency for Research on Cancer-World Health Organization. Estimated cancer incidence, mortality, prevalence and disability-adjusted life years (DALYs) worldwide. In: *Global cancer observatory*; 2018. https://gco.iarc.fr/. [Accessed 2 December 2019].
2. Ferlay J, Colombet M, Soerjomataram I, et al. Estimating the global cancer incidence and mortality in 2018: GLOBOCAN sources and methods. *Int J Cancer* 2019;**144**(8):1941–53. https://doi.org/10.1002/ijc.31937.
3. Department of Health-Xunta de Galicia. *Sistema de Información sobre Mortalidade por Cancro en Galicia. Sistema de Información sobre Mortalidade por Cancro en Galicia*; 2019 http://saudepublica.melisa.gal/. [Accessed 2 December 2019].
4. Araghi M, Soerjomataram I, Jenkins M, et al. Global trends in colorectal cancer mortality: projections to the year 2035. *Int J Cancer* 2019;**144**(12):2992–3000. https://doi.org/10.1002/ijc.32055.
5. Martínez ME. Primary prevention of colorectal cancer: lifestyle, nutrition, exercise. *Recent Results Cancer Res* 2005;**166**:177–211. https://doi.org/10.1007/3-540-26980-0_13.
6. Ahmed FE. Effect of diet, life style, and other environmental/chemopreventive factors on colorectal cancer development, and assessment of the risks. *J Environ Sci Health C Environ Carcinog Ecotoxicol Rev* 2004;**22**(2):91–148.
7. Chan AT, Giovannucci EL. Primary prevention of colorectal cancer. *Gastroenterology* 2010;**138**(6):2029–2043.e10. https://doi.org/10.1053/j.gastro.2010.01.057.
8. Coyle YM. Lifestyle, genes, and cancer. *Methods Mol Biol* 2009;**472**:25–56. https://doi.org/10.1007/978-1-60327-492-0_2.
9. Lieberman D. Colorectal cancer screening: practice guidelines. *Dig Dis* 2012;**30**(Suppl. 2):34–8. https://doi.org/10.1159/000341891.

Chapter 2

Prognosis and follow-up of CRC patients: Role of diagnostic and therapeutic delay

Sonia Pértega Díaz[a], Vanesa Balboa Barreiro[a], María Teresa Seoane Pillado[b], Mohammed Alhayek-Aí[c], Paloma González Santamaría[d], and Cristina González Martín[b]

[a]Research Support Unit, Institute for Biomedical Research of A Coruña (INIBIC), University Hospital Complex of A Coruña (CHUAC), SERGAS, University of A Coruña, A Coruña, Spain, [b]Department of Health Sciences, Faculty of Nursing and Podiatry, Universidade da Coruña, Ferrol, Spain, [c]Ophthalmology Department, Hospital de la Santa Creu i Sant Pau, Barcelona, Spain, [d]Meicende Health Centre, SERGAS, A Coruña, Spain

Survival and prognosis

Recent advances in the diagnosis and treatment of cancer patients have led to an overall increase in cancer patient survival. According to the latest report from the American Cancer Society (ACS), the 5-year relative survival after diagnosis for all tumors is around 65%.[1] Survival in cancer patients has thus improved substantially since the 1970s for all major sites except cervical and endometrial cancer.[2]

In Europe, EUROCARE-5 data show that cancer patient survival 5 years after diagnosis exceeds 50% for most tumor sites.[3] Thus about one-third of cancer patients have a 5-year survival probability of more than 80%, while a quarter of patients would have an estimated survival of less than 30%. The results for Spain in this report are similar to the European average for the main tumor sites, with 5-year survivals exceeding 80% for prostate (84.7%), nonmelanoma skin (84.6%), and breast cancers (82.8%).

As with most tumors, the life expectancy of patients with colorectal cancer has improved substantially in recent years. According to ACS data, the 5-year survival probability of patients diagnosed with colorectal cancer ranges from 14% to 90%, depending on the stage at diagnosis, and stands at 65% overall.[1] According to EUROCARE-5 data, the age-standardized 5-year survival rate for colon cancer is 57% and stands at 55.8% for rectal tumors. These overall figures at the European level are similar to those recorded for Spain in the same report (57.1% for colon cancer and 56.4% for rectal cancer). The 5-year survival figures thus increased from 54.2% in 1999–2001 to 58.1% in 2005–2007 for colon cancer, while rising from 52.1% to 57.6% for rectal cancer over the same period.[3]

This improvement in survival rates is undoubtedly linked to improvements in the early diagnosis of the disease and in its treatment, especially in terms of surgical techniques and adjuvant treatment of the disease. It should be noted, however, that this trend in survival figures may partly reflect the so-called lead time bias due to the implementation of screening programs for early detection of the disease.[1]

Pathological findings after potentially curative surgery are the most important information in determining the prognosis of the disease. Thus stage remains the prognostic factor with the greatest impact. In patients who receive neoadjuvant treatment prior to surgery, the posttreatment stage has even greater prognostic value.[4]

According to ACS data, 39% of patients with colorectal cancer are diagnosed at a local stage, with an estimated 5-year relative survival of 90%. These survival figures drop to 71% and 14% when patients are diagnosed at a regional or distant stage, respectively.[5]

Data from Eurocare-5, the first of its reports to make stage at diagnosis data available from several of the participating registries, provide figures that differ from previous ones, with only 13% of colon tumors and 19% of rectal tumors diagnosed at the local stage. In this case, the estimated 5-year relative survival decreases from 92.3% in local colon tumors to 12.0% in metastatic tumors for patients under the age of 70 at diagnosis, and from 87.6% to 6.2% among patients age 70 and older. In rectal tumors, these figures are 89.7% vs. 10.3% in those younger than 70, and 83.8% vs. 5.3% among those age 70 or older.[6]

In addition to stage, an increasing number of prognostic factors are being identified that may influence not only survival but also therapeutic decisions. Other important prognostic variables include the presence of extramural satellite nodules, lymphovascular and perineural invasion, histological grade, preoperative CEA values, microsatellite instability, and mutations in RAS and BRAF, among others.[4]

Formally, the terms "cured" or "survivors" are often used to refer to patients who have overcome a disease. In the case of oncological processes such as colorectal cancer, although controversial, a cancer patient is generally considered "cured" when he or she survives five years after diagnosis,[7] and these patients are often referred to as "great survivors" or "long-term survivors."

Despite the considerable percentage of colorectal cancer patients surviving more than 5 years after diagnosis, there are not many prospective studies assessing the recurrence rate or mortality in long-term survivors of colorectal cancer.[8,9] Some authors suggest that this "cure" is not guaranteed, estimating that among node-positive patients, more than one-third of deaths occurring after the first 5 years can be directly attributed to colorectal cancer, with this figure being 63% in cases with metastases.[8]

Although there are very few publications in this regard, some authors have highlighted the usefulness of cure models as an analysis strategy that can be incorporated into survival analysis, providing very useful information when quantifying colorectal cancer survival improvement figures.[10] Such models would allow a better estimation of the "cure" rate and analysis of the differences between cancer patients who are major survivors ("cured" patients) and those who are not. They also make it possible to identify covariates differently associated with short- or long-term prognosis. Studies to date generally show cure rates of around 50% of patients and a median survival of around 1 year of life after diagnosis for those not cured.[10–13] All of these papers have been based on population-based registries and do not incorporate clinical information other than age at diagnosis, gender, and disease stage, therefore the impact of other covariates on cure rate or survival time has not been explored through these models.

Follow-up strategies

Intensive follow-up programs after colorectal cancer surgery have been proposed, with the aim that early detection of asymptomatic recurrences would increase the proportion of patients potentially eligible for curative therapy. A more comprehensive follow-up strategy would also allow the detection and treatment of metachronous tumors in early stages, treat side effects from both surgery and adjuvant treatments, and monitor the radicality of surgical treatment and outcomes.[14]

There is great variability both in clinical practice and among the proposals of different scientific associations regarding the follow-up, frequency, and type of tests that should be performed after curative surgery for colorectal cancer.[15–19] There is also controversy about the impact that different follow-up strategies may have on the prognosis of these patients.[20,21] While some clinical trials[22,23] have demonstrated a significant reduction in the overall mortality of patients undergoing a more intensive follow-up strategy, others do not allow such an association[24–32] to be concluded.

The surveillance strategies proposed range from patient education to detect symptoms that may be an early indication of tumor recurrence, to periodic physical examinations, serological examinations (CEA determinations, liver function tests, hemogram, fecal occult blood in stool), colonoscopies or imaging tests (chest X-ray, ultrasound, CT).[15,16,33–35] Table 2.1 presents some of the follow-up strategies established by different medical institutions in the care of these patients.

A recent meta-analysis of data from 11 clinical trials concluded that a more intensive follow-up strategy after curative surgery for nonmetastatic colorectal cancer is associated with a significant improvement in the overall survival of these patients (hazard ratio HR = 0.75 [95% CI = 0.66–0.86]).[14,36] Similarly, it demonstrates that such follow-up is associated with increased detection of asymptomatic recurrences, increased likelihood of curative surgery for these recurrences, earlier diagnosis of recurrences, and increased survival after recurrence. In contrast, more intensive follow-up could not be associated with decreased tumor-specific mortality, nor with the detection of a higher number of local recurrences, partly due to the small number of studies that analyzed these outcome variables.[14,36]

This same study has shown that colonoscopies, imaging tests, CEA determinations, and regular medical visits are associated with an improvement in overall mortality[14,36] in relation to the diagnostic tests performed in the follow-up of patients undergoing CRC surgery. It stands to reason that any of these tests may allow earlier detection of recurrence and earlier intervention in these patients. However, the heterogeneity in the different follow-up schemes proposed, with very different combinations of tests to be performed and cadence of follow-up, makes it difficult to conclude an ideal follow-up scheme for these patients.

Finally, in terms of duration, standard follow-up strategies, including imaging, colonoscopy, and tumor marker testing, are usually performed periodically for 5 years after diagnosis.[24] Other authors, however, suggest that follow-up strategies should be expanded and modified according to the different risk factors presented by each patient.[8,37]

TABLE 2.1 Follow-up strategies after colorectal cancer surgery proposed by different institutions and medical associations.

Organization	Medical history and physical examination	CEA	CT	Endoscopy
ASCO and CCO	Every 3–6 months for 5 years	Every 3–6 months for 5 years	Abdomen and chest: annually for 3 years Pelvic: in case of rectal cancer, annually for the first 3–5 years	Colonoscopy 1 per year
NCCN	Every 3–6 months for 2 years	Every 3–6 months for 2 years, then every 6 months for 3 years	Abdomen/pelvis and chest: annually for 5 years Abdomen/pelvis and chest: every 3–6 months for 2 years, then every 6 months for another 3 years	Colonoscopy 1 per year
ESMO colon cancer	Every 3–6 months for 3 years	Every 3–6 months for 3 years, then every 6–12 months for 2 years	Abdomen and chest: every 6–12 months for 3 years	Colonoscopy 1 per year; every 3–5 years thereafter
ESMO rectal cancer	Every 6 months for 2 years	Not recommended	Not recommended	Colonoscopy every 5 years
New Zealand	Clinical assessment according to the risk of recurrence High-risk cancer: stage IIB or III. Every 6 to 12 months for 3 years, then once every 2 years thereafter. Low risk: stage I or IIA or with comorbidities that limit surgery. Once every 5 years or when symptoms occur.	For stage II and III: every 3–6 months for 3 years, once every 2 years thereafter	All individuals with stage I-III should be between 1 and 3 years old.	Colonoscopy 1 per year, every 6–12 months for 3 years for stages IIB and III, then annually for 5 years thereafter
British Columbia Medical Association	Every 3–6 months for 2 years	Every 3–6 months for 3 years, then every 6 months for 2 years	Every 6 months for 3 years, then annually for 2 years	Colonoscopy 1 per year, repeat 3 years later and, if normal, every 5 years

Source: Alhayek M. Systematic review and meta-analysis of follow-up strategies in patients operated on for colorectal cancer. [PhD thesis]. A Coruña, Spain: University of A Coruña. Published online, 2015.

Quality of life and long-term sequelae

Cancer survivors, despite overcoming the disease, may nevertheless suffer from other physical or psychosocial health problems, sometimes with a significant deterioration in their quality of life that may persist even decades after the initial treatment.[38–40] The reported improvement in survival figures for most cancer processes has led to increasing attention being paid to measuring the health-related quality of life of tumor disease survivors.[38, 41]

However, there are not much data evaluating long-term quality of life in major colorectal cancer survivors, with the most frequent studies being in the first 5 years after diagnosis.[42] A recent study assessed the prevalence of symptoms and health-related quality of life in a cohort of patients diagnosed with CRC 10 years after diagnosis. Their results suggest that these patients have a similar quality of life to the general population, although aspects such as bowel dysfunction contribute to deterioration even 15 years after diagnosis.[43]

These patients also have a higher prevalence of psychological disorders, particularly depressive symptoms.[44, 45] Thus, after the diagnosis of CRC and its treatment, patients are exposed to different problems that affect their daily life and quality of life. Complications arising from surgery, toxicity, and long-term effects of cancer treatment (including chemotherapy-induced peripheral neuropathy, postradiation proctitis, or fractures resulting from radiotherapy), comorbidities associated with progressive aging (some of which are more frequent in patients with CRC, such as diabetes and cardiovascular disease), bowel and anorectal problems, urinary symptoms, sexual dysfunction, and chronic fatigue are all determinants of the quality of life and psychological well-being of these patients.[46] Ostomy wearers also have to cope with all the economic, social, and occupational disadvantages and psychological effects.[47]

It is therefore important to incorporate aspects related to patients' psychosocial well-being and quality of life into the follow-up strategies proposed for them, as some associations are already doing.[48, 49] Other authors have shown that psychological therapies can contribute to improving prognosis in some types of tumors,[50, 51] as well as optimizing the quality of life of these patients.[52]

Impact of diagnostic and therapeutic delay on stage and prognosis

Colorectal cancer is usually diagnosed by clinical manifestations, as a result of a screening program or as a chance finding. There is a variable time interval between the onset of the disease and its diagnosis or treatment, known as a delay. This delay can be affected by the characteristics of the disease, the patient, or the healthcare system, and a general distinction is made between diagnostic delay and therapeutic delay.

There is great variability in the delay figures published in the literature.[53-57] Studies in Spain show a median delay time from the onset of symptoms attributable to colorectal cancer until diagnosis ranging from 49 days[58] to 128 days,[59, 60] with the median delay from the onset of symptoms to the start of treatment estimated at 155 days.[59]

Although there are different schemes to analyze the delay in cancer diagnosis and treatment,[61] the following time intervals can generally be considered[59, 62]:

(a) Patient Interval (PI): from the onset of the first symptoms to the first consultation with the primary care physician or other specialist.
(b) Symptom Diagnosis Interval (SDI): from the onset of the first symptoms to the pathological diagnosis of the tumor.
(c) Symptom Treatment Interval (STI): from the onset of first symptoms to the start of treatment.
(d) Health Services Interval (HSI): interval from first contact with a physician until diagnosis.
(e) Treatment interval (TI): interval from the date of diagnosis to the date of treatment is initiated.

Previous studies have shown that patient-attributable delay is generally higher than delay times attributable to the functioning of the healthcare system.[63] In a study of 1785 incident cases of colorectal cancer diagnosed in 22 Spanish hospitals, Zarcos-Pedrinaci et al.[64] found prolonged patient delay (>180 days) in up to 12.1% of cases, mainly associated with under-recognition of symptoms such as bowel rhythm disturbance.

Other authors, however, have reported shorter patient delays of around 19 days, compared to a median of 66 days between first contact with a doctor and definitive tumor diagnosis and a median of 22 days between diagnosis and surgical treatment or initiation of cancer treatment.[59] To minimize delays due to the health system, health authorities have implemented the practice of rapid diagnostic pathways[65] with the aim that any patient with suspected colorectal cancer will have a definitive diagnosis in less than 15 days.

It is clear that the variability in these ranges may be secondary to multiple reasons. It is thus difficult to compare the figures published in different studies due to the use of different definitions of delay, discrepancies in the inclusion and exclusion criteria used, as well as the differential characteristics of the patients analyzed and disparities between healthcare models in different countries. For example, poor agreement has been reported in the time intervals calculated from different data sources.[66] On the other hand, there are patient and environmental characteristics that can modify the diagnostic delay.[67] The severity of symptoms and the importance given to them may determine the delay.[68] On the other hand, variability in symptoms and physicians' attitudes toward the indication of diagnostic tests for symptoms that can be attributed to different processes and with low predictive values such as anemia also generate variability in diagnostic intervals.[69-71]

Several factors related to the patient or the healthcare system have been identified as modifiers of delay. Thus delay may be increased by failure to recognize the severity of symptoms, the location of the tumor, low socioeconomic status, errors in diagnosis, and the use of inappropriate tests or tests with previously negative results. Factors associated with shorter delays include associated comorbidity, going directly to the hospital, and the use of referral protocols. There is no clear evidence regarding other factors such as patient gender or age, fear of cancer diagnosis, presence of pain, educational level or family history, nor is there clear evidence for patient frequency or use of rapid access endoscopy.[72-75] A multicenter study conducted in five regions of Spain found that gender, perception of symptoms, and behavioral aspects related to help-seeking or emotional support are the main patient-related aspects influencing diagnostic delay in CRC.[59]

Intuitively, it seems reasonable to think that shortening the delay might allow diagnosis at less advanced stages of the disease and thus improve prognosis. However, the results are contradictory,[76–83] even pointing out that a longer delay is associated with a better prognosis[82] or that delay is not related to either stage or prognosis.[81] It is important to consider that the implementation of healthcare measures that reduce diagnostic delay, such as fast-track or high-resolution circuits, is based on reducing the uncertainty or social alarm suffered by the patient with symptomatic CRC and not so much on reducing mortality or costs which, together with early diagnosis, are objectives of population screening programs, among others.

The results of two reviews conducted to determine the role of delay in the prognosis of patients with colorectal cancer are inconclusive. Although the results of the meta-analysis suggest that longer delay is associated with longer survival, there was no clear certainty on the role of delay in both stage and survival.[84,85] The systematic review of the included studies allowed us to appreciate their main weaknesses and thus the considerations to be taken into account in future studies. The apparent beneficial effect of delay was observed more frequently in the colon than in the rectum, suggesting that this effect should be evaluated separately for each tumor. There was a need for larger samples of patients with sufficient power to detect significant differences, if any. The selection of samples restricted to certain patients in half of the studies was another limitation observed in the aforementioned reviews. Another problem observed was that most of the studies calculated survival from the date of diagnosis or intervention rather than the date of symptom onset, which could lead to an advance time bias.

Other more recent studies continue to point to no association between delay and stage at diagnosis, with a tendency toward even longer delays in earlier stages of disease.[60,66] However, controversy persists, as other publications report clear evidence of more advanced stages with longer diagnostic intervals.[86] In the same vein, most publications continue to find no clear association between delay and survival of patients with colorectal cancer[53,87] or even report a better prognosis in those patients with longer delays.[55] This paradox, according to which shorter delays would be associated with worse survival, may be explained by the impact of confounding variables that are not always taken into account in the analysis or may even be the result of what is known as confounding by indication bias.[56,57,88] Other authors have reported a different trend according to tumor location.[54,89]

Again, comparison of the results obtained in different series is not straightforward due to methodological differences, among other reasons. As concluded by several authors, patients with shorter delays are inherently different from those diagnosed with longer delays, likely in relation to other factors that condition their prognosis.[60] It should not be forgotten that, as in other oncological diseases, the symptomatic period of colorectal cancer represents only a small part of its natural history, with the asymptomatic phase playing a crucial role in prognosis, as demonstrated by the proven efficacy of screening programs in reducing mortality in these patients.[90]

References

1. Siegel RL, Miller KD, Jemal A. Cancer statistics, 2019. *CA Cancer J Clin* 2019;**69**(1):7–34. https://doi.org/10.3322/caac.21551.
2. Jemal A, Ward EM, Johnson CJ, et al. Annual report to the nation on the status of cancer, 1975–2014, featuring survival. *J Natl Cancer Inst* 2017;**109**(9). https://doi.org/10.1093/jnci/djx030.
3. De Angelis R, Sant M, Coleman MP, et al. Cancer survival in Europe 1999-2007 by country and age: results of EUROCARE—5-a population-based study. *Lancet Oncol* 2014;**15**(1):23–34. https://doi.org/10.1016/S1470-2045(13)70546-1.
4. Compton C, Tanabe KK, Savarese D. *Pathology and prognostic determinants of colorectal cancer. UpToDate*; 2019. Published online https://www.uptodate.com/contents/pathology-and-prognostic-determinants-of-colorectal-cancer.
5. American Cancer Society. *Colorectal cancer facts & figures 2017–2019*. Atlanta: American Cancer Society; 2017.
6. Minicozzi P, Innos K, Sánchez M-J, et al. Quality analysis of population-based information on cancer stage at diagnosis across Europe, with presentation of stage-specific cancer survival estimates: a EUROCARE-5 study. *Eur J Cancer* 2017;**84**:335–53. https://doi.org/10.1016/j.ejca.2017.07.015.
7. Hewitt M, Greenfield S, Stovall E, et al. *From cancer patient to cancer survivor*. National Academies Press; 2006. https://doi.org/10.17226/11468.
8. Abdel-Rahman O. Challenging a dogma: five-year survival does not equal cure in all colorectal cancer patients. *Expert Rev Anticancer Ther* 2018;**18**(2):187–92. https://doi.org/10.1080/14737140.2018.1409625.
9. O'Connell JB, Maggard MA, Ko CY. Colon Cancer survival rates with the new American joint committee on cancer sixth edition staging. *J Natl Cancer Inst* 2004;**96**(19):1420–5. https://doi.org/10.1093/jnci/djh275.
10. Gauci D, Allemani C, Woods L. Population-level cure of colorectal cancer in Malta: an analysis of patients diagnosed between 1995 and 2004. *Cancer Epidemiol* 2016;**42**:32–8. https://doi.org/10.1016/j.canep.2016.03.001.
11. Shack LG, Shah A, Lambert PC, Rachet B. Cure by age and stage at diagnosis for colorectal cancer patients in North West England, 1997-2004: a population-based study. *Cancer Epidemiol* 2012;**36**(6):548–53. https://doi.org/10.1016/j.canep.2012.06.011.
12. Lambert PC, Dickman PW, Osterlund P, Andersson T, Sankila R, Glimelius B. Temporal trends in the proportion cured for cancer of the colon and rectum: a population-based study using data from the Finnish Cancer Registry. *Int J Cancer* 2007;**121**(9):2052–9. https://doi.org/10.1002/ijc.22948.

13. Ito Y, Nakayama T, Miyashiro I, et al. Trends in "cure" fraction from colorectal cancer by age and tumour stage between 1975 and 2000, using population-based data, Osaka,Japan. *Jpn J Clin Oncol* 2012;**42**(10):974–83. https://doi.org/10.1093/jjco/hys132.
14. Alhayek M. *Systematic review and meta-analysis of follow-up strategies in patients operated on for colorectal cancer* [PhD thesis]. A Coruña, Spain: University of A Coruña; 2015. Published online.
15. Virgo KS, Vernava AM, Longo WE, McKirgan LW, Johnson FE. Cost of patient follow-up after potentially curative colorectal cancer treatment. *JAMA* 1995;**273**(23):1837–41. https://doi.org/10.1001/jama.1995.03520470045030.
16. Vernava AM, Longo WE, Virgo KS, Coplin MA, Wade TP, Johnson FE. Current follow-up strategies after resection of colon cancer. *Dis Colon Rectum* 1994;**37**(6):573–83. https://doi.org/10.1007/bf02050993.
17. Johnson FE, Longo WE, Vernava AM, Wade TP, Coplin MA, Virgo KS. How tumor stage affects surgeons' surveillance strategies after colon cancer surgery. *Cancer* 1995;**76**(8):1325–9. https://doi.org/10.1002/1097-0142(19951015)76:8<1325::aid-cncr2820760805>3.0.co;2-s.
18. Johnson FE, McKirgan LW, Coplin MA, et al. Geographic variation in patient surveillance after colon cancer surgery. *J Clin Oncol* 1996;**14**(1):183–7. https://doi.org/10.1200/JCO.1996.14.1.183.
19. Richert-Boe KE. Heterogeneity of cancer surveillance practices among medical oncologists in Washington and Oregon. *Cancer* 1995;**75**(10):2605–12. https://doi.org/10.1002/1097-0142(19950515)75:10<2605::aid-cncr2820751031>3.0.co;2-#.
20. Jeffery M, Hickey BE, Hider PN. Follow-up strategies for patients treated for non-metastatic colorectal cancer. *Cochrane Database Syst Rev* 2007;**1**. https://doi.org/10.1002/14651858.CD002200.pub2, CD002200.
21. Tjandra JJ, Chan MKY. Follow-up after curative resection of colorectal cancer: a meta-analysis. *Dis Colon Rectum* 2007;**50**(11):1783–99. https://doi.org/10.1007/s10350-007-9030-5.
22. Pietra N, Sarli L, Costi R, Ouchemi C, Grattarola M, Peracchia A. Role of follow-up in management of local recurrences of colorectal cancer: a prospective, randomized study. *Dis Colon Rectum* 1998;**41**(9):1127–33. https://doi.org/10.1007/bf02239434.
23. Secco GB, Fardelli R, Gianquinto D, et al. Efficacy and cost of risk-adapted follow-up in patients after colorectal cancer surgery: a prospective, randomized and controlled trial. *Eur J Surg Oncol* 2002;**28**(4):418–23. https://doi.org/10.1053/ejso.2001.1250.
24. Primrose JN, Perera R, Gray A, et al. Effect of 3 to 5 years of scheduled CEA and CT follow-up to detect recurrence of colorectal cancer: the FACS randomized clinical trial. *JAMA* 2014;**311**(3):263–70. https://doi.org/10.1001/jama.2013.285718.
25. Wang Z-D, Wu Z-Y, Li Y, Wu W-L, Lin F. Clinical efficacy comparison between laparoscopy and open radical resection for 191 advanced colorectal cancer patients. *Zhonghua Wei Chang Wai Ke Za Zhi* 2009;**12**(4):368–70. https://www.ncbi.nlm.nih.gov/pubmed/19598021.
26. Wattchow DA, Weller DP, Esterman A, et al. General practice vs surgical-based follow-up for patients with colon cancer: randomised controlled trial. *Br J Cancer* 2006;**94**(8):1116–21. https://doi.org/10.1038/sj.bjc.6603052.
27. Rodríguez-Moranta F, Saló J, Arcusa A, et al. Postoperative surveillance in patients with colorectal cancer who have undergone curative resection: a prospective, multicenter, randomized, controlled trial. *J Clin Oncol* 2006;**24**(3):386–93. https://doi.org/10.1200/JCO.2005.02.0826.
28. Grossmann EM, Johnson FE, Virgo KS, Longo WE, Fossati R. Follow-up of colorectal cancer patients after resection with curative intent-the GILDA trial. *Surg Oncol* 2004;**13**(2–3):119–24. https://doi.org/10.1016/j.suronc.2004.08.005.
29. Schoemaker D, Black R, Giles L, Toouli J. Yearly colonoscopy, liver CT, and chest radiography do not influence 5-year survival of colorectal cancer patients. *Gastroenterology* 1998;**114**(1):7–14. https://doi.org/10.1016/s0016-5085(98)70626-2.
30. Kjeldsen BJ, Kronborg O, Fenger C, Jorgensen OD. A prospective randomized study of follow-up after radical surgery for colorectal cancer. *Br J Surg* 1997;**84**(5):666–9. https://doi.org/10.1046/j.1365-2168.1997.02733.x.
31. Ohlsson B, Breland U, Ekberg H, Graffner H, Tranberg K-G. Follow-up after curative surgery for colorectal carcinoma. *Dis Colon Rectum* 1995;**38**(6):619–26. https://doi.org/10.1007/bf02054122.
32. Mäkelä JT, Laitinen SO, Kairaluoma MI. Five-year follow-up after radical surgery for colorectal cancer. Results of a prospective randomized trial. *Arch Surg* 1995;**130**(10):1062–7. https://doi.org/10.1001/archsurg.1995.01430100040009.
33. Johnson FE, Longo WE, Ode K, et al. Patient surveillance after curative-intent surgery for rectal cancer. *Int J Oncol* 2005;**27**(3):815–22. https://www.ncbi.nlm.nih.gov/pubmed/16077933.
34. Wichmann MW, Müller C, Hornung HM, Lau-Werner U, Schildberg F-W. Colorectal Cancer Study Group. Results of long-term follow-up after curative resection of Dukes A colorectal cancer. *World J Surg* 2002;**26**(6):732–6. https://doi.org/10.1007/s00268-002-6221-z.
35. Yamamoto S, Akasu T, Fujita S, Moriya Y. Postsurgical surveillance for recurrence of UICC stage I colorectal carcinoma: is follow-up by CEA justified? *Hepatogastroenterology* 2005;**52**(62):444–9. https://www.ncbi.nlm.nih.gov/pubmed/15816454.
36. Pita-Fernández S, Alhayek-Aí M, González-Martín C, López-Calviño B, Seoane-Pillado T, Pértega-Díaz S. Intensive follow-up strategies improve outcomes in nonmetastatic colorectal cancer patients after curative surgery: a systematic review and meta-analysis. *Ann Oncol* 2015;**26**(4):644–56. https://doi.org/10.1093/annonc/mdu543.
37. Macafee DAL, Whynes DK, Scholefield JH. Risk-stratified intensive follow up for treated colorectal cancer—realistic and cost saving? *Colorectal Dis* 2008;**10**(3):222–30. https://doi.org/10.1111/j.1463-1318.2007.01297.x.
38. Bloom JR, Petersen DM, Kang SH. Multi-dimensional quality of life among long-term (5+ years) adult cancer survivors. *Psychooncology* 2007;**16**(8):691–706. https://doi.org/10.1002/pon.1208.
39. Nord C, Mykletun A, Thorsen L, Bjøro T, Fosså SD. Self-reported health and use of health care services in long-term cancer survivors: Health and Use of Health Care Services in Cancer Survivors. *Int J Cancer* 2005;**114**(2):307–16. https://doi.org/10.1002/ijc.20713.
40. Bloom JR. Surviving and thriving? *Psychooncology* 2002;**11**(2):89–92. https://doi.org/10.1002/pon.606.
41. Llorca J, Delgado-Rodríguez M. Survival analysis in the presence of competing risks: event probability estimators. *Gac Sanit* 2004;**18**(5):391–7.

42. Jansen L, Koch L, Brenner H, Arndt V. Quality of life among long-term (≥5 years) colorectal cancer survivors—systematic review. *Eur J Cancer* 2010;**46**(16):2879–88. https://doi.org/10.1016/j.ejca.2010.06.010.
43. Hart TL, Charles ST, Gunaratne M, et al. Symptom severity and quality of life among long-term colorectal cancer survivors compared with matched control subjects: a population-based study. *Dis Colon Rectum* 2018;**61**(3):355–63. https://doi.org/10.1097/dcr.0000000000000972.
44. Lynch BM, Steginga SK, Hawkes AL, Pakenham KI, Dunn J. Describing and predicting psychological distress after colorectal cancer. *Cancer* 2008;**112**(6):1363–70. https://doi.org/10.1002/cncr.23300.
45. Clark CJ, Fino NF, Liang JH, Hiller D, Bohl J. Depressive symptoms in older long-term colorectal cancer survivors: a population-based analysis using the SEER-Medicare healthcare outcomes survey. *Support Care Cancer* 2016;**24**(9):3907–14. https://doi.org/10.1007/s00520-016-3227-x.
46. Haggstrom D, Cheung W. In: Nekhlyudov L, editor. *Approach to the long-term survivor of colorectal cancer*; 2020. Published online https://www.uptodate.com/contents/approach-to-the-long-term-survivor-of-colorectal-cancer.
47. Vonk-Klaassen SM, de Vocht HM, den Ouden MEM, Eddes EH, Schuurmans MJ. Ostomy-related problems and their impact on quality of life of colorectal cancer ostomates: a systematic review. *Qual Life Res* 2016;**25**(1):125–33. https://doi.org/10.1007/s11136-015-1050-3.
48. National Comprehensive Cancer Network (NCCN). *NCCN clinical practice guidelines in oncology*; 2019. Accessed September 9, 2019 https://www.nccn.org/professionals/physician_gls/default.aspx.
49. El-Shami K, Oeffinger KC, Erb NL, et al. American cancer society colorectal cancer survivorship care guidelines. *CA Cancer J Clin* 2015;**65**(6):428–55. https://doi.org/10.3322/caac.21286.
50. Renehan AG, Egger M, Saunders MP, O'Dwyer ST. Mechanisms of improved survival from intensive followup in colorectal cancer: a hypothesis. *Br J Cancer* 2005;**92**(3):430–3. https://doi.org/10.1038/sj.bjc.6602369.
51. Newell SA, Sanson-Fisher RW, Savolainen NJ. Systematic review of psychological therapies for cancer patients: overview and recommendations for future research. *J Natl Cancer Inst* 2002;**94**(8):558–84. https://doi.org/10.1093/jnci/94.8.558.
52. Kjeldsen BJ, Thorsen H, Whalley D, Kronborg O. Influence of follow-up on health-related quality of life after radical surgery for colorectal cancer. *Scand J Gastroenterol* 1999;**34**(5):509–15. https://doi.org/10.1080/003655299750026254.
53. Singh H, Shu E, Demers A, Bernstein CN, Griffith J, Fradette K. Trends in time to diagnosis of colon cancer and impact on clinical outcomes. *Can J Gastroenterol* 2012;**26**(12):877–80. https://doi.org/10.1155/2012/363242.
54. Pruitt SL, Harzke AJ, Davidson NO, Schootman M. Do diagnostic and treatment delays for colorectal cancer increase risk of death? *Cancer Causes Control* 2013;**24**(5):961–77. https://doi.org/10.1007/s10552-013-0172-6.
55. Terhaar sive Droste JS, Oort FA, RWM vdH, et al. Does delay in diagnosing colorectal cancer in symptomatic patients affect tumor stage and survival? A population-based observational study. *BMC Cancer* 2010;**10**:332. https://doi.org/10.1186/1471-2407-10-332.
56. Tørring ML, Frydenberg M, Hamilton W, Hansen RP, Lautrup MD, Vedsted P. Diagnostic interval and mortality in colorectal cancer: U-shaped association demonstrated for three different datasets. *J Clin Epidemiol* 2012;**65**(6):669–78. https://doi.org/10.1016/j.jclinepi.2011.12.006.
57. Tørring ML, Frydenberg M, Hansen RP, Olesen F, Vedsted P. Evidence of increasing mortality with longer diagnostic intervals for five common cancers: a cohort study in primary care. *Eur J Cancer* 2013;**49**(9):2187–98. https://doi.org/10.1016/j.ejca.2013.01.025.
58. Bernal Pérez M, Gómez Bernal FJ, Gómez Bernal GJ. Delay in the diagnosis of cancer. *Aten Primaria* 2001;**27**(2):79–85. https://doi.org/10.1016/s0212-6567(01)78778-1.
59. Esteva M, Leiva A, Ramos M, et al. Factors related with symptom duration until diagnosis and treatment of symptomatic colorectal cancer. *BMC Cancer* 2013;**13**:87. https://doi.org/10.1186/1471-2407-13-87.
60. Pita-Fernández S, González-Sáez L, López-Calviño B, et al. Effect of diagnostic delay on survival in patients with colorectal cancer: a retrospective cohort study. *BMC Cancer* 2016;**16**(1):664. https://doi.org/10.1186/s12885-016-2717-z.
61. Andersen BL, Cacioppo JT. Delay in seeking a cancer diagnosis: delay stages and psychophysiological comparison processes. *Br J Soc Psychol* 1995;**34**(Pt. 1):33–52. https://doi.org/10.1111/j.2044-8309.1995.tb01047.x.
62. Esteva M, Ramos M, Cabeza E, et al. Factors influencing delay in the diagnosis of colorectal cancer: a study protocol. *BMC Cancer* 2007;**7**(1):86. https://doi.org/10.1186/1471-2407-7-86.
63. Hansen RP, Vedsted P, Sokolowski I, Søndergaard J, Olesen F. Time intervals from first symptom to treatment of cancer: a cohort study of 2,212 newly diagnosed cancer patients. *BMC Health Serv Res* 2011;**11**(1):284. https://doi.org/10.1186/1472-6963-11-284.
64. Zarcos-Pedrinaci I, Téllez T, Rivas-Ruiz F, et al. Factors associated with prolonged patient-attributable delay in the diagnosis of colorectal cancer. *Cancer Res Treat* 2018;**50**(4):1270–80. https://doi.org/10.4143/crt.2017.371.
65. Ministerio de Sanidad y Consumo. *Estrategia En Cáncer Del Sistema Nacional de Salud. Líneas Estratégicas Priorizadas: Objetivos, Acciones e Indicadores (2006–2008)*; 2006.
66. Leiva A, Esteva M, Llobera J, et al. Time to diagnosis and stage of symptomatic colorectal cancer determined by three different sources of information: a population based retrospective study. *Cancer Epidemiol* 2017;**47**:48–55. https://doi.org/10.1016/j.canep.2016.10.021.
67. Ramos M, Arranz M, Taltavull M, March S, Cabeza E, Esteva M. Factors triggering medical consultation for symptoms of colorectal cancer and perceptions surrounding diagnosis. *Eur J Cancer Care* 2010;**19**(2):192–9. https://doi.org/10.1111/j.1365-2354.2008.00998.x.
68. Mitchell E, Macdonald S, Campbell NC, Weller D, Macleod U. Influences on pre-hospital delay in the diagnosis of colorectal cancer: a systematic review. *Br J Cancer* 2007;**98**(1):60–70. https://doi.org/10.1038/sj.bjc.6604096.
69. Jellema P, van der Windt DAWM, Bruinvels DJ, et al. Value of symptoms and additional diagnostic tests for colorectal cancer in primary care: systematic review and meta-analysis. *BMJ* 2010;**340**(mar31 3):c1269. https://doi.org/10.1136/bmj.c1269.
70. Adelstein B-A, Macaskill P, Chan SF, Katelaris PH, Irwig L. Most bowel cancer symptoms do not indicate colorectal cancer and polyps: a systematic review. *BMC Gastroenterol* 2011;**11**(1):65. https://doi.org/10.1186/1471-230X-11-65.

71. Astin M, Griffin T, Neal RD, Rose P, Hamilton W. The diagnostic value of symptoms for colorectal cancer in primary care: a systematic review. *Br J Gen Pract* 2011;**61**(586):e231–43. https://doi.org/10.3399/bjgp11X572427.
72. Macdonald S, Macleod U, Campbell NC, Weller D, Mitchell E. Systematic review of factors influencing patient and practitioner delay in diagnosis of upper gastrointestinal cancer. *Br J Cancer* 2006;**94**(9):1272–80. https://doi.org/10.1038/sj.bjc.6603089.
73. Porta M, Gallen M, Belloc J, Malats N. Predictors of the interval between onset of symptoms and first medical visit in patients with digestive tract cancer. *Int J Oncol* 1996;**8**(5):941–9. https://doi.org/10.3892/ijo.8.5.941.
74. Mariscal M, Llorca J, Prieto D, Delgado-Rodríguez M. Determinants of the interval between the onset of symptoms and diagnosis in patients with digestive tract cancers. *Cancer Detect Prev* 2000;**25**(5):420–9. https://www.ncbi.nlm.nih.gov/pubmed/11718448.
75. Maglinte DD, O'Connor K, Bessette J, Chernish SM, Kelvin FM. The role of the physician in the late diagnosis of primary malignant tumors of the small intestine. *Am J Gastroenterol* 1991;**86**(3):304–8. https://www.ncbi.nlm.nih.gov/pubmed/1998312.
76. McDermott FT, Hughes ES, Pihl E, Milne BJ, Price AB. Prognosis in relation to symptom duration in colon cancer. *Br J Surg* 1981;**68**(12):846–9. https://doi.org/10.1002/bjs.1800681206.
77. Smith C, Butler JA. Colorectal cancer in patients younger than 40 years of age. *Dis Colon Rectum* 1989;**32**(10):843–6. https://doi.org/10.1007/bf02554552.
78. Fegiz G, Barillari P, Ramacciato G, et al. Right colon cancer: long-term results after curative surgery and prognostic significance of duration of symptoms. *J Surg Oncol* 1989;**41**(4):250–5. https://doi.org/10.1002/jso.2930410412.
79. Arbman G, Nilsson E, Störgren-Fordell V, Sjödahl R. A short diagnostic delay is more important for rectal cancer than for colonic cancer. *Eur J Surg* 1996;**162**(11):899–904. https://www.ncbi.nlm.nih.gov/pubmed/8956960.
80. Mulcahy HE, O'Donoghue DP. Duration of colorectal cancer symptoms and survival: the effect of confounding clinical and pathological variables. *Eur J Cancer* 1997;**33**(9):1461–7. https://doi.org/10.1016/s0959-8049(97)00089-0.
81. Gonzalez-Hermoso F, Perez-Palma J, Marchena-Gomez J, Lorenzo-Rocha N, Medina-Arana V. Can early diagnosis of symptomatic colorectal cancer improve the prognosis? *World J Surg* 2004;**28**(7):716–20. https://doi.org/10.1007/s00268-004-7232-8.
82. Rupassara KS, Ponnusamy S, Withanage N, Milewski PJ. A paradox explained? Patients with delayed diagnosis of symptomatic colorectal cancer have good prognosis. *Colorectal Dis* 2006;**8**(5):423–9. https://doi.org/10.1111/j.1463-1318.2006.00958.x.
83. Korsgaard M, Pedersen L, Sørensen HT, Laurberg S. Delay of treatment is associated with advanced stage of rectal cancer but not of colon cancer. *Cancer Detect Prev* 2006;**30**(4):341–6. https://doi.org/10.1016/j.cdp.2006.07.001.
84. Ramos M, Esteva M, Cabeza E, Llobera J, Ruiz A. Lack of association between diagnostic and therapeutic delay and stage of colorectal cancer. *Eur J Cancer* 2008;**44**(4):510–21. https://doi.org/10.1016/j.ejca.2008.01.011.
85. Ramos M, Esteva M, Cabeza E, Campillo C, Llobera J, Aguiló A. Relationship of diagnostic and therapeutic delay with survival in colorectal cancer: a review. *Eur J Cancer* 2007;**43**(17):2467–78. https://doi.org/10.1016/j.ejca.2007.08.023.
86. Tørring ML, Murchie P, Hamilton W, et al. Evidence of advanced stage colorectal cancer with longer diagnostic intervals: a pooled analysis of seven primary care cohorts comprising 11 720 patients in five countries. *Br J Cancer* 2017;**117**(6):888–97. https://doi.org/10.1038/bjc.2017.236.
87. Dregan A, Møller H, Charlton J, Gulliford MC. Are alarm symptoms predictive of cancer survival?: population-based cohort study. *Br J Gen Pract* 2013;**63**(617):e807–12. https://doi.org/10.3399/bjgp13X675197.
88. Tørring ML, Frydenberg M, Hansen RP, Olesen F, Hamilton W, Vedsted P. Time to diagnosis and mortality in colorectal cancer: a cohort study in primary care. *Br J Cancer* 2011;**104**(6):934–40. https://doi.org/10.1038/bjc.2011.60.
89. Jullumstrø E, Lydersen S, Møller B, Dahl O, Edna T-H. Duration of symptoms, stage at diagnosis and relative survival in colon and rectal cancer. *Eur J Cancer* 2009;**45**(13):2383–90. https://doi.org/10.1016/j.ejca.2009.03.014.
90. Zauber AG, Lansdorp-Vogelaar I, Knudsen AB, Wilschut J, van Ballegooijen M, Kuntz KM. Evaluating test strategies for colorectal cancer screening: a decision analysis for the U.S. Preventive Services Task Force. *Ann Intern Med* 2008;**149**(9):659–69. https://doi.org/10.7326/0003-4819-149-9-200811040-00244.

Section B

Clinical manifestations and disease detection

Chapter 3

Primary prevention of CRC

José Luis Ulla Rocha[a], Raquel Sardina Ferreiro[b], Rosa Fungueiriño Suarez[c], and Juan Turnes Vázquez[a]
[a]Gastroenterology Department, Digestive Disease Unit, Pontevedra University Hospital Complex, Pontevedra, Spain, [b]Department of Internal Medicine, Arquitect Marcide Hospital, Ferrol, Spain, [c]Preventive Medicine Service, Pontevedra University Hospital Complex, Pontevedra, Spain

Introduction

Primary prevention of colorectal cancer (CRC) is considered the etiological study and analysis of the condition for the purpose of implementation of therapies that may reduce incidence of precursor adenomas or interrupt adenoma to carcinoma progression. Various epidemiological and experimental genetic studies suggest that CRC occurs as a result of complex interactions between genetic susceptibility factors and environmental factors. Hence, there are nonmodifiable factors, inherent to the individual, such as gender, race, age, and so on, on which, at present, intervention is not possible, and there are modifiable factors, on which intervention is possible. These are the subject of this chapter.

Triggers that cause normal colorectal epithelial cells to transform into tumor cells can be either genetic or epigenetic. Epigenetics consists of a series of chemical processes that modify the expression of deoxyribonucleic acid (DNA) without changing its sequence. These mechanisms include DNA methylation, acetylation, hydroxylation, histone protein modification, chromatin structure remodeling, and noncoded ribonucleic acid (RNA) or micro-RNA.[1] Epigenetic alterations can be induced by internal and external factors that could have similar effects to pathogenic mutations. Although epigenetic changes can be reversible, the majority remain stable through multiple cell division cycles, giving cells different identities, but maintaining the same genetic information. In other words, epigenetics studies why certain genes are expressed and others are not, and which environmental factors lead to those differences in different environmental circumstances.[2] Epigenetic alterations can be as significant as genetic modifications in causing neoplastic disease.[3]

Epidemiology and CRC

Geographical location plays a large part in the incidence of CRC, which has been found to be more widespread in developed societies with a Western lifestyle. This is why it has been called a disease of civilization.[4] The strongest evidence that CRC is a lifestyle-related disease, produced by environmental risk factors, stems from the fact that there is a relative increase in incidence of the condition in immigrant populations who have moved to countries with a Western lifestyle that differs from that of their country of origin. Accordingly, these changes in incidence within a single generation cannot be due to genetic changes.[5] We will briefly describe the mechanisms by which environmental factors can lead to onset of CRC. Epidemiology is a powerful tool for identifying potential etiological factors and has played a vital role in assembling our current knowledge of CRC.

We must take into consideration that when assessing the relationship between diet and cancer, the results must be interpreted with caution as there are a number of limitations to this type of study:

- On the one hand, observational studies can be inconsistent due to the possibility of finding elements causing confusion. Another problem may arise when the study populations simply do not remember their dietary intake or observers do not take notice of it.[5]
- Randomized clinical trials may also show erroneous results in relation to adherence to dietary guidelines, insufficient follow-up period, or the dosage itself or the form of administration of a particular nutrient may lead to different results. Studies tend to focus on an isolated nutrient, when dietary composition is more complex.[6]
- A very important aspect to be taken into account is whether factors that apparently prevent or lead to CRC are maintained throughout the general population or only have an effect on individuals with a high genetic risk. We have

TABLE 3.1 Level of evidence and mechanisms related to food and nutrients and CRC.

Nutrient or food	Level of evidence	Relative risk (RR)	Proposed mechanism
Red and processed meat	Convincing	RR: 1.16 (1.04–1.30) per 100 g/day increase intake	Carcinogenic effect of heme iron and heterocyclic amines
Milk	Probable	RR:0.91 (0.85–0.94) per 200 g/day increase of total milk intake	Antineoplastic effect of calcium, vitamin D, and butyric acid
Fruits	Limited—suggestive	RR:0–97 (0.94–0.99) per 100 g/day increase of fruit intake	Anticarcinogenic effects of folate, fiber, and flavonoids
Nonstarchy vegetables	Limited—suggestive	RR:0.98 (0.96–0.99) per 100 g/day increase of vegetable intake	Anticarcinogenic effects of folate, vitamins, and flavonoids
Whole grains	Convincing	RR.0.83 (0.78–0.89) per 3 servings/day increase	Anticancer properties of fiber and antioxidants
Omega-3 polyunsaturated fatty acids	Limited—no conclusion	RR:0.97 (0.86–1.10) comparing the highest to the lowest categories	Reduced inflammation through inhibition of Arachidonic acid
Total flat	Limited—no conclusion	RR:0.99 (0.89–1.09) comparing the highest to the lowest categories	Increased intestinal level of bile acids Increase of deoxycholic acid
Selenium	Limited—no conclusion	RR:0.81 (0.71–0.92) comparing the highest to the lowest categories	Antioxidative and inhibitor of cell proliferation
Vitamin A, C, E	Limited—no conclusion	RR: 0.93 (0.79–1.10) for total vitamin A RR: 0.86 (0.74–1.00) for total vitamin C	Antioxidative and inhibitor of cell proliferation
Folate	Limited—no conclusion	RR:0.99 (0.93–1.05) per 100 microg/day increase of dietary folate intake	Essential nutrient for DNA methylation and DNA synthesis
Fiber	Convincing	RR:0.90 (0.86–0.94) per 10 g/day increase of dietary fiber intake	Decreased transit time
Vitamin D	Limited—suggestive	RR:0.95 (00.93–0.98) per 100 IU/day increase of dietary vitamin D intake	Inhibition of invasion and metastases
Calcium	Probable	RR:0.92 (0.89–0.95) per 300 mg/day increase of total calcium intake	Suppression of cell proliferation

Adapted from Song M, Garrett WS, Chan AT. Nutrients, foods, and colorectal cancer prevention. Gastroenterology. 2015;148(6):1244–1260.e16. doi:10.1053/j.gastro.2014.12.035.

attempted to summarize and quantify the role of all CRC risk factors, as well as quantifying the dosage–response relationship for each, showing the most relevant in Table 3.1.[7] Thus we can readily refer to the CRC risk factors (RR > 1), the preventive factors (RR < 1), and the potentials of both, with the option of being able to intervene in each case. Relative risk (RR) is a measure used in cohort studies and clinical trials.

Environment and CRC

Various dietary and lifestyle factors have been identified affecting whether transcription takes place or not in certain genes, either leading to CRC or preventing onset; the process probably occurs through intricate metabolic and inflammatory mechanisms. Various studies estimate that approximately 50–60% of new cancer cases in the United States can be avoided with lifestyle changes. While the exact mechanism leading to cancer for each factor is likely to be different and a great deal more information is still needed, a number of research threads suggest gut microbiota could be a mechanism of action that may prove to be ultimately responsible. This information could establish the relationship between environmental factors and cancer.[8]

We outline epidemiological aspects, along with evidence linking diet and lifestyle to development of CRC in Fig. 3.1.[9] The role of microbiomes is discussed in another chapter.

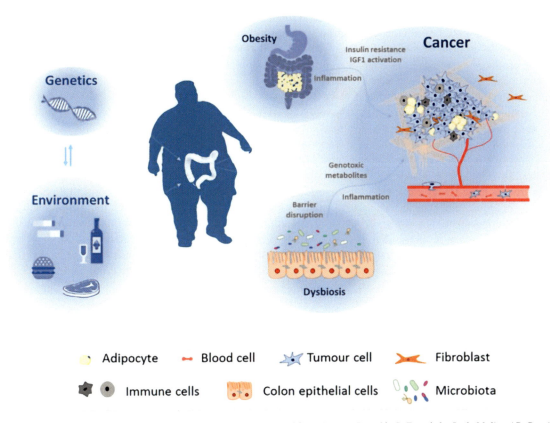

FIG. 3.1 Interactions between genetics/the environment and CRC. *(Adapted from Aguirre-Portolés C, Fernández L, de Molina AR. Precision nutrition for targeting lipid metabolism in colorectal cancer. Nutrients 2017;9(10):1076. doi:10.3390/nu9101076.)*

Diet

Diet is a key element affecting correct epigenetic cell regulation. A recent meta-analysis has used the dietary inflammatory index (DII) to show the relationship between the consumption of certain foods and increased CRC risk.[10] A high DII with proinflammatory potential produces increased CRC risk, while a low DII with antiinflammatory potential produces a low risk. Antiinflammatory dietary components include fiber; alcohol; mono- and polyunsaturated fatty acids; Omega 3 and 6; vitamins A, B6, C, D, and E; niacin; thiamine; zinc; magnesium; selenium; folic acid; beta-carotene; flavones; garlic; ginger; onion; thyme; rosemary; saffron; turmeric; caffeine; and tea. Proinflammatory dietary components include carbohydrates, proteins, total fat, cholesterol, saturated fatty acids, and iron.

Although analyzing the different epidemiological studies available with caution, due to the fact that diet is a complex phenomenon that includes production and cooking processes, we will methodically examine the key evidence available on the different dietary components.[11]

Foods recommended to be reduced

Processed and red meat

Meat that has been transformed by various means, such as smoking, salting, curing, or fermenting with the aim of improving flavor or for preservation, is referred to as processed meat. In 2015 the World Health Organization's International Agency for Research on Cancer (IARC)[12] reviewed the evidence linking the intake of processed meats and red meat to CRC, concluding that there was sufficient consistent information to consider the consumption of processed meat cancerogenic for humans (group 1). Although it has been established that the more processed meat is eaten, the greater the CRC risk will be, no exact guidelines for a safe level of consumption have been established. In relation to the consumption of red meat (e.g., beef, veal, pork, mutton, lamb, and horse) no such clear association has been found, since the part played by other dietary and lifestyle factors could not be excluded. Therefore red meat consumption was classified as a probable carcinogen for humans (group 2A). It has been suggested that the effects are due to the presence of the heme group, and also the fact that carcinogens are produced by cooking at high temperatures.[13] The production of toxic bile acids could also be a cause (cholecystectomized patients have a high risk of proximal CRC).

Data published in 2018 from the European Prospective Survey on Nutrition and Cancer (EPIC) found a positive association between high red and processed meat intake and CRC risk,[14] with an increase in risk of 35% in subjects who consumed more than 160 g per day, over those who consumed less than 20 g per day.[15] Cooking meat at high temperatures such as in barbecuing or frying is to be avoided, since compounds are produced that contribute to the creation of cancerogenic elements—mainly polycyclic aromatic hydrocarbons, although this has not yet been fully elucidated.

These claims are based on observational studies of dietary patterns in populations in various geographical locations and not all of the studies have found a consistent association between red and processed meat consumption and CRC risk. Therefore though consumption of processed meat products and red meat increases risk, the margin is small and relates to high daily consumption. It should also be taken into account that the risk may not be the same for all individuals. On the other hand, it needs to be borne in mind that consumption of red or processed meat at least once or twice a week forms part of a balanced diet.[16] For individuals over 70 years of age, it is not recommended that red meat intake be as restricted, as it is important in this age group to maintain adequate protein intake, and the development of CRC is a process occurring over a long period of time.[17]

Food and drink recommended to be increased

Dairy products

Intake of milk and other dairy products is associated with a decrease in CRC risk. The effect may be due to the fact that dairy products are one of the main sources of calcium and magnesium that have a preventive influence, although the presence of fatty acids such as linoleic and butyric acids can also have a preventive effect.[18]

In the EPIC study, multivariate analyses showed that milk consumption was inversely associated with the CRC risk (RR 0.93 per 200 g a day) and similar results were obtained for cheese and yoghurt[19]; however, in linear models this association was insignificant. These results strengthen the evidence that dairy products, regardless of fat content, can protect against CRC risk,

Dietary fiber and whole grain cereals

An inverse relationship between dietary fiber intake and CRC has been shown. High fiber content in the diet, especially fiber from whole grain cereals, has been shown to give protection from onset of both proximal and distal CRC.[20] The 2017 World Cancer Research Fund report on colorectal cancer concluded that whole grain and high fiber intake diets lower CRC risk.[21] On the other hand, refining cereals caused them to lose a large part of this preventive capability, as certain trace elements and vitamins, vitamin B6 in particular (pyridoxine), aiding in prevention are removed in the process.

In relation to recurring adenomas there is no evidence confirming that an increase in dietary fiber lessens the return of adenomatous polyps in patients who have previously had adenomas extirpated.[22] The exact mechanism by which fiber intervenes has not been established. Several theories have been raised suggesting that with a reduction in intestinal transit time, there is less time for contact between carcinogens and intestinal mucosa, and reduction in the production of biliary acids and short-chain fatty acids following fermentation of the fiber by microbiota.[11] The part played by starch in CRC prevention is still under study, but it seems that starch fermentation produces short-chain fatty acids, such as butyrate, which has been shown to have an antineoplastic effect on the colon, modulating immune response.

Fish

Fish is believed to have a moderate preventive effect against CRC, owing to its Omega 3 fatty acid content. Data from a recent meta-analysis found a modest association between high fish consumption and reduction in incidence of CRC.[23] Data published in a 2018 EPIC study found a 31% increase in CRC risk reduction in subjects consuming more than 80 g of fish a day over those who consumed less than 10 g a day.[24]

White and lean meat

White meat, mainly from poultry and rabbit, has little fat content. Epidemiological studies have not produced any evidence that high consumption of this kind of meat is related to an increase in CRC onset risk,[15] thus establishing that its consumption is associated with a reduction of CRC.

Coffee

The antioxidant effect brought about by drinking at least four cups of coffee a day, mainly due to coffee's polyphenols content, is associated with a decrease in CRC risk.[25] However, some authors believe that it's yet to be established if this negative consumption is a result of the fact that patients with CRC drink less coffee in an attempt to reduce symptoms of their neoplasia through changes in their bowel habits. In a recently published systematic review, no link was found between coffee consumption and CRC risk.[26] Stratification according to race showed a preventive effect in European men and Asian women. No relationship was found to rectal cancer. Ethnicity may explain the differences, but little is known about the relationship between genetics and CRC risk.

Garlic

Garlic was included as having a probable protective effect against CRC by the World Cancer Research Fund and American Institute for Cancer Research, but a review by the Food and Drug Administration of the United States (FDA) concluded that the level of evidence for this link was limited. More research is needed.[27]

Lifestyle behaviors

Several lifestyle behaviors have been suggested as risk factors for CRC onset, with varying degrees of reliability. These lifestyle factors can therefore present potential means for prevention.

Unhealthy practices

Excessive alcohol consumption

Several studies have shown a causal relationship between alcohol consumption and the incidence of CRC. In 2007 the International Agency for Research on Cancer (IARC) stated that there was sufficient evidence to suggest that alcohol is a causal factor in CRC.

In recent years, a number of studies have suggested that individuals consuming four or more alcoholic beverages a day (high consumption) increase CRC risk by 52% 28. This relationship was greater for Asian populations than white[28], most likely due to genetic factors that affect the metabolism of alcohol, dietary components, such as folate consumption, and body composition. The risk is greater for rectal cancer and with beer more than wine, although both beverages bring an increased risk.[29] Moderate alcohol consumption of less than 30 g a day (two alcoholic drinks) does not increase CRC risk in the general population.[30] Males are advised not to exceed this amount of alcohol, while women require a lower threshold of 20 g a day (one alcoholic drink) to minimize the risk.

It is suggested that bacteria in the large intestine or colon convert alcohol into large amounts of acetaldehyde, a chemical compound that causes cancer in laboratory animals. On the other hand, alcohol intake is associated with low levels of folic acid,[31] due to a decrease in its intake in heavy drinkers and also reduction in absorption. Folate is known to be necessary for DNA synthesis and repair; folate deficiency can contribute to alterations in genetic material leading to carcinogenesis. It is important for moderate drinkers to maintain an adequate intake of folate in their diet and a recommended weight to mitigate the effects of the alcohol.

Smoking

Early studies did not find higher CRC risk among smokers,[32] as it was not clear that smoking was a completely independent factor from other recognized risk factors and partly because of the long period of time between exposure to the effects of smoking and development of CRC. Later, the risk resulting from smoking was shown to persist even when other known CRC risk factors were controlled. As well, a dose–response relationship between smoking and cancer onset has been shown.[33] American studies show an increase of 50% in the prevalence of CRC in 20 or more a day smokers and a 40% increase in smokers of 35 packs a year. The increased risk is seen in both smokers and ex-smokers as opposed to nonsmokers.

Smoking is also a risk factor for the development of colonic polyps. Smokers increase risk of developing large adenomatous polyps two- to threefold.[34] Smoking causes the release of a large number of known carcinogenic components, including nitrosamines and aromatic amines that can reach the colorectal mucosa either via the circulatory system or by direct ingestion.[35]

Obesity

Obesity is a key factor due to the fact that in some western societies it has been identified as the most prevalent modifiable risk factor.[36] A meta-analysis that included 56 case–control and cohort studies[37] involving a total of 7,213,335 individuals showed increased colon cancer risk to be directly proportional to an increased body mass index. The trigger mechanism for cancer is not well established, but the mitogenic properties of insulin are known, and obesity causes insulin resistance, causing hyperinsulinemia that can lead to CRC pathogenesis. In addition, obesity is considered a proinflammatory state and inflammation has traditionally been linked to cancer development. Thus a recent study has shown an increase in overexpression of the proinflammatory enzyme, cyclo-oxygenase-2, in normal mucosa adjacent to colorectal tumors, in patients who are overweight, as compared to patients with a normal body mass index.[37] More recently, it has been established that obesity blocks the immune response of natural killer cells,[38] causing tumor cells to be unrecognized by immune system control, allowing tumor growth.

Recent studies have established that weight gain combined with lower levels of physical exercise is only associated with risk increase in approximately half of CRC patients.[39] In particular, this occurs in cancer developing via beta-catenin pathways (CTNNB1); this could be useful information when genetic studies are standardized in the future. Weight gain from youth to adulthood was shown to bring higher CRC risk than that from adulthood to advanced age.[40] A weight increase of 5 kg was shown to correlate with a 3% increase in CRC risk, with the increase somewhat greater for men than women. Abdominal obesity has been significantly linked to increased colon cancer risk for both sexes.[41] An increase in waist circumference and waist–hip ratio was linked to CRC. However, adult weight gain (abdominal obesity developing in the 50s) also increased the risk of colon cancer for both sexes.

Obesity is also associated with metabolic syndrome (obesity, hypertension, dyslipidemia, hyperglycemia), which is linked to increased CRC risk and mortality for both sexes. Hyperglycemia and/or increased body mass index and waist circumference are factors in the majority of risk associated with metabolic syndrome. Similarly, evidence suggests that diabetes increases CRC risk. A recent meta-analysis showed colon cancer risk in diabetics to be approximately 37% higher than for nondiabetics.[42] Weight reduction can improve all components of metabolic syndrome and insulin sensitivity.

Healthy practices

Physical exercise

Observational epidemiological studies provide evidence that regular physical exercise prevents CRC onset. This relationship occurs, regardless of the degree of individual obesity, suggesting the protective effect is greater than weight loss alone.[43] A dose–response relationship shows significant decrease in risk (25%), but there is no information suggesting one form of exercise is more beneficial than another. Physical exercise reduces the risk of colon cancer, specifically for ascending colon tumors[29] and for individuals of normal weight with a body mass index under 25.

In order to take advantage of primary prevention through daily exercise, an appropriate recommendation has been estimated as 30–60 min of moderate to vigorous exercise at least five days a week.[44] It is shown that physical exercise results in a lowering of circulating insulin levels,[45] improved immune function, alteration in bile acid metabolism, and increased intestinal transit.

Therapeutic strategies linked to lowering CRC risk

A number of therapies have been proposed with protective effects that help guard against CRC onset. We outline the most relevant as follows.

Salicylic acetyl acid (SAA)

Both long-term observational studies and randomized trials show that salicylic acetyl acid (SAA) has a preventive effect against CRC. It appears to act by inhibiting the effect of cyclooxygenase on nucleated cells in the intestinal epithelium[46]; high levels of the cyclooxygenase enzyme are found in colorectal neoplasms. SAA is known to increase prostaglandin which, in turn, promotes angiogenesis, cell proliferation, and migration, also preventing apoptosis. Trials have shown that a daily aspirin (SAA) reduces the risk of recurrence of colorectal adenomas by up to 17% and risk of advanced adenomas by up to 28%.[47] A daily aspirin taken over a five-year period reduces the CRC mortality rate after 20 years of follow-up by 30–40%. As well, the risk of the side effect of severe bleeding with the use of aspirin has also been found to lessen with prolonged use, suggesting a risk–benefit ratio favorable to the use of aspirin in CRC primary prevention. Assessment carried out on the result of intervention with aspirin concludes that the therapy could be cost effective, especially for patient populations at intermediate–high risk following polypectomy for colonic adenomas.[48]

In 2016 the US Preventive Services Task Force, after evaluating all the benefits of SAA for reducing cardiovascular events and CRC incidence, recommended a low dosage use of SAA for primary prevention of cardiovascular disease and CRC, within the 50–59 age group with cardiovascular disease risk over 10% and over 10 years life expectancy, for patients willing to take a low dose of SAA over at least a 10-year period.[49] In the 60- to 69-year-old cohort with similar characteristics and with no risk of intestinal bleeding, the decision should be made on an individual basis. However, the use of SAA is not recommended for those under 50 or over 70 years of age.

Nonsteroidal antiinflammatory drugs (NSAIDs)

Initial evidence of possible benefits from NSAIDS originated in the use of Sulindac over a number of years in patients with polyposis syndrome because of the capacity of the drug to slow down colon polyp development.[50] The mechanism of action also seems to occur through inhibition of the cyclooxygenase enzyme. A long-term cohort study that assessed intervention with a daily or weekly NSAID over a 10-year period on a population comprising thousands of individuals confirmed reduction in colon cancer incidence, but this did not apply to rectal adenocarcinoma.[51] However, in order to formulate effective recommendations for the use of NSAIDs, possible adverse side effects resulting from long-term use must be taken into account. Some NSAIDs have been associated with the development of cardiovascular or neurological ischemic events.[52]

In a recent meta-analysis examining the role of NSAIDs in populations aged 40 or over,[53] a protective effect was found for women (risk reduction 19%), for high dosages (18%), in distal colon cancer (22%), and for white populations (31–41%). It can be concluded from these results that NSAIDs could be used for CRC prevention in certain population groups. Accordingly, in individuals who have previously presented colorectal neoplasia, NSAIDs are the most effective agents for the prevention of metachronous advanced neoplasia, while SAA has the most effective risk–benefit profile.

Calcium and magnesium supplementation

Calcium

There are numerous studies showing calcium to have a protective role against CRC. A meta-analysis comprising 60 epidemiological studies showed that a high intake of calcium, either dietary or through supplements, has a significant protective effect against development of distal colon and rectal tumors.[54] Every 300-mg increase of calcium per day within total calcium intake is associated with approximately an 8% reduction in CRC risk. This protective effect was also linked to development of colorectal adenomas. The pathway by which calcium deficiency is thought to facilitate tumor progression relates to its importance in transcription of intracellular signals in the case of tumor cell apoptosis. The efficiency of calcium in lowering CRC risk greatly depends on the individual's vitamin D levels; optimum prevention requires adequate levels of this vitamin.

Magnesium

Based on previous studies undertaken on experimental animals, some authors believe that magnesium can also guard against CRC onset. In a study on a Swedish population[55] an inverse relationship was found between magnesium intake and CRC risk in women, for both colon and rectal cancer.

Vitamin D supplementation

The WHO has identified CRC as the cancer with the greatest likelihood of being associated with vitamin D deficiency. Both 25-hydroxyvitamin D and free vitamin D appear to play a protective role in CRC development. A 10-mg increase in circulating vitamin D levels was correlated with a 26% decrease in CRC risk. Some studies have shown this protective effect in women, but not in men. For CRC risk reduction, vitamin D levels need to be maintained at between 75 and 100 nmol/L.[56] It may be feasible for vitamin D to help prevent cancer biologically.[57] Vitamin D receptors are present in most tissue, and cell culture studies suggest that calcitriol or vitamin D can promote cell differentiation, inhibit proliferation of cancerogenic cells, and show proapoptotic properties.

In research on vitamin D, a number of considerations need to be taken into account that might cause bias in studies, relating to factors we have already seen: lack of physical activity results in less exposure to the sun, and this leads to lowering of synthesis of vitamin D from the skin, and obesity causes vitamin D to accumulate in adipose tissue. It should also be taken into account before prescribing massive vitamin D supplementation, that as with calcium, an excess can have a toxic effect on the body.

B-group vitamins

A meta-analysis evaluating the link between vitamin B6 and B2 levels and CRC found a moderate association in population groups with greater intake of foods rich in these vitamins with a reduction in cancer risk.[58] However, rather than supplementation, intake of foods containing these vitamins, namely vegetables, is recommended.

Folic acid and folate

Different studies have shown the role of folate as an inhibitor in colon cancer pathogenesis. It is believed to play a part in DNA stabilization and repair. Pharmacological supplementation of folic acid is frequently given to pregnant women to prevent neural tube defects in newborn babies. Folate is the natural form of the vitamin that is found profusely in nature, mainly in vegetables. Folic acid is the synthetic form used in food supplements. Due to different biochemical pathways, the effects of each can be different. Nevertheless, there is conflicting information on whether effective long-term folic acid supplementation from food is capable of preventing adenoma onset or colon adenocarcinoma. With the current information, it is not possible to recommend folic acid supplementation as a preventive measure for CRC onset even in individuals confirmed with increased risk (patients with polyposis).[59]

Antioxidants

Due to the fact that oxidative stress is involved in the mutagenesis process, the scientific community has debated whether antioxidant supplementation would be capable of reducing CRC incidence. However, not only did various studies fail to show a preventive effect but they also showed a possible increased risk,[60] due to mortality from unexplained causes. A systematic review by Cochrane on the subject found that antioxidant supplements significantly increased mortality; it also showed that beta-carotene increased mortality while vitamins A and C and selenium did not significantly affect mortality. No beneficial effects of antioxidants were found in primary or secondary prevention of colorectal adenomas.

Hormone replacement therapy

The incidence of CRC is higher in men than women, suggesting a possible preventive factor resulting from female sex hormones. Owing to this, the use of postmenopausal hormone therapy has been advocated as a means for decreasing CRC incidence. More recently, some authors have concluded that the estrogen preventive effect is limited to certain molecular subtypes that they labeled CRC-specific estrogen vitamin D receptors. Accordingly, the potential therapeutic part played by estrogen (or the soya product that provides it) would appear to depend on identification of target individuals or populations for whom this alternative might be used.[61]

Statins

Statins belong to a medication group used to reduce serum lipid levels and they have proven to be effective in reducing mortality from cardiovascular disease. However, in addition to their action in inhibiting cholesterol biosynthesis, they seem to have a pleiotropic effect that includes cell growth modulation, apoptosis, and inflammation. Through this mechanism they are capable of influencing a wide range of pathologies, including cancer. A meta-analysis of case–control studies showed a link between statin use and reduction of CRC risk.[62] Nevertheless, this information needs to be viewed with caution, since the use of statins is more prevalent in high socioeconomic populations that may also take NSAIDs and vitamin supplements. Hence, statins may be ineffective on their own in CRC chemoprevention, but, in combination with NSAIDs or aspirin, they may be considered a successful chemoprevention therapy. Notwithstanding, from a clinical practice perspective, there is now sufficient evidence to be able to recommend statins for CRC chemoprevention.[63]

Bisphosphonates

In some studies, the use of oral bisphosphonates, often prescribed for osteoporosis, over a period of a year or more, was linked to a decrease in the CRC risk. However, the mechanism for this to occur was not fully clarified—it was speculated that the information was relevant for populations with increased CRC risk, provided the risk–benefit equation was considered, due to the use of bisphosphonates being associated with serious adverse side effects[64] At the present time, the overall information linking bisphosphonates to CRC prevention is inconsistent.

Metformin

In recent years, this well-known oral antidiabetic medication has been gaining importance for CRC prevention. Observational studies suggest the use of metformin reduces the risk of adenomas and CRC.[2] The finding has been substantiated in numerous case–control and cohort studies and is not limited to CRC. Chemoprevention may extend to other kinds of neoplasia, especially hepatic carcinoma and breast cancer.[65] The biological plausibility for this is due to its direct effect on the reduction of circulating insulin levels, and also an indirect effect through other signaling pathways, such as the mTOR pathway.

Chondroitin sulfate and glucosamine

It is estimated that the combined use of both these two substances produces an antiinflammatory effect. The evidence comes from both in vitro and large-scale patient studies.[66] Prospective studies need to be carried out to substantiate this usage.

Angiotensin II inhibitors

Information is available from in vitro and in vivo studies on the role of angiotensin II in causing cancer, showing a link between inhibition of this enzyme and slowing growth of colon cancer cells. In a cohort study it was found that long-term taking of Lisinopril reduced the risk of advanced colorectal adenoma by 40%[67]; however, observational studies and data analysis from subsequent clinical trials showed contradictory results between the use of antihypertensive treatments using an angiotensin II converting enzyme inhibitor and CRC risk.

Dietary patterns and CRC

Different studies in recent years have assessed the association of healthy and unhealthy dietary patterns with cancer risk.[68] When compared to information on individual foods, information on dietary patterns can better describe the complexity of dietary intake and measure overall food intake. On the other hand, in relation to CRC risk, dietary patterns provide information beyond associations of individual food groups or nutrients. This allows the development of more practical dietary guidelines for the prevention and treatment of the disease. Healthy dietary patterns consist of fruit- and vegetable-based diets, whereas unhealthy patterns are referred to as Western, meat based, high fat, high salt, and refined. The study results show that healthy dietary patterns are negatively associated with CRC risk, as well as being correlated with healthier lifestyle factors that include increased physical exercise and nonsmoking. Studies carried out in Europe (MCC-Spain multiple case–control study and the cohort study on UK women) show that a Mediterranean diet is of greater benefit for CRC prevention.[69] These results can be attributed to its high content in oily fish, nuts, olives, and olive oil.

Conversely, Western dietary patterns characterized by greater intake of red and processed meats, sugar, and refined grains are associated with higher CRC risk—in particular, distal colon and rectal tumors. On the other hand, dietary patterns from a Mediterranean diet are associated with lower CRC risk in general, regardless of the anatomical or molecular site. Heterogeneity of CRC risk factors according to anatomical site indicates CRC, can exhibit different clinical and molecular characteristics, depending on its location, suggesting that in these cases therapies could be different.

Relationship between dietary index and CRC

Chronic inflammation contributes to onset and progression of many types of cancer, including CRC. Therefore measuring DII can be useful for comparing individuals with different dietary patterns. Proinflammatory dietary components typically include carbohydrates, protein, total fat, cholesterol, and saturated fatty acids. Antiinflammatory diets include fiber, polyunsaturated fatty acids, minerals, and vitamins. Several indices have been used to measure inflammatory potential and diet quality.[68] These include the Dietary Inflammatory Index (DII), the Healthy Eating Index (HEI), the Alternative Healthy Eating Index (AHEI), and the Nonenzymatic Antioxidant Activity Capacity (NEAC). Two meta-analyses found a positive association between the highest DII scores and higher CRC risk (>40%). Similar results were found in other regional epidemiological studies; the overall results of these were consistent with those examining dietary patterns. Thus determining DII scores for foods could make it possible to determine their quality and to design customized diets for prevention or treatment of CRC. However, there are limitations to this, as the inflammatory potential of most foods has not been established and further studies are needed.

Summary: Evidence and recommendations

In the light of current knowledge, we are able to strongly recommend to the general population, as a primary prevention measure for CRC, that healthy lifestyle practices be adopted. These include moderate daily physical exercise—not only for its own sake, but also for weight control—reduced alcohol consumption and nonsmoking. There are a number of dietary recommendations that seem to produce a beneficial effect on CRC prevention that include a diet rich in dairy foods, fresh fruit and vegetables, whole grain cereals, fish, and poultry. Moderate consumption of red meat, especially processed red meat is advised. Also recommended, but to a lesser degree, is a reduction in intake of overcooked or charred meat. Adequate intake of foods rich in beta-carotene, vitamins, and minerals is essential, but these antioxidants should not be administered as supplements. For particular populations at high risk of CRC, therapeutic strategies based on aspirin and other NSAIDs can be recommended. For individuals with a history of polyps, administration of calcium supplements for the prevention of adenoma relapse is recommended and, to a lesser extent, other medications that we have mentioned, either individually used or in combination. The risk–benefit equation must always be taken into account for each individual case. The present degree of recommendation for the different factors is presented in Table 3.2.[41]

TABLE 3.2 Recommendations for preventive intervention.

Recommendation	Quality of evidence	Strength of recommendation
Moderate intake of red and processed meats and overcooked or charred meat	Moderate	Weak
High-fiber diet (whole grain cereals, fruit, and vegetables)	Moderate	Weak
Dairy-rich diet	Moderate	Weak
Diet rich in fish and poultry	Low	Weak
Low-fat diet to avoid obesity	Low	Weak
Adequate folate, vitamin B, calcium, and vitamin D in dietary intake. Supplements *not* to be used	Moderate	Not recommended
Adequate dietary intake of foods rich in beta-carotene, vitamins, and minerals. Supplements *not* to be used	High	Not recommended
Calcium supplementation in cases of history of polyps to prevent adenoma recurrence	Low	Weak
Maintenance of healthy BMI and control of risk factors for metabolic syndrome (abdominal obesity, hyperinsulinemia)	Moderate	Strong
Regular physical exercise	Moderate	Strong
Nonsmoking	Moderate	Strong
Reduced alcohol consumption	Moderate	Strong
NSAIDs (including SAA) not to be administered as therapy. Risk–benefit allows administration of low-dose SAA in 50–59 age group with cardiovascular disease	Moderate	Mainly not recommended

Adapted from Cubiella J, Marzo-Castillejo M, Mascort-Roca JJ, et al. Clinical practice guideline. Diagnosis and prevention of colorectal cancer. 2018 update. Gastroenterol Hepatol 2018;41(9):585–596. doi:10.1016/j.gastrohep.2018.07.012.

References

1. Afanador CH, Muñetón Peña CM. *Epigenetics of colorectal cancer*. Rev Colomb Gastroenterol 2017; http://www.scielo.org.co/scielo.php?script=sci_arttext&pid=S0120-99572018000100032.
2. Umezawa S, Higurashi T, Komiya Y, et al. Chemoprevention of colorectal cancer: past, present, and future. Cancer Sci 2019;110(10):3018–26. https://doi.org/10.1111/cas.14149.
3. Menendez P, Villarejo P, Padilla D, Menéndez JM, Ja RM. *Epigenetics and colorectal cancer*. Cir Esp 2012;90(5):277–83. https://europepmc.org/article/med/22425513.
4. Watson AJM, Collins PD. Colon cancer: a civilization disorder. Dig Dis 2011;29(2):222–8. https://doi.org/10.1159/000323926.
5. Le Marchand L, Zhao LP, Quiaoit F, Wilkens LR, Kolonel LN. Family history and risk of colorectal cancer in the multiethnic population of Hawaii. Am J Epidemiol 1996;144(12):1122–8. https://doi.org/10.1093/oxfordjournals.aje.a008890.
6. Martínez ME, Marshall JR, Giovannucci E. Diet and cancer prevention: the roles of observation and experimentation. Nat Rev Cancer 2008;8(9):694–703. https://doi.org/10.1038/nrc2441.
7. Song M, Garrett WS, Chan AT. Nutrients, foods, and colorectal cancer prevention. Gastroenterology 2015;148(6):1244–1260.e16. https://doi.org/10.1053/j.gastro.2014.12.035.
8. Aituov B, Duisembekova A, Bulenova A, Alibek K. Pathogen-driven gastrointestinal cancers: time for a change in treatment paradigm? Infect Agent Cancer 2012;7(1):18. https://doi.org/10.1186/1750-9378-7-18.
9. Aguirre-Portolés C, Fernández L, de Molina AR. Precision nutrition for targeting lipid metabolism in colorectal cancer. Nutrients 2017;9(10):1076. https://doi.org/10.3390/nu9101076.
10. Shivappa N, Godos J, Hébert J, et al. Dietary inflammatory index and colorectal Cancer risk—a meta-analysis. Nutrients 2017;9(9):1043. https://doi.org/10.3390/nu9091043.
11. Thanikachalam K, Khan G. Colorectal cancer and nutrition. Nutrients 2019;11(1). https://doi.org/10.3390/nu11010164.
12. Bouvard V, Loomis D, Guyton KZ, et al. Carcinogenicity of consumption of red and processed meat. Lancet Oncol 2015;16(16):1599–600. https://doi.org/10.1016/S1470-2045(15)00444-1.
13. McCullough ML, Gapstur SM, Shah R, Jacobs EJ, Campbell PT. Association between red and processed meat intake and mortality among colorectal cancer survivors. J Clin Oncol 2013;31(22):2773–82. https://doi.org/10.1200/JCO.2013.49.1126.
14. Moskal A, Freisling H, Byrnes G, et al. Main nutrient patterns and colorectal cancer risk in the European prospective investigation into cancer and nutrition study. Br J Cancer 2016;115(11):1430–40. https://doi.org/10.1038/bjc.2016.334.
15. Norat T, Bingham S, Ferrari P, et al. Meat, fish, and colorectal cancer risk: the European prospective investigation into cancer and nutrition. J Natl Cancer Inst 2005;97(12):906–16. https://doi.org/10.1093/jnci/dji164.
16. Bilsborough S, Mann N. A review of issues of dietary protein intake in humans. Int J Sport Nutr Exerc Metab 2006;16(2):129–52. https://doi.org/10.1123/ijsnem.16.2.129.
17. Richi BE, Baumer B, Conrad B, Darioli R, Schmid A, Keller U. Health risks associated with meat consumption: a review of epidemiological studies. Int J Vitam Nutr Res 2015;85(1–2):70–8.
18. Aune D, Lau R, Chan DSM, et al. Dairy products and colorectal cancer risk: a systematic review and meta-analysis of cohort studies. Ann Oncol 2012;23(1):37–45. https://doi.org/10.1093/annonc/mdr269.
19. Barrubés L, Babio N, Becerra-Tomás N, Rosique-Esteban N, Salas-Salvadó J. Association between dairy product consumption and colorectal cancer risk in adults: a systematic review and meta-analysis of epidemiologic studies. Adv Nutr 2019;10(Suppl_2):S190–211. https://doi.org/10.1093/advances/nmy114.
20. He X, Wu K, Zhang X, et al. *Dietary intake of fiber, whole grains and risk of colorectal cancer: an updated analysis according to food sources, tumor location and molecular subtypes in two large US cohorts*. Int J Cancer 2019;145(11):3040–51. https://onlinelibrary.wiley.com/doi/abs/10.1002/ijc.32382.
21. Wiseman MJ. Nutrition and cancer: prevention and survival. Br J Nutr 2019;122(5):481–7. https://doi.org/10.1017/S0007114518002222.
22. Yao Y, Suo T, Andersson R, et al. Dietary fibre for the prevention of recurrent colorectal adenomas and carcinomas. Cochrane Database Syst Rev 2017;1:CD003430. https://doi.org/10.1002/14651858.CD003430.pub2.
23. Wu S, Feng B, Li K, et al. Fish consumption and colorectal cancer risk in humans: a systematic review and meta-analysis. Am J Med 2012;125(6):551–559.e5. https://doi.org/10.1016/j.amjmed.2012.01.022.
24. Aglago EK, Huybrechts I, Murphy N, et al. Consumption of fish and long-chain n-3 polyunsaturated fatty acids is associated with reduced risk of colorectal cancer in a large european cohort. Clin Gastroenterol Hepatol 2020;18(3):654–666.c6. https://doi.org/10.1016/j.cgh.2019.06.031.
25. Li G, Ma D, Zhang Y, Zheng W, Wang P. Coffee consumption and risk of colorectal cancer: a meta-analysis of observational studies. Public Health Nutr 2013;16(2):346–57. https://doi.org/10.1017/S1368980012002601.
26. Sartini M, Bragazzi NL, Spagnolo AM, et al. Coffee consumption and risk of colorectal cancer: a systematic review and meta-analysis of prospective studies. Nutrients 2019;11(3). https://doi.org/10.3390/nu11030694.
27. Kim JY, Kwon O. Garlic intake and cancer risk: an analysis using the food and drug administration's evidence-based review system for the scientific evaluation of health claims. Am J Clin Nutr 2009;89(1):257–64. https://doi.org/10.3945/ajcn.2008.26142.
28. Cai S, Li Y, Ding Y, Chen K, Jin M. Alcohol drinking and the risk of colorectal cancer death: a meta-analysis. Eur J Cancer Prev 2014;23(6):532–9.
29. Murphy N, Ward HA, Jenab M, et al. Heterogeneity of colorectal cancer risk factors by anatomical subsite in 10 European countries: a multinational cohort study. Clin Gastroenterol Hepatol 2019;17(7):1323–1331.e6. https://doi.org/10.1016/j.cgh.2018.07.030.

30. Klarich DS, Brasser SM, Hong MY. Moderate alcohol consumption and colorectal cancer risk. *Alcohol Clin Exp Res* 2015;**39**(8):1280–91. https://doi.org/10.1111/acer.12778.
31. Meyer F, White E. Alcohol and nutrients in relation to colon cancer in middle-aged adults. *Am J Epidemiol* 1993;**138**(4):225–36. https://doi.org/10.1093/oxfordjournals.aje.a116851.
32. Doll R, Peto R. Mortality in relation to smoking: 20 years' observations on male British doctors. *Br Med J* 1976;**2**(6051):1525–36. https://doi.org/10.1136/bmj.2.6051.1525.
33. Hannan LM, Jacobs EJ, Thun MJ. The association between cigarette smoking and risk of colorectal cancer in a large prospective cohort from the United States. *Cancer Epidemiol Biomarkers Prev* 2009;**18**(12):3362–7. https://doi.org/10.1158/1055-9965.EPI-09-0661.
34. Waluga M, Zorniak M, Fichna J, Kukla M, Hartleb M. Pharmacological and dietary factors in prevention of colorectal cancer. *J Physiol Pharmacol* 2018;**69**(3). https://doi.org/10.26402/jpp.2018.3.02.
35. Miller A. IARC monographs on the evaluation of the carcinogenic risk of chemicals to humans. Vol. 38. Tobacco smoking. *Food Chem Toxicol* 1987;**25**(8):627–8. https://doi.org/10.1016/0278-6915(87)90027-5.
36. Joshu CE, Parmigiani G, Colditz GA, Platz EA. Opportunities for the primary prevention of colorectal cancer in the United States. *Cancer Prev Res* 2012;**5**(1):138–45. https://doi.org/10.1158/1940-6207.CAPR-11-0322.
37. Ning Y, Wang L, Giovannucci EL. A quantitative analysis of body mass index and colorectal cancer: findings from 56 observational studies. *Obes Rev* 2010;**11**(1):19–30. https://doi.org/10.1111/j.1467-789X.2009.00613.x.
38. Michelet X, Dyck L, Hogan A, et al. Metabolic reprogramming of natural killer cells in obesity limits antitumor responses. *Nat Immunol* 2018;**19**(12):1330–40. https://doi.org/10.1038/s41590-018-0251-7.
39. Delage B, Rullier A, Capdepont M, Rullier E, Cassand P. The effect of body weight on altered expression of nuclear receptors and cyclooxygenase-2 in human colorectal cancers. *Nutr J* 2007;**6**:20. https://doi.org/10.1186/1475-2891-6-20.
40. Karahalios A, English DR, Simpson JA. Weight change and risk of colorectal cancer: a systematic review and meta-analysis. *Am J Epidemiol* 2015;**181**(11):832–45. https://doi.org/10.1093/aje/kwu357.
41. Cubiella J, Marzo-Castillejo M, Mascort-Roca JJ, et al. Clinical practice guideline. Diagnosis and prevention of colorectal cancer. 2018 update. *Gastroenterol Hepatol* 2018;**41**(9):585–96. https://doi.org/10.1016/j.gastrohep.2018.07.012.
42. Luo S, Li J-Y, Zhao L-N, et al. Diabetes mellitus increases the risk of colorectal neoplasia: an updated meta-analysis. *Clin Res Hepatol Gastroenterol* 2016;**40**(1):110–23. https://doi.org/10.1016/j.clinre.2015.05.021.
43. Hu FB, Willett WC, Li T, Stampfer MJ, Colditz GA, Manson JE. Adiposity as compared with physical activity in predicting mortality among women. *N Engl J Med* 2004;**351**(26):2694–703. https://doi.org/10.1056/NEJMoa042135.
44. Regensteiner JG, Mayer EJ, Shetterly SM, et al. Relationship between habitual physical activity and insulin levels among nondiabetic men and women: San Luis valley diabetes study. *Diabetes Care* 1991;**14**(11):1066–74. https://doi.org/10.2337/diacare.14.11.1066.
45. Friedenreich CM, Neilson HK, Lynch BM. State of the epidemiological evidence on physical activity and cancer prevention. *Eur J Cancer* 2010;**46**(14):2593–604. https://doi.org/10.1016/j.ejca.2010.07.028.
46. Patrono C, Patrignani P, García Rodríguez LA. Cyclooxygenase-selective inhibition of prostanoid formation: transducing biochemical selectivity into clinical read-outs. *J Clin Invest* 2001;**108**(1):7–13. https://doi.org/10.1172/JCI13418.
47. Rothwell PM. Aspirin in prevention of sporadic colorectal cancer: current clinical evidence and overall balance of risks and benefits. *Recent Results Cancer Res* 2013;**191**:121–42. https://doi.org/10.1007/978-3-642-30331-9_7.
48. Chan AT, Cook NR. Are we ready to recommend aspirin for cancer prevention? *Lancet* 2012;**379**(9826):1569–71. https://doi.org/10.1016/S0140-6736(11)61654-1.
49. Bibbins-Domingo K, on behalf of the U.S. Preventive Services Task Force. Aspirin Use for the primary prevention of cardiovascular disease and colorectal cancer: U.S. preventive services task force recommendation statement. *Ann Intern Med* 2016;**164**(12):836. https://doi.org/10.7326/m16-0577.
50. Labayle D, Fischer D, Vielh P, et al. Sulindac causes regression of rectal polyps in familial adenomatous polyposis. *Gastroenterology* 1991;**101**(3):635–9. https://doi.org/10.1016/0016-5085(91)90519-q.
51. Ruder EH, Laiyemo AO, Graubard BI, Hollenbeck AR, Schatzkin A, Cross AJ. Non-steroidal anti-inflammatory drugs and colorectal cancer risk in a large, prospective cohort. *Am J Gastroenterol* 2011;**106**(7):1340–50. https://doi.org/10.1038/ajg.2011.38.
52. Madka V, Rao C. Anti-inflammatory phytochemicals for chemoprevention of colon cancer. *Curr Cancer Drug Targets* 2013;**13**(5):542–57. https://doi.org/10.2174/15680096113139990036.
53. Tomić T, Domínguez-López S, Barrios-Rodríguez R. Non-aspirin non-steroidal anti-inflammatory drugs in prevention of colorectal cancer in people aged 40 or older: a systematic review and meta-analysis. *Cancer Epidemiol* 2019;**58**:52–62. https://doi.org/10.1016/j.canep.2018.11.002.
54. Huncharek M, Muscat J, Kupelnick B. Colorectal cancer risk and dietary intake of calcium, vitamin D, and dairy products: a meta-analysis of 26,335 cases from 60 observational studies. *Nutr Cancer* 2009;**61**(1):47–69. https://doi.org/10.1080/01635580802395733.
55. Larsson SC, Bergkvist L, Wolk A. Magnesium intake in relation to risk of colorectal cancer in women. *JAMA* 2005;**293**(1):86–9. https://doi.org/10.1001/jama.293.1.86.
56. McCullough ML, Zoltick ES, Weinstein SJ, et al. Circulating vitamin D and colorectal cancer risk: an international pooling project of 17 cohorts. *J Natl Cancer Inst* 2019;**111**(2):158–69. https://doi.org/10.1093/jnci/djy087.
57. Lee JE, Li H, Chan AT, et al. Circulating levels of vitamin D and colon and rectal cancer: the physicians' health study and a meta-analysis of prospective studies. *Cancer Prev Res* 2011;**4**(5):735–43. https://doi.org/10.1158/1940-6207.CAPR-10-0289.

58. Larsson SC, Orsini N, Wolk A. Vitamin B6 and risk of colorectal cancer: a meta-analysis of prospective studies. *JAMA* 2010;**303**(11):1077–83. https://doi.org/10.1001/jama.2010.263.
59. Nolfo F, Rametta S, Marventano S, et al. Pharmacological and dietary prevention for colorectal cancer. *BMC Surg* 2013;**13 Suppl 2**:S16. https://doi.org/10.1186/1471-2482-13-S2-S16.
60. Pais R, Dumitraşcu DL. *Do antioxidants prevent colorectal cancer? A meta-analysis*. *Rom J Intern Med* 2013;**51**(3–4):152–63. https://www.ncbi.nlm.nih.gov/pubmed/24620628.
61. Barzi A, Lenz AM, Labonte MJ, Lenz H-J. Molecular pathways: estrogen pathway in colorectal cancer. *Clin Cancer Res* 2013;**19**(21):5842–8. https://doi.org/10.1158/1078-0432.CCR-13-0325.
62. Taylor ML, Wells BJ, Smolak MJ. Statins and cancer: a meta-analysis of case–control studies. *Eur J Cancer Prev* 2008;**17**(3):259. https://doi.org/10.1097/CEJ.0b013e3282b721fe.
63. Lochhead P, Chan AT. *Statins and colorectal cancer*. *Clin Gastroenterol Hepatol* 2013;**11**(2):109–18. quiz e13-e14. https://doi.org/10.1016/j.cgh.2012.08.037.
64. Thosani N, Thosani SN, Kumar S, et al. Reduced risk of colorectal cancer with use of oral bisphosphonates: a systematic review and meta-analysis. *J Clin Oncol* 2013;**31**(5):623–30. https://doi.org/10.1200/JCO.2012.42.9530.
65. Singh S, Singh H, Singh PP, Murad MH, Limburg PJ. Antidiabetic medications and the risk of colorectal cancer in patients with diabetes mellitus: a systematic review and meta-analysis. *Cancer Epidemiol Biomarkers Prev* 2013;**22**(12):2258–68. https://doi.org/10.1158/1055-9965.EPI-13-0429.
66. Kantor ED, Zhang X, Wu K, et al. *Use of glucosamine and chondroitin supplements in relation to risk of colorectal cancer: results from the nurses' health study and health professionals follow-up study*. *Int J Cancer* 2016;**139**(9):1949–57. https://onlinelibrary.wiley.com/doi/abs/10.1002/ijc.30250.
67. Kedika R, Patel M, Pena Sahdala HN, Mahgoub A, Cipher D, Siddiqui AA. Long-term use of angiotensin converting enzyme inhibitors is associated with decreased incidence of advanced adenomatous colon polyps. *J Clin Gastroenterol* 2011;**45**(2):e12–6. https://doi.org/10.1097/MCG.0b013e3181ea1044.
68. Pan P, Yu J, Wang L-S. Diet and colon: what matters? *Curr Opin Gastroenterol* 2019;**35**(2):101–6. https://doi.org/10.1097/MOG.0000000000000501.
69. Obón-Santacana M, Romaguera D, Gracia-Lavedan E, et al. Dietary inflammatory index, dietary non-enzymatic antioxidant capacity, and colorectal and breast cancer risk (MCC-Spain Study). *Nutrients* 2019;**11**(6). https://doi.org/10.3390/nu11061406.

Chapter 4

Early onset of CRC

Andrés Dacal Rivas, Eva Martí Marqués, and Leopoldo López Rosés
Lucus Augusti University Hospital, Lugo, Spain

Introduction

Colorectal cancer is currently the third leading cause of cancer worldwide and, in terms of mortality, the second leading cause of death.[1]

In Spain it is the neoplasm with the third highest incidence in men and the second highest in women, but, in absolute numbers and considering both sexes, it is the most common cancer.[2]

The usual precursor lesion is the colorectal polyp, the sequence of progression to carcinoma is slow and, except for specific genetic syndromes, is estimated to occur in about 10–15 years (possibly faster if the precursor lesion is a serrated lesion). This carcinogenesis is due to an accumulation of genetic and epigenetic alterations that lead to the activation of oncogenes and inactivation of suppressor genes.

In developed countries, there has been evidence of a stabilization and even a downward trend in its incidence and mortality, which is related to the development and implementation of national population screening programs. However, current studies show that in the last three decades the number of cases diagnosed at young ages has been increasing.[3,4]

In line with most publications and also taking as a reference the age at which screening is initiated in the average-risk population, we can describe a colorectal cancer diagnosed at an early age as one that debuts before the age of 50 years. As a rule, cancer in the pediatric age group is not considered within this group.

Initially in the United States, and subsequently becoming a global phenomenon, an increase in diagnoses in patients between 20 and 49 years of age has been reported. At the European level, a study analyzing data from 20 countries shows a growth in the number of cases of between 1.6% and 7.9% per year, with a more marked increase in incidence in younger individuals, in the 20–29 age group.[4,5]

Exposure to environmental factors

The increased risk of colorectal cancer has been related in part to various lifestyle factors, which may condition a pro-inflammatory state at the colonic level and modify the microbiota.

These factors range from toxic habits such as smoking and alcohol consumption to a sedentary lifestyle and a high body mass index.

The consumption of a diet rich in saturated fats and red meat, as well as processed foods or the use of food additives (synthetic colorings, flavor enhancers such as monosodium glutamate, etc.), has also been associated with this increased risk.[3]

On the other hand, obesity is associated with a pro-inflammatory state and intestinal dysbiosis. In obese patients, factors such as a diet rich in fats and additives are also added, and it is believed that childhood obesity could condition a higher risk metabolic profile.[6]

However, this risk is not only conditioned by the body mass index, but it has been observed that physical inactivity, even in the absence of obesity, is a risk factor for the development of colorectal cancer at an early age.[6]

The effect of exposure to certain factors from the embryonic stage and birth is not yet known with sufficient evidence. For example, stress during pregnancy could be related to an increased risk of colorectal cancer in the offspring.[6]

Among the preventive measures to avoid the development of colorectal cancer, a diet rich in fiber, dairy products, and low in fat is recommended, together with regular physical activity and maintaining the body mass index in the normal range, controlling the factors that give rise to the metabolic syndrome.

Microbiota alteration

Although the balance of the interaction between the host and the intestinal microbiota is mostly beneficial, several organisms (such as *Fusobacterium nucleatum*, *Escherichia coli*, *Salmonella enterica*, *Bacteroides fragilis*) have been identified as playing a role in the carcinogenesis pathway.[6]

This microbiota is conditioned by multiple environmental factors, beginning from birth with breastfeeding and, subsequently, the type of diet and the consumption of food additives.

Exposure to antibiotics, especially at early ages, has been related to an increased risk of adenomas and colorectal cancer, with a higher risk associated with those of broad spectrum. The reasons are not fully known, but they are linked to an alteration in the individual's microbiota and gastrointestinal homeostasis.[6,7]

Hereditary factors

Approximately one-third of patients who develop colorectal cancer at an early age report at least one family history of this neoplasm, but although we actively search for hereditary syndromes that are related to the development of this pathology, in the majority of cancers diagnosed before the age of 50 years no pathogenic variant in genes involved in the development of cancer will be detected and in only 16%–20% of patients will we find these variants.[6,8] It should be clarified that these percentages are reached when using new generation sequencing methods with multigene panels and that not all the genes are of high penetrance, but up to half of those reported are unusual genes in colorectal cancer risk studies with moderate-high penetrance (ATM, CHECK2, BRCA1-2, TP53, etc.).[9]

The most frequently diagnosed hereditary colorectal cancer risk syndrome is Lynch syndrome. About 10%–20% of tumors diagnosed at an early age present microsatellite instability and, of this group, germline pathogenic variants of DNA repair genes are detected in most of them, which constitutes the diagnosis of Lynch syndrome. The biallelic or constitutive deficit of DNA repair genes is usually associated with the appearance of colorectal cancer in the 1st or 2nd decade of life, so it usually debuts in pediatric ages, and not later than 20 years of age.[10]

Other hereditary syndromes associated with colorectal cancer detected among these patients, although less frequently than Lynch syndrome, are Familial Adenomatous Polyposis. These are syndromes with a characteristic phenotype ranging from oligopolyposis forms (>20 but <100 adenomas) to classic forms (>100 and even thousands of adenomas), and which are related to pathogenic variants in the APC and MUTYH genes.

The genetic spectrum of predisposition to colorectal cancer is broad, ranging from genes of high penetrance to a polygenic influence constituted by multiple variants of low penetrance probably influenced by the environmental interaction previously described.[7]

Molecular differences

Three main molecular pathways of carcinogenesis have been described:

(1) Chromosomal instability (CIN) or suppressor pathway: the most frequent, accounting for 70%–90% of sporadic CRC. Carcinogenesis is promoted by inactivation of tumor suppressor genes (APC, SMAD4, TP53, KRAS, and BRAF). There are chromosomal abnormalities with allelic losses and gains (SCNA_somatic copy number alterations). The resulting cancer does not usually present microsatellite instability.
(2) Microsatellite instability (MSI) or mutator pathway: caused by a defect in the DNA repair system (MMR: MLH1, MSH2, MSH6, and PMS2), present in 2%–7% of sporadic CRC. The result is a CRC with microsatellite instability.
(3) Methylator phenotype or CIMP (CpG Island Methylator Phenotype): the phenomenon of hypermethylation of CpG islands in promoter regions of tumor suppressor genes (KRAS, BRAF, MLH1). Somatic inactivation of MLH1 by this mechanism is a frequent reason for CRC with microsatellite instability. This pathway of carcinogenesis is related to the serrated pathway and is considered to be responsible for 10%–20% of sporadic CRCs. It can give rise to stable or unstable tumors.

In 2015 a molecular classification was established that includes four different subtypes or CMS (Consensus Molecular Subtypes) based on the three previously described pathways.[11] This classification is presented in Table 4.1.

TABLE 4.1 Molecular classification established in four different subtypes or CMS (consensus molecular subtypes).

CMS1 MSI immune	CMS2 Canonical	CMS3 Metabolic	CMS4 Mesenchymal
MSI, CIMP high, hypermutation	SCNA high	Mixed MSI status, SCNA low, CIMP low	SCNA high
BRAF mutations		KRAS mutations	
Immune infiltration and activation	WNT an MYC activation	Metabolic deregulation	Stromal infiltration, TGFβ activation, angiogenesis
Worse survival after relapse			Worse relapse-free and overall survival

Thus the evidence currently available reveals that colorectal cancer at early ages presents certain different molecular characteristics.

A low prevalence of somatic mutation of APC and BRAF has been reported, and the serrated pathway of carcinogenesis is infrequent among these patients, so it is understood that serrated lesions do not constitute a cause that justifies the increase in incidence at these ages.[12]

The CMS1 subtype is more frequent than in older patients and the inflammatory-immune pathway may play an important role in the development of colorectal cancer below the age of 50 years.[3] We found a higher proportion of unstable tumors concerning the general population and that at these ages they are frequently related to germline mutations of the genes encoding the DNA repair system (Sd Lynch), while above 50 years of age the fundamental cause is the inactivation of MLH1 at the somatic level.[13]

The majority of CRCs at a young age are, however, stable tumors (not MSI) and, as in the general population, chromosomal instability remains the most frequent mechanism of carcinogenesis.[14] However, it is more frequent that they present a poor degree of cellular differentiation, as well as mucinous or signet ring cell differentiation. Synchronous and metachronous tumors are also more frequent.[6, 13]

Likewise, a subgroup of stable tumors without chromosomal instability has been described more frequently in these young patients, with a worse prognosis due to the fact that they are associated with advanced stages at diagnosis, a higher rate of recurrence, and lower survival.[13]

Clinical features

Colorectal cancer at early ages settles more frequently in distal sections of the intestine (especially in the sigmoid colon and rectum) than that diagnosed above the age of 50.[13, 15]

At diagnosis, it presents more advanced stages, with a prevalence of stages III-IV of 53%–72%.[7] This is related to a delay in diagnosis, since practically all patients are diagnosed when they are already symptomatic (86.4%–94%), with times from the onset of symptoms to diagnosis about 100 days longer than those patients over 50 years of age, resulting in a total delay of up to 6 months.[3, 15]

There is an excessive perception of benignity, both on the part of the physician and the patient, conditioned by the low probability of cancer diagnosis, which causes symptoms such as hematochezia or weight loss to go unnoticed.[7]

Alarm symptoms precede the diagnosis and the most frequent is rectorrhagia. Establishing a diagnostic algorithm for this presentation and not blaming the symptom on hemorrhoidal pathology could avoid diagnostic delays.[3] Some scientific societies recommend rectosigmoidoscopy or colonoscopy in the presence of rectorrhagia also at ages <50 years. Other strategies such as fecal immunochemical test (FIT) have been shown to have a good negative predictive value in the symptomatic patient scenario.[16]

Medical societies in other countries have recommended bringing forward the starting age for population screening to 45 years, basing this decision on mathematical predictive models, which show a favorable balance of benefit (in years of life gained) against the cost of the increase in the number of colonoscopies.[17] In our setting, we do not currently have scientific evidence to support this decision.

Conclusions

The incidence and mortality secondary to colorectal cancer below the screening age is increasing. Although the specific causes are unknown, it is probably due to an interaction of multiple factors: genetic, epigenetic, and environmental exposure.

Ruling out germline pathogenic variants is warranted, given the prevalence of hereditary syndromes in this population, but in the vast majority of cases, these studies will be uninformative.

Colorectal cancer diagnosed at an early age presents different molecular characteristics, which opens the door to identifying new therapeutic targets and individualized treatment.

The preventive strategies to reverse this situation have yet to be defined but, if the trend continues, it will generate the need to consider advancing the age of population screening to 45 years, as is already recommended in other countries. It would also be necessary to educate the general population about the importance of modifiable exposure factors.

Information and awareness-raising efforts should be made, both for the responsible physicians and the target population in this age range, on alarm symptoms or particular risk factors in order to avoid diagnostic delays.

References

1. Global Cancer Observatory. http://gco.iarc.fr/. [Accessed 9 March 2021].
2. Local index—HTTrack Website Copier. http://redecan.org/. [Accessed 9 March 2021].
3. Burnett-Hartman AN, Lee JK, Demb J, Gupta S. An update on the epidemiology, molecular characterization, diagnosis, and screening strategies for early-onset colorectal cancer. *Gastroenterology* 2021. https://doi.org/10.1053/j.gastro.2020.12.068. Published online January 5.
4. Siegel RL, Torre LA, Soerjomataram I, et al. Global patterns and trends in colorectal cancer incidence in young adults. *Gut* 2019;**68**(12):2179–85. https://doi.org/10.1136/gutjnl-2019-319511.
5. Vuik FE, Nieuwenburg SA, Bardou M, et al. Increasing incidence of colorectal cancer in young adults in Europe over the last 25 years. *Gut* 2019;**68**(10):1820–6. https://doi.org/10.1136/gutjnl-2018-317592.
6. Hofseth LJ, Hebert JR, Chanda A, et al. Early-onset colorectal cancer: initial clues and current views. *Nat Rev Gastroenterol Hepatol* 2020;**17**(6):352–64. https://doi.org/10.1038/s41575-019-0253-4.
7. Akimoto N, Ugai T, Zhong R, et al. Rising incidence of early-onset colorectal cancer—a call to action. *Nat Rev Clin Oncol* 2020. https://doi.org/10.1038/s41571-020-00445-1. Published online November 20.
8. Pearlman R, Frankel WL, Swanson B, et al. Prevalence and spectrum of germline cancer susceptibility gene mutations among patients with early-onset colorectal cancer. *JAMA Oncol* 2017;**3**(4):464–71. https://doi.org/10.1001/jamaoncol.2016.5194.
9. Stoffel EM, Koeppe E, Everett J, et al. Germline genetic features of young individuals with colorectal cancer. *Gastroenterology* 2018;**154**(4):897–905. e1. https://doi.org/10.1053/j.gastro.2017.11.004.
10. Wimmer K, Kratz CP, Vasen HFA, et al. Diagnostic criteria for constitutional mismatch repair deficiency syndrome: suggestions of the European consortium "care for CMMRD" (C4CMMRD). *J Med Genet* 2014;**51**(6):355–65. https://doi.org/10.1136/jmedgenet-2014-102284.
11. Guinney J, Dienstmann R, Wang X, et al. The consensus molecular subtypes of colorectal cancer. *Nat Med* 2015;**21**(11):1350–6. https://doi.org/10.1038/nm.3967.
12. Willauer AN, Liu Y, Pereira AAL, et al. Clinical and molecular characterization of early-onset colorectal cancer. *Cancer* 2019;**125**(12):2002–10. https://doi.org/10.1002/cncr.31994.
13. Silla IO, Rueda D, Rodríguez Y, García JL, de la Cruz VF, Perea J. Early-onset colorectal cancer: a separate subset of colorectal cancer. *World J Gastroenterol* 2014;**20**(46):17288–96. https://doi.org/10.3748/wjg.v20.i46.17288.
14. Mauri G, Sartore-Bianchi A, Russo A-G, Marsoni S, Bardelli A, Siena S. Early-onset colorectal cancer in young individuals. *Mol Oncol* 2019;**13**(2):109–31. https://doi.org/10.1002/1878-0261.12417.
15. Patel SG, Boland CR. Colorectal cancer in persons under age 50. *Gastrointest Endosc Clin N Am* 2020;**30**(3):441–55. https://doi.org/10.1016/j.giec.2020.03.001.
16. D'Souza N, Georgiou Delisle T, Chen M, Benton S, Abulafi M, NICE FIT Steering Group. Faecal immunochemical test is superior to symptoms in predicting pathology in patients with suspected colorectal cancer symptoms referred on a 2WW pathway: a diagnostic accuracy study. *Gut* 2020. https://doi.org/10.1136/gutjnl-2020-321956. Published online October 21.
17. Mannucci A, Zuppardo RA, Rosati R, Leo MD, Perea J, Cavestro GM. Colorectal cancer screening from 45 years of age: thesis, antithesis and synthesis. *World J Gastroenterol* 2019;**25**(21):2565–80. https://doi.org/10.3748/wjg.v25.i21.2565.

Chapter 5

Population-based universal screening for CRC: Secondary prevention

Loretta De Chiara[a,b], María Gallardo-Gómez[a,b], Vicent Hernández[c,d], and Oscar J. Cordero[e]

[a]*Department of Biochemistry, Genetics and Immunology, University of Vigo, Vigo, Spain,* [b]*Singular Research Center of Galicia (CINBIO), University of Vigo, Vigo, Spain,* [c]*Department of Gastroenterology, Xerencia Xestion Integrada de Vigo, SERGAS, Vigo, Spain,* [d]*Research Group in Digestive Diseases, Galicia Sur Health Research Institute (IIS Galicia Sur), SERGAS-UVIGO, Vigo, Spain,* [e]*University of Santiago de Compostela, Department of Biochemistry and Molecular Biology, Santiago de Compostela, Spain*

Secondary prevention and screening concept

In any community or human population there are two approaches to the problem of a disease or lack of health. On one hand, to prevent its occurrence, and on the other hand, if prevention was not successful, treat it and try to cure as many people as possible.

There are three levels of prevention: primary, secondary, and tertiary. Primary prevention consists in the control of risk factors or the causes of the disease. Secondary prevention consists in the early diagnosis, before clinical symptoms appear, or in the diagnosis and treatment of precursory lesions. Tertiary prevention consists in preventing complications once the disease has been diagnosed.[1,2]

Primary prevention avoids the development of the disease, that is, incidence decreases. With secondary prevention, treatment can be started earlier, when the disease is in an early stage and treatment is usually more effective or easier to apply. In the case of cancer, where tumors have a generally long natural history, early diagnosis is linked to higher cure rates, and therefore to greater survival.

The diagnosis of a pathology is achieved through a laboratory test, examination, or a procedure, which allows detecting the presence of the disease. When we apply this test to the population, the concept of screening emerges. Screening aims to identify in an apparently healthy population with a risk factor, such as age, those individuals who have the disease in a presymptomatic phase, with the objective of achieving a better prognosis.[1–3]

The concept of screening is intrinsic to a public health system or at least widespread in a population or community. Therefore since procedures or tests began to exist to identify diseases in these presymptomatic phases, especially due to the development of molecular biology since the 1950s, it was thought that these procedures should be fast, cheap, and accepted by the population. This may mean that in many cases the screening tests do not have to be the same as those used for diagnosis.[1,3]

Conditions for population screening: WHO general principles

As a result of both the creation of the UN and the development of the Welfare States in the years following the Second World War, the World Health Organization (WHO) was created: an agency specialized in managing prevention policies, promotion, and intervention in health worldwide. In 1968 WHO published the standards believed were necessary for a health system to propose a screening program[2] which are still practically valid (Table 5.1).[4]

Colorectal cancer screening

Colorectal cancer meets all the conditions required for the adoption of a screening policy. First, because it is an important public health problem since it constitutes one of the most common cancers. According to GLOBOCAN 2018 estimates for all men and women worldwide, CRC is the third cancer with the highest incidence (10.2%) and the second leading cause of cancer death (9.2%).[5] For 2018 more than 1.8 million new cases of CRC and 881,000 deaths were estimated, representing

TABLE 5.1 WHO standards for a screening program.

1. The disease must be a major health problem.
2. There should be a treatment for the disease.
3. Resources must be available for both diagnosis and treatment.
4. There must be a latent or preclinical state of the disease.
5. There must be a test or exam that recognizes it.
6. The test must be easily acceptable by the population.
7. The natural history of the disease must be properly understood.
8. There must be consensus on who to treat.
9. The total cost of the diagnosed cases and treated patients should be economically balanced with the total health budget.
10. The screening program should be a continuous process and not a one-time project.

approximately 1 in 10 cases of cancer and deaths. In Europe, 499,700 new cases and 242,500 deaths were estimated for 2018, with the highest incidence rates in Central Europe (Hungary, Slovakia and Slovenia).[6]

The natural history of CRC has been well known for many years. Currently, several pathways for colorectal carcinogenesis are recognized, including the chromosomal instability pathway (CIN), the microsatellite instability pathway (MSI), and the CpG island methylation phenotype pathway (CIMP), also known as the serrated neoplasia pathway.[7–10] The development of a colorectal tumor implies the evolution of a precancerous lesion to a carcinoma, a process that can last several decades.[11] Therefore, there is an opportunity to intervene in such sequence with the removal of the lesions (polypectomy). In a recent study comparing the initial CRC diagnosis by screening colonoscopy, diagnostic colonoscopy, and emergent surgery, it was observed that 38.5% of the tumors detected in the first group were stage I, compared with 7.2% and 0% with diagnostic colonoscopy and emergent surgery, respectively.[12] Hence, the understanding of the preclinical state and its molecular alterations contributes to the implementation of a screening program, highly recommended to reduce the incidence (due to resection of precursor lesions) and mortality (due to early stage diagnosis) of CRC.

There are also precise and reliable methods for diagnosis and confirmation, which allow the detection of the disease in early asymptomatic stages as well as in metastatic stages. The clinical guidelines of the European, Spanish, and American Society—ESMO (European Society for Medical Oncology), SEOM (Spanish Society of Medical Oncology), and the ACS (American Cancer Society)—collect all the information related to the levels of evidence for different diagnostic methods, patient management, treatment, and follow-up.[13–18]

In relation to the initial screening test, the framework document on population screening prepared by *Grupo de trabajo de la Ponencia de Cribado de la Comisión de Salud Pública*, part of the Spanish Ministry of Health,[19] indicates that the screening test should be simple, fast, easy to perform, safe, highly sensitive and specific, well accepted by professionals and patients, and with a good cost-effectiveness. There is currently no noninvasive CRC screening test that meets all the aforementioned characteristics. Nowadays, the noninvasive test recommended in the European guide for average-risk population is the fecal occult blood test (FOBT).[20] The next section includes information about this test and others.

Finally, although the implementation of a CRC screening program is a complex process that requires coordination between Primary Care, Specialized Care, and Public Health managers, in Spain and Galicia there are adequate resources for screening programs to be a continuous process. Thus, in Spain, the evolution of the implementation of screening programs has been exponential since 2000, and since 2017 all autonomous communities have developed screening programs targeting asymptomatic individuals between 50 and 69 years of age (*Red de programas de Cribado de Cáncer*, 2019).

For the early detection of cancer, the European Council[21] made available to its Member States explicit recommendations that in Spain were included in the Comprehensive Plan for Cancer of the Spanish Ministry of Health (*Plan Integral del Cáncer del Ministerio de Sanidad y Consumo*). Under these premises, the Spanish Society of Family and Community Medicine (SEMFYC), the Spanish Association of Gastroenterology (AEG), and the Ibero-American Cochrane Center (CCI) developed the CRC Prevention Clinical Practice Guide (March 2004) within the Program "Development of Clinical Practice Guidelines (CPG) in Digestive Diseases, from Primary to Specialized Care" (updated in 2009 and

2018). Recall that the clinical practice guidelines are a set of recommendations that guide professionals and patients in the decision-making process on the most appropriate health interventions for addressing a specific clinical condition. Recommendations are developed to be made systematically in specific health circumstances.

Colorectal cancer screening tests

A variety of strategies are currently available for the screening of average-risk population, that is, asymptomatic individuals 50 years or older without other known risk factors for the development of CRC. In general terms, the tests used for CRC screening can be classified into invasive tests, such as colonoscopy, and noninvasive tests, such as the FOBT. All methods have their advantages and disadvantages regarding their sensitivity, specificity, risk, accessibility, acceptance, and cost, but all of them have shown their utility to reduce the incidence and mortality of this neoplasm to some extent.[22]

Colonoscopy

Colonoscopy is a procedure that allows the visualization of the entire colon, requiring prior preparation for bowel cleansing. Colonoscopy is considered the reference test for the detection of cancer and colorectal precancerous lesions, so it is considered the definitive test when another screening test is positive.[23] Some of the advantages of colonoscopy are that it allows the accurate location of the lesion and to perform a biopsy, detect synchronous lesions, and eliminate polyps (polypectomy).

The use of colonoscopy in screening reduces both the incidence and mortality of CRC. It is estimated that the use of this screening strategy achieves a reduction in the incidence of CRC between 66% and 90%, while mortality is reduced between 31% and 88%.[24, 25]

One of the potential benefits of this test in the context of screening is a longer interval between exams, since colonoscopy can be performed every 10 years, as opposed to annual or biennial FOBT or sigmoidoscopy every 3–5 years.[3, 13, 15, 26] Likewise, the major drawbacks for the use of colonoscopy as a screening test are the complications due to its invasiveness, which include perforation (0.07 of every 1000 procedures) and major postcolonoscopy bleeding (0.8 of 1000).[27] Another drawback is the variation in the detection of polyps and the significant number of polyps not detected during the procedure. A recent meta-analysis based on 43 publications and more than 15,000 tandem colonoscopies reported a rate of unnoticed adenomas (miss rate) of 26%, 9% for advanced adenomas, and 27% for serrated polyps.[28] These results should be taken with caution given the geographical heterogeneity of the populations and the design of the studies. The authors emphasize that the rates of unnoticed lesions could be reduced with adequate cleansing of the colon prior to colonoscopy and the use of auxiliary techniques such as endoscopic accessories or improved imaging techniques.

In relation to auxiliary techniques, recent advances in artificial intelligence (AI) in the emerging field of "deep learning" have the potential to decrease the rates of unnoticed lesions, contributing to the increased detection of lesions, and improved optical diagnosis of colorectal lesions, managing to reduce unnecessary polypectomies of nonneoplastic lesions.[29, 30] Although today there are studies that report promising results, it will be necessary to obtain more evidence to promote the implementation of computer-assisted diagnosis in the practice of colonoscopy.

Flexible sigmoidoscopy

Flexible sigmoidoscopy allows the examination of a part of the descending colon, sigma, and rectum, typically reaching the splenic flexure that corresponds to the middle of the colon. The procedure is performed with a flexible 60-cm endoscope and the patient only requires an enema as preparation.

The use of sigmoidoscopy in screening is based on the fact that two out of three lesions are detected in the distal colon,[31] and on the fact that distal lesions can predict the presence of proximal lesions.[32] Therefore, when a finding is detected in sigmoidoscopy, the patient is referred to colonoscopy given the risk of presenting a clinically significant pathology in the proximal colon.[33]

Using flexible sigmoidoscopy as a screening test reduces the incidence of CRC between 18% and 23%, especially of tumors located in the distal colon, while mortality decreases between 22% and 31%.[31, 34] The effectiveness of sigmoidoscopy on the basis of age and gender has been analyzed in randomized studies: PLCO (US Prostate, Lung, Colorectal and Ovarian cancer screening trial), SCORE (Italian Screening for Colon and Rectum trial), and NORCCAP (Norwegian Colorectal Cancer Prevention trial). The study determined that sigmoidoscopy is a valid screening option for men (55–74 years) and for women under 60 years.[35]

Compared to colonoscopy, flexible sigmoidoscopy is safer, with a lower complication rate (0.005% perforation and 0.01% bleeding).[36, 37] The procedure is less expensive than colonoscopy and requires less time for execution. However, the main disadvantage is that it does not allow the visualization of the entire colon, so it cannot detect proximal lesions. On the other hand, one of the greatest advantages of flexible sigmoidoscopy over FOBT is its ability to examine a portion of the colon.

CT scan colonography

One of the most recent methods is the computed tomographic colonography (CTC), also called virtual colonoscopy. This technique was introduced in 1994 and involves the use of helical computerized tomography data in combination with graphic software that generates images of the colon in two and three dimensions. The test is only diagnostic and requires a good previous preparation of the colon. Although the individual is exposed to radiation, the dose is much lower than that used in other CT scans.[38]

Among the greatest advantages, the CTC allows the visualization of the anatomy of the colon from an endoluminal perspective. The patient's risk is minimal, the time to perform the test is 10–15 min, and allows the evaluation of segments proximal to obstructive cancers, where the colonoscopy could not reach. Currently, its clinical use includes the evaluation of patients in which a colonoscopy could not be performed and patients with obstructive cancer. However, it has the disadvantage of requiring additional time for the interpretation of the images obtained, does not allow biopsy or polypectomy, and still has a high cost.

Three European trials have provided evidence that supports the use of CTC for screening.[39] This review emphasizes the importance of choosing the appropriate option for clinical management based on the initial findings of the CTC. They also point out that the future of this test will be focused on the training of radiologists and the quality of the image evaluation protocol.

Regarding the comparison of CTC with colonoscopy, a recent review and meta-analysis reported participation rates of 29% for the first and 20% for the second.[40] The detection rates of advanced colorectal neoplasia were 5.7% for CTC and 8.5% for colonoscopy, so the authors conclude that virtual colonoscopy should not replace colonoscopy. However, a study that compared the cost-effectiveness of both tests in a screening context reported a higher participation rate for CTC, resulting in more cost-effective than screening colonoscopy.[41]

Fecal occult blood test (FOBT)

The effectiveness of a screening test depends not only on its diagnostic performance, but also to a large extent on the acceptance of the test by the target population.[25] The most commonly used noninvasive clinical test is the fecal occult blood.

The FOBT is based on the fact that polyps and tumors bleed, estimating that one-third of the lesions bleed, and blood loss increases with size.[36, 37] However, bleeding is intermittent and is unevenly distributed in the stool. In addition, blood in stool does not necessarily indicate the presence of a polyp or tumor, and may be due to other pathologies such as hemorrhoids or inflammatory processes.

Blood hemoglobin can be detected by chemical tests based on guaiac (gFOBT) or immunochemical (FIT, fecal immunochemical test). Given the nature of gFOBT, it is not specific to human blood, therefore meat, some raw fruits and vegetables, and the consumption of NSAIDs (nonsteroidal antiinflammatory drugs such as aspirin or ibuprofen) should be avoided between 3 and 5 days before the test to decrease the percentage of false positives. Currently, gFOBT does not constitute a relevant test for CRC screening.

Other tests based on the immunochemical detection of human hemoglobin have been developed, avoiding interactions with diet or drugs. These tests are based on a reaction with antibodies that detects human globin. Hemoglobin from the highest portions of the colon is generally degraded by bacteria and digestive enzymes, losing its ability to react with the chemical test, but still remains immunologically reactive. Therefore, these tests allow detecting significantly lower concentrations of fecal hemoglobin than gFOBT, which facilitates the detection not only of invasive CRC but also of advanced adenomas. Currently, quantitative FIT (OC-Light) is universally used, although there are also qualitative tests (OC-FIT-CHEK, OC-Sensor) available.[42]

FIT shows sensitivity between 73% and 88% to detect cancer cases, with 90%–94% specificity. However, its diagnostic capacity is more modest for the detection of advanced neoplasms, showing a sensitivity between 22% and 56%, depending on the cutoff selected.[26, 42–44] On the other hand, FIT detects distal lesions better than proximal lesions. This is due to the degradation of hemoglobin during its movement through the intestinal tract, and also to the different distribution of blood in the stool, which would be more superficial if it comes from distal lesions.[45]

Among all the methods recommended in clinical guidelines, FIT constitutes the noninvasive screening test with less risk, cheaper, and more accessible. However, although its sensitivity is adequate for CRC diagnosis, it is not suitable for premalignant lesions.

Other noninvasive tests: Stool or blood tests

Other noninvasive screening tests based on stool or blood samples have been developed, with the intention of improving the diagnostic performance of FIT. Until recently, stool DNA tests for screening were not satisfactory. However, Exact Sciences developed a multitarget test that includes several genetic mutations associated with CRC, including hypermethylation of BMP3 and NDRG4, and point mutations in KRAS, beta-actin gene as a reference gene, in addition to FIT.[42] Although there are no randomized prospective studies, a study with a large cohort reported a sensitivity of 92% for cancer, with 84% specificity (FIT: 72% sensitivity and 93% specificity). Regarding the detection of advanced adenomas, the multitarget test showed a sensitivity of 42% and 87% specificity (FIT: 24% sensitivity and 95% specificity).[46] These results indicate that the multitarget test has a higher rate of false positives, though sensitivity is better than that of FIT. In addition, its cost is very high compared to FIT and there is still no data on the optimal test interval or the impact of false positives.

On the other hand, serum markers are a good alternative compared to invasive tests and FIT, since obtaining a blood sample is simple, painless, inexpensive, and better accepted by the target population.[47] Another advantage of blood markers is their potential ability to detect lesions regardless of their location in the colorectal tract.[48]

A large number of biomarkers with potential utility for the detection of CRC and/or adenomas have been described in literature. These can be grouped mainly into genetic markers, epigenetic markers, protein markers, and antigenic-based markers. In a meta-analysis, 70 different blood biomarkers were evaluated, with sensitivities ranging between 18% and 65% for the detection of CRC.[49]

Many studies have reported changes in gene expression in the presence of CRC and adenomas. The main genetic markers include KRAS, p53, APC, BRAF, hMLH1, DcR3, TRAIL-R2, BANK1, BCNP1, and MS4A1, among others, with sensitivities for CRC between 50% and 94%, and specificities between 60% and 95%.[49–51] The most studied epigenetic marker is Septin-9 (SEPT9), approved by the FDA for the noninvasive detection of CRC. Plasma SEPT9 has a sensitivity for the detection of CRC between 36%–79% and a specificity of 86%–99%, with a reduced sensitivity for detecting advanced adenomas. Other notable methylation markers are IKZF1, BCAT1, ALX4,[52] VIM and SDC2, among others.[51,53]

Protein markers are also possible screening biomarkers. Within this group of markers we find antigens, antibodies, cytokines, and other proteins. One of the best known antigens related to CRC is CEA (carcinoembryonic antigen), with sensitivity between 43% and 69%, and 87%–90% specificity. Due to its low sensitivity and specificity, CEA has no utility for diagnosis or screening,[54] although it is considered the best prognostic and follow-up marker currently available in the clinic.

There are many studies focused on the search for protein biomarkers for the diagnosis and/or screening of CRC, some with quite promising results in terms of sensitivity and specificity: u-PA (76% and 96%, respectively), M2PK (69% and 90%), TPA-M (70% and 96%), sCD26 (90% and 90%), DR-70 (80% and 93%), prolactin (77% and 98%), laminin (89% and 88%), CCSA-2 (78%–97%).[49,55] Given the absence of a single marker that allows the correct classification of 100% of the target population (100% sensitivity, 100% specificity), the strategy is the combination of a panel of markers that can be measured in blood. Although it is difficult to establish which markers can be combined to be part of a panel, efforts of many research groups are focused on finding a panel with an optimal performance.[49,56,57] More detailed information on protein markers, in addition to genomic, epigenomic, and transcriptomic, in blood and feces, is included in Chapter 22 of this book.

Requirements for the implementation of a screening program

The Spanish framework document on population screening, prepared by the *Grupo de trabajo de la Ponencia de Cribado de la Comisión de Salud Pública*, depending on the Spanish Ministry of Health, includes the following issues among the criteria for strategic decision-making related to population screening programs[19]:

— Evidence of effectiveness: the effectiveness in reducing mortality or morbidity should be clearly demonstrated based on quality scientific studies, and should be evaluated by independent agencies or agencies specialized in health technologies.
— Benefit greater than the potential risks: before implementing a screening program it is necessary to quantify the benefit in terms of risk reduction and the impact of the intervention on the disease in the target population. Potential adverse effects such as psychological, procedural complications, overdiagnosis and overtreatment should also be assessed.

- Definition of the target population: the target population must be very well defined to identify and invite all individuals.
- Balanced cost: the economic impact of the entire screening program must be known in order to demonstrate that it is cost effective for the health system.
- Program acceptable from a clinical, social, and ethical point of view: equity in access must be promoted, confidentiality must be ensured, and autonomy must be respected.
- Evaluation and quality: it must be ensured that the final results related to the reduction of the disease burden are fixed in advance to measure the impact on health. Quality assurance should guide the actions of the program.
- Feasibility of the program within the National Health System: it will be necessary to evaluate the infrastructure, and material and human resources to determine the possibility of releasing certain activity and redirect it to the screening program. The beginning of a screening program may initially result in an increase in activity, waiting lists, and treatment costs. Initial and global investments over time should be considered, as screening can save more expensive tests and long-term treatments.

Although the criteria for establishing a population screening program are met, it cannot be initiated until the capacity of the system is ensured so that the program meets all of the following requirements related to its implementation:

- Population coverage and equity: screening programs can only be executed by the competent health administration. The implementation of the program must be done in a reasonably short time to reach the coverage of the entire target population, thereby avoiding inequality in access. To ensure cohesion and equity, the implementation must be coordinated between the different autonomous communities.
- Operational planning and coordination: screening begins with an initial test and ends with appropriate intervention in individuals in whom the diagnosis has been confirmed. Therefore, the comprehensive screening program must be planned and designed. Operational planning of the different aspects of the program must be carried out: management of the invitation to the target population, initial test, possible results, diagnostic confirmation and treatment, quality and evaluation, human and material resources, education, and information systems, among others.
- Program information system: it will be necessary to have an adequate information system that allows the integration of the program management tasks (personalized invitation, initial test), the coordination of the different actors in the entire screening process, the referral of the suspicious cases, the follow-up of the detected cases, the monitoring of the activity, the quality control, and the evaluation of results.
- Informed decision: according to the General Medical Council, for individuals to make a true informed decision about screening they should receive information about the purpose of the program; the possibility of positive and negative, false positive and false negative results; the uncertainties and risks associated with the entire process; any significant medical, social, or financial involvement; and the monitoring plan, including the availability of advisory and support services.[58]
- Protection of personal data and guarantee of confidentiality: throughout the entire screening process, the protection of personal privacy and the confidential treatment of personal data generated and stored in the documentation systems must be guaranteed, in accordance with current laws.
- Evaluation and quality plan: the program must have the evaluation and quality control mechanisms to ensure that the planned objectives are achieved. The evaluation must be planned and periodic. The program must be capable of updating mechanisms or making relevant changes according to health outcomes, cost-effectiveness, and cost-utility.
- Training of health professionals, social and media education: adequate training of health professionals, social and media education is necessary, with the involvement of scientific societies and health authorities explaining their limitations, benefits, and risks.

Evaluation of screening programs for their implementation: Efficacy, effectiveness, and efficiency

Before a screening program can be implemented, a detailed study of all the experimental work done internationally is usually carried out to know the best way to organize screening in a population.[2] The results that should be measured in screening programs are efficacy, effectiveness, and efficiency.[2]

By efficacy we understand that the screening program improves the health status of the population, for example, if the number of deaths due to colorectal cancer decreases in the population that has participated in the program over a period of 10 years. By effectiveness, we understand the improvement in the state of the target population or community screened. The difference with efficacy depends on several factors such as acceptance and adherence of the population, or the health professionals involved in the program. If the procedure or test is quick and convenient and the coverage is wide, acceptance is greater; the same if the cases detected as positive are diagnosed and treated by specialized services.

Directly related to the acceptance of the program is the positive predictive value (PPV) of the test, that is, its sensitivity and specificity. Simplifying, these concepts provide data for the detection of real cases avoiding the appearance of false positives (patients without the disease). This is particularly important if the subsequent diagnostic procedure is invasive, expensive, or involves a certain risk, such as colonoscopy. The performance of the program can also be improved by increasing the prevalence of the disease, that is, by limiting the population at risk (for example, those over 50 for colorectal cancer).

Finally, efficiency refers to the cost-benefit ratio of the program. It is an economic analysis comparing the cost of the program with other alternatives, such as the absence of screening (and therefore with a diagnosis and treatment of patients already with clinical symptoms), or with primary prevention, or between different programs. The ability of the system to carry out the program must also be taken into account. In the case of CRC in Galicia, each health area must have the number of equipment and specialized personnel for performing the screening test established in the target population, as well as perform the expected colonoscopies and pathological studies in a given period of time.

This type of study is not so easy to perform considering a community perspective, that is, in addition to collecting clinical costs such as drugs, therapies, human or administrative resources, infrastructure costs should be collected on one hand, and on the other the human capital of each of the CRC patients, as well as the indirect costs associated with their families.[59]

Problems related to a colorectal cancer screening program

The importance of the early detection of CRC and the role of screening programs is evident. However, there may be potential adverse effects in screening: a false sense of security can be created in the individual ("since they will do the screening test when the time comes, I do not pay attention now to clinical symptoms"). There may be misdiagnosis (wrong diagnosis) since the test is not perfect, although normally a wrong diagnosis is well understood because individuals have been previously informed about the benefits and harms of the test. Much more complicated is overdiagnosis.[4] In this case the test has worked correctly, but there are undesirable consequences related to the diagnosed individual.

One of the main problems in screening programs is participation. Several studies have identified factors that act as barriers or facilitators in the decision of the target individual to participate in a screening program. Participation in screening programs is usually low and tends to be lower among ethnic minorities, individuals with low socioeconomic status, and may vary by gender.[60]

Honein-AbouHaidar and collaborators[61] conducted a systematic review and meta-study to assess the problem of participation. Among the facilitators of participation in screening they included: awareness of the appropriate detection of CRC and its purpose, positive attitude toward the tests used for screening, motivation to be healthy, and peace of mind for not having CRC. On the other hand, among the barriers to participation in screening programs, they point out: lack of awareness regarding CRC screening and lack of understanding the purpose of screening, fear of cancer and the potential diagnosis derived from screening, fatalism regarding cancer and death; in the case of FIT, refusal to manipulate feces and store them in the kit; in relation to colonoscopy, the need to prepare for the exam, along with the pain, discomfort, and risk of perforation related to this test; additionally, lack of motivation and time.

They also describe specific barriers in certain cultures, such as cancer prevention based on natural remedies; the idea of certain ethnicities in India, China, and African-Caribbean people that their diet is enough to protect them against CRC; or the belief of losing masculinity in Latinos and African-Americans. Lastly, socioeconomic barriers including concerns about not being able to work and reducing income; concern about transportation and getting a person to accompany them after colonoscopy; lack of understanding of the terms "colon," "rectum," "percentage"; and language barriers that make it difficult to understand the doctor's instructions about sample collection.

Among the factors that can modify and influence the facilitator elements and barriers are education, promotion of awareness about CRC screening; and primary care doctors (GPs), who should talk with the asymptomatic patient about the importance of screening and its benefit. Participation rates can also be increased with patient management through the complex health system, including intensive reminder programs, distribution of the FIT, and decreased time between the initial invitation for screening and the final test.[62] It has also been described that telephone intervention carried out by primary care professionals significantly increases the population participation rate,[63] as well as personalized invitation letters.[64]

Situation of screening in Spain

A population-based CRC screening program is a complex process, which requires the participation and coordination of different levels of care (Public Health, Primary Care, Specialized Care), with very strict requirements on quality and information systems. Thus, the implementation of screening programs in Spain has represented a long road in which different actors have established the scientific, social, and political bases for their development.

Implementation of CRC screening in Spain

In 2003 the European Council recommended to the state members the screening of CRC through the FOBT (Recommendation 2003/878/CE, Official Journal of the European Union, 2003). In Spain, different pilot programs were established (Catalonia in 2000, Valencian Community in 2005 and Region of Murcia in 2006), which served as basis for the Spanish Ministry of Health in 2006 to elaborate a National Strategy Plan for CRC (*Estrategia en Cáncer del Sistema Nacional de Salud*). Among its objectives was the development of pilot studies for population screening based on FOBT, which would allow concluding, in the shortest possible time, what would be the best strategy for the implementation of a population-based program.[65]

In 2004 the CRC Prevention Clinical Practice Guide (*Guía de Práctica Clínica de Prevención del CCR*) was published. This guide, which was updated in 2009, was elaborated by the Spanish Society of Family and Community Medicine (SEMFYC), the Spanish Association of Gastroenterology (AEG), and the Ibero-American Cochrane Center (CCI). In the first version and in the update, the available evidence regarding the epidemiology and prevention of CRC was evaluated, establishing recommendations adapted to our environment, with special emphasis on the screening of average-risk population.

In 2008 the Alliance for the Prevention of Colon Cancer was created in Spain. This is an independent and nonprofit organization that integrates patient associations, nongovernmental altruistic organizations, and scientific societies, with the fundamental goal of disseminating the health and social importance of CRC in our country, and promoting screening, early detection and prevention measures.[66]

During that same year, the COLONPREV study started (ClinicalTrials.gov NCT00906997), promoted by the AEG, in which the efficacy of biennial FIT was compared with colonoscopy to reduce CRC mortality at 10 years. The study, which is carried out in 8 autonomous communities, showed that participation was higher among patients randomized to FIT than in those randomized to colonoscopy (results from the first screening round). On the other hand, the rate of CRC cases diagnosed with these methods was similar, although the rate of detection of advanced adenomas was higher in the colonoscopy arm.[67]

Thanks to the awareness campaigns and the social pressure from the Alliance, together with the scientific evidence accumulated in the COLONPREV study, other autonomous communities gradually initiated population screening programs. In July 2013 CRC screening through FOBT was added to the Basic Portfolio of Services of the National Health System by the Spanish Ministry of Health (*Cartera Básica de Servicios del Sistema Nacional de Salud*) in women and men between 50 and 69 years old.[68] At that time, 9 communities had screening programs at different stages of implementation.

After the incorporation of screening in the Basic Portfolio of Services, the rest of the autonomous communities implemented screening programs in their territories, so that in 2017 all the communities started their CRC screening program, with greater or lesser implementation in their territory.

Current situation of CRC screening in Spain: Cancer screening program network, 2019

Currently, the 17 autonomous communities have screening programs, and by the end of 2019 10 of them will have full coverage in their territory.

In the 2016–17 period, active screening programs covered 51% of the Spanish population between 50 and 69 years old. By communities, coverage exceeded 90% of the target population in Castilla y León, Catalonia, Navarra, Basque Country, and Valencian Community, and was between 40% and 50% in Castilla La Mancha and Galicia.

The average participation in Spain in the initial round was 42.5%. The communities with the highest participation are Navarra (75%) and the Basque Country (65%), while Andalusia, Canary Islands, Castilla La Mancha, and La Rioja have the lowest, with participation rates below 33%. On the other hand, participation in consecutive rounds reached an average of 90.3%, which implies that people who participated previously have good adherence to the program. Age and sex influence participation, with higher rates among women aged 60–69 and less in men aged 50–59. The colonoscopy acceptance rate for a positive FIT is 92% (range 83.4%–97.5%).

In the initial screening round, 20.94 medium- or high-risk adenomas were detected per 1000 individuals screened, and 2.63 CRC cases per 1000 individuals screened. Age and gender are also related to the presence of advanced lesions, with increased rates in men aged 60–69 (48.2‰ medium-high risk adenomas, and 7.5‰ CRC) and lower rates in 50–59 year women (11.08‰ medium-high risk adenomas and 1.53‰ CRC). The diagnosis of advanced lesions was lower in consecutive rounds (13.82‰ adenomas and 1.38‰ CRC).

A total of 3418 CRC cases were detected, of which there was no information on the tumor stage in 23.5%. Among the cases in which information was available, 66.4% were localized tumors (stages I and II) and 33.6% were tumors with regional or distant extension (stages III–IV).

Regarding the unwanted effects of a screening program, until December 2017, 218 serious complications were recorded, representing a rate of 0.31 complications per 100 patients who undergo the examination. Regarding interval cancers, information is available from 5 autonomous communities: 133 interval cancers were detected in more than 325,000 participants, representing a rate of 0.41‰ participants.

Current situation of CRC screening in Galicia: *Xunta de Galicia*, 2019

The CRC screening program in Galicia (*Programa Galego de Detección Precoz do Cancro Colorrectal*, PGDPCC) began in the area of Ferrol in 2013, followed by the area of Ourense in 2015; Pontevedra, Santiago and Lugo in 2016; and Vigo and Coruña in 2017. On December 2018 the coverage of the program reached 82% of the Galician population aged 50–69, and the program was expected to be fully implemented in all health areas on September 2019.

The PGDPCC is the only population program in Spain that includes in its planning the follow-up of patients with adenomas detected in the program, establishing follow-up indications (colonoscopy or FIT, with the corresponding time interval) and organizing the performance of the tests indicated.

The participation in initial rounds was 43%, with a greater participation among women than men. In consecutive rounds, participation reached 93%, with no notable differences between men and women. The acceptance rate of colonoscopy after a positive FIT is 97.1%.

From the beginning of the program until December 2017, 599 CRC cases (3.46‰ individuals screened) and 4256 medium- or high-risk adenoma cases (22.08‰) have been detected. Men had more frequent advanced lesions than women (4.65‰ vs. 1.89‰ CRC cases; 33.66‰ vs. 12.96‰ medium- or high-risk adenoma cases). As in the rest of Spanish programs, the detection of advanced lesions was lower in successive rounds.

Among the 560 cases with CRC diagnosed in the initial round, approximately 70% were stage I or II, in both men and women.

There were 19 serious complications (0.17%) and 21 interval cancers identified, representing 0.6 cases per 1000 participants with negative screening tests.

Strengths and areas for improvement in Spanish screening programs

At the present time, all the autonomous communities have initiated a population-based CRC screening program, so it can be considered that the implementation of screening in Spain is already accomplished. It is necessary to achieve the extension of the most recent programs to its entire target population to achieve full coverage in a short period of time.

As has been proven in other studies, invasive cancers detected in Spanish programs are more frequently found in localized stages (I or II), which implies that tumors are in a less advanced stage than tumors detected in symptomatic patients, and therefore have better prognosis. On the other hand, complication rates are low, within the levels accepted by the Spanish guidelines.[20]

However, participation in many areas is low, mainly among men, in which the probability of detecting advanced lesions is greater. Because of this, it is important to detect the factors behind this low participation and establish measures to improve it.

References

1. Marzo-Castillejo M, Vela-Vallespín C, Bellas-Beceiro B, et al. Recomendaciones de prevención del cáncer. Actualización PAPPS 2018. *Aten Primaria* 2018;**50**(Suppl 1):41–65. https://doi.org/10.1016/S0212-6567(18)30362-7.
2. Wilson JMG. Principles and practice of screening for diseases [by] J.M.G. Wilson [and] G. Jungner editor. World Health Organization; 1968. https://play.google.com/store/books/details?id=loF0NAAACAAJ.
3. Cubiella J, Marzo-Castillejo M, Mascort-Roca JJ, et al. Clinical practice guideline. Diagnosis and prevention of colorectal cancer. 2018 update. *Gastroenterol Hepatol* 2018;**41**(9):585–96. https://doi.org/10.1016/j.gastrohep.2018.07.012.
4. Smith RA. Can we improve on Wilson and Jungner's principles of screening for disease? *CMAJ* 2018;**190**(14):E414–5. https://doi.org/10.1503/cmaj.180330.

5. Bray F, Ferlay J, Soerjomataram I, Siegel RL, Torre LA, Jemal A. Global cancer statistics 2018: GLOBOCAN estimates of incidence and mortality worldwide for 36 cancers in 185 countries. *CA Cancer J Clin* 2018;**68**(6):394–424. https://acsjournals.onlinelibrary.wiley.com/doi/abs/10.3322/caac.21492@10.3322/%28ISSN%291542-4863.statistics.
6. Ferlay J, Colombet M, Soerjomataram I, et al. Cancer incidence and mortality patterns in Europe: estimates for 40 countries and 25 major cancers in 2018. *Eur J Cancer* 2018;**103**:356–87. https://doi.org/10.1016/j.ejca.2018.07.005.
7. IJspeert JEG, Medema JP, Dekker E. Colorectal neoplasia pathways. *Gastrointest Endosc Clin N Am* 2015;**25**(2):169–82. https://doi.org/10.1016/j.giec.2014.11.004.
8. Pino MS, Chung DC. The chromosomal instability pathway in colon cancer. *Gastroenterology* 2010;**138**(6):2059–72. https://doi.org/10.1053/j.gastro.2009.12.065.
9. Kawakami H, Zaanan A, Sinicrope FA. Microsatellite instability testing and its role in the management of colorectal cancer. *Curr Treat Options Oncol* 2015;**16**(7):30. https://doi.org/10.1007/s11864-015-0348-2.
10. Kim SY, Kim TI. Serrated neoplasia pathway as an alternative route of colorectal cancer carcinogenesis. *Intest Res* 2018;**16**(3):358–65. https://doi.org/10.5217/ir.2018.16.3.358.
11. Fearon ER, Vogelstein B. A genetic model for colorectal tumorigenesis. *Cell* 1990;**61**(5):759–67. https://doi.org/10.1016/0092-8674(90)90186-i.
12. Moreno CC, Mittal PK, Sullivan PS, et al. Colorectal cancer initial diagnosis: screening colonoscopy, diagnostic colonoscopy, or emergent surgery, and tumor stage and size at initial presentation. *Clin Colorectal Cancer* 2016;**15**(1):67–73. https://doi.org/10.1016/j.clcc.2015.07.004.
13. Glynne-Jones R, Wyrwicz L, Tiret E, et al. Rectal cancer: ESMO clinical practice guidelines for diagnosis, treatment and follow-up. *Ann Oncol* 2017;**28**:iv22–40. https://doi.org/10.1093/annonc/mdx224.
14. Van Cutsem E, Cervantes A, Adam R, et al. ESMO consensus guidelines for the management of patients with metastatic colorectal cancer. *Ann Oncol* 2016;**27**(8):1386–422. https://doi.org/10.1093/annonc/mdw235.
15. Labianca R, Nordlinger B, Beretta GD, et al. Early colon cancer: ESMO clinical practice guidelines for diagnosis, treatment and follow-up. *Ann Oncol* 2013;**24**:vi64–72. https://doi.org/10.1093/annonc/mdt354.
16. Gómez-España MA, Gallego J, González-Flores E, et al. SEOM clinical guidelines for diagnosis and treatment of metastatic colorectal cancer (2018). *Clin Transl Oncol* 2019;**21**(1):46–54. https://doi.org/10.1007/s12094-018-02002-w.
17. García-Carbonero R, Gómez España MA, Casado Sáenz E, et al. SEOM clinical guidelines for the treatment of advanced colorectal cancer. *Clin Transl Oncol* 2010;**12**(11):729–34. https://doi.org/10.1007/s12094-010-0587-4.
18. Vogel JD, Eskicioglu C, Weiser MR, Feingold DL, Steele SR. The American society of colon and rectal surgeons clinical practice guidelines for the treatment of colon cancer. *Dis Colon Rectum* 2017;**60**(10):999–1017. https://doi.org/10.1097/dcr.0000000000000926.
19. Poblacional PDEC. n.d.Documento marco sobre cribado poblacional. http://www.alianzaprevencioncolon.es/imagenesAdmin/prensa/Documento_marco_en%20CCR_2011.pdf.
20. Segnan N, Patnick J, von Karsa L. *European guidelines for quality assurance in colorectal cancer screening and diagnosis*. Publications Office of the European Union; 2010. https://play.google.com/store/books/details?id=1l2LZCYeDoAC.
21. Council recommendation. December 2003 on cancer screening. *Off J Eur Union* 2003;**16**(12):2003.
22. Edwards BK, Ward E, Kohler BA, et al. Annual report to the nation on the status of cancer, 1975-2006, featuring colorectal cancer trends and impact of interventions (risk factors, screening, and treatment) to reduce future rates. *Cancer* 2010;**116**(3):544–73. https://doi.org/10.1002/cncr.24760.
23. Kahi CJ, Imperiale TF, Juliar BE, Rex DK. Effect of screening colonoscopy on colorectal cancer incidence and mortality. *Clin Gastroenterol Hepatol* 2009;**7**(7):770–5. quiz 711 https://doi.org/10.1016/j.cgh.2008.12.030.
24. Vleugels JLA, van Lanschot MCJ, Dekker E. Colorectal cancer screening by colonoscopy: putting it into perspective. *Dig Endosc* 2016;**28**(3):250–9. https://doi.org/10.1111/den.12533.
25. Brenner H, Stock C, Hoffmeister M. Effect of screening sigmoidoscopy and screening colonoscopy on colorectal cancer incidence and mortality: systematic review and meta-analysis of randomised controlled trials and observational studies. *BMJ* 2014;**348**:g2467. https://doi.org/10.1136/bmj.g2467.
26. Lin JS, Piper MA, Perdue LA, et al. Screening for colorectal cancer: updated evidence report and systematic review for the US preventive services task force. *JAMA* 2016;**315**(23):2576–94. https://doi.org/10.1001/jama.2016.3332.
27. Vermeer NCA, Snijders HS, Holman FA, et al. Colorectal cancer screening: systematic review of screen-related morbidity and mortality. *Cancer Treat Rev* 2017;**54**:87–98. https://doi.org/10.1016/j.ctrv.2017.02.002.
28. Zhao S, Wang S, Pan P, et al. Magnitude, risk factors, and factors associated with adenoma miss rate of tandem colonoscopy: a systematic review and meta-analysis. *Gastroenterology* 2019;**156**(6):1661–1674.e11. https://doi.org/10.1053/j.gastro.2019.01.260.
29. Vinsard DG, Mori Y, Misawa M, et al. Quality assurance of computer-aided detection and diagnosis in colonoscopy. *Gastrointest Endosc* 2019;**90**(1):55–63. https://doi.org/10.1016/j.gie.2019.03.019.
30. Kudo S-E, Mori Y, Misawa M, et al. Artificial intelligence and colonoscopy: current status and future perspectives. *Dig Endosc* 2019;**31**(4):363–71. https://doi.org/10.1111/den.13340.
31. Atkin WS, Edwards R, Kralj-Hans I, et al. Once-only flexible sigmoidoscopy screening in prevention of colorectal cancer: a multicentre randomised controlled trial. *Lancet* 2010;**375**(9726):1624–33. https://doi.org/10.1016/S0140-6736(10)60551-X.
32. Levin TR, Palitz A, Grossman S, et al. Predicting advanced proximal colonic neoplasia with screening sigmoidoscopy. *JAMA* 1999;**281**(17):1611–7. https://doi.org/10.1001/jama.281.17.1611.
33. Pinsky PF, Schoen RE, Weissfeld JL, Bresalier RS, Hayes RB, Gohagan JK. Predictors of advanced proximal neoplasia in persons with abnormal screening flexible sigmoidoscopy. *Clin Gastroenterol Hepatol* 2003;**1**(2):103–10. https://doi.org/10.1053/cgh.2003.50017.

34. Stracci F, Zorzi M, Grazzini G. Colorectal cancer screening: tests, strategies, and perspectives. *Front Public Health* 2014;**2**:210. https://doi.org/10.3389/fpubh.2014.00210.
35. Holme Ø, Schoen RE, Senore C, et al. Effectiveness of flexible sigmoidoscopy screening in men and women and different age groups: pooled analysis of randomised trials. *BMJ* 2017;**356**:i6673. https://doi.org/10.1136/bmj.i6673.
36. Chorost MI, Datta R, Santiago RC, et al. Colon cancer screening: where have we come from and where do we go? *J Surg Oncol* 2004;**85**(1):7–13. https://doi.org/10.1002/jso.20008.
37. Huang CS, Lal SK, Farraye FA. Colorectal cancer screening in average risk individuals. *Cancer Causes Control* 2005;**16**(2):171–88. https://doi.org/10.1007/s10552-004-4027-z.
38. Johnson CD, Hara AK, Reed JE. Computed tomographic colonography (virtual colonoscopy): a new method for detecting colorectal neoplasms. *Endoscopy* 1997;**29**(6):454–61. https://doi.org/10.1055/s-2007-1004250.
39. Obaro AE, Burling DN, Plumb AA. Colon cancer screening with CT colonography: logistics, cost-effectiveness, efficiency and progress. *Br J Radiol* 2018;**91**(1090):20180307. https://doi.org/10.1259/bjr.20180307.
40. Duarte RB, Bernardo WM, Sakai CM, et al. Computed tomography colonography versus colonoscopy for the diagnosis of colorectal cancer: a systematic review and meta-analysis. *Ther Clin Risk Manag* 2018;**14**:349–60. https://doi.org/10.2147/TCRM.S152147.
41. van der Meulen MP, Lansdorp-Vogelaar I, Goede SL, et al. Colorectal cancer: cost-effectiveness of colonoscopy versus CT colonography screening with participation rates and costs. *Radiology* 2018;**287**(3):901–11. https://doi.org/10.1148/radiol.2017162359.
42. Rank KM, Shaukat A. Stool based testing for colorectal cancer: an overview of available evidence. *Curr Gastroenterol Rep* 2017;**19**(8):39. https://doi.org/10.1007/s11894-017-0579-4.
43. Cubiella J, Castro I, Hernandez V, et al. Diagnostic accuracy of fecal immunochemical test in average- and familial-risk colorectal cancer screening. *United European Gastroenterol J* 2014;**2**(6):522–9. https://doi.org/10.1177/2050640614553285.
44. Bailey JR, Aggarwal A, Imperiale TF. Colorectal cancer screening: stool DNA and other noninvasive modalities. *Gut Liver* 2016;**10**(2):204–11. https://doi.org/10.5009/gnl15420.
45. Haug U, Kuntz KM, Knudsen AB, Hundt S, Brenner H. Sensitivity of immunochemical faecal occult blood testing for detecting left- vs right-sided colorectal neoplasia. *Br J Cancer* 2011;**104**(11):1779–85. https://doi.org/10.1038/bjc.2011.160.
46. Imperiale TF, Ransohoff DF, Itzkowitz SH. Multitarget stool DNA testing for colorectal-cancer screening. *N Engl J Med* 2014;**371**(2):187–8. https://doi.org/10.1056/NEJMc1405215.
47. Ganepola GA, Nizin J, Rutledge JR, Chang DH. Use of blood-based biomarkers for early diagnosis and surveillance of colorectal cancer. *World J Gastrointest Oncol* 2014;**6**(4):83–97. https://doi.org/10.4251/wjgo.v6.i4.83.
48. Bresalier RS, Kopetz S, Brenner DE. Blood-based tests for colorectal cancer screening: do they threaten the survival of the FIT test? *Dig Dis Sci* 2015;**60**(3):664–71. https://doi.org/10.1007/s10620-015-3575-2.
49. Hundt S, Haug U, Brenner H. Blood markers for early detection of colorectal cancer: a systematic review. *Cancer Epidemiol Biomarkers Prev* 2007;**16**(10):1935–53. https://doi.org/10.1158/1055-9965.EPI-06-0994.
50. Vatandoost N, Ghanbari J, Mojaver M, et al. Early detection of colorectal cancer: from conventional methods to novel biomarkers. *J Cancer Res Clin Oncol* 2016;**142**(2):341–51. https://doi.org/10.1007/s00432-015-1928-z.
51. Singh MP, Rai S, Suyal S, et al. Genetic and epigenetic markers in colorectal cancer screening: recent advances. *Expert Rev Mol Diagn* 2017;**17**(7):665–85. https://doi.org/10.1080/14737159.2017.1337511.
52. Song L, Li Y. SEPT9: a specific circulating biomarker for colorectal cancer. *Adv Clin Chem* 2015;**72**:171–204. https://doi.org/10.1016/bs.acc.2015.07.004.
53. Worm Ørntoft M-B. Review of blood-based colorectal cancer screening: how far are circulating cell-free DNA methylation markers from clinical implementation? *Clin Colorectal Cancer* 2018;**17**(2):e415–33. https://doi.org/10.1016/j.clcc.2018.02.012.
54. Duffy MJ. Carcinoembryonic antigen as a marker for colorectal cancer: is it clinically useful? *Clin Chem* 2001;**47**(4):624–30. https://www.ncbi.nlm.nih.gov/pubmed/11274010.
55. De Chiara L, Rodríguez-Piñeiro AM, Rodríguez-Berrocal FJ, Cordero OJ, Martínez-Ares D, Páez de la Cadena M. Serum CD26 is related to histopathological polyp traits and behaves as a marker for colorectal cancer and advanced adenomas. *BMC Cancer* 2010;**10**:333. https://doi.org/10.1186/1471-2407-10-333.
56. Bünger S, Haug U, Kelly M, et al. A novel multiplex-protein array for serum diagnostics of colon cancer: a case-control study. *BMC Cancer* 2012;**12**:393. https://doi.org/10.1186/1471-2407-12-393.
57. Otero-Estévez O, De Chiara L, Rodríguez-Berrocal FJ, et al. Serum sCD26 for colorectal cancer screening in family-risk individuals: comparison with faecal immunochemical test. *Br J Cancer* 2015;**112**(2):375–81. https://doi.org/10.1038/bjc.2014.605.
58. Ruf M, Morgan. Diagnosis and screening. In: *Health knowledge public health textbook*. UK Department of Health; 2008. (Reviewed by Mackenzie, K, 2017). Published online https://www.healthknowledge.org.uk/public-health-textbook/disease-causation-diagnostic/2c-diagnosis-screening.
59. Carballo F, Muñoz-Navas M. Prevention or cure in times of crisis: the case of screening for colorectal cancer. *Rev Esp Enferm Dig* 2012;**104**(10):537–45. https://doi.org/10.4321/s1130-01082012001000006.
60. de Klerk CM, Gupta S, Dekker E, Essink-Bot ML. Expert working group "coalition to reduce inequities in colorectal cancer screening" of the world endoscopy organization. Socioeconomic and ethnic inequities within organised colorectal cancer screening programmes worldwide. *Gut* 2018;**67**(4):679–87. https://doi.org/10.1136/gutjnl-2016-313311.
61. Honein-AbouHaidar GN, Kastner M, Vuong V, et al. Systematic review and meta-study synthesis of qualitative studies evaluating facilitators and barriers to participation in colorectal cancer screening. *Cancer Epidemiol Biomarkers Prev* 2016;**25**(6):907–17. https://doi.org/10.1158/1055-9965.EPI-15-0990.

62. Dougherty MK, Brenner AT, Crockett SD, et al. Evaluation of interventions intended to increase colorectal cancer screening rates in the United States: a systematic review and meta-analysis. *JAMA Intern Med* 2018;**178**(12):1645–58. https://doi.org/10.1001/jamainternmed.2018.4637.
63. Luque Mellado FJ, Paino Pardal L, Condomines Feliu I, et al. Impact of a primary care intervention on the colorectal cancer early detection programme. *Gastroenterol Hepatol* 2019;**42**(6):351–61. https://doi.org/10.1016/j.gastrohep.2019.01.007.
64. Stratmann K, Bock H, Filmann N, et al. Individual invitation letters lead to significant increase in attendance for screening colonoscopies: results of a pilot study in northern Hesse, Germany. *United European Gastroenterol J* 2018;**6**(7):1082–8. https://doi.org/10.1177/2050640618769713.
65. Ministerio de Sanidad y Consumo. *Estrategia En Cáncer Del Sistema Nacional de Salud*; 2006 https://www.mscbs.gob.es/organizacion/sns/planCalidadSNS/pdf/excelencia/cancer-cardiopatia/CANCER/opsc_est1.pdf.
66. Andreu García M, Marzo M, Mascort J, et al. Prevención del cáncer colorrectal. *Gastroenterol Hepatol* 2009;**32**(3):137–9. https://medes.com/publication/49272.
67. Quintero E, Castells A, Bujanda L, et al. Colonoscopy versus fecal immunochemical testing in colorectal-cancer screening. *N Engl J Med* 2012;**366**(8):697–706. https://doi.org/10.1056/NEJMoa1108895.
68. Ministerio de Sanidad, Servicios Sociales e Igualdad. *Informe Del Grupo de Expertos Sobre Concreción de Cartera Común de Servicios Para Cribado de Cáncer*; 2013 https://www.mscbs.gob.es/profesionales/saludPublica/prevPromocion/docs/ResumenEjecutivoCribadoCancer.pdf.

Chapter 6

The role of primary care in early referral of CRC patients

Francisco Javier Maestro Saavedra[a], Carla Blanco Vázquez[b], Alain García de Castro[b], and Begoña Bravo Bueno[b]

[a]*Specialist in Family and Community Medicine, Elviña-Mesoiro Health Center, Xerencia Xestión Integrada de A Coruña, Galician Health Service, Santiago de Compostela, Spain,* [b]*Resident Physician in Family and Community Medicine, Elviña-Mesoiro Health Center, Teaching Unit of the Xerencia Xestión Integrada de A Coruña, Galician Health Service, Santiago de Compostela, Spain*

Primary health care is the primary and central element of any health system. The key representation of this element is the health professionals who develop their activity as a team of multidisciplinary activity and who perform their work at the level of health centers. This is an accessible and familiar place for patients, where they can feel safe and have very close contact with professionals who are concerned with preventing, promoting, and caring for their health.[1]

The World Health Organization—United Nations Children's Fund (WHO-UNICEF) conference of Alma-Ata defined Primary Health Care (PHC) as "the essential health care based on practical, scientifically sound and socially acceptable methods and technology made universally accessible to individuals and families in the community through its full participation and at a cost that the community and the country can afford to maintain at every stage of development in the spirit of self-reliance and self-determination. It is the first level of contact of individuals, the family, and community with the national health system bringing health care as close as possible to where people live and work, and constitutes the first elements of a continuing health care process."[2]

PHC is evolving from the traditional model and becoming more centered on the general practitioner. In addition, PHC is based on a curative conception of different pathologies toward a new conception of the care developed by doctors specialized in Family and Community Medicine and is more focused on the patient and the family and social context in which it operates. These concepts have made these doctors fundamental in the structure of the health system as they assume a wider range of responsibilities that can be classified in the following sections[3]:

- ✓ *Effective and efficient clinical care*
- ✓ *Integrated care throughout the disease process*
- ✓ *Oriented toward family health*
- ✓ *Oriented toward community health*
- ✓ *Development of prevention and health promotion programs for patient, family members, and the community*
- ✓ *Multidisciplinary teamwork*
- ✓ *Development of teaching activities*
- ✓ *Research functions*
- ✓ *Use of the scientific method*

The information provided to patients constitutes the fundamental axis that articulates a true sense of the process. Informed consent would be the last link in that right-to-user information.[4]

Finally, we must assess the role of biographical continuity in medical care, since your doctor will treat you from childhood until your death, where there will be acute processes and chronic diseases. They will also trust us in confidential matters requiring knowledge that may be relevant for good health care, hence contributing to confidentiality of the patient-doctor relationship.[5,6]

The role of primary care in the early release of patients with CRC[7,8]

In this document, we will focus on three pillars:

✓ Prevention and promotion of health
✓ Early diagnosis
✓ Follow-up of index or precursory lesions (polyps)

These pillars are important due to their relationship to healing potential and, if detected in the early stages, potential may be very high.

CRC is the third most common cancer, based on incidence, and the fourth in the world in terms of morbidity. In Spain, 30–40 new cases per 10,000 people are diagnosed each year, with CRC estimated to be the tumor of highest incidence in males and females in Spain. The relative five-year survival rate is 90% for those patients whose colorectal cancer is diagnosed and treated at the initial stage, although only 37% of the cases are detected during that period. In contrast, if the cancer has spread to neighboring organs or lymph nodes, the relative five-year survival rate decreases to 65%, in comparison to 9% for spread to distant parts of the body.

Prevention and promotion of health[9, 10]

Prevention through different actions could reduce the risk of developing colon cancer. These actions range from promoting lifestyle changes aimed at reducing risk factors and promoting protective factors, to early detection programs that allow the diagnosis of premalignant colonic lesions. In colon cancer, prevention includes measures of health promotion that fundamentally deal with lifestyle.

Most colon cancers begin in a staggered fashion, starting with growth of the cells of the intestinal mucosa, until they become a polyp. These adenomatous polyps, which are benign lesions, are involved in the process of colon cancer formation. However, with the passage of time, if they are not removed and remain in the intestine, some of them may become malignant and become cancerous. For this reason, one way to prevent colon cancer is to detect and remove these polyps before their transformation.

Within the prevention and promotion of health we must know what factors pose a risk for the development of colorectal cancer, since risk depends on partially modifiable factors, such as dietary, lifestyle, environmental and nonmodifiable (e.g., hereditary) factors. These include:

✓ Age: As the main risk factor, the majority of cases are detected in people older than 50 years.
✓ Adenomas: People who have had adenomatous polyps removed are more likely to develop new ones and colon cancer, therefore, this population requires regular monitoring.
✓ Chronic intestinal inflammatory disease (ulcerative colitis or Crohn's disease): The risk of cancer in patients with inflammatory bowel diseases becomes important eight years after presentation as a pancolitis or between 12 and 15 years after colitis on the left side.
✓ Personal history of colorectal cancer: People who have had colorectal cancer have a higher risk of developing a new one. In addition, women who have had ovarian or uterine cancer are more likely to develop colon cancer.
✓ Family history of colorectal cancer: People with first-degree relatives (parents, siblings, or children) who have been diagnosed with colon cancer have a higher risk of suffering from the disease. In 5% of patients, the cause of the disease is due to some hereditary genetic anomaly.
✓ Diet: Colon cancer is more common in upper-class people living in urban areas. The ingestion of animal fats increases the production of anaerobes in the intestinal microflora, transforming normal bile acids into carcinogens. There is a systematic increase in the risk of developing colorectal adenomas and carcinomas, associated with high levels of cholesterol and diets high in animal fat. The consumption of red, processed or cooked meat made or in direct contact with fire increases the risk of colon cancer, which promotes a diet rich in fish and poultry. A higher intake of vitamin D can prevent the genetic damage caused by lithocholic acid (one of the metabolites of bile acids). Fruits and vegetables have also been defined as protective.[9] There is a greater risk of developing colorectal cancer with a low consumption of dietary fiber.
✓ Obesity: The risk of developing colorectal cancer is increased in people who are overweight. The danger is even greater if excess fat at the surface of the waist is greater than that of the thighs or hips.
✓ Smoking: Some prospective cohort studies indicate that the probability of death from cancer increases in smokers approximately 30%–40%. Habitual smoking can be the cause of approximately 12% of fatal colorectal tumors.

✓ Ethnicity: Although not scientifically defined, it was discovered that the mutation of the "I1307K APC" gene may cause a greater proportion of colorectal cancer in Ashkenazi Jews.
✓ Diabetes mellitus: The risk of colorectal cancer in people with diabetes mellitus is increased by 30%–40%. Mortality rate is also higher after diagnosis.
✓ Bacteremia due to *Streptococcus bovis*: Patients with endocarditis or septicemia due to these fecal bacteria have a high incidence of hidden gastrointestinal tumors, such as colorectal cancer.
✓ Ureterosigmoidostomy: Patients who underwent surgery to correct a congenital bladder exstrophy have a 5%–10% incidence of colorectal cancer 15–30 years after surgery.
✓ Consumption of alcohol: Discrepancies exist regarding the attribution of the consumption of alcoholic beverages to an increase in the risk of contracting colorectal cancer. One hypothesis is that it may be due to the repercussions of alcohol on the metabolism of folic acid in the body.

Early diagnosis[11]

The associated symptoms and tests available for early detection of CRC are detailed as follows:

Symptomatology in colorectal cancer

The diagnostic evaluation of patients with low digestive symptoms requires a good anamnesis and a detailed physical examination, including anorectal examination. Patients with a rectal or abdominal mass suspected of CRC, palpable and/or visible by radiological imaging, should be referred without delay at the specialized level to confirm the diagnosis. Blood in the stool or rectal bleeding, independent of the deposition, may be present.

Changes in the usual way of making depositions

✓ Appearance of constipation or worsening of habitual constipation of more than three weeks of evolution
✓ Diarrhea
✓ Diarrhea, alternating with constipation
✓ Stools thinner than usual (taped stools)
✓ Rectal tenesmus

Abdominal discomfort (pain, inflammation, accumulation of gas, cramps) without known cause, repeated and of progressive intensity.
  Weight loss without known cause.
  Iron-deficiency anemia (lack of iron) not previously detected and without any other justifiable cause.

Tests for early detection of colorectal cancer

CRC is one of the clearest examples of cancer susceptible to screening. For early diagnosis of colon cancer, it is pertinent to regularly perform a test to find adenomatous polyps before they become cancer, or even cancer in its early stages when it is easier to treat and cure. In our country, according to the guidelines set by the European Union, colon cancer screening programs aimed at men and women from the age of 50 are being established in different, autonomous communities. In 2009, following these indications, the Spanish National Health System proposed population screening for those between 50 and 69 years of age, with biannual periodicity based on a fecal occult blood test.

According to the clinical practice guideline, the screening tests are of value in the population at medium risk. In the guideline's update in 2018, assessment of the population age ≥ 50 years without other risk factors is required for appropriate diagnosis and prevention of colorectal cancer.

The guideline includes

✓ Screening is recommended with a single determination of SOHi every two years between 50 and 75 years of age.
✓ Population screening with colonoscopy is not recommended.
✓ If the determination of SOHi is positive, colonoscopy is recommended.
✓ Before a complete and good quality previous colonoscopy without significant findings, a return to the screening program at 10 years is suggested.

- ✓ In the medium-risk population, it is not recommended to offer CT colonography or the colonic capsule as a CRC screening strategy. These options, however, are recommended for the evaluation of a positive SOH test in individuals in whom colonoscopy is contraindicated or with an incomplete colonoscopy for a cause other than a poor colonic cleaning.
- ✓ The use of biomarkers in peripheral blood is not recommended as a screening test for CRC.

The following tests should be performed for early detection of colon cancer:

Stool occult blood test

Patients with newly emerging low digestive symptoms who do not meet the criteria for referral without delay at a specialized level due to a high suspicion of CRC (rectal or abdominal mass, rectal bleeding, or iron-deficiency anemia) should undergo an immunological fecal occult blood test (SOHi), since CRC can appear without producing any symptoms and often bleeds in small, nonvisible quantities that are eliminated in the stool. The analysis of hidden blood in the stool (SOH) is used to detect the presence of these small amounts of blood, but not the cancer itself. Although it is not 100% accurate, it is the best studied early detection test for colon cancer. The SOH test is a simple test that consists of a collection of stool samples that the patient can perform in their own home.

There are two types of tests that can detect blood in the stool:

- ✓ Guaiac test: this test consists of a cardboard where a small amount of stool is deposited for three days in a row. The test expires if more than 14 days pass between the time the first sample is taken and its subsequent analysis.
- ✓ Immunological test: stool samples are collected from one or two different days. Preservation of the fresh material before analysis is required to maintain activity of the test.

If there is blood in the stool samples, the result of the test is considered positive (establishing the positive cutoff point at 10 μg/g of feces guarantees an optimal balance between sensitivity and specificity of the test). This does not indicate a positive diagnosis of colon cancer since the presence of blood may be due to causes other than cancer. In these cases, a colonoscopy is recommended.

Patients with a negative result (< 10 μg/g of feces) and persistence of symptoms of recent onset (2–4 weeks) should undergo a colonoscopy and/or referral at a specialized level for its completion and diagnostic confirmation. If the result of the test is negative, it is advisable to repeat it every two years, starting at age 50. This action may result in a one-third reduction in the possibility of death from colon cancer.

Sigmoidoscopy

Sigmoidoscopy is a test that involves the introduction of a short, thin, flexible tube with a lighting system through the anus that allows visualization of the rectum and lower part of the colon. It requires, as do other procedures that study the interior of the colon, proper diet and bowel preparation. Its advantage over colonoscopy is that it does not require sedation, but it has the disadvantage of not visualizing the right part of the intestine. In addition, if a polyp is detected, a colonoscopy is necessary since there may be another lesion in the nonvisualized part of the colon. An interval of 10 years between screening sigmoidoscopies is advised.

After detection by sigmoidoscopy of an adenomatous polyp or a distal serrated polyp with a size greater than 10 mm or with high-grade dysplasia, a complete colonoscopy is required, but not before the discovery of distal hyperplastic polyps.

Colonoscopy

Colonoscopy is a test that allows direct visualization of the interior of the large intestine by introducing a long, flexible fiber optic tube through the anus that allows inflation of air or CO_2. Scanned images are then collected for observation on a monitor. The colonoscope has one or two working channels that allow the introduction of devices to perform biopsies, remove polyps or tumors, and treat other types of injuries.

Indications

- ✓ A positive fecal occult blood test result
- ✓ Presence of symptoms that make it necessary to rule out colon cancer
- ✓ People with moderate or high risk factors for developing colon cancer

Preparation for tests

In order for the doctor performing the test to have a clear and complete vision, it is necessary to clean the intestine of all waste material prior to the examination. To achieve this status, patients are given instructions consisting of a special diet for one or two days, drinking plenty of fluids and taking laxatives, nonabsorbable, a few hours before the test. Patients are to inform their doctors about the medications they take, as there are some medications that are advisable to stop before the colonoscopy. Patients with diabetes, as well as those suffering from heart valve diseases, with a history of pacemaker implantation or taking anticoagulant pills, should inform their doctor before scheduling a colonoscopy since they may require a specific measure or additive medication. Patients should not eat or drink within six hours before the colonoscopy. Medications that do not interfere with test are allowable with sip of water at your usual time.

Technique of performing a colonoscopy

Most colonoscopies are performed under sedation. Before the procedure, a sedative or mild anesthetic is injected through a vein in the arm, which allows a patient to remain calm throughout the examination. The duration of the procedure is approximately 20–30 min, with a withdrawal time of not less than 6–8 min (up to one hour in therapeutic or screening colonoscopies), during which the patient will lie on the side, however position may be changed.

Most detected polyps are removed during the colonoscopy. To perform this, a metal loop is placed around the polyp and a special electric current (diathermy and/or coagulation) is applied. Due to sedation, the patient should not perceive. When it is suspected that a polyp has already transformed into a cancer, a small sample (biopsy) is taken for further study. After the examination, the patient is transported to the recovery room. Upon awakening, minimal discomfort may be observed, in relation to previous air insufflation. Losing a small amount of blood is not uncommon if a polyp has been removed or if a sample has been taken to analyze. Patients are advised to present for colonoscopy with a companion (due to sedation that will be used) and to avoid activities that may involve risks up to 24 h after the test.

Evaluation of results

Approximately 50% of colonoscopies are normal and the remainder are usually benign lesions, some of which, left to their evolution, could become malignant.

Normal colonoscopy

The absence of polyps or any carcinogenic process indicates a very low risk of colon cancer in the next 10 years.

If the colonoscopy has been performed (Table 6.1):

- ✓ to study intestinal symptoms: prevention measures are recommended based on risk of colon cancer,
- ✓ as a consequence of a positive fecal occult blood test (presence of blood in the samples): it is recommended to repeat the test 10 years after the colonoscopy, or
- ✓ as a screening test: subsequent colonoscopy will be performed based on the estimated risk of developing colon cancer.

Polyp detection on colonoscopy[14, 15]

A circumscribed, pedunculated, or sessile tumor that protrudes from the wall to the intestinal lumen is called a polyp. Normally, one in every 20 polyps evolves to CRC. Most importantly, the type of polyps should be determined. Based on histologic findings, they are classified as adenomatous (60%–70%), serrated (10%–30%), and the remainder (10%–20%) includes inflammatory, juvenile, hamartomatous, and other nonmucous lesions. The vast majority of CRCs develop from an adenomatous (70%–80%) or serrated polyp (20%–30%).

- ✓ Adenomatous polyps. Classifications include tubular (85%), tubulovillous (10%), and villous (5%, with more than 50% villous component). They can be of low or high degree of dysplasia (or carcinoma in situ). The presence of adenomas in advanced stages (defined by the presence of high-grade dysplasia, or villous component, or size ≥ 10 mm) or multiple adenomas (≥ 3 mm) are the most important risk factors for developing a CRC.
- ✓ Serrated polyps. Formerly called hyperplastic, these masses comprise a heterogeneous group of lesions with a common feature of the presence of a "sawtooth" architecture in the epithelium of the crypt, with or without dysplasia. Currently, the WHO divides them into three categories:

TABLE 6.1 Recommendations of scientific societies in patients without risk factors of CRC.[12, 13]

Screening test	SEMFYC[a], CCI[b], AEG[c]	ACS[d] (Choose test according to patient preference, giving at least 2 choices)
Occult blood in feces (SOH)	It should be done every two years to all individuals aged 50–75 years (Exception: up to 80 years in selected patients without comorbidities or risk factor) SOH-I[e] is recommended for medium-risk population, screening not recommended by DNA detection in feces	Annually for all individuals aged 45–75 years (Between 75 and 85 years depending on the patient's preferences, health status, life expectancy, and result of previous colonoscopies) or every 3 years if through DNA detection
Sigmoidoscopy	Dependent on acceptability and availability of resources	Every 5 years
SOH +sigmoidoscopy	Not proven to be more effective than sigmoidoscopy alone	
Colonoscopy	Indirect scientific tests samples that reduce incidence and mortality Protective effect for a period exceeding 10 years Not recommended in medium-risk populations	Every 10 years

[a] Sociedad Española de Medicina Familiar y Comunitaria.
[b] Centro Cochrane Iberoamericano.
[c] Asociación Española de Gastroenterología.
[d] American Cancer Society.
[e] SOH-I Sangre en heces realizado por test inmunoquímico.

- ○ Hyperplastic polyps (pH). Constituting 10%–30% of all colonic polyps, they represent approximately 75% of all serrated polyps. These masses are more prevalent in elderly people and are usually located in the distal part of the colon (sigmoid) and rectum. Specific monitoring of these masses outside the typical screening program is not required.
- ○ Sessile serrated adenomas (ASS). These masses represent 15%–25% of serrated polyps and are mainly found in the proximal colon. Since they may be dysplastic, they may malignify.
- ○ Traditional serrated adenomas (AST). These masses represent 1%–6% of serrated polyps and have dysplasia. They are more frequent in the left colon and are at risk of neoplastic transformation. The risk of malignization of a serrated polyp is determined by its histological characteristics (SSA with or without dysplasia and AST), quantity, size ≥ 10 mm, and proximal colon location.
- ✓ Inflammatory polyps. These masses are formed as a consequence of the regenerative process of an inflammatory focus and lack the potential for neoplastic degeneration.

(a) Colonoscopy detects nonadenomatous polyps (inflammatory, hyperplastic, etc.)

Since these types of polyps will not usually progress to cancer, the same indications as for a normal colonoscopy will be followed (with the exception of flat serrated polyps and S. of hyperplastic polyposis)

(b) Colonoscopy detects adenomas

Once complete removal is assured, follow-up by colonoscopy is required, depending on the risk for colon cancer (see table of risk factors) and/or the number of removed adenomas.

(c) Colonoscopy detects invasive cancer.

The most appropriate treatment according to the stage of diagnosis should be followed.
When it is not possible to perform a colonoscopy, a **ColonoTAC or Virtual Colonoscopy** can be performed as an alternative. The colon CT functions in that the CT scanner captures many images as it revolves around the person while lying on a table.
Endoscopic capsule or biomarkers in peripheral blood are not recommended for screening.
The usual screening in individuals without a family history of CRC and in those with a family history will be differentiated.

Monitoring of index or precursor injuries (polyps)

If polyps are resected during the colonoscopy, it should be determined if this basal colonoscopy meets the criteria of high quality to be considered valid.

These criteria include:

✓ adequate intubation and complete exploration with assessment of proximal, middle, and distal colon structures; accompanied by photographs of those areas,
✓ proper assessment of colon cleansing,
✓ thorough description of findings, including number of polyps, number removed and if removal is complete, morphology and localization, method of resection, result, and
✓ pathological anatomy report with histopathological diagnosis and degree of dysplasia.

These criteria allow appropriate classification of patients into different risk groups for the development of CRC (Table 6.2) and aids in the establishment of monitoring plan. This plan, for which control periods are evaluated, is established based on the findings of the last colonoscopy. This information will be summarized into a clinical management algorithm for clinical consultation in Primary Care, including recommendations of the scientific societies regarding surveillance after a first surveillance colonoscopy (Fig. 6.1).

Surveillance in special situations

Surveillance recommendations are always established after complete resection of lesions found during basal colonoscopy.

✓ In the case of incomplete resection, repeat colonoscopy until total exploration achieved and location is free of neoplastic lesions.
✓ In sessile or flat lesions of large size (≥ 20 mm) resected in fragmented manner, schedule colonoscopy within three to six months after basal colonoscopy.
✓ In view of loss or nonrecovery of the spent polyp, it is recommended to consider them as potential adenomas according to size and location: Thus, if it is about micropolyps in the rectum / sigmoid, they are considered hyperplastic; if <10 mm, they are considered tubular adenomas with low-grade dysplasia (nonadvanced adenomas); and if they are polyps >10 mm, they are considered advanced adenomas.
✓ In the event that adenomas and polyps are detected in the same exploration, the strategy that involves the shortest interval is advised.

TABLE 6.2 Risk groups and characteristics for development of CRC.

Risk groups and characteristics	
Low risk	One to two tubular adenomas <10 mm or Serrated polyps <10 mm without dysplasia
Intermediate risk	Three to four tubular adenomas <10 mm or One to four tubular adenomas 10–19 mm or One to four adenomas <20 mm with villous component and/or high-grade dysplasia and/or intramucosal carcinoma or Three to four serrated polyps <10 mm without dysplasia or One to four serrated polyps 10–19 mm without dysplasia or One to four serrated polyps <20 mm with dysplasia
High risk	Five or more adenomas/serrated polyps or Adenoma/serrated polyp >20 mm

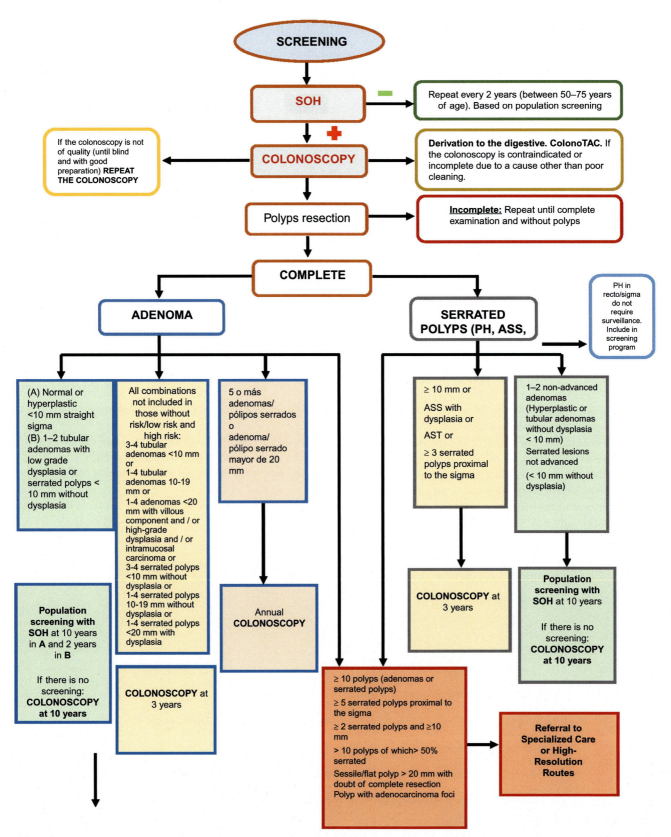

FIG. 6.1 Algorithm of surveillance of colorectal cancer for primary care.

Criteria for referral to specialized care

The criteria used to refer patients to specialized care include:

- ✓ patients with ≥10 adenomas, in a single examination or accumulated over time, to assess the possibility of a familial adenomatous polyposis syndrome;
- ✓ patients with multiple serrated polyps (≥ 10 in total or ≥5 proximal to the recto-sigma), to assess the possibility of a serrated polyposis syndrome;
- ✓ patients with polyps with adenocarcinoma focus;
- ✓ before unresectable polyps;
- ✓ in case of incomplete polypectomy;
- ✓ patients with a family history of the first degree of CRC that justify assessment in specialized units of high risk (genetic counseling);
- ✓ patients with a family history of hereditary syndromes (Lynch syndrome, familial adenomatous polyposis, serrated polyposis syndrome, polyposis associated with the MUTYH gene) for assessment in specialized high-risk units (genetic counseling);
- ✓ inability to request a colonoscopy; and
- ✓ doubts regarding postpolypectomy follow-up.

After two surveillance colonoscopies without advanced colorectal lesions, patients should be reincorporated into the general population screening programs.

CRC of family/hereditary characteristics[16, 17]

CRC is not hereditary in more than 90% of cases, and more than 75% of people who develop CRC do not have close relatives with a history. An increased risk of colon cancer must be suspected when two or more direct relatives in two or three generations (parents, siblings, or children) in a family are affected by colorectal cancer or, in some cases, diagnosed before age 50 years. Due to a family history, CRC screening in individuals at high risk requires a colonoscopy at shorter intervals and earlier age. Recommendations of the scientific societies of screening of family members with familial colorectal cancer (family CRC) are based on establishing a preventive strategy to determining if the family aggregation corresponds to any of the hereditary syndromes associated with known CRC:

- ✓ *Familial adenomatous polyposis (FAP)*. Characterized by the appearance of a large number of polyps of the adenomatous type (more than 100) in the colon and rectum from 20 to 30 years, these polyps have a high probability of becoming malignant in patients older than 30 years of age.
- ✓ *Polyposis associated with the MUTYH gene* is an autosomal recessive hereditary polyposis syndrome, which presents as attenuated polyposis, with fewer adenomas than in familial adenomatous polyposis, and occurring at more advanced age.
- ✓ *Syndrome associated with polymerase repair activity (SAARP)* involves two new genes associated with an increased risk of CRC and adenomatous polyposis, and *adenomatous polyposis associated with NTHL-1*, which is polyposis with an autosomal recessive pattern of inheritance that seems to be associated with an increase of risk of endometrial neoplasia.
- ✓ *Hamartomatous polyposis,* includes:
  - ○ Peutz-Jeghers syndrome (PJS): a condition of autosomal dominant inheritance that is characterized by the presence of hamartomatous polyps in the gastrointestinal tract; cutaneomucosal hyperpigmentation around the lips, buccal mucosa, and fingers; and high risk of cancer in various locations, in order of frequency: breast, colon, pancreas, stomach, ovary, and testicles. The overall risk of cancer in individuals with PJS is 93% at 65 years of age.
  - ○ Juvenile polyposis syndrome: a condition of autosomal dominant inheritance of variable penetrance that is characterized by the presence of multiple juvenile polyps throughout the entire gastrointestinal tract: colon and rectum (98%), stomach (14%), and small intestine (7%). The size of these polyps varies between a few millimeters and more than 3 cm in diameter. These lesions typically debut in the first decade of life and first symptoms usually occur within the first 2 decades (average age at diagnosis is 18.5 years). Symptoms include rectal bleeding with anemia, followed by diarrhea.
  - ○ Cowden syndrome: this condition is characterized by the development of multiple hamartoma-type polyps (in addition to other polyp types), cutaneous lesions, and increased risk of several tumors, most frequently, breast, thyroid, endometrium, kidney, colorectal, and melanoma.

- ✓ *Serial polyposis syndrome (PSS)* is characterized by the presence of numerous serrated polyps, which can be large in size, with a family history and an exceptionally high risk of CRC (estimated up to 70%). Patients should be recommended for inclusion in a program of close endoscopic surveillance. Annual total colonoscopy is currently advisable.
- ✓ *Hereditary nonpolyposis CRC (HNPCC)* refers primarily to the Lynch syndrome, although currently there are two additional related entities of interest in clinical practice: Lynch-like syndrome and familial CRC type X:
  - Lynch syndrome: this condition is of autosomal dominant inheritance and is characterized by the early development of CRC (usually before the age of 50), accelerated carcinogenesis, and an elevated tendency to present with synchronous or metachronous neoplasms, mainly of endometrial cancer. This syndrome also presents a higher risk for cancer of the stomach, pancreas, urinary system, ovary, biliary tract, and small intestine.
  - Lynch-like syndrome: the risk of CRC in these families appears to be lower than in patients with Lynch syndrome, but higher than in patients with sporadic CRC.
  - Familial CRC type X: this condition is of unknown etiology. The risk of developing CRC is lower than that observed in Lynch syndrome and there is no increased risk of extracolonic neoplasms.

The consensus would be based on the degree of kinship of the closest affected relative (Table 6.3, 6.4, 6.5).

These polyposis of familial characteristics should be studied by the specialized attention and by specific units of diagnosis, treatment, and follow-up.

The identification of families with possible hereditary cancers is important since its members can benefit from effective measures, not only in the early detection, but also in the prevention of tumors. In families with hereditary cancer, it is common to observe several cases of cancer, usually of the same type, appearing in one generation and the next and

TABLE 6.3 Individuals with first-degree relatives with CRC.

Number of family affected and age to diagnosis	Regimen
One family member > 60 years old	As medium-risk population, from age of 40: SOH biannual and colonoscopy every 10 years
One relative of < 60 years	Colonoscopy 5 years before the age of diagnosis of the affected family (or 10 years before the diagnosis age of the affected relative younger, whichever occurs first)
Two relatives of < 60 years	Colonoscopy every 5 years from 40 years of age (or 10 years before the diagnosis age of the affected relative younger, whichever occurs first)
Three or more relatives	Deriving to a specialized unit in the high risk of CRC

TABLE 6.4 Individuals with second-degree relatives with CRC.

Number of family affected	Regimen
One	As medium-risk population, from age of 50: SOH biannual and colonoscopy every 10 years.
Two or more	As medium-risk population, from age of 40: SOH biannual and colonoscopy every 10 years.

TABLE 6.5 Individuals with third-degree relatives with CRC.

Number of family affected	Regimen
One (With diagnosis at any age)	As medium-risk population, from age of 50: SOH biannual and colonoscopy every 10 years

occurring at an early age. This is most common with breast, ovarian, and colorectal cases among those aged 40–50 years, and bilateral involvement can be observed when organs are affected. In these families there are individuals who have also had more than one primary tumor or who, in addition to cancer, have developmental defects. When a doctor recognizes one or more of these signs in a family, it must be referred to a Hereditary Cancer Unit or, as an alternative, a Clinical Genetics or Clinical Oncology service for an individual risk assessment for each family member and appropriate genetic determinations. Clinical suspicion of a hereditary predisposition for cancer is based on the analysis of a personal or familial history of cancer. Genetic tests attempt to confirm the suspicion through a molecular study. After the process of genetic counseling, monitoring of the patient and high-risk family members by the UCG must be guaranteed in order to determine the occurrence of new cases of cancer, review follow-up results and other proposed medical interventions, and promote adherence to preventive and early detection measures. Imperative is to remember the importance of informing other relatives when a pathogenic mutation has been detected in family members and offer psychological help if necessary.

The realization of specific genetic tests is not possible in all cases, since there are entities in which the responsible gene is not yet known. The patient must be informed of the applicable genetic studies are most appropriate and what consequences can be expected from both positive and negative results.

Another aspect of hereditary cancer that has a significant impact on families is the monitoring of individuals at risk. In clinical practice, today, there are three main tools for handling these situations:

- ✓ periodic medical surveillance,
- ✓ chemoprevention and
- ✓ prophylactic surgery.

The most used tool is medical surveillance, mainly in medical oncology services. There are consensus recommendations for the care of individuals most frequently predisposed to cancer syndromes, such as Hereditary Breast and Ovarian Cancer Syndrome (CMOH), Lynch and Cowden Syndromes, Familial Adenomatous Polyposis (FAP), and Neoplasia Multiple Endocrine or Neurofibromatosis. For its part, chemoprevention in most situations is still under investigation. Prophylactic surgery allows primary prevention and has applications in PAF, CMOH and to a lesser extent, Lynch syndrome. ✓

The role of the family doctor in the approach to hereditary cancer is to:

- ✓ Identify people at risk for cancers in a context of family aggregation or hereditary cancer.
- ✓ Follow-up with people who, after assessment, identify themselves as low risk, and who will follow the recommendations of the general population.
- ✓ Inform the general population on general questions about hereditary cancer.
- ✓ Detect patients with higher risk and their referral to the corresponding level of care.
- ✓ Participate in action protocols for professionals and in dissemination campaigns aimed at patients.

References

1. Zurro AM, Cano Pérez JF. *Atención Primaria: Conceptos, Organización Y Práctica Clínica*; 2003. http://www.sidalc.net/cgi-bin/wxis.exe/?IsisScript=AGRIUAN.xis&method=post&formato=2&cantidad=1&expresion=mfn=030133.
2. Alma-Ata U. *Conferencia internacional sobre atención primaria de salud*. Alma-Ata URSS; 1978. p. 6–12. Published online https://www.academia.edu/download/45635225/APS_Alma_Ata-Declaracion-1978.pdf.
3. Gallo FJ. Perfil profesional del médico de familia. Bases conceptuales. In: *Medicina de Familia: La clave de un nuevo modelo España*. semFYC; 1997. p. 164–9. Published online.
4. de Galicia BO do P, de Galicia DO, del Estado BO. *Lei 3/2001, do 28 de maio, reguladora do consentimento informado e da historia clínica dos pacientes*. Boletín Oficial del Estado BOE; 2001. p. 23537–41. Published online.
5. Altisent R. Bioética y atención primaria: una relación de mutuas aportaciones. *Arch Med Fam* 2006;**8**(2):63–73. https://www.medigraphic.com/cgi-bin/new/resumen.cgi?IDARTICULO=8454. [Accessed 23 February 2021].
6. Trota RA, Espildora MNM, Moré DS. Ética y medicina de familia. In: *Atención Primaria: Principios, Organización Y Métodos En Medicina de Familia*. España: Elsevier; 2019. p. 154–71. https://dialnet.unirioja.es/servlet/articulo?codigo=7184890.
7. Fidler MM, Gupta S, Soerjomataram I, Ferlay J, Steliarova-Foucher E, Bray F. Cancer incidence and mortality among young adults aged 20–39 years worldwide in 2012: a population-based study. *Lancet Oncol* 2017;**18**(12):1579–89. https://doi.org/10.1016/S1470-2045(17)30677-0.
8. SEOM. *Las Cifras Del Cáncer En España*. Sociedad Española de Oncología Médica (SEOM); 2017. http://www.seom.org/seomcms/images/stories/recursos/Las_cifras_del_cancer_en_Esp_2017.pdf.
9. Cubiella J, Marzo-Castillejo M, Mascort-Roca JJ, et al. Guía de práctica clínica. Diagnóstico y prevención del cáncer colorrectal. Actualización 2018, *Gastroenterol Hepatol* 2018;**41**(9):585–96. https://doi.org/10.1016/j.gastrohep.2018.07.012.

10. de trabajo de la Guía G. de Práctica Clínica de Prevención del Cáncer Colorrectal. Actualización, *Guía de Práctica Clínica*. Barcelona: Asociación Espanola de Gastroenterología, Sociedad Espanola de Medicina de Familia y Comunitaria, y Centro Cochrane Iberoamericano; 2009. 2009. Programa de Elaboración de Guías de Práctica Clínica en Enfermedades Digestivas, desde la Atención Primaria a la Especializada: 4. Published online 2009.
11. Mascort J, Marzo M. Seguimiento de los pólipos intestinales. *AmFAR Rep* 2016;**12**(8):467–71.
12. American Cancer Society updates its colorectal cancer screening guideline. New recommendation is to start screening at age 45 years. *Cancer* 2018;**124**(18):3631–2. https://doi.org/10.1002/cncr.31742.
13. Vázquez ÁN, García BD, Gómez CA. Cribado del cáncer colorrectal. *Cadernos de atención primaria* 2010;**17**(1):24–9. https://www.agamfec.com/antiga2013/pdf/CADERNOS/VOL17/vol_1/04_Para_Saber_de.pdf.
14. Enríquez CF, Carot L, Bessa X. Seguimiento de los pólipos colorrectales. *FMC - Formación Médica Continuada en Atención Primaria* 2019;**26**(3):130–7. https://doi.org/10.1016/j.fmc.2018.12.001.
15. *Guía de Práctica Clínica Sobre Prevención Del Cáncer Colorrectal*. Ministerio de Sanidad n.d.https://portal.guiasalud.es/gpc/guia-de-practica-clinica-prevencion-del-cancer-colorrectal/?pdf=3919.
16. Robles L, Balmaña J, Barrel I, et al. Consenso en cáncer hereditario entre la Sociedad Española de Oncología Médica y las sociedades de atención primaria. *SEMERGEN—Medicina de Familia* 2013;**39**(5):259–66. https://doi.org/10.1016/j.semerg.2012.08.007.
17. Moreira L, Ferrández A. Protocolos de actuación conjunta entre médicos de familia y gastroenterólogos. In: *Revisión del paciente pospolipectomía AEGASTRUM Asociación Española de Gastroenterología (AEG)*. sociedad Española de Medicina Familiar y Comunitaria (semFYC); 2013. 1(2).

Chapter 7

Rapid diagnostic pathways for patients with suspected CRC

Alejandro Ledo Rodríguez[a] and Ismael Said-Criado[b]

[a]*Gastroenterology Department, Integrated Health Area of Pontevedra-O Salnés, Pontevedra, Spain,* [b]*Emergency Department, Integrated Health Area of Vigo, Health Research Institute Galicia Sur, Vigo, Spain*

Clinical pathways

Clinical pathways are programs of care that are used for patients with specific pathology and a predictable clinical course. Clinical pathways are also known as care maps, practical guidelines, coordinated care, and clinical case management. They bring together all areas of quality of care: these include more professional involvement and evaluation in regard to scientific and technical standards and communications between health professionals, keeping patients informed on processes and outcomes, and aspects of efficiency and cost management.

The most usual form of presentation used for clinical pathways is a temporal matrix, with divisions for days and hours of the day, in which all events and interventions are carefully distributed. The information is divided into categories entered in heading rows, with times in the columns. Zander innovated the method in the mid-1980s at the New England Medical Centre in Boston. It is often used for high-volume, high-risk, or high-cost medical procedures, such as aortocoronary bypasses, and knee or hip replacement. Today, more than 2500 clinical pathways have been established, in particular, in Anglo-Saxon countries. Compatible algorithms, protocols, and all care recommendations for patients with specific clinical diagnoses are used in order to provide an interdisciplinary approach. A decrease in unwarranted variability in patient care is one of the advantages of these elements, as the action to be taken in caring for each patient is established as a guideline. Thus inefficient action, unnecessary information, and delays in decision making are avoided. The responsibilities for each health professional are clearly defined, enhancing the work environment and providing legal security against malpractice suits (as with any protocol). The guidelines serve as valuable tools in training resident doctors and in communicating with patients and their families, keeping them informed of expectations on a day-to-day basis, and providing a commitment by the institution to their care and attention. A major difficulty with this form of clinical approach is the need for coordination between clinical and central services and different health areas.

Stages involved in creating a clinical pathway are as follows:

1. Bibliographic review.
2. Formation of a team of doctors and nursing staff from clinical services involved in diagnosed patient care, with the addition of a quality coordinator and a central management representative.
3. Deciding on a diagnosis strategy for creating the clinical pathway.
4. Creating a design plan—cutting and pasting models from other institutions can aid in the design phase and promote discussion between professionals involved in the pathway.
5. Obtaining acceptance of the plan through support from key professionals.
6. Carrying out of a pilot trial.
7. Preliminary analysis of trial results, using measurement unit indicators of effectiveness, efficiency, safety, and patient and professional satisfaction.
8. Revision and modification of the clinical pathway using established indicators.
9. Definitive Implementation.[1]

Rapid diagnosis pathways for colorectal cancer

Introduction

Colorectal cancer (CRC) was the most frequently diagnosed cancer in the Spanish population in 2017, making up 15% of the 34,331 total number of cases detected that year; this was followed by prostate cancer with 13%, lung cancer with 12%, and breast cancer with 11%. In terms of mortality, regardless of gender, CRC rates second, with 15,923 deaths a year, behind lung cancer, with 22,457 deaths a year, and in front of pancreatic cancer with close to 7000 deaths. In terms of gender, CRC ranks second behind lung cancer for men and second behind breast cancer for women. The main risk factor for colorectal cancer is known to be age, with 90% of cases diagnosed in people over 50. The incidence in Galicia for the 50–69 age group is 116 new diagnoses per every 100,000 inhabitants with 40 deaths per every 100,000 inhabitants.[2]

The stage of the disease at time of diagnosis is the most important factor in prognosis for the CRC patient.[3-5] Performing rapid diagnosis in patients suspected of CRC and reducing the wait time between diagnosis and treatment—from clinical suspicion of cancer to definite diagnosis, and then to initiation of treatment—is a major objective of health systems.[6,7] The main priority in caring for patients diagnosed with cancer is to improve survival and reduce mortality rates; this has become a greater concern in cancers with high incidence, such as CRC, due to the significant level of social and economic impact.[8,9]

Spain and countries such as England, Canada, and the United States have introduced preventive and improved care plans in their oncology programs.[5,10,11] These provide clinical rapid diagnosis and treatment pathways for symptomatic patients.[6] A delayed diagnosis can have significant consequences for the patient's well-being, especially psychologically, due to anxiety caused by the diagnostic process. However, within the large number of studies published on the subject, it has not been possible to establish a link between diagnostic delay and the stage of neoplastic disease at diagnosis.[12,13] The same applies to survival.[14-18] Although these benefits have not been demonstrated, early commencement of the first treatment decreases anxiety in the patient with suspected cancer and improves the quality of ongoing care,[8] by reducing variability in referrals from primary care (PC) and standardizing the specialist care (SC) process.[19]

Implementation

The great majority of studies published on CRC rapid diagnostic pathways (RDP) are of British origin. Concerns about delays in diagnosing CRC patients prompted the British government, in 2000, under the endorsement of the National Institute for Health and Care Excellence (NICE), to publish guidelines on rapid referral procedures for patients with suspected CRC. The guidelines sought to identify patients with new symptoms indicative of a high probability of the presence of CRC. In this way, the aim was established to diagnose 90% of CRC cases within RDPs.[20] The patients would need to be seen by a specialist in a hospital setting within two weeks of referral by a PC doctor (two-week rule) and diagnosed and treated within 62 days from the time of the initial referral. This rapid diagnosis and treatment method was established to reduce morbidity and mortality.[5] The guidelines have been updated several times, the last update being in July 2017.[21] In this latest update it was recommended that patients meeting any of the criteria for suspected CRC referral presented in Table 7.1 be seen within the following two-week period. The 2017 update excluded patients, previously included in the RDP, whose results were only positive for the fecal occult blood test. This test was recommended as a means of separating out patients in PC for suspected CRC referral who did not have rectorrhagia or symptoms that met the referral criteria.

TABLE 7.1 CRC rapid referral criteria.

Abdominal or rectal mass
Aged over 40 with unexplained weight loss and abdominal pain
Aged over 50 with unexplained rectal bleeding
Aged over 60 with iron deficiency anemia or changes in bowel habits
Aged under 50 with rectorrhagia and abdominal pain, changes in bowel habits, weight loss or iron deficiency anemia

Adapted from NICE guideline [NG12] Suspected cancer: recognition and referral 2015 [updated July 2017]. Available from: https://www.nice.org.uk/guidance/ng12/chapter/1-Recommendations-organised-by-site-of-cancer#lower-gastrointestinal-tract-cancers.

As mentioned, Spain has introduced preventive strategies and improved cancer care programs promoting rapid diagnosis and treatment pathways for symptomatic patients.[6, 7, 9] In addition, rapid care pathways have been developed for CRC diagnosis and treatment in the autonomous community of Galicia.[22–28]

The Galician Health Service (SERGAS) aimed to design care pathways to be applied throughout the Galician health care system. These would guarantee patients with suspected CRC (or other tumors) that care for their pathology would be developed and streamlined to ensure that the wait time from suspicion of diagnosis to the first treatment would not be more than 30 days. To achieve this, groups of professionals from different disciplines and areas in the region came together to identify specific responsibilities for each of them. These working groups held their first meetings in October and November of 2009. As well, inclusion criteria for the program and care and administrative practices for the coordinated pathways were established. Coordination approaches, databases for monitoring practices, and means for control and assessment of time and resource management were outlined.[29]

The most up-to-date CCR rapid diagnostic pathway in the Galician Health Service, with the latest update occurring in January 2017, after an earlier update in October 2012, is that at the Integrated Healthcare Area of Pontevedra e O Salnés. The aim of its design was to streamline care of suspected CRC patients to give a standardized or shortened wait time. Start and finish, and entry and exit criteria were defined, as were developmental resources. The suspicion criteria established for going straight to colonoscopy are shown in Fig. 7.1. The colonoscopy appointment would be administered by admissions, following assessment of referral criteria fulfillment by the gastrointestinal service. Once diagnosis is established, staging of the tumor would take place. Each case would then be discussed by the tumor team to decide on a treatment plan. This is outlined in the RDP flowchart, with six biannual or annual evaluation indicators. Among these was the target of performing colonoscopies in over 95% of cases in less than two weeks after their order.[22]

Published research

The results, mainly from British studies, on the most relevant studies published relating to implementation of RDPs are now summarized.

Meeting referral criteria

The implementation of the two-week rule in British hospital settings has shown benefits. Most patients referred to the CRC rapid diagnostic program are seen by a specialist within two weeks. However, appropriate fulfillment of the guideline criteria by primary care doctors has been variable, with percentages, according to a number of studies, ranging from 41% to 96%.[30–40] Nonetheless, increased percentages in fulfillment have not been linked to improvement in CRC diagnostic performance for patients referred to the rapid pathway.[31]

Among research papers from Spain, Valentin-Lopez et al.[7] published a retrospective study that included 252 patients referred to CRC rapid diagnostic pathways who were immediately ordered colonoscopies in primary care without previous assessment by a hospital specialist. The percentage for meeting at least one of the high-risk criteria for rapid referral according to their protocol was 79.2%. Other Spanish researchers have demonstrated higher rates of criteria fulfillment of up to 100%,[9] while in other cases fulfillment has not been as high,[41, 42] with figures lower than 50%. These lower percentages for meeting referral criteria could be explained by the PC doctors using their own judgment, as some doctors have admitted to doing in a study by Redaniel et al.[43] for the purpose of ensuring that patients were seen early, rather than concern about a CRC diagnosis. There are studies that demonstrate that at times there has been a discrepancy between the symptoms for which the PC doctor sends the patient for rapid diagnosis and those the patient describes to the specialist.[44]

In regard to the possibility of rereferral of a patient to an RDP who had previously been referred to the same pathway without a CRC diagnosis being established, Vaughan-Shaw et al.[45] found no difference between diagnostic performance for these patients and first-time referral patients. This information shows that previous passage through the pathway does not exclude a rereferral. However, the probability of a CRC diagnosis in patients previously passing through a CRC rapid diagnostic pathway appears to be low. In a study by Patel et al.[39] no patient assessed via RDP and not diagnosed with CRC received a CRC diagnosis at their 5-year follow-up.

Cancer detection rates

Recently, Mozdiak et al.[46] published a systematic review and meta-analysis of all British studies published from 2000 to 2017 that had analyzed the results of RDPs that included patients with suspected CRC. In total, 49 studies were included covering 93,655 patients: 37 were retrospective and 12 prospective cohort studies. Of the studies included, 19 were

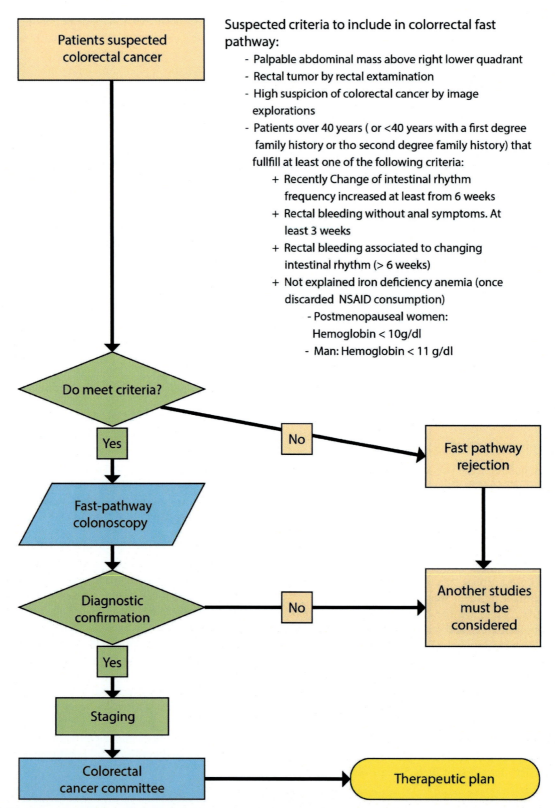

FIG. 7.1 Flowchart showing CRC rapid diagnosis pathway at Pontevedra e O Salnés Integrated Healthcare Area.

considered of acceptable quality, 30 of low quality, and none of high quality. This was mainly due to insufficient information on the characteristics of the study population or inadequate descriptions of the results. In addition, there was a high degree of heterogeneity for all the results assessed. The percentage rates for CRC detection were analyzed in 38 of the studies, with the combined rate proving very low at just 7.7% (95% CI: 6.9–8.5). Some Spanish studies have shown similar CRC detection rates.[47] Other Spanish studies have shown higher percentages but still within a low range.[9] Vallverdú Cartié et al. ccarried out a retrospective study that included 316 patients who entered CRC rapid diagnostic pathways in 2008–2009. Of these, 19.3% eventually received CRC diagnoses. Valentin-Lopez et al.[7] found a slightly higher percentage, at 20.6%. In a Spanish study by Prades et al.[19] far exceeding any other in terms of the number of patients, a quantitative and qualitative analysis was undertaken of the Catalan RDP program for different types of tumor, from 2006 to 2009. In regard to CRC, in the final year of the analysis (2009), 8077 patients entered an RDP. In 28.7% of cases, CRC was diagnosed, showing a progressive decrease in the detection rate compared with the first year of the analysis (2006), when it was 40.7%.

Percentages of diagnosed cancer cases

The most recent data published in the British studies show that approximately half of the CRC diagnoses resulted via RDPs, if considering only patients referred from PC consultations. In 2018 Zhou et al.[48] published data from the National Cancer Data Repository (NCDR) containing diagnostic pathway information on all the patients diagnosed with cancer in England during the period 2006–2010, who had been exclusively referred from a PC consultation. The study did not include cancer cases diagnosed through any other diagnostic means. Of the 46,012 patients referred from PC and diagnosed with colon cancer during the period of the study, 48.4% of diagnoses were via RDPs and of the 35,542 patients diagnosed with rectal cancer, 56.1% were diagnosed via RDPs.

A systematic review and meta-analysis by Mozdiak et al.[46] did not analyze this information, but did refer to the results for the UK from the National Cancer Registration and Analysis Service for 2016, which showed that approximately 31% of all CRC diagnoses were via RDP. Ten percent of the diagnoses resulted from screening programs, 23% were referred from PC for clinical diagnosis other than via RDPs, 23% presented urgently, and the remaining cases resulted from other less usual means. The Spanish studies show that percentage rates for patients diagnosed with CRC via RDPs were variable, with a percentage of 54.3% out of the total number CRC diagnoses in a study by Prades et al.[19] Similar results appear in a study by Alburquerque Miranda et al.[41] Nevertheless, these researchers suggested that when knowledge of RDPs became more widespread in PC, the percentage rate may increase.

Are RDPs sensitive and specific?

Different British researchers have demonstrated low specificity in the guidelines for rapid referral.[31, 49, 50] Jones et al. found a positive predictive value of only 30% for RDP referral criteria.[51] Cark et al. observed that even if the RDP criteria were correctly applied, a significant percentage of patients with CRC would be missed.

In the systematic review and metaanalysis by Mozdiak et al.[46] 54.6% of patients who had been included in RDPs showed completely normal colonoscopies. In other words, when patients are referred to CRC rapid diagnostic pathways, it is more probable that they do not present any colonic pathology than that they do, and among the positive cases, most will be nonmalignant. This data reflects the fact that minor gastrointestinal symptoms are nonspecific and have a low predictive value, as they form part of the usual symptoms experienced by patients with functional gastrointestinal disorders. For this reason, the authors concluded that a preferentially based referral system, depending on the presence of certain symptoms, was unlikely to improve early detection of early stage CRC.

Effect on care timing

Successful implementation of the two-week pathway in the UK has shortened the wait time for patients in almost all the immediate steps for diagnosis and treatment, as compared with patients referred via other pathways.[33, 52, 53] In the 14 studies assessing wait time for a specialist appointment included in the meta-analysis by Mozdiak et al.,[46] the two-week target was met for the great majority of patients. In some clinics the reduction in wait time meant patients not diagnosed via RDPs had to wait longer, possibly up to twice as long, for treatment.[54] This is particularly significant, due to the fact that as we have seen in British records (with a great number of cases) most CRCs are diagnosed by means other than RDPs.

It has been shown that the key points in the diagnosis and treatment process at which long delays occur relate to radiology, endoscopy, and oncology, followed by the wait for treatment to begin. By making changes in these areas the time from referral to treatment can be reduced by up to 32%.[55] An indication of the actual wait time for the various stages in diagnosis and treatment can be found in studies from Spain.[9] Vallverdú Cartié, et al. found the following results after implementation of a clinical pathway: the average wait time for a specialist appointment was 3 days (range 1–8); then the wait for the first colonoscopy, 11.5 days (range 1–41); followed by the wait for the first consultation with the colorectal surgeon, 20 days (1–48); and then 13 days wait (range 3–46) for the intervention.[6] Guzmán, et al. carried out a study comparing 156 patients diagnosed with CRC through an RDP program with a control group of 156 patients diagnosed with CRC in the conventional way. The average wait time from diagnosis to treatment was 39.2 days for the RDP group and 63.4 days for the control group ($p < 0.001$). In the RDP group 28% of patients waited less than 30 days (the target of the CRC rapid diagnosis and treatment protocol implemented in Catalonia in 2005), and in the control group the percentage was 10%. In the study by Valentin-Lopez et al.[7] the wait time for a colonoscopy was significantly shorter for the RDP group of patients (13.8 days) than for the traditional pathway patients (33.8 days). The percentage of RDP patients who met the program target of colonoscopies in two weeks or less was 64.7%, versus 43% for the control group ($p = 0.004$).

In relation to results for the Autonomous Community of Galicia, Piñón Gamallo analyzed the RDP at the Integrated Healthcare Area of Pontevedra e O Salnés. The case histories of all the patients, totaling 105, who entered the RDP in the final quarter of 2015 were collected. Ninety-six of these patients eventually underwent colonoscopies. These were performed after a mean wait of 17.8 days and a median of 7 days. Out of 10 patients diagnosed with CRC, 8 underwent surgical procedures after a mean wait of 56.3 days and a median of 38 days.

Effect on tumor stage and survival

In ten of the studies analyzed in the meta-analysis by Mozdiak et al.,[46] Dukes staging information at diagnosis was shown. According to this, percentages of patients diagnosed for each stage were as follows: Stage A: 11.2% (95% CI: 7.4–15.6), Stage B: 36.7% (95% CI: 30.8–42.8), Stage C: 35.7% (95% CI: 30.8–40.8), Stage D: 11.1% (95% CI: 7.3–15.5). These results were compared with information from the National Cancer Registration and Analysis Service, without finding significant differences in Dukes staging at diagnosis between RDP patients and those diagnosed by other means. The percentage of patients diagnosed early, at Stage A, was low. The researchers found the data confirmed that by sending patients via one or other diagnostic pathway on the basis of symptoms (as occurred in the studies analyzed) independently of how rapidly they were seen, did not improve CRC detection capacity, nor diagnosis in the earliest stages. Information on survival was only provided in seven of the studies, and not in a consistent way, making it impossible to undertake a meta-analysis. However, none of the studies showed higher survival rates for RDP patients. In three of them data were given on survival at 5-year follow-up without finding statistically significant differences between RDP patients and those diagnosed by other means.

In a study from Spain, did not find significant differences in tumor staging between the RDP and control groups.[6] Conversely, Valentin-Lopez et al.[7] found that, with rapid referral, 26% of cancers were diagnosed at Stage A (Astler-Coller), compared with only 11.6% of those diagnosed the traditional way ($p = 0.030$). Although, in most of the studies published no link seems to have been found between CRC diagnosis via RDP and disease prognosis,[56] in an analysis of seven population-based studies in five countries, assessing patients diagnosed in PC centers, but not specifically RDP patients, it was found that longer wait times for diagnosis corresponded to more advanced CRC cases. However, an acceptable wait time between the initial PC appointment and diagnosis could not be established.

Possibilities for increasing effectiveness of CRC rapid pathways

Taking into account the poor results from establishing rapid diagnostic and treatment pathways due to low capacity for tumor detection and lack of benefit for long-term survival, several options have been proposed in an attempt to increase effectiveness.[43] Redaniel et al. carried out the first study assessing the response of health professionals to RDPs. Based on the information obtained, four main courses of action on which to focus for possible improvement were established. These included (1) promoting patient behavior in relation to seeking help when symptoms are first noticed, (2) reviewing RDP referral criteria, (3) optimizing resources, and (4) facilitating coordination of different care areas.

In regard to the first proposal on patient behavior, the study proposes distribution of positive messages about cancer treatment, investment in early screening and diagnosis, and promotion of patient education campaigns, such as are carried out in the UK.[18] More attention may need to be paid to socioeconomic groups with lower education levels as a link has been made between those groups and long delays before initiation of CRC treatment.[57]

In regard to the second proposal on referral, on the basis of available evidence, there is considerable debate about the validity of a referral system, preferentially based on the presence of certain symptoms, as a way of improving early detection of the earliest stages of CRC. The introduction of a referral system based on objective laboratory data to stratify the risks for symptomatic patients could increase the sensitivity and specificity of RDPs.[46] In relation to this, there are several recently published studies confirming the superior diagnostic accuracy of the immunological fecal occult blood test compared with symptom-based referral criteria, either alone or associated with other predictive score variables.[58–60] There is also evidence showing the presence of anemia (Hb < 120 g/L for women and Hb < 130 g/L for men) multiplies by three of the CRC diagnostic probabilities for patients referred with suspicion of tumor.[61]

In relation to the other two other proposals by Redaniel et al.,[43] the available evidence focuses on the preferential referral system from primary care to CRC diagnosis. In shortening wait time to diagnosis for patients meeting the rapid pathway criteria, the straight-to-test system involves bypassing the initial specialist appointment in the hospital setting and carrying out endoscopies and imaging directly. There are studies showing the straight-to-colonoscopy strategy shortens the time between the first PC visit and diagnosis[62, 63] and is cost effective compared with the traditional system of referral to specialist care.[32, 64] Similar results have been obtained with the system, straight to CT colonography, although only a group of patients over 60 years of age with alternating bowel rhythm and iron deficiency anemia were evaluated.[65] In addition to specific referral systems, improvement in PC resources is fundamental, including ongoing training programs in CRC and coordination with hospital care.[9] The role of the PC doctor at the beginning of the sequence is also important as it has been indicated that continuity of health care and especially trust in the doctor are the best predictors of early stage CRC.[66, 67] It is the responsibility of health systems to provide administrative support, diagnostic and treatment services, and availability of qualified personnel. For this to occur, the presence of a responsible figure appears essential, for CRC patient management in follow-up, for optimization of timing and proper management of financial resources, avoiding unnecessary delays attributable to the process, coordinating processes, and acting as a liaison with the patient and other health care units.[9]

The use of new technologies could be useful for improving RDPs. One possibility in this regard would be to design pathways using results from process mining in each care environment. The aims of process mining are the discovery, monitoring, and improvement of real processes, not virtual processes, by being able to make use of a wide range of health procedures.[68–71] Process mining could therefore be used to determine what the real CRC patient flow process is and to define a new RDP based on the results obtained from the use of these techniques.

Conclusions

Available evidence suggests CRC rapid diagnostic pathways have not achieved one of their objectives, if not the main one, for which they were created—to diagnose CRC in the early stages when probabilities for patient survival are greater. In addition, the CRC detection rate was shown to be low. Consequently, the standard preferential referral system via RDP, based on the presence of certain symptoms, independently of how rapidly the patient is diagnosed, does not improve the CRC detection rate, nor provide diagnosis in the early stages.[46]

Problems in meeting referral criteria and lack of sensitivity and specificity of these cause most CRC cases to be diagnosed apart from the programs. The progressive increase in patients referred to RDPs may result in patients not referred this way waiting longer for diagnostic tests and treatments they need. Even patients referred via RDPs may not fall within the guideline wait time targets; this could be due to an increase in RDP referrals without the diagnostic and treatment resources to cope with them.[50] The straight-to-colonoscopy strategy has been shown to shorten wait times throughout the whole process and to be reliable and cost effective; hence, it could be appropriate to include it in all rapid diagnosis programs. In addition, there are several recently published papers confirming the superiority of the diagnostic accuracy of the immunological fecal occult blood test (alone or associated with other variables) over symptom-based referral criteria, such as currently used in RDPs. Therefore the complementary use of objective laboratory data could increase the diagnostic effectiveness of current RDPs, by better stratifying CRC risk and preventing unnecessary colonoscopies.

Although efforts for early diagnosis have not demonstrated their effectiveness in improving survival, we cannot say that diagnostic delay does not matter.[9] Rapid diagnosis of any kind of cancer is a priority in cancer treatment plans, as stipulated in the WHO guidelines.[66] Reducing wait times for diagnosis and treatment means improvement in the quality of care for the cancer patient, by reducing distress for the patient and their family. In relation to health systems, it means enhanced clinical organization and coordination between different care areas.[6]

Symptoms of CRC are frequently nonspecific and can present in numerous nontumorous gastrointestinal pathologies; this accounts for the low effectiveness of RDPs in which the referral system is fundamentally based on the presence of certain symptoms. In spite of the benefits of implementation of the programs we have seen, but because they have mainly not been shown to have an impact on patient survival, it could be necessary to promote investment in screening program development for the asymptomatic population.[9, 66] On the other hand, as this is a priority issue in health care systems and in order to continue serving the symptomatic population, for nonparticipants and those outside the screening age range, it will be necessary to implement new diagnostic advances in information technology to be able to update and optimize the rapid diagnostic process in order to diagnose more CRC cases at earlier stages and with affordable costs.

References

1. Rodríguez Pérez MP, Grande AM. Calidad asistencial: Conceptos, dimensiones y desarrollo operativo. In: *Gestión clínica: Desarrollo e instrumentos Ed Luis Angel Oteo Edición Díaz de Santos*; 2006. Published online.
2. Informe-incidencia-colon.pdf. https://www.aecc.es/sites/default/files/content-file/Informe-incidencia-colon.pdf.
3. Ciccolallo L, Capocaccia R, Coleman MP, et al. Survival differences between European and US patients with colorectal cancer: role of stage at diagnosis and surgery. *Gut* 2005;**54**(2):268–73. https://doi.org/10.1136/gut.2004.044214.
4. Gonzalez-Hermoso F, Perez-Palma J, Marchena-Gomez J, Lorenzo-Rocha N, Medina-Arana V. Can early diagnosis of symptomatic colorectal cancer improve the prognosis? *World J Surg* 2004;**28**(7):716–20. https://doi.org/10.1007/s00268-004-7232-8.
5. Service GNH. *The NHS Cancer Plan: A Plan for Investment, A Plan for Reform*. Department of Health; 2000.
6. Guzmán Laura KP, Bolíbar Ribas I, Alepuz MT, Gonzalez D, Martín M. Impacto en el tiempo asistencial y el estadio tumoral de un programa de diagnóstico y tratamiento rápido del cáncer colorrectal. *Rev Esp Enferm Dig* 2011;**103**(1):13–9. http://scielo.isciii.es/scielo.php?script=sci_arttext&pid=S1130-01082011000100003. [Accessed 25 February 2021].
7. Valentin-Lopez B, Ferrandiz-Santos J, Blasco-Amaro J-A, Morillas-Sainz J-D, Ruiz-Lopez P. Assessment of a rapid referral pathway for suspected colorectal cancer in Madrid. *Fam Pract* 2012;**29**(2):182–8. https://doi.org/10.1093/fampra/cmr080.
8. Brown ML, Lipscomb J, Snyder C. The burden of illness of cancer: economic cost and quality of life. *Annu Rev Public Health* 2001;**22**:91–113. https://doi.org/10.1146/annurev.publhealth.22.1.91.
9. Vallverdú Cartié H, Comajuncosas Camp J, Orbeal Sáenz RA, et al. Resultados de la implementación del circuito de diagnostico rápido de cáncer colorrectal. *Rev Esp Enferm Dig* 2011;**103**(8):402–7. https://doi.org/10.4321/S1130-01082011000800003.
10. Freund KM, Battaglia TA, Calhoun E, et al. National Cancer Institute patient navigation research program. *Cancer* 2008;**113**(12):3391–9. https://doi.org/10.1002/cncr.23960.
11. Paterson WG, Depew WT, Paré P, et al. Canadian consensus on medically acceptable wait times for digestive health care. *Can J Gastroenterol* 2006;**20**(6):411–23. https://doi.org/10.1155/2006/343686.
12. Ramos M, Esteva M, Cabeza E, Llobera J, Ruiz A. Lack of association between diagnostic and therapeutic delay and stage of colorectal cancer. *Eur J Cancer* 2008;**44**(4):510–21. https://doi.org/10.1016/j.ejca.2008.01.011.
13. Tiong J, Gray A, Jackson C, Thompson-Fawcett M, Schultz M. Audit of the association between length of time spent on diagnostic work-up and tumour stage in patients with symptomatic colon cancer. *ANZ J Surg* 2017;**87**(3):138–42. https://doi.org/10.1111/ans.12804.
14. Aslam MI, Chaudhri S, Singh B, Jameson JS. The "two-week wait" referral pathway is not associated with improved survival for patients with colorectal cancer. *Int J Surg* 2017;**43**:181–5. https://doi.org/10.1016/j.ijsu.2017.05.046.
15. Helewa RM, Turner D, Park J, et al. Longer waiting times for patients undergoing colorectal cancer surgery are not associated with decreased survival. *J Surg Oncol* 2013;**108**(6):378–84. https://doi.org/10.1002/jso.23412.
16. Pruitt SL, Harzke AJ, Davidson NO, Schootman M. Do diagnostic and treatment delays for colorectal cancer increase risk of death? *Cancer Causes Control* 2013;**24**(5):961–77. https://doi.org/10.1007/s10552-013-0172-6.
17. Schneider C, Bevis PM, Durdey P, Thomas MG, Sylvester PA, Longman RJ. The association between referral source and outcome in patients with colorectal cancer. *Surgeon* 2013;**11**(3):141–6. https://doi.org/10.1016/j.surge.2012.10.004.
18. Thornton L, Reader H, Stojkovic S, Allgar V, Woodcock N. Has the "Fast-Track" referral system affected the route of presentation and/or clinical outcomes in patients with colorectal cancer? *World J Surg Oncol* 2016;**14**(1). https://doi.org/10.1186/s12957-016-0911-8.
19. Prades J, Espinàs JA, Font R, Argimon JM, Borràs JM. Implementing a Cancer fast-track Programme between primary and specialised care in Catalonia (Spain): a mixed methods study. *Br J Cancer* 2011;**105**(6):753–9. https://doi.org/10.1038/bjc.2011.308.
20. Mangion D, Brennan A. Relationship between the "two-week rule" and colorectal cancer diagnosis. *Br J Nurs* 2014;**23**(12):660–7. https://doi.org/10.12968/bjon.2014.23.12.660.
21. NICE guideline [NG12] Suspected cancer: recognition and referral 2015 [updated July 2017]. Available from: https://www.nice.org.uk/guidance/ng12/chapter/1-Recommendations-organised-by-site-of-cancer#lower-gastrointestinal-tract-cancers.
22. Circuito Asistencial Rápido Para Pacientes Con Cáncer Colorrectal. Sergas. *Estrutura Organizativa de Xestión Integrada (EOXI) Pontevedra e O Salnés*; 2017.
23. Camiño Clínico Cancro Colorrectal. Sergas. *Área Sanitaria de Ferrol. Área de Xerencia de Procesos*; 2010.
24. Dirección de Procesos Asistenciais. *Procedemento continuidade asistencial en Dixestivo. Vía Rápida de Cancro Colorrectal*. SERGAS. Xerencia de Xestión Integrada de Santiago de Compostela. Sistema de Xestión da Calidade; 2011.
25. *Vía Rápida Cáncer Colorrectal. Sergas*. Complexo Hospitalario Universitario de Vigo; 2011.
26. *Vía Rápida de Cáncer Colorrectal. Sergas*. Xerencia de Xestión Integrada de Ourense Verín e O Barco de Valdeorras; 2012.
27. *Derivación Especial Desde Atención Primaria a Consulta Especializada Hospitalaria. Derivación Especial Para Colonoscopia. Sergas*. Hospital Universitario Lucus Augusti. Xerencia de Atención Primaria Lugo; 2012.
28. *Guía Indicacións Colonoscopia área A Coruña. Sergas*. Xerencia Xestión Integrada A Coruña; 2013.
29. *Mellora de Accesibilidade Dos Pacientes Con Sospeita de Cancro ós Dispositivos Asistencias. Santiago de Compostela. Sergas*. Dirección de asistencia sanitaria; 2011. http://intranet/DOCUMENTOSXerencia%20de%20atencin%20integrada%20de%20Santiag/201112/Presentaci%c3%b3n%20Circuitos%20Cancro_20111216_102314_0154.pdf.
30. Ballal M, Hodder R, Ameh V, Selvachandran S, Cade D. Guideline compliance—do we maintain the standards? *Colorectal Dis Suppl* 2003;**5**.
31. Barwick TW, Scott SB, Ambrose NS. The two week referral for colorectal cancer: a retrospective analysis. *Colorectal Dis* 2004;**6**(2):85–91. https://doi.org/10.1111/j.1463-1318.2004.00589.x.

32. Beggs AD, Bhate RD, Irukulla S, Achiek M, Abulafi AM. Straight to colonoscopy: the ideal patient pathway for the 2-week suspected cancer referrals? *Ann R Coll Surg Engl* 2011;**93**(2):114–9. https://doi.org/10.1308/003588411X12851639107917.
33. Bevis PM, Donaldson OW, Card M, et al. The association between referral source and stage of disease in patients with colorectal cancer. *Colorectal Dis* 2008;**10**(1):58–62. https://doi.org/10.1111/j.1463-1318.2007.01222.x.
34. Chohan DPK, Goodwin K, Wilkinson S, Miller R, Hall NR. How has the "two-week wait" rule affected the presentation of colorectal cancer? *Colorectal Dis* 2005;**7**(5):450–3. https://doi.org/10.1111/j.1463-1318.2005.00821.x.
35. Choudhary R, Debnath D, Gunning K. A 3 year study of compliance to the guidelines for urgent referral of suspected colorectal cancer: P057. *Colorectal Dis Suppl* 2005;**7**.
36. Debnath D, Choudhary R, Dielehner N, Gunning K. Rapid access colorectal cancer referral: a significant association between cancer diagnosis and compliance with the guidelines. *Colorectal Dis Suppl* 2003;**5**:41–2.
37. Eccersley AJ, Wilson EM, Makris A, Novell JR. Referral guidelines for colorectal cancer—do they work? *Ann R Coll Surg Engl* 2003;**85**(2):107–10. https://doi.org/10.1308/003588403321219885.
38. Leung E, Grainger J, Bandla N, Wong L. The effectiveness of the "2-week wait" referral service for colorectal cancer: effectiveness of 2-week wait service. *Int J Clin Pract* 2010;**64**(12):1671–4. https://doi.org/10.1111/j.1742-1241.2010.02505.x.
39. Patel RK, Sayers AE, Seedat S, Altayeb T, Hunter IA. The 2-week wait service: a UK tertiary colorectal centre's experience in the early identification of colorectal cancer. *Eur J Gastroenterol Hepatol* 2014;**26**(12):1408–14. https://doi.org/10.1097/MEG.0000000000000206.
40. Warwick M, Zeiderman M, Watkinson A. Referral of patients to a rapid access clinic for suspected colorectal cancer in the absence of "high-risk"-symptoms. *Colorectal Dis Suppl* 2003;**vol. 5**.
41. Miranda MA, Alburquerque Miranda M, Fernández Alvarez A, et al. Diagnóstico rápido de cáncer colorectal: ¿justifica la sobrecarga de las salas de endoscopia? *Endoscopy* 2013;**45**(11). https://doi.org/10.1055/s-0033-1354660.
42. Vega-Villaamil P, Salve-Bouzo M, Cubiella J, et al. Evaluación de la implantación de las indicaciones y niveles de prioridad del Servizo Galego de Saude para la colonoscopia en pacientes sintomáticos: estudio prospectivo y transversal. *Rev Esp Enferm Dig* 2013;**105**(10):600–8. http://scielo.isciii.es/scielo.php?script=sci_arttext&pid=S1130-01082013001000005.
43. Redaniel MT, Ridd M, Martin RM, Coxon F, Jeffreys M, Wade J. Rapid diagnostic pathways for suspected colorectal cancer: views of primary and secondary care clinicians on challenges and their potential solutions. *BMJ Open* 2015;**5**(10). https://doi.org/10.1136/bmjopen-2015-008577, e008577.
44. Chohan D, Goodwin K, Wilkinson S, Miller R, Hall N. How has the "2-week wait rule" affected colorectal cancer presentation? *Colorectal Dis Suppl* 2003;**5**:40–1.
45. Vaughan-Shaw PG, Cutting JE, Borley NR, Wheeler JMD. Repeat 2-week wait referrals for colorectal cancer. *Colorectal Dis* 2013;**15**(3):292–7. https://doi.org/10.1111/j.1463-1318.2012.03173.x.
46. Mozdiak E, Weldeselassie Y, McFarlane M, et al. Systematic review with meta-analysis of over 90 000 patients. Does fast-track review diagnose colorectal cancer earlier? *Aliment Pharmacol Ther* 2019;**50**(4):348–72. https://doi.org/10.1111/apt.15378.
47. Casamitjana M, Pozuelo A, López M, Segura JM, Peris M. *Evaluación de un programa piloto de diagnóstico rápido de cáncer entre seis áreas básicas y su hospital de referencia*. http://www.postermedic.com/parcdesalutmar/npimas062099/pdfbaja/npimas062099.pdf. [Accessed 25 February 2021].
48. Zhou Y, Mendonca SC, Abel GA, et al. Variation in "fast-track" referrals for suspected cancer by patient characteristic and cancer diagnosis: evidence from 670 000 patients with cancers of 35 different sites. *Br J Cancer* 2018;**118**(1):24–31. https://www.nature.com/articles/bjc2017381.
49. Jones R, Rubin G, Hungin P. Is the two week rule for cancer referrals working? *BMJ* 2001;**322**(7302):1555–6. https://doi.org/10.1136/bmj.322.7302.1555.
50. Rai S, Kelly MJ. Prioritization of colorectal referrals: a review of the 2-week wait referral system. *Colorectal Dis* 2007;**9**(3):195–202. https://doi.org/10.1111/j.1463-1318.2006.01107.x.
51. Cark J, Williams A, Steger A, et al. Missing colorectal Cancer—the 2-week wait. *Colorectal Dis* 2002;**4**(suppl 1):49. PO27.
52. Aryal K, Sverrisdottir A. Treatment of colorectal Cancer in a district general hospital; where do we stand in terms of waiting times? *Colorectal Dis Suppl* 2003;**5**, A39.
53. Foster P, Scott S, Ambrose N. Waiting times for colorectal cancer. *Colorectal Dis Suppl* 2003;**5**, A39.
54. Pullyblank A, Silavant M, Cook T. Failure to recognize high risk symptoms of colorectal cancer in standard referral letters leads to a delay in initiation of treatment. *Br J Surg* 2003;**90**(suppl I):133.
55. Lloyd T, Sutton C, Marshall LJ, Marshall D, Beach M, Kelly M. Application of cancer collaborative initiatives. *Colorectal Dis* 2002;**4**(suppl I):63.
56. Tørring ML, Murchie P, Hamilton W, et al. Evidence of advanced stage colorectal cancer with longer diagnostic intervals: a pooled analysis of seven primary care cohorts comprising 11 720 patients in five countries. *Br J Cancer* 2017;**117**(6):888–97. https://doi.org/10.1038/bjc.2017.236.
57. Zarcos-Pedrinaci I, Fernández-López A, Téllez T, et al. Factors that influence treatment delay in patients with colorectal cancer. *Oncotarget* 2017;**8**(22):36728–42. https://doi.org/10.18632/oncotarget.13574.
58. Cubiella J, Salve M, Díaz-Ondina M, et al. Diagnostic accuracy of the faecal immunochemical test for colorectal cancer in symptomatic patients: comparison with NICE and SIGN referral criteria. *Colorectal Dis* 2014;**16**(8):O273–82. https://doi.org/10.1111/codi.12569.
59. Herrero J-M, Vega P, Salve M, Bujanda L, Cubiella J. Symptom or faecal immunochemical test based referral criteria for colorectal cancer detection in symptomatic patients: a diagnostic tests study. *BMC Gastroenterol* 2018;**18**(1):155. https://doi.org/10.1186/s12876-018-0887-7.
60. Rodríguez-Alonso L, Rodríguez-Moranta F, Ruiz-Cerulla A, et al. An urgent referral strategy for symptomatic patients with suspected colorectal cancer based on a quantitative immunochemical faecal occult blood test. *Dig Liver Dis* 2015;**47**(9):797–804. https://doi.org/10.1016/j.dld.2015.05.004.
61. Mashlab S, Large P, Laing W, et al. Anaemia as a risk stratification tool for symptomatic patients referred via the two-week wait pathway for colorectal cancer. *Ann R Coll Surg Engl* 2018;**100**(5):350–6. https://doi.org/10.1308/rcsann.2018.0030.

62. Gregory C. Improving colorectal cancer referrals. *BMJ Open Qual* 2018;**7**(1). https://doi.org/10.1136/bmjoq-2017-000280, e000280.
63. Mukherjee S, Fountain G, Stalker M, et al. The "straight to test" initiative reduces both diagnostic and treatment waiting times for colorectal cancer: outcomes after 2 years. *Colorectal Dis* 2010;**12**(10Online):e250–4. https://onlinelibrary.wiley.com/doi/abs/10.1111/j.1463-1318.2009.02182.x.
64. Banerjea A, Voll J, Chowdhury A, et al. Straight-to-test colonoscopy for 2-week-wait referrals improves time to diagnosis of colorectal cancer and is feasible in a high-volume unit. *Colorectal Dis* 2017;**19**(9):819–26. https://doi.org/10.1111/codi.13667.
65. Stephenson JA, Pancholi J, Ivan CV, et al. Straight-to-test faecal tagging CT colonography for exclusion of colon cancer in symptomatic patients under the English 2-week-wait cancer investigation pathway: a service review. *Clin Radiol* 2018;**73**(9). https://doi.org/10.1016/j.crad.2018.05.013, 836.e1-e836.e7.
66. Bixquert JM. Early diagnosis of colorectal cancer. Diagnostic delay reduction or rather screening programs? *Rev Esp Enferm Dig* 2006;**98**(5):315–21. https://doi.org/10.4321/s1130-01082006000500001.
67. Mainous 3rd AG, Kern D, Hainer B, Kneuper-Hall R, Stephens J, Geesey ME. The relationship between continuity of care and trust with stage of cancer at diagnosis. *Fam Med* 2004;**36**(1):35–9. https://www.ncbi.nlm.nih.gov/pubmed/14710327.
68. Homayounfar P. Process mining challenges in hospital information systems. In: *2012 Federated Conference on Computer Science and Information Systems (FedCSIS)*; 2012. p. 1135–40. https://ieeexplore.ieee.org/abstract/document/6354456/.
69. Lenz R, Reichert M. IT support for healthcare processes—premises, challenges, perspectives. *Data Knowl Eng* 2007;**61**(1):39–58. https://doi.org/10.1016/j.datak.2006.04.007.
70. Mans RS, Schonenberg MH, Song M, van der Aalst WMP, Bakker PJM. Application of process mining in healthcare—a case study in a dutch hospital. In: *International Joint Conference on Biomedical Engineering Systems and Technologies*. Springer; 2008. p. 425–38. https://link.springer.com/chapter/10.1007/978-3-540-92219-3_32.
71. van der Aalst WMP. Process mining. In: *Encyclopedia of Database Systems*; 2009. p. 2171–3. https://doi.org/10.1007/978-0-387-39940-9_1477. Published online.

Section C

Diagnosis and staging

Section C.1

Current diagnosis methods in colorectan cancer

Chapter 8

Colon capsule endoscopy

Ramiro Manuel Macenlle García
Digestive System Service Ourense University Hospital Complex, Ourense, Spain

Introduction

The scientific progress experienced in the last two decades in fields such as electronics has allowed the design of small devices equipped with advanced technology. The capsule endoscopy is a clear example of this. It consists of a device that once ingested is able to take multiple images for several hours as it passes through the digestive tract and transmit them to an external recorder.

At the World Congress of Gastroenterology held in Los Angeles in 1994, Gabriel Iddan and Paul Swain first presented this technique, and Swain's first two publications on this date from 1996 and 1997.[1, 2]

Initially the capsule endoscopy was developed for the study of the small intestine (Given Imaging, Yoqneam, Israel), and subsequently a model for the study of the esophagus and a model for the study of the colon were developed. Recently the second generation of colon capsule endoscopy (CCE), called Pillcam COLON 2, has been released, which compared to the previous model offers technical improvements.

Technical characteristics of the second-generation colon capsule

The CCE Pillcam COLON 2 is a double head capsule (see Fig. 8.1). It consists of a double camera of low consumption with CMOS technology, two light sources with four emitters of white LED light each, two optics of eight magnifications, a radio frequency transmitter (see Fig. 8.2) in the UHF band as well as control electronics and batteries power, which are button batteries containing silver oxide. The autonomy is 10 h and the size is 31.5×11.6 mm. It has three lenses by optics and an automatic light control system. The viewing angle is 172 degrees, greater than in the previous model, and the depth of field is 0–30 mm.

This device is capable of taking up to 17.5 images per second in each camera, that is, 35 images per second in total, a notable increase with respect to the 4 images per second captured by the first-generation CEC. These images are transmitted to the receiver / transmitter that the patient carries during the entire duration of the study, where they will be stored. This receiver/transmitter is placed in the pocket of a special belt that adapts to the patient's waist and must be connected to seven mini-antennae that receive the capsule signals and that are placed in the patient's abdomen in the specified positions by the manufacturer. In addition, another antenna also placed in the abdomen will serve to transmit signals from the transmitter/receiver to the capsule. There is permanent communication of the capsule with the transmitter/receiver, so that it is able to order an increase or decrease in the frequency of taking pictures depending on whether the capsule is moving or static. In addition, the transmitter/receiver is able to analyze the images to determine if the capsule remains in the stomach or has already passed into the small intestine, sending alarm signals to the patient as appropriate for the intake of laxatives or prokinetics.

Once the study is finished, the data stored in the transmitter/receiver must be transmitted from a cradle connected to a computer equipped with the software that will allow the visualization of the images obtained. One of the novelties of the Pillcam COLON 2 software is that it includes a function to estimate the size of the lesions detected.[3, 4] Technical characteristics of the Pillcam COLON 2 are presented in Table 8.1.

Preparation of the patient

Intestinal cleansing as exhaustive as possible is essential for the CCE to be able to clearly identify the different lesions that may exist in the mucosa. It is recommended to follow a diet with clear liquids the day before the exploration. The beneficial role of a low-residue diet the previous days has not been clearly demonstrated.

FIG. 8.1 Colon capsule Pillcam COLON 2.

In the studies performed with the first-generation CEC and with the second generation, the cleaning method has varied; patients have usually been recommended to take 4 L of polyethylene glycol before the examination, two or three liters the day before and the rest the day of the test, together with additional doses of sodium phosphate laxative solution once the capsule is ingested.

In addition, the administration of an oral prokinetic and a bisacodyl suppository has been employed. With the Pillcam COLON 2, if the transmitter/receiver detects that the capsule remains in the stomach after 1 h from its ingestion, an alarm signal is activated indicating that the patient should take a prokinetic. And once the passage to the small intestine has been detected, another alarm signal will indicate to the patient that the additional dose of sodium phosphate should be taken.[5, 6]

In an extensive review on the subject, Singhal et al.[7] consider that the ideal method to achieve a perfect intestinal cleansing has not yet been standardized and that more studies are needed to optimize it.

Comparative studies between colon capsule endoscopy and colonoscopy

The vast majority of the scientific evidence accumulated in recent years with the CCE has been using the first-generation model, with a growing and progressive information available with the use of the second-generation model after its commercialization. There are no direct comparative studies between both models.

Almost all studies have included adult patients with different indications for performing a colonoscopy, and the diagnostic accuracy of the CCE has been compared with the colonoscopy as a gold standard. Generally, the detection of three or more polyps of any size and the detection of polyps of size ≥ 6 mm or masses have been considered significant findings.

With the first-generation capsule the sensitivity for the detection of polyps of size ≥ 6 mm has been low, from 58% to 64%, with the specificity being 73%–84%.[8, 9] In the multicenter study by Van Gossum et al.,[10] in the 328 patients studied, 19 cases of colorectal cancer (CRC) were detected with colonoscopy, of which 14 were also identified with the capsule, which means a sensitivity of 74% for this finding.

In this last study, 7.9% of the patients reported mild-moderate adverse effects, almost all related to bowel preparation. Sacher-Huvelin et al.,[11] also conducted a multicenter study with 545 patients with medium or increased risk of CRC. Of the

FIG. 8.2 Transmitter/receiver on its cradle.

TABLE 8.1 Technical characteristics of the Pillcam COLON 2 (courtesy of Endo Técnica S.L.).

Dimensions	31.5 mm (L) × 11.6 mm (D)
Weight	2.87 ± 0.3 g
Optic characteristics	3 lentes
Field of vision	172 degrees (ISO_8600_3) by camera
Optic illumination	4 LEDs in each camera
Magnification	1:8 approximately
Depth of field	0–30 mm
Automatic light control	Yes
Minimum detectable object	0,09 mm
Transmission frequency	434,1 MHz
Type of transmitter modulation	MSK
Transmission modulation signal	Digital
Operational characteristics	
Type of battery	Silver oxide 399
Autonomy	10 h
Storage temperature	0–30°
Regulations	
FDA, FCC, UL 2601, ISO- 8600-3	IEC 60601, CE, TGA

five cases of CRC detected with colonoscopy, two were not appreciated with the CCE, this last test also showing a low sensitivity for the detection of polyps equal to or greater than 6 mm, of only 39%, the specificity being of 88%. The authors concluded that CCE could not replace colonoscopy as a CCR screening technique.

There have been published several prospective studies in which the second-generation CCE has been compared with colonoscopy for the detection of polyps. In the study by Eliakim et al.,[8] in 98 patients studied the sensitivity and specificity of the CCE for the detection of polyps of size ≥6 mm have been 89% and 76%, respectively, and for polyps of size equal to or greater than 10 mm, 88% and 89%, respectively. In the study by Spada et al.,[6] including 109 patients, the sensitivity and specificity of CCE for the detection of polyps of size ≥6 mm have been 84% and 64%, respectively, and for polyps of size equal or greater than 10 mm, 88% and 95%, respectively. In this study there were three patients with CRC, all detected with the capsule. In both studies the adverse effects have been 8% and 6.8%, respectively, all mild-moderate and almost always related to bowel preparation.

In the multicenter study by Rex et al.[12] with the inclusion of 884 subjects, the sensitivity and specificity for the finding of polyps equal to or greater than 6 mm have been 88% and 82%, respectively, and for polyps equal to or greater than 10 mm, 92% and 95%, respectively.

Spada et al.[13] analyzed in a systematic review and meta-analysis the diagnostic accuracy of the first (CCE-1) and second generation (CCE-2) in the detection of polyps. They identified 14 studies with 2420 patients in whom a colonoscopy and a CEC-1 or a CEC-2 had been performed. For the detection of polyps larger than 6 mm, the sensitivity of CEC-2 was 86% and 58% with the CEC-1. For the detection of polyps of 10 mm or greater, the sensitivity and specificity of the CEC-1 were 54% and 97.4%, while with the CEC-2 the sensitivity was 87%, with a specificity of 95.3%.

Depending on the results obtained, and in the absence of comparative studies with both models of CCE, a greater sensitivity is observed with the second-generation model in the detection of polyps of size ≥6 mm. The low specificity found with this model is due, in the majority of cases, to the detection of polyps whose size is considered significant by the CCE and that in the colonoscopy are shown ≤6 mm in size. It is possible that this overestimation can be solved in the future with adjustments in the computer program.

Comparative studies between CCE and colonography by computerized axial tomography

One of the techniques used for colonic evaluation is colonography by computerized axial tomography or colonoCT. Spada et al.[14] conducted a comparative study between the two and found no differences in the rate of complete examination of the colon between them (98%) and also in the level of intestinal cleansing. In the detection of polyps of 6–10 mm, the CCE was superior to the colonoCT, and also showed greater diagnostic capacity for flat / sessile lesions. In contrast, Rondonotti et al.[15] found no differences between the two techniques in the detection rate of polyps larger than 6 or 10 mm.

Current indications and contraindications of the colon capsule endoscopy: Future perspectives

Currently this technique does not have clearly defined indications. It could be an alternative to colonoscopy in patients who do not wish to have this test or have any contraindication to it, as well as a complement if the colonoscopy is incomplete and thus allow the study of the rest of the colon.[16–18]

In a Spanish multicenter study, 96 patients with incomplete colonoscopy underwent a study with CCE. In 69 of them (71.9%), the entire colon could be examined, finding new lesions in 58 patients (60.4%), mainly polyps, detected in 41 subjects, and 2 cases of cancer (2.1%). In 43 cases, the attitude taken consisted of performing a new colonoscopy using a pediatric colonoscope, a balloon enteroscope, or a standard colonoscope with change in the type of sedation, achieving a complete exploration in 40 patients.[19]

In another multicenter study, Baltes et al.[20] performed CCE to 81 patients after an incomplete colonoscopy, using two protocols of intestinal cleansing, achieving the visualization of colon segments not explored by colonoscopy in 90% and 97% of cases, respectively.

The main interest of the CCE manufacturing company is to position this diagnostic technique as an alternative to colonoscopy in CRC screening in patients at medium risk. In some countries this screening is performed by colonoscopy, and generally a low population participation is observed, not greater than 25%.[16] A noninvasive technique such as CCE, despite lower diagnostic accuracy than colonoscopy, may be better accepted and may increase the number of patients diagnosed with CRC or advanced adenomas.

Hassan et al.[21] conducted a cost-effectiveness study in which they have compared first-generation CCE with colonoscopy as a CRC screening technique, showing that CCE is more cost effective if adherence greater than 30% is achieved with respect to that achieved with colonoscopy. Since the second-generation CCE seems to offer better results than its predecessor, its application in a screening program could be cost effective with a difference in adhesion less than that 30%. To date, there has been no prospective, randomized, large population-based study with a medium risk of suffering from CRC, comparing colonoscopy and CCE as screening techniques.

In other countries, such as Spain, CRC screening programs in medium-risk population are carried out by means of immunological fecal occult blood (FOB), and colonoscopy is performed in those with positive results. We also lack studies comparing this strategy with the CCE.

In the Spanish guideline for diagnosis and prevention of CRC updated in 2018, it is advised to perform colonoCT in patients with contraindication for colonoscopy or with an incomplete colonoscopy for reasons other than poor intestinal cleansing. As for the CCE, they suggest performing this technique in the same cases as the colonoCT.[22]

An Italian study has been carried out with 110 patients from a CRC screening program, all with positive FOB, who were offered the study by CEC after having rejected colonoscopy. Only 10 patients agreed, which means that possibly less than 10% of the subjects can be rescued with a less invasive technique than colonoscopy.[23]

As contraindications for performing CCE, they are considered the same as for the small intestine capsule. The use of sodium phosphate as a stimulant of intestinal transit should be avoided in patients at risk of toxicity from this product. The risk of capsule retention is low and if it occurs, endoscopy or surgery must be considered according to the clinical situation.[16]

Technological advances are expected in the coming years. The application of artificial intelligence to the CCE could allow automated reading of the captured images as well as the identification of lesions. New capsules are also being developed, such as a magnetically controlled endoscopic capsule or Check-cap, which obtains images using a low dose of radiation without requiring prior intestinal cleansing.[24]

Conclusions

CCE offers the possibility of studying the large intestine safely and noninvasively. A thorough intestinal cleansing is an essential requirement so that the visualization of the intestinal mucosa is optimal; however, the ideal method to achieve it has not yet been found.

With the first-generation CCE a low diagnostic accuracy has been observed, which significantly improves with the second-generation CCE, so that this technique could be an alternative to colonoscopy in certain situations. Automated reading could contribute in the near future to the expansion of the use of this diagnostic technique.

Other types of capsules for the study of the colon are in development and their results are expected with great interest.

References

1. Swain CP, Gong F, Mills TN. Wireless transmission of a colour television moving image from the stomach using a miniature CCD camera, light source and microwave transmitter. *Gut* 1996;**39A**:26. https://doi.org/10.1016/S0016-5107(97)80063-6.
2. Swain CP, Gong F, Mills TN. Wireless transmission of a colour television moving image from the stomach using a miniature CCD camera, light source and microwave transmitter. *Gastrointest Endosc* 1997;**45**(4):AB40. https://doi.org/10.1016/S0016-5107(97)80063-6.
3. González-Suárez B, Llach J. Nueva generación de cápsula endoscópica colónica: ¿una opción no invasiva en el cribado del cáncer colorrectal? *Gastroenterol Hepatol* 2011;**34**(5):346–51. https://doi.org/10.1016/j.gastrohep.2011.03.011.
4. Spada C, De Vincentis F, Cesaro P, et al. Accuracy and safety of second-generation PillCam COLON capsule for colorectal polyp detection. *Ther Adv Gastroenterol* 2012;**5**(3):173–8. https://doi.org/10.1177/1756283X12438054.
5. Eliakim R, Fireman Z, Gralnek IM, et al. Evaluation of the PillCam Colon capsule in the detection of colonic pathology: results of the first multicenter, prospective, comparative study. *Endoscopy* 2006;**38**(10):963–70. https://doi.org/10.1055/s-2006-944832.
6. Spada C, Hassan C, Munoz-Navas M, et al. Second-generation colon capsule endoscopy compared with colonoscopy. *Gastrointest Endosc* 2011;**74**(3):581–589.e1. https://doi.org/10.1016/j.gie.2011.03.1125.
7. Singhal S, Nigar S, Paleti V, Lane D, Duddempudi S. Bowel preparation regimens for colon capsule endoscopy: a review. *Ther Adv Gastroenterol* 2014;**7**(3):115–22. https://doi.org/10.1177/1756283X13504730.
8. Eliakim R, Yassin K, Niv Y, et al. Prospective multicenter performance evaluation of the second-generation colon capsule compared with colonoscopy. *Endoscopy* 2009;**41**(12):1026–31. https://doi.org/10.1055/s-0029-1215360.
9. Schoofs N, Devière J, Van Gossum A. PillCam colon capsule endoscopy compared with colonoscopy for colorectal tumor diagnosis: a prospective pilot study. *Endoscopy* 2006;**38**(10):971–7. https://doi.org/10.1055/s-2006-944835.
10. Van Gossum A, Munoz-Navas M, Fernandez-Urien I, et al. Capsule endoscopy versus colonoscopy for the detection of polyps and cancer. *N Engl J Med* 2009;**361**(3):264–70. https://doi.org/10.1056/NEJMoa0806347.
11. Sacher-Huvelin S, Coron E, Gaudric M, et al. Colon capsule endoscopy vs. colonoscopy in patients at average or increased risk of colorectal cancer. *Aliment Pharmacol Ther* 2010;**32**(9):1145–53. https://doi.org/10.1111/j.1365-2036.2010.04458.x.
12. Rex DK, Adler SN, Aisenberg J, et al. Accuracy of PillCam COLON 2 for detecting subjects with adenomas ≥6 mm. *Gastrointest Endosc* 2013;**77**(S5):AB703.
13. Spada C, Pasha SF, Gross SA, et al. Accuracy of first- and second-generation colon capsules in endoscopic detection of colorectal polyps: a systematic review and meta-analysis. *Clin Gastroenterol Hepatol* 2016;**14**(11):1533–1543.e8. https://doi.org/10.1016/j.cgh.2016.04.038.
14. Spada C, Hassan C, Barbaro B, et al. Colon capsule versus CT colonography in patients with incomplete colonoscopy: a prospective, comparative trial. *Gut* 2015;**64**(2):272–81. https://doi.org/10.1136/gutjnl-2013-306550.
15. Rondonotti E, Borghi C, Mandelli G, et al. Accuracy of capsule colonoscopy and computed tomographic colonography in individuals with positive results from the fecal occult blood test. *Clin Gastroenterol Hepatol* 2014;**12**(8):1303–10. https://doi.org/10.1016/j.cgh.2013.12.027.
16. Riccioni ME, Urgesi R, Cianci R, Bizzotto A, Spada C, Costamagna G. Colon capsule endoscopy: advantages, limitations and expectations. Which novelties? *World J Gastrointest Endosc* 2012;**4**(4):99–107. https://doi.org/10.4253/wjge.v4.i4.99.
17. Spada C, Hassan C, Galmiche JP, et al. Colon capsule endoscopy: European Society of Gastrointestinal Endoscopy (ESGE) guideline. *Endoscopy* 2012;**44**(5):527–36. https://doi.org/10.1055/s-0031-1291717.
18. Spada C, Barbaro F, Andrisani G, et al. Colon capsule endoscopy: what we know and what we would like to know. *World J Gastroenterol* 2014;**20**(45):16948–55. https://doi.org/10.3748/wjg.v20.i45.16948.
19. Nogales Ó, García-Lledó J, Luján M, et al. Therapeutic impact of colon capsule endoscopy with PillCam™ COLON 2 after incomplete standard colonoscopy: a Spanish multicenter study. *Rev Esp Enferm Dig* 2017;**109**(5):322–7. https://doi.org/10.17235/reed.2017.4369/2016.
20. Baltes P, Bota M, Albert J, et al. PillCamColon2 after incomplete colonoscopy—a prospective multicenter study. *World J Gastroenterol* 2018;**24**(31):3556–66. https://doi.org/10.3748/wjg.v24.i31.3556.
21. Hassan C, Zullo A, Winn S, Morini S. Cost-effectiveness of capsule endoscopy in screening for colorectal cancer. *Endoscopy* 2008;**40**(5):414–21. https://doi.org/10.1055/s-2007-995565.
22. Cubiella J, Marzo-Castillejo M, Mascort-Roca JJ, et al. Guía de práctica clínica. Diagnóstico y prevención del cáncer colorrectal. Actualización 2018. *Gastroenterol Hepatol* 2018;**41**(9):535–610. https://doi.org/10.1016/j.gastrohep.2018.07.012.
23. Mussetto A, Triossi O, Gasperoni S, Casetti T. Colon capsule endoscopy may represent an effective tool for colorectal cancer screening: a single-centre series. *Dig Liver Dis* 2012;**44**(4):357–8. https://doi.org/10.1016/j.dld.2011.11.004.
24. Yung DE, Rondonotti E, Koulaouzidis A. Review: capsule colonoscopy-a concise clinical overview of current status. *Ann Transl Med* 2016;**4**(20):398–411. https://doi.org/10.21037/atm.2016.10.71.

Chapter 9

Endoscopic ultrasound in CRC

José Luis Ulla Rocha[a], Raquel Sardina Ferreiro[b], and Juan Turnes Vázquez[a]

[a]*Gastroenterology Department, Digestive Disease Unit, Pontevedra University Hospital Complex, Pontevedra, Spain,* [b]*Department of Internal Medicine, Arquitect Marcide Hospital. Ferrol, Spain*

Outline of technique

Endoscopic ultrasonography (EUS) or echoendoscopy was first seen in 1985 as an extension of conventional endoscopy. The new echoendoscope was formed by fitting an echographic probe to the distal end of the endoscope. For transmission of ultrasound, the echographic probe is covered with a distensible balloon filled with still water.[1] Very small echography probes (miniprobes) that could be inserted into the endoscope work channel were also developed but are now scarcely used. Rectal EUS usually uses frequencies of 7.5–12 MHZ, which allows for detailed examination of the gastrointestinal tract wall and adjacent delimiting structures, demonstrating great usefulness for staging neoplasms found there. The main drawback is operator dependence, with the subjectivity this entails.[2] Currently, both radial and sectorial echoendoscopes are available—the radial gives circumferential echographic visualization and is the most used for tumoral staging, while the sectorial linear echoendoscope has an elevator and provides sectorial visualization, allowing puncturing for extraction of histopathological samples and performance of therapeutic interventions.[3] Frontal view echoendoscopes are also available, but at present only in specialized clinics.[4] The echoendoscopy technique requires a learning curve to be highly skilled, especially for identification of malignant lymph nodes. At present the technique can only be applied to localized tumors in the rectum or at the recto-sigmoidal junction. This is due to the fact that the radial EUS equipment with an oblique lateral view has difficulty advancing through the sigma.

Patients referred to us for EUS staging have usually already undergone a full colonoscopy during which biopsy samples have been taken from the tumor area. Enema is the only intestinal preparation required for rectal EUS due to the fact that the exploration distance is limited to 30 cm. Sedation is not usually given except in the case of adenopathy aspiration puncture or therapeutic procedures.

EUS for rectal cancer staging

For management of rectal adenocarcinoma, it is essential to know the exact state of the tumor at diagnosis in order to choose the most appropriate therapeutic option, such as local excision, radical resection, radiotherapy, and others. Examination using EUS provides information on the rectum wall and adjacent tissue, so is useful for loco-regional TN staging. It provides a high degree of accuracy in ascertaining depth of infiltration and affectation of adenopathies. If EUS is carried out for pretherapeutic staging using the TNM system, it is established practice to place a U (ultrasonography) before the T or N. For rectal examination, the technique is initially carried out while the patient is in the left lateral decubitus position, after which the apparatus is moved to provide optimum visualization of the tumor until it is seen in its entirety. The rectum may be filled with water in some cases to optimize the quality of the image. Care is taken not to put any or as little pressure as possible on the tumor. To accurately establish the distal margin of a tumor, the apparatus is usually inserted to a distance of 30 cm to be able to assess adenopathy presence in the aorta-iliac area and then the rectum is checked for invasion of the anal sphincter complex, However, despite the accuracy of the technique in determining TN staging, a CT scan is needed to rule out the presence of distant metastasis. At present, the staging classification system most widely used is that of the American Joint Cancer Committee[5] (Table 9.1).

Degree of invasion of intestinal wall (uT)

Using EUS, five concentric layers of the normal rectal wall can be visualized, showing echogenic differences according to the acoustic impedance of the tissue. These five layers in order of distance from the transducer are the hyperechoic

TABLE 9.1 TNM staging.

TNM staging categories
Tx: Primary tumor cannot be assessed
T0: No evidence of primary tumor
Tis: Carcinoma in situ: intraepithelial or invasion of lamina propria
T1: Tumor invades submucosa
T2: Tumor invades muscularis propria
T3: Tumor invades perirectal tissue through muscularis propria
T4a: Tumor penetrates visceral peritoneum surface
T4b: Tumor directly invades or adheres to other organs and structures
N0: No regional lymph nodes metastasis
N1: Metastasis in one to three regional lymph nodes
N1a: Metastasis in one regional lymph node
N1b: Metastasis in 2–3 regional lymph nodes
N1c: Tumor deposit(s) in the subserosa or mesentery without regional node metastasis
N2: Metastasis present in four or more regional lymph nodes
N2a: metastasis in 4–6 regional lymph nodes
N2b: metastasis in 7 or more regional lymph nodes
M0: No evidence of distant metastasis
M1a: Metastasis confined to one organ or site (liver, lung, ovary or nonregional ganglia)
M1b: Metastasis in more than one organ or site or in the peritoneum

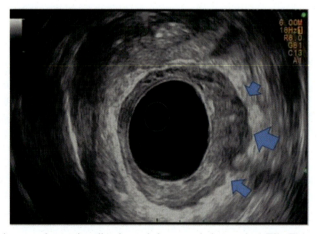

FIG. 9.1 Arrows indicate hypoechoic tumoral mass invading beyond the muscularis propria (uT3). The normal rectal wall can be seen on the opposite side.

balloon-mucosa interface, the hypoechoic muscularis mucosa, the hyperechoic submucosa, the hypoechoic muscularis propria, and the hyperechoic serosa (Fig. 9.1). Rectal adenocarcinomas appear as hypoechogenic masses that deeply invade the different layers of the rectum wall (Fig. 9.1). Early studies on EUS produced T-staging accuracy of 80%–95% in relation to the correspondence between the degree of invasion established by echoendoscopy, and that determined by analysis of histopathological samples from surgery; this was a higher percentage than produced by CT scans and nuclear magnetic resonance (MRI). Some authors, however, have subsequently expressed concern about possible overestimation of some tumors (uT3) causing unnecessary treatment for some patients whose tumors were in fact found to be in early stages (T1-T2). Due to this, it has been suggested that the accuracy obtained in routine clinical practice in an endoscopy unit may not be as high as that formerly obtained in specifically designed studies in focused clinics performing a high volume of cases (>30 cases of EUS rectal staging per year). Despite this, EUS is considered a technique with a favorable cost-benefit ratio that if judiciously used in combination with other techniques, such as CT scanning and MRI, will allow the most apt treatment to be chosen for each patient.

A recent systematic review has shown that EUS demonstrates greater suitability for T1-T2 staging in cases of early adenocarcinomas than for T3-T4 advanced cases. A possible reason is that EUS does not discriminate accurately if irregularities observed in the serosa are the result of tumoral invasion or peritumoral inflammation. Conversely, MRI provides more accurate staging results for advanced adenocarcinomas, as will be discussed later. Finally, we must take into account that the presence of stenosis and polypoid morphology of tumoral lesions is inversely associated with the accuracy of EUS.

EUS provides more accurate staging for early stage T1-T2 tumors with a low degree of invasion. This has particular relevance for tumors in the distal rectum, both for those affecting the mucosa alone and those affecting the submucosa close to the surface. Either endoscopic or surgical endoanal excision can be performed for these, providing the decision is made on an individual basis with the risk-benefit equation in mind, and close follow-up of the local treatment is carried out, all with the aim of maintaining continence with the anal sphincter complex remaining intact.

Affectation of local adenopathies (uN)

The possibility of the tumor affecting the lymph nodes can be visualized using EUS. As we have shown in the TNM classification system, it is important to enumerate the affected lymph nodes. Some authors have established criteria for identifying adenopathic malignancy (Table 9.2).

These predictive indicators for malignancy—hypoechoic aspect, rounded shape, peritumoral localization, and size greater than 5 mm—are a guide only, with the percentage of overall correspondence between lymphatic tumor affectation detected using EUS and using a surgical sample (sN) being around 70%–80%. The reasons for this discrepancy are varied. On the one hand, adenopathic metastasis can occur in ganglia of less than 5 mm (underestimation) and, on the other hand, the coexistence of peritumoral inflammatory adenopathies greater than 10 mm is possible (overestimation). Consequently, use of another technique such as a CT scan or MRI in combination with EUS is recommended for lymphatic tumoral staging. The possibility of carrying out EUS using puncture aspiration of the lymph nodes (EUS-FNA) will be discussed later.

Significance of the circumferential resection margin

The circumferential resection margin (CRM) was initially described postoperatively as the surgical plain created by tumoral resection. Later it was realized that the CRM could be assessed preoperatively in such a way that it was considered the least distance between the tumor and the mesorectal fascia, which is a sheath of connective tissue that envelops the rectum and perirectal fat, including the lymphatic vessels and ganglia, acting as a natural barrier against the spread of the cancer. The CRM is considered cancer free if the tumor is localized at a distance of more than 1 mm from the mesorectal fascia. If the tumor has entered the mesorectal fascia or if the distance that exists between the adenocarcinoma and the CRM is less than 1 mm, this indicates infiltration of the resection margin and is equivalent to stage T4.

In patients with anterior rectal adenocarcinomas, EUS can clearly demarcate possible invasion of the mesorectal fascia—called rectoprostatic, rectovaginal, or Denonvilliers' fascia—and can predict the distance between the tumor and the CRM. Conversely, in patients with posterior or posterolateral tumors, EUS cannot estimate the distance between the tumor and the CRM, due to the absence of neighboring structures that facilitate it. Therefore, for these, NMR is the most

TABLE 9.2 Characteristics of lymph node malignancy.

Hypoechoic aspect
Rounded shape
Peritumoral localization
Size greater than 5 mm

efficacious technique for assessing the CRM. The distance between the tumor and the CRM has prognostic significance and can determine which patients are candidates for neoadjuvant chemo-radiotherapy.

A positive CRM result is defined as that in which the tumor invades or is less than 1 mm from the mesorectal fascia and is the most significant factor in postsurgery tumor recurrence. In such cases, when the tumor invades the mesorectal fascia or is found <1 mm from it, there is a high probability that the CRM will be seen to be affected in rectal surgery for complete excision of the mesorectal fascia.

EUS for assessment of neoadjuvant therapy response

Up to 90 percent of rectal tumors that have initially been considered unresectable have become resectable following neoadjuvant therapy. At the moment, none of the imaging techniques available—EUS, CT scan, MRI—can verify complete remission of the tumoral lesion after the therapy. The use of EUS can more accurately demarcate tumors not affecting the sphincter complex and therefore the surgeon can leave the complex intact. Although reduction of tumor size can be determined with these methods, the correspondence between stage sT (surgical sample) and the percentage for a histopathological regression outcome is low with a specificity of 37.5%. Some authors have attempted to create a regression grading system for tumors following neoadjuvant treatment but have been unable to produce an independent reproducible value. MRI including diffusion imaging[6] seems to have a better capability to assess tumoral response after neaoadjuvant therapy.

Interventional EUS

EUS can also be used for intervention, due to the fact that diagnostic and therapeutic maneuvering can be achieved with the technique in addition to echographic visualization. It is possible to take tissue samples and drain fluid collections using echoendoscopic guidance.

EUS-FNA initial diagnosis and recurrence

As previously referred to, the insertion of an elevator and the sectorial quality of the ultrasonographic image, together with access to highly facilitative work channels, allows the insertion of puncture needles, which are becoming more and more sophisticated, for taking histopathological samples.

Differential diagnosis and diagnosis of neoplastic affectation of adenopathies

EUS-FNA can also allow the differential diagnosis to be made with other benign pathologies that affect the pelvic area and simulate neoplasms, which would otherwise be impossible to diagnose at a time prior to surgery such as endometriosis,[7] that, on the other hand, don't need a surgical management.

Diagnostic accuracy for N staging can be improved with EUS or EUS-FNA, provided its usage is limited to adenopathies that are not adjacent to the tumor, in order to prevent contamination of the sample with neoplastic cells. The majority of authors do not consider that EUS-FNA gives more accurate staging than conventional EUS for perirectal lymphatic ganglia and it is only recommended for the early stages. They believe EUS-FNA does not add significant value to conventional EUS because rectal lymph nodes are too small to be seen with EUS unless they are metastatic. Hence, EUS-FNA would not be expected to improve the echographic criteria for malignancy in adenopathies. On the contrary, some literature suggests EUS-FNA provides more accurate results in determining malignancy than echography and therefore can be used on a wider spectrum of patients for different treatments with better final outcomes, primarily in relation to expenditure to reduce the incidence of tumor recurrence. Clinically, a practice of not routinely using EUS-FNA seems to be imposed and it is only used for a very small group of patients with T1 tumors. In the case of EUS or EUS-FNA used for these patients, treatment by means of local resection for pathological lymph nodes would be withheld.

Diagnosing tumor recurrence

A set of follow-up guidelines has been established for post rectal tumor that broadly includes trimestral determination of CEA levels and colonoscopies at one and three years.[8] Local extra-luminal recurrence may frequently develop, making diagnosis by colonoscopy difficult. On the other hand, a CT scan is not capable of distinguishing inflammatory changes or fibrosis in the tumoral recurrence itself or artifacts such as surgical clips may be seen as well as postradiotherapy alterations. Thus the recent incorporation of the new technique, PET-CT scanning, can add information to images derived from a

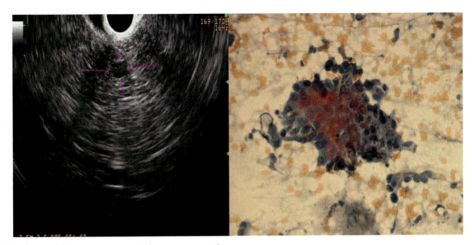
FIG. 9.2 EUS-FNAP adenopathic recurrence extrinsic to anastomosis.

combination of morphological and functional examinations; however, the technique still has the disadvantage that it is unable to give clear anatomopathological confirmation of tumor recurrence. As opposed to the limitations of the techniques mentioned, EUS-FNA can histopathologically confirm tumor recurrence. This recurrence may appear as extra-luminal masses (Fig. 9.2), or as regional or distant adenopathies with a sensitivity and specificity in relation to the adenopathies of 98% and 100%, respectively.[9] In this way, the use of EUS-FNA, along with the findings of EUS itself or the suspected recurrence appearing with other techniques (CT or PET-CT scans), can become the final step for confirming or ruling out tumoral recurrence, even being able to attribute the possible recurrence to a different type of neoplasm from a different origin the patient had previously presented with[10] (Table 9.2). Nevertheless, we need to take into account that, as with any operator-dependent technique, results improve correspondingly with the greater degree of experience of the cytopathologist, whose presence when puncture occurs is advisable, and the echoendoscopist.

Drainage of postsurgical fluid collections

As part of the continuing development of EUS, the opportunity has arisen for performing therapeutic procedures via the wider work channels it provides, with the possibility of insertion of a wide variety of instruments. Thus when guides, dilators, diathermic catheters, or prostheses are inserted, several therapeutic modes are presented. These include post rectal surgery drainage of perirectal fluid collections enabled by the insertion of trans-rectal prostheses. This provides outstanding long-term results. In this way, internal collection drainage of fluid collections into the gastrointestinal tract can be achieved, obviating the need for reintervention or implantation of larger percutaneous catheters with consequent avoidance of damage to adjacent organs.[11] The prosthesis insertion technique is a simple one, whether performed by means of previous use of a cystostoma or by single step stent insertion. Another therapeutic possibility, as we have recently been able to observe in our clinic, consists of drainage of perihepatic fluid collections resulting from resection surgery for single site hepatic metastasis stemming from a rectal adenocarcinoma. Using EUS, the perihepatic fluid collection localized close to the gastric wall can be drained via the upper gastrointestinal tract through the stomach, by insertion of a transgastric drainage-enabling stent, using a single step insertion technique.[12]

Extrahepatic bile duct drainage

Tumoral recurrence of CRC frequently occurs as adenopathies around the hepatic hilum or the bile duct.[13] The employment of EUS-FNA can give early definitive diagnostic confirmation of these as we have seen. Once the diagnosis is established, antitumor chemotherapy plans require the total bilirubin serum values to habitually be less than 5 mg/dL. After that, the usual procedure consists of performing bile duct drainage by means of stent insertion through the Vater papilla, employing the technique known as endoscopic retrograde cholangiopancreatography (ERCP). In a case when it is not technically possible to drain the bile duct with ERCP, it can be drained using EUS, either by directly inserting a trans-duodenal or transgastric stent or inserting a guide that allows a later meeting with ERCP, known as the rendezvous technique.[14]

Conclusion

Echoendoscopic ultrasound alone or in conjunction with FNA is a very useful and cost-effective tool for rectal adenocarcinoma staging.[15] Nonetheless, latest advances in nuclear magnetic resonance (MRI) have made it the technique of choice for tumor staging,[16] while echoendoscopy is limited to staging of early lesions (T1-T2) or for use when confirmation is needed with puncture aspiration (EUS-FNA). Another increasing usage of EUS-FNAP is that of minimally invasive drainage of postsurgical fluid collections. In clinical follow-up of postintervention CRC patients it is the only technique that provides diagnostic confirmation of extra-luminal tumor recurrence, in conjunction with the techniques, CT and PET-CT scanning. The new EUS equipment and instruments can provide clearance of bile duct obstruction in the case of recurring adenopathy, sometimes in conjunction with ERCP. More recently, the introduction of three-dimensional EUS,[17] and, moreover, the appearance of forward-viewing echoendoscope[18] have achieved a better view and an easier accessibility to colorectal tumors.

References

1. Hildebrandt U, Feifel G. Preoperative staging of rectal cancer by intrarectal ultrasound. *Dis Colon Rectum* 1985;**28**(1):42–6. https://doi.org/10.1007/bf02553906.
2. Garcia-Aguilar J, Pollack J, Lee S-H, et al. Accuracy of endorectal ultrasonography in preoperative staging of rectal tumors. *Dis Colon Rectum* 2002;**45**(1):10–5. https://doi.org/10.1007/s10350-004-6106-3.
3. Sasaki Y, Niwa Y, Hirooka Y, et al. The use of endoscopic ultrasound-guided fine-needle aspiration for investigation of submucosal and extrinsic masses of the colon and rectum. *Endoscopy* 2005;**37**(2):154–60. https://doi.org/10.1055/s-2004-826152.
4. Iwashita T, Nakai Y, Lee JG, Park DH, Muthusamy VR, Chang KJ. Newly-developed, forward-viewing echoendoscope: a comparative pilot study to the standard echoendoscope in the imaging of abdominal organs and feasibility of endoscopic ultrasound-guided interventions. *J Gastroenterol Hepatol* 2012;**27**(2):362–7. https://doi.org/10.1111/j.1440-1746.2011.06923.x.
5. Jin M, Frankel WL. Lymph node metastasis in colorectal cancer. *Surg Oncol Clin N Am* 2018;**27**(2):401–12. https://doi.org/10.1016/j.soc.2017.11.011.
6. Juchems MS, Wessling J. Rational staging and follow-up of colorectal cancer: do guidelines provide further help? *Radiologe* 2019;**59**(9):820–7. https://doi.org/10.1007/s00117-019-0578-6.
7. Kishimoto K, Kawashima K, Moriyama I, et al. Sigmoid endometriosis diagnosed preoperatively using endoscopic ultrasound-guided fine-needle aspiration. *Clin J Gastroenterol* 2020;**13**(2):158–63. https://doi.org/10.1007/s12328-019-01046-x.
8. Schmoll HJ, Van Cutsem E, Stein A, et al. ESMO consensus guidelines for management of patients with colon and rectal cancer. A personalized approach to clinical decision making. *Ann Oncol* 2012;**23**(10):2479–516. https://doi.org/10.1093/annonc/mds236.
9. Chen VK, Eloubeidi MA. Endoscopic ultrasound-guided fine needle aspiration is superior to lymph node echofeatures: a prospective evaluation of mediastinal and peri-intestinal lymphadenopathy. *Am J Gastroenterol* 2004;**99**(4):628–33. https://doi.org/10.1111/j.1572-0241.2004.04064.x.
10. Hünerbein M, Totkas S, Moesta KT, Ulmer C, Handke T, Schlag PM. The role of transrectal ultrasound-guided biopsy in the postoperative follow-up of patients with rectal cancer. *Surgery* 2001;**129**(2):164–9. https://doi.org/10.1067/msy.2001.110428.
11. Varadarajulu S, Drelichman ER. Effectiveness of EUS in drainage of pelvic abscesses in 25 consecutive patients (with video). *Gastrointest Endosc* 2009;**70**(6):1121–7. https://doi.org/10.1016/j.gie.2009.08.034.
12. Ulla-Rocha JL, Vilar-Cao Z, Sardina-Ferreiro R. EUS-guided drainage and stent placement for postoperative intra-abdominal and pelvic fluid collections in oncological surgery. *Therap Adv Gastroenterol* 2012;**5**(2):95–102. https://doi.org/10.1177/1756283X11427420.
13. Warshaw AL, Welch JP. Extrahepatic biliary obstruction by metastatic colon carcinoma. *Ann Surg* 1978;**188**(5):593–7. https://doi.org/10.1097/00000658-197811000-00002.
14. Sarkaria S, Lee H-S, Gaidhane M, Kahaleh M. Advances in endoscopic ultrasound-guided biliary drainage: a comprehensive review. *Gut Liver* 2013;**7**(2):129–36. https://doi.org/10.5009/gnl.2013.7.2.129.
15. Harewood GC, Wiersema MJ. Cost-effectiveness of endoscopic ultrasonography in the evaluation of proximal rectal cancer. *Am J Gastroenterol* 2002;**97**(4):874–82. https://doi.org/10.1111/j.1572-0241.2002.05603.x.
16. Tombazzi CR, Loy P, Bondar V, Ruiz JI, Waters B, Tombazzi CR. Accuracy of endoscopic ultrasound in staging of early rectal cancer. *Fed Pract* 2019;**36**(Suppl. 5):S26–9. https://www.ncbi.nlm.nih.gov/pubmed/31507310.
17. Kim JC, Kim HC, Yu CS, et al. Efficacy of 3-dimensional endorectal ultrasonography compared with conventional ultrasonography and computed tomography in preoperative rectal cancer staging. *Am J Surg* 2006;**192**(1):89–97. https://doi.org/10.1016/j.amjsurg.2006.01.054.
18. Larghi A, Ibrahim M, Fuccio L, et al. Forward-viewing echoendoscope versus standard echoendoscope for endoscopic ultrasound-guided tissue acquisition of solid lesions: a randomized, multicenter study. *Endoscopy* 2019;**51**(5):444–51. https://doi.org/10.1055/a-0790-8342.

Chapter 10

CT-Colonography (CTC): Technical requirements, indications and current status

Concepción Crespo García, María Jose Martinez-Sapiña Llanas, and Susana A. Otero Muinelo
Department of Radiology, A Coruña University Hospital, A Coruña, Spain

Introduction

Optical colonoscopy (OC) is the technique of choice for the diagnosis of colorectal cancer and its precursor adenoma, but it is invasive and not without risk (perforation, bleeding, and anesthetic complications). Furthermore, 10%–15% of examinations are incomplete.[1,2] Since the first CT colonography (CTC) images were published in 1994,[3] and thanks to significant improvements in image quality, numerous successive studies have demonstrated this technique's ability to detect colonic lesions, mainly polyps and masses. In 2006 the American Gastrointestinal Association's Clinical Practice and Economics Committee accepted CTC as the method of choice in the case of incomplete OC, replacing barium enema.[4] In 2008 the American Cancer Society accepted CTC as a screening technique for colorectal cancer.[5] It is a technique that, when performed according to high-quality standards, has proven to be reliable and robust in the study of the colon, with a sensitivity similar to OC in the detection of colorectal cancer and polyps larger than 1 cm.[4] Another strength of CTC is the ability to detect extracolonic abnormalities as noncontrast images of the entire abdomen and pelvis are acquired.[6,7]

CTC has few contraindications, but it is necessary to be aware of them. They consist of all those situations in which there is a risk of perforation when introducing the gas necessary to distend the colon (Table 10.1).

Technical requirements

To perform a CTC, a multidetector computed tomography (CT) scanner with at least 8 rows of detectors, software with specific CTC tools, and an experienced radiologist are required for correct interpretation. There are also two technical aspects that are fundamental for an optimal study of the colon: correct preparation and adequate distention.

Preparation of the colon

Optimal preparation of the colon is decisive for adequate CTC interpretation since the presence of residue can simulate or hide lesions. It is currently widely accepted and recommended by the IVIRCO group (Virtual Imaging of the Colon) to prepare the colon without cathartic agents, with a residue-free diet and oral administration of iodinated contrast (Diatrizoate) which, in addition to tagging the contents of the intestine, helps to homogenize it and has a cathartic effect.[8–11] The use of cathartics is reserved for patients with habitual constipation, avoiding polyethylene glycol due to the abundant residual fluid it produces in the colon.[8,12]

Distension of the colon

Adequate colonic distension is essential for the correct evaluation of the colon. The technique recommended by the ESGAR (European Society of Gastrointestinal and Abdominal Radiology) is automatic low-pressure CO_2 insufflation through a flexible rectal cannula, which reduces spasms and improves comfort.[11] During insufflation, the patient's postural changes prevent inadequate distention of any segment (Fig. 10.1) due to the presence of stool, loops, or angulations of the colon.[10,13]

TABLE 10.1
CTC contraindications

- Acute symptomatic colitis
- Acute-phase diverticulitis
- Recent colorectal surgery (<3 months)
- Hernia of the abdominal wall involving any segment of the colon
- Recent biopsy or recent deep endoscopic polypectomy
- Inflammatory bowel disease in active phase
- Intestinal obstruction with suspected perforation

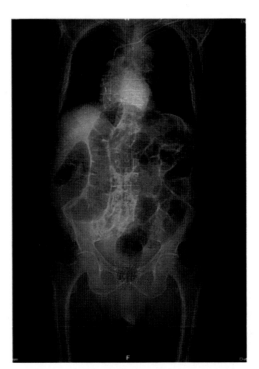

FIG. 10.1 Topogram with the patient in the supine position showing optimal distension of the colon prior to the acquisition of the helix.

Administration of antispasmodics 30 min before the examination is recommended[8, 9, 11] to facilitate colonic distention, provided there are no contraindications. This procedure is very well tolerated by the patient and does not require sedation.

Colon perforation secondary to barotrauma is virtually nonexistent (0.005%) with the use of automatic CO_2 injectors. Most of the cases reported were symptomatic patients with acute inflammatory processes or stenosis undergoing manual insufflation of ambient air through rigid cannulas and these cases were usually resolved with conservative management.[9]

Image acquisition

Two helices are always acquired, one in the supine position and another in the prone position. In this way, lesions hidden by stools will be visible in one of the two series[14]. A third helix in lateral decubitus may be necessary if a segment of the colon has not been distended in either of the previous positions.[10, 13] If colon cancer is diagnosed, one of the helices is performed by administrating intravenous iodinated contrast to complete the extension study. Since one of the major limitations of CTC is exposure to ionizing radiation, low-dose acquisition protocols should be sought using automatic dose modulation systems that adapt the current intensity to the patient's anatomy and iterative reconstruction systems, if available, which allow working with doses of 1–1.5 mSv (an effective dose of less than 5–7 mSv is recommended). A low dose is considered when the effective dose is lower than 5–7 mSv.[15, 16]

Recommended acquisition parameters:
- Minimum slice thickness of 1.25 mm
- Low-dose technique: maximum 140 mA (recommended 50 mA) and 100 kV with dose modulation and iterative reconstruction
- Standard reconstruction algorithm
- Rotation time 0.5 s

Software and tools

Acquired data are sent to a workstation or platform with specific CTC software for review and postprocessing.[17] The images can be displayed in 2D (axial and multiplanar) and 3D. It is recommended to start the interpretation using 3D navigation through the colonic lumen in an anterograde and retrograde direction to avoid blind spots hidden by prominent haustra. The 2D images are used to complete the information and solve diagnostic problems. For correct interpretation, both forms of visualization have to be combined as they are complementary.[17]

Postprocessing tools (Fig. 10.2) that facilitate interpretation and increase diagnostic efficiency are as follows:

- Virtual dissection: 360 degree view presenting the colon open and unfolded, eliminating blind spots and allowing a unidirectional 3D assessment.
- Virtual biopsy or translucency: assigns colors to tissues according to Hounsfield units, which help to differentiate polyps from stool or fatty lesions.
- Computer-aided diagnosis (CAD) software: facilitates detection of polyps, although with a high false-positive rate and limitations for detection of flat lesions.
- Electronic stool and fluid removal: allows removal of stool and fluid stained with oral contrast to facilitate visualization of the mucosa.

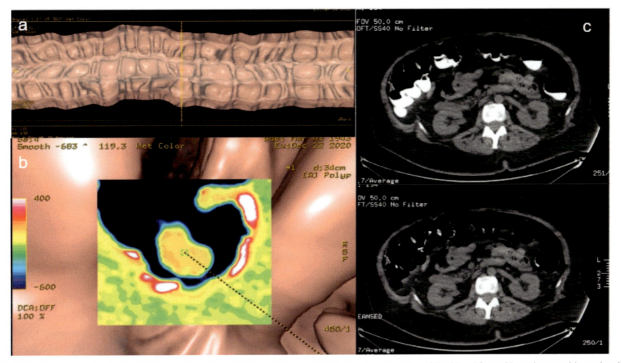

FIG. 10.2 Postprocessing tools. (A) Virtual dissection. (B) Virtual biopsy. (C) Electronic subtraction of liquid and feces. Upper image without cleaning and lower image with subtraction.

Indications for virtual colonoscopy

This imaging technique has universally accepted and clearly established indications.[4, 17] In recent years, new indications have been developed and gained positions, mainly in the preoperative study of patients with colorectal carcinoma, with promising results for differentiating low- and high-risk tumors.

The *indications* are as follows:

1. **Incomplete OC, the most important and frequent indication, which may be due to:**
   - intolerance
   - redundant colon or exhaustion of the endoscope
   - angulations
   - spasm
   - stenosis
2. **Contraindications or refusal of OC:**
   - frail patients with severe cardiac or pulmonary disease
   - risk of bleeding
   - risk for sedation
   - history of incomplete OC
   - patient's refusal of OC
3. **Colorectal cancer screening**: Allows a complete examination of the colon in a safe way, with good tolerance and patient adherence, and also detects extracolonic findings
4. **Study of chronic uncomplicated diverticular disease**
5. **Characterization of submucosal lesions detected in OC**
6. **Follow-up after surgery for colorectal cancer**
7. **Follow-up of lesions previously detected in CTC** as intermediate-sized polyps not removed, if less than 3
8. **Precise localization of the tumor and detection and localization of synchronous lesions** before laparoscopic surgery for colorectal cancer
9. **To evaluate the extent of colon involvement in inflammatory bowel disease** in the chronic fibro-stenotic phase

Clinically significant findings in CTC

Most colorectal cancers develop from benign adenomatous polyps.[10, 14, 18] They are slow-growing lesions and the target lesion for CRC screening. The aim of CTC screening is to detect these premalignant lesions noninvasively,[10, 18] not looking for just any polyp, but specifically advanced adenoma, which is a polyp that is at risk of progressing to cancer and is equal to or larger than 1 cm in size with a significant villous component or high-grade dysplasia. The risk of malignancy of a polyp increases with size, which determines its clinical management.[18]

Polyps

Polyp is a nonspecific term that defines any lesion that protrudes into the mucosa. In CTC, it is a homogeneous lesion with soft-tissue attenuation and well-defined contours. It may be surrounded by contrast or marked fluid, has no gas within it and is not mobile, except for pedunculated polyps or polyps that settle in a mobile segment of the colon.

Polyps are classified according to their morphology and size.

Types of polyps according to their morphology:

- **Sessile**: they present a wide implantation base in the colon wall, so they do not change position when the patient does (Fig. 10.3). They have a "mushroom cap" appearance and are the most common.
- **Pediculate**: they have a well-defined head and pedicle (Fig. 10.4).
- **Flat**: a subgroup of sessile polyps no more than 3 mm in height or half their length (Fig. 10.5). They are more difficult to detect in CTC and require excellent technique in terms of preparation, distension, and interpretation. A thin layer of contrast material covering them is helpful in their detection. They are much less frequent in the western population and the likelihood of malignancy is also lower.[19] The term carpet lesion refers to a lesion that is extensive in surface area (usually larger than 3 cm) but only slightly elevated. They are characteristically located in the cecum and rectum, two segments that are usually well distended in CTC and they are therefore often detected.[8]

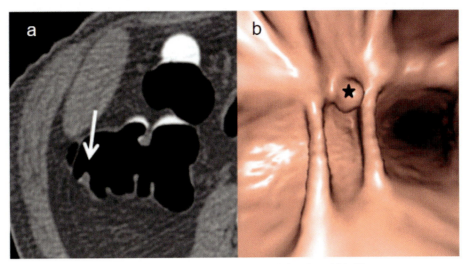

FIG. 10.3 Small sessile polyp in 2D axial image *(arrow* in A) and 3D endoluminal view *(asterisk* in B).

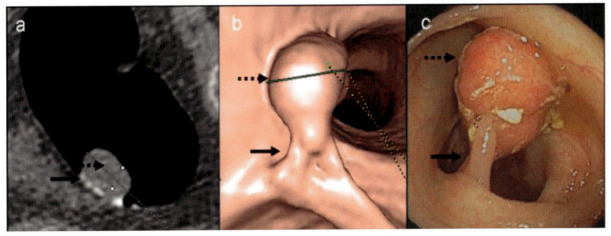

FIG. 10.4 1-cm pedunculated polyp. (A) 2D image showing the head of the polyp *(dashed black arrow)* and how the ingested contrast is introduced between the colon wall and the polyp *(continuous black arrow)*. (B) 3D endoluminal image with the head of the polyp *(dashed arrow)* and the pedicle *(continuous arrow)*. (C) Correlation with the optical colonoscopy image: head *(dashed arrow)* and pedicle *(continuous arrow)*.

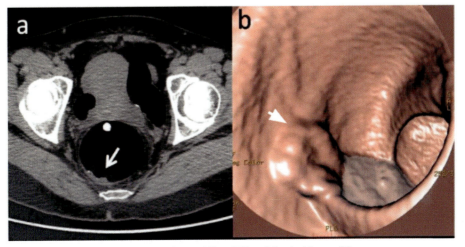

FIG. 10.5 Flat polyp in rectum *(white arrows)* in 2D axial image (A) and 3D endoluminal image (B).

Types of polyps according to their size:

- **Small**: equal to or less than 5 mm
- **Medium**: 6–9 mm
- **Large**: equal to or larger than 10 mm

The sensitivity of CTC for the detection of polyps is directly proportional to their size, being close to 100% for large polyps and high for medium polyps.[20] Sensitivity decreases for lesions smaller than 6 mm. However, polyps of this size have a minimal risk of progressing to cancer and there is controversy among radiologists and gastroenterologists as to whether these lesions should be cited.[21] There is currently a consensus within ESGAR (European Society of Gastrointestinal and Abdominal Radiology): all lesions less than 5 mm with a high probability of being polyps should be recorded, especially if there are more than three.[11] It is not possible to differentiate inflammatory polyps from hyperplastic polyps in CTC, so a histological examination of all detected polyps 1 cm or more in size is necessary.

Cancers

Colon tumors present in CTC in three main patterns[22, 23]:

(1) **Broad-based endoluminal mass** (Fig. 10.6), the most frequent presentation.
(2) **Polypoid or fungoid lesion** (Fig. 10.7).
(3) **Annular lesion** presenting with the classic "apple core" or "napkin ring" sign (Fig. 10.8) and usually causing stenosis.

Small tumors may present as a focal polypoid thickening of a haustra, like a sessile or even flat lesion. This is the usual appearance of tumors detected in asymptomatic patients and synchronous tumors that occur in 1.5%–9% of patients diagnosed with colorectal cancer.[24] For this reason, CTC plays a fundamental role in these patients, since up to 26% of OC is incomplete and 58% of these cases are due to stenosing tumors that do not allow passage of the colonoscope.[8, 25] The usefulness of CTC in the surgical planning of patients with colorectal cancer is widely accepted. In addition to the detection of tumors and synchronous adenomatous polyps (present in 27%–55% of these patients), CTC is more accurate than OC in the segmental localization of tumors and polyps, which is essential for optimal surgical planning in laparoscopic colectomies.[26]

In recent years, several publications have also evaluated the efficacy of CTC in the T and N staging of colon tumors, excluding those located in the rectum for which MRI is more sensitive. Although CTC does not allow differentiation of the layers of the bowel wall, it does allow the assessment of the external and internal surface of the bowel wall. A **simplified T-staging system** has been proposed for CTC reports with three categories.[26–28]

- **T1/T2**: confined to the bowel wall.
- **T3**: invading subserosal fat.
- **T4**: adjacent organ involvement.

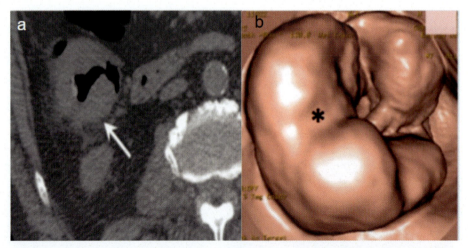

FIG. 10.6 Right colon carcinoma presenting as an endoluminal mass in 2D view (*arrow* in A) and 3D view (*asterisk* in B).

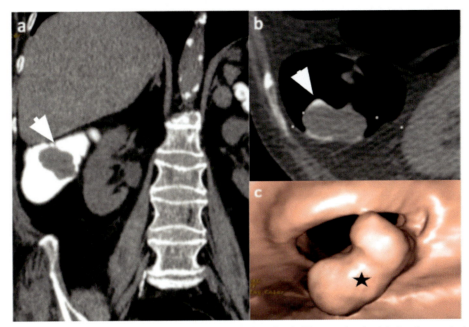

FIG. 10.7 Carcinoma of the hepatic angle of the colon presenting as a polypoid lesion in 2D coronal and axial view (*arrows* in A and B) and 3D endoluminal view (*asterisk* in C).

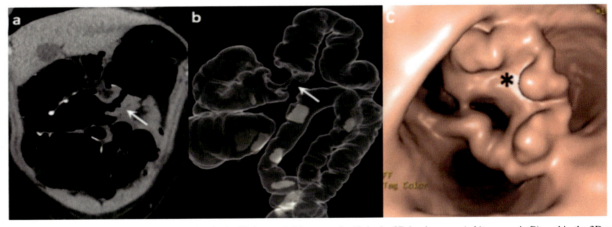

FIG. 10.8 Transverse colon cancer: annular lesion in the 2D image (*white arrow* in A), in the 3D luminogram (*white arrow* in B), and in the 3D endoluminal view (*asterisk* in C).

A morphological classification that evaluates the deformity of the bowel wall and the external contours of the tumor assessed in 3D surface images allows T1/T2 tumors to be distinguished from T3 tumors. For this purpose, tumors are classified into three basic morphological types: arciform, trapezoid, and "apple core" (Fig. 10.9). The first two are associated with T1/T2 tumors and the third with T3/T4.[29] Following this classification, the accuracy of CTC is 76%–79% in most series.[26] Sensitivity and specificity are increased by classifying them into two broad T1/T2 and T3/T4 categories: low- and high-risk tumors.[27, 30]

A nodal staging system with CTC has also been proposed[31]:

- **N0**: no nodes or less than 3, all smaller than 10 mm.
- **N1**: 3 nodes regardless of size or less than 3, but 1 larger than 10 mm.
- **N2**: more than 3 nodes of any size.

The reliability of CTC in determining the N stage is 70%–85%.[26, 30, 31]

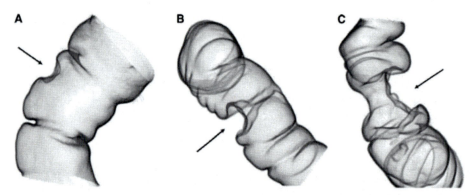

FIG. 10.9 Morphological classification of bowel wall deformity in 3D surface images. (A) Aciform. (B) Trapezoid. (C) "Appel core" type.

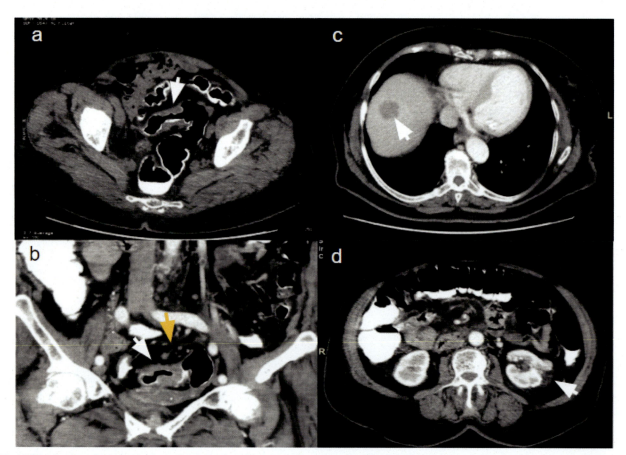

FIG. 10.10 Staging of colorectal cancer. Stenosing sigmoid tumor (*white arrows* in A and B) whose contours and morphology are compatible with a locally advanced tumor, T3/T4. Accumulation of regional lymph nodes (*yellow arrow* in B). Liver metastasis (*white arrow* in C) in the supine CTC study with intravenous contrast. In addition, as an incidental finding, an angiomyolipoma in the left kidney (*white arrow* in D).

In conclusion, CTC is a reliable diagnostic tool in the extension study of patients with colorectal carcinoma (Fig. 10.10), together with contrast enhanced CT necessary for the evaluation of distant disease (M).

Submucosal lesions

Another of the utilities of CTC is the characterization of lesions detected in OC, such as submucosal lesions.[32] A submucosal lesion is any lesion that simulates a tumor, projecting into the intestinal lumen covered by normal mucosa.

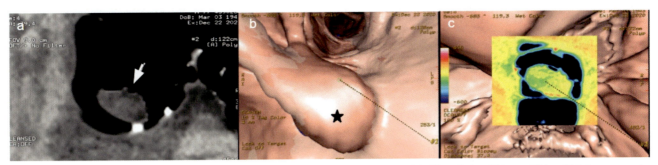

FIG. 10.11 Submucosal lesion: lipoma in 2D image (*arrow* in A), 3D endoluminal (*asterisk* in B), and virtual biopsy (C) in which the *green color* assigned to the fat in our software predominates.

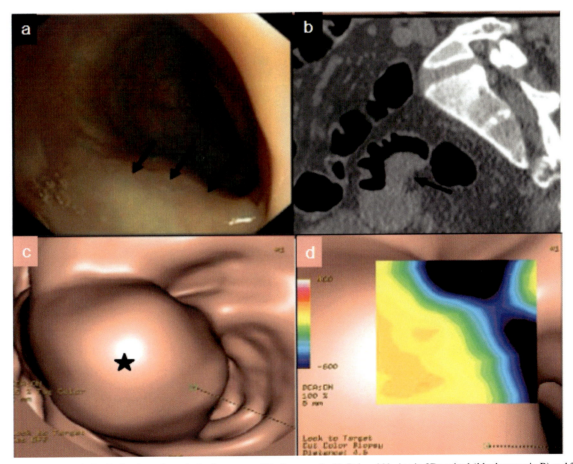

FIG. 10.12 Submucosal lesion in the sigma in optical colonoscopy (*black arrows* in A). Polypoid lesion in 2D sagittal (*black arrow* in B) and 3D endoluminal (*asterisk* in C). In the virtual biopsy (D) the *yellow color* predominates, assigned to soft tissues in our software. Histological diagnosis: endometriosis.

Characteristically, they are lesions with a broad base, obtuse angles to the bowel wall, and a smooth surface. They can be **intramural**, such as lipomas (Fig. 10.11), vascular lesions, carcinoid tumors, cystic lesions, pneumatosis, etc., or **extramural**, such as endometriosis (Fig. 10.12), developmental cystic lesions, extrinsic tumors, etc.[8, 33]

Chronic diverticular disease

Diverticulosis is not an indication for CTC, but it is the most frequent cause of incomplete OC. As a consequence of recurrent diverticular processes that cause mural fibrosis, stenosis of the lumen and resistance to distension of the colon

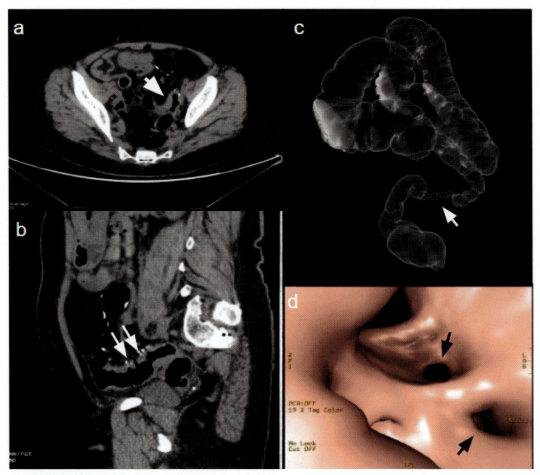

FIG. 10.13 Diverticulosis: 2D axial and sagittal images showing a smooth and uniform thickening of a long segment of the sigma (*white arrows* in A and B). The 3D surface image (C) shows the luminal stricture *(white arrows)* and 3D endoluminal image (d) showing the diverticula *(black arrows)*.

occurs, which impedes the passage of the colonoscope. The greatest difficulty in the diagnosis of chronic diverticulitis is to exclude the possibility of colorectal cancer, as both entities are frequent in elderly patients and can coexist in the same patient. CTC makes it possible to complete the study of the entire colon and study the stricture to differentiate inflammatory strictures from neoplastic strictures.[34] In general, neoplastic strictures are those that are short (less than 50 mm), with abrupt margins, mural thickening greater than 15 mm, and marked distortion of the mucosal pattern. Strictures secondary to chronic diverticulitis (Fig. 10.13) are long, the mural thickening is less than 15 mm, the transition with the rest of the colon is conical in morphology, and the folds are partially distorted.[35, 36]

Other

CT colonography is also useful in chronic stages of inflammatory bowel diseases, in benign strictures of inflammatory (Fig. 10.14), ischemic or actinic origin, or in anatomical variants, diaphragmatic hernias and postsurgical status, which are a frequent cause of incomplete optical colonoscopy.[4, 37]

Finally, it should be noted that a systematic protocol for reports: C-RADS (Colonography Reporting and Data System) is available for radiologists. It standardizes and classifies the findings and proposes their management to compare results between programs and guarantee the quality of the studies[38] (Table 10.2).

Conclusions

CT colonography is a diagnostic technique with high sensitivity in the detection of clinically significant lesions, provided it is performed according to high-quality standards. The British Society of Gastrointestinal and Abdominal Radiology and the

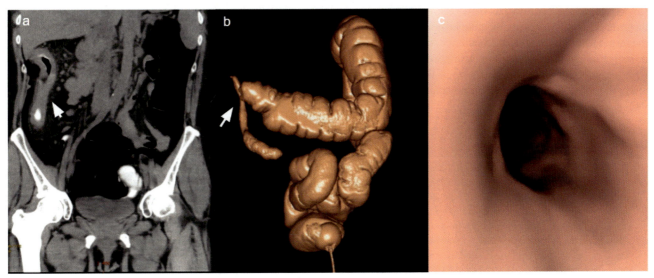

FIG. 10.14 Other causes of stenosis: Intestinal tuberculosis. Severe stenosis of almost the entire right colon due to a smooth and uniform mural thickening in 2D coronal view (*arrow* in A). The 3D opaque surface image shows a right "microcolon" (*white arrow* in B) and a normal mucosal pattern in the endoluminal view (C).

TABLE 10.2 C-RADS classification of findings and management recommendations.[38]

C0	Inadequate study/comparison with previous CTC needed
	Inadequate preparation: lesions <1 cm cannot be excluded
	Inadequate insufflation: one or more collapsed segments in both series performed
	Previous studies to compare are needed
C1	Normal colon or benign lesions
	Recommendation: resume screening routine (every 5–10 years)
	No abnormalities are observed in the colon
	There are no polyps larger than 6 mm
	Lipoma or inverted diverticula
	Nonneoplastic lesions (such as diverticula)
C2	Medium-sized polyps or indeterminate findings
	Recommendation: follow up or optical colonoscopy
	Less than 3 medium-sized polyps (6–9 mm)
	Indeterminate findings, polyps ≥6 mm cannot be excluded
C3	Polyp, probable advanced adenoma
	Recommendation: optical colonoscopy
	Polyps ≥10 mm
	More than 3 polyps, each 6–9 mm in size
C4	Mass in colon, suspicious of malignancy
	Recommendation: surgery consultation
	Lesion that involves the colonic lumen or shows extracolonic extension

Royal College of Radiologists have recently published a consensus guideline document that sets out the standards for performing CTC.[39] The main objective is the detection of precursor lesions of colorectal cancer, with advanced adenoma being the target lesion. It is also an excellent diagnostic tool that complements optical colonoscopy when it is incomplete or contraindicated. Most patients do not require cathartics; it is performed without sedation, without intravenous contrast, and with a low radiation dose. The future of this noninvasive, accessible, well-tolerated, and virtually complication-free technique is promising, especially in the staging and subsequent therapeutic approach to colorectal cancer.

References

1. Shah HA, Paszat LF, Saskin R, Stukel TA, Rabeneck L. Factors associated with incomplete colonoscopy: a population-based study. *Gastroenterology* 2007;**132**(7):2297–303. https://doi.org/10.1053/j.gastro.2007.03.032.
2. Aslinia F, Uradomo L, Steele A, Greenwald BD, Raufman J-P. Quality assessment of colonoscopic cecal intubation: an analysis of 6 years of continuous practice at a university hospital. *Am J Gastroenterol* 2006;**101**(4):721–31. https://doi.org/10.1111/j.1572-0241.2006.00494.x.
3. Vining D. Technical feasibility of colon imaging with helical CT and virtual reality. In: *Proc ann meeting of Amer Roentgen Ray Soc.* vol. 104; 1994. https://ci.nii.ac.jp/naid/10009003962/. [Accessed 26 February 2021].
4. Laghi A, Rengo M, Graser A, Iafrate F. Current status on performance of CT colonography and clinical indications. *Eur J Radiol* 2013;**82**(8):1192–200. https://doi.org/10.1016/j.ejrad.2012.05.026.
5. Levin B, Lieberman DA, McFarland B, et al. Screening and surveillance for the early detection of colorectal Cancer and adenomatous polyps, 2008: a joint guideline from the American Cancer Society, the US multi-society task force on colorectal Cancer, and the American College of Radiology. *Gastroenterology* 2008;**134**(5):1570–95. https://doi.org/10.1053/j.gastro.2008.02.002.
6. Martínez-Sapiña Llanas MJ, Rodriguez Antuña J, Rois Siso A, Romeu Villar D, Fraga Manteiga D, López RD. *Incidentally clinically important extracolonic findings at CT-colonography: prevalence in 600 consecutive examinations.* Published online; 2013. https://doi.org/10.1594/ECR2013/C-2491.
7. Veerappan GR, Ally MR, Choi J-HR, Pak JS, Maydonovitch C, Wong RKH. Extracolonic findings on CT colonography increases yield of colorectal Cancer Screening. *Am J Roentgenol* 2010;**195**(3):677–86. https://doi.org/10.2214/ajr.09.3779.
8. Martínez-Sapiña Llanas MJ, Otero Muinelo SA, Crespo GC. Patología del recto: hallazgos en la colonografía-TC. *Radiologia* 2018;**60**(3):208–16. https://doi.org/10.1016/j.rx.2017.10.005.
9. Mang T, Graser A, Schima W, Maier A. CT colonography: techniques, indications, findings. *Eur J Radiol* 2007;**61**(3):388–99. https://doi.org/10.1016/j.ejrad.2006.11.019.
10. Pickhardt PJ, Screening CT. Colonography: how I do it. *Am J Roentgenol* 2007;**189**(2):290–8. https://doi.org/10.2214/ajr.07.2136.
11. Neri E, ESGAR CT. Colonography working group, Halligan S, et al. the second ESGAR consensus statement on CT colonography. *Eur Radiol* 2013;**23**(3):720–9. https://doi.org/10.1007/s00330-012-2632-x.
12. Zueco C, Zueco CZ, Sampedro CS, Corroto JD, Fernández PR, Fontanillo MF. CT colonography without cathartic preparation: positive predictive value and patient experience in clinical practice. *Eur Radiol* 2012;**22**(6):1195–204. https://doi.org/10.1007/s00330-011-2367-0.
13. Buchach CM, Kim DH, Pickhardt PJ. Performing an additional decubitus series at CT colonography. *Abdom Imaging* 2011;**36**(5):538–44. https://doi.org/10.1007/s00261-010-9666-9.
14. Pickhardt PJ, Kim DH. Colorectal cancer screening with CT colonography: key concepts regarding polyp prevalence, size, histology, morphology, and natural history. *Am J Roentgenol* 2009;**193**(1):40–6. https://doi.org/10.2214/ajr.08.1709.
15. Liedenbaum MH, Venema HW, Stoker J. Radiation dose in CT colonography–trends in time and differences between daily practice and screening protocols. *Eur Radiol* 2008;**18**(10):2222–30. https://doi.org/10.1007/s00330-008-0994-x.
16. Geyer LL, Schoepf UJ, Meinel FG, et al. State of the art: iterative CT reconstruction techniques. *Radiology* 2015;**276**(2):339–57. https://doi.org/10.1148/radiol.2015132766.
17. Pagés Llinás M, Darnell Martín A, Ayuso Colella JR. Colonografía por TC. Lo que el radiólogo debe conocer. *Radiologia* 2011;**53**(4):315–25. https://doi.org/10.1016/j.rx.2011.01.009.
18. Winawer SJ, Zauber AG. The advanced adenoma as the primary target of Screening. *Gastrointest Endosc Clin N Am* 2002;**12**(1):1–9. https://doi.org/10.1016/s1052-5157(03)00053-9.
19. Pickhardt PJ. Missed lesions at CT colonography: lessons learned. *Abdom Imaging* 2013;**38**(1):82–97. https://doi.org/10.1007/s00261-012-9897-z.
20. Chaparro M, Gisbert JP, del Campo L, Cantero J, Maté J. Accuracy of computed tomographic colonography for the detection of polyps and colorectal tumors: a systematic review and meta-analysis. *Digestion* 2009;**80**(1):1–17. https://doi.org/10.1159/000215387.
21. Pickhardt PJ, Hassan C, Laghi A, et al. Small and diminutive polyps detected at screening CT colonography: a decision analysis for referral to colonoscopy. *Am J Roentgenol* 2008;**190**(1):136–44. https://doi.org/10.2214/ajr.07.2646.
22. Martínez-Sapiña Llanas MJ, Rois Siso A, Rodriguez AJ. *La colonografía por TC (CTC) en el diagnóstico de las lesiones estenosantes del colon.* Published online; 2014. https://doi.org/10.1594/SERAM2014/S-1274.
23. Kim DH, Pickhardt PJ, Taylor AJ, et al. CT Colonography versus colonoscopy for the detection of advanced neoplasia. *N Engl J Med* 2007;**357**(14):1403–12. https://doi.org/10.1056/nejmoa070543.
24. Leksowski K, Rudzinska M, Rudzinski J. Computed tomographic colonography in preoperative evaluation of colorectal tumors: a prospective study. *Surg Endosc* 2011;**25**(7):2344–9. https://doi.org/10.1007/s00464-010-1566-0.

25. Yang J, Peng J-Y, Chen W. Synchronous colorectal cancers: a review of clinical features, diagnosis, treatment, and prognosis. *Dig Surg* 2011;28(5–6):379–85. https://doi.org/10.1159/000334073.
26. Sali L, Falchini M, Taddei A, Mascalchi M. Role of preoperative CT colonography in patients with colorectal cancer. *World J Gastroenterol* 2014;20(14):3795–803. https://doi.org/10.3748/wjg.v20.i14.3795.
27. Nerad E, Lahaye MJ, Maas M, et al. Diagnostic accuracy of CT for local staging of Colon Cancer: a systematic review and meta-analysis. *Am J Roentgenol* 2016;207(5):984–95. https://doi.org/10.2214/ajr.15.15785.
28. Horvat N, Raj A, Liu S, et al. CT colonography in preoperative staging of colon cancer: evaluation of FOxTROT inclusion criteria for neoadjuvant therapy. *Am J Roentgenol* 2019;212(1):94–102. https://doi.org/10.2214/ajr.18.19928.
29. Utano K, Endo K, Togashi K, et al. Preoperative T staging of colorectal cancer by CT colonography. *Dis Colon Rectum* 2008;51(6):875–81. https://doi.org/10.1007/s10350-008-9261-0.
30. Llanas MM-S, Chaves AV, García CC, Lopez LA, Sanchez MC, Martinez CF, et al. *CT colonography in preoperative T and N staging of colorectal cancer.* ESGAR; 2019. https://doi.org/10.5444/esgar2019/SE-28. Published online.
31. Filippone A, Ambrosini R, Fuschi M, Marinelli T, Genovesi D, Bonomo L. Preoperative T and N staging of colorectal cancer: accuracy of contrast-enhanced multi–detector row CT colonography—initial experience. *Radiology* 2004;231(1):83–90. https://doi.org/10.1148/radiol.2311021152.
32. Martinez-Sapiña Llanas MJ, Álvarez Devesa L, Pazos Silva V, Rois Siso A, Díaz Angulo C, Crespo GC. *CT colonography (CTC) in the diagnosis of polypoid lesions of the colon: impact of findings on the therapeutic management.* Published online; 2013. https://doi.org/10.1594/ECR2013/C-0244.
33. Pickhardt PJ, Kim DH, Menias CO, Gopal DV, Arluk GM, Heise CP. Evaluation of submucosal lesions of the large intestine. *Radiographics* 2007;27(6):1681–92. https://doi.org/10.1148/rg.276075027.
34. Flor N, Rigamonti P, Ceretti AP, et al. Diverticular disease severity score based on CT colonography. *Eur Radiol* 2013;23(10):2723–9. https://doi.org/10.1007/s00330-013-2882-2.
35. Gryspeerdt S, Lefere P. Chronic diverticulitis vs. colorectal cancer: findings on CT colonography. *Abdom Imaging* 2012;37(6):1101–9. https://doi.org/10.1007/s00261-012-9858-6.
36. Martínez-Sapiña Llanas MJ. *Chronic diverticulitis vs. colorectal cancer: role of the CT-colonography after incomplete optical colonoscopy.* Published online; 2016. https://doi.org/10.1594/ECR2016/C-1426.
37. Martinez-Sapiña Llanas MJ, Rois Siso A, Rodriguez Antuña J, Romeu Villar D, Fraga Manteiga D, López RD. *CT colonography in the diagnosis of stenotic colonic lesions: impact of the findings in subsequent therapeutic management.* RSNA Chicago; 2013. Published online.
38. Zalis ME, Barish MA, Richard Choi J, et al. CT Colonography reporting and data system: a consensus proposal. *Radiology* 2005;236(1):3–9. https://doi.org/10.1148/radiol.2361041926.
39. *Standards of practice for computed tomography colonography (CTC) Joint guidance from the British Society of Gastrointestinal and Abdominal Radiology and The Royal College of Radiologists.* https://www.rcr.ac.uk/publication/standards-practice-computed-tomography-colonography-ctc-joint-guidance-british-society. [Accessed 26 February 2021].

Chapter 11

Rectal pathology: Findings at CT-Colonography

María Jose Martinez-Sapiña Llanas, Susana A. Otero Muinelo, and Concepción Crespo García
Department of Radiology, A Coruña University Hospital Complex, A Coruña, Spain

Introduction

Rectal pathology is varied and prevalent and, although the most serious lesion is the carcinoma, in most cases the lesion is usually benign. The optical colonoscopy (OC) is the standard imaging modality for its study since it entirely evaluates the rectum in most cases. Nonetheless, the OC procedure is an invasive imaging modality with associated risks such as perforation, bleeding, and complications following sedation.[1] Whenever the OC is contraindicated (Table 11.1) or is incomplete (10%–15% of the times[2]), the computed tomography-colonography (CTC)[1,2] is indicated. In other occasions, the CTC is conducted as the first imaging modality for the screening of colorectal cancer.[3–7]

The CTC is a quick, noninvasive emerging imaging modality developed for the screening of colorectal cancer and approved by the American Cancer Society back in 2008.[8] It is usually implemented as an alternative to the incomplete or contraindicated OC and is considered the most suitable radiological imaging modality for the screening of colorectal cancer and polyps. Its diagnostic performance for the detection of cancer is similar to that of the OC and clearly superior to the barium enema.[2]

The CTC allows us to perform easy, well-tolerated, and almost risk-free[2] 2D and 3D examinations of the colon, and it is also capable of showing extracolonic findings[9,10] using low doses and no IV contrast. The CTC has different indications (Table 11.2)[3,11] and very few contraindications (Table 11.3).[12]

The assessment of the anorectal region using the CTC is especially problematic due to a wide range of unique pathologies in this area, the presence of a rectal balloon catheter, the possible artifacts, and the particular funicular morphology of the anal canal, which all may lead to false positive findings or conceal serious pathologies.[13–15] The rectum is the most common location of hidden cancers in the CTC.[16]

Technical considerations

Conducting one CTC requires one 8-row multidetector CT machine,[17] the adequate preparation and distension of the colon, and specific software.[1]

The preparation of the colon (Table 11.4) is essential here, since the residual fecal matter can simulate or hide lesions, and an inadequate distension will not let us assess the colonic wall or surface.[3,15,17,18]

One moderately inflated balloon catheter is inserted into the rectum after an optional, although recommended, digital examination. Distension can be manual, using ambient air, or preferably automatic with CO_2. The whole process starts in the right lateral decubitus position and different series are acquired both in the supine decubitus and prone positions[12–17] without IV contrast. It is advisable to partially deflate the balloon in its helix in the prone position so that no adjacent lesions are hidden.[12] If a segment is found that remains persistently collapsed, then a third helix should be acquired in the lateral decubitus position.[17,18] If available, protocols with a low dose of radiation[17] and iterative reconstruction are used. In the presence of a known tumor, the staging process with the use of contrast in one of the series is optional.

The analysis of the images obtained allows 2D (axial images and multiplanar reconstructions) and 3D endoluminal views with anterograde and retrograde navigation visualizations.[16] Postprocessing tools are virtual dissection, virtual biopsy or translucency, second readings, and the electronic subtraction of fluid and feces.

We should remember that for the adequate assessment of the rectal region, an excellent colonic preparation and distension are needed. The balloon should be moderately deflated in its helix in the decubitus prone position.

TABLE 11.1 Contraindications of the optical colonoscopy procedure.

Absolute contraindications

- Severe pulmonary or heart disease
- Diathesis, bleeding, or treatment with anticoagulants
- Risks due to sedation
- Patient refusing to undergo the procedure

Relative contraindications

- Prior history of incomplete optical colonoscopy
- Weak patient and with mobility issues

TABLE 11.2 Indications of the CT-colonography.

- Contraindicated optical colonoscopy
- Incomplete optical colonoscopy
- Patient refusal to undergo the optical colonoscopy procedure
- Assessment of diverticular disease (after the acute phase)
- Assessment of patients with colonic stoma
- Other indications:
    - Screening of colorectal cancer
    - Controls after colorectal cancer surgery or polypectomy

TABLE 11.3 Contraindications of the CT-colonography.

- Acute bowel inflammatory disease
- Acute diverticulitis
- Recent surgery (<3 months)
- Inguinal hernia with colonic content

Rectal pathology

Rectal pathology includes processes of very different origin: congenital, acquired, tumors, inflammatory, vascular, or artifactual. Although the most severe lesions are carcinomas and lymphomas, we may find a wide variety of benign lesions in the rectum.

TABLE 11.4 Colonic preparation for the CT-colonography.

- Diet without fiber 3 days prior to the examination
- Complete diet with liquid food supplement (Isosource) 1 day prior to the examination
- Oral iodinated contrast (diatriazoate): 3 doses of 7 cc diluted in water 2 days prior to the examination, and 5 doses of 7 cc diluted in water 1 day prior to the examination
- Microenema of local action (Micralax) first time in the morning of the examination day; immediately prior to the CTC, evacuation of the rectal ampulla
- Take 2 L of water a day as a complement to the whole preparation
- Optional: prescription of intramuscular bowel muscle relaxants (Buscopan) 1 h prior to the test; they are contraindicated in cases of glaucoma, prostatic hypertrophy, heart disease, severe myasthenia gravis, or porphyria

TABLE 11.5 Classification of colonic polyps.

Based on their morphology	Based on their size
Sessile: wide base of implantation	Tiny: <6 mm
Pedunculated: with stalk or pedicle	Intermediate: 6–9 mm
Flat: protrude <3 mm over the mucosa; carpet lesions are flat lesions >3 cm in size that usually affect the cecum and the rectum	Large: ≥10 mm

Polyps

They are homogeneous attenuation structures of soft tissues that originate in the mucosa and project toward the lumen. They may be found anywhere in the colon and are common in the rectum, where the rectal catheter can end up masking them.[13]

They are classified based on their morphology and size (Table 11.5),[17] being this the criterion that stratifies its malignant potential.

The goal of the CTC is to detect advanced adenomas: polyps ≥10 mm, villous component >25%, or high-grade dysplasia. Size should be assessed in both helixes through 2D and 3D visualizations[13] and also in the plane that better shows its actual dimension.[16]

Rectal polyps can be single or multiple polyps, be part of polyposis syndromes, and coexist with other conditions (Fig. 11.1).

For screening purposes, polyps ≥6 mm identified through the CTC should appear in the radiological report, the endoscopic polypectomy being the recommended procedure here.[2] CTC monitoring is an alternative in patients where the polypectomy is risky and with one or two polyps of intermediate size. Polyps ≥10 mm should undergo endoscopic polypectomy procedures.[2] Polyps <5 mm are difficult to detect on the CTC, grow slowly, and have a low malignancy risk; however, the European Society of Gastrointestinal and Abdominal Radiology recommends reporting polyps >3 mm when they have been safely detected.[2]

We should remember that the target lesion of the CTC is the advanced adenoma: polyp ≥10 mm, villous component >25%, or high-grade dysplasia. There is a direct correlation between size and malignancy risk.

Villous tumors

They are rare in the rectum and represent 5% of all colorectal neoplasms. They are large in size, and have a lobular appearance on the CTC, which is consistent with the dense appearance seen on the OC. They have a higher risk of degeneration (Fig. 11.2). The diagnosis should be confirmed through OC and biopsy.

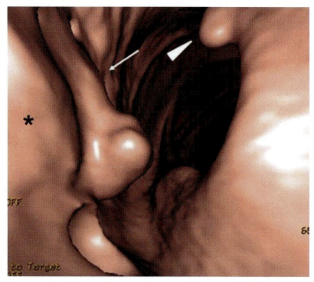

FIG. 11.1 Rectal polyps. Coexistence of different polypoid lesions in the rectum: pedunculated polyp *(white arrow)*, sessile polyp *(arrowhead)*, rectal balloon *(asterisk)*.

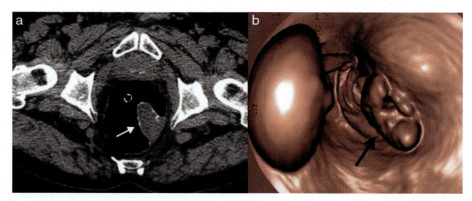

FIG. 11.2 Villous tumor. (A) Axial CT image. Lobulated lesion based on the lateral wall of the rectum *(white arrow)*. The soft cover of its surface after the administration of oral contrast shows its dense appearance. (B) Match with 3D view. The anatomopathological finding was villous adenoma.

Malignant neoplasms

Of variable morphology, they may present as small or big size stenosing or polypoid lesions. The most difficult cancers to detect are the small ones, since they can remain kind of hidden by the balloon catheter and look like polypoid focal thickenings (Fig. 11.3), which is why it is advisable to slightly deflate the balloon in its helix in the decubitus prone position. Between 1.5% and 6% of all colonic neoplasms associate synchronic lesions (Fig. 11.4). The CTC is especially useful if the distal lesion is oclusive.[1, 9, 19]

One rectal lesion suspicious of malignancy on the CTC should be biopsied with CO.[2]

We should remember that the biggest problem when it comes to the anorectal region is misdiagnosing low malignant lesions hidden by the balloon catheter or darkened by the artifacts.

Submucosal lesions

There is a wide variety of benign and malignant conditions (Table 11.6). They originate in deep areas (intramural or extramural), protrude toward the intestinal lumen, make up obtuse angles with the wall, and displace the folds without interrupting them.[12]

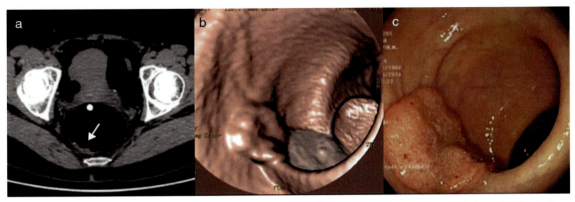

FIG. 11.3 Carcinoma. (A) This axial CTC image corresponds to one patient studied due to anemia and shows one flat mural lesion in the rectum discretely protruding toward the lumen *(arrow)* that turned out to be an adenocarcinoma in the cylinder biopsy. (B) 3D virtual colonoscopy. (C) Optical colonoscopy.

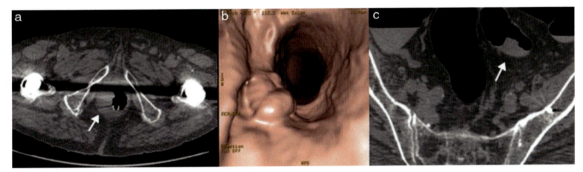

FIG. 11.4 Synchronic neoplasms. CTC of a 75-year-old woman after an incomplete optical colonoscopy due to occlusive stenosing lesion in her rectum. (A) Axial CT image. Lesion inside the distal rectum *(arrow)* corresponding with one polypoid elevated lesion (B, 3D endoluminal view) consistent with one carcinoma. The CTC was good for the detection of another lesion of malignant appearance in the sigma (C). Note the beam hardening artifact caused by the metallic prosthetic material in both hips (A), which makes the assessment of the rectum even more difficult.

TABLE 11.6 Submucosal lesions.

Of intramural origin	Of extramural origin
• Leiomyoma	• Endometriosis
• Lipoma	• Developmental retrorectal cystic lesions:
• Neuroendocrine tumor	– Retrorectal cystic hamartoma
• Gastrointestinal stromal tumor	– Rectal duplication
• Schwannoma	– Epidermoid cyst
• Lymphoma	– Dermoid cyst
• Melanoma	• Rectal invasion by other
• Other primary tumors	
• Metastasis	

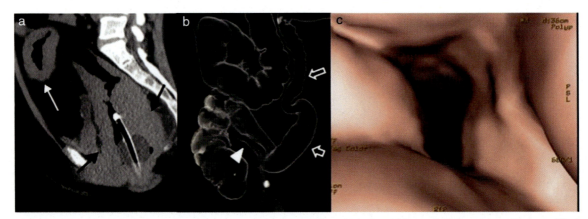

FIG. 11.5 Lymphoma. (A) CT image reconstruction in the sagittal plane. Diffuse thickening of the wall of the rectum and sigma *(black arrows)*. (B) Virtual luminogram. Loss of distension in the damaged segments *(arrowheads)*. Note the loss of haustration of the descending colon *(hollow arrows)* relative to ulcerative colitis in chronic stage. (C) Virtual colonoscopy. Stenotic appearance of submucosal masses in intestinal lymphoma.

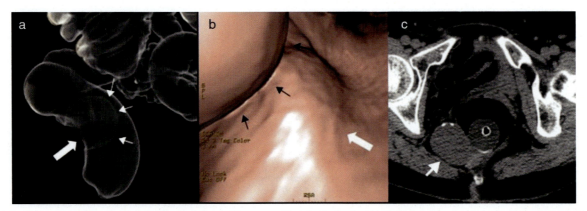

FIG. 11.6 Extramural submucosal lesion. Sixty-eight-year-old woman with abdominal pain. (A) CTC, virtual luminogram. Repletion defect in the lateral wall of the rectum *(thick arrow)* that looks like an extrinsic compression on the 3D endoluminal image (B) *(thick arrow* in B). (C) The axial CT image shows one extramural submucosal fluid density lesion *(white arrow)* exerting that extrinsic compression and consistent with a developing cyst. The *thin arrows* in (A) and (B) point at the rectal balloon.

The most common of all are lipomas that can be easily identified by their fat density. When it comes to malignant lesions, the primary rectal lymphoma is relatively rare compared to the small intestine gastric lymphoma. Almost all of them are non-Hodgkin lymphomas type B associated with immunosuppression and bowel inflammatory disease. On the CTC they appear as one big polylobulated or multifocal single mass (Fig. 11.5).

The CTC assessment of suspected submucosal lesions found on the OC is useful to be able to distinguish an intramural process from an extramural extrinsic compression, identify its true nature, and study the spread of the disease[13,14] (Fig. 11.6).

Management can vary. Fat density characterizes lipomas and is diagnostic on the 2D images. In other submucosal lesions, other imaging modalities (MRI, transrectal ultrasound, etc.) may help us characterize these lesions.

We should remember that the CTC allows us to distinguish intramural from extramural submucosal lesions, identify their true nature, and study the spread of the disease.

Vascular lesions

Internal hemorrhoids

It is the most common rectal pathology. It consists of the dilation of the veins of the superior plexus that are covered by the mucosa[13] over the dentate line.[14] They have a typical appearance on the CTC, of anorectal location, and on circumferential disposition around the catheter, giving the appearance of one submucosal lesion, or a wrinkled appearance of the mucosa

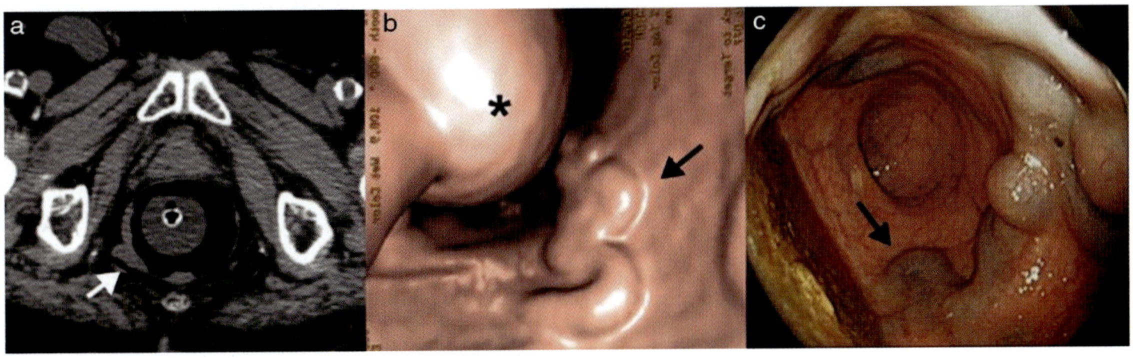

FIG. 11.7 Rectal varicose vein. Eighty-year-old male with anemia and rectal bleeding. (A) axial CTC image. Parietal lesion in his rectum of soft tissue density *(white arrow)*. (B) 3D CTC. The *black arrow* shows the tubular and winding morphology of the lesion seen in (A). Rectal catheter *(asterisk)*. (C) Rectal varicose veins as seen on the optical colonoscopy *(arrow)*.

around the rectal tube. When hemorrhoids become thrombosed they may look like a tumor.[14] A digital rectal exam, instead of the OC, may help confirm the diagnosis.

Rectal varices

They are less common than internal hemorrhoids, are associated with portal hypertension, and have a winding and tubular morphology (Fig. 11.7). Diagnosis is achieved through the OC.

Venous malformations

They are rare. They may be part of the Blue Rubber Bleb Nevus Syndrome (BRBNS) or appear in isolation on the CTC simulating one polyp. On the OC they show a characteristic blue color. On the MRI, their hyperintensity on the T2-weighted sequences and their spread into the mesorectal fat are specific characteristics.[20]

Hypertrophied anal papilla

They are focal fibrous protrusions on the dentate line.[13] They may look like polyps, but their location in the anorectal junction is pathognomonic, and almost always in contact with the catheter (Fig. 11.8). The OC is diagnostic.

Inflammatory conditions

The rectum is affected in the ulcerative colitis and Crohn's disease whenever there is associated perianal disease.[13] The radiation therapy-related iatrogenia in the pelvis affects the rectum in the form of actinic proctitis. These conditions appear as one diffuse, circumferential thickening of the wall of the rectum that causes variable stenoses, with important frequencies, but with signs of benignity. The patient's personal history facilitates the diagnosis (Fig. 11.9).

The CTC outside the acute episode allows us to assess the degree of stenosis and plan the course of treatment. The solitary rectal ulcer consists of an intense inflammatory reaction around an ulcer that conditions one mass effect that can be interpreted as a malignant tumor in a patient with rectal bleeding and painful defecation.[13] Both the OC and the biopsy are indicated to achieve the diagnosis.

Postoperative changes

The surgical clips placed on colorectal anastomosis usually appear on the 3D images as irregularities of the mucosa that can be taken for tumor relapses. The 2D images are key here since they reveal their metallic density (Fig. 11.10).

Recto-colonic anastomosis is a common cause of incomplete OC, but they rarely cause significant stenosis.

The signs of recurrence of neoplastic disease are irregularity, wall thickening, and distortion of the mucosal pattern compared to common postoperative findings such as small size inflammatory polyps located in the anastomotic line.[9]

In the presence of suspicious images and suspicion of relapse, we should try to biopsy with the OC, or else, with a surgical biopsy in cases of impassable stenosis.

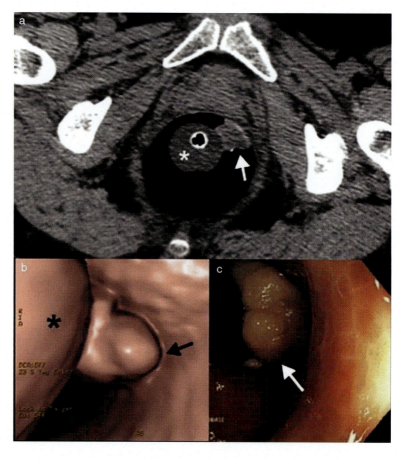

FIG. 11.8 Hypertrophied anal papilla. Fifty-seven-year-old-woman. CTC after incomplete optical colonoscopy. (A) 2D image on the axial plane. Lesion in the rectal lumen stained in its periphery after the administration of oral contrast *(white arrow)*. (B) 3D endoluminal view. The *arrow* points at the same lesion in contact with the rectal balloon *(asterisk)* and close to the anorectal junction (C) The optical colonoscopy confirms it is consistent with one hypertrophied anal papilla *(white arrow)*.

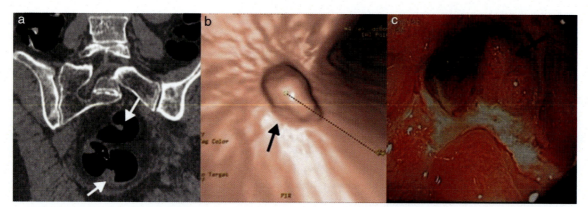

FIG. 11.9 Actinic proctitis. CTC in one patient with cervical carcinoma treated with radiation therapy after incomplete optical colonoscopy due to impassable stenosis. (A) CT image reconstruction in the coronal plane that reveals protrusions *(arrows)* and stenosis (not shown). (B) The 3D image shows one of the polypoid lesions that looks ulcerative in the optical colonoscopy (C).

Pseudolesions and artifacts

With an optimal fecal marking and colonic distention,[13] most artifacts are easily recognizable:

(a) The rectal catheter: constant finding in the anorectal region.[13, 14] Its tip can have a polypoid appearance on the 3D views or cause compression on an adjacent rectal fold. Both the partial balloon deflating in the decubitus prone position[14] and the verification of its presence on the 2D images (Fig. 11.11) are of great help.
(b) Stained feces: they may appear as polyps or masses based on their size on the 3D endoluminal images,[14] but they can be easily identified on the 2D images after contrast staining.[13, 17]

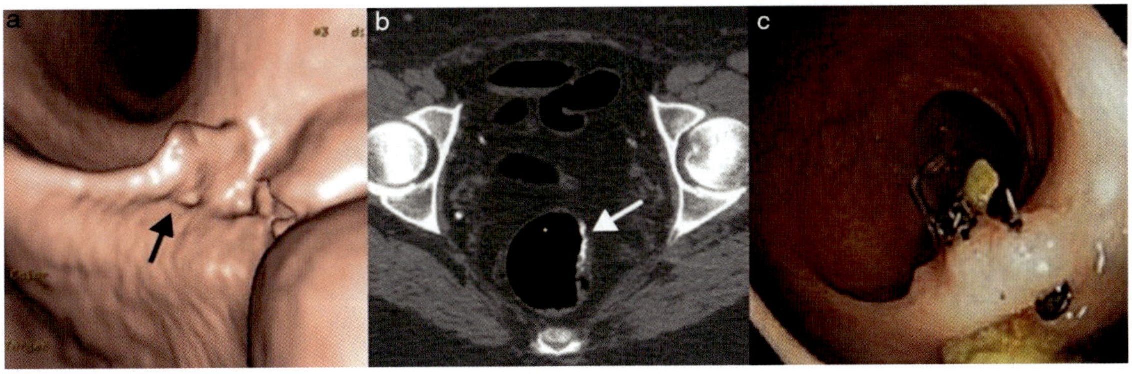

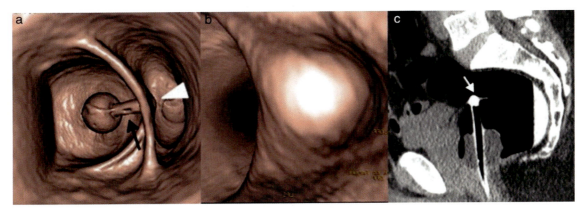

FIG. 11.10 Surgical material. (A) The 3D image shows a significant mucosal irregularity inside one rectal valve *(black arrow)*. The CTC image on the axial plane (B) shows material of metallic density *(white arrow)*, and eventually tumor pathology is ruled out after the OC confirms the presence of surgical clips (C).

FIG. 11.11 Rectal balloon catheter. (A) The distal end of the rectal catheter *(black arrow)* can compress the rectal valves *(arrowhead)* and create an image of submucosal extrinsic compression. (B) 3D image. Appearance of the catheter protrusion over the rectal valve. (C) The position of the balloon catheter needs to be confirmed on the multiplanar reconstructions.

(c) Unstained feces: they can be a problem if they are small.[14] Internal air foci and supine-to-prone position changes are characteristic here.[13, 17]

(d) Fecalomas: they are common in the rectal ampulla. The 3D endoluminal image shows one lobulated irregular mass that simulates a tumor or cancer. The 2D image is diagnostic and shows the heterogeneous composition that is typical of unstained feces.

Other lesions

The condyloma acuminatum can present as a polypoid lesion, although it is rare. The diagnosis of anorectal lesions is achieved through anoscopy examination or rectal digital exam.

The diverticula are exceptional in the rectum. They look exactly the same in all colonic locations, and their finding does not require follow-up or diagnostic confirmation.

Conclusion

Although rectal lesions can go misdiagnosed on the CTC because it is particularly difficult to assess this anatomical region, and even though it is not the modality of choice for the study of rectal pathologies, it is indicated in cases of incomplete or contraindicated OC. For this reason, it is essential to know the rectal pathology and its semiology on the CTC, have an excellent command while performing the technique when it comes to preparation and distension, perform moderate balloon insufflations, and careful 2D and 3D navigations.

References

1. Bouzas SR. Optical colonoscopy and virtual colonoscopy: the current role of each technique. *Radiologia* 2015;**57**(2):95–100. https://doi.org/10.1016/j.rx.2014.04.003.
2. Spada C, Stoker J, Alarcon O, et al. Clinical indications for computed tomographic colonography: European Society of Gastrointestinal Endoscopy (ESGE) and European Society of Gastrointestinal and Abdominal Radiology (ESGAR) guideline. *Eur Radiol* 2015;**25**(2):331–45. https://doi.org/10.1007/s00330-014-3435-z.
3. Laghi A. Computed tomography colonography in 2014: an update on technique and indications. *World J Gastroenterol* 2014;**20**(45):16858–67. https://doi.org/10.3748/wjg.v20.i45.16858.
4. Johnson CD. CT colonography: coming of age. *AJR Am J Roentgenol* 2009;**193**(5):1239–42. https://doi.org/10.2214/AJR.08.1859.
5. de Haan MC, Halligan S, Stoker J. Does CT colonography have a role for population-based colorectal cancer screening? *Eur Radiol* 2012;**22**(7):1495–503. https://doi.org/10.1007/s00330-012-2449-7.
6. Pickhardt PJ, Kim DH. Colorectal cancer screening with CT colonography: key concepts regarding polyp prevalence, size, histology, morphology, and natural history. *AJR Am J Roentgenol* 2009;**193**(1):40–6. https://doi.org/10.2214/AJR.08.1709.
7. Ho W, Broughton DE, Donelan K, Gazelle GS, Hur C. Analysis of barriers to and patients' preferences for CT colonography for colorectal cancer screening in a nonadherent urban population. *AJR Am J Roentgenol* 2010;**195**(2):393–7. https://doi.org/10.2214/AJR.09.3500.
8. Levin B, Lieberman DA, McFarland B, et al. Screening and surveillance for the early detection of colorectal cancer and adenomatous polyps, 2008: a joint guideline from the American Cancer Society, the US multi-society task force on colorectal Cancer, and the American College of Radiology. *CA Cancer J Clin* 2008;**58**(3):130–60. https://doi.org/10.3322/CA.2007.0018.
9. Hong N, Park SH. CT colonography in the diagnosis and management of colorectal cancer: emphasis on pre- and post-surgical evaluation. *World J Gastroenterol* 2014;**20**(8):2014–22. https://doi.org/10.3748/wjg.v20.i8.2014.
10. *ACR-SAR-SCBT-MR practice parameter for the performance of computed tomography (CT) colonography in adults.* https://www.acr.org/media/A81531ACA92F45058A83B5281E8FE826.pdf. [Accessed 25 July 2015].
11. Pooler BD, Kim DH, Pickhardt PJ. Potentially important extracolonic findings at screening CT colonography: incidence and outcomes data from a clinical screening program. *AJR Am J Roentgenol* 2016;**206**(2):313–8. https://doi.org/10.2214/AJR.15.15193.
12. Siewert B, Kruskal JB, Eisenberg R, Hall F, Sosna J. Quality initiatives: quality improvement grand rounds at Beth Israel Deaconess Medical Center: CT colonography performance review after an adverse event. *Radiographics* 2010;**30**(1):23–31. https://doi.org/10.1148/rg.301095125.
13. Silva AC, Vens EA, Hara AK, Fletcher JG, Fidler JL, Johnson CD. Evaluation of benign and malignant rectal lesions with CT colonography and endoscopic correlation. *Radiographics* 2006;**26**(4):1085–99. https://doi.org/10.1148/rg.264055166.
14. Pickhardt PJ, Kim DH. CT colonography: pitfalls in interpretation. *Radiol Clin North Am* 2013;**51**(1):69–88. https://doi.org/10.1016/j.rcl.2012.09.005.
15. Pickhardt PJ. Missed lesions at CT colonography: lessons learned. *Abdom Imaging* 2013;**38**(1):82–97. https://doi.org/10.1007/s00261-012-9897-z.
16. Pickhardt PJ, Hassan C, Halligan S, Marmo R. Colorectal cancer: CT colonography and colonoscopy for detection—systematic review and meta-analysis. *Radiology* 2011;**259**(2):393–405. https://doi.org/10.1148/radiol.11101887.
17. Pagés Llinás M, Darnell Martín A, Ayuso Colella JR. CT colonography: what radiologists need to know. *Radiologia* 2011;**53**(4):315–25. https://doi.org/10.1016/j.rx.2011.01.009.
18. Taylor SA, Halligan S, Goh V, et al. Optimizing colonic distention for multi-detector row CT colonography: effect of hyoscine butylbromide and rectal balloon catheter. *Radiology* 2003;**229**(1):99–108. https://doi.org/10.1148/radiol.2291021151.
19. Sali L, Falchini M, Taddei A, Mascalchi M. Role of preoperative CT colonography in patients with colorectal cancer. *World J Gastroenterol* 2014;**20**(14):3795–803. https://doi.org/10.3748/wjg.v20.i14.3795.
20. Yoo S. GI-associated hemangiomas and vascular malformations. *Clin Colon Rectal Surg* 2011;**24**(3):193–200. https://doi.org/10.1055/s-0031-1286003.

Chapter 12

Applications of [18F] FDG PET and PET/CT in colorectal carcinoma

Sofía Rodríguez Martínez de Llano, Fernando Zelaya Reinquet, Shirly Margarita Nieves Maldonado, Sara Seijas Marcos, and Paulino Pais Silva

Department of Nuclear Medicine, Oncology Centre of Galicia, A Coruña, Spain

Introduction

Colorectal cancer (CRC) is the fourth most frequently diagnosed cancer and the second leading cause of cancer death in the United States. In 2020 an estimated 104 610 new cases of colon cancer and 43 340 cases of rectal cancer will occur. An estimated 53 200 people will die from CRC in 2020, including 3640 men and women younger than age 50. It is estimated that the incidence rates for colon and rectal cancers will increase by 90.0% and 124.2%, respectively, for patients 20 to 34 years of age by 2030. Approximately 85% of CRCs are adenocarcinomas (not otherwise specified), 10% are mucinous adenocarcinomas, and the remaining are rare histologic types, such as papillary carcinoma, adenosquamous carcinoma, and signet cell carcinoma. The risk of developing CRC is influenced by both environmental and genetic factors, with the majority of cases being sporadic. Approximately 20% of cases of colon cancer are associated with familial clustering, and first-degree relatives of patients with colorectal adenomas or invasive CRC are at increased risk for CRC.[1]

Once the diagnosis is established, the local and distant extent of disease spread will provide the scaffold for therapy planning and prognosis, with resection being the mainstay of treatment for curative intent. Following surgical exploration of the abdomen, histopathologic staging is performed. This will frequently include evaluation of the grade of the tumor; depth of penetration, and extension to adjacent structures (T); number of regional lymph nodes evaluated and number involved (N); status of proximal, distal, and radial margins; lymphovascular invasion; perineural invasion; and assessment of distant metastases to other organs, peritoneum, or nonregional lymph nodes (M).[2]

Modern imaging techniques play a critical role in the management of patients with CRC and help in selection of patients who would benefit most from surgical or invasive approaches. Postoperatively, surveillance by imaging is usually performed in tandem with serum carcinoembryonic antigen (CEA) assays. When CEA levels rise during postoperative surveillance, imaging is performed in an attempt to identify the presence and location of recurrence to facilitate prompt treatment.[3]

[18F] FDG PET imaging is a noninvasive diagnostic tool that provides tomographic images and can be used to obtain quantitative parameters concerning the metabolic activity of target tissues. 18F is a cyclotron-produced radioisotope of fluorine that emits positrons and has a short half-life (109.7 min). It allows labeling of numerous molecular tracers that can be imaged within a few hours (typically <3 h) after injection. FDG is an analogue of glucose and is taken up by living cells via cell membrane glucose transporters and subsequently incorporated into the first step of the normal glycolytic pathway.

[18F] FDG PET/CT has become one of the cornerstones of patient management in oncology. PET is a tomographic technique that measures the three-dimensional distribution of positron-emitting labeled radiotracers. PET allows noninvasive quantitative assessment of biochemical and functional processes. The most commonly used tracer at present is the 18F-labeled glucose analogue FDG. FDG accumulation in tissue is proportional to the amount of glucose utilization. Increased consumption of glucose is characteristic of most cancers and is in part related to overexpression of the GLUT glucose transporters and increased hexokinase activity. [18F] FDG PET/CT has been proven to be a sensitive imaging modality for detection, staging, and restaging and therapy response assessment in oncology. [18F] FDG PET/CT provides essential information for radiation treatment planning, helping with critical decisions when delineating tumor volumes.[4]

CT uses a combined X-ray transmission source and detector system rotating around the subject to generate tomographic images. CT allows not only attenuation correction but also the visualization of morphological and anatomical structures with a high spatial resolution. Anatomical and morphological information derived from CT can be used to improve the localization, extent, and characterization of lesions detected by FDG PET. Recently, combined or integrated PET and MRI systems (PET/MRI) have come onto the market.

Patient preparation and precautions include reducing tracer uptake in normal tissue (kidneys, bladder, skeletal muscle, myocardium, brown fat) while maintaining and optimizing tracer uptake in the target structures (tumor tissue) and keeping patient radiation exposure levels as low as reasonably possible (ALARA). Nondiabetic patients should not consume any food, simple carbohydrates or liquids other than plain (unflavored) water for at least 4 h prior to the start of the FDG PET/CT study (i.e., with respect to the time of injection of FDG). Those scheduled for an afternoon FDG PET/CT study may have a light breakfast at least 4 h prior to the time of their PET/CT examination appointment. Medication can be taken as prescribed.

The main objectives of patient preparation with at least 4 h of fasting are to ensure low blood glucose and low insulinemia, as insulin is directly responsible for glucose uptake by nontumor cells.

If the plasma glucose level is lower than 11 mmol/L (about 200 mg/dL), the FDG PET/CT study can be performed. If the plasma glucose level is higher than or equal to 11 mmol/L (about 200 mg/dL), the FDG PET/CT study should be rescheduled.

Indications for FDG PET/CT include, but are not limited to, the following[4]:

- Differentiation of benign from malignant lesions.
- Searching for an unknown primary tumor when metastatic disease is discovered as the first manifestation of cancer or when the patient presents with a para-neoplastic syndrome.
- Staging patients with known malignancies.
- Monitoring the effect of therapy on known malignancies.
- Determining whether residual abnormalities detected on physical examination or on other imaging studies following treatment represent tumor or posttreatment fibrosis or necrosis.
- Detecting tumor recurrence, especially in the presence of elevated tumor markers.
- Selection of the region of tumor most likely to yield diagnostic information for biopsy.
- Guiding radiation therapy planning.

Staging

Accurate initial staging of CRC is mandatory for optimal therapeutic planning. As a whole-body imaging technique, fluorodeoxyglucose (FDG) positron emission tomography (PET) and PET/CT have the unique capability of providing staging for the tumor (T) stage, nodal (N) stage, and metastatic (M) stage in a single imaging session.[5]

In the assessment of the primary tumor, PET with FDG has presented high values of sensitivity, 95%–100% of primary intraluminal known cancers are visible with [18F] FDG PET (Fig. 12.1). However, it has limitations, such as the presence of false negatives that can be found in patients with mucinous tumors[5] with small tumor foci in tubulovillous adenomas,[6,7] as well as the false positive, conditioned by benign polyps, precancerous lesions, previous polypectomies, and intestinal inflammatory processes.[5,7,8]

T characterization of the tumor can usually be performed with CT and RM (magnetic resonance), since they have a high precision rate of up to 80%.[9] A recent study showed a diagnostic accuracy of PET/ceCT for T staging of 82%.[10] The MR characterization of the ratio of tumor with the mesorectal fascia seems likewise adequate.[11]

Innovative fused positron emission tomography–magnetic resonance (PET/MR) outperformed PET/CT in CRC staging. PET/MR might allow accurate local and distant staging of CRC patients during both at the time of diagnosis and during follow-up.[12]

Lymphatic involvement is an important prognostic indicator in patients with CCR. CT and MRI show sensitivity values of 55%–66%[9] in the detection of nodal disease. Metabolic changes sometimes precede morphological ones, so [18F] FDG PET can demonstrate involvement of normal-sized lymph nodes. There are limitations in the spatial resolution of the equipment, in detection of lymph nodes close to the primary tumor that cannot be differentiated from the primary tumor, and in the detection of disease in lymph nodes with low tumor burden, which reduces sensitivity to values of 29%–37% in N staging, with high specificity of up to 88%.[5–7] A meta-analysis determined that PET/CT studies with [18F] FDG show a sensitivity of 42.9% and 87.9% specificity in pretherapeutic lymph node detection in patients with CRC.[13]

PET/CT with [18F] FDG shows a sensitivity of 81% in the detection of liver metastases in these patients and a specificity of 91% as well as in the detection of extrahepatic disease of 84% and 70%, respectively.[14] In addition, it has shown that FDG PET scans 120 min after injection found hepatic lesions that were missed in 17% of images obtained 90 min after

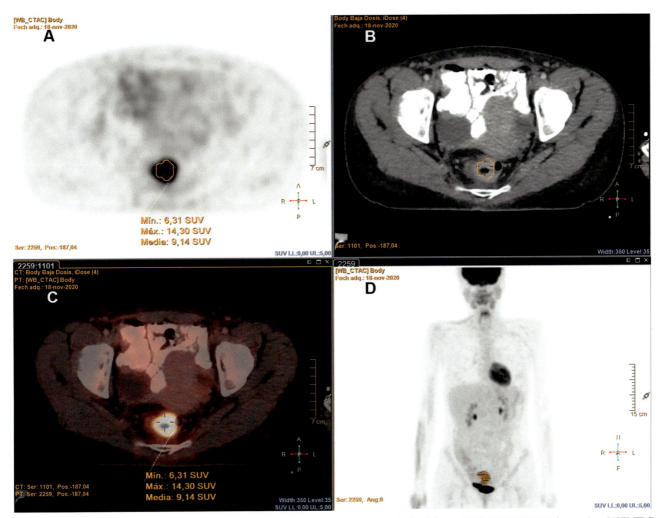

FIG. 12.1 Images from a 40-year-old woman diagnosed with rectal adenocarcinoma, submitted for an [18F] FDG PETTAC for staging. [18F] FDG PETTAC images allowed to identify an hypermetabolic focus in pelvis (SUVmax 14.3 g/mL) suggestive of the primary lesion at the lower rectum as shown in the transaxial PET views (A), fused (C), and MIP (Maximum Intensity Projection) (D). CT images (B) showed thickening of the walls at the level of the rectal ampulla, hyperdense, with an irregular and anfractuous lumen, compatible with a malignant neoplastic lesion at this level.

injection, so this practice should be considered, if necessary.[5] PET images fused with MRI showed a sensitivity of 98.3% in the detection of metastatic hepatic lesions, detecting smaller lesions (<1 cm) than fused PET studies with CT.[15]

Regarding the detection of brain metastases, FDG-PET has limitations, MRI with gadolinium contrast would be recommended.

Peritoneal metastases, if they present FDG uptake, are visualized as isolated or multiple focal lesions (Fig. 12.2) or as a diffuse pattern in patients with advanced carcinomatosis, its positivity depends on the amount of volume tumor, finding false negatives if the nodules are small.

Recent studies using combined PET/CT colonography have reported that, in staging colon cancer, combined PET/CT colonography delivers accuracies superior to CT alone and to CT plus PET performed separately.[5] It has been reported an 80% sensitivity for N stage and a 100% sensitivity for M detection in a pilot study of 47 patients. They also reported that PET/CT colonography affected the therapy decisions in four patients (9%). PET/CT colonography was able to localize synchronous colon cancers proximal to the obstruction precisely and in conjunction with optical colonoscopy may be suitable strategy for staging colorectal cancer.[16]

The use of [18F] FDG PET/CT in the initial staging of patients with CRC is therefore reserved in high-risk patients, for example those with elevation of CEA (carcinoembryonic antigen) values > 10 ng/mL, locally advanced disease, or inconclusive results with other image techniques. Its incorporation may lead to a change in patient management in up to 24% of cases,[17] demonstrating unexpected metastasis or clarifying the nature of indeterminate lesions.[18]

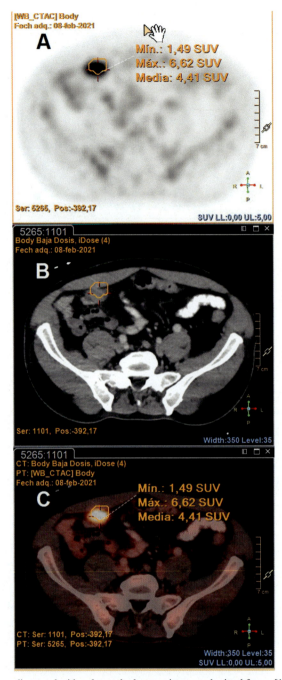

FIG. 12.2 Images from a 62-year-old man diagnosed with colorectal adenocarcinoma, submitted for an [18F] FDG PETTAC for staging. [18F] FDG PETTAC images allowed to identify hypermetabolic foci in the greater omentum adjacent to abdominal wall, in relation to areas of tissue thickening in the CT scan (transaxial CT view B), presenting lesion with highest SUVmax (6.62 g/mL) in the right iliac fossa, as shown in the transaxial PET views (A), and fused (C).

Restaging

Recurrent disease detection

Recurrence occurs approximately in 30% of patients with colorectal cancer within the first couple of years after surgery. Despite potentially curative surgery and the use of modern adjuvant chemotherapy and/or radiation therapy (RT), more than 40% of patients who present with stage II or III disease will have a disease recurrence following primary therapy.[3]

An increase of serum carcinoembryonic antigen (CEA) level is considered as an early marker in 60% of them,[19] although this is not exclusive of CRC, but also appears in some other malignancies and benign processes. The CEA is routinely monitored in follow-up, and its appearance could precede the clinical symptoms in 4.5–8 months,[20] but does not localize the disease site. If it is normal, it does not exclude the recurrent disease and it is essential to have an exhaustive clinical and imaging follow-up.

Approximately 25% of recurrent disease patients will reach remission. The relapse might be localized loco-regionally (25%) or could be distant metastases (75%), and they can be treated by resection to reach remission. Intensive postoperative surveillance programs have been justified in the hope that early detection of asymptomatic recurrences will increase the proportion of patients who are potentially eligible for curative therapy. A survival benefit from such an approach has in fact been shown in several meta-analyses. Furthermore, periodic imaging can detect early, potentially resectable recurrences.

Colonoscopy is useful for detection of relapse at the surgical anastomosis level, with a limited role in cases of extraintestinal disease. Computed tomography (CT) of the chest, abdomen, and pelvis is the most used tool, with high precision for localizing disease, limited to liver, peritoneum, mesentery, lymph nodes, and postsurgical changes.

Detection of recurrent disease and restaging in CRC patients is the most frequent recommendation for [18F] FDG PET/CT. Patients with CEA high levels and with negative radiologic imaging studies or equivocal show [18F] FDG PET/CT sensitivity of about 67%–75% in recurrence detection, with an NPV from 95% to 100%.[21,22] The sensitivity depends on the histological type of the carcinoma: lower sensitivity, 58%, is noted for the detection of mucinous cancer, while for non-mucinous cancer sensitivity reaches 92%.[23] [18F] FDG PET/CT evaluation of these patients could indicate potentially curative surgery if distant metastases are ruled out; this window of opportunity could be overpassed if we take only into account morphological changes (CT).

[18F] FDG PET/CT scans have no role in routine surveillance, and their use is specifically reserved for certain cases in guidelines from expert groups, including ASCO and the NCCN. If the PET/CT result is negative, the actions will depend on the existence or not of morphological lesions at other image modalities. If FDG PET/CT is pathological, findings must be compared to previous imaging studies, and when lesions are less than 1 cm, metastatic disease could not be discarded, because of the system's spatial resolution and/or the metabolic behavior. If other morphological images are not pathological and the suspicion is based on high levels of tumor markers, a follow-up and a new PET/CT could be planned at 3 months.

The likelihood of false-negative findings depends on the size of metastases (for lesions <1 cm, sensitivity is only 25%) and the histological type.[24] [18F] FDG PET/CT is a useful tool in the recognition of local recurrences: differentiation between postoperative or postradiation therapy scar/fibrotic tissue and vital tumor in rectal cancer is difficult or impossible when using morphology-based imaging modalities.[25]

The [18F] FDG PET/CT must be considered in cases of suspected recurrence due to high CEA levels and with non-conclusive results in conventional image techniques, as well as in characterization of equivocal or doubtful lesions seen on them.

Preoperative evaluation of potentially resectable disease

Recto-sigmoidal colon double venous drainage explains the variety of bone and pulmonary metastases through venous dissemination and liver metastases through the portal venous system. Liver is the most frequent site of metastasis in colorectal cancer patients, and up to 25% of patients show them from the moment of diagnosis.[26] Some authors recommend to delay the beginning of the acquisition of FDG PET/CT images until 90–120 min postinjection, in order to evaluate liver lesions, given the normal background to tumor contrast ratio which increases within time.[14,27]

Medium survival rate between 30% and 40% in 5 years could be achieved in adequate patients for surgical treatment (hepatectomy) and with localized metastatic disease.[24] Nevertheless, it has been demonstrated that even 50% of patients with apparent localized metastatic disease show inoperability findings at laparotomy.[28] Many practitioners may accept that more accurate staging will lead to a better choice of treatment plan, thereby avoiding overtreatment and sparing patients the unnecessary risks or side effects of therapy or avoiding undertreatment when patients might otherwise benefit from aggressive curative-intent therapy.[29]

PET/CT sensitivity for detection of liver metastases decreases after adjuvant chemotherapy administration. It is recommended to perform a baseline scan and planned treatment even when there is no evidence of lesions after chemotherapy.[30] FDG PET/CT presents higher sensibility than CT for identifying extrahepatic metastatic disease,[31] changing therapeutic decision (nonsurgical) even up to 40% of cases.[32]

In addition, pre-surgery PET/CT assessment could be used to identify accurately adequate candidates with unique pulmonary metastatic disease.[33]

Bone metastatic disease could be identified with FDG PET/CT, but is less frequent, especially if there are no other metastatic sites of disease; its findings may be useful to plan surgical or palliative treatment.[34] This imaging technique is particularly useful to detect mostly osteolytic bone lesions.[27]

Others: Incidental diagnosis of colorectal cancer. Assessment of treatment response

Incidental diagnosis of colorectal cancer

Physiological bowel uptake is seen in [18F] FDG PET/CT, sometimes there is a diffuse or a patchy uptake with the [18F] FDG, without morphological changes in CT. There is also high bowel uptake in diabetic patients treated with metformin, performing late images may help to verify a change in location or intensity of bowel uptake, decreasing the false-positive rate.[35]

Unexpected focal colonic or rectal radiotracer activity is a usual finding in patients referred for a PET study. Three percent of patients who undergo a PET/CT study can have focal bowel uptake. In 60% of patients the focal uptake means a precancerous lesion or a carcinoma.[36] So according to these results, it is recommended to perform colonoscopy and biopsy, in all patients with unexpected focal FDG activity found in colon or rectum during a 18F-FDG PET/CT examination.[37]

Although [18F] FDG PET/CT has high precision in detecting polyps larger than 13 mm, the sensibility in detecting small polyps is low (24%), so [18F] FDG PET/CT is not good for screening in this cases.[8]

Assessment of treatment response

It is well known that morphological images have some limitations to assess treatment response (adjuvant chemotherapy and palliative treatments).[38] [18F] FDG PET/CT provides a valuable tool to differentiate and compare the effects of different therapeutic approaches.

[18F] FDG PET/CT is an asset for improving patient care by reducing the effort, costs, and morbidity associated with ineffective treatment in nonresponder CRC patients (Figs. 12.3 and 12.4).[39]

There is evidence that suggest that the assessment of the metabolic response using [18F] FDG PET/CT may allow prediction of long-term responders, by demonstrating a reduction of the SUVmax by the tumor.[40,41]

[18F] FDG PET/CT is recommended after neoadjuvant treatment in the preoperative setting in primary rectal cancer and it has a better correlation with pathology than morphologic imaging modalities. Therefore this technique may be used to guide chemotherapy for rectal cancer after neoadjuvant and local treatments.

Many patients with CRC have presacral fibrosis after surgery or radiotherapy. Sometimes it is difficult to differentiate between local recurrence and fibrosis with some imaging modalities (ERUS, MRI, CT). In these cases performing the [18F] FDG PET/CT may change management. A sensitivity and specificity of detection of local recurrences is relatively high with this technique (sensitivity 100%; specificity 96%).[42]

The possibility of false-positive findings should be taken into account in the assessment of these studies, especially if the patient has had infectious /postoperative inflammatory diseases (abscesses in the scar line) or if the PET study is being performed within 6 months after completion of surgical treatment and/or radiotherapy.[27] Presacral metabolism after radiotherapy normally shows a diffuse pattern that decreases over time.

All therapies can alter images interpretation. Reduction in FDG uptake may be a direct consequence of the lower volume in tumor cells, but also due to possible cellular stunning up to 1 month after chemotherapy. For this reason, it is recommended to wait at least 3–5 weeks after chemotherapy to minimize false negatives. Conventional imaging studies also have difficulties in detecting residual or recurrent disease in sites treated by radiofrequency. Theoretically, treated cells may be less susceptible to 18F-FDG uptake, so hybrid imaging equipment (with CT) could help in detection of recurrences after liver and/or lung ablation[43,44] when there is an anatomical alteration that makes it difficult to interpretation of the study.

18F-FDG PET/CT also seems to have a role in assessing treatment response in patients with CRC and unresectable liver metastases, treated with radioembolization (Yttrium-90 microspheres).[45]

New perspectives

18F-FDG-PET may be useful to evaluate response to chemotherapy, although the optimum timing of the assessment of metabolic response remains unsettled. Moreover, new drugs targeted to angiogenesis or tyrosine kinase have opened new frontiers to the use of [18F] FDG PET/CT in evaluating response because of their cytostatic rather than cytoreductive

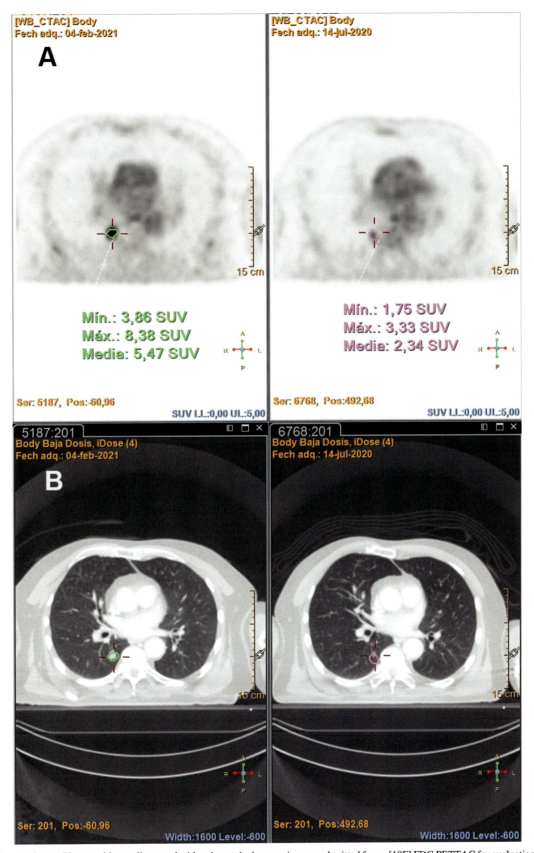

FIG. 12.3 Images from a 72-year-old man diagnosed with colorectal adenocarcinoma, submitted for an [18F] FDG PETTAC for evaluation of treatment response. [18F] FDG PETTAC images allowed to identify an hypermetabolic focus in the lower right lung lobe of 14 mm (previous 10 mm) with an increase in SUVmax (8.35, previous 3.32 g/mL), as shown in the transaxial PET views (A), and CT views (B), compatible with a malignant neoplastic lesion in progression.

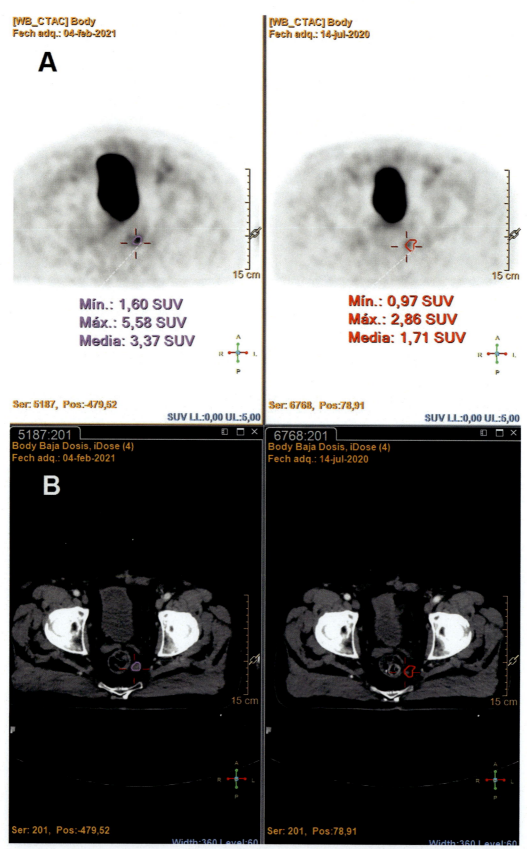

FIG. 12.4 Images from a 72-year-old man diagnosed with colorectal adenocarcinoma (same patient as Fig. 12.3), submitted for an [18F] FDG PETTAC for evaluation of treatment response. [18F] FDG PETTAC images allowed to identify an hypermetabolic focus in a left mesorectal adenopathy with an increase in size (11 mm, previous 5 mm) and in metabolic activity SUVmax (5.58, previous 2.86 g/mL) as shown in the transaxial PET views (A), and CT views (B), compatible with a malignant neoplastic lesion in progression.

effect. In rectal cancer it is often difficult to evaluate radiotherapy response by anatomic imaging due to residual tissue mass, but 18F-FDG-PET/CT may detect residual tumor with an increase in metabolic activity.[46] In addition, [18F] FDG PET/CT has been proposed in the evaluation of local treatment response in liver and lung metastases by radiofrequency ablation (RFA).

New imaging techniques are being used in CRC, if available, such as PET/MRI. The biggest advantage of combined PET/MRI is expected in the assessment of therapeutic response and tumor relapse. MRI was shown to be reliable for the assessment of tumor-free distance to the mesorectal fascia, with reported sensitivity of 77% and specificity of 94%. MR DWI may predict the tumor clearance of the mesorectal fascia in locally advanced rectal cancer and was reported to provide an imaging biomarker for tumor invasiveness. Lower apparent diffusion coefficient values correlated significantly with more aggressive tumor profiles, including high grades, high frequency of lymph node metastases, and invasion of the mesorectal fascia. Residual tumor or locally recurrent tumors are sometimes difficult to identify on the basis of morphologic criteria, partially because of potential postsurgical changes or alterations related to chemoradiation that can lead to scar tissue or desmoplastic reactions. In these cases, metabolic information of 18F-FDG activity should be considered for correct restaging. It is also known that regeneration tissue and inflammation may present increased 18F-FDG uptake. This is a real limitation of [18F] FDG PET/CT in prediction of histopathologic tumor response after chemoradiation when scanning is performed shortly after therapy. PET/MRI integrates the advantages of MRI and [18F] FDG PET/CT and thus may evolve as the first-line restaging modality in CRC patients with suspected tumor relapse or newly developed metastases.[47]

New PET (non-FDG) radiotracers are also being developed. Choline is a marker of phospholipid synthesis. 11C-Choline has been considered as a potential PET radiopharmaceutical for tumor detection. 11C-Choline PET may be a reasonable approach for detection of CRC. Even though further studies are necessary, the authors hypothesized that 11C-Choline PET/CT might be useful for the detection of primary lesions. Because of the high level of physiologic 11C-Choline activity in the liver, visualization of hepatic metastasis is challenging. However, 11C-Choline PET might be useful for detecting other metastases of CRC such as brain, lung, and bone metastases, with less interference from background activity.[48]

In patients with unresectable liver metastases and/or advanced burden of liver disease, transarterial radioembolization with microspheres labeled with 90Y is becoming a valid therapeutic alternative to chemoembolization and RFA.[49]

References

1. NCCN Guidelines. *Colon Cancer Version 2.2021*; 2021. Available from: https://www.nccn.org/professionals/physician_gls [Accessed 21 January 2021].
2. Amin MB, Edge SB, Greene FL, et al. Cancer, AJCo. In: *AJCC cancer staging manual*; 2017.
3. Siegel RL, Miller KD, Fuchs HE, Jemal A. Cancer statistics, 2021. *CA Cancer J Clin* 2021;**71**(1):7–33. https://doi.org/10.3322/caac.21654.
4. Boellaard R, Delgado-Bolton R, Oyen WJG, et al. FDG PET/CT: EANM procedure guidelines for tumour imaging: version 2.0. *Eur J Nucl Med Mol Imaging* 2015;**42**(2):328–54. https://doi.org/10.1007/s00259-014-2961-x.
5. Lonneux M. FDG-PET and PET/CT in colorectal cancer. *PET Clin* 2008;**3**(2):147–53. https://doi.org/10.1016/j.cpet.2008.08.004.
6. Kantorová I, Lipská L, Bêlohlávek O, Visokai V, Trubač M, Schneiderová M. Routine 18F-FDG PET preoperative staging of colorectal cancer: comparison with conventional staging and its impact on treatment decision making. *J Nucl Med* 2003;**44**(11):1784–8. https://jnm.snmjournals.org/content/44/11/1784.short. [Accessed 22 February 2021].
7. Abdel-Nabi H, Doerr RJ, Lamonica DM, et al. Staging of primary colorectal carcinomas with fluorine-18 fluorodeoxyglucose whole-body PET: correlation with histopathologic and CT findings. *Radiology* 1998;**206**(3):755–60. https://doi.org/10.1148/radiology.206.3.9494497.
8. Yasuda S, Fujii H, Nakahara T, et al. 18F-FDG PET detection of colonic adenomas. *J Nucl Med* 2001;**42**(7):989–92. https://www.ncbi.nlm.nih.gov/pubmed/11438616.
9. Bipat S, Glas AS, Slors FJM, Zwinderman AH, Bossuyt PMM, Stoker J. Rectal cancer: local staging and assessment of lymph node involvement with endoluminal US, CT, and MR imaging—a meta-analysis. *Radiology* 2004;**232**(3):773–83. https://doi.org/10.1148/radiol.2323031368.
10. Kunawudhi A, Sereeborwornthanasak K, Promteangtrong C, Siripongpreeda B, Vanprom S, Chotipanich C. Value of FDG PET/contrast-enhanced CT in initial staging of colorectal cancer—comparison with contrast-enhanced CT. *Asian Pac J Cancer Prev* 2016;**17**(8):4071–5. https://www.ncbi.nlm.nih.gov/pubmed/27644663.
11. Klessen C, Rogalla P, Taupitz M. Local staging of rectal cancer: the current role of MRI. *Eur Radiol* 2007;**17**(2):379–89. https://doi.org/10.1007/s00330-006-0388-x.
12. Catalano OA, Coutinho AM, Sahani DV, et al. Colorectal cancer staging: comparison of whole-body PET/CT and PET/MR. *Abdom Radiol (NY)* 2017;**42**(4):1141–51. https://doi.org/10.1007/s00261-016-0985-3.
13. Lu Y-Y, Chen J-H, Ding H-J, Chien C-R, Lin W-Y, Kao C-H. A systematic review and meta-analysis of pretherapeutic lymph node staging of colorectal cancer by 18F-FDG PET or PET/CT. *Nucl Med Commun* 2012;**33**(11):1127. https://doi.org/10.1097/MNM.0b013e328357b2d9.

14. Fuster D, Lafuente S, Setoain X, et al. Dual-time point images of the liver with 18F-FDG PET/CT in suspected recurrence from colorectal cancer. *Rev Esp Med Nucl Imagen Mol* 2012;**31**(3):111–6. https://doi.org/10.1016/j.remnie.2012.05.005.
15. Yong TW, Yuan ZZ, Jun Z, Lin Z, He WZ, Juanqi Z. Sensitivity of PET/MR images in liver metastases from colorectal carcinoma. *Hell J Nucl Med* 2011;**14**(3):264–8. https://www.ncbi.nlm.nih.gov/pubmed/22087447.
16. Kijima S, Sasaki T, Nagata K, Utano K, Lefor AT, Sugimoto H. Preoperative evaluation of colorectal cancer using CT colonography, MRI, and PET/CT. *World J Gastroenterol* 2014;**20**(45):16964–75. https://doi.org/10.3748/wjg.v20.i45.16964.
17. Park IJ, Kim HC, Yu CS, et al. Efficacy of PET/CT in the accurate evaluation of primary colorectal carcinoma. *Eur J Surg Oncol* 2006;**32**(9):941–7. https://doi.org/10.1016/j.ejso.2006.05.019.
18. Llamas-Elvira JM, Rodríguez-Fernández A, Gutiérrez-Sáinz J, et al. Fluorine-18 fluorodeoxyglucose PET in the preoperative staging of colorectal cancer. *Eur J Nucl Med Mol Imaging* 2007;**34**(6):859–67. https://doi.org/10.1007/s00259-006-0274-4.
19. Valk PE, Abella-Columna E, Haseman MK, et al. Whole-body PET imaging with [18F] fluorodeoxyglucose in management of recurrent colorectal cancer. *Arch Surg* 1999;**134**(5):503–11. discussion 511-513 https://doi.org/10.1001/archsurg.134.5.503.
20. McCall JL, Black RB, Rich CA, et al. The value of serum carcinoembryonic antigen in predicting recurrent disease following curative resection of colorectal cancer. *Dis Colon Rectum* 1994;**37**(9):875–81. https://doi.org/10.1007/BF02052591.
21. Flamen P, Stroobants S, Van Cutsem E, et al. Additional value of whole-body positron emission tomography with fluorine-18-2-fluoro-2-deoxy-D-glucose in recurrent colorectal cancer. *J Clin Oncol* 1999;**17**(3):894–901. https://doi.org/10.1200/JCO.1999.17.3.894.
22. Flanagan FL, Dehdashti F, Ogunbiyi OA, Kodner IJ, Siegel BA. Utility of FDG-PET for investigating unexplained plasma CEA elevation in patients with colorectal cancer. *Ann Surg* 1998;**227**(3):319–23. https://doi.org/10.1097/00000658-199803000-00001.
23. Whiteford MH, Whiteford HM, Yee LF, et al. Usefulness of FDG-PET scan in the assessment of suspected metastatic or recurrent adenocarcinoma of the colon and rectum. *Dis Colon Rectum* 2000;**43**(6):759–67. discussion 767-770 https://doi.org/10.1007/BF02238010.
24. Fong Y, Saldinger PF, Akhurst T, et al. Utility of 18F-FDG positron emission tomography scanning on selection of patients for resection of hepatic colorectal metastases. *Am J Surg* 1999;**178**(4):282–7. https://doi.org/10.1016/s0002-9610(99)00187-7.
25. Kunikowska J, Królicki L. Clinical applications of PET/CT in oncology: digestive tract tumours. EANM. Part 2; 2008. p. 89–101 (chapter 7).
26. Scheele J, Stangl R, Altendorf-Hofmann A. Hepatic metastases from colorectal carcinoma: impact of surgical resection on the natural history. *Br J Surg* 1990;**77**(11):1241–6. https://doi.org/10.1002/bjs.1800771115.
27. Cabrera Villegas A, Gámez Cenzano C, Martín Urreta JC. Positron emission tomography (PET) in clinical oncology. Part II. *Rev Esp Med Nucl* 2002;**21**(2):131–47. quiz 149-151 https://doi.org/10.1016/s0212-6982(02)72051-x.
28. Hughes KS, Simon R, Songhorabodi S, et al. Resection of the liver for colorectal carcinoma metastases: a multi-institutional study of patterns of recurrence. *Surgery* 1986;**100**(2):278–84. https://www.ncbi.nlm.nih.gov/pubmed/3526605.
29. Chan K, Welch S, Walker-Dilks C, Raifu AO. PET imaging in colorectal cancer. Toronto (ON): Cancer Care Ontario; 2009 [updated 2010 Nov 30]. Program in Evidence-based Care Recommendation Report No.: 1 Version 2. *Citation: Chan K, Welch S, Walker-Dilks C, Raifu A Evidence-based guideline recommendations on the use of positron emission tomography imaging in colorectal cancer. Clin Oncol* 2011;**1**:2. https://doi.org/10.1016/j clon 2011 11 008. *Epub 2011 Dec 20* https://www.cancercareontario.ca/sites/ccocancercare/files/guidelines/full/pebcpet-1f_0.pdf.
30. Tan MCB, Linehan DC, Hawkins WG, Siegel BA, Strasberg SM. Chemotherapy-induced normalization of FDG uptake by colorectal liver metastases does not usually indicate complete pathologic response. *J Gastrointest Surg* 2007;**11**(9):1112–9. https://doi.org/10.1007/s11605-007-0218-8.
31. Truant S, Huglo D, Hebbar M, Ernst O, Steinling M, Pruvot FR. Prospective evaluation of the impact of [18F] fluoro-2-deoxy-D-glucose positron emission tomography of resectable colorectal liver metastases. *Br J Surg* 2005;**92**(3):362–9.
32. Desai DC, Zervos EE, Arnold MW, Burak WE, Mantil J, Martin EW. Positron emission tomography affects surgical Management in Recurrent Colorectal Cancer Patients. *Ann Surg Oncol* 2003;**10**(1):59–64. https://doi.org/10.1245/aso.2003.05.006.
33. Inoue M, Ohta M, Iuchi K, et al. Benefits of surgery for patients with pulmonary metastases from colorectal carcinoma. *Ann Thorac Surg* 2004;**78**(1):238–44. https://doi.org/10.1016/j.athoracsur.2004.02.017.
34. Onesti JK, Mascarenhas CR, Chung MH, Davis AT. Isolated metastasis of colon cancer to the scapula: is surgical resection warranted? *World J Surg Oncol* 2011;**9**(1):137. https://doi.org/10.1186/1477-7819-9-137.
35. Miyake KK, Nakamoto Y, Togashi K. Dual-time-point 18F-FDG PET/CT in patients with colorectal cancer: clinical value of early delayed scanning. *Ann Nucl Med* 2012;**26**(6):492–500. https://doi.org/10.1007/s12149-012-0599-y.
36. Kamel EM, Thumshirn M, Truninger K, et al. Significance of incidental 18F-FDG accumulations in the gastrointestinal tract in PET/CT: correlation with endoscopic and histopathologic results. *J Nucl Med* 2004;**45**(11):1804–10. https://www.ncbi.nlm.nih.gov/pubmed/15534047.
37. Fuertes J, Montagut C, Bullich S, et al. Incidental focal uptake in colorectal location on oncologic FDG PET and PET/CT studies: histopathological findings and clinical significances. *Rev Esp Med Nucl Imagen Mol* 2015;**34**(2):95–101. https://doi.org/10.1016/j.remn.2014.07.008.
38. Eisenhauer EA, Therasse P, Bogaerts J, et al. New response evaluation criteria in solid tumours: revised RECIST guideline (version 1.1). *Eur J Cancer* 2009;**45**(2):228–47. https://doi.org/10.1016/j.ejca.2008.10.026.
39. de Geus-Oei L-F, Vriens D, HWM vL, WTA vdG, WJG O. Monitoring and predicting response to therapy with 18F-FDG PET in colorectal cancer: a systematic review. *J Nucl Med* 2009;**50**(Suppl_1):43S–54S. https://doi.org/10.2967/jnumed.108.057224.
40. Capirci C, Rubello D, Chierichetti F, et al. Long-term prognostic value of 18F-FDG PET in patients with locally advanced rectal Cancer previously treated with neoadjuvant Radiochemotherapy. *Am J Roentgenol* 2006;**187**(2):W202–8. https://doi.org/10.2214/ajr.05.0902.
41. Mertens J, De Bruyne S, Van Damme N, et al. Standardized added metabolic activity (SAM) IN ^{18}F-FDG PET assessment of treatment response in colorectal liver metastases. *Eur J Nucl Med Mol Imaging* 2013;**40**(8):1214–22. https://doi.org/10.1007/s00259-013-2421-z.

42. Even-Sapir E, Parag Y, Lerman H, et al. Detection of recurrence in patients with rectal cancer: PET/CT after abdominoperineal or anterior resection. *Radiology* 2004;**232**(3):815–22. https://doi.org/10.1148/radiol.2323031065.
43. Kuehl H, Antoch G, Stergar H, et al. Comparison of FDG-PET, PET/CT and MRI for follow-up of colorectal liver metastases treated with radio-frequency ablation: initial results. *Eur J Radiol* 2008;**67**(2):362–71. https://doi.org/10.1016/j.ejrad.2007.11.017.
44. Lafuente S, Arguis P, Fuster D, Vilana R, Lomeña F, Pons F. Assessment of radiofrequency ablation of lung metastasis from colorectal cancer using dual time-point PET/CT. *Clin Nucl Med* 2011;**36**(7):603–5. https://doi.org/10.1097/rlu.0b013e318217741b.
45. Miller FH, Keppke AL, Reddy D, et al. Response of liver metastases after treatment with Yttrium-90 microspheres: role of size, necrosis, and PET. *Am J Roentgenol* 2007;**188**(3):776–83. https://doi.org/10.2214/ajr.06.0707.
46. Pelosi E, Deandreis D, Cassalia L, Penna D. Diagnostic applications of nuclear medicine: colorectal cancer. Nucl Oncol. Published online 2016:1-21. doi:https://doi.org/10.1007/978-3-319-26067-9_19-1.
47. Aktolun C, Goldsmith SJ. Nucl Oncol. Published online 2015. https://jnm.snmjournals.org/content/jnumed/56/9/1465.full-text.pdf.
48. Terauchi T, Tateishi U, Maeda T, et al. A case of colon cancer detected by Carbon-11 choline positron emission tomography/computed tomography: an initial report. *Jpn J Clin Oncol* 2007;**37**(10):797–800. https://doi.org/10.1093/jjco/hym102.
49. Shackett P. *Nuclear medicine technology: procedures and quick reference.* Lippincott Williams & Wilkins; 2000. https://play.google.com/store/books/details?id=rPp_zfv3eKEC.

Chapter 13

Magnetic resonance imaging (MRI) in staging and restaging after neoadjuvant therapy

Cristina Méndez Díaz, Esther Rodríguez García, and Rafaela Soler Fernández
Department of Radiology, A Coruña University Hospital Complex, A Coruña, Spain

Introduction

Rectal cancer (RC) is one of the neoplasms with the highest incidence in our environment and one of the oncological diseases in which advances in individualized treatment strategies, based on the characteristics of the patient and the tumor, have achieved an improvement in both local and systemic control of the disease.[1]

Correct preoperative staging helps to define prognosis, identify patients with disseminated disease and a higher risk of local recurrence, select candidates for preoperative neoadjuvant treatment, and plan surgical treatment appropriately.[2]

The factors that have the most influence on the decrease in local recurrence are accurate preoperative staging by imaging techniques, the introduction of the surgical concept of total excision of the mesorectum,[3] and the use of preoperative radiotherapy and chemoradiotherapy.

Endorectal ultrasound accurately defines the degree of penetration of the tumor into the thickness of the rectal wall, discriminating among T1, T2, and T3 tumor stages. However, in the case of T3 tumors, ultrasound cannot determine the distance of the tumor to the mesorectal fascia or accurately identify the mesorectal lymph nodes because the field of view is limited by the transducer.[2–4]

Magnetic resonance imaging (MRI) now plays a key role in the pre- and posttreatment assessment of CR, helping the multidisciplinary team to select the most appropriate treatment option.[4, 5]

The prognosis of RC is directly related to tumor infiltration of the mesorectum and the ability to surgically achieve negative circumferential resection margins and treatment with neoadjuvant chemotherapy and radiotherapy in patients with locally advanced rectal cancers,[1] diagnosed on the basis of MRI findings.

The main objective of MRI in RC is to define whether total excision of the mesorectum is possible or whether the tumor is advanced. Both factors are essential for the multidisciplinary team to define the most appropriate treatment, at the most appropriate time for the patients.

Total mesorectal excision

Total mesorectal excision (TME) consists of excision of the lymphatic tissue and perirectal fat including the lateral and circumferential margins.

The mesorectum is an anatomical term that refers to the surface of the surgically resected specimen and corresponds only to the part of the rectum not covered by peritoneum. The anterior peritoneal reflection (APR) delimits the transition between the peritoneum-lined and non-peritoneum-lined portions of the rectum (Fig. 13.1).

Based on this anatomical definition, the mesorectum is only circumferential for rectal tumors below the peritoneal reflection. In tumors of the upper rectum, the mesorectum is located posteriorly and in tumors of the upper-middle third, the mesorectum is posterolateral. For this reason, the concept of mesorectum should not be used in anterior and anterolateral tumors above the APR, where the rectum is covered by peritoneum.

As circumferential mesorectum depends on the extent of surgical resection, which cannot be predicted by MRI, the most appropriate term for MR staging of RC is mesorectal fascia (MRF). Similar to circumferential mesorectum, MRF is

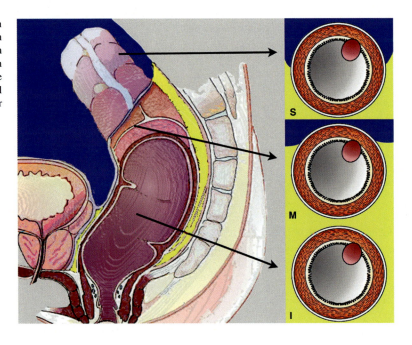

FIG. 13.1 Diagram of the relationship between the rectum and the mesorectum, and the anterior peritoneal reflection (lateral and axial views). The anterior peritoneal reflection covers the anterior and lateral wall of the superior rectum and the anterior wall of the superior portion of the middle rectum. The lower rectum is entirely covered by perirectal fat. S = superior rectum, M = middle rectum, I = inferior rectum.

circumferential only in rectal tumors below the APR and is not included in the anterior and anterolateral rectal surface above the APR not covered by peritoneum.

Currently, TME is the technique of choice in RC surgery. In clinical practice, TME is advised for middle and lower third tumors and removal of the mesorectum with a 5 cm margin distal to the tumor. In lower third tumors, surgery is accepted from an oncological point of view if 1–2 cm of rectum between the caudal end of the tumor and the sphincter complex remains free, as infiltration in the RC occurs mainly circumferentially and not longitudinally.[6]

The success of TME depends on staging, tumor extension, and the minimum distance between the tumor and the MRF.

Key anatomical references

The rectum is approximately 15 cm long from the anal margin and is traditionally divided into upper (10–15 cm), middle (5–10 cm), and lower (<5 cm).[1] This classification is the one commonly used for digital rectal examination and endoscopy. However, the anal margin (junction of perianal skin with hair and perianal skin without hair) is obviously not a reference point in MRI studies. Instead, it is more accurate to measure the distance from the puborectalis muscle, which is identified on sagittal T2-weighted images as a muscular thickening behind the anorectal angulation (Fig. 13.2).

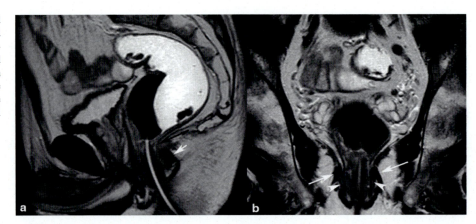

FIG. 13.2 Puborectalis muscle and anorectal angulation. Sagittal (A) and coronal (B) TSE-T2 images show the relationship among a cancer in the lower third of the rectum and the anorectal angulation (*short arrow*) and the puborectalis muscle (*long arrows*) that continues with the levator ani muscle and the external anal sphincter (*white arrowheads*).

Mesorectal fascia and anterior peritoneal reflection

The mesorectal fascia and the anterior peritoneal reflection are two of the most important structures for proper staging of RC.

The MRF represents the visceral fascia of the extraperitoneal portion of the rectum and envelops the mesorectal fat. On MRI, the mesorectum is seen as a hyperintense structure on T2-weighted sequences surrounding the rectum, bounded by a hypointense line corresponding to the MRF and corresponding to the surgical margin of circumferential resection when TME is performed (Fig. 13.3).[5–7]

The intraperitoneal and extraperitoneal portions of the rectum are separated by the APR. The APR is identified on most rectal MRI scans on sagittal T2-weighted slices as a thin hypointense line extending from the upper wall of the bladder (men) (Fig. 13.4A) or uterus (women) (Fig. 13.4B) to the anterior wall of the rectum. On axial images, the APR adopts a "V" shape (Fig. 13.3).[1, 8, 9]

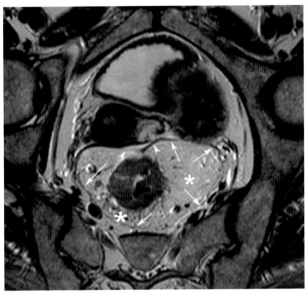

FIG. 13.3 Mesorectal fascia and anterior peritoneal reflection. Oblique axial TSE-T2 image perpendicular to a cancer in the middle third of the rectum. The hypointense line of mesorectal fascia (*arrows*) completely surrounds the fat of the mesorectum (*asterisks*). The *arrowheads* point to the anterior peritoneal reflection that joins the anterior wall of the rectum with a "V" morphology.

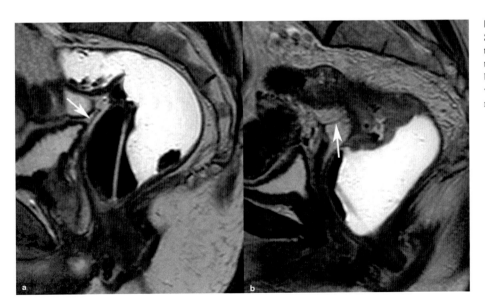

FIG. 13.4 Anterior peritoneal reflection. Sagittal TSE-T2 images show the hypointense line of the anterior peritoneal reflection (arrows) from the upper wall of the bladder in a man (A) and the uterus in a woman (B) to the anterior wall of the rectum.

From an oncological point of view, rectal cancer is cancer affecting the extraperitoneal rectum, which lies below the peritoneal reflection, while neoplasms affecting the intraperitoneal rectum behave in the same way as those of the left colon. The location of a tumor above or below the APR is probably the most important anatomical reference that helps the surgeon to choose the type of intervention.[8, 9]

Upper rectum

The anterior and lateral wall of the upper rectum is covered by the APR (Fig. 13.1). In this location, there is a higher risk of peritoneal perforation during surgery and a higher incidence of transcellular tumor spread.[1] For more advanced tumors treated with radiation, effective radiotherapy would be difficult and potentially morbid, because the tumor may be in a mobile mesentery, surrounded by small bowel, exposing the patient to increased morbidity due to the possibility of actinic enteritis.[10]

Middle rectum

The middle third of the rectum is located below the peritoneal reflection, is completely surrounded by the mesorectum, and is therefore the ideal segment for TME. The uppermost portion of the middle third of the rectum is covered anteriorly by the peritoneum (Fig. 13.1).[7]

Lower rectum

The lower rectum is entirely extraperitoneal and is clinically defined as the portion between 0 and 5 cm from the anal margin (Fig. 13.1). In this segment, the mesorectum thins progressively caudally until it merges with the mesorectal fascia at the upper end of the internal sphincter, making surgical approach difficult.[8]

Tumors of the lower rectum are classified according to MRI into two categories based on the relationship with the superior border of the puborectalis muscle. This muscle appears as a muscular thickening on either side of the levator ani muscle in the coronal plane (Fig. 13.2B) and behind the anorectal angle in the sagittal plane (Fig. 13.2A). The puborectalis muscle is continued by the external anal sphincter which surrounds the internal sphincter and is continued by the wall of the rectum (Fig. 13.2B).[7, 8]

The length of the normal rectum above the junction of the levator ani muscle with the puborectalis muscle is the key to determining whether the sphincters can be preserved during surgery. A resection margin distal to the RC of more than 2 cm is considered optimal to avoid tumor recurrence.[7, 11] In tumors below the superior border of the puborectalis muscle, abdominoperineal amputation is necessary.

The sagittal and coronal planes are the most useful to show the relationship among tumor, levator ani muscle, and anal sphincter, and allow measurement of the distance between the lower margin of the carcinoma and the upper margin of the external sphincter, which is essential in sphincter-sparing surgery with adequate tumor margins.[8]

Lymph nodes

The lymphatic drainage of the middle and upper third of the rectum is via the mesorectal nodes and the inferior mesenteric chain, while the distal third drains together with the middle rectus vessels to the internal iliac chain. Distal tumors infiltrating the anal canal also drain to the inguinal lymph nodes.[3, 7]

Local staging

The TNM classification, described by the International Union Against Cancer (IUAC) and the American Joint Committee on Cancer (AJCC), is currently used for staging of RC.[12]

T-staging

On MRI, the mucosal and muscularis propria layers are hypointense, while the submucosa and perirectal fat are hyperintense. The wall of the rectum often shows focal interruptions on its surface, corresponding to vessels penetrating its wall, which should not be confused with tumor.[5, 7, 11]

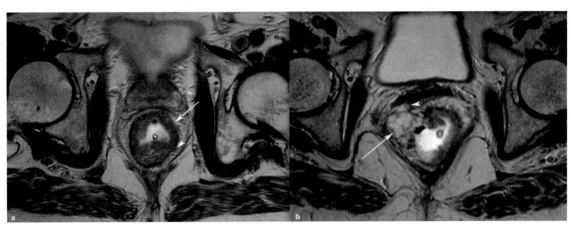

FIG. 13.5 Tumor signal. (A) Axial TSE-T2 image of an undifferentiated adenocarcinoma in the lower third of the rectum. The hyposignal of the muscularis is replaced by the hyperintense tumor reaching the surface of the rectal wall (*arrow*). The *arrowhead* points to small vessels penetrating the rectal wall. (B) Axial TSE-T2 image of the middle third of the distended rectum with ultrasound gel via the rectal route shows a very hyperintense mucinous differentiated tumor (*arrow*) contacting the anterior mesorectal fascia (*arrowhead*).

RC is usually more hypointense than submucosa, but more hyperintense than muscle (Fig. 13.5A). An exception is tumors with mucinous differentiation, which may show varying degrees of high signal on T2-weighted sequences (Fig. 13.5B).[5, 11]

T1 and T2 tumors

T1 tumors invade the submucosa and T2 tumors extend into the muscularis propria (Fig. 13.6A).[13] Low-grade T1 tumors with <1 mm of submucosal invasion have virtually no risk of lymph node metastasis.[13] This means that transanal endoscopic microsurgery can provide similar results to TME in these patients, with greatly reduced morbidity and mortality.

The specificity of MRI to discriminate between T1 and T2 tumors is low (69%), except in some patients with T1 tumors when it is possible to identify a preserved submucosal layer (hyperintense signal) beneath the lesion. In the remaining patients with T1 stage tumors who are to undergo local excision, endorectal ultrasound is the technique of choice for preoperative staging because its specificity (86%) is superior to that of MRI.[3]

A tumor is defined on MRI as T2 stage when the hyposignal of part or all the muscle layer thickness is replaced by the intermediate or hyperintense signal of the tumor (Fig. 13.6B).

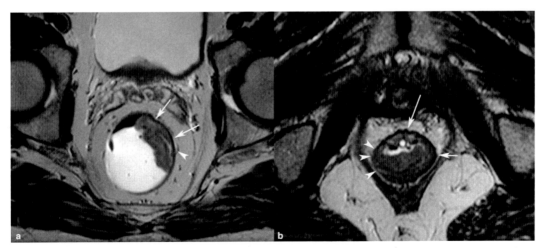

FIG. 13.6 Difference between stage T1 and T2 rectal cancer. Axial TSE-T2 images of the lower third of the rectum of stage T1 (A) and T2 (B) rectal carcinoma. In stage T1, the hyperintense tumor is surrounded by undamaged hypointense muscle (*arrows*) with a small interruption in its surface corresponding to vessels penetrating the wall (*arrowhead*). In stage T2, the hyperintense tumor invades part of the muscle layer (*arrowheads*) which loses its normal hypointense signal (*arrows*).

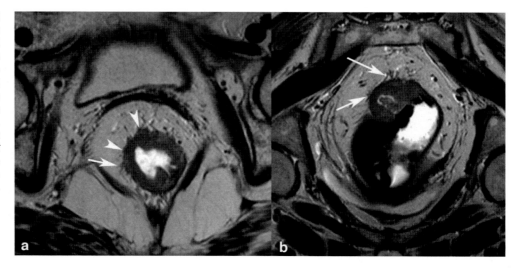

FIG. 13.7 Stage T3 (A) and T2 rectal cancer with fine spicules on its surface (B). TSE-T2 axial images. In stage T3, the outline of the tumor is nodular (*arrowheads*) and invades the perirectal fat and there are hypointense fine spicules extending from the rectal wall into the fat (*arrow*). In stage T2, the tumor does not invade the hypointense line of muscle and the hypointense spicules extending into the perirectal fat (*arrows*) represent peritumoral fibrous reaction.

T3 tumors

On MRI, protrusion or nodular configuration of the tumor beyond the muscle contour is defined as a T3 stage (Fig. 13.7A).[13] The diagnostic accuracy of MRI in defining a T3 stage tumor varies in the literature from 71% to 94%.[4]

The decision to include wall contour spiculation in the tumor area as a T3 stage is controversial in the literature because it is often caused by fibrosis or desmoplastic reaction (Fig. 13.7B). Most failures in MRI T-staging occur because of the difficulty in distinguishing between minimal T2 and minimal T3, i.e., between isolated fibrosis (T2) or fibrosis containing tumor cells (T3).[5]

In the assessment of stage T3, the most important parameters are the depth of extramural tumor extension and the minimum distance between the tumor or metastatic lymph nodes and the MRF (Fig. 13.8). Survival of patients with T3 tumors and extramural extension on MRI greater than 5 mm is lower than that of patients with extramural tumor extension equal to or less than 5 mm. The benefit of preoperative neoadjuvant treatment is related to extramural tumor extension of less than 5 mm.[14] Extramural tumor extension is a parameter that should be included in the MRI report.

The current criterion used in the literature as a free circumferential resection margin is the distance between the tumor and the MRF of at least 1 mm.[12–15] This measure is essential to classify cases based on the potential risk of recurrence after surgery and should always be defined, even if there is no transmural infiltration of the rectum.

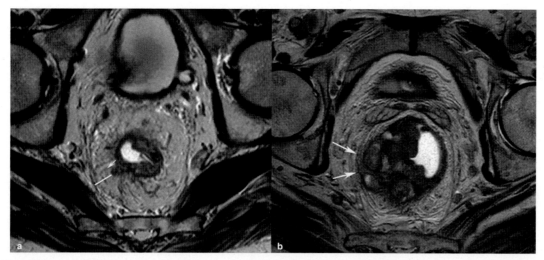

FIG. 13.8 Extramural extension of the tumor and distance to the mesorectal fascia. (A) Axial TSE-T2 image of a tumor with free circumferential resection margin. The tumor invades the mesorectal fat (*arrow*) and the distance to the mesorectal fascia (*white line*) is greater than 5 mm. (B) In the axial TSE-T2 image of a carcinoma with invaded circumferential resection margin, the tumor contacts the mesorectal fascia (*arrows*).

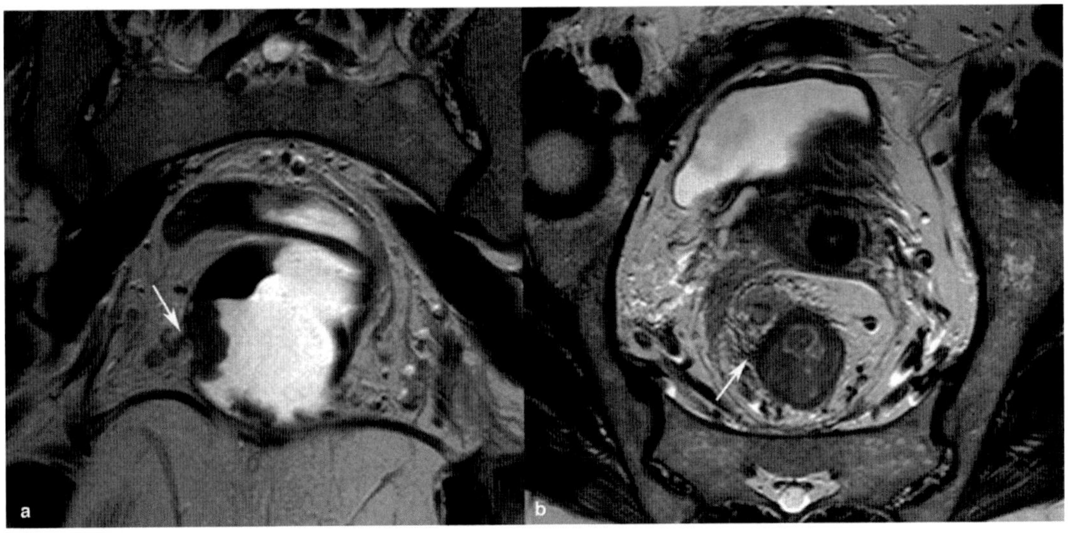

FIG. 13.9 Infiltration of extramural vessels. Coronal (A) and axial oblique (B) TSE-T2 images of a stage T3 carcinoma. The *arrows* show the expansion of a superficial vein infiltrated by the tumor.

The accuracy of MRI in defining the circumferential resection margin is greater than 13, 92.5%.[16, 17] In the anterior plane, only the circumferential margin below the peritoneal reflection is considered circumferential.[2, 6]

The usefulness of MRI in the evaluation of the circumferential resection margin may be limited in tumors located in the anterior wall of the middle third of the rectum and in those of the lower third, since in these locations the mesorectal fat is scarce and the rectum is remarkably close to the mesorectal fascia, which results in poor visualization of the latter. Even in T2 tumors located in these regions and affecting the full thickness of the rectal wall, even if they do not infiltrate the perirectal fat, they may be very close to the mesorectal fascia and will often have a threatened radial margin.[13, 15–17]

Invasion of extramural veins is determined by the presence of tumor cells in the endothelium of the vessels outside the muscular layer of the rectal wall. Their presence raises suspicion of locally advanced tumor (T3/T4) and distant disease.[13, 14, 18] Although vascular invasion does not influence therapeutic decisions, it is one of the prognostic factors in RC. MRI diagnosis is obvious when the vein penetrating the wall in the area of tumor infiltration is expanded, and its contours are nodular (Fig. 13.9). These findings have a diagnostic sensitivity and specificity of 62% and 88%, respectively.[18]

T4 tumors

Stage T4 implies that the tumor has either overgrown the mesorectal fascia, extends to the peritoneal surface (stage T4a), or infiltrates adjacent organs or structures (stage T4b).[13, 14]

Tumor extension to the peritoneal surface is identified by an anterior nodular infiltration through the peritoneal reflection, at or above its attachment to the anterior aspect of the rectum (Fig. 13.10).[14, 19]

N-staging

Preoperative assessment of lymph node status in patients with RC is important because both the absence and number of infiltrated nodes are independent predictors of patient survival, and because the presence of tumor nodes near the MRF increases the risk of local and systemic recurrence.[5, 17, 20]

None of the currently available imaging techniques allow accurate preoperative detection of lymph node involvement. Endorectal ultrasound is not reliable in detection because it cannot analyze lymph nodes located outside the field of view of the transducer. Despite the false negatives and positives of MRI, this technique is currently considered the most effective for locating lymph nodes within the mesorectum, particularly in relation to MRF.[17, 20]

The diagnostic accuracy of MRI in characterizing lymph nodes varies in the literature from 47% to 89.5%.[20] This variability is due to the different criteria used for diagnosis.

The size criterion alone is unreliable for discriminating between benign and malignant lymph nodes. In RC, nodes with a short axis smaller than 5 mm may be infiltrated and larger nodes may be reactive.[20] Using a nodal size of 5 mm (short axis) as a criterion, the sensitivity and specificity of MRI for predicting nodal infiltration are 66% and 76%, respectively.[21]

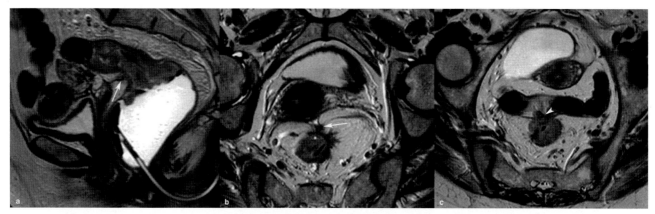

FIG. 13.10 Stage T4 rectal cancer. Sagittal (A) and oblique axial (B, Cc) TSE-T2 images. The tumor extends through the muscularis and invades the anterior peritoneal reflection (*arrows*). The uppermost portion of the tumor infiltrates the wall of the ileum (*arrowhead*).

FIG. 13.11 Carcinoma in the lower third of the rectum stage T3 with suspicious mesorectal adenopathies. Axial oblique (A) and sagittal (B) TSE-T2 images show tumor extension into the perirectal fat, infiltration of the external anal sphincter, and an adenopathy in the perirectal fat larger than 5 mm with irregular contour and heterogeneous signal probably infiltrated by tumor (*arrows*).

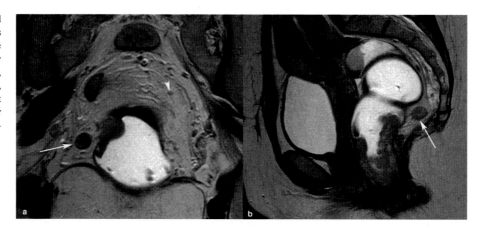

Irregular contour and heterogeneous signal (Fig. 13.11) in the node have been shown to be better predictors of nodal infiltration, with sensitivity and specificity figures of 85% and 97%, respectively.[21]

The European Society of Gastrointestinal and Abdominal Radiology (ESGAR) has proposed a consensus guideline to analyze the status of mesorectal and extramesorectal lymph nodes based on a combination of size and morphological criteria. In this guideline, nodes with a short axis greater than or equal to 9 mm are considered malignant; if the short axis is between 5 and 8 mm, the node must have at least two morphological criteria to be considered malignant (round morphology, irregular contour, heterogeneous signal); for nodes smaller than 5 mm, only those with all three morphological criteria (round morphology, irregular contour, heterogeneous signal) are considered malignant. All nodes of a mucinous tumor are considered malignant, regardless of their size.[22]

When infiltrating nodes are suspected, it is important to analyze the number to differentiate between N1 and N2 stages.[12] It is important to remember that infiltrating nodes are usually located at the same height or above the tumor, as the lymphatic drainage of the tumor is cranial.

Based on TNM staging criteria,[12] internal iliac lymph nodes are considered regional lymph nodes, while external iliac lymph nodes, obturator, and retroperitoneal nodes are considered nonregional lymph nodes and therefore compatible with distant disease M (metastases).

Restaging after neoadjuvant treatment

Restaging after neoadjuvant treatment is usually done 8–10 weeks after completion of the last course of treatment and is performed on endoscopy and MRI.

MRI restaging after neoadjuvant treatment may modify surgical planning, especially in cases of complete tumor response and reevaluation of the circumferential resection margin.[23]

T-restaging

The ability of imaging techniques in general, and MRI in particular, in restaging RC after neoadjuvant treatment is limited. The diagnostic accuracy of this technique in T-restaging is 50%[23] and in predicting circumferential resection margin involvement is 77%.[17, 24]

The major limitation of MRI in T-restaging is due to overstaging of T2 stage tumors. Complete disappearance of the tumor and hypointense thickening of the rectal wall with or without hypointense spiculations extending into the perirectal fat is classified as stage T2-T0 (Fig. 13.12).[25] The fibrosis present after neoadjuvant treatment causes thickening of the rectal wall in most cases, making it difficult to differentiate residual tumor from fibrosis, desmoplastic reaction, or inflammation.[25, 26]

Understaging may be due to the absence of tumor on MRI, which does not always represent tumor stage T0, or due to the presence of viable tumor nests within fibrotic areas, which are impossible to identify.[5] Understaging could be a problem when considering a change in the initial surgical strategy for some patients, especially when the tumor response is apparently complete.

Nodular growth of intermediate or hyperintense signal, with invasion of perirectal fat or adjacent organs or structures, is classified as residual tumor T3 or T4, respectively (Fig. 13.13) (Table 13.1). Thin, hypointense spiculations extending from the tumor into the mesorectal fat or adjacent organs and structures correspond to fibrosis and represent tumor response (Fig. 13.13B).[25]

Some tumors treated with neoadjuvant therapy may appear very hyperintense on T2-weighted sequences because they respond by increasing mucin production.[25, 26] This mucin response (Fig. 13.14)[13] cannot be differentiated from inflammatory changes associated with residual tumor.[27] Furthermore, this high signal should not be confused with the high signal of untreated mucinous tumors (Fig. 13.5B).[25]

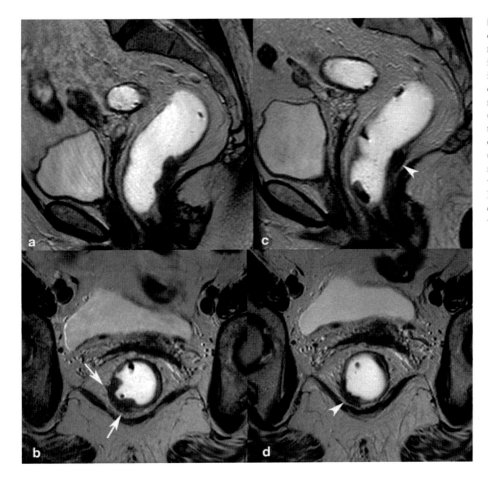

FIG. 13.12 Carcinoma in the lower third of the rectum stage T3 with complete response after neoadjuvant treatment. Sagittal (A) and oblique axial (B) TSE-T2 images before neoadjuvant radiotherapy of a carcinoma in the lower third of the rectum invading the perirectal fat (*arrows*). Images obtained after radiotherapy (C, D) show the decrease in size of the lesion showing hypointense signal (*arrowheads*) and small hypointense spicules in the perirectal fat. The tumor was classified on MRI as stage T2-T0. Histological study after total mesorectal excision showed intense fibrosis with no viable tumor cells.

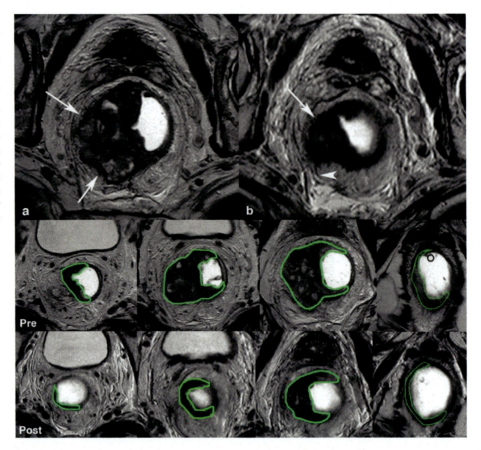

FIG. 13.13 Carcinoma in the middle third of the rectum stage T3 with good response after neoadjuvant treatment. Axial TSE-T2 image obtained before neoadjuvant treatment (A) shows a carcinoma in the middle third of the rectum (*arrows*) reaching the mesorectal fascia. The axial TSE-T2 image obtained after neoadjuvant treatment (B) shows the decrease in size and signal of the tumor (*arrow*) and hypointense fine spicules extending from the periphery of the tumor (*arrowhead*). Volumetric images obtained before (pre) and after (post) neoadjuvant treatment show a 70% decrease in tumor volume, indicating a good response to treatment.

TABLE 13.1 Morphological criteria for TNM restaging by MRI.[25]

Criteria	Description
Viable tumor	Intermediate signal larger than muscle
Tumor response	Decrease or increase in signal intensity compared to the MRI study before neoadjuvant therapy
T2 or less (T0-T2)	Normal rectal wall. Hypointense thickening of the rectal wall with or without spiculations in perirectal fat
T3	Nodular tumor with extension into mesorectal fat, larger than muscle signal
T4	Nodular tumor with extension to adjacent organs, signal larger than muscle

Volumetric analysis by manual tumor volume contouring provides more realistic and reproducible data than conventional measurements in tumor restaging. A tumor volume reduction of 65% represents a partial response (Fig. 13.13).[28] The correlation between MRI and histological findings of tumor volume and shrinkage after neoadjuvant treatment is good, although it does not allow differentiation between complete histological regression and residual tumor.[23] Furthermore, volumetric analysis is time consuming and may therefore be difficult to implement routinely in clinical practice. As a practical alternative, the most recent ESGAR guidelines suggest using tumor length before and after neoadjuvant treatment as a reproducible quantification to estimate the change in tumor size–tumor volume after treatment.[22]

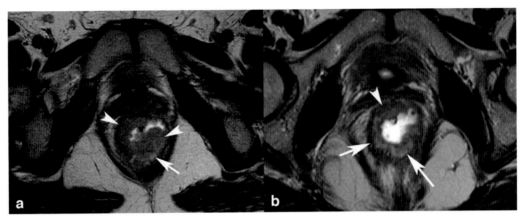

FIG. 13.14 Carcinoma in the lower third of the rectum stage T3 with moderate response and mucinous areas after neoadjuvant therapy. (A) Axial TSE-T2 image before neoadjuvant radiotherapy shows the tumor with hyperintense signal (*arrowheads*) with some small hyperintense foci (*arrow*). (b) Axial TSE-T2 image obtained after radiotherapy shows acellular areas of highly hyperintense mucin (*arrows*) and hyperintense areas of residual tumor (*arrowhead*).

Some authors state that the circumferential resection margin is more effectively predicted in patients who have not received neoadjuvant or only a short course, whereas if they receive a long course, posttreatment fibrosis prevents a correct assessment.[24, 29] However, others[30] conclude that all patients with mesorectal fat infiltration greater than 5 mm should be identified because they have a much worse prognosis than infiltration less than 5 mm. A resection margin less than or equal to 2 mm is associated with tumor infiltration of the MRF in 90% of cases treated with neoadjuvant therapy and should be considered as a potentially invaded circumferential resection margin.[24–26, 30]

Extramural vascular invasion, which is usually associated with stage T3 or T4, may disappear completely with neoadjuvant treatment. Identification of linear or spiculated structures in the extramural veins penetrating the rectal wall signifies a good response to treatment.[26]

Several studies have demonstrated the ability of diffusion-enhanced sequences in the assessment of therapeutic response. A significant increase in the ADC map value of the tumor has been demonstrated (Fig. 13.15), which can appear

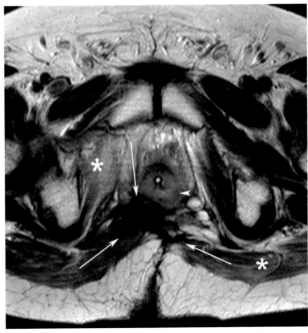

FIG. 13.15 Postradiotherapy fibrosis. Axial TSE-T2 image of the lower third of the rectum demonstrates a mass with angulated contours and homogeneous hypointense signal (*arrows*) representing postradiotherapy fibrosis and high signal in the rectal wall (*arrowhead*) and in the right obturator internus and left gluteus muscles due to secondary edema (*asterisks*).

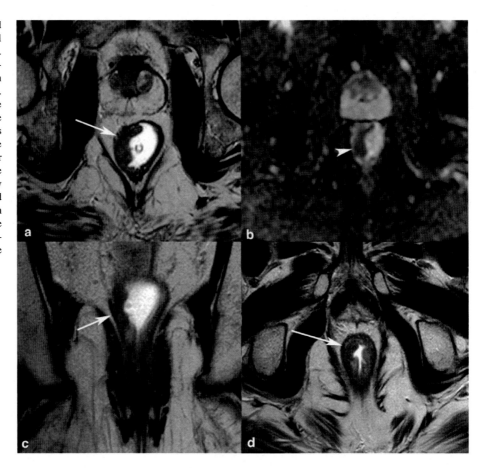

FIG. 13.16 Carcinoma in the lower third of the rectum with complete histological response after neoadjuvant treatment. (A) Axial TSE-T2 image before neoadjuvant treatment demonstrates a tumor in the lower third of the rectum (*arrows*). (B) Diffusion-enhanced image shows the marked hyposignal of the tumor on the ADC map. In the axial TSE-T2 images (C) and in the ADC map of the diffusion-weighted sequence (D) after neoadjuvant treatment, the hypointense thickening of the rectal wall secondary to posttreatment fibrosis (*arrows*) and the increased signal of the tumor bed in the ADC map (*arrowhead*) can be observed. Histological study of the surgical specimen demonstrated a complete response.

as early as 1 week after the start of treatment and is more durable after radiotherapy (due to persistent edema). The response to antiangiogenic drugs, on the other hand, causes transient decreases in these values, secondary to decreased flow, cellular edema, and reduction of the extracellular space.[15]

A low pretreatment ADC map value correlates with a good posttreatment response (Fig. 13.16). This could be explained by the fact that tumors with a high ADC value often have necrosis and necrosis is associated with a poor response to treatment. The possibility of early detection of nonresponders may allow for treatment intensification or a change in the initial treatment decision.[31]

There is growing evidence in the literature suggesting that the addition of diffusion-weighted to T2-weighted imaging (Fig. 13.17) improves diagnostic accuracy to 80%–85% in cases of complete response, with a prediction of absence of mesorectal fascia infiltration of 89%–93% with the added value of excellent interobserver correlation.[25, 32, 33]

N-restaging

Neoadjuvant treatment reduces the size and number of both benign and malignant. The diagnostic accuracy of MRI in N-restaging is 65% using size as a criterion and 85% using irregular contour and heterogeneous signal as criteria.[23, 26]

Some authors state that even though diffusion-enhanced sequences help to detect the number of nodes, they do not differentiate between benign and malignant nodes and do not improve the efficacy of MRI in nodal restaging.[22, 34] However, most authors agree that the absence of signal in the nodes on diffusion sequences has a high negative predictive value for differentiating infiltrated nodes from fibrosis, improving diagnostic accuracy in restaging.[23, 32]

In N-restaging, MRI is more accurate than in N-staging. After neoadjuvant treatment, most lymph nodes decrease in size and approximately 44% of nodes smaller than 4mm disappear. In the absence of mesorectal and extramesorectal lymph nodes at restaging with diffusion sequences, MRI has a high negative predictive value.[32]

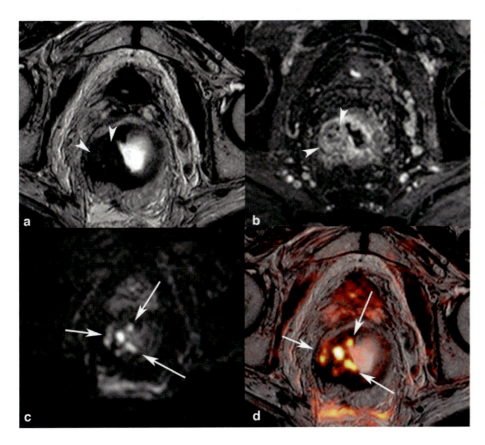

FIG. 13.17 Carcinoma in the middle third of the rectum stage T3 with good response after neoadjuvant treatment. Axial TSE-T2 (A) and perfusion (B) images show the tumor bed with hyperintense areas (*arrowheads*) of residual tumor. The diffusion-enhanced image with b-factor of 1000 (C) and the false color map obtained from the fusion of the diffusion image ($b=1000$) and TSE-T2 (D) more easily demonstrate the areas of residual tumor (*arrows*) and their relationship with the mesorectal fascia.

The structured radiology report

One of the key aspects of RC treatment planning is the clear and accurate communication of preoperative staging provided by the radiologist to the multidisciplinary team. There is often great variability between radiological reports issued in different hospitals, and even between radiologists in the same department, so they need to be properly structured and standarized.[35]

A standardized radiology report is important because it ensures uniform language and improves the clarity and accuracy of what the radiologist communicates to the multidisciplinary team. The ultimate goal is that patients receive the right diagnostic information and the right treatment at the right time, with minimal negative consequences and maximum effectiveness.[36]

The structured RC report should include the location and extent of the tumor, T-staging, presence of suspicious nodes, and response to neoadjuvant treatment.

Location

The location of the tumor and its cranio-caudal length is especially important for surgeons because it helps to decide the type of surgical intervention.

In all cases the distance from the lower edge of the tumor to the anal margin should be described for upper and middle third tumors. In lower third tumors, it is important to define the relationship of the lower edge of the tumor to the upper edge of the puborectalis muscle.

Morphology

Differentiating between infiltrative and polypoid morphology is important because polypoid tumors are usually less invasive, extend into the rectal wall through their pedicle, and may require local resection.

T-staging

In all cases it is essential to describe the T stage because it conditions the treatment. In cases where a T stage cannot be assigned with certainty, it is advisable to describe a range of T categories (e.g., superficial T2-T3). The accuracy of T-staging is limited by the MRI technology itself, so describing a range of categories simply emphasizes that there is some uncertainty in the diagnosis and that the multidisciplinary team must take this into account when making treatment decisions.

To improve the consistency of the MRI report, it is recommended that spiculations of mesorectal fat be described as superficial T2-T3.

In stage T4 tumors, it is important to describe the organs or structures involved, because they may change the management of the patient.

Extramural extension of the tumor

The extramural extent of the tumor is described for stages T3 and T4 and should be measured from the edge of the tumor excluding spiculations in the perirectal fat.

For stage T1 and T2 tumors, the extramural extent of the tumor should be defined as 0 mm.

Distance to mesorectal fascia

The minimum distance between the tumor and the MRF should be described for all stage T2 tumors or higher. If it is not possible to clearly see the MRF, the minimum distance should be defined as not assessable.

For tumors located above the APR and involving the peritonealized rectal wall, this distance should be described as not applicable.

For tumors invading adjacent structures (stage T4), the distance to the MRF should be described as 0 mm.

The minimum distance from the spiculations adjacent to the tumor to the MRF should be described separately from the minimum distance from the tumor.

In anterior tumors close to the APR, the tumor may have a T3 component above the APR and a T2 component below. In these cases, the minimum distance to the MRF of the most penetrating part of the tumor or T3 component above the APR should be described as not applicable. However, the minimum distance between the MRF and the T2 component below the APR should be accurately measured because tumors in this location will often have a threatened radial margin even if they do not infiltrate the perirectal fat. In these cases, despite the T2 stage, preoperative neoadjuvant treatment is often considered.

Adenopathies and mesorectal tumor deposits

Any mesorectal lymph node or tumor deposit with an irregular border, heterogeneous signal, and/or short axis larger than 5 mm should be described as suspicious.

Most infiltrating nodes are usually located close to the tumor. Although infiltrating nodes caudal to the tumor are rare, they may condition the extent of the field of radiotherapy or surgery. It is therefore important to define the location of lymph nodes above, at the same level or below the tumor.

Extramesorectal adenopathies

Extramesorectal nodes with irregular border, heterogeneous signal, and/or short axis greater than 10 mm should be described as suspicious.

Restaging after neoadjuvant treatment

After neoadjuvant treatment in patients with locally advanced RC, MRI is useful to verify the effectiveness of the treatment after having considered what the tumor was like previously in order not to confuse posttreatment changes with tumor persistence, as well as colloid degeneration with a mucinous tumor.

The presence of mucin in tumors is seen as an area of high signal on T2-weighted sequences. After chemotherapy, colloid or mucinous degeneration may occur in nonmucinous tumors, demonstrating a response to treatment and a better

TABLE 13.2 Grades of tumor regression by MRI.[22]

Criteria	Description
Full answer	Completely normal rectal wall, or minimal linear or crescentic scarring or no alteration in signal intensity at the initial tumor site
Almost complete answer	Rectal wall thickening without restriction in diffusion at the initial tumor site
Tumor persistence	Residual tumor with diffusion restriction at the initial tumor site

prognosis. A mucinous tumor that was already mucinous at initial staging and shows no significant response after chemotherapy is associated with a higher risk of local recurrence and worse prognosis.

It is therefore important to visualize both the tests performed before and after chemotherapy treatment to differentiate colloid degeneration from a mucinous tumor.

T-restaging after neoadjuvant treatment is analyzed according to the criteria described in Table 13.1.

The degree of tumor regression is quantified according to the morphological response and changes in signal intensity in diffusion sequences (Table 13.2). In the literature, 5 groups are considered, which, for practical purposes, in the ESGAR consensus are grouped into 3: complete response, near complete response, and tumor persistence.[22]

Locoregional recurrence

The incidence of locoregional recurrence of rectal cancer has decreased since the introduction of TME and the use of preoperative neoadjuvant treatment. Most recurrences occur in the first 2 years after surgery for the primary tumor.

A recent multicenter study by the MERCURY group shows that the accuracy of MRI in assessing the circumferential resection margin is similar to histological accuracy. In this study, they conclude that preoperative assessment of the circumferential resection margin status by MRI is superior to the criteria used in the TNM classification for predicting the risk of recurrence and overall and disease-free survival of patients.[37]

Many patients are asymptomatic and locoregional recurrence is detected in routine postoperative follow-up studies. Although pelvic MRI is not used in the routine follow-up of rectal cancer patients, this technique can help confirm the suspicions of locoregional recurrence.[38]

The most accurate sign on MRI to differentiate fibrosis from recurrence is the morphology of the lesion margins. A mass with angulated contours, homogeneous and hypointense signal on T2-weighted sequences, and with less than 40% enhancement on perfusion studies is diagnostic of fibrosis, regardless of the time elapsed since surgery (Fig. 13.15).[39]

The accuracy of MRI in the diagnosis of locoregional recurrence depends on the time that has elapsed since surgery. When the time since surgery is greater than 1 year, the detection of a mass of nodular morphology with hyperintense areas on T2-weighted sequences and enhancement greater than 40% on intravenous contrast perfusion study (Fig. 13.18)

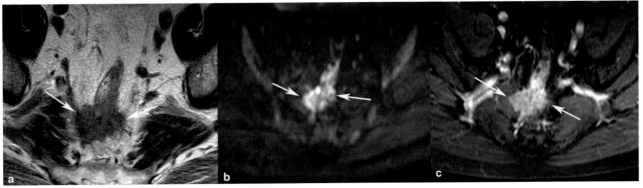

FIG. 13.18 Locoregional recurrence of a carcinoma in the upper third of the rectum. Axial TSE-T2 images (A), enhanced in diffusion with b-factor 1000 (B) and perfusion (C). Tumor with infiltrative and nodular contours (*arrows*) invading the sigma, with heterogeneous hyperintense signal in the T2-weighted sequences, with restriction in the diffusion-weighted sequences and with enhancement greater than 40% in the perfusion study.

has a sensitivity of 100%, a specificity of 85%, and an accuracy of 92% in the diagnosis of recurrence.[39,40] However, these findings are not specific and can be observed in immature fibrosis, when the time since surgery is less than 1 year.[13,39–41] Diffusion-enhanced sequences help to detect recurrences with more confidence.[41]

References

1. Salerno G, Daniels IR, Moran BJ, Wotherspoon A, Brown G. Clarifying margins in the multidisciplinary management of rectal cancer: the MERCURY experience. *Clin Radiol* 2006;**61**(11):916–23. https://doi.org/10.1016/j.crad.2006.06.005.
2. Engstrom PF, Arnoletti JP, Benson 3rd AB, et al. NCCN clinical practice guidelines in oncology: rectal cancer. *J Natl Compr Canc Netw* 2009;**7**(8):838–81. https://doi.org/10.6004/jnccn.2009.0057.
3. Bipat S, Glas AS, Slors FJM, Zwinderman AH, Bossuyt PMM, Stoker J. Rectal cancer: local staging and assessment of lymph node involvement with endoluminal US, CT, and MR imaging—a meta-analysis. *Radiology* 2004;**232**(3):773–83. https://doi.org/10.1148/radiol.2323031368.
4. Bianchi PP, Ceriani C, Rottoli M, et al. Endoscopic ultrasonography and magnetic resonance in preoperative staging of rectal cancer: comparison with histologic findings. *J Gastrointest Surg* 2005;**9**(9):1222–7. discussion 1227-8 https://doi.org/10.1016/j.gassur.2005.07.008.
5. Beets-Tan RGH, Beets GL. Rectal cancer: review with emphasis on MR imaging. *Radiology* 2004;**232**(2):335–46. https://doi.org/10.1148/radiol.2322021326.
6. Van Cutsem E, Dicato M, Haustermans K, et al. The diagnosis and management of rectal cancer: expert discussion and recommendations derived from the 9th World Congress on Gastrointestinal Cancer, Barcelona, 2007. *Ann Oncol* 2008;**19**:vi1–8. https://www.academia.edu/download/48193863/vi1.pdf.
7. Brown G, Kirkham A, Williams GT, et al. High-resolution MRI of the anatomy important in total mesorectal excision of the rectum. *AJR Am J Roentgenol* 2004;**182**(2):431–9. https://doi.org/10.2214/ajr.182.2.1820431.
8. Dujovny N, Quiros RM, Saclarides TJ. Anorectal anatomy and embryology. *Surg Oncol Clin N Am* 2004;**13**(2):277–93. https://doi.org/10.1016/j.soc.2004.01.002.
9. Gollub MJ, Maas M, Weiser M, et al. Recognition of the anterior peritoneal reflection at rectal MRI. *AJR Am J Roentgenol* 2013;**200**(1):97–101. https://doi.org/10.2214/AJR.11.7602.
10. Birgisson H, Påhlman L, Gunnarsson U, Glimelius B. Swedish rectal cancer trial group. Adverse effects of preoperative radiation therapy for rectal cancer: long-term follow-up of the Swedish rectal Cancer trial. *J Clin Oncol* 2005;**23**(34):8697–705. https://doi.org/10.1200/JCO.2005.02.9017.
11. Kim CK, Kim SH, Chun HK, et al. Preoperative staging of rectal cancer: accuracy of 3-tesla magnetic resonance imaging. *Eur Radiol* 2006;**16**(5):972–80. https://doi.org/10.1007/s00330-005-0084-2.
12. Amin M, Edge S, Greene F, et al. *AJCC cancer staging manual*. 8th ed. Springer; 2017.
13. Horvat N, Carlos Tavares Rocha C, Clemente Oliveira B, Petkovska I, Gollub MJ. MRI of rectal cancer: tumor staging, imaging techniques, and management. *Radiographics* 2019;**39**(2):367–87. https://doi.org/10.1148/rg.2019180114.
14. Curvo-Semedo L. Rectal cancer: staging. *Magn Reson Imaging Clin N Am* 2020;**28**(1):105–15. https://doi.org/10.1016/j.mric.2019.09.003.
15. Taylor FGM, Quirke P, Heald RJ, et al. Preoperative high-resolution magnetic resonance imaging can identify good prognosis stage I, II, and III rectal cancer best managed by surgery alone: a prospective, multicenter, European study. *Ann Surg* 2011;**253**(4):711–9. https://doi.org/10.1097/SLA.0b013e31820b8d52.
16. Strassburg J. Magnetic resonance imaging in rectal cancer: the MERCURY experience. *Tech Coloproctol* 2004;**8**(S1):s16–8. https://doi.org/10.1007/s10151-004-0100-6.
17. MERCURY Study Group. Diagnostic accuracy of preoperative magnetic resonance imaging in predicting curative resection of rectal cancer: prospective observational study. *BMJ* 2006;**333**(7572):779. https://doi.org/10.1136/bmj.38937.646400.55.
18. Smith NJ, Shihab O, Arnaout A, Swift RI, Brown G. MRI for detection of extramural vascular invasion in rectal cancer. *AJR Am J Roentgenol* 2008;**191**(5):1517–22. https://doi.org/10.2214/AJR.08.1298.
19. Brown G, Radcliffe AG, Newcombe RG, Dallimore NS, Bourne MW, Williams GT. Preoperative assessment of prognostic factors in rectal cancer using high-resolution magnetic resonance imaging. *Br J Surg* 2003;**90**(3):355–64. https://doi.org/10.1002/bjs.4034.
20. Koh DM, Brown G, Husband JE. Nodal staging in rectal cancer. *Abdom Imaging* 2006;**31**(6):652–9. https://doi.org/10.1007/s00261-006-9021-3.
21. Brown G, Richards CJ, Bourne MW, et al. Morphologic predictors of lymph node status in rectal cancer with use of high-spatial-resolution MR imaging with histopathologic comparison. *Radiology* 2003;**227**(2):371–7. https://doi.org/10.1148/radiol.2272011747.
22. Beets-Tan RGH, Lambregts DMJ, Maas M, et al. Magnetic resonance imaging for clinical management of rectal cancer: updated recommendations from the 2016 European Society of Gastrointestinal and Abdominal Radiology (ESGAR) consensus meeting. *Eur Radiol* 2018;**28**(4):1465–75. https://doi.org/10.1007/s00330-017-5026-2.
23. Sclafani F, Brown G, Cunningham D, et al. Comparison between MRI and pathology in the assessment of tumour regression grade in rectal cancer. *Br J Cancer* 2017;**117**(10):1478–85. https://doi.org/10.1038/bjc.2017.320.
24. Vliegen RFA, Beets GL, Lammering G, et al. Mesorectal fascia invasion after neoadjuvant chemotherapy and radiation therapy for locally advanced rectal cancer: accuracy of MR imaging for prediction. *Radiology* 2008;**246**(2):454–62. https://doi.org/10.1148/radiol.2462070042.
25. Kalisz KR, Enzerra MD, Paspulati RM. MRI evaluation of the response of rectal cancer to neoadjuvant chemoradiation therapy. *Radiographics* 2019;**39**(2):538–56. https://doi.org/10.1148/rg.2019180075.
26. Patel UB, Blomqvist LK, Taylor F, et al. MRI after treatment of locally advanced rectal cancer: how to report tumor response—the MERCURY experience. *AJR Am J Roentgenol* 2012;**199**(4):W486–95. https://doi.org/10.2214/AJR.11.8210.

27. Allen SD, Padhani AR, Dzik-Jurasz AS, Glynne-Jones R. Rectal carcinoma: MRI with histologic correlation before and after chemoradiation therapy. *AJR Am J Roentgenol* 2007;**188**(2):442–51. https://doi.org/10.2214/AJR.05.1967.
28. Therasse P, Arbuck SG, Eisenhauer EA, et al. New guidelines to evaluate the response to treatment in solid tumors. European Organization for Research and Treatment of Cancer, National Cancer Institute of the United States, National Cancer Institute of Canada. *J Natl Cancer Inst* 2000;**92**(3):205–16. https://doi.org/10.1093/jnci/92.3.205.
29. Kulkarni T, Gollins S, Maw A, Hobson P, Byrne R, Widdowson D. Magnetic resonance imaging in rectal cancer downstaged using neoadjuvant chemoradiation: accuracy of prediction of tumour stage and circumferential resection margin status. *Colorectal Dis* 2008;**10**(5):479–89. https://doi.org/10.1111/j.1463-1318.2007.01451.x.
30. MERCURY Study Group. Extramural depth of tumor invasion at thin-section MR in patients with rectal cancer: results of the MERCURY study. *Radiology* 2007;**243**(1):132–9. https://doi.org/10.1148/radiol.2431051825.
31. Kim SH, Lee JY, Lee JM, Han JK, Choi BI. Apparent diffusion coefficient for evaluating tumour response to neoadjuvant chemoradiation therapy for locally advanced rectal cancer. *Eur Radiol* 2011;**21**(5):987–95. https://doi.org/10.1007/s00330-010-1989-y.
32. Maas M, Dijkhoff RAP, Beets-Tan R. Rectal cancer: assessing response to neoadjuvant therapy. *Magn Reson Imaging Clin N Am* 2020;**28**(1):117–26. https://doi.org/10.1016/j.mric.2019.09.004.
33. Engin G, Sharifov R, Güral Z, et al. Can diffusion-weighted MRI determine complete responders after neoadjuvant chemoradiation for locally advanced rectal cancer? *Diagn Interv Radiol* 2012;**18**(6):574–81. https://doi.org/10.4261/1305-3825.DIR.5755-12.1.
34. Lambregts DMJ, Maas M, Riedl RG, et al. Value of ADC measurements for nodal staging after chemoradiation in locally advanced rectal cancer-a per lesion validation study. *Eur Radiol* 2011;**21**(2):265–73. https://doi.org/10.1007/s00330-010-1937-x.
35. Schwartz LH, Panicek DM, Berk AR, Li Y, Hricak H. Improving communication of diagnostic radiology findings through structured reporting. *Radiology* 2011;**260**(1):174–81. https://doi.org/10.1148/radiol.11101913.
36. Mirnezami R, Nicholson J, Darzi A. Preparing for precision medicine. *N Engl J Med* 2012;**366**(6):489–91. https://doi.org/10.1056/NEJMp1114866.
37. Sinaei M, Swallow C, Milot L, Moghaddam PA, Smith A, Atri M. Patterns and signal intensity characteristics of pelvic recurrence of rectal cancer at MR imaging. *Radiographics* 2013;**33**(5):E171–87. https://doi.org/10.1148/rg.335115170.
38. Schaefer O, Langer M. Detection of recurrent rectal cancer with CT, MRI and PET/CT. *Eur Radiol* 2007;**17**(8):2044–54. https://doi.org/10.1007/s00330-007-0613-2.
39. Markus J, Morrissey B, de Gara C, Tarulli G. MRI of recurrent rectosigmoid carcinoma. *Abdom Imaging* 1997;**22**(3):338–42. https://doi.org/10.1007/s002619900203.
40. Kinkel K, Tardivon AA, Soyer P, et al. Dynamic contrast-enhanced subtraction versus T2-weighted spin-echo MR imaging in the follow-up of colorectal neoplasm: a prospective study of 41 patients. *Radiology* 1996;**200**(2):453–8. https://doi.org/10.1148/radiology.200.2.8685341.
41. Lambregts DMJ, Cappendijk VC, Maas M, Beets GL, Beets-Tan RGH. Value of MRI and diffusion-weighted MRI for the diagnosis of locally recurrent rectal cancer. *Eur Radiol* 2011;**21**(6):1250–8. https://doi.org/10.1007/s00330-010-2052-8.

Chapter 14

Histopathological diagnosis of CRC

Álvaro Gómez Castro[a] and Héctor Lázare Iglesias[b]
[a]Head of the Pathology Department, Pontevedra Hospital Complex, Pontevedra, Spain, [b]Specialist in Anatomic Pathology, Santiago de Compostela University Hospital Complex, Santiago de Compostela, Spain

Introduction

Histopathology is the gold standard test in the diagnostic process of colorectal cancer.[1] It is a process that is routinely performed in all Anatomic Pathology departments worldwide. It can be performed preoperatively by taking biopsies from a colonoscopy, or once the patient has undergone surgery with bowel resection.

Histopathological diagnosis of CRC in biopsies and endoscopic polypectomies

The most common entry procedure for histopathological diagnosis is endoscopic biopsy or polypectomy from colonoscopy (Fig. 14.1). The primary role of the pathologist is to prepare a histopathological report that has a dual purpose: to provide a basic description of the specimen being submitted and to draw a final conclusion based on the pathological findings for diagnostic purposes. When the situation so requires, it should also guide the clinician to adopt specific therapeutic attitudes (malignant polyps requiring complete excision, or assessment of prognostic factors or behavior of the neoplasm).

There are a number of general concepts to consider in the endoscopic diagnosis of colorectal cancer. The accepted consensus is that neoplasms that infiltrate at least the submucosa are referred to as adenocarcinoma.[2] Intramucosal or intraepithelial carcinomas may be termed high-grade dysplasia, given their virtually zero risk of metastasis. The distinction between high-grade dysplasia and intramucosal adenocarcinoma is actually based on the fact that the latter infiltrates the lamina propria or muscularis of the mucosa, which can sometimes be difficult to distinguish histologically. In any case, unlike in other locations in the gastrointestinal tract, in the colon, infiltration of the lamina propria is biologically equivalent to high-grade dysplasia, so this distinction is not as important.

In order to diagnose invasive carcinoma in endoscopic biopsies, infiltration of the submucosal tissue should ideally be observed, but in most cases, it is not possible to obtain samples where the submucosa is clearly identified. Nevertheless, the presence of atypical glands and loose cells associated with desmoplastic stroma almost invariably indicates the presence of an underlying invasive carcinoma.[3]

Most primary colorectal carcinomas are well or moderately differentiated (low-grade) adenocarcinomas, so confirmatory immunohistochemistry is not necessary. However, for those cases where it is necessary, these tumors almost always show immunohistochemical positivity for cytokeratin 20, CDX2, and SATB2, with negative staining for cytokeratin 7. In fact, SATB2 positivity in combination with cytokeratin 20 identifies more than 95% of colorectal carcinomas.[4]

An anatomopathological report should be issued following the recommendations of the AJCC[5] (American Joint Committee on Cancer), **8th edition**. The variables that refer to **excisional biopsy (polypectomy)** samples are summarized as follows:

— **Location:** cecum/right colon (ascending)/hepatic angle/transverse colon/left colon (descending)/sigmoid colon (sigma)/rectum/other locations/unspecified
— **Completeness of sample:** intact (complete)/fragmented
— **Size of polyp**: largest diameter (cm)/other measurements (cm)/cannot be measured/comments
— **Polyp configuration**: pedunculated/pedicle size (cm)/sessile
— **Size of infiltrating carcinoma:** area of greatest invasion (cm)/cannot be determined
— **Histological type:** adenocarcinoma/mucinous adenocarcinoma/signet ring cell carcinoma/medullary carcinoma/micropapillary carcinoma/serrated adenocarcinoma/large cell neuroendocrine carcinoma/small cell neuroendocrine carcinoma/neuroendocrine carcinoma (poorly differentiated)/squamous cell carcinoma/adenosquamous

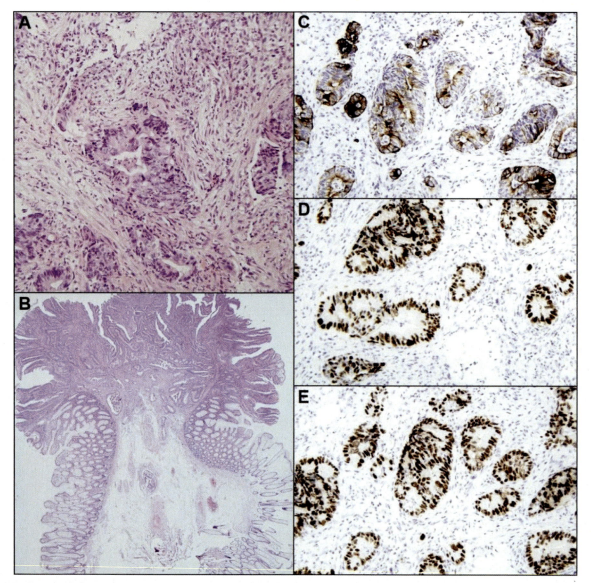

FIG. 14.1 Colorectal adenocarcinoma. Digestive endoscopic biopsy of an ulcerated mass, showing an atypical glandular proliferation associated with a desmoplastic stromal response (A). Endoscopic excision of a pedunculated polyp showing adenocarcinoma over tubulovillous adenoma infiltrating the submucosa (B). Typical immunophenotype of intestinal adenocarcinoma with cytoplasmic positivity for CK20 (C) and nuclear positivity for CDX2 (D) and SATB2 (E).

carcinoma/spindle cell carcinoma/mixed adenoneuroendocrine carcinoma/undifferentiated carcinoma/other histological types (specify)/cannot be determined/not applicable
- **Histological grade**: G1: well differentiated/G2: moderately differentiated/G3: poorly differentiated/G4: undifferentiated/other (specify)/cannot be determined/not applicable
- **Level of infiltration**: lamina propria/muscularis mucosae/submucosa/muscularis propria/cannot be determined
- **Deep margin (pedicle margin):** cannot be determined/not affected/affected by carcinoma (delimit distance)
- **Mucosal margins (lateral margins):** cannot be determined/not affected/affected by carcinoma/affected by adenoma
- **Lymphovascular invasion:** not identified/present (small vessel or large vessel)/unable to determine
- **Tumor budding** (number of tumor buds in 1 hotspot field): low score (0–4)/intermediate score[5–9]/high score (10 or more)/cannot be determined
- **Type of polyp in which the carcinoma originates:** tubular adenoma/villous adenoma/tubulovillous adenoma/traditional serrated adenoma/sessile serrated adenoma/hamartomatous polyp/other (specify)

- **Additional pathological findings:** not identified/ulcerative colitis/Crohn's disease/other (specify)
- **Complementary studies:** immunohistochemistry of DNA repair proteins/KRAS/NRAS/BRAF/PIK3CA/PTEN

In patients with adenocarcinoma on a polyp, "high risk of residual disease" is considered: the presence of poorly differentiated or high-grade carcinoma, tumor in contact with or within 1 mm of the resection border of the polypectomy, incomplete endoscopic excision (punch resection), lymphovascular invasion present, sessile polyp (not pedunculated), and budding tumor present.[6]

To assign a level of adenocarcinoma invasion within the polyp, one of the most widely used systems corresponds to that described by Haggitt et al..[7] In this proposal, four levels of submucosal infiltration are distinguished. Level 1 corresponds to invasive adenocarcinoma limited to the head of the polyp, level 2 implies involvement of the neck of the polyp, level 3 indicates infiltration of the polyp stalk, and level 4 indicates invasion of the submucosa at the level of the adjacent intestinal wall. This system is especially useful for pedunculated polyps, since an adenocarcinoma on a sessile polyp is by definition assigned level 4 and other systems have been designed for these particular cases, such as those proposed by Kikuchi et al.[8] or Kitajima et al.[9] which are based on the depth of submucosal invasion (sm).[10]

Anatomopathological diagnosis of CRC in surgical specimens

The pathology report is the final product of the diagnostic process. It must be drawn up in accordance with the three basic quality standards: safety, speed, and complete information.[11] In addition to stating all the essential findings that characterize the process under study, it is necessary to make reference to prognostic variables, therapeutic indications, and predictive factors for both response to treatment and risk of recurrence.

Standardization[12] is an important quality factor since it makes the report reproducible, homologous, homogeneous, and more objective by eliminating variables susceptible to individual interpretation. In this sense, the anatomopathological report protocols are of great help in the homogenization and standardization process, contributing to the improvement of the quality of the report.

For the preparation of the report, as mentioned before, it should be issued following the recommendations of the **AJCC**,[5] **8th edition**. The variables that refer to samples corresponding to **surgical specimens of resection of the colon and rectum** (Fig. 14.2) are summarized as follows:

- **Procedure:** right hemicolectomy/transverse colectomy/left hemicolectomy/sigmoidectomy/lower anterior resection/total abdominal colectomy/abdomino-perineal resection/transanal disc excision/other (specify)/no referenced procedure
- **Tumor localization:** cecum/right colon (ascending)/hepatic angle/transverse colon/splenic angle/left colon (descending)/sigmoid colon (sigmoid)/rectum-sigmoid junction/rectum/ileocecal valve/unspecified region/unidentifiable
- **Tumor location (only applicable to rectal primaries):** completely above the anterior peritoneal reflection/completely below the anterior peritoneal reflection/above the anterior peritoneal reflection/unspecified
- **Size of the lump:** largest size (cm)/other measurements/cannot be determined
- **Macroscopic perforation:** unidentified/present/cannot be determined
- **Macroscopic assessment of the mesorectum:** complete/nearly complete/incomplete/cannot be ascertained
- **Histological type:** adenocarcinoma/mucinous adenocarcinoma/signet ring cell carcinoma medullary carcinoma/micropapillary carcinoma/serrated adenocarcinoma/large cell neuroendocrine carcinoma/small cell neuroendocrine carcinoma/poorly differentiated neuroendocrine carcinoma/squamous cell (epidermoid) carcinoma/adenosquamous carcinoma/undifferentiated carcinoma/other (specify)/carcinoma of undetermined type
- **Histological grade:** G1: well differentiated/G2: moderately differentiated/G3: poorly differentiated/G4: undifferentiated/other (specify)/cannot be determined/not applicable
- **Tumor extension:** no tumor observed/no invasion observed (high-grade dysplasia)/tumor invades lamina propria or muscularis mucosae (intramucosal adenocarcinoma)/tumor invades submucosa/tumor invades muscularis propria/tumor invades pericolorectal tissues/tumor invades visceral peritoneum/tumor directly invades adjacent structures (specify)/cannot be determined
- **Proximal margin:** cannot be determined/not affected by invasive carcinoma (distance from the tumor to the margin in mm or cm)/affected by invasive carcinoma
- **Distal margin:** cannot be determined/not affected by invasive carcinoma (distance from tumor to margin in mm or cm)/affected by invasive carcinoma

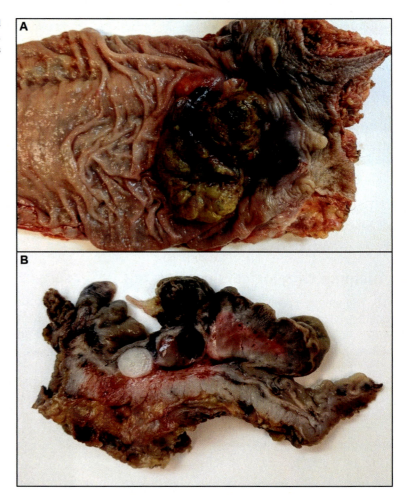

FIG. 14.2 Surgical resection specimen. (A) Mameloned exophytic mass in the rectum, close to the pectineal line. (B) Cross-section showing the lesion and the different layers that make up the intestinal wall.

- **Radial or mesenteric margin** (for rectum only): cannot be determined/not affected by invasive carcinoma (distance from tumor to margin in mm or cm)/affected by invasive carcinoma (tumor present within 0–1 mm of margin)
- **Noninvasive tumor status at margins:** affected by intramucosal adenocarcinoma/affected by high-grade dysplasia/affected by adenoma
- **Other margins (specify):** cannot be ascertained/not affected by invasive carcinoma/affected by invasive carcinoma
- **Response to neoadjuvant therapy:** unknown/absence of residual neoplastic cells (complete response, tumor regression grade 0)/loose cells or occasional small clusters of neoplastic cells (near complete response, tumor regression grade 1)/residual neoplasm with evidence of tumor regression, but more than single cells or occasional small clusters of neoplastic cells (partial response, tumor regression grade 2)/extensive residual neoplasm with no evidence of tumor regression (poor or no response, tumor regression grade 3)/cannot be determined
- **Lymphovascular invasion:** not identified/present (small vessel or large vessel)/unable to determine
- **Perineural invasion:** not identified/present/cannot be determined
- **Tumor budding** (number of tumor buds in 1 hotspot field): low score (0–4)/intermediate score (5–9)/high score (10 or more)/cannot be determined
- **Type of polyp in which the carcinoma originates:** tubular adenoma/villous adenoma/tubulovillous adenoma/traditional serrated adenoma/sessile serrated adenoma/hamartomatous polyp/other (specify)
- **Tumor pathological stage: primary tumor (pT):** pTX: cannot be determined/pT0: no primary tumor recognized/pTis: carcinoma in situ, intramucosal carcinoma (infiltration of the lamina propria without extension through the muscularis mucosae)/pT1: tumor invades submucosa/pT2: tumor invades muscularis propria/pT3: tumor invades through muscularis propria into pericolorectal tissues/pT4a: tumor invades through visceral peritoneum/pT4b: tumor invades directly or adheres to neighboring organs or structures

- **Tumor pathological stage: regional lymph nodes (pN):** pNX: cannot be determined/pN0: no lymph node metastases observed/pN1a: metastases in 1 lymph node/pN1b: metastases in 2–3 lymph nodes/pN1c: no lymph node metastases observed, but there are tumor deposits in the subserosa, mesentery, pericolonic nonperitonealized tissues or perirectal/mesorectal tissues/pN2a: metastases in 4–6 lymph nodes/pN2b: metastases in 7 or more lymph nodes
- **Tumor pathological stage: distant metastasis (pM):** pM1a: metastasis in one organ, without evidence of peritoneal metastasis/pM1b: metastasis in two organs, without evidence of peritoneal metastasis/pM1c: metastasis on the peritoneal surface alone or accompanied by metastasis to other organs
- **Additional pathological findings:** not identified/ulcerative colitis/Crohn's disease/other (specify)

Histopathological types of colorectal neoplasms

The different types and subtypes of colorectal neoplasms are described as follows, analyzing adenocarcinoma in greater detail, which is the most common and the subject of this chapter, without forgetting that we must also be aware of other types of neoplasms in order to make a proper differential diagnosis.

1. Adenocarcinomas: the majority of malignant colorectal tumors are adenocarcinomas. Adenocarcinoma is defined as a malignant epithelial tumor that has penetrated the muscularis mucosae and infiltrates the submucosa. Three grades are distinguished based on glandular differentiation percentage: well differentiated, moderately differentiated and poorly differentiated. To simplify, two grades can be established: low grade, which combines well and moderately differentiated forms, and high grade, corresponding to poorly differentiated forms.

Although most cases are diagnosed as adenocarcinoma NOS, several histological subtypes with characteristic clinical and molecular features can be distinguished, most of which were included in the latest **WHO (World Health Organization) classification of tumors of the digestive system in 2019.**[13] The most relevant ones are briefly outlined as follows:

Mucinous adenocarcinoma: Accounts for approximately 15% of colorectal carcinomas.[14] It requires at least 50% of the neoplasm to be composed of extracellular mucin lakes with the presence of malignant epithelial structures.

Signet ring cell carcinoma: For a tumor to be called signet ring carcinoma, more than 50% of the tumor cells must consist of intracytoplasmic mucin, which often displaces the nucleus toward the periphery. These are particularly aggressive tumors that usually occur in advanced stages.

Medullary carcinoma: A rare variant of colorectal carcinoma consisting of sheets of large cells with vesicular nuclei, prominent nucleoli, and abundant eosinophilic cytoplasm,[15] often showing a conspicuous peritumoral lymphocytic infiltrate.

Serrated adenocarcinoma: This histological subtype is so named because of its histological resemblance to serrated polyps, showing glandular serration, which may be accompanied by areas of mucinous component.

Micropapillary adenocarcinoma: This is a rare form of adenocarcinoma characterized by the presence of small aggregates of malignant cells in relation to lacunar spaces simulating vascular channels. This subtype is often associated with unfavorable prognostic factors such as lymphovascular invasion.

Although rare, there are also variants of colorectal carcinoma that exhibit components of sarcomatoid habitus, as well as morphologically, immunohistochemically and molecularly undifferentiated forms, which generally entail a worse prognosis.

2. Neuroendocrine neoplasms: Neuroendocrine neoplasms of the colon and rectum constitute a group of epithelial tumors with neuroendocrine differentiation that include well-differentiated neuroendocrine tumors (NETs), which generally have a good prognosis; poorly differentiated neuroendocrine carcinomas (NECs), which can be large or small cell and have a worse prognosis; and finally, mixed tumors. These types of neoplasms can be recognized not only by their morphology, but also because they often have a positive immunophenotypic profile for neuroendocrine markers such as synaptophysin, chromogranin, and CD56.

3. Lymphomas: Lymphomas are more frequent in the small intestine than in the colon. Primitive colorectal lymphomas account for only 0.2% of all colorectal malignancies, although their frequency increases when analyzed in the context of patients with immunodeficiency. Primary colorectal lymphomas are mostly non-Hodgkin B lymphomas, of which diffuse large B-cell lymphoma (60%) is the most frequent in the colon, followed by MALT lymphoma (15%) and Burkitt lymphoma (15%), with other variants being much less frequent.[16] They may offer differential diagnosis with other tumors, although they are generally recognized by their immunohistochemical positivity for pan-lymphoid markers such as CD45 and absence of expression for cytokeratins.

4. Mesenchymal tumors: There are essentially three types of primary malignant mesenchymal tumors of the colorectum: gastrointestinal stromal tumors (GIST), leiomyosarcomas, and Kaposi's sarcoma. GIST presumably originate from specialized cells of Cajal found in the colorectal wall. They are composed of predominantly spindle cellularity, although epithelioid forms exist. They frequently show immunohistochemical positivity for Vimentin, CD34, DOG-1, and CD117. At the molecular level, they show mutations specific to different exons of the c-kit gene and can be treated with tyrosine kinase inhibitors (Glivec). Most malignant colorectal mesenchymal tumors belong to this category. Leiomyosarcomas are malignant mesenchymal tumors with muscle differentiation and, therefore, usually have immunostaining for smooth muscle actin and desmin. Kaposi's sarcoma is a relatively frequent tumor in patients with acquired immunodeficiency syndrome[17] but is very rare outside this context.

5. Melanomas: It is important to bear in mind that most melanomas diagnosed in the gastrointestinal tract are metastases of cutaneous or ocular primaries, although the anorectal region is the gastrointestinal tract site where primary mucosal melanomas originate most frequently. These tumors can offer a broad histological differential diagnosis, although they are often distinguished by their characteristic immunohistochemical profile with positivity for S100 and more specific melanocytic markers such as Melan A and HMB 45.

6. Metastatic tumors: Colorectal metastases of primary carcinomas from other sites represent a rare and often unexpected histological finding in a biopsy. They usually involve the submucosa, with the mucosa normally remaining intact and often nonulcerated. Such infiltration often occurs in the context of patients with a known primary tumor, most commonly lung, ovarian, breast, prostate, kidney, skin, stomach, or hepatobiliary.[18]

Biomarkers in CRC

Colorectal carcinoma is the third most common cause of cancer-related death worldwide, and therefore the development of new diagnostic and prognostic biomarkers is of vital importance in order to prevent deaths associated with this neoplasm. Colorectal carcinoma is a sporadic tumor in the vast majority of cases (75%–80%), although genetic mutations such as BRAF, KRAS, P53, and microsatellite instability, as well as epigenetic alterations, such as DNA methylation in promoter regions with CpG islands, may play a key role in cancer development.[19]

Some of the most important findings in the treatment of colorectal carcinoma include the possibility of detecting the absence of mutations in the KRAS, NRAS, BRAF, and PIK3CA genes, with the intention of administering effective anti-EGFR therapies such as Cetuximab and Panitumumab. In addition, patients with this type of neoplasm may also benefit from microsatellite instability studies and detection of loss of heterozygosity of chromosome 18q, which are useful when making decisions regarding therapy with 5-fluorouracil.

Some 7.5% of colorectal carcinomas in Spain present alterations in the DNA mismatch repair (MMR) system. Although most cases are sporadic, about 20% develop in patients with hereditary nonpolyposis colorectal carcinoma (HNPCC) or Lynch syndrome. Clinical, molecular, histopathological, and immunohistochemical criteria are used to identify these patients.[20]

Among the clinical criteria, the Amsterdam criteria published in 1990, which have been extended over time, are particularly well known. From an anatomopathological perspective, it is also important to be familiar with the Bethesda criteria, whose recommendations include immunohistochemistry for DNA repair proteins (MLH1, PMS2, MSH2, and MSH6) for histological types frequently associated with high microsatellite instability, such as those with lymphocytic infiltration, mucinous differentiation, signet ring cells, or medullary growth pattern. Therefore the immunohistochemical study performed by the pathologist, in cases that require it, allows the search to be directed to the affected gene, thus increasing the chances of detecting Lynch syndrome patients and reducing the costs of the genetic study.

For a correct immunohistochemical interpretation, it is necessary to know that a tumor has high microsatellite instability (MSI-H) if there is loss of nuclear expression in two or more markers (Fig. 14.3), and low instability (MSI-L) if it shows loss of nuclear expression in only one marker. If there is conserved expression in all the markers studied, then it is a tumor with microsatellite stability (MSS). It should be noted that a loss of immunohistochemical expression should only be considered real if no neoplastic cell staining is observed and internal controls (such as stromal cells or positive lymphocytes) are available in the sample. It is also important to know that loss of MLH1 expression is usually associated with secondary loss of PMS2, as is the case with MSH2 and MSH6. Tumors showing loss of MSH2 and MSH6, as well as those with isolated loss of PMS2 or MSH6, have a high probability of being associated with Lynch syndrome, so germline analysis would be necessary in these cases. However, if loss of MLH1 and PMS2 expression is observed, performing a methylation analysis of the MLH1 promoter combined with the mutational determination of BRAF can help us to discriminate between a sporadic tumor and one associated with Lynch syndrome.

For all of the earlier mentioned reasons, the pathologist plays a fundamental role in both the identification of patients with hereditary nonpolyposis colorectal carcinoma or Lynch syndrome as well as in the analysis of mutations in the KRAS, NRAS, or BRAF genes, among others, which are useful when deciding on the administration of certain oncological therapies, especially in advanced stages of colorectal carcinoma.

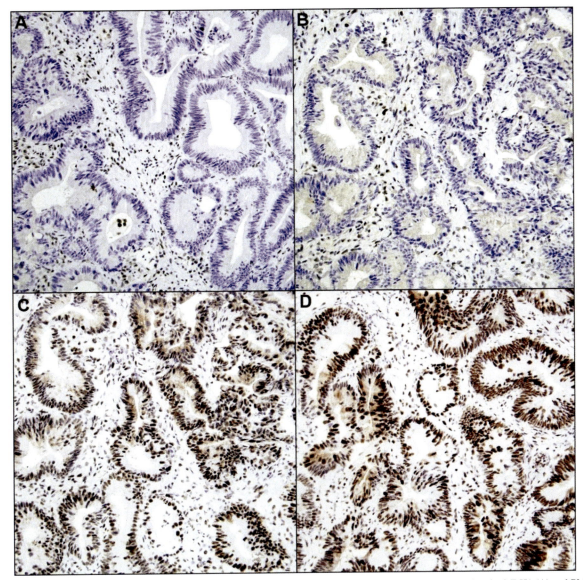

FIG. 14.3 Colorectal carcinoma with high microsatellite instability. Note the loss of immunohistochemical expression for MLH1 (A) and PMS2 (B), with conserved expression for MSH2 (C) and MSH6 (D).

References

1. Goldblum JR, Lamps LW, JK MK, Myers JL. *Rosai and Ackerman's surgical pathology*. Elsevier Health Sciences; 2018.
2. Odze RD, Goldblum JR. *Surgical pathology of the GI tract, liver, biliary tract and pancreas*. Elsevier Health Sciences; 2015.
3. Montgomery EA, Voltaggio L. *Biopsy interpretation of the gastrointestinal tract mucosa*. Wolters Kluwer; 2018.
4. Magnusson K, de Wit M, Brennan DJ, et al. SATB2 in combination with cytokeratin 20 identifies over 95% of all colorectal carcinomas. *Am J Surg Pathol* 2011;**35**(7):937–48. https://doi.org/10.1097/PAS.0b013e31821c3dae.
5. Amin MB, Edge SB, Greene FL, et al, editors. *AJCC cancer staging manual*. Springer International Publishing; 2017.
6. Aarons CB, Shanmugan S, Bleier JIS. Management of malignant colon polyps: current status and controversies. *World J Gastroenterol* 2014;**20**(43):16178–83. https://doi.org/10.3748/wjg.v20.i43.16178.
7. Haggitt RC, Glotzbach RE, Soffer EE, Wruble LD. Prognostic factors in colorectal carcinomas arising in adenomas: implications for lesions removed by endoscopic polypectomy. *Gastroenterology* 1985;**89**(2):328–36. https://doi.org/10.1016/0016-5085(85)90333-6.
8. Kikuchi R, Takano M, Takagi K, et al. Management of early invasive colorectal cancer. Risk of recurrence and clinical guidelines. *Dis Colon Rectum* 1995;**38**(12):1286–95. https://doi.org/10.1007/bf02049154.

9. Kitajima K, Fujimori T, Fujii S, et al. Correlations between lymph node metastasis and depth of submucosal invasion in submucosal invasive colorectal carcinoma: a Japanese collaborative study. *J Gastroenterol* 2004;**39**(6):534–43. https://doi.org/10.1007/s00535-004-1339-4.
10. Bujanda L, Cosme A, Gil I, Arenas-Mirave JI. Malignant colorectal polyps. *World J Gastroenterol* 2010;**16**(25):3103–11. https://doi.org/10.3748/wjg.v16.i25.3103.
11. Valenciana G. Oncoguide of colorectal cancer in the Valencian community. Cancer Plan Office, Conselleria de Sanitat. Published online 2007.
12. Talbot I, Price A, Salto-Tellez M. *Biopsy pathology in colorectal disease*. 2nd ed. CRC Press; 2006. https://play.google.com/store/books/details?id=HMgmSyaFRlsC.
13. Nagtegaal ID, Odze RD, Klimstra D, et al. The 2019 WHO classification of tumours of the digestive system. *Histopathology* 2019;**76**(2):182–8. https://doi.org/10.1111/his.13975.
14. Sasaki O, Atkin WS, Jass JR. Mucinous carcinoma of the rectum. *Histopathology* 1987;**11**(3):259–72. https://doi.org/10.1111/j.1365-2559.1987.tb02631.x.
15. Day DW, Jass JR, Price AB, et al. *Morson and Dawson's gastrointestinal pathology*. John Wiley & Sons; 2008. https://play.google.com/store/books/details?id=OzZaT2AZhIwC.
16. Times M. Colorectal lymphoma. *Clin Colon Rectal Surg* 2011;**24**(3):135–41. https://doi.org/10.1055/s-0031-1285997.
17. Parente F, Cernuschi M, Orlando G, Rizzardini G, Lazzarin A, Bianchi PG. Kaposi's sarcoma and AIDS: frequency of gastrointestinal involvement and its effect on survival. A prospective study in a heterogeneous population. *Scand J Gastroenterol* 1991;**26**(10):1007–12. https://doi.org/10.3109/00365529109003949.
18. Galanopoulos M, Gkeros F, Liatsos C, et al. Secondary metastatic lesions to colon and rectum. *Ann Gastroenterol Hepatol* 2018;**31**(3):282–7. https://www.ncbi.nlm.nih.gov/pmc/articles/pmc5924850/.
19. Vacante M, Borzì AM, Basile F, Biondi A. Biomarkers in colorectal cancer: current clinical utility and future perspectives. *World J Clin Cases* 2018;**6**(15):869–81. https://doi.org/10.12998/wjcc.v6.i15.869.
20. Payá A, Alenda C, Jover R, Aranda FI. Mismatch-repair deficiency colorrectal carcinoma. Identification keys and clinical relevance. *Rev Esp Patol* 2006;**39**(4):201–8.

Chapter 15

Colonoscopy: Technique and quality factors

Ramón Vázquez Dourado, Leticia García Diéguez, and Javier Castro Alvariño
Gastroenterology Department, Ferrol University Hospital Complex, A Coruña, Spain

Historical introduction

Great strides were made in the endoscopic examination of the digestive tract made when image transmission through fiber bundles was first described by Hopkins and Kapany in a paper published in the journal *Nature* in 1954.[1] Three years later, Hirschowitz presented the first prototype of a flexible gastroscope at an American Gastroscopy Society meeting.

The first fiber colonoscope was demonstrated by Overholt to the American Society for Gastrointestinal Endoscopy at the University of Michigan in 1967 and was quickly adopted due to previous experience with gastroscopes and the wide repertoire of diagnostic indications in gastrointestinal bleeding, infectious, inflammatory, and tumor disease. It progressively replaced indirect diagnostic techniques such as the barium enema as well as the rectosigmoidoscopy using rigid instruments, as it offered the possibility to examine the entire colon.

The development and miniaturization of CCD image sensors facilitated their inclusion in flexible endoscopy equipment, significantly improving the color and resolution of the images. The first models of videoendoscopes appeared in the United States in 1983 and were later improved thanks to the advances in technological research in Japan. Nowadays, we can perform a detailed evaluation of the mucosa in high definition using essentially the same controls and channels as the ones in the earliest fiberscopes, while keeping their strength and reliability intact. The reduction of the caliber and the greater flexibility of the distal end facilitate examination in acute angulations, insertion into the terminal ileum, and retroflexion in the rectum and ascending colon.

Colorectal cancer

Colorectal cancer (CRC) is considered to be one of the tumors that cause the highest mortality in humans worldwide.

According to data from the GLOBOCAN project (http://globocan.iarc.fr), it is estimated that in 2018, there were about 880 792 deaths from CRC worldwide.[2] Its natural history is widely known, as the sequential biomolecular changes that lead to its onset require at least 10 years to develop into a malignant tumor. The basic premalignant lesion from which CRC develops is the adenomatous polyp. In recent years, management of these lesions has included those originating in the serrated pathway, with specific endoscopic and histologic characteristics and responsible for up to 20%–30% of CRC. If the polyp presents a size greater than 1 cm or some histologic characteristics such as a villous configuration or high-grade dysplasia (HGD), then it presents a greater potential for malignancy. In most cases, it is a lesion that can be effectively treated through endoscopic resection, which has been shown to reduce the incidence and mortality of CRC.[3]

CRC screening procedures based on immunologic fecal occult blood tests have been shown to be cost effective and to reduce the incidence and mortality in medium-risk populations in which they have been implemented.[4]

Colonoscopy technique

Colonoscopy is currently the standard technique for the examination of the colon and part of the ileum and offers the advantage of facilitating the performance of biopsies for the diagnosis of different diseases, allowing early or premalignant lesions to be treated in most cases. For this reason, it is taking up increasingly more of the schedules of Gastroenterology departments all over the world. However, it can be a complex technique and is not free of a certain risk of complications.

The colonoscope has a working length of approximately 130 cm. The small-caliber or "pediatric" 11-mm model is sometimes used instead as it makes it easier to overcome strictures and the examination of a tortuous sigmoid colon or one with severe angulations. We currently make use of a wide range of accessories, although the basic ones are biopsy forceps, which can be large, polypectomy loops, Roth nets for polyp retrieval, needles for submucosal injection or tagging, and argon plasma coagulation (APC) probes.

We also have different options to improve the final results of the colonoscopy, depending on the performance of the endoscope itself or the use of auxiliary elements:

— High-definition image.
— Auxiliary water channel.
— Digital or electronic chromoendoscopy (NBI, FICE, iScan).
— Colonoscopes with increased viewing angles.
— Optical magnification.
— Conventional chromoendoscopy: instillation of stains such as indigo carmine through the working channel of the colonoscope to improve visualization and characterization of polypoid lesions.
— Caps or other devices attached to the distal end of the colonoscope to separate the folds of the colon and provide an enhanced view of the entire mucosal surface.

Endoscopic description

Prior cleansing of the colon is essential to perform optimal examination. For this purpose, it is advisable to stick to a liquid diet the day before, suspend oral iron supplements 5 days before the colonoscopy and drink polyethylene glycol-based electrolyte solutions (the only ones indicated in chronic renal and cardiac insufficiency) or combinations of magnesium citrate and sodium picosulfate. It is now standard practice for the second half of the chosen preparation to be taken four to 6 h prior to colonoscopy to ensure optimal cleansing.[5] If the patient reports previous chronic constipation, consideration should be given to intensifying these preparations. If the patient has diabetes on insulin therapy, a dose reduction will be advised in accordance with the caloric intake of that day's diet.

In certain therapeutic procedures such as polypectomy, endoscopic balloon dilatation for stenosis, and echoendoscopy with puncture, the suspension of antiplatelet or anticoagulant treatment should be considered, taking into account the thromboembolic risk of the patient.[6] In general, acenocoumarol should be suspended for 3 days and the new oral anticoagulants for 2 days, using bridging therapy with low-molecular-weight heparin. In the case of antiplatelet agents such as clopidogrel, they should be suspended 7 days before and replaced with 100 mg of acetylsalicylic acid unless they wear a vasoactive stent, in which case therapeutic maneuvers would be contraindicated.

Insufflation and traction of the mesocolon make colonoscopy an uncomfortable exam. In addition to using a refined technique, the administration of premedication with a sedative–analgesic combination of fentanyl and midazolam or even propofol anesthesia should be considered.[7]

The patient starts out in the left lateral decubitus position with both knees flexed. The examination begins with an inspection of the perianal region that can detect various lesions such as skin flaps, scarring, fistulae and fissures, hemorrhoids, and prolapses. After performing a rectal examination with lubrication, the insertion of the endoscope begins, which can be done by auxiliary personnel who will handle the endoscope shaft or by the endoscopists themselves, with the assistance of auxiliary personnel in this case for abdominal compression and position changes. The tip of the endoscope is usually supported at the anal margin to facilitate sphincter relaxation.

The anal canal cannot be fully examined on initial insertion due to its short length and the anal sphincter. Once the endoscope has been withdrawn at the level of the distal rectum, retroflexion should be performed, which provides an opportunity to explore this area, which is difficult to observe during insertion. It is easy to move the colonoscope through the rectum because it is attached to the retroperitoneal wall and has almost no mobility. Houston's valves are characteristically observed as we insert using left and right turns. A general rule is to move the tip of the endoscope away from the area where the most light is reflected and toward the darker area or the center of the arches formed by the mucosal folds, in order to follow the intestinal lumen. In most cases, at the rectosigmoid junction, the lumen opens toward the left axis forming a sharp curve. To avoid the formation of a loop, it is advisable to advance the colonoscope while applying a left torque with mild pushing.

The examination of the sigmoid colon and its junction with the descending colon presents a variable degree of difficulty that depends on the length of the mesentery, the degree of redundancy of intestinal loops, the presence of diverticular disease, and the history of previous surgery or inflammatory processes. Approximately two-thirds of the time spent on insertion will be taken up by the examination of the left colon. It is essential to ensure that all loops are resolved and that the colonoscope is kept straight before proceeding to more proximal sections.[8,9] To this end, it is advisable to follow a series of recommendations:

- Keep the endoscope shaft straight so that the torquing movements exerted by the hand holding the control unit are faithfully transmitted to the end of the endoscope.
- Properly match these torquing movements with the angulation of the distal end of the endoscope.
- Any polyps discovered upon insertion should be resected, especially smaller ones.
- Push forward and pull back along short segments.
- Insufflate as little as possible, suctioning frequently and keeping the endoscope in the center of the lumen of the colon. This will make it easier for the colon to make folds over the endoscope, which can also be achieved by flexing the distal end behind a mucosal fold and withdrawing the endoscope 5–10 cm.
- In case of diverticular disease, position perpendicularly with respect to the orifices and irrigate with water to facilitate intubation.
- If no progress can be made in proximal sections despite insertion, it means that a loop has formed. In this case, rectify by pulling back and applying a clockwise torque.
- To overcome angulations, keep pushing forward and pulling back, together with clockwise rotation, which are essential to avoid the formation of loops in the sigmoid colon (Fig. 15.1), especially when the descending colon, which is retroperitoneal, has already been reached. If there are sharp angulations, we will position the distal end in such a way that its axis is at 12 or 6 o'clock because flexion of the distal end is easier to manage and wider.
- Avoid pushing through angulations that are not easily overcome using the previous method. Colonoscopy is currently the standard technique for the exploration of the colon and part of the ileum and offers us the advantage of facilitating the taking of biopsies for the diagnosis of different pathologies, also allowing in most cases the treatment of early or premalignant lesions. For this reason it occupies more and more time and space in the examination room agendas of Digestive Departments all over the world. However, it can be a complex technique and is not free of a certain risk of complications.
- Manual compression on the left iliac fossa and changing the patient's position to supine may be necessary.

Sometimes there has already been a history of laborious or incomplete colonoscopy. If the reason is the presence of a narrow or sharply angulated sigmoid colon, the use of a pediatric colonoscope with frequent suctioning and warm water irrigation is recommended. In case of a redundant colon, a variable stiffness colonoscope should be used and adjusted to the maximum level when kinking is evident, inserting the scope with abdominal compression and shortening the colon as soon as possible to avoid loop formation. If there has been an unsuccessful attempt to perform a complete colonoscopy, the use of a gastroscope with a distal cap[10] or even a double-balloon enteroscope should be considered.

FIG. 15.1 Clockwise torque maneuver and withdrawal of the colonoscope.

After passing the sigmoid-descending junction, we access a retroperitoneal section with concentric haustra that will be examined using straightforward insertion. Afterward, we will see the splenic flexure and we will advance the colonoscope using an up and down movement of the distal end. Changing the patient's position and abdominal compression on the sigmoid colon might help. If there are considerable difficulties in overcoming it, we may be facing an inverted splenic flexure associated with a mobile descending colon, in which case we will have to perform an anticlockwise torque and compress the area. At this point, the inserted length of the endoscope should be 45–50 cm.

Access to the transverse colon will be confirmed by observing the typical triangular configuration determined by the arrangement of the taenia coli and haustra. By pushing forward and using air suction, progress is generally smooth. Occasionally a loop may form in the middle due to a long mesocolon, which is reduced by performing a clockwise torque, using air suction and abdominal compression at the umbilical level, sometimes combined with that of the left hypochondrium and iliac fossa.

The hepatic flexure is easy to identify because the hepatic spot can be visualized through the colon wall and will be located at about 70–80 cm with the colonoscope. The lumen of the ascending colon is usually located to the right, so repeatedly pushing forward and pulling back and using air suction, combined with gently twisting to the right and left are usually enough to overcome the flexure, all of which is made even easier if the patient is in the left lateral decubitus position. Access to the ascending colon and cecum is facilitated by air suction, abdominal compression on the transverse colon, and placing the patient in the right lateral decubitus position, if necessary. It should always be documented with an image of the cecum and appendiceal orifice.

Intubation of the terminal ileum should be considered according to the indications of the examination and previous findings. For this purpose, the end of the endoscope is placed in the cecum and the ileocecal valve is placed in the lower left quadrant of the image, for which it may be necessary to change the patient's position to the supine decubitus position. A slight suctioning is performed and the distal end of the colonoscope is withdrawn until the lower lip of the valve is reached by bending it in the presumed direction of the orifice. With anticlockwise twisting and insufflation, we can access the ileum, observing granular mucosa with visible villi or pseudopolypoid due to the presence of Peyer's patches.

There are a series of anatomical references that must be taken into account to determine the position of the tip of the endoscope such as visualizing the spleen and liver in the respective angles, the arrangement of the taenia coli in the transverse colon that makes it take on a triangular shape, the crescent shape of the appendiceal orifice, the length corresponding to the ileocecal valve that is found before reaching the blind-ending sac of the cecum, and visualizing the mucosa with ileal villi. The degree of colon mobilization after digital palpation and transillumination can help us. It is always advisable to record images of the cecum and ileum to confirm that a complete examination has been performed.

A uniform approach must be taken at all times with regard to the lesions that are identified during the insertion of the endoscope: their shape, size, appearance, and location with respect to the anatomical references will be described. Biopsies will be taken or excised according to their size and consistency.

The colonoscope will be removed by slowly suctioning the liquid remains that we find and making a systematic effort to explore the proximal part of the folds, haustra, and ileocecal valve to detect mucosal irregularities as well as submucosal circulation.

Magnification techniques

Although colonoscopy represents the best tool to reduce CRC mortality, the technique providing conventional images may not be sufficient to evaluate subtle or incipient changes at the level of the colorectal mucosa. Therefore, in recent decades, new techniques have been developed to deliver a better diagnosis of colorectal mucosal abnormalities: optical chromoendoscopy with dyes or optical filters such as Narrow Band Imaging (NBI Olympus) or FICE (Fujinon) are among the most common. The use of indigo carmine or methylene blue solutions makes it possible to show new neoplastic patterns not detectable using conventional white light, thus increasing the diagnosis of lesions with high-grade dysplasia and polypoid, flat, or depressed carcinomas. Therefore it should be used routinely to facilitate the early diagnosis of CRC, while chromoendoscopy with magnification in experienced hands also represents a significant breakthrough in endoscopic practice to improve the individual diagnosis of each lesion, increasing the therapeutic efficacy for colorectal tumors.[11]

Chronic inflammation, additionally, is a well-known risk factor for the development of neoplastic mucosal changes. In long-standing ulcerative colitis and colonic Crohn's disease, there is a significant increase in the incidence of CRC, although with wide variations among studies. Conventional colonoscopy is effective in detecting polypoid changes but the general consensus is that in these diseases, neoplastic lesions are generally flat or depressed, which a routine colonoscopy may not be able to detect. The introduction of chromoendoscopy combined with magnification techniques has made a breakthrough in defining, as well as limiting, the multiple biopsies typically recommended in cancer surveillance protocols for these patients. The cumulative evidence of data following the implementation of this technique demonstrates its effectiveness. However, the introduction of chromoendoscopy in surveillance programs requires meticulous training and further comparative studies, especially between different magnification techniques.[12]

Classification of colorectal polyps

The macroscopic characteristics of colorectal polyps allow us to classify these lesions and therefore establish indications for their endoscopic treatment. Currently, the Paris classification is used for all lesions and while Kudo's classification (Fig. 15.2) is used for laterally spreading tumors or LSTs. The risk of submucosal invasion varies with endoscopic morphology so these classifications have prognostic value and are the ones used to determine the most appropriate treatment in terms of an en bloc endoscopic resection as cancer treatment or it may contraindicate any type of endoscopic treatment.

The Paris classification (Fig. 15.3)[13] applies to superficial polyps limited to mucosa and submucosa (type 0 lesions) which correspond to lesions that can be treated endoscopically. This classification divides polyps into two groups:

1. Polypoid or protruded lesions (type 0–I lesions):
   — Type 0-Ip: pedunculated polyps.
   — Type 0-Is: sessile polyps, without pedicle.

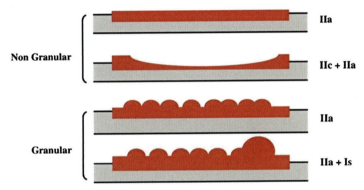

FIG. 15.2 Kudo's classification.

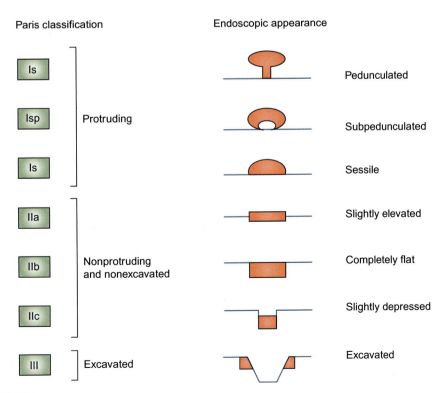

FIG. 15.3 Paris classification.

2. Nonpolypoid or flat lesions (lesions type 0-II):
   - Type 0-IIa: slightly elevated flat
   - Type 0-IIb: completely flat (they are not usually found in the colon).
   - Type 0-IIc: slightly depressed flat lesions

   Some lesions may have several components:
   - Type 0-Is+IIc.
   - Type 0-IIa+IIc.
   - Type 0-IIc+IIa.

Flat lesions larger than 1 cm are classified as laterally spreading tumors (LST).
   The classification described by Kudo (Fig. 15.2) divides these lesions into two groups:

1. Lesions with a granular surface (LST-G), traditionally known as "carpet lesions":
   - Homogeneous LST-G, with nodules of similar size
   - Mixed LST-G, with nodules of different sizes
2. Lesions with a nongranular surface (LST-NG):
   - LST-NG elevated type, with a slightly elevated surface
   - LST-NG pseudodepressed type

The general consensus is that the larger the size of a polypoid lesion, the greater the risk of deep invasion, a criterion that does not influence 0-IIa and 0-IIb lesions so much. In superficial depressed 0-IIc lesions, the risk is higher in smaller lesions. On the other hand, in LST-G lesions—which are the majority—the risk of submucosal invasion increases when the lesion is not homogeneous and presents nodules of than 3 mm in size or depressed areas, with the latter criterion also applicable to nongranular lesions.

The next step is to assess the pattern of glandular crypts and vascularization on the surface of the lesion to consider the most likely histology and determine the risk of deep invasion, which will help us to plan the type of treatment (Fig. 15.4). One of the most practical classifications is the NICE classification,[14] which can be used with conventional colonoscopes featuring narrow-band imaging electronic chromoendoscopy and which evaluates color, vascularization, and surface pattern (Fig. 15.5). It allows lesions to be divided into three categories: nonadenomatous or nonneoplastic lesions (type1), adenomatous lesions (type 2), and lesions with suspected deep submucosal invasive carcinoma (type 3). In addition, there are a series of characteristics that will make us suspect submucosal invasion: a size greater than 20 mm, presence of white spots around the lesion, circumscribed areas of erythema, firm consistency, extensive appearance, convergence of folds, and deep depression.[15]

Serrated pathway in CRC

It is generally accepted that CRC is a heterogeneous disease. The serrated pathway is an evolutionary pathogenic concept that has been clearly accepted in CRC carcinogenesis and it has been postulated that up to 30% of cancers develop through this Figs. 15.6–15.18 alternative pathway. Since its original description two decades ago, our morphological, histological,

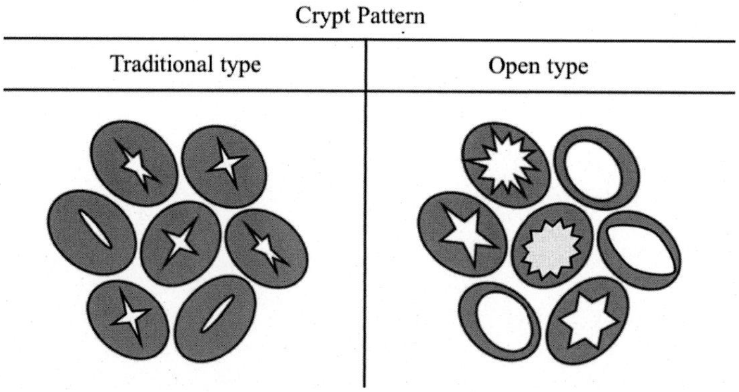

FIG. 15.4 Crypt pattern most common in serrated polyps.

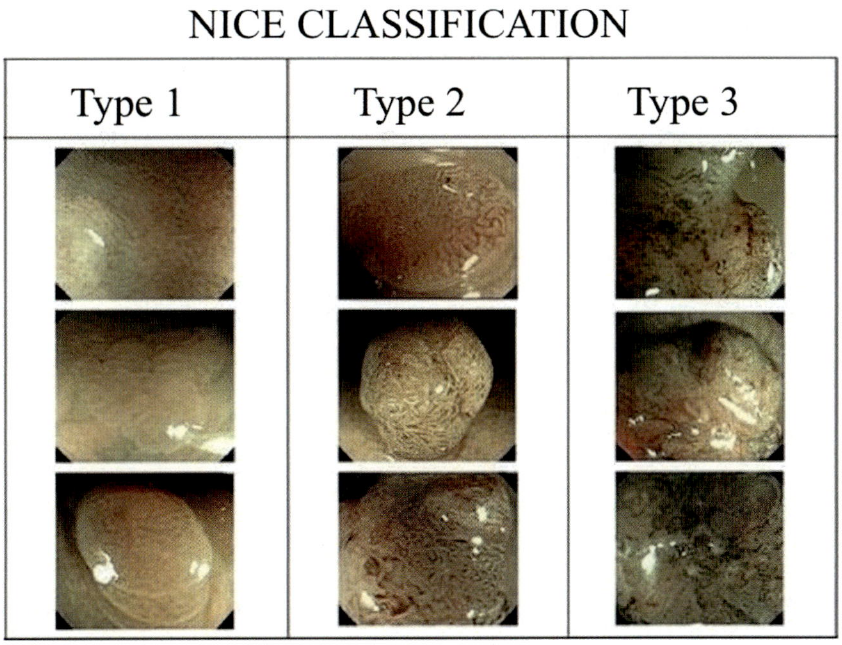

FIG. 15.5 NICE classification.

and molecular knowledge of serrated lesions has increased, likewise gaining more recognition from the endoscopic point of view. That said, knowledge of their natural history, optimal treatment, and surveillance criteria that allow us to avoid interval cancers still remains to be transferred to clinical practice.[16]

Sessile serrated adenomas are slightly elevated polyps with indistinct borders and a cloud-like surface using white-light endoscopy.[17] Unlike hyperplastic polyps, they are usually larger than 5 mm, are frequently covered with a mucus layer, and are more common in the proximal colon, sometimes mimicking a thickened fold. Their characteristics make them more difficult to detect and completely resect, so that advanced magnification and chromoendoscopy techniques are often necessary to recognize typical mucosal structural patterns (pit patterns) for diagnosis and treatment. A recent study using magnification endoscopy has described a new open-shape pattern (pit type II-O) with a high predictive value for serrated adenomas (sensitivity, 65.5%; specificity, 97.3%). Certain endoscopic features such as semipedunculated morphology, double elevation, central depression, and redness may represent an increased risk of having high-grade dysplasia or invasive submucosal cancer.[18] Serrated adenocarcinomas arising from these lesions are predominantly located in the cecum (52%) and rectum (33%). It is estimated that 16% of proximal CRC are serrated adenocarcinomas while the proportion in the distal colon is only 6%. One study highlights that women comprise progressively higher proportions of individuals with more advanced serrated lesions (they accounted for 53%, 57%, 69%, and 76% of patients with nondysplastic sessile serrated adenomas, with low-grade dysplasia, high-grade dysplasia, and with invasive adenocarcinoma, respectively).[16]

Colonoscopy in the diagnosis of CRC in medium-risk populations

There are wide variations in the performance of colonoscopy among different examiners. The rate of missed adenomas increases with smaller polyps, and colon cancers may even be missed during a routine colonoscopy. The occurrence of cancer early after colonoscopy (interval cancer) may occur primarily in proximal sections in patients with a history of advanced adenomas, although most would probably be preventable as a result of previously missed or incompletely resected polyps. However, some interval cancers may appear as a result of rapid tumor growth secondary to genetic alterations due to CIMP/LINE-1 methylation or microsatellite instability without representing a failure of colonoscopy.[19] Operator variability and circumstances (inadequate technique, incomplete colonoscopy or polypectomy, personal characteristics such as degree of concentration or motivation, etc.) are important factors in diagnostic errors. Some lesions may also be missed due to poor bowel preparation or funding-related factors.

The detection rates of adenomas and polypectomies are inversely related to the development of new or nonvisualized cancers after diagnostic colonoscopy and can be used as quality indicators.[20,21]

After the quality indicators initially disclosed by the ASGE-ACG, Lieberman et al. in 2007 published a study by the Quality Assurance Task Group of the National Colorectal Cancer Roundtable on quality criteria in colonoscopy and reported a database.[22]

Colonoscopy is a powerful tool for CRC prevention, but its quality is operator dependent. Although clinicians may be hesitant with regard to comprehensive quality controls, the implementation of quality control programs is mandatory to enhance its value. There are, however, important differences among the strategies established in Western countries, mainly in Europe. European quality guidelines should tend to unify, taking advantage of the introduction and generalization of screening programs.[23] In our setting, a clinical practice guideline on quality in colonoscopy screening has been in place since 2011. It was developed by the working group of the Spanish Association of Gastroenterology and the Spanish Society of Digestive Endoscopy and it was updated in 2018.

Colonoscopy in the diagnosis of CRC in high-risk populations

In individuals with a family history of familial adenomatous polyposis (FAP) or Gardner syndrome, a genetic study should be recommended and once the pathogenic mutation in the APC gene has been confirmed, annual sigmoidoscopy should be performed from the age of 10 to 12 years and an annual colonoscopy once the phenotype is confirmed. This endoscopic surveillance will be delayed in cases of attenuated FAP to 18 years. A complete colectomy should be considered for these patients at around 25 years of age because of the near 100% probability of developing CRC.

The diagnosis of hereditary nonpolyposis CRC (HNPCC), also called Lynch syndrome, should be considered in populations with several family members with CRC, particularly if one or more of them develop cancer under the age of 50. It is an autosomal dominant disorder caused by mutations in DNA repair genes with an approximate 60%–80% lifetime risk of developing CRC. After confirmation of risk with the corresponding germline mutation study, these patients should undergo regular screening with colonoscopy every 1–2 years, starting at 20–25 years of age. This strategy will be modified according to personal history and familial aggregation of cases of Lynch-like syndrome and familial CRC type X, which will start 10 years before the age of diagnosis of the youngest relative.

Individuals with two first-degree relatives diagnosed with CRC should undergo screening by means of colonoscopy every 5 years starting at an age 10 years younger than the youngest affected relative or at age 40 years.

The follow-up schedule for CRC screening varies in patients with inflammatory bowel disease (IBD). Patients with extensive colitis, presence of pseudopolyps, and first-degree relatives with CRC at age 50 or older should undergo surveillance at 2–3 years of disease progression. This interval will be shortened to annual or biannual in cases of extensive colitis with severe activity, a history of strictures or dysplasia, first-degree relatives with CRC before age 50, a reservoir with dysplasia or cancer, and the presence of primary sclerosing cholangitis. Ideally, because of the difficulty in differentiating inflammatory from premalignant changes, follow-up colonoscopy should not be performed during periods of active colitis, and biopsies from areas with less inflammation would be preferable. It has been suggested in the past, perhaps exaggeratedly, that up to 64 serial biopsies of the colon may be necessary to achieve 95% sensitivity when screening for dysplasia in IBD. As discussed, the implementation of new technologies including chromoendoscopy, magnification techniques, and narrow-band imaging (NBI) may improve the detection of dysplasia in these circumstances and allow endoscopists to obtain fewer samples.

Colonoscopy quality indicators

1. Availability of appropriate facilities and equipment, waiting room, toilets and monitoring systems or pulse oximeter.
2. Adherence to standard disinfection recommendations.
3. Documentation of medical history to identify risk factors for sedation and management of the anticoagulated patient.
4. Documentation of an appropriate indication for colonoscopy and follow-up criteria.
5. Obtaining informed consent. In most cases it would be beneficial in addition to a prior consultation to explain the cleansing preparation, to examine the clinical history together with the treatments followed by the patient, and to provide them with knowledge on the examination.[24]
6. Documentation of adequate bowel preparation essential for visualization of the entire colon and allowing the detection of lesions larger than 5 mm. Adherence to dietary instructions has a significant impact on quality. If preparation is deficient in more than 10% of scans, the method or type of products used should be reviewed.[25]
7. Cecal intubation rate not less than 90%–95%, passing the end of the endoscope beyond the ileocecal valve, visualizing the appendiceal orifice and the entire cecum. Graphic documentation is recommended from a medicolegal perspective.
   The percentage could be adjusted or not by excluding colonoscopies with poor preparation or clinical situations that do not allow the proximal colon to be reached.

8. Colonoscope withdrawal time (not including the necessary time for biopsies and polypectomies). The detection of adenomas is increased if it is longer than 6 min.[26] It is a controversial indicator because of the difficulty of measurement or the possibility of delay in the rectum. In addition, the length or individual anatomy of the colon is variable, as well as the viewing angles of the endoscopes, but it is essential to use a minimum withdrawal time to suction out residues, visualize the details of the mucosa, and perform position changes, especially for the right colon.
9. The adenoma detection rate (ADR) is an important indicator of quality in screening programs and should be at least 20% if colonoscopy is the primary screening strategy or 40% if it is secondary, after confirmation of fecal occult blood. The measurement of the individual ADR of the endoscopists performing CRC screening is imperative to establish quality criteria. Lower rates express poor colonoscopies and probably some interval cancers are related to nonvisualized adenomas.[26,27]
10. Polypectomy should be performed in all polyps detected except in multiple micropolyps (smaller than 5 mm) that appear hyperplastic in the rectosigmoid that can be removed using biopsy forceps.[28] The use of chromoendoscopy or NBI with high-resolution endoscopy can help in the characterization of the lesions. All polyps smaller than 2 cm should be resected by conventional polypectomy or endoscopic mucosal resection (mucosectomy). Larger sessile polyps should be resected by experienced endoscopists. Complete removal preferably en bloc is necessary to prevent CRC. At least 90% of resected polyps should be retrieved for a pathological study.[29]
11. Malignant polyps are those containing cancer with submucosal invasion. Occasionally, a complete resection of early CRC is possible (curative polypectomy), but the presence of affected nodes is associated with deep submucosal invasion. Several histological factors are associated with a high risk of lymph node metastasis and local recurrence after polypectomy. Polypectomy can be considered complete if the resection margins (lateral and deep) are free of cancer, if there is no histologic evidence of vascular or lymphatic invasion, and if the pathological anatomy is not very indistinct. The risk of residual or recurrent cancer postpolypectomy is lower than the surgical risk for complications and can be cost effective.[20] However, recurrence after resection in fragments of large adenomas is high, so it is desirable to mark the area where the polypectomy was performed and to perform an eschar biopsy 3–6 months to 1 year after resection.[30]
12. Record of adverse effects. Perforation rates greater than 1/1000 in screening and postpolypectomy bleeding rates greater than 1/100 should be taken into account. At least 90% of bleedings should be treated endoscopically.[31]
13. Standardized endoscopic report. Pathological findings should be evaluated with the pathologist and clear follow-up recommendations based on clinical guidelines should be established.
14. Interval cancers. Their monitoring is recommended as an important indicator to determine the quality of CRC screening programs. Although a proportion may be due to aggressive tumor biology, undetected lesions or incompletely resected polyps are notable factors to be taken into account. They are defined as the occurrence of CRC between screening periods, in the context of adequate screening according to clinical practice guideline recommendations. Their frequency has been estimated to be between 0 and 6 cases per 1000 patients with a previous colonoscopy without lesions or with complete resection of lesions.[32,33]

Endoscopic iconography

See Figs. 15.6–15.18.

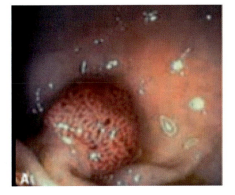

FIG. 15.6 Polypoid adenoma.

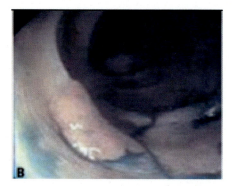

FIG. 15.7 Moderate high.

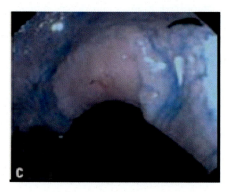

FIG. 15.8 Flat adenoma.

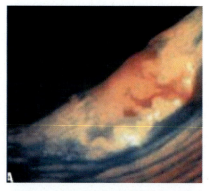

FIG. 15.9 Image of depressed lesion using indigo carmine in high-grade intraepithelial neoplasia.

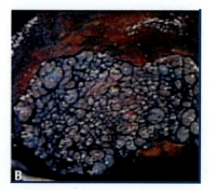

FIG. 15.10 Image of serrated adenoma using indigo carmine in high-grade intraepithelial neoplasia.

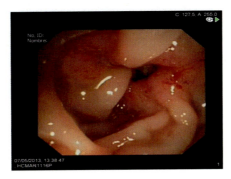

FIG. 15.11 Impassable neoplastic stenosis.

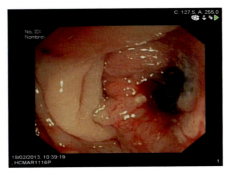

FIG. 15.12 Ulcerated stenosing carcinoma. Sigmoid colon.

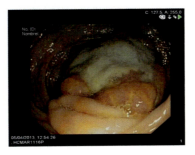

FIG. 15.13 Polypoid carcinoma of the cecal pole.

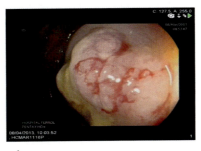

FIG. 15.14 Ulcerative carcinoma of the ascending colon.

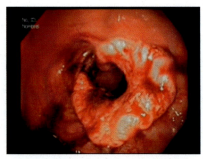

FIG. 15.15 Rectosigmoid carcinoma "in napkin ring."

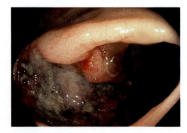

FIG. 15.16 Extensively ulcerated proximal colon neoplasia.

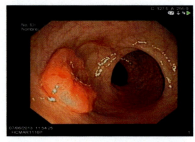

FIG. 15.17 Incipient polypoid carcinoma of the rectum.

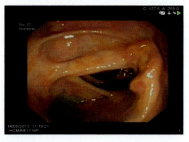

FIG. 15.18 Neoplastic excavated flat lesion of the right colon.

References

1. Hopkins HH, Kapany NS. A flexible fibrescope, using static scanning. *Nature* 1954;**173**(4392):39–41.
2. Bray F, Ferlay J, Soerjomataram I, Siegel RL, Torre LA, Jemal A. Global cancer statistics 2018: GLOBOCAN estimates of incidence and mortality worldwide for 36 cancers in 185 countries. *CA Cancer J Clin* 2018;**68**(6):394–424. https://doi.org/10.3322/caac.21492.
3. Zauber AG, Winawer SJ, O'Brien MJ, et al. Colonoscopic polypectomy and long-term prevention of colorectal cancer deaths. *Obstet Gynecol Surv* 2012;**366**(8):687–96. https://doi.org/10.1097/ogx.0b013e31825bc1f5.
4. Towler BP, Irwig L, Glasziou P, Weller D, Kewenter J. Screening for colorectal cancer using the faecal occult blood test, hemoccult. *Cochrane Database Syst Rev* 2000;**2**, CD001216.

5. Lewis JR, Cohen LB. Update on colonoscopy preparation, premedication and sedation. *Exp Rev Gastroenterol Hepatol* 2013;**7**(1):77–87.
6. Veitch AM, Vanbiervliet G, Gershlick AH, et al. Endoscopy in patients on antiplatelet or anticoagulant therapy, including direct oral anticoagulants: British Society of Gastroenterology (BSG) and European Society of Gastrointestinal Endoscopy (ESGE) guidelines. *Gut* 2016;**65**(3):374–89.
7. Antolín SM, Da Silva BAM, Santamarta FS, et al. Severe cardiorespiratory complications derived from propofol sedation monitored by an endoscopist. *Rev Esp Enferm Dig* 2018;**110**(4):237–9. https://doi.org/10.17235/reed.2018.5282/2017.
8. Berzin TM. Colonoscopic tips and tricks—advice from 3 master endoscopists. *Gastrointest Endosc* 2009;**70**(2):370–1.
9. González-Huix Lladó F, Figa Francesch M, Huertas NC. Criterios de calidad que deben exigirse en la indicación y en la realización de la colonoscopia. *Gastroenterol Hepatol* 2010;**33**(1):33–42.
10. Shida T, Takano S, Kaiho M, Miyazaki M. Transparent hood attached to a gastroscope: a simple rescue technique for patients with difficult or incomplete colonoscopy. *Endoscopy* 2008;**40**(Suppl 2):E139.
11. Backes Y, Moss A, Reitsma JB, Siersema PD, Moons LMG. Narrow band imaging, magnifying chromoendoscopy, and gross morphological features for the optical diagnosis of T1 colorectal cancer and deep submucosal invasion: a systematic review and meta-analysis. *Am J Gastroenterol* 2017;**112**(1):54–64.
12. Thorlacius H, Toth E. Role of chromoendoscopy in colon cancer surveillance in inflammatory bowel disease. *Inflamm Bowel Dis* 2007;**13**(7):911–7.
13. Group ECR, Endoscopic Classification Review Group. Update on the Paris classification of superficial neoplastic lesions in the digestive tract. *Endoscopy* 2005;**37**(6):570–8. https://doi.org/10.1055/s-2005-861352.
14. Hayashi N, Tanaka S, Hewett DG, et al. Endoscopic prediction of deep submucosal invasive carcinoma: validation of the narrow-band imaging international colorectal endoscopic (NICE) classification. *Gastrointest Endosc* 2013;**78**(4):625–32.
15. Matsuda T, Parra-Blanco A, Saito Y, Sakamoto T, Nakajima T. Assessment of likelihood of submucosal invasion in non-polypoid colorectal neoplasms. *Gastrointest Endosc Clin N Am* 2010;**20**(3):487–96.
16. Lash RH, Genta RM, Schuler CM. Sessile serrated adenomas: prevalence of dysplasia and carcinoma in 2139 patients. *J Clin Pathol* 2010;**63**(8):681–6.
17. Hazewinkel Y, López-Cerón M, East JE, et al. Endoscopic features of sessile serrated adenomas: validation by international experts using high-resolution white-light endoscopy and narrow-band imaging. *Gastrointest Endosc* 2013;**77**(6):916–24.
18. Murakami T, Sakamoto N, Ritsuno H, et al. Distinct endoscopic characteristics of sessile serrated adenoma/polyp with and without dysplasia/carcinoma. *Gastrointest Endosc* 2017;**85**(3):590–600.
19. Nishihara R, Wu K, Lochhead P, et al. Long-term colorectal-cancer incidence and mortality after lower endoscopy. *N Engl J Med* 2013;**369**(12):1095–105.
20. Bressler B, Paszat LF, Chen Z, Rothwell DM, Vinden C, Rabeneck L. Rates of new or missed colorectal cancers after colonoscopy and their risk factors: a population-based analysis. *Gastroenterology* 2007;**132**(1):96–102.
21. Hewett DG, Kahi CJ, Rex DK. Does colonoscopy work? *J Natl Compr Canc Netw* 2010;**8**(1):67–77.
22. Lieberman D, Nadel M, Smith RA, et al. Standardized colonoscopy reporting and data system: report of the Quality Assurance Task Group of the National Colorectal Cancer Roundtable. *Gastrointest Endosc* 2007;**65**(6):757–66.
23. European Colorectal Cancer Screening Guidelines Working Group, von Karsa L, Patnick J, et al. European guidelines for quality assurance in colorectal cancer screening and diagnosis: overview and introduction to the full supplement publication. *Endoscopy* 2013;**45**(1):51–9.
24. Rex DK, Bond JH, Winawer S, et al. Quality in the technical performance of colonoscopy and the continuous quality improvement process for colonoscopy: recommendations of the U.S. Multi-Society Task Force on Colorectal Cancer. *Am J Gastroenterol* 2002;**97**(6):1296–308. https://doi.org/10.1111/j.1572-0241.2002.05812.x.
25. Froehlich F, Wietlisbach V, Gonvers J-J, Burnand B, Vader J-P. Impact of colonic cleansing on quality and diagnostic yield of colonoscopy: the European Panel of Appropriateness of Gastrointestinal Endoscopy European multicenter study. *Gastrointest Endosc* 2005;**61**(3):378–84.
26. Barclay RL, Vicari JJ, Doughty AS, Johanson JF, Greenlaw RL. Colonoscopic withdrawal times and adenoma detection during screening colonoscopy. *N Engl J Med* 2006;**355**(24):2533–41.
27. Sanchez W, Harewood GC, Petersen BT. Evaluation of polyp detection in relation to procedure time of screening or surveillance colonoscopy. *Am J Gastroenterol* 2004;**99**(10):1941–5.
28. Peluso F, Goldner F. Follow-up of hot biopsy forceps treatment of diminutive colonic polyps. *Gastrointest Endosc* 1991;**37**(6):604–6. https://doi.org/10.1016/s0016-5107(91)70863-8.
29. Hurlstone DP. Colonoscopic resection of lateral spreading tumours: a prospective analysis of endoscopic mucosal resection. *Gut* 2004;**53**(9):1334–9. https://doi.org/10.1136/gut.2003.036913.
30. Robertson DJ, Greenberg ER, Beach M, et al. Colorectal cancer in patients under close colonoscopic surveillance. *Gastroenterology* 2005;**129**(1):34–41.
31. Sorbi D, Norton I, Conio M, Balm R, Zinsmeister A, Gostout CJ. Postpolypectomy lower GI bleeding: descriptive analysis. *Gastrointest Endosc* 2000;**51**(6):690–6. https://doi.org/10.1067/mge.2000.105773.
32. Heresbach D, Barrioz T, Lapalus M, et al. Miss rate for colorectal neoplastic polyps: a prospective multicenter study of back-to-back video colonoscopies. *Endoscopy* 2008;**40**(04):284–90. https://doi.org/10.1055/s-2007-995618.
33. Imperiale TF, Glowinski EA, Lin-Cooper C, Larkin GN, Rogge JD, Ransohoff DF. Five-year risk of colorectal neoplasia after negative screening colonoscopy. *N Engl J Med* 2008;**359**(12):1218–24. https://doi.org/10.1056/nejmoa0803597.

Chapter 16

Endoscopic diagnosis of preneoplastic and neoplastic lesions

Beatriz Romero-Mosquera[a,b], Alfonso Martínez-Turnes[a,b], and Vicent Hernández[a,b]
[a]Department of Gastroenterology, Xerencia Xestion Integrada de Vigo, SERGAS, Vigo, Spain, [b]Research Group in Digestive Diseases, Galicia Sur Health Research Institute (IIS Galicia Sur), SERGAS-UVIGO, Vigo, Spain

Introduction

The fundamental objective of colonoscopy is the diagnosis of neoplastic lesions, especially in the early stages.[1]

According to the adenoma-carcinoma sequence, it is well established that the premalignant lesion from which colorectal cancer develops is generally an adenomatous polyp. These lesions are potentially treatable by endoscopy and their removal reduces the incidence of colorectal cancer.[2]

In recent years, and with the implementation of screening programs, great interest has been placed on improving techniques and developing quality standards with the purpose of increasing the detection of adenomas.[3]

The improvement in endoscopy equipment and the implementation of new technologies have allowed the assessment of colonic mucosa to become increasingly accurate, both in morphological characterization and in the identification of mucosal patterns. This translates not only into a greater number of lesions detected, but also into the possibility of predicting the histology of the lesion during colonoscopy and, based on this, determining the appropriate therapeutic approach in each case.[4]

Identification of lesions

The detection of adenomas and the characterization of lesions during colonoscopy can be influenced by various factors and can be modified by the application of different techniques or auxiliary devices.

Quality of colonoscopy

The quality of colonoscopy is a fundamental factor in identifying neoplastic lesions in the colon. A complete colonoscopy up to the caecal fundus, good bowel cleansing and a detailed examination of the mucosa with a withdrawal time of more than 6 min have been associated with a higher detection rate of adenomas.[5,6]

High-definition colonoscopy

Although colonoscopy is the gold standard for the diagnosis of neoplastic lesions, conventional imaging methods are unable to detect subtle or incipient changes at the level of the colonic mucosa. High-definition colonoscopy has demonstrated greater detection of adenomas than conventional colonoscopy[7] and therefore the use of high-definition colonoscopy is now routinely recommended.

Conventional chromoendoscopy

Chromoendoscopy consists in using dyes during colonoscopy to improve the detection and characterization of lesions. Of the dyes available, the two most commonly used are indigo carmin and methylene blue. The difference between observing a

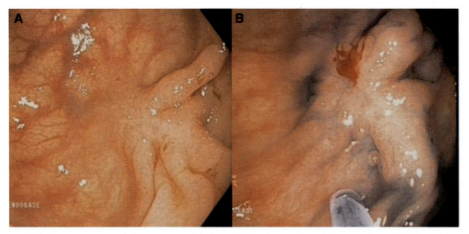

FIG. 16.1 Lat lesion. We can see in panel (A) the lesion only with light, while in panel (B) we have performed chromoendoscopy with indigo carmin improving the delimitation of the *upper margin. (Photos from the archive of the Department.)*

lesion without and with indigo carmin chromoendoscopy is shown in Fig. 16.1. To ensure that they are properly distributed, adequate intestinal cleansing is important, and the prior application of mucolytics can be useful.[4]

Their application on a previously detected lesion enhances mucosal crypt patterns and irregular areas of the intestinal mucosa, improving the delimitation of lesions.

In addition, dyes can also be used to stain the entire colonic mucosa (panchromoendoscopy), a technique that has been shown to be effective for increased lesion detection in patients at high risk for CRC, such as inflammatory bowel disease, patients with Lynch syndrome or patients with serrated polyposis.[8]

Virtual chromoendoscopy

Virtual chromoendoscopy has emerged as a fast and simple alternative to conventional chromoendoscopy. It allows the mucosal pattern to be enhanced by modifying the light of the colonoscope by applying optical light filters (Olympus® NBI system) or by digital processing of the captured image (Pentax® iScan system and Fuji® FICE system).

Its great advantage is its ease of use, as it is activated by simply pressing a button on the endoscope, doing away with the need for more cumbersome staining.

Its use in assessing the entire mucosa has not been shown to improve the rate of adenomas,[9] but it does improve the characterization of lesions detected with white light, as will be seen later.

Use of ancillary equipment and new colonoscopes

Various endoscopic devices have been developed that, when added to the colonoscope, are intended to improve the detection of adenomas, especially by facilitating the identification of lesions "hidden" behind the folds or bends of the colon.

These devices include the cap, the EndoCuff®, and the EndoRing®. Multiple studies have shown an increase in adenoma detection with their use, especially for the cap and EndoCuff.[10]

With the same objective of improving the lesion detection rate, new colonoscopes have appeared on the market. This is the case of the FUSE® endoscope, which increases the usual field of view from 170 to 330 degrees, or the G-EYE® endoscope, which has a balloon attached to its distal part to improve visualization of the folds of the colon.

However, these endoscopes are not commonly available and more clinical studies evaluating their real benefit are still needed.[11]

Characterization of lesions

Once a lesion has been detected, it is essential to adequately characterize the lesion. Although colonoscopy will allow the removal of a large part of the lesions detected, not all neoplastic lesions can be approached from the endoscopic point of view. The morphological evaluation as well as the glandular pattern will help us to determine the most adequate therapeutic method in each case.

Before delving into the endoscopic characterization of neoplastic lesions of the colon, it is important to keep several key concepts in mind.

The concept of superficial neoplasia refers to those lesions whose deep invasion is limited at most to the submucosa, i.e., without involvement of the muscularis propria. Only superficial lesions can be removed during a colonoscopy using the usual techniques.

However, a complete endoscopic resection of a superficial lesion does not necessarily imply its cure. Only lesions that have a low risk for lymph node and lymphatic metastasis after histopathological analysis can be considered cured by endoscopy. This situation is known as early cancer.

The criteria for cure after endoscopic treatment are dealt with in greater depth in another chapter of the book, but for a start, and in order to understand what follows, it is important to bear in mind that superficial lesions with deep submucosal invasion cannot be considered cured after endoscopic resection due to the risk of distant metastasis.

Therefore the objective when characterizing a neoplastic lesion is to distinguish between lesions that can be treated endoscopically (i.e., with mucosal or superficial submucosal involvement) and lesions that cannot be cured by endoscopic techniques (superficial with deep submucosal involvement or advanced neoplasms).[12]

Morphological evaluation

With the acceptance of the Paris Classification[12] (Fig. 16.2), the morphological characterization of neoplastic lesions began to take on greater importance because of its ability to predict histology.

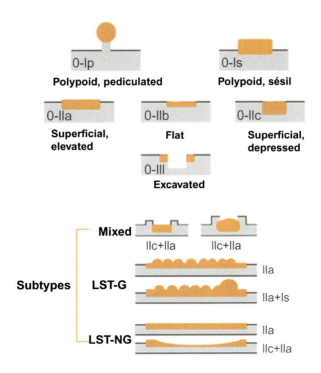

FIG. 16.2 Paris classification.[12]

The Paris classification[12] divides superficial neoplasms into polypoid or nonpolypoid, depending on their elevation above the surrounding mucosa; those with a height greater than 2.5 mm are considered polypoid.

In turn, polypoid lesions (0-I) can be divided into pedunculated (0-Ip) or sessile (0-Is) and nonpolypoid lesions into slightly elevated (0-IIa), flat (0-IIb), slightly depressed (0-IIc), and excavated (0-III).

There are also lesions showing a combined morphology (e.g., 0-IIa–Is).

Laterally spreading tumors (LST) refer to lesions of more than 1 cm in which, although the height sometimes exceeds 2.5 mm, the growth of the lesion is clearly horizontal, the base being much larger than its height. Depending on whether they have a nodular surface or not, they are classified as granular (LST-G) or nongranular (LST-NG), respectively.[13]

Some examples of polypoid lesions are shown in Fig. 16.3.

The relevance of these classifications, as previously mentioned, lies in their relationship with the histology of the lesion.

Thus, in polypoid lesions, the probability of deep infiltration is lower and increases with the size of the lesion, while nonpolypoid lesions, especially flat and excavated lesions, are associated with a higher probability of advanced histology, even when small in size.[13]

Granular homogeneous LSTs present a low risk of deep invasion, while mixed LSTs with nodules larger than 1 cm may present deep submucosal invasion. The highest risk of deep invasion is found in nongranular LSTs, especially if there is a depressed component.[14]

There are other morphological aspects not included in the Paris classification such as circumscribed areas of erythema, fold convergence, and "chicken skin" appearance, which have also been associated with the existence of deep infiltration.[15] As for the nonlifting of the lesion after injection of submucosal solutions (nonlifting sign), traditionally considered a sign of deep invasion, it is less reliable for diagnosing deep invasion than the endoscopic aspect.[16]

Evaluation of the glandular pattern

The evaluation of the glandular pattern of the lesions, together with the morphological evaluation described before, makes it possible to characterize them. It is performed using a chromoendoscopy technique, either conventional or virtual, on the

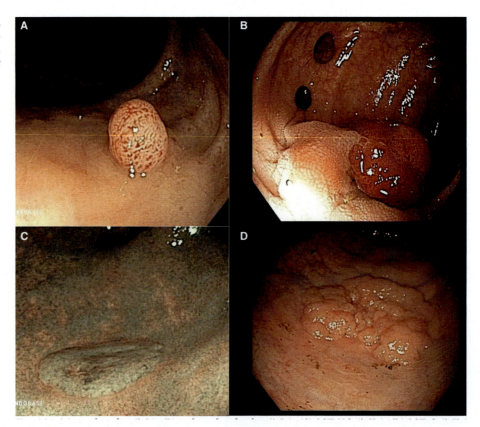

FIG. 16.3 (A) sessile polyp (0-Is), (B) pedunculated polyp (0-Ip), (C) - LST-NG (0-IIa), (D) LST-G (0-IIa). *(Photos from the archive of the Digestive System Department.)*

lesions detected, thus highlighting the glandular pattern in order to differentiate between neoplastic and nonneoplastic lesions. Among the neoplastic lesions, it aims to determine the degree of deep involvement.

Kudo's classification

Kudo's classification (see Fig. 16.4) establishes a correlation between the crypt pattern and the histology of the lesion, on which conventional chromoendoscopy has been previously performed.[17,18]

Its application requires the availability of magnifying colonoscopy, a high-definition colonoscopy that optically magnifies images up to ×100.

Subsequent to Kudo's classification, other classifications evaluating the glandular pattern using magnifying endoscopes were also developed. However, they made use of virtual chromoendoscopy instead of conventional chromoendoscopy. Among these, Sano's classification is the most important.[19]

Nevertheless, the main limitation for both classifications is that the implementation of magnifying colonoscopy is not widespread in Western countries, so there is less experience in this regard than that reported by Asian countries.

NICE classification

The advent of virtual chromoendoscopy led to the development of a new classification based on vascular and glandular pattern, using endoscopes without optical magnification (i.e., with commonly available high-definition endoscopes). This

Type	Schematic	Endoscopic	Description	Suggested Pathology	Ideal Treatment
I	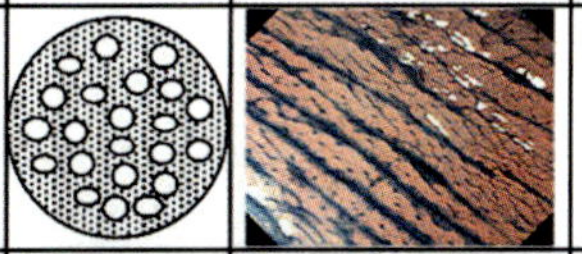		Round pits.	Non-neoplastic.	Endoscopic or none.
II	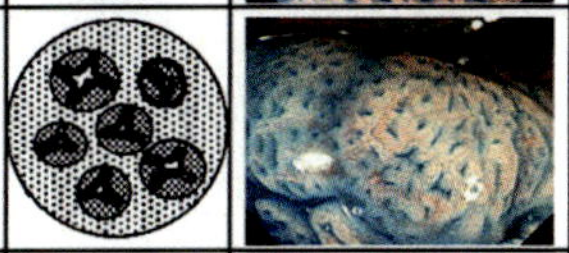		Stellar or papillary pits.	Non-neoplastic.	Endoscopic or none.
III$_S$	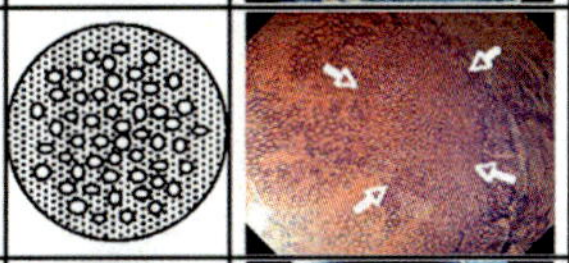		Small tubular or round pits that are smaller than the normal pit	Neoplastic.	Endoscopic.
III$_L$	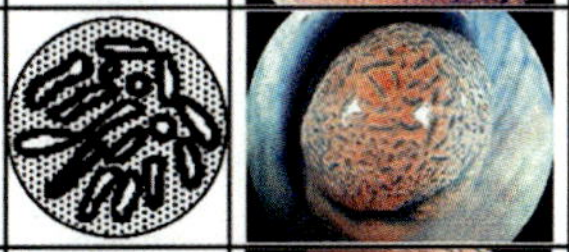		Tubular or roundish pits that are larger than the normal pits.	Neoplastic.	Endoscopic.
IV	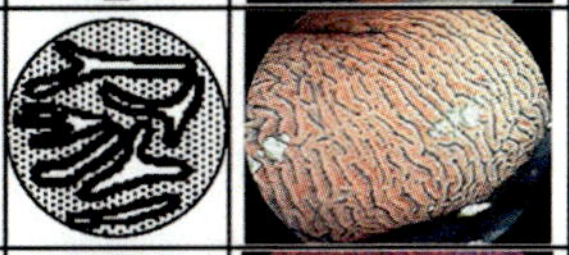		Branch-like or gyrus-like pits.	Neoplastic.	Endoscopic.
V$_I$	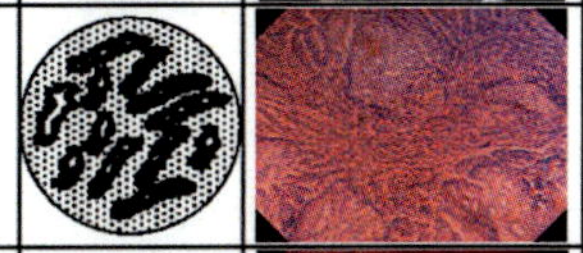		Irregularly arranged pits with type III$_S$, III$_L$, IV type pit patterns.	Neoplastic (invasive).	Endoscopic or surgical.
V$_N$	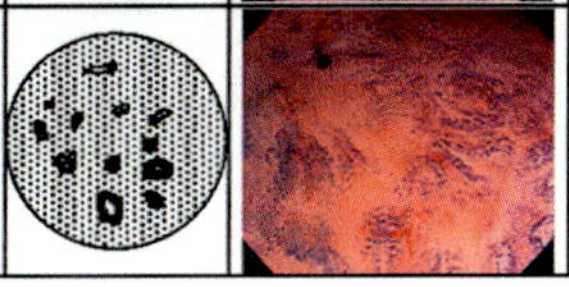		Non-structural pits.	Neoplastic (massive submucosal invasive).	Surgical.

FIG. 16.4 Schema of Kudo's classification.[17]

	TYPE 1	TYPE 2	TYPE 3
Color	Same or lighter than background	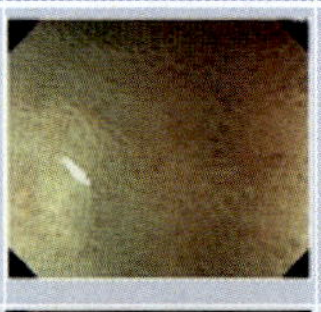 Browner relative to background	 Brown to dark brown relative to background; sometimes patchy whiter areas 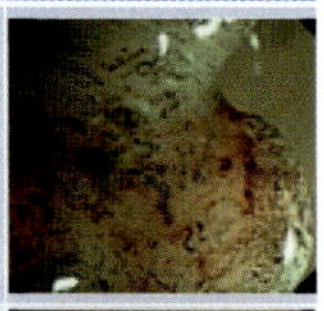
Vessels	None, or isolated lacy vessels may be present coursing across the lesion	 Brown vessels surrounding white structures	 Has area(s) of disrupted or missing vessels 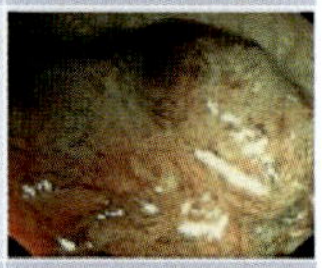
Surface pattern	Dark or white spots of uniform size, or homogeneous absence of pattern	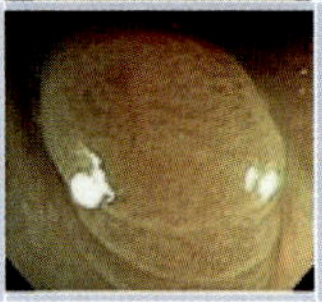 Oval, tubular or branched white structures surrounded by brown vessels	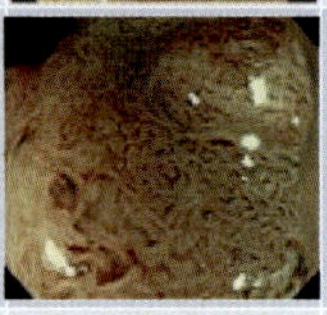 Amorphous or absent surface pattern 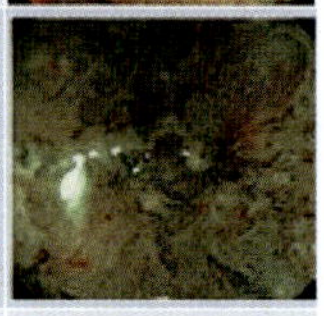
Most likely pathology	Hyperplastic	Adenoma	Deep submucosal invasive cancer
Treatment	Follow up	Polypectomy / EMR / ESD	Surgery

FIG. 16.5 NICE classification. *(Adapted from Iwatate M, Ikumoto T, Hattori S, Sano W, Sano Y, Fujimori T. NBI and NBI combined with magnifying colonoscopy.* Diagn Ther Endosc *2012;2012:173269.)*

classification, called the NICE (NBI International Colorectal Endoscopic) classification (see Fig. 16.5), is based on three criteria to predict the risk of deep invasion: colors, vessels, and surface patterns.[4]

Due to its ease of application, it has been widely accepted among endoscopists and its use has become widespread among endoscopy units. Numerous studies have shown it to have good diagnostic accuracy, especially in the differentiation of neoplastic from nonneoplastic lesions.[20,21] In Fig. 16.6 the same lesion is displayed with white light and with NBI, we find NICE 3 patterns, suggestive of invasive adenocarcinoma, later confirmed by histology.

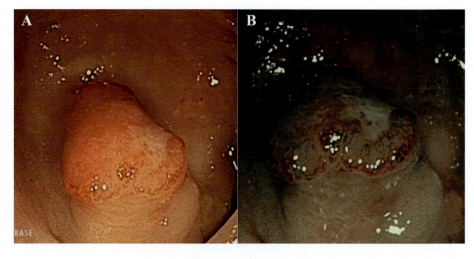

FIG. 16.6 (A) lesion with white light. (B) Lesion with NBI, with NICE 3 patterns, suggestive of invasive adenocarcinoma, later confirmed by histology. *(Photos from the archive of the Digestive System Department.)*

WASP classification

Traditionally, adenomas have been considered the only preneoplastic lesions in colon cancer. However, with the discovery of the serrated pathway of carcinogenesis, it is estimated that approximately 15%–30% of cancers have their origin in serrated lesions (traditional serrated adenomas and sessile serrated adenomas).

It is difficult to detect these lesions during colonoscopy because they are flat lesions with poorly defined margins. In addition, the previously described glandular pattern classification systems have not been developed for these lesions, so they are often misclassified as hyperplastic polyps or adenomas.

The WASP (Workgroup Serrated Polyps and Polyposis) classification was developed with the aim of differentiating hyperplastic lesions, adenomas, and serrated lesions. Lesions with a high probability of being serrated are considered to be those that fulfill two of these four characteristics: cloud-like appearance, poorly defined borders, irregular shape, and presence of black dots within the crypts.[22]

Example of a polyp with poorly defined borders and cloud-like appearance is shown in Fig. 16.7.

Therefore, to summarize, when performing an examination, we should use the best colonoscope available (high definition if possible) to evaluate the colonic mucosa. After detection of a lesion, its characterization should be supported by performing conventional or virtual chromoendoscopy, thus improving the morphological characterization and allowing evaluation of the glandular pattern (with magnifying endoscopes if available). Based on this, we will be able to establish the probability of deep invasion of the lesion that will determine the most appropriate therapeutic approach in each case.

Role of colonoscopy in the diagnostic of advanced colonic lesions

The diagnostic role of colonoscopy in advanced lesions basically consists, in addition to detecting the lesion, of taking biopsies for histopathological diagnosis. It also allows tattooing of the colonic mucosa to make it easier to locate the lesion during surgery. Different types of colon neoplasms are shown in Fig. 16.8.

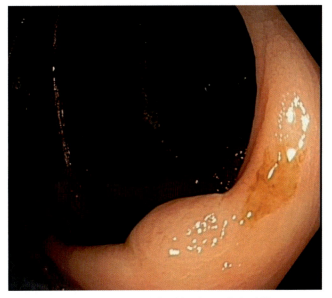

FIG. 16.7 Elevated flat polyp (0-IIa) in the right colon with poorly defined borders and cloud-like appearance, confirming the histology that it is a serrated adenoma. *(Photos from the archive of the Digestive System Department.)*

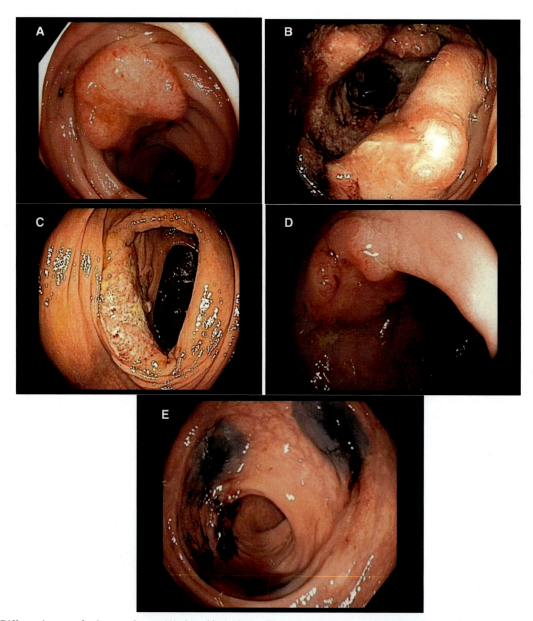

FIG. 16.8 Different images of colon neoplasms: (A) sigmoid neoplasm, (B) stenosing neoplasm of rectum, (C) ascending colon neoplasm, (D) early cancer in splenic flexure, (E) endoscopic tattoo distal to sigmoid neoplasm. *(Photos from the archive of the Digestive System Department.)*

References

1. Levin B, Lieberman DA, McFarland B, et al. Screening and surveillance for the early detection of colorectal cancer and adenomatous polyps, 2008: a joint guideline from the American Cancer Society, the US Multi Society Task Force on Colorectal Cancer, and the American College of Radiology. *Gastroenterology* 2008;**134**:1570–95.
2. Winawer SJ, Zauber AG, Ho MN, et al. Prevention of colorectal cancer by colonoscopic polypectomy. *N Engl J Med* 1993;**329**(27):1977–81. https://doi.org/10.1056/nejm199312303292701.
3. Grupo de trabajo AEG-SEED. *Guía de práctica clínica de calidad en la colonoscopia de cribado del cáncer colorectal*; 2011. https://www.aegastro.es/documents/pdf/guia_clinica_-_calidad_en_la_colonoscopia.pdf.
4. Iwatate M, Ikumoto T, Hattori S, Sano W, Sano Y, Fujimori T. NBI and NBI combined with magnifying colonoscopy. *Diagn Ther Endosc* 2012;**2012**:173269.
5. Froehlich F, Wietlisbach V, Gonvers J-J, Burnand B, Vader J-P. Impact of colonic cleansing on quality and diagnostic yield of colonoscopy: the European panel of appropriateness of gastrointestinal endoscopy European multicenter study. *Gastrointest Endosc* 2005;**61**(3):378–84.

6. Barclay RL, Vicari JJ, Doughty AS, Johanson JF, Greenlaw RL. Colonoscopic withdrawal times and adenoma detection during screening colonoscopy. *N Engl J Med* 2006;**355**(24):2533–41.
7. Subramanian V, Mannath J, Hawkey C, Ragunath K. High definition colonoscopy vs. standard video endoscopy for the detection of colonic polyps: a meta-analysis. *Endoscopy* 2011;**43**(06):499–505. https://doi.org/10.1055/s-0030-1256207.
8. Kamiński MF, Hassan C, Bisschops R, et al. Advanced imaging for detection and differentiation of colorectal neoplasia: European Society of Gastrointestinal Endoscopy (ESGE) guideline. *Endoscopy* 2014;**46**(5):435–49.
9. Nagorni A, Bjelakovic G, Petrovic B. Narrow band imaging versus conventional white light colonoscopy for the detection of colorectal polyps. *Cochrane Database Syst Rev* 2012;**1**, CD008361.
10. Gkolfakis P, Tziatzios G, Spartalis E, Papanikolaou IS, Triantafyllou K. Colonoscopy attachments for the detection of precancerous lesions during colonoscopy: a review of the literature. *World J Gastroenterol* 2018;**24**(37):4243–53.
11. Ishaq S, Siau K, Harrison E, et al. Technological advances for improving adenoma detection rates: the changing face of colonoscopy. *Digest Liver Dis* 2017;**49**(7):721–7. https://doi.org/10.1016/j.dld.2017.03.030.
12. Participants in the Paris Workshop. The Paris endoscopic classification of superficial neoplastic lesions: esophagus, stomach, and colon: November 30 to December 1, 2002. *Gastrointest Endosc* 2003;**58**(6 Suppl):S3–S43.
13. Group ECR, Endoscopic Classification Review Group. Update on the Paris classification of superficial neoplastic lesions in the digestive tract. *Endoscopy* 2005;**37**(6):570–8. https://doi.org/10.1055/s-2005-861352.
14. Tutticci N, Bourke MJ. Advanced endoscopic resection in the colon: recent innovations, current limitations and future directions. *Exp Rev Gastroenterol Hepatol* 2014;**8**(2):161–77. https://doi.org/10.1586/17474124.2014.866894.
15. Matsuda T, Parra-Blanco A, Saito Y, Sakamoto T, Nakajima T. Assessment of likelihood of submucosal invasion in non-polypoid colorectal neoplasms. Gastrointest Endosc Clin N Am. 2010;20(3):487-496.
16. Kobayashi N, Saito Y, Sano Y, et al. Determining the treatment strategy for colorectal neoplastic lesions: endoscopic assessment or the non-lifting sign for diagnosing invasion depth? *Endoscopy* 2007;**39**(08):701–5. https://doi.org/10.1055/s-2007-966587.
17. Tanaka S, Kaltenbach T, Chayama K, Soetikno R. High-magnification colonoscopy (with videos). *Gastrointest Endosc* 2006;**64**(4):604–13. https://doi.org/10.1016/j.gie.2006.06.007. 16996357.
18. Kudo S, Rubino C, Teixeira C, Kashida H, Kogure E. Pit pattern in colorectal neoplasia: endoscopic magnifying view. *Endoscopy* 2001;**33**(04):367–73. https://doi.org/10.1055/s-2004-826104.
19. Sano Y, Horimatsu T, Fu KI, Katagiri A, Muto M, Ishikawa H. Magnifying observation of microvascular architecture of colorectal lesions using a narrow-band imaging system. *Dig Endosc* 2006;**18**(s1):S44–51. https://doi.org/10.1111/j.1443-1661.2006.00621.x.
20. Hewett DG, Kaltenbach T, Sano Y, et al. Validation of a simple classification system for endoscopic diagnosis of small colorectal polyps using narrow-band imaging. *Gastroenterology* 2012;**143**(3). https://doi.org/10.1053/j.gastro.2012.05.006. 599-607.e1.
21. Puig I, López-Cerón M, Arnau A, et al. Accuracy of the narrow-band imaging international colorectal endoscopic classification system in identification of deep invasion in colorectal polyps. *Gastroenterology* 2019;**156**(1):75–87.
22. IJspeert JEG, Bastiaansen BAJ, van Leerdam ME, et al. Development and validation of the WASP classification system for optical diagnosis of adenomas, hyperplastic polyps and sessile serrated adenomas/polyps. *Gut* 2016;**65**(6):963–70.

Chapter 17

Oligometastatic disease

José Luis Ulla Rocha[a] and Paloma Sosa Fajardo[b]
[a]Gastroenterology Department, Digestive Disease Unit, Pontevedra University Hospital Complex, Pontevedra, Spain, [b]Radiation Oncology Unit, Universitary Hospital Complex of Santiago de Compostela, Santiago de Compostela, Spain

The increasing development of diagnostic tools in recent years has led to a new scenario of early diagnosis of distant colorectal metastasis, even before the appearance of any symptoms. This new stage of metastatic disease differs from the traditional Stage IV adenocarcinoma and represents a paradigm shift in the classification, treatment, and prognosis of metastatic colorectal cancer, showing improved outcomes in metastatic local control and survival rates. A new term was then required to describe this spectrum of a low burden metastatic disease: OLIGOMETASTASIS. This term was first proposed by Hellman and Weichselbaum in 1995,[1] and its defined as a condition with a few metastases arising from tumors that have not acquired a potential for widespread metastases.[2]

The oligometastatic disease implies that tumoral cells migrate through the body and colonize and nest in few locations, generating new tumors in those organs. Oligometastasis depicts an intermediate state between a localized disease and advanced metastatic stage and implies a clear difference in therapeutic approach, since it refers to less biologically aggressive tumors whose metastatic potential is believed to be more limited.

Currently, when a colorectal cancer is detected, a complete disease staging is mandatory. PET-CT and MRI can detect foci of tumoral cell groups in unexpected areas, apart from the locoregional site of disease.

Based on the different outcomes of the oligometastatic disease, many authors have described a subclassification that is now widely accepted and used, and includes different terms like oligorecurrence, oligoprogression, and oligopersistence, considering whether the oligometastatic disease is diagnosed during the treatment or the follow-up and whether or not an oligometastatic lesion is progressing on current imaging.[3] These terms will be deeply reviewed in next chapters.

So far, there is no established biomarker for the prediction of oligometastatic disease, but new hypothesis about an inherent susceptibility to develop a less invasive disease is currently being investigated, in addition to define which patient subgroups were expected to benefit most from local treatment.

A minimal invasive approach of metastases by sequential surgery was established and surgical teams have been developed in order to deal with liver and lung metastases, currently being this scheme the standard of care in metastatic colorectal cancer treatment.[4] The major limitation of this technique is the preservation of the FLR (Future Liver Remnant), a viable remnant of functioning liver that preserves the hepatic function in the future and avoids an acute hepatic failure.

Several groups have studied more aggressive therapeutic approaches in the oligometastatic setting of colorectal cancer, using a new concept "Metastasis-directed therapy (MDT)-local therapy."[5] Treatment of oligometastatic disease is expected to achieve long-term local control and to improve survival.

Among them, Rees[6] and colleagues and Fong[7] and colleagues described a couple of series of almost 2000 patients who underwent hepatic metastasectomy. Ten years cancer-specific survival and overall survival rates were 23% and 22%, respectively. In the setting of hepatic metastasectomy for colorectal primaries, numerous factors were associated with inferior outcome.[8] These included positive nodal status, extrahepatic disease, short metastasis disease-free interval (DFI), number of tumors (>1 or >3 depending on the study), preoperative carcinoembryonic antigen level (>60 or >200 ng/mL depending on the study), and size of largest tumor >5 cm. In the lung metastasectomy scenario, two series with more than 6000 patients achieved a 5-year actuarial survival rate of 50%: Casiraghi[9] and colleagues demonstrated underlying histology and DFI but not number of metastases as predictors of outcome, whereas in the larger series by Pastorino and colleagues, not only DFI but also number of metastases were found to be important.[10]

Disease-free interval (DFI), also known as disease-free survival (DFS) is defined as the interval between resection of the colorectal cancer and the diagnosis of metastases, and it is considered an important prognostic indicator. Several groups demonstrated the impact of a short DFI in predicting disease recurrence, but its paper in survival prediction is more controverted.[11,12]

Traditionally, radiotherapy played a solely palliative role in the therapeutic approach of metastatic stages due to the lack of efficient ways to deliver RT doses while limiting the adjacent organs (organs at risk) toxicity.

However, in the past years, solid advances in technology allowed to safely deliver high doses of RT per fraction in to a small target using SABR[13] (Stereotactic Ablative Radiotherapy). Oligometastases can now be treated in a less invasive local way. A recent randomized phase II study, SABR-COMET,[14] proved that SBRT was associated with an improvement in overall survival. Following the success of stereotactic radiosurgery, SABR has been widely accepted for treating extra-cranial metastases, considering its efficacy and minimum invasiveness.[15] SABR provides high level of local control with minimal associated toxicity.

In addition, in recent past years a new method of primary diagnosis known as liquid biopsy was described. Liquid biopsy[16] extracts normal cell and tumoral DNA and RNA from peripheral blood in a minimal invasive procedure of venepuncture. DNA and RNA can be useful tools used in a near future for molecular diagnosis of any occult neoplasia, tumor relapse, or their body oligometastases.

In the long term it is expected that a top notch immunotherapy will allow to curb even an advanced disease, thus the current difference between oligometastatic and advanced disease can be ignored.

References

1. Hellman S, Weichselbaum RR. Oligometastases. *J Clin Oncol* 1995;**13**(1):8–10. https://doi.org/10.1200/JCO.1995.13.1.8.
2. Otake S, Goto T. Stereotactic radiotherapy for oligometastasis. *Cancers* 2019;**11**(2). https://doi.org/10.3390/cancers11020133.
3. Guckenberger M, Lievens Y, Bouma AB, et al. Characterisation and classification of oligometastatic disease: a European Society for Radiotherapy and Oncology and European Organisation for Research and Treatment of Cancer consensus recommendation. *Lancet Oncol* 2020;**21**(1): e18–28. https://doi.org/10.1016/S1470-2045(19)30718-1.
4. Jarabo JR, Gómez AM, Calatayud J, et al. Combined hepatic and pulmonary metastasectomies from colorectal carcinoma. Data from the prospective Spanish registry 2008-2010. *Arch Bronconeumol* 2018;**54**(4):189–97. https://doi.org/10.1016/j.arbres.2017.10.003.
5. Beckham TH, Yang TJ, Gomez D, Tsai CJ. Metastasis-directed therapy for oligometastasis and beyond. *Br J Cancer* 2021;**124**(1):136–41. https://doi.org/10.1038/s41416-020-01128-5.
6. Rees JR, Rees M, McNair AGK, et al. The prognostic value of patient-reported outcome data in patients with colorectal hepatic metastases who underwent surgery. *Clin Colorectal Cancer* 2016;**15**(1). https://doi.org/10.1016/j.clcc.2015.07.003. 74–81.e1.
7. Fong Y, Fortner J, Sun RL, Brennan MF, Blumgart LH. Clinical score for predicting recurrence after hepatic resection for metastatic colorectal cancer: analysis of 1001 consecutive cases. *Ann Surg* 1999;**230**(3):309–18. discussion 318–321 https://doi.org/10.1097/00000658-199909000-00004.
8. Engstrand J, Nilsson H, Strömberg C, Jonas E, Freedman J. Colorectal cancer liver metastases—a population-based study on incidence, management and survival. *BMC Cancer* 2018;**18**(1). https://doi.org/10.1186/s12885-017-3925-x.
9. Casiraghi M, De Pas T, Maisonneuve P, et al. A 10-year single-center experience on 708 lung metastasectomies: the evidence of the international registry of lung metastases. *J Thorac Oncol* 2011;**6**(8):1373–8. https://doi.org/10.1097/jto.0b013e3182208e58.
10. Pastorino U, Buyse M, Friedel G, et al. Long-term results of lung metastasectomy: prognostic analyses based on 5206 cases. *J Thorac Cardiovasc Surg* 1997;**113**(1):37–49. https://doi.org/10.1016/s0022-5223(97)70397-0.
11. Höppener DJ, Nierop PMH, van Amerongen MJ, et al. The disease-free interval between resection of primary colorectal malignancy and the detection of hepatic metastases predicts disease recurrence but not overall survival. *Ann Surg Oncol* 2019;**26**(9):2812–20. https://doi.org/10.1245/s10434-019-07481-x.
12. Zhao L, Wang J, Li H, Che J, Cao B. Meta-analysis comparing maintenance strategies with continuous therapy and complete chemotherapy-free interval strategies in the treatment of metastatic colorectal cancer. *Oncotarget* 2016;**7**(22):33418–28. https://doi.org/10.18632/oncotarget.8644.
13. Kobiela J, Spychalski P, Marvaso G, et al. Ablative stereotactic radiotherapy for oligometastatic colorectal cancer: systematic review. *Crit Rev Oncol Hematol* 2018;**129**:91–101. https://doi.org/10.1016/j.critrevonc.2018.06.005.
14. Palma DA, Olson R, Harrow S, et al. Stereotactic ablative radiotherapy versus standard of care palliative treatment in patients with oligometastatic cancers (SABR-COMET): a randomised, phase 2, open-label trial. *Lancet* 2019;**393**(10185):2051–8. https://doi.org/10.1016/S0140-6736(18)32487-5.
15. Gutiontov SI, Pitroda SP, Weichselbaum RR. Oligometastasis: past, present, future. *Int J Radiat Oncol Biol Phys* 2020;**108**(3):530–8. https://doi.org/10.1016/j.ijrobp.2020.02.019.
16. Stelcer E, Konkol M, Głęboka A, Suchorska WM. Liquid biopsy in oligometastatic prostate cancer—a biologist's point of view. *Front Oncol* 2019;**9**. https://doi.org/10.3389/fonc.2019.00775.

Chapter 18

Hereditary nonpolyposis CRC

M. Lidia Vazquez-Tunas
Medical Oncology Services, Álvaro Cunqueiro Hospital, Vigo University Hospital Complex, Vigo, Pontevedra, Spain

Introduction

Colorectal cancer (CRC) is the second most common cancer in women (after breast cancer) and the third most common cancer diagnosed in men (after lung and prostate cancer).[1] Familial CRC comprises approximately 20%–30% of all diagnosed cases. However, the genetic variant responsible for most of them is currently unknown. In addition, nongenetic factors such as exposure to certain environmental carcinogens may contribute to familial aggregation of CRC cases without an underlying intergenerationally transmissible genetic cause.

Hereditary nonpolyposic colorectal cancer (HNPCC), also known as Lynch syndrome, is the most common of the hereditary CRC syndromes, accounting for at least 2%–3% of all CRC and 3% of endometrial cancer cases.[2,3] Clinically, it corresponds to a genetic predisposition to the appearance of colon cancer at an early age, but it is also associated, although with a lower risk of occurrence, with extracolonic neoplasms: endometrial, ovarian, stomach, small intestine, hepatobiliary tract, upper urinary tract, skin, and brain.

It is an inherited disease with an autosomal dominant pattern of inheritance, due to germline mutation in genes involved in the repair of missing DNA bases (mismatch repair genes or *MMR* genes).[4,5]

An individual carrying a mutation in one of these repair genes has a cumulative lifetime risk of developing colorectal cancer of approximately 80%, 60% for endometrial cancer, between 10% and 15% for ovarian or stomach tumors, and a higher risk than the general population for tumors of the urinary tract, small intestine, bile duct and pancreas, and sebaceous tumors of the skin.[6] This makes the identification of subjects predisposed to HNPCC essential, since it allows the adoption of prevention measures of recognized efficacy, both for colorectal cancer and for other types of associated tumors.

We must differentiate Lynch syndrome from other hereditary familial colorectal cancer syndromes (i.e., familial adenomatous polyposis, etc.), as both surveillance and treatment may be different.

Genetic basis

The molecular genetic characteristics of Lynch syndrome were established in 1993 following genome sequencing of large families.[7,8] It presents an autosomal dominant inheritance pattern and 3% is due to germline mutations in the *MLH1*, *MSH2*, *MSH6*, *PMS2*, or *EPCAM* genes. The lifetime risk of developing CRC in carriers of a mutation varies between 10% and 75%.[1]

The DNA replication process, a prerequisite for cell division, is prone to errors. These errors, if not repaired, result in the development of tumors. The mismatch repair (*MMR*) gene complex is one of the DNA base pairing error repair systems in humans and its function is to maintain genomic integrity by correcting nucleotide change and pairing errors in small insertions–deletions that are generated during DNA duplication prior to mitosis.

MMR genes associated with Lynch syndrome include:

— *MSH2*: MutS homolog 2, located on chromosome 2p16.
— *MLH1*: MutL homolog 1, located on chromosome 3p21
— *PMS1*: postmeiotic segregation 1, located on chromosome 2q31
— *PMS2*: postmeiotic segregation 2, located on chromosome 7p22
— *MSH6*: MutS homolog 6, located on chromosome 2p16.[4]

Mutations in *MLH1* and *MSH2* comprise about 32% and 39% of HNPCC cases, both of which have the highest frequency of occurrence; mutations in *MSH6* account for about 15%, and mutations in *PMS2* are very sporadic (less than 14%).[9] There are other genes that may also be involved in HNPCC, such as *MSH3*, *EXO1*, and *TGFbR2*, although their clinical significance is not well established. The existence of de novo mutations is anecdotal and generation skipping is rarely observed. Even so, between one-third and one-half of hereditary colorectal cancer cases are caused by as yet unknown mutations.[10–12]

The *MMR* system recognizes base pair mismatches and repairs them, and requires inactivation of both alleles of some *MMR* gene to malfunction. Patients with HNPCC have a germline mutation in one allele of an *MMR* gene and the second allele is inactivated by mutation, loss of heterozygosity, or epigenetic silencing by promoter hypermethylation. This biallelic inactivation results in an increased mutation rate (genomic instability) due to the inability to repair mismatches that normally occur during DNA synthesis. DNA mismatches commonly occur in regions of short repetitive nucleotide sequences (microsatellites). Therefore a characteristic feature of loss of mismatch repair in tumors is the expansion or contraction of these microsatellite sequence regions in the tumor compared with normal tissue. This genetic alteration is termed microsatellite instability and is a molecular signature of HNPCC-associated tumors.[13]

The resulting increased mutation rate leads to alterations in nucleotide repeats in many genes, including some that control cell growth (TGF beta and IGF receptors) or regulate cell death by apoptosis (caspase 5, Bax), and some in the repair genes themselves (*MSH3*, *MSH6*).[14] It is believed that the accumulation of mutations in these genes induces the process of carcinogenesis in Lynch syndrome.

However, microsatellite instability (MSI) is not unique to Lynch syndrome and is also observed in tumors of sporadic origin as a consequence of biallelic somatic mutation of *MMR* genes, with up to 15% of CRC tumors estimated to have MSI. Tumors with elevated MSI (MSI-H) generally have a higher level of DNA methylation, no mutation in *KRAS* but mutation in *BRAF*, and a low frequency of mutations in *APC* and *TP53*.[1]

There are two other possible scenarios, but less frequent, that cause Lynch syndrome and that are not a direct consequence of mutations in *MMR* genes, although they do determine its malfunction. One of them is the presence of large deletions at the 3′ end of the *EPCAM* gene, which determines the transcriptional read-through and subsequent downstream epigenetic silencing of the neighboring *MSH2* gene.[15] *MSH2* is only inactivated in cells in which the *EPCAM* locus is active, so the tumor shows a mosaic pattern of *MSH2* inactivation.

The other situation is constitutive inactivation of *MLH1* as a consequence of hypermethylation of the gene in the embryonic stage.[13]

Molecular identification of Lynch syndrome

Lynch syndrome should be suspected in patients with synchronous or metachronous CRC, CRC diagnosed at an early age (before 50 years of age), presentation of several tumors of the Lynch syndrome spectrum (CRC, endometrial, ovarian, gastric, renal, etc.), and in cases with familial association. The study of the tumor tissue provides rapid information to select the cases in which to carry out a subsequent complete genetic study to confirm the diagnosis. This study is performed through the analysis of microsatellite instability and/or determination of protein expression of *MMR* genes. The absence of MSI and the correct expression of MMR proteins rule out most cases of HNPCC.

Microsatellite instability analysis

It consists in the study of the length of tumor DNA fragments containing microsatellites, which is evidence of malfunction of the *MMR* genes. In order to avoid false negatives, this test must be performed on a paraffin-embedded tumor block when the area corresponding to the tumor is correctly marked. It is advisable to include healthy tissue samples for analysis (in the same piece of paraffin or separately, colorectal mucosa adjacent to the tumor or blood).

By performing PCR (polymerase chain reaction) techniques, we can establish microsatellite instability on a standard panel of DNA sequences. The American National Cancer Institute has defined a standard panel of five microsatellite markers (BAT25, BAT26, D2S123, D5S346, and D17S250).[16,17] High instability (MSI-H) is defined as two or more of the five markers showing instability (or 30% or more of the markers if more than five are analyzed). Low microsatellite instability (MSI-L) is defined when one of the five markers shows microsatellite instability (or less than 30% of the markers), and we speak of microsatellite stability (MSS) when there is an absence of instability.[18]

The presence of MSI-H in tumor tissue suggests the existence of a defect in one of the *MMR* genes responsible for the DNA matching repair system.[19] More than 90% of tumor tissue from patients with Lynch syndrome show high levels of MSI-H, but despite this high sensitivity, its specificity is low and up to 7%–15% of sporadic colorectal tumors have MSI-H,[20] determined by hypermethylation of the *MLH1* promoter region.

The two groups of MSI-H tumors can be differentiated by direct measurement of *MLH1* methylation status in the tumor, or more simply by genetic analysis of the *BRAF* gene. Somatic mutation of exon 15 of the *BRAF* oncogene (V600E) in tumor tissue occurs almost universally in sporadic MSI-H cancers, but is rare in Lynch syndrome-associated cancers. Identification of a *BRAF* mutation in MSI-H colorectal cancer essentially rules out Lynch syndrome.[21–24]

Immunohistochemistry for reparative proteins in tumor tissue

Mutations in the *MMR* genes that cause Lynch syndrome often result in a truncated or missing protein product; therefore, immunohistochemistry (IHC) techniques for identification of MMR proteins can be used to establish colon tumors that have lost these products.[17,22] Loss of *MSH2* or *MSH6* expression by IHC in colorectal cancer is highly specific to Lynch syndrome, whereas loss of *MLH1* expression can also occur in sporadic colorectal cancers due to hypermethylation of the *MLH1* promoter.[18,25] The determination of expression by IHC allows us to find out, by means of a rapid and inexpensive technique, the gene affected, as well as the importance of the defect that has occurred.[10] The sensitivity and specificity values for detecting MMR deficiency by IHC are 83% and 89%, respectively.

The correlation between loss of *MMR* gene expression and microsatellite instability is very high,[26] such that only 0.2% loss of protein expression has been demonstrated in tumors with microsatellite stability or low instability. It should be noted that up to 5% of tumors with high instability show no loss of protein expression.[27,28] The loss of protein expression can guide subsequent genetic study as it is associated with a germline mutation in one of the genes that code for them.

IHC techniques have also proven effective in endometrial cancer for the identification of Lynch syndrome, *MLH1* methylation can cause epigenetic MSI-H endometrial tumors, but *BRAF* V600E mutations are not common in endometrial tumors. In other Lynch syndrome-associated tumors, IHC techniques have been less rigorously studied, and while abnormal tumor studies may be indicative of Lynch syndrome, normal tumor studies do not necessarily rule it out.[18]

Hypermethylation of *MLH1*

The most frequent mechanism of MMR deficiency in sporadic tumors is methylation of the *MLH1* promoter, leading to blockage of *MLH1* gene expression and loss of function.

The presence of *MLH1* methylation in the tumor suggests a sporadic origin, with two exceptions:

- The possibility of constitutive methylation or epimutation of *MLH1*.
- The possibility that *MLH1* methylation occurs as a second inactivating event in cases with germline mutation. This situation has been described in up to 15% of colorectal tumors in Lynch syndrome.[1]

Detection of germline mutations in genomic DNA

The detection of point mutations in the involved repair genes can be performed by various techniques depending on the preferences and availabilities of the laboratory (SSCP [Single Strand Conformation Polymorphism], PTT [Protein Truncation Test], DHPLC [Denaturing High Performance Liquid Chromatography], DGGE [Denaturing Gradient Gel Electrophoresis], etc.).[18,22,29] These techniques should cover the entire coding sequence of genes and adjacent intronic regions.

When an alteration in the DNA strand is detected, it is necessary to characterize it by direct Sanger sequencing which, due to its sensitivity, is considered the gold standard for mutation characterization.

Detection of large genomic rearrangements in the repair genes can also be performed prior to screening for point mutations, since these alterations appear with some frequency in families with HNPCC. In case of detection of a genomic rearrangement, point mutation screening could be bypassed.[30–32]

Clinical features

The first description of a family with HNPCC was made by Warthin in 1913, he called it "family G," and in its family tree, there were numerous cases of patients diagnosed with CRC in the absence of polyposis, and also patients with endometrial and gastric cancer.[33] Subsequently, Lynch described two other large families in which a tumor association similar to that described by Warthin was present.[34] The characteristics of HNPCC are summarized in Table 18.1.

Patients with Lynch syndrome have a significantly increased risk of colorectal cancer, endometrial cancer, and several other types of cancer such as cancer of the ovary, ureter and renal pelvis, stomach, small intestine, bile duct and pancreas, skin (sebaceous adenomas and carcinomas and keratoacanthomas), and brain tumors. Over time, a variation in the spectrum of tumors presented in families with Lynch syndrome has been observed, thus the family described by Warthim[33] presented mainly gastric and endometrial tumors, while the families described by Henry Lynch[34,35] had CRC as the tumor that most

TABLE 18.1 Characteristics of Lynch syndrome or HNPCC.

Autosomal dominant inheritance
Association of multiple tumors: CRC, gastric, endometrial, ovarian, ureter and renal pelvis, small intestine, biliary tract and pancreas, skin
Early age of onset
Synchronous or metachronous multiple tumors
Colorectal cancer characteristics: proximal colon, diploid tumors, poorly differentiated, presence of lymphocytes around the tumor, signet ring cell differentiation and Crohn-like reaction
Rapid progression of the adenoma-carcinoma sequence (3 years)
Frequent microsatellite instability
Loss of *MLH1, MSH2, MSH6, PMS2* protein expression

commonly appeared. This variation is due to the fact that environmental factors are also important contributors to carcinogenesis. Table 18.2 presents the risk of developing different types of cancer in patients with HNPCC.[17,22,36,37]

The early age of presentation and the multiplicity of cancers have been considered characteristic of Lynch syndrome, as is the case in other hereditary cancer syndromes.

The median age at diagnosis of CRC is between 44 and 52 years,[38–40] in contrast to sporadic CRC which presents at a median age of 71 years. Similarly, the median age of diagnosis of endometrial cancer in families with HNPCC is 10 years lower than that of sporadic endometrial cancer onset (50 and 60 years, respectively).[22] Although Lynch syndrome is characterized by an early age of CRC onset, in families with known mutation when probands were excluded and all diagnoses and absence of diagnoses were ascertained, the mean age at CRC diagnosis was found to be 61 years,[41] suggesting that this syndrome should also be considered in families with colorectal cancer expression in older patients.

Patients with HNPCC often have synchronous colorectal tumors (7%–10% of patients) and metachronous tumors (20%–40% of patients) if subtotal colectomy is not performed at diagnosis,[42] and the risk of second neoplasm increases with time. In addition, they are at higher risk of developing colon adenomas and the onset of these appears to occur at an earlier age than in individuals not carrying *MMR* gene mutations from the same families, the mean age at diagnosis of adenoma in carriers was 43.3 years (range, 23–63.2 years) and the mean age at diagnosis of carcinoma was 45.8 years (range, 25.2–57.6 years).[43] Similar to sporadic cancers, HNPCC originate from adenomas, but are usually larger, flat in morphology, have high-grade dysplasia, or have villous histology. It is believed that the adenoma-carcinoma sequence progresses much more rapidly in Lynch syndrome, so that it is possible to observe a new neoplasm and within 2–3 years after what appeared to be a negative colonoscopy.[44,45]

TABLE 18.2 Lifetime risk of developing cancer in Lynch syndrome.

CCR (male)	30%–80%
CCR (female)	25%–70%
Endometrial cancer	30%–70%
Ovarian cancer	3%–13%
Gastric cancer	2%–19%
Urinary tract cancer	1%–12%
Brain tumor	1%–4%
Bile duct and biliary tract cancer	2%–7%
Small bowel cancer	1%–7%

Unlike patients with sporadic cancers, which most often develop on the left side of the colon, approximately two-thirds of cancers in Lynch syndrome arise on the right side of the colon, proximal to the splenic flexure.[22] In general, tumors in patients with HNPCC resemble sporadic CRC, but have a higher frequency of lymphocytic infiltration, Crohn-like reaction, and increased presence of peritumoral lymphocytes. Colorectal tumors with MSI-H are more likely to present two or more cellular subclones, such that we find areas of different differentiation or different tumor type (e.g., mucinous and nonmucinous).[19,46] Although their histologic features may appear more aggressive, the 5-year survival rate is better than that observed in sporadic colorectal cancer, suggesting that the biology and natural history of Lynch syndrome-associated tumors is fundamentally different from that of sporadic colorectal cancer.[47,48]

Lynch syndrome patients and their family members are at risk for a wide variety of synchronous or metachronous associated tumors (Table 18.2), not just colorectal cancer. The most common is endometrial adenocarcinoma. The cumulative lifetime risk of a woman with HNPCC for endometrial cancer varies depending on the mutated DNA repair gene and is estimated at 44%–54% if they are carriers of mutation in *MLH1* and *MSH2*. If the mutation is in the *MSH6* or *PMS2* genes, risk decreases to around 25%–25%.[37,41,49] Most endometrial cancers are of the endometrioid type, although clear cell carcinoma, uterine papillary serous carcinoma, and mixed malignant Müllerian tumors can also be found.[50]

Different studies have shown that patients with Lynch syndrome are also at risk of developing transitional cell carcinoma of the ureters and renal pelvis and cancer of the stomach, small bowel, liver and biliary tract, brain, breast, and ovary.[51–53]

Muir-Torre syndrome is currently considered a variant of Lynch syndrome and includes in its phenotype the presence of multiple cutaneous neoplasms (including sebaceous adenomas, sebaceous carcinomas, and keratoacanthomas) and visceral carcinomas. Sebaceous tumors are rare and represent a good diagnostic marker for HNPCC.[1] Mutations in the *MLH1* and *MSH2* genes have been found in Muir-Torre families,[54–57] although alterations have also been described in MUTYH-associated polyposis.

Turcot syndrome associates brain tumors, predominantly gliomas, but also glioblastomas, astrocytomas, and oligodendrogliomas.[22] The relative risk of developing a brain tumor in patients with HNPCC and their first-degree relatives is estimated to be 6 times higher than in the general population,[58] but with low cumulative lifetime risk, so screening for brain tumors in these families is not recommended. The term Turcot syndrome is also used in families with adenomatous polyposis due to *APC* gene mutation presenting with central nervous system tumors, most typically medulloblastomas.

Diagnosis of Lynch syndrome (HNPCC) and criteria for referral to a genetic counseling unit (GCU)

In 1989 the international collaborative group on HNPCC (now the International Society of Gastrointestinal Hereditary Tumors—InSiGHT) was established to promote research projects in this field, and in 1991, the Amsterdam criteria were published.[59] These criteria provided general guidance for identifying families with HNPCC, allowed international studies and unified the terminology in use (Table 18.3). In 1999 a second set of criteria, Amsterdam II, was proposed to include extracolonic cancers (endometrial, ovarian, urinary tract, etc.).[60]

In 1996, following the discovery of microsatellite instability, the Bethesda criteria were proposed to improve sensitivity in the identification of affected families; these criteria were revised in 2004 (Table 18.4) [61,62]. These criteria are less stringent in identifying families with germline mutations in one of the *MMR* genes.

Tumors associated with HNPCC: colorectal tumors, endometrial, stomach, ovarian, urinary tract, pancreas, small intestine, biliary tract and brain tumors, sebaceous gland adenomas, and keratoacanthomas.

These suspicion criteria (Amsterdam and Bethesda) allow a strict selection of the individuals on whom we will perform the genetic analysis, which will provide us with the definitive diagnosis of HNPCC in case a mutation in the repair genes is found. The Amsterdam criteria have low sensitivity, which is a limitation for their use as a diagnostic tool in clinical practice. The Bethesda criteria can be used to identify individuals who are candidates for microsatellite instability and repair protein expression techniques.

The indications for referring an individual with suspected HNPCC to a Genetic Counseling Unit are:

— Individual affected by colorectal cancer if they meet the Bethesda criteria.
— Healthy individual or individual with CRC or associated tumor if they belong to a family that meets the Amsterdam criteria.

The indication criteria for the performance of a molecular study are patients with colorectal cancer who meet the Amsterdam or Bethesda criteria.

TABLE 18.3 Amsterdam I and II criteria (all criteria must be met).

AMSTERDAM CRITERIA I

There must be at least three affected family members with CRC:

- One of the affected parties must be a first-degree relative of the other two
- At least two successive generations must be affected
- At least one tumor must be diagnosed before the age of 50 years old
- Familial adenomatous polyposis should be excluded
- Tumors must be confirmed by anatomopathological study

AMSTERDAM CRITERIA II

At least three family members with CRC or a cancer associated with Lynch syndrome (endometrial, small bowel, ureter, or renal pelvis cancer)

- One of the affected parties must be a first-degree relative of the other two
- At least two successive generations must be affected
- At least one tumor must be diagnosed before the age of 50 years old
- Familial adenomatous polyposis should be excluded in case of CRC.
- Tumors must be confirmed by anatomopathological study

TABLE 18.4 Revised Bethesda criteria (at least one must be met).

CRC diagnosed in a patient younger than 50 years old

Presence of synchronous, metachronous, or other colorectal tumors associated with HNPCC, regardless of age

CRC with MSI-H histology diagnosed in a patient younger than 60 years

Patient with CRC and a first-degree relative presenting with tumor associated with HNPCC, with one of the tumors diagnosed under 50 years of age

CRC patient with two or more first- or second-degree relatives with a tumor associated with HNPCC, regardless of age

Technological advances for the determination of MSI and the study of DNA mutations revealed the existence of families that met the Amsterdam criteria for Lynch syndrome, but in which it was not possible to detect a pathogenic mutation, either because it occurs in regions of DNA not analyzed or because it was classified as a variant of uncertain significance. In these families, there is a lower incidence of colorectal cancer than in families with Lynch syndrome, the age of diagnosis is later, and they do not have a higher incidence of other types of cancers, so the surveillance and follow-up protocol is also less strict. These families should not be described or classified as HNPCC, so it is advisable to refer to them as "familial colorectal cancer type X."[63]

Molecular study strategy

The detection of germline *MMR* gene mutations in families at high risk for HNPCC is desirable because it improves the efficiency of cancer prevention and early diagnosis measures. Family members who do not carry the mutation causing the spectrum of tumors in the family will not be subjected to intensive surveillance or follow-up beyond that recommended for the general population. We must perform an appropriate selection of at-risk families prior to mutation screening, since it represents not only an economic cost but also an added psychological cost.

Different algorithms have been proposed for the evaluation of individuals with suspected Lynch syndrome, mostly based on the identification of high-risk individuals and the probability of isolating a mutation.[64] In families meeting the Amsterdam I and II criteria, the individual with the most clinical factors suggestive of HNPCC will be selected for testing and MSI and IHC testing of repair proteins (*MLH1*, *MSH2*, *MSH6*) will be performed. The advantage of the IHC study over microsatellite instability is that it allows targeting the mutation analysis, since the staining pattern suggests the underlying genetic defect, therefore, most authors recommend using MSI and IHC in a complementary manner. If the tumor shows instability or loss of expression of any protein, germline mutation analysis of the corresponding *MMR* gene should be performed. Otherwise, it is not recommended to continue with the genetic study given the low probability of identifying a pathogenic mutation. If IHC and MSI are not assessable or no viable tumor is available, germline study of *MLH1* and *MSH2* should be performed.

Exceptionally, in the absence of a living tumor-affected relative, mutation testing can be considered in healthy individuals at risk of being carriers, provided that there is an orientation of the gene to be studied and that they have received genetic counseling.[18]

The results of the study can be:

- IHQ/MSI:
  - IHQ with conserved repair protein expression and microsatellite stability (MSS): it is not appropriate to perform mutational study.
  - MSI-H or altered IHQ: mutation study in repair genes.
- Study of mutations in repair genes:
  - Informative germline study. Assumes the finding of a mutation in *MMR* genes, deletion/insertion catalogued as a pathogenic mutation. It is necessary to perform a carrier study in family members.
  - Noninformative study. No pathogenic alteration is found. Study of relatives is not appropriate.
  - Variant of uncertain significance. It is a germline mutation finding, but its pathogenic potential is unknown. A joint assessment between geneticists and clinicians is recommended and research studies should be carried out to determine the pathogenicity of the variant and whether or not to extend the study to relatives.

Predictive models

Several computer predictive models have been proposed to quantify the risk of having a germline mutation of reparative genes, incorporate data from personal and family history of cancer to calculate the probability of mutations in *MMR* genes, and recommend or not the genetic study. However, these models do not replace the clinician's judgment in the suspicion of a hereditary syndrome.[65]

These models include MMRPredict, MMRPro, and Prediction of genetic mismatch repair mutations in *MLH1*, *MSH2*, and *MSH6* (PREMM1,2,6).[66–69]

Clinical controls and surveillance

Surveillance of individuals who belong to families with HNPCC should be different from what we suggest in the general population, since tumor onset occurs at an earlier age than for sporadic cancers (estimated risk of colorectal cancer at age 40 years is 31% for males and 32% for females).[7] A higher proportion of CRC occurs in the right colon, so sigmoidoscopy alone is not a good follow-up strategy and a complete colonoscopy should be performed. Moreover, since adenoma-carcinoma progression is accelerated, early detection tests should be performed more frequently than in the general population (every 1–2 years).

We should also not forget that individuals belonging to families with NSCLC HNPCC are at high risk for other types of tumors, especially endometrial and ovarian carcinoma in women.

Follow-up of healthy individuals in families with HNPCC

Not all members of a family are to be monitored equally. Individuals are considered at risk:

- Asymptomatic healthy individuals carrying pathogenic mutation in *MMR* genes.
- Asymptomatic first-degree relatives of a patient with HNPCC in whom it has not been possible to demonstrate the pathogenic mutation responsible for the disease.[18]

Colorectal cancer screening

The different guidelines recommend starting screening at the age of 20–25 years (or 2–5 years before the age of the youngest diagnosis if younger than 25 years), although the diagnosis of CRC before the age of 25 years is usually exceptional.[1,18,70–72] The risk of CRC after the age of 80 years is considered low, so individualized assessment is recommended.

Complete colonoscopy should be performed biannually until the age of 40 years; after this age, the recommendation is to perform a colonoscopy every year.[18] Follow-up colonoscopy has been shown to offer a benefit in terms of reducing the incidence rate by 62% and the mortality rate by 65% in these families.[73]

Chromoendoscopy added to conventional colonoscopy as a screening technique appears to be more effective in patients diagnosed with Lynch syndrome.[72]

Endometrial and ovarian cancer screening

Given the cumulative lifetime risk of women in these families for endometrial and ovarian cancer (25%–50% and 8%–12%, respectively), annual transvaginal ultrasound and endometrial aspiration cytology are recommended from the age of 30 years and, optionally, determination of serum CA 125 levels also on an annual basis.[18,72] However, the sensitivity and specificity of these tests is low and more studies are needed to confirm the efficacy of these surveillance protocols.

Prophylactic hysterectomy and bilateral salpingo-oophorectomy are an option to be discussed with the patient once she has fulfilled her reproductive desires or has reached menopause.

Control of other neoplasms

Screening for gastric cancer (annual gastric endoscopy), pancreatic tumor (MRI and/or echoendoscopy), or urinary tract tumors (sediment, cytology, and annual renal ultrasound) is recommended when we encounter families with these tumors in their spectrum of neoplasms.[18,72]

Follow-up in patients with colorectal cancer and HNPCC

Mutation carriers diagnosed with a neoplasm related to this syndrome, after their cure, must also follow appropriate controls, both at the colonic and extracolonic levels, given that they are at high risk of metachronous tumors.[42]

In cases of colorectal cancer diagnosed in patients carrying the MMR mutation, the risk of presenting a new colon neoplasm if a segmental resection has been performed is estimated at around 16% after 10 years, and it is therefore recommended that endoscopies continue to be performed every 1–3 years depending on age, associated pathology, and the type of surgery performed.[74] To avoid second tumors it is also advisable to consider subtotal colectomy, especially in young patients.[1] The results of a study on life expectancy in patients with CRC according to the type of surgery performed (subtotal or segmental colectomy) have been presented, indicating that subtotal colectomy increases life expectancy by 2.3 years.[75]

General recommendations

The patient should be advised to seek medical consultation in case of onset or long duration (more than 6 weeks) of any of the following symptoms:[18]

- Dyspepsia, disorders of gastric secretion, motility or sensitivity that disturb digestion
- Continuous epigastric pain or colicky abdominal pain
- Repeated unexplained vomiting
- Unexplained weight loss
- Unexplained anemia
- Dysphagia
- Upper abdominal mass
- Jaundice
- Change in bowel habits (frequency/consistency).
- Rectal tenesmus
- Presence of blood in stool repeatedly or blood in urine
- Any persistent symptom not explained by another cause

Chemoprevention

The Colorectal Adenoma/Carcinoma Prevention Program (CAPP2) was a double-blind, placebo-controlled, randomized trial aimed at determining the role of acetylsalicylic acid in the prevention of CRC in patients with Lynch syndrome who were in surveillance programs at various international centers. Published in 2011, the authors conclude that 600 mg of aspirin daily for a mean of 25 months substantially reduces the incidence of colon cancer in patients with HNPCC.[76] Optimal duration of treatment and exact dose of ASA are currently unknown and the CAPP3 study with different doses of ASA is ongoing.

Systemic treatment

The presence of high MSI is a prognostic factor in patients diagnosed with CRC. MMR-deficient status may be useful in determining the group of patients with stage II CRC who are at low risk of recurrence and in whom treatment with adjuvant chemotherapy may not be accurate, as they are patients resistant to 5-fluorouracil-based chemotherapy.[77]

Several studies have shown that MMR-deficient tumors have high mutational burden and express numerous neoantigens that make them sensitive to immunotherapy. Two checkpoint inhibitors have demonstrated efficacy in patients with advanced CRC and MMR protein deficiency, pembrolizumab for any MMR-deficient solid tumor and nivolumab for MMR-deficient colorectal tumors.[78,79]

Other entities

Colorectal cancer type X

Families have been identified that meet the Amsterdam I/II criteria for HNPCC, but in which no defective DNA repair gene is detected and no microsatellite instability in the tumor is present. These families are termed "colorectal cancer type X families" and have clinical features different from the usual Lynch syndrome.[63]

Studies that have directly compared the age of onset between HNPCC families and type X families suggest a slightly higher age of presentation in type X families, but a lower colorectal cancer risk and no increased incidence of other HNPCC-related tumors have been found.[80] In addition, tumors arising within type X families also appear to have a different pathologic phenotype, with a lower number of tumor-infiltrating lymphocytes than those in HNPCC families.[81]

In these cases, given the lower incidence of colorectal cancer and other tumor types, surveillance recommendations are less intensive; colonoscopy will be performed every 3 years starting at age 45 years or beginning 10 years before the youngest diagnosis.[36]

Lynch-like syndrome

Lynch-like syndrome is called when the tumor shows a deficiency of MMR protein expression or elevated MSI (excluding cases of *MLH1* hypermethylation) but no pathogenic germline mutation is observed. When DNA mutation study is performed, it is observed that 50%–70% of cases present biallelic somatic mutations, which would explain the absence of protein expression in IHC and/or elevated MSI.[82] In these cases, since it is a sporadic somatic biallelic inactivation, surveillance and follow-up of at-risk relatives would not be necessary.

Conclusions

Within the daily clinical practice of oncology, it is important to identify families at risk of hereditary cancer, since these individuals are the ones who benefit the most from cancer screening and early detection programs.

In the field of colorectal cancer, we have multiple clinical and laboratory tools for the categorization and study of these families at our disposal. Until relatively recently, only the Amsterdam and Bethesda criteria were available for clinical diagnosis. The introduction of immunohistochemistry and the detection of microsatellite instability are major advances prior to the identification of mutations in DNA repair genes.

Genetic testing has an approximately 70% probability of detecting deleterious mutations in one of the four *MMR* genes involved. The joint interpretation of the results of the laboratory tests will allow us to offer family members at risk a series of surveillance and follow-up recommendations appropriate to each case.

We must not forget that this is multidisciplinary work, involving geneticists, pathologists, gastroenterologists, general surgeons, and medical oncologists alike, in which promoting adherence to the follow-up protocols is important for them to be truly effective. Guidelines for the management of hereditary cancer syndromes, such as the *I Guía de Cáncer Hereditario de Galicia*[18] or the *Libro de Cáncer Hereditario de SEOM (Sociedad Española De Oncología Médica)*,[1] should be promoted so that medical professionals have all the necessary tools to recognize these syndromes and refer patients to genetic counseling units when indicated within our reach.

References

1. Cáncer hereditario. *Sociedad Española de Oncología Médica*. Tercera Edición; 2019. https://seom.org/publicaciones/publicaciones-seom/cancer-hereditario/207453-3-edicion-libro-seom-de-cancer-hereditario.
2. Jemal A, Bray F, Center MM, Ferlay J, Ward E, Forman D. Global cancer statistics. *CA Cancer J Clin* 2011;**61**(2):69–90. https://doi.org/10.3322/caac.20107.
3. Lynch HT, de la Chapelle A. Hereditary colorectal cancer. *N Engl J Med* 2003;**348**(10):919–32.
4. Chung DC, Rustgi AK. The hereditary nonpolyposis colorectal cancer syndrome: genetics and clinical implications. *Ann Intern Med* 2003;**138**(7):560–70.
5. Peltomäki P, Vasen H. Mutations associated with HNPCC predisposition—update of ICG-HNPCC/INSiGHT mutation database. *Disease Markers* 2004;**20**(4–5):269–76. https://doi.org/10.1155/2004/305058.
6. Vasen HF. Clinical diagnosis and management of hereditary colorectal cancer syndromes. *J Clin Oncol* 2000;**18**(21 Suppl):81S–92S.
7. Peltomäki P, Aaltonen LA, Sistonen P, et al. Genetic mapping of a locus predisposing to human colorectal cancer. *Science* 1993;**260**(5109):810–2.
8. Lindblom A, Tannergård P, Werelius B, Nordenskjöld M. Genetic mapping of a second locus predisposing to hereditary non-polyposis colon cancer. *Nat Genet* 1993;**5**(3):279–82.
9. Palomaki GE, McClain MR, Melillo S, Hampel HL, Thibodeau SN. EGAPP supplementary evidence review: DNA testing strategies aimed at reducing morbidity and mortality from Lynch syndrome. *Genet Med* 2009;**11**(1):42–65.
10. Wu Y, Berends MJ, Sijmons RH, et al. A role for MLH3 in hereditary nonpolyposis colorectal cancer. *Nat Genet* 2001;**29**(2):137–8.
11. Peltomaki P. Role of DNA mismatch repair defects in the pathogenesis of human cancer. *J Clin Oncol* 2003;**21**(6):1174–9.
12. Win AK, Jenkins MA, Dowty JG, et al. Prevalence and penetrance of major genes and polygenes for colorectal cancer. *Cancer Epidemiol Biomarkers Prev* 2017;**26**(3):404–12.
13. Lázaro C, Feliubadaló L, del Valle J. Genetic testing in hereditary colorectal cancer. In: Valle L, Gruber SB, Capellá G, editors. *Hereditary colorectal cancer: genetic basis and clinical implications*. Springer International Publishing; 2018. p. 209–32.
14. Kraus C, Kastl S, Günther K, Klessinger S, Hohenberger W, Ballhausen WG. A proven de novo germline mutation in HNPCC. *J Med Genet* 1999;**36**(12):919–21.
15. Kempers MJE, Kuiper RP, Ockeloen CW, et al. Risk of colorectal and endometrial cancers in EPCAM deletion-positive Lynch syndrome: a cohort study. *Lancet Oncol* 2011;**12**(1):49–55.
16. Tannergård P, Lipford JR, Kolodner R, Frödin JE, Nordenskjöld M, Lindblom A. Mutation screening in the hMLH1 gene in Swedish hereditary nonpolyposis colon cancer families. *Cancer Res* 1995;**55**(24):6092–6.
17. Boland CR, Thibodeau SN, Hamilton SR, et al. A National Cancer Institute workshop on microsatellite instability for cancer detection and familial predisposition: development of international criteria for the determination of microsatellite instability in colorectal cancer. *Cancer Res* 1998;**58**(22):5248–57.
18. autores M. I Guía de Cáncer Hereditario de Galicia. Sociedad Oncológica de Galicia. Published online 2011. https://sog-galicia.org/wp-content/uploads/2019/02/guia_cancer_hereditario.pdf.
19. Weissman SM, Bellcross C, Bittner CC, et al. Genetic counseling considerations in the evaluation of families for Lynch syndrome—a review. *J Genet Couns* 2011;**20**(1):5–19.
20. Clendenning M, Senter L, Hampel H, et al. A frame-shift mutation of PMS2 is a widespread cause of Lynch syndrome. *J Med Genet* 2008;**45**(6):340–5.
21. Jenkins MA, Hayashi S, O'Shea A-M, et al. Pathology features in Bethesda guidelines predict colorectal cancer microsatellite instability: a population-based study. *Gastroenterology* 2007;**133**(1):48–56.
22. Lynch HT, Lynch JF, Lynch PM. Toward a consensus in molecular diagnosis of hereditary nonpolyposis colorectal cancer (Lynch syndrome). *J Natl Cancer Inst* 2007;**99**(4):261–3. https://doi.org/10.1093/jnci/djk077.
23. Cáncer Hereditario. *Sociedad Española de Oncología Médica*. II edición; 2010. https://seom.org/es/publicaciones/publicaciones-seom/103484-iiedicion-libro-seom-de-cancer-hereditario.
24. Raptis S, Mrkonjic M, Green RC, et al. MLH1- -93G > A promoter polymorphism and the risk of microsatellite-unstable colorectal cancer. *J Natl Cancer Inst* 2007;**99**(6):463–74.
25. Mrkonjic M, Raptis S, Green RC, et al. MSH2- 118T > C and MSH6- 159C > T promoter polymorphisms and the risk of colorectal cancer. *Carcinogenesis* 2007;**28**(12):2575–80.
26. Shia J. Immunohistochemistry versus microsatellite instability testing for screening colorectal cancer patients at risk for hereditary nonpolyposis colorectal cancer syndrome. Part I. the utility of immunohistochemistry. *J Mol Diagn* 2008;**10**(4):293–300.
27. Lindor NM, Burgart LJ, Leontovich O, et al. Immunohistochemistry versus microsatellite instability testing in phenotyping colorectal tumors. *J Clin Oncol* 2002;**20**(4):1043–8.
28. Baudhuin LM, Burgart LJ, Leontovich O, Thibodeau SN. Use of microsatellite instability and immunohistochemistry testing for the identification of individuals at risk for Lynch syndrome. *Fam Cancer* 2005;**4**(3):255–65.

29. Halvarsson B, Lindblom A, Rambech E, Lagerstedt K, Nilbert M. Microsatellite instability analysis and/or immunostaining for the diagnosis of hereditary nonpolyposis colorectal cancer? *Virchows Arch* 2004;**444**(2):135–41.
30. Aaltonen LA, Salovaara R, Kristo P, et al. Incidence of hereditary nonpolyposis colorectal cancer and the feasibility of molecular screening for the disease. *N Engl J Med* 1998;**338**(21):1481–7.
31. Gille JJP, Hogervorst FBL, Pals G, et al. Genomic deletions of MSH2 and MLH1 in colorectal cancer families detected by a novel mutation detection approach. *Br J Cancer* 2002;**87**(8):892–7.
32. Giardiello FM, Allen JI, Axilbund JE, et al. Guidelines on genetic evaluation and management of Lynch syndrome: a consensus statement by the US multi-society task force on colorectal cancer. *Am J Gastroenterol* 2014;**109**(8):1159–79.
33. Warthin AS. Heredity with reference to carcinoma: as shown by the study of the cases examined in the pathological laboratory of the University of Michigan, 1895-1913. *Arch Intern Med* 1913;**12**(5):546–55.
34. Lynch HT, Shaw MW, Magnuson CW, Larsen AL, Krush AJ. Hereditary factors in cancer. Study of two large midwestern kindreds. *Arch Intern Med* 1966;**117**(2):206–12.
35. Lynch HT, Krush AJ. Cancer family "G" revisited: 1895-1970. *Cancer* 1971;**27**(6):1505–11.
36. Vasen HFA, Möslein G, Alonso A, et al. Guidelines for the clinical management of Lynch syndrome (hereditary non-polyposis cancer). *J Med Genet* 2007;**44**(6):353–62.
37. Bonadona V, Bonaïti B, Olschwang S, et al. Cancer risks associated with germline mutations in MLH1, MSH2, and MSH6 genes in Lynch syndrome. *JAMA* 2011;**305**(22):2304–10.
38. Hampel H, Frankel WL, Martin E, et al. Screening for the Lynch syndrome (hereditary nonpolyposis colorectal cancer). *N Engl J Med* 2005;**352**(18):1851–60.
39. Hampel H, Frankel WL, Martin E, et al. Feasibility of screening for Lynch syndrome among patients with colorectal cancer. *J Clin Oncol* 2008;**26**(35):5783–8.
40. Vasen HFA. Clinical description of the Lynch syndrome [hereditary nonpolyposis colorectal cancer (HNPCC)]. *Fam Cancer* 2005;**4**(3):219–25.
41. Hampel H, Stephens JA, Pukkala E, et al. Cancer risk in hereditary nonpolyposis colorectal cancer syndrome: later age of onset. *Gastroenterology* 2005;**129**(2):415–21.
42. Lin KM, Shashidharan M, Ternent CA, et al. Colorectal and extracolonic cancer variations in MLH1/MSH2 hereditary nonpolyposis colorectal cancer kindreds and the general population. *Dis Colon Rectum* 1998;**41**(4):428–33.
43. De Jong AE, Morreau H, Van Puijenbroek M, et al. The role of mismatch repair gene defects in the development of adenomas in patients with HNPCC. *Gastroenterology* 2004;**126**(1):42–8.
44. Jass JR, Stewart SM, Stewart J, Lane MR. Hereditary non-polyposis colorectal cancer—morphologies, genes and mutations. *Mutat Res/Fundam Mol Mech Mutag* 1994;**310**(1):125–33.
45. Vasen HF, Nagengast FM, Khan PM. Interval cancers in hereditary non-polyposis colorectal cancer (Lynch syndrome). *Lancet* 1995;**345**(8958):1183–4.
46. Jass JR, Walsh MD, Barker M, Simms LA, Young J, Leggett BA. Distinction between familial and sporadic forms of colorectal cancer showing DNA microsatellite instability. *Eur J Cancer* 2002;**38**(7):858–66.
47. Sankila R, Aaltonen LA, Järvinen HJ, Mecklin JP. Better survival rates in patients with MLH1-associated hereditary colorectal cancer. *Gastroenterology* 1996;**110**(3):682–7.
48. Gryfe R, Kim H, Hsieh ET, et al. Tumor microsatellite instability and clinical outcome in young patients with colorectal cancer. *N Engl J Med* 2000;**342**(2):69–77.
49. Quehenberger F, Vasen HFA, van Houwelingen HC. Risk of colorectal and endometrial cancer for carriers of mutations of the hMLH1 and hMSH2 gene: correction for ascertainment. *J Med Genet* 2005;**42**(6):491–6.
50. Broaddus RR, Lynch HT, Chen L-M, et al. Pathologic features of endometrial carcinoma associated with HNPCC: a comparison with sporadic endometrial carcinoma. *Cancer* 2006;**106**(1):87–94.
51. Vasen HF, Offerhaus GJ, den Hartog Jager FC, et al. The tumour spectrum in hereditary non-polyposis colorectal cancer: a study of 24 kindreds in the Netherlands. *Int J Cancer* 1990;**46**(1):31–4.
52. Watson P, Lynch HT. Extracolonic cancer in hereditary nonpolyposis colorectal cancer. *Cancer* 1993;**71**(3):677–85.
53. Aarnio M, Mecklin JP, Aaltonen LA, Nyström-Lahti M, Järvinen HJ. Life-time risk of different cancers in hereditary non-polyposis colorectal cancer (HNPCC) syndrome. *Int J Cancer* 1995;**64**(6):430–3.
54. Lynch HT, Lynch PM, Pester J, Fusaro RM. The cancer family syndrome. Rare cutaneous phenotypic linkage of Torre's syndrome. *Arch Intern Med* 1981;**141**(5):607–11.
55. Bapat B, Xia L, Madlensky L, et al. The genetic basis of Muir-Torre syndrome includes the hMLH1 locus. *Am J Hum Genet* 1996;**59**(3):736–9.
56. Suspiro A, Fidalgo P, Cravo M, et al. The Muir-Torre syndrome: a rare variant of hereditary nonpolyposis colorectal cancer associated with hMSH2 mutation. *Am J Gastroenterol* 1998;**93**(9):1572–4.
57. South CD, Hampel H, Comeras I, Westman JA, Frankel WL, de la Chapelle A. The frequency of Muir-Torre syndrome among Lynch syndrome families. *J Natl Cancer Inst* 2008;**100**(4):277–81.
58. Vasen HF, Sanders EA, Taal BG, et al. The risk of brain tumours in hereditary non-polyposis colorectal cancer (HNPCC). *Int J Cancer* 1996;**65**(4):422–5.
59. Vasen HF, Mecklin JP, Khan PM, Lynch HT. The international collaborative group on hereditary non-polyposis colorectal Cancer (ICG-HNPCC). *Dis Colon Rectum* 1991;**34**(5):424–5.

60. Vasen HF, Watson P, Mecklin JP, Lynch HT. New clinical criteria for hereditary nonpolyposis colorectal cancer (HNPCC, Lynch syndrome) proposed by the international collaborative group on HNPCC. *Gastroenterology* 1999;**116**(6):1453–6.
61. Rodriguez-Bigas MA, Boland CR, Hamilton SR, et al. A National Cancer Institute workshop on hereditary nonpolyposis colorectal Cancer syndrome: meeting highlights and Bethesda guidelines. *J Natl Cancer Inst* 1997;**89**(23):1758–62.
62. de la CA Ruschoff J, et al. UABCTJSS. Revised Bethesda guidelines for hereditary nonpolyposis colorectal cancer (Lynch syndrome) and microsatellite instability. *J Natl Cancer Inst* 2004;**96**(5):261–8.
63. Lindor NM, Rabe K, Petersen GM, et al. Lower cancer incidence in Amsterdam-I criteria families without mismatch repair deficiency: familial colorectal cancer type X. *JAMA* 2005;**293**(16):1979–85.
64. Lindor NM, Petersen GM, Hadley DW, et al. Recommendations for the care of individuals with an inherited predisposition to Lynch syndrome: a systematic review. *JAMA* 2006;**296**(12):1507–17.
65. Kastrinos F, Balmaña J, Syngal S. Prediction models in Lynch syndrome. *Fam Cancer* 2013;**12**(2):217–28.
66. Barnetson RA, Tenesa A, Farrington SM, et al. Identification and survival of carriers of mutations in DNA mismatch-repair genes in colon cancer. *N Engl J Med* 2006;**354**(26):2751–63.
67. Chen S, Wang W, Lee S, et al. Prediction of germline mutations and cancer risk in the Lynch syndrome. *JAMA* 2006;**296**(12):1479–87.
68. Balmaña J, Stockwell DH, Steyerberg EW, et al. Prediction of MLH1 and MSH2 mutations in Lynch syndrome. *JAMA* 2006;**296**(12):1469–78.
69. Kastrinos F, Steyerberg EW, Mercado R, et al. The PREMM1, 2, 6 model predicts risk of MLH1, MSH2, and MSH6 germline mutations based on cancer history. *Gastroenterology* 2011;**140**(1):73–81.
70. Wood DE. National Comprehensive Cancer Network (NCCN) clinical practice guidelines for lung Cancer screening. *Thorac Surg Clin* 2015;**25**(2):185–97.
71. Stoffel EM, Mangu PB, Gruber SB, et al. Hereditary colorectal cancer syndromes: American Society of Clinical Oncology clinical practice guideline endorsement of the familial risk-colorectal cancer: European Society for Medical Oncology clinical practice guidelines. *J Clin Oncol* 2015;**33**(2):209–17.
72. Stjepanovic N, Moreira L, Carneiro F. Hereditary gastrointestinal cancers: ESMO Clinical Practice Guidelines for diagnosis, treatment and follow-up. Ann Oncol. Published online 2019. 30(10):1558-1571 https://www.annalsofoncology.org/article/S0923-7534(19)60977-4/abstract.
73. Järvinen HJ, Mecklin JP, Sistonen P. Screening reduces colorectal cancer rate in families with hereditary nonpolyposis colorectal cancer. *Gastroenterology* 1995;**108**(5):1405–11.
74. de Vos tot Nederveen Cappel WH, Buskens E, van Duijvendijk P, et al. Decision analysis in the surgical treatment of colorectal cancer due to a mismatch repair gene defect. *Gut* 2003;**52**(12):1752–5.
75. Lynch HT, Lynch PM, Lanspa SJ, Snyder CL, Lynch JF, Boland CR. Review of the Lynch syndrome: history, molecular genetics, screening, differential diagnosis, and medicolegal ramifications. *Clin Genet* 2009;**76**(1):1–18.
76. Burn J, Gerdes A-M, Macrae F, et al. Long-term effect of aspirin on cancer risk in carriers of hereditary colorectal cancer: an analysis from the CAPP2 randomised controlled trial. *Lancet* 2011;**378**(9809):2081–7.
77. Tejpar S, Saridaki Z, Delorenzi M, Bosman F, Roth AD. Microsatellite instability, prognosis and drug sensitivity of stage II and III colorectal cancer: more complexity to the puzzle. *J Natl Cancer Inst* 2011;**103**(11):841–4.
78. Le DT, Uram JN, Wang H, et al. PD-1 blockade in tumors with mismatch-repair deficiency. *N Engl J Med* 2015;**372**(26):2509–20.
79. Overman MJ, McDermott R, Leach JL, et al. Nivolumab in patients with metastatic DNA mismatch repair-deficient or microsatellite instability-high colorectal cancer (CheckMate 142): an open-label, multicentre, phase 2 study. *Lancet Oncol* 2017;**18**(9):1182–91.
80. Llor X, Pons E, Xicola RM, et al. Differential features of colorectal cancers fulfilling Amsterdam criteria without involvement of the mutator pathway. *Clin Cancer Res* 2005;**11**(20):7304–10.
81. Valle L, Perea J, Carbonell P, et al. Clinicopathologic and pedigree differences in Amsterdam I—positive hereditary nonpolyposis colorectal cancer families according to tumor microsatellite instability status. *J Clin Oncol* 2007;**25**(7):781–6.
82. Haraldsdottir S, Hampel H, Tomsic J, et al. Colon and endometrial cancers with mismatch repair deficiency can arise from somatic, rather than germline, mutations. *Gastroenterology* 2014;**147**(6). 1308–1316.e1.

Chapter 19

Hereditary polyposis CRC

José Manuel Mera Calviño and Enrique González de la Ballina González
Gastroenterology Service. Pontevedra University Hospital Complex, Pontevedra, Spain

The importance of the HPCC study is reflected in Table 19.1.

Adenomatous polyposis syndromes: Familial adenomatous polyposis and MUTYH-associated polyposis

Adenomatous polyposis includes classic and attenuated familial adenomatous polyposis (FAP and AFAP) and MUTYH-associated polyposis. These diseases are the most common forms of gastrointestinal polyposis.

Genetics

Classic FAP is an autosomal dominant disease caused by germline mutations in the APC gene (5q21-q22). The penetrance of the disease is greater than 95%,[1] with a prevalence of 1 in 10000 to 1 in 2000 [2] and is estimated to be responsible for 1% of all cases of colorectal cancer (CRC), making it the second most common hereditary predisposition syndrome for this neoplasm.

The discovery of the genetic defect associated with FAP has allowed syndromes previously considered as independent entities to be included within the same genetic condition. It is now known that Gardner syndrome and two-thirds of patients with Turcot syndrome have genetic abnormalities in the APC gene.

Up to 30% of cases are associated with de novo mutations,[3] meaning that the germline mutation originated in the sperm or egg of an unaffected individual and was transmitted to their offspring; therefore, one-third of those affected will have no family history of the disease.

The clinical manifestations of FAP and even the risk of cancer are related to the location of the mutation. If the mutation is located at the ends of the gene or in certain areas of exon 9, the patient develops a less aggressive form of the disease called attenuated FAP.

MAP is an autosomal recessive disease caused by germline biallelic mutations in the mutY homolog (MUTYH) gene located on chromosome 1p34.3-1p32.1. Patients with MAP have great clinical variability, most have an attenuated polyposis phenotype while a small percentage suffer from a clinical picture indistinguishable from classic FAP.[4]

A mutation in APC is found in up to 80% of patients with more than 1000 adenomas. This frequency decreases to 56% in patients with between 100 and 999 adenomas; 10% in individuals with 20–99 polyps and only 5% in cases with less than 20 adenomas.

MUTYH mutations are only found in 2% of patients with more than 1000 adenomas; in 7% of those with between 999 and 100 polyps and also the group with between 99 and 20 adenomas and only in 4% of those with fewer than 20 polyps.[5]

In the group of attenuated adenomatous polyposis, the genetic cause is much more heterogeneous than in the classical form, as in addition to mutations of APC and MUTYH, mutations of other genes such as POLE, POLD1, or NTLHL1 are involved, while no mutation is identified in up to one-third of the cases.[6]

Clinical symptoms

Classic FAP is characterized by the progressive development of hundreds to thousands of adenomatous polyps in the large intestine—with all varieties of adenomas, including tubular, tubulovillous, and villous—and to a lesser extent along other regions of the gastrointestinal tract.

TABLE 19.1 Risk Cancer in different pathologies.

Syndrome	Inheritance	Gene	Polyp type	Cancer risk (%)
Familial adenomatous polyposis	A.D.	APC	Adenoma	100
MUTYH-associated polyposis	A.R.	MUTYH	Adenoma	80
Peutz–Jeghers syndrome	A.D.	STK11	Hamartoma	39
Juvenile polyposis	A.D.	SMAD4, BMPR1a	Hamartoma	40–60
Cowden syndrome	A.D.	PTEN	Hamartoma	16
Serrated polyposis syndrome	Unknown	Unknown	Serrated	50

The disease debuts in childhood with the presence of polyps smaller than 5 mm in the left colon. The number of lesions and their size increase in adolescence, spreading to the entire surface of the colon.

Practically half of patients with FAP have adenomas at age 15 and 95% have adenomas at age 35 which is the median age of diagnosis of colorectal cancer.[7]

Extracolonic manifestations are present in 40% of cases,[8] such as congenital hypertrophy of the retinal pigment epithelium, which is present in 58%–92% of cases; other common manifestations would be gastroduodenal lesions such as fundic gland hyperplasia (20%–56%), hyperplastic polyps (8%–44%), gastric adenomas (2%–13%), gastric cancers (<1%), duodenal adenomas (24%–100%), and duodenal and periampullary carcinomas (<1%), as well as other alterations such as mandibular and cranial osteomas (14%–93%), epidermoid cysts and fibromas (50%), desmoid tumors (4%–15%), hepatobiliary tumors (<1%), papillary thyroid carcinoma (2%), central nervous system tumors (<1%), and dental anomalies.

Attenuated FAP (AFAP) is a variant of classic FAP characterized by the presence of multiple polyps, usually more than 15 and less than 100, preferentially located in the right colon. The colorectal cancer usually appears 10 years later than in the classic variant, is not accompanied by congenital hypertrophy of the retinal pigment epithelium but may present duodenal adenomas, gastric polyps, increased risk of papillary thyroid cancer and, rarely, desmoid tumors. It is vitally important to recognize these extracolonic manifestations, especially in the attenuated form, as they may precede the colonic manifestations.

Diagnostics

Patients with polyposis may debut with nonspecific symptoms, such as hematochezia, diarrhea, and abdominal pain. However, the key to the diagnosis and management of the disease is to identify the presymptomatic individual, and this goal is achieved by continuous diagnostic workup in the relatives of affected patients.

Classic FAP is defined by the presence of more than 100 adenomatous polyps, distributed throughout the colon, and appears at an early age. The most frequent location of the cancer is the left colon.

Attenuated FAP should be suspected in an individual with 20 or more colorectal adenomas accumulated over a lifetime, multiple colorectal adenomas (more than 10) at an age younger than 40 years or in the presence of more than 10 adenomas, and a family history of colon cancer before the age of 60 years or attenuated adenomatous polyposis.[7]

If a diagnosis of suspected FAP is established, the study should be completed in an attempt to discover possible extracolonic manifestations, including, in addition to pancolonoscopy:

- Gastroduodenoscopy, adequately assessing the periampullary region (assessing the need to use a side-view endoscope, if necessary).
- Abdominal ultrasound or abdominopelvic CT scan (to rule out desmoid tumor).
- Clinical thyroid evaluation.

In case of diagnostic doubt, when supporting the request for a molecular study, we could consider retinoscopy (congenital hypertrophy of the retinal pigment epithelium is detected in 2% of the general population and in FAP, in 58%–92%) and orthopantomography, which could be omitted if the diagnosis is clear.

Genetic counseling and testing should be provided:

- To individuals who present an increased risk of FAP due to family history.
- To asymptomatic healthy individuals belonging to a family with known APC mutation.
- Whenever there is a clinical diagnosis of adenomatous polyposis (by colonoscopy) regardless of family history.

In familial colonic polyposis, the study is recommended when the maturity of the patient allows it and the clinical endoscopic status indicates that prophylactic colectomy may be necessary in the short term.

Monitoring of healthy individuals with classic PAF

This monitoring is advisable for asymptomatic individuals carrying the pathogenic mutation in the APC gene and first-degree relatives of a FAP sufferer for whom it has not been possible to identify the genetic mutation responsible for the disease. It is recommended to perform:

Annual colonoscopy from 10 to 12 years of age. When adenomatous polyps are detected, colonoscopies should be performed every year until surgery.[9,10] If no polyps are evident in the endoscopies performed, the recommended frequency is.[11]

- Every year from 10 to 12 years of age to 40 years of age.
- Every 5 years from the age of 40 to 50–60 years.

Baseline upper gastrointestinal endoscopy for gastric, duodenal, and ampullar assessment every 5 years from 25 to 30 years of age. If polyps are detected in the duodenum or ampulla, they will be biopsied and followed up according to the Spigelman classification, which is based on the number, size, and characteristics of the polyps detected (see Tables 19.2–19.4).[12]

TABLE 19.2 Spigelman scores classification.

Score	1	2	3
Number	1–4	5–20	>20
Size (mm)	1–4	5–10	>10
Histological type	Tubular	Tubulovillous	Villous
Dysplasia	Mild	Moderate	Severe

TABLE 19.3 Spigelman stage classification.

Stage	0	I	II	III	IV
Score	0	1–4	5–6	7–8	9–12

TABLE 19.4 Recommendations for endoscopic monitoring of possible upper gastrointestinal involvement according to Spigelman stages.

Stage 0	Endoscopies every 5 years, with multiple biopsies of mucosal folds
Stage I	Endoscopies every 5 years
Stage II	Endoscopies every 3 years
Stage III	Endoscopies every ½–2 years
Stage IV	Endoscopies every 3 months and evaluate duodenal surgery

- Annual clinical evaluation, including thyroid palpation.[13,14]
- Determination of AFP and abdominal ultrasound in children of patients with FAP from birth to 7 years of age due to the risk of hepatoblastoma (1.6%).[13,14]
- Regular clinical evaluation is recommended and in case of suspicion of desmoid tumor, screening by CT and/or abdominal ultrasound if there is a history of abdominal surgery, since desmoid tumors are a common cause of morbidity and mortality.[13,14]

In patients with FAP for whom colectomy has been performed:

- Desmoid tumors are the second cause of death after periampullary carcinoma. Regular clinical evaluation is recommended and, in case of suspicion, screening by CT and/or abdominal ultrasound.
- In individuals with ileal reservoir or ileo-rectal anastomosis, endoscopic follow-up is recommended after surgery,[13,14] rectoscopy is performed every 6–12 months or reservoroscopy every 2 years.[13,14]
- Follow-up of extracolonic manifestations as previously indicated.[7, 13, 14]

Follow-up of healthy individuals with AFAF

At-risk individuals are considered to be healthy, asymptomatic, mutation carriers and healthy, asymptomatic individuals belonging to families with AFAP in whom the pathogenic mutation has not been identified. It is advisable to perform:

- Complete colonoscopy at 18–20 years of age, repeating it every 2 years until 74 years of age. Subsequently, each case will be individualized according to the associated pathology and the patient's PS.
- If adenomatous polyps are detected in the colonoscopy and these can be resected by polypectomy, the test should be repeated annually.[13, 14]

Follow-up of healthy individuals with MAP

A colonoscopy is recommended every 1–2 years in patients affected by MUTYH-associated polyposis starting at 25–30 years of age.

Gastroduodenoscopy will be initiated upon the appearance of colonic polyps or at 30–35 years of age and repeated every 5 years if there are no adenomas or according to Spigelman's protocol if there are any.

Annual cancer screening by examination and ultrasound has been suggested, although there is no data to support this recommendation.[15]

In monozygous MYH gene mutation carriers there is a small increased risk of developing CRC, with disparity among studies as to the clinical relevance and need for follow-up. Screening by colonoscopy every 5 years from the age of 40 years or from 10 years before the age of diagnosis of the youngest affected relative, whichever is earlier, has been suggested.[16, 17]

Treatment of FAP

Surgery is the only reasonable management option for classic FAP, and the timing and extent of surgery are the primary clinical considerations. Treatment of the FAP patient is aimed at avoiding the most common causes of morbidity and mortality in FAP and desmoid tumors.

Colonic involvement of FAP should be treated with prophylactic colectomy, generally recommended before the age of 25 years, with the size and histology of the polyps detected determining the ideal time to perform colectomy. Sometimes surgery is delayed in women who have not yet had offspring.

As for the type of surgery, it has evolved considerably, seeking not only to remove the entire colonic mucosa, but also trying to ensure normal fecal evacuation, anal continence and bladder and sexual function.[13,14]

The choice of the surgical technique depends on different factors, such as the age at diagnosis of FAP, its phenotype, family history, rectal involvement, subsequent follow-up and, above all, the decision of the affected individual.[18]

The surgical options are as follows:

- Total proctocolectomy with ileal reservoir and ileoanal anastomosis (of choice), although it presents the greatest operative morbidity.
- Total colectomy with ileo-rectal anastomosis presents a high risk of recurrence in the mucosa of the rectal stump, indicated in young patients with few polyps in the rectum and a family history of mild forms of FAP or attenuated FAP, and who accept stricter follow-up.

It has been shown that the cumulative risk of rectal cancer at 10, 20, and 40 years is 4%, 12%, and 32%, respectively, and it is advisable to inform the patient of a possible subsequent surgical rescue.

As for upper gastrointestinal polyps, the overall approach is usually conservative. Biopsies of gastric polyps should be taken to determine whether they are adenomas or, on the contrary, hyperplastic or fundic glands; the latter two do not usually require treatment.

In the duodenum, villous adenomas, those with high-grade dysplasia, large and symptomatic adenomas should be resected regardless of histology. Endoscopic ablation of periampullary adenomas can be performed relatively safely by experienced endoscopists. In selected patients with a large number of polyps in complex locations, surgical excision of the duodenum may be necessary, either by local exeresis or cephalic duodenopancreatectomy.[13,14]

Desmoid tumors of the abdominal wall and extremities can be treated with wide surgical excision. Mesenteric or retroperitoneal desmoid tumors are difficult to treat surgically due to the high rate of complications, intestinal obstructions, intestinal ischemia, etc. Surgical treatment is generally reserved for a third or fourth therapeutic option after progression with other treatments such as NSAIDs associated with antioestrogens, imatinib, sorafenib, or radiotherapy.[13,14]

Treatment of AFAP

Prophylactic colonic surgery will be performed in case of multiple adenomas in which an adequate follow-up by colonoscopy and polypectomies cannot be performed; subtotal colectomy with ileo-rectal anastomosis will be the treatment of choice.[13,14]

Regarding extracolonic involvement, the same therapeutic approach as in classic FAP will be used.[13,14]

Treatment of FAP

With regard to FAP associated with the MYH gene (MAP), the risk of CRC is 19% at 50 years of age and 43% at 60 years of age. The occurrence of duodenal adenomas has been described less frequently in adenomatous polyposis associated with APC mutations.

If it is susceptible to control by endoscopic polypectomy this will be the first option. Otherwise, surgery will be necessary.

Chemoprevention

Sulindac and celecoxib have been shown to reduce the number and cause regression of established adenomas, so both the FDA and the EMEA have authorized the use of these drugs with the intention of delaying the need for surgery. Their administration is not justified in the primary prevention of FAP (IA).[13,14] They are only accepted as adjuvant therapy to surgery in case of recurrence or in patients who cannot undergo surgery.[7,13,14]

Other inherited adenomatous polyposis syndromes

Gardner syndrome

In the past, intestinal polyposis with certain benign extraintestinal growths was considered Gardner syndrome. This is a familial disease composed of gastrointestinal polyps and osteomas, associated with a variety of benign soft tissue tumors (lipomas, fibromas, desmoid tumors) and other extraintestinal manifestations (congenital hypertrophy of the retinal pigment epithelium, thyroid tumors, adrenal tumors, dental impactions, supernumerary teeth).

FAP and Gardner syndrome are variable manifestations of a disease localized to a single genetic locus, the APC gene.

Turcot syndrome (glioma-polyposis)

It is an autosomal dominant inherited disease that was initially described as an association of FAP and primary tumors of the central nervous system: malignant gliomas and medulloblastomas.[17, 19] However, the presence of germline mutations of both the APC gene and genes responsible for DNA repair (MLH1, MSH2, PMS2) and MUTYH has recently been described.[20]

The number of adenomas is usually less than in cases of classic familial adenomatous polyposis and the treatment is the same for this disease. There are no guidelines on how often or how to screen for brain tumors in these patients.

Hamartomatous polyposis syndrome

Peutz–Jeghers syndrome (PJS)

Peutz and Jeghers described a familial syndrome characterized by mucocutaneous pigmentation and gastrointestinal polyposis in 1921 and 1949, respectively.

It is an autosomal dominantly inherited syndrome caused by germline mutation in the STK11/LKB1 gene,[21,22] but not all families with PJS are related to that gene locus, suggesting genetic heterogeneity.

The incidence is estimated at 1 in 200 000 live births.

It is characterized by the association of multiple hamartomatous gastrointestinal polyps with melanic pigmentation of the buccal mucosa, lips, hands, feet, and perineal region. The typical mucocutaneous pigmentation of PJS can be observed during early infancy, with brown to greenish-black colored macular melanin deposits that tend to disappear during puberty, except for the pigmentation of the mouth. It is important to remember that these skin lesions do not have malignant potential.

Polypoid lesions are of different sizes and are profusely distributed throughout the digestive tract, with a predominance in the small intestine (60%–90%) and colon (50%–64%). Histologically, they are characterized by a fine, highly branched connective axis which gives them an arborescent appearance.

The predominant symptom is recurrent abdominal pain of colic type due to episodes of intestinal invagination, with the average age of diagnosis of this syndrome 23–26 years.

Treatment is endoscopic polypectomy; due to the diffuse distribution of polyps, surgical treatment is not indicated. However, segmental bowel resection is sometimes necessary due to complications associated with polyps, such as chronic bleeding with secondary anemia, invagination, and intestinal obstruction.

A higher incidence of cancer has been described in this syndrome with respect to that observed in the general population. Different studies have described an overall lifetime risk of cancer of up to 93%, the most common being cancers of the breast (54%) and colon (39%), followed by pancreas (36%), stomach (29%), ovaries (21%), and testes (9%).[23] The risk of cancer of the cervix (10%), endometrium (9%), and lung (15%) is also increased.[24]

It is advisable to begin screening in infancy, looking for characteristic phenotypic features. The first gastrointestinal examination is recommended at 8 years of age, with upper gastrointestinal endoscopy and capsule endoscopy. They will be repeated every 3 years if there are polyps and at 18 years of age if there are none, starting with colonoscopy, gastroduodenoscopy, and capsule endoscopy or barium swallow every 2–3 years, as well as gynecological examinations (monthly breast self-examination, annual cervical cytology and pelvic examination). From the age of 25 onward, it is recommended to add an annual mammogram or MRI. For pancreatic cancer screening, annual EUS is recommended from the age of 30 years (or 10 years before the age of diagnosis of the youngest relative).[25]

Juvenile polyposis syndrome

Juvenile polyps are differentiated hamartomas that are usually solitary and are located mostly in the rectum of children.

Juvenile polyposis is usually diagnosed during childhood and is characterized by the presence of hamartomatous polyps in the gastrointestinal tract, defined by the presence of five or more juvenile polyps in the colon, juvenile polyps throughout the gastrointestinal tract or any number of juvenile polyps associated with a history of juvenile polyposis.[26,27]

It is a disease with autosomal dominant inheritance with variable penetrance, with an incidence of 1 in 100 000 births. Several genes are involved in its pathogenesis, including SMAD4 and BMPR1, and to a lesser extent ENG and PTEN.

Although these polyps have no intrinsic malignant potential, a high percentage (10%–60%) is associated with digestive neoplasms: CRC 20% at 35 years of age, 68% at 60 years of age, and gastric and duodenal cancer to a lesser extent. Treatment consists of endoscopic polypectomy, due to the tendency to hemorrhage and obstruction. Surgical resection is indicated in cases of multiple polyposis or when there is severe hemorrhage, intestinal obstruction, or invagination.[11]

Screening endoscopy of the upper and lower gastrointestinal tract should be performed every 1–2 years. As screening measures for family members, a colonoscopy is recommended from the age of 15, with associated gastroduodenal endoscopy and a barium swallow or capsule endoscopy from the age of 25.[11]

Cowden syndrome

It is a rare disease, with an estimated incidence of 1 in 300 000 inhabitants, with autosomal dominant inheritance and variable penetrance.

In terms of clinical manifestations, 90% of patients have characteristic skin lesions, facial trichilemmomas, consisting of skin-colored papules, pink or brown, with the appearance of flat warts located in the central part of the face, periorbital areas, lips, corners of the mouth, and ears, translucent punctate keratosis on palms and soles, and flat hyperkeratotic papules on the back of the hands and forearms. The mucosal lesions are characteristic: papules on the gums, lips and palate are paler in color than the adjacent mucosa and give a cobblestone appearance, being easy to confuse with papillomas of the buccal mucosa and tongue.

At gastrointestinal level they present multiple polyps (juvenile, adenomas, inflammatory) without an increased risk of colorectal cancer, but there is an increased risk of thyroid, endometrial, and mainly breast cancer. They may also present cranial alterations in the form of macrocephaly or dolichocephaly.[28, 29]

A colonoscopy is recommended every 3–5 years from the age of 35, annual mammography/MRI from the age of 30 and annual thyroid ultrasound from the age of 18.

Bannayan–Ruvalcaba–Riley syndrome

This is a very rare autosomal dominantly inherited entity. It is characterized by the presence of hamartomatous polyps located in the gastrointestinal tract, mental retardation, macrocephaly, lipomas, and pigmented macules on the penis.[30]

Rare hereditary hamartomatous polyposis syndromes

Finally, here are the hereditary syndromes associated with hamartomatous polyps. We are only going to name four very infrequent entities observed in some isolated families, which are as follows:

- Hereditary mixed polyposis syndrome
- Intestinal ganglioneuromatosis and neurofibromatosis
- Devon family syndrome
- Basal cell nevus syndrome

Serrated polyposis syndrome (SPS)

PS are a group of lesions that can progress to CRC by the so-called serrated pathway of carcinogenesis. There are three types of serrated polyps:

1. hyperplastic polyps
2. adenomas or sessile serrated polyps
3. traditional serrated adenomas or polyps

Today we know that up to 30% of colon cancers develop by the serrated route.

SPS is the most frequent polyposis syndrome, with multiple PS that can be large and with a high risk of transformation to CRC. The recently revised diagnostic criteria for SPS are as follows:

1. presence of five or more PS proximal to the rectum, all >5 mm and at least 2 of them ≥10 mm
2. presence of more than 20 PS distributed throughout the entire colon.

Clinical features

SPS is considered a high-risk form of CRC; the risk estimate has dropped in recent studies to a 5-year cumulative incidence of less than 2%.[31]

No specific mutation has been identified. SPS is a disease with a complex etiology involving an interaction between genetics and environmental factors (especially smoking).

Endoscopic surveillance is currently recommended for first-degree relatives from the age of 25 and will be repeated every 1–2 years if serrated polyps are present and every 5 years if they are not.

These recommendations are currently under revision and it is likely that in the future the presence of a single serrated lesion will not require annual/biannual monitoring and that the age of onset of colonoscopies will be set at 10 years before the age of diagnosis of the index case.

In affected patients it is advisable to remove all polyps larger than 5 mm and biannual surveillance. There is no evidence of increased risk of extracolonic cancer so there is no indication for gastroscopy or nondigestive cancer screening. In cases where endoscopic management is not possible, due to the number or location of polyps, surgical treatment including segmental resection or total colectomy with ileo-rectal anastomosis will be considered. Surgery is required in about 30% of cases.[32]

References

1. Bisgaard ML, Fenger K, Bülow S, Niebuhr E, Mohr J. Familial adenomatous polyposis (FAP): frequency, penetrance, and mutation rate. *Hum Mutat* 1994;**3**(2):121–5. https://doi.org/10.1002/humu.1380030206.
2. Galiatsatos P, Foulkes WD. Familial adenomatous polyposis. *Am J Gastroenterol* 2006;**101**(2):385–98. https://doi.org/10.1111/j.1572-0241.2006.00375.x.
3. Aretz S, Uhlhaas S, Goergens H, et al. MUTYH-associated polyposis: 70 of 71 patients with biallelic mutations present with an attenuated or atypical phenotype. *Int J Cancer* 2006;**119**(4):807–14. https://doi.org/10.1002/ijc.21905.
4. Nielsen M, Hes FJ, Nagengast FM, et al. Germline mutations in APC and MUTYH are responsible for the majority of families with attenuated familial adenomatous polyposis. *Clin Genet* 2007;**71**(5):427–33. https://doi.org/10.1111/j.1399-0004.2007.00766.x.
5. Grover S, Kastrinos F, Steyerberg EW, et al. Prevalence and phenotypes of APC and MUTYH mutations in patients with multiple colorectal adenomas. *JAMA* 2012;**308**(5):485–92. https://doi.org/10.1001/jama.2012.8780.
6. Carballal S, Leoz ML, Moreira L, Ocaña T, Balaguer F. Hereditary colorectal cancer syndromes. *Colorectal Cancer* 2014;**3**(1):57–76. https://doi.org/10.2217/crc.13.80.
7. Cubiella J, Marzo-Castillejo M, Mascort-Roca JJ, et al. Clinical practice guideline. Diagnosis and prevention of colorectal cancer. *Gastroenterol Hepatol (Engl Ed)* 2018;**41**(9):585–96. https://doi.org/10.1016/j.gastre.2018.07.008.
8. Sener SF, Miller HH, DeCosse JJ. The spectrum of polyposis. *Surg Gynecol Obstet* 1984;**159**(6):525–32. https://www.ncbi.nlm.nih.gov/pubmed/6505938.
9. Al-Tassan N, Chmiel NH, Maynard J, et al. Inherited variants of MYH associated with somatic somatic G:C−>T:A mutations in colorectal tumors. *Nat Genet* 2002;**30**(2):227–32. https://doi.org/10.1038/ng828.
10. Lipton L, Tomlinson I. The multiple colorectal adenoma phenotype and MYH, a base excision repair gene. *Clin Gastroenterol Hepatol* 2004;**2**(8):633–8. https://doi.org/10.1016/s1542-3565(04)00286-1.
11. Dunlop MG, British Society for Gastroenterology, Association of Coloproctology for Great Britain and Ireland. Guidance on gastrointestinal surveillance for hereditary non-polyposis colorectal cancer, familial adenomatous polypolis, juvenile polyposis, and Peutz-Jeghers syndrome. *Gut* 2002;**51**(Suppl. 5):V21–7. https://doi.org/10.1136/gut.51.suppl_5.v21.
12. Brosens LAA, Keller JJ, Offerhaus GJA, Goggins M, Giardiello FM. Prevention and management of duodenal polyps in familial adenomatous polyposis. *Gut* 2005;**54**(7):1034–43. https://doi.org/10.1136/gut.2004.053843.
13. de Catalunya G. Onco Guia del consejo y asesoramiento genético en el cáncer hereditario. *Versión breve para la aplicación en la práctica clínica Barcelona: Agència d'Avaluació de Tecnologia i Recerca Mèdiques Barcelona: Departament de Salut Generalitat de Catalunya*. Published online 2006.
14. Pérez Segura P. Cáncer colorrectal hereditario. Unidades de Consejo Genético. Cáncer colorrectal. Published online 2009.
15. Syngal S, Brand RE, Church JM, et al. ACG clinical guideline: genetic testing and management of hereditary gastrointestinal cancer syndromes. *Am J Gastroenterol* 2015;**110**(2):223–62. quiz 263. https://doi.org/10.1038/ajg.2014.435.
16. Provenzale D, Gupta S, Ahnen DJ, et al. Genetic/familial high-risk assessment: colorectal version 1.2016, NCCN clinical practice guidelines in oncology. *J Natl Compr Canc Netw* 2016;**14**(8):1010–30. https://doi.org/10.6004/jnccn.2016.0108.
17. Turcot J, Després J-P, St. Pierre F. Malignant tumors of the central nervous system associated with familial polyposis of the colon. *Dis Colon Rectum* 1959;**2**(5):465–8. https://doi.org/10.1007/bf02616938.
18. King JE, Dozois RR, Lindor NM, Ahlquist DA. Care of patients and their families with familial adenomatous polyposis. *Mayo Clin Proc* 2000;**75**(1):57–67. https://doi.org/10.4065/75.1.57.
19. Baughman Jr FA, List CF, Williams JR, Muldoon JP, Segarra JM, Volkel JS. The glioma-polyposis syndrome. *N Engl J Med* 1969;**281**(24):1345–6. https://doi.org/10.1056/NEJM196912112812407.
20. Barel D, Cohen IJ, Mor C, et al. Mutations of the adenomatous polyposis coli and p53 genes in a child with Turcot's syndrome1This work is in partial fulfillment of the requirements for the Ph.D. degree of D. Barel, Sackler Faculty of Medicine, Tel Aviv University, Tel Aviv, Israel.1. *Cancer Lett* 1998;**132**(1–2):119–25. https://doi.org/10.1016/s0304-3835(98)00167-0.
21. Hemminki A, Markie D, Tomlinson I, et al. A serine/threonine kinase gene defective in Peutz-Jeghers syndrome. *Nature* 1998;**391**(6663):184–7. https://doi.org/10.1038/34432.
22. Jenne DE, Reimann H, Nezu J, et al. Peutz-Jeghers syndrome is caused by mutations in a novel serine threonine kinase. *Nat Genet* 1998;**18**(1):38–43. https://doi.org/10.1038/ng0198-38.
23. Beggs AD, Latchford AR, Vasen HFA, et al. Peutz-Jeghers syndrome: a systematic review and recommendations for management. *Gut* 2010;**59**(7):975–86. https://doi.org/10.1136/gut.2009.198499.
24. Giardiello FM, Brensinger JD, Tersmette AC, et al. Very high risk of cancer in familial Peutz-Jeghers syndrome. *Gastroenterology* 2000;**119**(6):1447–53. https://doi.org/10.1053/gast.2000.20228.
25. Giardiello FM, Trimbath JD. Peutz-Jeghers syndrome and management recommendations. *Clin Gastroenterol Hepatol* 2006;**4**(4):408–15. https://doi.org/10.1016/j.cgh.2005.11.005.

26. Beacham CH, Shields HM, Raffensperger EC, Enterline HT. Juvenile and adenomatous gastrointestinal polyposis. *Am J Dig Dis* 1978;**23**(12): 1137–43. https://doi.org/10.1007/bf01072892.
27. Sachatello CR, Pickren JW, Grace Jr JT. Generalized juvenile gastrointestinal polyposis. A hereditary syndrome. *Gastroenterology* 1970; **58**(5):699–708. https://www.ncbi.nlm.nih.gov/pubmed/5444174.
28. Salem OS, Steck WD. Cowden's disease (multiple hamartoma and neoplasia syndrome). A case report and review of the English literature. *J Am Acad Dermatol* 1983;**8**(5):686–96. https://doi.org/10.1016/s0190-9622(83)70081-2.
29. Fitzpatrick TB, et al, editors. *Color atlas and synopsis of clinical dermatology*. 4th ed. McGraw-Hill Publishing; 2001.
30. Ruvalcaba RH, Myhre S, Smith DW. Sotos syndrome with intestinal polyposis and pigmentary changes of the genitalia. *Clin Genet* 1980; **18**(6):413–6. https://doi.org/10.1111/j.1399-0004.1980.tb01785.x.
31. IJspeert JEG, Rana SAQ, Atkinson NSS, et al. Clinical risk factors of colorectal cancer in patients with serrated polyposis syndrome: a multicentre cohort analysis. *Gut* 2017;**66**(2):278–84. https://doi.org/10.1136/gutjnl-2015-310630.
32. Patel R, Hyer W. Practical management of polyposis syndromes. *Frontline Gastroenterol* 2019;**10**(4):379–87. https://doi.org/10.1136/flgastro-2018-101053.

Section C.II

New tools of diagnosis in CRC

Chapter 20

Diagnostic, prognostic, predictive and therapeutic molecular biomarkers in CRC: Understanding the present and foreseeing the future

Alberto Veiga[a], Francisco Queipo[b], Germán Bou[c], Alfonso Cepeda-Emiliani[d], and Ángel Concha[e]

[a]Department of Anatomical Pathology, University Hospital Complex of Ferrol (CHUF), A Coruña, Spain, [b]Department of Anatomical Pathology, University Hospital Complex of A Coruña (CHUAC), A Coruña, Spain, [c]Department of Microbiology, University Hospital Complex of A Coruña (CHUAC). Microbiology Research Group, Biomedical Research Institute of A Coruña (INIBIC), A Coruña, Spain, [d]Department of Morphological Sciences, School of Medicine and Dentistry, University of Santiago de Compostela, A Coruña, Spain, [e]Department of Anatomical Pathology, University Hospital Complex of A Coruña (CHUAC). Biobank of A Coruña, Biomedical Research Institute of A Coruña (INIBIC), A Coruña, Spain

Introduction

Colorectal carcinoma (CRC) is one of the most frequent malignant tumors and one of the main causes of mortality in developed countries. Histologically, CRCs are usually adenocarcinomas of different varieties (conventional, serrated, with signet ring cells, mucinous, etc.).[1]

Three fundamental types of CRCs can be distinguished at the molecular level: (i) those with chromosomal instability producing numerous alterations in the number of copies (amplifications/deletions) of a small number of genes; (ii) those with microsatellite instability (MSI) that cause high mutation rates in numerous genes (hypermutants); and (iii) those with polymerase defects, which give rise to a large number of mutations in multiple genes (ultramutants).[1]

Many oncogenes are involved in the carcinogenesis of the large intestine, but due to their frequency and importance we highlight the following: *APC*, *K-RAS*, *N-RAS*, *B-RAF*, *p53*, *SMAD4*, *PI3KCA*, *POLE*, and *PTEN*.[1] However, only some of them are identified as biomarkers when selecting specific targeted treatments.

Precision medicine is based on the concept that molecular alterations associated with the origin of disease can be neutralized through therapeutic strategies specifically aimed to block or minimize the deleterious effect of the alterations. Although this concept is applicable to many areas of medical practice, it is in the field of oncology/hematology where it has shown the most evident development.

This is a relatively new concept with different names since its inception, such as personalized or individualized medicine, and represents an evolution of classical medicine in which treatments were empirically administered based on clinical judgment, determining the cause of disease subsequently (*diagnosis ex juvantibus*), or in which the patient was placed in a diagnostic/prognostic category according to clinical parameters, and standard treatments were administered based on them.

The beginnings of personalized precision medicine occurred at the end of the last century when the FDA approved the use of Trastuzumab (Herceptin, Genetech) for the treatment of HER-2 positive breast cancer.[2] Later, the use of Imatinib mesylate was approved for the treatment of gastrointestinal stromal tumors (GIST) with *c-KIT* activating mutations.[3,4]

The next example was the use of other specific tyrosine kinase inhibitors (TKIs), such as Gefitinib in nonsmall cell lung carcinomas (NSCLC) presenting activating mutations of epidermal growth factor receptor (*EGFR*) gene in exons 18–21.[5,6]

Subsequently, the use of Cetuximab (a monoclonal anti-EGFR antibody) was approved for treatment of CRCs without mutations in *RAS* (wild type),[7] and later Vemurafenib began to be used for treatment of metastatic melanomas with *B-RAF* gene mutations.[8]

On the basis of these advances, new molecular targets and specific drugs have been incorporated, generally humanized monoclonal antibodies directed against cellular antigens or TKIs.

What are biomarkers? Concept and classification

With the sequencing of the human genome, biotechnological developments in different areas (computer science, genetic engineering, and gene editing) and progress in disciplines such as artificial intelligence or big data, drugs and biological products are being developed (humanized monoclonal antibodies, cell therapies, T cells with chimeric antigen receptors or T-CARs) designed to act against cells with molecular alterations (mutations) or against specific cellular antigens, with relevant clinical results.

The concept of *biomarker* arises in this scenario of personalized precision medicine. Although multiple definitions exist, biomarkers can be defined as molecules that can be detected by laboratory tests or histophenotypic studies and which provide information about the sensitivity or resistance that a patient, carrier of a specific disease, may present to specific targeted therapies. That is, biomarkers have prognostic *and* predictive value. Therefore, as science progresses, many molecules now considered strictly prognostic markers will become biomarkers if they are associated with a target-specific therapeutic strategy.

In the field of molecular pathology/onco-hematology, biomarkers studied in daily clinical practice can be of different types:

Specific to a distinct type of tumor originating from a specific organ

This is the most frequent situation, as is the case of NSCLC, in which mutations of exons 18–21 of the *EGFR* gene are studied to decide if the patient is treated with TKIs, as previously mentioned.

It is noteworthy that with increases in knowledge and technological developments, biomarkers initially characteristic of a given tumor expand their scope of application to other tumor types. Thus, *K-RAS*, which classically has been studied in CRCs is now also being explored in NSCLCs.[9] *B-RAF*, initially determined if it had the V600E mutation only in melanomas, is now being analyzed in other tumors like CRCs,[10] as large intestine adenocarcinomas with *B-RAF* mutations can be treated with specific inhibitors such as Encorafenib.

Something similar occurs with the *HER-2* gene, whose activation state (overexpression/amplification) was determined in breast carcinomas, later in stomach adenocarcinomas (mainly of the gastroesophageal junction) and is now being studied in other tumors, CRCs in particular, to explore possibilities of administering anti-HER-2 treatments. Curiously, the method to evaluate HER-2 expression is different in breast cancer, as 10% positive tumor cells are required, compared to the 50% required to consider positive a case of CRC. However, to decide if there is HER-2 amplification criteria are similar in both cases.[11]

Likewise, rearrangements in *ALK* and *ROS1* genes in aggressive forms of CRC have been recently described, similar to those occurring in some cases of NSCLC, and are treated with specific inhibitors such as Crizotinib, a finding that opens new scenarios in CRC treatment.[12]

Sometimes the same gene can suffer different types of alterations associated with specific tumors. Returning to *HER-2*, which tends to present amplifications in breast and stomach carcinomas and in some cases of CRC, it can also undergo mutations in exon 20 in NSCLCs and in a small percentage of CRCs. Therefore, its study should be included in the molecular diagnostic panel of NSCLC[9] and CRC.[13,14] These different alterations may have relevant therapeutic implications that should be explored.

On the other hand, in some situations the same gene presents almost identical mutations but with small differences, as is the case with *B-RAF*. This gene usually shows the V600E mutation (exon 15) in most tumors where alterations of the gene are detected, but there are variants and mutations are classified as class 1, 2, and 3. Only class 1 mutations determine good response to Encorafenib in CRC cases.[15]

Contrary to the foregoing, a phenomenon to consider with therapeutic purposes is the different relevance that the same molecular alteration may have depending on the type of tumor on which it arises. For example, *K-RAS* has mutations in exons 2, 3, and 4 in multiple tumors. At least 90% of pancreatic adenocarcinomas present *K-RAS* mutations[16] and, however, there are no positive clinical results to date with specific anti-K-RAS inhibitor treatments, contrary to what occurs in the lung case.

NSCLC patients presenting the G12C mutation in exon 2 of *K-RAS* (13%) can be treated with Sotorasib with good results.[17] Additionally, CRC patients with this same mutation (1%–3%) might receive this treatment in the future, although this indication does not currently exist.

Finally, there are families of genes that can present similar alterations in any of their components and , regardless of the organ where the tumor originates, patients can be successfully treated. This is the case of the *N-TRK* 1, 2, and 3 gene family.

Generic "pan-cancer" biomarkers

The other large group of biomarkers studied in molecular oncological pathology are the so-called pan-cancer biomarkers. This term refers to genes that participate in the neoplastic transformation process in different organs. In this context, in principle the type of tumor or organ in which it originates does not matter, as what is important is to recognize if these genes have molecular alterations that can be inhibited by a specific treatment.

An example is the family of *N-TRK* genes (1, 2, and 3) which frequently present rearrangements in rare tumors such as childhood fibrosarcoma or breast and salivary gland carcinomas of the "secretory" type, but which in rare occasions suffer activating alterations in common tumors such as lung carcinomas or CRCs, in which frequencies of such rearrangements have been detected in less than 5% of them.[18] Patients with tumors presenting *N-TRK* rearrangements can be treated with drugs such as Entrectinib or Larotrectinib, regardless of tumor type.

Biomarkers can also be classified attending to their biological characteristics:

Cell surface membrane antigens

Tumor cells can present shared (sometimes altered) antigens with nonneoplastic cells or neoantigens associated with tumor transformation (tumor antigens). This group includes the leukocyte differentiation or cluster of differentiation antigens (CDs) present in normal hematologic cells and neoplastic cells, such as CD20 or CD30. Monoclonal antibodies have been designed to kill cells that express these surface antigens. This is also the case of CD19, which is being used to treat lymphomas and leukemias (lymphoblastic lymphoma, diffuse large cell lymphoma) employing CAR-T technology.[19]

Another example is the *HER-2* oncogene that encodes a membrane protein, specifically a receptor for the growth factor corresponding to the HER family, expressed at very low levels in nontumor cells. In neoplastic cells undergoing molecular alterations of the gene (mainly amplification) there is increased transcription and subsequent cell surface protein overexpression. This overexpressed receptor can be attacked with humanized monoclonal antibodies such as Trastuzumab or Pertuzumab, with monoclonal antibodies conjugated with toxic molecules such as TDM-1, or it can also be blocked by TKIs such as Lapatinib.[20]

Molecular alterations of oncogenic genes (oncogenes)

Mutations in oncogenes eventually lead to formation of anomalous or chimeric proteins with transforming power that can in turn act as tumor neoantigens. These mutations can be of different types such as those that alter the reading frame, point mutations (in a certain codon), insertions, deletions, or rearrangements.[21]

Although there are many data obtained regarding the "mutational landscape" in most human malignant tumors, there are currently very few target-specific drugs developed with approval for clinical use. It should also be remembered that it is necessary to differentiate key mutations in the neoplastic transformation process (*drivers*), from those with no relevant roles in carcinogenesis (*transient*). Furthermore, mutations that may represent a therapeutic target are called *actionable*.

Most oncogenes suffer activating mutations and alteration of one allele is sufficient to initiate/collaborate in oncogenic transformation, while others (tumor suppressor genes) present mutations that silence their activity, as is the case of *p53*, *APC*, *pRB*, *PTEN*, or *BRCA1* and *2*. These genes have a protective role and when their function is altered the cell transformation process is favored. In general both alleles must undergo mutations, as occurs in many examples of childhood or family cancer, where one altered allele can be inherited and the inactivation of the second one contributes to the generation of cancer.[22]

Furthermore, these "recessive oncogenes" can not only be inactivated by mutations but also by epigenetic processes such as methylation of their promoters. This phenomenon is highly relevant in the case of silencing of genes that encode repair proteins of DNA mismatches (mismatch repair proteins), such as *MLH-1* in CRC, since promoter hypermethylation of *MLH-1* decreases its activity and associates with the genesis of CRC.[23]

Sometimes biomarkers are studied to apply a specific therapy against an altered molecule that matches the biomarker, as happens in most cases. However, on other occasions, biomarkers are studied to apply a therapy aimed directly against another molecule different from the one studied as a biomarker, but whose inhibition by the treatment is conditioned by the functional state of the studied biomarker. Examples are the genes *K-RAS* and *N-RAS* (also now *B-RAF*) that need to be free of mutations (wild type) so that monoclonal anti-EGFR antibodies (Cetuximab or Panitumumab) can be successfully applied in the treatment of CRC, even if intrinsic and acquired resistance mechanisms exist.[24]

Other biomarkers

Apart from these biomarkers that can be defined as classical, there are other situations to be considered that also help to predict the sensitivity or resistance that certain patients may present to specific therapies.

For example, monoclonal antibodies have been recently developed against cellular antigens of the immune system that modulate the immune response by inhibiting the activation of certain immune cells. These are the so-called checkpoint antigens that include cytotoxic T-lymphocyte antigen-4 (CTLA-4), programmed death-1 (PD-1), or its ligand PD-L1.[25]

Initially Ipilimumab (directed against CTLA-4) and later Pembrolizumab and Nivolumab (directed against PD-1), and more recently Atezolizumab, Avelumab, and Durvalumab (designed against PD-L1) are being used successfully for treatment of various solid tumors.[26]

At present, immunohistochemical (IHC) techniques are performed in tissue to evaluate the expression level of PD-L1, either in tumor cells, in inflammatory cells, or in both, depending on the case. However, selection of patients who can receive this type of treatment is very heterogeneous as it depends on the type of tumor (melanoma, NSCLC, CRC, bladder carcinoma, etc.), the type of monoclonal antibody to be administered (Pembrolizumab, Nivolumab, Atezolizumab, etc.), the clone of the antibody to be used in the immunohistochemical study (28.8, 22C3, SP142, SP263, etc.), and the system to evaluate immunohistological results: percentage of positive tumor cells, of inflammatory/stromal cells, of the relationship between them, etc. All of these variables increase confusion and lack of intra and interinstitutional reproducibility, which is why a normalization of PD-L1 testing is convenient.[27] Automated digital image analysis systems help and improve the evaluation of PD-L1 expression in tissues.[28]

It has recently been postulated that other biomarkers can predict the response to treatment with these monoclonal antibodies. One is the total mutational burden (TMB), defined as the number of mutations in a tumor per megabase of DNA studied. TMB is derived from next-generation sequencing (NGS) studies and can also be determined in liquid biopsy (LB).[29,30] It has acquired great importance, since it is assumed that tumors with high mutational rate generate more neoantigens and therefore, induce greater antitumor immune response, at least in theory.

For similar reasons, regulatory agencies have already approved that CRC patients with high microsatellite instability (MSI-H) or with polymerase epsilon (*POLE*) mutations may also receive these immunotherapeutic treatments successfully.[31]

Also related to antitumor immunosurveillance/immunotherapy, the study of inflammatory tumor infiltrates or Immunoscore has been developed, based on the count of tumor infiltrating lymphocytes (TILs) CD3+/CD8+ located in the central portion and leading edge (margin) of the tumor. This system has been internationally validated for its application in clinical practice in CRC cases, since it not only has prognostic and chemotherapy response value,[32] but also important predictive value, especially in patients undergoing immunotherapy.[33]

In this area of nonclassical biomarkers the study of the digestive tract microbiome must also be mentioned. The presence or absence of certain microbial agents in the digestive tract, especially in the large intestine, can indicate sensitivity to classical therapies such as chemotherapy or radiotherapy. There is also solid scientific evidence indicating that the intestinal microbiota composition can be a magnificent biomarker of response to strategies based on immunotherapy of CRC and other solid tumors or even in hematologic processes.[34]

But biomarkers are not only analyzed to choose target-specific drugs or biological therapies but also to evaluate the response to other conventional therapeutic strategies such as radiotherapy in CRC.[35]

Determination of biomarkers in CRC

An interesting point to highlight in this field refers to the excellent quality that determinations and laboratory tests performed for diagnosis of oncological biomarkers must have. There are guidelines and recommendations from national and international scientific societies that address this issue in general, and in particular with regard to CRC.[10,13,14]

These studies must be carried out by professionals with sufficient knowledge and training to perform them within facilities of the health systems where cancer patients are cared for, as part of the portfolio of services of these healthcare laboratories.

In addition, approved in vitro diagnostic kits and procedures should be used and only methods designed and validated for clinical practice should be used. This is of special interest as it facilitates that results of studies are reproducible and provide sufficient guarantees to patients diagnosed and treated based on these studies. Similarly, it is very important that diagnostic clinical services are subject to quality assurance programs and have certification/accreditation to perform such studies.[10]

Other important aspects highlighted in the consensus guidelines and recommendations previously mentioned are the indications and the *what* (type of biomarker), *when* (moment of patient's evolution), *how* (methodology to be applied), and *where* (in primary tumor, in metastasis, in LB) the determinations of biomarkers in general and in CRC in particular should be carried out.[13,14] Intrinsic characteristics of malignant neoplasms such as their heterogeneity and the dynamic plasticity they present throughout their progression and dissemination should be especially considered.

According to current consensus guidelines and recommendations, biomarkers used in daily clinical practice in CRC patients are mutations in exons 2, 3, and 4 of *K-RAS* and *N-RAS*, the V600 mutation of *B-RAF*, the MSI state, and HER-2. In the immediate future, other biomarkers such as PI3K, ROS-1, ALK, RET, PTEN, N-TRK, or PD-L1 will possibly be incorporated.

Due to the continuous advancement of research and the number and type of determinations that must be carried out according to the different types of tumors, the offer of target-specific drugs or personalized therapeutic strategies is increasingly important. The way to study these biomarkers depends on the type of target molecules, their alterations, their cellular or tissue location, or the type of drug to be used.

For the study of cell surface membrane receptors or antigens of solid tumors such as CRC, IHC techniques are generally used, especially if they have to be done on biopsy samples or routinely processed surgical specimens, i.e., formalin-fixed paraffin-embedded tissues. As examples, HER-2 expression, DNA repair proteins, and PD-L1 are routinely studied by IHC.

Immunofluorescence (IF) or flow cytometry (FC) techniques have multiple applications but require fresh tissue or isolated cells, which is why they are used mainly in oncohematology and not so much in pathology. Fluorescent in situ hybridization (FISH) techniques have been traditionally used in cytogenetics and oncohematology but are increasingly used in pathology, as they have been optimized for use in the study of solid paraffin-embedded tumors, and even more adequate adaptations have been made, such as chromogenic in situ hybridization (CISH) and silver in situ hybridization (SISH).

These techniques are especially useful for detection of variation in the number of copies of certain genes (e.g. amplification of the *HER-2* gene) or for the study of translocations and rearrangements as occurs in *ALK*, *ROS-1*, or *N-TRK*. The advantage of techniques that use tissue samples on a slide, i.e., IHC or ISH techniques, is to topographically locate cellular alterations, which is highly relevant in some cases.

Molecular techniques are increasingly used in clinical practice, fundamentally those based on amplification by polymerase chain reaction (PCR) in different modalities such as qPCR, RT-PCR, or digital PCR. They are based on the amplification of small DNA or RNA sequences of target genes. Currently they are also used in fixed and paraffin-embedded tissue, and their use has been generalized in the study of solid tumor biomarkers.

These PCR techniques are designed to detect known mutations of a certain gene, either in one or in different exons (hot spots), although currently there are multiple diagnostic kits validated for in vitro clinical diagnosis (IVD) that examine multiple sequences of different genes in so-called multiplex assays.

The sensitivity and specificity of these kits are very high and are undergoing constant improvements. Thus, ddPCR (droplet digital PCR) or BEAMing (Beads, Emulsion Amplification and Magnetics) techniques can have a mutation detection threshold of 001%.[36]

Finally, in recent years NGS techniques are being introduced in different formats, either as specific panels for certain tumors such as NSCLC or CRC, or in so-called pan-cancer panels that study numerous genes in their hot spots and that can be applied to many different types of solid or hematologic tumors. Likewise, techniques are beginning to be applied to study the DNA methylation pattern, both in tissue and free tumor DNA in plasma by LB.[37]

Blood and stool biomarkers for CRC screening

The diagnosis of cancer in general and CRC in particular has been and continues to be carried out fundamentally by histopathological study of biopsies and surgical resection pieces. In recent years, the so-called liquid biopsy (LB) is emerging with great interest.[38] LB is the study of cells or cellular products (extracellular vesicles, nucleic acids, proteins) either to establish a diagnosis or to monitor patients suffering from this disease,[39] although LB is increasingly being directed toward early diagnosis and cancer screening.[40]

There are different types of LB. On the one hand, the type of liquid in which target cells or molecules are studied is relevant. The most widely used is blood and mainly plasma, although determinations can also be made in serum, in the cellular fraction, or in whole blood. Other biological fluids such as urine, pleural fluid, ascitic fluid, or even cerebrospinal fluid are being used to perform specific studies in specific oncologic pathologies.[41] However, peripheral blood biomarker studies are the most widely accepted both in research and clinical practice.[42]

On the other hand, referring specifically to solid tumors, different types of cellular and subcellular structures or molecules can be studied in LBs, among which we highlight:

Circulating tumor cells (CTCs)

CTCs can be detected in peripheral blood with different techniques, some based on immunological methods and others based on physical or chemical properties of the CTCs themselves, that allow them to be differentiated and isolated from the rest of the hematologic cells found in the bloodstream.[43]

Among immunological or phenotypic isolation methods are those such as EPCam that separate CTCs by detection (immunoaffinity) of surface antigens associated with epithelial cells. In some technologies they are complemented with PCR studies for detection of cytokeratins (CKs), as is the case of Cell Search, a method approved by the Food and Drugs Administration (FDA) for clinical use for detecting CTCs in breast, prostate, and colorectal cancer.[44]

Other methods based on physicochemical properties of CTCs are often used in research but have no significant presence in the clinical field. Lately, specific procedures like DEP Array are being developed to isolate individualized CTCs, and they can provide additional information of interest on tumor heterogeneity.[45]

The study of CTCs in CRC has shown that it can have important prognostic, predictive, and diagnostic value, both in early stages of disease and when there is progression of it. Specifically, using Cell Search technology to study the molecular profile of CTCs, it has been shown that depending on the presence of numerous tumor cells identified in 7.5 mL of peripheral blood, cutoff points can be established to determine prognosis based on the evolution of the disease, and predict the response to chemotherapy and target-specific drugs.[46]

Exosomes

Another way to study molecular components of tumor cells by LB is through extracellular vesicles in general, and specifically exosomes, which are 40–100 nm in size and present several membrane antigens such as CD9, CD63, CD81, or tetraspanins.[47] They can thus be identified by immunophenotypic methods, and as they are covered by cell membrane, they constitute an ideal system for the preservation of cell components.[48]

Previously exosomes were considered a kind of "garbage dump" that cells used to dump waste products of cellular metabolism into the extracellular space, but now it is known that exosomes have very relevant biological functions, such as those related to local intercellular and long-distance communication or with immunomodulation. Regarding tumor cells, exosomes are known to have a very relevant role in the conditioning of tumor niches and in metastatic spread, as well as in immune regulation of antitumor response.[49,50]

As exosomes are a faithful reflection of the constitution of cells in which they originate, they can transport proteins and nucleic acids such as DNA, mRNA, and even microRNAs (miRNAs) and keep them in good condition. Thus, from the isolation of exosomes in peripheral blood (or in other bodily fluids) we can obtain an important source of information.[48]

Previously, isolation and identification of exosomes was technically complicated and required special centrifugation methods, but now there are easier isolation methods (commercial kits) that allow them to be used in clinical settings.

One of the best applications of exosome isolation is to be able to perform studies of mRNA, since this molecule is very labile, and its degradation is rapid due to the action of blood RNases. However, in exosomes mRNA is well preserved and can be extracted for study.[41] Similarly, other molecules such as miRNAs, DNA, or proteins can be isolated from exosomes. However, although study of exosomes and especially mRNA is a very promising field, currently it has no important role in clinical practice of molecular pathology diagnosis of solid tumors.[51]

Cancer-educated platelets (CEPs)

Something similar happens with the so-called cancer-educated platelets (CEPs). As with exosomes, platelets can be an important source of molecular information. Today we know that they are directly involved in the process of tumor progression and metastatic dissemination and have an important, intimate relationship with tumor cells.[52]

Platelets trap molecules derived from tumor cells via different mechanisms (alternative splicing of pre-mRNA or direct phagocytosis)[53] and are capable of transporting and conserving in perfect conditions tumor cell products such as proteins, mRNAs, miRNAs, or DNA. In the future they may also be an important source for obtaining biological products of tumor origin such as mRNA or miRNAs. Although it is a very promising field, at present its application in oncological care practice is also not developed.[54]

Messenger ribonucleic acid (mRNA)

As we mentioned before, the study of free mRNA in plasma by LB is conditioned by the fragility of this molecule and its sensitivity to the action of RNAases. This is the reason why LB studies are not generally used despite the important information they provide.

Small noncoding ribonucleic acids (miRNAs)

miRNAs are small noncoding RNA molecules of about 21–25 bases that are not translated into proteins but are coupled to sequences of different target mRNAs to modulate (generally inhibit) their function. In this way, they regulate multiple cellular processes of great relevance, e.g., immunological modulation, among many others.[55]

miRNAs are adopting an important protagonism in different areas of biology and medicine and more specifically in the field of diagnosis, prognosis, and predictive value in oncopathology.[56] Due to their small size, they are easier to obtain free in plasma, although it is advisable that their study be carried out either in tissue or in exosomes and CEPs, for their better preservation. There are kits that permit the isolation of miRNAs from both tissue samples and peripheral blood with high quality.

At present, it is considered that some types of miRNAs may have important value in early diagnosis of CRC and in follow-up of patients, as is the case of miR-21,[57] although others such as miR-200c, miR, miR-1290, miR-29c, miR-122, or miR-203 could also have prognostic value.[46]

Proteins

The so-called classical tumor markers such as carcinoembryonic antigen (CEA), prostate specific antigen (PSA), or CA 125, CA 15.3, or CA 19.9, among others, are determined in peripheral blood in daily clinical routine, but their sensitivity and specificity are not high. CEA and sometimes also CA 19.9 have been used as markers for monitoring patients with CRC, although their sensitivity and especially their specificity are not high. CEA can also be elevated in different types of tumors (generally adenocarcinomas) of different organs, and even in situations of nonneoplastic diseases (ulcerative colitis, liver disease, etc.). CA 19.9 is generally associated with pancreatic and mucosecretory adenocarcinomas, CA15.3 with breast carcinoma, and CA 125 with ovarian and endometrial tumors, but their diagnostic value is also relatively low. These markers can also be detected in CRC tissue samples, but their diagnostic value is limited.[58]

Protein profiles in LB can also be an important source of oncological diagnostic information using appropriate technology, as occurs in procedures like Cancer Seek, which uses a double methodology. It studies mutations in "hot spots" of 16 genes (1933 distinct genomic positions) but also the expression of 8 protein markers such as CEA, CA19.9, or CA125 and has a high sensitivity and specificity for different solid tumors in early stage (surgically resectable), including CRC.[59]

Free tumor deoxyribonucleic acid (ctDNA)

It is, on the other hand, the study of ctDNA that is generally used in clinical practice for diagnosis and monitoring of cancer patients, being relatively stable and present in sufficient quantity to be detected with different technologies.[60]

ctDNA concentration varies markedly with respect to the total amount of free circulating DNA (cfDNA) and can present levels ranging between 10 and 100 ng/mL.[61] ctDNA fragments are usually small (about 180–200 bp)[62] but sufficient to be studied using different molecular techniques. The amount of ctDNA can vary depending on different factors but is intimately related to the tumor mass that the patient presents. Logically, small tumors in early stages of disease contribute less DNA to the bloodstream than those with disseminated disease and significant tumor burden. In consequence, the proportion of ctDNA with respect to total cfDNA can vary between 0.1% and 10%.[63]

Furthermore, the percentage of ctDNA with respect to total cfDNA can vary throughout the disease and depending on the effect of therapies administered to the patient or on intrinsic phenomena of the neoplasm. In fact, tumors that may present a higher rate of apoptosis or larger areas of necrosis in theory are those that can contribute a greater amount of ctDNA to the bloodstream.[64] On the contrary, those cell clones that do not undergo apoptosis or necrosis may represent a significant contingent of the neoplasm but their contribution to ctDNA may be scarce, although it is known that there may be an "active export" of ctDNA into the bloodstream, which may have a role in tumor progression.[65]

The study of cfDNA provides prognostic data when it is quantified and also predictive data when it is used to detect biomarkers. In general, mutational profiles in solid tumors tend to have good correlation with those detected by LB, so LB may be used when there is no tumor tissue available to carry out molecular studies.

Furthermore, although in the specific case of CRC the concordance between detection of K-RAS mutations in tissue presents a correlation higher than 90% when compared with LB results,[66] there are dynamic biological processes of great clinical interest that LB can reveal. These include the high heterogeneity of K-RAS mutational profiles described in CRC,[67] which may partly explain the different response that patients may present to the same anti-EGFR therapy.

Specifically, in CRC the presence of *RAS* mutations is studied to monitor response to EGFR inhibitors in CRC patients with no *RAS* mutations in the primary tumor. It is possible that small cell clones with *K-RAS* mutations in the primary tumor can expand after pressure with anti-EGFR treatment and be detected with LB.[68]

Additionally, a mutation in *EGFR* (S492R) has been described that can be detected in LB after treatment with anti-EGFR antibodies and which induces resistance to treatment with Cetuximab but not so much with Panitumumab, although this aspect has yet to be elucidated with prospective studies.[69]

We now know that some of these patients receiving anti-EGFR therapy develop acquired *RAS* mutations secondary to therapy, but that they are reversible, since upon withdrawal of the specific anti-EGFR treatment, the tumor ctDNA once again shows a wild profile type.[70]

ctDNA methylation profiles also provide very relevant information about the neoplastic process that a patient undergoes. Methodologies such as CancerLocator not only provide information on the percentage of circulating tumor ctDNA with respect to the total, but also indicate the origin of the tumor,[71] whereas other methods establish useful molecular profiles in the exploration of new biomarkers.[37]

Therefore LB has important advantages and complements the classical biopsy (without replacing it) being easy to obtain, noninvasive, cheap, and can be repeated as many times as necessary to correctly monitor and treat patients with CRC. Notably, LB has advantages in the detection of mechanisms of primary or acquired resistance,[72] or to detect minimal residual disease in patients.[73]

It is important to consider that despite the important development that molecular biology techniques applied to LB are having in the diagnosis and monitoring of patients with CRC, there are today many drawbacks for applying this technology in clinical practice with all the guarantees. In particular, regarding the role of LB in CRC, there is solid evidence that LB provides relevant information in early diagnosis, establishment of prognostic categories, risk of recurrence, identification of biomarkers and useful parameters in the follow-up, and monitoring of patients with CRC, whether CTCs such as cfDNA (mutations and methylation patterns), mRNA, or miRNAs are used.[46]

Perhaps the biggest problem is the lack of normalization or standardization of procedures and the establishment of normal or pathological values or cutoff points from which we can consider that the result of the study is negative or positive.

There are many rigorous scientific articles describing different methodologies and orientations of LB in cancer patients, but there is great intra and interinstitutional variability, and sometimes results obtained by research groups are not reproducible or are performed on case series with small numbers of patients. These are aspects that must be addressed and improved in the future so that LB has the clinical importance that we assume it should have.[74]

Regarding the early diagnosis of CRC, we must also review the studies carried out in screening programs generally applied in people over 50 years of age. In some countries endoscopy is the test initially performed in the general population for this purpose, although in countries like Spain a fecal occult blood test is carried out prior to performing endoscopy, which is only indicated if this test is considered positive.

The tests generally used are based on detecting hemoglobin (cutoff 20 μg hemoglobin/g of feces) by immunochemical methods, to which detection of calprotectin can be added complementarily. However, there is evidence that addition of this second marker does not provide relevant information, which is why it is recommended that the study of hemoglobin in feces is enough to make an adequate selection of patients who can be subjected to endoscopic study.[75]

It is well known that although these tests have a high negative predictive value, they nevertheless have a low specificity since there are many patients who test positive for hemoglobin in feces and subsequently when an endoscopic study is performed do not present neoplastic lesions, not even precursor lesions such as adenomas.

For this reason, other complementary studies are proposed such as the detection of oncogenic molecules, miRNAs, or volatile organic compounds, which can be detected in fluids such as urine and help to improve selection of patients who should be subjected to endoscopy.[76]

Recently, performance of a bacterial signature of the intestinal microbiota to reduce the rate of false positives in these tests was proposed.[77]

RAS: K-RAS and N-RAS, the current key for precision medicine in CRC

Aberrant activation of EGFR and RAS/RAF signaling pathways in CRC is primarily associated with activating mutations of genes in the mitogen-activated protein kinase (MAPK) and phosphatidylinositol-3-kinase (PI3K) pathways. *RAS* is the most frequently mutated oncogene family in cancer. It's a critical driver of oncogenesis making proteins involved in cell signaling pathways that control cell growth and death. Alterations in *RAS* may cause cancer cells to grow and spread through the body. There are three *RAS* genes in the human genome: *K-RAS*, *N-RAS*, and *H-RAS*. The first two are closely related, their mutations are mutually exclusive, and they are the most frequently mutated in CRC, being found in more than 50% of cases.[78] *H-RAS* mutations are very rare in this type of tumor.

The essential role of *RAS* in tumor genesis and progression together with the high incidence of mutations seen in CRC justify the important role that the study and research of this gene family have acquired as targets of so-called precision medicine. Cetuximab and panitumumab are monoclonal antibodies that inhibit signal transduction through their binding to the extracellular domain of EGFR. Both are essential treatment options in patients with metastatic CRC.[10]

The crucial role of RAS status as a predictive biomarker of response to treatment with EGFR inhibitors has been demonstrated in several relevant trials. In the CRYSTAL study, administration of Cetuximab with FOLFIRI showed an increase in the response rate and median overall survival in those patients with "KRAS wt/native" tumors.[79] Similar results were seen in trials that included therapeutic regimens combining FOLFOX and EGFR inhibitors, as OPUS (FOLFOX+Cetuximab) and PRIME (FOLFOX+panitumumab) studies.[80–83]

In agreement with evidence demonstrating that *RAS* mutations are associated with resistance to EGFR inhibitor therapies, a mutational study of the *RAS* gene should be performed at the time of diagnosis in all patients with metastatic CRC. If possible, this study should be performed on representative tissue of any metastatic lesion.

At first, only mutations in exon 2 of the *K-RAS* gene were studied. From results obtained in the retrospective analysis of the PRIME trial[10] it was concluded that, in addition to the known evidence regarding this exon, there are other mutations in *K-RAS* and *N-RAS* related to a lack of response to anti-EGFR therapies.

In the meta-analysis published by Sorich et al.,[78] tumors with "new" mutations were compared with those "nonmutated" and those harboring *K-RAS* exon 2 mutations, to evaluate the possible impact of anti-EGFR therapy. With use of anti-EGFR mAbs, no differences in progression-free survival or overall survival were observed between tumors with any *R-AS* mutation. Thus, actually, mutational analysis should include an "expanded" panel with *K-RAS* and *N-RAS* codons 12 and 13 of exon 2, codons 59 and 61 of exon 3, and codons 117 and 146 of exon 4, following recommendations from the ASCO, ESMO, NCCN, and SEOM-SEAP guidelines.[10,13,14,84]

B-RAF mutation in colorectal cancer

B-RAF is another gene directly involved in the RAS/MAPK intracellular signaling pathway. The most frequent *B-RAF* mutation is the substitution of glutamic acid for valine in codon V600E, which produces constitutive activation of the MAPK pathway. *B-RAF* activating mutations occur in a significant subset of patients with CRC, being present in approximately 8%–10% of tumors in patients with advanced disease[85,86] and in about 14% of patients with stage II and III CRC.[87] Mutated B-RAF tumors are associated with female sex, predominance in the right colon, poorly differentiated or mucinous morphology, advanced stage, and MSI. Patients with advanced CRC who possess a *B-RAF* mutation have significantly poorer outcomes as measured by progression-free and overall survival and have a decreased response rate to anti-EGFR therapy relative to those with nonmutated *B-RAF*. It is important to know the *B-RAF* mutation status since standard therapy is inadequate for patients with metastatic disease and *B-RAF* mutation.[10] The value of the *B-RAF* mutation as a negative prognostic factor and its relationship with MSI was assessed in an analysis of four clinical trials including more than 3000 patients with metastatic CRC.[88] Statistically significant differences were found in progression-free (6.2 vs 7.7) and overall survival (11.4 vs 17.2) between the group of patients with mutated B-RAF tumors and patients without *B-RAF* mutations. Analysis of the relationship between B-RAF and MSI revealed no differences in terms of progression-free or overall survival in the population with MSI-H tumors regardless of B-RAF status. However, a significant decrease in both parameters was observed in patients with stable B-RAF mutated tumors, reinforcing the negative prognostic role of this mutation.

On the basis of this evidence, international clinical guidelines include the mutational analysis of *B-RAF* both for prognostic stratification purposes and for assessing the risk of Lynch syndrome in tumors with loss of MLH1, favoring the presence of a mutation of sporadic origin of the deficiency.

Finally, data from the BEACON study support the possible predictive value of *B-RAF* mutationanalysis.[89] In this phase III study a combination of Encorafenib, Cetuximab, and Binimetinib resulted in significantly longer overall survival and a higher response rate than standard therapy in patients with metastatic colorectal cancer with the *B-RAF* V600E mutation. Further studies will be necessary to establish definitely the predictive value of the *B-RAF* mutation.

EGFR/HER FAMILY/TP53-APC/β-CATENINS/C-MET as prognostic indicators

HER-2

HER-2 is a well-known therapeutic target in breast carcinoma and gastric cancer. In CRC its role as a prognostic biomarker is yet to be developed. However, its usefulness as a therapeutic target and negative predictive factor of response to anti-EGFR treatment is increasing. *HER-2* amplification in CRC carcinoma has been studied employing molecular techniques, with results ranging from 1.8% to 22% of the cases studied.[90] Results from several relatively recent studies strengthen the hypothesis that HER-2 must be considered an essential therapeutic target in this disease and a mechanism of resistance to anti-EGFR therapies. Similarly, results observed in relation to high levels of *ERBB3* amplification as a negative prognostic factor in colorectal carcinoma suggest that HER-3 may also play an important therapeutic role in this type of tumor.[91] Data from the HERACLES and MyPathway studies showed objective response rates varying between 30% in patients with KRAS native and HER-2 positive CRC treated with Lapatinib and Trastuzumab,[92] and 38% in CRC patients with overexpression of HER-2 treated with dual inhibition by Trastuzumab+Pertuzumab.[93] Protein overexpression and genomic amplification are now easily detectable by immunohistochemical methods and FISH. The validated criteria in the aforementioned HERACLES study define as HER-2 positive those tumors with immunohistochemical staining of 3+ intensity in more than 50% of neoplastic cells or, alternatively, those staining with moderate 2+ intensity with a HER-2 ratio: CEP>2 objectifiable by FISH, in more than 50% of the cells.

In this context, determination of HER-2 may be of clinical utility in advanced colorectal carcinoma. Its determination seems more than justified in native RAS tumors that show resistance to treatment with anti-EGFR drugs. The potential role that other receptors of the same family may have and the frequency with which high expression of HER-3[94,95] has been observed in some studies make it necessary to consider the possibility of designing therapies directed against both receptors.

P53

Most CRCs develop as a result of sequential inactivation of suppressor genes (*APC*, *TP53*, *SMAD4*) and activation of oncogenes such as *RAS* and *B-RAF*. p53 is mutated in multiple types of cancer. It regulates many genes involved in DNA repair, cell senescence, and apoptosis. Mutations in *p53* play a fundamental role in determining the biological behavior of CRC and are related to depth of invasion, metastatic capacity, lymphatic invasion in the proximal colon, and invasion of both lymphatic and venous vessels in the distal colon.[96] Therefore, mutations in *p53* not only play a fundamental role in the transformation to cancer cells, but are also crucial in their aggressiveness and invasiveness. Research in this regard has been based on attempts to modulate gene activity (MDM2 inhibitors, Nutlinas, RITA), activation of other members of the family (e.g. p67 and p73), and reactivation of the mutant p53.[97] In this sense, there is no doubt that reactivation and restoration of p53 activity could be key points in the therapeutic arsenal against CRC. However, most molecules potentially capable of this have only been tested in cell lines and animal models, and there are no reliable data from clinical trials. Furthermore, the most critical mechanisms of action of these potential therapeutic molecules are complex and currently not fully understood.. Thus, the subject should be explored in greater depth. Recent studies[98,99] have investigated the role that TRIM67 protein could have in activation of p53 and its possible inhibitory effects on tumor proliferation and distant metastasis.

APC/β-CATENINS

A very high percentage of CRCs (~80%) have mutations in the adenomatous polyposis coli (APC) gene. Mutations in APC result in elevated β-catenin mediated transcription.[100] Half of the cases without this mutation present alterations in the

Wnt/β-catenin signaling pathway as a result of mutations in the β-catenin gene. Furthermore, more than 94% of CRCs present alterations in proteins directly involved in some of the known Wnt signaling cascades.[101] Despite extensive knowledge and the crucial role it has in CRC, it remains difficult to predict to what extent it will be feasible to successfully intervene this pathway for therapeutic purposes. The issue is very complex and there are important difficulties to fully understand its relationship with other key pathways in the cell signaling cascade such as AKT/PI3K, NOTCH, or mTOR. The potential importance of its study in patient selection is beyond doubt. At the moment, most drugs are in early stages of development.[102,103]

C-MET

Hepatocyte growth factor receptor (C-MET) is a cell surface tyrosine kinase. Activation of the signaling cascade occurs when c-MET binds to its ligand, hepatocyte growth factor (HGF). The c-MET/HGF signaling pathway is involved in multiple key biological processes, including the ability of cells to proliferate and migrate, angiogenesis, and the epithelium–mesenchyme transition. In CRC, as in other types of neoplasms, abnormal expression of c-MET has been associated with tumors with high cell proliferation rate, cell dissociation, tumor aggressiveness, invasiveness and, ultimately, worse prognosis.[104] Thus it is clear that pharmacological intervention of this signaling pathway can constitute a good therapeutic alternative when trying to prevent tumor progression both at the locoregional and at a distance level.[105] To the difficulties in using this pathway as a therapeutic target, the frequent interactions it presents with other signaling pathways directly involved in tumor genesis and development must be added. c-MET inhibitors used to treat CRC are not widely used in a clinical context.[106]

MSI, its role in CRC

Carcinogenesis in colon cancer is a multistage model.[107] The main pathway is inactivation of the APC gene, followed by mutation of the *K-RAS* gene and loss of heterozygosity in 18q. Inactivation of the *p53* gene is also implicated in the adenoma–carcinoma transition. However, another carcinogenetic pathway related to MSI and loss of mismatch repair (MMR) proteins has also been identified.[107]

Microsatellites are repetitive DNA sequences of 1–6 nucleotides distributed in a nonrandom way throughout the genome and are a preferred site of DNA replication failures.[108,109]

MMR proteins are responsible for correcting replicative errors. These proteins are a system of heterodimers where MLH1 binds to PMS2, PMS1, or MLH3 to form MutLα, MutLβ, and MutLγ, respectively; or on the contrary MSH2 binds to MSH6 or MSH3 to form MutSα or MutSβ, respectively.[108,110] A deficiency in such heterodimers is called deficient MMR (dMMR).

MSI arises from a deficiency of MMR proteins, from promoter methylation (MLH1 and to a lesser extent MSH2) or from somatic or germline mutations of genes involved in them.[108,111] Regardless of origin, the result is an inability of cells to recognize and repair spontaneous mutations, originating tumors with a high mutational load and alteration in microsatellite sequences.[112,113]

MSI is estimated to be present in 15% of colon carcinomas (especially in proximal colon), approximately 10%–15% of sporadic carcinomas and 90% of those associated with Lynch syndrome.[112,114] Eighty to 90% of sporadic cases are due to methylation of CpG islands of the MLH1 promoter.[115] Most hereditary cases (~70%) are due to germline mutations of MLH1 and MSH2.[112]

In approximately 3% of cases with MSI, neither germline mutations nor MLH1 methylations are detected.[107] In these cases, double somatic mutations have been identified, accompanied by a higher frequency of somatic *PI3K* mutations.[107,116]

MSI can be studied by PCR technology or NGS.[107,117] The most common strategies for its study are direct determination in the tumor or after loss of immunohistochemical expression of repair proteins, although in the face of strong family burden it may be indicated even with preserved repair proteins.[108,117]

In addition to being useful for identifying possible Lynch syndromes, MSI also has predictive and prognostic utility.[107,117] Tumors with MSI or dMRR show less sensitivity to chemotherapy regimens with 5 FU, without benefit in terms of survival.[107,116,117] Its better prognosis and less sensitivity to the usual treatments help to make the decision whether or not to indicate adjuvant therapies in stage II.[107] On the contrary, tumors with MSI have greater sensitivity to PD1 inhibitors, improvements in the response rate, and progression-free and global survival being observed in phase III trials compared to chemotherapy.[107,112,117,118]

Its favorable prognostic value is more established in stage II, with better survival rates in patients without adjuvant therapy than with it. In stage III, the prognostic value is more controversial, suggesting that it is limited to patients treated with oxaliplatin regimens.

miRNAs, small players with large potential

Dysregulation in CRC is broad and includes genetic and nongenetic changes. One of the main and most studied mechanisms in carcinogenesis relates to miRNAs.

miRNAs are short (19–24 nucleotic) noncoding RNA sequences. They influence many biological processes (development, proliferation, differentiation, apoptosis, signal transduction, etc.) and alteration of their expression pattern associates with numerous diseases including cancer.[119–124] They act by binding posttranscriptional products.[122–124] It is estimated that they represent 3%–4% of the genome.[125] More than 25,000 miRNAs have been described and their number continues to increase.[121]

miRNAs regulate numerous signaling processes, pathways, and cascades. A miRNA can interact with >1 mRNA and alter or regulate various functions, which change depending on cell types, behaving sometimes as tumor promoters or suppressors.[125] They stand out for their stability, which allows their study in different types of samples, including LB.[124]

In colon cancer more than 500 miRNAs have been described. The most frequent are miR21, 143, and 145.[122,126] They are involved in a multitude of processes related to progression, metastases, and resistance to chemotherapy and radiotherapy:

Proliferation: they act on APC, Notch, and Wnt/Bcatenin.[119]

Metastatic process: regulation of mesenchymal–epithelial transition (MET), cadherin switching, AKT/mTOR, NFKB, regulation of MMPs, and the generation of a pro-inflammatory microenvironment more prone to metastasis.[119]

Tumor angiogenesis: they act on the MAPK, PI3K, and MET/ERK/ELK1/HIF1a/VEGFA pathways. Some act directly on the VEGFA pathway. Generally, their alteration increases tumor angiogenesis.[119]

Apoptosis: the target of most miRNAs involved is activation of caspases and induction of apoptosis.[119]

Resistance to radiotherapy: they can help increase or reduce sensitivity to radiotherapy.[119,124]

Resistance to chemotherapy: they can be related to an increase or decrease in sensitivity to 5 FU, cisplatin, oxaliplatin, or combinations of them. Many of these resistances are due to activation of mesenchymal–epithelial transition (MET).[119,123]

In clinical practice miRNAs may have several applications, mainly as predictive and prognostic factors.[124] A main application may be to identify tumors with a more prometastatic or noninvasive profile, even in early stages. Another promising use, associated to specific tumor miRNAs, is in scenarios of metastasis of unknown origin, helping to recognize the primary neoplasia.

Aberrant DNA methylation in CRC

During the last decades knowledge of epigenetic alterations in neoplasias has increased, and one of the most important is aberrant DNA methylation.

The incorporation of the methyl group (methylation) into DNA is an epigenetic, frequent, dynamic, and reversible process, controlled by regulators such as methyltransferase and demethylase.[127–132] This alters gene expression but without permanent changes in the DNA sequence.[127]

Aberrant methylation patterns are involved in carcinogenesis, including CRC, and regulate other mechanisms that modulate gene expression such as miRNAs.[127,133]

Hypo and hypermethylation phenomena are seen in carcinogenesis with predominance of a global hypomethylation pattern contributing to genomic instability, together with activation of silenced oncogenes and inactivation of suppressor genes by hypermethylation of their promoter (methylation in CpG islands).[127,129,134]

In the case of CRC, hypomethylation is seen in early stages of carcinogenesis.[127] Hypomethylation coupled with activation of proto-oncogenes is observed at three levels: in promoter regions, which can cause loss of gene imprinting (IGF2) or direct activation of proto-oncogenes (*MYC* or *H-RAS*); in distant regulatory regions, as superactivators (gene encoding B catenin); and in antisense promoters.[127,135,136]

Currently there are many technologies and commercial kits to study DNA methylation. PCR methods are sensitive, scalable, specific, reliable, and consume fewer resources than other options, especially methylation-specific PCR.[137] DNA methylation studies can be performed on a wide range of samples, including LB.[119]

The utility of methylation studies is broad. Tools such as Epi ProColon® have a moderate to high sensitivity and a high specificity in blood.[132,138,139] Although its sensitivity decreases in asymptomatic cases, it may be an alternative to occult blood test in screening or a less invasive alternative than colonoscopy for the diagnosis of CRC.[132,138,139]

Colonoscopies are invasive procedures not without complications. Methylation studies can help refine the indication for colonoscopy, increasing performance and avoiding it in patients at higher risk.[132,137] Furthermore, commercial kits are available that double the sensitivity of CEA to detect relapses in patients operated on for CRC.[132,140,141]

Microbial markers in CRC prognosis and therapeutics

The set of viruses, bacteria, fungi, archaea, and protozoa that populate our body is called the microbiota. In recent years, there has been growing interest in the study of the microbiota, especially of the digestive tract, and its relationship with the development of diseases, including cancer, as well as its participation in the success or failure of antineoplastic therapies.[142]

The study of the microbiota is complex as it is difficult to carry out cultures under strictly anaerobic conditions to be able to make a complete and functional description of the microbiota of a certain patient. For this reason, other strategies are used such as the so-called metagenomics, in which bacterial DNA/RNA in a sample is studied, in order to characterize the main types of bacteria that make up the microbiota of a specific individual.[143]

This methodology has advantages and disadvantages, being valid as a research method but far from being useful in daily healthcare practice. Therefore, new technologies and more affordable methodologies must be developed and optimized so that they can have clinical use.

Today we know that the intestinal microbiota (particularly of the large intestine) changes over time from birth to old age. Many factors are involved in this process, such as diet. Balance between microbiota composition and the functioning of the body is called *eubiosis*, whereas when this balance is broken is called *dysbiosis*.

In a situation of eubiosis the intestinal microbiota plays physiological roles of great importance, and very especially in the regulation of the immune system.[144] On the contrary, when there is dysbiosis, alterations occur in multiple processes, including immunomodulation, and carcinogenesis processes can also be induced.[145]

The relationship between the microbiota and the host organism is much more complex than might seem at first. Today we know that bacteria interact with cellular elements and immunomodulatory molecules with an innate and adaptive response through complex mechanisms. Bacteria can cross the mucin layer of the apical surface of epithelial cells and be phagocytosed by macrophages and antigen presenting cells, and even being alive and undegraded, they can be transported to regional lymph nodes where they perform an important immunomodulatory function, or they can even be transferred to other distant places in the body.[146]

In fact, bacteria associated with CRC can not only be detected on the surface of the tumor but also within the neoplastic tissue itself and even in metastases that occur at a distance from the tumor. It has been speculated that in these places they may be modulating or blocking the antitumor immune system response to favor tumor growth and dissemination of metastases.[147]

Today some of the mechanisms by which the microbiota participates in development of diseases such as cancer are known, and it can also contribute to success or failure of different antitumor treatments,[148] and even modulate the toxicity of these therapeutic strategies.

The microbiota can play a relevant role in CRC genesis not only due to modulation of the immune system and the aforementioned stromal microenvironment, but also by promoting tissue inflammation, altering cell metabolism, inducing damage to the genome of epithelial cells, or by altering intercellular communication mechanisms mediated by molecules such as E-cadherin or β-catenin.[142]

Thus there are beneficial bacteria such as lactobacilli or bifidobacteria that produce positive effects on the organism, either directly or through their metabolites, and negative or harmful bacteria that cause severe functional alterations or develop diseases. Specifically, *Fusobacterium nucleatum* is related not only to CRC carcinogenesis,[149] but its detection is a factor of poor prognosis in patients with this type of cancer.[150]

The composition and quantity of the microbiota can also play a fundamental role in the response to conventional treatments such as chemotherapy or those based on immunotherapy. While presence of *L. murinus* or *L. johnsonii* increases the efficacy of cyclophosphamide, the presence of *E. coli* or *M. hyorhinis* increases resistance to gemcitabine. The presence of *B. longum* or *E. faecium* improves the efficacy of anti-PD-1 treatments, as occurs with *Bifidobacterium* and anti-PD-L1 treatments or with *B. fragilis* and anti-CTLA-4 treatments.[34]

Therefore, the study of the microbiome, i.e., of DNA/RNA genetic material from the microbiota, is important for knowledge of etiopathogenesis and development of diseases, and also as a biomarker that could determine prognosis and response to classical treatments such as chemotherapy. Especially, it may have predictive value for selecting immunological or target-specific treatments in patients with CRC.

In this sense, treatment with antibiotics, probiotics, prebiotics, or synbiotics, or repopulating the microbiota with beneficial bacteria or even fecal transplantation can be interesting therapeutic initiatives, complementary to specific treatments for CRC.[151]

Thus, in the future, when in the framework of personalized precision medicine different biomarkers of a CRC patient have to be studied, it is necessary not only to analyze classical biomarkers (mutations and rearrangements of driver genes), but also intrinsic factors of the patient as the HLA typing, cytochrome P450 polymorphisms, and very especially their intestinal microbiota.[152]

New insights in CRC biomarkers, recent advances, and future challenges

"Undruggable" RAS

More than 30 years have passed since we became aware of the benefits of patient selection based on *RAS* mutational status. This knowledge allows to define a group of "responding" patients and a group that will not benefit from potential effects of targeted therapies focused on *RAS*. During all this time, failure of therapeutic attempts in mutated RAS tumors has given the latter patient subgroup the status of "intractable."[153] Mutations in *K-RAS* are by far the most common in the group of mutated RAS tumors, and many show the specific G12C mutation. Efforts to develop inhibitory molecules targeting this mutated cysteine have shown that it may be possible to redefine the usual criteria for selecting patients currently considered untreatable with RAS-targeted drugs.[154] In this way, new therapeutic horizons are opened that can be tremendously promising.

The role of mass sequencing techniques

Absence of therapeutic alternatives in these mutated *RAS* cases makes it necessary to develop and continuously search for other targets to act upon. Generalized use of sequencing techniques will probably guide development of a large part of the treatments designed in the near future. Furthermore, based on their potential ability to detect genetic alterations, they will facilitate development of new CRC classifications that must consider the different behaviors of tumors and the heterogeneity of therapeutic responses. In this way, design of increasingly individualized treatments will be facilitated, and patient selection criteria will be improved, thus improving response rates.

The continuing value of the anatomopathological report

Advances and invaluable information derived from the potential of massive sequencing techniques should not overlook the always relevant role of anatomopathological studies. Protocolized and systematic determination of the budding tumor (BT) has been added to well-established prognostic factors obtained from the study of tissue by means of classical methods (tumor type and differentiation, stage, venous/lymphatic/perineural invasion, lymph node metastases and/or discontinuous tumor deposits, presence of intratumoral inflammatory infiltrate). BT is defined as the presence of single tumor cells or small groups smaller than five elements at the leading edge of the neoplasm. Criteria for its measurement and grading have been agreed by international consensus, being classified in a system of three degrees (low, medium, and high) defined by the number of buds in a determined area in mm^2. BT has been shown to be a predictive factor of risk of lymph node metastasis in tumors with submucosal layer invasion[155] and risk of recurrence in stage II CRC.[156] Also, budding is a strong prognostic predictor of survival in rectal cancer patients after neoadjuvant therapy.[157]

Molecular staging of CRC

Molecular staging of CRC has revealed in several studies the high rate of false negatives resulting from classical evaluation of periintestinal nodes using hematoxylin–eosin. The prospective study by Aldecoa et al.[158] showed that in stage I-II patients, detection of CK19 mRNA in lymph nodes correlated with classical high-risk factors, establishing that the sum of all copies of CK19 mRNA from each of the metastasized nodes (total tumor load or TTL) may improve the ability to stage patients with early stage tumors.

Advantages of liquid biopsy

Peripheral blood ctDNA study is the best known and most developed option in LB. Several studies have observed the high concordance of results obtained from these samples with those obtained from tissue samples.[66] Advantages of LB are indisputable in terms of its ability to provide information on tumor heterogeneity and the possibilities it offers to learn about the evolutionary dynamics of the neoplasia and to detect minimal residual diseases after surgery.

Study of NTRK rearrangements

In CRC, rearrangements in the *NTRK1*, *NTRK2*, and *NTRK3* genes are predictive factors of response to TKIs.[18] Their presence associates with highly unstable native Ras tumors[23] and worse survival data, as occurs in cases with rearrangement in *ALK* and *ROS-1*.[12] In the algorithm recently proposed by Penault-Llorca et al.[159] for the identification of these patients use of immunohistochemical techniques is recommended as screening followed by subsequent confirmation of positive cases using molecular techniques. TRK inhibitors Larotrectinib and Entrectinib were approved in November 2018 and in mid-2019 for treatment of tumors with *NTRK* fusions.

TRIM67/P53 axis

In addition to its known role in innate immunity, the tripartite motif (TRIM) superfamily is involved in a wide variety of biological processes including key processes in cell cycle regulation (proliferation, differentiation, apoptosis) and tumor development. At present more than 70 members of this family are known. In CRC, the expression of TRIM67 has been associated with degree of invasion and tumor size, presence of metastases in regional lymph nodes and, ultimately, with the clinical and pathological stage.[98] In CRC, TRIM67 is frequently silenced and its downregulation is associated with worse overall survival data. Reactivation of TRIM67, in addition to restoring the activity of p53, has been shown to be effective in sensitizing tumor cells to chemotherapy. Its suppressive role in CRC makes TRIM67 a potential target in the search for better response rates with these treatments.[99] In addition to its potential role as a predictive response factor, TRIM67 has a significant role as a therapeutic target as it is directly involved in both inhibition of tumor cell proliferation and metastatic capacity through regulation of mitogen-activated protein kinase 11 (MAPK11).[98]

Immunotherapy in colon cancer

Colon carcinoma is the second leading cause of cancer mortality worldwide.[160,161] In metastatic disease, 80%–90% of cases are unresectable, and combination of chemotherapy with VEGF or EGFR inhibitors is considered first-line treatment.[162]

Despite this, its effectiveness is limited and overall survival in these patients is approximately 30 months. This underscores the search for new drugs and approaches.[162]

An important novelty is the use of immunotherapy. The FDA has authorized Pembrolizumab and Nivolumab, although only in patients with loss of repair proteins or MSI.[160]

Signals of the PD1/PDL1 axis in tumor microenvironment are involved in resistance to adaptive antitumor immunity.[160] Activation of immunity leads to expression of PD1 and an increase in inflammatory cytokines such as interferon γ, which induces expression of PDL1, inhibiting T antitumor response.[160,163] CTLA-4 may act in the initial phase of the immune response inhibiting activation of T lymphocytes, while PD1 in later phases may act disconnecting the antitumor T response.[160,164]

Tumors associated with loss of repair proteins/MSI are characterized by having a higher proportion of TILs, in which Th1 (interferon γ producers) and cytotoxic CD8 predominate. These are preclinical findings that explain the rationale for the use of immunotherapy in these tumors.[160]

In recent years many clinical trials have been performed for immunotherapy in the PD1/PDL1 axis, with best responses observed in tumors with loss of repair proteins/MSI.[112,160,165,166]

Considering that approximately 30% of patients treated with Pembrolizumab have primary resistance, other strategies are necessary in this field.[112,160]

One of the most promising options is the double blockade of PD1 and CTLA-4. Currently only results of a phase II clinical trial are available. The benefit is limited to tumors with loss of repair proteins/MSI, although with considerable toxicity. In consequence, the FDA authorized in July of 2018 the treatment with Nivolumab and Ipilimumab, as second line after chemotherapy in these patients.[160,166]

PD1/PDL1 interaction is not the only regulatory pathway for activation of T lymphocytes in the tumor microenvironment. Other implicated molecules have been described which in the future might be pharmacological targets, TIM3, TIGIT, and LAG3 being the most prominent.[166–168] Their use in clinical trials is very incipient, with some positive experiences in phase II trials.[169]

Immunotherapy can be combined to enhance its effect with other therapeutic tools such as radiation therapy.[170] In other tumors such as NSCLC, the synergistic effect of radiotherapy and immunotherapy has been demonstrated.[171,172] In CRC, experience is very scarce, and doubts exist about the time of use and radiotherapy parameters.[173,174]

The role of immunotherapy in the treatment of colon cancer is not limited to metastatic patients, but is expanding to neoadjuvant and adjuvant scenarios, especially in patients with loss of repair proteins/MSI.[160,175,176]

Immunotherapy drugs are expensive and not free from side effects, which is why the search for biomarkers is important to select patients who will benefit the most from these treatments. The cheapest and most widely available biomarker in daily practice is the loss of repair proteins/MSI.[160]

An emerging factor is TMB, defined as the number of mutations per coding area of tumor genome.[160,177] Greater benefit of immunotherapy has been observed in tumors with high tumor mutational burden (TMB-H). This led the FDA in June of 2020 to authorize Pembrolizumab in tumors with TMB-H. Caution is warranted, since the proportion of colon carcinomas in the approval study was low and the benefit is greater in tumors with loss of repair proteins/MSI than TMB-H.[177,178]

Another possible biomarker is *POLE* status. *POLE* is a gene involved in maintaining DNA replication fidelity and in preventing mutagenesis. Somatic mutations appear in 1%–3% of colon carcinomas, they are mutually exclusive with loss of repair proteins and endow tumors with a somatic hypermutation profile, with increased CD8+ TILs.[113,179]

Tumor microenvironment in CRC

Understanding and management of CRC has evolved from an approach based on radiological and pathology management to a more comprehensive one that includes molecular characteristics of the tumor and the tumor microenvironment.[180–182] The tumor microenvironment is composed of mesenchymal cells, cells infiltrating the tumor, endothelial cells, extracellular matrix, and inflammatory mediators.[183]

Evidence suggests that tumor progression and metastasis are due to alterations of intratumoral immunity, but this is not the only factor. For metastasis to occur, genetic and molecular alterations must converge in the tumor and in the microenvironment.[184] The tumor stroma is dynamic and heterogeneous. Different types of inflammatory cells, fibroblasts, and even cells from the bone marrow are observed, influenced on many occasions by paracrine secretions.

In the tumor microenvironment, the role of cancer-associated fibroblasts (CAFs) stands out, although CAFs are a heterogeneous population with different functions.[185–187] In a desmoplastic stroma, CAFs interact with the extracellular matrix and other cells, helping tumor cells. Effects produced by this interaction are increase in tumor proliferation and migration, immune suppression, and appearance of foci of tumor stem cells (chemoradioresistant).[185] In addition, CAFs constitute a physical barrier that hinders or prevents diffusion of antineoplastic drugs.[188]

One of the aspects that best shows the importance of the tumor microenvironment is the stem tumor cells, a subpopulation of tumor cells with self-renewal and tumor initiation capacity, long-term clonal growth and resistant to treatments.[189–191] These cells reside in favorable tumor microenvironments, escaping immune surveillance, apoptosis and maintaining their plasticity, which favors metastasizing capacity.[192,193]

Stem tumor cells modulate secretion and behavior of tumor microenvironments through exosomes.[194] The role of exosomes is very important, since they allow distant tissues to be site of metastasis, e.g., with the delivery of miRNAs.[195] Stem tumor cells also secrete mediators such as chemokines, which are chemotactic for cells that secrete mediators that favor or promote stem tumor cell functions.[196]

Immune checkpoints in relation to the microenvironment influence prognosis of colon carcinoma, especially TIM3, LAG3, PD1, and PDL1, and in particular their location. Their presence in tumor cells indicates bad prognosis, while their presence in stromal immune cells increases cancer-specific survival.[180]

High tumor expression of PD1 is associated with poorer survival, either due to its pro-inflammatory effect at the systemic level or due to decrease of T lymphocytes (total, cytotoxic, and activated regulatory T cells).[196,197] However, its localization in stromal immune cells has the opposite effect.[196,197]

High expression of TIM3 and LAG3 in stromal immune cells is associated with improvement in survival, probably due to their effect on the local lymphocytic infiltrate, but in the case of LAG3 the effect is more powerful.[180]

PDL1 expression does not have a homogeneous prognostic influence, depending on its location. Its expression in tumor cells associates with worse prognosis, whereas if observed in immune cells prognosis is better, perhaps because of greater infiltration of immune cells in the tumor microenvironment.[197–202]

In CRC the value of the tumor microenvironment begins to be transferred to daily practice. Two good examples of this are the tumor/stroma ratio and the type of desmoplastic reaction on the front of tumor advance.[203–205] They are fast morphological features, inexpensive, with low interobserver variability and with prognostic value, although their use is not yet massive.[203–205]

In CRC the importance of the microenvironment is not limited to the primary tumor, but also extends to metastases.[184] Differences in composition (approximately 75%–80% of the tumor mass is made up of CAFs), differences in interstitial pressure that hinder drug diffusion, and a defective immune response help to understand drug resistance.[185] Finally, as in primary tumors, composition of the stroma is a dynamic process.

Concluding remarks

Although there is extensive knowledge of molecular mechanisms, genes, and epigenetic events involved in the neoplastic transformation process of CRC, there is currently no extensive offer of specific personalized treatments for CRC.

This situation is different from that observed in other types of tumors, especially NSCLC, which has undergone a major revolution in its diagnosis and treatment in recent years due to the identification of various molecular targets, and the generation of drugs and specific biological therapies, leading to an important qualitative change in the prognosis of this type of tumor.

Since the application of monoclonal antibodies (Cetuximab and Panitumumab) to inhibit the action of EGFR in patients without mutations in *K-RAS*, *N-RAS*, and *B-RAF*, there have been no important incorporations to the therapeutic armamentarium to treat advanced CRC. In contrast, in recent years new drugs have begun to be incorporated to treat CRC patients with *B-RAF* mutations (V600E), such as Encorafenib, or anti-HER-2 therapies (Trastuzumab and Lapatinib) to treat tumors with activation of this gene. In the near future possibly the use of Sotorasib will be approved in CRCs with *K-RAS* (C12G) mutations, as in NSCLC.

Moreover, other types of studies are being incorporated into clinical practice, such as evaluation of the antitumor inflammatory infiltrate (Immunoscore), determination of the expression level of MMR and the MSI status, in order to apply new treatments such as immunotherapy.

Even so, there is still a long way to go contemplating different emerging scenarios, both in research and technological developments. The study of the intestinal microbiota, the development of high-throughput technologies, the introduction of NGS and molecular profile studies (in tissue and LB), advanced digital imaging and application of new therapeutic strategies such as CAR-Ts for treating solid tumors including CRC open new expectations of great interest.

Along with these initiatives, we must put our effort into what really improves the lives of patients, facilitating their total cure: *the early diagnosis of CRC*. To achieve this, a social, political, scientific, and welfare effort must be made, improving programs already underway which are giving very satisfactory results.

Early diagnosis of CRC not only improves patient health but is also very efficient from a sociosanitary perspective and helps the sustainability of health systems. For this reason, as improvement of advanced CRC treatment continues, it is very important to direct a large part of the resources to processes and procedures for the analysis of diagnostic biomarkers in preclinical initial phases of CRC.

References

1. Nagtegaal ID, Odze RD, Klimstra D, et al. The 2019 WHO classification of tumours of the digestive system. *Histopathology* 2020;**76**(2):182–8.
2. Pegram M, Hsu S, Lewis G, et al. Inhibitory effects of combinations of HER-2/neu antibody and chemotherapeutic agents used for treatment of human breast cancers. *Oncogene* 1999;**18**(13):2241–51. https://doi.org/10.1038/sj.onc.1202526.
3. Miettinen M, Sarlomo-Rikala M, Lasota J. Gastrointestinal stromal tumors: recent advances in understanding of their biology. *Hum Pathol* 1999;**30**(10):1213–20. https://doi.org/10.1016/s0046-8177(99)90040-0.
4. Hirota S, de Ilarduya CT, Barron LG, Szoka Jr FC. Simple mixing device to reproducibly prepare cationic lipid-DNA complexes (lipoplexes). *Biotechniques* 1999;**27**(2):286–90. https://doi.org/10.2144/99272bm16.
5. Lynch TJ, Bell DW, Sordella R, et al. Activating mutations in the epidermal growth factor receptor underlying responsiveness of non–small-cell lung cancer to gefitinib. *N Engl J Med* 2004;**350**(21):2129–39. https://doi.org/10.1056/NEJMoa040938.
6. Paez JG, Jänne PA, Lee JC, et al. EGFR mutations in lung cancer: correlation with clinical response to gefitinib therapy. *Science* 2004;**304**(5676):1497–500. https://doi.org/10.1126/science.1099314.
7. Van Cutsem E, Köhne C-H, Hitre E, et al. Cetuximab and chemotherapy as initial treatment for metastatic colorectal cancer. *N Engl J Med* 2009;**360**(14):1408–17. https://doi.org/10.1056/NEJMoa0805019.
8. Bollag G, Hirth P, Tsai J, et al. Clinical efficacy of a RAF inhibitor needs broad target blockade in BRAF-mutant melanoma. *Nature* 2010;**467**(7315):596–9. https://doi.org/10.1038/nature09454.

9. Lindeman NI, Cagle PT, Aisner DL, et al. Updated molecular testing guideline for the selection of lung cancer patients for treatment with targeted tyrosine kinase inhibitors: guideline from the College of American Pathologists, the International Association for the Study of Lung Cancer, and the Association for Molecular Pathology. *Arch Pathol Lab Med* 2018;**142**(3):321–46. https://meridian.allenpress.com/aplm/article-abstract/142/3/321/103064.
10. Sepulveda AR, Hamilton SR, Allegra CJ, et al. Molecular biomarkers for the evaluation of colorectal cancer: guideline from the American society for clinical pathology, college of American pathologists, association for molecular pathology, and American society of clinical oncology. *Arch Pathol Lab Med* 2017;**141**(5):625–57. https://mayoclinic.pure.elsevier.com/en/publications/molecular-biomarkers-for-the-evaluation-of-colorectal-cancer-guid-2.
11. Valtorta E, Martino C, Sartore-Bianchi A, et al. Assessment of a HER2 scoring system for colorectal cancer: results from a validation study. *Mod Pathol* 2015;**28**(11):1481–91. https://doi.org/10.1038/modpathol.2015.98.
12. Pietrantonio F, Di Nicolantonio F, Schrock AB, et al. ALK, ROS1, and NTRK Rearrangements in Metastatic Colorectal Cancer. *J Natl Cancer Inst* 2017;**109**(12). https://doi.org/10.1093/jnci/djx089.
13. García-Alfonso P, García-Carbonero R. Update of the recommendations for the determination of biomarkers in colorectal carcinoma: National Consensus of the Spanish Society of Medical Oncology. *Clin Transl Oncol* 2020. Published online https://link.springer.com/content/pdf/10.1007/s12094-020-02357-z.pdf.
14. García-Alfonso P, Díaz-Rubio E, Abad A, et al. First-line biological agents plus chemotherapy in older patients with metastatic colorectal cancer: a retrospective pooled analysis. *J Clin Oncol* 2020;**38**(15):4017. https://doi.org/10.1007/s40266-021-00834-w.
15. Schirripa M, Biason P, Lonardi S, et al. Class 1, 2, and 3 BRAF-mutated metastatic colorectal cancer: a detailed clinical, pathologic, and molecular characterization. *Clin Cancer Res* 2019;**25**(13):3954–61. https://doi.org/10.1158/1078-0432.CCR-19-0311.
16. Almoguera C, Shibata D, Forrester K, Martin J, Arnheim N, Perucho M. Most human carcinomas of the exocrine pancreas contain mutant c-K-ras genes. *Cell* 1988;**53**(4):549–54. https://doi.org/10.1016/0092-8674(88)90571-5.
17. Hong DS, Fakih MG, Strickler JH, et al. KRASG12C inhibition with sotorasib in advanced solid tumors. *N Engl J Med* 2020;**383**(13):1207–17. https://doi.org/10.1056/NEJMoa1917239.
18. Cocco E, Scaltriti M, Drilon A. NTRK fusion-positive cancers and TRK inhibitor therapy. *Nat Rev Clin Oncol* 2018;**15**(12):731–47. https://doi.org/10.1038/s41571-018-0113-0.
19. Huang R, Li X, He Y, et al. Recent advances in CAR-T cell engineering. *J Hematol Oncol* 2020;**13**(1):86. https://doi.org/10.1186/s13045-020-00910-5.
20. Voigtlaender M, Schneider-Merck T, Lapatinib TM. In: Martens U, editor. *Small molecules in oncology.recent results in cancer research.* vol. 211. Springer; 2018. p. 19–44. https://doi.org/10.1007/978-3-319-91442-8_2.
21. Weinberg RA. Oncogenes and tumor suppressor genes. *CA Cancer J Clin* 1994;**44**(3):160–70. https://doi.org/10.3322/canjclin.44.3.160.
22. Kontomanolis EN, Koutras A, Syllaios A, et al. Role of oncogenes and tumor-suppressor genes in carcinogenesis: a review. *Anticancer Res* 2020;**40**(11):6009–15. https://doi.org/10.21873/anticanres.14622.
23. Cocco E, Benhamida J, Middha S, et al. Colorectal carcinomas containing Hypermethylated MLH1 promoter and wild-type BRAF/KRAS are enriched for targetable kinase fusions. *Cancer Res* 2019;**79**(6):1047–53. https://doi.org/10.1158/0008-5472.CAN-18-3126.
24. Bray SM, Lee J, Kim ST, et al. Genomic characterization of intrinsic and acquired resistance to cetuximab in colorectal cancer patients. *Sci Rep* 2019;**9**(1):15365. https://doi.org/10.1038/s41598-019-51981-5.
25. Ribas A, Wolchok JD. Cancer immunotherapy using checkpoint blockade. *Science* 2018;**359**(6382):1350–5. https://doi.org/10.1126/science.aar4060.
26. Hargadon KM, Johnson CE, Williams CJ. Immune checkpoint blockade therapy for cancer: an overview of FDA-approved immune checkpoint inhibitors. *Int Immunopharmacol* 2018;**62**:29–39. https://doi.org/10.1016/j.intimp.2018.06.001.
27. Ionescu DN, Downes MR, Christofides A, Tsao MS. Harmonization of PD-L1 testing in oncology: a Canadian pathology perspective. *Curr Oncol* 2018;**25**(3):e209–16. https://doi.org/10.3747/co.25.4031.
28. Humphries MP, Hynes S, Bingham V, et al. Automated tumour recognition and digital pathology scoring unravels new role for PD-L1 in predicting good outcome in ER-/HER2+ breast Cancer. *J Oncol* 2018;**2018**:2937012. https://doi.org/10.1155/2018/2937012.
29. Khagi Y, Goodman AM, Daniels GA, Patel SP. Hypermutated circulating tumor DNA: correlation with response to checkpoint inhibitor–based immunotherapy. *Clin Cancer Res* 2017. Published online https://clincancerres.aacrjournals.org/content/23/19/5729.
30. Khagi Y, Kurzrock R, Patel SP. Next generation predictive biomarkers for immune checkpoint inhibition. *Cancer Metastasis Rev* 2017;**36**(1):179–90. https://doi.org/10.1007/s10555-016-9652-y.
31. Wang C, Gong J, Tu TY, Lee PP, Fakih M. Immune profiling of microsatellite instability-high and polymerase ε (POLE)-mutated metastatic colorectal tumors identifies predictors of response to anti-PD-1 therapy. *J Gastrointest Oncol* 2018;**9**(3):404–15. https://doi.org/10.21037/jgo.2018.01.09.
32. Pagès F, Mlecnik B, Marliot F, et al. International validation of the consensus immunoscore for the classification of colon cancer: a prognostic and accuracy study. *Lancet* 2018;**391**(10135):2128–39. https://doi.org/10.1016/S0140-6736(18)30789-X.
33. Angell HK, Bruni D, Barrett JC, Herbst R, Galon J. The immunoscore: colon cancer and beyond. *Clin Cancer Res* 2020;**26**(2):332–9. https://doi.org/10.1158/1078-0432.CCR-18-1851.
34. Panebianco C, Andriulli A, Pazienza V. Pharmacomicrobiomics: exploiting the drug-microbiota interactions in anticancer therapies. *Microbiome* 2018;**6**(1):92. https://doi.org/10.1186/s40168-018-0483-7.
35. Wen Y, Zhao S, Holmqvist A, et al. Predictive role of biopsy based biomarkers for radiotherapy treatment in rectal cancer. *J Pers Med* 2020;**10**(4). https://doi.org/10.3390/jpm10040168.

36. Hindson CM, Chevillet JR, Briggs HA, et al. Absolute quantification by droplet digital PCR versus analog real-time PCR. *Nat Methods* 2013;**10**(10):1003–5. https://doi.org/10.1038/nmeth.2633.
37. Huang J, Wang L. Cell-free DNA methylation profiling analysis—technologies and bioinformatics. *Cancer* 2019;**11**(11):1741. https://doi.org/10.3390/cancers11111741.
38. Poulet G, Massias J, Taly V. Liquid biopsy: general concepts. *Acta Cytol* 2019;**63**(6):449–55. https://doi.org/10.1159/000499337.
39. Siravegna G, Marsoni S, Siena S, Bardelli A. Integrating liquid biopsies into the management of cancer. *Nat Rev Clin Oncol* 2017;**14**(9):531–48. https://doi.org/10.1038/nrclinonc.2017.14.
40. Chen M, Zhao H. Next-generation sequencing in liquid biopsy: cancer screening and early detection. *Hum Genomics* 2019;**13**(1):34. https://doi.org/10.1186/s40246-019-0220-8.
41. Palmirotta R, Lovero D, Cafforio P, et al. Liquid biopsy of cancer: a multimodal diagnostic tool in clinical oncology. *Ther Adv Med Oncol* 2018;**10**:1–24. https://doi.org/10.1177/1758835918794630.
42. Crowley E, Di Nicolantonio F, Loupakis F, Bardelli A. Liquid biopsy: monitoring cancer-genetics in the blood. *Nat Rev Clin Oncol* 2013;**10**(8):472–84. https://doi.org/10.1038/nrclinonc.2013.110.
43. Alix-Panabières C, Pantel K. Challenges in circulating tumour cell research. *Nat Rev Cancer* 2014;**14**(9):623–31. https://doi.org/10.1038/nrc3820.
44. de Wit S, van Dalum G, Terstappen LWMM. Detection of circulating tumor cells. *Scientifica* 2014;**2014**:819362. https://doi.org/10.1155/2014/819362.
45. Abonnenc M, Manaresi N, Borgatti M, et al. Programmable interactions of functionalized single bioparticles in a dielectrophoresis-based microarray chip. *Anal Chem* 2013;**85**(17):8219–24. https://doi.org/10.1021/ac401296m.
46. Normanno N, Cervantes A, Ciardiello F, De Luca A, Pinto C. The liquid biopsy in the management of colorectal cancer patients: current applications and future scenarios. *Cancer Treat Rev* 2018;**70**:1–8. https://doi.org/10.1016/j.ctrv.2018.07.007.
47. Théry C, Zitvogel L, Amigorena S. Exosomes: composition, biogenesis and function. *Nat Rev Immunol* 2002;**2**(8):569–79. https://doi.org/10.1038/nri855.
48. De Rubis G, Rajeev Krishnan S, Bebawy M. Liquid biopsies in cancer diagnosis, monitoring, and prognosis. *Trends Pharmacol Sci* 2019;**40**(3):172–86. https://doi.org/10.1016/j.tips.2019.01.006.
49. Weidle UH, Birzele F, Kollmorgen G, Rüger R. The multiple roles of exosomes in metastasis. *Cancer Genomics Proteomics* 2017;**14**(1):1–15. https://doi.org/10.21873/cgp.20015.
50. Tucci M, Mannavola F, Passarelli A, Stucci LS, Cives M, Silvestris F. Exosomes in melanoma: a role in tumor progression, metastasis and impaired immune system activity. *Oncotarget* 2018;**9**(29):20826–37. https://doi.org/10.18632/oncotarget.24846.
51. Vaidyanathan R, Soon RH, Zhang P, Jiang K, Lim CT. Cancer diagnosis: from tumor to liquid biopsy and beyond. *Lab Chip* 2018;**19**(1):11–34. https://doi.org/10.1039/c8lc00684a.
52. Kanikarla-Marie P, Lam M, Menter DG, Kopetz S. Platelets, circulating tumor cells, and the circulome. *Cancer Metastasis Rev* 2017;**36**(2):235–48. https://doi.org/10.1007/s10555-017-9681-1.
53. Sol N, Wurdinger T. Platelet RNA signatures for the detection of cancer. *Cancer Metastasis Rev* 2017;**36**(2):263–72. https://doi.org/10.1007/s10555-017-9674-0.
54. In't Veld S, Wurdinger T. Tumor-educated platelets. *Blood* 2019;**133**(22):2359–64. https://doi.org/10.1182/blood-2018-12-852830.
55. Schickel R, Boyerinas B, Park S-M, Peter ME. MicroRNAs: key players in the immune system, differentiation, tumorigenesis and cell death. *Oncogene* 2008;**27**(45):5959–74. https://doi.org/10.1038/onc.2008.274.
56. Eichmüller SB, Osen W, Mandelboim O, Seliger B. Immune modulatory microRNAs involved in tumor attack and tumor immune escape. *J Natl Cancer Inst* 2017;**109**(10). https://doi.org/10.1093/jnci/djx034.
57. Peng S-B, Van Horn RD, Yin T, et al. Distinct mobilization of leukocytes and hematopoietic stem cells by CXCR4 peptide antagonist LY2510924 and monoclonal antibody LY2624587. *Oncotarget* 2017;**8**(55):94619–34. https://doi.org/10.18632/oncotarget.21816.
58. Krasinskas AM, Goldsmith JD. Immunohistology of the gastrointestinal tract. In: Dabbs D, editor. *Diagnostic immunohistochemistry*. 3rd ed. Elsevier; 2010. p. 500–40.
59. Cohen JD, Li L, Wang Y, et al. Detection and localization of surgically resectable cancers with a multi-analyte blood test. *Science* 2018;**359**(6378):926–30. https://doi.org/10.1126/science.aar3247.
60. Heitzer E, Haque IS, Roberts CES, Speicher MR. Current and future perspectives of liquid biopsies in genomics-driven oncology. *Nat Rev Genet* 2019;**20**(2):71–88. https://doi.org/10.1038/s41576-018-0071-5.
61. Fleischhacker M, Schmidt B. Circulating nucleic acids (CNAs) and cancer—a survey. *Biochim Biophys Acta* 2007;**1775**(1):181–232. https://doi.org/10.1016/j.bbcan.2006.10.001.
62. Mouliere F, Robert B, Arnau Peyrotte E, et al. High fragmentation characterizes tumour-derived circulating DNA. *PLoS One* 2011;**6**(9). https://doi.org/10.1371/journal.pone.0023418, e23418.
63. Diehl F, Schmidt K, Choti MA, et al. Circulating mutant DNA to assess tumor dynamics. *Nat Med* 2008;**14**(9):985–90. https://doi.org/10.1038/nm.1789.
64. Jahr S, Hentze H, Englisch S, et al. DNA fragments in the blood plasma of cancer patients: quantitations and evidence for their origin from apoptotic and necrotic cells. *Cancer Res* 2001;**61**(4):1659–65. https://www.ncbi.nlm.nih.gov/pubmed/11245480.
65. Bergsmedh A, Szeles A, Henriksson M, et al. Horizontal transfer of oncogenes by uptake of apoptotic bodies. *Proc Natl Acad Sci U S A* 2001;**98**(11):6407–11. https://doi.org/10.1073/pnas.101129998.
66. García-Foncillas J, Tabernero J, Élez E, et al. Prospective multicenter real-world RAS mutation comparison between OncoBEAM-based liquid biopsy and tissue analysis in metastatic colorectal cancer. *Br J Cancer* 2018;**119**(12):1464–70. https://doi.org/10.1038/s41416-018-0293-5.

67. Kondo Y, Hayashi K, Kawakami K, Miwa Y, Hayashi H, Yamamoto M. KRAS mutation analysis of single circulating tumor cells from patients with metastatic colorectal cancer. *BMC Cancer* 2017;**17**(1):311. https://doi.org/10.1186/s12885-017-3305-6.
68. Diaz Jr LA, Williams RT, Wu J, et al. The molecular evolution of acquired resistance to targeted EGFR blockade in colorectal cancers. *Nature* 2012;**486**(7404):537–40. https://doi.org/10.1038/nature11219.
69. Price T, Ang A, Boedigheimer M, et al. Frequency of S492R mutations in the epidermal growth factor receptor: analysis of plasma DNA from patients with metastatic colorectal cancer treated with panitumumab or cetuximab monotherapy. *Cancer Biol Ther* 2020;**21**(10):891–8. https://www.tandfonline.com/doi/abs/10.1080/15384047.2020.1798695.
70. Siravegna G, Mussolin B, Buscarino M, et al. Clonal evolution and resistance to EGFR blockade in the blood of colorectal cancer patients. *Nat Med* 2015;**21**(7):795–801. https://doi.org/10.1038/nm.3870.
71. Kang S, Li Q, Chen Q, et al. CancerLocator: non-invasive cancer diagnosis and tissue-of-origin prediction using methylation profiles of cell-free DNA. *Genome Biol* 2017;**18**(1):53. https://doi.org/10.1186/s13059-017-1191-5.
72. Klein-Scory S, Maslova M, Pohl M, et al. Significance of liquid biopsy for monitoring and therapy decision of colorectal cancer. *Transl Oncol* 2018;**11**(2):213–20. https://doi.org/10.1016/j.tranon.2017.12.010.
73. Yamada T, Matsuda A, Koizumi M, et al. Liquid biopsy for the management of patients with colorectal cancer. *Digestion* 2019;**99**(1):39–45. https://doi.org/10.1159/000494411.
74. Rossi G, Ignatiadis M. Promises and pitfalls of using liquid biopsy for precision medicine. *Cancer Res* 2019;**79**(11):2798–804. https://doi.org/10.1158/0008-5472.CAN-18-3402.
75. Widlak MM, Thomas CL, Thomas MG, et al. Diagnostic accuracy of faecal biomarkers in detecting colorectal cancer and adenoma in symptomatic patients. *Aliment Pharmacol Ther* 2017;**45**(2):354–63. https://doi.org/10.1111/apt.13865.
76. Malagón M, Ramió-Pujol S, Serrano M, et al. New fecal bacterial signature for colorectal cancer screening reduces the fecal immunochemical test false-positive rate in a screening population. *PLoS One* 2020;**15**(12). https://doi.org/10.1371/journal.pone.0243158, e0243158.
77. Widlak MM, Neal M, Daulton E, et al. Risk stratification of symptomatic patients suspected of colorectal cancer using faecal and urinary markers. *Colorectal Dis* 2018;**20**(12):O335–42. https://doi.org/10.1111/codi.14431.
78. Sorich MJ, Wiese MD, Rowland A, Kichenadasse G, McKinnon RA, Karapetis CS. Extended RAS mutations and anti-EGFR monoclonal antibody survival benefit in metastatic colorectal cancer: a meta-analysis of randomized, controlled trials. *Ann Oncol* 2015;**26**(1):13–21. https://doi.org/10.1093/annonc/mdu378.
79. Van Cutsem E, Kohne C-H, Láng I, et al. Cetuximab plus irinotecan, fluorouracil, and leucovorin as first-line treatment for metastatic colorectal cancer: updated analysis of overall survival according to tumor KRAS and BRAF mutation status. *J Clin Oncol* 2011;**29**(15):2011–9. https://pdfs.semanticscholar.org/86d6/c0146a6e40a4d22a4a3dfeda066ba6d44f43.pdf.
80. Douillard J-Y, Oliner KS, Siena S, et al. Panitumumab–FOLFOX4 treatment and RAS mutations in colorectal cancer. *N Engl J Med* 2013;**369**(11):1023–34. https://doi.org/10.1056/NEJMoa1305275.
81. Bokemeyer C, Bondarenko I, Hartmann JT, et al. Efficacy according to biomarker status of cetuximab plus FOLFOX-4 as first-line treatment for metastatic colorectal cancer: the OPUS study. *Ann Oncol* 2011;**22**(7):1535–46. https://doi.org/10.1093/annonc/mdq632.
82. Douillard JY, Siena S, Cassidy J, et al. Final results from PRIME: randomized phase III study of panitumumab with FOLFOX4 for first-line treatment of metastatic colorectal cancer. *Ann Oncol* 2014;**25**(7):1346–55. https://doi.org/10.1093/annonc/mdu141.
83. Douillard JY, Siena S, Cassidy J. Randomized, phase III trial of panitumumab with infusional fluorouracil, leucovorin, and oxaliplatin (FOLFOX4) versus FOLFOX4 alone as first-line treatment in patients with previously untreated metastatic colorectal cancer: the PRIME study. *J Clin Oncol* 2010. Published online https://www.researchgate.net/profile/Gyoergy_Bodoky/publication/47336663_Randomized_Phase_III_Trial_of_Panitumumab_With_Infusional_Fluorouracil_Leucovorin_and_Oxaliplatin_FOLFOX4_Versus_FOLFOX4_Alone_As_First-Line_Treatment_in_Patients_With_Previously_Untreated_Metastatic_/links/568641a408ae197583971f17/Randomized-Phase-III-Trial-of-Panitumumab-With-Infusional-Fluorouracil-Leucovorin-and-Oxaliplatin-FOLFOX4-Versus-FOLFOX4-Alone-As-First-Line-Treatment-in-Patients-With-Previously-Untreated-Metastat.pdf.
84. Van Cutsem E, Cervantes A, Adam R, et al. ESMO consensus guidelines for the management of patients with metastatic colorectal cancer. *Ann Oncol* 2016;**27**(8):1386–422. https://doi.org/10.1093/annonc/mdw235.
85. Yuan Z-X, Wang X-Y, Qin Q-Y, et al. The prognostic role of BRAF mutation in metastatic colorectal cancer receiving anti-EGFR monoclonal antibodies: a meta-analysis. *PLoS One* 2013;**8**(6):e65995. https://doi.org/10.1371/journal.pone.0065995.
86. Tran B, Kopetz S, Tie J, et al. Impact of BRAF mutation and microsatellite instability on the pattern of metastatic spread and prognosis in metastatic colorectal cancer. *Cancer* 2011;**117**(20):4623–32. https://doi.org/10.1002/cncr.26086.
87. Forbes SA, Bhamra G, Bamford S, et al. The catalogue of somatic mutations in cancer (COSMIC). *Curr Protoc Hum Genet* 2008. https://doi.org/10.1002/0471142905.hg1011s57. Chapter 10(1):Unit 10.11.
88. Venderbosch S, Nagtegaal ID, Maughan TS, et al. Mismatch repair status and BRAF mutation status in metastatic colorectal cancer patients: a pooled analysis of the CAIRO, CAIRO2, COIN, and FOCUS studies. *Clin Cancer Res* 2014;**20**(20):5322–30. https://doi.org/10.1158/1078-0432.CCR-14-0332.
89. Kopetz S, Grothey A, Yaeger R, et al. Encorafenib, binimetinib, and cetuximab in BRAF V600E–mutated colorectal cancer. *N Engl J Med* 2019;**381**(17):1632–43. https://doi.org/10.1056/NEJMoa1908075.
90. Siena S, Sartore-Bianchi A, Marsoni S, et al. Targeting the human epidermal growth factor receptor 2 (HER2) oncogene in colorectal cancer. *Ann Oncol* 2018;**29**(5):1108–19. https://doi.org/10.1093/annonc/mdy100.
91. Ross JS, Fakih M, Ali SM, et al. Targeting HER2 in colorectal cancer: the landscape of amplification and short variant mutations in ERBB2 and ERBB3. *Cancer* 2018;**124**(7):1358–73. https://doi.org/10.1002/cncr.31125.

92. Sartore-Bianchi A, Trusolino L, Martino C, et al. Dual-targeted therapy with trastuzumab and lapatinib in treatment-refractory, KRAS codon 12/13 wild-type, HER2-positive metastatic colorectal cancer (HERACLES): a proof-of-concept, multicentre, open-label, phase 2 trial. *Lancet Oncol* 2016;**17**(6):738–46. https://www.sciencedirect.com/science/article/pii/S1470204516001509.
93. Hurwitz H, Raghav KPS, Burris HA, et al. Pertuzumab + trastuzumab for HER2-amplified/overexpressed metastatic colorectal cancer (mCRC): interim data from MyPathway. *J Clin Orthod* 2017;**35**(4_suppl):676. https://doi.org/10.1200/JCO.2017.35.4_suppl.676.
94. Styczen H, Nagelmeier I, Beissbarth T, et al. HER-2 and HER-3 expression in liver metastases of patients with colorectal cancer. *Oncotarget* 2015;**6**(17):15065–76. https://doi.org/10.18632/oncotarget.3527.
95. Conradi L-C, Spitzner M, Metzger A-L, et al. Combined targeting of HER-2 and HER-3 represents a promising therapeutic strategy in colorectal cancer. *BMC Cancer* 2019;**19**(1):880. https://doi.org/10.1186/s12885-019-6051-0.
96. Russo A, Bazan V, Iacopetta B, Kerr D, Soussi T, Gebbia N. The TP53 colorectal cancer international collaborative study on the prognostic and predictive significance of p53 mutation: influence of tumor site, type of mutation, and adjuvant treatment. *J Clin Oncol* 2005;**23**(30):7518–28. https://ascopubs.org/doi/abs/10.1200/jco.2005.00.471.
97. Li X-L, Zhou J, Chen Z-R, Chng W-J. P53 mutations in colorectal cancer—molecular pathogenesis and pharmacological reactivation. *World J Gastroenterol* 2015;**21**(1):84–93. https://doi.org/10.3748/wjg.v21.i1.84.
98. Liu Y, Wang G, Jiang X, et al. TRIM67 inhibits tumor proliferation and metastasis by mediating MAPK11 in colorectal cancer. *J Cancer* 2020;**11**(20):6025–37. https://doi.org/10.7150/jca.47538.
99. Wang S, Zhang Y, Huang J, et al. TRIM67 activates p53 to suppress colorectal cancer initiation and progression. *Cancer Res* 2019;**79**(16):4086–98. https://doi.org/10.1158/0008-5472.CAN-18-3614.
100. Morin PJ, Sparks AB, Korinek V, et al. Activation of beta-catenin-Tcf signaling in colon cancer by mutations in beta-catenin or APC. *Science* 1997;**275**(5307):1787–90. https://doi.org/10.1126/science.275.5307.1787.
101. Cheng X, Xu X, Chen D, Zhao F, Wang W. Therapeutic potential of targeting the Wnt/β-catenin signaling pathway in colorectal cancer. *Biomed Pharmacother* 2019;**110**:473–81. https://doi.org/10.1016/j.biopha.2018.11.082.
102. Wu L, Zhou Z, Han S, et al. PLAGL2 promotes epithelial–mesenchymal transition and mediates colorectal cancer metastasis via β-catenin-dependent regulation of ZEB1. *Br J Cancer* 2020;**122**(4):578–89. https://doi.org/10.1038/s41416-019-0679-z.
103. Chen J, Zhao J, Chen X, Ding C, Lee K, Jia Z. Hyper activation of β-catenin signalling induced by IKK ε inhibition thwarts colorectal cancer cell proliferation. *Cell Prolif* 2017. Published online https://onlinelibrary.wiley.com/doi/abs/10.1111/cpr.12350.
104. Birchmeier C, Birchmeier W, Gherardi E, Vande Woude GF. Met, metastasis, motility and more. *Nat Rev Mol Cell Biol* 2003;**4**(12):915–25. https://doi.org/10.1038/nrm1261.
105. Lee SJ, Lee J, Park SH, et al. C-MET overexpression in colorectal cancer: a poor prognostic factor for survival. *Clin Colorectal Cancer* 2018;**17**(3):165–9. https://doi.org/10.1016/j.clcc.2018.02.013.
106. Parizadeh SM, Jafarzadeh-Esfehani R, Fazilat-Panah D, et al. The potential therapeutic and prognostic impacts of the c-MET/HGF signaling pathway in colorectal cancer. *IUBMB Life* 2019;**71**(7):802–11. https://iubmb.onlinelibrary.wiley.com/doi/abs/10.1002/iub.2063.
107. Jin Z, Sinicrope FA. Prognostic and predictive values of mismatch repair deficiency in non-metastatic colorectal cancer. *Cancer* 2021;**13**(2). https://doi.org/10.3390/cancers13020300.
108. Diao Z, Han Y, Chen Y, Zhang R, Li J. The clinical utility of microsatellite instability in colorectal cancer. *Crit Rev Oncol Hematol* 2021;**157**:103171. https://doi.org/10.1016/j.critrevonc.2020.103171.
109. Baretti M, Le DT. DNA mismatch repair in cancer. *Pharmacol Ther* 2018;**189**:45–62. https://doi.org/10.1016/j.pharmthera.2018.04.004.
110. Jiricny J. The multifaceted mismatch-repair system. *Nat Rev Mol Cell Biol* 2006;**7**(5):335–46. https://doi.org/10.1038/nrm1907.
111. Luchini C, Bibeau F, Ligtenberg MJL, et al. ESMO recommendations on microsatellite instability testing for immunotherapy in cancer, and its relationship with PD-1/PD-L1 expression and tumour mutational burden: a systematic review-based approach. *Ann Oncol* 2019;**30**(8):1232–43. https://www.sciencedirect.com/science/article/pii/S0923753419312694.
112. André T, Shiu K-K, Kim TW, et al. Pembrolizumab in microsatellite-instability–high advanced colorectal cancer. *N Engl J Med* 2020;**383**(23):2207–18. https://doi.org/10.1056/NEJMoa2017699.
113. Cancer Genome Atlas Network. Comprehensive molecular characterization of human colon and rectal cancer. *Nature* 2012;**487**(7407):330–7. https://doi.org/10.1038/nature11252.
114. Carr PR, Alwers E, Bienert S, et al. Lifestyle factors and risk of sporadic colorectal cancer by microsatellite instability status: a systematic review and meta-analyses. *Ann Oncol* 2018;**29**(4):825–34. https://doi.org/10.1093/annonc/mdy059.
115. Miyakura Y, Sugano K, Akasu T, et al. Extensive but hemiallelic methylation of the hMLH1 promoter region in early-onset sporadic colon cancers with microsatellite instability. *Clin Gastroenterol Hepatol* 2004;**2**(2):147–56. https://doi.org/10.1016/s1542-3565(03)00314-8.
116. Cohen SA, Turner EH, Beightol MB, et al. Frequent PIK3CA mutations in colorectal and endometrial tumors with 2 or more somatic mutations in mismatch repair genes. *Gastroenterology* 2016;**151**(3):440–447.e1. https://doi.org/10.1053/j.gastro.2016.06.004.
117. Lee C-T, Chow N-H, Chen Y-L, et al. Clinicopathological features of mismatch repair protein expression patterns in colorectal cancer. *Pathol Res Pract* 2021;**217**:153288. https://doi.org/10.1016/j.prp.2020.153288.
118. Le DT, Uram JN, Wang H, et al. PD-1 blockade in tumors with mismatch-repair deficiency. *N Engl J Med* 2015;**372**(26):2509–20.
119. Zhang N, Hu X, Du Y, Du J. The role of miRNAs in colorectal cancer progression and chemoradiotherapy. *Biomed Pharmacother* 2021;**134**:111099. https://doi.org/10.1016/j.biopha.2020.111099.
120. Valeri N, Croce CM, Fabbri M. Pathogenetic and clinical relevance of microRNAs in colorectal cancer. *Cancer Genomics Proteomics* 2009;**6**(4):195–204. https://www.ncbi.nlm.nih.gov/pubmed/19656996.

121. Moridikia A, Mirzaei H, Sahebkar A, Salimian J. MicroRNAs: potential candidates for diagnosis and treatment of colorectal cancer. *J Cell Physiol* 2018;**233**(2):901–13. https://doi.org/10.1002/jcp.25801.
122. Javed Z, Javed Iqbal M, Rasheed A, et al. Regulation of hedgehog signaling by miRNAs and nanoformulations: a possible therapeutic solution for colorectal cancer. *Front Oncol* 2021;**10**:607607. https://www.researchgate.net/profile/Javad_Sharifi-Rad/publication/348297898_Regulation_of_Hedgehog_Signaling_by_miRNAs_and_Nanoformulations_A_Possible_Therapeutic_Solution_for_Colorectal_Cancer/links/5ff7100ea6fdccdcb83a068b/Regulation-of-Hedgehog-Signaling-by-miRNAs-and-Nanoformulations-A-Possible-Therapeutic-Solution-for-Colorectal-Cancer.pdf.
123. Escalante PI, Quiñones LA, Contreras HR. Epithelial-mesenchymal transition and microRNAs in colorectal cancer chemoresistance to FOLFOX. *Pharmaceutics* 2021;**13**(1). https://doi.org/10.3390/pharmaceutics13010075.
124. Hibner G, Kimsa-Furdzik M, Francuz T. Relevance of microRNAs as potential diagnostic and prognostic markers in colorectal cancer. *Int J Mol Sci* 2018;**19**(10). https://doi.org/10.3390/ijms19102944.
125. Lu TX, Rothenberg ME. MicroRNA. *J Allergy Clin Immunol* 2018;**141**(4):1202–7. https://doi.org/10.1016/j.jaci.2017.08.034.
126. Cekaite L, Eide PW, Lind GE, Skotheim RI, Lothe RA. MicroRNAs as growth regulators, their function and biomarker status in colorectal cancer. *Oncotarget* 2016;**7**(6):6476–505. https://doi.org/10.18632/oncotarget.6390.
127. Jung G, Hernández-Illán E, Moreira L, Balaguer F, Goel A. Epigenetics of colorectal cancer: biomarker and therapeutic potential. *Nat Rev Gastroenterol Hepatol* 2020;**17**(2):111–30. https://doi.org/10.1038/s41575-019-0230-y.
128. Kanai Y, Hirohashi S. Alterations of DNA methylation associated with abnormalities of DNA methyltransferases in human cancers during transition from a precancerous to a malignant state. *Carcinogenesis* 2007;**28**(12):2434–42. https://doi.org/10.1093/carcin/bgm206.
129. Fu B, Du C, Wu Z, et al. Analysis of DNA methylation-driven genes for predicting the prognosis of patients with colorectal cancer. *Aging* 2020;**12**(22):22814–39. https://doi.org/10.18632/aging.103949.
130. Bestor TH, Edwards JR, Boulard M. Notes on the role of dynamic DNA methylation in mammalian development. *Proc Natl Acad Sci U S A* 2015;**112**(22):6796–9. https://doi.org/10.1073/pnas.1415301111.
131. Ramchandani S, Bhattacharya SK, Cervoni N, Szyf M. DNA methylation is a reversible biological signal. *Proc Natl Acad Sci U S A* 1999;**96**(11):6107–12. https://doi.org/10.1073/pnas.96.11.6107.
132. Constâncio V, Nunes SP, Henrique R, Jerónimo C. DNA methylation-based testing in liquid biopsies as detection and prognostic biomarkers for the four major cancer types. *Cell* 2020;**9**(3). https://doi.org/10.3390/cells9030624.
133. Liu J, Li H, Sun L, et al. Epigenetic alternations of microRNAs and DNA methylation contribute to liver metastasis of colorectal cancer. *Dig Dis Sci* 2019;**64**(6):1523–34. https://doi.org/10.1007/s10620-018-5424-6.
134. McCabe MT, Brandes JC, Vertino PM. Cancer DNA methylation: molecular mechanisms and clinical implications. *Clin Cancer Res* 2009;**15**(12):3927–37. https://doi.org/10.1158/1078-0432.CCR-08-2784.
135. Baba Y, Nosho K, Shima K, et al. Hypomethylation of the IGF2 DMR in colorectal tumors, detected by bisulfite pyrosequencing, is associated with poor prognosis. *Gastroenterology* 2010;**139**(6):1855–64. https://doi.org/10.1053/j.gastro.2010.07.050.
136. Antelo M, Balaguer F, Shia J, et al. A high degree of LINE-1 hypomethylation is a unique feature of early-onset colorectal cancer. *PLoS One* 2012;**7**(9):e45357. https://doi.org/10.1371/journal.pone.0045357.
137. Zhan Y-X, Luo G-H. DNA methylation detection methods used in colorectal cancer. *World J Clin Cases* 2019;**7**(19):2916–29. https://doi.org/10.12998/wjcc.v7.i19.2916.
138. Lamb YN, Dhillon S. Epi proColon® 2.0 CE: a blood-based screening test for colorectal cancer. *Mol Diagn Ther* 2017;**21**(2):225–32. https://doi.org/10.1007/s40291-017-0259-y.
139. Zhao G, Liu X, Liu Y, et al. Aberrant DNA methylation of SEPT9 and SDC2 in stool specimens as an integrated biomarker for colorectal cancer early detection. *Front Genet* 2020;**11**:643. https://doi.org/10.3389/fgene.2020.00643.
140. Young GP, Pedersen SK, Mansfield S, et al. A cross-sectional study comparing a blood test for methylated BCAT 1 and IKZF 1 tumor-derived DNA with CEA for detection of recurrent colorectal cancer. *Cancer Med* 2016;**5**(10):2763–72. https://onlinelibrary.wiley.com/doi/abs/10.1002/cam4.868.
141. Murray DH, Symonds EL, Young GP, et al. Relationship between post-surgery detection of methylated circulating tumor DNA with risk of residual disease and recurrence-free survival. *J Cancer Res Clin Oncol* 2018;**144**(9):1741–50. https://doi.org/10.1007/s00432-018-2701-x.
142. Garrett WS. Cancer and the microbiota. *Science* 2015;**348**(6230):80–6. https://doi.org/10.1126/science.aaa4972.
143. Lloyd-Price J, Arze C, Ananthakrishnan AN, et al. Multi-omics of the gut microbial ecosystem in inflammatory bowel diseases. *Nature* 2019;**569**(7758):655–62. https://doi.org/10.1038/s41586-019-1237-9.
144. Jandhyala SM, Talukdar R, Subramanyam C, Vuyyuru H, Sasikala M, Nageshwar RD. Role of the normal gut microbiota. *World J Gastroenterol* 2015;**21**(29):8787–803. https://doi.org/10.3748/wjg.v21.i29.8787.
145. Roy S, Trinchieri G. Microbiota: a key orchestrator of cancer therapy. *Nat Rev Cancer* 2017;**17**(5):271–85.
146. Macpherson AJ, Uhr T. Induction of protective IgA by intestinal dendritic cells carrying commensal bacteria. *Science* 2004;**303**(5664):1662–5. https://doi.org/10.1126/science.1091334.
147. Bullman S, Pedamallu CS, Sicinska E, et al. Analysis of fusobacterium persistence and antibiotic response in colorectal cancer. *Science* 2017;**358**(6369):1443–8.
148. Brandi G, Frega G. Microbiota: overview and implication in immunotherapy-based cancer treatments. *Int J Mol Sci* 2019;**20**(11). https://doi.org/10.3390/ijms20112699.
149. Yu T, Guo F, Yu Y, et al. Fusobacterium nucleatum promotes chemoresistance to colorectal cancer by modulating autophagy. *Cell* 2017;**170**(3):548–563.e16. https://doi.org/10.1016/j.cell.2017.07.008.

150. Mima K, Nishihara R, Qian ZR, et al. Fusobacterium nucleatum in colorectal carcinoma tissue and patient prognosis. *Gut* 2016;**65**(12): 1973–80. https://doi.org/10.1136/gutjnl-2015-310101.
151. Dzutsev A, Goldszmid RS, Viaud S, Zitvogel L, Trinchieri G. The role of the microbiota in inflammation, carcinogenesis, and cancer therapy. *Eur J Immunol* 2015;**45**(1):17–31. https://doi.org/10.1002/eji.201444972.
152. Blanco-Calvo M, Concha Á, Figueroa A, Garrido F, Valladares-Ayerbes M. Colorectal cancer classification and cell heterogeneity: a systems oncology approach. *Int J Mol Sci* 2015;**16**(6):13610–32. https://doi.org/10.3390/ijms160613610.
153. Cox AD, Fesik SW, Kimmelman AC, Luo J, Der CJ. Drugging the undruggable RAS: mission possible? *Nat Rev Drug Discov* 2014;**13**(11): 828–51. https://doi.org/10.1038/nrd4389.
154. Ni D, Li X, He X, Zhang H, Zhang J, Lu S. Drugging K-RasG12C through covalent inhibitors: mission possible? *Pharmacol Ther* 2019;**202**: 1–17. https://doi.org/10.1016/j.pharmthera.2019.06.007.
155. van Wyk HC, Park J, Roxburgh C, Horgan P, Foulis A, McMillan DC. The role of tumour budding in predicting survival in patients with primary operable colorectal cancer: a systematic review. *Cancer Treat Rev* 2015;**41**(2):151–9. https://doi.org/10.1016/j.ctrv.2014.12.007.
156. Nakamura T, Mitomi H, Kanazawa H, Ohkura Y, Watanabe M. Tumor budding as an index to identify high-risk patients with stage II colon cancer. *Dis Colon Rectum* 2008;**51**(5):568–72. https://doi.org/10.1007/s10350-008-9192-9.
157. Trotsyuk I, Sparschuh H, Müller AJ, et al. Tumor budding outperforms ypT and ypN classification in predicting outcome of rectal cancer after neoadjuvant chemoradiotherapy. *BMC Cancer* 2019;**19**(1):1033. https://doi.org/10.1186/s12885-019-6261-5.
158. Aldecoa I, Atares B, Tarragona J, et al. Molecularly determined total tumour load in lymph nodes of stage I–II colon cancer patients correlates with high-risk factors. A multicentre prospective study. *Virchows Arch* 2016;**469**(4):385–94. https://doi.org/10.1007/s00428-016-1990-1.
159. Penault-Llorca F, Rudzinski ER, Sepulveda AR. Testing algorithm for identification of patients with TRK fusion cancer. *J Clin Pathol* 2019;**72** (7):460–7. https://doi.org/10.1136/jclinpath-2018-205679.
160. Ooki A, Shinozaki E, Yamaguchi K. Immunotherapy in colorectal cancer: current and future strategies. *J Anus Rectum Colon* 2021;**5**(1): 11–24. https://doi.org/10.23922/jarc.2020-064.
161. Sung H, Ferlay J, Siegel RL, et al. Global cancer statistics 2020: GLOBOCAN estimates of incidence and mortality worldwide for 36 cancers in 185 countries. *CA Cancer J Clin* 2021. https://doi.org/10.3322/caac.21660. Published online February 4.
162. National Comprehensive Cancer Network (NCCN). *NCCN clinical practice guidelines in oncology*; 2020. Accessed September 9, 2019 https://www.nccn.org/professionals/physician_gls/default.aspx.
163. Chen DS, Mellman I. Oncology meets immunology: the cancer-immunity cycle. *Immunity* 2013;**39**(1):1–10. https://doi.org/10.1016/j.immuni.2013.07.012.
164. Zappasodi R, Merghoub T, Wolchok JD. Emerging concepts for immune checkpoint blockade-based combination therapies. *Cancer Cell* 2018;**34** (4):690. https://doi.org/10.1016/j.ccell.2018.09.008.
165. Le DT, Kim TW, Van Cutsem E, et al. Phase II open-label study of pembrolizumab in treatment-refractory, microsatellite instability-high/mismatch repair-deficient metastatic colorectal cancer: KEYNOTE-164. *J Clin Oncol* 2020;**38**(1):11–9. https://doi.org/10.1200/JCO.19.02107.
166. Overman MJ, Lonardi S, Wong KYM, et al. Durable clinical benefit with nivolumab plus ipilimumab in DNA mismatch repair-deficient/ microsatellite instability-high metastatic colorectal cancer. *J Clin Oncol* 2018;**36**(8):773–9. https://doi.org/10.1200/JCO.2017.76.9901.
167. Llosa NJ, Cruise M, Tam A, et al. The vigorous immune microenvironment of microsatellite instable colon cancer is balanced by multiple counter-inhibitory checkpoints. *Cancer Discov* 2015;**5**(1):43–51. https://doi.org/10.1158/2159-8290.CD-14-0863.
168. Koyama S, Akbay EA, Li YY, et al. Adaptive resistance to therapeutic PD-1 blockade is associated with upregulation of alternative immune checkpoints. *Nat Commun* 2016;**7**:10501. https://doi.org/10.1038/ncomms10501.
169. Rodriguez-Abreu D, Johnson ML, Hussein MA, et al. Primary analysis of a randomized, double-blind, phase II study of the anti-TIGIT antibody tiragolumab (tira) plus atezolizumab (atezo) versus placebo plus atezo as first-line (1L) treatment in patients with PD-L1-selected NSCLC (CITYSCAPE). *J Clin Orthod* 2020;**38**(15_suppl):9503. https://doi.org/10.1200/JCO.2020.38.15_suppl.9503.
170. Twyman-Saint Victor C, Rech AJ, Maity A, et al. Radiation and dual checkpoint blockade activate non-redundant immune mechanisms in cancer. *Nature* 2015;**520**(7547):373–7. https://doi.org/10.1038/nature14292.
171. Antonia SJ, Villegas A, Daniel D, et al. Durvalumab after chemoradiotherapy in stage III non–small-cell lung cancer. *N Engl J Med* 2017;**377** (20):1919–29. https://doi.org/10.1056/NEJMoa1709937.
172. Shaverdian N, Lisberg AE, Bornazyan K, et al. Previous radiotherapy and the clinical activity and toxicity of pembrolizumab in the treatment of non-small-cell lung cancer: a secondary analysis of the KEYNOTE-001 phase 1 trial. *Lancet Oncol* 2017;**18**(7):895–903. https://doi.org/10.1016/S1470-2045(17)30380-7.
173. Segal NH, Kemeny NE, Cercek A, et al. Non-randomized phase II study to assess the efficacy of pembrolizumab (Pem) plus radiotherapy (RT) or ablation in mismatch repair proficient (pMMR) metastatic colorectal cancer (mCRC) patients. *J Clin Orthod* 2016;**34**(15_suppl):3539. https://doi.org/10.1200/JCO.2016.34.15_suppl.3539.
174. Kabiljo J, Harpain F, Carotta S, Bergmann M. Radiotherapy as a backbone for novel concepts in cancer immunotherapy. *Cancer* 2019;**12**(1). https://doi.org/10.3390/cancers12010079.
175. Sinicrope FA, Ou F-S, Shi Q, et al. Randomized trial of FOLFOX alone or combined with atezolizumab as adjuvant therapy for patients with stage III colon cancer and deficient DNA mismatch repair or microsatellite instability (ATOMIC, Alliance A021502). *J Clin Orthod* 2017;**35**(15_suppl): TPS3630. https://doi.org/10.1200/JCO.2017.35.15_suppl.TPS3630.
176. Lau D, Kalaitzaki E, Church DN, et al. Rationale and design of the POLEM trial: avelumab plus fluoropyrimidine-based chemotherapy as adjuvant treatment for stage III mismatch repair deficient or POLE exonuclease domain mutant colon cancer: a phase III randomised study. *ESMO open* 2020;**5**(1):e000638. https://www.sciencedirect.com/science/article/pii/S205970292030020X.

177. Yarchoan M, Hopkins A, Jaffee EM. Tumor mutational burden and response rate to PD-1 inhibition. *N Engl J Med* 2017;**377**(25):2500. https://www.ncbi.nlm.nih.gov/pmc/articles/pmc6549688/.
178. Samstein RM, Lee C-H, Shoushtari AN, et al. Tumor mutational load predicts survival after immunotherapy across multiple cancer types. *Nat Genet* 2019;**51**(2):202–6. https://doi.org/10.1038/s41588-018-0312-8.
179. Domingo E, Freeman-Mills L, Rayner E, et al. Somatic POLE proofreading domain mutation, immune response, and prognosis in colorectal cancer: a retrospective, pooled biomarker study. *Lancet Gastroenterol Hepatol* 2016;**1**(3):207–16. https://doi.org/10.1016/S2468-1253(16)30014-0.
180. Al-Badran SS, Grant L, Campo MV, et al. Relationship between immune checkpoint proteins, tumour microenvironment characteristics, and prognosis in primary operable colorectal cancer. *Hip Int* 2021;**7**(2):121–34. https://doi.org/10.1002/cjp2.193.
181. Hutchins G, Southward K, Handley K, et al. Value of mismatch repair, KRAS, and BRAF mutations in predicting recurrence and benefits from chemotherapy in colorectal cancer. *J Clin Oncol* 2011;**29**(10):1261–70. https://doi.org/10.1200/JCO.2010.30.1366.
182. Park JH, Richards CH, McMillan DC, Horgan PG, Roxburgh CSD. The relationship between tumour stroma percentage, the tumour microenvironment and survival in patients with primary operable colorectal cancer. *Ann Oncol* 2014;**25**(3):644–51. https://doi.org/10.1093/annonc/mdt593.
183. Hanahan D, Coussens LM. Accessories to the crime: functions of cells recruited to the tumor microenvironment. *Cancer Cell* 2012;**21**(3):309–22. https://doi.org/10.1016/j.ccr.2012.02.022.
184. Van den Eynde M, Mlecnik B, Bindea G, Galon J. Multiverse of immune microenvironment in metastatic colorectal cancer. *Onco Targets Ther* 2020;**9**(1):1824316. https://doi.org/10.1080/2162402X.2020.1824316.
185. Garcia-Vicién G, Mezheyeuski A, Bañuls M, Ruiz-Roig N, Molleví DG. The tumor microenvironment in liver metastases from colorectal carcinoma in the context of the histologic growth patterns. *Int J Mol Sci* 2021;**22**(4). https://doi.org/10.3390/ijms22041544.
186. Erkan M, Michalski CW, Rieder S, et al. The activated stroma index is a novel and independent prognostic marker in pancreatic ductal adenocarcinoma. *Clin Gastroenterol Hepatol* 2008;**6**(10):1155–61. https://doi.org/10.1016/j.cgh.2008.05.006.
187. Calon A, Lonardo E, Berenguer-Llergo A, et al. Stromal gene expression defines poor-prognosis subtypes in colorectal cancer. *Nat Genet* 2015;**47**(4):320–9. https://doi.org/10.1038/ng.3225.
188. Sabath DE. Minimal residual disease. In: Maloy S, Hughes K, editors. *Brenner's encyclopedia of genetics*. 2nd ed. Academic Press; 2013. p. 417–9. https://doi.org/10.1016/B978-0-12-374984-0.00951-7.
189. O'Brien CA, Pollett A, Gallinger S, Dick JE. A human colon cancer cell capable of initiating tumour growth in immunodeficient mice. *Nature* 2007;**445**(7123):106–10. https://doi.org/10.1038/nature05372.
190. Todaro M, Alea MP, Di Stefano AB, et al. Colon cancer stem cells dictate tumor growth and resist cell death by production of interleukin-4. *Cell Stem Cell* 2007;**1**(4):389–402. https://doi.org/10.1016/j.stem.2007.08.001.
191. Brabletz T. To differentiate or not—routes towards metastasis. *Nat Rev Cancer* 2012;**12**(6):425–36. https://doi.org/10.1038/nrc3265.
192. Sainz Jr B, Carron E, Vallespinós M, Machado HL. Cancer stem cells and macrophages: implications in tumor biology and therapeutic strategies. *Mediators Inflamm* 2016;**2016**:9012369. https://doi.org/10.1155/2016/9012369.
193. Oskarsson T, Batlle E, Massagué J. Metastatic stem cells: sources, niches, and vital pathways. *Cell Stem Cell* 2014;**14**(3):306–21.
194. Raposo G, Stoorvogel W. Extracellular vesicles: exosomes, microvesicles, and friends. *J Cell Biol* 2013;**200**(4):373–83. https://doi.org/10.1083/jcb.201211138.
195. Rana S, Malinowska K, Zöller M. Exosomal tumor microRNA modulates premetastatic organ cells. *Neoplasia* 2013;**15**(3):281–95. https://doi.org/10.1593/neo.122010.
196. Hwang W-L, Lan H-Y, Cheng W-C, Huang S-C, Yang M-H. Tumor stem-like cell-derived exosomal RNAs prime neutrophils for facilitating tumorigenesis of colon cancer. *J Hematol Oncol* 2019;**12**(1):10. https://doi.org/10.1186/s13045-019-0699-4.
197. Melling N, Simon R, Mirlacher M, et al. Loss of RNA-binding motif protein 3 expression is associated with right-sided localization and poor prognosis in colorectal cancer. *Histopathology* 2016;**68**(2):191–8. https://doi.org/10.1111/his.12726.
198. González-Trejo S, Carrillo JF, Carmona-Herrera DD, et al. Baseline serum albumin and other common clinical markers are prognostic factors in colorectal carcinoma: a retrospective cohort study. *Medicine* 2017;**96**(15):e6610. https://doi.org/10.1097/MD.0000000000006610.
199. Boussiotis VA. Molecular and biochemical aspects of the PD-1 checkpoint pathway. *N Engl J Med* 2016;**375**(18):1767–78. https://doi.org/10.1056/NEJMra1514296.
200. Calik I, Calik M, Turken G, et al. Intratumoral cytotoxic T-lymphocyte density and PD-L1 expression are prognostic biomarkers for patients with colorectal cancer. *Medicina* 2019;**55**(11). https://doi.org/10.3390/medicina55110723.
201. Berntsson J, Eberhard J, Nodin B, Leandersson K, Larsson AH, Jirström K. Expression of programmed cell death protein 1 (PD-1) and its ligand PD-L1 in colorectal cancer: relationship with sidedness and prognosis. *Onco Targets Ther* 2018;**7**(8):e1465165. https://doi.org/10.1080/2162402X.2018.1465165.
202. Lee LH, Cavalcanti MS, Segal NH, et al. Patterns and prognostic relevance of PD-1 and PD-L1 expression in colorectal carcinoma. *Mod Pathol* 2016;**29**(11):1433–42. https://www.nature.com/articles/modpathol2016139.
203. van Pelt GW, Sandberg TP, Morreau H, et al. The tumour-stroma ratio in colon cancer: the biological role and its prognostic impact. *Histopathology* 2018;**73**(2):197–206. https://doi.org/10.1111/his.13489.
204. Ueno H, Kanemitsu Y, Sekine S, et al. A Multicenter study of the prognostic value of desmoplastic reaction categorization in stage II colorectal Cancer. *Am J Surg Pathol* 2019;**43**(8):1015–22. https://doi.org/10.1097/PAS.0000000000001272.
205. Nearchou IP, Kajiwara Y, Mochizuki S, Harrison DJ, Caie PD, Ueno H. Novel internationally verified method reports desmoplastic reaction as the most significant prognostic feature for disease-specific survival in stage II colorectal cancer. *Am J Surg Pathol* 2019;**43**(9):1239–48. https://doi.org/10.1097/PAS.0000000000001304.

Chapter 21

Bioinformatic tools for research in CRC

Virginia Mato-Abad[a], Alejandro Pazos[a], Cristian R. Munteanu[a], Jose Liñares-Blanco[a], Sara Alvarez-Gonzalez[a], José M. Vázquez-Naya[a], Nieves Pedreira[a], Jorge Amigo[b], and Carlos Fernandez-Lozano[a]

[a]Department of Computer Science and Information Technologies, Faculty of Computer Science, CITIC-Research Center of Information and Communication Technologies, Universidade da Coruna, Biomedical Research Institute of A Coruña (INIBIC), A Coruña, Spain, [b]Galician Public Foundation of Genomic Medicine, A Coruña, Spain

Introduction

Each human cell has a DNA sequence of 3000 million base pairs distributed in 23 pairs of chromosomes, where a total of approximately 20,000 genes are found. These genes, which will become proteins, produce between 10^5 and 10^6 mRNA molecules, which will later give rise to around 100,000 different proteins, specifically expressed in each cell and tissue. The different alterations that may occur during this process of genetic information are incalculable. In addition, the interaction that the cellular metabolism has with the environment (microorganisms, environmental exposures, nutrition, etc.) generates greater complexity, offering a much greater search space than we already had.

This makes cancer diagnosis a tremendously complex field, and a multidisciplinary approach is essential, combining the best techniques developed in the different research fields (biotechnological sequencing platforms, computer data storage and analysis techniques, complex mathematical models, etc.).

To assimilate and process all these data from heterogeneous sources in order to molecularly characterize the multitude of tumor types would be unattainable without the computational capacity we have today. That is why, at present, any progress in the field of biomedicine, and specifically in the field of oncology, is accompanied by solid support from bioinformatics and computational techniques in general.

Bioinformatics has proved to be the discipline capable of integrating different types of data and information and making predictions and virtual models that can be very orientative and applicable in the clinical field. Bioinformatics is present in many of the steps of oncological research. From the output of data from large biotechnological platforms, to its final analytical processing to extract new knowledge. In this way, there are algorithms and bioinformatic methodologies for the alignment of genomes, the annotation of variants, the study of the metagenomic profile, as well as a specific analysis pipeline for each of the omics to be studied. In addition, the field of bioinformatics is increasingly encompassing processes and protocols from the world of data analysis, such as the execution of automatic learning algorithms or even artificial vision. Thus there are a multitude of experts within the world of bioinformatics specialized in extremely specific tasks.

These advances, as well as the specialization of the experts, have provided a technological impulse and decisive knowledge in cancer research, offering an opportunity to understand the biological processes that underlie the appearance and evolution of the disease throughout its different stages. The evolution of these technologies and knowledge about the disease is considered as a promising source of new diagnostic and therapeutic tools through the identification of new targets.[1]

Emphasizing the type of tumor on which this book focuses, colorectal cancer (CRC) is the third most common type of cancer in the world and the second leading cause of death worldwide. Each year, approximately 1,000,000 new cases of CRC are diagnosed and 50,000 disease-related deaths occur.[2]

Survival rates have improved markedly when diagnosed early, but limited access to colonoscopy remains a problem. This is why the identification and characterization of new biomarkers is crucial for better diagnosis, prognosis, and knowledge about the molecular nature of the disease.

Recent advances in genomics and proteomics have contributed to the molecular understanding of CRC by evaluating the expression profiles of genes and proteins in cancerous and noncancerous tissues. The identification of genes and/or proteins that are characteristic of CRC development can help define potential biomarkers that facilitate early detection of CRC.

The achievement of all these objectives, to improve the quality of treatment of colorectal cancer patients, will inevitably come hand in hand with bioinformatics. Because the correct and efficient diagnosis of cancer will be obtained thanks to the correct processing and understanding of the so-called omic data.

Next, and in an effort to delve into the possibilities offered by the world of bioinformatics, an exhaustive review will be made of both the bioinformatics tools currently developed and the databases available for their own use.

Methodologies for obtaining primary data in cancer

All the advances that have been taking place in the last two decades in the field of molecular biology have gone in one direction: the reduction of time and costs, increasing the precision and, above all, the ability of the technique. It is precisely this enormous increase in data generation capacity that constantly pointed toward an urgent need to make all these data quickly and easily meaningful to transform them into useful information through bioinformatics. From the beginning of computational chemistry in the mid-twentieth century, through the sequencing of proteins and then directly from DNA, to the consultation of genomic markers with high-performance technologies or the massive parallelization of sequencing, it has always been tried to interrogate in various ways the genomic information that characterizes living beings.

If something characterizes the change in order of magnitude in obtaining biological data, it is the miniaturization and systematization of interrogation processes. The growing ability to massively obtain metrics from a biological system allowed for accurate and rapid characterization. This, together with the reduction of costs and the robotization of the processes, has made the complete functional study of a sample in a matter of hours viable. In the last 20 years we have undergone a constant paradigm shift as new interrogation techniques emerged. Thus, at the end of the 20th century, we were able to evaluate thousands of expression values of a cancer in a single experiment, at the beginning of the century we were able to consult hundreds of thousands of markers in a single sample, and we are currently able to sequence all genes associated to the CRC in just a few hours. Because of its relevance in creating possibilities for obtaining cancer information, two technological milestones are described as follows:

- High-performance screening technologies, such as high-throughput genotyping.
- The massive parallelization of sequencing, also called ultra-sequencing or next-generation sequencing.

High-throughput genotyping (HTG)

The genotype of a sample was determined through known sequencing methods which, although highly reliable, were tedious when the number of variants or samples to be handled ceased to be counted on the fingers. The appearance of more direct methods, which allowed working with dozens of samples at a time and that could determine the content of dozens of specific genome positions, allowed molecular biology to jump from the analysis of simple traits or disorders to study complex phenotypes and diseases, something that before was practically impossible because of the low efficiency in process cost.

Genotyping technologies have even evolved exponentially over time. Since its appearance 20 years ago, the capacities both in handling the number of samples and in genome positions investigated have increased dramatically. It soon began to discriminate between techniques based on their performance, and what 10 years ago was considered high performance (many hundreds or a few thousand genotyped variants) soon had to be categorized as medium performance to distinguish it from the different generations of massive genotyping chips that companies like Affymetrix or Illumina have been positioning as benchmark standards in the market. In fact, if there is a possibility, it is common practice to include more than one platform in a genotyping project, since very high-performance technologies cover a wide range of possibilities but are not as flexible and adaptable as low and medium performance. These can be used to cover the gaps that the earlier mentioned leave in the areas of greatest interest for a particular project.

At present there are many high-performance genotyping technologies, although they all try to determine the allelic content of very specific positions or regions present within the genome to be investigated. Although not entirely, most of the work of these technologies, and their great strength in order to increase their performance, is based on the discrimination of binary alleles, so that the genotyping platform will have to choose between one of the two known forms of the investigated allele. As analysis technologies, they differ both in the allelic discrimination reaction and in the method of detecting the products of said reaction, even in the format in which the reaction takes place, and its application is determined not only by the number of markers and samples to be analyzed, but also by the type of sample, the flexibility of the design, or the possibility of automation.

Next-generation sequencing (NGS)

Since the late 1980s, the sequencing today called traditional or Sanger sequencing has been improving both the quality and quantity of the sequence obtained from a sample, being able to obtain several hundreds of nucleotides in each reaction. Given that the greatest limitation to improve this technique was due to the increasing difficulty in distinguishing DNA fragments whose size differs by a single nucleotide, at the beginning of this century it was decided to massively parallelize every process and mechanize it to the extent of the possible in ultra-sequencing stations. Although the preparation of the samples is more complex and expensive, and that the analysis of the results requires an important bioinformatics dedication using high-performance computing, the appearance of this technology allowed the enormous expansion of the study regions in a genome within the same reaction, as well as the drastic reduction of sequencing costs.

As in genotyping, there are many ultra-sequencing techniques such as the older Roche's 454 and LifeTech's SOLiD or the newer Ion S5 and all the Illumina Seq family, although they all work essentially the same: the DNA is divided into millions of small pieces or fragments, which are then amplified and read in the sequencing platform to finally contextualize them by means of bioinformatics techniques of alignment and assembly. It is not a direct method such as traditional sequencing, but a massive reading exercise in which the more a particular region is read the more accurate the variant call will be. The whole process involves multiple steps, from the sample preparation and the proper sequencing process in the wet lab where the costs have been exponentially reduced in the last 15 years, to the data processing and analysis in the dry lab where the computing and archiving costs have increased in time. Overall, the power of this methodology is so large that it is very useful and economically very advantageous. In this way, all the genomic information related to the CRC of a sample can be obtained in just a few hours, which allows to develop diagnostic and preventive tests in a simple and barely invasive way.

Bioinformatics tools for cancer research: Applications in colorectal cancer

Traditionally, biological research typically studies one or a few genes at a time. However, genomics, proteomics, and bioinformatics exploration approaches allow researchers to simultaneously measure the changes that occur in the genome under certain biological conditions. As a result, these high-performance technologies result in a list of "interesting" genes. Biological interpretation of these lists of genes, which range in size from hundreds to thousands, is a difficult and daunting task.[3] In recent times, bioinformatics has made it possible to systematically dissect and interpret large lists of genes/proteins through the use of different methods.

The following points will describe some of the fundamental bioinformatics tools in cancer research and their applications in CRC. These tools can be grouped into two large groups:

- Databases that contain vast amounts of data and are a potential source of useful information. Some incorporate data mining techniques that allow sophisticated analysis.
- Analysis through Complex Networks, which have demonstrated a high efficacy in problems of classification and prediction in different types of cancer.

Databases

There are several databases containing oncogenomic research data, all aimed at helping to improve cancer prevention, diagnosis, and treatment through a better understanding of the molecular bases of this disease. Being able to distinguish between different types of databases we can distinguish:

- Ontological databases, which aim to address the need for a common and consistent description of genes.
- Databases for oncogenomic research: these contain cancer-specific biological data, are dedicated to oncogenomic research, and offer a certain level of online data mining analysis and techniques.

A more comprehensive description of the two types of databases is presented as follows.

Ontological databases

In computer science, an ontology is a kind of controlled vocabulary, but written in a way that can be both human-readable and machine-readable. In other words, an ontology is an artifact that defines the basic terms and relations comprising the vocabulary of a topic area as well as the rules for combining terms and relations to define extensions to the vocabulary.[4]

One of the objectives of the ontologies is to describe data accurately, relating the data with the corresponding terms of a given ontology. This process is known as "annotate" data and can be an arduous task, but it allows to later integrate data from different sources or to make precise analyses.

Ontologies are used in many fields, but in domains such as biomedicine, where a very large number of terms are handled, and where it is very important to use these terms precisely, ontologies have proved to be a very useful tool.

Ontologies that deal with biomedical concepts are known as "biomedical ontologies" or "bio-ontologies."[5] A proof of the importance that biomedical ontologies have gained in recent years is the creation of centers dedicated exclusively to maintaining and developing them. Thus there is the National Center for Biomedical Ontology (NCBO)[6] in the United States or the European Center for Ontological Research (ECOR) in Europe.[7]

At the time of writing, there are a great number of relevant ontologies. For example, if we use the browser available Bioportal website (a resource of the NCBO), we can find more than 800 biomedical ontologies.[8] Some of these ontologies are of a general purpose, others are specific to cancer, and some are specific to a type of cancer, such as the Prostate Cancer Ontology or the Breast Cancer Staging Ontology. In the case of colorectal cancer, no specific ontology is found. However, we can find terms related to colorectal cancer in a multitude of existing ontologies.

The following are some of the most relevant ontologies in the biomedical field, cancer and, more specifically, colorectal cancer.

- *Gene Ontology (GO)*. It is the most accepted and successful example of bio-ontology. Its aim is to address the need for consistent descriptions of gene products across species and databases.[9] GO is composed of three ontologies that describe gene products in terms of their related biological processes, cellular components, and molecular functions in a species-independent way.[10]
- *NCI Thesaurus (NCIt)*. It is one of the most important ontologies in the area of cancer. NCIt is NCI's reference terminology and core biomedical ontology, covering some 140,000 key biomedical concepts with a rich set of terms, codes, 120,000 textual definitions, and over 500,000 interconcept relationships. The aim of the NCI Thesaurus is to integrate molecular and clinical cancer-related information with a controlled terminology, providing a structured representation of key cancer-related concepts in fields such as drugs, therapies, pathways, cellular processes, etc.[11]
- *Unified Medical Language System (UMLS)*. It is a popular and widely used ontological resource, which is a large compilation of names, relationships, and associated information from a variety of bio-ontologies (including the abovementioned Gene Ontology and the NCI Thesaurus).[12]

As we have seen in the previous examples, one of the main purposes of ontologies is as controlled vocabularies. But ontologies are much more than that. The knowledge they represent is encoded in formal language and can be processed by computers to infer new knowledge. Thus, for example, in,[13] the authors explain like ontologies are useful not only as controlled vocabularies, but as making possible automatic reasoning, in this case for colorectal cancer drug target prediction. And, it is to be expected that in the coming years, new research works will appear in this direction, at the same time as new ontologies appear and the existing ontologies evolve.

Databases for oncogenomic research

Access to the large amount of genomic data that researchers currently have offers the possibility of generating new methodologies and algorithms capable of extracting new knowledge from them. In this section we are going to take a look at the most important repositories in terms of oncological databases. It is in this field of research where there is a greater amount of data, not only because of the prevalence of the disease and the interest it arouses in the population, but also because of the difficulty of molecularly characterizing this multifactorial disease, which encompasses more than 100 different types and subtypes of tumors. In order to do this, and in order to offer precision oncology treatment in the future, the appearance and development of the tumor in all its stages and individuals must be analyzed by means of omic variables.

It is obvious that the figures that handle large initiatives are not acceptable to most research groups but allowing these data to be freely accessible offers a new alternative to smaller and/or more technical research groups in order to offer their point of view and provide new methodologies, approaches, or ideas for the study of cancer in general.

It is important to highlight the great variety of data that can be generated from a single tumor sample. At the beginning, it was considered that if the genetic sequence of a tumor sample was obtained, it was sufficient to infer all the aspects underlying a specific type of tumor in a patient. Subsequently, it was observed that the genetic sequence, although storing a large amount of information, was not sufficient to infer, for example, the causes of the appearance of the tumor, and even less to predict its prognosis. It is true that some types of tumors, with a very marked hereditary load, are good candidates to be studied only by genomic sequence, but in the vast majority of them, there is a very predominant environmental factor. It is in this context that, with the advent of ultrasequencing techniques, biotechnological platforms began to specialize not only in genetic sequence, but also in the quantification and characterization of many other molecular variables, which we now named as omic variables. From this moment on, tumors were characterized by their genome, proteome, transcriptome, epigenome, exposome, and many more and more "-omics," Unfortunately, this was not the end of cancer research, and as is often the case, every time an unknown door is opened in research, many more are found unopened. In this case, and what is more topical, is that it is being observed that it is not possible to characterize the tumor solely from a sample. Recent work has shown that, for the complete characterization of a specific tumor tissue, it is necessary to catalogue each of its different cells. It should be understood that the expression "each of its cells" does not refer to all the cells of the tumor (there may be millions of them, so sequencing costs would not be affordable), but to the different clusters or populations of cells with different alterations. It is convenient to think of a tumor as an ecosystem, if we only consider one species as all the variety that exists within the ecosystem, we are losing a lot of information about it. In the same way, in the tumor there are different cell populations with different alterations. For this reason, the tumor is beginning to be studied using techniques known as single cell genomics.

Next, and referring to the above, we will explore in depth those initiatives and repositories that present databases of free access that can be used by any research group and/or individual researcher with interest in the field of oncogenomics. We will also focus on the data presented by each initiative in relation to the CRC, and how this can support any type of research in this regard.

- COSMIC: Catalogue of somatic mutations in cancer

In the United Kingdom, the Wellcome Trust Sanger Institute started The Cancer Genome Project, with the aim of identifying sequences of important variants in the development of human cancers. From the results of the experiments, the tool COSMIC (catalogue of somatic mutations in cancer) was developed, where it is possible to navigate along the different variants and obtain detailed information on the results obtained in the project.

COSMIC comprises the COSMIC database and the Cell Lines Project, two separate but related resources. COSMIC database combines two main types of data. On one hand, high precision data, manually curated by experts which include targeted gene-screening panels, over 25,000 peer-reviewed papers, metadata from environment and patient history and full details of the curation process and data captured. On the other hand, genome-wide screen data, including over 32,000 genomes consisting in peer-reviewed large-scale genome screening data and other databases such as TCGA and ICGC.

The Cell Lines Project, based on Wellcome Sanger Institute, started with the aim of improving the utility of cancer cell lines through standardization, also provide a systematic characterization of the genetics and genomics of large numbers of cancer cell lines and data from the full exome sequencing of 1020 cancer cell lines. In addition, the molecular characterization includes at the moment CNV and Gene expression data. The cancer cell lines are from major publicly accessible repositories from around the world and a few lines which are not publicly available.

In terms of CRC, the COSMIC repository houses a total of 384 works that have sequenced and identified mutations in colorectal tumors. The web tool allows the user to navigate in each of the works, showing the ratio of mutations in each of the genes, being very useful for researchers focused on the different nucleotide variants that present the different types of cancer, as can be seen in Fig. 21.1, which shows a screenshot of the COSMIC tool. Among the different variants catalogued are point mutations, variants in the number of copies, gene fusion mutations, as well as all kinds of possible mutations. In addition, their data also include gene expression data and methylation data.

FIG. 21.1 Screenshot showing the COSMIC web-tool. *Image obtained from https://cancer.sanger.ac.uk/cosmic/browse/tissue?wgs=off&sn=large_intestine&ss=colon&hn=carcinoma&sh=adenocarcinoma&in=t&src=tissue&all_data=n.*

- InSiGHT

The International Society for Gastrointestinal Hereditary Tumors (InSiGHT) is an international multidisciplinary scientific organization that aims to improve the quality of care of patients and families with any hereditary condition resulting in gastrointestinal tumors. Fig. 21.2 shows a screenshot of the InSiGHT web-tool. InSiGHT was formed in 2005 by the merger of the Leeds Castle Polyposis Group (LCPG) and the International Collaborative Group on Hereditary Non-Polyposis Colorectal Cancer (ICG-HNPCC). In 2010, it became an incorporated charity in England and Wales, and today it houses and curates the most comprehensive database of DNA variants that contribute to gastrointestinal cancer. This database, along with the ones provided by a number of other institutions, is supported by curators and panels of experts who review the pathogenicity assignments; therefore, most of the databases are considered to present the most authoritative interpretation of the variants, based on defined criteria for interpretation.

- TCGA—The Cancer Genome Atlas

In order to lay the foundations and achieve great advances in prevention, early detection, stratification, and success in the treatment of cancer, it is necessary to identify the complete changes generated by each type of cancer in its genome and to understand how these changes interact with its environment, intra- and intercellular, so that the disease manifests itself. Thus The Cancer Genome Atlas (TCGA) emerges as a collaboration between the National Cancer Institute (NCI) and the National Human Genome Research Institute (NHGRI) of the United States, with the aim of obtaining comprehensive multidimensional genomic maps of all key changes in most types and subtypes of cancer. A first pilot project in 2006 confirmed that an atlas of these changes could be created specifically for different types of cancer. Since that time, TCGA has been collecting tissues from more than 11,000 cancer patients and healthy patients, allowing the study of more than 33 types and subtypes of cancer, including 10 rare cancers. The most interesting aspect of this initiative is that all the information is free and accessible to any researcher who wants to focus their efforts on this disease. The development of new bioinformatics techniques is progressively making it easier to interpret multiple and different sets of data at the genomic, epigenomic, transcriptomic, metabolomic, and other scales.

Specifically, the TCGA project currently has different types of data ranging from DNA sequences, different types of RNA, copy number mutations, expression data, methylation data, and protein data, among others. If we refer to colon cancer, the TCGA currently presents data from 629 patients with clinical data, 616 samples analyzed by SNP6 Copy Num, 104 LowPass DNASeq Copy number, 223 MAF, 622 Methylation, 549 miRSeq, 222 mRNA, 623 mRNASeq, 489 raw Mutation Annotation File, 491 Reverse Phase Protein Array, and 28 images. Fig. 21.3 shows a screenshot of the Genomic Data Commons Data Portal, where different data can be downloaded.

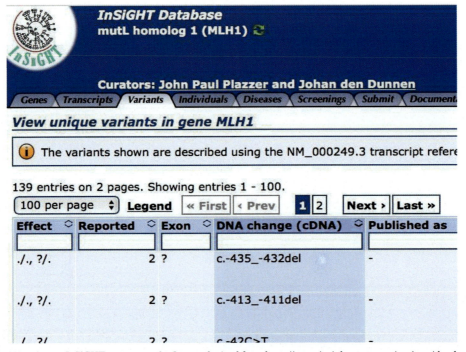

FIG. 21.2 Screenshot showing an InSiGHT gene example. *Image obtained from https://www.insight-group.org/variants/databases/.*

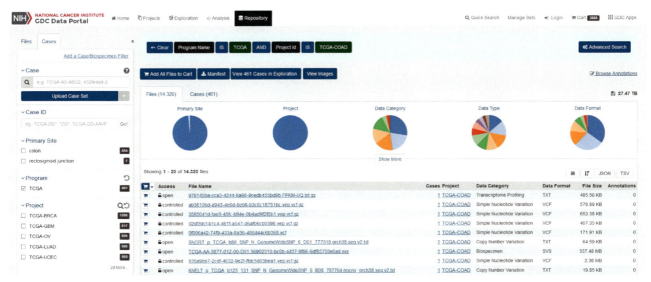

FIG. 21.3 Screenshot of Genomic Data Commons Data Portal, where TCGA data are stored. Specifically can be seen all different data types from COADREAD cohort. *Image obtained from https://portal.gdc.cancer.gov/repository.*

The consortium, thanks to the integrative analysis of its data, made large contributions to the CRC. In 2012 they published in Nature a complete characterization of the genetic alterations present in colorectal adenocarcinoma.[14] To do so, they performed a genomic-scale analysis of 276 samples, analyzing their exoma, DNA copy number, promoter methylation, and messenger RNA and microRNA expression. It was found that 16% of colorectal carcinomas were hypermutalized: three quarters of them had the expected high instability of microsatellites, generally with hypermethylation and silencing with MLH1, and a quarter had somatic mutations to repair missing gene and polymerase θ (POLE). In addition, they observed that cases of colon and rectal cancer that are not hypermutated had similar modification patterns. They also identified 24 genes with significant mutations, finding frequent mutations in ARID1A, SOX9, and FAM123B. Finally, the integrative analyses they performed suggested new markers for aggressive colorectal carcinoma with an important role for MYC-directed transcriptional activation and repression.

- HCA—Human Cell Atlas

Finally, in relation to the genomic data of single cells, we find a project that has recently begun its journey, the Human Cell Atlas (HCA). This project was conceived in 2016 in London, when a group of pioneering scientists in their fields of research met and discussed how to build a human cell atlas, in order to describe and define the cellular bases of health and disease.

As has already been mentioned, research into complex diseases, in this case cancer, has evolved into increasingly sensitive and specific technology, to the point where it is necessary to molecularly define each of the cells that participate in the development of the disease. Thus the HCA, still in development, will be composed of complete reference maps of all human cells as a basis for understanding fundamental human biological processes and for diagnosing, monitoring, and treating diseases. In this way, it will help scientists understand how genetic variants impact disease risk, define drug toxicities, discover better therapies, and advance regenerative medicine.

Currently, the project already has an open-access repository where scientists can download the data generated by the project (https://data.humancellatlas.org/explore/projects). Fig. 21.4 shows a screenshot of the HCA Data Portal, where different data can be downloaded. Mostly, they are sequence data without being preprocessed, so a large storage capacity is needed. However, there are other types of platforms such as JingleBells (http://jinglebells.bgu.ac.il/) where you can download standardized scRNASeq databases and import them directly as an R object.

- ICGC—The International Cancer Genome Consortium

As can be seen, most omic data repositories in tumors are closely related and benefit from each other. This approach lays the foundation for standardization and integration of data using ontologies, as discussed in the previous section. The reference project in this regard is The International Cancer Genome Consortium (ICGC), which is an integration of different cancer-related projects. It is a scientific organization that provides a forum for collaboration among the world's leading cancer researchers. It was initiated in 2008 to coordinate large-scale cancer genomic studies. It currently has the participation of 86 projects from different countries that study around 25,000 patients belonging to 22 types of cancer. Of these, three projects are dedicated to the study of Colorectal Cancer. In addition, they not only present genomic data but also host data

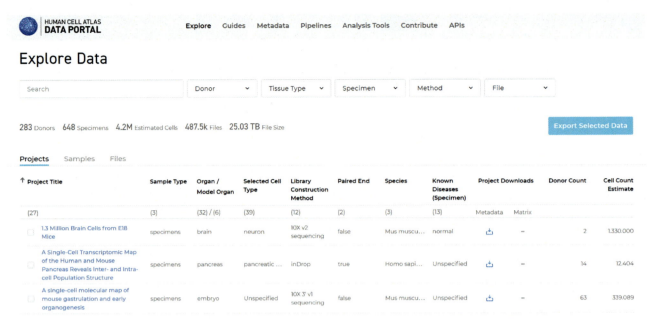

FIG. 21.4 Screenshot from Human Cell Atlas Data Portal, where it can download open access data. *Image obtained from https://data.humancellatlas.org/explore/projects.*

FIG. 21.5 Screenshot of ICGC data repository. *Image obtained from https://dcc.icgc.org/repositories.*

from different biotechnological platforms, capable of inferring levels of genetic expression, protein, methylation profiles, annotation of mutations, or even levels of different transcripts. All these data are available through the ICGC web portal (https://dcc.icgc.org/projects/details?projects=%7B%22from%22:1%7D&filters=%7B%22project%22:%7B%22primarySite%22:%7B%22is%22:%5B%22Colorectal%22%5D%7D%7D%7D). A total of 1779 samples belonging to 1041 donors are currently stored. Fig. 21.5 shows a screenshot from ICGC data repository.

- Other repositories
  - IntOGen: developed by the BBGLab of the Institute for Research Biomedicine (IRB) in Barcelona, which integrates multidimensional oncogenomic data and shows the different drivers in each type of cancer (see https://www.intogen.org/search).

- cBioPortal: is a tool that allows exploratory analysis of data and where it is possible to download small segments of data (user-defined genes and sample sets) without having to download complete datasets (see https://www.cbioportal.org/).
- GEO: is a public functional genomics data repository that contained array- and sequence-based data (see https://www.ncbi.nlm.nih.gov/geo/).

Complex networks

The large amount of data and information generated by genomic and proteomic techniques requires that in some cases it is necessary to use complex systems that can analyze and connect all this information. Complex Networks or Graph Theory is very useful for tackling these problems, using information about the connections or relationships that exist between the different parts of a system.

With these methods it is possible to address systems with a large number of dimensions, from small molecules to social networks: a drug molecule formed by atoms linked by chemical bonds, a protein containing amino acids, a nucleic acid molecule with nucleotides linked by phosphate bonds, a metabolic pathway composed of several types of molecules, or a population linked by the same type of activity.

A network is a collection of connected objects. In mathematics and computer science, it is an abstract representation of a real network and is made up of links that connect pairs of nodes. These nodes are used to model the relationships between pairs of objects in a given collection.

In biology these graphs are used to encode the structure of proteins or for other properties such as DNA/RNA nucleotide sequences, microarrays, etc. A biological example of a complex network is the graphical representation of a protein where the nodes or vertexes are the amino acids connected by peptide bonds.

Recent work using complex networks of proteins or proteomics in human serum has contributed to the creation of theoretical models applicable to the diagnosis of cancer and the detection of molecules related to CRC. The proteins involved in the CRC processes are of great importance to find new molecular targets in the research of new drugs against this type of cancer. Proteins related to CRC have very diverse forms, locations, and functions within the body. Thus the task of obtaining a classification model that can discriminate this class of proteins from others represents a very difficult task. When the molecular characteristics are unknown in order to carry out a direct classification of molecules with specific functions, it is possible to use an indirect method that uses the information encoded in the structure or topology of the molecule.[15]

Consequently, the methodology of complex networks or graphs can be used to encode the internal topological information of protein molecules into specific topological indices. These indices will be the basis for the search for classification models of a specific biological function, such as the CRC. Several publications have presented classification models for biological functions such as enzymes,[16] antioxidants,[17] enzymes[18], or lipid complexes.[19] In the case of cancer, there are also publications that use information from molecular graphs.[20, 21]

In the case of CRC, data from the primary protein structure have been used in Ref. 22 to generate a specific classification model that can evaluate whether or not a protein is related to CRC. The mathematical model used is based on the following steps:

1. The primary structures (sequences of peptides/protein chains) are transformed into star-like graphs, where each node is an amino acid.
2. A series of specific topological indices are calculated for star graphs, as a series that characterizes only a protein chain. The S2SNet application[23] is used for this purpose.
3. With these indices, we look for the best model that can discriminate between proteins that are related to the CRC or not, using Machine Learning techniques.

The best model obtained to predict new peptides related to CRC is a linear model and has been implemented in an online tool called Human Colon Cancer Prediction (HCC-Pred, http://bio-aims.udc.es/HCCPred.php). This model is based on five topological indices of the protein star graphs and correctly classifies 89.5% of the proteins, being very useful to predict the relationship with the CRC of new protein sequences. The flow diagram in Fig. 21.6 describes the methodology used by the tool, which only requires from the user the protein sequence whose relationship with the CRC is intended to predict.

In addition, there are more publications about complex networks and CRC. Reference 24 proposes a new classification model for CRC using amino acid sequences of proteins and spiral-like molecular graphs with a predictive capacity of 90.92% of new CRC-related proteins. A complete review of the work done on the classification and prediction of CRC using complex networks can be found in Ref. 10.

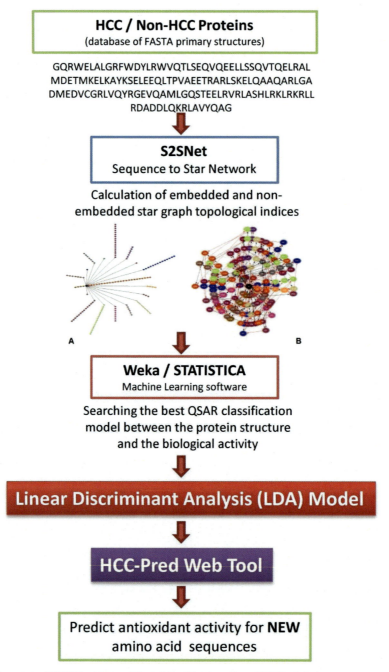

FIG. 21.6 Methodology of the web tool HCC-Pred (accessible at: http://bio-aims.udc.es/HCCPred.php).

Biomedical tools for assessing the risk of suffering from colorectal cancer

There are individuals who, based on their family and/or personal history, are at high risk for the development of CRC. For these people it is feasible to establish specific, individualized, and highly effective screening and surveillance measures in order to prevent or detect early this disease. There are different risk prediction models of suffering from CRC:

- The absolute risk of cancer is the probability that a person with a certain age and certain risk factors will develop cancer in a defined period of time. Some examples of these risk factors include age, gender, race, genetics, body mass index, family history of cancer, history of cigarette smoking, physical activity, the use of hormone replacement therapies, reproductive factors, dietary factors, etc. The development of statistical models to estimate the probability of

developing cancer over a defined period of time helps physicians to identify the individuals most at risk of suffering specific cancers. By using this, you can advise and monitor the individual earlier and more frequently to reduce the risk. Different models of absolute CRC risk prediction have been developed,[25] mainly for use in research. The methodology and results of these models have been evaluated by qualified scientists and doctors and published in scientific and medical journals.
- There are other models that estimate the probability of detecting a mutation in a gene susceptible to CRC in a family or individual.
- There are also CRC risk assessment tools more focused on the orientation than on the research, which may be useful for obtaining a greater knowledge of the risk to develop CCR. Using easily obtainable information (e.g., personal and family medical history, lifestyle behaviors, age, etc.), these tools provide an estimate of an individual's risk of developing CCR over certain time periods.

Some examples of these tools focused on the orientation are the Cleveland Clinic's colon cancer risk assessment,[26] the "My CancerIQ" website designed by "Cancer Care Ontario"[27] that helps to understand the risk for different types of cancer including CRC and what to do to help lower that risk. The downside of this tool is that the assessment is calculated relative to the population of Ontario because it only uses Ontario-specific data. Another one is the Colorectal Cancer RISk Prediction tool (CRISP) developed by researchers at the University of Melbourne in collaboration with general practitioners, gastroenterologists, geneticists, policy makers, IT developers, and epidemiologists.[28] The CRISP software tool is for use in primary care with the aim of increase risk-appropriate CRC screening so it is not available online. Definitely, the most important of these tools, due to the wide range of populations and ages that it includes, is the Colorectal Risk Assessment Tool[29] developed by the National Cancer Institute (NCI) that is described next.

The Colorectal Risk Assessment Tool (CCRAT) is based on the first absolute risk model for CRC incidence. The model was developed by the NCI using data from two large population studies of colon and rectal cancer in the United States, incidence data from NCI Surveillance, Epidemiology, and End Results (SEER) registries, and mortality rates. The tool uses the information about risk and preventive factors to calculate a patient's absolute risk of CRC for a specific time period. The model was tested in a large population of non-Hispanic whites and has been shown to be accurate in predicting this absolute risk. The risk-assessment model[29] and its validation[30] are published in the Journal of Clinical Oncology.

The CCRAT was designed for doctors and other health care providers to use with their patients. The tool estimates the risk of colorectal cancer over the next 5 years and the lifetime risk for men and women who are between the ages of 45 and 85, white, African American, Asian American, or Hispanic. This tool cannot accurately estimate risk of colorectal cancer for people who have the following health conditions: ulcerative colitis, Crohn's disease, Familial adenomatous polyposis (FAP), Hereditary Nonpolyposis Colorectal Cancer (HNPCC), also known as Lynch Syndrome or personal history of CRC.

The risk calculator takes about 5 min to complete 13 questions from 4 different categories: demographics, diet and physical activity, medical history, and family history. It is available on the NCI Web site at www.cancer.gov/colorectalcancerrisk. People using it should work with their health care providers to interpret the results. The tool is updated periodically as new data or research becomes available. The current incidence data are from SEER 18 in the period 2010 through 2015. In addition, it may prove useful to researchers who are designing clinical intervention studies.

In addition to the existence of statistical models, mutation estimation, and tools that can serve as guidance for early detection of CRC, we can also find free software platforms and online execution that, without having hardly any notions of bioinformatic analysis, allow us to process genomic data in order to extract useful information from the data obtained in different tests on patients. A clear example of this would be Galaxy, a biomedical analysis platform for anyone.[31]

Conceptualized as a web-based platform for scientific analysis, since its creation in 2005 it has been used by hundreds of thousands of scientists worldwide. Framed under three key milestones for data-driven biomedical science (accessibility, reproducibility, and transparent communication), the Galaxy team has achieved substantial enhancements throughout the years that enable the analysis of extensive datasets and more than 5500 tools that are already available in the Galaxy ToolShed. It has numerous high quality tutorials focused on analysis of common types of genomics, in addition, the user has mainly two modalities to analyze the data, by exploratory execution or pipeline, being able to carry them out simultaneously. Automatically, reusable and generalizable workflows can be generated from an ad hoc analysis, and there is also an interactive workflow editor to modify or generate them from scratch. One characteristic that makes this platform especially remarkable is that it enables you to perform complex analyses without having any type of knowledge in bioinformatics programming, despite the fact that for complete genomic data processing it is required to customize the scripts, counting with an updated infrastructure in its Galaxy Interactive Environments.[31, 32]

This project consists of four complementary components: the Galaxy public server (https://usegalaxy.org/), composed of a diverse toolset for large-scale genomic analysis, terabytes of public data to use, and a large number of shared analyses, workflows, as well as interactive publications; the Galaxy framework and software ecosystem (https://github.com/gal axyproject) which is an open-source software package that anyone can use to run the Galaxy server on a Unix-based operating system; the Galaxy ToolShed (https://toolshed.g2.bx.psu.edu/), available to "AppStore" users where they can share Galaxy tools, as can be Gemini, to explore genetic variations,[33] QIIME for quantitative analysis of the microbiome for raw DNA sequencing data[33, 34] or deepTolls for analysis of deeply sequence data,[33–36] in addition to workflows and visualizations; finally it provides the Galaxy Community (https://galaxyproject.org/community/), place where all the aspects within this project.

Deep learning for medical imaging in colorectal cancer

Deep Learning (DL)[37] is a specific type of Machine Learning techniques that become the state of the art in Medical Imaging for skin cancer,[10, 38] lung cancer,[39] breast cancer,[39, 40] Alzheimer,[41] etc.

Several studies are using deep neural networks (DL) to help the diagnosis of CRC. Thus DL was applied for detection of colon cancer in histology images,[42] tumor images,[43] or epithelial colon cells.[44]

The colon polyps detection represents a great opportunity to localize future CRC by estimating the risk of CRC or to help the surveillance of patients. Korbar et al.[45] has used DL and 239 for polyp detection in histological images with an accuracy of 93%.

Ribeiro et al.[46] proposed a texture-based method with Convolutional Neural Networks (the state of the art in imaging)[47] to automatically classify the colonic mucosa in colonoscopy images with an accuracy up to 91%.

The missing of a public application/script for polyp detection with high accuracy determined my group to propose two different tools for the identification of colon polyps in colonoscopy images using Keras (based on Tensorflow) and Fastai (based on PyTorch).

The first project, CNN4Polyps, Munteanu C.R. used a public dataset downloaded from http://site.uit.no/deep healthresearch/2017/07/19/polyp-detection-using-deep-learning/ and a CNN topology to find the best model that can detect and localize the polyp into a colonoscopy image (see Fig. 21.7).

All the python code is presented as jupyter notebooks into a public GitHub repository: https://github.com/muntisa/Colo noscopy-polyps-detection-with-CNNs. Fig. 21.8 shows the flow of the methodology, script by script:

– From the original dataset, different images are cropped with and without polyps (606 images for each class);
– A specific structure of folders with the images has created as inputs for the DL classifier;
– The first models used small CNNs;
– In the second models, transfer learning was used by using pretrained VGG16 networks;
– The last type of models used a fine tuning of the VGG16 networks;
– The localization step was done by window slide method.

CNN4Polyps results showed that it is possible to obtain 98% accuracy to detect polyps in colonoscopy images:

– If we are using small CNN with only 2 or 3 convolutional blocks, it is possible to obtain in only 2 min a classifier with over 90% accuracy (CPU i7, 16G RAM, GPU Nvidia Titan Xp).
– If transfer learning is used with VGG16 (training only the fully connected layers), the results are similar with the small CNNs. This could be explained by the training of VGG16 with the Imagenet dataset for very different types of images

FIG. 21.7 Detection and localization of colon polyps in colonoscopy images using deep learning methods.

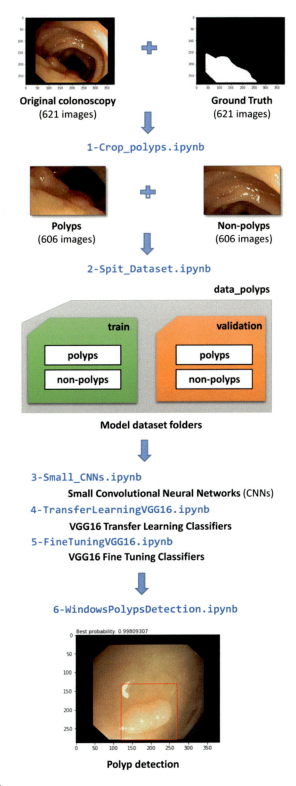

FIG. 21.8 CNN4Polyps methodology.

compared with polyps (VGG16 contains features that are not the best ones for our classification). In addition, we used the original dataset, without data augmentation because of the transfer learning advantage.
- If fine tuning is applied with the same VGG16 (training the fully connected layers with the last convolutional block), the classifier is extracting better features to classify the colon polyps. Thus it is possible to obtain an accuracy over 98% (learning rate = 0.0002, momentum = 0.9, batch size = 64).

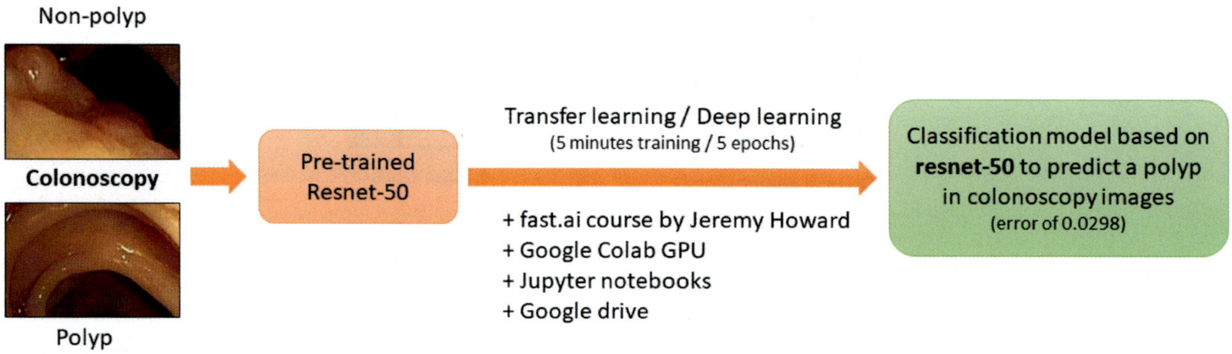

FIG. 21.9 Fastai for classification of images with colon polyps.

The classifier could be improved by using additional hyperparameter optimization that includes different drop rate, optimizer, or the use of a different base model such as Resnet or Inception.

The second project (Fastai-Colon-Polyps) is available as a free GitHub repository at https://github.com/muntisa/Fastai-Colon-Polyps and it is based on the Fastai python package,[48] jupyter notebooks executed for free in Google Colab with GPU support (Fig. 21.9). This project is using a pretrained Resnet-50 network for transfer learning and fine tuning.

Using the minimum code from Fastai, fine tuning of Resnet-50 with different learning rate for convolutional blocks and fully connected layers, an accuracy of 99% was obtained in only 2 min of training with the free GPU support from Google. All the projects can be reproduced with the repository code.

In the same repository, the best Fastai model was implemented as a public docker image for the Web app for colon polyp detection using deep learning with colonoscopy images (Fig. 21.10). The docker image is available at Docker Hub.[49] Run this docker locally with the following command: `docker run -p 5000:5000 colon-polyps-fastai`.

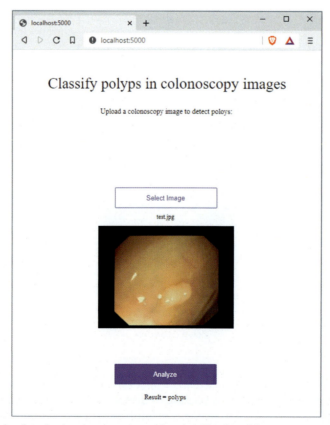

FIG. 21.10 Web application for polyp detection in colonoscopy images based on Fastai model.

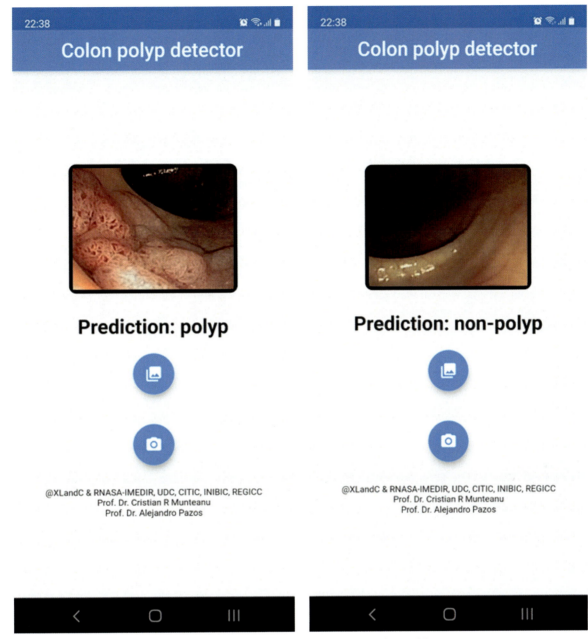

FIG. 21.11 Android application for polyp detection in colonoscopy images using deep learning models.

In addition to the GitHub repository with scripts to obtain all the prediction models, the Web app implementation and docker option, RNASA-IMEDIR with REGICC and XLandC lab[50] implemented a free Android mobile app (see Google Play[51]) of the polyp detector (Fig. 21.11) by using a Tensorflow Lite model and Flutter development kit (tools from Google).

Our DL models demonstrated again the power of CNNs for any type of image, with excellent accuracy and free calculation support. Similar tools will become the new computer-aided support for helping the medical diagnosis.

Conclusions

This chapter has reviewed the different bioinformatics tools currently used in cancer research, the public databases that can be consulted and analyzed, as well as the different applications of these in the CRC. These tools include a multitude of different techniques and methodologies, such as oncogenomic research databases, which incorporate data mining techniques, or analyses with complex networks and graphs, which have shown high effectiveness in problems of classification and prediction in different types of cancer.

It is undeniable that the emergence of mass sequencing techniques led to the emergence and rapid rise and development of bioinformatics algorithms and methodologies. This change of paradigm at the time of confronting such a complex disease as cancer requires the collaboration of big international research consortiums to unite efforts in the research tasks. The high level of funding that these consortia have, mentioned throughout this chapter, cannot be completed by small research groups that seek to provide new ideas, tools, or solutions for the possible diagnosis or treatment of the disease. The fact that these large consortiums share the data obtained with the scientific community favors the participation of these small groups in the fight against cancer. This action highlights the need for collaboration in order to reach such a desired objective.

It is in this context that a significant increase has been seen in the development of new bioinformatics tools, increasingly modeled on data obtained from biotechnological platforms. In addition, as discussed in this chapter, data analysis techniques, specifically Machine Learning, are increasingly present in bioinformatics methodologies. All this favors the generation of new knowledge, although good practice in the development and execution of all these pipelines is mandatory, focusing on standardization in order to generate robust and quality scientific results that can be replicated and contrasted by different research groups. It is at this point that emphasis should be placed on laying the foundations for future findings and research.

The ultimate goal of all these tools and research is to reach the patient in a personalized way. The role played by bioinformatics is to facilitate the work of clinicians when making decisions, as well as the automation of processes in which there may be some human failure. Finally, making a more precise medicine will help in the treatment and diagnosis of colorectal cancer and other pathologies.

References

1. Lopez-Campos GH, Romera-Lopez A, Seoane JA, Perez-Villamil B, Lopez-Alonso V, Martin-Sanchez F. Microarrays and colon cancer in the road for translational medicine. *Curr Bioinform* 2011;**6**(2):145–62. https://doi.org/10.2174/157489361106020145.
2. García-Bilbao A, Armañanzas R, Ispizua Z, et al. Identification of a biomarker panel for colorectal cancer diagnosis. *BMC Cancer* 2012;**12**(1). https://doi.org/10.1186/1471-2407-12-43.
3. Huang DW, Sherman BT, Lempicki RA. Bioinformatics enrichment tools: paths toward the comprehensive functional analysis of large gene lists. *Nucleic Acids Res* 2009;**37**(1):1–13. https://doi.org/10.1093/nar/gkn923.
4. Neches R, Fikes RE, Finin T, et al. Enabling technology for knowledge sharing. *AI Mag* 1991;**12**(3):36. https://www.aaai.org/ojs/index.php/aimagazine/article/view/902.
5. Blake J. Bio-ontologies—fast and furious. *Nat Biotechnol* 2004;**22**(6):773–4. https://doi.org/10.1038/nbt0604-773.
6. National Center for Biomedical Ontology. *BioPortal website*. https://www.bioontology.org/. Accessed October 19, 2019.
7. European Center for Ontological Research (ECOR). http://www.ecor.uni-saarland.de/home.html. Accessed October 19, 2019.
8. *Browse ontologies. NCBO bioportal.* http://bioportal.bioontology.org/ontologies. Accessed October 19, 2019.
9. Ashburner M, Ball CA, Blake JA, et al. Gene ontology: tool for the unification of biology. The Gene Ontology Consortium. *Nat Genet* 2000;**25**(1):25–9. https://doi.org/10.1038/75556.
10. Martínez-Romero M, Vázquez-Naya JM, Rabuñal JR, et al. Artificial intelligence techniques for colorectal cancer drug metabolism: ontology and complex network. *Curr Drug Metab* 2010;**11**(4):347–68. https://www.ncbi.nlm.nih.gov/pubmed/20446907.
11. National Cancer Institute. *Enterprise vocabulary services*. https://evs.nci.nih.gov/. Accessed October 18, 2019.
12. Bodenreider O. The unified medical language system (UMLS): integrating biomedical terminology. *Nucleic Acids Res* 2004;**32**(Database issue):D267–70. https://doi.org/10.1093/nar/gkh061.
13. Tao C, Sun J, Zheng WJ, Chen J, Xu H. Colorectal cancer drug target prediction using ontology-based inference and network analysis. *Database* 2015;**2015**. https://doi.org/10.1093/database/bav015.
14. Cancer Genome Atlas Network. Comprehensive molecular characterization of human colon and rectal cancer. *Nature* 2012;**487**(7407):330–7. https://doi.org/10.1038/nature11252.
15. González-Díaz H, Pérez-Montoto LG, Duardo-Sanchez A, et al. Generalized lattice graphs for 2D-visualization of biological information. *J Theor Biol* 2009;**261**(1):136–47. https://doi.org/10.1016/j.jtbi.2009.07.029.
16. Concu R, DS Cordeiro MN, Munteanu CR, González-Díaz H. PTML model of enzyme subclasses for mining the proteome of biofuel producing microorganisms. *J Proteome Res* 2019;**18**(7):2735–46. https://doi.org/10.1021/acs.jproteome.8b00949.
17. Fernández-Blanco E, Aguiar-Pulido V, Munteanu CR, Dorado J. Random Forest classification based on star graph topological indices for antioxidant proteins. *J Theor Biol* 2013;**317**:331–7. https://doi.org/10.1016/j.jtbi.2012.10.006.
18. Munteanu CR, González-Díaz H, Magalhães AL. Enzymes/non-enzymes classification model complexity based on composition, sequence, 3D and topological indices. *J Theor Biol* 2008;**254**(2):476–82. https://doi.org/10.1016/j.jtbi.2008.06.003.

19. González-Díaz H, Munteanu CR, Postelnicu L, Prado-Prado F, Gestal M, Pazos A. LIBP-Pred: web server for lipid binding proteins using structural network parameters; PDB mining of human cancer biomarkers and drug targets in parasites and bacteria. *Mol Biosyst* 2012;**8**(3):851–62. https://doi.org/10.1039/c2mb05432a.
20. González-Díaz H, Ferino G, Prado-Prado FJ, et al. Protein graphs in cancer prediction. In: *An omics perspective on cancer research*. Springer; 2010. p. 125–40. https://doi.org/10.1007/978-90-481-2675-0_7. Published online.
21. Munteanu CR, Vázquez JM, Dorado J, et al. Complex network spectral moments for ATCUN motif DNA cleavage: first predictive study on proteins of human pathogen parasites. *J Proteome Res* 2009;**8**(11):5219–28. https://doi.org/10.1021/pr900556g.
22. Munteanu CR, Magalhães AL, Uriarte E, González-Díaz H. Multi-target QPDR classification model for human breast and colon cancer-related proteins using star graph topological indices. *J Theor Biol* 2009;**257**(2):303–11. https://doi.org/10.1016/j.jtbi.2008.11.017.
23. Munteanu C, Magalhaes A, Duardo-Sanchez A, Pazos A, Gonzalez-Diaz H. S2SNet: a tool for transforming characters and numeric sequences into star network topological indices in chemoinformatics, bioinformatics, biomedical, and social-legal sciences. *Curr Bioinform* 2013;**8**(4):429–37. https://doi.org/10.2174/1574893611308040005.
24. Aguiar-Pulido V, Munteanu CR, Seoane JA, et al. Naïve Bayes QSDR classification based on spiral-graph Shannon entropies for protein biomarkers in human colon cancer. *Mol Biosyst* 2012;**8**(6):1716–22. https://doi.org/10.1039/c2mb25039j.
25. Colorectal cancer risk prediction models. https://epi.grants.cancer.gov/cancer_risk_prediction/colorectal.html. Accessed October 1, 2019.
26. Cleveland clinic colon cancer risk assessment. https://digestive.ccf.org/. Accessed October 2, 2019.
27. My CancerIQ. *My CancerIQ*. https://www.mycanceriq.ca/Cancers/Risk. Accessed October 3, 2019.
28. Walker JG, Bickerstaffe A, Hewabandu N, et al. The CRISP colorectal cancer risk prediction tool: an exploratory study using simulated consultations in Australian primary care. *BMC Med Inform Decis Mak* 2017;**17**(1):13. https://doi.org/10.1186/s12911-017-0407-7.
29. Freedman AN, Slattery ML, Ballard-Barbash R, et al. Colorectal cancer risk prediction tool for white men and women without known susceptibility. *J Clin Oncol* 2009;**27**(5):686–93. https://doi.org/10.1200/JCO.2008.17.4797.
30. Park Y, Freedman AN, Gail MH, et al. Validation of a colorectal cancer risk prediction model among white patients age 50 years and older. *J Clin Oncol* 2009;**27**(5):694–8. https://doi.org/10.1200/JCO.2008.17.4813.
31. Afgan E, Baker D, Batut B, et al. The galaxy platform for accessible, reproducible and collaborative biomedical analyses: 2018 update. *Nucleic Acids Res* 2018;**46**(W1):W537–44. https://doi.org/10.1093/nar/gky379.
32. Grüning BA, Rasche E, Rebolledo-Jaramillo B, et al. Jupyter and galaxy: easing entry barriers into complex data analyses for biomedical researchers. *PLoS Comput Biol* 2017;**13**(5):e1005425. https://doi.org/10.1371/journal.pcbi.1005425.
33. Paila U, Chapman BA, Kirchner R, Quinlan AR. GEMINI: integrative exploration of genetic variation and genome annotations. *PLoS Comput Biol* 2013;**9**(7):e1003153. https://doi.org/10.1371/journal.pcbi.1003153.
34. Caporaso JG, Kuczynski J, Stombaugh J, et al. QIIME allows analysis of high-throughput community sequencing data. *Nat Methods* 2010;**7**(5):335–6. https://doi.org/10.1038/nmeth.f.303.
35. Ramírez F, Dündar F, Diehl S, Grüning BA, Manke T. deepTools: a flexible platform for exploring deep-sequencing data. *Nucleic Acids Res* 2014;**42**(Web Server issue):W187–91. https://doi.org/10.1093/nar/gku365.
36. Ramírez F, Ryan DP, Grüning B, et al. deepTools2: a next generation web server for deep-sequencing data analysis. *Nucleic Acids Res* 2016;**44**(W1):W160–5. https://doi.org/10.1093/nar/gkw257.
37. LeCun Y, Bengio Y, Hinton G. Deep learning. *Nature* 2015;**521**(7553):436–44. https://doi.org/10.1038/nature14539.
38. Esteva A, Kuprel B, Novoa RA, et al. Dermatologist-level classification of skin cancer with deep neural networks. *Nature* 2017;**542**(7639):115–8. https://doi.org/10.1038/nature21056.
39. Sun W, Zheng B, Qian W. Computer aided lung cancer diagnosis with deep learning algorithms. In: *Medical imaging 2016: computer-aided diagnosis*. SPIE Digital Library; 2016. Published online.
40. Antropova N, Huynh BQ, Giger ML. A deep feature fusion methodology for breast cancer diagnosis demonstrated on three imaging modality datasets. *Med Phys* 2017;**44**(10):5162–71. https://doi.org/10.1002/mp.12453.
41. Puente-Castro A, Munteanu CR, Fernandez-Blanco E. System for automatic assessment of Alzheimer's disease diagnosis based on deep learning techniques. In: *Multidisciplinary digital publishing institute proceedings*. 21; 2019. p. 28. https://doi.org/10.3390/proceedings2019021028. 1.
42. Sirinukunwattana K, Ahmed Raza SE, Tsang Y-W, Snead DRJ, Cree IA, Rajpoot NM. Locality sensitive deep learning for detection and classification of nuclei in routine colon cancer histology images. *IEEE Trans Med Imaging* 2016;**35**(5):1196–206. https://doi.org/10.1109/tmi.2016.2525803.
43. Bychkov D, Linder N, Turkki R, et al. Deep learning based tissue analysis predicts outcome in colorectal cancer. *Sci Rep* 2018;**8**(1). https://doi.org/10.1038/s41598-018-21758-3.
44. Chen CL, Mahjoubfar A, Tai L-C, et al. Deep learning in label-free cell classification. *Sci Rep* 2016;**6**:21471. https://doi.org/10.1038/srep21471.
45. Korbar B, Olofson AM, Miraflor AP, et al. Deep learning for classification of colorectal polyps on whole-slide images. *J Pathol Inform* 2017;**8**:30. https://doi.org/10.4103/jpi.jpi_34_17.
46. Ribeiro E, Uhl A, Hafner M. Colonic polyp classification with convolutional neural networks. In: *2016 IEEE 29th international symposium on computer-based medical systems (CBMS)*; 2016. https://doi.org/10.1109/cbms.2016.39. Published online.
47. Krizhevsky A, Sutskever I, Hinton GE. ImageNet classification with deep convolutional neural networks. *Commun ACM* 2017;**60**(6):84–90. https://doi.org/10.1145/3065386.
48. fastai fastai/fastai. GitHub. https://github.com/fastai/fastai. Accessed September 12, 2019.
49. Docker Hub. https://hub.docker.com/repository/docker/muntisa/colon-polyps-fastai. Accessed December 25, 2020.
50. XLandC. XLandC Technologies. http://xlandc.eu. Accessed March 22, 2021.
51. Polyp detector app. Google Play. https://play.google.com/store/apps/details?id=com.xlandc.polypdetect. Accessed March 21, 2021.

Chapter 22

Omics-based biomarkers for CRC

María Gallardo-Gómez[a,b], Paula Álvarez-Chaver[c], Alberto Cepeda[d], Patricia Regal[d], Alexandre Lamas[d], and Loretta De Chiara[a,b]

[a]Department of Biochemistry, Genetics and Immunology, University of Vigo, Vigo, Spain, [b]Singular Research Center of Galicia (CINBIO), University of Vigo, Vigo, Spain, [c]Proteomics Unit, Service of Structural Determination, Proteomics and Genomics, Center for Scientific and Technological Research Support (CACTI), University of Vigo, Vigo, Spain, [d]Department of Analytical Chemistry, Nutrition and Bromatology, University of Santiago de Compostela, Lugo, Spain

Introduction

Given the growing understanding of the molecular fingerprint of colorectal cancer (CRC), it is clear that colorectal tumorigenesis is a multistep process, consequence of the progressive accumulation of genetic and epigenetic alterations. This leads to dysregulation of gene expression, aberrant proteome, and impaired homeostatic functions, resulting in neoplastic transformation that displays both intra- and intertumoral heterogeneity.

This broad range of alterations sets an open door for the discovery of novel biomarkers for screening/diagnosis, prognosis, and treatment monitoring using different omics approaches. Genome- and epigenome-wide studies, proteomics, transcriptomics, and micro RNA (miRNA) profiling, together with other emerging omics such as metabolomics or lipidomics, can improve the delivery of personalized precision medicine.

According to the specific application, ideal biomarkers for CRC should have certain characteristics. A screening biomarker should be a clinically feasible test with high patient acceptance and is expected to detect patients with early neoplastic transformation. Prognosis biomarkers should identify cancer progression independently of conventional classifications, such as TNM staging, to improve patient outcome. On the other hand, predictive biomarkers should predict the response to specific treatments. The discovery of new biomarkers for CRC still remains a challenge in clinical practice.

The evolution of the omics fields for biomarker discovery and validation should be oriented toward minimally invasive or noninvasive tests; thus, enabling an early diagnosis or CRC screening, pre- and postoperative monitoring, and assistance for selecting the most suitable therapeutic strategy and follow-up.

Biological samples for the discovery of biomarkers and their clinical application

In general, case-control studies are performed for biomarker discovery, in which samples of individuals with the disease (cases) and samples of healthy individuals (controls) are used. Candidate markers are analyzed in the samples and statistical measurements are used to determine differences between the two groups. Such case-control studies result in a filtered list of potential candidates, which will later need to be validated to assess their clinical utility.

The strategy and design of a study for the search of biomarkers must be consistent with the intended use of the marker: diagnosis, prognosis, or follow-up. Another very important consideration is the estimation of the sample size, which is closely related to the type of sample. In the discovery of candidate CRC tumor markers, tissue is widely used, although in most studies the candidate found is usually validated in blood (serum or plasma) since it is the most appropriate clinical sample given its availability and noninvasiveness.

The analysis of *tissue samples (biopsy)* allows the comparison of the molecular or protein profile of the tumor with that of the adjacent healthy mucosa, thus enabling the identification of biomarkers. The samples may be fresh frozen tissue or tissue included in paraffin and fixed with formalin (formalin-fixed, paraffin-embedded, FFPE). In case of FFPE samples, it is important to note that DNA can be degraded and the analysis in this type of samples may be difficult.

Tissue samples can be used to detect gene mutations, microsatellite instability (MSI), gene expression, and miRNA profiles, among others. This approach has several advantages over serum/plasma analyses. First, not all molecules (DNA, RNA, proteins) altered in the tumor are secreted to the blood by tumor cells. Second, the analysis of the tumor itself does not call into question the origin of the candidate biomarker. Third, its concentration is higher in tumor than in blood.

However, tumor studies also have limitations. Tissue recovery is invasive and sample availability is scarce, in addition to intra- and intertumoral heterogeneity, and the fact that tumors are not only made up of tumor cells but also surrounding stromal cells.

In the last years, *liquid biopsy* has appeared as a noninvasive alternative to solid and invasive samples. Any body fluid that can be used as a biological sample is known as a liquid biopsy. Unlike tumor tissue biopsy, liquid biopsy has the advantage of being easily available, less expensive and less invasive for the patient, allowing to obtain samples more frequently, and therefore providing information on the prognosis of the disease and the response to therapy. In liquid biopsy we can find proteins, DNA, RNA, and tumor miRNA, both free (cell-free DNA: cfDNA; cell-free RNA: cfRNA) and within circulating tumor cells (CTCs), or extracellular vesicles such as exosomes. All of them constitute potential sources of biomarkers.

The presence in blood and other biological fluids of *CTCs* may be indicative of metastasis. Current methods for the identification and capture of CTCs in CRC are based on the detection of epithelial markers such as EpCAM (epithelial cell surface antigen) and cytokeratins 19 and 20, whose expression is also altered in inflammatory processes and leads to false positives.[1,2] The effectiveness of different methods such as label-free chips based on size and deformity of CTCs is being assessed, allowing the capture of these cells with an efficiency of 70%.[3]

Exosomes are small 30–200nm membrane extracellular vesicles, released by cells into circulating body fluids, including blood. Nowadays they are thought to have a very important role in intercellular communication, participating in metastasis and resistance to treatments.[4] These vesicles not only carry proteins, but also nucleic acids (DNA, mRNA, and miRNA), that may participate in CRC progression.

In the case of *saliva*, not much studies based on this type of sample have been published for the search of CRC biomarkers since this sample is distant from the tumor location. However, advantages include it is easy to obtain and noninvasive for the patient.

Recently, low levels of several cytosine-derived nucleosides have been found in *urine* samples from CRC patients.[5] These authors conclude that epigenetic markers could be very useful as noninvasive biomarkers for the early detection and prognosis of CRC.

The use of *feces* for biomarker discovery has the advantage of being a readily available sample that is obtained in a noninvasive way, although for proteome analysis there is the disadvantage of proteolysis caused by the microbial flora. Feces also offer the possibility of analyzing the gut *microbiome*, which is sensitive to the presence of lesions. Under the assumption that CRC patients have an altered microbiome compared to healthy individuals, several bacteria such as *Streptococcus bovis*, *Fusobacterium nucleatum*, and *Helicobacter pylori* have been related to the presence of this neoplasm (reviewed in Refs. 6, 7).

As an alternative to clinical samples, *cell lines* derived from colorectal tumors have also been used for biomarker discovery. Cell lines have advantages over clinical samples because they are homogeneous cell populations, easy to maintain in culture, and their availability is virtually unlimited. In addition, they can be used to analyze changes caused by overexpression or silencing of genes, or drug response. However, the use of cell lines also has some drawbacks. In the case of colon cancer cell lines, all of them come from tumors and there are no cell lines derived from normal mucosa epithelial cells, which makes it impossible to analyze changes associated with the sequence of the malignant transformation: normal mucosa-adenoma-adenocarcinoma. In addition, cell cultures do not fully represent the in vivo situation since the interaction between tumor cells and stromal cells with the immune system is lacking. Therefore, it is essential that biomarkers identified in cell lines be subsequently validated in clinical samples of CRC patients.

Finally, *animal models* may be another alternative to the use of clinical samples since they eliminate the genetic and physiological variations that exist between individuals. In the case of CRC, there are only a small number of animal models that have some of the genetic alterations that give rise to this pathology. Among them are Apc/Min mice that have a mutation in the APC gene and may develop tumors.

Genomics approaches for colorectal cancer biomarker discovery and validation

Genomics refers to the study of the whole genome of an organism, including all the genes, their interrelations, and influence on the organism. Genomics implicates the sequencing and analysis of genomes through DNA sequencing. Advances in genomics have triggered a revolution in discovery-based research. In parallel, bioinformatics has become crucial for sequence analysis, assembly, and annotation of entire genomes.

For genomics studies, once the type of sample is defined, the next step is to select the platform. In recent years, the technologies available to carry out genomic studies and their potential applications have increased, opening new study possibilities in cancer research.

De novo identification of genetic biomarkers

The development of *next-generation sequencing (NGS)* technologies has revolutionized genomic assays. NGS is an excellent tool for biomedical research since it allows the analysis of the whole genome or exome of a sample in one assay. In whole genome sequencing (WGS) the entire genome is sequenced, while whole exome sequencing (WES) is focused in the sequencing of exons, which are the coding sequences of proteins. Though WES has less coverage than WGS (only ~2% of the genome corresponds to the exome), this approach is widely used to save money and time. However, WGS generates much more uniform coverage of the genome compared to WES and has the advantage of longer reads that allow a better determination of copy number variants (CNV), rearrangements, and other structural variations.

NGS has the ability to detect a wide range of mutations in a wide range of genes at the same time, in a single assay. Therefore when using NGS it is not necessary to develop specific protocols for each gene mutation, as in PCR-based technologies. Although its application in research studies is not subject to discussion, its use in clinical routine is still limited due to the need for highly trained personnel, complex data interpretation, quality control of sample processing, and high cost. However, the companies that commercialize these technologies are developing more simple protocols and specific software analysis to simplify the analysis of samples, to facilitate the use in clinic.

NGS technology is very useful in discovery studies due to its ability to search for unknown candidate biomarkers associated with CRC. By comparing samples of patients with different CRC subtypes and control patients, it is possible to find potential mutations associated with this neoplasia. A comprehensive molecular characterization of CRC was carried out by The Cancer Genome Atlas Network by using the potential of NGS to analyze the exome sequence of 276 CRC samples. This study found that 24 genes were significantly mutated, among them APC, TP53, SMAD4, PIK3CA, KRAS, ARID1A, SOX9, and FAM123B. In relation to disease monitoring, there are commercially available kits to evaluate mutations in 14 genes (AKT1, BRAF, CTNNB1, EGFR, ERBB2, FBXW7, GNAS, KRAS, MAP2K1, NRAS, PIK3CA, MAD4, TP53, APC) with >240 hotspots covered in cfDNA from patients with CRC. The gradual introduction of NGS in the clinic will be very helpful to personalize treatment options in CRC patients.

Approaches for targeted validation of genomic biomarkers

One of the most common and extended technique in molecular biology is the *Polymerase Chain Reaction (PCR)* and its quantitative real-time PCR version. This technique is based on the use of specific primers targeting a defined region of interest. For example, conventional PCR can be used to analyze microsatellite instability (MSI), which is a prognostic marker for CRC.

PCR can also be used for the detection of punctual gene mutations in combination with Sanger sequencing. For example, this method is used to detect mutations in KRAS gene exons 2 and 3, or BRAF gene. Briefly, the gene exon that is analyzed is amplified by conventional PCR and then the amplicon is analyzed by Sanger sequencing. This technology has the advantage of detecting all the mutations in the analyzed sequence in the same analysis. However, Sanger sequencing has poor sensitivity for low levels of mutation. Some studies have determined that at least 10–25% of mutant DNA in the sample is necessary for a reliable detection, limiting its application in clinical settings.

Another alternative to conventional PCR is *COLD-PCR*. This method is based on the enrichment of variant alleles from a mixture of wild-type and mutation-containing DNA. This method is based on the different melting temperatures of wild and mutated alleles and the use of denaturing temperatures. The ability to preferentially amplify and identify minority alleles and low-level somatic DNA mutations in the presence of excess wild-type alleles is useful for the detection of mutations. This method was successfully used to detect mutations of KRAS in liquid biopsy samples from CRC patients.

Real-time or quantitative PCR (qPCR) is another technique that can be used for detecting mutations in CRC-related genes. It is important to note that real-time PCR-based assays are considered one of the most sensitive methods. There are two common methods that are used in qPCR to detect punctual mutations. The first is based on the use of allelic-specific probes complementary to the wild allele and to the mutated allele. This method is used to study well-known punctual mutations of specific genes, as previous knowledge of the gene sequence with mutation is necessary to design the specific probes. However, it is not possible to discover new mutations or to analyze genes with mutations in several exons. For this first option, fluorophores such as SybrGreen can be used, although the wild-type allele and the mutant cannot be detected in the same assay since the SybrGreen binding is not sequence specific. This fluorophore binds to double-stranded DNA, so the design of the primers is a critical point. Alternatively, qPCR probes can be used allowing multiplexing. The second option is High Resolution Melting real-time PCR (HRM PCR), which is based on a qPCR followed by a melting curve step. The melting temperature is the temperature in which the two strands of the DNA amplicon separate or "melt" apart and depends on the nucleotide composition of the amplicon. This method allows the detection of punctual mutations in a specific gene exon.

Digital droplet PCR (ddPCR) emerged as a refinement of real-time PCR offering an alternative method for absolute quantification and rare allele detection. Explained in a simple way, ddPCR works by dividing each sample of DNA or cDNA into many individual parallel PCR reactions in nano-liter sized droplets. Some reactions contain the target molecule and others do not. After PCR amplification cycles, the fluorescence of each individual reaction is recorded with a binary readout of 0 for no fluorescence or 1 for fluorescence. These features increase the quantification capability of ddPCR compared to real-time PCR, as smaller fold-change differences can be detected. Because of this, ddPCR is an ideal tool for liquid biopsy samples. In this sense, this technique was used to detect common point mutations in the KRAS and BRAF oncogenes in cfDNA from CRC patients, showing that it is highly specific and useful for mutation detection.[8] In addition, ddPCR was used in combination with a microfluidic cell filter for isolation of circulating tumor cells (CTCs) to detect mutations that can confirm that CTCs were from colorectal origin. Thus, ddPCR was able to detect APC mutation in CTCs from metastatic CRC patients. These studies show the high potential of this technology, to be used routinely in clinic.[3]

Another option for the analysis of mutations in CRC is *DNA-based microarrays* with labeled oligonucleotide probes specific for mutant and wild-type sequences. For example, this technology was used to detect KRAS mutations in codon 12–13 in liquid biopsies of CRC patients. Although microarrays showed positive results, the comparison with PCR-based methods showed more sensitivity. Microarrays have a great utility in Genome-wide association studies (GWAS). In this type of studies, the genotyping of the human genome is carried out searching for associations between genetic variants and a phenotypic characteristic. Microarrays are an adequate technology as it allows the analysis of all the known variants of a sample in a single assay. Finally, microarrays can also be used to determine CNV of genes related with CRC. For example, this technology allows the detection of recurrent copy-number alterations and their relation to variations with specific gene expression changes in samples from CRC patients.[9]

Proposed genetic biomarkers for colorectal cancer

Many mutations in genes related to MSI and chromosomal instability (CIN), such as APC, TP53, KRAS, BRAF, PIK3CA, and genes related to signaling pathways like WNT and TGF-β, among others,[10] have been evaluated as potential biomarkers in many studies. The National Cancer Institute recommended a panel of five microsatellite loci (three dinucleotides: D2S123, D5S346, and D17S250; two mononucleotide repeats: BAT-25 and BAT-26) for prognosis.

Currently, FDA-approved and CE-IVD-certified diagnostic devices such as TherascreenKRAS RGQ PCR Kit (Qiagen, Hilden, Germany) and the Cobas KRAS Mutation Test (Roche Molecular Systems, New Jersey, USA) are available for detecting KRAS mutations in tissue biopsy. KRAS mutations are estimated to be present in 30%–40% of CRC cases, and it is well established that the presence of any activating mutation in KRAS predicts resistance to anti-EGFR therapy in metastatic CRC (reviewed in Ref. 11).

Regarding potential biomarkers in liquid biopsy,[12] analyzing a panel of 55 genes using TEC-Seq (targeted error correction sequencing) mutations were found in 50 and 89% of cfDNA samples from stage I and II CRC patients, respectively.

Many research groups are focusing on the analysis of the phenotype and the molecular profile of CTCs in liquid biopsy, with the aim of performing personalized therapy for each patient. An example of this is the recent proposal of a panel of genetic markers (GAPDH, VIL1, CLU, TIMP1, TLN1, LOXL3, and ZEB2) to detect CTCs in blood as a prognostic tool and response to therapy in patients with metastatic CRC.[13]

Transcriptomics approaches for colorectal cancer biomarker discovery and validation

Logically, the advances in genomic analysis techniques were fundamental for the development of transcriptomics, since the principle of the technology is the same for both. As a matter of fact, the goal of transcriptomics is to identify differentially expressed genes under different conditions. Transcriptomics allow a comprehensive characterization of the whole transcriptome. For this purpose, there are two main techniques that can be used: *microarrays* and *next-generation RNA sequencing* (RNA-seq).

Identification of transcriptomic biomarkers

RNA-seq is based on the combination of a high-throughput sequencing technology and bioinformatics tools which capture and quantify the transcripts present in an RNA extract. One of the main advantages of RNA-seq is that no previous knowledge of gene sequence is necessary, and therefore allows the detection of new transcript sequences. In addition, this technology is very useful to detect RNA sequence variants, isoforms, unknown splice variants, and point mutations related to the development of cancer. On the other hand, RNA-seq protocols normally have several steps and require highly trained laboratory technicians. Also, due to the high amount of raw data generated in RNA-seq experiments, bioinformatics is essential for data processing and analysis.

RNA-seq has a great utility to evaluate genes differentially expressed in CRC samples. For example, RNA-seq is used in combination with GWAS and *trans* and *cis*-Expression Quantitative Trait Loci (eQTL) to search for associations between genome variants and the expression of genes related to CRC.[14] As in the case of mutations, there are also commercially available kits especially designed to evaluate gene expression in solid samples.

RNA-seq and microarray studies are also useful to evaluate new drug targets and drugs for colorectal cancer.[15] In addition, transcriptomic technologies can be used to search and validate biomarkers to stratify CRC into molecular subtypes based on the up- or downregulation of important genes related to progression. Additionally, RNA-seq and microarrays can be combined to study the relation between gene expression variations and mutations or CNV, also analyzed by NGS and microarrays. For example, a study that evaluated 276 samples using transcriptomic technologies observed upregulation of gene *IGF2* related to the gain of 100–150 kb region of the chromosome arm11p15.5.[9] Despite the huge potential of transcriptomics technologies, *real-time PCR* is still the gold standard to study gene expression and validate transcriptional differences of genes observed in RNA-seq and microarrays studies.

Proposed mRNAs as colorectal cancer biomarkers

In the last years, gene-prognostic profiles based on the levels of mRNA have shown to provide a great accuracy in cancer prognosis. In addition, these gene expression profiles enable a better individualized and more effective therapy. By using RNA sequencing data from more than 1000 CRC patients it was possible to obtain a 6-gene signature (EPHA6, TIMP1, IRX6, ART5, HIST3H2BB, and FOXD1) predicting prognosis in colorectal cancer patients.[16] The expression profiles of these genes allow to discriminate between high- and low-risk patients, implying poor and good outcomes, respectively. In another study, EEF2K was significantly downregulated at both mRNA and protein levels in tumors of CRC patients, resulting prognostic marker.[17] On the other hand, high expression of SLC17A9 and MN1 genes has been correlated with reduced survival rates in CRC patients.[18,19] An interesting study used the expression of hypoxia-related genes (HSPA1L, PUM1, UBE2D2, and HSP27) to develop a prognostic nomogram predicting overall survival of colorectal cancer patients.[20] mRNA analysis can also be used to determine the predictive value of chemotherapy efficacy. Patients with lower expression of ERCC1 and TYMS had better 3-year survival rates than patients with higher expression.[21]

Proposed miRNAs as colorectal cancer biomarkers

RNA-seq and microarrays technologies are also used to study noncoding RNA molecules such as long noncoding RNAs (lncRNAs) or micro RNAs (miRNAs). The main characteristic of these RNA molecules is that they do not encode proteins but act as posttranscriptional regulators.

Specifically, miRNAs are noncoding single-stranded short RNAs (18–22 bp) involved in the posttranscriptional regulation of gene expression. Due to their important role in physiopathological processes and their direct or indirect role in gene regulation of tumor cells, miRNAs can be considered potential biomarkers and therapeutic targets for cancer treatment. In this sense, it has been extensively reported that miRNAs are stable in body fluids such as serum and stool, and thus provide another source of biomarkers for CRC.

High-throughput miRNA profiling has been widely performed to reveal CRC-related miRNA fingerprints. Several miRNAs have been proposed as biomarkers for screening and early detection, prognosis and prediction, both in blood (serum/plasma) and stool samples. miR-92a and miR-29a were identified as potential plasma biomarkers for CRC and advanced adenomas (reviewed in Ref. 22). In serum samples, miR-194 and miR-29b have been proposed as potential biomarkers for early diagnosis.

Stool-based miRNAs such as miR-21, miR-92a, and the cluster miR-17-92 and miR-223 have been suggested for prognosis (reviewed in Refs. 11, 22–24). In relation to saliva samples, Sazanov et al.[25] also analyzed miR-21 in saliva.

Differentially expressed miRNAs have strong correlation with treatment response in CRC. For example, increased levels of miR-143 in serum are associated with improved progression-free survival in 5-FU-based therapy in metastatic CRC patients. The combination of the upregulation of miR-320e and downregulation of miR-148 and miR-150 is a predictive indicator of poor response to FOLFOX chemotherapy (reviewed in Refs. 11, 22–24).

A comprehensive review of miRNA profiles in CRC patients indicates that 164 miRNAs are differentially expressed in CRC patients, of which two-thirds are upregulated. For example, it was observed that miR-21 acts as an oncogene by regulating the expression of ITGb4, PDCD4, PTEN, SPRY2, RECK and other downstream target genes. Other examples of upregulated miRNAs that act as oncogenes are miR-182, miR-301a, miR-96, and miR-50. They may promote proliferation and differentiation, leading to the occurrence, development, and multiple therapeutic resistance of CRC.

Finally, miRNAs can be used as potential biomarkers for CRC diagnosis and disease monitoring using qPCR. As an example, a study successfully developed a panel of seven miRNAs (miR-103a-3p, miR-127-3p, miR-151a-5p, miR-17-5p, miR-181a-5p, miR-18a-5p, and miR-18b-5p) for CRC diagnosis.[26] Additionally, some companies offer miRNA-based screening tests using noninvasive samples to identify individuals at initial CRC stages.

Epigenomics approaches for colorectal cancer biomarkers discovery and validation

The term "epigenetics" is defined as modifications in the genome and its structure that regulate gene expression without altering the nucleotide sequence, which is reversible and can be inherited. The epigenetic regulation involves four mechanisms: DNA methylation of cytosines, histone posttranslational modifications, nucleosome positioning, and posttranscriptional regulation through noncoding RNAs, such as micro RNAs (miRNAs, detailed in "Transcriptomics" section). Aberrations in these regulatory processes can lead to impaired gene function and thus promote the transformation of normal cells into malignant cells.[24, 27] This section will focus on DNA methylation since this epigenetic mark is widely studied in CRC. This modification consists in the enzymatic addition of a methyl group to the 5′ of cytosine residues in CG dinucleotides, known as CpG sites.

The aberrant methylation occurring during colorectal neoplasia is characterized by promoter hypermethylation and transcriptional silencing of tumor suppressor or DNA repair genes, coexisting with a global loss of methylation. Global hypomethylation mainly occurs in repetitive transposable DNA sequences such as LINE-1 and SINE, that leads to chromosomal and microsatellite instability and oncogene activation. On the other hand, hypermethylation mainly occurs in CG-rich regions (CpG islands), which are found in promoters of 40%–60% human genes. Cancer-specific DNA methylation is a particularly interesting source of biomarkers.[22, 28, 29]

The main approach that can differentiate methylated from unmethylated cytosines in DNA is the chemical modification with sodium bisulfite that converts unmethylated cytosines to uracils, while methylated cytosines remain intact. Methods based on sodium bisulfite treatment are considered to be the most robust and provide simple and straightforward solutions both in research and clinical scenarios.[30]

DNA obtained from almost any body fluid or tissue can be used for DNA methylation analysis, given the stability of this epigenetic mark during sample processing and DNA extraction. Regarding DNA methylation analysis techniques, the methods can be divided into three main categories depending on the purpose of the study: (A) assays used for the discovery of novel biomarkers, in which the CpG sites available for analysis are not predefined; (B) targeted approaches focused on the analysis of DNA methylation within particular known and defined CpG sites or regions of interest; and (C) methods for the assessment of genome global methylation (Fig. 22.1).

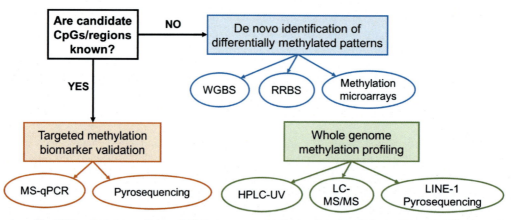

FIG. 22.1 Overview of the DNA methylation analysis methods that are covered in this section, classified according to the objectives of a specific project. *(Adapted from Kurdyukov S, Bullock M. DNA methylation analysis: choosing the right method. Biology 2016; 5 (1): 3. https://doi.org/10.3390/biology5010003.)*

De novo identification of differentially methylated patterns

Several methodologies are suitable for high-throughput methylation biomarker discovery, including whole and reduced bisulfite sequencing and microarray-based technology.[30, 31] Bisulfite sequencing is considered the gold standard for DNA methylation analysis. The combination of bisulfite treatment with NGS technologies is known as *whole genome bisulfite sequencing (WGBS)*. Total DNA is modified with sodium bisulfite, sequenced and compared to the reference genome enabling a precise quantification of the methylation at single-nucleotide resolution in a genome-wide scenario. Thus WGBS is the most comprehensive and widely used approach to study DNA methylation and to detect aberrant methylation patterns.[28, 30]

A variant of WGBS that allows to reduce costs is *reduced representation bisulfite sequencing (RRBS)*, where only a fraction of the genome, enriched in CpG sites, is sequenced. RRBS ensures isolation of approximately 85% of CpG islands in the human genome, achieving a high coverage of potentially differentially methylated regions while greatly reducing sequencing read requirement. Therefore this approach provides a time- and cost-effective identification of methylation status in cancer.[30, 31]

The search of differentially methylated patterns across the genome can also be achieved with *methylation-specific microarrays*, such as Illumina Infinium Methylation technology. After bisulfite conversion, DNA is amplified on whole genome level and hybridized to a chip consisting of CpG locus-specific oligonucleotides immobilized to bar-coded beads. The Infinium MethylationEPIC BeadChip array (EPIC) is the most comprehensive methylation microarray platform available to date and interrogates more than 850,000 CpG sites distributed across the genome.[32, 33] Infinium technology allows a precise quantification of the methylation level of interrogated cytosines and generates data that are more quickly and easily analyzable than sequencing data. This translates into a powerful cost-effective tool, notably attractive for high-throughput biomarker discovery studies.

Analysis of targeted methylation biomarkers

For the assessment of methylation in specific biomarkers, bisulfite conversion is still the first step in many downstream methods. *Methylation-specific quantitative PCR (MS-qPCR)* is a classical method that uses bisulfite-treated DNA. Two pairs of primers are designed, to target both the methylated and unmethylated sequences, and two parallel qPCR reactions are carried out for each sample using fluorescent probes or dyes and, relative methylation is calculated.[30, 31]

In *pyrosequencing*, following bisulfite conversion, PCR products for the region of interest are obtained and short-read sequencing reactions can be performed. With this technique, the amounts of C and T at individual sites are measured based on the amounts of pyrophosphate (PPi) released when nucleotides are incorporated by the polymerase. Such PPi is accurately quantified bioluminometrically. The main advantage of pyrosequencing is the accurate quantification of individual CpG methylation levels. Despite being an easy procedure, equipment specifically designed for this method is needed.[30, 31]

Whole genome methylation profiling

Global DNA methylation has an impact in CRC pathogenesis and progression. Indeed, global hypomethylation associates with poor prognosis in CRC patients, which justifies the interest in profiling the levels of whole genome methylation. *High performance liquid chromatography-ultraviolet (HPLC-UV)* is considered the current gold standard assay for the quantification of global methylation in hydrolyzed DNA samples. The main limitation of this method is the requirement of specialized laboratory equipment and large DNA input (3–10 μg, hard to achieve depending on the sample type). *Liquid chromatography coupled with tandem mass spectrometry (LC-MS/MS)* is an alternative with higher sensitivity than HPLC-UV, though the need for specialized equipment and expertise limits its use.

The methylation levels of LINE-1 retrotransposons, which compose ~17% of the human genome, are considered a surrogate measure for global DNA methylation, as it has been reported they reflect global DNA methylation changes. Methylation levels of LINE-1 can be assessed by bisulfite conversion followed by PCR amplification and pyrosequencing.

Proposed DNA methylation biomarkers for CRC

In the last years it has been shown that cfDNA present in body fluids (serum, plasma, feces) reflects the aberrant methylation patterns present in tumor cells, thus representing a source of potential noninvasive biomarkers for the detection of CRC.[34, 35] Several studies have evaluated the methylation status of several candidates such as SEPT9, APC, HLTF, NEUROG1, or ALX4 in plasma and serum samples.[28, 36] Recently it has been reported that the analysis of DNA methylation in serum cfDNA pooled samples represents an affordable and effective strategy for the discovery of noninvasive methylation biomarkers for CRC.[37]

Nowadays, the only epigenetic biomarker approved by the FDA for the detection of CRC in plasma is Epi proColon 2.0CE, which is based on the quantification of methylation in the promoter of SEPT9.[38] The results on the diagnostic performance of this biomarker are variable and inconsistent. In an asymptomatic average-risk population, SEPT9 showed lower sensitivity and specificity than FIT (Sens: 68% vs. 79%; Esp: 80% vs. 94%, respectively).[39, 40] Cologuard is a stool-based epigenetic test, also approved by the FDA. This is a noninvasive test that combines FIT with the detection of mutations in KRAS and aberrant methylation of NDGR4 and BMP3.[41] The test is indicated for the screening of CRC in asymptomatic average-risk individuals and has a 92.3% sensitivity and 86.6% specificity.

Regarding prognostic noninvasive epigenetic biomarkers, the BCAT1/IKZF1 test shows more sensitivity for prediction of both local and distant recurrence (75% and 66.7%, respectively) compared to CEA (50% and 29.2%).[42] Other methylation biomarkers associated with a worse prognosis are hypomethylation of LINE-1 or MGMT, hypermethylation of CDKN2A, IGFBP3, HTLF, or HPP1, and the methylation status of CIMP (CpG Island Methylator Phenotype), classically evaluated as a panel including CDKN2A, MINT1, MINT2, MINT31, and MLH1 (reviewed in Refs. 22, 24).

Proteomics approaches for colorectal cancer biomarkers discovery and validation

The analysis of protein expression changes between case and control samples is known as differential expression proteomics and is fundamental in the discovery of potential biomarkers for CRC. Currently, the proteomic strategies used can be divided into two main areas: discovery proteomics, which aims to find proteins whose expression is altered in the tumor; and targeted proteomics, which confirms or validates the alteration of these proteins.

Identification of proteomic biomarkers

During many years *two-dimensional electrophoresis (2-DE)* has been one of the most used techniques in studies of protein expression patterns in cancer. 2-DE enables both a separation prior to protein characterization and/or identification by mass spectrometry (MS). Despite being a laborious technique that requires a relatively high amount of sample, it has a good resolution and allows the separation of thousands of proteins in a single analysis. A representative scheme of the main proteomic techniques for the discovery and validation of CRC biomarkers is presented in Fig. 22.2.

Among the many 2-DE studies carried out in recent years to search for new markers for CRC, we can highlight that of Chen et al.[43] These researchers found overexpression of alpha-enolase proteins, HSP27 (Heat Shock Protein 27) and macrophage migration inhibitor factor (MIF) in tumor tissue of patients with low preoperative serum CEA levels. In another

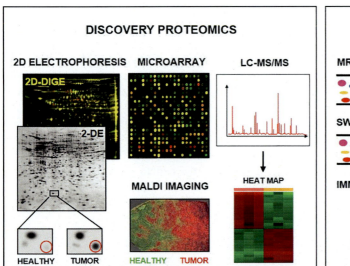
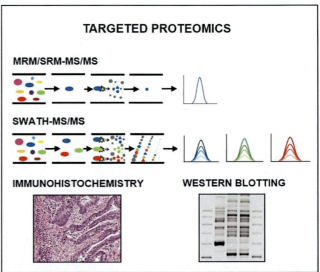

FIG. 22.2 Schematic representation of the principal techniques in proteomics for CRC biomarker discovery and validation. *Adapted from Álvarez-Chaver P, Otero-Estévez O, Páez de la Cadena M, Rodríguez-Berrocal FJ, Martínez-Zorzano VS. Proteomics for discovery of candidate colorectal cancer biomarkers.* World J Gastroenterol *2014;20(14):3804–24. https://doi.org/10.3748/wjg.v20.i14.3804.*

study, the carbonic anhydrase II enzyme (CA II) was identified as a potential biomarker for CRC diagnosis, exhibiting decreased tumor expression that was corroborated by RT-PCR, western blotting, and immunohistochemistry.[44] Various studies have used serum samples from patients with and without metastases, among which we can highlight that of Zhao et al.,[45] who proposed transthyretin protein (TTR) as a specific metastasis marker for CRC. Other tumor markers that have been extensively studied in both serum and tissue using 2-DE technology are clusterin protein (CLU)[46] and nucleoside diphosphate kinase A.[47]

The *two-dimensional differential gel electrophoresis (DIGE, Differential In Gel Electrophoresis)* is an improvement of the 2-DE. It is based on the previous labeling of samples with fluorescent dyes, minimizing the gel-to-gel variability. Through this technique, Strohkamp et al.[48] proposed clusterin and glutathione synthetase as potential biomarkers for CRC diagnosis. Using the Caco-2 cell line, which consists of an in vitro model for the study of colorectal carcinogenesis, potential biomarker candidates were found, including TCTP or TGFβIp.[49] 2D-DIGE has great sensitivity, it decreases the analysis time and allows precise quantification. However, so far it does not allow comparisons on the same gel of more than three different samples.

The use of *protein microarrays* allows the simultaneous analysis of multiple proteins in a single experiment, as well as their identification, quantification, study of interactions with other proteins, and their function. Through this technology, greater expression of the nucleoside diphosphate kinase A (NDKA) and prohibitin proteins was found in tumors compared to adjacent healthy tissue.[50,51] In another study, several candidate serum biomarkers (Apo A1 and C9) were selected by liquid chromatography coupled to MS, and subsequently validated using protein microarrays.[52] This methodology is widely applied to cancer research, since very small sample quantity is used and allows many samples to be analyzed and quantified in a single experiment. Also, it has a lot of interest in the clinical field since it is a fast and easily automatable technique. In addition, there is currently SPR (Surface Plasmon Resonance) equipment that allows the analysis of protein microarrays to study protein interactions and, subsequently, identify the ligands of interest by MS.

A variant of the protein matrix used in the search for new biomarkers is the *SELDI-TOF-MS (Surface-Enhanced Laser Desorption/Ionization Time-Of-Flight Mass Spectrometry)* technique. Although it can be applied to tissue samples, it has been more used in serum and plasma studies. It combines two very powerful techniques, liquid chromatography (LC) and MS, and consists of a solid aluminum or steel support coated with a specific chromatographic surface (reverse phase, anion exchange, etc.). For a review of this type of studies we recommend the work of Gemoll et al.[53]

Most of the recent studies in differential expression proteomics have used *liquid nanochromatography coupled to tandem mass spectrometry (nLC-MS/MS)*. This methodology allows a previous separation of the peptides from the sample in a reverse phase chromatographic column. As the LC equipment is connected online to the mass spectrometer, the fractions from the chromatography enter the mass analyzer, which allows the peptides present in the sample to be fragmented one by one. This analysis can be performed with tryptic digested bands of a 1-DE gel, spots of a 2-DE gel, or with a complex sample of proteins not previously separated. This last type of analysis is known as shotgun proteomics.

In the search for new CRC biomarkers, the aim is to find proteins that are easily detectable and with large expression changes. Therefore proteomic techniques based on nLC-MS/MS without prior labeling of the samples (label-free proteomics) are currently used, although they involve much longer analysis time than isobaric tag methods (iTRAQ, TMT) and not all peptides are detected equally due to ionic competition, dynamic range, and sensitivity of the equipment. Applying this methodology, a change in the expression of NG1, OLFM4, and Sec24C proteins in early staged CRC tissue was found compared to preneoplastic and healthy tissue, which was corroborated by immunohistochemistry.[54]

Targeted validation of protein biomarkers

Once the candidate biomarkers have been identified through discovery strategies, especially shotgun proteomics, the next step is to validate the alteration in their expression. Classically, antibody-based molecular biology techniques are used, such as *immunohistochemistry* or *Western blotting*. However, there are currently novel strategies that allow quantification and validation of hundreds of candidate biomarkers in a single MS analysis with great sensitivity and specificity, and without the need for antibodies. Among them are the techniques called *MRM (Multiple Reaction Monitoring)*, *SRM (Selected Reaction Monitoring)*, and *SWATH (Sequential Window Acquisition of all THeoretical spectra)*. These are methods used in tandem mass spectrometry (MS/MS) that consist in selecting specific (proteotypic) peptides from a list of certain precursor ions, then fragmentation and selection of one of their product ions. This is detected and used to make a relative or absolute quantification of the peptide and, by extension, of the protein to which it belongs in the analyzed sample.[55] However, these techniques require highly qualified personnel and sophisticated equipment. In the recent work of Atak et al.,[56] the iTRAQ labeling and MRM were used to validate a panel of 9 CRC diagnostic biomarkers. HDGF (hepatoma-derived growth factor), LDHA (L-lactate dehydrogenase A chain), PKM (pyruvate kinase), S100A8,

S100A9, and S100A11 decreased in tumor tissue expression, while EHD2 (EH domain-containing protein 2), LUM (Lumican), and AOC3 (membrane primary amine oxidase) were significantly overexpressed.

Despite the enormous progress made in recent years, still one of the great limitations of proteomics in the search for biomarkers is the lack of information in databases. Many of the proteins that could be useful as markers are coded by mutated genes in the tumor, so their sequence will also be truncated. However, if those mutated peptide sequences are not deposited in the databases, such aberrant proteins cannot be identified or quantified by MS. On the other hand, as not all peptides are ionized in the same way, in many cases good sequence coverage is not obtained and the MS equipment is not able to detect variants of the same protein (isoforms). In addition, if the protein has posttranslational modifications (PTMs) there are also important limitations.

Proposed protein biomarkers for CRC

The detection of protein biomarkers in liquid biopsy presents some difficulties. One of them is that we found a heterogeneous mixture of proteins from different tissues, which makes it difficult to establish a direct relationship between the alteration of a protein and the disease. This may result in the identification of nonspecific candidate protein biomarkers. On the other hand, majority proteins mask minority proteins that might be of interest as biomarkers. Despite these limitations, numerous proteomic studies have been conducted with the aim of finding serum markers that can distinguish CRC patients from healthy individuals.

Different authors have identified the same proteins using different techniques. Among them we can mention apolipoproteins, alpha-1 antitrypsin, beta-2 microglobulin, cathepsins, chaperones of the HSP family, clusterin (CLU), gelsolin C3 and C9 from the complement system, transferrin, transthyretin, defensins, prolactin, isoenzyme M2 of pyruvate kinase, metalloproteases 7 and 9, CCSA 2, 3, and 4 (reviewed in Refs. 57–59).

Analysis of the protein profile of exosomes may represent a useful method for identifying diagnostic markers for CRC. Studying exosomes from CRC patients, the high expression of ECM1 has been related to the presence of liver metastasis.[60]

Among the candidate protein biomarkers discovered in lysates or secretomes of colon tumor cell lines, subsequently validated in serum or tissue of patients, we can mention CRMP-2 protein (protein 2 mediating the response to collapsin), growth factor/differentiation GDF15, and "trefoil factor 3" (TFF3), which is a peptide associated with the mucins of the gastrointestinal tract.[57] De Wit et al., analyzing the secretome of healthy and tumor tissue of patients with CRC by SDS-PAGE followed by LC-MS/MS, found 76 candidate biomarkers, of which 21 are useful for early tumor detection.[61]

Regarding the use of stool samples, it has been shown that the panel formed by the proteins S100A12, TIMP1, hemoglobin, haptoglobin, calprotectin (S100A8/A9), and CEA presents greater sensitivity than the fecal occult blood test, identifying 74% of early staged patients (reviewed in Ref. 62). It should be mentioned that many of these proteins identified in CRC patients are related to inflammation and are nonspecific.

A good approximation to detect less abundant proteins is the isolation of subcellular fractions, such as membrane proteins, and the subproteome characterization. Most of the biomarkers currently used in the clinic, such as CEA, and approximately 70% of all known therapeutic targets, are membrane proteins. Subproteomes of the membrane fraction and soluble fraction of colorectal tumors and adjacent healthy mucosa have been analyzed. Among the altered membrane proteins, we find cytoskeleton proteins, chaperones, and two isoforms of the calcium binding protein S100A6[63]; while in the soluble fraction, proteins 14-3-3-zeta/delta, DJ-1, RBBP-4, and NDKA were identified.[47, 64, 65]

Other omics technologies for the discovery of biomarkers

The study of the *metabolite profile (metabolomics)* in a biological system is currently an omic technology in expansion, with special interest in cancer research. Through the use of advanced analytical techniques and bioinformatics tools, the metabolome provides key information that has been included in several reviews, such as that of Zhang et al.,[66] which assessed the potential of studying these small molecules for the discovery of new CRC biomarkers. Some examples are lactate, a product of anaerobic glycolysis that was found in greater amounts in both tissue and serum of patients with CRC; fumarate, intermediate product of the Krebs cycle, whose quantity is reduced in colorectal tumors; and glucose, with decreased levels in serum, feces, and colorectal tumors.

The imaging technique *MALDI-MSI (matrix-assisted laser desorption/ionization mass spectrometry imaging)* allows the analysis of histological sections by mass spectrometry, with the advantage that it retains the spatial integrity of the molecules in the tissue, providing information that is lost in homogenization. Tumor biology not only depends on tumor cells but also on their interaction with stromal cells, blood vessels, and the immune system. Therefore the MALDI-MSI technique is not only useful for the discovery of CRC biomarkers but also for tumor characterization. Histological sections are deposited on a glass surface similar to a slide that has conductivity, so it is usually coated with titanium oxide and indium. Next, the MALDI matrix is applied to the tissue. The matrix is a solution of an organic molecule that cocrystallizes with the analytes present in the tissue, so that when the laser strikes the sample it allows them to be extracted and ionized to reach the mass spectrometer detector. The result of the MS analysis is an image of the histological section with a color code (2D and 3D density map) that allows differentiating the areas of the tissue based on the detected analytes. This technique is especially useful for the analysis of metabolites, peptides, and low molecular weight proteins. However, nowadays it has several limitations, such as the fact that the matrix has multiple signals in the range of low molecular weight masses that interfere with analyte signals. Therefore one of the current challenges is to find better matrixes for the MALDI-MSI analysis, which in the future will probably be very useful in the clinical setting.[67]

Currently, the most important clinical application of MALDI-MSI is the histological characterization of tumors, since it allows differentiating tissue adjacent to the tumor with changes at the level of proteins, peptides, lipids, carbohydrates, and other small molecules, which are not differentiated by classical histology.[68] MALDI-MSI is also used to determine how certain drugs enter and distribute in tumor cells,[69] in addition to its use in the discovery of biomarkers.[70,71] This method offers a great advantage for the identification of prognostic markers since it allows retrospective studies to be carried out using cohort samples from patients whose clinical follow-up data are already available.

Phases for the development of clinical biomarkers

The field of biomarker discovery has considerably expanded due to the evolution and development of new omics technologies detailed in this chapter. This progress has allowed the transition from targeted studies in which candidate genes or proteins are analyzed, to massive studies in which hundreds or thousands of nucleotide sequences, proteins, or metabolites are interrogated. Although association studies such as GWAS or EWAS do not allow inferring causal relationship, the differential patterns associated with the study phenotype, once validated, result in useful biomarkers.

However, there is still a period of time between the initial discovery of biomarkers and their clinical implementation. The diagnostic, prognostic, or predictive capacity of candidate biomarkers is not always reproduced in subsequent studies due to failures in experimental design, technical variability between platforms and assays, or sample availability. These factors result in success rates of 0.1% in the translation of new cancer biomarkers into clinical practice (reviewed in Refs. 28, 36). A complete experimental development prior to the beginning of the study is crucial, as well as the exhaustive definition of the characteristics of the specimens (type of samples, patients, and cohort size) involved in the development of new biomarkers. Therefore the development of biomarkers is challenging.

The general biomarker development process can be divided into four phases (Fig. 22.3). Each phase pursues specific objectives and the accomplishment from one to another must add evidence in favor of the candidate biomarkers.

Biomarker discovery: Exploratory analysis

The initial phase of biomarker development generally consists in conducting an exploratory preclinical study on a cohort of cases and controls, with the objective of obtaining a prioritized list of candidate biomarkers. One of the most relevant aspects to consider in the early stages is the choice of the study samples, which should be in the final biomarker application format (plasma, serum, feces, etc.), as well as the definition of the target population. In an optimal context, samples should be collected prospectively, based on clinical inclusion and exclusion criteria previously defined and specified in the study protocol. The posterior choice of "samples of convenience" should be avoided, since this practice may introduce unknown confounding factors that contribute to the appearance of false positives and false negatives. The robustness and validity of the methods and algorithms used in the bioinformatic processing and analysis of data is another factor that influences the reproducibility of biomarker studies, as well as their validation and transfer to clinical practice.

In order to prioritize candidate biomarkers, this discovery phase must include a validation in the same study context using a set of samples independent of the ones used for the discovery. As this is generally not possible (sample availability, increased costs), it is often replaced by cross validation in the discovery cohort.

Development of the validation assay: Technical/analytical validation

The objective of this phase is to adapt the candidate biomarkers to platforms commonly used in clinical practice. Typically, the same samples used in the previous exploratory analysis are analyzed using the clinical platform of choice. This determines the reproducibility of the measurements between the omic platform used in the initial discovery and the assay chosen for validation. In addition, this technical validation allows evaluating the analytical validity of candidate biomarkers and the clinical test in the desired target sample format (such as serum, plasma, or stool).

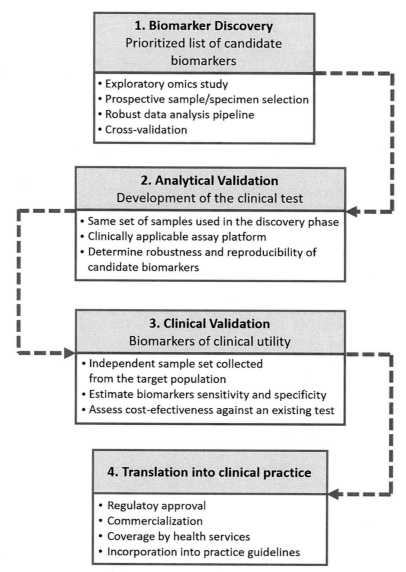

FIG. 22.3 Summary of the phases for the process of discovery and development of new biomarkers. *(Adapted from Goossens N, Nakagawa S, Sun X, Hoshida Y. Cancer biomarker discovery and validation. Transl Cancer Res 2015; 4 (3): 256–69. https://doi.org/10.3978/j.issn.2218-676X.2015.06.04.)*

Validation of the clinical utility

After the analytical validation, the diagnostic or predictive performance of the candidate biomarkers should be evaluated and confirmed in a patient cohort independent of that used for the discovery phase. This phase consists of a prospective study and the validation cohort must have the characteristics of the target population. It must have adequate statistical power to determine the sensitivity and specificity of biomarkers and to define the criteria for the interpretation of results.

A very relevant aspect, especially in the case of screening biomarkers, is the choice of controls as this will affect the estimation of specificity. To avoid the selection of biomarkers that will later be discarded due to lack of specificity, the control group must have the same sex and age as the patient group. In addition, controls must be colonoscopically confirmed, instead of self-declared healthy.

The validation phase of the clinical utility of the candidate biomarkers must also include an analysis of the benefits of the new test compared to the standard or existing procedure, together with a cost-effectiveness analysis of the introduction of the biomarker in clinical practice.

Clinical translation

The process of translation of a new biomarker analytically and clinically validated includes aspects that vary widely at the regional level: intellectual property, approval for clinical use by competent organisms or agencies (FDA, EMA, AEMPS), commercialization, coverage by health services, and incorporation into clinical practice guidelines.

Once a new biomarker is translated to the clinical practice and after a period of actual use, prospective studies are carried out. Thus it is possible to evaluate aspects such as the feasibility of the test, real cost-effectiveness, and acceptance by the patients. Other factors that should also be evaluated are the impact on the reduction of incidence or mortality, on the choice of treatment, and on patients' quality of life.

References

1. Hardingham JE, Grover P, Winter M, Hewett PJ, Price TJ, Thierry B. Detection and clinical significance of circulating tumor cells in colorectal cancer—20 years of progress. *Mol Med* 2015;**21**(S1):S25–31. https://doi.org/10.2119/molmed.2015.00149.
2. Normanno N, Cervantes A, Ciardiello F, De Luca A, Pinto C. The liquid biopsy in the management of colorectal cancer patients: current applications and future scenarios. *Cancer Treat Rev* 2018;**70**:1–8. https://doi.org/10.1016/j.ctrv.2018.07.007.
3. Ribeiro-Samy S, Oliveira MI, Pereira-Veiga T, et al. Fast and efficient microfluidic cell filter for isolation of circulating tumor cells from unprocessed whole blood of colorectal cancer patients. *Sci Rep* 2019;**9**(1):8032. https://doi.org/10.1038/s41598-019-44401-1.
4. Siveen KS, Raza A, Ahmed EI, et al. The role of extracellular vesicles as modulators of the tumor microenvironment, metastasis and drug resistance in colorectal cancer. *Cancers* 2019;**11**(6):746. https://doi.org/10.3390/cancers11060746.
5. Guo C, Xie C, Chen Q, et al. A novel malic acid-enhanced method for the analysis of 5-methyl-2′-deoxycytidine, 5-hydroxymethyl-2′-deoxycytidine, 5-methylcytidine and 5-hydroxymethylcytidine in human urine using hydrophilic interaction liquid chromatography-tandem mass spectrometry. *Anal Chim Acta* 2018;**1034**:110–8. https://doi.org/10.1016/j.aca.2018.06.081.
6. Conte B, Kopetz S. Challenges and strategies for identifying biomarkers for colorectal cancer. *Colorectal Cancer* 2013;**2**(6):487–9. https://doi.org/10.2217/crc.13.65.
7. Toma SC, Ungureanu BS, Patrascu S, Surlin V, Georgescu I. Colorectal cancer biomarkers—a new trend in early diagnosis. *Curr Health Sci J* 2018;**44**(2):140–6. https://doi.org/10.12865/CHSJ.44.02.08.
8. Liebs S, Keilholz U, Kehler I, Schweiger C, Haybäck J, Nonnenmacher A. Detection of mutations in circulating cell-free DNA in relation to disease stage in colorectal cancer. *Cancer Med* 2019;**8**(8):3761–9. https://doi.org/10.1002/cam4.2219.
9. Cancer Genome Atlas Network. Comprehensive molecular characterization of human colon and rectal cancer. *Nature* 2012;**487**(7407):330–7. https://doi.org/10.1038/nature11252.
10. Fearon ER. Molecular genetics of colorectal cancer. *Annu Rev Pathol: Mech Dis* 2011;**6**(1):479–507. https://doi.org/10.1146/annurev-pathol-011110-130235.
11. Goel G. Molecular characterization and biomarker identification in colorectal cancer: toward realization of the precision medicine dream. *Cancer Manag Res* 2018;**10**:5895–908. https://doi.org/10.2147/cmar.s162967.
12. Phallen J, Sausen M, Adleff V, et al. Direct detection of early-stage cancers using circulating tumor DNA. *Sci Transl Med* 2017;**9**(403). https://doi.org/10.1126/scitranslmed.aan2415.
13. Insua YV, de la Cámara J, Vázquez EB, et al. Predicting outcome and therapy response in mCRC patients using an indirect method for CTCs detection by a multigene expression panel: a multicentric prospective validation study. *Int J Mol Sci* 2017;**18**(6):1265. https://doi.org/10.3390/ijms18061265.
14. Lu Y, Kweon S-S, Tanikawa C, et al. Large-scale genome-wide association study of east Asians identifies loci associated with risk for colorectal cancer. *Gastroenterology* 2019;**156**(5):1455–66. https://doi.org/10.1053/j.gastro.2018.11.066.
15. Pacheco MP, Bintener T, Ternes D, et al. Identifying and targeting cancer-specific metabolism with network-based drug target prediction. *EBioMedicine* 2019;**43**:98–106. https://doi.org/10.1016/j.ebiom.2019.04.046.
16. Zuo S, Dai G, Ren X. Identification of a 6-gene signature predicting prognosis for colorectal cancer. *Cancer Cell Int* 2019;**19**(1). https://doi.org/10.1186/s12935-018-0724-7.
17. Ng TH, Sham KWY, Xie CM, et al. Eukaryotic elongation factor-2 kinase expression is an independent prognostic factor in colorectal cancer. *BMC Cancer* 2019;**19**(1):649. https://doi.org/10.1186/s12885-019-5873-0.
18. Ho Y-J, Liu F-C, Chang J, et al. High expression of meningioma 1 is correlated with reduced survival rates in colorectal cancer patients. *Acta Histochem* 2019;**121**(5):628–37. https://doi.org/10.1016/j.acthis.2019.05.006.
19. Yang L, Chen Z, Xiong W, et al. High expression of SLC17A9 correlates with poor prognosis in colorectal cancer. *Hum Pathol* 2019;**84**:62–70. https://doi.org/10.1016/j.humpath.2018.09.002.
20. Lee J-H, Jung S, Park WS, et al. Prognostic nomogram of hypoxia-related genes predicting overall survival of colorectal cancer—analysis of TCGA database. *Sci Rep* 2019;**9**(1). https://doi.org/10.1038/s41598-018-38116-y.
21. Jiang H, Li B, Wang F, Ma C, Hao T. Expression of ERCC1 and TYMS in colorectal cancer patients and the predictive value of chemotherapy efficacy. *Oncol Lett* 2019. https://doi.org/10.3892/ol.2019.10395.
22. Zamani M, Hosseini SV, Mokarram P. Epigenetic biomarkers in colorectal cancer: premises and prospects. *Biomarkers* 2018;**23**(2):105–14. https://doi.org/10.1080/1354750x.2016.1252961.

23. Wang X, Kuang Y-Y, Hu X-T. Advances in epigenetic biomarker research in colorectal cancer. *World J Gastroenterol* 2014;**20**(15):4276–87. https://doi.org/10.3748/wjg.v20.i15.4276.
24. Singh MP, Rai S, Suyal S, et al. Genetic and epigenetic markers in colorectal cancer screening: recent advances. *Expert Rev Mol Diagn* 2017;**17**(7):665–85. https://doi.org/10.1080/14737159.2017.1337511.
25. Sazanov AA, Kiselyova EV, Zakharenko AA, Romanov MN, Zaraysky MI. Plasma and saliva miR-21 expression in colorectal cancer patients. *J Appl Genet* 2017;**58**(2):231–7. https://doi.org/10.1007/s13353-016-0379-9.
26. Zhang H, Zhu M, Shan X, et al. A panel of seven-miRNA signature in plasma as potential biomarker for colorectal cancer diagnosis. *Gene* 2019;**687**:246–54. https://doi.org/10.1016/j.gene.2018.11.055.
27. Esteller M. Cancer epigenomics: DNA methylomes and histone-modification maps. *Nat Rev Genet* 2007;**8**(4):286–98. https://doi.org/10.1038/nrg2005.
28. Worm Ørntoft M-B. Review of blood-based colorectal cancer screening: how far are circulating cell-free DNA methylation markers from clinical implementation? *Clin Colorectal Cancer* 2018;**17**(2):e415–33. https://doi.org/10.1016/j.clcc.2018.02.012.
29. Berdasco M, Esteller M. Clinical epigenetics: seizing opportunities for translation. *Nat Rev Genet* 2019;**20**(2):109–27. https://doi.org/10.1038/s41576-018-0074-2.
30. Kurdyukov S, Bullock M. DNA methylation analysis: choosing the right method. *Biology* 2016;**5**(1):3. https://doi.org/10.3390/biology5010003.
31. Soozangar N, Sadeghi MR, Jeddi F, Somi MH, Shirmohamadi M, Samadi N. Comparison of genome-wide analysis techniques to DNA methylation analysis in human cancer. *J Cell Physiol* 2018;**233**(5):3968–81. https://doi.org/10.1002/jcp.26176.
32. Moran S, Arribas C, Esteller M. Validation of a DNA methylation microarray for 850,000 CpG sites of the human genome enriched in enhancer sequences. *Epigenomics* 2016;**8**(3):389–99. https://doi.org/10.2217/epi.15.114.
33. Pidsley R, Zotenko E, Peters TJ, et al. Critical evaluation of the Illumina MethylationEPIC BeadChip microarray for whole-genome DNA methylation profiling. *Genome Biol* 2016;**17**(1). https://doi.org/10.1186/s13059-016-1066-1.
34. Galanopoulos M, Tsoukalas N, Papanikolaou IS, Tolia M, Gazouli M, Mantzaris GJ. Abnormal DNA methylation as a cell-free circulating DNA biomarker for colorectal cancer detection: a review of literature. *World J Gastroint Oncol* 2017;**9**(4):142–52. https://doi.org/10.4251/wjgo.v9.i4.142.
35. Krishnamurthy N, Spencer E, Torkamani A, Nicholson L. Liquid biopsies for cancer: coming to a patient near you. *J Clin Med* 2017;**6**(1):3. https://doi.org/10.3390/jcm6010003.
36. Goossens N, Nakagawa S, Sun X, Hoshida Y. Cancer biomarker discovery and validation. *Transl Cancer Res* 2015;**4**(3):256–69. https://doi.org/10.3978/j.issn.2218-676X.2015.06.04.
37. Gallardo-Gómez M, Moran S, Páez de la Cadena M, et al. A new approach to epigenome-wide discovery of non-invasive methylation biomarkers for colorectal cancer screening in circulating cell-free DNA using pooled samples. *Clin Epigenetics* 2018;**10**:53. https://doi.org/10.1186/s13148-018-0487-y.
38. Lamb YN, Dhillon S. Epi proColon® 2.0 CE: a blood-based screening test for colorectal cancer. *Mol Diagn Ther* 2017;**21**(2):225–32. https://doi.org/10.1007/s40291-017-0259-y.
39. Song L, Li Y. Progress on the clinical application of the SEPT9 gene methylation assay in the past 5 years. *Biomark Med* 2017;**11**(6):415–8. https://doi.org/10.2217/bmm-2017-0091.
40. Wang Y, Chen P-M, Liu R-B. Advance in plasma SEPT9 gene methylation assay for colorectal cancer early detection. *World J Gastroint Oncol* 2018;**10**(1):15–22. https://doi.org/10.4251/wjgo.v10.i1.15.
41. Imperiale TF, Ransohoff DF, Itzkowitz SH. Multitarget stool DNA testing for colorectal-cancer screening. *N Engl J Med* 2014;**371**(2):187–8. https://doi.org/10.1056/NEJMc1405215.
42. Young GP, Pedersen SK, Mansfield S, et al. A cross-sectional study comparing a blood test for methylated BCAT1 and IKZF1 tumor-derived DNA with CEA for detection of recurrent colorectal cancer. *Cancer Med* 2016;**5**(10):2763–72. https://doi.org/10.1002/cam4.868.
43. Chen WT-L, Chang S-C, Ke T-W, Chiang H-C, Tsai F-J, Lo W-Y. Identification of biomarkers to improve diagnostic sensitivity of sporadic colorectal cancer in patients with low preoperative serum carcinoembryonic antigen by clinical proteomic analysis. *Clin Chim Acta* 2011;**412**(7):636–41. https://doi.org/10.1016/j.cca.2010.12.024.
44. Zhou R, Huang W, Yao Y, et al. CA II, a potential biomarker by proteomic analysis, exerts significant inhibitory effect on the growth of colorectal cancer cells. *Int J Oncol* 2013;**43**(2):611–21. https://doi.org/10.3892/ijo.2013.1972.
45. Zhao L, Liu Y, Sun X, Peng K, Ding Y. Serum proteome analysis for profiling protein markers associated with lymph node metastasis in colorectal carcinoma. *J Comp Pathol* 2011;**144**(2–3):187–94. https://doi.org/10.1016/j.jcpa.2010.09.001.
46. Rodríguez-Piñeiro AM, de la Cadena MP, López-Saco Á, Rodríguez-Berrocal FJ. Differential expression of serum clusterin isoforms in colorectal cancer. *Mol Cell Proteomics* 2006;**5**(9):1647–57. https://doi.org/10.1074/mcp.m600143-mcp200.
47. Álvarez-Chaver P, Rodríguez-Piñeiro AM, Rodríguez-Berrocal FJ, García-Lorenzo A, de la Cadena MP, Martínez-Zorzano VS. Selection of putative colorectal cancer markers by applying PCA on the soluble proteome of tumors: NDK A as a promising candidate. *J Proteomics* 2011;**74**(6):874–86. https://doi.org/10.1016/j.jprot.2011.02.031.
48. Strohkamp S, Gemoll T, Humborg S, et al. Protein levels of clusterin and glutathione synthetase in platelets allow for early detection of colorectal cancer. *Cell Mol Life Sci* 2018;**75**(2):323–34. https://doi.org/10.1007/s00018-017-2631-9.
49. García-Lorenzo A, Rodríguez-Piñeiro A, Rodríguez-Berrocal F, Cadena M, Martínez-Zorzano V. Changes on the Caco-2 secretome through differentiation analyzed by 2-D differential in-gel electrophoresis (DIGE). *Int J Mol Sci* 2012;**13**(12):14401–20. https://doi.org/10.3390/ijms131114401.
50. Oliveira LA, Artigiani-Neto R, et al. NM23 protein expression in colorectal carcinoma using TMA (tissue microarray): association with metastases and survival. *Arq Gastroenterol* 2010;**47**(4):361–7. https://doi.org/10.1590/s0004-28032010000400008.

51. Chen D, Chen F, Lu X, Yang X, Xu Z, Pan J, et al. Identification of prohibitin as a potential biomarker for colorectal carcinoma based on proteomics technology. *Int J Oncol* 2010;**37**(2). https://doi.org/10.3892/ijo_00000684.
52. Murakoshi Y, Honda K, Sasazuki S, et al. Plasma biomarker discovery and validation for colorectal cancer by quantitative shotgun mass spectrometry and protein microarray. *Cancer Sci* 2011;**102**(3):630–8. https://doi.org/10.1111/j.1349-7006.2010.01818.x.
53. Gemoll T, Roblick UJ, Auer G, Jörnvall H, Habermann JK. SELDI-TOF serum proteomics and colorectal cancer: a current overview. *Arch Physiol Biochem* 2010;**116**(4–5):188–96. https://doi.org/10.3109/13813455.2010.495130.
54. Quesada-Calvo F, Massot C, Bertrand V, et al. OLFM4, KNG1 and Sec24C identified by proteomics and immunohistochemistry as potential markers of early colorectal cancer stages. *Clin Proteomics* 2017;**14**:9. https://doi.org/10.1186/s12014-017-9143-3.
55. Chauvin A, Boisvert F-M. Clinical proteomics in colorectal cancer, a promising tool for improving personalised medicine. *Proteomes* 2018;**6**(4):49. https://doi.org/10.3390/proteomes6040049.
56. Atak A, Khurana S, Gollapalli K, et al. Quantitative mass spectrometry analysis reveals a panel of nine proteins as diagnostic markers for colon adenocarcinomas. *Oncotarget* 2018;**9**(17):13530–44. https://doi.org/10.18632/oncotarget.24418.
57. de Wit M, Fijneman RJA, Verheul HMW, Meijer GA, Jimenez CR. Proteomics in colorectal cancer translational research: biomarker discovery for clinical applications. *Clin Biochem* 2013;**46**(6):466–79. https://doi.org/10.1016/j.clinbiochem.2012.10.039.
58. Álvarez-Chaver P, Otero-Estévez O, Páez de la Cadena M, Rodríguez-Berrocal FJ, Martínez-Zorzano VS. Proteomics for discovery of candidate colorectal cancer biomarkers. *World J Gastroenterol* 2014;**20**(14):3804–24. https://doi.org/10.3748/wjg.v20.i14.3804.
59. Wang K, Huang C, Nice EC. Proteomics, genomics and transcriptomics: their emerging roles in the discovery and validation of colorectal cancer biomarkers. *Expert Rev Proteomics* 2014;**11**(2):179–205. https://doi.org/10.1586/14789450.2014.894466.
60. Santasusagna S, Moreno I, Navarro A, et al. Proteomic analysis of liquid biopsy from tumor-draining vein indicates that high expression of exosomal ECM1 Is associated with relapse in stage I-III colon cancer. *Transl Oncol* 2018;**11**(3):715–21. https://doi.org/10.1016/j.tranon.2018.03.010.
61. de Wit M, Kant H, Piersma SR, et al. Colorectal cancer candidate biomarkers identified by tissue secretome proteome profiling. *J Proteomics* 2014;**99**:26–39. https://doi.org/10.1016/j.jprot.2014.01.001.
62. Ang C-S, Phung J, Nice EC. The discovery and validation of colorectal cancer biomarkers. *Biomed Chromatogr* 2011;**25**(1-2):82–99. https://doi.org/10.1002/bmc.1528.
63. Alvarez-Chaver P, Rodríguez-Piñeiro AM, Rodríguez-Berrocal FJ, Martínez-Zorzano VS, Páez de la Cadena M. Identification of hydrophobic proteins as biomarker candidates for colorectal cancer. *Int J Biochem Cell Biol* 2007;**39**(3):529–40. https://doi.org/10.1016/j.biocel.2006.10.001.
64. Otero-Estévez O, De Chiara L, Barcia-Castro L, et al. Evaluation of serum nucleoside diphosphate kinase A for the detection of colorectal cancer. *Sci Rep* 2016;**6**(1). https://doi.org/10.1038/srep26703.
65. Álvarez-Chaver P, De Chiara L, Martínez-Zorzano VS. Proteomic profiling for colorectal cancer biomarker discovery. *Methods Mol Bio* 2018;241–69. https://doi.org/10.1007/978-1-4939-7765-9_16.
66. Zhang F, Zhang Y, Zhao W, et al. Metabolomics for biomarker discovery in the diagnosis, prognosis, survival and recurrence of colorectal cancer: a systematic review. *Oncotarget* 2017;**8**(21):35460–72. https://doi.org/10.18632/oncotarget.16727.
67. Baker TC, Han J, Borchers CH. Recent advancements in matrix-assisted laser desorption/ionization mass spectrometry imaging. *Curr Opin Biotechnol* 2017;**43**:62–9. https://doi.org/10.1016/j.copbio.2016.09.003.
68. Kriegsmann J, Kriegsmann M, Casadonte R. MALDI TOF imaging mass spectrometry in clinical pathology: a valuable tool for cancer diagnostics (Review). *Int J Oncol* 2015;**46**(3):893–906. https://doi.org/10.3892/ijo.2014.2788.
69. Bianga J, Bouslimani A, Bec N, et al. Complementarity of MALDI and LA ICP mass spectrometry for platinum anticancer imaging in human tumor. *Metallomics* 2014;**6**(8):1382–6. https://doi.org/10.1039/c4mt00131a.
70. Meding S, Balluff B, Elsner M, et al. Tissue-based proteomics reveals FXYD3, S100A11 and GSTM3 as novel markers for regional lymph node metastasis in colon cancer. *J Pathol* 2012;**228**(4):459–70. https://doi.org/10.1002/path.4021.
71. Gemoll T, Strohkamp S, Schillo K, Thorns C, Habermann JK. MALDI-imaging reveals thymosin beta-4 as an independent prognostic marker for colorectal cancer. *Oncotarget* 2015;**6**(41):43869–80. https://doi.org/10.18632/oncotarget.6103.

Section D

Treatment

Section D.1

Endoscopic treatment

Chapter 23

Multidisciplinary committee for a comprehensive approach to CRC patients

Ramón Vázquez Dourado, Pedro Carpintero Briones, and Javier Castro Alvariño
Gastroenterology Department, Ferrol University Hospital Complex, Ferrol, Spain

Introduction

The development of oncological care in Spain has been characterized by the difficulty of articulating, in the traditional organizational structure of hospital care services, the multidisciplinary approach to care for cancer patients. The complexity of the diagnostic and therapeutic process involves the clinical care of a large number of professionals. There is, therefore, a possibility that coordination and communication problems may arise.

The evidence points out the role played by these aspects in contributing to greater survival rates and improving the quality of life for patients. The implementation of a developed model of multidisciplinary care is one of the greatest challenges to further care quality improvement for cancer patients.

Different reviews of the literature associate the multidisciplinary approach to cancer care with better adherence to clinical practice guidelines, greater patient access to clinical trials, and better coordination between different hospital services. These results, along with the role assigned to multidisciplinary care in various cancer-related plans, reveal its great importance for health systems in general and for the quality of care in particular. This aspect has already been highlighted at the round table organized by the Portuguese Presidency of the European Union in 2007.[1]

The Cancer Care Strategy of the Spanish Health System[2] was developed by the Ministry of Health and Social Policy, the Scientific Societies, the Autonomous Communities and the Patient Associations, and updated in the Inter-territorial Council of the Spanish Health System held in October 2009. It proposes that diagnosed patients be treated within the framework of a multidisciplinary and integrated team, with a professional who acts as a reference point for the patient. The dynamics of care led to the organizational changes of this type.

Clinical oncology committees

Oncology patient care poses a number of challenges, including the establishment of effective clinical coordination of surgical, medical, and radiotherapeutic treatments. The patient should be provided clear information in this regard. This step involves different professionals, different levels of care, a long duration of treatment and follow-up and, in general, complex care processes.

The growing number of cases and the emergence of increasingly specific treatments involving different specialties have allowed professionals' specialization according to tumor pathology because no one can meet the needs of the patient on their own. This is how the need for specialization against a specific type of cancer arises and also for collaboration among the different professionals.

The development of multidisciplinary cancer care[3] therefore involves a number of main aspects:

- Specialization of the professional by disease and in diverse diagnostic techniques.
- Standardization of clinical processes and criteria in clinical care guidelines and processes.
- Redistribution of tasks toward a multidisciplinary team level.
- Tendency toward the identification and specific allocation of resources according to the disease or organ (colon, breast, lung, prostate).

Within the scope of the Spanish Health System, the oncology committees are the main instrument for discussion and decision making on the diagnosis and therapeutic plan of the cancer patient. They were created in the 1980s to discuss care processes with high clinical complexity. Advances in the field of science and disease management made the initial function of assessing clinical cases by the committee the starting point for a series of changes in the process of caring for cancer patients.

During the 1970s, cancer became an issue in the hospitals, thus the process of creating oncology committees represented the starting point for providing more and better care for patients whose lives depended on surgery alone. The existence of different therapeutic options determines which surgeons, medical oncologists, radiotherapists, and various specialists come together with the objective of considering the best possible treatment for a specific patient. Better knowledge of oncologic diseases involves more and more professionals, modifies their functions and calls for cross-disciplinary care within a health system with deficiencies in the integration.

The 2000s included the effective development of the committees and the beginning of providing care, which can already be described as multidisciplinary. A process of **specialization according to tumor pathology** is promoted among the different medical specialties that question the care structures which, up to that moment, had limited the possibilities of coordination, communication, research, and teaching, all as a specific response to an increasingly complex oncology care.

In the field of colorectal cancer, the committees are made up of specialists in the Digestive System, Surgery, Medical Oncology and Radiation Therapy, Pathological Anatomy, and Radiodiagnosis; in many cases, simultaneously with the development of specific or high-resolution consultations, as is the case of our hospital.

The fact that the effective administration of the combination therapy has prognostic value in the survival of the patient[4] demonstrates the importance of prior decision making and how this is done. The emergence of new potentially effective treatments also implies choices about the therapeutic options and how they should be presented to patients.[5] Although the initial clinical assessment is limited to the therapy decision, many teams subsequently transcend that scope, considering the processes as a whole, including diagnostic and nursing specialties.

Currently, in Spain there are three models (Table 23.1) for the establishment of oncology committees: mutual adaptation, advisory committee, and comprehensive care.

The model of mutual adaptation is one of the most widely used in the Spanish Health System and consensus plays a key role in it. The team acts as the frame of reference for professionals who share their views on the diagnosis, treatment, and follow-up of a specific type of cancer. The meeting is open to all specialists involved in patient management, with the coordinator or team leader playing a key role. The fundamental factor driving this type of approach is the agreement

TABLE 23.1 Models for establishing oncology committees.

	Advisory committee	Mutual adaptation	Comprehensive care
Cases referred	Complex or out-of-protocol cases 10%–50%	All possible cases 50%–80%	Primary source of clinical opinion 90%–100%
Patient access phase	Treatment (ongoing or not)	Diagnosis or treatment	Suspicion or histological diagnosis (early access)
Decisions	Recommendations for the responsible clinician	Decisions by consensus which are not always respected	Binding decisions
Role of professionals	Negative perception	Temporary nurse coordinator	Nurse coordinator
Impact on clinical process	Minor changes	Impact on segments of attention	Comprehensive diagnostic and therapeutic process
Participation of specialists	Treatment	Diagnosis and treatment (occasional absences due to unavailability)	The meetings are part of the care-giving work
Doctors in training and nursing staff	Absent	Occasional presence which is encouraged	Compulsory participation
Role of the hospital management	Lack of interest	Recognition without express support	Explicit recognition
Presence in the health system	40%	50%	10%

on the need for joint decision making prior to the administration of any treatment, and for all cases to be dealt with in the multidisciplinary meeting. Both aspects are hindered, however, by the inertia of hospital services when it comes to disease management.[3]

Another committee model adopted in up to 40% of the centers is that of an "advisory committee." This is a group of professionals made up of specialists in different therapeutic areas who meet regularly on an informal basis to discuss cases considered clinically complex. Since patients may have received some of the treatments (usually surgery), the multidisciplinary meeting is aimed at subjecting them to the assessment of other professionals for further treatment. This approach involves rigorous respect for the autonomy of the physician and the overlapping of boundaries between the patient's treatment team and the multidisciplinary meeting: patient assessment is conducted without further consideration of health care performance.

Teams working under the comprehensive action model share a broader view of patient management, including clinical coordination, research, and economic assessment of the treatment. Since this model provides teams with early access to patients, knowledge of their preferences, comorbidities, and their psychosocial context are incorporated into the multidisciplinary discussions. The role of the professional team has an impact on improving clinical care, although this model is currently seldom used.

Factors influencing the performance of committee activity

Many physicians recognize the existence of *variability in clinical practice* as a result of diagnosing and treating patients who, sharing the same symptomatology, may receive different initial therapy because their access to the hospital is carried out through different departments. When protocols are established and it is agreed to submit the patient to the meeting of the multidisciplinary team, this variability becomes limited, which is further reduced with the *unification of access routes* into a single department of the hospital, thus providing the necessary quality in terms of standardization of criteria and clinical pathways. A clear example of this type of organizational change is the unification of the admission to the Digestive System consultation of patients who exhibit alarming symptoms or have risk factors for colorectal cancer. More high resolution consultations are emerging in different hospitals that attempt to centralize these cases. When agreement is reached on common access to suspicious cases, three consequences are identified: providing early access to patients to the multidisciplinary team, reducing the feeling that they are "owned" by a particular doctor or department, and establishing a referral area for Primary Care where high-risk patients should be referred. Given that the department hosting the unification process is a diagnostic unit, this implies that it also plays a relevant role within the multidisciplinary team.

The *paucity of common guidelines* for the entire country and the lack of coordination strategies for the implementation of existing ones result in their reduced use and a lack of systematic assessment of compliance with them. Due to this situation, hospital clinical protocols are often based on guidelines from other countries, with different consequences. Consequently, hospitals that refer complex cases have protocols based on different nonstandardized guidelines throughout the health system. Second, different levels of development occur in multidisciplinary care, thus hospital patients may be referred to a specific department and not to the oncology committee. On the other hand, there is a lack of predefined criteria for the derivation of levels of attention. The consequence of all this is that some decisions are made without the scientific consensus of a multidisciplinary committee. Decisions can vary widely, causing patients' confusion and lack of trust.

In addition, professionals participating in committees often consider that meeting time is not usually recognized as effective working time and identify two *clear priorities for hospital management teams*: protecting these meetings and their working time, and promoting the participation of new professionals, such as case management nurses or support clerks.

On the other hand, health managers must face the transition from a model based on macroservices, which addresses delimited processes, to one that encourages multidisciplinary work based on transversal groups, which sometimes do not appear in the organization chart, but play an important role in the quality of health care and allow integrated processes to be projected.

An experience which was quite successful in several health care services and became the basis for the assessment of each multidisciplinary team was the creation and implementation of a *fast track* for diagnosis and a treatment program, which promoted integration between services and multidisciplinary teams. Its implementation demonstrated the fundamental role that health policy can play in improving the organization of oncology care.

It is also essential that there be easy access to the decisions made in multidisciplinary teams and their grounds. The *recording of decisions* reflects the result of consensus, establishing the final point of the decision-making process and generating a positive perception among the hospital staff. The main shortcomings in this respect arise from the lack of standardization and identification of those responsible for the records as well as the lack of administrative support. The most glaring example of the importance of recording decisions is when facing a case in which there is no a priori consensus, a situation with clear medical and legal implications.

In conclusion, we can define *five basic points* that should help to transform decision-making processes in order to avoid problems in terms of quality of oncology care services and to project the best possible care:

(1) All cancer patients should be presented to an oncology committee.
(2) Patient access to the committee should take place at the diagnostic stage (suspicion or histological diagnosis).
(3) Agreements between specialists must take the form of consensual and binding decisions.
(4) The clinical results of this process should be systematically assessed.
(5) Health managers should protect committee time and promote a type of organization that is open to the development of multidisciplinary teams, focusing first on the transversal roles (coordinator, case management nurse) that represent them.

Effect of the multidisciplinary approach on colon and rectal cancer

When dealing with cancer care, it is increasingly difficult to attribute clinical results to the work of individual professionals: the functioning of the teams and the organizational context in which they operate have an important impact on them.

It is widely believed that an accurate and comprehensive preoperative assessment, and the adoption of active strategies for adjuvant therapy planning carried out by the members of the colorectal cancer committee are major factors in improving clinical outcomes.[6]

The study conducted by Wood et al. published in 2008[7] prospectively evaluated compliance with all decisions made by the colorectal cancer committee by analyzing hospital records and identifying the reasons for modifying them. They found that the compliance rate was 90% and that in the rest of the cases the presence of comorbidity was the reason why the therapeutic plan was unsuitable or could not be applied, opting for more conservative attitudes. At other times, the decision to use palliative radiotherapy or chemotherapy was changed to a more comprehensive treatment. This points out the need to obtain up-to-date information about the patient's general health and their preferences prior to the committee meeting. Such information could include relevant details on cardiorespiratory or psychosocial pathology. Whichever means is used to include more information on patient-related factors in meetings requires time, and evidence suggests that when patients are consulted about the treatment decision, compliance is better.[8]

The multidisciplinary approach in rectal cancer is associated with a significantly lower rate of circumferential resection margins with tumor infiltration due to a complete staging (involving Magnetic Resonance Imaging and Echoendoscopy) and a correct assessment of the need for neoadjuvant chemoradiotherapy.[9] Despite this, its implementation still has room for improvement.[10]

Approximately 15 years ago, the diagnosis of stage IV colorectal cancer was associated with a five-year survival rate of less than 1%, and surgical resection of liver metastases isolated from a minority of patients was considered, reaching a five-year survival rate of 30%–40%. Currently, about 20% of patients with liver metastases are candidates for curative surgery and the five-year survival rate increased to 50%. Advances in chemotherapy have made it easier for many patients with initially unresectable disease to become clear candidates for surgery. The assessment of these patients by a liver surgery specialist included in the multidisciplinary team makes it easier for patients to not be excluded from a potentially curative treatment[11] determining significant differences in the average survival rate, although they are not so clear in disease-free survival.[12]

References

1. Gouveia J, Coleman MP, Haward R, et al. Improving cancer control in the European Union: conclusions from the Lisbon round-table under the Portuguese EU presidency, 2007. *Eur J Cancer* 2008;**44**(10):1457–62. https://doi.org/10.1016/j.ejca.2008.02.006.
2. SNS. *Plan de Calidad Para El Sistema Nacional de Salud*. Madrid: Ministerio de Sanidad y Consumo; 2010.
3. Prades J, Borràs JM. Multidisciplinary cancer care in Spain, or when the function creates the organ: qualitative interview study. *BMC Public Health* 2011;**11**(1):141. https://doi.org/10.1186/1471-2458-11-141.
4. Smith TJ, Hillner BE. Ensuring quality cancer care by the use of clinical practice guidelines and critical pathways. *J Clin Oncol* 2001;**19**(11): 2886–97. https://doi.org/10.1200/JCO.2001.19.11.2886.
5. Fleissig A, Jenkins V, Catt S, Fallowfield L. Multidisciplinary teams in cancer care: are they effective in the UK? *Lancet Oncol* 2006;**7**(11): 935–43. https://doi.org/10.1016/S1470-2045(06)70940-8.
6. Cervantes A, Roselló S, Rodríguez-Braun E, et al. Progress in the multidisciplinary treatment of gastrointestinal cancer and the impact on clinical practice: perioperative management of rectal cancer. *Ann Oncol* 2008;**19**(Suppl. 7). https://doi.org/10.1093/annonc/mdn438. vii266–72.

7. Wood JJ, Metcalfe C, Paes A, et al. An evaluation of treatment decisions at a colorectal cancer multi-disciplinary team. *Colorectal Dis* 2008;**10**(8):769–72. https://doi.org/10.1111/j.1463-1318.2007.01464.x.
8. Solomon MJ, Pager CK, Keshava A, et al. What do patients want? *Dis Colon Rectum* 2003;**46**(10):1351–7. https://doi.org/10.1007/s10350-004-6749-0.
9. Burton S, Brown G, Daniels IR, et al. MRI directed multidisciplinary team preoperative treatment strategy: the way to eliminate positive circumferential margins? *Br J Cancer* 2006;**94**(3):351–7. https://doi.org/10.1038/sj.bjc.6602947.
10. Swellengrebel HAM, Peters EG, Cats A, et al. Multidisciplinary discussion and management of rectal cancer: a population-based study. *World J Surg* 2011;**35**(9):2125–33. https://doi.org/10.1007/s00268-011-1181-9.
11. Jones RP, Vauthey J-N, Adam R, et al. Effect of specialist decision-making on treatment strategies for colorectal liver metastases. *Br J Surg* 2012;**99**(9):1263–9. https://doi.org/10.1002/bjs.8835.
12. Lordan JT, Karanjia ND, Quiney N, Fawcett WJ, Worthington TR. A 10-year study of outcome following hepatic resection for colorectal liver metastases - the effect of evaluation in a multidisciplinary team setting. *Eur J Surg Oncol* 2009;**35**(3):302–6. https://doi.org/10.1016/j.ejso.2008.01.028.

Chapter 24

Endoscopic treatment of preneoplastic or early lesions

Leticia García Diéguez, Ramón Vázquez Dourado, and Javier Castro Alvariño
Gastroenterology Services, Ferrol University Hospital Complex, A Coruña, Spain

Introduction

Most colorectal cancers arise from previously benign polyps, so their endoscopic resection reduces the incidence and mortality of colon cancer and is considered an essential basic technique for the endoscopist performing a colonoscopy.[1–3]

Colonic polyps can be divided into two main groups: neoplastic (adenomas and carcinomas) and nonneoplastic. Adenomas and carcinomas share an inexorable characteristic: the presence of dysplasia. The knowledge that serrated polyps also have malignant potential now allows them to be classified as neoplastic polyps.[4]

Surgical treatment (open or laparoscopic) of early colorectal neoplasms involves a higher cost and a higher rate of complications compared to local treatments, so considering that the vast majority of these lesions lack invasive potential, endoscopic treatment is the treatment of choice and is considered curative in most cases.[5–8]

Endoscopic polypectomy

Prior to making decisions during an adequate polypectomy, criteria must be established in relation to the characteristics of the polyp, location, or size: there are several clinical guidelines that help us to establish the technique of choice, but previously a series of classifications must be known that will help us to characterize these lesions.

The Paris classification of superficial neoplastic lesions allows us to predict advanced histology or invasive lesions that would have an impact on our approach toward the technique to be performed[9] (Table 24.1).

Another of the patterns to consider would be contemplated by Kudo's classification published in 1993.[10] Lateral spreading neoplastic lesions (LST) described in this classification are not reflected in the Paris classification. Lateral spreading tumors can be divided into granular (homogeneous or mixed) and nongranular (raised or pseudodepressed) and there are differences in the risk of invasive cancer.[11] In addition to the shape and pattern of Kudo's crypts, another factor to take into account in predicting invasive risk is the size of the polyp.

Basically, the European Society for Gastrointestinal Endoscopy (ESGE) recommends characterizing the polyp prior to excision based on morphology (using the Paris classification) and size in millimeters.

The ESGE provided specific recommendations through its 2017 clinical guidelines on colorectal polypectomy and endoscopic mucosal resection (Fig. 24.1).

In summary, in routine clinical practice:

- All polyps should be removed whenever technically possible, except for tiny polyps (<6mm) in the rectum or rectosigmoid that are expected to be hyperplastic.[12]
- All polyps removed should be studied histologically. In expert centers where high-resolution technology methods are available to make an optical diagnosis with an advanced degree of certainty, the "resect and discard" strategy may be considered for tiny polyps.[13]
- The currently preferred technique for the resection of tiny polyps (< 6mm) is cold snare polypectomy (see Fig. 24.2) (without using a cutting and/or coagulation current). In the case of excessive technical difficulty or impossibility due to minimal size (polyps between 1 and 3mm), cold forceps polypectomy is recommended.[14]
- For sessile polyps measuring 6–9mm, the use of hot snare polypectomy (with a cutting and/or coagulation current) is recommended. The use of biopsy forceps is discouraged due to high rates of incomplete resection.[15]

TABLE 24.1 Paris classification: Colorectal neoplasms.

Endoscopic appearance	Paris classification	Scheme	Features
Protruding lesions (> 2.5 mm)	0-Ip		Classic pedunculated polyp
	0-Isp	Mixed 0-Ip/0-Is	Semipedunculated polyp
	0-Is		Sessile polyp
Flat "slightly elevated" lesions (<2.5 mm)	0-IIa		Flat lesion with a slight elevation
	0-IIa/c	Mix 0-IIa/0-IIc	Flat lesion with a slight elevation at the margins and a discrete central depression
Flat lesions	0-IIb		Mucosal change "without any elevation"
	0-IIc		Mucosal depression
	0-IIc/a	Mix 0-IIc/0-IIa	Mucosal depression with a very slight elevation at the margins

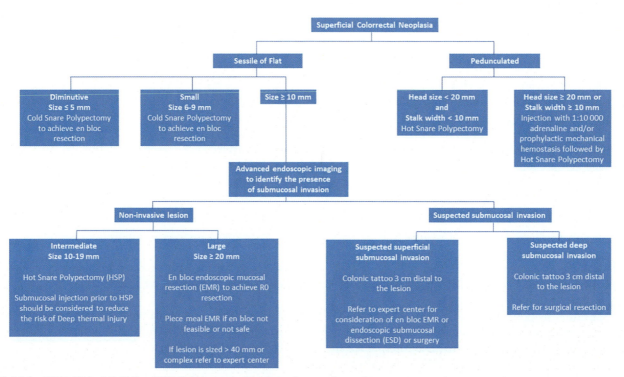

FIG. 24.1 ESGE Clinical Guideline 2017: Polypectomy and colorectal mucosal resection.

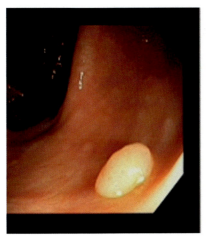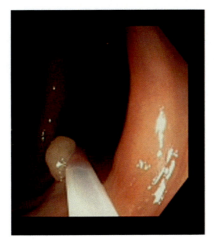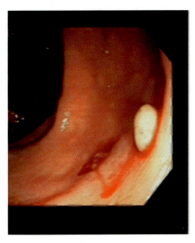

FIG. 24.2 Cold snare polypectomy of a 5-mm polyp, leaving a good eschar after complete resection with almost no bleeding.

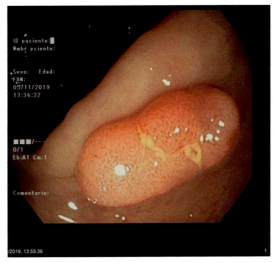

FIG. 24.3 Eleven millimeters sessile polyp.

- The ESGE always recommends hot snare polypectomy (with or without submucosal injection of substances) for the resection of polyps measuring between 10 and 19 mm.[15] We can observe an example for an 11-mm sessile polyp in Fig. 24.3 and its hot snare polypectomy in Fig. 24.4.
- Most pedunculated lesions are easily removed with a hot snare. The most common side effect is bleeding. The following are established as risk factors: cephalic size greater than 10 mm, pedicle greater than 5 mm in diameter, location in the right colon, or suspicion of malignancy.[16, 17]
- The ESGE recommends the use of hot snares in pedunculated polyps, in polyps with a head larger than 20 mm or a pedicle with a diameter larger than 10 mm as a rule. To prevent bleeding, pretreatment with a diluted adrenaline injection in the pedicle or placement of mechanical hemostasis devices (endoloops or clips) could be performed.[15]
- Most colorectal lesions can be removed by standard polypectomy or mucosal resection (EMR).[15] Endoscopic mucosal resection is safe, efficient, and cost effective compared to surgery or other more complex endoscopic techniques.[18]
- The ESGE recommends that en bloc resection techniques (surgery, endoscopic mucosal resection, and submucosal dissection (SMD)) should be considered for lesions with suspected carcinoma with superficial invasion.[15] En bloc resection by SMR is generally limited to lesions no larger than 20 mm or up to 25 mm in the rectum where the risk of perforation is lower. Larger lesions, in order to achieve an adequate en bloc resection, normally require surgery or SMD.[19] Otherwise the alternative will be complete resection in fragments (piecemeal resection).

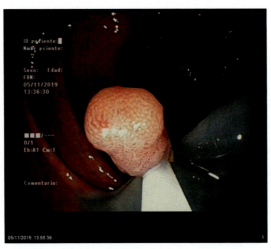

FIG. 24.4 Hot snare polypectomy of an 11-mm sessile polyp after submucosal injection of a solution of saline and indigo carmine.

- Lesions with characteristics of deep submucosal invasion should not be removed by endoscopic techniques but instead by surgery. If there is a lack of data on deep submucosal invasion before the procedure, it would be appropriate to request prior assessment by an expert endoscopy centre.[15]
- Lesions (excised or not) to be located by surgery or future endoscopies should be tattooed during endoscopy.[15] A sterile, safe, and easy-to-use charcoal particle suspension is recommended for tattooing.[20] Although it is not free of occasional complications (peritonitis due to transmural injection or submucosal fibrosis), the risks can be reduced by tattooing in an appropriate place (marginal area) or by previously performing a saline bleb in the submucosal layer. Usually 2 or 3 injections of at least 1–1.5 mL each are used, 3 cm away from the lesion.[15, 21, 22]
- The ESGE establishes that the final objective of EMR is always to achieve a complete snare resection, with adequate margins, in the smallest number of fragments and if possible without the need for associated ablative techniques such as argon plasma coagulation.[15, 23, 24]

Main advanced techniques for endoscopic treatment of colorectal tumors

Endoscopic mucosal resection (EMR)

The basic differences between EMR and other resection techniques arise from, on the one hand, the elevation of the lesion by submucosal injection, which is not performed during a simple polypectomy, and the use of a snare for resection, which would be the main difference with endoscopic submucosal dissection (ESD).[25]

Lesions should be positioned in the lower field of view of the endoscope, between 5 and 6 h. This is because the working channel of the colonoscope is located in this axis, therefore any instrument or device will come out at this level, facilitating the approach and the resection technique.[26]

Changes in the patient's position make it possible to vary the view of the lesions. Sometimes moving the patient so that the lesion is in an antigravity position (so that blood, fragments of the lesion, liquid from the colon, etc., do not accumulate in the lesion, preventing proper visualization) can be advantageous. In other cases, as in the underwater resection technique ("underwater EMR"), the modification of the patient's position to locate the lesion in favor of gravity allows the water to accumulate over it, making it easier to perform the procedure.[25, 27]

Submucosal injection allows elevation of the lesion and facilitates resection while reducing the risk of perforation. The absence of elevation of a lesion after submucosal injection raises suspicion of deep invasion and can therefore also be considered a diagnostic technique, although the lack of elevation may also be due to other causes.[25, 28]

The addition of adrenaline has shown a decrease in immediate bleeding, but not in delayed bleeding.[29] Thermal ablation of the edges with argon primarily in large lesions requiring resection in fragments may be necessary to reduce adenoma recurrence rates.

Indigo carmine is recommended as a method of submucosal staining (submucosal chromoendoscopy). It can be used in different concentrations mixed with the solution to be injected. It facilitates the identification of losses of continuity in the muscular layer indicating perforation.[2]

Material

There are different materials that can be used for the same technique depending on the needs and experience of the endoscopist:

- **Caps:** Caps are cylindrical transparent plastic devices that are placed at the tip of the endoscope creating a ridge 2–4 mm in front of the distal end. They stabilize the endoscope and prevent the mucosa from sticking to the lens, improving visibility and increasing its diagnostic and therapeutic capacity.[25, 30, 31]
- **Snares:** There is a great variety of loops with different sizes and degrees of stiffness that must be used according to the type of lesion and that confer different properties for different resection needs.[25]
- **Cutting current:** there is no uniformity or standardization in the recommended settings for colon resection, although an ASGE review suggests some adjustments according to the type of generator used. With the available data it seems prudent to avoid pure cutting currents due to the greater risk of immediate bleeding and pure coagulation current due to the risk of delayed bleeding.[25, 32]
- **Endoloops:** Their usefulness is limited to the prevention of hemorrhage, either by placing them before or after hemorrhage in pedunculated polyps. They are particularly effective in those with wide and long pedicles.[25, 33, 34] A second use of the endoloop, with greater application in EMR, is the closure of perforations made by resection.[25]
- **Clips:** Endoclips can be used in two contexts: prevention of bleeding and treatment of complications.[25] A prospective randomized comparative study of endoclips with endoloops for postpolypectomy bleeding prevention found no difference between the two techniques in the incidence of immediate or delayed bleeding.[25, 34] The over-the-scope-clip (OTSC, Ovesco Endoscopy®, Tübingen, Germany) is especially useful in the closure of larger perforations because it achieves a stronger transmural closure.[25]
- **Electrocoagulation forceps:** there is currently a method called "hot avulsion" which is performed with hot forceps biopsy. It is safe and effective for completing resections of polyps with small areas that are difficult to elevate or do not elevate; it reduces recurrence rates.[35]

EMR variants

- **Simple or en bloc mucosectomy:** consists in the injection of a solution in the submucosa in order to elevate it and proceed with the resection by means of electrocoagulation with a snare in a single piece or fragment.[36] This technique is used for the resection of flat or sessile lesions smaller than 20 mm[11] as the percentage of en bloc resections decreases drastically for larger lesions.[37]
- **Mucosectomy in fragments (piecemeal):** This technique is performed on lesions larger than 20 mm, ideally starting at the distal margin to ensure proper visualization. Once completed, the edges and the center of the resected area should be carefully studied. No visible debris should remain; debris between 3 and 10 mm should be excised with a 10-mm snare. Debris smaller than 3 mm can be removed with jumbo forceps, argon plasma, or monopolar electrocoagulation.[25, 38] Thermal ablation of the edges, especially in large laterally spreading lesions in fragments, is necessary to reduce adenoma recurrence rates[39] (Fig. 24.5).
- **EMR with bands:** it is not recommended in the colon as it is not considered safe, except in the rectum due to the risk of perforation.[40]
- **Mucosectomy using the "underwater" immersion technique:** a technical variant that allows resection to be performed without submucosal injection. Once the lesion has been identified, air is suctioned out and between 500 and

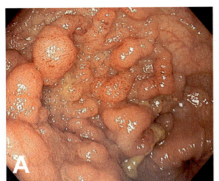 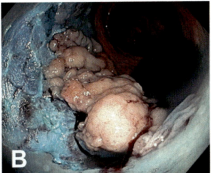 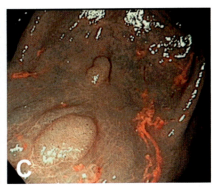

FIG. 24.5 Lateral spreading tumor before (A), during (B) and after (C) EMR.

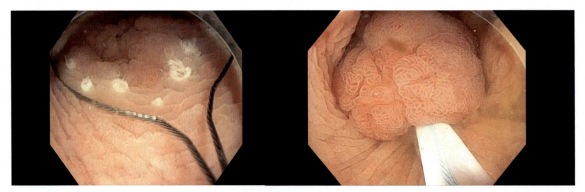

FIG. 24.6 Mucosal resection "underwater."

1000 mL of water is injected until the lumen of the colon is completely full. The margins of the lesion are marked underwater using argon plasma. Mucosal resection is performed in fragments with a 15-mm snare.[25] It is recommended for the resection of residual lesions on the scars of a previous resection[40] and as an alternative to simple resection or resection by fragments[41] (Fig. 24.6).

Endoscopic submucosal dissection (ESD)

ESD is a technique developed in the East and currently emerging in the West that has attracted the interest of many expert endoscopists. ESD achieves high en bloc resection rates, but its learning curve is complex, requiring a considerable training period that is difficult to complete in Western countries. Its percentage of serious complications is higher at the beginning of the learning curve and requires longer execution times.[25, 42]

This technique commonly used for lateral spreading tumors (LST) consists in creating a submucosal "cushion" using an injection catheter followed by controlled dissection within the submucosal plane below the lesion, allowing en bloc resection.[43, 44]

ESD could offer advantages in the treatment of neoplasms larger than 20 mm for which a higher risk of infiltration of the submucosal layer is foreseeable. More data on the effectiveness of ESD in our setting is needed before establishing its role in the treatment of early colorectal neoplasms in the West.[25] Nevertheless, the Spanish EMR Group proposes an endoscopic treatment algorithm for colorectal lesions, as set out in the following table according to morphological subtype and size (Fig. 24.7, Table 24.2).

Other techniques

There are other advanced endoscopic possibilities such as full-thickness endoscopic parietal resection ("full-thickness resection," EFTR) used mainly for GIST-type submucosal tumors or lesions with excessive submucosal fibrosis still in development, which would allow make it possible to avoid surgery in a certain number of cases.[45]

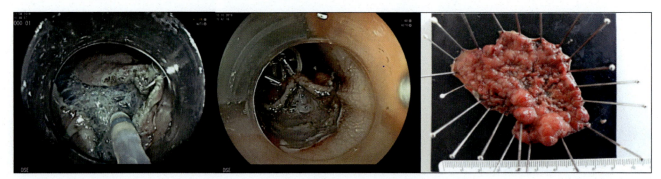

FIG. 24.7 DSM. Mixed LST-G 70 × 45 mm in cecum. *(Image provided by Dr. F Gallego, H del Poniente.)*

TABLE 24.2 Strategy for endoscopic treatment for colorectal lesions (Spanish Endoscopic Mucosal Resection Group).

Morphological subtype	Size <20–30 mm	Size >20–30 mm
LST-G	EMR	RME (ESD on an exceptional basis)
LST-G NM	EMR	Case × case (EMR/ESD: if EMR, separate drying/analysis of large modules)
LST-NG FE	EMR	Case × case (EMR/ESD)
LST-NG PD	EMR	DSE
0Is	Polypectomy/EMR	Case × case (EMR/ESD)
No elevation symbol	Case × case (EMR, hybrid EMR, underwater EMR, EMR + avulsion or APC. ESD)	
0Ip	Polypectomy	Polypectomy/EMR (large pedicles or semipedunculated lesions)

References

1. Zauber AG, Winawer SJ, O'Brien MJ, et al. Colonoscopic polypectomy and long-term prevention of colorectal-cancer deaths. *N Engl J Med* 2012; **366**(8):687–96. https://doi.org/10.1056/NEJMoa1100370.
2. Bretthauer M, Kaminski MF, Løberg M, et al. Population-based colonoscopy screening for colorectal cancer: a randomized clinical trial. *JAMA Intern Med* 2016;**176**(7):894–902. https://doi.org/10.1001/jamainternmed.2016.0960.
3. Brenner H, Stock C, Hoffmeister M. Effect of screening sigmoidoscopy and screening colonoscopy on colorectal cancer incidence and mortality: systematic review and meta-analysis of randomised controlled trials and observational studies. *BMJ* 2014;**348**(apr09 1). https://doi.org/10.1136/bmj.g2467. g2467.
4. Itzkowitz SH, Potack J. Colonic polyps and polyposis syndromes. In: *Sleisenger and Fordtran's Gastrointestinal and Liver Disease*. Elsevier; 2010. https://doi.org/10.1016/b978-1-4160-6189-2.00122-0. 2155–2189.e7.
5. Puli SR, Kakugawa Y, Saito Y, Antillon D, Gotoda T, Antillon MR. Successful complete cure en-bloc resection of large nonpedunculated colonic polyps by endoscopic submucosal dissection: a meta-analysis and systematic review. *Ann Surg Oncol* 2009;**16**(8):2147–51. https://doi.org/10.1245/s10434-009-0520-7.
6. Masci E, Viale E, Notaristefano C, et al. Endoscopic mucosal resection in high- and low-volume centers: a prospective multicentric study. *Surg Endosc* 2013;**27**(10):3799–805. https://doi.org/10.1007/s00464-013-2977-5.
7. Moss A, Bourke MJ, Williams SJ, et al. Endoscopic mucosal resection outcomes and prediction of submucosal cancer from advanced colonic mucosal neoplasia. *Gastroenterology* 2011;**140**(7):1909–18. https://doi.org/10.1053/j.gastro.2011.02.062.
8. Moss A, Williams SJ, Hourigan LF, et al. Long-term adenoma recurrence following wide-field endoscopic mucosal resection (WF-EMR) for advanced colonic mucosal neoplasia is infrequent: results and risk factors in 1000 cases from the Australian colonic EMR (ACE) study. *Gut* 2015;**64**(1):57–65. https://doi.org/10.1136/gutjnl-2013-305516.
9. Endoscopic Classification Review Group. Update on the Paris classification of superficial neoplastic lesions in the digestive tract. *Endoscopy* 2005; **37**(6):570–8. https://doi.org/10.1055/s-2005-861352.
10. Kudo S. Endoscopic mucosal resection of flat and depressed types of early colorectal cancer. *Endoscopy* 1993;**25**(7):455–61. https://doi.org/10.1055/s-2007-1010367.
11. Kudo S, Lambert R, Allen JI, et al. Nonpolypoid neoplastic lesions of the colorectal mucosa. *Gastrointest Endosc* 2008;**68**(Suppl. 4). https://doi.org/10.1016/j.gie.2008.07.052. S3–47.
12. Kamiński MF, Hassan C, Bisschops R, et al. Advanced imaging for detection and differentiation of colorectal neoplasia: European Society of Gastrointestinal Endoscopy (ESGE) Guideline. *Endoscopy* 2014;**46**(5):435–49. https://doi.org/10.1055/s-0034-1365348.
13. Rex DK, Kahi C, O'Brien M, et al. The American Society for Gastrointestinal Endoscopy PIVI (preservation and incorporation of valuable endoscopic innovations) on real-time endoscopic assessment of the histology of diminutive colorectal polyps. *Gastrointest Endosc* 2011;**73**(3):419–22. https://doi.org/10.1016/j.gie.2011.01.023.
14. Kim JS, Lee B-I, Choi H, et al. Cold snare polypectomy versus cold forceps polypectomy for diminutive and small colorectal polyps: a randomized controlled trial. *Gastrointest Endosc* 2015;**81**(3):741–7. https://doi.org/10.1016/j.gie.2014.11.048.
15. Ferlitsch M, Moss A, Hassan C, et al. Colorectal polypectomy and endoscopic mucosal resection (EMR): European Society of Gastrointestinal Endoscopy (ESGE) clinical guideline. *Endoscopy* 2017;**49**(3):270–97. https://doi.org/10.1055/s-0043-102569.
16. Dobrowolski S, Dobosz M, Babicki A, Głowacki J, Nałecz A. Blood supply of colorectal polyps correlates with risk of bleeding after colonoscopic polypectomy. *Gastrointest Endosc* 2006;**63**(7):1004–9. https://doi.org/10.1016/j.gie.2005.11.063.
17. Buddingh KT, Herngreen T, Haringsma J, et al. Location in the right hemi-colon is an independent risk factor for delayed post-polypectomy hemorrhage: a multi-center case-control study. *Am J Gastroenterol* 2011;**106**(6):1119–24. https://doi.org/10.1038/ajg.2010.507.

18. Swan MP, Bourke MJ, Alexander S, Moss A, Williams SJ. Large refractory colonic polyps: is it time to change our practice? A prospective study of the clinical and economic impact of a tertiary referral colonic mucosal resection and polypectomy service (with videos). *Gastrointest Endosc* 2009; **70**(6):1128–36. https://doi.org/10.1016/j.gie.2009.05.039.
19. Repici A, Pellicano R, Strangio G, Danese S, Fagoonee S, Malesci A. Endoscopic mucosal resection for early colorectal neoplasia: pathologic basis, procedures, and outcomes. *Dis Colon Rectum* 2009;**52**(8):1502–15. https://doi.org/10.1007/DCR.0b013e3181a74d9b.
20. Askin MP, Waye JD, Fiedler L, Harpaz N. Tattoo of colonic neoplasms in 113 patients with a new sterile carbon compound. *Gastrointest Endosc* 2002;**56**(3):339–42. https://doi.org/10.1067/mge.2002.126905.
21. Moss A, Bourke MJ, Pathmanathan N. Safety of colonic tattoo with sterile carbon particle suspension: a proposed guideline with illustrative cases. *Gastrointest Endosc* 2011;**74**(1):214–8. https://doi.org/10.1016/j.gie.2011.01.056.
22. Park JW, Sohn DK, Hong CW, et al. The usefulness of preoperative colonoscopic tattooing using a saline test injection method with prepackaged sterile India ink for localization in laparoscopic colorectal surgery. *Surg Endosc* 2008;**22**(2):501–5. https://doi.org/10.1007/s00464-007-9495-2.
23. Burgess NG, Bahin FF, Bourke MJ. Colonic polypectomy (with videos). *Gastrointest Endosc* 2015;**81**(4):813–35. https://doi.org/10.1016/j.gie.2014.12.027.
24. Holt BA, Bourke MJ. Wide field endoscopic resection for advanced colonic mucosal neoplasia: current status and future directions. *Clin Gastroenterol Hepatol* 2012;**10**(9):969–79. https://doi.org/10.1016/j.cgh.2012.05.020.
25. Arbizu EA, Urquiza MP. Guía clínica para resección endoscópica de pólipos de colon y recto. *Sociedad Española de Endoscopia Digestiva* 2017. Published online 2017.
26. Mönkemüller K, Neumann H, Malfertheiner P, Fry LC. Advanced colon polypectomy. *Clin Gastroenterol Hepatol* 2009;**7**(6):641–52. https://doi.org/10.1016/j.cgh.2009.02.032.
27. Binmoeller KF, Weilert F, Shah J, Bhat Y, Kane S. "Underwater" EMR without submucosal injection for large sessile colorectal polyps (with video). *Gastrointest Endosc* 2012;**75**(5):1086–91. https://doi.org/10.1016/j.gie.2011.12.022.
28. ASGE TECHNOLOGY COMMITTEE, Kantsevoy SV, Adler DG, et al. Endoscopic mucosal resection and endoscopic submucosal dissection. *Gastrointest Endosc* 2008;**68**(1):11–8. https://doi.org/10.1016/j.gie.2008.01.037.
29. Hsieh YH, Lin HJ, Tseng GY, et al. Is submucosal epinephrine injection necessary before polypectomy? A prospective, comparative study. *Hepatogastroenterology* 2001;**48**(41):1379–82. https://www.ncbi.nlm.nih.gov/pubmed/11677969.
30. Sanchez-Yague A, Kaltenbach T, Yamamoto H, Anglemyer A, Inoue H, Soetikno R. The endoscopic cap that can (with videos). *Gastrointest Endosc* 2012;**76**(1). https://doi.org/10.1016/j.gie.2012.04.447. 169–78.e1–2.
31. Tada M, Inoue H, Yabata E, Okabe S, Endo M. Feasibility of the transparent cap-fitted colonoscope for screening and mucosal resection. *Dis Colon Rectum* 1997;**40**(5):618–21. https://doi.org/10.1007/bf02055390.
32. ASGE Technology Committee, Tokar JL, Barth BA, et al. Electrosurgical generators. *Gastrointest Endosc* 2013;**78**(2):197–208. https://doi.org/10.1016/j.gie.2013.04.164.
33. Iishi H, Tatsuta M, Narahara H, Iseki K, Sakai N. Endoscopic resection of large pedunculated colorectal polyps using a detachable snare. *Gastrointest Endosc* 1996;**44**(5):594–7. https://doi.org/10.1016/s0016-5107(96)70015-9.
34. Ji J-S, Lee S-W, Kim T, et al. Comparison of prophylactic clip and endoloop application for the prevention of postpolypectomy bleeding in pedunculated colonic polyps: a prospective, randomized, multicenter study. *Endoscopy* 2014;**46**(09):817. https://doi.org/10.1055/s-0034-1367670.
35. Veerappan SG, Ormonde D, Yusoff IF, Raftopoulos SC. Hot avulsion: a modification of an existing technique for management of nonlifting areas of a polyp (with video). *Gastrointest Endosc* 2014;**80**(5):884–8. https://doi.org/10.1016/j.gie.2014.05.333.
36. Soetikno RM, Gotoda T, Nakanishi Y, Soehendra N. Endoscopic mucosal resection. *Gastrointest Endosc* 2003;**57**(4):567–79. https://doi.org/10.1067/mge.2003.130.
37. Kim HH, Kim JH, Park SJ, Park MI, Moon W. Risk factors for incomplete resection and complications in endoscopic mucosal resection for lateral spreading tumors. *Dig Endosc* 2012;**24**(4):259–66. https://doi.org/10.1111/j.1443-1661.2011.01232.x.
38. Zlatanic J, Waye JD, Kim PS, Baiocco PJ, Gleim GW. Large sessile colonic adenomas: use of argon plasma coagulator to supplement piecemeal snare polypectomy. *Gastrointest Endosc* 1999;**49**(6):731–5. https://doi.org/10.1016/s0016-5107(99)70291-9.
39. Klein A, Tate DJ, Jayasekeran V, et al. Thermal ablation of mucosal defect margins reduces adenoma recurrence after colonic endoscopic mucosal resection. *Gastroenterology* 2019;**156**(3). https://doi.org/10.1053/j.gastro.2018.10.003. 604–613.e3.
40. Albéniz E, Pellisé M, Gimeno García AZ, et al. Guía clínica para la resección mucosa endoscópica de lesiones colorrectales no pediculadas. *Gastroenterol Hepatol* 2018;**41**(3):175–90. https://doi.org/10.1016/j.gastrohep.2017.08.013.
41. Curcio G, Granata A, Ligresti D, et al. Underwater colorectal EMR: remodeling endoscopic mucosal resection. *Gastrointest Endosc* 2015;**81**(5):1238–42. https://doi.org/10.1016/j.gie.2014.12.055.
42. Bourke MJ, Neuhaus H. Colorectal endoscopic submucosal dissection: when and by whom? *Endoscopy* 2014;**46**(8):677–9. https://doi.org/10.1055/s-0034-1377449.
43. Soetikno RM, Inoue H, Chang KJ. Endoscopic mucosal resection. Current concepts. *Gastrointest Endosc Clin N Am* 2000;**10**(4):595–617. vi https://doi.org/10.1016/s1052-5157(18)30100-4.
44. Kitajima K, Fujimori T, Fujii S, et al. Correlations between lymph node metastasis and depth of submucosal invasion in submucosal invasive colorectal carcinoma: a Japanese collaborative study. *J Gastroenterol* 2004;**39**(6):534–43. https://doi.org/10.1007/s00535-004-1339-4.
45. Marín-Gabriel JC, Díaz-Tasende J, Rodríguez-Muñoz S, del Pozo-García A, Ibarrola-Andrés C. Colonic endoscopic full-thickness resection (EFTR) with the over-the-scope device (FTRD): a short case series. *Rev Esp Enferm Dig* 2017;**109**. https://doi.org/10.17235/reed.2017.4259/2016.

Chapter 25

Endoscopic surveillance

Nereida Fernández Fernández[a,b], Antonio Rodríguez-D'Jesús[a,b], and Arantza Germade Martínez[a,b]
[a]Department of Gastroenterology, Xerencia de Xestión Integrada de Vigo, SERGAS, Vigo, Spain, [b]Research Group on Digestive Diseases, Galicia Sur Health Research Institute (IIS Galicia Sur), SERGAS-UVIGO, Pontevedra, Spain

Surveillance of preneoplastic colonic lesions

Currently, 20%–25% of colonoscopies performed in people over 50 years of age are for follow-up endoscopic surveillance. After an initial colonoscopy with resection of lesions, approximately 30%–50% of patients will require follow-up endoscopies. After 5 years of follow-up, less than 1% of patients in surveillance programs will develop colorectal cancer (CRC). Quality colonoscopy in the previous 10 years can reduce CRC incidence and mortality by about 60%.[1]

The implementation of CRC screening programs in all autonomous communities is leading to a considerable increase in the number of colonoscopies, as a result of the positivity of the fecal occult blood test (FOBT) and the indications for follow-up. It is estimated that 22% of the population undergoing CRC screening will require endoscopic follow-up.[2]

Follow-up endoscopies are resource intensive and increase the complications inherent to the technique. It is important to direct endoscopic surveillance to patients who really benefit from it, to achieve the greatest possible prevention of CRC with the minimum frequency necessary so as not to increase complications, and to limit unnecessary examinations. To this end, patients are divided into risk groups, which we will define later on, thereby calling for differentiated follow-up strategies.

The aim of this chapter is to establish risk stratification for metachronous CRC following polyp removal in order to determine appropriate surveillance intervals based on current evidence.

The characteristics of previously resected adenomas and the quality of the initial colonoscopy are the two factors that determine the likelihood of diagnosing adenomas and neoplastic lesions during follow-up. The occurrence of neoplastic lesions during follow-up, known as interval cancer, is mainly due to procedure-related factors, nondetection of lesions, and incomplete resections. Patient-related biological factors play a minor role.[3]

Initial colonoscopy

Colonoscopy is the gold standard test for CRC prevention and diagnosis; it is widely established in clinical practice, safe and effective but not always perfect. Therefore, in recent years it has become increasingly important to measure the quality parameters that ensure the performance of a high-quality colonoscopy to meet the follow-up intervals of clinical guidelines.

The follow-up recommendations of the Scientific Societies are based on an initial quality colonoscopy that should meet certain parameters, including complete and thorough inspection of the colonic mucosa, adequate colonic cleansing, definition and complete resection of the lesions found, and histology report. These requirements are defined in Table 25.1.[4] If any of the earlier mentioned factors were not be fulfilled, it is advisable to repeat the examination within at most 1 year.[5]

Characteristics of basal polyps

Polyps can be flat, sessile, or pedunculated. According to their morphology they are classified according to the aforementioned Paris Classification. Histologically they are classified as adenomas (>60%), serrated, and others (hyperplastic, hamartomas, inflammatory, etc.).

Hyperplastic polyps (Fig. 25.1) account for >50% of colorectal polyps. They are between 1 and 5 mm in size, sessile, and are usually many. These polyps smaller than 10 mm are not malignant and rarely have dysplasia and therefore do not require follow-up. Nowadays, with proper optical diagnosis and in the hands of expert endoscopists, these rectal lesions can even be left unresected due to their benign characteristics.

TABLE 25.1 Colonoscopy Quality Criteria.

1. Complete inspection of the colon: cecum, including ileocecal valve and appendicular orifice
   - It is recommended that photographs be attached

2. Degree of colonic cleanliness according to a valid scale. (Boston scale)

3. Colonoscopy report including:
   - Number of polyps removed and recovered
   - Size, location, morphology, and predictive histology of each polyp
   - Method of removal of each polyp
   - Complete, fragmented, or "en bloc" resection

4. Anatomical Pathology Report:
   - Number of adenomas and serrated polyps
   - Hairy component and degree of dysplasia

Adapted from Grupo de trabajo AEG-SEED. Guía de práctica clínica de calidad en la colonoscopia de cribado del cáncer colorrectal. (Jover R, editor). Editores Médicos SA; 2011. https://www.aegastro.es/publicaciones/guia-clinica-de-calidad-en-la-colonoscopia-de-cribado-del-cancer-colorrectal/.

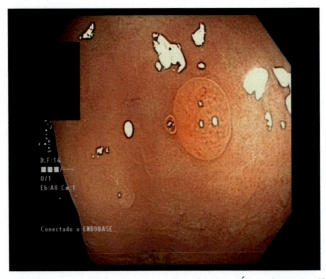

FIG. 25.1 Hyperplastic polyp. *Taken from the archives of the Gastroenterology Services. Álvaro Cunqueiro Hospital.*

Adenomas are the most common preneoplastic colonic lesions with malignant potential. They are classified by glandular histology and degree of dysplasia, which determines the potential for malignancy. Tubular lesions account for more than 80% of the total (Fig. 25.2) and tubulovillous lesions, those with a villous component >20% (Fig. 25.3). Although all adenomas are dysplastic, they are classified as low- or high-grade dysplasia.[6]

Lesions with a villous component or high-grade dysplasia (HGD) have an increased risk of CRC.[7]

The size and number of adenomas are also directly related to the risk of metachronous CRC. Patients with the presence of three or more adenomas are at increased risk of advanced lesion or synchronous and metachronous CRC. A size larger than 10 mm increases the risk of advanced lesion or CRC at follow-up.

An advanced adenomatous lesion is therefore considered to be one that is larger than 10 mm and has a villous component or HGD.

Serrated lesions are less common, but it is estimated that approximately 20%–30% of current CRCs arise from these lesions (serrated pathway of carcinogenesis). This group includes hyperplastic polyps (70%–95%) (see earlier), sessile serrated adenoma (5%–25%), and traditional serrated adenoma (2%–4%). These lesions are histologically difficult to classify,

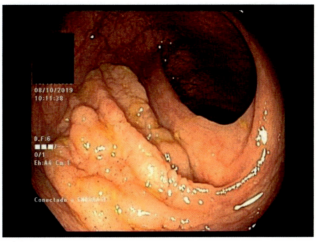

FIG. 25.2 Tubular adenoma polyp. *Taken from the archives of the Gastroenterology Services. Álvaro Cunqueiro Hospital.*

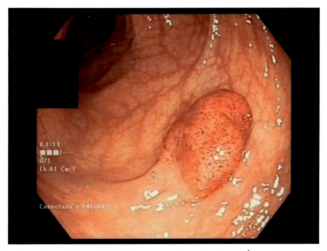

FIG. 25.3 Tubulovillous polyp. *Taken from the archives of the Gastroenterology Services. Álvaro Cunqueiro Hospital.*

so the role of the pathologist is crucial. In addition, these lesions are difficult to visualize during colonoscopy because of their flattened appearance and because they are often mucus covered, making them a diagnostic challenge (Fig. 25.4).

Serrated lesions that confer the highest risk of CRC are those larger than 10mm and with dysplasia, so these are considered advanced.

Patients with nonadvanced adenomas have a slightly higher risk of advanced adenomas at the first follow-up colonoscopy than those without adenomas.[8] However, this incidence is not clinically relevant. It has also been shown that it does not increase the risk of CRC—the incidence of which is similar to that of the general population—so these patients will not require endoscopic follow-up. However, the protective effect of performing initial polypectomy of these lesions is evident, with a 25% risk reduction of CRC compared with the general population in these patients.[9]

With respect to adenomas, when at least 1 adenoma ≥10mm is present during the initial colonoscopy, the likelihood of advanced lesions at follow-up increases by 2–5 times, and villous histology is identified as a risk factor for the development of metachronous advanced lesions. Patients with high-risk colorectal lesions (at least one advanced adenoma or serrated polyp or more than two nonadvanced adenomas) are 5–7 times more likely to develop advanced colorectal lesions at follow-up than patients without adenomas.[10] Therefore these patients will require appropriate endoscopic follow-up.

In terms of CRC risk, there is evidence in favor of a relationship between advanced adenomatous lesions and the presence of CRC. Patients with high-risk adenomas have a higher risk of CRC mortality. Endoscopic surveillance has been shown to reduce the risk of CRC in these patients and is therefore clearly beneficial.[11]

Serrated lesions, which have been increasingly studied in recent years, also pose an increased risk of advanced lesions. Location in the right colon, size >10mm, and the presence of dysplasia may be risk factors for synchronous and

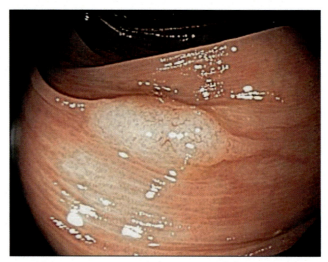

FIG. 25.4 Serrated polyp. *Taken from the archives of the Gastroenterology Services. Álvaro Cunqueiro Hospital.*

metachronous advanced lesions.[12, 13] In addition, several studies have linked the presence of these lesions to an increased risk of metachronous CRC, mainly interval cancer in the right colon.[14]

Risk groups

Based on the initial colonoscopy findings, patients are classified into risk groups. There are several clinical practice guidelines, published over time with some differences among them, which propose different follow-up algorithms for these patients. The two most widely used are the European Guidelines published in 2010[15] and the Clinical Practice Guideline on the Diagnosis and Prevention of Colorectal Cancer published in 2018.[16]

In the 2010 European Guidelines, patients are classified into three risk groups (low, intermediate, and high) with different follow-up strategies, depending on the risk of adenoma and CRC development.[17]

The high-risk group includes patients with more than five adenomas or with at least one 20-mm adenoma and 1-year follow-up is recommended, compared to a 3-year follow-up in patients with 3–4 adenomas, adenomas larger than 10 mm or adenomas with a villous component. A study conducted in Spain[18] found that the incidence of advanced lesions was 16% in the high-risk group compared to 12% in the intermediate-risk group and that there was no difference in the incidence of CRC. Other studies comparing the two groups found no difference in the presence of CRC. For this reason, in the update to the Spanish Guidelines for the Diagnosis and Prevention of Colorectal Cancer,[16] the follow-up groups established in the European Guidelines were changed:

- *Low-risk patients*: these are patients with 1–2 lesions considered nonadvanced. These patients have a small increased risk of advanced colorectal lesions but lower CRC mortality, so the benefit of endoscopic follow-up is residual.

  This group includes patients with 1–2 adenomatous lesions with DBG or <10 mm and patients with serrated lesions without dysplasia <10 mm regardless of number. Also patients with hyperplastic polyps in the rectum and sigmoid colon <10 mm.

  It is recommended that all these patients should be included in population screening programs again. A FOBT or colonoscopy should be performed at 10 years if no prevention program is in place.

- *Intermediate/high-risk patients*: This group requires endoscopic follow-up. This includes patients with at least three adenomatous lesions, one adenomatous lesion, or advanced serrated lesions. These patients have an increased risk of developing advanced lesions and CRC mortality, so endoscopic follow-up at 3 years is proposed.

It is very important to bear in mind that surveillance recommendations are always made after a high-quality baseline colonoscopy and complete excision of lesions.

Sometimes, due to different reasons, lesions are resected and cannot be recovered for histological analysis. If these lesions measure more than 10 mm, they are considered advanced. If they measure <10 mm, they are considered not advanced, and should be included in the final count to determine our patient's risk group.

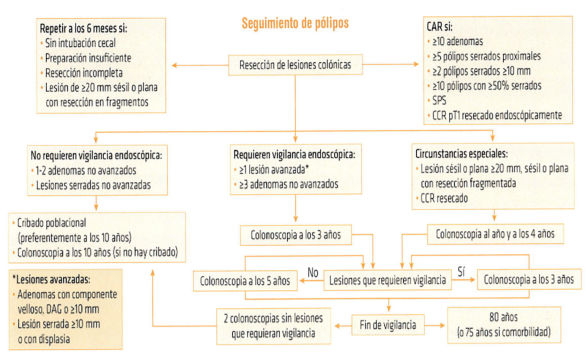

FIG. 25.5 Follow-up Endoscopy. *Adapted from Cubiella J, Marzo-Castillejo M, Mascort-Roca JJ, et al. Clinical practice guideline. Diagnosis and prevention of colorectal cancer. 2018 Update. Gastroenterol Hepatol 2018;41(9):585–596. https://doi.org/10.1016/j.gastrohep.2018.07.012.*

There are two special situations at baseline colonoscopy that influence endoscopic surveillance:

1. *Incomplete resection of lesions:* Complete resection of the lesions must be ensured to initiate follow-up. Incomplete resection occurs more frequently than previously thought. In some studies, these figures are as high as 10%. If a complete resection of the visualized lesions has not been performed, another colonoscopy should be done until all lesions have been removed. Incomplete resection increases the risk of interval cancer and CRC.[19]
2. *Fragmented lesion resection*: Fragmented resection of polyps is associated with incomplete resection, mainly in flat lesions >20mm. Fragmented resection increases the risk of residual tissue and thus of early recurrence (3–6months) and interval CRC.[20, 21] The overall risk of recurrence after fragmented resection exceeds 20%, with more than 90% occurring within the first 6months.

It is recommended that before including the patient in a follow-up program, they should undergo an early endoscopic checkup at 3–6months where eschar biopsy is performed to rule out recurrence. A negative biopsy result is predictive of freedom from recurrence in subsequent follow-up colonoscopies in 97% of cases.[22] Once the absence of residual tissue is confirmed at the first examination, the first follow-up colonoscopy should be performed after 1year.

Once endoscopic surveillance is initiated, all studies suggest that the findings during the second colonoscopy depend more on the findings during the first surveillance endoscopy than during the baseline endoscopy. However, in patients with advanced lesions during the first colonoscopy, the risk of lesions during the second colonoscopy is 10% and a second surveillance colonoscopy is recommended. This second colonoscopy would be at 5years in the case of low-risk lesions or absence of lesions at the first follow-up endoscopy and at 3years in the case of advanced lesions. It is advisable—despite the lack of robust scientific evidence—to refer the patient to the screening program after two normal colonoscopies or with lesions that do not require surveillance. All indications for follow-up are summarized in Fig. 25.5.

When to end follow-up

There is significant debate about the appropriate time to end endoscopic follow-up. Life expectancy has increased and patients are getting older, aside from having multiple diseases, and now these crucial tests are at the center of the debate.

The risk of CRC in patients older than 80 years is three times higher than in patients aged 50–59 years and the risk of developing CRC at 8 years in patients with intermediate-risk adenomas is 3.3%.[23] The benefit of endoscopic surveillance when life expectancy is less than 10 years is questionable due to the risk to which the patient is subjected and considering the long natural history of the adenoma-carcinoma sequence. Although it is true that not all patients are the same and the recommendation needs to be individualized, it seems reasonable to stop endoscopic follow-up at 80 years of age, and it is suggested to stop follow-up between 75 and 80 years of age in patients with multiple diseases and to individually assess each case in case of doubt.

Nonhereditary polyposis: Nonhereditary attenuated polyposis and serrated polyposis

Polyposis is a group of syndromes characterized by the appearance of multiple polyps in the large intestine. Depending on the histology of the polyps they can be classified as adenomatous, hyperplastic, or hamartomatous.

Adenomatous polyposis is classified as classic, when more than 100 adenomas are detected, and attenuated, when fewer than 100 lesions are present.

Hereditary polyposis syndromes (adenomatous and hamartomatous) have been described in other chapters, so this section focuses on the characteristics and management of patients with polyposis in whom a genetic basis has not been identified or described.

Nonhereditary attenuated polyposis

Definition

This group includes patients with 10–100 polyps in whom no germline mutation has been identified and has been reported to account for up to one-third of cases of attenuated polyposis.[24] It is the most common type of polyposis and is often challenging as there is no evidence on clinical management. The presentation is extremely varied, ranging from the appearance of a relatively low number of adenomas (25–30) at an advanced age, to a high number in a young individual or with a history of colorectal cancer (CRC), which may raise the possibility of an unidentified germline mutation.

Management

These patients are usually diagnosed after an average-risk screening colonoscopy, as they are very often patients with no family history.

Resection of adenomatous lesions is usually performed endoscopically, although surgical treatment may occasionally be considered.

Once the diagnosis of attenuated polyposis has been established, the possibility of a patient with a germline mutation should be considered, and genetic testing is recommended in the following cases.[16]

- 20 or more colorectal adenomas in an individual, regardless of age.
- 10 or more colorectal adenomas before the age of 40.
- 10 or more adenomas when there is a personal or family history of CRC before the age of 60 years.
- 10 or more adenomas when there is a family history of attenuated adenomatous polyposis.

Surveillance after resection of colorectal lesions

After complete resection of all lesions, it is recommended to perform a colonoscopy after 1 year and then continue surveillance according to endoscopic findings based on current recommendations. In individuals with two or more normal surveillance colonoscopies, endoscopic follow-up may be continued every 5 years (Fig. 25.6).

Surveillance for extracolonic lesions

Although evidence is scarce, gastroscopy is recommended at the time of diagnosis of colorectal disease to exclude gastroduodenal involvement. If duodenal polyps are observed, the frequency of surveillance examinations should be based on the Spigelman staging system. Screening for extraintestinal neoplasms is not necessary for this group of patients.

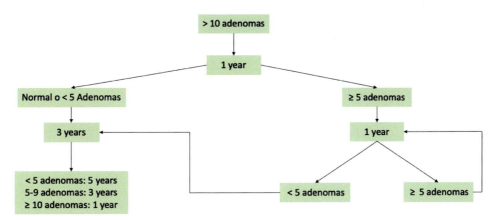

FIG. 25.6 Surveillance Recommendations in patients with attenuated adenomatous polyposis with no identified genetic cause/oligopolyposis. *Adapted from Cubiella J, Marzo-Castillejo M, Mascort-Roca JJ, et al. Clinical practice guideline. Diagnosis and prevention of colorectal cancer. 2018 Update. Gastroenterol Hepatol 2018;41(9):585–596. htttps://doi.org/10.1016/j.gastrohep.2018.07.012.*

Recommendations for family screening

In individuals with attenuated adenomatous polyposis without an identified genetic cause, in the absence of an identified mutation to direct screening, recommendations are directed at first-degree relatives at risk (parents, siblings, and children).

For them, it is recommended to start screening with a complete colonoscopy at 40 years of age or 10 years before the diagnosis of the youngest affected relative.

Earlier screening could be considered at 20–25 years of age in those cases in which there is a high index of suspicion that the origin of the polyposis is hereditary, even if no germline mutation has been identified: index case with >40 polyps or a history of CRC, and/or a family history and/or extracolonic manifestations. The frequency will be established according to the findings of each examination following the current surveillance guidelines for colorectal polyps.[16]

Serrated polyposis

Definition

Serrated polyposis syndrome (SPS) is characterized by the presence of multiple serrated lesions, with the potential to transform into CRC through the expression of an alternative carcinogenesis, the so-called serrated pathway. They appear to progress to CRC through a molecular pathway involving increased methylation of CpG islands resulting in decreased gene expression that may lead to microsatellite instability.

Approximately up to 30% of tumors develop through this pathway, and of these, 10% are BRAF-mutated (proximally located), 15% are KRAS-mutated (preferentially distally located), and 5% are BRAF- and KRAS-native. This fact may be related to the existence of an increased risk for developing foci of dysplasia in serrated polyps (SP).[25]

Endoscopic types and characteristics

Serrated lesions account for approximately one-third of all colonic polyps, with a prevalence in the general population of 20%–35%[26]. These lesions can be difficult to detect with conventional white light colonoscopy, as they are typically flat, located in the right colon, covered by a mucous layer, and have poorly defined margins.

This group of SP comprises a heterogeneous group of lesions with a common feature: the presence of a "sawtooth" architecture in the crypt epithelium due to an accumulation of colonocytes in the crypt due to inhibition of apoptosis.

The WHO currently divides the histologically into three subtypes: (1) hyperplastic polyps, (2) sessile serrated adenomas or SSA, and (3) traditional serrated adenomas or TSA.

SPS is a condition currently defined by clinical-endoscopic criteria as characterized by a tendency to develop multiple or large SP, which is associated with a high CRC risk.

SPS is defined by the presence of one of the two criteria established by the WHO. These criteria are applied to the total number of polyps diagnosed during the patient's lifetime and have recently been updated.[27]

FIG. 25.7 Differences between serrated polyposis syndrome diagnostic criteria 2010 and 2019. *Adapted from Dekker E, Bleijenberg A, Balaguer F, Dutch-Spanish-British Serrated Polyposis Syndrome Collaboration. Update on the World Health Organization criteria for diagnosis of serrated polyposis syndrome. Gastroenterology 2020;158(6):1520–1523. https://doi.org/10.1053/j.gastro.2019.11.310.*

I. Presence of five or more polyps or serrated lesions proximal to the rectum, all ≥5 mm and at least two of them ≥1 cm
II. Presence of more than 20 polyps or serrated lesions of any size distributed throughout the colon, with at least five of them proximal to the rectum.

The differences between the 2010 criteria and the current criteria are summarized in Fig. 25.7. Although these criteria were established arbitrarily, they are very useful for standardizing diagnosis, treatment, and follow-up.[28]

Epidemiology and risk factors

Although SPS was described a few decades ago and was considered a rare entity, the most recent evidence places it as the most common polyposis syndrome today.

Its prevalence is not well known. Nowadays, increased awareness and improved diagnostic techniques have led to a clear increase in the detection of these lesions. Prevalence is higher in populations that undergo population-based CRC screening colonoscopy, following the detection of a positive fecal occult blood test, with figures between 0.34% and 0.66%.[29, 30]

The average age at diagnosis is 55 years, with no gender differences.

The etiology of SPS is unknown, but features involving an interaction between genetic predisposition and environmental factors have been described. Several studies link smoking and being overweight to the development of serrated lesions and SPS.[31–33]

Regarding the risk of developing CRC, the association of SPS with an increased prevalence/incidence of CRC (25%–70%) has been classically described. However, recent studies appear to reduce the estimated risk to around 15%, with a cumulative 5-year interval cancer risk ranging from 2% to 7%.[32]

Certain risk-modifying characteristics related to lesion location and phenotype (number of polyps, size, and certain histologic features) have been described and could be used to stratify recommended endoscopic surveillance intervals, but further prospective studies are needed to confirm these findings.

The presence of at least two sessile SP proximal to the splenic flexure or the presence of high-grade dysplasia in any of the polyps is associated with an increased risk of CRC. In a case–control study,[14] patients with serrated lesions without dysplasia had a 10-year CRC risk of 2.56%, a result very similar to that of patients with conventional adenomas. In contrast, the risk of advanced serrated lesions is higher, reaching a CRC risk of 4.43% in patients with serrated lesions with dysplasia (sessile SP with dysplasia and traditional serrated adenomas).

There is no evidence of an increased risk of extracolonic neoplasms in patients with SPS or their relatives.

Endoscopic monitoring

Given the high risk of CRC in patients meeting WHO criteria for SPS and that effective surveillance appears to reduce the risk of CRC, annual colonoscopy screening is recommended.

Treatment is based on endoscopic polypectomy, if technically possible.

Polypectomy of all serrated lesions >3 mm should be performed to give the protective effect of colonoscopy.

Although some authors recommend a minimum withdrawal time of 6 min, a major study has highlighted the importance of extending the withdrawal time up to 9 min to achieve up to 30% higher detection rates of serrated lesions.[34] It is suggested that the use of chromoendoscopy may improve the detection of serrated lesions.[35]

In cases where endoscopic management is not possible either because of the number or unresectability of the serrated lesions, or when CRC is detected, a surgical approach (total colectomy with ileo-rectal anastomosis or segmental colectomy in selected cases) is recommended.

After surgery, annual endoscopic surveillance is recommended.

Family screenings

It has been observed that more than 30% of first-degree relatives of patients with SPS diagnostic criteria are themselves diagnosed with SPS. The risk of CRC in FPG of patients with SPS is estimated to be five times higher than in the general population.

It is currently recommended that all first-degree relatives undergo colonoscopy screening from the age of 40 years, or 10 years before the youngest SPS index case. Thereafter, surveillance with colonoscopy is recommended every 5 years, unless there are lesions that modify management.[36]

There is insufficient evidence of increased incidence of extracolonic neoplasms in patients with SPS or FPG, so screening is not called for.[37]

Follow-up of pT1 neoplastic lesions after endoscopic treatment

Colorectal cancer (CRC) is one of the most common tumors in the world. However, of all malignant lesions of the gastrointestinal tract, it has the highest cure rate, as long as it is detected in the early stages as a premalignant lesion (e.g., adenoma) or before lymph node involvement.[16, 38]

Malignant polyps are defined as polyps with invasive cancer that overlaps the muscularis mucosae, affecting the submucosa without overlapping it (pT1).[16] In European population-based screening programs, they account for 17% of detected colorectal cancers. These lesions can have lymph node involvement in 8%–16%.[39, 40] The decision to perform endoscopic follow-up or perform an adjunctive oncological segmental resection is a balance between the surgical risk (mortality between 1.9% and 9.8%)[40] and the risk of recurrence (1%–5%) which can have a mortality rate of around 40%.[41, 42] There are risk factors that are associated with the presence of lymph node metastases or a worse prognosis, with poorly differentiated histology and the presence of lymphatic involvement being the most widely agreed risk factors in clinical practice guidelines.[16, 38, 43] Other risk factors include resection margins or budding, among others. A brief review of each of these is given as follows.

Histology

Poorly differentiated lesions (grade 3) have a worse prognosis compared to grade 1 or well-differentiated lesions. In a systematic review, this factor was associated with a relative risk (RR) of 4.8 times the likelihood of lymph node involvement.[40] Furthermore, poorly differentiated carcinoma was associated with a higher likelihood of distant metastasis and cancer-associated mortality.[44]

Resection margin after polypectomy

A resection margin of less than 1 mm was associated with residual cancer in 16% of colectomies performed[39] or a 21%–33% local recurrence in patients followed up.[41] In this regard, current European guidelines consider the margin to be affected if the distance to the lesion is less than or equal to 1 mm.[38]

Although a positive margin was associated with a similar rate of lymph node involvement to patients with negative margins, the former had a higher frequency of hematological metastases and cancer-related death.[44]

Submucosal infiltration

The Haggitt classification divides pedunculated polyps into four levels according to the depth of pedicle involvement, with Haggitt 4 being associated with a poorer prognosis.[42,45] All sessile lesions, regardless of the degree of submucosal involvement according to this classification, would be equivalent to Haggitt 4. The Kikuchi classification further defines submucosal (sm) involvement of sessile malignant polyps by dividing the submucosal space into three levels (sm 1–3), with involvement of the lower third (sm3) being associated with more than 23% risk of lymphatic involvement. However, this classification can be difficult for pathologists to apply if the submitted material does not have a good proportion of submucosa. Simplifying this concept, it has been shown that involvement of more than 1 mm of the submucosa is associated with a significant risk of lymph node involvement.[46] In the case of pedunculated polyps, the 2018 clinical practice guideline of the Spanish Association of Gastroenterology reports SM involvement <3 mm (or Haggitt 1 or 2) as a good prognostic criterion.[16]

Tumor budding

This refers to small clusters of undifferentiated cells ahead of the invasive front of the lesion. However, it is not a parameter routinely analyzed by all pathologists, although there is increasing evidence that the determination of the degree of budding reflects the clinical aggressiveness of the tumor.

In this regard, some authors consider a high-grade budding, i.e., more than 10 clusters of tumor cells in a high-magnification microscopic field (200–400 ×) as a poor prognostic criterion, with an RR of 5.1 (95% CI 3.6–7.3) for lymph node involvement.[40]

Lymphovascular infiltration

Due to the absence of lymphatic vessels in the mucosal layer, intramucosal neoplastic lesions behave like a benign adenoma. However, in the submucosa, the presence of submucosal lymphatic vessels opens the way for metastasis through lymphatic vessels and blood vessels.[44] Lymphatic involvement is defined as the presence of tumor cells, covered by endothelial cells without the presence of erythrocytes. Vascular involvement is described as the presence of tumor cells in a space bounded by smooth muscle cells or endothelium with the presence of erythrocytes or fibrin clots.[40] It is important to report vascular and/or lymphatic invasion independently as each has a different prognosis. In a meta-analysis by Bosch et al.,[40] lymphatic tumor invasion was the strongest predictor of lymph node involvement, with an RR of 5.2 (95% CI 4.0–6.8), much higher than the relative risk if the involvement is only vascular (RR 2.2 95% CI 1.4–3.2).

Interaction between factors

The sum of unfavorable factors is associated with an increased risk of nodal metastasis. It has been shown that combining unfavorable histologic factors, such as poor tumor differentiation, lymphovascular invasion, and budding, results in a lymph node involvement rate of 0.7%, 20.7%, and 36.4% in the absence, presence of a single factor, or multiple factors, respectively.[46]

In summary, surgery should be considered in patients who are amenable to surgery and have poor prognostic factors: poorly differentiated lesions, extensive lymphovascular involvement, deep submucosal invasion (>1 mm or SM2 and 3), resection margin of less than 1 mm, and fragmented excision (if resection margins cannot be determined) (see Table 25.1).

Follow-up

For patients without poor prognostic criteria, the consensus is that endoscopic polypectomy is appropriate as a treatment, with endoscopic surveillance recommended.[16,38,45] However, there is currently no standard guideline for surveillance after endoscopic resection. The European guidelines recommend follow-up with colonoscopy at 1 year.[38] Subsequent follow-up remains even more variable and a matter of debate.

References

1. Robertson DJ, Lieberman DA, Winawer SJ, et al. Colorectal cancers soon after colonoscopy: a pooled multicohort analysis. *Gut* 2014;**63**(6):949–56. https://doi.org/10.1136/gutjnl-2012-303796.
2. Albéniz E, Pellisé M, Gimeno-García AZ, et al. Clinical guidelines for endoscopic mucosal resection of non-pedunculated colorectal lesions. *Rev Esp Enferm Dig* 2018;**110**(3):179–94.
3. Dekker E, Rex DK. Advances in CRC prevention: screening and surveillance. *Gastroenterology* 2018;**154**(7):1970–84. https://doi.org/10.1053/j.gastro.2018.01.069.

4. Grupo de trabajo AEG-SEED. In: Jover R, editor. *Guía de práctica clínica de calidad en la colonoscopia de cribado del cáncer colorrectal*. Editores Médicos SA; 2011. https://www.aegastro.es/publicaciones/guia-clinica-de-calidad-en-la-colonoscopia-de-cribado-del-cancer-colorrectal/.
5. Hassan C, Quintero E, Dumonceau J-M, et al. Post-polypectomy colonoscopy surveillance: European Society of Gastrointestinal Endoscopy (ESGE) Guideline. *Endoscopy* 2013;**45**(10):842–51. https://doi.org/10.1055/s-0033-1344548.
6. Short MW, Layton MC, Teer BN, Domagalski JE. Colorectal cancer screening and surveillance. *Am Fam Physician* 2015;**91**(2):93–100. https://www.ncbi.nlm.nih.gov/pubmed/25591210.
7. Martínez ME, Baron JA, Lieberman DA, et al. A pooled analysis of advanced colorectal neoplasia diagnoses after colonoscopic polypectomy. *Gastroenterology* 2009;**136**(3):832–41. https://doi.org/10.1053/j.gastro.2008.12.007.
8. Lieberman DA, Weiss DG, Harford WV, et al. Five-year colon surveillance after screening colonoscopy. *Gastroenterology* 2007;**133**(4):1077–85. https://doi.org/10.1053/j.gastro.2007.07.006.
9. Løberg M, Kalager M, Holme Ø, Hoff G, Adami H-O, Bretthauer M. Long-term colorectal-cancer mortality after adenoma removal. *N Engl J Med* 2014;**371**(9):799–807. https://doi.org/10.1056/NEJMoa1315870.
10. de Jonge V, Sint Nicolaas J, van Leerdam ME, Kuipers EJ, van Zanten SJO V. Systematic literature review and pooled analyses of risk factors for finding adenomas at surveillance colonoscopy. *Endoscopy* 2011;**43**(7):560–72. https://doi.org/10.1055/s-0030-1256306.
11. Cottet V, Jooste V, Fournel I, Bouvier A-M, Faivre J, Bonithon-Kopp C. Long-term risk of colorectal cancer after adenoma removal: a population-based cohort study. *Gut* 2012;**61**(8):1180–6. https://doi.org/10.1136/gutjnl-2011-300295.
12. Schreiner MA, Weiss DG, Lieberman DA. Proximal and large hyperplastic and nondysplastic serrated polyps detected by colonoscopy are associated with neoplasia. *Gastroenterology* 2010;**139**(5):1497–502. https://doi.org/10.1053/j.gastro.2010.06.074.
13. Holme Ø, Bretthauer M, Eide TJ, et al. Long-term risk of colorectal cancer in individuals with serrated polyps. *Gut* 2015;**64**(6):929–36. https://doi.org/10.1136/gutjnl-2014-307793.
14. Erichsen R, Baron JA, Hamilton-Dutoit SJ, et al. Increased risk of colorectal cancer development among patients with serrated polyps. *Gastroenterology* 2016;**150**(4):895–902.e5. https://doi.org/10.1053/j.gastro.2015.11.046.
15. European Commission. In: Segnan N, Patnick J, Von Karsa L, editors. *European guidelines for quality assurance in colorectal cancer screening and diagnosis*. Luxembourg: Publications Office of the European Union; 2010. https://op.europa.eu/en/publication-detail/-/publication/e1ef52d8-8786-4ac4-9f91-4da2261ee53.
16. Cubiella J, Marzo-Castillejo M, Mascort-Roca JJ, et al. Clinical practice guideline. Diagnosis and prevention of colorectal cancer. 2018 Update. *Gastroenterol Hepatol* 2018;**41**(9):585–96. https://doi.org/10.1016/j.gastrohep.2018.07.012.
17. Atkin WS, Valori R, Kuipers EJ, et al. European guidelines for quality assurance in colorectal cancer screening and diagnosis. First edition—colonoscopic surveillance following adenoma removal. *Endoscopy* 2012;**44**(Suppl 3):SE151–63. https://doi.org/10.1055/s-0032-1309821.
18. Cubiella J, Carballo F, Portillo I, et al. Incidence of advanced neoplasia during surveillance in high- and intermediate-risk groups of the European colorectal cancer screening guidelines. *Endoscopy* 2016;**48**(11):995–1002. https://doi.org/10.1055/s-0042-112571.
19. le Clercq CMC, Bouwens MWE, Rondagh EJA, et al. Postcolonoscopy colorectal cancers are preventable: a population-based study. *Gut* 2014;**63**(6):957–63. https://doi.org/10.1136/gutjnl-2013-304880.
20. Pohl H, Srivastava A, Bensen SP, et al. Incomplete polyp resection during colonoscopy-results of the complete adenoma resection (CARE) study. *Gastroenterology* 2013;**144**(1):74–80.e1. https://doi.org/10.1053/j.gastro.2012.09.043.
21. Oh DM, Lee JK, Kim H, et al. Local recurrence and its risk factor after incomplete resection of colorectal advanced adenomas: a single center, retrospective study. *Korean J Gastroenterol* 2017;**70**(1):33–8. https://doi.org/10.4166/kjg.2017.70.1.33.
22. Khashab M, Eid E, Rusche M, Rex DK. Incidence and predictors of "late" recurrences after endoscopic piecemeal resection of large sessile adenomas. *Gastrointest Endosc* 2009;**70**(2):344–9. https://doi.org/10.1016/j.gie.2008.10.037.
23. Atkin W, Wooldrage K, Brenner A, et al. Adenoma surveillance and colorectal cancer incidence: a retrospective, multicentre, cohort study. *Lancet Oncol* 2017;**18**(6):823–34. https://doi.org/10.1016/S1470-2045(17)30187-0.
24. Nielsen M, Hes FJ, Nagengast FM, et al. Germline mutations in APC and MUTYH are responsible for the majority of families with attenuated familial adenomatous polyposis. *Clin Genet* 2007;**71**(5):427–33. https://doi.org/10.1111/j.1399-0004.2007.00766.x.
25. Chan TL, Zhao W, Leung SY, Yuen ST. Cancer Genome Project. BRAF and KRAS mutations in colorectal hyperplastic polyps and serrated adenomas. *Cancer Res* 2003;**63**(16):4878–81. https://www.ncbi.nlm.nih.gov/pubmed/12941809.
26. Burgess NG, Pellise M, Nanda KS, et al. Clinical and endoscopic predictors of cytological dysplasia or cancer in a prospective multicentre study of large sessile serrated adenomas/polyps. *Gut* 2016;**65**(3):437–46. https://doi.org/10.1136/gutjnl-2014-308603.
27. Rosty C, Brosens LAA, Dekker E, Nagtegaal ID. Serrated polyposis. In: Bosman FT, Carneiro F, Hruban RH, Theise ND, editors. *WHO classification of tumours of the digestive system*. vol. 5. International Agency for Research on Cancer; 2019. p. 532–4.
28. Dekker E, Bleijenberg A, Balaguer F. Dutch-Spanish-British Serrated Polyposis Syndrome Collaboration. Update on the World Health Organization criteria for diagnosis of serrated polyposis syndrome. *Gastroenterology* 2020;**158**(6):1520–3. https://doi.org/10.1053/j.gastro.2019.11.310.
29. Moreira L, Pellisé M, Carballal S, et al. High prevalence of serrated polyposis syndrome in FIT-based colorectal cancer screening programmes. *Gut* 2013;**62**(3):476–7. https://doi.org/10.1136/gutjnl-2012-303496.
30. Biswas S, Ellis AJ, Guy R, Savage H, Madronal K, East JE. High prevalence of hyperplastic polyposis syndrome (serrated polyposis) in the NHS bowel cancer screening programme. *Gut* 2013;**62**(3):475. https://doi.org/10.1136/gutjnl-2012-303233.
31. IJspeert JEG, Bossuyt PM, Kuipers EJ, et al. Smoking status informs about the risk of advanced serrated polyps in a screening population. *Endosc Int Open* 2016;**4**(1):E73–8. https://doi.org/10.1055/s-0034-1393361.
32. Carballal S, Rodríguez-Alcalde D, Moreira L, et al. Colorectal cancer risk factors in patients with serrated polyposis syndrome: a large multicentre study. *Gut* 2016;**65**(11):1829–37. https://doi.org/10.1136/gutjnl-2015-309647.

33. Walker RG, Landmann JK, Hewett DG, et al. Hyperplastic polyposis syndrome is associated with cigarette smoking, which may be a modifiable risk factor. *Am J Gastroenterol* 2010;**105**(7):1642–7. https://doi.org/10.1038/ajg.2009.757.
34. Butterly L, Robinson CM, Anderson JC, et al. Serrated and adenomatous polyp detection increases with longer withdrawal time: results from the New Hampshire Colonoscopy Registry. *Am J Gastroenterol* 2014;**109**(3):417–26. https://doi.org/10.1038/ajg.2013.442.
35. East JE, Atkin WS, Bateman AC, et al. British Society of Gastroenterology position statement on serrated polyps in the colon and rectum. *Gut* 2017;**66**(7):1181–96. https://doi.org/10.1136/gutjnl-2017-314005.
36. Boparai KS, Mathus-Vliegen EMH, Koornstra JJ, et al. Increased colorectal cancer risk during follow-up in patients with hyperplastic polyposis syndrome: a multicentre cohort study. *Gut* 2010;**59**(8):1094–100. https://doi.org/10.1136/gut.2009.185884.
37. Syngal S, Brand RE, Church JM, et al. ACG clinical guideline: genetic testing and management of hereditary gastrointestinal cancer syndromes. *Am J Gastroenterol* 2015;**110**(2):223–62. quiz 263 https://doi.org/10.1038/ajg.2014.435.
38. Hassan C, Wysocki PT, Fuccio L, et al. Endoscopic surveillance after surgical or endoscopic resection for colorectal cancer: European Society of Gastrointestinal Endoscopy (ESGE) and European Society of Digestive Oncology (ESDO) Guideline. *Endoscopy* 2019;**51**(3):266–77. https://doi.org/10.1055/a-0831-2522.
39. Bartel MJ, Brahmbhatt BS, Wallace MB. Management of colorectal T1 carcinoma treated by endoscopic resection from the Western perspective. *Dig Endosc* 2016;**28**(3):330–41. https://doi.org/10.1111/den.12598.
40. Bosch SL, Teerenstra S, de Wilt JHW, Cunningham C, Nagtegaal ID. Predicting lymph node metastasis in pT1 colorectal cancer: a systematic review of risk factors providing rationale for therapy decisions. *Endoscopy* 2013;**45**(10):827–34. https://doi.org/10.1055/s-0033-1344238.
41. Ciocalteu A, Gheonea DI, Saftoiu A, Streba L, Dragoescu NA, Tenea-Cojan TS. Current strategies for malignant pedunculated colorectal polyps. *World J Gastrointest Oncol* 2018;**10**(12):465–75. https://doi.org/10.4251/wjgo.v10.i12.465.
42. Aarons CB, Shanmugan S, Bleier JIS. Management of malignant colon polyps: current status and controversies. *World J Gastroenterol* 2014;**20**(43):16178–83.
43. Benson 3rd AB, Venook AP, Cederquist L, et al. Colon Cancer, version 1.2017, NCCN clinical practice guidelines in oncology. *J Natl Compr Canc Netw* 2017;**15**(3):370–98. https://doi.org/10.6004/jnccn.2017.0036.
44. Hassan C, Zullo A, Risio M, Rossini FP, Morini S. Histologic risk factors and clinical outcome in colorectal malignant polyp: a pooled-data analysis. *Dis Colon Rectum* 2005;**48**(8):1588–96. https://doi.org/10.1007/s10350-005-0063-3.
45. Watanabe T, Muro K, Ajioka Y, et al. Japanese Society for Cancer of the Colon and Rectum (JSCCR) guidelines 2016 for the treatment of colorectal cancer. *Int J Clin Oncol* 2018;**23**(1):1–34. https://doi.org/10.1007/s10147-017-1101-6.
46. Ueno H, Mochizuki H, Hashiguchi Y, et al. Risk factors for an adverse outcome in early invasive colorectal carcinoma. *Gastroenterology* 2004;**127**(2):385–94. https://doi.org/10.1053/j.gastro.2004.04.022.

Chapter 26

Advanced endoscopy in colorectal cancer: Colorectal prostheses

María Teresa Vázquez Rey, Benito González Conde, Ignacio Couto Worner, and Pedro A. Alonso Aguirre
Department of Gastroenterology, University Hospital Complex of A Coruña (CHUAC), SERGAS, A Coruña, Spain

Introduction

Intestinal obstruction is the form of presentation of colorectal cancer (CRC) in 8%–13% of cases,[1] with tumors of the splenic angle being the most frequent to debut in this manner (50%), followed by those of the descending colon (25%) and rectum-sigmoid (6%).[2] This situation represents a medical emergency that should be resolved in the shortest possible time to avoid complications secondary to colonic wall distress, such as ischemia or intestinal perforation.

In this context, emergency surgery usually involves the performance of an unloading colostomy and resection of the tumor (Hartmann's technique) to attempt to reconstruct the transit in a second stage. This technique yields high morbimortality rates (10%–36% according to a previous series[3]) and has important deleterious effects on the quality of life of patients, as in up to 50% of cases, the colostomy becomes definitive when transit reconstruction is rejected owing to several factors, including the patient's condition and stage of the tumor.

In 1991 Dohmoto et al.[4] described the use of endoscopically placed prostheses to manage colonic strictures of neoplastic origin for the first time based on their experience with the use of prostheses for the treatment of esophageal malignant pathology. This technique became very popular, and by the end of the 1990s, the indication for prosthesis placement in patients with acute colonic obstruction secondary to CRC was already fully established.

There are multiple studies on the efficacy of this technique, with two recent meta-analyses concluding that in the short term, colonic prosthesis placement is associated with a lower morbidity, similar postoperative mortality, and higher proportion of primary anastomoses.[5,6]

However, there is controversy regarding whether the placement of a prosthesis as a bridge to surgery could increase loco-regional recurrence,[7] although this seems to be observed in a series with high perforation rates during the procedure (>10%).[8] In the absence of evidence of this association, the ESGE in its latest recommendations considers colonic prosthesis placement as an option to be considered in this scenario.[9]

Indications and contraindications

Indications

The indications for the use of colonic prostheses in tumor pathology could be summarized as follows:

- *"Bridge" treatment to surgery*[10]: In right colonic tumors, colonic prostheses should be used for palliative purposes, as urgent surgery in a single stage is possible in most cases.
- *Palliative treatment*: It is not uncommon for cases that debut with obstruction to be advanced tumors with disseminated disease. In patients who are inoperable owing to comorbidities or nonsurgical tumors considering their stage, the placement of a prosthesis avoids the need for surgery and has a palliative purpose, with a lower rate of stoma and a shorter duration of hospital stay, considerably improving the quality of life of these patients.
- *Treatment of postoperative complications*: In recent years, prostheses have been used for the treatment of anastomotic complications in surgeries for CRC or Crohn's disease. Stenosis of colonic anastomoses occurs in approximately 8% of cases, being more frequent in the distal rectum. Although the initial treatment is usually endoscopic balloon dilatation, the use of coated or biodegradable stents has been described as an alternative to surgery in refractory strictures in selected cases.[11–15] Similarly, case series wherein esophageal overlay prostheses were used as a treatment for surgical anastomotic leaks have also been described.[16]

Contraindications

There are two aspects that must be taken into account in cases of colonic obstruction, as they may contraindicate the placement of an expandable prosthesis.

- *Perforation/abscess*: In these cases, placement of a prosthesis increases the risk of peritonitis and is of no benefit.[17]
- *Distance to the anal margin*: If the tumor is located very close to the anal margin (<5 cm), placement of a prosthesis should be avoided, as secondary tenesmus would require removal of the device.

Endoscopic technique

Although the diagnosis of colonic obstruction can be established with clinical findings and a simple abdominal radiography, a complementary imaging technique (computed tomography [CT]) must be performed to confirm the existence of the lesion and its location and dimensions.[18] Complications (e.g., abscess or perforation) must also be excluded, and the existence of distant disease or synchronous lesions must be assessed. Furthermore, prior to the procedure, it is necessary to inform the patients regarding the technique and its benefits and risks in their specific case and obtain their consent.

The placement of a colonic prosthesis, as shown later, is not free of complications and should therefore be performed by experienced medical personnel. Studies estimate that 10–20 procedures, as well as experience in the use of guidewires and radiology, are required to achieve the highest success rate (96%) and the lowest number of complications (10%).[19]

Scopic equipment should be available to allow radiological vision during the procedure. Although not essential, especially in cases of more distal obstruction, sedation improves patient cooperation and tolerance.

Different stents for the endoscopic treatment of tumoral colonic strictures, with different sizes, materials, and delivery mechanisms, are currently available in the market. A recent meta-analysis compared the efficacy between coated and uncoated stents.[20] Generally, uncoated prostheses were associated with fewer complications, including migration, longer duration of patency, and lesser need for reinsertion. The main disadvantage of uncoated stents is stent occlusion owing to intratumoral growth, which can be addressed either by placing a second stent coaxially to the initial stent (Fig. 26.1) or by endoscopic ablative treatments.

The length of the stent is selected according to the length of the stenosis visualized on CT scan or with the administration of contrast during the procedure. An endoscope with a therapeutic working channel is usually used so that the stent can be introduced through it. Progress is taken using as minimal insufflation as possible until the lesion is reached, and biopsies are performed for histological confirmation.

The next step is to introduce a guidewire through the stricture with the aid of radiological control and once in place, the prosthesis must slide over it (Fig. 26.2). Most of the technical failures when placing a prosthesis are due to the impossibility of progressing with the guidewire, usually because of the angulation of the lesion and difficulty in correctly facing it with the endoscope. To solve this problem, several studies have described the use of a rotating sphincterotome, which allows the

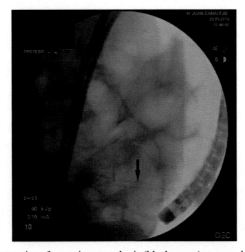

FIG. 26.1 Scopic image showing how the obstruction of a previous prosthesis (black arrow) was resolved by fitting a second prosthesis coaxially to it (red arrow).

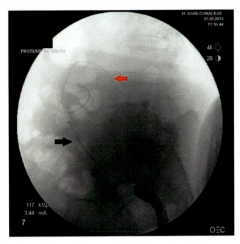

FIG. 26.2 Image showing the guidewire passing through the tumor stenosis (red arrow) and the not yet released prosthesis sliding over it (black arrow).

tip of the guidewire to be angled and oriented toward the small lumen of the tumor.[21–23] In addition, as already mentioned, small amounts of contrast instilled through the endoscope channel can also help draw the morphology of the stricture.[18]

Once the guidewire is in place, the prosthesis is advanced over the guidewire and introduced through the endoscope channel until it is positioned inside the tumor under direct endoscopic vision and radiological control. The prostheses have radiopaque markings that help locate the ends. Notably, these markings must be at least 2 cm above and below the lesion to ensure adequate expansion once the device is released.

When the prosthesis is correctly positioned, it must be released, while considering that at this point, it might move proximally (toward the "interior" aspect of the tumor); thus, its position must be continuously corrected by pulling on the introducer that contains it.

Once released, the guidewire and the introducer sheath are removed, and whether both ends are well expanded and whether the prosthesis exceeds the area of tumor stenosis on both sides are checked using the scope (Fig. 26.3). With endoscopic vision, its correct functioning can be checked by visualizing the exit of feces (Fig. 26.4).

As an alternative to the usual procedure, and when the diameter of the stricture permits, a fine-gauge endoscope can be used to pass through the tumor and leave the guidewire at the other end of the lesion. Once this is performed, the endoscope is removed, leaving the guidewire in place; with the aid of a biopsy forceps, the end of the guidewire is reinserted in a retrograde manner through the working channel of the operating endoscope, and from this point onward, the usual procedure is followed.

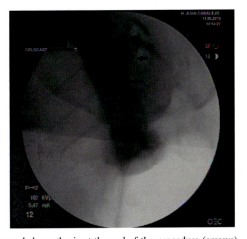

FIG. 26.3 Scopic view of the released and expanded prosthesis at the end of the procedure (arrows).

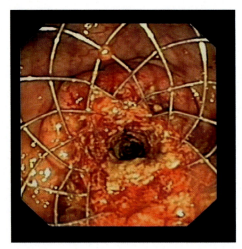

FIG. 26.4 Endoscopic view of the prosthesis once in place across the tumor stenosis.

Results and complications

In general, the technical success rate is 93% (varies from 83% to 100% according to a series).[24–26] The causes of failure are the impossibility of reaching the stricture owing to angulation of the colon secondary to the process itself or in most cases, the impossibility of progressing the guidewire through the lesion. Notably, technical success is not always synonymous with clinical success, which is achieved in 77%–100% of cases. Although a prosthesis is correctly placed, the obstructive condition may not resolve owing to poor expansion of the prosthesis (e.g., frozen pelvis or excessive angulation of the tumor) or the existence of stenosis at other levels[18] (e.g., synchronous tumors or stenosis of the intestine due to carcinomatosis).

Therefore strict clinical observation of the patient (gradual improvement of abdominal distension and pain must be checked) or even radiological monitoring to detect cases in which the prosthesis does not work is necessary, as they will require urgent surgery in the same manner.

Although the mortality rate of the procedure is very low (<1%), it is not without risk. The univariate analysis of some previous studies[26] described factors that appear to increase the risk of complications, such as the presence of complete obstruction, previous dilatation, male sex, lack of experience of the operator, or treatment with antiangiogenic drugs (bevacizumab).

The rate of complications varies between 25% and 10%,[27] and the most frequent complications observed are as follows:

- *Perforation (0%–12%)*: This complication can occur both at the level of the tumor itself during the placement of the prosthesis (with the end of the prosthesis or the guidewire) and at a distance as a result of colonic distension added with insufflation or the nonresolution of the obstructive condition. It results in urgent surgery and is more likely in very angulated lesions, e.g., lesions in the sigmoid colon, or if dilatation of the previous stricture is performed[18, 28] (procedure currently in disuse).
- *Migration (1%–10%)*: This complication more frequently occurs in the first week and distally, expelling through the anus, and seems to be more likely if the prosthesis is covered (in cases of late migration) after treatment with CRT or if dilatation has been performed. It can sometimes be solved by fitting a new prosthesis.
- *Tenesmus*: As already mentioned, when the tumor is located in the rectum and the prosthesis is close to the pectineal line, pain, tenesmus, or incontinence may occur.
- *Reobstruction (7.3%)*: When the prosthesis is used for long-term palliative purposes (median: 24 weeks), tumor tissue may grow through the prosthesis strands, causing reobstruction. This is less likely to occur if it is a coated prosthesis, and placement of a second prosthesis through the first may resolve the problem.[29, 30]

In summary, endoscopic prosthesis placement is currently an effective alternative for selected patients with colonic obstruction secondary to CRC, either as a palliative treatment or as a bridge to surgery.

References

1. Watson AJM, Shanmugam V, Mackay I, et al. Outcomes after placement of colorectal stents. *Colorectal Dis* 2005;**7**(1):70–3. https://doi.org/10.1111/j.1463-1318.2004.00727.x.
2. Adler DG, Baron TH. Endoscopic palliation of colorectal cancer. *Hematol Oncol Clin North Am* 2002;**16**(4):1015–29. https://doi.org/10.1016/s0889-8588(02)00034-5.
3. Hennekinne-Mucci S, Tuech J-J, Bréhant O, et al. Emergency subtotal/total colectomy in the management of obstructed left colon carcinoma. *Int J Colorectal Dis* 2006;**21**(6):538–41. https://doi.org/10.1007/s00384-005-0048-7.
4. Dohmoto M. New method-endoscopic implantation of rectal stent in palliative treatment of malignant stenosis. *Endoscopia Digest* 1991;**3**:1507–12. https://ci.nii.ac.jp/naid/10007034916/. [Accessed 22 February 2021].
5. Foo CC, Poon SHT, Chiu RHY, Lam WY, Cheung LC, Law WL. Is bridge to surgery stenting a safe alternative to emergency surgery in malignant colonic obstruction: a meta-analysis of randomized control trials. *Surg Endosc* 2019;**33**(1):293–302. https://doi.org/10.1007/s00464-018-6487-3.
6. Yang P, Lin X-F, Lin K, Li W. The role of stents as bridge to surgery for acute left-sided obstructive colorectal cancer: meta-analysis of randomized controlled trials. *Rev Invest Clin* 2018;**70**(6):269–78. https://doi.org/10.24875/RIC.18002516.
7. Sloothaak DAM, Van Den Berg MW, Dijkgraaf MGW, et al. Oncological outcome of malignant colonic obstruction in the Dutch stent-in 2 trial. *Br J Surg* 2014;**101**(13):1751–7. https://www.academia.edu/download/44889512/Oncological_outcome_of_malignant_colonic20160419-19960-rkgscb.pdf.
8. Amelung FJ, Burghgraef TA, Tanis PJ, et al. Critical appraisal of oncological safety of stent as bridge to surgery in left-sided obstructing colon cancer; a systematic review and meta-analysis. *Crit Rev Oncol Hematol* 2018;**131**:66–75. https://doi.org/10.1016/j.critrevonc.2018.08.003.
9. van Hooft JE, Veld JV, Arnold D, et al. Self-expandable metal stents for obstructing colonic and extracolonic cancer: European Society of Gastrointestinal Endoscopy (ESGE) guideline—update 2020. *Endoscopy* 2020;**52**(5):389–407. https://doi.org/10.1055/a-1140-3017.
10. Ng KC, Law WL, Lee YM, Choi HK, Seto CL, Ho JWC. Self-expanding metallic stent as a bridge to surgery versus emergency resection for obstructing left-sided colorectal cancer: a case-matched study. *J Gastrointest Surg* 2006;**10**(6):798–803. https://doi.org/10.1016/j.gassur.2006.02.006.
11. Keränen I, Lepistö A, Udd M, Halttunen J, Kylänpää L. Outcome of patients after endoluminal stent placement for benign colorectal obstruction. *Scand J Gastroenterol* 2010;**45**(6):725–31. https://doi.org/10.3109/00365521003663696.
12. Geiger TM, Miedema BW, Tsereteli Z, Sporn E, Thaler K. Stent placement for benign colonic stenosis: case report, review of the literature, and animal pilot data. *Int J Colorectal Dis* 2008;**23**(10):1007–12. https://doi.org/10.1007/s00384-008-0518-9.
13. Small AJ, Young-Fadok TM, Baron TH. Expandable metal stent placement for benign colorectal obstruction: outcomes for 23 cases. *Surg Endosc* 2008;**22**(2):454–62. https://doi.org/10.1007/s00464-007-9453-z.
14. García-Cano J. Dilation of benign strictures in the esophagus and colon with the polyflex stent: a case series study. *Dig Dis Sci* 2008;**53**(2):341–6. https://doi.org/10.1007/s10620-007-9864-7.
15. Repici A, Pagano N, Rando G, et al. A retrospective analysis of early and late outcome of biodegradable stent placement in the management of refractory anastomotic colorectal strictures. *Surg Endosc* 2013;**27**(7):2487–91. https://doi.org/10.1007/s00464-012-2762-x.
16. DiMaio CJ, Dorfman MP, Gardner GJ, et al. Covered esophageal self-expandable metal stents in the nonoperative management of postoperative colorectal anastomotic leaks. *Gastrointest Endosc* 2012;**76**(2):431–5. https://doi.org/10.1016/j.gie.2012.03.1393.
17. Keymling M. Colorectal stenting. *Endoscopy* 2003;**35**(3):234–8. https://doi.org/10.1055/s-2003-37265.
18. Mosca S. How can we improve the implementation of new endoscopic techniques? Concerning colonic stenting. *Endoscopy* 2003;**35**(8):709–10. author reply 711-2 https://doi.org/10.1055/s-2003-41525.
19. Geraghty J, Sarkar S, Cox T, et al. Management of large bowel obstruction with self-expanding metal stents. A multicentre retrospective study of factors determining outcome. *Colorectal Dis* 2014;**16**(6):476–83. https://doi.org/10.1111/codi.12582.
20. Mashar M, Mashar R, Hajibandeh S. Uncovered versus covered stent in management of large bowel obstruction due to colorectal malignancy: a systematic review and meta-analysis. *Int J Colorectal Dis* 2019;**34**(5):773–85. https://doi.org/10.1007/s00384-019-03277-3.
21. Vázquez-Iglesias JL, Gonzalez-Conde B, Vázquez-Millán MA, Estévez-Prieto E, Alonso-Aguirre P. Self-expandable stents in malignant colonic obstruction: insertion assisted with a sphincterotome in technically difficult cases. *Gastrointest Endosc* 2005;**62**(3):436–7. https://doi.org/10.1016/j.gie.2005.04.028.
22. Armstrong EM, Fox BM. Assistance of colorectal stent insertion by sphincterotome. *Dis Colon Rectum* 2007;**50**(3):399–400. https://doi.org/10.1007/s10350-006-0715-y.
23. Rosés L, González Ramírez A, Lancho Seco A, Soto Iglesias S, Santos Blanco E, Avila S. A new use for the rotatable sphincterotome as an aid for stenting malignant gastrointestinal tract stenoses. *Endoscopy* 2004;**36**(12):1132. https://doi.org/10.1055/s-2004-825981.
24. Cases-Baldó MJ, García-Marín JA, Aguayo-Albasini JL, Pellicer-Franco E, Soria-Aledo V, Pérez-Cuadrado E. Colorectal stents: efficacy and complications in our center. *Cir Cir* 2012;**80**(6):523–7. https://www.medigraphic.com/cgi-bin/new/resumenI.cgi?IDREVISTA=10&IDARTICULO=38725&IDPUBLICACION=4094.
25. Tirosh D, Perry Z, Walfisch S, et al. Endoscopic self-expanding metal stents for acute colonic obstruction. *Am Surg* 2013;**79**(1):30–4. https://www.ncbi.nlm.nih.gov/pubmed/23317598.
26. Small AJ, Coelho-Prabhu N, Baron TH. Endoscopic placement of self-expandable metal stents for malignant colonic obstruction: long-term outcomes and complication factors. *Gastrointest Endosc* 2010;**71**(3):560–72. https://doi.org/10.1016/j.gie.2009.10.012.
27. Gianotti L, Tamini N, Nespoli L, et al. A prospective evaluation of short-term and long-term results from colonic stenting for palliation or as a bridge to elective operation versus immediate surgery for large-bowel obstruction. *Surg Endosc* 2013;**27**(3):832–42. https://doi.org/10.1007/s00464-012-2520-0.

28. Baron TH, Rey JF, Spinelli P. Expandable metal stent placement for malignant colorectal obstruction. *Endoscopy* 2002;**34**(10):823–30. https://doi.org/10.1055/s-2002-34271.
29. Ding X-L, Li Y-D, Yang R-M, Li F-B, Zhang M-Q. A temporary self-expanding metallic stent for malignant colorectal obstruction. *World J Gastroenterol* 2013;**19**(7):1119–23. https://doi.org/10.3748/wjg.v19.i7.1119.
30. Zhang Y, Shi J, Shi B, Song CY, Xie WF, Chen YX. Comparison of efficacy between uncovered and covered self-expanding metallic stents in malignant large bowel obstruction: a systematic review and meta-analysis. *Colorectal Dis* 2012;**14**(7):e367–74. https://doi.org/10.1111/j.1463-1318.2012.03056.x.

Section D.II

Surgical treatment

Chapter 27

Emergency surgery for CRC

Alejandra García Novoa[a], Alba Gómez Dovigo[a], Tatiana María Civeira Taboada[b], and José Francisco Noguera Aguilar[c]

[a]*General Surgery Service, Quironsalud A Coruña Hospital, A Coruña, Spain,* [b]*Coloproctology Unit—CHUAC, A Coruña, Spain,* [c]*Coloproctology Unit—CHUAC, Head of the General Surgery Department of the University Hospital of A Coruña, A Coruña, Spain*

Colon cancer is the third leading type of cancer worldwide and fourth leading cause of cancer death. It is the third leading type of cancer in men (after prostate and lung cancers) and the second in women (after breast cancer).[1]

In Spain, the absolute number of colorectal cancers diagnosed has been increasing for decades. This is probably related to two main factors: (1) characteristics of the population itself, such as aging; exposure to risk factors, including tobacco use, alcohol consumption, and pollution; and obesity or a sedentary lifestyle and (2) development of early detection programs in some types of cancer, such as colorectal, breast, cervical, or prostate cancer. In 2017 more than 34,000 new cases of colorectal cancer were detected, representing 15% of all solid tumors; this type of cancer was the second leading cause of death from cancer, after lung cancer.[2]

Despite greater access to healthcare and the application of screening programs, approximately 30% of colorectal carcinomas are diagnosed in emergency owing to complications of the tumor.[3,4] These tumors are usually more aggressive and have a worse prognosis. Obstruction due to colorectal carcinoma accounts for 80% of these emergencies, and intestinal perforation accounts for the remaining 20% of colon carcinoma emergencies.[1]

Despite the standardization of the treatment of patients with colorectal carcinoma, the ideal treatment in the emergency department remains controversial.[5] The aim of this chapter is to discuss the surgical options in different emergency scenarios for colorectal tumor complications.

Diagnosis in the emergency department

There are three frequent causes of consultation in the emergency department secondary to a colonic tumor: bleeding from the tumor, bowel obstruction, and intestinal perforation. The first may present as rectorrhagia or melena, which destabilizes the patient and requires urgent intervention on rare occasions.[5]

Bowel obstruction is the most frequent complication of colorectal tumors in the emergency department. Most obstructing tumors are found in the left colon/sigmoid colon; in fact, in 75% of patients, the tumor causing the obstruction is distal to the splenic flexure of the colon.[1] Clinically, patients present to the emergency room with colicky abdominal pain, absence of gas (90%) and bowel movements (80%), and abdominal distention (65%).[1]

Intestinal perforation secondary to a colonic tumor can be at the level of or proximal to the tumor. In up to 30% of patients, perforation occurs proximal to the tumor secondary to the obstruction generated at the level of the tumor.[1] Patients debuting with perforation proximal to the tumor usually have fecaloid peritonitis and secondary septic shock. Mortality in these patients is 19%–65% compared to 0%–24% in patients who debut with collections contained by perforations at the level of the tumor.[1]

Diagnostic tests

Abdominal radiography has very low sensitivity and specificity to diagnose perforation and/or obstruction of the colon. However, when there is clinical suspicion of intestinal obstruction, intestinal radiography is a good, easy, and inexpensive resource for a first diagnostic approach.

When colon cancer is suspected, bedside ultrasound can help in the diagnosis. However, the radiological test with the highest sensitivity and specificity for the diagnosis of a complicated colonic tumor is computed tomography (CT) (S: 95%; E: 90%).[1] However, in patients with clinical symptoms of diffuse peritonitis or hemodynamic instability, treatment (medical and surgical) should not be delayed to perform CT. CT allows staging of the tumor, ruling out distant disease (hepatic) or peritoneal carcinomatosis, which may lead to modifications in the therapeutic approach. In addition, CT helps the surgeon establish a surgical strategy and the required material.

Chest CT for staging does not need to be performed in the emergency department. The sensitivity of this test is very high (100%), but with false negative rates as high as 34%. In contrast, chest radiography has low sensitivity (30%–64%) but has a specificity of 90%. Therefore, in the emergency department setting, chest radiography will be sufficient for thoracic staging of a patient with suspected complicated colon cancer.[1]

Colonoscopy is not mandatory; however, in stable patients with colorectal obstruction, diagnostic colonoscopy can be considered to rule out other causes of obstruction or even to evaluate endoscopic treatment (prosthesis/stent).

Treatment of complicated colorectal carcinoma

The main objective of urgent treatment of patients with colorectal cancer is to resolve the life-threatening cause. However, obstruction and/or perforation occurs in an oncologic setting. In this context, the maximum oncologic safety that allows urgency must be provided. Therefore the obstruction or perforation must be managed following the oncologic principles of tumor resectability and lymphadenectomy as long as the patient is hemodynamically stable. In addition, the treatment that provides the least morbidity and the best quality of life, without modifying the safety of the intervention and avoiding stomas, must be provided to the patient.

Several studies have shown that patients who undergo emergency surgery have a higher rate of loco-regional and distant relapses,[5] and the simple fact of debuting with obstruction or perforation increases the possibility of requiring adjuvant chemotherapy. For this reason, surgical intervention must solve the complication generated by the tumor, ensuring early recovery that avoids delay in the start of systemic treatment, which will allow distant control of the disease and therefore prolong the overall survival of the patient.

In this context, the UK clinical guidelines[6] suggest that all patients with colorectal carcinoma should be operated on by a colorectal surgeon as part of a multidisciplinary team. However, in the emergency setting, it is not always possible. For this reason, some groups have proposed using the strategy of damage control surgery in patients with colon cancer in the emergency department. Obviously, there are situations in which an emergency resection is mandatory; however, in most cases, definitive treatment can be safely postponed following the same principles of damage control surgery for trauma. Thus, for example, in cases of obstructing tumors, the emergency can be solved by using a shunt (ileostomy, colostomy, or stent); in cases of perforated tumors, resection of the perforated bowel can be performed; and in cases of poor general conditions, drainage and proximal shunting can be considered.[5]

Thus the objectives of urgent surgery in colorectal cancer, in order of priority, are as follows:

1. To solve the intestinal obstruction or perforation that has generated the need for urgent surgery
2. To perform an intervention with oncological criteria (e.g., free margins or lymphadenectomy) as long as it does not increase morbidity
3. To perform quality surgery with the least morbidity necessary, allowing early recovery. If the surgeon's skill and the patient's conditions allow it, a laparoscopic intervention can be offered, which will allow early recovery.
4. To avoid a stoma, if the infectious and systemic context of the patient allows it, particularly a definitive stoma.
5. To increase overall survival. The goal of any treatment and even more so a surgical intervention should be to increase the patient's overall survival or palliative care, which increases the quality of life.

Type of surgery

There are three main possibilities in the urgent surgery of patients with colorectal tumor:

- Tumor resection and anastomosis, with or without a protective stoma.
- Tumor resection and terminal stoma.
- Stoma proximal to the tumor, without tumor resection.

The decision of the type of intervention to be performed will depend on different factors, including the following:

- *Surgeon's experience*: A colorectal surgeon will be able to perform interventions safely with quality and oncological criteria in an emergency. On the contrary, a young surgeon not specialized in coloproctology will have more difficulty in solving the emergency.

- *Patient's age*: The treatment proposed to the patient will depend on the patient's life expectancy. Sometimes, more important than age will be the patient's overall condition and comorbidities. Depending on the patient's life expectancy, an aggressive curative intervention or palliative measures may be proposed.
- *Tumor staging and localization*: Obviously, tumor staging will change the approach strategy. A patient with obstructive colorectal carcinoma with distant metastasis will probably benefit from surgery with minimal morbidity, as tumor resection will not modify the survival, and early recovery is needed to receive adjuvant systemic treatments to prolong the overall survival. Therefore, in patients with stage IV colorectal carcinoma, a stoma (ileostomy or colostomy) proximal to the point of obstruction will be an optimal solution that will resolve the obstruction with minimal morbidity. Even in patients with intestinal obstruction, a colonic stent should be considered. The only controversial point of a stent is that it presents a higher rate of complications (perforation) in patients who require treatment with antiangiogenic agents.[7]

Similarly, in patients with locally advanced tumors without distant metastases, a proximal stoma allows treatment with neoadjuvant chemotherapy and salvage surgery if the tumor mass responds to systemic treatment. This strategy is particularly beneficial in rectal tumors.

- *Patient stability*: In a hemodynamically unstable patient, we must always give priority to resolving the urgent tumor complication, followed by the oncological criteria. Therefore, in these cases, most of which will correspond to patients with fecaloid peritonitis due to perforations proximal to the tumor, the sepsis must be resolved, and anastomosis should not be considered.

Each scenario is a different context, in which the surgeon must decide the strategy based on the resources and the patient's presentation. In patients debuting with intestinal perforation, it is advisable to resect the tumor and thus control the focus of peritonitis. In patients with unresectable tumors, a proximal stoma without resection can be considered; however, control of the focus of peritonitis will be more tortuous.[1,5]

The decision for an anastomosis will depend mainly on two factors[1]: the surgeon's experience and[2] the patient's conditions and peritoneal environment during surgery. A surgeon with minimal experience in colorectal surgery in a center with few cases should not consider anastomosis. On the contrary, an expert surgeon should consider anastomosis as long as the patient is stable and there is no fecaloid peritonitis. In cases of rectal tumors, anastomosis and proximal ileostomy can be considered.

It is possible to perform surgery in two stages. A first intervention for damage control and then an appropriate oncologic intervention can be performed by an expert colorectal surgeon. However, the current literature regarding this topic is controversial, and there is no consensus on whether it is better for an on-call surgeon to perform the emergency oncologic intervention or whether it is better to perform the intervention in two stages, sometimes allowing neoadjuvant treatment. In patients with operable disease, both surgical approaches may be acceptable. The 2019 National Comprehensive Cancer Network (NCCN) clinical guidelines[8] consider tumor resection, bypass surgery, or stenting as a valid strategy. Some meta-analyses[9,10] have reported similar short- and long-term results of resection surgery between stenting and deferred surgery. These results are similar to those reported in the multicenter clinical trial ESCO,[11] which included Spanish centers. Another recent meta-analysis[12] demonstrated similar 30-day morbimortality between bypass surgery and resection, with a lower rate of permanent stomas in the bypass surgery group, establishing that stents are a safe strategy for resolving the emergency.

Another notable strategy, which was also contemplated in the NCCN guidelines, is neoadjuvant chemotherapy in patients with locally unresectable tumors or in patients who are medically inoperable at debut. Thus, in some patients, it is possible to transform unresectable tumors into potentially resectable lesions.

In conclusion, the surgical strategy will depend on the surgeon's skills and the patient's condition. In emergencies, the oncologic context is different, and performance of two-stage surgery may reduce morbidity.

References

1. Pisano M, Zorcolo L, Merli C, Cimbanassi S, Poiasina E, Ceresoli M, et al. 2017 WSES guidelines on colon and rectal cancer emergencies: obstruction and perforation. *World J Emerg Surg* 2018;**13**:36.
2. Sociedad Española de Oncología Médica (SEOM). 2019. Las cifras del Cáncer en España 2019. (www.seom.org). https://seom.org/dmcancer/wp-content/uploads/2019/Informe-SEOM-cifras-cancer-2019.pdf (Consultado por última vez: 13/10/2019).
3. Gunnarsson H, Jennische K, Forssell S, Granström J, Jestin P, Ekholm A, et al. Heterogeneity of colon cancer patients reported as emergencies. *World J Surg* 2014;**38**(7):1819–26.

4. Renzi C, Lyratzopoulos G, Card T, Chu TP, Macleod U, Rachet B. Do colorectal cancer patients diagnosed as an emergency differ from non-emergency patients in their consultation patterns and symptoms? A longitudinal data-linkage study in England. *Br J Cancer* 2016;**115**(7):866–75.
5. Tebala GD, Natili A, Gallucci A, Brachini G, Khan AQ, Tebala D, et al. Emergency treatment of complicated colorectal cancer. *Cancer Mang Res* 2018;**10**:827–38. https://doi.org/10.2147/CMAR.S158335.
6. Moran B, Cunningham C, Singh T, Sagar P, Bradbury J, Geh I, et al. Association of Coloproctology of Great Britain and Ireland (ACPGBI): guidelines for the management of cancer of the colon, rectum and anus (2017). Surgical management. *Colorectal Dis* 2017;**19**(Suppl. 1):18–36.
7. Pacheco-Barcia V, Mondéjar R, Martínez-Sáez O, Longo F, Moreno JA, Rogado J, et al. Safety and oncological outcomes of bevacizumab therapy in patients with advanced colorectal Cancer and self-expandable metal stents. *Clin Colorectal Cancer* 2019;**18**(3):e287–93. https://doi.org/10.1016/j.clcc.2019.05.009.
8. Guía clínica de cáncer de colon de NCCN (National Compnhesive Cancer Network), 2019. https://www.nccn.org/professionals/physician_gls/pdf/colon.pdf (Consultado por última vez: 13/10/2019).
9. Huang X, Lv B, Zhang S, Meng L. Preoperative colonic stents versus emergency surgery for acute left-sided malignant colonic obstruction: a meta-analysis. *J Gastrointest Surg* 2014;**18**(3):584–91. https://doi.org/10.1007/s11605-013-2344-9.
10. Matsuda A, Miyashita M, Matsumoto S, Matsutani T, Sakurazawa N, Takahashi G, et al. Comparison of long-term outcomes of colonic stent as "bridge to surgery" and emergency surgery for malignant large-bowel obstruction: a meta-analysis. *Ann Surg Oncol* 2015;**22**(2):497–504. https://doi.org/10.1245/s10434-014-3997-7.
11. Arezzo A, Balague C, Targarona E, Borghi F, Giraudo G, Ghezzo L, et al. Colonic stenting as a bridge to surgery versus emergency surgery for malignant colonic obstruction: results of a multicentre randomised controlled trial (ESCO trial). *Surg Endosc* 2017;**31**(8):3297–305. https://doi.org/10.1007/s00464-016-5362-3.
12. Amelung FJ, Mulder CL, Verheijen PM, Draaisma WA, Siersema PD, Consten EC. Acute resection versus bridge to surgery with diverting colostomy for patients with acute malignant left sided colonic obstruction: systematic review and meta-analysis. *Surg Oncol* 2015;**24**(4):313–21. https://doi.org/10.1016/j.suronc.2015.10.003.

Chapter 28

Innovation and new technologies in colorectal cancer UNIVEC device development experience

Alba Gómez Dovigo[a], Alejandra García Novoa[a], Javier Aguirrezabalaga González[b], José Francisco Noguera Aguilar[c], and Alberto Centeno Cortés[d]

[a]*General Surgery Service, Quironsalud A Coruña Hospital, A Coruña, Spain*, [b]*HBP Unit—CHUAC, A Coruña, Spain*, [c]*General Surgery Service of CHUAC, A Coruña, Spain*, [d]*CTF-XXIAC, A Coruña, Spain*

Introduction

In the history of medicine, the existence of innovative surgeons has been fundamental to the development and diffusion of technology, just as the context, which includes the support of the scientific community, is equally important.

Innovation is defined as the creation or modification of a product and its introduction into the market. To innovate is to use knowledge to generate value.[1] Currently, *"surgical value"* includes not only the quality and safety of the procedure, but also refers to the ratio between the results (clinical effectiveness + perceived quality) and the adverse effects (costs and invasiveness).

Technological innovation applied to colorectal cancer (CRC) surgery is in continuous evolution and progress with the aim of improving oncologic outcomes and the quality of life of CRC patients.[2]

Minimally invasive surgery

It has always been considered that the benefits of surgical treatment of diseases are obtained with a certain acceptable level of pain and trauma to the patient. Minimizing this detrimental effect of any surgical procedure has been the driving force behind advances in surgery in the search for the least invasive, yet maintaining the same functional and oncologic outcomes.

With the improvement of technologies and the advancement of Minimally Invasive Surgery (MIS), the concept of "scarless surgery" arose in an attempt to treat certain diseases by obviating the need for incisions to access the peritoneal cavity. This evolution culminated in *Laparoendoscopic Single Site Surgery* (LESS) and *Natural Orifice Transluminal Endoscopic Surgery* (NOTES).[3]

- **LESS**: *Laparoendoscopic Single Site Surgery or* **SINGLE PORT**: encompasses the concept of any minimally invasive surgical procedure, performed through a single incision, using both conventional laparoscopic instruments and instruments with a certain degree of deflection (curved), viewing cameras and multivalvular platforms specially designed to facilitate the performance of LESS surgery.
- **NOTES**: *Natural Orifice Transluminal Endoscopic Surgery*: arises to further minimize access trauma to the abdominal wall, which could lead to better and faster recovery from surgery, less pain, postoperative complications, and long-term problems such as hernias. It is based on the possibility of performing surgical techniques through natural orifices (stomach, vagina, colon, rectum, bladder, urethra) by perforating the organ allowing direct entry into the peritoneal or thoracic cavity. The technical and ethical challenges involved in the perforation and closure of a healthy organ induced multiple and reasonable doubts and controversies at this procedure.

It is necessary that NOTES surgery complies with basic surgical principles: adequate visualization of the surgical field, adequate manipulation of organs and tissues, hemostatic control capacity, as well as minimizing peritoneal contamination. In 2006 the *Society of American Gastrointestinal and Endoscopic Surgeons* (SAGES) and the *American Society of*

Gastrointestinal Endoscopists (ASGE) established guidelines for performing NOTES surgery in centers that meet a series of requirements prior to its development in humans. The NOSCAR (*Natural Orifice Surgery Consortium for Assessment and Research*) was created to coordinate the development of NOTES. Recommendations for performing NOTES procedures included[4]:

- To have a multidisciplinary team, with the participation of gastroenterologists and surgeons with advanced therapeutic and laparoscopic endoscopic skills.
- To have previous training in experimental models.
- All clinical activities or publications must be carried out under the supervision of the ethics committee of the hospital where the intervention is performed.
- To be a member of SAGES and/or ASGE.
- Share the results with other members of the NOSCAR group in biannual meetings and make a registry of results.

The operating room should have double equipment, a rigid endoscopy tower, and a flexible endoscopy tower.

Of great importance for the consolidation of the NOTES technique is to achieve secure closure of the entry orifices. Transcolonic NOTES procedures have been almost entirely restricted to hybrid interventions with laparoscopic support. The use of the anastomotic site as access to the peritoneal cavity does not carry an additional risk of infection or complications apart from the risk of anastomosis. Besides, it has the advantage that when the anastomosis is performed, the access orifice is safely closed.

Transanal Total Mesorectal Excision (TaTME) is a *hybrid* NOTES approach, in which transanal access is used in combination with conventional laparoscopy for treatment of lower-middle rectal cancer with a distance ≤ 10 cm from the anal margin, providing advantages in men with narrow pelvis, obesity, or bulky uterus. The contraindications for this approach are involvement of the circumferential CRM+ margin, T4 or long anal canal tumors, pelvis with previous interventions, anal canal stenosis, fecal incontinence, and perianal sepsis.[5,6]

The future of NOTES surgery will depend on advances in medical engineering, the development and application of computerized and robotic technologies that facilitate optimal visual control, full movement of endoscopic instruments that allow for safe, responsible, and reproducible application. Like all emerging surgical procedures, it requires clinical practice and continuous evaluation.

Instruments, innovative devices, and robotic surgery

Highly skilled laparoscopic instruments

Conventional handheld laparoscopic instruments are long, straight, with rigid jaws at the tips and 4 degrees of movement, which offer limited dexterity inside the patient, making laparoscopic surgery difficult. Robotic systems offer greater range of motion but at a substantially increased cost. This has prompted the development of mechanically articulated (nonrobotic) devices that can offer some of the advantages of surgical robots without significantly increasing costs[7]:

- *Articulated laparoscopic instruments with pistol grips*: RealHand (Novare Surgical Systems, Inc.), Autonomy Laparo-Angle (Cambridge Endoscopic Devices, Inc.), SILS Hand Instrument (Covidien): These are 5-mm disposable instruments.
- *The Radius Surgical System* (Tuebingen Scientific): is a 10-mm reusable instrument that uses a mixed control method.
- *MAESTRO* (Vanderbilt University): Nonrobotic Dexterous Laparoscopic Instrument with a Wrist providing seven degrees of freedom.
- The FlexDex surgical instrument (FlexDex Surgical, Brighton, Michigan, EE. UU.): The device is connected by a frame to the forearm. It transmits the surgeon's hand, wrist, and arm movements from outside the patient to an end effector inside the patient's body. It allows suturing in difficult areas by offering greater precision and a 360° range of motion[8] (Fig. 28.1).
- *ArtiSential*: has a double-jointed structure that allows the instrument to move 360°. It has bipolar forceps, a needle holder, and a clip applicator.
- Others: *DragonFlex, MiFlex, Intuitool, Easy Grasp*.

There are a number of partially motorized JAiMY instruments (Endocontrol Company, Grenoble, France) that use motors within the handle of the device to actuate the wrist, although the incorporation of motors increases the cost compared to purely mechanical ones.

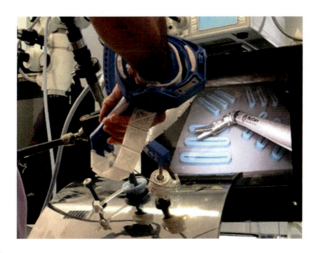

FIG. 28.1 FlexDex Surgical System.

Innovation in imaging systems

3D-HD display technology (4th generation): The conventional Laparoscopic Surgery (LS) is limited by a 2D vision that does not allow the perception of the operative field as in open surgery. To solve this problem, three-dimensional (3D) imaging technology was developed with the aid of a 3D-HD screen and polarized glasses. The transition from 2D to 3D vision requires an initial adaptation period. Once adaptation to the 3D view is achieved, it offers greater depth and spatial orientation allowing greater speed and precision of movements, improves intraoperative performance and postoperative results of colorectal cancer procedures, especially for procedures involving laparoscopic suturing or meticulous laparoscopic dissection as required to respect oncological principles.[9]

4K image systems: Ultra high-definition (4K) imaging systems are a new two-dimensional technology with 4 times the number of pixels of HD. It offers more detailed colors and images with greater depth perception and avoids the side effects of 3D systems (the need to use polarized glasses, eye fatigue, blurred vision, and difficulty in focusing).[10]

Fluorescence imaging with indocyanine green (ICG)[11]: is a simple, easily reproducible, relatively inexpensive, and useful technique for the following indications:

- *Evaluation of blood perfusion in anastomoses in colorectal surgery*: This allows real-time assessment of intestinal perfusion and irrigation, which would favor the prevention of anastomotic leakage. The PILAR II study showed a change in the surgical plan in 7.9% of patients.[12]
- *ICG lymphography* (Fig. 28.2):
  - One of the main reasons for 30% of early stage CRC recurrences is a pathologic underestimation related to an undetected lymph node. Fluorescence lymphography allows the identification of aberrant lymphatic vessels outside the planned resection with the potential to improve lymphadenectomy outcomes in colorectal cancer[11] by reducing the recurrence rate, more accurate disease staging, and appropriate adjuvant treatment.

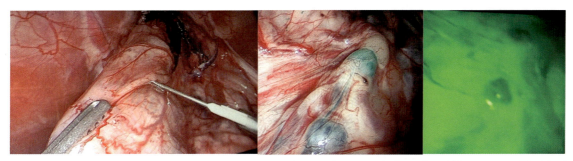

FIG. 28.2 ICG-guided GC identification procedure in a porcine model [performed at CTF-XXIAC (Centro Tecnológico de Formación de la Xerencia de Xestión integrada de A Coruña)].

- Although colorectal surgery currently includes lymphadenectomy as a routine with at least 12 LNs obtained, the possibility of ICG-guided sentinel node detection in CRC in combination with OSNA technology is being studied and would allow stratification of the extent of surgery based on the information obtained without compromising the prognostic or therapeutic value (e.g., change from transanal surgery to radical surgery). Further studies are needed to establish exact protocols limiting surgical dissection based on this technology.[13,14]
- *Assessment of CRC liver metastases:* Useful in the detection of additional metastatic lesions to those identified by previous diagnostic tests and for obtaining adequate resection margins, which would improve their staging and guarantee R0, resulting in a clear improvement in the prognosis of these patients.
- *Assessment of peritoneal carcinomatosis:* The use of ICG for in vivo localization of carcinomatosis during cytoreduction surgery and intraperitoneal hyperthermia is under study.[15]
- *Guide for total mesorectal excision (TME):* It has been used for the identification of the adequate plane in endoanal surgery, thus avoiding injury to neighboring structures (urethra, ureters, presacral fascia, prostate, vessels).[16]
- *Visualization of the ureter*: Avoids iatrogenic injury to the ureter, especially in complex rectal resections and reinterventions.
- *Tumor tattooing*: Endoscopic marking of the tumor may be useful for perioperative identification of the location of the tumor (as a substitute for Chinese ink).

The 3D reconstruction for surgical planning in colorectal surgery: The 3D technology makes it easier for radiologists to quantify or evaluate the extent of the tumor objectively and to recognize infiltration of neighboring structures (as in cases of low anterior rectal tumors). This allows surgeons to plan surgery more accurately: it evaluates complex anatomical regions such as the pelvis or the vascular regions of the abdomen, and allows oncologists to opt for more precise treatment pathways.[17]

Three-dimensional (3D) printing: is an emerging technology with a wide spectrum of possible applications in colorectal surgery: it allows the production of patient-specific anatomical replicas from 2D images in order to improve preoperative planning, as a tool to facilitate perioperative dissection, as a complement to current imaging techniques, and facilitates surgical training of the surgeon by improving surgical skills and patient understanding of their disease as well as training for stoma care. Initial results are satisfactory and further funding and research are needed for updating and maintenance.[18]

5G-assisted telementored surgery: Fifth generation (5G) wireless networks offer the possibility of working with more stable and high-speed data transmission, which is a major advantage of 5G. The main advantage of 5G is ***telementoring*** which allows complex surgical procedures to be performed safely and efficiently through telemedicine and remote surgery in real time, with a very high degree of satisfaction of the surgical team, as well as ***telestration*** (possibility of performing indications on monitors) with hardly any structures with the probable reduction of costs (since the system does not depend on a robot). Thus it would be an optimal resource for areas with a shortage of specialized surgeons.[19,20]

Google Glass: is a wearable device mounted on a pair of glasses, designed to display information similar to that of a smartphone on a screen, while allowing users to be hands-free. It has proven useful in various fields of surgery primarily as a teaching and training tool.[21]

Robotic surgery

For the past 20 years, the predominant robot in laparoscopic surgery has been the **Da Vinci Surgical System** (Intuitive Surgical, Sunnyvale, California, United States). This monopoly situation has led to rising costs and relatively slow innovation. New robotic platforms have been developed to compete in the market:

- **Da Vinci Surgical System** (Intuitive Surgical, Sunnyvale, California, United States). There are 4 versions of the system: ***Standard*** (2000–07), ***modelo S*** (2006), ***modelo Si HD*** (2011) access to the "new technologies" such as the single port, and the ***Xi HD*** (2014) which offers Firefly imaging for the visualization of tissues following infrared injection of indocyanine green ICG. The robotic system provides a three-dimensional view, allowing manipulation of miniaturized articulated instruments at a distance that offer human wrist-like movements with 7 degrees of freedom (**EndoWrist instruments**) with virtually no tremors. It consists of two parts: the master console for the surgeon with HD-3D imaging and 4 robotic arms. The movements of the surgeon's fingers are reproduced by the robot arms and instruments. Despite the advantages, the robotic system is used in only 2.8% of surgical cases. There are several obstacles that have limited the advancement of this technology: one of the main reasons is the cost of the robot, its instruments and maintenance, the size of the system, the impossibility of changing instruments quickly during the procedure, and the lack of tactile feedback. Another reason is the lack of evidence of the superiority of robots over laparoscopy. The ROLARR study has shown that robotic surgery is at least equivalent to other minimally invasive techniques. All this conditions its cost-effectiveness and widespread use.[22]

- **The Senhance Surgical System** (TransEnterix, Morrisville, North Carolina), available since 2016, uses 5-mm instruments, which is the standard size in laparoscopy, provides tactile feedback and HD-3D technology coupled with an eye-tracking camera control system that focuses the image on the point at which the surgeon is looking.[23]
- **Versius Surgical Robotic System**: It has an open console that allows you to sit or stand. The surgeon controls the robotic arms through a joystick on the console while wearing HD-3D glasses and viewing the monitor. It is one of the main robotic systems that will compete with the Da Vinci System with studies underway to evaluate its efficacy and safety.[24, 25]
- **Revo-i robotic surgical system** (Meere Company Inc.): is a Korean robotic system, which will compete in the future with the Da Vinci market. The first preclinical studies and clinical cases in humans demonstrate its safety and viability. It uses reusable endoscopic instruments.[26, 27]

Robotic camera controller[28]

- *FreeHand surgical robotic* (FreeHand 2010 Ltd., Cardiff, United Kingdom): is a mechanical robotic arm for holding the laparoscopic camera (no assistant required). It allows the surgeon to control the camera hands-free; it can move in three dimensions controlled by the operator's head and guided by laser. The surgeon wears a surgical cap containing an infrared transmitter that sends a signal to a receiver on the monitor. To select the direction of movement the surgeon moves his head in the desired direction.
- *SoloAssist II* (Aktormed, Barbing, Germany): It is a robotic camera control system with 6 joints and is installed on the side of the table. It contributes to shorten the operation time and reduces the number of assistant surgeons.
- **Other**: Robot-assisted colonoscope (***Invendoscopy E200 system*** (Invendo Medical GmbH, Germany), ***Flex Robotic System Technology*** (Medrobotics Corp., Raynham, MAI), Computer-assisted colonoscopy (the **NeoGuide Endoscopy System**), **SPIDER** (*Single-Port Instrument Delivery Extended Research* system, TransEnterix Durham).[29]

At present, the number of robotic laparoscopic colectomies performed in the world does not exceed 50% of the interventions for colon cancer, and at the national level it does not exceed 40%. The adoption of robotic colectomies requires studies that confirm safety, feasibility, and at least equality of results.[30]

Flexible single incision surgery

In Spain, the flexible endoscope is a device used by the digestive physician. At present, the flexible endoscope is rarely used in the operating room and is relegated to some specific procedures, such as exploration of the biliary tract using a choledochoscope, intraoperative localization of lesions, and certain emergency situations: extraction of foreign bodies, exploration of the upper and lower digestive tract (revision of anastomoses to demonstrate AF, assessment of bleeding, etc.), but the scarce use by surgeons means that on most occasions we depend on endoscopists for its management.[31] The training of surgeons dedicated to colorectal surgery will make flexible intraoperative endoscopy routinely available within the arsenal of operating room instruments. In addition to specific training programs in the use of the flexible endoscope for endoscopic problem solving, additional equipment in the operating room is indispensable.[31, 32]

Laparoendoscopic fusion surgery is a "hybrid" approach that takes advantage of previous experience with the use of the flexible endoscope and new minimally invasive surgical approaches. This technique known as Flexible Single Incision Surgery (FSIS) is performed using the flexible endoscope as the optical camera, light source, and working channel, which means adding 2 additional working channels and the standard laparoscopic surgery instruments and those of LESS surgery, thus transforming it into a surgery with 4 working channels.[33]

The choice of the platform for performing surgery remains a matter of debate. Currently, single-port platforms or devices show limitations (angulation, visualization, force transmission, etc.) that are limited to the pelvis below the promontory since they only allow the use of rigid laparoscopic optics.

The **UNI-VEC** device is a new multichannel transanal access platform that allows the maintenance of the pneumorectum and the introduction of flexible and surgical endoscopic instrumentation. It has a silicone introducer ring, a closing piece, and a rigid plastic top cover containing 3 working channels: a main channel with pneumatic sealing system through which the optical instrument is introduced (flexible endoscope or rigid laparoscope) and two secondary working channels for the introduction of minimally invasive surgical instrumentation (5 and 11–12 mm cannulas); the maintenance of the

FIG. 28.3 UNI-VEC device.

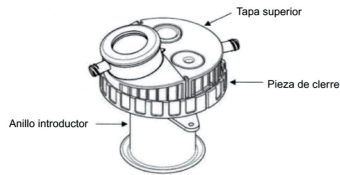

pneumorectum is performed by insufflation of CO_2 through its Luer-Lock connection with a working pressure of 12–15 mmHg (Fig. 28.3).

It will facilitate the performance of endoscopic and minimally invasive surgical procedures both in endoscopy rooms and in the operating room. It is especially indicated for **Endoscopic Mucosal Resection (EMR)**, for superficial nonneoplastic lesions; **Endoscopic Submucosal Dissection (ESD)**, for superficial benign and malignant lesions; **Endoscopic full thickness resection (EFTR)**, for thicker malignant lesions; **Transanal Minimally Invasive Surgery (TAMIS)** of rectal lesions of all types, benign and malignant, from mucosal excision to full-thickness resection; **Transanal Total Mesorectal Excision (TaTME)** endoscopic resection of malignant neoplastic lesions that are treated by excision of the rectum and mesorectum by combined transanal and abdominal route.

In the preclinical trials, a validation of the functional design of the device and the experimental evaluation in an in vivo animal model was performed at the CTF-XXIAC, with 2 main phases: *Conceptual Validation Phase* where it was demonstrated that the device maintains the pneumorectum with a minimal global leak with a CO_2 consumption rate of less than 0.5 L/min and the *Animal Validation Phase* where the device design was modified until reaching the final design with all the definitive components and which has allowed validating the use of the device in local transanal resection (EMR, ESD and EFTR) in porcine models, obtaining very satisfactory results, and proving that the device meets the minimum performance requirements.

A prospective multicenter clinical trial with 6 participating centers and an expected minimum number of 20 cases has been approved. The coordinating investigator is Dr. José Francisco Noguera of the Complexo Hospitalario Universitario de A Coruña (CHUAC) center in collaboration with *VecMedical Spain S.L.* (Figs. 28.4–28.6).

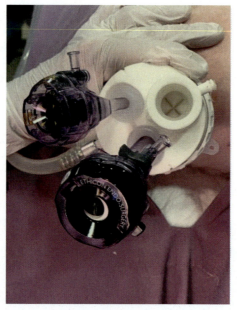

FIG. 28.4 UNIVEC device. Animal validation phase in a porcine model at CTF-XXIAC for the performance of a transanal ESD.

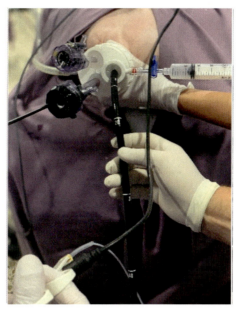

FIG. 28.5 UNIVEC device. Animal validation phase in a porcine model at CTF-XXIAC for the performance of a transanal ESD.

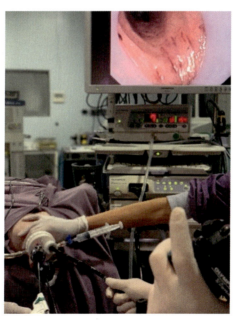

FIG. 28.6 UNIVEC device. Animal validation phase in a porcine model at CTF-XXIAC for the performance of a transanal ESD.

The UNIVEC multichannel surgical device will facilitate procedures not only in the surgical setting but also in the endoscopy suite. Close collaboration with endoscopic gastroenterologists is essential, while colorectal surgeons are being trained to work together.

References

1. DeJong CHC, Earnshaw JJ. Surgical innovation. *Br J Surg* 2015;**102**:e8–9.
2. Mayol J. Innovación en cirugía. *Cir Esp* 2016;**94**(4):207–9.
3. Noguera JF, Cuadrado A. NOTES, MANOS, SILS and other new laparoendoscopic techniques. *World J Gastrointest Endosc* 2012;**4**(6):212–7. https://doi.org/10.4253/wjge.v4.i6.212.

4. Rattner D, Kalloo A. ASGE/SAGES Working Group on Natural Orifice Translumenal Endoscopic Surgery. White Paper October 2005. *Gastrointest Endosc* 2006;**63**:199–203.
5. de Lacy AM, Rattner DW, Adelsdorfer C, Tasende MM, Fernández M, Delgado S, et al. Transanal natural orifice transluminal endoscopic surgery (NOTES) rectal resection: "down-to-up" total mesorectal excision (TME)—short-term outcomes in the first 20 cases. *Surg Endosc* 2013;**27**(9): 3165–72. https://doi.org/10.1007/s00464-013-2872-0. Epub 2013 Mar 22 23519489.
6. Díaz-López C. Escisión transanal mesorrectal ¿futuro o presente? *Cirugía Andaluza* 2018;**29**(4):520–4.
7. Anderson PL, Lathrop RA, Webster III RJ. Robot-like dexterity without computers and motors: a review of hand-held laparoscopic instruments with wrist-like tip articulation. *Expert Rev Med Devices* 2016;**13**(7):661–72. https://doi.org/10.1586/17434440.2016.1146585. 26808896. PMC5927586.
8. García Jiménez ML, Castro Diez L, Aguirrezabalaga González J, Noguera Aguilar JF. Sutura laparoscópica mecanizada con FlexDex Surgical System para ubicaciones anatómicamente difíciles. *Cir Esp* 2020;**14**. https://doi.org/10.1016/j.ciresp.2020.10.005.
9. Pantalos G, Patsouras D, Spartalis E, Dimitroulis D, Tsourouflis G, Nikiteas N. Three-dimensional versus two-dimensional laparoscopic surgery for colorectal cancer: systematic review and meta-analysis. *In Vivo* 2020;**34**(1):11–21. https://doi.org/10.21873/invivo.11740. 31882458. PMC6984079.
10. Abdelrahman M, Belramman A, Salem R, Patel B. Acquiring basic and advanced laparoscopic skills in novices using two-dimensional (2D), three-dimensional (3D) and ultra-high definition (4K) vision systems: a randomized control study. *Int J Surg* 2018;**53**:333–8. https://doi.org/10.1016/j.ijsu.2018.03.080. Epub 2018 Apr 12 29656132.
11. Barreiro Dominguez EM, Montero Zorrilla C, Ortín Navarro M. Test del verde de indocianina ¿es realmente útil en cirugía de colon y recto? *Archivos de Coloproctología* 2020;**3**(2):6–40. https://doi.org/10.26754/ojs_arcol/archcolo.202024567.
12. Jafari MD, Wexner SD, Martz JE, McLemore EC, Margolin DA, Sherwinter DA, et al. Perfusion assessment in laparoscopic left-sided/anterior resection (PILLAR II): a multi-institutional study. *J Am Coll Surg* 2015;**220**(1):82–92.e1. https://doi.org/10.1016/j.jamcollsurg.2014.09.015. Epub 2014 Sep 28 25451666.
13. Yeung TM, Wang LM, Colling R, Kraus R, Cahill R, Hompes R, et al. Intraoperative identification and analysis of lymph nodes at laparoscopic colorectal cancer surgery using fluorescence imaging combined with rapid OSNA pathological assessment. *Surg Endosc* 2018;**32**(2):1073–6. https://doi.org/10.1007/s00464-017-5644-4. Epub 2017 Jun 22 28643063. PMC5772117.
14. Noguera J, Castro L, Garcia L, Mosquera C, Gomez A. Lymphadenectomy guided by indocyanine-green (ICG) in colorectal cancer: a pilot study. *J Surg Tech Proced* 2019;**3**(1):1023.
15. Lieto E, Auricchio A, Cardella F, Mabilia A, Basile N, Castellano P, et al. Fluorescence-guided surgery in the combined treatment of peritoneal carcinomatosis from colorectal cancer: preliminary results and considerations. *World J Surg* 2018;**42**(4):1154–60. https://doi.org/10.1007/s00268-017-4237-7. 28929277.
16. Dapri G, Cahill R, Bourgeois P, Liberale G, Galdon Gomez M, Cadière GB. Peritumoural injection of indocyanine green fluorescence during transanal total mesorectal excision to identify the plane of dissection—a video vignette. *Colorectal Dis* 2017;**19**(6):599–600. https://doi.org/10.1111/codi.13698. 28467625.
17. Garcia-Granero A, Pellino G, Giner F, Frasson M, Fletcher-Sanfeliu D, Romaguera VP, et al. A mathematical 3D-method applied to MRI to evaluate prostatic infiltration in advanced rectal cancer. *Tech Coloproctol* 2020;**24**(6):605–7. https://doi.org/10.1007/s10151-020-02170-4. Epub 2020 Feb 28 32107687.
18. Papazarkadas X, Spartalis E, Patsouras D, Ioannidis A, Schizas D, Georgiou K, et al. The role of 3D printing in colorectal surgery: current evidence and future perspectives. *In Vivo* 2019;**33**(2):297–302. https://doi.org/10.21873/invivo.11475. 30804106. PMC6506312.
19. Lacy AM, Bravo R, Otero-Piñeiro AM, Pena R, De Lacy FB, Menchaca R, et al. 5G-assisted telementored surgery. *Br J Surg* 2019;**106**(12):1576–9. https://doi.org/10.1002/bjs.11364. Epub 2019 Sep 4 31483054.
20. Rodríguez García JI, Contreras Sáiz E, García Munar M, García Flórez L, Granero TJ. Telemedicine, telementoring and telematic evaluation in surgery. Is it your time after COVID-19? *Cir Esp* 2020;**1**. https://doi.org/10.1016/j.ciresp.2020.05.022. S0009-739X(20)30193–7. English, Spanish. Epub ahead of print 32591110. PMC7262517.
21. Iqbal MH, Aydin A, Lowdon A, Ahmed H, Muir GH, Khan MS, et al. The effectiveness of Google GLASS as a vital signs monitor in surgery: a simulation study. *Int J Surg* 2016;**36**(Pt A):293–7. https://doi.org/10.1016/j.ijsu.2016.11.013. Epub 2016 Nov 7 27833004.
22. Jayne D, Pigazzi A, Marshall H, Croft J, Corrigan N, Copeland J, et al. Effect of robotic-assisted vs conventional laparoscopic surgery on risk of conversion to open laparotomy among patients undergoing resection for rectal cancer: the ROLARR randomized clinical trial. *JAMA* 2017;**318** (16):1569–80. https://doi.org/10.1001/jama.2017.7219. 29067426. PMC5818805.
23. Montlouis-Calixte J, Ripamonti B, Barabino G, Corsini T, Chauleur C. Senhance 3-mm robot-assisted surgery: experience on first 14 patients in France. *J Robot Surg* 2019;**13**(5):643–7. https://doi.org/10.1007/s11701-019-00955-w. Epub 2019 Apr 5 30953270.
24. Morton J, Hardwick RH, Tilney HS, Gudgeon AM, Jah A, Stevens L, et al. Preclinical evaluation of the Versius surgical system, a new robot-assisted surgical device for use in minimal access general and colorectal procedures. *Surg Endosc* 2020. https://doi.org/10.1007/s00464-020-07622-4. Epub ahead of print 32405893.
25. Kelkar D, Borse MA, Godbole GP, Kurlekar U, Slack M. Interim safety analysis of the first-in-human clinical trial of the Versius surgical system, a new robot-assisted device for use in minimal access surgery. *Surg Endosc* 2020. https://doi.org/10.1007/s00464-020-08014-4. Epub ahead of print 32989548.
26. Chang KD, Abdel Raheem A, Choi YD, Chung BH, Rha KH. Retzius-sparing robot-assisted radical prostatectomy using the Revo-i robotic surgical system: surgical technique and results of the first human trial. *BJU Int* 2018;**122**(3):441–8. https://doi.org/10.1111/bju.14245. Epub 2018 May 27 29645348.

27. Lim JH, Lee WJ, Park DW, et al. Robotic cholecystectomy using Revo-i Model MSR-5000, the newly developed Korean robotic surgical system: a preclinical study. *Surg Endosc* 2017;**31**:3391–7. https://doi.org/10.1007/s00464-016-5357-0.
28. Holländer SW, Klingen HJ, Fritz M, Djalali P, Birk D. Robotic camera assistance and its benefit in 1033 traditional laparoscopic procedures: prospective clinical trial using a joystick-guided camera holder. *Surg Technol Int* 2014;**25**:19–23. 25419950.
29. Peters BS, Armijo PR, Krause C, Choudhury SA, Oleynikov D. Review of emerging surgical robotic technology. *Surg Endosc* 2018;**32**(4):1636–55. https://doi.org/10.1007/s00464-018-6079-2. Epub 2018 Feb 13 29442240.
30. Díaz Pavón JM, Jiménez Rodriguez R, De la Portilla de Juan F. Cirugía robótica en el cáncer colorrectal. *Cirugía Andaluza* 2018;**29**(4). 2018.
31. Noguera Aguilar JF, Cuadrado A, Olea JM, García JC, Sanfeliu G. Integración del endoscopio flexible en cirugía digestiva. *Cir Esp* 2012;**90**(9):558–63.
32. Noguera J, Tejada S, Tortajada C, Sánchez A, Muñoz J. Prospective, randomized clinical trial comparing the use of a single-port device with that of a flexible endoscope with no other device for transumbilical cholecystectomy: LLATZER-FSIS pilot study. *Surg Endosc* 2013;**27**(11):4284–90. https://doi.org/10.1007/s00464-013-3044-y. Epub 2013 Jun 29 23812286.
33. Noguera JF, Dolz C, Cuadrado A, Olea J, García J. Flexible single-incision surgery: a fusion technique. *Surg Innov* 2013;**20**(3):256–9. https://doi.org/10.1177/1553350612451355. Epub 2012 Jun 19 22717701.

Chapter 29

Colon cancer surgery

Fernando Fernández López[a], Jesús P. Paredes Cotoré[a], and Manuel Bustamante Montalvo[b]

[a]Coloproctology Unit of the General Surgery Service, Associate Professor in Health Sciences, Department of Surgery, Faculty of Medicine, University of Santiago de Compostela, A Coruña, Spain, [b]Head of the Department of General and Digestive Surgery, University Hospital Complex of Santiago de Compostela, A Coruña, Spain

Surgical treatment of nonmetastatic colon cancer

The treatment with curative intent for colon cancer is an en bloc resection of the entire tumor, with histologically free margins and absence of residual lymph node metastases or distant organ metastases; otherwise, it would be a palliative resection.[1]

The letter R describes the absence or presence of residual tumor after surgical treatment. The term R0 resection is used when the tumor has been excised en bloc with disease-free histological margins, R1 resection when the histological margins are positive, and R2 resection when macroscopic residual tumor has been left. If all these data cannot be provided, the term Rx resection is used. The prognosis according to the TNM classification presupposes complete tumor resection, and for adjuvant treatment to be curative, R0 resection is also required.[2]

The commonly accepted rate of curative resection is 60%. However, if the proportion of patients undergoing advanced-stage surgery is high, this percentage will decrease.

Although it may be controversial, a recent review of the literature suggests that the greater the hospital and surgeon volume and training, the better the results in postoperative morbidity/mortality and long-term survival.[3,4] However, although these data seem to point in this direction, they should be examined with caution for several reasons, including the following: The results are influenced by a large number of variables not always reflected in the studies, such as patient groups not always having the desired homogeneity.

General surgical technique

The most commonly used access route is via median laparotomy, which allows both adequate exploration of the abdominal cavity and its extension, as needed, while leaving the flanks free in case an unplanned stoma is required.[5]

The first gesture after laparotomy should be intraoperative assessment by meticulous exploration of the entire abdominal cavity: liver, ovaries, peritoneum, omentum, mesenteric and retroperitoneal adenopathies, and biopsies of lesions suspected of metastasis. All these data should be recorded in the operative report.

There is currently no evidence that the "do not touch" technique during curative resection has any influence on tumor recurrence or survival.[6]

Although debated, it seems that 5–10 cm on both sides of the tumor is sufficient to remove the epicolic and paracolic nodes and avoid local recurrence.[3] However, it will depend on the extent of lymphovascular resection. More extensive resections have not been shown to improve survival. The length of the resected ileum does not affect recurrence but should be minimized to avoid malabsorptive problems. Patients with established colon cancer who have another synchronous tumor or are from a family with hereditary nonpolyposis colorectal cancer (CRC) are candidates for total colectomy. In cases of ulcerative colitis with severe dysplasia or cancer, proctocolectomy is performed, either restorative or nonrestorative with definitive ileostomy.[1]

Lymphovascular resection has prognostic and therapeutic implications. Patients with apical lymph node involvement have a significantly reduced 5-year survival.[4] Conversely, the greater the number of lymph nodes analyzed, the higher the percentage of lymph nodes with micrometastases identified, which conditions staging and treatment. Nevertheless, prospective studies have not shown that high ligation of the vessels or extended lymphadenectomy improves survival. Lymphadenectomy should be performed from the root of the vessels and extended to nodes outside the resection field if infiltration is suspected.

Leaving positive nodes indicates an incomplete resection (R2). A minimum of 12 negative nodes must be examined to establish stage II; this has been shown to be positively associated with survival in both stage II and stage III.[7]

How should locally advanced tumors be treated?

Tumors are attached to the wall or neighboring organs in 15% of patients. In these cases, if possible, en bloc resection is performed with the organs or structures affected and with histologically free margins to consider curative resection. This is the only method to achieve survival rates similar to those of standard resection, although burdened by greater morbidity owing to the extent of the resection. Adhesions of the tumor to neighboring structures are neoplastic in 40%–50% of cases but are difficult to observe macroscopically "a priori." For this reason, en bloc resection is recommended, as if the tumor is detached from its adhesions, tumor dissemination may occur, which can worsen survival, and the surgery will be considered incomplete.[8] Further, the only factors associated with survival in this type of resection are lymph node metastases and transfusion. Conversely, if tumor perforation occurs during dissection, the risks of local recurrence and decreased survival increase. The perforated tumor will be classified as T4 and the resection as R1. In any of these locally advanced cases, it is advisable to mark the surgical site using metal clips in case postoperative radiotherapy is considered in the future. There are insufficient data to support routine prophylactic oophorectomy, especially in women of childbearing age. Bilateral oophorectomy is debatable when the tumor is attached to the ovary or when one or both ovaries are enlarged.[9]

Surgical technique according to specific locations and special situations

Basically, the surgical technique consists of a colectomy resecting the segment where the tumor is located with the corresponding mesocolon and the omentum adjacent to the lymph node territory. The segment of the resected colon will depend on the height of the vascular section performed to remove the entire draining lymph node territory. The anastomosis can be performed manually or mechanically, according to the surgeon's preference, as there are no advantages between the two,[10] but considering that it must be performed with well vascularized and tension-free ends; this will allow the angles of the colon to be freed on certain occasions.

- *Right colon*: More or less extensive right hemicolectomy depending on the location of the tumor, ligating the ileocolic vessels, right colic, and right branch of the middle colic and with ileotransverse anastomosis.
- *Transverse colon*: Segmental colectomy, ligating the middle colic vessels and with colocolic anastomosis. In cases of transverse tumors located to the right, the current trend is to perform a right hemicolectomy extended to the left with ileocolic anastomosis. If the tumor is located close to the splenic angle, the technique described for splenic angle tumors is used.
- *Splenic angle*: High left colectomy, ligating the upper left colic vessels and with colocolic anastomosis. However, a very widespread approach is the extended right hemicolectomy with ileosigmoid anastomosis, sectioning the ascending branch of the inferior mesenteric artery.
- *Descending colon*: Left hemicolectomy, ligating the inferior mesenteric artery at its root and the vein at the lower edge of the pancreas, performing colorectal anastomosis.
- *Sigmoid colon*: Lower segmental colectomy, ligating the mesenteric artery above or below the left colic and the vein at the inferior border of the pancreas, with colorectal anastomosis.

When and why should radical surgery be performed after endoscopic removal of a colon polyp?

Endoscopic polypectomy is an appropriate treatment for adenocarcinoma in situ or intramucosal adenocarcinoma and in certain cases of early carcinoma (T1 Nx Mx). Early carcinoma is invasive, as it breaks the muscularis mucosa, and there may be lymph node or metastatic invasion, which induces the need to determine parameters to consider whether polypectomy is sufficient or whether oncological resection should be performed. Radical surgery should be performed after endoscopic excision of a colon polyp if any of the following criteria are evident[1]: tumor at the resection margin or less than 1 mm (incomplete excision),[2] vascular or lymphatic infiltration, and[3] poorly differentiated tumor. Oncological resection is advised in the presence of one of these criteria because of the high likelihood of residual disease.

Treatment of colon cancer presenting as a surgical emergency

Between 15% and 20% of colonic tumors present as a surgical emergency, and the most frequent urgent manifestation is obstruction. CRC is the leading cause of colon occlusion and is the most frequent cause of splenic angle tumors, followed by left-sided tumors.[11] Another complication is perforation, the incidence of which is lower, and it can occur both at the tumor level and diastasis, proximal to the tumor and secondary to occlusion; in these cases, morbidity and mortality are greater owing to intestinal ischemia, peritonitis, and septic shock, which are frequently associated conditions. Conversely, massive hemorrhage as a complication of CRC is rare.

Urgent presentation appears to have a negative impact on prognosis. Prospective, multicenter studies document poorer long-term survival in patients undergoing urgent surgery, as well as more advanced staging. A lower resection rate has even been found in emergency surgery than in elective surgery.[12] The literature contains highly variable values for postoperative mortality in patients undergoing emergency surgery. The range is 8%–21% for emergency surgery compared to 3%–9% for elective surgery, with significant differences. In any case, the overall mortality in emergency surgery should be less than 20%.[12]

Patients with occlusion should be carefully prepared for surgery with adequate fluid intake, monitoring of blood pressure, and diuresis. Antibiotic and antithrombotic prophylaxis should be administered. Massive hemorrhage, perforation, and obstruction with competent ileocecal valves are nondifferentiable emergencies. However, obstruction without danger of perforation can be performed as a deferred emergency, whenever possible, by experienced surgeons and anesthetists, attempting definitive curative resections.

Surgical options for urgent CRC

The surgical options for CRC depend mainly on the location of the tumor, extent of the disease, clinical status of the patient, and associated comorbidity. Once general support has been optimized to allow control of the life-threatening situation, the first debate centers on the possibility of resection of the diseased colon segment. Several prospective studies[12] have sought to determine whether primary resection has advantages over staged resection in terms of morbidity, mortality, and survival. None of them have shown any difference in these parameters, and they have only commonly demonstrated a significantly shorter hospital stay in primary resection. Although a Cochrane review in 2004[12] did not draw any clear conclusions in this regard owing to the lack of larger randomized clinical trials, the idea that except in palliative or extremely serious circumstances, urgent tumor resection is mandatory is now well established. This resection should be oncologically radical, applying the same oncological principles that guide elective surgery.

Recommended resections depending on the location of the tumor

It is widely accepted that the strategy for the treatment of occlusive lesions and perforations proximal to the splenic angle of the colon is to perform an extended right or right hemicolectomy with primary ileocolic anastomosis. However, the treatment of left colon emergencies remains controversial, and the literature shows limited consensus on the optimal treatment for a particular patient.

The Hartmann procedure yields satisfactory results with a mortality rate ranging from 2.6% to 9% and has been the technique of choice for the management of urgent pathology of the left colon. Its main disadvantage is the need for a second operation, which increases morbidity and mortality, leading to a low rate of reconstructions. Currently, one-stage surgery is considered safe in selected patients even in the presence of peritonitis, and the decision to perform primary anastomosis should be based primarily on the patient's general condition rather than on intraoperative findings.

Palliative procedures

Several systematic reviews have established that the two main indications for prostheses in CRC are colonic decompression as a bridge to elective surgery and permanent decompression as palliative treatment of unresectable stenosing neoplasms. Successful placement (technical success) is achieved in 92% of patients, with a clinical success rate (resolution of occlusion) of 88%. The major complications are perforation (4%), migration of the prosthesis (9.8%), reobstruction (9.9%), and death (1%). The only absolute contraindication is the presence of colonic perforation. The route of prosthesis placement can be radiologically controlled or combined (radiological and endoscopic), with no differences in the percentage of technical success or failure and the dose of radiation administered. When the endoscopic route is used, the prosthesis is introduced through the endoscope for neoplasms located more than 20 cm from the anal margin, taking into account that the recommended limit from the distal end should not be less than 5 cm from the pectineal line. The placement of the prosthesis is not an impediment to subsequent surgery or to the insertion of a new prosthesis in case of obstruction of the prosthesis owing to tumor growth or migration when the intention is palliative.[13,14]

In terms of cost-effectiveness, some authors have shown that placement of self-expandable prostheses is associated with a significant cost benefit in patients with curative intent over conventional emergency surgery. However, the overall cost is similar in both groups when compared with palliative intent.

Laparoscopic approach to colon cancer

It is technically feasible to perform colon resection using a laparoscopic approach. In cases of resection with oncological criteria, several prospective and randomized studies in patients with CRC have shown no difference in the extent of resection, number of lymph nodes assessed, and length of the primary vascular pedicle between open surgery and the laparoscopic approach.[3]

Prospective, randomized studies have found no significant differences in the rate and severity of intraoperative and postoperative complications and postoperative mortality. Laparoscopic surgery offers greater postoperative comfort than does open surgery. Patients who underwent surgery with the laparoscopic approach reported a significant decrease in pain at rest and during coughing spells and mobilization after surgery and a significant decrease in the use of narcotics and oral analgesia.

The postoperative length of stay was significantly shorter in patients who underwent surgery with the laparoscopic approach than in those who underwent open surgery. In terms of long-term disease-free survival, different studies have detected no significant differences between the laparoscopic approach and open surgery (86%–75% vs. 85%–78%, respectively).

However, as in any new surgical procedure, adequate systematic learning and training are necessary to acquire the necessary experience to perform adequate oncological resection of the colon via the laparoscopic approach.[15]

Contraindications for laparoscopic surgery

An absolute contraindication for the laparoscopic approach to colon cancer is locally advanced neoplasia and/or infiltration of adjacent organs (T4). In these cases, the manipulation will presumably be greater, so there will be higher risks of release of neoplastic cells and their dissemination by the effect of carbon dioxide. For the same reason, perforated colon cancer will also be an absolute contraindication. Other complicated forms, such as severe obstruction, as well as tumor sites that are difficult to access, may be relative contraindications owing to the increased technical difficulty. Similarly, once the laparoscopic approach has been initiated, the impossibility of performing the operation with oncological principles should be a reason for early conversion to open surgery.

References

1. ACPGBI (The association of coloproctology of Great Britain and Ireland). *Guidelines for the Management of Colorectal Cancer*. Royal College of Surgeons London; 2001. http://www.opengrey.eu/item/display/10068/605329.
2. Yarbro JW, Page DL, Fielding LP, Partridge EE, Murphy GP. American Joint Committee on Cancer prognostic factors consensus conference. *Cancer* 1999;**86**(11):2436–46. https://doi.org/10.1002/(sici)1097-0142(19991201)86:11%3C2436::aid-cncr35%3E3.0.co;2-#.
3. Hida J-I, Okuno K, Yasutomi M, et al. Optimal ligation level of the primary feeding artery and bowel resection margin in colon cancer surgery: the influence of the site of the primary feeding artery. *Dis Colon Rectum* 2005;**48**(12):2232–7. https://doi.org/10.1007/s10350-005-0161-2.
4. Chapuis PH, Dent OF, Bokey EL, Newland RC, Sinclair G. Adverse histopathological findings as a guide to patient management after curative resection of node-positive colonic cancer. *Br J Surg* 2004;**91**(3):349–54. https://doi.org/10.1002/bjs.4389.
5. Smith JAE, King PM, Lane RHS, Thompson MR. Evidence of the effect of "specialization" on the management, surgical outcome and survival from colorectal cancer in Wessex. *Br J Surg* 2003;**90**(5):583–92. https://doi.org/10.1002/bjs.4085.
6. Harris GJC, Church JM, Senagore AJ, et al. Factors affecting local recurrence of colonic adenocarcinoma. *Dis Colon Rectum* 2002;**45**(8):1029–34. https://doi.org/10.1007/s10350-004-6355-1.
7. Chang GJ, Rodriguez-Bigas MA, Skibber JM, Moyer VA. Lymph node evaluation and survival after curative resection of colon cancer: systematic review. *J Natl Cancer Inst* 2007;**99**(6):433–41. https://doi.org/10.1093/jnci/djk092.
8. Otchy D, Hyman NH, Simmang C, et al. Practice parameters for colon cancer. *Dis Colon Rectum* 2004;**47**(8):1269–84. https://doi.org/10.1007/s10350-004-0598-8.
9. Banerjee S, Kapur S, Moran BJ. The role of prophylactic oophorectomy in women undergoing surgery for colorectal cancer. *Colorectal Dis* 2005;**7**(3):214–7. https://doi.org/10.1111/j.1463-1318.2005.00770.x.
10. Lustosa SA, Matos D, Atallah AN, Castro AA. Stapled versus handsewn methods for colorectal anastomosis surgery. *Cochrane Database Syst Rev* 2001;**3**. https://doi.org/10.1002/14651858.CD003144, CD003144.
11. Anderson JH, Hole D, McArdle CS. Elective versus emergency surgery for patients with colorectal cancer. *Br J Surg* 1992;**79**(7):706–9. https://doi.org/10.1002/bjs.1800790739.

12. De Salvo GL, Gava C, Pucciarelli S, Lise M. Curative surgery for obstruction from primary left colorectal carcinoma: primary or staged resection? *Cochrane Database Syst Rev* 2004;**2**. https://doi.org/10.1002/14651858.CD002101.pub2, CD002101.
13. Martinez-Santos C, Lobato RF, Fradejas JM, Pinto I, Ortega-Deballón P, Moreno-Azcoita M. Self-expandable stent before elective surgery vs. emergency surgery for the treatment of malignant colorectal obstructions: comparison of primary anastomosis and morbidity rates. *Dis Colon Rectum* 2002;**45**(3):401–6. https://doi.org/10.1007/s10350-004-6190-4.
14. Khot UP, Lang AW, Murali K, Parker MC. Systematic review of the efficacy and safety of colorectal stents. *Br J Surg* 2002;**89**(9):1096–102. https://doi.org/10.1046/j.1365-2168.2002.02148.x.
15. Veldkamp R, Kuhry E, Hop WCJ, et al. Laparoscopic surgery versus open surgery for colon cancer: short-term outcomes of a randomised trial. *Lancet Oncol* 2005;**6**(7):477–84. https://doi.org/10.1016/S1470-2045(05)70221-7.

Chapter 30

Rectal cancer surgery

Jesús P. Paredes Cotoré[a], FernandoFernández López[b], and Manuel Bustamante Montalvo[c]

[a]Colorectal Unit, General Surgery Department, Department of Surgery, University of Santiago de Compostela, Santiago de Compostela, A Coruña, Spain,
[b]Coloproctology Unit of the General Surgery Service, Department of Surgery, Faculty of Medicine, University of Santiago de Compostela, Santiago de Compostela, A Coruña, Spain, [c]Department of General and Digestive Surgery, University Hospital Complex of Santiago de Compostela, Santiago de Compostela, A Coruña, Spain

Surgical treatment of rectal cancer

Rectal cancer has a high incidence in the West. In the EU, 125,000 cases are diagnosed each year[1] (Glynne-Jones). It accounts for approximately 35% of all cases of colorectal cancer, which is the most common tumor in men and women combined. The forecasts for 2020 are far more than 14,000 cases in Spain[1] (SEOM data).

As for most solid tumors, surgery is the main form of curative treatment. The various advances in diagnostic imaging and staging; population screening and early diagnosis programs; selection of the best therapeutic modality assessed by multidisciplinary committees; neo and adjuvant treatment based on radiotherapy, chemotherapy, and molecular biology; and precision of the current surgery to meet well-defined oncological objectives have improved the prognosis and treatment of rectal cancer. This indicates greater attention to the patient's quality of life, which is assessed using tools, such as the EORTC QLQ-C30 and EORTC QLQ-C29. In addition, the increase in the geriatric population, with many octogenarian patients, makes it necessary to tailor treatment to their biological frailty, comorbidity, and lower functional reserve. For an elderly patient, cure may not be the priority objective, but rather achieving the least functional repercussions after surgery.

The experience of the surgeon and the volume of cases treated by the surgeon and the hospital are of particular importance to obtain the best results[2] (Iversen). The fundamental objective of surgery is local control of the disease, reducing local recurrence, which, in addition to compromising cure, has a very negative effect on the patient's quality of life.

Currently, surgeons can select from a wide range of surgical procedures, approaches, and complex and expensive technology. For this reason, it is particularly necessary to carefully select the surgical technique required for each patient.

Preoperative assessment and staging

Confirmation of the diagnosis and adequate preoperative staging are necessary. The objectives are to determine the risk of recurrence and the existence of metastases to select the therapeutic modality and to establish the prognosis.

Different aspects of the tumor, such as the degree of parietal invasion, possible lymph node involvement, and radial margin of resection, as well as its morphological characteristics, including size, distance to the anal canal, and location (height and number of affected quadrants), must be considered in this assessment.[3]

In the physical examination, in addition to a complete general examination, it is essential to perform a rectal examination, which provides information on the morphology of the lesion, degree of mobility or fixation to the wall, distance to the anal margin, and extent of circumferential involvement.

At consultation and prior to any treatment, the surgeon must perform a rigid rectoscopy, which is the only valid procedure for the exact measurement of the distance between the lower edge of the tumor and the anal margin. The following complementary diagnostic and staging tests are necessary[4]:

- Complete colonoscopy, with biopsy of the tumor (assessment of the degree of differentiation) and exclusion of synchronous tumors or polyps. In the absence of a complete examination, a barium enema or computed tomography (CT) colonography may be selected.
- Pretreatment baseline CEA level determination.
- Thoraco-abdomino-pelvic CT scan for the detection of distant metastases (TNM M).

- Specific evaluation of TNM T and N performed via echo-endoscopy and magnetic resonance imaging (MRI). It is probably not necessary to perform both in all cases. Echo-endoscopy can be performed for the assessment of more superficial tumors. Perfect characterization of the tumor via MRI is very important: distance of the tumor from the anal and puborectal margin, relationship with peritoneal reflection, cranial-caudal length of the tumor, rectal circumference involvement, T stage of parietal invasion, extramural extension (subclassify T3), nodal staging, MRC risk, extramural vascular invasion, and invasion of neighboring organs. If there is a tumor at the level of the puborectalis muscle, the level of involvement (submucosal or internal sphincter [partial or total], intersphincteric plane, or external sphincter) should be specified. The response to neoadjuvant treatment (T4 tumors and MRC risk) should also be selectively assessed. Tumor contact or proximity to the mesorectal fascia (margin: <1 mm) is a risk factor for positive circumferential margins at surgical resection and is an independent prognostic factor for local failure.

Vascular invasion on MRI is a prognostic factor for systemic recurrence.[3]

Therefore the surgeon must have precise knowledge of the location, stage, and characteristics of the tumor to plan for the type of surgical intervention.

The rectum includes the last 15 cm of the digestive tract. It is classically divided into lower, middle, and upper thirds, corresponding, respectively, to 0–5 cm, 6–10 cm, and 11–15 cm from the anal margin. The upper limit of the rectum has also been considered to be the line from the pubis to the sacral promontory. Recently, the rectum has been defined using MRI criteria[5] and is considered to be up to 12 cm from the anal margin. The tumors that pose the greatest difficulty for correct surgical treatment are precisely those located in the middle and lower thirds. Upper rectal cancer is more similar in its behavior to colon cancer and should be treated according to the same technical principles.

Preoperative sphincter function and continence should be taken into account when deciding, whenever possible, whether to have bowel continuity restorative surgery or to opt for a definitive stoma. The patient's general condition and lifestyle should also be considered. The surgeon should discuss with the patient which operation is to be performed, what the possible complications and risks are, how it may affect the patient's lifestyle, and what operative eventualities that may lead to changes in previous decisions may occur. It is essential to establish understanding and trust. The informed consent signed by the patient is only a legal formality that signifies all of the earlier.

As with all surgical patients, a general preoperative examination and preanesthetic consultation should be performed. The patient's usual medical treatment should be reviewed and adjusted and prescribed or modified as necessary. The patient should be referred to a stoma-therapist, and the location of possible stomas (colostomy and ileostomy) should be marked. Instructions should be given on the preoperative preparation of the colon, and antibiotic and antithrombotic prophylaxis should be prescribed.

Oncological and technical principles

Rectal cancer resection must comply with the principles of oncological surgery: atraumatic manipulation of the tumor; complete resection with adequate proximal, distal, and radial margins; en bloc removal of the tumor and possible adjacent affected organs, without opening or penetrating the tumor; and complete regional lymphadenectomy.[4] It is debated whether in rectal cancer, in contrast to colon cancer, 12 should also be the minimum number of isolated nodes in the resection specimen. Whether neoadjuvant radiochemotherapy has an influence on a smaller specimen remains unclear. In any case, it is important to make a special effort to search for nodes in the operative specimen.

The macroscopic distal tumor margin should be 1–2 cm. A margin of <1 cm (5 mm) may be acceptable in patients undergoing neoadjuvant radiotherapy.[6]

A debated issue is the level of ligation of the inferior mesenteric artery (IMA). Several studies have shown that there is no difference in survival between ligation of the IMA at the level of its origin in the aorta and distal to the origin of the left colic.[7] However, ligation at a higher level may be advisable for technical reasons, as it allows greater mobility in the colon to reach the pelvic floor.[8]

The mesorectum, a lympho-fatty, vascular and neural tissue that circumferentially surrounds the rectum, is enveloped by the fascia propria. Between this structure and the presacral fascia, there is a layer of relatively avascular lax areolar tissue through which the hypogastric nerves run. Laterally, there are condensations of connective tissue that attach the mesorectum to the lateral walls of the pelvis and contain the pelvic autonomic plexus. The posterior mesorectum is thick and bilobed in appearance. The anterior mesorectum is thinner and is limited to the extraperitoneal portion of the rectum. At this level, the fascia propria of the rectum is more tenuous and is separated from the urogenital organs by Denonvilliers' fascia. Adequate excision of the mesorectum is the treatment of choice for rectal cancer.[2,9] In upper third tumors, partial excision including 5 cm distal to the tumor is indicated, while in lower and middle third tumors, total mesorectal excision

(TME) is indicated. TME refers to sharp or cutting dissection in the plane between the fascia propria of the rectum and the presacral fascia with complete removal of the mesorectum down to the pelvic floor. Several studies[10, 11] have shown a close association of the quality of mesorectal excision (optimal, suboptimal, and unsatisfactory) with local recurrence. Because of the particular difficulty in achieving a good circumferential margin owing to anatomical issues, lower third tumors have a worse prognosis. Cylindrical abdomino-perineal amputation can be attempted to solve this problem.

In the colonic lumen of patients with colorectal cancer, owing to surgical maneuvers, there are exfoliated malignant cells, which could implant distally or at the anastomosis and cause metastasis. Irrigation of the rectal stump with saline or other agents prior to anastomosis eliminates such cells and should be a routine practice to reduce the risk of local recurrence.

It is advisable to check the tightness of the colorectal anastomosis during the operation. In the air test (insufflation via the transanal route with the suture immersed in water looking for absence of bubbling), a dye, such as methylene blue, or flexible endoscopy can be used. The circular integrity of the tissue impingements of the mechanical suture must also be verified.

The routine use of pelvic drains is controversial. Proponents argue that it allows the outflow of fluids (lymph) that could become infected and predispose to anastomotic complications. It will allow early recognition and perhaps control of anastomotic dehiscence. Others relate them to increased incidence of fistula. Many surgeons continue to drain anastomoses below the peritoneal reflection. Taken together, there is no scientific evidence of a difference between the use and nonuse of drains.

Another issue debated in recent years is pelvic lymphadenectomy extended to the lateral nodes (along the common and internal iliac arteries). It is a common practice in Eastern countries, with a very different patient biotype from that in Europe or America. In the absence of evident clinical involvement of the lateral pelvic lymph nodes in the postneoadjuvant control MRI, it is not necessary to perform this procedure. It does not improve oncological evolution and increases the occurrence of sexual and urinary dysfunctions.[8]

Surgical techniques for rectal cancer

After oncological resection of the tumor, digestive continuity should be reestablished whenever possible or desirable. After removal of a tumor in the upper third of the rectum, colorectal anastomosis is usually performed without difficulty. At the level of the middle third, depending on the tumor and pelvic anatomy (e.g., large tumor, obesity, male sex, or narrow pelvis), anastomosis may be difficult to perform. Tumors in the lower third were previously classically treated with amputation of the rectum and a stoma creation. Currently, in many cases, some form of anastomosis can be performed in the most favorable cases from anatomical and functional points of view. Different surgical techniques are available, including the following:

Anterior resection

Anterior resection is a sphincter-preserving procedure that involves partial or total resection of the rectum and mesorectum (depending on the location of the tumor) and anastomosis between the colon and rectum or anal canal, thus avoiding a definitive stoma.

There are different techniques for reestablishing intestinal transit depending on the height of the resection:

- Anterior rectal resection with direct colorectal anastomosis, manual (exceptional) or mechanical (double stapling technique) and high or low, preserving a rectal stump larger than 2 cm. It is mostly performed for resections of tumors located in the middle and upper thirds of the rectum.
- Anterior rectal resection with very low colorectal anastomosis, with colorectal anastomosis to the apex of the anal canal. The rectal stump is smaller than 2 cm. This anastomosis is sometimes referred to as colo-supraanal. Modifications of the procedure have been designed to minimize the functional disturbances called anterior resection syndrome.
    - Colonic "J" reservoir: It aims to increase the reservoir function of the colon. Prospective, randomized studies have found an improvement in the quality of life of patients reconstructed with a J-colon reservoir compared to direct end-to-end anastomosis in low resections. These advantages have been demonstrated mainly during the first postoperative year.
    - Latero-terminal anastomosis (with results comparable to the "J" reservoir)
    - Transverse coloplasty

- Anterior resection of the rectum with coloanal anastomosis: The entire rectum is removed up to the anorectal junction. The colon is excised through the anus to whose mucosa it is sutured manually. Functionally, it yields worse results.
- Anterior resection of the rectum, with delayed coloanal anastomosis (Turnbull-Cutait technique). The left colon is exteriorized through the anal canal by approximately 10 cm by fixing the plasty at the level of the anal canal. The coloanal anastomosis is performed at a second stage (between the fifth and tenth postoperative day) to achieve adhesion between the colic stump and the anal canal.[12] This avoids the need for a defunctionalizing stoma.

After sphincter-preserving resection, patients often present with the so-called *postproctectomy or anterior resection syndrome (LARS)*. It encompasses a number of functional disturbances, such as diarrhea, urgency, fragmentation, imperfect continence, and increased stool frequency. Its cause is not well known; however, factors, such as sphincter dilatation, partial or total sphincter resection, loss of reservoir and accommodative function in the neorectum, lack of transitional sensory zone epithelium, and radiotherapy play a role. Postoperative rectal function is better when some form of colonic reservoir is performed as described. There are scales to quantify LARS[13] and its evolution and response to treatment.

Intersphincteric resection

In intersphincteric resection, abdominal resection of the rectum is combined with a transanal approach through the intersphincter space with total or partial removal of the internal sphincter. This procedure should be considered in young, well-selected patients with good sphincter function and patient acceptance of a suboptimal functional outcome.[14, 15]

Rullier[16] has classified low tumors with practical orientation according to the type of reconstruction possible:

- Type I: supra-anal tumors (>1 cm from the anorectal junction); coloanal anastomosis.
- Type II: juxta-anal tumors (<1 cm from the anorectal junction); intersphincteric anastomosis with partial resection of the internal sphincter.
- Type III: intra-anal tumors (invasion of the internal sphincter); intersphincteric anastomosis with total resection of the internal sphincter.
- Type IV: transanal tumors (invasion of the external sphincter); abdomino-perineal amputation.

Transanal mesorectal resection (TME, "down-to-up" approach)

The evolution of laparoscopic surgery and surgical technique has given rise to natural orifice specimen extraction and laparoscopy-assisted natural orifice surgery or *minilaparoscopy-assisted natural orifice surgery*. The fusion of the concepts of TME, integration of transanal access platforms, and mobilization of the distal rectum using the *bottom-to-up* or *down-to-up* transabdominal and transanal approach have allowed the development of this new approach to rectal cancer.[17]

This technique has become popular as an attractive alternative in patients with obesity and/or a narrow pelvis, where mobilization of the distal rectum and adequate visualization are technically difficult. Its main advantage is establishment of a correct distal margin under direct vision, as well as sectioning of the rectum from its lumen, avoiding an inadequate abdominal stapling owing to its difficult access.

A circumferential incision is made over the anorectal ring (or intersphincteric resection depending on the height of the tumor) to access the mesorectal plane, combining its dissection with that of the laparoscopic abdominal time. Laparoscopic and transanal dissections can be performed simultaneously, optimizing the surgical time. It is a technique that represents a new anatomical vision, which must be performed after an adequate and demanding training. Its misuse can lead to new and serious postoperative complications.

Dysfunctionalizing stoma

When the colorectal anastomosis is low, ultra-low, or coloanal, despite its perfect technical execution, there is a risk of dehiscence, a serious complication that can endanger the life of the patient and the final functional result. The risk factors for anastomotic failure include distance of the suture from the anal margin, neoadjuvant chemo-/radiotherapy, total removal of the mesorectum, immunosuppression, malnutrition, and anatomical or surgical technique factors that hinder the construction of an optimal anastomosis.

To minimize the consequences of such a complication, surgeons create a dysfunctionalizing stoma, in the form of a *colostomy* or much more frequently, a temporary *lateral ileostomy*. Once the patient has stabilized and the tightness of the anastomosis has been verified using imaging techniques, the digestive transit is reestablished in a second surgical stage. In addition to reducing the morbidity of dehiscence, the stoma may reduce its incidence.

However, it is not in itself a procedure free of surgical and metabolic complications and must therefore be performed in a selective and individualized manner.

Abdomino-perineal amputation

Abdomino-perineal amputation consists of the combined abdominal and perineal removal of the entire rectum, anus, and sphincteric apparatus, creating a definitive terminal colostomy in the left iliac fossa.

It is indicated in the following cases:

- Tumor infiltration of the external sphincter, which is currently considered the only absolute indication.
- Tumors of the anal canal or <1 cm of the anorectal ring when intersphincteric resection is not indicated.
- Incontinence/poor sphincter function (with intersphincteric amputation or Hartmann-type resection as an alternative).

The *"classic" abdomino-perineal amputation* is performed following the mesorectal plane up to the levators. Owing to the anatomy of the mesorectum in the lower third, it can have oncological disadvantages related to frequent positive circumferential margins or perforation of the rectum during dissection maneuvers. Recently, the so-called *extra-elevator or "cylindrical" abdomino-perineal amputation* has been developed, in which the dissection is performed on the outside, at the level of the insertion of the levator muscles in the lateral wall of the pelvis. The abdominal dissection only extends until the mesorectum is "coned" and stops at that level. It has the disadvantage of causing a more problematic perineal wound in terms of closure, as the defect created is larger. This can be resolved with the use of a mesh or muscle flap. A very important advantage, when the perineal part of surgery is performed with the patient in the prone position (*jack-knife* position), is the excellent view of all the structures of the anterior pelvic region (recto-prostate/vaginal dissection plane). Studies comparing the two approaches are currently underway; however, to date, no clear oncological advantages have been demonstrated.

Hartmann-type resection

In Hartmann-type resection, rectal resection is performed without reestablishing intestinal continuity. A terminal colostomy is made with the proximal colon in the left iliac fossa, and the distal rectum is left closed in the pelvis.

It is indicated in the following cases:

- Advanced age or presence of a comorbidity ("frail") and high surgical risk (Jonker).
- Preoperative anal incontinence in tumors that do not require rectal amputation (alternative: intersphincteric amputation).
- Cases when anastomosis would not be safe owing to local tissue conditions (e.g., edema or fibrosis). Reconstruction of the digestive tract may be possible via a subsequent operation.

Local resection

The use of local treatment for rectal cancer is justified by the problems associated with radical resections: stomas, poor functional results (defecatory, sexual, or urinary dysfunction), and considerable morbidity.

Local treatment of rectal cancer is curative in the absence of tumors that have spread to regional lymph nodes. However, preoperative identification of lymph node metastases is difficult. It is related to the depth of tumor invasion, degree of differentiation, and presence of lymphatic or venous invasion. As no lymphadenectomy is performed in local resection, cases with a low risk of lymph node spread should be selected. Therefore this intervention is restricted to T1 tumors with favorable histology. It may also be indicated as a palliative option in patients with metastatic involvement and in patients at a high surgical risk, including elderly patients, in the event of stoma rejection.

In cases of a more advanced tumor stage after examination of the surgical specimen or local recurrence, radical salvage surgery is necessary as a second step.[18, 19]

The approaches for local resection of rectal tumors are as follows:

- Parks' "open or classic" transanal route for the most distal lesions, which are approached under direct vision, with the aid of Lone-Star-type leaflets or retractors.
- Endoscopic surgical approach: Specific platforms, such as transanal endoscopic microsurgery or transanal endoscopic operation, are used. Recently, flexible and disposable devices, such as SILS (initially designed for single-port laparoscopic surgery) or Gelpoint, have been incorporated. They have given rise to transanal minimally invasive surgery.

Pneumorectal and optical equipment and conventional laparoscopic instruments are used. Taken together, this has led to a major development of this approach.

The currently accepted indications for local treatment of rectal cancer include a number of tumor conditions[3]:

- Location: distal rectum (last 6–8 cm).
- Tumor size: 3–4 cm.
- Location: posterior or lateral aspect.
- Involvement: 30%–40% of rectal circumference.
- Staging: T1.
- Mobile.
- No lymph node involvement.
- Well-differentiated with no lymphatic or vascular or perineural invasion.
- Free margin of >3 mm on the resection specimen.

The technical principles to be followed in surgical resection are diathermy marking of the excision line, perilesional margin of ≥1 cm, full wall thickness resection (down to the perirectal fat), and optional suturing of the parietal defect. Submucosal resection is incorrect in suspicious or malignant lesions. It is essential not to destroy the architecture of the tumor for its correct pathological study.

Laparoscopic surgery for rectal cancer

Laparoscopic surgery is a technically difficult surgery, which requires a skilled surgeon and the correct selection of patients. Laparoscopic surgery for rectal cancer must carefully comply with the aforementioned principles of oncological surgery. The advantages of this surgery are an excellent view of the pelvic structures and a more comfortable and faster postoperative recovery of the patient.

There are well-designed trials conducted by highly qualified technical and specialized groups that demonstrate that this surgery can be performed with results comparable to those of open surgery.[20]

Robotic surgery

Owing to the wide acceptance of the use of robots in prostate cancer surgery, there is interest in studying its applicability in rectal cancer surgery. It is under evaluation and it is to be hoped that technological development will improve its implementation. Studies such as the ROLARR conclude that it does not offer a clear advantage over other approaches.[21]

Nonsurgical treatment

Some rectal cancers treated with radiotherapy or neoadjuvant chemoradiotherapy disappear completely (15%–20%). This is called a pathological complete response (pCR). Examination of the surgical specimen by a pathologist must confirm the absence of tumors in the rectum. Tumors with a pCR have a better prognosis (local recurrence of <1% and 5-year survival of >95%). Whether it is necessary to subject affected patients with a poorer quality of life to radical oncological surgery remains unclear. In Sao Paulo, Habr-Gama proposed not to operate in such cases, giving rise to a groundbreaking and novel approach, which is the so-called *wait & see* or *wait & watch* (W&W).

Patients with a complete clinical response (pCR) need to be accurately identified via assessment:

- Clinical: no symptoms and normal rectal examination findings.
- Endoscopic (rectoscopy, colonoscopy): no lesion or a whitish scar and telangiectasia.
- MRI: regression of the tumor.

Very close monitoring of the patient is required, including the performance of the following procedures: digital rectal examination, rectoscopy, and CEA level determination to detect tumor regrowth and consider salvage surgery. According to the international registry, the long-term oncological results seem to be similar between W&W and radical surgery; meanwhile, the functional results seem to be better with the former.

However, this intervention is not yet a proven option outside of rigorous trials and registries and perhaps should only be offered in very selected or high-surgical risk cases.

References

1. Glynne-Jones R, Wyrwicz L, Tiret E, et al. Rectal cancer: ESMO clinical practice guidelines for diagnosis, treatment and follow-up. *Ann Oncol* 2017;**28**(Suppl. 4):22–40. https://doi.org/10.1093/annonc/mdy161.
2. Iversen LH, Harling H, Laurberg S, Wille-Jørgensen P, Group DCC. Influence of caseload and surgical speciality on outcome following surgery for colorectal cancer: a review of evidence part 2: long-term outcome. *Colorectal Dis* 2007;**9**(1):38–46. https://doi.org/10.1111/j.1463-1318.2006.01095.x.
3. Daniels IR, Fisher SE, Heald RJ, Moran BJ. Accurate staging, selective preoperative therapy and optimal surgery improves outcome in rectal cancer: a review of the recent evidence. *Colorectal Dis* 2007;**9**(4):290–301. https://doi.org/10.1111/j.1463-1318.2006.01116.x.
4. National Comprehensive Cancer Network (NCCN). n.d., NCCN Clinical practice guidelines in oncology. Accessed 9 September 2019. https://www.nccn.org/professionals/physician_gls/default.aspx.
5. D'Souza N, Balyasnikova S, Tudyka V, et al. Variation in landmarks for the rectum: an MRI study. *Colorectal Dis* 2018;**20**(10):O304–9. https://doi.org/10.1111/codi.14398.
6. Kang DW, Kwak HD, Sung NS, et al. Oncologic outcomes in rectal cancer patients with a ≤1-cm distal resection margin. *Int J Colorectal Dis* 2017;**32**(3):325–32. https://doi.org/10.1007/s00384-016-2708-1.
7. Hida J-I, Okuno K, Yasutomi M, et al. Optimal ligation level of the primary feeding artery and bowel resection margin in colon cancer surgery: the influence of the site of the primary feeding artery. *Dis Colon Rectum* 2005;**48**(12):2232–7. https://doi.org/10.1007/s10350-005-0161-2.
8. Monson JRT, Weiser MR, Buie WD, et al. Practice parameters for the management of rectal cancer (revised). *Dis Colon Rectum* 2013;**56**(5):535–50. https://doi.org/10.1097/DCR.0b013e31828cb66c.
9. Otchy D, Hyman NH, Simmang C, et al. Practice parameters for colon cancer. *Dis Colon Rectum* 2004;**47**(8):1269–84. https://doi.org/10.1007/s10350-004-0598-8.
10. Harris GJC, Church JM, Senagore AJ, et al. Factors affecting local recurrence of colonic adenocarcinoma. *Dis Colon Rectum* 2002;**45**(8):1029–34. https://doi.org/10.1007/s10350-004-6355-1.
11. Yarbro JW, Page DL, Fielding LP, Partridge EE, Murphy GP. American Joint Committee on Cancer prognostic factors consensus conference. *Cancer* 1999;**86**(11):2436–46. https://doi.org/10.1002/(sici)1097-0142(19991201)86:11%3C2436::aid-cncr35%3E3.0.co;2-#.
12. Biondo S, Trenti L, Espín E, et al. Complicaciones y mortalidad postoperatorias tras anastomosis coloanal en dos tiempos según técnica de Turnbull-Cutait. *Cir Esp* 2012;**90**(4):248–53. https://doi.org/10.1016/j.ciresp.2011.12.006.
13. Emmertsen KJ, Laurberg S. Low anterior resection syndrome score: development and validation of a symptom-based scoring system for bowel dysfunction after low anterior resection for rectal cancer. *Ann Surg* 2012;**255**(5):922–8. https://doi.org/10.1097/SLA.0b013e31824f1c21.
14. Aly EH. Colorectal surgery: current practice & future developments. *Int J Surg* 2012;**10**(4):182–6. https://doi.org/10.1016/j.ijsu.2012.02.016.
15. Chua TC, Chong CH, Liauw W, Morris DL. Approach to rectal cancer surgery. *Int J Surg Oncol* 2012;**2012**:247107. https://doi.org/10.1155/2012/247107.
16. Rullier E, Denost Q, Vendrely V, Rullier A, Laurent C. Low rectal cancer: classification and standardization of surgery. *Dis Colon Rectum* 2013;**56**(5):560–7. https://doi.org/10.1097/DCR.0b013e31827c4a8c.
17. Arroyave MC, DeLacy FB, Lacy AM. Transanal total mesorectal excision (TaTME) for rectal cancer: step by step description of the surgical technique for a two-teams approach. *Eur J Surg Oncol* 2017;**43**(2):502–5. https://doi.org/10.1016/j.ejso.2016.10.024.
18. Chang GJ, Rodriguez-Bigas MA, Skibber JM, Moyer VA. Lymph node evaluation and survival after curative resection of colon cancer: systematic review. *J Natl Cancer Inst* 2007;**99**(6):433–41. https://doi.org/10.1093/jnci/djk092.
19. Chapuis PH, Dent OF, Bokey EL, Newland RC, Sinclair G. Adverse histopathological findings as a guide to patient management after curative resection of node-positive colonic cancer. *Br J Surg* 2004;**91**(3):349–54. https://doi.org/10.1002/bjs.4389.
20. Veldkamp R, Kuhry E, Hop WCJ, et al. Laparoscopic surgery versus open surgery for colon cancer: short-term outcomes of a randomised trial. *Lancet Oncol* 2005;**6**(7):477–84. https://doi.org/10.1016/S1470-2045(05)70221-7.
21. Jayne D, Pigazzi A, Marshall H, et al. Effect of robotic-assisted vs conventional laparoscopic surgery on risk of conversion to open laparotomy among patients undergoing resection for rectal cancer. *JAMA* 2017;**318**(16):1569–80.

Chapter 31

Liver metastases from CRC: A treatment paradigm

Manuel Bustamante Montalvo[a], Francisco Javier González Rodríguez[b], and Sergio Manuel Estévez Fernández[c]

[a]University Hospital Complex of Santiago de Compostela, Santiago de Compostela, Spain, [b]Santiago de Compostela University Hospital Complex, Santiago de Compostela, Spain, [c]Álvaro Cunqueiro Hospital of Vigo, Vigo, Spain

Introduction

Approximately 22,000 cases of colorectal cancer are diagnosed annually in Spain; one-third of patients die as a direct consequence of the disease owing to the development of metastases. The liver is the universal receptor of metastases, and in one-third of patients, it is the only target organ. There are multiple possibilities for local treatment of metastatic disease, such as surgical resection, local tumor ablation with radiofrequency or cryotherapy, intra-arterial liver chemotherapy, chemoembolization, selective liver irradiation with Ytrium90-loaded microspheres implanted through the hepatic artery, and radiotherapy. Of these, the only treatment that has yielded a significant increase in survival is surgical treatment.

In recent years, there have been important surgical advances; the increase in survival in affected patients has been a consequence of an increased indication for resection of metastatic disease in selected patients and more effective chemotherapy. Old dogmas in liver surgery that prevented resection of metastases in patients with more than three lesions in one lobe or when it was not possible to obtain a resection margin of more than 1 cm have been abandoned, allowing a much more aggressive approach to the treatment of liver metastases of colorectal origin.

The natural history of this disease is well documented; no patient survives without treatment. This changes dramatically with surgery, with advances in chemotherapy and with the development of new biologic drugs. In fact, there are patients with liver metastases who are not candidates for hepatectomy in whom liver resection is indicated after several cycles of chemotherapy. After surgical treatment and chemotherapy with curative intent, patients who presented hepatic or extrahepatic recurrence, but who received new drugs, have had very prolonged disease survival; patients whose primary tumors and liver disease have been treated with disease-free survival of several years and who had subsequently required rehepatectomy or even resection of primary tumor recurrence through pelvic exenterations or localized peritonectomies have survived for many years with the disease. For this reason, strict clinical, biochemical, and radiological monitoring of each patient is very important because their evolution is uncertain (cure or chronification of the disease with a good quality of life).

Treatment of metastatic disease

Since 1940, when Cattell performed the first liver resection for metastatic disease, liver surgery has increased exponentially, especially in the last three decades, representing the only opportunity for cure for patients with metastases of colorectal cancer located in the liver. The demonstrated survival in surgical series ranges from 24% to 58%, with a median of 40%, and the operative mortality was less than 5%. One-third of patients who survive up to 5 years die of a cancer-related cause, and those who survive up to 10 years are considered cured (Table 31.1).

Liver resection has a clear role in increasing survival and is therefore the treatment of choice; unfortunately, only 20% of patients with metastases located in the liver are candidates for surgical intervention owing to the size, number, and location of metastases or inadequate hepatic reserve.

Some patients with initially unresectable disease become surgical candidates owing to the success of chemotherapy; the percentage at which this phenomenon occurs is variable and depends on several factors, including the liver surgeon's subjective perception of resectability (5%–15%). Clinical guidelines (National Comprehensive Cancer Network [NCCN][17]) suggest that initially unresectable metastatic liver disease of colorectal origin should be classified as potentially resectable

TABLE 31.1 Outcomes of liver resection for metastatic colorectal cancer.

Author and year	Number of patients	Survival at 5 years	Average survival, months
Hughes K; 1986[1]	607	33	–
Scheele J; 1995[2]	434	33	40
Nordlinger B; 1996[3]	1568	28	–
Jamison RL; 1997[4]	280	27	33
Fong Y; 1999[5]	1001	37	42
Iwatsuki S; 1999[6]	305	32	–
Choti M; 2002[7]	133	58	–
Abdalla E; 2004[8]	190	58	–
Fernandez FG; 2004[9]	100	58	–
Wei AC; 2006[10]	423	47	–
Rees M; 2008[11]	929	36	42.5
de Jong M; 2009[12]	1669	47	36
Morris EJ; 2010[13]	3116	44	–
Pawlik; 2017[14]	2505	44	–
Imai; 2017[15]	553	57	–
Adam R; 2018[16]	12,406	49	–

or unresectable so that patients considered to have unresectable disease receive palliative chemotherapy, as opposed to potentially resectable patients who would receive chemotherapy with neoadjuvant intent.

The key to obtaining good oncological results and avoiding postoperative complications is correct patient selection, considering factors related to patient characteristics, tumor biology, and anatomical relationships between the tumor and existing metastatic lesions.

Patient-related factors

Hepatectomy is a surgery that causes significant physiological stress; therefore, in patients in whom the metastatic liver disease is technically resectable, but in whom significant comorbidity (e.g., advanced age, cardiopulmonary insufficiency, and renal insufficiency) exists, a high perioperative risk will contraindicate surgical resection.

Tumor-related factors

Tumor biology is one of the most important factors in predicting recurrence and survival. In more than half of patients, the disease will progress, with a significant decrease in overall survival. A number of studies have been conducted in an attempt to identify these, including the establishment of preoperative metrics with four to five poor prognostic factors. Risk metrics that predict which patients will benefit most from surgical resection have very limited clinical utility, especially among patients receiving neoadjuvant treatment.

The following are poor prognostic factors related to the primary tumor: poorly differentiated, mucinous or high-grade tumor type, vascular infiltration, presence of adenopathy in the hepatic pedicle, extrahepatic involvement, metastasis size greater than 5 cm, more than four metastases, satellitosis, bilobarity, synchronicity, a short time interval in the development of metachronous metastases, elevated CEA level before and after chemotherapy, radiological response to chemotherapy (probably one of the most relevant prognostic factors), and pathological complete response to chemotherapy (absence of tumor cells in the anatomopathological study of the resection specimen has a very favorable prognosis). Meanwhile, the following are the factors related to surgery: a positive surgical margin and a very extensive resection. The embryological origin of the colonic tumor also influences the prognosis of patients with liver metastases; patients with tumors located in the right colon and transverse colon have a worse pathological response to chemotherapy and lower postresection survival than patients with tumors located in the left colon and sigmoid colon, independently of the RAS oncogene mutation.

Anatomical factors

The classic clinical guidelines for the treatment of patients with metastatic disease determined the performance of liver resection in relation to the number and size of lesions and the possibility of performing tumor removal with a tumor-free margin of more than 1 cm. Currently, the concept of resectability has been simplified to the point that liver resection can be indicated provided that a complete removal of R0 tumors can be conducted while preserving a sufficient residual liver volume. The following directions are recommended:

- Avoid resection of liver metastases with a positive surgical margin owing to the higher local recurrence and lower overall survival.
- Do not contraindicate liver resection when a priori, the expected tumor-free margin is less than 1 cm, which would be optimal.
- Determine the resectability of metastatic liver disease of colorectal origin based on the ability to perform complete tumor resection with negative margins to be able to preserve two contiguous liver segments, preserve adequate liver inflow and outflow with good biliary drainage, and preserve sufficient remaining liver volume (>20% in a healthy liver, >30% in a liver that has received neoadjuvant chemotherapy).
- Do not contraindicate liver resection, even if there is an extrahepatic metastatic disease as long as it is resectable or even if lymph node metastases exist in the hepatic pedicle (the oncological results are much better when lymphadenopathies are limited to the portal without the involvement of the common hepatic artery).

In some cases, in addition to the purely anatomical factors that allow hepatectomy, it is necessary to consider whether it is worthwhile to perform large liver resection, with the sacrifice of a healthy liver that this entails for single small lesions deep in the liver parenchyma in patients with no risk factors, assuming life-threatening complications, when local treatment of these lesions is possible (e.g., tumor ablation or radiotherapy). This is a difficult decision to make by multidisciplinary committees.

For indicating resection of liver metastases, resection of the primary tumor and any extrahepatic disease must be guaranteed, which is why imaging tests are necessary to obtain vital information for planning a hepatectomy, e.g., number, location, and vascular relationships of existing metastatic lesions, and to rule out the presence of unresectable extrahepatic metastatic disease.

Abdominal computed tomography (CT)

Abdominal CT with contrast has a sensitivity and specificity of 95% for the diagnosis of liver metastases; however, the percentage of false-negative results is higher than 10% when these are less than 1 cm in size, and this procedure has a low sensitivity for the diagnosis of intraperitoneal extrahepatic metastatic disease (peritoneum, Glisson's capsule, or hepatic pedicle). Nevertheless, it is very useful in the planning of liver resection because it allows the estimation of volumes before and after portal embolization to accurately quantify the future liver remnant.

Magnetic resonance imaging (MRI)

MRI with hepatospecific contrast (Primovist®) has a sensitivity and specificity of over 97% in the diagnosis of liver metastases compared to abdominal CT with contrast and is especially useful for the detection of subcentimetric lesions in livers with fatty steatosis after neoadjuvant chemotherapy. However, it is less accessible and difficult to perform in patients with claustrophobia or metal implants.

Positron emission tomography

Although not universally used, positron emission tomography can be useful in ruling out extrahepatic metastatic disease before indicating liver resection in a patient with liver metastases of colorectal origin.

Colonoscopy

A colonoscopy is necessary for the 6-month period before resection of liver metastases to rule out a second colorectal neoplasm or a recurrence in the anastomosis.

Three-dimensional (3D) planning

The integration of CT/MRI data into the 3D planning software allows the composition of a 3D image that can be rotated 360° in all directions, providing a better interpretation of the image and a much more accurate perception of the tumor and its relationship to adjacent vital structures and allowing the performance of virtual resections. All data are processed and transferred to a 3D printer for the creation of 1:1 scale models, which are sterilized and incorporated into the operating field during surgery along with the rest of the instruments, improving the surgical conditions and reducing the risks involved.

The reconstruction of 3D images and the creation of models at the original size allow the observation of anatomical details impossible to explore on CT or MRI, so that surgeons can increase their objectivity in deciding the resectability of a tumor, develop an appropriate surgical approach knowing perfectly the location of the tumor and its anatomical relationships, and explain to the patient the intervention and its risks and complications in a real, clear, and sincere manner.

Surgical technique

Liver metastases can be removed via anatomical resection based on the hepatic segmental anatomy described by Coinaud (Fig. 31.1) or via nonanatomical resection in which priority is given to the preservation of the liver parenchyma. The type of liver resection performed (anatomical vs nonanatomical) does not influence the resection margin positivity, recurrence, or survival. However, in patients who have received neoadjuvant chemotherapy, parenchymal-sparing surgery allows resection of the disease without the need for a major hepatectomy which, in many cases, the patient would not tolerate; this consequently allows reoperation in cases of recurrence because the liver parenchyma would be sufficient, unlike a first surgery in which an anatomical hepatectomy would be performed.

The type of hepatectomy to be performed depends on the size, location, and number of metastases and the relationship with vascular and biliary structures; therefore, atypical resection or metastasectomy is generally accepted for small and superficial metastases, while anatomical resection is performed for large or multiple metastases located in the same lobe.

Intraoperative ultrasound has been used in hepatobiliary surgery since the 1980s and has become an essential procedure in all liver resections because it allows the identification of small nonpalpable lesions 2–3 mm in size and defines the

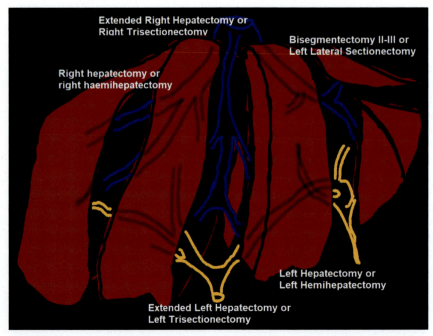

FIG. 31.1 Types of hepatectomy.

relationship of the liver tumor with the vascular and biliary structures within the liver parenchyma; thus, it determines the resectability of the tumor and defines the most appropriate surgical procedure, which is modified in a third of patients during the operation. This intervention has a sensitivity and specificity comparable only to those of the latest generation of liver MRI.

Treatment is different depending on the time of diagnosis of liver metastases: simultaneous with the diagnosis of the colorectal tumor (synchronous metastases) or after the diagnosis of the tumor (metachronous metastases).

In patients with metachronous metastases, the indication for direct surgery or neoadjuvant chemotherapy depends on the number of lesions, bilobar hepatic involvement, and existence of vascular infiltration. Chemotherapy is usually indicated in cases where there are multiple bilateral lesions or vascular involvement to facilitate surgical resection, make lesions that were not resectable a priori, and avoid unnecessary surgery in patients with tumor progression during chemotherapy.

In patients with synchronous liver metastases of colorectal origin, the treatment of the primary tumor and metastases is multidisciplinary and multimodal (chemotherapy, radiotherapy, surgery, and locoregional therapy). The timing of the indication for surgical intervention and the treatment sequence into which the treatment is integrated depend on the existence of symptoms and the tumor burden in each patient.

- In patients with symptoms related to their primary colorectal tumor (e.g., hemorrhage, obstruction, or perforation), removal of the primary tumor is a priority.
- In patients with no symptoms related to their colorectal primary tumor, whether to perform a simultaneous resection of the entire disease or a two-stage resection depends mainly on the existing tumor burden in the liver:
  - If the location of the primary tumor is favorable (right colon or transverse colon), and the liver disease is limited, a one-stage resection of the colonic tumor and liver metastases could be performed in most patients, with many of them benefiting from a laparoscopic approach. Some publications have shown that simultaneous resection of the entire disease has morbidity and mortality similar to those of patients undergoing a two-stage treatment, provided that resection of the liver disease does not involve a major hepatectomy, because in that case, morbidity would double (36.1% vs 15.1%), and mortality would become unacceptable (8.3% vs 1.4%).
- In patients with multiple and bilateral liver metastases, it is often necessary to perform major anatomical hepatectomies associated with atypical resections of metastases in the contralateral lobe; if this is added to the resection of the primary tumor, the risk of liver failure and dehiscence of the colorectal anastomosis becomes very high. Therefore, in these cases, a good option is to resect the primary tumor and, in the same surgical intervention, to perform atypical metastasectomies in the future liver remnant associated with ligation of the portal branch of the hepatic lobe in which there is a greater tumor load employing a laparoscopic approach, for subsequent resection at a later stage once hypertrophy of the future liver remnant has been achieved, as described in the following section. For patients with metastatic liver disease, what is known as reverse therapy has been developed in recent years, which consists of resecting the liver metastases to determine the prognosis and then resecting the colorectal tumor; thus, if neoadjuvant chemotherapy is decided, the first resection of the liver disease may be performed, followed by resection of the colorectal tumor with or without chemotherapy in the period between the two surgical interventions after the patient's recovery. If a laparoscopic approach can be used in the first surgery, there would be fewer intraperitoneal adhesions, which facilitate a second surgery, which, in many patients, can be performed again using a laparoscopic technique. Leaving excision of the colorectal tumor for a second time may increase the risk of developing complications in up to 20% of patients. In patients with a rectal neoplasm with an indication for neoadjuvant chemoradiotherapy (e.g., T3 or N1), there is the possibility of resection of the liver disease in the time interval between neoadjuvant treatment and excision of the rectal tumor (8–12 weeks).

The only opportunity for achieving long-term survival is surgical resection. Incomplete resection of the tumor should be avoided because the oncological results are similar to those of nonsurgical resection interventions. Failure to remove the entire tumor while preserving a sufficient amount of functional liver parenchyma (at least 30% of the initial liver volume) is the main cause of unresectability owing to the high risk of postoperative liver failure. There are different onco-surgical strategies to increase resectability in tumors in which it would be impossible to achieve R0 resection a priori guaranteeing the survival of the patient postoperatively.

Neoadjuvant chemotherapy

Neoadjuvant chemotherapy has two aims: provide knowledge of the evolution of the metastatic disease to avoid conducing liver resection in patients in whom this does not modify their survival (fundamentally in patients with synchronous liver metastases) and reduce the size of the liver metastases to facilitate their resection.

There is still no universal criterion for the selection of patients who would benefit from neoadjuvant chemotherapy, for the choice of the treatment schedule, and for determining the treatment duration and how to integrate it with surgery for the primary colorectal tumor and liver metastases. However, most authors accept that in potentially resectable low-risk patients (healthy with no comorbidities, four or fewer liver metastases, metachronous presentation, and metastatic involvement only in the liver), resection of the initial liver disease should be performed, and then adjuvant chemotherapy should be considered; in patients with unresectable or difficult to resect metastatic disease, neoadjuvant chemotherapy should be performed with the intention of rescuing such patients for surgery (10%–15%). Regardless of the treatment schedule selected, the duration of chemotherapy should be limited to the shortest possible time, evaluating the radiological response to treatment every 6–8 weeks, to indicate resection of liver metastases as soon as a clear response is evident and as long as they are resectable. This is because prolonged chemotherapy treatment with oxaliplatin, irinotecan, and bevacizumab can lead to hepatic and extrahepatic complications (e.g., steatohepatitis, vascular lesions, portal hypertension, thromboembolic events, intestinal perforation, and delayed tissue healing), making liver resection of a technically resectable lesion difficult.

There are multiple possibilities for neoadjuvant chemotherapy using different combinations of drugs; most guidelines recommend FOLFOX, FOLFIRI, and XELOX, in combination or not with cetuximab, panitumumab, or bevacizumab, depending on the status of the RAS oncogene (wild-type or mutated).

Preoperative portal embolization and associating liver partition and portal vein ligation for staged hepatectomy (ALPPS)

There are three levels of risk of postoperative liver failure depending on the remaining liver volume: (1) low risk when it is greater than 40%, (2) medium risk when it is between 40% and 25%, and (3) high risk when the functional liver parenchymal remnant is less than 25%. Preoperative portal embolization of the liver to be resected induces compensatory hypertrophy of the future liver remnant, allowing a larger resection of liver tissue with a much clearer tumor-free margin and decreasing the risk of postoperative liver failure. Portal embolization is routinely recommended in patients with an estimated future liver tissue remnant volume of less than 30% or 40% in cases where neoadjuvant chemotherapy has been administered. After portal embolization, liver resection is performed once the highest degree of hypertrophy of the remaining liver has been reached, as calculated by a volumetric study performed on CT (4–6 weeks after embolization). With preoperative embolization, it has been estimated that 20% more extended hepatectomies are possible in initially unresectable tumors. It is also accepted for use in major hepatectomies to increase the safety of the procedure.

Sequential hepatectomy, described in the Anglo-Saxon literature with the acronym ALPPS, is a novel technique that offers rapid and effective growth of the remaining liver volume and allows surgical resection of liver lesions initially considered unresectable. Initially described by Schnitzbauer et al., it consists of performing portal occlusion of the hemiliver to be resected together with hepatic parenchymal transection and, if necessary, resection of contralateral lesions, and completing the resection of the embolized liver (Fig. 31.2). The most recognized advantage of this technique is its proven ability to achieve rapid and effective liver regeneration in a short period. Conversely, the main limitation is its high morbidity and

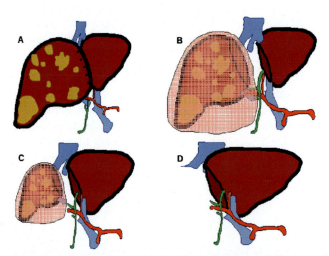

FIG. 31.2 Schematic diagram of trisectionectomy associated with hepatic partitioning and ligation of the right portal branch (ALPPS). (A) Cholecystectomy and ligation of the right portal branch. (B) Transection of the hepatic parenchyma and introduction of the right liver, which will be subsequently removed in a plastic bag. (C) Hypertrophy of the future hepatic remnant and atrophy of the right liver. (D) Hepatectomy is completed by section of the right hepatic artery, right bile duct, and right and middle suprahepatic veins.

mortality, with values usually higher than those published in series of liver resections in recent years, making it a prohibitive technique for some authors. ALPPS is probably one of the surgical techniques that has aroused the most controversy in surgical forums in recent years, even though there are no large series available, and the opinions in favor are almost as numerous as those against. The advantages described with this technique are as follows:

- Sufficient hypertrophy of the contralateral liver remnant achieved in a short period; this growth is variable and ranges from 40% to 80% in 7–13 days, with an average of 74% in 9 days.
- Accurate knowledge of the extent of the disease, which can be assessed from the outset with exploratory laparotomy and intraoperative ultrasound, with consequent reduction in the percentage of false-negative results of preoperative imaging techniques, allowing the treatment strategy to be modified to the patient's benefit.
- Possibility of extemporaneous liver biopsy to assess parenchymal involvement, which will facilitate the decision on the indication for resection.
- With portal embolization, the recommended period until the sufficient remaining liver volume is available is usually between 4 and 8 weeks, and in 18%–40% of cases, resection cannot be completed owing to disease progression. ALPPS can reduce the waiting time and shorten the interval between the two surgical resections, which, in theory, would reduce the progression of the disease.

Local treatment for metastatic disease

There are patients with liver metastases who are not candidates for surgical treatment with curative intent owing to the size, number, and location of metastases or inadequate hepatic reserve; however, nonsurgical treatment options are available.

Regional chemotherapy via the hepatic artery

Hepatic metastases have a vascularization that depends fundamentally on arterial flow, unlike normal hepatocytes in which portal flow is a priority; theoretically, the administration of chemotherapy selectively through the hepatic artery would yield a predominant effect on the tumor with less repercussion on healthy hepatic tissue, less exposure of the other organs to chemotherapy with mainly hepatic metabolization, and therefore, a theoretically smaller number of systemic complications. However, although multiple studies and research have been conducted, these benefits have not been demonstrated.

Radiotherapy

It has been estimated that the dose of radiotherapy necessary for the destruction of liver metastases of colorectal origin is greater than 70 Gy, which is double the maximum dose that a healthy liver can tolerate; this is why radiotherapy has not formed part of the therapeutic arsenal for the treatment of metastases until the appropriate technological development has taken place (stereotactic radiotherapy or selective internal radiotherapy), which has allowed the administration of high doses of radiotherapy selectively on the tumor, respecting vascular and biliary structures and the surrounding hepatic parenchyma.

Stereotactic radiotherapy selectively targets the tumor while sparing normal liver tissue. It has been used successfully in small unresectable metastases with local disease control of more than 50% at 3 years.

Selective internal radiotherapy with Yttrium spheres can also achieve local control of the disease, utilizing their application through terminal branches of the hepatic artery selectively in the liver tissue where the lesion is located, given that they migrate to the arterioles surrounding the tumor, releasing radiation doses with a maximum penetration of 1 cm and thus respecting healthy liver tissue.

Tumor ablation

Hyperthermic tumor ablation by radiofrequency or microwave is a widely used technique in patients who are not candidates for surgical resection with less than three liver metastases and less than 3-cm metastatic size. It can be applied during laparotomy, laparoscopic surgery, or percutaneous surgery. Its main limitation is the proximity of the tumor to vascular or biliary structures. Survival is disappointing at 3 years (20%–40%) and virtually nonexistent at 5 years. A tumor size of more than 3 cm is not applicable even if the needle is repositioned and several applications are made in an attempt to completely destroy the lesion. A location of the metastasis close to the hepatic hilum prevents its application owing to the risk of injuring the biliary tract and causing a stricture or biliary fistula. Electrical burns (diaphragmatic and colonic perforations), pleural effusion, bleeding, subcapsular hematomas, biliomas, abscesses, arterio-portal fistulas, and portal thrombosis (irreversible and potentially fatal lesions) can occur. Local ablation techniques alone or in combination with surgery yield superior survival to nonsurgical resection (40% vs 10% survival at 3 years) but significantly inferior survival to surgical resection. However, it may have a role in patients with incomplete macroscopic resection of metastases, when there are small lesions that are surgically inaccessible, in patients with unresectable recurrence after hepatectomy, and in patients who are not candidates for surgery owing to associated comorbidity. After portal embolization, there is a large growth of metastases from the nonembolized liver; thus, local destructive techniques are often used in conjunction to prevent their rapid progression.

Two-stage hepatectomy

Two-stage hepatectomy is a sequential resective strategy with curative intent that is applied in the treatment of multiple bilobar metastases where resection with a single procedure would be impossible. Thus, during the first hepatectomy, the aim is to remove as many metastatic lesions as possible, with the goal of making a second hepatectomy potentially curative. After the first resection, the liver remnant is hypertrophied, and systemic chemotherapy limits the growth and spread of any residual tumor deposits. This postoperative chemotherapy, which usually uses the same treatment regimen as neoadjuvant chemotherapy, is started 3 weeks after the first resection and does not interfere with early liver regeneration. A second hepatectomy will only be performed if it is potentially curative, i.e., no significant tumor progression has occurred and there is adequate parenchymal hypertrophy to reduce the risk of postoperative liver failure. Multinodularity represents the most frequent cause of the inability to resect metastases owing to a technical problem.

Schematically, seven situations can occur, to which different surgical solutions can be applied depending on the number, size, and location of the lesions (Fig. 31.3):

1. Resectable metachronous metastases requiring resection of <60% of parenchyma: resection
2. Resectable synchronous metastases in the same surgical intervention and less liver resection: one-stage resection

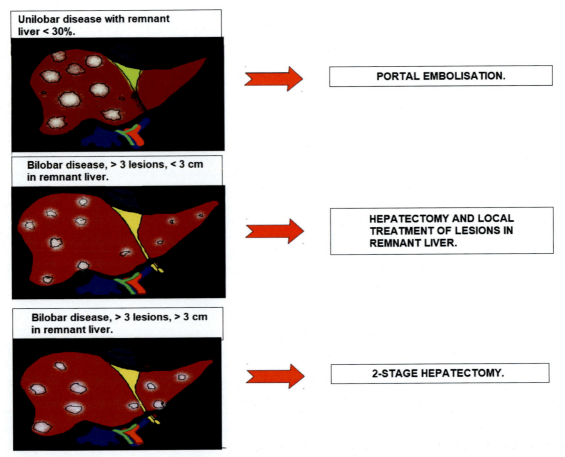

FIG. 31.3 Surgical strategies.

3. Synchronous cases that cannot be approached in the same surgical intervention and/or major resection: primary surgery, chemotherapy, and subsequent hepatectomy
4. Multiple unilobar metastases requiring complete resection of an extended hepatectomy with a functional parenchymal resection volume greater than 60%–70%: preoperative portal embolization vs ALPPS (Fig. 31.4). The hypertrophy induced in the future liver remnant allows for curative resection with minimal risk of postoperative liver failure.
5. Multiple bilobar metastases whose resection would leave three or fewer nodules of less than 3 cm in the remnant liver: radiofrequency/microwave-associated liver resection of the unresected nodules
6. Multiple bilobar metastases whose resection would leave more than three nodules larger than 3 cm in the remaining liver: two-stage hepatectomy
7. R0 with very poor prognostic factors: intra-arterial catheter for postoperative chemotherapy or aggressive postoperative chemotherapy

In recent years, there has been an expansion of the resectability criteria for patients with liver metastases of colorectal origin. Underlying this change is the development and application of modern chemotherapy, which has increased the response of lesions to treatment, allowing tumors initially considered unresectable to benefit from treatment with curative intent. Tumor growth over the suprahepatic veins or the vena cava is the second cause of the inability to resect for technical reasons, owing to the high risk of intraoperative hemorrhage. Surgeons have various resources at their disposal to deal with this type of risky resection; these include intermittent clamping of the portal pedicle (Pringle maneuver), hepatic vascular exclusion with or without preservation of the vena cava flow, extracorporeal venovenous bypass, vascular reconstruction of the vena cava or hepatic veins, and hypothermic perfusion of the liver with in situ or ex vivo resection. The type of vascular control depends on its complexity and the estimated duration of the hepatectomy, with the aim of minimizing bleeding and avoiding prolonged ischemia time. The use of these and other techniques originally developed for liver transplantation or living donor grafts within the arsenal of cancer resection surgery is what is known as extreme liver surgery, with a mortality in most series close to 30% and a 5-year survival of less than 10%.

Liver transplantation

Liver transplantation is considered the standard treatment for other unresectable neoplasms, such as hepatocarcinomas and low-grade neuroendocrine tumor metastases. Therefore, in patients with unresectable colorectal metastases, liver transplantation has emerged as a curative alternative in selected patients. In the 1990s, Mühlbacher et al. published the first experience with 17 patients transplanted for unresectable colorectal metastases, with an overall 5-year survival of 12% and a recurrence rate of 60%. Penn et al. published a retrospective study in which eight patients with colorectal metastases were transplanted, with a recurrence rate of 70% and a postoperative mortality of 11%. Owing to the poor results obtained, the idea of transplantation in patients with colorectal metastases was abandoned until the Norwegian SEcondary CAncer I study was published in 2013. Norway is a country where the supply of livers exceeds the demand, so it was possible to conduct this study with expanded criteria in which 21 patients were transplanted, obtaining an overall survival at 1 and

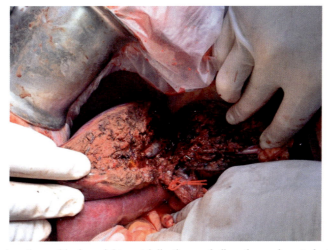

FIG. 31.4 First stage of ALPPS, hepatic partitioning, right portal ligation, and dissection and control of the right hepatic artery and right suprahepatic vein.

5 years of 95% and 65%, respectively. Despite the initial enthusiasm for the publication of these results, the following directions are necessary before universal acceptance:

- Demonstrate that liver transplantation yields superior survival to chemotherapy as the sole treatment.
- Define strict patient selection criteria.
- Define a neoadjuvant and adjuvant chemotherapy protocol.

Liver transplantation is a victim of its own success and is already exhausted before including a new indication, its main problem being the shortage of donors and the high mortality on the waiting list for a compatible graft; adding a new indication would make this situation even more critical. However, this situation could be reversed in the future with NECMO-controlled asystole donation and new two-stage liver transplantation techniques using the left lateral segments associated with liver splitting and ligation of the right portal vein with living or cadaveric donors (RAPID, LD-RAPID). Therefore clinical trials are ongoing.

Treatment of recurrence

Despite correct surgical and chemotherapy with curative intent, approximately 70% of patients have a recurrence of the disease, 28% of which is localized exclusively in the liver after liver resection of metastases of colorectal origin. In these cases, rehepatectomy has the same indication as the first liver resection, yielding a significant increase in survival, with a similar morbimortality, especially if laparoscopic resections are performed to facilitate subsequent hepatectomies (Table 31.2). In their Paul Brousse series, Adam et al. have obtained a 5-year survival of 30% after the third hepatectomy, indicating that any liver recurrence should be resected whenever technically possible. In this series, the interval between initial resection and diagnosis of recurrence was a very important predictor of survival, which means that when the disease-free interval is short, the tumor is much more aggressive and has a greater potential for dissemination and recurrence.

In 20% of cases, there is localized recurrence in the lung, and resection is indicated owing to the good results obtained in terms of long-term survival.

Recurrent disease at the level of the primary tumor resection bed or at the mesenteric lymph node level occurs in less than 10% of patients, in whom resection is also indicated.

TABLE 31.2 Outcomes of rehepatectomy for recurrent metastases of colorectal origin.

Author and year	Number of patients	Survival at 5 years	Average survival, months
Fong Y; 1994[18]	25	44	30
Nordlinger B; 1994[19]	116	33	24
Fernandez-Trigo V; 1995[20]	170	32	34
Adam R; 1997[21]	64	41	46
Yamamoto J; 1999[22]	90	31	31
Petrowsky H; 2002[23]	126	34	37
Nagakura S; 2002[24]	28	42	27
Tanaka K; 2004[25]	26	48	–
Pessaux P; 2006[26]	42	55	41
Ishiguro S; 2006[27]	111	41	43
Cunha A; 2007[28]	40	31	–
Jones N; 2012[29]	52	34	22
Adam R; 2012[30]	195	29	–
Demoux; 2016	447	56	
Hamada; 2020[31]	99	60	

Chemotherapy after resection

It is unclear whether adjuvant chemotherapy is necessary after R0 resection. In 1998[32] Lorenz showed no benefit with intra-arterial chemotherapy with 5-fluorouracil in a large multicenter study. However, in 1999[33] and 2002[34], Kemeny showed that intra-arterial plus intravenous therapy is effective in delaying the time to intrahepatic recurrence and in preventing recurrence. Studies on irinotecan and oxaliplatin with promising results are underway, and the use of chemoimmunotherapy, chemoradiotherapy, vasopressors, angiogenesis inhibitors, radionuclides, and isotopes, such as 90 Yttrium, is also being tested.

Aggressive and early recurrence of the metastatic disease is an undesirable event usually observed in patients with a high tumor burden. It seems that sometimes, surgery "awakens" small inactive foci or reactivates the disease, and for this reason, a more aggressive intervention, such as postoperative chemotherapy with more powerful drugs or intra-arterial chemotherapy, would be a good idea in certain patients with risk factors.

Clinical guidelines (NCCN) recommend the administration of systemic chemotherapy for 6 months after resection of liver metastases from colorectal cancer, with the following treatment regimens being possible:

- FOLFOX or CAPOX or FOLFIRI with or without bevacizumab.
- FOLFOX or CAPOX or FOLFIRI plus cetuximab (only if K-ras-negative).

Follow-up after resection

Clinical guidelines (NCCN) recommend the following:

- CEA level determination every 3 months for 2 years and then every 6 months for 3 years.
- Thoraco-abdomino-pelvic CT scan every 3 months for 2 years and every 6 months for the following 3 years.
- Colonoscopy once per year. If adenomas are present, repeat at 3 years and then at 5 years. If there is an adenoma, and it is completely resected, colonoscopy should be repeated at 1 year.

Conclusions

The efficacy of surgery in metastatic liver disease is already established, although no randomized, controlled, prospective study has been performed to date. Liver metastases secondary to colorectal cancer are the most frequently treated metastases owing to their frequency, form of presentation, and good response to chemotherapy. In the past two decades, affected patients have gone from being considered terminal, with no possibility of treatment, to being surgical candidates, with a very significant increase in survival and even achieving a definitive cure. Advances in liver surgery, improvements in chemotherapy, and the interdisciplinary approach to the disease have undoubtedly been a major step in this direction.

References

1. Hughes KS, Simon R, Songhorabodi S, et al. Resection of the liver for colorectal carcinoma metastases: a multi-institutional study of patterns of recurrence. *Surgery* 1986;**100**(2):278–84. https://www.ncbi.nlm.nih.gov/pubmed/3526605.
2. Scheele J, Stang R, Altendorf-Hofmann A, Paul M. Resection of colorectal liver metastases. *World J Surg* 1995;**19**(1):59–71. https://doi.org/10.1007/BF00316981.
3. Nordlinger B, Guiguet M, Vaillant JC, et al. Surgical resection of colorectal carcinoma metastases to the liver. A prognostic scoring system to improve case selection, based on 1568 patients. Association Française de Chirurgie. *Cancer* 1996;**77**(7):1254–62. https://www.ncbi.nlm.nih.gov/pubmed/8608500.
4. Jamison RL, Donohue JH, Nagorney DM, Rosen CB, Harmsen WS, Ilstrup DM. Hepatic resection for metastatic colorectal cancer results in cure for some patients. *Arch Surg* 1997;**132**(5):505–10. discussion 511 https://doi.org/10.1001/archsurg.1997.01430290051008.
5. Fong Y, Fortner J, Sun RL, Brennan MF, Blumgart LH. Clinical score for predicting recurrence after hepatic resection for metastatic colorectal Cancer. *Ann Surg* 1999;**230**(3):309. https://doi.org/10.1097/00000658-199909000-00004.
6. Iwatsuki S, Dvorchik I, Madariaga JR, et al. Hepatic resection for metastatic colorectal adenocarcinoma: a proposal of a prognostic scoring system. *J Am Coll Surg* 1999;**189**(3):291–9. https://doi.org/10.1016/s1072-7515(99)00089-7.
7. Choti MA, Sitzmann JV, Tiburi MF, et al. Trends in long-term survival following liver resection for hepatic colorectal metastases. *Ann Surg* 2002;**235**(6):759–66. https://doi.org/10.1097/00000658-200206000-00002.
8. Abdalla EK, Vauthey J-N, Ellis LM, et al. Recurrence and outcomes following hepatic resection, radiofrequency ablation, and combined resection/ablation for colorectal liver metastases. *Ann Surg* 2004;**239**(6):818–25. discussion 825-827 https://doi.org/10.1097/01.sla.0000128305.90650.71.

9. Fernandez FG, Drebin JA, Linehan DC, Dehdashti F, Siegel BA, Strasberg SM. Five-year survival after resection of hepatic metastases from colorectal cancer in patients screened by positron emission tomography with F-18 fluorodeoxyglucose (FDG-PET). *Ann Surg* 2004;**240**(3):438–47. discussion 447-450 https://doi.org/10.1097/01.sla.0000138076.72547.b1.
10. Wei AC, Greig PD, Grant D, Taylor B, Langer B, Gallinger S. Survival after hepatic resection for colorectal metastases: a 10-year experience. *Ann Surg Oncol* 2006;**13**(5):668–76. https://doi.org/10.1245/ASO.2006.05.039.
11. Rees M, Tekkis PP, Welsh FKS, O'Rourke T, John TG. Evaluation of long-term survival after hepatic resection for metastatic colorectal cancer. *Ann Surg* 2008;**247**(1):125–35. https://doi.org/10.1097/sla.0b013e31815aa2c2.
12. de Jong MC, Pulitano C, Ribero D, et al. Rates and patterns of recurrence following curative intent surgery for colorectal liver metastasis: an international multi-institutional analysis of 1669 patients. *Ann Surg* 2009;**250**(3):440. https://doi.org/10.1097/SLA.0b013e3181b4539b.
13. Morris EJA, Forman D, Thomas JD, et al. Surgical management and outcomes of colorectal cancer liver metastases. *Br J Surg* 2010;**97**(7):1110–8. https://doi.org/10.1002/bjs.7032.
14. Pawlik TM. Intrahepatic cholangiocarcinoma: from diagnosis to treatment. *Hepatobiliary Surg Nutr* 2017;**6**(1):1. https://doi.org/10.21037/hbsn.2017.01.04.
15. Imai K, Allard M-A, Castro Benitez C, et al. Long-term outcomes of radiofrequency ablation combined with hepatectomy compared with hepatectomy alone for colorectal liver metastases. *Br J Surg* 2017;**104**(5):570–9. https://doi.org/10.1002/bjs.10447.
16. Adam R, Karam V, Cailliez V, et al. 2018 annual report of the European liver transplant registry (ELTR)—50-year evolution of liver transplantation. *Transpl Int* 2018;**31**(12):1293–317. https://doi.org/10.1111/tri.13358.
17. NCCN - evidence-based cancer guidelines, oncology drug compendium, oncology continuing medical education. Accessed February 18, 2021. http://www.nccn.org.
18. Fong Y, Blumgart LH, Cohen A, Fortner J, Brennan MF. Repeat hepatic resections for metastatic colorectal cancer. *Ann Surg* 1994;**220**(5):657–62. https://doi.org/10.1097/00000658-199411000-00009.
19. Nordlinger B, Vaillant JC, Guiguet M, et al. Survival benefit of repeat liver resections for recurrent colorectal metastases: 143 cases. Association Francaise de Chirurgie. *J Clin Oncol* 1994;**12**(7):1491–6. https://doi.org/10.1200/JCO.1994.12.7.1491.
20. Fernandez-Trigo V, Shamsa F, Vidal-Jové J, Kern DH, Sugarbaker PH. Prognostic implications of chemoresistance-sensitivity assays for colorectal and appendiceal cancer. *Am J Clin Oncol* 1995;**18**(5):454–60. https://doi.org/10.1097/00000421-199510000-00020.
21. Adam R, Bismuth H, Castaing D, et al. Repeat hepatectomy for colorectal liver metastases. *Ann Surg* 1997;**225**(1):51–62. https://doi.org/10.1097/00000658-199701000-00006.
22. Yamamoto J, Kosuge T, Shimada K, Yamasaki S, Moriya Y, Sugihara K. Repeat liver resection for recurrent colorectal liver metastases. *Am J Surg* 1999;**178**(4):275–81. https://doi.org/10.1016/s0002-9610(99)00176-2.
23. Petrowsky H, Gonen M, Jarnagin W, et al. Second liver resections are safe and effective treatment for recurrent hepatic metastases from colorectal cancer: a bi-institutional analysis. *Ann Surg* 2002;**235**(6):863–71. https://doi.org/10.1097/00000658-200206000-00015.
24. Nagakura S, Shirai Y, Suda T, Hatakeyama K. Multiple repeat resections of intra- and extrahepatic recurrences in patients undergoing initial hepatectomy for colorectal carcinoma metastases. *World J Surg* 2002;**26**(2):141–7. https://doi.org/10.1007/s00268-001-0196-z.
25. Tanaka K, Shimada H, Ohta M, et al. Procedures of choice for resection of primary and recurrent liver metastases from colorectal cancer. *World J Surg* 2004;**28**(5):482–7. https://doi.org/10.1007/s00268-004-7214-x.
26. Pessaux P, Lermite E, Brehant O, Tuech J-J, Lorimier G, Arnaud J-P. Repeat hepatectomy for recurrent colorectal liver metastases. *J Surg Oncol* 2006;**93**(1):1–7. https://doi.org/10.1002/jso.20384.
27. Ishiguro S, Akasu T, Fujimoto Y, et al. Second hepatectomy for recurrent colorectal liver metastasis: analysis of preoperative prognostic factors. *Ann Surg Oncol* 2006;**13**(12):1579–87. https://doi.org/10.1245/s10434-006-9067-z.
28. Cunha AS. A second liver resection due to recurrent colorectal liver metastases. *Arch Surg* 2007;**142**(12):1144. https://doi.org/10.1001/archsurg.142.12.1144.
29. Jones NB, McNally ME, Malhotra L, et al. Repeat hepatectomy for metastatic colorectal cancer is safe but marginally effective. *Ann Surg Oncol* 2012;**19**(7):2224–9. https://doi.org/10.1245/s10434-011-2179-0.
30. Adam R, Karam V, Delvart V, et al. Evolution of indications and results of liver transplantation in Europe. A report from the European liver transplant registry (ELTR). *J Hepatol* 2012;**57**(3):675–88. https://doi.org/10.1016/j.jhep.2012.04.015.
31. Hamada T, Ishizaki H, Haruyama Y, et al. Neutrophil-to-lymphocyte ratio and intratumoral CD45RO-positive T cells as predictive factors for longer survival of patients with colorectal liver metastasis after hepatectomy. *Tohoku J Exp Med* 2020;**251**(4):303–11. https://doi.org/10.1620/tjem.251.303.
32. Lorenz M, Müller HH, Schramm H, et al. Interim analysis of a prospective, randomized multi-center study by the "Liver Metastases" Study Group: adjuvant intra-arterial chemotherapy after curative liver resection of colorectal metastases. In: *Langenbecks Archiv Fur Chirurgie. Supplement. Kongressband. Deutsche Gesellschaft Fur Chirurgie. Kongress*. vol. 115; 1998. p. 523–8. europepmc.org. https://europepmc.org/article/med/14518310.
33. Kemeny N, Huang Y, Cohen AM, et al. Hepatic arterial infusion of chemotherapy after resection of hepatic metastases from colorectal cancer. *N Engl J Med* 1999;**341**(27):2039–48. https://doi.org/10.1056/nejm199912303412702.
34. Kemeny MM, Adak S, Gray B, et al. Combined-modality treatment for resectable metastatic colorectal carcinoma to the liver: surgical resection of hepatic metastases in combination with continuous infusion of chemotherapy—an intergroup study. *J Clin Oncol* 2002;**20**(6):1499–505. https://doi.org/10.1200/JCO.2002.20.6.1499.

Chapter 32

Pulmonary metastasectomy for CRC

Rodrigo A.S. Sardenberg[a,b] and Diego Gonzalez-Rivas[c,d,e,f]
[a]Thoracic Surgeon Hospital Alemão Oswaldo Cruz and United Health Group, São Paulo, Brazil, [b]Center for Advanced Research at University of Great Lakes São José do Rio Preto, São Paulo, Brazil, [c]Uniportal VATS Training Program, Shanghai Pulmonary Hospital, Shanghai, People's Republic of China, [d]Ernst Von Bergmann Hospital, Berlin, Germany, [e]Department of Thoracic Surgery and Lung Transplantation, A Coruña University Hospital, A Coruña, Spain, [f]Minimally Invasive Thoracic Surgery Unit (UCTMI), A Coruña, Spain

Introduction

Colorectal cancer (CRC) is the third most commonly diagnosed cancer among men and women in the United States.[1] Though colon and rectum cancer are often referred to as colorectal cancer, they are actually distinct disease entities. They have a different inherent prognosis, different potential for treatment response (and indeed different standard treatment options), and may have different metastatic patterns.[1] Previous studies had generated insight into metastatic patterns and showed that different primary cancers tended to metastasize with different frequencies and to different sites.[2] At the time of diagnosis, about 20% of CRC patients have already developed metastatic diseases.

It is well known that the most common metastatic site for CRC patients is liver, followed by lung.[1,2] A nationwide retrospective review of 5817 pathological records of CRC patients showed that rectum cancer patients more often had metastasis at extraabdominal sites while patients with colon cancer had a higher rate of abdominal metastases.[3]

Pulmonary metastases are the harbinger of systemic spreading of malignant tumor. Although primary tumors can be locally controlled with surgery and/or radiotherapy, the treatment of choice for metastatic disease is chemotherapy or targeted therapy. Although lung metastases often represent widespread systemic disease, selected groups of patients present with disease localized solely in the lungs and may have distinct biological behavior. Therefore such patients can be candidates for surgical treatment.

Surgical resection of pulmonary metastases was introduced as an important form of treatment for a variety of solid tumors.[4] Several series evaluating salvage surgery for metastatic disease showed survival rates of 5 years ranging from 5% to 85%, with acceptable morbidity and mortality.[5] These data are presented in Table 32.1.

However, most studies in the medical literature reported results of pulmonary metastasectomy (PM) in a patient population typically collected over a long period of time when treatment, diagnosis, and management have changed but have not resulted in any convincing conclusions. None of the prospective studies included a control group or randomization, comparing surgery with other standard systemic treatment. All data are susceptible to the limitations of retrospective analysis and patient selection bias.[6]

A quarter of patients with colorectal cancer (CRC) have metastatic lesions at diagnosis and in nearly half of them, metastases will develop, often in liver or lung or both.[7] Surgery has been consistently reported as a potentially curative option for liver-limited disease, with 5-year survival rate of 30%–40%; in 10%–15% of cases, lung metastases are documented at advanced disease and are diagnosed mostly as multiple or bilateral metastases, with only 2%–7% as single lesion.[7]

Advanced stage colorectal cancer remains one of the most common causes of pulmonary metastases and currently no studies directly comparing surgical intervention with supportive care have been reported.[8] Stage IV colorectal cancer without chemotherapy or surgical intervention notably carries a dismal prognosis, with a 5-year survival rate of less than 10% and a median survival time of 5 months even with optimal supportive care.[8]

Results

The results of PM should be analyzed from the point of view according to the critical factors that can affect survival rate. Such analysis results should also be based on reviews of primary studies in histology (breast, colon, melanoma) or in patients with similar histologies (soft tissue sarcomas) in a sufficient number of patients. The results of the International

TABLE 32.1 Overall survival rate estimated at 5 years in patients undergoing pulmonary metastasectomy.

Type of tumor	Five-year overall survival rate after pulmonary metastasectomy (%)
Soft tissue sarcomas	18–35
Colorectal carcinomas	25–42
Germ cell tumors	30–84
Malignant melanoma	4–14
Kidney carcinoma	13–55
Breast carcinoma	14–36

Registration of Lung Metastases confirm that metastasectomy is a form of potential therapy for healing that can be safe with low mortality.[9]

According to the principles of oncological surgery, complete removal of all pulmonary foci is associated with increased survival rate. The data also suggest that the preoperative radiological workup evaluation, however, has low accuracy and the intraoperative staging by experienced surgeons is essential for the resection of all metastases. In this context, video-assisted thoracoscopic surgery (VATS) cannot provide security for the removal of all metastases, particularly if more than one lesion, micro and/or deeper nodules are identified intraoperatively.

A limitation lies in the fact that there is no evaluation of variables related to the biological behavior of different histological types, which may explain the evolution of clinical metastases. Currently, VATS may be indicated for the diagnosis or staging to evaluate the extent of disease, or resection of metastasis in patients with single nodules in peripheral topography, after evaluation by helical chest CT. Complications of VATS include incomplete resection and pleural implantation after extraction of lung metastases.[10, 11]

The role of PM is less evident in patients with breast tumors and melanoma, which require further definition in future prospective studies. Another area of uncertainty is based on the effectiveness of surgical staging of hidden metastases by contralateral disease detected by median sternotomy—or bilateral thoracotomy—in some tumor histology, not only sarcomas. The actual role in the contribution of induction or adjuvant chemotherapy in some specific types of tumors deserves further investigations.[11]

Discussion

The first surgical resection of metastatic lung is credited to Weinlechner, who, in 1882 as part of a surgical resection of primary tumor of the chest wall, incidentally resected metastatic pulmonary nodule as a complementary procedure. However, pulmonary resection for metastatic disease was first reported as independent procedure in the medical literature by Division in Europe in 1926 and Torek in the United States in 1930.[12] In 1927 Tudor held in London sublobar resection for metastatic nodule from lower limb sarcoma. This patient remained alive 18 years after metastasectomy. In 1933 in North America, Barney and Churchill resected a solitary mass by lobectomy, and subsequently proven metastatic renal cell carcinoma. The patient then underwent nephrectomy to remove the primary tumor and survived disease-free for 23 years.[12, 13]

In 1947 Alexander and Haight reviewed 24 patients (16 carcinomas and 8 sarcomas) who underwent pulmonary metastasectomy, and initially discussed the selection criteria. This series also contains the first report of metachronous metastasectomy, and the patient remained disease-free 14 years after the second resection.

Interestingly at this time, the vast majority (80%) of patients were submitted to lobectomy or pneumonectomy for resection of pulmonary metastasis, when rates of postoperative mortality were around 10%.[12] Simultaneous resections of multiple synchronous lung metastasis were described by Mannix in 1953. This time it was the first reported resection of lesions by means of a six lingulectomies and basal segmentectomy, held in the same surgical procedure. The review of the Mayo Clinic between 1942 and 1962, based on 205 patients, highlighted the economic value of resection and the limitations of preoperative staging by Thomford et al. In this series, the probability of survival rate at 5 years was 30%, and there were no differences between patients with single or multiple metastases.[12, 14]

In 1979 McCormack et al. reported the experience of Memorial Sloan-Kettering Cancer Center through the results obtained in 402 patients and 622 thoracotomies, with very reasonable survival rate, according to selection criteria established in advanced disease.[12]

The most frequent sites of distant metastases from colorectal cancer are the liver and lung. After curative resection for colorectal cancer, hepatic metastases are detected in 8%–30% of patients and pulmonary metastases in 10%–20%.[15]

Patient selection

The patient selection was extracted from The International Registry of Lung Metastases (IRLM), published in 1997. It counts with retrospective and prospective data, experience based on 5206 patients from 18 centers in the United States, Europe, and Canada. With median follow-up of 46 months observed in patients with complete resection, they found a survival rate of 36% in 5 years, 26% in 10 years, and 22% in 15 years. The survival rate for patients with incomplete resection was 13% in 5 years and 7% in 10 years, with a median of 15 months.[9] Data analysis from this important study also allowed the development of a prognostic model with four groups:

Group I: resectable metastases, disease-free interval >36 months and single metastases (or no risk factor);
Group II: resectable metastases, free interval disease <36 months or multiple metastases (or one risk factor);
Group III: unresectable metastases, disease-free interval <36 months and multiple metastases (or two risk factors);
Group IV: unresectable metastases.

Thus the prognostic factors of greatest impact were related to disease-free interval greater than 36 months, presence of a single lesion, and the possibility of complete resection of the lesions. The overall survival rate was different for the four groups and is presented in Table 32.2.

Downey et al. emphasized the importance of surgical reoperation in patients with new pulmonary recurrence, where survival for patients operated at least twice was 42.8 months, with an estimated 36% in 5 years. In this series, the most important factors correlated with better survival were complete re-resection, less than three nodules resected and greater resected nodule <2 cm ($P = .003$).[6]

The selection criteria for surgical resection of lung metastases are currently considered as standard treatment in select cases and routinely performed in specialized centers. In some tumors such as sarcomas and germ cell tumors, a large proportion of patients (>50%) may be candidates for metastasectomy.[16] With regard to patients with epithelial tumors, the minority of patients may benefit from the procedure. The decision to refer the patient for surgical treatment should be multi-disciplinary, and the most relevant criteria for surgical indication include:

(1) Controlled or controllable primary tumor;
(2) Completely resectable lesions;
(3) Absence of extrapulmonary disease;
(4) Lack of effective systemic therapy;
(5) Clinical condition to tolerate the planned procedure.

Still, it is worth noting the presence of criteria extended or exception for patients with lung metastases in specific situations:

(1) Need for diagnosis;
(2) Removal of residual nodules after chemotherapy;
(3) Symptomatic metastases;
(4) Obtaining tissue for tumor markers and immunohistochemistry.

The current data available suggest that PM can increase survival rate of patients with low morbidity and mortality, and the 5-year survival rate ranges between 20% and 40%, figures stronger than chemotherapy or radiation therapy only.[17]

TABLE 32.2 Survival according to prognostic group of the International Registry of Lung Metastases.

Group	Median survival (months)
I	61
II	34
III	24
IV	14

Recently, the Pulmonary Metastasectomy in Colorectal Cancer (PulMiCC) randomized controlled trial was stopped because of poor and worsening recruitment. The small number of participants in the trial ($N = 65$) precludes a conclusive answer to the research question given the large overlap in the confidence intervals in the proportions still alive at all time points. A widely held belief is that the 5-year absolute survival benefit with metastasectomy is about 35%: 40% after metastasectomy compared to <5% in controls. The estimated survival rate in this study was 38% (23%–62%) for metastasectomy patients and 29% (16%–52%) in the well-matched controls. That is the new and important finding of this RCT.[18]

Pathophysiology

The spread of cancer cells with subsequent appearance of metastatic disease is the single factor of worse negative prognostic factor in cancer treatment. As an example of what happens with colorectal tumors, where overall survival rate is around 90% in localized disease, but in the presence of liver or lung metastases, survival rate drops to 10%.[19]

The purpose of the follow-up of cancer patients treated thus aims to ultimately identify the development of metastases. When the tumor is locally controlled, chemotherapy can eradicate occult metastases or micrometastases. Even with the advancement of multidisciplinary treatment, however, metastatic disease remains the leading cause of mortality in patients with cancer. Determining which patients and what types of tumors have a higher risk of dissemination is a major goal of the oncologist. The histological heterogeneity of certain tumors is related to the propensity to develop lung metastases. These observations raise some questions: (1) What causes the tumor to become metastatic? (2) Why some types of tumors have the pulmonary parenchyma as the preferred target for metastases?

Many patients developed more than one site of metastatic diseases. About 10.01% of colon cancer and 7.66% of rectum cancer patients had only liver metastases. Rare CRC patients had only brain or bone metastases at the time of diagnosis. The most common two-site metastases combination is liver and lung (2.24% for colon cancer and 2.84% for rectal cancer). Rectum cancer is more likely to have two-site metastasis than colon cancer, especially for lung and liver, lung and bone, as well as liver and bone combination. Qiu et al. found that patients with lung metastasis had a higher risk of bone (10.0% vs 4.5%, $P < .001$) or brain metastases (3.1% vs 0.1%, $P < .001$) than patients without it.[20] Though a similar phenomenon was noted for liver metastases, with higher risk of bone (4.7% vs 0.3%, $P < .001$) and brain metastases (1.0% vs 0.1%, $P < .001$) for liver metastases patients than those without, CRC patients with lung metastases had a higher incidence rate of bone or brain metastases than patients with liver metastases.[20]

Usually, malignant tumors metastasize through hematogenous routes, lymphatic, or by direct invasion. Hematogenous metastases are most often found in the lungs, liver, brain, and bones. Tumor cells that metastasize to the lungs can become trapped in the capillary endothelium, and many of these tumor emboli die. However, others survive and ultimately proliferate. Tumor cells can also be transported through capillaries and occupy a specific location in the lung or spread throughout the parenchyma (e.g., lymphangitis in breast tumors).

Until the 1970s, Paget's theory of choice of target organs by the primary tumor to metastatic cell installation, proposed in 1889, was followed. Such theory stablished the hypothesis of random dissemination cancer, the most widely accepted so far. The theory became known as the "seed and soil theory," in which the tumor would find affinity for the installation environment and proliferation in certain organs.[6]

More recently, improvements in technologies of molecular biology, cellular and images, highlighted the events that occurred between the primary tumor growth and metastases of radiological appearance—also known as cascade of metastases. Given the complexity of this process, the events within this cascade are postulated as the primary target of intervention for the prevention of distant metastases.

Symptoms

Normally the pulmonary nodules are identified during routine follow-up of the primary tumor after treatment in radiography or chest computed tomography. Respiratory symptoms such as cough and hemoptysis may be present in cases where there is hilar involvement or bronchial invasion (Fig. 32.1). In endobronchial metastases, there may be persistent dry cough or massive hemoptysis. In some patients (especially in breast, melanoma, and renal tumors), endobronchial metastases may cause segmental atelectasis or even of the correspondent entire lobe.[6]

The peripheral location of tumors can invade the parietal pleura or chest wall, in which cases the patient usually referred chest pain. More rarely, peripheral tumors can rupture and cause pneumothorax, a fact more common in patients with sarcomas. When there are multiple nodes (or lymphangitis carcinomatosis), especially in bilateral topography, patients also may develop a dry cough or dyspnea (Fig. 32.2).

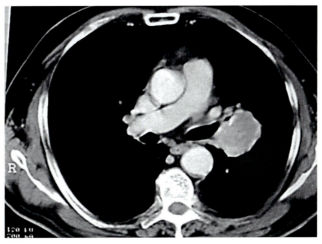

FIG. 32.1 Chest CT, metastatic CRC tumor with invasion of hilar left bronchus.

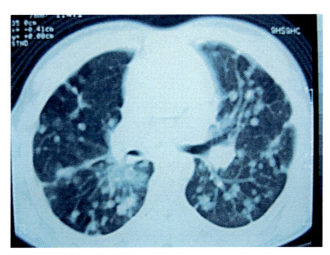

FIG. 32.2 Chest CT showing multiple nodules with lymphangitis.

Preoperative evaluation

As the main objective is to assess which patients will benefit from PM, the appropriate selection is the key to success for this procedure. The radiological evaluation is mandatory in sensitive preoperative. Many times during the follow-up it is possible to observe the pulmonary lesions in chest X-ray, either due to size or due to the multiplicity of lesions (Fig. 32.3). There is no specific radiological finding of pulmonary metastasis, however, in patients with a history of treated tumor and present with well-circumscribed nodule(s), mainly in the lower lobe(s), should be strongly investigated in this direction (Figs. 32.4 and 32.5).

In chest CT, the lung metastases can be diagnosed as single or multiple lesions, and may still present as diffuse lesions (Figs. 32.6 and 32.7).

This examination allows a consistent anatomical idea, highly important with reference to intrathoracic structures and for planning resection and extent of surgery.[21] Other test that can be part of the preoperative evaluation to get the levels of tumor markers is CEA, especially in CRC.

More recently, positron emission tomography (PET-CT) with fluorine-18 deoxyglucose has been shown to be a new modality in the evaluation of patients with lung metastases. It is known that PET-scan has limited sensitivity in nodules <1 cm, which is not higher with integrated PET-CT.[22] On the other hand, the higher value of PET-CT is its high sensitivity to detect extrathoracic recurrence, information that can alter the surgical plan. PET-CT can also better assess the extent of intrathoracic disease, such as the presence of mediastinal lymph nodes (Fig. 32.8). In this context, it is important to stablish that the PET-CT negative for the lung node should not influence the conduct of state to surgical resection, provided that

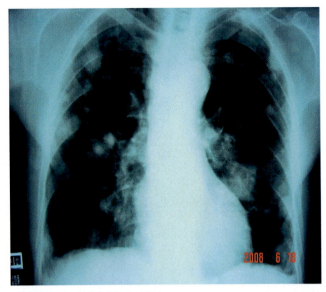

FIG. 32.3 Plain chest radiograph of a patient with multiple bilateral lesions.

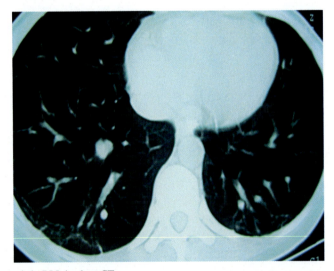

FIG. 32.4 Patient with single metastasis in RLL in chest CT.

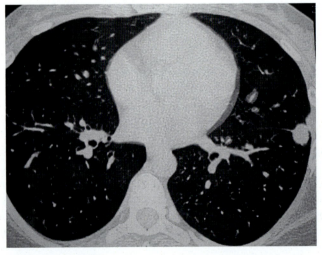

FIG. 32.5 Patient with lung metastasis in LLL in chest CT.

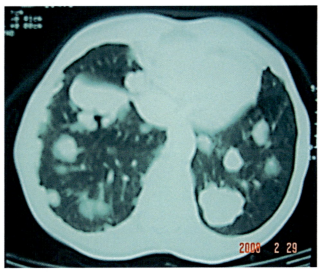

FIG. 32.6 Patients with multiple bilateral pulmonary metastases in chest CT.

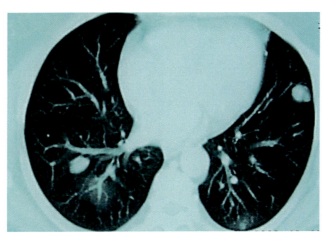

FIG. 32.7 CT chest showed bilateral pulmonary metastases (resectable).

there is strong suspicion for this, especially in nodules growth during the follow-up.[23] The PET-CT has a sensitivity of 50%, specificity of 98%, and accuracy of 87% for lung metastases, compared with helical CT of 75%, 100%, and 94%, respectively.[23]

Factors such as evaluation of the anatomic extent; characterization of solitary pulmonary nodule—in squamous cell carcinomas—and metastatic primary lung tumors; evaluation of extrathoracic disease and regional lymph nodes, should help as additional clinical information which is required for treatment planning.[24]

Prognostic factors

Survival prognostic factors are related to patient characteristics (e.g., age, sex, symptoms, diagnosis, and location) or to tumor-related characteristics itself, and its treatment (tumor size, neoadjuvant and adjuvant chemotherapy, disease-free interval) and factors related to lung recurrence (number of metastases, the laterality of nodes, number of thoracotomies, the last resection, histopathological aspects, molecular markers). The search for characteristics that allow to predict clinical behavior of certain disease is of great importance, especially for treatment planning.

Younes et al. reported a retrospective study of 529 patients undergoing PM of different histologies, that for those with disease-free interval (DFI) less than 12 months, there was a significant worsening of overall survival rate, but this factor did not show statistical difference in the multivariate analysis.[25]

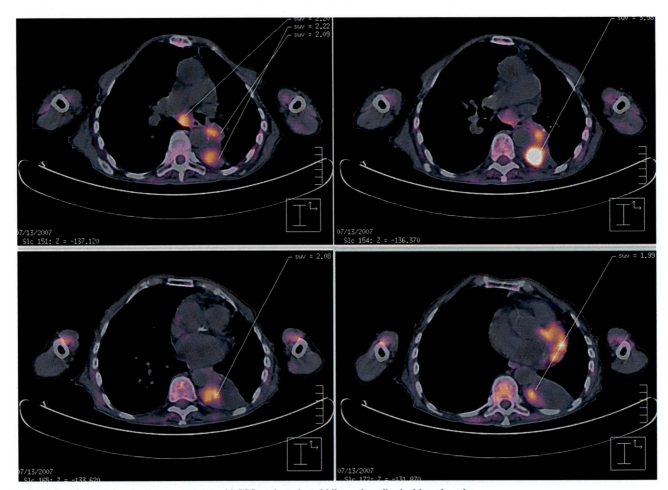

FIG. 32.8 PET-CT showing lung metastases with LLL atelectasis and hilar and mediastinal lymph nodes.

Pastorino et al. analyzed 5206 cases of tumors of various histologies and showed that 31% of patients had 11 months DFI, 36% had 12–35 months, and 31% with 36 months or more. Different survival rates were observed: for patients with DFI < 11 months, survival rate was 33% in 5 years and a median of 29 months; for those with DFI between 12 and 35 months, survival rate was 31% in 5 years and a median of 30 months; those who had DFI > 36 months, survival rate was 45% in 5 years and a median of 49 months.[9]

Although there is a consensus in the literature that the DFI is a prognostic factor, some authors disagree with such information. Although not confirmed in previous reports, probably, the research methods and interpretation are different in each report.[26]

Some authors make their calculations for a fixed DFI, while others look for a DFI with the best statistical difference. The best results are obtained with DFI > 12 months.

Pulmonary metastases can be accompanied by hilar or mediastinal lymphatic involvement; it is believed, though unproven, that the disease may spread to lymph nodes by lung. Retrospective series and autopsy revealed nodal metastatic disease in 5%–33% of cases, depending on the histological type. Thus they are more common in carcinomas than in sarcomas. Although the presence of thoracic lymph node metastasis is a negative prognostic factor, systematic lymphadenectomy has no defined role.[27–29]

Prognosis of patients with metastatic disease appears to be largely determined by the number of metastatic nodules, according to several authors.[21–24] However, a subgroup analysis of patients reported multiple metastatic nodules treated with complete resection, nothing of that approach was a strong positive predictor of survival rate, reducing the impact of the high number of metastases on survival rate. Rusch et al. reported that patients with pulmonary nodule solitary metastasis had better survival rate.[5] Younes et al. showed in a prospective study of pulmonary metastasectomy in 529 patients with tumors of various histologies, that the number of resected nodules greater than two unfavorable correlated with survival rate, a significant feature in the univariate ($P = .0007$) and multivariate ($P = .016$), both overall survival rate and disease-free survival rate ($P = .003$ and $P = .004$) in univariate and multivariate analysis, respectively.[25]

Several authors have emphasized that patients undergoing pulmonary metastasectomy and complete resection cancer were prognostic factors of greater impact, and associated with a longer survival rate. Patients treated with complete resection can present 5-year survival rate between 30% and 40%, and patients with incomplete resection, a survival rate much lower, below 10%.[5, 30, 31]

Previous studies reported the following independent prognostic factors after lung metastasectomy: number of pulmonary metastases, hilar or mediastinal lymph node metastasis, time when pulmonary metastases were detected, prethoracotomy CEA level, distribution of pulmonary metastases, liver metastasectomy before thoracotomy.[15] Surgical resection of both liver and lung metastases from colorectal cancer is thought to prolong survival rate in highly selected patients.[15] Therefore prospective data analysis is needed to establish operative indications.

Surgical technique

In selected patients, resectable lung metastases in the absence of extrapulmonary disease, complete resection is associated with increased overall survival rate. In those selected which extrathoracic metastases were operated (e.g., liver), the indication of pulmonary resection aims to remove all disease at first detected. In these cases, the surgeon has in its favor the knowledge of tumor biology such as CRC, in which the dissemination is made to preferred organs (e.g., lung and liver).

The decision between minimally invasive and conventional thoracotomy or sternotomy depends on the number, location, and size of metastases. The main objective is to remove all node(s) with a gross margin of at least 1 cm. The open or conventional surgery, with more limited incisions in the skin, associated with muscle-sparing thoracotomy, have gradually replaced the classic posterolateral thoracotomy—with the intention to reduce the trauma to muscle function and pain after surgery—thereby promoting early mobilization of the patient.

Other authors recommend further staged axillary thoracotomy for bilateral metastases as an alternative to conventional thoracotomy.[32] There is considerable controversy about patients with unilateral disease diagnosed by chest CT, with respect to unilateral or bilateral approach. Since CT scans can miss up to 24% of multiple lesions, some patients are operated upon bilateral approach (sternotomy or thoracotomy). However, improvements in image quality by helical CT and positron emission tomography with FDG PET-CT in preoperative evaluation increased the accuracy of staging and selection of patients for decision on unilateral or bilateral approach.[23] The video-assisted thoracic surgery (VATS) has been proposed as a procedure with less morbidity for PM. This approach has obvious appeal because of shorter hospital stay and less postoperative pain when compared to sternotomy or thoracotomy. The VATS often has been used as a diagnostic procedure—especially in cases with a single nodule, and as a curative potential in PM—especially for lesions located in the outer third of the lung with up to 3 cm.[10, 28] The deeper lesions, or smaller, cannot be detected by thoracoscopy. In addition, for patients with multiple metastases, palpation of the lesions during the stapling offers greater security with regard to the resection margins.[32, 33] Currently, after evaluation by chest CT, VATS may be indicated for the diagnosis or staging to evaluate the extent of disease, or for resection of metastases in patients with single nodules in peripheral topography. Complications of VATS include incomplete resection and pleural implantation after extraction of lung metastases.[10, 11]

The possible incisions for resection of metastases include thoracotomy, staged bilateral thoracotomy, sternotomy, bilateral thoracotomy (Clamshell incision), and minimally invasive VATS. The incision of choice should not influence the survival rate of all patients when the main goal is complete resection. There are advantages and disadvantages for any choice in access for bilateral lesions (Table 32.3).

New perspectives

The current drive is toward stereotactic (ablative) body radiotherapy (abbreviated as SABR/ SBRT) for metastases.[18] There has already been large investment: practitioners, for-profit health providers, and the devices industry all expect a return on this investment.[34] However, it is probably more rational to treat systemic cancer with the now more effective systemic treatments.[35]

In colorectal cancer the evidence from a metaanalysis of 16 RCTs showed no survival benefit from detecting metastases 1–2 years earlier, indicating that the growing practice of metastasectomy may not improve survival rate. These findings were regarded as "bleak nihilism" by the *British Journal of Surgery*'s editor who wrote: "it is counterintuitive that earlier identification of metastatic disease does not improve survival."[36] The findings were confirmed by a Cochrane review.[37]

TABLE 32.3 Characteristics of each surgical approach—advantages and disadvantages.

Surgical technique	Advantages	Disadvantages
Posterolateral thoracotomy	Excellent exposure	Major pain; need for reoperation and more morbidity
Median sternotomy	Simultaneous bilateral exposure; less pain	Difficult access to the hilum, posterior segments, and lower lobes
Anterior bilateral thoracotomy (Clamshell)	Excellent exposure, less pain than conventional thoracotomy	Possibility of internal mammary vessels lesion and possible difficult access to posterior segments or lobes
VATS	Good visualization; less pain and morbidity	No palpation; difficult access to deeper and central lesions

Conclusions

The radical surgical resections of PM in appropriately selected patients may offer increased disease-free survival rate without systemic treatment. Retrospective studies with large samples have shown that survival rates at 5 and 10 years can reach 35% and 25%, respectively.

Surgical resection for lung metastases alone or for both liver and lung metastases produced good outcomes. However, the survival rate after both liver and lung metastasectomies was worse than the survival rate after lung metastasectomy alone. The number of metastatic organs was significantly higher after resection of hepatic and pulmonary metastases than after lung metastasectomy alone. Thus the different survival rates may be due to differential patterns of recurrence after pulmonary resection. Tumor recurrence after pulmonary metastasectomy is high. Therefore multimodality therapy with systemic chemotherapy may provide a survival benefit.

For patients with single pulmonary nodule, even in previously treated tumor, the possibility of primary cancer should be considered, especially in those patients with previous history of smoking, when lobectomy with lymph node dissection should be taken into account.

The long-term follow-up chest CT every 3–4 months is mandatory, and all nodes of recent onset must be regarded as new metastases.

The results published in the literature confirm that lung metastasectomy is a potentially curative treatment. The resectability, the DFI, the CEA level, and the number of resected nodules are the most important predictors of survival.

References

1. Lee YC, Lee YL, Chuang JP, Lee JC. Differences in survival between colon and rectal cancer from SEER data. *PLoS One* 2013;**8**:e78709.
2. Cook AD, Single R, McCahill LE. Surgical resection of primary tumors in patients who present with stage IV colorectal cancer: an analysis of surveillance, epidemiology, and end results data, 1988 to 2000. *Ann Surg Oncol* 2005;**12**:637–45.
3. Hugen N, van de Velde CJ, de Wilt JH, Nagtegaal ID. Metastatic pattern in colorectal cancer is strongly influenced by histological subtype. *Ann Oncol* 2014;**25**:651–7.
4. Treasure T, Milošević M, Fiorentino F, Macbeth F. Pulmonary metastasectomy: what is the practice and where is the evidence for effectiveness? *Thorax* 2014;**10**:1136–9.
5. Rusch VW. Pulmonary metastasectomy. Current indications. *Chest* 1995;**107**(6)322S–31S.
6. Van Raemdonck D, Friedel G. The European society of thoracic surgeons lung metastasectomy project. *J Thorac Oncol* 2010;**5**(6 Suppl 2):S127–9.
7. Zampino MG, Maisonneuve P, Ravenda PS, et al. Lung metastases from colorectal cancer: analysis of prognostic factors in a single institution study. *Ann Thorac Surg* 2014;**98**:1238–45.
8. Cheung FPY, Alam NZ, Wright GM. The past, present and future of pulmonary metastasectomy: a review article. *Ann Thorac Cardiovasc Surg* 2019;**25**:129–41.
9. Pastorino U. The development of an international registry. *J Thorac Oncol* 2010;(6 Suppl 2):S196–9.
10. Landreneau RJ, Pigula F, Luketich JD. Acute and chronic morbidity differences between muscle-sparing and standard thoracotomies. *J Thorac Cardiovasc Surg* 1996;**112**(5):1346–51.
11. Cerfolio RJ, Bryant AS, McCarty TP, Minnich DJ. A prospective study to determine the incidence of non-imaged malignant pulmonary nodules in patients who undergo metastasectomy by thoracotomy with lung palpation. *Ann Thorac Surg* 2011;**91**:1696–700 [discussion 1700–1].
12. Pastorino U. History of the surgical management of pulmonary metastases and development of the international registry. *Semin Thorac Cardiovasc Surg* 2002;**14**:18–28.

13. Sihag S, Muniappan A. Lymph node dissection and pulmonary metastasectomy. *Thorac Surg Clin* 2016;**26**(3):315–23.
14. Dai J, Jiang G. Lymphadenectomy in pulmonary metastasectomy: for diagnostic purpose or therapeutic effect? *Ann Thorac Surg* 2017;**103**(2) 688-73.
15. Hattori N, Kanmitsu Y, Komori K, Shimizu Y, Sano T, Senda Y, Fukui T. Outcomes after hepatic and pulmonary metastasectomies compared with pulmonary metastasectomy alone in patients with colorectal Cancer metastasis to liver and lungs. *World J Surg* 2013;**37**(6):1315–21.
16. Downey RJ. Surgery for colorectal and sarcomatous pulmonary pulmonary metastases- history, current management, and future directions. *Thorac Surg Clin* 2006;133–9.
17. Pastorino U, Gasparini M, Tavecchio L, Azzarelli A, Zucchi V, et al. The contribution of salvage surgery to the management of childhood osteosarcoma. *J Clin Oncol* 1991;**9**:1357–62.
18. Treasure T, Farewell V, Macbeth F, Monson K, Williams NR, Brew-Graves C, Lees B, Grigg O, Fallowfield L. Pulmonary metastasectomy versus continued active monitoring in colorectal Cancer (PulMiCC): a multicentre randomised clinical trial. *Trial* 2019;**20**:718.
19. Ries L, Eisner M, Kosary C, et al. *SEER cancer statistics review, 1975–2002*. Bethesda, MD: National Cancer Institute; 2005.
20. Qiu M, Hu J, Yang D, Cosgrove DP, Xu R. Pattern of distant metastases in colorectal cancer: a SEER based study. *Oncotarget* 2015;**36**:3658–66.
21. Margaritora S, et al. Pulmonary metastases: can accurate radiological avoid thoracotomy approach? *Eur J Cardiothorac Surg* 2002;**21**:1111–5.
22. Franzius C, Daldrup-Link HE, Sciuk J, et al. FDG-PET for detection of pulmonary metastases from malignant primary bone tumors: comparison with spiral CT. *Ann Oncol* 2001;**12**:479–82.
23. Pastorino U, Veronesi G, Landoni C, et al. Fluorodeoxyglicose positron emisssion tomography improves staging of resectable lung metastases. *J Thorac Cardiovasc Surg* 2003;**126**:1906.
24. Woodard PK, Dehdashi F, Putman CE. Radiologic diagnosis of extrathoracic metastases to the lung. *Oncology* 1998;**12**:431.
25. Younes RN, Gross JL, Taira AM, Martins AAC, Neves GS. Surgical resection of lung metastases: results from 529 patients. *Clinics* 2009;**64**(6):535–41.
26. Su X, Ma G, Zhang X, Long H, Rong TH. Surgical approach and outcomes for treatment of pulmonary metastases. *Ann Thorac Med* 2013;**8**(3):160–4.
27. Pastorino U, Buyse M, Frielder G. Long- term results of lung metastasectomy: prognostic analyses based on 5206 cases. *J Thorac Cardiovasc Surg* 1997;**113**(1):31–49.
28. Eckardt J, Licht PB. Thoracoscopic or open surgery for pulmonary metastasectomy: an observer blinded study. *Ann Thorac Surg* 2014;**98**(2):466–9.
29. Berry MF. Role of segmentectomy for pulmonary metastases. *Ann Cardiothorac Surg* 2014;**3**(2):176–82.
30. Rehders A, Hosch SB, Scheuneman P, Stoecklein NH, Knoefel WT, Peiper M. Benefit of surgical of lung metastases in soft tissue sarcoma. *Arch Surg* 2007;**142**(1):70–5.
31. Caushi F, Xhemalaj D, Skenduli I, Hafizi H, Pupla Z, Hatibi A, Shima E, Selenica E, Gjerazi J, Gega B. Which is the most common malignancy that profits from the lung metastasectomy? Ten years experience. *Ann Transl Med* 2016;**4**(6):124.
32. Downey RJ, McComarck P, LoCicero III J. Dissemination of malignant tumors after video-assisted thoracic surgery: a report of 21 cases. The video-assisted thoracic surgery group. *J Thorac Cardiovasc Surg* 1996;**111**:954–9.
33. Deo SS, Shukla NK, Khanna P, Jha D, Pandit A, Thulkar S. Pulmonary metastasectomy: review of experience at a tertiary cancer care center. *J Cancer Res Ther* 2014;**10**(3):535–9.
34. Palma DA, Olson R, Harrow S, Gaede S, Louie AV, Haasbeek C, et al. Stereotactic ablative radiotherapy versus standard of care palliative treatment in patients with oligometastatic cancers (SABR-COMET): a randomised, phase 2, open-label trial. *Lancet* 2019;**393**(10185):2051–8 40.
35. Lewis SL, Porceddu S, Nakamura N, Palma DA, Lo SS, Hoskin P, et al. Definitive stereotactic body radiotherapy (SBRT) for extracranial oligometastases: an international survey of > 1000 radiation oncologists. *Am J Clin Oncol* 2017;**40**(4):418–22.
36. Jeffery M, Hickey BE, Hider PN, See AM. Follow-up strategies for patients treated for non-metastatic colorectal cancer. *Cochrane Database Syst Rev* 2016;**11**:CD002200.
37. Mokhles S, Macbeth F, Farewell V, Fiorentino F, Williams NR, Younes RN, et al. Meta-analysis of colorectal cancer follow-up after potentially curative resection. *Br J Surg* 2016;**103**(10):1259–68.

Section D.III

Pharmacological and radiotherapeutic treatment

Chapter 33

Molecularly targeted therapy in metastatic CRC

Juan Ruiz-Bañobre[a,b], Elena Brozos-Vázquez[a,b], Francisca Vázquez-Rivera[a,b], Yolanda Vidal-Ínsua[a,b], Rafael López-López[a,b,*], and Sonia Candamio-Folgar[a,b,*]

[a]Medical Oncology Department, University Clinical Hospital of Santiago de Compostela, University of Santiago de Compostela (USC), Santiago de Compostela, Spain, [b]Translational Medical Oncology Group (Oncomet), Health Research Institute of Santiago (IDIS), University Clinical Hospital of Santiago de Compostela, University of Santiago de Compostela (USC), CIBERONC, Santiago de Compostela, Spain

Introduction

Over the past decades, the survival of patients with metastatic colorectal cancer (mCRC) has gradually increased[1] because of the implementation of combined chemotherapy regimens as well as targeted molecular therapies against epidermal growth factor receptor (EGFR); different angiogenic factors such as vascular endothelial growth factor (VEGF) A and B, and placental growth factor (PlGF) 1 and 2; and more recently, against V-raf murine sarcoma viral oncogene homolog B (BRAF).[1–3] Moreover, immunotherapy has recently become a reality for patients with microsatellite instability-high (MSI-H) or DNA deficient mismatch repair (dMMR) mCRC.[4] Other treatment options already available but with a more modest impact on survival are regorafenib and trifluridine/tipiracil (TAS-102).[5]

In the next sections, we will describe not only the current landscape of molecularly targeted therapy but also the future perspectives in the field of mCRC. Antiangiogenic therapy is beyond the scope of this chapter and will not be further discussed in the next sections.

Anti-EGFR antibodies

The EGFR is one of the four members of the ErbB protein family.[6] The family encompasses four members, all of which are receptor tyrosine kinases, including EGFR, ErbB2/Neu/Her2, ErbB3/Her3, and ErbB4/Her4.[6] These proteins consist of an extracellular ligand-binding domain connected by a short transmembrane stretch with an intracellular tyrosine kinase domain.[6] Their binding induces structural changes of the receptors inducing homodimer and heterodimer formation, followed by an increase of the kinase activity and subsequent phosphorylation of intracellular tyrosine residues. These phosphotyrosines recruit specific partner proteins that trigger intracellular signaling pathways.[7] Signaling pathways activated by ErbB family receptors include phosphatidyl-inositol-3 kinase (PI3K) and mitogen-activated protein kinase (MAPK) cascades.[6,8]

Although different strategies to inhibit the EGFR pathway have been developed, to date only monoclonal antibodies targeting this receptor have shown positive results in mCRC. In 2004 and 2006 the US Food and Drug Administration (FDA), and in 2004 and 2007 the European Medicines Agency approved cetuximab and panitumumab, respectively, to the treatment of EGFR-expressing mCRC. Later, in view of the emerging evidence that EGFR immunohistochemistry (IHC) was not a reliable predictor of response to EGFR-targeted therapy, all patients irrespective of EGFR status were eligible for been treated with these two anti-EGFR antibodies in the mCRC scenario. In addition to those trials successfully conducted in mCRC patients refractory to fluorouracil and irinotecan, several randomized clinical trials have established the effectiveness of cetuximab and panitumumab in combination with fluorouracil (5-FU) plus irinotecan (FOLFIRI), and 5-FU plus oxaliplatin (FOLFOX) in the first and second line of therapy for metastatic disease.[9–12] Later, based on

---

*These authors contributed equally.

retrospective analyses of tumor samples from prospective clinical trials, RAS mutations were identified and validated as negative predictive markers for anti-EGFR-based therapies.

The search for predictive biomarkers for anti-EGFR blockade was initially directed toward EGFR expression, which varies around 48%–82% of mCRC specimens.[13,14] While initially EGFR-positive expression assessed by IHC was a inclusion criterion among the first anti-EGFR antibody-based clinical trials, later it was confirmed as not predictive for treatment efficacy.[15,16] Data from the BOND study (NCT00063141) indicated that the intensity of immunohistochemical EGFR staining in colorectal tumor cells did not correlate with the objective response rate (ORR) to cetuximab.[17] Subsequently, mutations conferring resistance to anti-EGFR therapies were identified in codons 12 and 13 of exon 2 of the KRAS gene, which result in constitutive activation of the RAS/RAF/MEK/ERK pathway.[3,18,19] Activating mutations in KRAS are detected in approximately 40% of mCRC,[20] with good concordance between the primary tumors and matched distant metastases.[21,22] More recently, several studies have found that resistance to anti-EGFR therapy can also be mediated by lower-frequency mutations in KRAS exon 3 or 4, or in NRAS exon 2, 3, or 4.[11,21,23] Exclusion of patients with any RAS mutation identifies a population that is more likely to benefit from anti-EGFR therapies.[24] In the PRIME trial, 17% of patients without mutations in KRAS exon 2 did have mutations in KRAS exon 3 or 4, or in NRAS exon 2, 3, or 4.[11] All of these RAS mutations predicted lack of response to panitumumab, and in fact, their presence was associated with inferior progression-free survival (PFS) and overall survival (OS) in patients receiving panitumumab plus FOLFOX compared with FOLFOX alone. Median OS was 25.8 months vs 20.2 months (HR = 0.77, 95% CI, 0.64–0.94, $P = 0.009$) in wild-type RAS populations, in favor of the combination of panitumumab and FOLFOX. Similar results were presented for all RAS genotypes in the CRYSTAL (NCT00154102) and OPUS (NCT00125034) trials, in which randomized patients received first-line cetuximab in combination with FOLFIRI or FOLFOX, respectively.[9,25] In addition, a meta-analysis of nine randomized controlled trials of anti-EGFR antibodies for mCRC demonstrated the predictive value of RAS mutational profiles for both PFS and OS benefits.[24] Altogether, these results indicate that therapy with anti-EGFR antibodies should be restricted to RAS wild-type mCRC patients.

Inmmune checkpoint inhibitors

In 2017 the US FDA approved the anti-programmed cell death 1 (PD-1) therapies pembrolizumab and nivolumab for the treatment of patients with microsatellite instability-high (MSI-H) or deficient DNA mismatch repair (dMMR) mCRC for whom the disease has progressed after treatment with fluoropyrimidine, oxaliplatin, and irinotecan.[4] In 2018 nivolumab plus ipilimumab combination regimen was approved, bringing three novel treatment options for ~5% of all patients with mCRC (those with MSI-H or dMMR tumors).[26] Although MSI-H mCRC patients have worse prognosis, it is hypothesized that they derive clinical benefit from anti-PD-1 therapy because of a large proportion of lymphocytic infiltration and the presence of mutation-associated neoantigens.[27–31]

Later, in 2020, the US FDA approved pembrolizumab as first-line therapy for patients with MSI-H/dMMR mCRC. This approval was based on the results of the KEYNOTE-177 study (NCT02563002),[32] a multicenter, international, open-label, active-controlled, randomized trial that compared the efficacy of first-line therapy with pembrolizumab versus chemotherapy in 307 patients with MSI-H/dMMR mCRC. This study demonstrated a statistically significant improvement in PFS, with a median PFS of 16.5 months vs 8.2 months for pembrolizumab compared to standard-of-care chemotherapy. Importantly, health-related quality of life (HRQoL) results were reported, demonstrating a clinically meaningful improvement in HRQoL favoring pembrolizumab versus standard of care chemotherapy.[33] Although results are promising, longer-term analysis is needed to assess the effect on OS. Moreover, in the same year the US FDA granted accelerated approval to pembrolizumab for the treatment of patients with any unresectable or metastatic solid tumor with high mutational burden (as determined by the FDA-approved test, the FoundationOneCDx assay) whose cancer has progressed after previous treatment and has no satisfactory alternative treatment options.[34]

With the aim of improving on results from previous studies and further extending survival and quality of life of these patients, several clinical trials evaluating the combination of anti-PD-1 blockade with other different therapeutic strategies such as chemotherapy, molecularly targeted therapy, or radiotherapy are ongoing for previously untreated MSI-H/dMMR mCRC patients. Meanwhile, other different immunotherapeutic approaches are being evaluated for treatment of microsatellite stable colorectal cancer (CRC), which is less responsive to immune checkpoint inhibition than MSI-H mCRC.[35] Although the results presented in this section represent substantial therapeutic advances in the treatment of mCRC, they also emphasize the growing need for more precise predictive biomarkers to develop more rational immunotherapeutic approaches.[36]

BRAF tyrosine kinase inhibitors

BRAF mutations occur in 10%–15% of all CRCs and in ~7% of all mCRC.[37,38] Although most *BRAF* mutations occur in codon 600 (mainly *BRAF* V600E), which leads to constitutive BRAF kinase activity and sustained MAPK pathway signaling, 2% of mCRCs have atypical *BRAF* mutations that are outside of codon 600, usually in codon 594.[39] Surprisingly, although monotherapy with BRAF inhibitors (BRAFi) has proven effective in the treatment of *BRAF*-mutant melanoma, it was ineffective in *BRAF* V600E-mutant CRCs. Preclinical evidence demonstrated that despite transient inhibition of pERK by BRAFi such as vemurafenib, rapid ERK reactivation occurs through EGFR-mediated activation of RAS and CRAF.[40] Furthermore, the fact that *BRAF* V600E-mutant CRCs express higher levels of pEGFR than do *BRAF*-mutant melanomas positions them for EGFR-mediated resistance.[40] Collectively, these findings provided rationale to test dual BRAF and EGFR blockade. Results from preclinical studies and early phase clinical trials have demonstrated this strategy is feasible and safe and can potentially improve therapeutic efficacy of BRAFi.[36] Moreover, preclinical studies have suggested that combined inhibition of BRAF and MEK was more effective than dual BRAF and EGFR blockade. This strategy was tested in subsequent phase 1 and phase 2 clinical trials that combined BRAF inhibitors with both anti-EGFR monoclonal antibodies and MEK inhibitors.[40–42] In April 2020 results of these trials led to US FDA approval of encorafenib, a BRAF tyrosine kinase inhibitor, used in combination with cetuximab for the treatment of adult patients with *BRAF* V600E-mutated mCRC. The efficacy of this combination of drugs was evaluated in the BEACON CRC study, a phase 3 randomized, active-controlled, open-label, multicenter trial (NCT02928224).[43] In this trial, encorafenib plus cetuximab demonstrated a clinical and statistically significant OS and PFS benefit compared to the control arm of either irinotecan or FOLFIRI plus cetuximab in patients with *BRAF* V600E-mutated mCRC who had progressed on one or two prior regimens. The median OS in the encorafenib plus cetuximab group was 8.4 months (HR = 0.60, 95% CI, 45–79).[43] The PFS and ORR in the doublet-therapy group were 4.2 months (95% CI, 3.7–5.4) and 20% (95% CI, 13–29), respectively, both significantly higher than those in the control arm.[43] This trial also evaluated the efficacy of triple-therapy with encorafenib, binimetinib (a MEK inhibitor [MEKi]), and cetuximab in a second experimental arm, but although this regimen showed an improved OS and PFS compared to the control arm, it was more toxic than the dual BRAF and EGFR blockade and had similar efficacy.

Based on the positive results of triple therapy with encorafenib, binimetinib, and cetuximab in the second or third line of therapy, an open-label, single arm, two-stage design, phase 2 study, the ANCHOR CRC (NCT03693170), is currently ongoing in patients with *BRAF* V600E-mutated mCRC who did not receive any prior systemic therapy for metastatic disease.[44]

Although with more modest clinical activity, vemurafenib, another BRAFi, was recently included in the National Comprehensive Cancer Network (NCCN) guidelines as a treatment option for patients with *BRAF* V600E-mutated mCRC when used in combination with cetuximab/panitumumab plus irinotecan irinotecan.[5,45] Inclusion in the guidelines was based on results of the randomized phase 2 Southwest Oncology Group (SWOG) 1406 trial (NCT02164916), in which the triple therapy (vemurafenib, cetuximab, and irinotecan) demonstrated improved efficacy as compared with cetuximab plus irinotecan.[45] PFS, the primary endpoint of this trial, was improved with the addition of vemurafenib to cetuximab and irinotecan (HR = 0.50, $P = 0.001$). The response rate was 17% vs 4%, with a DCR of 65% vs 21%, both statistically favorable to the triple therapy.[45]

In addition to the previously described regimens, based on the results of a phase 1 study (NCT01750918),[42] the NCCN Panel has recommended the combination of dabrafenib (BRAFi) plus trametinib (MEKi) plus either cetuximab or panitumumab as another treatment option beyond the first-line setting for *BRAF* V600E-mutated mCRC.[5]

HER-2 blockade

A large body of evidence, accrued primarily from breast and gastric cancer patients, supports the role of HER-2 amplification or overexpression as a predictive biomarker for anti-HER-2-based therapies.[3,36] Therefore there is renewed interest in evaluating HER-2 as a clinically actionable target in mCRC,[46] which is presented in ~5% of patients with KRAS wild-type mCRC. While initial mCRC clinical trials interrogating the anti-HER-2 monoclonal antibody trastuzumab in combination with other chemotherapeutic agents (either FOLFOX or irinotecan) closed early due to lack of patient accrual, mechanistic insights gained from preclinical analyses of HER-2-amplified mCRC patient-derived xenografts have led to improved design of new clinical trials.[47–49]

The fact that HER2 amplification can be a driving genetic alteration in mCRC was first documented in the HERACLES phase 2 trial (NCT03225937), in which patients with *KRAS* exon 2 (codons 12 and 13) wild-type and HER-2-positive mCRC refractory to standard of care therapy (including cetuximab or panitumumab) were treated with trastuzumab and lapatinib. Twenty-seven patients were included, eight of them achieved an objective response (30%, 95% CI, 14–50), yielding the predefined activity threshold.[50] Moreover, the responses were durable and the combination regimen was well tolerated. These results suggested that the combination of trastuzumab and lapatinib is an active regimen in heavily pretreated patients with metastatic disease.[50] Additionally, results from MyPathway trial (NCT02091141) have contributed to support the use of dual HER-2 targeted therapy. MyPathway is an ongoing phase 2a, multiple basket study in which 57 patients with HER-2-amplified, RAS wild-type, heavily pretreated mCRC received trastuzumab plus pertuzumab combination. In the HERACLES study, the primary endpoint was ORR. Among the 57 evaluable patients enrolled, one (2%) patient achieved a complete response and 17 (30%), partial responses; thus overall 18 of 57 patients achieved an objective response (32%, 95% CI, 20%–45%). Furthermore, at the time of data cutoff, estimated median OS was 11.5 months (95% CI 7.7—not estimable).[51] Recently in the context of the DESTINY-CRC01 clinical trial (NCT03384940), trastuzumab deruxtecan, a HER2-directed antibody drug conjugate, has demonstrated promising results as a treatment option in HER-2-expressing, RAS wild-type mCRC patients who had progressed on two or more prior regimens of treatment.[52] DESTINY-CRC01 is a phase 2, open-label, multicenter study designed to test the safety and efficacy of trastuzumab deruxtecan in three cohorts of mCRC based on the expression of HER-2: cohort A, HER-2 IHC +++ or IHC ++/ in situ hybridization (ISH) +; cohort B, IHC ++/ISH −; and cohort C, IHC +). The primary endpoint was confirmed ORR by independent central review in cohort A; secondary endpoints included disease control rate (DCR), duration of response, PFS, OS, and ORR in cohorts B and C. At data cutoff, 78 patients (53 in cohort A, 7 in cohort B, and 18 in cohort C) were included and received trastuzumab deruxtecan. The confirmed ORR was 45.3% in cohort A, including 1 complete response and 23 partial responses; median duration of response was not reached (95% CI, 4.2 months—not estimable). The ORR in patients with prior anti-HER2 treatment was 43.8% (95% CI, 19.8%–70.1%). The DCR was 83% (95% CI, 70.2%–91.9%), median PFS was 6.9 months (95% CI, 4.1 months—not estimable), and median OS was not reached. Unexpectedly, no responses were observed in cohorts B or C. On the other hand, the safety profile was acceptable and consistent with previous studies.[52]

These data, which demonstrate the potential of anti-HER-2 therapy as a promising treatment option for patients with advanced HER-2-positive mCRC, have paved the path for the development of ongoing phase 2 clinical trials evaluating the efficacy of new anti-HER-2 agents, such as S1613 (NCT03365882), trastuzumab emtansine (NCT03418558), or tucatinib (NCT03043313) in this clinical scenario. On the other hand, determining the utility of ctDNA analyses in monitoring therapeutic efficacy and in identifying mechanisms of resistance to dual HER-2 blockade is also an attractive area of study.[53] Moreover, due to the fact that HER-2 amplification is a well-described mechanism of resistance to anti-EGFR therapy, determining HER-2 status not only at diagnosis but also after anti-EGFR therapy progression can help to detect an increasing number of HER-2-positive candidates to be treated with anti-HER-2-based strategies.[54]

Tyrosine kinase inhibitors

A pan-cancer analysis of the transcriptomes of nearly 7000 tumors from The Cancer Genome Atlas detected several low-frequency, pan-cancer kinase fusion events that drive tumorigenesis in a small fraction of multiple cancers regardless of tissue type. The neurotrophic tyrosine receptor kinases NTRK1, NTRK2, and NTRK3 genes were among those with different fusion events.[55] Following this discovery, new drugs that target tyrosine kinase fusions in genes such as NRTK1/2/3, RET, ALK, and ROS1 have been developed and tested, showing promising preliminary results in phase 1 and 2 clinical trials that include patients with CRC. One agent, LOXO 101 (larotrectinib), is a selective tropomyosin receptor kinase (TRK) inhibitor that has demonstrated rapid, potent, and durable tumor-agnostic activity in children and adult patients with NTRK fusion-positive malignancies (including four patients with CRC who achieved a partial response).[56] A second agent, entrectinib, an ALK, ROS1, TRKA, TRKB, and TRKC selective inhibitor, exhibited clinical activity in patients who had fusions in the previously described tyrosine kinase genes.[57] Patients who responded to entrectinib included two patients whose mCRC harbored CAD-ALK or LMNA-NTRK1 gene fusions.[58,59] Anticipating potential resistance mechanisms to larotrectinib based on evidence from other pan-tyrosine kinase inhibitors, Drilon et al. developed LOXO-195 (selitrectinib), a potent and selective TRK kinase inhibitor designed to have a molecular structure that would overcome typical TRK resistance mutations.[60] LOXO-195 was initially evaluated in a mCRC patient whose cancer had an LMNA-NTRK1 rearrangement with a G595R larotrectinib resistance mutation. This patient successfully achieved a durable partial response.[60] Although the prevalence of rearrangements in tyrosine kinase genes in mCRC patients may be low (1.5%), the accelerated development of tyrosine kinase inhibitors offers new hope for some heavily pretreated mCRC patients who may otherwise be devoid of alternative treatment options.[61]

Given the promising results reported by trials conducted to date, the US FDA granted accelerated approval to larotrectinib and entrectinib in November 2018 and August 2019, respectively, for patients with solid tumors that have NTRK gene fusion without a known resistance mutation, are metastatic or where resection is likely to result in severe morbidity, and have no satisfactory alternative treatments or have progression following treatment. Furthermore, the Committee for Medicinal Products for Human Use of the European Medicines Agency recently has also recommended the granting of a conditional marketing authorization for larotrectinib and entrectinib for the same indication.

KRAS inhibitors

One of the most commonly altered oncogenes in human cancers, *KRAS*, was long considered an undruggable target because of the small size of altered KRAS proteins; the presence of few binding sites; and the rapid, tight binding of active KRAS to GTP. However, recent data have suggested that KRAS might be targetable. For example, preliminary data on the activity of AMG510 (sotorasib), a small covalent inhibitor, have shown that it rapidly and irreversibly occupies *KRAS* G12C and extinguishes its activity through a unique interaction with the P2 pocket.[62] The *KRAS* G12C mutation occurs in ~4% of CRC.[63] In a recent phase 1 trial, sotorasib showed encouraging antitumor activity in heavily pretreated patients who had advanced, *KRAS* G12C-mutated solid tumors.[64] A total of 129 patients were included in this study, 42 of whom had CRC. Within CRC patients, sotorasib treatment yielded an ORR and DCR of 7.1% and 73.8%, respectively. The median duration of stable disease was 5.4 months and the median PFS was 4.0 months. Although sotorasib showed promising anticancer activity in patients with heavily pretreated solid tumors bearing the *KRAS* G12C mutation, inconsistency was seen in tumor response between patients with non-small cell lung cancer and those with CRC, which the authors suggested indicated either that *KRAS* G12C is not the dominant oncogenic driver for CRC or that other pathways, such as the WNT or EGFR pathways, mediate oncogenic signaling beyond *KRAS*. These hypotheses are supported by solid preclinical evidence,[65–67] and therefore, clinical trials that combine sotorasib with other agents that block additional pathways have already been initiated (i.e., NCT04185883 and NCT04303780). Although many *KRAS* G12C inhibitors in addition to sotorasib are under development, to date only adagrasib, an irreversible covalent inhibitor, has shown promising antitumor activity in *KRAS* G12C-mutated CRC. Furthermore, inhibitors for mutations other than *KRAS* G12C are being developed. For example, initial preclinical data for MRTX1133, a new, first-in-class *KRAS* G12D inhibitor, have demonstrated significant tumor regression in preclinical animal models.[68] Thus, through development of a range of inhibitors, effective means of targeting KRAS are emerging.[36]

Conclusion

Step by step, management of mCRC patients is moving from approaches where *"all comers are welcome"* to a more personalized medicine strategy, which based on the presence of different driver molecular alterations in tumors, select the more appropriate therapy for each patient. With this concept in mind, biomarker research and clinical trial design are moving forward, bringing together many different types of data and therapeutic strategies to more efficiently drive CRC clinical management.

Acknowledgments

Juan Ruiz-Bañobre is supported by Río Hortega fellowship from the Institute of Health Carlos III (CM19/00087) and 2020 TTD Research Grant from the Spanish Cooperative Group for the Treatment of Digestive Tumors (TTD).

References

1. Siegel RL, Miller KD, Goding Sauer A, et al. Colorectal cancer statistics, 2020. *CA Cancer J Clin* 2020;**70**(3):145–64. https://doi.org/10.3322/caac.21601.
2. Kopetz S, Chang GJ, Overman MJ, et al. Improved survival in metastatic colorectal cancer is associated with adoption of hepatic resection and improved chemotherapy. *J Clin Oncol* 2009;**27**(22):3677–83. https://doi.org/10.1200/JCO.2008.20.5278.
3. Ruiz-Bañobre J, Kandimalla R, Goel A. Predictive biomarkers in metastatic colorectal cancer: a systematic review. *JCO Precis Oncol* 2019;**3**:1–17. https://doi.org/10.1200/PO.18.00260.
4. Ruiz-Bañobre J, Goel A. DNA mismatch repair deficiency and immune checkpoint inhibitors in gastrointestinal cancers. *Gastroenterology* 2019. https://doi.org/10.1053/j.gastro.2018.11.071.
5. National Comprehensive Cancer Network. *Colon Cancer (Version 2.2021)*; 2021 https://www.nccn.org/professionals/physician_gls/pdf/colon.pdf. Accessed February 21, 2021.
6. Irmer D, Funk JO, Blaukat A. EGFR kinase domain mutations—functional impact and relevance for lung cancer therapy. *Oncogene* 2007;**26**(39):5693–701. https://doi.org/10.1038/sj.onc.1210383.

7. Pawson T, Gish GD, Nash P. SH2 domains, interaction modules and cellular wiring. *Trends Cell Biol* 2001;**11**(12):504–11. https://doi.org/10.1016/s0962-8924(01)02154-7.
8. Yarden Y. The EGFR family and its ligands in human cancer: signalling mechanisms and therapeutic opportunities. *Eur J Cancer* 2001;**37**(Suppl 4): S3–8. https://doi.org/10.1016/s0959-8049(01)00230-1.
9. Van Cutsem E, Köhne C-H, Hitre E, et al. Cetuximab and chemotherapy as initial treatment for metastatic colorectal cancer. *N Engl J Med* 2009; **360**(14):1408–17. https://doi.org/10.1056/NEJMoa0805019.
10. Peeters M, Price TJ, Cervantes A, et al. Randomized phase III study of panitumumab with fluorouracil, leucovorin, and irinotecan (FOLFIRI) compared with FOLFIRI alone as second-line treatment in patients with metastatic colorectal cancer. *J Clin Oncol Off J Am Soc Clin Oncol* 2010;**28** (31):4706–13. https://doi.org/10.1200/JCO.2009.27.6055.
11. Douillard JY, Siena S, Cassidy J, et al. Randomized, phase III trial of panitumumab with infusional fluorouracil, leucovorin, and oxaliplatin (FOLFOX4) versus FOLFOX4 alone as first-line treatment in patients with previously untreated metastatic colorectal cancer: The PRIME study. *J Clin Oncol* 2010;**28**(31):4697–705. https://doi.org/10.1200/JCO.2009.27.4860.
12. Bokemeyer C, Bondarenko I, Makhson A, et al. Fluorouracil, leucovorin, and oxaliplatin with and without cetuximab in the first-line treatment of metastatic colorectal cancer. *J Clin Oncol Off J Am Soc Clin Oncol* 2009;**27**(5):663–71. https://doi.org/10.1200/JCO.2008.20.8397.
13. Antonacopoulou AG, Tsamandas AC, Petsas T, et al. EGFR, HER-2 and COX-2 levels in colorectal cancer. *Histopathology* 2008;**53**(6): 698–706. https://doi.org/10.1111/j.1365- 2559.2008.03165.x.
14. McKay JA, Murray LJ, Curran S, et al. Evaluation of the epidermal growth factor receptor (EGFR) in colorectal tumours and lymph node metastases. *Eur J Cancer* 2002;**38**(17):2258–64. https://doi.org/10.1016/s0959-8049(02)00234-4.
15. Chung KY, Shia J, Kemeny NE, et al. Cetuximab shows activity in colorectal Cancer patients with tumors that do not express the epidermal growth factor receptor by immunohistochemistry. *J Clin Oncol* 2005;**23**(9):1803–10. https://doi.org/10.1200/JCO.2005.08.037.
16. Hecht JR, Mitchell E, Neubauer MA, et al. Lack of correlation between epidermal growth factor receptor status and response to Panitumumab monotherapy in metastatic colorectal cancer. *Clin Cancer Res* 2010;**16**(7):2205–13. https://doi.org/10.1158/1078-0432.CCR-09-2017.
17. Cunningham D, Humblet Y, Siena S, et al. Cetuximab monotherapy and cetuximab plus irinotecan in irinotecan-refractory metastatic colorectal cancer. *N Engl J Med* 2004;**351**(4):337–45. https://doi.org/10.1056/NEJMoa033025.
18. Wadlow RC, Hezel AF, Abrams TA, et al. Panitumumab in patients with KRAS wild-type colorectal cancer after progression on cetuximab. *Oncologist* 2012;**17**(1):14. https://doi.org/10.1634/theoncologist.2011-0452.
19. Graham DM, Coyle VM, Kennedy RD, Wilson RH. Molecular subtypes and personalized therapy in metastatic colorectal cancer. *Curr Colorectal Cancer Rep* 2016;**12**:141–50. https://doi.org/10.1007/s11888-016-0312-y.
20. Amado RG, Wolf M, Peeters M, et al. Wild-type KRAS is required for panitumumab efficacy in patients with metastatic colorectal cancer. *J Clin Oncol* 2008;**26**(10):1626–34. https://doi.org/10.1200/JCO.2007.14.7116.
21. Peeters M, Kafatos G, Taylor A, et al. Prevalence of RAS mutations and individual variation patterns among patients with metastatic colorectal cancer: a pooled analysis of randomised controlled trials. *Eur J Cancer* 2015;**51**(13):1704–13. https://doi.org/10.1016/J.EJCA.2015.05.017.
22. Han C-B, Li F, Ma J-T, Zou H-W. Concordant KRAS mutations in primary and metastatic colorectal cancer tissue specimens: a meta-analysis and systematic review. *Cancer Invest* 2012;**30**(10):741–7. https://doi.org/10.3109/07357907.2012.732159.
23. Loupakis F, Ruzzo A, Cremolini C, et al. KRAS codon 61, 146 and BRAF mutations predict resistance to cetuximab plus irinotecan in KRAS codon 12 and 13 wild-type metastatic colorectal cancer. *Br J Cancer* 2009;**101**(4):715–21. https://doi.org/10.1038/sj.bjc.6605177.
24. Sorich MJ, Wiese MD, Rowland A, Kichenadasse G, McKinnon RA, Karapetis CS. Extended RAS mutations and anti-EGFR monoclonal antibody survival benefit in metastatic colorectal cancer: a meta-analysis of randomized, controlled trials. *Ann Oncol* 2015;**26**(1):13–21.
25. Bokemeyer C, Bondarenko I, Hartmann JT, et al. Efficacy according to biomarker status of cetuximab plus FOLFOX-4 as first-line treatment for metastatic colorectal cancer: the OPUS study. *Ann Oncol* 2011;**22**(7):1535–46. https://doi.org/10.1093/annonc/mdq632.
26. Braun MS, Richman SD, Quirke P, et al. Predictive biomarkers of chemotherapy efficacy in colorectal cancer: results from the UK MRC FOCUS trial. *J Clin Oncol* 2017;**26**(16). https://doi.org/10.1200/JCO.2007.15.5580.
27. Le DT, Uram JN, Wang H, et al. PD-1 blockade in tumors with mismatch-repair deficiency. *N Engl J Med* 2015;**372**(26):2509–20. https://doi.org/10.1056/NEJMoa1500596.
28. Le DT, Durham JN, Smith KN, et al. Mismatch repair deficiency predicts response of solid tumors to PD-1 blockade. *Science* 2017;**357**(6349): 409–13. https://doi.org/10.1126/science.aan6733.
29. Overman MJ, Lonardi S, Leone F, et al. Nivolumab in patients with DNA mismatch repair deficient/microsatellite instability high metastatic colorectal cancer: update from CheckMate 142. *J Clin Oncol* 2017;**35**(4_suppl):519. https://doi.org/10.1200/JCO.2017.35.4_suppl.519.
30. Overman MJ, Lonardi S, Wong KYM, et al. Durable clinical benefit with nivolumab plus Ipilimumab in DNA mismatch repair–deficient/microsatellite instability–high metastatic colorectal cancer. *J Clin Oncol* 2018;**36**(8):773–9. https://doi.org/10.1200/JCO.2017.76.9901.
31. Giannakis M, Mu XJ, Shukla SA, et al. Genomic correlates of immune-cell infiltrates in colorectal carcinoma. *Cell Rep* 2016;**15**(4):857–65. https://doi.org/10.1016/j.celrep.2016.03.075.
32. André T, Shiu K-K, Kim TW, et al. Pembrolizumab in microsatellite-instability–high advanced colorectal cancer. *N Engl J Med* 2020;**383**(23): 2207–18. https://doi.org/10.1056/NEJMoa2017699.
33. André T, Amonkar M, Norquist J, et al. 396O Health-related quality of life (HRQoL) in patients (pts) treated with pembrolizumab (pembro) vs chemotherapy as first-line treatment in microsatellite instability-high (MSI-H) and/or deficient mismatch repair (dMMR) metastatic colorectal cancer (mCRC): phase III KEYNOTE-177 study. *Ann Oncol* 2020;**31**:S409. https://doi.org/10.1016/j.annonc.2020.08.507.
34. Bersanelli M. Tumour mutational burden as a driver for treatment choice in resistant tumours (and beyond). *Lancet Oncol* 2020;**21**(10):1255–7. https://doi.org/10.1016/S1470-2045(20)30433-2.

35. Jung G, Benítez-Ribas D, Sánchez A, Balaguer F. Current treatments of metastatic colorectal cancer with immune checkpoint inhibitors-2020 update. *J Clin Med* 2020;**9**(11). https://doi.org/10.3390/jcm9113520.
36. Ruiz-Bañobre J, Goel A. Genomic and epigenomic biomarkers in colorectal cancer: From diagnosis to therapy. In: *Advances in cancer research*. Academic Press, Elsevier; 2021. ISBN 0065-230 https://doi.org/10.1016/bs.acr.2021.02.008.
37. Davies H, Bignell GR, Cox C, et al. Mutations of the BRAF gene in human cancer. *Nature* 2002;**417**(6892):949–54. https://doi.org/10.1038/nature00766.
38. Clarke CN, Kopetz ES. BRAF mutant colorectal cancer as a distinct subset of colorectal cancer: clinical characteristics, clinical behavior, and response to targeted therapies. *J Gastrointest Oncol* 2015;**6**(6):660–7. https://doi.org/10.3978/j.issn.2078-6891.2015.077.
39. Jones JC, Renfro LA, Al-Shamsi HO, et al. (Non-V600) BRAF mutations define a clinically distinct molecular subtype of metastatic colorectal cancer. *J Clin Oncol* 2017;**35**(23):2624–30. https://doi.org/10.1200/JCO.2016.71.4394.
40. Corcoran RB, Ebi H, Turke AB, et al. EGFR-mediated reactivation of MAPK signaling contributes to insensitivity of BRAF-mutant colorectal cancers to RAF inhibition with vemurafenib. *Cancer Discov* 2012;**2**(3):227–35. https://doi.org/10.1158/2159-8290.CD-11-0341.
41. Corcoran RB, Dias-Santagata D, Bergethon K, Iafrate AJ, Settleman J, Engelman JA. BRAF gene amplification can promote acquired resistance to MEK inhibitors in cancer cells harboring the BRAF V600E mutation. *Sci Signal* 2010;**3**(149):ra84. https://doi.org/10.1126/scisignal.2001148.
42. Corcoran RB, Andre T, Atreya CE, et al. Combined BRAF, EGFR, and MEK inhibition in patients with BRAFV600E-mutant colorectal cancer. *Cancer Discov* 2018;**8**:428–43.
43. Kopetz S, Grothey A, Yaeger R, et al. Encorafenib, binimetinib, and cetuximab in BRAFV600E–mutated colorectal cancer. *N Engl J Med* 2019;**381**(17):1632–43. https://doi.org/10.1056/NEJMoa1908075.
44. Grothey A, Yaeger R, Paez D, et al. ANCHOR CRC: a phase 2, open-label, single arm, multicenter study of encorafenib (ENCO), binimetinib (BINI), plus cetuximab (CETUX) in patients with previously untreated BRAF V600E-mutant metastatic colorectal cancer (mCRC). *Ann Oncol* 2019;**30**:iv109. https://doi.org/10.1093/annonc/mdz155.399.
45. Kopetz S, Guthrie KA, Morris VK, et al. Randomized trial of irinotecan and cetuximab with or without vemurafenib in BRAF-mutant metastatic colorectal cancer (SWOG S1406). *J Clin Oncol* 2020. https://doi.org/10.1200/JCO.20.01994. JCO.20.01994.
46. Valtorta E, Martino C, Sartore-Bianchi A, et al. Assessment of a HER2 scoring system for colorectal cancer: results from a validation study. *Mod Pathol* 2015;**28**(11):1481–91. https://doi.org/10.1038/modpathol.2015.98.
47. Clark JW, Niedzwiecki D, Hollis DP. Phase-II trial of 5-fluororuacil (5-FU), leucovorin (LV), oxaliplatin (Ox), and trastuzumab (T) for patients with metastatic colorectal cancer (CRC) refractory to initial therapy. *Onkologie* 2003;**26**:13–46.
48. Ramanathan RK, Hwang JJ, Zamboni WC, et al. Low overexpression of HER-2/neu in advanced colorectal cancer limits the usefulness of trastuzumab (Herceptin) and irinotecan as therapy. A phase II trial. *Cancer Invest* 2004;**22**(6):858–65.
49. Bertotti A, Migliardi G, Galimi F, et al. A molecularly annotated platform of patient-derived xenografts ("xenopatients") identifies HER2 as an effective therapeutic target in cetuximab-resistant colorectal cancer. *Cancer Discov* 2011;**1**(6):508–23. https://doi.org/10.1158/2159-8290.CD-11-0109.
50. Sartore-Bianchi A, Trusolino L, Martino C, et al. Dual-targeted therapy with trastuzumab and lapatinib in treatment-refractory, KRAS codon 12/13 wild-type, HER2-positive metastatic colorectal cancer (HERACLES): a proof-of-concept, multicentre, open-label, phase 2 trial. *Lancet Oncol* 2017;**17**(6):738–46. https://doi.org/10.1016/S1470-2045(16)00150-9.
51. Hainsworth JD, Meric-Bernstam F, Swanton C, et al. Targeted therapy for advanced solid tumors on the basis of molecular profiles: results from MyPathway, an open-label, phase IIa multiple basket study. *J Clin Oncol* 2018;**36**(6):536–42. https://doi.org/10.1200/JCO.2017.75.3780.
52. Siena S, Di Bartolomeo M, Raghav KPS, et al. A phase II, multicenter, open-label study of trastuzumab deruxtecan (T-DXd; DS-8201) in patients (pts) with HER2-expressing metastatic colorectal cancer (mCRC): DESTINY-CRC01. *J Clin Oncol* 2020;**38**(15_suppl):4000. https://doi.org/10.1200/JCO.2020.38.15_suppl.4000.
53. Siravegna G, Lazzari L, Crisafulli G, et al. Radiologic and genomic evolution of individual metastases during HER2 blockade in colorectal cancer. *Cancer Cell* 2018;**34**(1):148–162.e7. https://doi.org/10.1016/j.ccell.2018.06.004.
54. Bregni G, Sciallero S, Sobrero A. HER2 amplification and anti-EGFR sensitivity in advanced colorectal cancer. *JAMA Oncol* 2019;**5**(5):605–6. https://doi.org/10.1001/jamaoncol.2018.7229.
55. Stransky N, Cerami E, Schalm S, Kim JL, Lengauer C. The landscape of kinase fusions in cancer. *Nat Commun* 2014;**5**(1):4846. https://doi.org/10.1038/ncomms5846.
56. Drilon A, Laetsch TW, Kummar S, et al. Efficacy of larotrectinib in TRK fusion–positive cancers in adults and children. *N Engl J Med* 2018;**378**(8):731–9. https://doi.org/10.1056/NEJMoa1714448.
57. Drilon A, Siena S, Ou S-HI, et al. Safety and antitumor activity of the multitargeted pan-TRK, ROS1, and ALK inhibitor entrectinib: combined results from two phase I trials (ALKA-372-001 and STARTRK-1). *Cancer Discov* 2017;**7**(4):400–9.
58. Amatu A, Somaschini A, Cerea G, et al. Novel CAD-ALK gene rearrangement is drugable by entrectinib in colorectal cancer. *Br J Cancer* 2015;**113**:1730.
59. Sartore-Bianchi A, Ardini E, Bosotti R, et al. Sensitivity to entrectinib associated with a novel LMNA-NTRK1 gene fusion in metastatic colorectal cancer. *J Natl Cancer Inst* 2016;**108**(1):djv306.
60. Drilon A, Nagasubramanian R, Blake JF, et al. A next-generation TRK kinase inhibitor overcomes acquired resistance to prior TRK kinase inhibition in patients with TRK fusion-positive solid tumors. *Cancer Discov* 2017;**7**(9):63–972.
61. Pietrantonio F, Di Nicolantonio F, Schrock AB, et al. ALK, ROS1, and NTRK rearrangements in metastatic colorectal cancer. *J Natl Cancer Inst* 2017;**109**(12):djx089.

62. Janes MR, Zhang J, Li L-S, et al. Targeting KRAS mutant cancers with a covalent G12C-specific inhibitor. *Cell* 2018;**172**(3):578–589.e17. https://doi.org/10.1016/j.cell.2018.01.006.
63. Neumann J, Zeindl-Eberhart E, Kirchner T, Jung A. Frequency and type of KRAS mutations in routine diagnostic analysis of metastatic colorectal cancer. *Pathol Res Pract* 2009;**205**(12):858–62. https://doi.org/10.1016/j.prp.2009.07.010.
64. Hong DS, Fakih MG, Strickler JH, et al. KRASG12C inhibition with sotorasib in advanced solid tumors. *N Engl J Med* 2020;**383**(13):1207–17. https://doi.org/10.1056/NEJMoa1917239.
65. Lee S-K, Jeong W-J, Cho Y-H, et al. β-Catenin-RAS interaction serves as a molecular switch for RAS degradation via GSK3β. *EMBO Rep* 2018;**19**(12). https://doi.org/10.15252/embr.201846060.
66. Amodio V, Yaeger R, Arcella P, et al. EGFR blockade reverts resistance to KRASG12C inhibition in colorectal cancer. *Cancer Discov* 2020;**10**(8):1129–39. https://doi.org/10.1158/2159-8290.CD-20-0187.
67. Xue JY, Zhao Y, Aronowitz J, et al. Rapid non-uniform adaptation to conformation-specific KRAS(G12C) inhibition. *Nature* 2020;**577**(7790):421–5. https://doi.org/10.1038/s41586-019-1884-x.
68. https://ir.mirati.com/news-releases/news-details/2020/Mirati-Therapeutics-Reports-Investigational-Adagrasib-MRTX849-Preliminary-Data-Demonstrating-Tolerability-and-Durable-Anti-Tumor-Activity-as-well-as-Initial-MRTX1133-Preclinical-Data/default.aspx. Accessed February 22, 2021.

Chapter 34

A roadmap for medical treatment of metastatic CRC

Gala Martínez-Bernal, Julia Martínez-Pérez, and Manuel Valladares-Ayerbes
Medical Oncology Unit, Virgen del Rocío University Hospital, Sevilla, Spain

Background

It is estimated that 1,931,590 new cases of colorectal cancer (CRC) have been diagnosed in the world in 2020, representing 10% of all malignant tumors and being the third in incidence.[1,2] The estimate of the number of new cases in Spain for the year 2021 (both sexes) is more than 43,000, being the first in incidence in both sexes. The total prevalence of CRC in Spain was 126,000 in men and 100,900 in women, by 2020. In Spain, by 2020 it has been estimated that CRC was the cause of death in 16,470 cases (14.6% of cancer deaths), being the first in women and second in men. The most common sites of metastasis include lymph nodes, liver, lung, and peritoneum. In metastatic CRC, the 5-year survival rate is less than 20%. Median age at onset is 67 years.[3]

In the last decades, the treatment of advanced CRC has undergone substantial changes that have provided a consistent benefit in terms of rates of disease control, progression-free survival (PFS), and overall survival (OS) of patients, allowing new paradigms, such as decision-making in multidisciplinary teams and the resection of metastases that were initially unresectable.[4] Resection of metastatic CRC achieves long-term cure for less than 20% of metastatic CRC patients. Although only a limited fraction of patients are candidates for multimodal treatments with a "curative" intention, there is high-quality scientific evidence that demonstrates the benefit of new pharmacological strategies, based on combinations of cytostatic and biological agents, in the treatment of most patients with metastatic CRC cancer.

The first incorporation of irinotecan and oxaliplatin in regimens based on 5FUluorouracil (5FU), the demonstration of the therapeutic efficacy of capecitabine, the addition of monoclonal antibodies (Ab Mo) against specific molecular targets (bevacizumab, aflibercept, ramucirumab, cetuximab, panitumumab), and the appearance of new active drugs (regorafenib, TAS102) are clear advances, based on prospective randomized trials, the results of which guide our recommendations in the treatment of metastatic CRC. Lastly, we have biological markers that determine the effectiveness of drugs, especially cetuximab and panitumumab, based on the status of the RAS oncogenes (KRAS and NRAS), antitarget drugs in specific subgroups (encorafenib in mutated BRAF V600E). Although in a limited group (3%–5%), immunotherapy has a clear role in microsatellite instable (MSI-H) or mismatch repair deficient (dMMR) tumors.

Roadmap for treatment

To define the roadmap for treatment of patients with advanced CRC, it must be deemed what objectives are pursued and what prognostic factors are present. Different subgroups of patients have been defined in which the therapeutic objectives are clearly different, from the patient with exclusive but currently unresectable liver disease to patients with various metastatic sites, practically asymptomatic and with an indolent course of the disease.[5] Thus, in the selection of the initial, first-line treatment, the following factors are needed to be considered:

1. Factors related to the patient, including biological age, comorbidities, performance status, physical and psychological capacities, presence of symptoms, and their expectations and preferences.
2. Factors related to the tumor and its biology, such as side of the primary tumor, location, and number of metastases, current and potential resectability in case of response, "natural history" or dynamics of the disease, or potential serious complications derived of disease growth. Biological markers especially RAS oncogenes (KRAS and NRAS) sequence variants, mutated BRAF V600E, and microsatellite status (MSI).
3. Toxicity profile and potential efficacy of drugs.

4. Objectives of treatment, such as obtaining the maximum response for a potential surgery, avoiding a rapid and fatal progression, or maintaining maximum survival with minimum side effects, in an indolent and asymptomatic disease.
5. Possibilities of access to the different active drugs, for example, economic, social, or accessibility reasons.

There are different prognosis models[6, 7] in advanced CRC, which essentially include: (i) decreased functional capacity (ECOG $\geq$2); (ii) biochemical parameters such as elevated levels of carcinoembryonic antigen (CEA > 50 µg/l), alkaline phosphatase $\geq$300 U/l, platelets $\geq 400 \times 10^9$/l, hemoglobin <11 g/dL, leukocytes $\geq 10 \times 10^9$/l, high levels of LDH and low levels of albumin; (iii) disease related, such as the number of metastatic sites, extensive peritoneal carcinomatosis, and the number of circulating tumor cells; (iv) molecular, such as RAS/BRAF mutations and microsatellite instability status.

A consensus molecular classification of CRC has been developed based on gene expression data and linked to the tumor phenotype and its clinical behavior. After joint analysis of the transcriptional profiles of more than 4000 CRC samples, four consensus molecular subtypes (CMS1–4) were established.[8] Most of the tumors with MSI high are found in the CMS1 or immune group (14%) and includes hypermethylated tumors, greater mutational tumor burden, a high prevalence of BRAF sequence variations, and an abundant peritumoral immune infiltrate. From a clinical point of view, it is characterized by being more frequent in older women, with tumors located in the right colon, a high degree of histological differentiation, and a poor prognosis after tumor recurrence. The benefit of immunotherapy in these MSI high tumors has been demonstrated. However, a specific drug approach for most of the CRC patients based on the CMS is not feasible currently but is under investigation.

The OS of the patients will depend on these prognostic factors and to a great extent on the integration, not only of the different active drugs[4] but also of other therapeutic modalities, such as surgery[9, 10] and even other forms of local treatment, such as radiotherapy, or ablation specially in oligometastatic disease.

Since the end of the 1990s and with the use of different forms of administration of 5FU, it is considered that the choice of the most effective regimens as first-line treatment can have a positive and long-term impact on patient survival.[11] In a meta-analysis with individual data from more than 3700 patients included in 25 clinical trials,[12] OS was directly correlated with first-line response rates, with a hazard ratio of 0.90 ($P = 0.003$). OS will depend not only on the initial treatment but also on the efficacy of successive therapeutic interventions. A new definition of objectives is necessary[13] that collect with greater precision the effects of successive or alternative interventions or of different strategies and that allow a better evaluation of the benefit in the clinic. Finally, the efficacy of successive lines of treatment after the progression of the disease to a previous scheme and how many of our patients will be able to receive these second- or third-line treatments will condition the results. Thus defining a strategy or a treatment trajectory so that most of our patients can receive all active drugs, within their defined indications, that is, a strategy of "continuity of care," will have a substantial impact on the survival of our patients.[3, 14]

Median survival with treatment is approximately 30 months for patients with RAS/BRAF wild-type metastatic CRC.[15] A pooled analysis showed a median overall survival of 21.0 months for patients with KRAS-mutated tumors and 11.7 months for those with BRAF V600E mutation.[16] Median survival was 19 months for right-sided primary tumors and 34 months for left-sided primary tumors.

First-line chemotherapy

The selection of the first treatment for metastatic disease will depend on the clinical situation of the patient, the potential objectives in the specific patient, and the global strategy that we can define. Mutation analysis (sequence variants) for RAS genes (including KRAS and NRAS) and BRAF and pathologic testing for microsatellite instability or mismatch repair protein expression should be performed. These are usually done on primary tumor or metastatic biopsy. Recently, dihydropyrimidine dehydrogenase (DPYD) testing prior to fluoropyrimidine treatment is recommended.[17]

In the early 2000s, the addition of oxaliplatin or irinotecan to the backbone of FU/LV resulted in an improvement of median OS to nearly 20 months, median PFS up to 8 months, and an overall response rate around 45%–50%. Different multicenter randomized trials have evaluated the noninferiority of capecitabine with respect to continuous infusion of 5FU (modulated with FA or not) in combination regimens with oxaliplatin.[18–20] A meta-analysis[21] that included 3494 patients has demonstrated the noninferiority of the regimens based on capecitabine and oxaliplatin compared to 5FU in infusion and oxaliplatin, in terms of PFS or OS. However, the objective response rate was lower with capecitabine and oxaliplatin (ratio probability, 0.85; $P = 0.02$). In two phase III studies, combination of capecitabine and irinotecan resulted in greater toxicity, especially diarrhea and dehydration, and a greater need for dose reduction with capecitabine regimens. In the BICC study,[22] the CapeIri scheme produced greater toxicity, including nausea and vomiting, diarrhea, dehydration, and hand-foot syndrome. Likewise, the PFS was lower than that obtained with FOLFIRI. A meta-analysis of 7 RCTs compared FOLFIRI (fluorouracil, leucovorin, and irinotecan) and CAPIRI (capecitabine and irinotecan) and showed no difference in PFS and OS. High incidence of diarrhea was found with capecitabine and irinotecan combination.[23]

A doublet of chemotherapy plus one of the monoclonal antibodies is the first-line choice for most patients with mCRC, with adequate performance status and no comorbidities. The oxaliplatin regimens with fluoropyrimidines and the FOLFIRI regimen are considered equivalent, although with a different toxicity profile. The superiority of FOLFOX/XELOX with bevacizumab over FOLFOX/XELOX in PFS has been demonstrated. Table 34.1 presents the results of the main comparative studies that include randomized bevacizumab regimens. The VEGF inhibitor bevacizumab is indicated for both RAS native and RAS mutant tumors. The toxicity profile observed with bevacizumab consists of hypertension, bleeding, gastrointestinal perforations, impaired wound healing, and arterial-venous thrombotic events. Different biosimilars of bevacizumab have been approved by the regulatory agencies.

In patients with native RAS tumors, both the combination of FOLFIRI with cetuximab[29] and FOLFOX with panitumumab[30] are superior to FOLFIRI and FOLFOX, respectively, in OS, PFS, and response rate. The monoclonal antibodies cetuximab and panitumumab bind to EGFR, in different binding domains with the ligands. This binding of Ab Mo to EGFR is of higher affinity than that of endogenous ligands and inhibits receptor autophosphorylation. Main toxicities associated with anti-EGFR antibodies are diarrhea, magnesium-wasting syndrome, infusion reaction, and skin toxicity.

In general, combinations of capecitabine and oxaliplatin or irinotecan with cetuximab or panitumumab are not considered appropriate due to toxicity. Some studies have shown the inferiority of these schemes.[31]

In selected patients, with poor performance status or comorbidities, disease not subsidiary to surgery in any case, asymptomatic, with favorable prognostic factors and without risk of rapid deterioration in the event of disease progression, an initial sequential treatment could be considered, with a first line based on a fluoropyrimidine, alone, or preferably, in combination with bevacizumab. Older patients could be also selected for less intensive treatment. In a randomized phase III

TABLE 34.1 First-line Bevacizumab and chemotherapy comparative trials.

Author	Schedules	Patients	Response (%)	PFS (months)	P	OS (months)	P
Kabbinavar[24]	Fu+Lv	105	15	5,5		12,9[a]	
	Fu+Lv+ Bevacizumab	104	26	9,2	0.0002	16,6[a]	NS
Hurwitz[25]	IFL	411	35	6,2		15,5[a]	
	IFL+Bevacizumab	402	45	10,6	<0.001	20,3[a]	<0.001
Fuchs[a,22]	mIFL+ Bevacizumab	60	53,3	8,3[a]		19,2	
	FOLFIRI+Beva	57	57,9	11,2[a]	NS	28	0.037
Saltz[26]	FOLFOX4/XELOX	701	38	8[a]		19,9	
	FOLFOX4/XELOX + Bevacizumab	699	38	9,4[a]	0,0023	21,1	NS
Tebbutt[27]	Capecitabine	142	30,3	5,7[a]		18,9	
	Capecitabine-Mitomycin-Bevacizumab	146	45,9	8,4[a]		16,4	
	Capecitabine-Bevacizumab	147	38,1	8,5[a]	<0.0001	18,9	NS
Cunningham[28]	Capecitabine	140	10	5,1[a]		16,8	
	Capecitabine-Bevacizumab	140	19	9,1[a]	<0.0001	20,7	NS

NS: no significant differences.
[a]Primary objective.

study, which specifically included 280 patients over 70 years of age,[28] the combination of capecitabine with bevacizumab resulted in a PFS significantly higher (9.1 vs 5.1 months; HR: 0.53; $P < 0.0001$). More severe adverse events occurred in the capecitabine and bevacizumab group (40% vs 22%). The most relevant adverse effects were hand-foot syndrome (16% vs 7%), diarrhea (7% in both arms), and venous thromboembolic events (8% vs 4%). Bleeding (of any grade, although mild in most cases) occurred in 25% of patients allocated to capecitabine and bevacizumab versus 7% of those allocated to capecitabine alone. This study demonstrates the efficacy and safety of the combination of capecitabine and bevacizumab in the treatment of elderly patients with metastatic CRC.

Dihydropyrimidine dehydrogenase (DPD) is the fundamental enzyme in the metabolism of fluorouracil, with an activity subject to interindividual variability and genetic polymorphism in its DPYD gene. Deficiency in the activity of this enzyme is rare, estimating that between 3% and 8% of the European Caucasian population presents a partial deficiency, being completely absent in up to 0.5%. Treatment with fluoropyrimidines carries a significant risk of serious, life-threatening adverse reactions for patients who are partially or completely deficient in DPD activity. For this reason, the European Medicines Agency (EMA) recommends carrying out genotype and/or phenotype tests for DPD deficiency before starting therapeutic regimens containing fluoropyrimidines in all patients. These recommendations were officially presented by the Spanish Agency for Medicines and Health Products (AEMPS) in May 2020, consequently being adopted by the different scientific societies of Medical Oncology. The administration of fluoropyrimidines is contraindicated in patients with complete DPD deficiency and in those with partial deficiency, treatment with dose reduction should be started and fluorouracil levels should be monitored during treatment if possible (this last recommendation is not applicable to capecitabine because there is no good correlation between plasma levels and toxicity). Alternative options such as raltitrexed or irinotecan-oxaliplatin combination can be recommended in the case of high-risk DPYD variants.[17,32,33]

Selection of bevacizumab or anti-EGFR in combination with chemotherapy

The randomized trials that directly compare chemotherapy combined with bevacizumab or with anti-EGFR are presented in Tables 34.1 and 34.2.

In the PEAK study,[34] the PFS of the FOLFOX scheme combined with panitumumab versus the FOLFOX-bevacizumab scheme has been estimated with a phase II design with random assignment, in patients with metastatic CRC in the first line, native KRAS (exon 2). One hundred and forty-two and 143 patients have been included in the regimens of FOLFOX combined with panitumumab or bevacizumab, respectively. The PFS was similar 10.9 versus 10.1 months (HR 0.84; $P = 0.22$). The median OS was 34.2 months versus 24.3 months, in favor of the panitumumab arm (HR 0.62, $P = 0.009$). The response rates were 58% and 54%. In a preset biomarker analysis[34] (LS), additional mutations in KRAS (exons 3 and 4) and NRAS (exons 2, 3, and 4) were analyzed. In patients without RAS activating mutations (KRAS and NRAS wild types) assigned to

TABLE 34.2 Bevacizumab or anti-EGFR in first line: Randomized clinical trials.

Author and trial	Arms	Phase	RAS wild-type (n)	Median survival (months)	CI	Hazard ratio	CI	P
Schwartzberg LS[34]	CT/Pan	II	88	41.3	28.8–41.3			
PEAK	CT/Beva		82	28.9	23.9–31.3	0.63	0.39–1.02	0.058
Stintzing S[35]	CT/Cet	III	199	33.1	24.5–39.4			
FIRE-3	CT/Beva		201	25.0	23.0–28.1	0.70	0.54–0.90	0.0059
Venook AP[15]	CT/Beva	III	559	30.0	NR			
ALLIANCE	CT/Cet		578	29.0	NR	0.88	0.77–1.01	0.08

FOLFOX-cetuximab, the efficacy results were PFS 13 months, SG 41.3 months, and 64% of responses. In the group assigned to chemotherapy with bevacizumab: PFS 10.1 months, SG 28.9 months, and 60% responses.

The results of the FIRE3 study in patients with native KRAS tumors have been communicated.[36] Five hundred and ninety-two patients have been included. In the intention-to-treat analysis, there were no differences in the main objective, which was the response rate: 62% in the group assigned to FOLFIRI-cetuximab (with 297 patients) and 58% in the group (with 295 patients) assigned to FOLFIRI-bevacizumab ($P = 0.183$). PFS was 10 months with cetuximab and 10.3 months with bevacizumab (hazard ratio, 1.06; $P = 0.547$). However, OS (one of the secondary endpoints of the study) was higher in the cetuximab group (28.7 months vs. 25 months, hazard ratio 0.77, $P = 0.017$). There were no differences in hematological side effects. Nausea and vomiting, hypertension, bleeding, abscesses, and fistulas were more frequent with bevacizumab, while hand-foot syndrome, skin-mucosa toxicity, hypocalcemia, hypomagnesemia, and infusion reactions were more frequent with cetuximab. Adverse events with a fatal outcome were more frequent with bevacizumab (1.7% vs 0%). Results based on the presence or absence of additional sequence variations in the KRAS and NRAS genes have recently been reported.[35] KRAS exons 3 and 4 and NRAS exons 2, 3, and 4 were analyzed, in addition to BRAF. Tumors from 488 patients were evaluable. Additional mutations in KRAS and NRAS (RAS mutated) were detected in 16%, which were all native to KRAS exon 2 mutations. In the intention-to-treat analysis, there were no differences in the response rate: 65.5% in the native RAS group assigned to FOLFIRI-cetuximab and 51.9% in the native RAS group assigned to FOLFIRI-bevacizumab ($P = 0, 32$). PFS was 10.4 months with cetuximab and 10.2 months with bevacizumab (hazard ratio 0.93; $P = 0.54$). PFS was lower ($P = 0.085$) in mutated patients assigned to cetuximab. OS was 7.5 months higher in patients with native RAS assigned to cetuximab (33.1 months vs 25.6 months, hazard ratio 0.70, $P = 0.011$).

The Alliance CALGB/SWOG 80405 study[15] analyzed, with OS as the main objective, the comparative efficacy of the combination of FOLFOX or FOLFIRI with cetuximab versus FOLFOX or FOLFIRI combined with bevacizumab in 1137 randomized KRAS wild-type (codons 12 & 13) patients. It is concluded that there are no significant differences for the primary endpoint, OS, between the two groups: the OS of 29.9 months in the cetuximab group, compared to 29.04 months in the bevacizumab group (HR = 0.92, $P = 0.34$).

The location of the primary tumor has been gaining importance as a prognostic factor, as well as a predictive factor of response to drugs. In a meta-analysis published in 2016,[37] 66 studies were reviewed, with a total of 1,437,846 patients included and with a median follow-up of 65 months. The left side was associated with a better prognosis, with a significant reduction in the risk of death of 19% in absolute terms, behaving as an independent prognostic factor in both localized and advanced disease (HR = 0.82). Subgroup analyses have been reported from the FIRE3 and CALGB/SWOG 80405 trials in which it has been suggested that anti-EGFR therapy would have less benefit in right-side mCRC. In the CALGB/SWOG 80405 study, retrospective analyses were performed based on the location of the primary tumor. Differences in OS were observed, being 33.3 months for the left side and 19.4 months for the right side ($P < 0.0001$), regardless of the treatment. These differences in OS were more evident in the subgroup of patients treated with chemotherapy and cetuximab, being 36.0 months on the left side and 16.7 months on the right side ($P < 0.0001$). Similarly, retrospective analyses were performed in the FIRE3 study based on the location of the primary, finding significant differences in OS as a function of laterality, especially in the subgroup of "all RAS native" patients treated with chemotherapy and cetuximab, being the OS of 38.3 months for the left side and 18.3 months for the right side ($P < 0.00001$). Thus the authors conclude that the laterality of the primary tumor constitutes a prognostic and predictive factor for first-line treatment when using biological agents.

The influence of the location of the primary tumor in patients with RAS wild-type mCRC was retrospectively analyzed, including six randomized studies (CRYSTAL, FIRE 3, CALGB80405, PRIME, PEAK, and 20,050,181), as a prognostic and predictive factor, comparing treatment with chemotherapy plus anti-EGFR (experimental arm) with treatment with chemotherapy alone or with bevacizumab (control arm).[38] A worse prognosis was observed for patients with primary on the right side in the control arm and the experimental arm for OS (HRs 2.03 and 1.38 respectively, significantly) and PFS (HRs = 1.59 and 1.25, respectively, significant). Regarding the predictive value, a significant benefit was found for treatment with chemotherapy plus anti-EGFR in patients with a primary on the left side (HR = 0.75 for OS and 0.78 for PFS), compared with a nonsignificant benefit in patients with a primary on the right side (HR = 1.12 for OS and 1.12 for PFS). For response rate (ORR), a trend toward greater benefit from chemotherapy plus anti-EGFR was observed in patients with left-sided tumors (OR = 2.12) compared to those with right-sided tumors (OR = 1.47). Thus the authors conclude that in patients with RAS native (BRAF native) left-sided tumors, the treatment of choice should be chemotherapy doublet plus anti-EGFR, regardless of treatment objective. In the case of patients with RAS native right-sided tumors, elective treatment would be a triplet (FOLFOXIRI) plus bevacizumab if our goal is cytoreduction; if our goal is disease control, the treatment of choice would be a doublet of chemotherapy with or without bevacizumab.

Triplet combinations of 5-fluorouracilo, oxaliplatin, and irinotecan

Two randomized studies have evaluated the triple combination of 5FU, oxaliplatin, and irinotecan versus FOLFIRI as first-line treatment in patients with advanced CRC. In a first trial,[39] with 283 patients, there were no differences in OS, PFS, or response rate. There was an increase in metastases resection (4%–10%) in patients assigned to FOLFOXIRI. In the second study,[40] with 244 patients, the FOLFOXIRI scheme (with higher doses of oxaliplatin and irinotecan compared to the previous study) was associated with a higher response rate (34% vs 60%; $P < 0.0001$), higher PFS (6.9 vs. 9.8 months; HR 0.63; $P = 0.0006$), and higher OS (16.7 vs. 22.6 months; HR: 0.70; $P = 0.032$). Likewise, the R0 metastasis surgery rate was higher (6% vs. 15%; $P = 0.033$) among all randomized patients and especially among patients with only liver metastases: 12% vs. 36% ($P = 0.017$). The treatment with the three drugs in both studies was associated with greater toxicity, especially alopecia, diarrhea, neutropenia, and neurotoxicity.

The results of the phase III TRIBE study[41] evaluating the efficacy and safety of the FOLFOXIRI-bevacizumab regimen have been reported. It included 508 patients who were randomly assigned to the control regimen with FOLFIRI plus bevacizumab or the experimental regimen with FOLFOXIRI plus bevacizumab. The main target was the PFS. In the FOLFOXIRI-bevacizumab arm, PFS was significantly higher (12.2 months vs. 9.7 months; HR 0.71; $P = 0.0006$). The response rate was also significantly higher in the experimental Scheme (65% vs. 53%; $P = 0.006$). Likewise, an improvement in OS was observed when updating its data[41]: 29.8 months for the FOLFOXIRI-bevacizumab group vs 25.8 months for the control group (HR = 0.8, $P = 0.03$). Molecular subgroups were also established and differences in OS were observed, with the median OS of 37.1 months in the subgroup of native patients for RAS and BRAF, 25.6 months in the group with mutations in RAS, and 13.4 months in the subgroup for BRAF mutated (HR = 2.79, $P < 0.0001$). The most common serious side effects with FOLFOXIRI-bevacizumab were diarrhea, stomatitis, neutropenia, and neurotoxicity. There was no increase in febrile neutropenia, serious adverse effects, or treatment-related deaths. There were also no significant differences in the incidence of bevacizumab-related toxicities such as hypertension, thromboembolic events, or bleeding. The authors conclude that the FOLFOXIRI-bevacizumab triplet would constitute a feasible treatment alternative in patients with good performance status and who meet the inclusion criteria of the trial, regardless of the molecular characteristics.

Data from the TRIBE 2 study[42] have recently been published. The primary endpoint of this trial was PFS, beginning with randomization and ending with death or progression after second line. The FOLFOXIRI-bevacizumab scheme following maintenance with 5FU and bevacizumab with reintroduction of irinotecan and oxaliplatin at progression was compared with sequential FOLFOX-bevacizumab followed by FOLFIRI-bevacizumab at the progression. A total of 679 patients have been randomized. The median PFS was 19.2 months in the experimental group and 16.4 months in the control group (HR = 0.74, $P = 0.0005$). OS benefits were observed for the experimental group (27.4 months vs 22.5 months in the control group (HR = 0.82, $P = 0.03$).

Equally recent are the results of VISNU 1,[43] in which the FOLFOXIRI-bevacizumab scheme is compared with FOLFOX-bevacizumab in patients with metastatic CRC and ≥ 3 circulating tumor cells, a marker of poor prognostic. With 349 patients, differences in PFS are observed, 12.4 months for FOLFOXIRI-bevacizumab and 9.3 months for FOLFOX-bevacizumab (HR = 0.64, $P = 0.0006$). Adverse effects $\geq$G3 were more common in the experimental group. The authors conclude that, in patients with ≥ 3 CTCs, the FOLFOXIRI-bevacizumab scheme significantly improves PFS.

Duration and intensity of first-line treatment

Defining the optimal duration of first-line treatment in patients with advanced CRC is extremely important in clinical practice. In patients not subsidiary to an eventual metastasis surgery, the objectives are to obtain maximum survival, maintaining or improving quality of life, and controlling symptoms and with the least possible toxicity. Different strategies have been evaluated to improve tolerance without compromising its effectiveness. These strategies would include the complete suspension of treatment, intermittent chemotherapy, the concept of "stop and go" ("stop and go"), and different forms of maintenance treatment, of less intensity and potentially better tolerated.

When oxaliplatin is included in first-line chemotherapy, the appearance of cumulative neurotoxicity makes it necessary to consider interrupting its administration in a significant number of patients, after a variable cumulative dose. Likewise, this cessation in the administration of oxaliplatin, in many cases before a clinical progression of the disease and apparently without developing resistance to the drug, would allow its eventual reintroduction into the continued care of the patient.

In the OPTIMOX2 study[44] the duration of disease control was similar, although the PFS was higher (8.3 versus 6.7 months; $P = 0.04$) in the control group, with "maintenance" chemotherapy.

The noninferiority of the strategy of complete withdrawal of treatment (intermittent chemotherapy) could not be formally demonstrated in the COIN study either.[45] Median OS in "intention-to-treat" patients was 15.8 months in the control arm and 14.4 months in the experimental arm (HR 1.084, 80% CI, 1.008–1.165). Thus although the noninferiority of the discontinuation of all chemotherapy cannot be formally demonstrated, in the clinic the small benefit in OS in favor of continued treatment must be taken into account against the potential advantages of intermittent treatment: improvement in certain aspects of quality of life, less time with treatment, and fewer hospital visits.

The CONCEPT study[46] demonstrated an increase in time to treatment failure in the subgroup of patients who received intermittent oxaliplatin (median 25 weeks) compared to the continuous group (median 18 weeks), with a 42% risk reduction ($P = 0.0025$). PFS was 7.3 months in the continuous treatment group versus 12 months with the intermittent strategy.

Maintenance chemotherapy combined with bevacizumab significantly improved PFS and showed a trend toward prolonged OS in a recent meta-analysis.[47]

Second-line treatment strategies

The disease progression after a first therapeutic scheme determines the need for a second line of treatment. Sometimes the previous treatment or part of it can be reintroduced (for example, after a complete interruption without disease progression and without residual toxicity, reintroduce oxaliplatin). It is often necessary and appropriate to maintain some of the drugs used in the first line (for example, 5FU or capecitabine, and sometimes bevacizumab). In the second line, the fundamental objective will be to control the progression of the disease and increase survival, but surgery will rarely be proposed again. The functional situation of the patient, tolerance to previous treatment, possible accumulated toxicities (for example, neurotoxicity due to oxaliplatin), and the presence of symptoms will determine the selection of the second therapy.

Different regimens have been shown to increase overall survival, regardless of KRAS: (i) FOLFOX plus bevacizumab in patients who have been treated only with 5FU and irinotecan,[48] (ii) continue bevacizumab and fluoropyrimidine by switching from oxaliplatin to irinotecan or vice versa,[49] (iii) FOLFIRI plus aflibercept in patients previously treated with FOLFOX with or without bevacizumab,[50] (iv) FOLFIRI plus ramucirumab in patients previously treated with FOLFOX and bevacizumab.[51] Median OS with chemotherapy and antiangiogenics in second-line setting is around 11–13,5 months, with a median PFS of 6–8 months. Overall response rate is less than 20%.

Combinations of FOLFOX or FOLFIRI with second-line bevacizumab have been the control scheme in different randomized trials and in observational studies.[28, 52] The OS for these combinations has ranged between 14.1 and 15.7 months, with PFS between 6.4 and 7.8 months. The second-line combination of FOLFIRI and bevacizumab has not been evaluated in phase III trials. In a systematic analysis, with 435 patients treated in different nonrandomized studies[53] the OS was 17.2 months, with a PFS of 8.3 months and a response rate of 26%.

In relation to EGFR antibodies, different studies have shown, in patients with RAS native tumors, an increase in the response rate and PFS. The combination of FOLFIRI plus panitumumab[54] has been shown to be superior to FOLFIRI alone, in patients with native KRAS (exon 2) in both PFS (HR: 0, 73; $P = 0.004$), with a median of 5.9 months for panitumumab-FOLFIRI and response rate (35% versus 10%). However, the difference in OS was not statistically significant (14.5 versus 12.5 months, in favor of the panitumumab-FOLFIRI arm).

The SPIRITT study[55] explores, with a phase II design, the efficacy (with PFS as the main objective of the study) of FOLFIRI plus panitumumab versus FOLFIRI plus bevacizumab, in second line, in patients with native KRAS exon 2 tumors previously treated with a regimen based on oxaliplatin and bevacizumab. Ninety-one patients were included in each of the treatment arms. The PFS for the panitumumab arm was 7.7 months, with an OS of 18 months, and a response rate of 28%. The PFS for the bevacizumab arm was 9.2 months, with an OS of 21.4 months, and a response rate of 16%. None of the risk ratios (HR) were significant. The results of this study cannot be considered definitive, but they suggest that, in the second line, switching to anti-EGFR plus chemotherapy is not superior compared to continuing antiangiogenic plus chemotherapy, in patients with wild-type tumors.

Exploratory analyses suggest a trend toward improved OS (36.8 months) for first-line panitumumab plus chemotherapy followed by second-line anti-VEGF, compared with first-line bevacizumab followed by second-line EGFR inhibitors (27.8 months) in patients with RAS native mCRC with a HR of 0.65 (95% CI 0.42–1.03).[56]

Treatment-refractory metastatic CRC

It is estimated that about 30%–50% of all patients with mCRC will be candidates for treatments beyond the second line. The overall survival (OS) after progression to two or more previous lines of treatment ranges from 4 to 6 months when they receive exclusively symptomatic care.

The selection of treatment after the second line will depend on the functional capacity, comorbidities, and organ dysfunction; the possible residual toxicity; and the previous treatments received, as well as the response to these.[57] Especially in this context of advanced disease refractory to previous treatments, the essential objectives would be the improvement of symptoms and the maintenance or improvement of the quality of life, with the least possible toxicity and, if possible, the prolongation of the SG. The wishes and expectations of the patient, properly informed, will be essential. Patients with ECOG ≥ 2, organ dysfunction, comorbidity, or limited life expectancy (generally less than 3 months) will be candidates for symptomatic care. On the other hand, we must always consider inclusion in a clinical trial (CT), especially in refractory patients.[58,59]

In patients in whom oxaliplatin was discontinued without progression to it and who do not maintain residual toxicity, reintroducing a regimen with FP and oxaliplatin could be an option beyond the second line. A better response to this reintroduction has been associated with an oxaliplatin-free interval of at least 6 months and a prior objective response. In these patients, reintroduction can obtain up to 22% of responses with PFS of 5.5 months and OS of 16 months.[60]

For patients with native RAS who have not previously received EGFR antibodies, third-line treatment options include cetuximab plus irinotecan or monotherapy with cetuximab or panitumumab.[58] A randomized comparative study[61] in native KRAS exon 2 patients, refractory to chemotherapy, evaluated the efficacy of cetuximab ($n = 500$) and panitumumab ($n = 499$). The results in OS (10.4 months with panitumumab and 9.9 months with cetuximab), in PFS (4.2 and 4.4 months, respectively), and response rate (22% and 19.8%) show no differences. In small phase II study, the ORR with irinotecan and panitumumab was 15.2%, with median PFS of 3.8 months and median OS of 12.5 months. Wild-type BRAF patients showed a 13% response rate.[62]

For patients with mCRC who have progressed after chemotherapy and biologics, there is only evidence of efficacy based on phase III trials for regorafenib[63] and TAS-102.[64] Regorafenib is an oral drug that acts as an inhibitor of protein kinases involved in tumor angiogenesis (VEGFR, TIE2), oncogenesis (KIT, RET, RAF-1, BRAF), and the tumor stroma or microenvironment (PDGFR, FGFR). TAS-102 is a combination of trifluorothymidine (trifluridine or TFT), a thymidine analogue antineoplastic, and tipiracil hydrochloride (TPI), a thymidine phosphorylase (TPase) inhibitor. There are no direct comparative studies of regorafenib and TAS-102. The toxicity profile is different, with more hematological toxicity with TAS-102 and a higher incidence of nonhematological toxicity with regorafenib, especially asthenia and hand-foot syndrome.

Both regorafenib and TAS-102 have been shown to increase OS compared to placebo in both phase III CTs 7, 8. The magnitude of this benefit is around 1.4 to 2 months. The toxicity profile of both drugs is different. In both studies, only patients with ECOG 0-1, without organ dysfunction and with a life expectancy of at least 3 months were included.

BRAF V600E-variant metastatic CRC

The BRAF V600E variants constitute therapeutically actionable molecular targets. These mutations are found in 5%–10% of all metastatic CRC and confer a worse prognosis.[65] In a meta-analysis[66] that groups the individual data of the patients treated in the first-line trials with random assignment between FOLFOXIRI and bevacizumab versus doublet and bevacizumab, 115 patients with mutated BRAF were included. The median OS was 14.5 months for the control arm, with doublet and 13.6 months with triplet (HR 1.14). Regarding PFS, the difference was not significant either, with a risk ratio of 0.84. Mutations in BRAF confer resistance to anti-EGFR. In the Alliance trial,[67] OS was numerically higher with bevacizumab (15 months) than with cetuximab (11.7 months; HR, 0.67; $P = 0.176$).

In a pooled analysis[68] includes 129 patients with BRAF V600E treated in the second line, an advantage in overall survival was shown in those patients ($n = 82$) who received antiangiogenics in combination with chemotherapy, with a HR of 0.50 ($P = 0.01$).

In the phase 3 BEACON study[69] 665 patients with metastatic CRC with BRAF V600E mutations, who had progressed to at least a first line, were randomized to receive the triplet combination of encorafenib (BRAF inhibitor), binimetinib (MEK inhibitor) and cetuximab, the doublet encorafenib, and cetuximab, or the investigator's choice control group with irinotecan/ cetuximab or FOLFIRI/cetuximab. Statistically significant differences were observed in the primary endpoint OS, in both experimental groups, with OS of 9.0 months in the triplet group versus 5.4 months in the control group (HR = 0.52, $P < 0.001$). OS in the encorafenib and cetuximab group was 8.4 months (HR = 0.6, $P > 0.001$). Likewise, differences were observed in the objective response rates, being 26% in the triplet group, compared to 2% in the control group. Grade 3 or higher adverse effects were less frequent in the encorafenib and cetuximab subgroup (50%) compared to the triplet group (58%). The most common side effects were fatigue, anemia, and acneiform dermatitis. Updated OS data shows 9.3 months in the triplet or doublet groups, compared to 5.9 months in the control group (HR = 0.6).[70] The treatment with encorafenib and cetuximab is considered as a new standard of second- and third-line treatment in V600E-mutated BRAF patients.

Immunotherapy

One of the most significant advances in the treatment of cancer has been the introduction of immunotherapy with checkpoint inhibitors (CPI). Pembrolizumab and nivolumab are two humanized monoclonal antibodies that bind to the programmed cell death receptor-1 (PD-1) present on T lymphocytes, thus blocking their interaction with the PD-L1 and PD-L2 ligands that are expressed on T cells and antigen presenting cells. The PD-1 receptor is a negative regulator of the activity of T lymphocytes, so its blockade allows the antitumor immune response to be enhanced by T cells. Ipilimumab is a human monoclonal antibody that blocks CTLA-4. It has an indirect mechanism of action since CTLA-4 prevents the activation of T lymphocytes and thus, ipilimumab activates the T lymphocytes and therefore the antitumor immune response. In mCRC, the CPI have demonstrated efficacy only in patients with MSI-H/dMMR tumors.[71–73]

In different phase II studies (Table 34.3), in patients with mCCR with MSI-H/dMMR, both pembrolizumab and nivolumab, both PD-1 inhibitors, have shown objective response rates in pretreated patients of up to 30%–40% in monotherapy. In most patients with objective response, it is long-lasting. Landmark analysis showed a 31%–74% PFS rate at 24 months and a 31%–79% of patients alive at 24 months.

The efficacy of nivolumab and ipilimumab in combination, in pretreated CRC MSI-H, has also been evaluated. The reported objective response rate has been 55% (95% CI 45.2%–63.8%) and the disease control rate at 12 weeks 80%. The observed responses were durable, with 71% of patients responding at 12 months, regardless of tumor PDL-1 expression. The PFS and OS at 12 months in this cohort were 71% and 85%, respectively.[74, 77] The presence of Lynch syndrome, RAS/BRAF mutations, and PD-L1 expression has not been identified as predictors of response with CPI in mCCR. Based on these studies, the use of pembrolizumab, nivolumab, and the combination of nivolumab + ipilimumab has been approved for the treatment of CRC MSI after failure of standard chemotherapy treatment (Food and Drug Administration).

In the first-line treatment[75] the phase III trial Keynote 177 evaluates pembrolizumab against standard chemotherapy treatment in the first line of treatment of mCRC-MSI-H/dMMR. With a total of 307 patients, pembrolizumab showed a statistically significant improvement in PFS compared to chemotherapy (16.5 vs 8.2 months, respectively, HR: 0.6, $P = 0.0002$). Furthermore, almost half of the patients (48.3%) in the pembrolizumab arm had not progressed at the time of analysis at 24 months. The objective response rate was also higher with pembrolizumab (43.8% vs 33.1% with chemotherapy), while grade 3–4 toxicity was lower than the control arm (22% with immunotherapy vs 66% with chemotherapy). OS, the other primary endpoint along with PFS, is still under study and has yet to be reported.

TABLE 34.3 Checkpoint Inhibitors in MSI-H/dMMR mCRC.

	Pembrolizumab KEYN 164/A n = 61	Pembrolizumab KEYN 164/B n = 63	Nivolumab CHK-M 142 n = 74	NIVO IPI CHK-M 142 n = 119	NIVO IPI 1ªL CHK-M 142 n = 45
ORR (%)	33 (21–46)	33 (22–46)	25 (23–46)	55 (45–64)	60 (44–64)
PFS, median, months 95% CI	2.3 (2.1–8.1)	4.1 (2.1–18.9)	6,6 (3-NR)	Not Reach	NR (14,1-not estimated)
PFS % Landmark	31% 24m	34% 24m	44%[37–59] 18m	71%[65–83] 12m	74% (57.2–84.5) 24m
OS median, months 95% CI	31.4 (21.4-NE)	NR (19,2-NE)	NR (19,6-NE)	NR (18-NE)	NE
OS % Landmark	31% 24m	63% 24m	67%[59–81] 18m	85%[81–94] 12m	79% (64.1–88.7) 24m

Data extracted from several studies: Le DT, Uram JN, Wang H, et al. PD-1 Blockade in tumors with mismatch-repair deficiency. N Engl J Med. 2015;372 (26):2509-2520; Le DT, Kim TW, Van Cutsem E, et al. Phase II open-label study of pembrolizumab in treatment-refractory, microsatellite instability-high/ mismatch repair-deficient metastatic colorectal cancer: KEYNOTE-164. J Clin Oncol. 2020;38(1):11-19. https://doi.org/10.1200/JCO.19.02107; Overman MJ, Lonardi S, Wong KYM, et al. Durable clinical benefit with nivolumab plus ipilimumab in DNA mismatch repair-deficient/microsatellite instability-high metastatic colorectal cancer. J Clin Oncol. 2018;36(8):773-779. https://doi.org/10.1200/JCO.2017.76.9901; André T, Shiu K-K, Kim TW, et al. Pembrolizumab in microsatellite-instability–high advanced colorectal cancer. N Engl J Med. 2020;383(23):2207-2218. https://doi.org/10.1056/NEJMoa2017699; Lenz H-J, Lonardi S, Zagonel V, et al. Nivolumab (NIVO) + low-dose ipilimumab (IPI) as first-line (1L) therapy in microsatellite instability-high/mismatch repair-deficient (MSI-H/dMMR) metastatic colorectal cancer (mCRC): Two-year clinical update. J Clin Orthod. 2020;38(15_suppl):4040-4040. https://doi.org/10.1200/JCO. 2020.38.15_suppl.4040; Lenz H-JJ, Van Cutsem E, Limon ML, et al. Durable clinical benefit with nivolumab (NIVO) plus low-dose ipilimumab (IPI) as first-line therapy in microsatellite instability-high/mismatch repair deficient (MSI-H/dMMR) metastatic colorectal cancer (mCRC). Ann Oncol. 2018;29(suppl_8):viii714. https://doi.org/10.1093/annonc/mdy424.019.

On the other hand, another cohort from Phase II CheckMate 142[76] studied the combination of nivolumab with ipilimumab in the first line of treatment for CRC-MSI. The results were published in ASCO 2020 and after a median follow-up of 29 months, the objective response rate was 69%, the PFS rate at 24 months was 74%, and the OS rate was 79%.

Currently, there are multiple trials under development that include combinations of checkpoint inhibitors with traditional chemotherapy regimens for mCRC, or even with other treatment modalities such as radiotherapy, also in tumors with MSS. In stable CRC, there are mechanisms (genomic, epigenetic, transcriptional) of primary resistance to CPI.[78,79] No meaningful objective responses have been obtained in phase II trials with pembrolizumab, nivolumab, or the nivolumab-ipilimumab combination. In randomized studies, atezolizumab[80] was not superior to regorafenib in refractory patients. In a phase II[81] trial suggests that combined immune checkpoint inhibition with durvalumab (anti-PD-L1) plus tremelimumab (anti-CTL4) may be associated with prolonged OS in patients with advanced refractory CRC and elevated tumor mutational burden (TMB). Further research is needed to expand the immune therapeutic horizon of mCRC. In addition, it is a priority to identify other possible predictive biomarkers of response beyond the determination of microsatellite instability such as TMB.

Chemotherapy and surgery of metastases

This topic has been dealt with in greater length and depth in the corresponding chapter. We can clearly differentiate two situations. When metastatic disease (mainly liver) is resectable, we can opt for an upfront surgery or a perioperative strategy. Systemic treatment before and after the metastasis surgery has the aims of avoiding systemic progression and treating the minimal residual disease that may exist either in the liver or in other potential locations, although not detectable. In this context, the administration of the FOLFOX pre- and postsurgery[82,83] has shown an increase in PFS. In this context, no Mo Ab has shown benefit in randomized studies with an adequate number of patients.

For metastatic disease not amenable to initial surgery, the medical treatment objectives and options are different. The potential resectability of the lesions should be assessed in a multidisciplinary team. The objective in this case will be to obtain an adequate reduction in tumor size (radiological response, but trying to avoid complete disappearance), ideally accompanied by a pathological response after surgery and with minimal damage to the healthy liver parenchyma, mainly by limiting the number of cycles of chemotherapy (+/− biological agents) administered preoperatively. It is needed to avoid VEGF inhibitors for 6 to 8 weeks before and after major surgery.

Different meta-analyses have been published[83,84] that evaluate the efficacy of adding agents against EGFR (cetuximab or panitumumab) to first-line chemotherapy, in patients with tumors RAS wild type, with an increased response rate, an improvement in R0 curative intent resection of liver metastases, and consistently an improvement in PFS. Exploratory analyses suggest that the obtention of an early tumor shrinkage (ETS) and the depth of response (DpR) in patients with RAS wild-type (WT) mCRC are relevant goals to improve resection rates and to improve OS.[85,86]

Recently, a prospective randomized study with 138 patients with native KRAS mCRC and unresectable liver metastatic disease, carried out in China,[87] has shown an increase in R0 "curative" surgery in patients assigned to chemotherapy (FOLFOX or FOLFIRI) plus cetuximab (25.7%) compared to those assigned only to chemotherapy (7.4%; $P < 0.01$). Likewise, patients who received cetuximab had a higher percentage of survival at 3 years (41% vs. 18% $P = 0.013$) and a higher median survival (30.9 vs. 21 months; P = 0.013). In a Spanish cooperative study in patients with native-KRAS mCRC and liver limited disease, first-line panitumumab combined either with FOLFOX4 or FOLFIRI resulted in high ORR (67%–74%) allowing potentially curative resection up to 37%–69% in KRAS and NRAS wild-type tumors.[88]

It is reasonable, and thus advised by various expert recommendations, to consider in patients with native RAS the combination of chemotherapy and anti-EGFR as preoperative treatment in patients with potentially resectable liver metastases, specially, in left-side tumors. In patients with mutated RAS, the options would be the combination of chemotherapy (FOLFOX or FOLFIRI) with bevacizumab, the triple combination of chemotherapy (FOLFOXIRI) or even, the combination of FOLFOXIRI plus bevacizumab.

Management of colorectal cancer and peritoneal metastases

Up to 10% of the patients with colorectal cancer have synchronous peritoneal metastases at the time of first diagnosis. In addition, more than half the patients with recurrent disease will present with metachronous peritoneal metastases. However, it is estimated the peritoneum is the only dissemination site in about 5% of CRC cases. Several factors are associated with peritoneal disease. It is more frequent for colon tumors (5.7%), especially in right primary location than for rectal tumors (1.7%). Other risk factors are pT4 stage[89] (in 17%–50%), mucinous and poor-differentiated histology, age less than 70–75 years, emergency surgery because of obstructive or perforated tumor at diagnosis, lymph node metastases, and non-radical resection during the first surgery. Molecular alterations linked to peritoneal metastases are MSI-high, BRAF

V600E, and KRAS mutations. All of these have been related to worse prognosis. A recent study shows that three-quarters of the peritoneal metastases analyzed were classified as consensus molecular subtypes (CMS) 4.

Treatment options[90] for colorectal cancer and peritoneal metastases included systemic chemotherapy and biological agents and, for selected patients with limited peritoneal disease, cytoreductive surgery (CRS) in most cases combined with hyperthermic intraperitoneal chemotherapy (HIPEC). The goal is to remove all macroscopic metastases surgically, while eliminating any residual microscopic disease by HIPEC. Patient selection is of paramount relevance for success of CRS. Characteristics of peritoneal metastases, presence of other site of metastases, and clinical factors such as fitness for surgery, performance status, age, body mass index, comorbidities, frailty, previous interventions are needed to be into account.

No definitive consensus concerning the ideal peritoneal carcinomatous index (PCI) cutoff at which to perform CRS has been adopted. The most widely accepted value is 20 or less for well- or moderately differentiated colorectal adenocarcinoma. A PCI greater than 20 is associated with a poor prognosis.

Nonrandomized studies[91,92] suggest that CRS and different HIPEC modalities could improve the prognosis of CRC patients with peritoneal metastases. In a retrospective multicenter study[93] that included 506 patients, the overall 1-year, 3-year, and 5-year disease-free survival rates were 40%, 16%, and 10%, respectively. The overall median survival was 19.2 months. Patients in whom cytoreductive surgery was complete had a median survival of 32.4 months. In a recent systematic review,[89] the median survival ranged from 14.6 to 60.1 months. The 5-year overall survival ranged from 23.4% to 52%. For those patients in whom a complete cytoreduction was obtained, the median survival reached 25 to 49 months. Major morbidity and mortality ranged from 15.1% to 47.2% and 0% to 4.5%, respectively.

Only two randomized clinical trials (RCT), the Netherlands Cancer Institute trial[94] and the PRODIGE 7,[95] have been fully published in this setting and they conform the level 1 evidence on the use of CRS and HIPEC for the treatment of colorectal peritoneal carcinomatosis.

In the first one, 105 patients were randomly assigned to receive either standard treatment consisting of systemic chemotherapy (fluorouracil-leucovorin) with or without palliative surgery or experimental therapy based on the combination of CRS and HIPEC. Initial results show the median survival was 12.6 months in the standard therapy arm and 22.3 months in the experimental therapy arm (log-rank test, $P=0.032$). After 8-year median follow-up, the median progression-free survival was 7.7 months in the control arm and 12.6 months in the HIPEC arm ($P=0.020$). The median disease-specific survival was 12.6 months in the control arm and 22.2 months in the CRS-HIPEC arm ($P=0.028$). The 5-year survival was 45% for those patients in whom a R1 resection was achieved. These results show that CRS and HIPEC are better than old systemic chemotherapy with palliative resection. The exact role of CRS, HIPEC, or the combination of both cannot be addressed in this study. Moreover, current standard chemotherapy and targeted agents are more efficacious than 5FU-LV, including in patients with peritoneal metastases, resulting in median survivals of 15 to 23 months.

Recently, in the PRODIGE 7 trial, the addition of HIPEC to CRS was evaluated in 265 patients with a PCI score up to 25 and a complete or optimal cytoreduction. Control arm involved CRS without HIPEC. All individuals received systemic chemotherapy with or without targeted therapy before or after surgery, or both. The results showed no difference in 5-year overall survival (39.4% compared with 36.7%; HR: 1.00; 95% CI: 0.73–1.37) or DFS (14.8% compared with 13.1%; HR: 0.908; 95% CI: 0.69–1.19). Median survival in this highly selected cohort was 41 months in both experimental and control arms. A subgroup analysis, although not preplanned, suggested an improvement in survival when HIPEC is added to CRS in patients with an intermediate PCI between 11 and 15. Late complications were significantly more frequent in the HIPEC group. The PRODIGE 7 study suggests that systemic chemotherapy and cytoreductive surgery without oxaliplatin-based HIPEC should be the cornerstone of therapeutic strategies with curative intent for colorectal peritoneal metastases.

Actual guidelines and expert's opinions recommend that the management of patients with colorectal cancer and peritoneal metastases should be led by a multidisciplinary team carried out in experienced centers but many questions about the optimal management approach for such patients remain.

References

1. European Union. *2020 Cancer incidence and mortality in EU-27 countries*; 2020. An official website of the European Union https://ec.europa.eu/jrc/en/news/2020-cancer-incidence-and-mortality-eu-27-countries. [Accessed 18 February 2021].
2. SEOM, 2021. Cancer figures in Spain 2021. Sociedad Española de Oncología Médica. Accessed 18 February 2021. https://seom.org/images/Cifras_del_cancer_en_Espnaha_2021.pdf.
3. Biller LH, Schrag D. Diagnosis and treatment of metastatic colorectal cancer: A review. *JAMA* 2021;**325**(7):669–85. https://doi.org/10.1001/jama.2021.0106.
4. Grothey A, Sargent D, Goldberg RM, Schmoll H-J. Survival of patients with advanced colorectal cancer improves with the availability of fluorouracil-leucovorin, irinotecan, and oxaliplatin in the course of treatment. *J Clin Oncol* 2004;**22**(7):1209–14. https://doi.org/10.1200/JCO.2004.11.037.

5. Schmoll HJ, Van Cutsem E, Stein A, et al. ESMO Consensus Guidelines for management of patients with colon and rectal cancer. a personalized approach to clinical decision making. *Ann Oncol* 2012;**23**(10):2479–516.
6. Köhne CH, Cunningham D, Di Costanzo F, et al. Clinical determinants of survival in patients with 5-fluorouracil-based treatment for metastatic colorectal cancer: results of a multivariate analysis of 3825 patients. *Ann Oncol* 2002;**13**(2):308–17. https://doi.org/10.1093/annonc/mdf034.
7. Chibaudel B, Bonnetain F, Tournigand C, et al. Simplified prognostic model in patients with oxaliplatin-based or irinotecan-based first-line chemotherapy for metastatic colorectal cancer: a GERCOR study. *Oncologist* 2011;**16**(9):1228–38. https://doi.org/10.1634/theoncologist.2011-0039.
8. Guinney J, Dienstmann R, Wang X, et al. The consensus molecular subtypes of colorectal cancer. *Nat Med* 2015;**21**(11):1350–6.
9. Folprecht G, Grothey A, Alberts S, Raab H-R, Köhne C-H. Neoadjuvant treatment of unresectable colorectal liver metastases: correlation between tumour response and resection rates. *Ann Oncol* 2005;**16**(8):1311–9. https://doi.org/10.1093/annonc/mdi246.
10. Poston GJ, Figueras J, Giuliante F, et al. Urgent need for a new staging system in advanced colorectal cancer. *J Clin Oncol* 2008;**26**(29):4828–33. https://doi.org/10.1200/JCO.2008.17.6453.
11. Thirion P, Wolmark N, Haddad E, Buyse M, Piedbois P. Survival impact of chemotherapy in patients with colorectal metastases confined to the liver: A re-analysis of 1458 non-operable patients randomised in 22 trials and 4 meta-analyses. *Ann Oncol* 1999;**10**(11):1317–20. https://doi.org/10.1023/a:1008365511961.
12. Buyse M, Thirion P, Carlson RW, Burzykowski T, Molenberghs G, Piedbois P. Relation between tumour response to first-line chemotherapy and survival in advanced colorectal cancer: a meta-analysis. *Lancet* 2000;**356**(9227):373–8. https://doi.org/10.1016/s0140-6736(00)02528-9.
13. Allegra C, Blanke C, Buyse M, et al. End points in advanced colon cancer clinical trials: a review and proposal. *J Clin Oncol* 2007;**25**(24):3572–5. https://doi.org/10.1200/JCO.2007.12.1368.
14. Goldberg RM, Rothenberg ML, Van Cutsem E, et al. The continuum of care: a paradigm for the management of metastatic colorectal cancer. *Oncologist* 2007;**12**(1):38–50. https://doi.org/10.1634/theoncologist.12-1-38.
15. Venook AP, Niedzwiecki D, Lenz H-J, et al. Effect of first-line chemotherapy combined with cetuximab or bevacizumab on overall survival in patients with KRAS wild-type advanced or metastatic colorectal cancer. *JAMA* 2017;**317**(23):2392. https://doi.org/10.1001/jama.2017.7105.
16. Modest DP, Ricard I, Heinemann V, et al. Outcome according to KRAS-, NRAS- and BRAF-mutation as well as KRAS mutation variants: pooled analysis of five randomized trials in metastatic colorectal cancer by the AIO colorectal cancer study group. *Ann Oncol* 2016;**27**(9):1746–53. https://doi.org/10.1093/annonc/mdw261.
17. Wörmann B, Bokemeyer C, Burmeister T, et al. Dihydropyrimidine Dehydrogenase Testing prior to Treatment with 5-Fluorouracil, Capecitabine, and Tegafur: a consensus paper. *Oncol Res Treat* 2020;**43**(11):628–36. https://doi.org/10.1159/000510258.
18. Porschen R, Arkenau H-T, Kubicka S, et al. Phase III study of capecitabine plus oxaliplatin compared with fluorouracil and leucovorin plus oxaliplatin in metastatic colorectal cancer: A final report of the AIO colorectal study group. *J Clin Oncol* 2007;**25**(27):4217–23. https://doi.org/10.1200/jco.2006.09.2684.
19. Díaz-Rubio E, Tabernero J, Gómez-España A, et al. Phase III study of capecitabine plus oxaliplatin compared with continuous-infusion fluorouracil plus oxaliplatin as first-line therapy in metastatic colorectal cancer: Final report of the Spanish cooperative group for the treatment of digestive tumors trial. *J Clin Oncol* 2007;**25**(27):4224–30. https://doi.org/10.1200/jco.2006.09.8467.
20. Cassidy J, Clarke S, Díaz-Rubio E, et al. Randomized phase III study of capecitabine plus oxaliplatin compared with fluorouracil/folinic acid plus oxaliplatin as first-line therapy for metastatic colorectal cancer. *J Clin Oncol* 2008;**26**(12):2006–12. https://doi.org/10.1200/JCO.2007.14.9898.
21. Arkenau H-T, Arnold D, Cassidy J, et al. Efficacy of oxaliplatin plus capecitabine or infusional fluorouracil/leucovorin in patients with metastatic colorectal cancer: a pooled analysis of randomized trials. *J Clin Oncol* 2008;**26**(36):5910–7. https://doi.org/10.1200/JCO.2008.16.7759.
22. Fuchs CS, Marshall J, Mitchell E, et al. Randomized, controlled trial of irinotecan plus infusional, bolus, or oral fluoropyrimidines in first-line treatment of metastatic colorectal cancer: results from the BICC-C Study. *J Clin Oncol* 2007;**25**(30):4779–86. https://doi.org/10.1200/JCO.2007.11.3357.
23. Ding H-H, Wu W-D, Jiang T, et al. Meta-analysis comparing the safety and efficacy of metastatic colorectal cancer treatment regimens, capecitabine plus irinotecan (CAPIRI) and 5-fluorouracil/leucovorin plus irinotecan (FOLFIRI). *Tumour Biol* 2015;**36**(5):3361–9. https://doi.org/10.1007/s13277-014-2970-1.
24. Kabbinavar F, Hurwitz HI, Fehrenbacher L, et al. Phase II, randomized trial comparing bevacizumab plus fluorouracil (FU)/leucovorin (LV) with FU/LV alone in patients with metastatic colorectal cancer. *J Clin Oncol* 2003;**21**(1):60–5. https://doi.org/10.1200/JCO.2003.10.066.
25. Hurwitz H, Fehrenbacher L, Novotny W, et al. Bevacizumab plus irinotecan, fluorouracil, and leucovorin for metastatic colorectal cancer. *N Engl J Med* 2004;**350**(23):2335–42. https://doi.org/10.1056/NEJMoa032691.
26. Saltz LB, Clarke S, Díaz-Rubio E, et al. Bevacizumab in combination with oxaliplatin-based chemotherapy as first-line therapy in metastatic colorectal cancer: a randomized phase III study. *J Clin Oncol* 2008;**26**(12):2013–9. https://doi.org/10.1200/JCO.2007.14.9930.
27. Tebbutt NC, Wilson K, Gebski VJ, et al. Capecitabine, bevacizumab, and mitomycin in first-line treatment of metastatic colorectal cancer: results of the Australasian gastrointestinal trials group randomized phase III MAX study. *J Clin Oncol* 2010;**28**(19):3191–8. https://doi.org/10.1200/JCO.2009.27.7723.
28. Cunningham D, Lang I, Marcuello E, et al. Bevacizumab plus capecitabine versus capecitabine alone in elderly patients with previously untreated metastatic colorectal cancer (AVEX): an open-label, randomised phase 3 trial. *Lancet Oncol* 2013;**14**(11):1077–85. https://doi.org/10.1016/s1470-2045(13)70154-2.
29. Van Cutsem E, Lang I, D'haens G, et al. KRAS status and efficacy in the first-line treatment of patients with metastatic colorectal cancer (mCRC) treated with FOLFIRI with or without cetuximab: The CRYSTAL experience. *J Clin Oncol* 2008;**26**(Suppl. 15). 2-2 https://ascopubs.org/doi/abs/10.1200/jco.2008.26.15_suppl.2.

30. Douillard J-Y, Siena S, Cassidy J, et al. Randomized, phase III trial of panitumumab with infusional fluorouracil, leucovorin, and oxaliplatin (FOLFOX4) versus FOLFOX4 alone as first-line treatment in patients with previously untreated metastatic colorectal cancer: the PRIME study. *J Clin Oncol* 2010;**28**(31):4697–705. https://doi.org/10.1200/JCO.2009.27.4860.
31. Maughan TS, Adams RA, Smith CG, et al. Addition of cetuximab to oxaliplatin-based first-line combination chemotherapy for treatment of advanced colorectal cancer: results of the randomised phase 3 MRC COIN trial. *Lancet* 2011;**377**(9783):2103–14. https://doi.org/10.1016/S0140-6736(11)60613-2.
32. Amstutz U, Henricks LM, Offer SM, et al. Clinical Pharmacogenetics Implementation Consortium (CPIC) guideline for dihydropyrimidine dehydrogenase genotype and fluoropyrimidine dosing: 2017 update. *Clin Pharmacol Ther* 2018;**103**(2):210–6. https://doi.org/10.1002/cpt.911.
33. Goldberg RM, Sargent DJ, Morton RF, et al. A randomized controlled trial of fluorouracil plus leucovorin, irinotecan, and oxaliplatin combinations in patients with previously untreated metastatic colorectal cancer. *J Clin Oncol* 2004;**22**(1):23–30. https://doi.org/10.1200/JCO.2004.09.046.
34. Schwartzberg LS, Rivera F, Karthaus M, et al. PEAK: a randomized, multicenter phase II study of panitumumab plus modified fluorouracil, leucovorin, and oxaliplatin (mFOLFOX6) or bevacizumab plus mFOLFOX6 in patients with previously untreated, unresectable, wild-type KRAS exon 2 metastatic colorectal cancer. *J Clin Oncol* 2014;**32**(21):2240–7. https://doi.org/10.1200/JCO.2013.53.2473.
35. Stintzing S, Modest DP, Rossius L, et al. FOLFIRI plus cetuximab versus FOLFIRI plus bevacizumab for metastatic colorectal cancer (FIRE-3): a post-hoc analysis of tumour dynamics in the final RAS wild-type subgroup of this randomised open-label phase 3 trial. *Lancet Oncol* 2016;**17**(10):1426–34. https://doi.org/10.1016/s1470-2045(16)30269-8.
36. Heinemann V, von Weikersthal LF, Decker T, et al. FOLFIRI plus cetuximab versus FOLFIRI plus bevacizumab as first-line treatment for patients with metastatic colorectal cancer (FIRE-3): a randomised, open-label, phase 3 trial. *Lancet Oncol* 2014;**15**(10):1065–75. https://doi.org/10.1016/S1470-2045(14)70330-4.
37. Petrelli F, Tomasello G, Borgonovo K, et al. Prognostic Survival Associated With Left-Sided vs Right-Sided Colon Cancer: A Systematic Review and Meta-analysis. *JAMA Oncol* 2016;**3**(2):211–9. https://doi.org/10.1001/jamaoncol.2016.4227.
38. Arnold D, Lueza B, Douillard J-Y, et al. Prognostic and predictive value of primary tumour side in patients with RAS wild-type metastatic colorectal cancer treated with chemotherapy and EGFR directed antibodies in six randomized trials. *Ann Oncol* 2017;**28**(8):1713–29. https://doi.org/10.1093/annonc/mdx175.
39. Souglakos J, Androulakis N, Syrigos K, et al. FOLFOXIRI (folinic acid, 5-fluorouracil, oxaliplatin and irinotecan) vs FOLFIRI (folinic acid, 5-fluorouracil and irinotecan) as first-line treatment in metastatic colorectal cancer (MCC): a multicentre randomised phase III trial from the Hellenic Oncology Research Group (HORG). *Br J Cancer* 2006;**94**(6):798–805. https://doi.org/10.1038/sj.bjc.6603011.
40. Falcone A, Ricci S, Brunetti I, et al. Phase III trial of infusional fluorouracil, leucovorin, oxaliplatin, and irinotecan (FOLFOXIRI) compared with infusional fluorouracil, leucovorin, and irinotecan (FOLFIRI) as first-line treatment for metastatic colorectal cancer: the Gruppo Oncologico Nord Ovest. *J Clin Oncol* 2007;**25**(13):1670–6. https://doi.org/10.1200/JCO.2006.09.0928.
41. Cremolini C, Loupakis F, Antoniotti C, et al. FOLFOXIRI plus bevacizumab versus FOLFIRI plus bevacizumab as first-line treatment of patients with metastatic colorectal cancer: updated overall survival and molecular subgroup analyses of the open-label, phase 3 TRIBE study. *Lancet Oncol* 2015;**16**(13):1306–15. https://doi.org/10.1016/S1470-2045(15)00122-9.
42. Cremolini C, Antoniotti C, Rossini D, et al. Upfront FOLFOXIRI plus bevacizumab and reintroduction after progression versus mFOLFOX6 plus bevacizumab followed by FOLFIRI plus bevacizumab in the treatment of patients with metastatic colorectal cancer (TRIBE2): a multicentre, open-label, phase 3, randomised, controlled trial. *Lancet Oncol* 2020;**21**(4):497–507. https://doi.org/10.1016/S1470-2045(19)30862-9.
43. Aranda E, Viéitez JM, Gómez-España A, et al. FOLFOXIRI plus bevacizumab versus FOLFOX plus bevacizumab for patients with metastatic colorectal cancer and ≥3 circulating tumour cells: the randomised phase III VISNÚ-1 trial. *ESMO Open* 2020;**5**(6). https://doi.org/10.1136/esmoopen-2020-000944. e000944.
44. Chibaudel B, Maindrault-Goebel F, Lledo G, et al. Can chemotherapy be discontinued in unresectable metastatic colorectal cancer? The GERCOR OPTIMOX2 Study. *J Clin Oncol* 2009;**27**(34):5727–33. https://doi.org/10.1200/JCO.2009.23.4344.
45. Adams RA, Meade AM, Seymour MT, et al. Intermittent versus continuous oxaliplatin and fluoropyrimidine combination chemotherapy for first-line treatment of advanced colorectal cancer: results of the randomised phase 3 MRC COIN trial. *Lancet Oncol* 2011;**12**(7):642–53. https://doi.org/10.1016/S1470-2045(11)70102-4.
46. Grothey A, Hart LL, Rowland KM, et al. Intermittent oxaliplatin (oxali) administration and time-to-treatment-failure (TTF) in metastatic colorectal cancer (mCRC): Final results of the phase III CONcePT trial. *J Clin Oncol* 2008;**26**(15_suppl):4010. https://doi.org/10.1200/jco.2008.26.15_suppl.4010.
47. Ma H, Wu X, Tao M, et al. Efficacy and safety of bevacizumab-based maintenance therapy in metastatic colorectal cancer: A meta-analysis. *Medicine (Baltimore)* 2019;**98**(50). https://doi.org/10.1097/MD.0000000000018227, e18227.
48. Giantonio BJ, Catalano PJ, Meropol NJ, et al. Bevacizumab in combination with oxaliplatin, fluorouracil, and leucovorin (FOLFOX4) for previously treated metastatic colorectal cancer: results from the Eastern Cooperative Oncology Group Study E3200. *J Clin Oncol* 2007;**25**(12):1539–44. https://pdfs.semanticscholar.org/7d66/db10695c406aaeba4f430732c0734bee7217.pdf.
49. Kubicka S, Greil R, André T, et al. Bevacizumab plus chemotherapy continued beyond first progression in patients with metastatic colorectal cancer previously treated with bevacizumab plus chemotherapy: ML18147 study KRAS subgroup findings. *Ann Oncol* 2013;**24**(9):2342–9. https://doi.org/10.1093/annonc/mdt231.
50. Van Cutsem E, Tabernero J, Lakomy R, et al. Addition of aflibercept to fluorouracil, leucovorin, and irinotecan improves survival in a phase III randomized trial in patients with metastatic colorectal cancer previously treated with an oxaliplatin-based regimen. *J Clin Oncol* 2012;**30**(28):3499–506. https://doi.org/10.1200/JCO.2012.42.8201.

51. Tabernero J, Yoshino T, Cohn AL, et al. Ramucirumab versus placebo in combination with second-line FOLFIRI in patients with metastatic colorectal carcinoma that progressed during or after first-line therapy with bevacizumab, oxaliplatin, and a fluoropyrimidine (RAISE): a randomised, double-blind, multicentre, phase 3 study. *Lancet Oncol* 2015;**16**(5):499–508. https://doi.org/10.1016/S1470-2045(15)70127-0.
52. Moriwaki T, Bando H, Takashima A, et al. Bevacizumab in combination with irinotecan, 5-fluorouracil, and leucovorin (FOLFIRI) in patients with metastatic colorectal cancer who were previously treated with oxaliplatin-containing regimens: a multicenter observational cohort study (TCTG 2nd-BV study). *Med Oncol* 2012;**29**(4):2842–8. https://doi.org/10.1007/s12032-011-0151-2.
53. Beretta GD, Petrelli F, Stinco S, et al. FOLFIRI + bevacizumab as second-line therapy for metastatic colorectal cancer pretreated with oxaliplatin: a pooled analysis of published trials. *Med Oncol* 2013;**30**(1):486. https://doi.org/10.1007/s12032-013-0486-y.
54. Peeters M, Price TJ, Cervantes A, et al. Randomized phase III study of panitumumab with fluorouracil, leucovorin, and irinotecan (FOLFIRI) compared with FOLFIRI alone as second-line treatment in patients with metastatic colorectal cancer. *J Clin Oncol* 2010;**28**(31):4706–13. https://doi.org/10.1200/JCO.2009.27.6055.
55. Hecht JR, Cohn AL, Dakhil SR, et al. SPIRITT (study 20060141): A randomized phase II study of FOLFIRI with either panitumumab (pmab) or bevacizumab (bev) as second-line treatment (tx) in patients (pts) with wild-type (WT) KRAS metastatic colorectal cancer (mCRC). *J Clin Orthod* 2013;**31**(4_suppl):454. https://doi.org/10.1200/jco.2013.31.4_suppl.454.
56. Peeters M, Forget F, Karthaus M, et al. Exploratory pooled analysis evaluating the effect of sequence of biological therapies on overall survival in patients with RAS wild-type metastatic colorectal carcinoma. *ESMO Open* 2018;**3**(2). https://doi.org/10.1136/esmoopen-2017-000297, e000297.
57. Arnold D, Prager GW, Quintela A, et al. Beyond second-line therapy in patients with metastatic colorectal cancer: a systematic review. *Ann Oncol* 2018;**29**(4):835–56. https://doi.org/10.1093/annonc/mdy038.
58. Van Cutsem E, Cervantes A, Adam R, et al. ESMO consensus guidelines for the management of patients with metastatic colorectal cancer. *Ann Oncol* 2016;**27**(8):1386–422. https://doi.org/10.1093/annonc/mdw235.
59. National Comprehensive Cancer Network (NCCN), n.d. Colon Cancer, Clinical practice guidelines in oncology. Accessed February 2017. https://www.nccn.org/professionals/physician_gls/pdf/colon.pdf.
60. Chibaudel B, Tournigand C, Bonnetain F, et al. Platinum-sensitivity in metastatic colorectal cancer: towards a definition. *Eur J Cancer* 2013;**49**(18):3813–20. https://doi.org/10.1016/j.ejca.2013.07.150.
61. Price T, Kim TW, Li J, et al. Final results and outcomes by prior bevacizumab exposure, skin toxicity, and hypomagnesaemia from ASPECCT: randomized phase 3 non-inferiority study of panitumumab versus cetuximab in chemorefractory wild-type KRAS exon 2 metastatic colorectal cancer. *Eur J Cancer* 2016;**68**:51–9. https://doi.org/10.1016/j.ejca.2016.08.010.
62. Elez E, Pericay C, Valladares-Ayerbes M, et al. A phase 2 study eof panitumumab with irinotecan as salvage therapy in chemorefractory KRAS exon 2 wild-type metastatic colorectal cancer patients. *Br J Cancer* 2019;**121**(5):378–83. https://doi.org/10.1038/s41416-019-0537-z.
63. Grothey A, Van Cutsem E, Sobrero A, et al. Regorafenib monotherapy for previously treated metastatic colorectal cancer (CORRECT): an international, multicentre, randomised, placebo-controlled, phase 3 trial. *Lancet* 2013;**381**(9863):303–12. https://doi.org/10.1016/S0140-6736(12)61900-X.
64. Mayer RJ, Van Cutsem E, Falcone A, et al. Randomized trial of TAS-102 for refractory metastatic colorectal cancer. *N Engl J Med* 2015;**372**(20):1909–19. https://doi.org/10.1056/NEJMoa1414325.
65. Bläker H, Alwers E, Arnold A, et al. The association between mutations in BRAF and colorectal cancer-specific survival depends on microsatellite status and tumor stage. *Clin Gastroenterol Hepatol* 2019;**17**(3). https://doi.org/10.1016/j.cgh.2018.04.015. 455–462.e6.
66. Cremolini C, Antoniotti C, Stein A, et al. Individual patient data meta-analysis of FOLFOXIRI plus bevacizumab versus doublets plus bevacizumab as initial therapy of unresectable metastatic colorectal cancer. *J Clin Oncol* 2020;**38**(28). https://doi.org/10.1200/JCO.20.01225. JCO2001225.
67. Lenz H-J, Ou F-S, Venook AP, et al. Impact of consensus molecular subtype on survival in patients with metastatic colorectal cancer: Results from CALGB/SWOG 80405 (alliance). *J Clin Oncol* 2019;**37**(22):1876–85. https://doi.org/10.1200/JCO.18.02258.
68. Gelsomino F, Casadei-Gardini A, Rossini D, et al. The role of anti-angiogenics in pre-treated metastatic BRAF-mutant colorectal cancer: A pooled analysis. *Cancers (Basel)* 2020;**12**(4):1022. https://doi.org/10.3390/cancers12041022.
69. Kopetz S, Grothey A, Yaeger R, et al. Encorafenib, binimetinib, and cetuximab in BRAF V600E-mutated colorectal cancer. *N Engl J Med* 2019;**381**(17):1632–43. https://doi.org/10.1056/NEJMoa1908075.
70. Tabernero J, Grothey A, Van Cutsem E, et al. Encorafenib plus cetuximab as a new standard of care for previously treated BRAF V600E-mutant metastatic colorectal cancer: Updated survival results and subgroup analyses from the BEACON study. *J Clin Oncol* 2021;**39**(4):273–84. https://doi.org/10.1200/JCO.20.02088.
71. Le DT UJN, Wang H, et al. PD-1 blockade in tumors with mismatch-repair deficiency. *N Engl J Med* 2015;**372**(26):2509–20.
72. Le DT DJN, Smith KN, et al. Mismatch repair deficiency predicts response of solid tumors to PD-1 blockade. *Science* 2017;**357**(6349):409–13. https://doi.org/10.1126/science.aan6733.
73. Le DT KTW, Van Cutsem E, et al. Phase II open-label study of pembrolizumab in treatment-refractory, microsatellite instability-high/mismatch repair-deficient metastatic colorectal cancer: KEYNOTE-164. *J Clin Oncol* 2020;**38**(1):11–9. https://doi.org/10.1200/JCO.19.02107.
74. Overman MJ, Lonardi S, KYM W, et al. Durable clinical benefit with Nivolumab plus Ipilimumab in DNA mismatch repair-deficient/microsatellite instability-high metastatic colorectal cancer. *J Clin Oncol* 2018;**36**(8):773–9. https://doi.org/10.1200/JCO.2017.76.9901.
75. André T, Shiu K-K, Kim TW, et al. Pembrolizumab in microsatellite-instability-high advanced colorectal cancer. *N Engl J Med* 2020;**383**(23):2207–18. https://doi.org/10.1056/NEJMoa2017699.
76. Lenz H-J, Lonardi S, Zagonel V, et al. Nivolumab (NIVO) + low-dose ipilimumab (IPI) as first-line (1L) therapy in microsatellite instability-high/mismatch repair-deficient (MSI-H/dMMR) metastatic colorectal cancer (mCRC): Two-year clinical update. *J Clin Orthod* 2020;**38**(15_suppl):4040. https://doi.org/10.1200/JCO.2020.38.15_suppl.4040.

77. Overman MJ, McDermott R, Leach JL, et al. Nivolumab in patients with metastatic DNA mismatch repair-deficient or microsatellite instability-high colorectal cancer (CheckMate 142): an open-label, multicentre, phase 2 study. *Lancet Oncol* 2017;**18**(9):1182–91. https://doi.org/10.1016/S1470-2045(17)30422-9.
78. Grasso CS, Giannakis M, Wells DK, et al. Genetic mechanisms of immune evasion in colorectal cancer. *Cancer Discov* 2018;**8**(6):730–49. https://doi.org/10.1158/2159-8290.CD-17-1327.
79. Becht E, de Reyniès A, Giraldo NA, et al. Immune and stromal classification of colorectal cancer is associated with molecular subtypes and relevant for precision immunotherapy. *Clin Cancer Res* 2016;**22**(16):4057–66. https://doi.org/10.1158/1078-0432.CCR-15-2879.
80. Eng C, Kim TW, Bendell J, et al. Atezolizumab with or without cobimetinib versus regorafenib in previously treated metastatic colorectal cancer (IMblaze370): a multicentre, open-label, phase 3, randomised, controlled trial. *Lancet Oncol* 2019;**20**(6):849–61. https://doi.org/10.1016/S1470-2045(19)30027-0.
81. Chen EX, Jonker DJ, Loree JM, et al. Effect of combined immune checkpoint inhibition vs best supportive care alone in patients with advanced colorectal cancer: The Canadian cancer trials group CO.26 study. *JAMA Oncol* 2020;**6**(6):831–8. https://doi.org/10.1001/jamaoncol.2020.0910.
82. Nordlinger B, Sorbye H, Glimelius B, et al. Perioperative chemotherapy with FOLFOX4 and surgery versus surgery alone for resectable liver metastases from colorectal cancer (EORTC Intergroup trial 40983): a randomised controlled trial. *Lancet* 2008;**371**(9617):1007–16. https://doi.org/10.1016/S0140-6736(08)60455-9.
83. Petrelli F, Barni S. Anti-EGFR agents for liver metastases. Resectability and outcome with anti-EGFR agents in patients with KRAS wild-type colorectal liver-limited metastases: a meta-analysis. *Int J Colorectal Dis* 2012;**27**(8):997–1004. https://doi.org/10.1007/s00384-012-1438-2.
84. Vale CL, Tierney JF, Fisher D, et al. Does anti-EGFR therapy improve outcome in advanced colorectal cancer? A systematic review and meta-analysis. *Cancer Treat Rev* 2012;**38**(6):618–25. https://doi.org/10.1016/j.ctrv.2011.11.002.
85. Heinemann V, Stintzing S, Modest DP, Giessen-Jung C, Michl M, Mansmann UR. Early tumour shrinkage (ETS) and depth of response (DpR) in the treatment of patients with metastatic colorectal cancer (mCRC). *Eur J Cancer* 2015;**51**(14):1927–36. https://doi.org/10.1016/j.ejca.2015.06.116.
86. Taieb J, Rivera F, Siena S, et al. Exploratory analyses assessing the impact of early tumour shrinkage and depth of response on survival outcomes in patients with RAS wild-type metastatic colorectal cancer receiving treatment in three randomised panitumumab trials. *J Cancer Res Clin Oncol* 2018;**144**(2):321–35. https://doi.org/10.1007/s00432-017-2534-z.
87. Ye L-C, Liu T-S, Ren L, et al. Randomized controlled trial of cetuximab plus chemotherapy for patients with KRAS wild-type unresectable colorectal liver-limited metastases. *J Clin Oncol* 2013;**31**(16):1931–8. https://doi.org/10.1200/JCO.2012.44.8308.
88. Carrato A, Abad A, Massuti B, et al. First-line panitumumab plus FOLFOX4 or FOLFIRI in colorectal cancer with multiple or unresectable liver metastases: A randomised, phase II trial (PLANET-TTD). *Eur J Cancer* 2017;**81**:191–202. https://doi.org/10.1016/j.ejca.2017.04.024.
89. Flood M, Narasimhan V, Waters P, et al. Survival after cytoreductive surgery and hyperthermic intraperitoneal chemotherapy for colorectal peritoneal metastases: A systematic review and discussion of latest controversies. *Surgeon* 2020. https://doi.org/10.1016/j.surge.2020.08.016. Published online October 3, 2020.
90. van Oudheusden TR, Razenberg LG, van Gestel YR, Creemers GJ, Lemmens VE, de Hingh IH. Systemic treatment of patients with metachronous peritoneal carcinomatosis of colorectal origin. *Sci Rep* 2015;**5**:18632. https://doi.org/10.1038/srep18632.
91. Zani S, Papalezova K, Stinnett S, Tyler D, Hsu D, Blazer 3rd DG. Modest advances in survival for patients with colorectal-associated peritoneal carcinomatosis in the era of modern chemotherapy: advances in CRC-associated PC survival. *J Surg Oncol* 2013;**107**(4):307–11. https://doi.org/10.1002/jso.23222.
92. Chua TC, Morris DL, Saxena A, et al. Influence of modern systemic therapies as adjunct to cytoreduction and perioperative intraperitoneal chemotherapy for patients with colorectal peritoneal carcinomatosis: a multicenter study. *Ann Surg Oncol* 2011;**18**(6):1560–7. https://doi.org/10.1245/s10434-010-1522-1.
93. Glehen O, Kwiatkowski F, Sugarbaker PH, et al. Cytoreductive surgery combined with perioperative intraperitoneal chemotherapy for the management of peritoneal carcinomatosis from colorectal cancer: a multi-institutional study. *J Clin Oncol* 2004;**22**(16):3284–92. http://nlp.case.edu/public/data/TargetedToxicity_JCOFullText/SVM_text_classifier_training/training/positive/1_128.html.
94. Verwaal VJ, van Ruth S, de Bree E, et al. Randomized trial of cytoreduction and hyperthermic intraperitoneal chemotherapy versus systemic chemotherapy and palliative surgery in patients with peritoneal carcinomatosis of colorectal cancer. *J Clin Oncol* 2003;**21**(20):3737–43. https://doi.org/10.1200/JCO.2003.04.187.
95. Quénet F, Elias D, Roca L, et al. Cytoreductive surgery plus hyperthermic intraperitoneal chemotherapy versus cytoreductive surgery alone for colorectal peritoneal metastases (PRODIGE 7): a multicentre, randomised, open-label, phase 3 trial. *Lancet Oncol* 2021;**22**(2):256–66. https://doi.org/10.1016/S1470-2045(20)30599-4.

Chapter 35

Adjuvant chemotherapy for colorectal cancer

Marta Covela Rúa, Silvia Varela Ferreiro, and Begoña Campos Balea
Medical Oncology Service, Lucus Augusti Hospital, Lugo, Spain

Introduction

At the time of diagnosis of colon cancer, the tumor is located in the bowel wall, without regional node infiltration (stages I and II) or with regional node invasion (stage III), in 75%–80% of patients.[1] Table 35.1 presents the probability of being in the different stages at the time of diagnosis.

Surgery is the most important part of the curative treatment of colon cancer, and the quality of surgical treatment cannot be replaced by any other treatment, including chemotherapy.

Although most patients undergo surgery, approximately half of those who are treated with curative intent will die from a relapse of their cancer. The percentage of patients with distant metastases after surgery increases with more advanced stage.

Therefore the prognosis depends fundamentally on the anatomopathological stage. Depending on the stage of the disease, the 5-year survival in patients treated with surgery alone is established as follows: stage I: 74%–78%, stage II: 38%–67%, and stage III: 28%–78%.[2]

Patients who debut with stage IV cancer with distant metastases constitute the remaining 20%–25%; in this group, complete resection of liver and/or lung metastases can achieve cure.

Rectal tumors are classified according to their distance from the anal margin: low (4–5 cm), medium (5–10 cm), and high (10–15 cm). When defining a therapeutic strategy, it is essential to determine mesorectal involvement or the presence of extramural or venous invasion, in addition to the location and existence of lymph node involvement (N).

Surgical resection is also the cornerstone of curative treatment for rectal cancer. After potential curative resection, the 5-year survival is 80%–90% in patients with stage I disease but drops to around 70% in patients with stage II and III disease (AJCC, seventh edition).

Adjuvant chemotherapy at the colorectal level is administered with the aim of eradicating any micrometastatic disease that may remain after surgery, with the consequent increase in disease-free survival and overall survival of the patient.[3]

The drugs that have been studied in the framework of adjuvant treatment of colon cancer with observed benefits are 5-fluorouracil (5-FU) alone or modulated with leucovorin (LV), oral fluoropyrimidine (FPO), and oxaliplatin.

Studies on irinotecan (CALGB-89803 and PETACC-3) in the adjuvant setting in targeted therapy and studies on NSABP C-08 and AVANT with bevacizumab and NCCTG NOA147 and PETACC-0 with cetuximab have been negative probably owing to the different biological characteristics of the localized disease and advanced disease.

The efficacy of adjuvant treatment will depend on tumor staging. Given their excellent prognosis, patients with stage I disease do not require adjuvant treatment. In patients with stage II disease, adjuvant treatment with 5-FU as a single drug reduces the risk of death from colon cancer after resection of the primary tumor by 3%–5%. In patients with stage III disease, chemotherapy regimens combining 5-FU and oxaliplatin reduce the risk of death by 15%–20%. Given that in stage II disease, around 80% of patients are cured by surgery alone, and given that in stage III disease, this percentage is reduced to almost 60%, the recommendations must be adjusted separately for each patient group, while explaining in detail the baseline situation and the possible benefits and complications to the patient.[4]

TABLE 35.1 Stage at diagnosis.

Stage I	15%
Stage II	20%–30% (20% poor prognosis)
Stage III	30%–40%
Stage IV	20%–25%

Adjuvant treatment of colon cancer

The decision for adjuvant treatment in patients diagnosed with colon cancer should be based on patient-dependent factors, such as PS, age, comorbidities, and preferences, as well as tumor-related factors, such as tumor staging and grade, which contribute to the overall risk of relapse.[4] The guidelines for treating colon cancer are shown in Fig. 35.1.

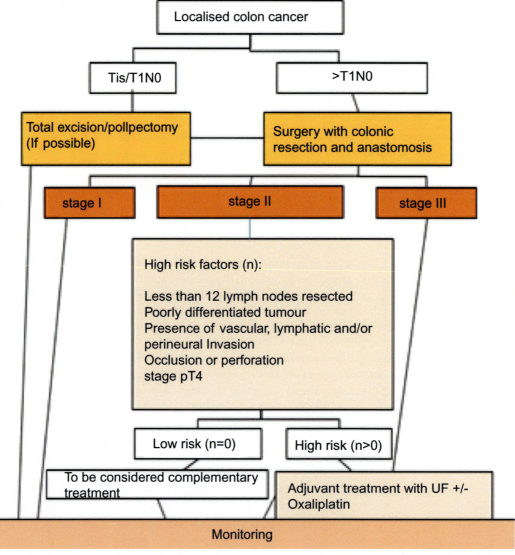

FIG. 35.1 Algorithm for colon cancer treatment.

Stage II

Adjuvant treatment should only be recommended for patients with stage II disease considered to be at a high risk, for which one of the following factors must be present[4–8]:

- Number of resected lymph nodes: <12
- Poorly differentiated tumor
- Presence of vascular, lymphatic, or perineural invasion
- Stage pT4
- Clinical presentation with bowel obstruction or perforation

In these cases, adjuvant treatment with 5-FU-based chemotherapy should be considered, either intravenously combined with LV or orally (capecitabine). The results of most published studies offer improvement in progression-free survival (PFS). Specifically, the results from a 2015 meta-analysis of 25 studies showed that the 5-year PFS in chemotherapy-naive patients with stage II disease was 84.4% compared with 79.3% in those treated with chemotherapy. In contrast, the QUASAR trial demonstrated a small but significant survival benefit for patients with stage II disease treated with adjuvant 5-FU/LV.

The benefit of adding oxaliplatin to adjuvant therapy is more controversial, as several studies have shown negative results (MOSAIC, C-07). In very high-risk patients (T4 or more than one poor prognostic factor), 5-FU with oxaliplatin could be considered.

Decision-making on the use of adjuvant treatment in patients with stage II disease should be individualized, taking into account both the clinical and molecular characteristics of the disease, as well as possible associated toxicities. Observation and participation in clinical trials are options to be considered.

Microsatellite instability (MSI)

MSI is critical in adjuvant treatment decision-making because of its prognostic implications. This is of particular value in patients with stage II disease. The PETACC-3 study showed that tumors with MSI are more frequent in stage II than in stage III (22% vs 12%). This means that these types of neoplasms are less likely to metastasize and are characterized as tumors with a better prognosis. Notably, some studies have suggested that MSI may be a predictive biomarker; thus, there could be a detrimental effect of FPO monotherapy in stage II. However, a recent study published by Sargent et al., in which almost 2000 patients with stage III colon cancer, of whom around 50% had received adjuvant chemotherapy, were evaluated showed that although MSI was a prognostic factor, it was not a predictive factor and did not imply a detrimental effect of chemotherapy.[9] This conclusion was corroborated by CLGB 9581 and CALGB 89803. The latest version of the National Comprehensive Cancer Network (NCCN) guidelines recommends determining MSI in all patients.

Stage III

For the past two decades, 5-FU has been the mainstay adjuvant treatment for colon cancer, first as a single agent, then in combination with levamisole, and later with LV. The latter combination proved to be superior and has been the standard adjuvant treatment for stage III colon cancer since 1998. Treatment with high-dose continuous-infusion 5-FU in the adjuvant setting has been provided after several clinical trials have demonstrated its better tolerability profile without affecting its impact on survival.[3] It has been accepted as the standard treatment in the phase III clinical trials MOSAIC and PETACC-3. Oxaliplatin was added to the combination of continuous-infusion 5-FU+LV in the FOLFOX scheme as standard treatment following the results of the MOSAIC trial in 2004 (Fig. 35.2), which confirmed a 23% decrease in the risk of recurrence, 6% improvement in EFS, and 2.6% improvement in overall survival in the combination group after more than 6 years of follow-up. This European study compared the efficacy of FOLFOX and 5-FU/LV in adjuvant colon cancer after randomizing both regimens in 2246 patients with stage II and III disease after complete resection of their colon cancer. The main toxicity of the FOLFOX scheme was peripheral neurotoxicity, which affected 12.4% of patients in the FOLFOX arm to grade 3 compared with 0.2% in the 5-FU/LV arm. This neuropathy remained in 15.4% of patients in the first arm at 4 years, with the majority classified under grade 1.[8]

The NSABP C-07 study compared the efficacy of the FLOX schedule between the combination of oxaliplatin and bolus 5-FU and the administration of bolus 5-FU and LV. The results of this study confirm the benefit of the combination but at the expense of increased digestive and neurological toxicities; thus, this scheme is not currently used in the adjuvant setting.[4,8]

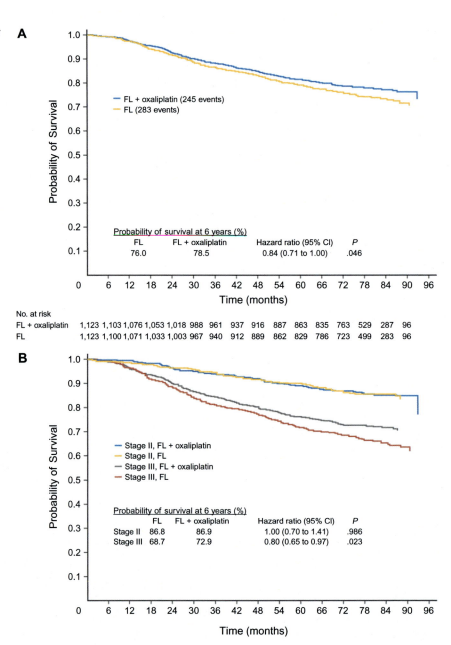

FIG. 35.2 Graphs from the MOSAIC study (data from SLE).

As for other drugs, such as capecitabine, there have been studies assessing its role in this context, either in monotherapy or in combination with oxaliplatin (XELOX). In the first case, the X-ACT study, with almost 2000 patients diagnosed with stage III colon cancer, showed the equivalence between bolus 5-FU and capecitabine in monotherapy for 6 months. Its combination with oxaliplatin has been compared with the combination of 5-FU and LV, revealing similar results to those obtained with the FOLFOX scheme.[8]

An observational analysis of patients from five databases included in the SEER-Medicare and NCCN showed that the addition of oxaliplatin to 5-FU/LV provides a benefit regarding overall survival in patients with stage III colon cancer.

Duration of treatment

The International Duration Evaluation of Adjuvant Chemotherapy (IDEA) collaboration aimed to conduct a meta-analysis of six international clinical trials (CALGB/SWOG 80702, IDEA FRANCE, SCOT, ACHIEVE, TOSCA, and HORG) on the duration of adjuvant chemotherapy in colon cancer.[10] The objective was to assess whether 3 months of treatment with oxaliplatin and FPO (CAPOX or FOLFOX) is noninferior to 6 months of treatment for stage III colon cancer. The objective of the pooled analysis was EFS; it ultimately included 12,834 patients and after a median follow-up of 4.8 months, noninferiority of 3-month to 6-month therapy was not confirmed ($P = 0.11$). The short treatment was better tolerated, with a rate of grade 2 or higher neurotoxicity of 16.6% for FOLFOX and 14.2% for CAPOX in the 3-month arm compared with 47.7% for FOLFOX and 44.9% for CAPOX in the 6-month arm.

Although the noninferiority hypothesis was not confirmed in the overall analysis, the study contained prespecified subgroup analyses based on the treatment type (CAPOX or FOLFOX) and tumor stages T and N. Patients receiving FOLFOX for 3 months had an increased risk of relapse compared to those receiving treatment for 6 months; however, in the CAPOX-treated subgroup, the noninferiority criterion was met for the 3-month therapy over the 6-month therapy.

Regarding the interpretation of the subgroup analyses based on tumor stage, the 3-month duration had an inferior efficacy over the 6-month duration in the T4 tumors. Finally, an exploratory analysis combining stages T1-3N1 (low risk) and T4 or N2 (high risk) was performed; the noninferiority criterion was met in the low-risk tumors, but not in the high-risk tumors. The study authors proposed a risk-based adjuvant treatment strategy. For T1-3N1 tumors, they suggested that 3 months of treatment may be sufficient, especially when CAPOX is used. In contrast, for T4 or N2 tumors, they suggested a longer adjuvant treatment (6 months), especially if FOLFOX is to be used.

The SCOT study[11] evaluated the noninferiority of adjuvant therapy with oxaliplatin and FPO for 3 months over that for 6 months in high-risk patients with stage II and stage III colon cancers.

After a median follow-up of 37 months, the 3-year EFS was 76.7% in the 3-month arm and 77.1% in the 6-month arm. The 3-year overall survival was 90% in the 3-month arm and 89.6% in the 6-month arm; the study met the prespecified noninferiority criterion. The authors performed a post hoc (exploratory) analysis, combining T1-3N1 and T4 or N2 stages. The noninferiority criterion was met for the T1-3N1 tumors but not for the T4 or N2 tumors and was achieved with CAPOX but not with FOLFOX. Severe sensory neuropathy during treatment was more frequent in the prolonged treatment group.

Both the IDEA and SCOT studies raise questions on the benefits and risks in relation to the adjuvant treatment duration. The final decision on the duration of treatment has to be individualized on a case-by-case basis. Table 35.2 lists different adjuvant treatment recommendations depending on the stage of the patient.

TABLE 35.2 Adjuvant treatment recommendations according to the latest update of the European Society for Medical Oncology guidelines (2020).

Stage III	Definition	Scheme and duration
Low risk	pT1-3 N1, MSI or MSS	FOLFOX 6 months (I, A), CAPOX 3 months (I, A)
High risk	pT4 and/or N2, MSI or MSS	FOLFOX 6 months (I, A), CAPOX 6 months (I, A)
Stage II	**Definition**	**Scheme and duration**
Low risk	No risk factors, MSI or SMS	Monitoring
Medium risk	Lymphatic invasion Perineural invasion Vascular invasion Histological grade 3 Tumor obstruction Preoperative CEA level of >5 ng/mL	MSI: monitoring SMS: De Gramont 6 months (I, B), capecitabine 6 months (V)
High risk	pT4 <12 isolated lymph nodes, multiple intermediate risk factors	MSI or MSS: FOLFOX 6 months (I, B), CAPOX 6 months (I, B), CAPOX 3 months (I, B)

MSI, microsatellite instability.

Resected stage IV

In cases of potentially resectable metastatic disease, perioperative chemotherapy could be considered, either neoadjuvant for 3 months and then complementary for another 3 months or adjuvant after metastasectomy if this is feasible from the outset.

The benefit of neoadjuvant treatment is based on the early treatment of micrometastatic disease, identification of patients who respond to treatment, and surgical management. The main disadvantages are disease progression during treatment, which may make it unresectable, or conversely, achievement of a complete radiological response that may make it difficult to identify the area to resect.[8]

A recent meta-analysis identified three clinical studies evaluating 642 patients with colon cancer liver metastases randomized to surgery alone and surgery plus chemotherapy. The results confirmed the benefit in PFS (HR: 0.75; CI: 0.62–0.91; $P = 0.003$) but not in overall survival.[8]

The regimen to be used depends on several factors, such as previous treatment and potential toxicity. If it is possible to administer FOLFOX for 6 months, it would be the standard treatment based on indirect evidence, because no studies have yet reported supporting evidence to date.[4]

In cases of induction treatment, an early reevaluation should be performed after 2–3 months to avoid hepatotoxicity that contraindicates or reduces the opportunity for liver resection.[8]

Elderly patients

In this patient subpopulation, it is more difficult to assess the efficacy of adjuvant therapy, as it is underrepresented in available clinical trials.[8] When the results of the MOSAIC and NSABP C-07 studies were pooled and patients were divided according to age into those older and younger than 70 years, the overall survival efficacy was shown to be higher in the group of patients younger than 70 years, with an HR of 1.18 vs 0.81, respectively.

As for the role of oxaliplatin, the NSABP C-07 subgroup analysis showed that the addition of oxaliplatin to 5-FU/LV did not provide any survival benefit in patients with stage II and III colon cancer aged 70 years or older, with a downward trend in survival. Similarly, in an analysis of the MOSAIC trial, 315 patients with stage II and III colon cancer aged 70–75 years did not benefit from the addition of oxaliplatin. In the phase III XELOXA study comparing capecitabine plus oxaliplatin with bolus 5-FU/LV for stage III disease, the subgroup of patients older than 70 years appeared to have benefitted in terms of EFS in the experimental arm, although this benefit is 8% lower than that in patients younger than 70 years. A recent pooled analysis of individual patient data from the NSABP C-08, XELOXA, X-ACT, and AVANT trials found that EFS and overall survival improved by treatment with CAPOX or FOLFOX over 5-FU/LV in patients aged 70 years and older.

The benefit and toxicities of 5-FU/LV and capecitabine as adjuvant therapy appear similar in older and younger patients. However, notably, no clear benefit has been demonstrated for the addition of oxaliplatin to 5-FU/LV in patients with stage II and III colon cancer aged 70 years or older; thus, in this patient population, treatment selection has to be individualized, taking into account comorbidities, treatment toxicity, and disease risk.

Conclusions

In summary, the schemes of adjuvant treatment for colon cancer that have shown benefits are those based on 5-FU, either in monotherapy or in combination with oxaliplatin; the schemes of the combination of both drugs have shown greater benefits in terms of PFS and overall survival.[4-8] Thus the following can be concluded:

- 5-FU infusion is preferred over bolus 5-FU schedules owing to its better tolerance, although this implies the need to place a venous device for its administration, with potential associated complications, such as thrombosis or infections.
- It is also possible to prescribe orally administered 5-FU derivatives, which avoid the need for venous access.
- Treatment should be started as soon as possible after surgical treatment, up to a maximum of 8 weeks, beyond which it should be considered on an individual basis.
- The duration of treatment should take into account the stage of the disease, expected toxicities, and patient preferences.

Neoadjuvant/adjuvant treatment of rectal cancer

Total excision of the mesorectum (TME) is a standard procedure in rectal cancer surgery; it is recommended for all patients with tumors located in the middle and lower thirds of the rectum. Meanwhile, partial resection is appropriate for tumors located in the upper third of the rectum because it reduces morbidity. Abdomino-perineal resection (APR) is the indicated surgical approach in cases of involvement of the anorectal junction and anal sphincter or as salvage treatment after local relapse (LR).

The management of tumors of the upper rectum (>10cm from the anal margin) is the same as that of colonic tumors, with the exception of large tumors with extension to adjacent structures or to the peritoneal reflection, which require preoperative chemotherapy (QT) and radiotherapy (RTP).

Neoadjuvant treatment

The goals of neoadjuvant treatment are to reduce the risk of LR, improve resectability to achieve complete resection in patients with mesorectal involvement or T4 disease, and preserve the sphincter function in low tumors by avoiding colostomy.

Adjuvant chemoradiotherapy has been the standard treatment for resected rectal cancer until a German trial[12] firmly established the role of neoadjuvant QT. The CAO/ARO/AIO-9 study compared between preoperative and postoperative chemoradiotherapies with 5-FU and demonstrated no difference in disease-free interval (DFI) and overall survival but with a significant reduction in local recurrence at 5 and 10 years (6% vs 13% and 7% vs 10%) and in short- and long-term toxicities. Based on these study findings, neoadjuvant QT-RTP has been established as standard treatment in resectable locally advanced rectal cancer (LARC).

Several randomized trials and a meta-analysis have addressed the issue of simultaneous administration of QT with conventional RTP fractionation. The largest trial, EORTC 22921, examined the benefit of concurrent chemoradiotherapy (bolus 5-FU and LV during weeks 1 and 5 of RTP) vs preoperative RTP alone (45 Gy over 5 weeks) and the contribution of postoperative adjuvant QT (four cycles of bolus 5-FU and LV). Patients undergoing preoperative chemoradiotherapy had a significantly higher pathological complete response rate (14% vs 5%); significantly less advanced tumors; fewer cases with venous, perineural, and lymphatic invasions; and better local control than those receiving RTP alone.

The results of previous randomized trials suggest that presurgical combination therapy reduces the local recurrence rate, yields lower short- and long-term toxicities, and allows a higher rate of sphincter-sparing surgery, thus improving the functional outcome of low tumors; however, the rate of distant relapse and overall survival are similar with both approaches. Further, combined therapy does not seem to increase the rate of perioperative complications of surgical resection.

Indications for neoadjuvant treatment

- Based on the results of randomized trials, the only definitive indication for neoadjuvant QT+RTP is the presence of T3–4 tumors. Patients with these tumors, if operated on initially, will require postoperative RTP.
- The optimal management of cT3N0 tumors is unclear, as many of affected patients have a favorable prognosis; hence, questions regarding the need for subsequent adjuvant treatment after excision of the mesorectum arise. Conversely, a percentage of these patients (1/5) may be understaged with presurgical imaging. Therefore, given the limitations of current imaging modalities, all patients with cT3N0 rectal cancer diagnosed on transrectal ultrasound or magnetic resonance imaging (MRI) are candidates for neoadjuvant chemoradiotherapy.
- The relative indications for neoadjuvant chemoradiotherapy include T1–2 tumors diagnosed on MRI or transrectal ultrasound, with clinically positive nodes; distal tumors for which APR is thought as necessary; and tumors appearing to have invaded the mesorectal fascia on preoperative imaging.

There are two modalities of radiotherapy delivery: the short course, based on the Swedish trial,[13] wherein 25 Gy is delivered in five fractions in 1 week, and classical radiotherapy, wherein 45–50.4 Gy is delivered in 25–28 fractions. Two randomized studies (one Polish study[14] and one Australian[15] study) have directly compared short-course RTP with "extended" RTP concurrent with 5-FU. Although the studies were not sufficiently powered to demonstrate equivalence between the two treatments, they have reported no significant differences in the rate of local control, sphincter preservation, or short- or long-term complications. Notably, both trials showed a nonsignificant trend toward worse local control with the short course.

The long course of radiotherapy should always be combined with FPO (5-FU)-based chemotherapy in continuous infusion or bolus form (continuous infusion is preferred owing to the more favorable side effect profile).

Another alternative could be to use oral 5-FU prodrugs (capecitabine) concurrent with radiotherapy. Substituting capecitabine for infusional 5-FU is an attractive option that has been studied in several clinical trials, the most representative of which are the NSABP R-04[16] and AIO,[17] showing equal efficacy between the two drugs (noninferiority studies). Therefore, although definitive survival data are not available, capecitabine can be considered as an alternative option to 5-FU infusion in the setting of chemoradiotherapy (either pre- or postsurgery), especially as it avoids central venous access. If this is selected, it is reasonable to administer a dose of 825 mg/m² twice daily during RTP.

The role of oxaliplatin in the neoadjuvant setting has been studied in several studies with the aim of improving the results obtained with standard neoadjuvant therapy. A meta-analysis of the main studies concluded that this combination increases grade 3 and 4 toxicities, with a small benefit in the rate of pathological complete responses that, in general, does not reach statistical significance.

Targeted therapies in neoadjuvant therapy have not shown any benefit in the studies conducted (studies on cetuximab, panitumumab, and bevacizumab).

Currently, the combination of FPO with radiotherapy remains the standard neoadjuvant treatment for rectal cancer. The use of other drugs outside of clinical trials is not indicated.

Induction and consolidation chemotherapies associated with neoadjuvant treatment

For LARC cT3-4 ± N+, several studies have been conducted in recent years combining various strategies to improve local control and survival outcomes.

The theoretical benefits of administering induction or consolidation chemotherapy before surgery are to treat the micrometastatic disease early, increase adherence to postoperative therapy, assess disease response, and allow early closure of the ileostomy.

Total neoadjuvant treatment (TNT) refers to administering all treatment before surgery. Of the various studies in this field, the phase II study by the Spanish GEMCAD group,[18] which compared in one arm the long course of RTP plus QT followed by surgery and adjuvant QT with the same adjuvant QT schedule, but which was administered before the long course of QT-RTP, is worth highlighting. The analysis showed a higher rate of grade 3 and 4 toxicities in the adjuvant arm and higher dose intensity in the TNT arm, with no differences in pathologic complete response (pCR), LR, and DFI at 5 years.

Another alternative is consolidation chemotherapy, which is based on administering chemotherapy after QT-RTP. With this strategy, a possible greater adherence to treatment is maintained by avoiding the sequelae of surgery that could lead to a delay attributed to a short postoperative period, early administration of the entire treatment, and lengthening the interval between QT-RTP and surgery, aiming to obtain a greater pCR. In 2016 the POLISH-2 study compared a short course of radiotherapy and FOLFOX for three cycles with the branch considered standard (long course of RTP with QT with oxaliplatin and adjuvant 5-FU). No differences were observed in radical surgery, which was the main objective of the study; however, there were less acute toxicity and a significant increase in overall survival for the short course of radiotherapy, followed by neoadjuvant chemotherapy.

Another study in this regard is the study conducted by Bujko et al.[19]; it is a phase III study on cT4a-b-cN2 tumors. Therein, a short course of radiotherapy (5 × 5 Gy) was provided, followed by chemotherapy (six cycles of CAPOX or nine cycles of FOLFOX4) and then ETM vs standard treatment. The experimental arm achieved a higher complete response rate and a lower percentage of distant metastases.

Another notable study is the phase III PRODIGE 23 study,[20] which compared FOLFIRINOXm (3 months) provided before neoadjuvant QT (capecitabine)-RTP, surgery, and at 3 months (a scheme that adds irinotecan, in addition to oxaliplatin +5-FU) with the standard regimen. The experimental arm showed increased complete response rate, EFS, and metastasis-free survival.

Currently, the NCCN guidelines (version 2021) already contemplate the indication of induction or consolidation chemotherapy (FOLFOX or XELOX scheme) combined with a short or long course of radiotherapy before surgery for LARC.

Adjuvant treatment

While adjuvant treatment strategies for resected colon cancer have focused on QT alone, RTP has become an important component in adjuvant therapy for rectal cancer owing to the different patterns of failure after surgical resection. In contrast to colon cancer, where failure occurs predominantly at a distance, failure occurs equally at a distance (liver and lung), as well as locally, in LARC.

Although a local recurrence rate of less than 10% was achieved with neoadjuvant treatment, patients still have a distant metastasis rate of around 30%, indicating the role of adjuvant chemotherapy.

The survival benefit following adjuvant QT after potential curative resection in rectal cancer was demonstrated in a meta-analysis[21] of 21 trials comparing outcomes between 4367 patients who did not receive adjuvant chemotherapy and 4854 who did. All trials used FPO-based QT. The use of adjuvant QT was associated with a significant reduction in the risk of recurrence (HR: 0.75, 95% CI: 0.68–0.83) and death (HR: 0.83, 95% CI: 0.76–0.91). There is no overall survival benefit; however, notably, a high percentage of patients did not complete treatment.

In a previous meta-analysis including the main studies (PETACC6, CAO/ARO/AIO-04, and ADORE) associating oxaliplatin with adjuvant therapy together with FPO[22], there was no reported evidence of benefit in overall survival; however, there was an observed evidence of benefit in ILE in the last two trials. As in the 5-FU studies, a high number of patients did not complete treatment; in general, 27% of patients who were candidates for adjuvant treatment did not receive it, and of those who did, only 50% received full doses. There is also the well-known loss of benefit of adjuvant treatment over time after surgery; the median duration in rectal cancer can be as long as 20 weeks.

In adjuvant studies, there is no stratification by subgroups. As known, the overall benefit of adjuvant treatment in stage II colon cancer is low. In rectal studies, stage II and III diseases are included together. In the retrospective study by Maas et al.[23] conducted among 3133 patients, there was evidence of a benefit in patients with LARC treated with adjuvant therapy depending on the pathological stage. Thus patients with ypT1–2 or ypT3–4 tumors benefited more from adjuvant treatment than did patients with pT0N0 tumors.[24] In patients with pathological stage III and II type diseases with risk factors, even though the benefit is likely to be less than that in colon cancer.

Adjuvant chemotherapy with FPO alone or in combination with oxaliplatin (FOLFOX or XELOX regimen) should be individualized according to the risk of recurrence and expected toxicity. Currently, the main treatment guidelines (NCCN and ESMO) advise postoperative chemotherapy in stage II and III diseases, according to the recommendations established for colon carcinoma. Patients with resected stage I disease have an excellent prognosis with surgery alone and do not need adjuvant therapy.

Perioperative treatment should last for 6 months.

Elderly patients

In this subpopulation, there is no age limit for treatment, as long as the patient's comorbidities allow. Dose reduction may be considered in frail patients.

Conclusions

The standard treatment of rectal cancer, except for early stages, is RTP concomitant with 5-FU, followed by TME. The role of adjuvant 5-FU ± oxaliplatin should be assessed on an individual basis depending on the risk of recurrence and expected toxicity. In locally advanced tumors, especially T4N2 tumors, TNT may be considered before surgery. Table 35.3 presents the therapeutic regimens for adjuvant colorectal cancer.

TABLE 35.3 Therapeutic regimens for adjuvant colorectal cancer.[4]

Regimen	Drugs/dose/scheme	Periodicity
Monotherapy		
Capecitabine	1250 mg/m^2 every 12 h days 1–14	22 days
De Gramont	5-FU 400 mg/m^2 iven bolo and LV 200 mg/m^2 iv followed by 5-FU 600 mg/m^2 iv as a 22 h continuous infusion, days 1 & 2	15 days
Polychemotherapy		
XELOX	Capecitabine 1000 mg/m^2 every 12 h days 1–15, Oxaliplatin 130 mg/m^2 day 1	22 days
mFOLFOX6	5-FU 400 mg/m^2 iven bolus and LV 400 mg/m^2 iv followed by 5-FU 2400 mg/m^2 iv continuous infusion over 46 h, Oxaliplatin 85 mg/m^2 day 1	15 days
FOLFOX4	5-FU 400 mg/m^2 iven bolus and LV 200 mg/m^2 iv followed by 5-FU 600 mg/m^2 iv as a 22-h continuous infusion, days 1 and 2, Oxaliplatin 85 mg/m^2 day 1	15 days

5-FU; 5-fluorouracil; LV, leucovorin.

References

1. Ahmedin Jemal DA, Tiwari RC, Murray T. Cancer statistics. *CA Cancer J Clin* 2004. Published online 2004 https://www.academia.edu/download/46786197/pdf.pdf.
2. Labianca R, Nordlinger B, Beretta GD, et al. Early colon cancer: ESMO Clinical Practice Guidelines for diagnosis, treatment and follow-up. *Ann Oncol* 2013;**24**(Suppl 6):vi64–72. https://doi.org/10.1093/annonc/mdt354.
3. Grávalos C, Ghamen I, Malón D. Tratamiento adyuvante del cáncer de colon. In: LR SJ, editor. *Cáncer colorrectal*. vol. 97. You&Us; 2009.
4. Schmoll HJ, Van Cutsem E, Stein A, et al. ESMO consensus guidelines for management of patients with colon and rectal cancer. A personalized approach to clinical decision making. *Ann Oncol* 2012;**23**(10):2479–516. https://doi.org/10.1093/annonc/mds236.
5. Kuebler JP, Wieand HS, O'Connell MJ, et al. Oxaliplatin combined with weekly bolus fluorouracil and leucovorin as surgical adjuvant chemotherapy for stage II and III colon cancer: results from NSABP C-07. *J Clin Oncol* 2007;**25**(16):2198–204. https://doi.org/10.1200/JCO.2006.08.2974.
6. André T, Boni C, Mounedji-Boudiaf L, et al. Oxaliplatin, fluorouracil, and leucovorin as adjuvant treatment for colon cancer. *N Engl J Med* 2004;**350**(23):2343–51. https://doi.org/10.1056/NEJMoa032709.
7. André T, Boni C, Navarro M, et al. Improved overall survival with oxaliplatin, fluorouracil, and leucovorin as adjuvant treatment in stage II or III colon cancer in the MOSAIC trial. *J Clin Oncol* 2009;**27**(19):3109–16. https://doi.org/10.1200/JCO.2008.20.6771.
8. *About the NCCN clinical practice guidelines in oncology (NCCN guidelines®)*; 2021. Accessed February 17, 2021 https://www.nccn.org/professionals/default.aspx.
9. Sargent DJ, Marsoni S, Monges G, et al. Defective mismatch repair as a predictive marker for lack of efficacy of fluorouracil-based adjuvant therapy in colon cancer. *J Clin Oncol* 2010;**28**(20):3219–26. https://doi.org/10.1200/JCO.2009.27.1825.
10. André T, Meyerhardt J, Iveson T, et al. Effect of duration of adjuvant chemotherapy for patients with stage III colon cancer (IDEA collaboration): final results from a prospective, pooled analysis of six randomised, phase 3 trials. *Lancet Oncol* 2020;**21**(12):1620–9. https://doi.org/10.1016/S1470-2045(20)30527-1.
11. Iveson TJ, Kerr RS, Saunders MP, et al. 3 versus 6 months of adjuvant oxaliplatin-fluoropyrimidine combination therapy for colorectal cancer (SCOT): an international, randomised, phase 3, non-inferiority trial. *Lancet Oncol* 2018;**19**(4):562–78. https://doi.org/10.1016/S1470-2045(18)30093-7.
12. Sauer R, Becker H, Hohenberger W, et al. Preoperative versus postoperative chemoradiotherapy for rectal cancer. *N Engl J Med* 2004;**351**(17):1731–40. https://doi.org/10.1056/nejmoa040694.
13. Improved survival with preoperative radiotherapy in resectable rectal cancer: Swedish rectal cancer. *N Engl J Med* 1997;**336**:980–7. Cancer/Radiothérapie. 1997;1(4):361–362 https://doi.org/10.1016/S1278-3218(97)81507-7.
14. Bujko K, Nowacki MP, Nasierowska-Guttmejer A, Michalski W, Bebenek M, Kryj M. Long-term results of a randomized trial comparing preoperative short-course radiotherapy with preoperative conventionally fractionated chemoradiation for rectal cancer. *Br J Surg* 2006;**93**(10):1215–23. https://doi.org/10.1002/bjs.5506.
15. Ngan SY, Burmeister B, Fisher RJ, et al. Randomized trial of short-course radiotherapy versus long-course chemoradiation comparing rates of local recurrence in patients with T3 rectal cancer: Trans-Tasman Radiation Oncology Group trial 01.04. *J Clin Oncol* 2012;**30**(31):3827–33. https://doi.org/10.1200/JCO.2012.42.9597.
16. Hofheinz R-D, Wenz F, Post S, et al. Chemoradiotherapy with capecitabine versus fluorouracil for locally advanced rectal cancer: a randomised, multicentre, non-inferiority, phase 3 trial. *Lancet Oncol* 2012;**13**(6):579–88. https://doi.org/10.1016/S1470-2045(12)70116-X.
17. Roh MS, Yothers GA, O'Connell MJ, et al. The impact of capecitabine and oxaliplatin in the preoperative multimodality treatment in patients with carcinoma of the rectum: NSABP R-04. *J Clin Oncol* 2011;**29**(15_suppl):3503. https://doi.org/10.1200/jco.2011.29.15_suppl.3503.
18. Fernández-Martos C, Pericay C, Aparicio J, et al. Phase II, randomized study of concomitant chemoradiotherapy followed by surgery and adjuvant capecitabine plus oxaliplatin (CAPOX) compared with induction CAPOX followed by concomitant chemoradiotherapy and surgery in magnetic resonance imaging-defined, locally advanced rectal cancer: Grupo cancer de recto 3 study. *J Clin Oncol* 2010;**28**(5):859–65. https://doi.org/10.1200/JCO.2009.25.8541.
19. Bujko K, Wyrwicz L, Rutkowski A, et al. Long-course oxaliplatin-based preoperative chemoradiation versus 5×5 Gy and consolidation chemotherapy for cT4 or fixed cT3 rectal cancer: results of a randomized phase III study. *Ann Oncol* 2016;**27**(5):834–42. https://doi.org/10.1093/annonc/mdw062.
20. Conroy T, Lamfichekh N, Etienne P-L, et al. Total neoadjuvant therapy with mFOLFIRINOX versus preoperative chemoradiation in patients with locally advanced rectal cancer: final results of PRODIGE 23 phase III trial, a UNICANCER GI trial. *J Clin Orthod* 2020;**38**(15_suppl):4007. https://doi.org/10.1200/JCO.2020.38.15_suppl.4007.
21. Petersen SH, Harling H, Kirkeby LT, Wille-Jørgensen P, Mocellin S. Postoperative adjuvant chemotherapy in rectal cancer operated for cure. *Cochrane Database Syst Rev* 2012;**3**. https://doi.org/10.1002/14651858.CD004078.pub2, CD004078.
22. Bujko K, Glimelius B, Valentini V, Michalski W, Spalek M. Postoperative chemotherapy in patients with rectal cancer receiving preoperative radio(chemo)therapy: a meta-analysis of randomized trials comparing surgery ± a fluoropyrimidine and surgery + a fluoropyrimidine ± oxaliplatin. *Eur J Surg Oncol* 2015;**41**(6):713–23. https://doi.org/10.1016/j.ejso.2015.03.233.
23. Maas M, Nelemans PJ, Valentini V, et al. Adjuvant chemotherapy in rectal cancer: defining subgroups who may benefit after neoadjuvant chemoradiation and resection: a pooled analysis of 3,313 patients. *Int J Cancer* 2015;**137**(1):212–20. https://doi.org/10.1002/ijc.29355.
24. Yang Y-J, Cao L, Li Z-W, et al. Fluorouracil-based neoadjuvant chemoradiotherapy with or without oxaliplatin for treatment of locally advanced rectal cancer: an updated systematic review and meta-analysis. *Oncotarget* 2016;**7**(29):45513–24. https://doi.org/10.18632/oncotarget.9995.

Chapter 36

Oral administration of cytostatic drugs in the treatment of CRC

Carmen Álvarez Lorenzo[a], Martina Lema Oreiro[b], and Ángel Concheiro Nine[a]
[a]*Department of Pharmacology, Pharmacy and Pharmaceutical Technology, Faculty of Pharmacy, University of Santiago de Compostela, A Coruña, Spain,* [b]*Pharmaceutical Services Management Service, S.X. de Farmacia, General Directorate of Health Care, Consellería de Sanidade, SERGAS, A Coruña, Spain*

Treatment of colorectal cancer (CRC) via the oral route

The three pillars of CRC treatment are surgery, radiotherapy, and drug therapy. These three treatment methods can be applied in combination or sequentially; thus, the approach to CRC is multidisciplinary. In general, drug therapy may be indicated as a complementary (adjuvant) treatment to surgery in earlier stages of the disease, in resectable metastatic disease to reduce the size and number of metastases so that surgical resection can be conducted, or in unresectable metastatic disease for mainly palliative purposes.

The mainstay systemic treatment of CRC is fluoropyrimidine (FP) alone or in combination with other interventions. For more than 40 years, 5-fluorouracil (5-FU) has been the most widely used active substance. Among the various strategies to improve its activity and reduce the toxicity of its intravenous bolus administration, the prolongation of the infusion time and its combination with leucovorin (LV; as calcium folinate), which enhances its antitumor efficacy, stands out. Thus, dosing schedules of 5-FU in combination with LV in continuous infusion have become the most widely used strategy. Another strategy for improvement was the development of oral FPs, such as capecitabine. Capecitabine is another antimetabolite, a precursor of 5-FU, whose activity mimics that obtained by continuous infusion of 5-FU. It is administered orally in the morning and evening for 14 days, followed by a week off. Its safety profile is generally more favorable than that of 5-FU, except for hand-foot syndrome.

Subsequently, the combination of 5-FU/LV with other chemotherapeutic agents, such as irinotecan (FOLFIRI) or oxaliplatin (FOLFOX), has led to better response rates, better progression-free survival values, and higher overall survival than 5-FU/LV alone. Both schemes are widely used, have similar activity, and differ in their toxicity profile, with neuropathy predominating for oxaliplatin and severe diarrhea for irinotecan. The combination of capecitabine with oxaliplatin (CAPOX) is an alternative to 5-FU/LV continuous infusion. Another option available is the combination of all of the earlier mentioned drugs, i.e., 5-FU/LV, irinotecan, and oxaliplatin, that make up the FOLFOXIRI or FOLFIRINOX regimens.[1]

These therapeutic regimens have been a major advance in the treatment of CRC but are based on nonspecific agents that destroy not only tumor cells but also normal cells in the body, producing adverse effects that have a negative impact on patients' quality of life.[2] Advances in the molecular understanding of cancer have led to the development of molecules that act on specific targets of the pathways involved in the control of cell survival, cell cycle progression, and angiogenesis. Among other therapeutic alternatives, this has led to the development of antiangiogenic monoclonal antibodies that act against the vascular endothelial growth factor (VEGF), such as bevacizumab and ramucirumab, and proteins, such as aflibercept, although there is no validated molecular marker predictive of their indication. Another example focuses on the RAS gene, whose activation by the epidermal growth factor receptor (EGFR) in tumors contributes to increased proliferation, survival, and production of pro-angiogenic factors. Mutations in the RAS gene are detected in approximately 40% of CRC cases and have been identified as biomarkers of tumor cell resistance to EGFR-targeted monoclonal antibodies, such as cetuximab and panitumumab. For this reason, these two drugs are indicated only in patients without RAS mutation.[3] This therapeutic strategy is a good example of precision medicine. Depending on the authorized indication, these drugs can be used first-line or successively and as monotherapy or in combination with previous regimens, giving rise to an infinite number of possibilities, such as FOLFOX-bevacizumab, FOLFIRI-cetuximab, FOLFIRI-aflibercept, or CAPEOX-bevacizumab.

Other targeted therapies licensed in recent years use regorafenib, an inhibitor of multiple kinases involved in tumor angiogenesis (VEGFR receptor), which is indicated as monotherapy in patients who have already been treated or are not candidates for available therapies, including FPO-based chemotherapy, oxaliplatin, irinotecan, and anti-VEGF and anti-EGFR agents.

There is also evidence that microsatellite instability (high microsatellite instability [MSI-H]) may serve as a predictive factor for treatment with checkpoint inhibitor antibodies or immune checkpoints (nivolumab, pembrolizumab, or ipilimumab). MSI-H corresponds to a pathogenic molecular pathway that originates when the DNA mismatch repair system, known as the MMR (mismatch repair deficient) system, is dysfunctional, resulting in the accumulation of mutations originating from DNA polymerase. Although it affects only 3%–5% of CRC cases, a high mutational tumor burden, high expression of programmed death ligand-1 (PD-L1), and high neoantigen load are observed. PD-L1 and PD-L2 are expressed in tumor cells and can suppress the body's immune response by binding to the programmed cell death protein 1 (PD-1) receptor in effector T cells. Nivolumab and pembrolizumab are monoclonal antibodies that have a high affinity for the PD-1 receptor, preventing its binding to PD-L1 and PD-L2 and allowing tumor cell recognition and immune response.[4] The Food and Drug Administration (FDA) has already approved the use of these anti-PD-1 monoclonal antibodies for immunotherapy in patients with CRC previously treated with FPO, oxaliplatin, and irinotecan. Nivolumab can be used as monotherapy or in combination with ipilimumab, an inhibitor of the immune checkpoint cytotoxic T-lymphocyte-associated antigen 4, which is a key regulator of T-cell activity.

Other strategies are related to overexpression or amplification of human EGFR 2 (HER-2), which is seen in 2%–4% of CRC cases but is more prevalent in patients with nonmutated RAS/BRAF. Potential anti-HER-2 drug combinations (trastuzumab with lapatinib or with pertuzumab) are being studied but have not yet been licensed for the treatment of CRC. Between 5% and 10% of CRCs have a mutation in the BRAF gene (V600), which is associated with a worse prognosis in this tumor type. Affected patients could be treated with BRAF inhibitors, such as vemurafenib or encorafenib, and orally administered tyrosine kinase inhibitors. However, unlike in melanoma, where this mutation is also present, it has been shown that in CRC, BRAF inhibitors are not effective as monotherapy; thus, they are used in combination with other drugs directed at other targets in the same pathway.[5]

Despite the growing development of increasingly targeted therapies, chemotherapy (e.g., FOLFOX or FOLFIRI), in combination or not with other drug therapies, remains the backbone of CRC treatment. Moreover, some of the targeted therapies can only be used in patients with a certain biomarker, which significantly reduces the number of candidates. Therefore it can be stated that the predominant route of treatment administration for CRC is intravenous, with a few exceptions, such as treatment with capecitabine and regorafenib or the trifluridine-tipiracil combination. The combined formulation of the latter two drugs in one tablet was authorized in 2016 for a similar indication as regorafenib, i.e., for patients who have received one of the above-mentioned intravenous regimens.

Intravenous chemotherapy requires the insertion of a central or peripheral venous access in the patient, which can lead to complications, such as infection and extravasation. Administration takes place in a hospital, requiring the patient to travel to the center for each session.[6] It also requires the involvement of specialized staff and is generally resource intensive.

Conversely, the oral route is more convenient for the patient, as it allows for self-administration of treatment at home, which is particularly important for long-term treatment. Patients can maintain their daily routine and, in some cases, their work activity, which improves their quality of life considerably. Several studies report a preference for the oral route.[7,8] This has led to a growing interest in the development of drugs that can be administered orally and that provide plasma profiles similar to those achieved when intravenous infusion is used. Thus, for example, 19 of the 29 (66%) innovative drugs approved by the FDA for the treatment of cancer between 2015 and 2017 are orally administered.[9] However, it is not the treatment of CRC that has benefited the most from this trend, especially when compared to the treatment of other tumors, such as renal cancer or chronic myeloid leukemia.

The oral route also has certain disadvantages. Oral administration does not avoid the toxicity inherent to each molecule. Further, patients often believe that it is less effective. However, the main limitation associated with this route is undoubtedly the lack of adherence, as it has been shown that nonadherence to the dosing regimen can compromise the efficacy of treatment. A multitude of factors that can negatively influence adherence, including poor doctor-patient communication, certain demographic and psychosocial characteristics of the patient, and complexity of the treatment or its adverse effects, have been identified. Despite the importance of adherence to oral treatments and their increasing use in oncology, the number of published studies on this topic is still scarce.[10]

From a technological point of view, the development of oral dosage forms is a challenge owing to the physicochemical characteristics of antitumor agents and the significant physiological barriers that drugs have to overcome to reach their site of action. In the particular case of CRC, the oral route can be used with a dual aim: (1) to achieve systemic absorption, so that the drug can access tumor cells from the bloodstream, and (2) to locally treat tumor cells in the affected areas of the colon and rectum (Fig. 36.1).

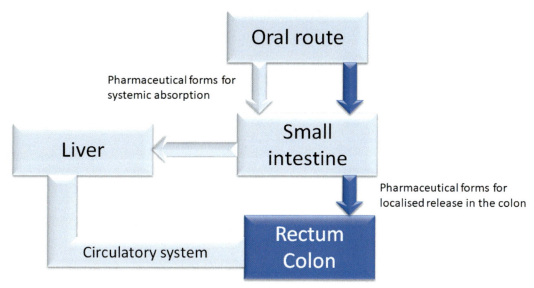

FIG. 36.1 Schematic diagram of the route taken by a drug to access tumor cells in the colon or rectum when administered in a systemic absorption dosage form (in light gray) and in a localized colon delivery dosage form (in dark gray).

Systemic treatment

Antitumor drugs have a low oral bioavailability owing to various causes, such as low water solubility, poor intestinal permeability, or intense presystemic metabolization in the intestinal wall and liver. As a result, only 5%–20% of the administered dose reaches the bloodstream. For 5-FU, early oral products yielded very erratic intestinal absorption owing to interindividual differences in the levels of dihydropyrimidine dehydrogenase, an enzyme that catabolizes the drug in the intestinal mucosa. This limitation was overcome by administering the drug together with an enzyme inhibitor or by substituting it with a prodrug. Tegafur (5-FU prodrug) administered orally with uracil, which inhibits dihydropyrimidine dehydrogenase, yields similar response times and median survival levels to those yielded by parenteral administration of 5-FU.

Capecitabine, another 5-FU prodrug, is absorbed and remain unchanged through the intestinal wall and, once absorbed, is transformed into an active molecule by a series of successive enzymatic conversions. Oral administration of twice-daily doses of 1250 mg/m^2 of capecitabine (14 days then every 3 weeks) yields a therapeutic response similar to that achieved with the standard monthly bolus intravenous regimen of 5-FU associated with LV, but with fewer side effects.

Trifluridine, an antineoplastic thymidine nucleoside analog, is rapidly degraded by thymidine phosphorylase (TPase) and undergoes a strong first-pass effect when administered orally. The combination of trifluridine with the TPase inhibitor, tipiracil, which is marketed in Spain as Lonsurf in tablet form, significantly increases its oral bioavailability. This combination is effective against 5-FU-sensitive and resistant colorectal cancer cell lines.

Fig. 36.2 summarizes the main technological strategies that have been developed to overcome the obstacles limiting the oral bioavailability of drugs used in the treatment of CRC.

Physicochemical properties of antitumor drugs

For drugs that are not affected by biotransformation processes in the intestinal wall, the likelihood that the administered dose will be absorbed depends on the solubility and permeability of the drug.[11] This is because the flux of the drug through the intestinal wall is directly proportional to the concentration of the drug in the intestinal lumen and the permeability coefficient. Based on these two variables, the Biopharmaceutical Classification System for orally administered drugs has been established (Fig. 36.3).

Class I drugs, which are characterized by their high solubility and permeability, are the most suitable drugs for oral absorption. Capecitabine could be considered a class I drug, as its oral absorption is complete if it is protected from contact with the acidic environment of the stomach in which it is degraded. This instability in an acidic environment is avoided by formulating the drug in film-coated tablets.

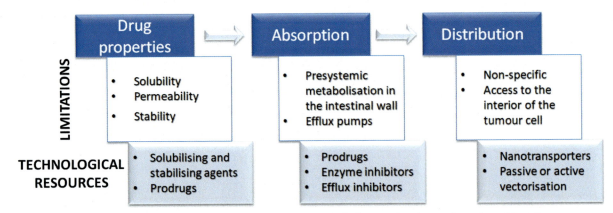

FIG. 36.2 Challenges in systemic oral delivery of antitumor drugs and technological approaches that can be used to address them.

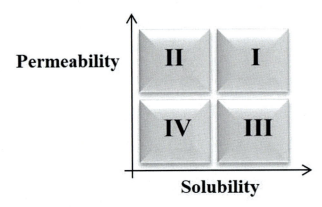

FIG. 36.3 Biopharmaceutical classification system for orally administered drugs.

The vast majority of antitumor agents do not fit under class I. For example, regorafenib, tamoxifen, rubitecan, sorafenib, and gefitinib are categorized under class II, that is, their low solubility or dissolution rate prevents the dose from being fully dissolved in the volume of gastrointestinal media available in the period that the formulation remains in the digestive tract. As a result, complete absorption will not occur unless technological means are applied to increase solubility, such as preparation of solid dispersions.[12] Oxaliplatin, cyclophosphamide, anastrozole, letrozole, doxorubicin, and methotrexate have low permeability and have difficulty crossing biological barriers owing to their high hydrophilicity and are therefore categorized under class III. It is also common for class II and III drugs to serve as substrates for efflux pumps, which makes it difficult for drugs to accumulate inside cells.[11]

Absorption

Efflux is a cellular protection mechanism involving membrane transporter proteins (efflux pumps) that expel from the interior of cell substances that the cells recognize as harmful.[13] Transport occurs against a concentration gradient, i.e., from the inside (low concentration) to the outside of the cell (higher concentration), and therefore requires an energy input. Efflux pumps belong to the ABC family of transporters, so named because they are encoded by the ATP Binding Cassette (ABC) genes. The first efflux pump identified as responsible for incomplete drug absorption was permeability glycoprotein (P-gp). P-gp is identified as the ABCB1 transporter, although it is also called PGY1 or multidrug resistance protein 1, because it is involved in multidrug resistance mechanisms. Tumor cells overexpress several efflux pumps in the membrane, making them less sensitive to antitumor drugs than healthy cells.

Similar with other efflux pumps, P-gp is widely distributed in (1) the apical side of healthy enterocytes; (2) the canalicular membrane of hepatocytes; and (3) certain tissues, such as the brain, lymphocytes, and testis. As a consequence, drugs that are P-gp substrates exhibit low absorption, high elimination into bile and urine, and impaired distribution to particularly sensitive tissues.

Efflux can be reduced by coadministering drugs with pump-inhibiting substances that act by any of the following mechanisms: alteration of membrane fluidity, interference with energy generation (ATPase inhibition or ATP depletion), blockade of the drug binding site, or decreased pump expression. For reducing efflux pump activity and increasing intestinal wall permeability, P-gp inhibitors, such as verapamil or cyclosporin A, can be used, which can increase the oral bioavailability of antitumor agents, such as paclitaxel or docetaxel, by up to 10-fold. Newer-generation inhibitors, such as elacridar, zosuquidar, and tariquidar, have the advantage of lacking pharmacological activity but may cause suppression of the immune system. In addition to playing a technologically relevant role in facilitating the incorporation of the drug into the dosage form, certain excipients attenuate the effect of efflux mechanisms. This is the case for some natural polymers, such as dextrans, flavonoids (quercetin), anionic polysaccharides (xanthan gum, gellan gum, or alginates), or polyphenols (curcumin or silymarin), and certain synthetic polymers with polyethylene glycol (PEG) or poly(ethylene oxide) (PEO) chains or with thiol groups in their structure, which are effective when they reach concentrations of more than 0.05% in the intestinal contents.[13]

PEO-poly(propylene oxide) (PPO) block copolymers from the poloxamer family (Pluronic or Lutrol) are multifunctional excipients that can act simultaneously as solubilizing and stabilizing agents for drugs, as nanotransporters capable of targeting them to tumor tissues, and as inhibitors of efflux pumps. The PPO block is more hydrophobic than PEO; thus, when it exceeds a certain concentration (critical micellar concentration), it tends to minimize its contact with the aqueous medium, associating with the PPO blocks of other polymeric chains. This phenomenon leads to the formation of supramolecular aggregates, which are called polymeric micelles because of their similarity to the aggregates (micelles) formed by conventional low molecular weight surfactants. Polymeric micelles have a hydrophobic core-hydrophilic corona structure and can accommodate large amounts of low-solubility drugs in the core, protecting them from agents that may cause their destabilization, while the corona facilitates the solubilization of the aggregate in the aqueous medium and inhibits its recognition by the mononuclear phagocytic system. Unlike conventional surfactant micelles that disintegrate very easily, polymeric micelles form at lower concentrations and are much more stable against dilution in biological fluids. These properties make it possible to formulate effective doses of low-solubility drugs in small-volume aqueous liquid formulations. Once the micellar system is administered orally, the micelles progressively disassemble, and the drug is released in the intestine. The individualized polymer chains (unimers) have the ability to inhibit efflux pumps in both healthy and tumor tissues, even at very low concentrations. Micelles that remain assembled (with the drug encapsulated) in the intestinal lumen can be taken up as such via intestinal enterocytes (by an endocytosis mechanism that bypasses efflux pumps) or M-cells (specialized phagocytic enterocytes) found in the lymph nodes of the ileum (Peyer's patches) that can translocate nanoparticles from the gastrointestinal tract to the systemic circulation.

Cyclodextrins (CDs) are also useful as solubilizing agents and promoters of oral absorption in solid (tablets or capsules) or liquid (solutions or suspensions) dosage forms. They are cyclic oligosaccharides consisting of rings of six (αCD), seven (βCD), eight (γCD), or even more glucose units, adopting a hollow truncated cone shape with a maximum diameter between 0.47 and 0.83 nm and a height of 0.79 nm. The outer face is hydrophilic, while the inner face is hydrophobic. CDs can form reversible inclusion complexes with a wide variety of drugs capable of fully or partially penetrating their cavity. Because of their large size and hydrophilicity, CDs and CD-drug complexes do not cross biological membranes but serve as reservoirs that provide a very high apparent drug concentration in the intestinal lumen, facilitating the passage of free drug via diffusion. As the drug passes through the intestinal wall, progressive decomplexation occurs. CDs can protect drugs by preventing their degradation at the site of absorption, while acting as absorption promoters and removing lipophilic components from membranes. In addition, complexes can be used to mask unpleasant organoleptic properties and prevent the incidence of local side effects (mainly gastrointestinal irritation).[11]

Self-emulsifying solid systems (SMEDDSs) based on mixtures of surfactants from the sorbitan and polyoxyethylene fatty acid ester group (known as Tween), polyglycolized glycerides (Labrasol) or poloxamers, with oily excipients, usually modified or hydrolyzed vegetable oils, are also of great interest. SMEDDSs can be formulated and packaged in a variety of forms, including soft and hard gelatin capsules and pellets. Under the temperature and agitation conditions of the digestive tract, SMEDDSs result in the spontaneous formation of microemulsions that increase drug solubility and promote drug absorption.[14]

Formulation strategies that promote the passage of hydrophilic antitumor drugs, such as oxaliplatin, through the intestinal wall are required to improve the oral absorption of these drugs. Recently, the formation of complexes with a lipophilic cationic derivative of a bile acid (DCK) has been tested. This derivative can utilize specific transporters at the intestinal level, which facilitates the passage of the complex and increases the bioavailability of the drug. In preclinical trials, the coformulation of the oxaliplatin-DCK complex and 5-FU in a multiple water-in-oil-in-water nanoemulsion has been shown to increase the oral absorption of both drugs and improve therapeutic efficacy.[15]

Distribution and elimination

Substances that penetrate through the intestinal wall are carried by mesenteric blood flow to the portal vein, from where they enter the liver. Liver and intestinal enzymes, such as glucuronosyl- and sulfo-transferases and enzymes of the cytochrome P450 family, may be involved in degradative processes that affect the drug before it enters the general circulation (first-pass effect). Presystemic losses associated with the activity of efflux pumps and cytochrome P450 enzymes can be greatly reduced by administering substances that inhibit them, e.g., paclitaxel with cyclosporin A or docetaxel with ritonavir, simultaneously with the antitumor drug. This practice has the limitation of suppressing the immune system. The drug that "escapes" the first-pass effect is distributed to the tissues in quantity and at a rate dependent on the degree of irrigation and tissue affinity and thus gains undifferentiated access to healthy and tumor cells in the target tissue. As efflux pumps are overexpressed in tumor cells, access of the drug into the cells to exert its therapeutic effect is more difficult than in healthy cells, which negatively affects the efficacy/safety profile. Efflux pumps also facilitate renal excretion. For the purpose of increasing the residence time of the drug in the bloodstream and selectively target (vectorize) it to tumor cells, facilitating its penetration, different strategies have been developed based on covalent binding (conjugation) of the drug to PEG chains (PEGylation) or incorporation of the drug into nanocarriers.

PEGylation

PEG is a highly hydrophilic synthetic polymer, consisting of repeating ethylene oxide units, which does not give rise to an antigenic response (although repeated administrations may result in the formation of specific antibodies). It is approved by regulatory agencies for use as an excipient for oral, parenteral (including intravenous), and topical dosage forms. The immediate consequence of PEGylation is a marked increase in the solubility of the drug, which facilitates its oral absorption.[16] In addition, PEG conjugation results in increased residence time in the bloodstream, as the conjugated drug cannot be extravasated through the endothelium of healthy blood vessels. This also makes it more difficult for the drug to reach healthy tissue cells. In tumors, blood vessels in the formation phase have poorly aligned endothelial cells, leaving gaps of 100–1000 nm between them, through which macromolecules and nanocarriers can be extravasated. This high permeability of the vascular wall and the slowness with which macromolecules leave the tumor tissue result in an increased permeability and retention known as the enhanced permeability and retention (EPR) effect, which manifests itself in all tumors, except hypovascular tumors (pancreas and prostate).

The conjugation of a drug to PEGs can be accomplished by establishing permanent bonds or biodegradable bonds. The former leads to the resulting molecule having to be considered, for regulatory purposes, as a new chemical entity different from the original drug, which must pass the corresponding clinical trials before it can be used therapeutically. For this application, PEGs with a molecular weight of less than 1000 Da are used, so that the polymer chain does not cause steric hindrance affecting the activity of the original drug. Conjugation through biodegradable linkages is in the field of prodrugs, and the PEGylated drug gives rise in the body to the original molecule by the effect of a chemical or enzymatic reaction. In this case, high-molecular-weight PEGs (1000–60,000 Da) can be used to achieve more significant increases in solubility and result in a longer residence time in the bloodstream and significant improvements in the biodistribution profile. The PEG chain must not be cleaved before the conjugate reaches the target tissue, so that the drug can be released into the tumor cells, thus mitigating adverse effects and improving treatment efficacy. PEGs with a branched structure contain numerous hydroxyl groups to which drug molecules can bind to form bulky conjugates, even if the molecular weight of the PEG is not excessively high.

A PEGylated irinotecan conjugate (NKTR-102, Nektar, USA) is currently in clinical phase III/II for second-line treatment of CRC, which allows a 300-fold increase in the concentration of drug in the tumor compared to the levels achieved by administering it free and prolongs the duration of effects by up to 50 days. Other doxorubicin conjugates, such as camptothecin, 7-ethyl-10-hydroxy-camptothecin (SN38), docetaxel, cisplatin, gemcitabine, wortmannin, pemetrexed, lamellarin-D, methotrexate, or gambogic acid, have also been evaluated or are in various stages of development for the treatment of solid tumors.[16]

The EPR effect is mostly seen in small tumors, and its efficacy is compromised by substances that reduce capillary permeability and high intratumoral pressure. As a result, assessing the opportunities for successful passive targeting requires a thorough understanding of the pathophysiology of the tumor to be treated. For the purpose of a more effective targeting, components that act as active targeting elements are incorporated. Certain markers that are overexpressed on the surface of tumor cells can be used as targets for recognition. The recognition component that is incorporated into the nanocarrier can consist of different structures, such as peptides, antibodies, aptamers, or small ligands (e.g., folic acid or sialic acid).[2]

Nanostructures

Nanoparticulate systems can improve the oral bioavailability of antitumor drugs via several mechanisms: increasing solubility and dissolution rate, promoting mucoadhesion, opening tight junctions between cells of the intestinal epithelium, and favoring uptake of transporters by enterocytes through receptor-mediated endocytosis and transcytosis, phagocytosis by M cells of Peyer's patches and other lymphoid tissues associated with the intestinal mucosa, and lymphatic uptake by a mechanism similar to that involved in the uptake of chylomicrons by enterocytes (in the case of lipid transporters).[17]

Nanocrystals

Highly efficient technologies to prepare crystalline nanoparticles are currently available, which can significantly increase the bioavailability of drugs by improving their solubility and dissolution rate. This can start from larger crystals that undergo fragmentation processes in special milling equipment (top-down approach). Nanocrystals can also be formed from drug solutions that are subjected to appropriate conditions to promote crystallization into very small structures (bottom-up approach).

The performance of nanocrystals can be improved by incorporating mucoadhesive agents or functional excipients that act as efflux pump inhibitors. For example, nanocrystals of paclitaxel alone or in combination with camptothecin surface stabilized with poloxamer (Pluronic F127) have been prepared, resulting in significant improvements in antitumor efficacy. For the purpose of mitigating the side effects on the gastrointestinal tract caused by the drug when administered free in solution, nanocrystals coated with hydrophobic excipients and absorption promoters are being evaluated, so that the intact nanocrystals can be absorbed by mechanisms that are specific to nanotransporters.[18]

Nanotransporters

Drugs can be encapsulated in polymeric systems between 100 and 500 nm in size to provide them with a physically stable shell that protects them from degradative processes in the gastrointestinal tract. The ability of the nanocarriers or "nanomedicines" to pass through the intestinal mucosa depends largely on their size and surface properties. In general, those with a size of approximately 200 nm and a nonionic or an anionic surface are the ones that diffuse best through the mucus. In addition, they can be "decorated" with ligands that promote interaction with the epithelium and their penetration via para- or trans-cellular pathways. Nanotransporters prevent premature drug release into the bloodstream and vector the drug to its site of action by EPR and active recognition mechanisms. A relatively large number of biocompatible and biodegradable or bioeliminable polymers are available, and the diversity of nanometric structures to which they can give rise makes this approach highly versatile in terms of the nature of the drugs to which the technology can be applied and the therapeutic targets that can be achieved.[19]

For a nanocarrier to be effective, it must remain for a long time in the bloodstream without being recognized by the mononuclear phagocytic system. A particle with a hydrophilic coating that makes it undetectable (stealth) can circulate for several hours in the bloodstream, making it more likely to gain access to tumor tissue. PEG derivatives are often used to prepare stealth particles, and the longer the chain length, the more effective they are. When administered intravenously, silent systems reach concentrations in the tumor that are 10 to 2000 times higher than plasma levels and 10 times higher than those in healthy tissues. Retention in the tumor can last for weeks or even months.

To ensure that the nanocarriers reach their destination without losing significant cargo, they are designed with a matrix or shell that is very stable in any environment other than the target site. In or around tumor cells, the nanocarriers must disaggregate to deliver the drug at a rate fast enough to allow therapeutic levels to be reached. In addition to being ineffective, subtherapeutic levels can lead to the development of resistance phenomena. For ensuring that the release is triggered in the tumor site, it is useful to incorporate components (polymers, lipids, or metal particles) into the nanocarriers that are sensitive to the physiological or pathological conditions of the target site (pH and redox potential) or to external variables (heat, light radiation, and magnetic field). Activatable (also called smart) nanocarriers rapidly release the drug when they experience the effect of a local or external stimulus.

Several types of nanocarriers have been tested for oral delivery of antitumor agents:

(a) Polymeric nanocarriers, including solid polymeric nanoparticles, which have high physical stability and in which the drug is embedded in a matrix structure (nanospheres) or inside a reservoir (nanocapsules). Other useful nanocarriers are polymeric micelles with a hydrophobic core-hydrophilic corona structure and polymersomes with a vesicular structure consisting of layers of polymers arranged in a palisade, which result from the spontaneous self-association of amphiphilic copolymers and which allow combining functionalities, such as hydrosolubilization, vectorization, and efflux pump inhibition. For the latter function, some varieties of Pluronic already present in various commercially available products are useful, as are polymers that respond to changes in pH, temperature, or redox potential characteristics of tumor cells. Dendrimers or multifunctional branched polymers that can conjugate several antitumor drug molecules through labile bonds are also included in this group. Dendrimers facilitate the passage of drugs across various epithelial barriers to specifically deliver them via enzymatic hydrolysis of covalent bonds into tumor cells.[20]

(b) Lipid nanocarriers, including solid lipid nanoparticles composed of fatty acids, fatty acid esters or waxes, and liposomes, which are vesicular systems with concentric phospholipid bilayers. For example, docetaxel-loaded lipid nanoparticles coated with a conjugate of glycolic acid and chondroitin sulfate show increased uptake in the ileum via interaction with bile acid transporters and result in sustained plasma docetaxel levels over 24 h.[21]

(c) Hybrid nanocarriers, which combine polymers or lipids with magnetite or gold particles or quantum dots. These systems can allow diagnosis, treatment, and monitoring of treatment efficacy and are therefore called theranostics.[22-24] The presence of metal particles allows their localization to be monitored using physical procedures (e.g., radiography, confocal imaging, or NMR). The therapeutic effect can be achieved by incorporating drugs to be released once the transporter reaches the tumor cells or by generating local hyperthermia through the application of near-infrared (NIR) laser light or oscillating magnetic fields. Although there is very limited information available on the efficacy of oral delivery of theranostic systems, there are approved products and a large number in the clinical phase that are useful parenterally.

In the specific case of CRC, PEGylated hyaluronic acid nanoparticles containing irinotecan and a dye that emits fluorescence when excited by a radiation source in the NIR have been developed.[25] When injected intravenously, they passively accumulate in tumor tissue, and the effect of the drug on tumor volume can be monitored in real time via fluorescence imaging analysis. Incorporation of ligands on the surface of the particles that recognize specific CRC tumor markers allows for more effective active targeting.

Magnetic core nanoparticles incorporating drugs and antibodies can accumulate in the tumor through a dual vectorization mechanism involving specific recognition and targeting promoted by a continuous external magnetic field. Once localized in the tumor, the application of an oscillating magnetic field causes local hyperthermia that contributes to the destruction of tumor cells, which are more sensitive to heat than healthy cells, increasing the efficacy of the treatment.[26] It is also possible to coat the magnetic cores with a gold film that increases the stability of the magnetite and makes the system sensitive not only to the magnetic field but also to NIR radiation.

Drug dosage forms for colonic delivery

Colon-specific drug delivery has great potential in the local treatment of diseases of the large intestine, such as ulcerative colitis, Crohn's disease, amebiasis, and colon cancer, and in the systemic delivery of peptides and proteins. The aim when designing oral dosage forms for colon-specific delivery is to provide effective protection for labile drugs during the transit of the drug through the gastrointestinal tract, avoiding exposure to adverse agents in the stomach and small intestine, and to ensure delivery once the drug reaches the colon.

Colon-specific drug delivery may be useful for the treatment of stage II, III, and IV CRCs after surgery, combining local tissue exposure with systemic absorption of drugs, such as 5-FU, oxaliplatin, capecitabine, and irinotecan. In the particular case of 5-FU, bioactivation to 5-fluoro-2′-deoxyuridine, which disrupts RNA synthesis and interferes with thymidylate synthase activity and thus kills cancer cells, is more intense in colonic tumor tissue than in healthy tissue (as opposed to bioinactivation to dihydro-5-fluoruracil). In the case of oxaliplatin, local treatment in stage I disease minimizes systemic exposure, reduces side effects, and allows reaching of higher levels in the mucosal layers.

The use of colon-targeted dosage forms that also increase the permeability of the intestinal epithelium allows systemic absorption of oxaliplatin in the treatment of cancer that has spread to deeper tissues or lymph nodes (stages II, III, and IV after surgery). In addition to antitumor agents, oral administration of nonsteroidal antiinflammatory drugs (NSAIDs) can reduce the incidence of colonic polyps and decrease CRC mortality. NSAIDs that have been shown to be useful in this regard include acetylsalicylic acid, naproxen, celecoxib, and sulindac.[27] The incorporation of these drugs into colon-specific delivery forms can significantly reduce the local side effects that occur in the upper digestive tract. Formulations developed to treat inflammatory processes at the colonic level could also be useful for the prevention and local treatment of CRC.

Different strategies have been implemented to design dosage forms for colon-specific drug release based on the peculiarities of this area of the intestine, in particular the transit time, pH, and metabolic activity of the more than 400 species of microorganisms that make up the microbiota.[28]

Time-dependent release

The time that elapses between the oral administration of a drug and its arrival in the colon depends on the rate of gastric emptying and the transit time through the small intestine. While the passage through the small intestine takes place in a relatively constant time (3–4 h), the gastric emptying time is highly variable owing to numerous causes that result in significant inter- and intraindividual differences. This indicates that the arrival in the colon may require 4–12 h. The transit time through the colon is also highly variable (1–60 h).

For enabling the drug to be released only when the formulation reaches the colon, tablets and capsules may be coated with a polymer that is impermeable under acidic pH conditions (enteric coating) and dissolves upon reaching the small intestine. The system is designed in such a manner that the drug is released within 5 h of the disappearance of the enteric coating.

pH-dependent release

There is a significant pH gradient in the gastrointestinal tract. The pH of the empty stomach varies between 1.5 and 3 and increases to 4 or 5 in the presence of food. In the duodenum, the pH ranges between 1.7 and 4.3 under fast conditions and between 3 and 6 with food; in the jejunum, between 6 and 7.5; and in the colon, between 6.4 and 7.

Various solid dosage forms have been developed, such as tablets, capsules, pellets, microspheres, or nanoparticles, which release the drug under certain pH conditions. Most of these formulations use acrylic polymers with pH-dependent solubility, which are incorporated into coatings. The impossibility of accurately predicting the pH in different areas of the digestive tract means that drug release may be initiated in the small intestine. Nevertheless, medicines showing pH-dependent release are already marketed for the treatment of ulcerative colitis with antiinflammatory agents. Osmotic systems coated with polymer films that dissolve at pH greater than 7 (e.g., OROS-CT system) prevent prematurely triggered release and can provide constant release rates in the colon for 12 h.

Enzyme-activated release

There is an abundance of aerobic and anaerobic microbiota in the digestive tract. Bacteria in the colon are primarily anaerobic and engage in an intense metabolic activity involving reactions as varied as hydrolysis of glycosides, esters, amides, nitrates, and sulfamates; reduction of C=C bonds and azo, aldehyde, ketone, alcohol, and nitrogen oxide groups; dehydroxylation; decarboxylation; dealkylation; dehalogenation; deamination; heterocyclic ring cleavage; acetylation; and esterification.

Reduction and enzymatic hydrolysis are particularly useful for activating release at colon. Prodrug strategies involve conjugating the active substance to a water-soluble polymer to produce a molecule that is not absorbed in the stomach or small intestine. In the colon, the conjugation bond is broken by bacterial enzymes, leaving the drug free to trigger its local effect and to be absorbed. One of the most tested polymers for conjugation is poly(N-(2-hydroxypropyl)methacrylamide), which can bind via spacers with azo or peptide groups to various antitumor drugs. These conjugates are highly stable in the gastric contents but rapidly release the drug upon reaching the colon, favoring its accumulation in the mucosa.[29]

Another approach is incorporation into the matrix or in the coating of solid dosage forms (tablets or pellets) of polymers that are degradable only by enzymes present in the colon. Studies have investigated the usefulness of cross-linked networks, such as hydrogel particles, in which the drug of interest is encapsulated so that it remains retained inside until azoreductases break the chains or the bonds between the polymer chains.

Along with synthetic polymers, numerous polysaccharides that are not digested in the stomach or small intestine but serve as substrates for enzymatic degradation processes in the colon have received considerable attention. Pectins, guar gums, chitosan, inulin, alginate, amylose, and chondroitin have been proven to be useful. An important advantage of these natural polymers is their excellent safety profile.

To improve the specificity of release, colonic enzyme-responsive materials have been combined with polymers with pH-dependent solubility and time-dependent release systems. For example, pellets and pectin microparticles cross-linked with calcium chloride and coated with enteric polymers are useful for colonic 5-FU binding. Another advanced system is CODES™, which consists of a core of the drug, polysaccharides (lactulose), and other excipients, coated with a film of an acrylic polymer soluble at pH below 5 and a second outermost enteric film. The dosage form passes through the stomach unchanged, and the outer coating dissolves in the small intestine. This allows the inner coating to swell and become permeable. Once in the colon, the polysaccharides are enzymatically degraded, and the acidic degradation products cause the pH to decrease, which dissolves the inner film, allowing the drug to be released.[30]

Encapsulation of oxaliplatin in liposomes decorated with folic acid and incorporated into alginate particles coated with Eudragit S-100 has also been tested. Liposome delivery occurs at the level of the colon when the alginate is exposed to enzymes characteristic of this area, facilitating direct interaction with tumor cells.[31]

In summary, the available information on colon-specific drug delivery for the treatment of CRC, which is the result of the activity developed in various research centers, shows that this approach has considerable clinical potential and that it could be implemented using relatively simple dosage forms that are easily adaptable to the requirements of industrial production of medicines. In any case, the implementation of this mode of administration requires an additional effort focused on clinical trials to confirm the efficacy and safety of the treatments.

References

1. Van Cutsem E, Cervantes A, Adam R, et al. ESMO consensus guidelines for the management of patients with metastatic colorectal cancer. *Ann Oncol* 2016;**27**(8):1386–422. https://doi.org/10.1093/annonc/mdw235.
2. Matos AI, Carreira B, Peres C, et al. Nanotechnology is an important strategy for combinational innovative chemo-immunotherapies against colorectal cancer. *J Control Release* 2019;**307**:108–38. https://doi.org/10.1016/j.jconrel.2019.06.017.
3. Gómez-España MA, Gallego J, González-Flores E, et al. SEOM clinical guidelines for diagnosis and treatment of metastatic colorectal cancer (2018). *Clin Transl Oncol* 2019;**21**(1):46–54. https://doi.org/10.1007/s12094-018-02002-w.
4. *About the NCCN clinical practice guidelines in oncology (NCCN guidelines®)*. Accessed February 17, 2021 https://www.nccn.org/professionals/default.aspx.
5. Kim SY, Kim TW. Current challenges in the implementation of precision oncology for the management of metastatic colorectal cancer. *ESMO Open* 2020;**5**(2). https://doi.org/10.1136/esmoopen-2019-000634, e000634.
6. Shahiwala A, Qawoogha SS, Tambuwala MM. Oral adjuvant therapy for colorectal cancer: recent developments and future targets. *Ther Deliv* 2019;**10**(10):659–69. https://doi.org/10.4155/tde-2019-0067.
7. Aurilio G, Gori S, Nolè F, et al. Oral chemotherapy and patient perspective in solid tumors: a national survey by the Italian association of medical oncology. *Tumori* 2016;**102**(1):108–13. https://doi.org/10.5301/tj.5000383.
8. Bassan F, Peter F, Houbre B, et al. Adherence to oral antineoplastic agents by cancer patients: definition and literature review. *Eur J Cancer Care* 2014;**23**(1):22–35. https://doi.org/10.1111/ecc.12124.
9. Siden R, Modlin J, Lee-Gabel L, Redic KA. Handout for research subjects receiving investigational oral chemotherapy. *Am J Health Syst Pharm* 2019;**76**(24):2009–12. https://doi.org/10.1093/ajhp/zxz239.
10. Barillet M, Prevost V, Joly F, Clarisse B. Oral antineoplastic agents: how do we care about adherence? *Br J Clin Pharmacol* 2015;**80**(6):1289–302. https://doi.org/10.1111/bcp.12734.
11. Thanki K, Gangwal RP, Sangamwar AT, Jain S. Oral delivery of anticancer drugs: challenges and opportunities. *J Control Release* 2013;**170**(1):15–40. https://doi.org/10.1016/j.jconrel.2013.04.020.
12. Sawicki E, Schellens JHM, Beijnen JH, Nuijen B. Inventory of oral anticancer agents: pharmaceutical formulation aspects with focus on the solid dispersion technique. *Cancer Treat Rev* 2016;**50**:247–63. https://doi.org/10.1016/j.ctrv.2016.09.012.
13. Alvarez-Lorenzo C, Sosnik A, Concheiro A. PEO-PPO block copolymers for passive micellar targeting and overcoming multidrug resistance in cancer therapy. *Curr Drug Targets* 2011;**12**(8):1112–30. https://doi.org/10.2174/138945011795906615.
14. Zhang L, Zhu W, Yang C, et al. A novel folate-modified self-microemulsifying drug delivery system of curcumin for colon targeting. *Int J Nanomedicine* 2012;**7**:151–62. https://doi.org/10.2147/IJN.S27639.
15. Pangeni R, Choi SW, Jeon O-C, Byun Y, Park JW. Multiple nanoemulsion system for an oral combinational delivery of oxaliplatin and 5-fluorouracil: preparation and in vivo evaluation. *Int J Nanomedicine* 2016;**11**:6379–99. https://doi.org/10.2147/ijn.s121114.
16. Li W, Zhan P, De Clercq E, Lou H, Liu X. Current drug research on PEGylation with small molecular agents. *Prog Polym Sci* 2013;**38**(3–4):421–44. https://doi.org/10.1016/j.progpolymsci.2012.07.006.
17. Etheridge ML, Campbell SA, Erdman AG, Haynes CL, Wolf SM, McCullough J. The big picture on nanomedicine: the state of investigational and approved nanomedicine products. *Nanomed Nanotechnol Biol Med* 2013;**9**(1):1–14. https://doi.org/10.1016/j.nano.2012.05.013.
18. Fan M, Geng S, Liu Y, et al. Nanocrystal technology as a strategy to improve drug bioavailability and antitumor efficacy for the cancer treatment. *Curr Pharm Des* 2018;**24**(21):2416–24. https://doi.org/10.2174/1381612824666180515154109.
19. You X, Kang Y, Hollett G, et al. Polymeric nanoparticles for colon cancer therapy: overview and perspectives. *J Mater Chem B* 2016;**4**(48):7779–92. https://doi.org/10.1039/c6tb01925k.
20. Sadekar S, Ghandehari H. Transepithelial transport and toxicity of PAMAM dendrimers: implications for oral drug delivery. *Adv Drug Deliv Rev* 2012;**64**(6):571–88. https://doi.org/10.1016/j.addr.2011.09.010.
21. Kim KS, Youn YS, Bae YH. Immune-triggered cancer treatment by intestinal lymphatic delivery of docetaxel-loaded nanoparticle. *J Control Release* 2019;**311–312**:85–95. https://doi.org/10.1016/j.jconrel.2019.08.027.
22. Carvalho MR, Reis RL, Oliveira JM. Dendrimer nanoparticles for colorectal cancer applications. *J Mater Chem B* 2020;**8**(6):1128–38. https://doi.org/10.1039/c9tb02289a.
23. Lacava Z, da Paz C, Santos A, et al. Anti-CEA loaded maghemite nanoparticles as a theragnostic device for colorectal cancer. *Int J Nanomedicine* 2012;5271. https://doi.org/10.2147/ijn.s32139. Published online.

24. Rampado R, Crotti S, Caliceti P, Pucciarelli S, Agostini M. Nanovectors design for theranostic applications in colorectal cancer. *J Oncol* 2019;**2019**: 1–27. https://doi.org/10.1155/2019/2740923.
25. Choi KY, Jeon EJ, Yoon HY, et al. Theranostic nanoparticles based on PEGylated hyaluronic acid for the diagnosis, therapy and monitoring of colon cancer. *Biomaterials* 2012;**33**(26):6186–93. https://doi.org/10.1016/j.biomaterials.2012.05.029.
26. Lima SAC, Costa Lima SA, Gaspar A, Reis S, Durães L. Multifunctional nanospheres for co-delivery of methotrexate and mild hyperthermia to colon cancer cells. *Mater Sci Eng C* 2017;**75**:1420–6. https://doi.org/10.1016/j.msec.2017.03.049.
27. Mohammed A, Yarla NS, Madka V, Rao CV. Clinically relevant anti-inflammatory agents for chemoprevention of colorectal cancer: new perspectives. *Int J Mol Sci* 2018;**19**(8):2332. https://doi.org/10.3390/ijms19082332.
28. Bak A, Ashford M, Brayden DJ. Local delivery of macromolecules to treat diseases associated with the colon. *Adv Drug Deliv Rev* 2018;**136–137**: 2–27. https://doi.org/10.1016/j.addr.2018.10.009.
29. Gao S-Q, Lu Z-R, Petri B, Kopečková P, Kopeček J. Colon-specific 9-aminocamptothecin-HPMA copolymer conjugates containing a 1,6-elimination spacer. *J Control Release* 2006;**110**(2):323–31. https://doi.org/10.1016/j.jconrel.2005.10.004.
30. Krishnaiah YSR, Khan MA. Strategies of targeting oral drug delivery systems to the colon and their potential use for the treatment of colorectal cancer. *Pharm Dev Technol* 2012;**17**(5):521–40. https://doi.org/10.3109/10837450.2012.696268.
31. Bansal D, Gulbake A, Tiwari J, Jain SK. Development of liposomes entrapped in alginate beads for the treatment of colorectal cancer. *Int J Biol Macromol* 2016;**82**:687–95. https://doi.org/10.1016/j.ijbiomac.2015.09.052.

Chapter 37

Neoadjuvant, adjuvant, and intraoperative radiotherapy for rectal cancer

Ana María Carballo Castro[a], Paula Peleteiro Higuero[a], Begoña Taboada Valladares[a], Patricia Calvo Crespo[a], Jesús Paredes Cotoré[b], Roberto García Figueiras[c], and Antonio Gómez Caamaño[a]

[a]*Department of Radiation Oncology, Complexo Hospitalario Universitario de Santiago de Compostela, A Coruña, Spain,* [b]*Department of Surgery, Complexo Hospitalario Universitario de Santiago de Compostela, A Coruña, Spain,* [c]*Department of Radiology, Complexo Hospitalario Universitario de Santiago de Compostela, A Coruña, Spain*

Introduction

The efficacy of surgery in rectal cancer is conditioned by two fundamental factors: the absence of serosa in the lower portion of the rectum and the difficulty in obtaining wide resection margins owing to the presence of the pelvic bone structure; thus, the fundamental problem when it comes to rectal cancer lies in optimal surgery of the mesorectum.[1] Several studies have shown that macroscopic examination of the resection specimen is associated with microscopic radial resection margin, local recurrence, distant metastasis, and overall survival.[2] Therefore an intact mesorectum reflects the quality of surgery and predicts the patient's prognosis. In a review of the literature including surgical specimens from 17,500 patients with rectal cancer, this margin is presented as one of the main predictors of local recurrence, distant metastasis, and survival, mainly in patients receiving neoadjuvant treatment.[2]

In contrast to colon cancer where the predominant pattern of failure is distant metastases, in patients with rectal cancer, the first site of relapse after surgery is equally distributed locally and distantly. Of particular clinical concern is the local recurrence rate ranging from 10% to 45% in the surgical arm of randomized multicenter trials.[3] The risk of recurrence is related to the depth of tumor extension into the bowel wall and the presence of lymph node involvement, ranging from 5% to 10% in stage I to 30%–50% in stage III.[3] Local relapse substantially influences the prognosis, with 5-year survival decreasing from 85% in patients without recurrence to 20% in patients with recurrence.[3] Conversely, the development of local recurrence causes profound morbidity with severe and disabling symptoms that compromise the quality of life and are difficult to manage therapeutically (pain, perforation, obstruction, or fistulas).

The management of rectal cancer requires a multidisciplinary approach that allows individualization of treatment and the selection of patients who are candidates for combined therapies based on the location, stage, and resectability of the tumor. This strategy associated with the optimization of surgery, chemotherapy, and radiotherapy has allowed local recurrence to decrease from 50 to less than 10% and overall survival to increase from 50 to nearly 80% in patients with non-metastatic rectal cancer in recent decades.[1,4]

Adjuvant treatment

With the aim of improving both local control and survival, several North American cooperative groups performed multiple randomized trials in the 1980s and 1990s, investigating the role of radiotherapy and chemotherapy in adjuvant treatment of rectal cancer.[5–10] Of all these studies, the GITSG 7175 (radiochemotherapy vs. monotherapy) and MAYO/NCCTG 79-47-51 (radiochemotherapy vs. monotherapy) stand out for their impact on clinical practice, as well as the MAYO/NCCTG 85-47-51 (continuous infusion of 5-fluorouracil [5-FU] during radiotherapy vs. bolus), INT 0114 (modulation of 5-FU), NSBPA-R02 (radiochemotherapy vs. chemotherapy), and INT 0144 (continuous infusion of 5-FU throughout adjuvant treatment). The main conclusions can be summarized as follows:

– Exclusive surgery has a high local recurrence rate.
– Postoperative radiotherapy increases local control but does not increase survival.
– Postoperative chemotherapy increases survival but does not increase local control.

- Five-FU-based radiochemotherapy increases local control and survival when compared with postoperative radiotherapy alone or control.
- The addition of radiotherapy to chemotherapy does not increase overall survival or disease-free survival but decreases the rate of local recurrence.
- Administration of 5-FU as a continuous infusion during radiotherapy is superior to bolus in terms of overall survival, disease-free survival, and risk of distant metastases.
- There is no advantage to the addition of leucovorin (LV) or levamisole to 5-FU.
- Administration of 5-FU as a continuous infusion throughout adjuvant treatment is not superior to bolus in terms of the locoregional recurrence rate, disease-free survival, and overall survival, although grade 3–4 hematological toxicity is lower (4% vs. 50%).

Neoadjuvant treatment

The shift from adjuvant to neoadjuvant radiotherapy is attributed to the theoretical superiority, in terms of efficacy and toxicity, of administering this treatment before surgery. Its main potential advantages can be summarized as follows:

- Increased radiochemosensitivity: Well-oxygenated tumor cells are more radio- and chemosensitive than hypoxic cells, and surgery alters the vascularization and oxygenation of the tumor bed.
- Sterilization of the operative field with decreased risk of surgical seeding.
- Decrease in tumor size with a consequent increase in resectability.
- Decreased acute and chronic toxicity: There is no fixation of the intestinal loops in the pelvis, no need to radiate a surgical anastomosis, and no need to cover the perineal region with radiation in cases of abdomino-perineal amputation.
- Possibility of sphincter preservation mediated by a decrease in tumor volume.
- Higher degree of adherence to treatment: After surgery, chemotherapy and radiotherapy are often delayed or not even administered owing to postoperative complications.

In contrast to these advantages, the detractors of neoadjuvant treatment allege the disadvantage of patient selection with the risk of overtreatment in cases of T1-2N0 tumors or metastatic disease. However, imaging techniques that are highly accurate in determining the tumor stage, such as endorectal ultrasound, magnetic resonance imaging (MRI), and positron emission tomography, are available. Moreover, MRI can predict the circumferential resection margin with high reliability and consistency, allowing preoperative identification of patients at a high risk of recurrence, who would benefit from more aggressive treatments.

All the theoretical advantages of neoadjuvant treatment in rectal cancer have been demonstrated in phase III trials, except for the ability to achieve sphincter preservation.

Neoadjuvant radiotherapy

Early studies in neoadjuvant radiotherapy focused on investigating whether preoperative radiotherapy improved the results of surgery for rectal cancer. The doses and schedules used varied widely between studies, making interpretation of the results difficult. Finally, the data obtained from a Swedish study[11] (Swedish Rectal Cancer Trial), a large study randomizing 1168 patients with resectable rectal cancer (T1–T3) in the pretotal mesorectal excision (TME) era to surgery vs. radiotherapy (25 Gy in five fractions of 5 Gy) followed by surgery within 1 week, suggest that radiotherapy at 5 years increases local control (89% vs. 73%; $P < .001$) and survival (58% vs. 48%; $P = .004$). However, this trial and its radiation schedule have a number of limitations, such as high morbidity and mortality, inadequate patient selection (including stage I, although the advantage in local control is also evident in this group), lack of downstaging (a 7-day radiotherapy-surgery interval does not allow for a marked decrease in tumor size), and inability to integrate chemotherapy in such a short treatment period. The 10-year results of the Swedish study confirmed that preoperative radiotherapy confers a significant benefit on the risk of local recurrence (9% vs. 26%; $P < .001$), cancer-specific survival (72% vs. 62%; $P = .03$), and overall survival (38% vs. 30%; $P = .008$).[12]

Once TME was shown to decrease local recurrence to the same extent as preoperative radiotherapy did in the Swedish study, the next major contribution was made with the publication of a phase III study sponsored by the Dutch Colorectal Cancer Group,[13] which demonstrated the benefit of preoperative radiotherapy (25 Gy in five fractions) even in the presence of optimal surgery. The study randomized 1861 patients with resectable rectal cancer (T1-T3) to receive radiotherapy (25 Gy in five fractions) followed by surgery with TME or TME alone. Preoperative radiotherapy was found to reduce the local recurrence rate at both 2 and 5 years (5.6% vs. 10.9%; $P < .001$), mainly in stage III disease and tumors located less than 10 cm from the anal margin, although it did not impact overall survival. In the group treated with preoperative radiotherapy, more healing problems were observed after abdomino-perineal amputation (29% vs. 18%; $P = .05$), fecal incontinence (62% vs. 38%; $P < .001$), and sexual dysfunction and dissatisfaction with bowel function. At the 12-year follow-up, the differences in local control were maintained (5% vs. 11%; $P < .0001$), with no benefit in terms of overall survival (48% vs. 49%; $P = .86$). While overall survival was similar in both groups, preoperative radiotherapy significantly improved the 10-year survival (50% vs. 40%; $P = .032$) in those with stage III disease with negative circumferential margins.[4] The benefit of radiotherapy was found to be greater as the distance of the tumor from the anal margin increased; however, when patients with circumferentially resected margin involvement were excluded, this benefit became independent of the distance from the anal margin as was the increase in cancer-specific survival.

Neoadjuvant radiotherapy vs. adjuvant radiochemotherapy

The MRC/NCIC 07 trial[14] randomized 1350 patients with resectable rectal cancer to receive preoperative radiotherapy (25 Gy in five fractions) followed by surgery vs. surgery followed by postoperative radiochemotherapy (45 Gy in 180 cGy fractions with 5-FU) in patients with positive circumferential resection margins; adjuvant chemotherapy was allowed in both arms in cases of positive nodes or circumferential margin involvement. After a median follow-up of 4 years, the conclusions of the study are that preoperative radiotherapy decreases local recurrence ($P = .001$; HR: 0.39) and increases cancer-specific survival ($P = .013$; HR: 0.76), with no difference in terms of survival. The benefit in favor of preoperative radiotherapy involves all tumor sites and disease stages, although its impact is more pronounced in tumors located between 10 and 15 cm from the anal margin and in stage III disease. Published results on quality of life indicate that in men, sexual dysfunction is the main adverse effect and that surgery rather than radiotherapy is the main cause of such effect, although preoperative radiotherapy may affect both sexual and bowel functions.[15, 16]

Therefore it is currently accepted that preoperative radiotherapy reduces the rate of local recurrence; however, there is no agreement on its potential benefit on overall survival. Although this technique is commonly used in the Nordic countries and the UK, radiochemotherapy is often given in conjunction with radiochemotherapy in the rest of Europe and the USA.

Neoadjuvant radiochemotherapy

Since the late 1990s and despite the absence of studies comparing the efficacy between neoadjuvant and adjuvant therapies, the practice of preoperative radiochemotherapy has spread both in Europe and in the USA based on the extrapolation of the results of postoperative radiochemotherapy studies that demonstrated the superiority of combined treatment over radiotherapy alone. Its fundamental aim is to maximize tumor response, so that the elimination of local disease by radiochemotherapy and of micrometastases by chemotherapy will result in increased local control and survival. Several phase II trials in patients with resectable rectal cancer treated with conventional doses of radiotherapy concomitant with bolus or continuous infusion 5-FU demonstrated good results in terms of pathological complete responses (pCRs) (10%–30%), local control (95%–100%), and 5-year survival (80%–95%), as well as acceptable toxicity (grade 3: 20%–25%).[17]

Following the publication of several clinical trials (CAO/ARO, EORTC, and FFCD), neoadjuvant radiochemotherapy is currently considered the standard treatment for stage II and III rectal cancers in many countries. This indication is limited to tumors located between 0 and 16 cm from the anal margin and to patients with acceptable clinical condition (ECOG 0-1). In patients with ECOG 2, a rigorous evaluation of radiation dose and volumes and the need for concomitant chemotherapy should be performed.

Classical schemes use doses ranging from 45 to 50 Gy (fraction: 180 cGy) combined with 5-FU-based chemotherapy. The delineation of radiation volumes is based on findings derived from digital rectal examination, endoscopy, barium enema, echoendoscopy, computed tomography (CT), and MRI. If available, fusion of CT and MRI is recommended for treatment planning, as well as advice from a radiodiagnostic specialist for correct delineation of areas of the highest risk of recurrence.

Interval between neoadjuvant treatment and surgery

The optimal time from the end of neoadjuvant treatment to the surgical intervention is controversial.[1] The ideal interval to achieve the maximum effect of neoadjuvant therapy and to allow safe surgery with optimal results remains to be established, with wide intervals being considered in the different consensus guidelines.[18]

Several randomized studies and at least three meta-analyses have attempted to establish the most appropriate timing of surgical intervention.

In 1999 the Lyon R90-01 study[19] showed that a long interval between the end of neoadjuvant treatment and surgery (6–8 weeks) was associated with a higher rate of response to treatment (53.1% vs. 71.7%, $P = .007$) and a higher rate of downstaging than a short interval (26% vs. 10.3%, $P = .0054$). In 2016[20] the results of this study were reviewed after a 15-year follow-up; the long interval to surgery group demonstrated superiority in terms of pCRs (26% vs. 10.3%, $P = .015$). pCR was associated with better survival outcomes for patients ($P = .0048$). No significant differences were observed between the two study arms in relation to local recurrence or survival.

In 2017[21] and 2019[22] the results of two substudies of the Stockholm III trial (multicenter phase III trial with 3 arms: short-cycle RT and early surgery, short-cycle RT and delayed surgery [4–8 weeks], and long-cycle RT with delayed surgery [4–8 weeks]) were published; therein, no significant differences were observed in terms of local recurrence among the three study arms, and the short cycle with delayed surgery had a higher rate of complete responses, with no differences in postoperative complications.

The GRECCAR-6[23] study was a randomized study whose results indicate that a surgical waiting interval after neoadjuvant treatment of more than 11 weeks does not increase the rate of pCRs but can lead to greater morbidity and surgical difficulty. This finding of the GRECCAR-6 study has not been confirmed in other prospective studies, such as those of García Aguilar[24] or Fokas,[24] in which a longer interval between radiochemotherapy and consolidation chemotherapy as total neoadjuvant treatment did not increase surgical morbidity.

In 2018 Kim et al[25] analyzed the outcomes of 249 patients with locally advanced rectal cancer operated at different time intervals after completion of neoadjuvant treatment. The majority of patients (113 patients, 45.4%) were operated on in the interval between the 7th and 9th weeks. The pCR rate was higher in the 9–11th-week interval (3 patients, 8.6%; $P = .886$), although the difference was not significant. The highest downstaging was seen in the 7–9th-week interval (52.9%; $P = .087$); however, this difference was also not significant.

Several meta-analyses have investigated the optimal timing of surgical intervention, concluding that the rate of pCR and downstaging is higher at an interval of ≥ 8 weeks after neoadjuvant surgery, with no differences in the rates of R0 resection, local recurrence, sphincter preservation, disease-free survival, overall survival, or postoperative complications.[26]

Neoadjuvant radiochemotherapy vs. adjuvant radiochemotherapy

Two studies have compared between the administration of neoadjuvant radiochemotherapy and adjuvant radiochemotherapy.

The NSABP R-03[27] study compared the administration of radiotherapy with 5-FU/LV before and after surgery. A total of 267 patients with cT3-4 and/or N+ rectal tumors were randomized to receive pre- or postoperative radiochemotherapy with 5-FU/LV (50.4 Gy in 28 fractions). In both cases, treatment was completed with four cycles of 5-FU/LV, and TME was not mandatory. Recruitment was slower than expected and was closed after including 267 patients. Neoadjuvant treatment achieved a pCR rate of 15% and increased the 5-year disease-free survival (64.7% vs. 53.4%; $P = .01$), with no difference in the 5-year overall survival (74.5% vs. 65.6%; $P = .065$). Sphincter-sparing surgery could be performed more frequently in patients treated with neoadjuvant treatment (47.8% vs. 39.2%; $P = .227$).

The German Rectal Cancer Group (CAO/ARO/AIO 94)[28] randomized 823 patients with stage II and III rectal cancers to two arms that are exactly the same in design, except for the timing of radiochemotherapy, i.e., 4–6 weeks before surgery vs. 4–6 weeks after surgery. An important peculiarity with respect to other studies is that it includes the combination of an optimal surgery (TME) with a combined treatment considered standard (50 Gy of radiotherapy with concomitant continuous infusion 5-FU). The conclusions are that preoperative radiochemotherapy is significantly superior to postoperative radiochemotherapy in terms of local control (6% vs. 13%; $P = .006$), acute grade 3–4 toxicity (27% vs. 40%; $P = .001$), chronic toxicity (14% vs. 24%; $P = .01$), and possibility of sphincter preservation in patients who were initially considered to require abdomino-perineal amputation (39% vs. 19%; $P = .004$). However, no difference was found in the 5-year survival (76% vs. 74%; $P = .080$). Other interesting findings were the pCR rate of 8%, lack of impact of neoadjuvant therapy on operative morbidity, and difficulty in completing therapy in the adjuvant arm (only 50% of patients received the full prescribed radiotherapeutic and chemotherapeutic treatment). The 11-year results confirm the significant benefit of preoperative radiochemotherapy in local control.[29]

A substudy of the German Rectal Cancer Group evaluated possible prognostic factors in patients treated in the neoadjuvant radiochemotherapy arm; it was found that disease-free survival is related to the degree of histological differentiation, ypTN, degree of tumor regression, vascular invasion, and lymphatic invasion, while both metastasis-free survival and local failure-free survival are exclusively related to ypN.[30]

A recent post hoc analysis by the German Rectal Cancer Group (CAO/ARO/AIO 94)[31] studied the long-term impact of surgical complications. Patients with postoperative complications had worse oncological outcomes at 10 years after treatment, with lower overall survival (46.6% vs. 63.8%; $P < .001$), lower distant metastasis-free survival (63.2% vs. 72.0%; $P = .030$), and higher local recurrence rate (15.5% vs. 6.4%; $P < .001$). Postoperative complications were found to be an independent predictor of poorer overall survival.

In summary, preoperative radiochemotherapy is associated with a 50% reduction in the risk of local recurrence in cT3-4 rectal cancer, the possibility of sphincter preservation, and less toxicity compared to postoperative radiochemotherapy, with no difference found in terms of survival.

Neoadjuvant radiochemotherapy vs. neoadjuvant radiotherapy

Chemotherapy has been integrated into irradiation schedules, with the aim of enhancing the effect of radiotherapy and achieving better local control of the disease. The association with radiosensitizing drugs has been shown to enhance these phenomena. The possibility of adding chemotherapy to preoperative radiotherapy has been evaluated in four European clinical trials: three in resectable disease (EORTC 22921, FFCD 9203, and Polish Trial) and one in unresectable disease (LARC Sweden Trial).

The EORTC 22921[32] study was a randomized study with a four-arm 2 × 2 factorial design that aimed to establish the value of neoadjuvant radiochemotherapy vs. radiotherapy alone and the value of adjuvant chemotherapy vs. observation. It included 1011 patients with T3–T4 tumors located below 15 cm from the anal margin. The conclusion of the trial is that in patients with stage II and III rectal cancers receiving preoperative radiotherapy, adding 5-FU-based chemotherapy pre- or postoperatively has no impact on survival. However, regardless of whether it is provided before or after surgery, chemotherapy confers a significant benefit on the risk of local recurrence (9% vs. 17%; $P = .002$), rate of pCR (14% vs. 5%; $P < .001$), and degree of perineural (14.3% vs. 7.6%; $P = .001$) and lymphatic (17.4% vs. 11.4%; $P = .008$) invasions. Conversely, adding chemotherapy to radiotherapy increases grade 3–4 acute toxicity (14% vs. 7%), mainly at the expense of diarrhea, but does not compromise compliance with radiotherapy or surgery.

The 10-year results of the study have been published, and the authors concluded that adjuvant 5-FU-based chemotherapy after preoperative radiotherapy (with or without chemotherapy) has no impact on overall survival and disease-free survival, not supporting the use of adjuvant chemotherapy after preoperative radiotherapy with or without chemotherapy in locally advanced rectal cancer.[33]

The second study comparing preoperative radiotherapy with radiochemotherapy was conducted by the Fédération Francophone de Cancerologie Digestive (FFCD).[26] The FFCD randomized 762 patients with T3-4 rectal cancer accessible to digital rectal examination to receive neoadjuvant radiotherapy vs. radiotherapy +5-FU/LV during weeks 1 and 5. After a median follow-up of 69 months, neoadjuvant radiochemotherapy yielded a higher rate of pCRs (11.4% vs. 3.6%; $P < .001$), with a decrease in local recurrences (8.1% vs. 16.5%; $P = .004$), but no difference in the sphincter preservation rate (41.7% vs. 42.3%; $P = .837$) or 5-year overall survival (67.9% vs. 67.4%; $P = .684$). However, a higher acute grade 3–4 toxicity was observed in the radiochemotherapy arm (14.9% vs. 2.9%; $P < .001$).

Therefore, in patients with T3-4 rectal tumors, the combination of chemotherapy with preoperative radiotherapy was associated with a relative risk reduction of local recurrence by approximately 50% compared to preoperative radiotherapy alone. This significant difference in local recurrence did not translate into a significant difference in the overall survival or in the rate of sphincter-sparing surgery.

The LARCS[11] study randomized 207 patients with unresectable or recurrent rectal cancer to preoperative radiotherapy (50 Gy in 2-Gy fractions) vs. neoadjuvant radiochemotherapy (50 Gy with bolus 5-FU/LV) followed by adjuvant 5-FU-based chemotherapy. Local control (82% vs. 67%; $P = .03$), cancer-specific survival (72% vs. 55%; $P = .02$), and overall survival (66% vs. 53%; $P = .09$) all favored the combination arm at the expense of higher grade 3–4 toxicities (29% vs. 6%; $P = .001$). After 10 years of follow-up, the updated study results were published, and although the difference in overall survival between the two groups was still 8%, it was not significant (HR: 1.27, CI: 0.87–01.84; $P = .21$); however, the significant advantage in local control in the neoadjuvant radiochemotherapy group remained.

Short-course neoadjuvant radiotherapy vs. long-course neoadjuvant radiochemotherapy

Randomized studies have shown that exclusive preoperative radiotherapy (25 Gy in five fractions) or preoperative radiochemotherapy (50.4 Gy in 28 fractions with concomitant chemotherapy) results in increased local control, less toxicity, and better adherence than does adjuvant treatment.[10]

The best results of exclusive preoperative radiotherapy have been obtained in the large Swedish trial,[11] wherein radiotherapy was found to improve not only local control but also disease-free survival. It is logical to compare this short-course schedule with conventional fractionation radiochemotherapy schedules as used in the EORTC[34] and FFCD[35] studies.

These competing approaches evolved in parallel, and it is not clear whether one is better than the other; the short course was developed in northern Europe and the UK and the long course in the US and some European countries. The proponents of each approach based their enthusiasm on the results of several randomized trials published in recent years.

The Polish study[36] randomized 316 patients with resectable T3 and/or T4 rectal tumors to receive short-course preoperative radiotherapy (25 Gy in five fractions) and surgery at 7 days or radiochemotherapy (50.4 Gy in 28 fractions of 1.8 Gy, 5-FU bolus, and LV) and delayed surgery at 4–6 weeks. They concluded that the combined treatment is superior to radiotherapy alone in terms of the pCR rate (16% vs. 1%; $P < .001$), positive lymph node rate (31.6% vs. 47.6%; $P = .007$), and positive circumferential margin (12.9% vs. 4.4%; $P = .017$). Conversely, adding chemotherapy to radiotherapy increased grade 3–4 acute toxicity (18.2% vs. 3.2%; $P < .001$), with no difference in chronic toxicity (10.1% vs. 7.1%; $P = .360$), sphincter preservation (61.2% vs. 58.0%; $P = .570$), local control (9% vs. 14.2%; $P = .170$), and survival (67.2% vs. 66.2%; $P = .960$).

In 2016 this same working group (Polish Colorectal Study Group) published the results of a phase III study in which 541 patients with cT4 or cT3 rectal cancer "fixed by rectal touch" were randomized to short-course 5×5-Gy radiotherapy and FOLFOX x4 (group A) or radiotherapy 50.4 Gy in 28 fractions combined with two bolus cycles of 5-FU 325 mg/m^2/day and LV 20 mg/m^2/day at weeks 1 and 5 of radiation and oxaliplatin infusion 50 mg/m^2 weekly (group B). There were no differences found in the pCR rates, R0 resection, disease-free survival, local recurrence rates, and distant metastases between the two groups. There were higher preoperative acute toxicity in group B (long course) and lower adherence to radiotherapy. There was no difference in the chronic toxicity. The overall survival at 3 years was better in group A (short course): 73% vs. 65%, $P = .04666$. Their updated results have recently been published[37]; after a median follow-up of 7 years, the initially detected differences in the overall survival in favor of the short-course arm disappeared in the long term (HR: 0.90, 95% CI: 0.70–1.15; $P = .38$); this concluded that superiority in the oncological outcomes of short-course radiotherapy vs. radiochemotherapy could not be demonstrated.

The Australian study (Trans-Taman Radiation Oncology Group Trial 01.04)[38] randomized 326 patients with T3N0-2 rectal tumors located below 12 cm from the anal margin to short-course radiotherapy (25 Gy in 5-Gy fractions) followed by surgery plus six cycles of adjuvant chemotherapy or radiotherapy (50.4 Gy in 28 fractions) with 5-FU chemotherapy (225 mg/m^2 continuous infusion) followed by surgery at 4–6 weeks. Their conclusion is that there was no difference in terms of local control (7.5% vs. 4.4% at 3 years; $P = .24$) and survival (74% vs. 70% at 5 years; $P = .62$) between short-course radiotherapy and radiochemotherapy. However, in tumors located below 5 cm from the anal margin, a lower risk of local recurrence was observed with the radiochemotherapy arm, although no significant differences were found.

The Swedish study (STOCKHOLM III trial) evaluated the optimal fractionation and timing of surgery of different preoperative schedules. Patients with resectable rectal cancer were randomized to receive short-course radiotherapy (25 Gy in five fractions) followed by surgery within 1 week (group 1) or after 4–8 weeks (group 2) or radiochemotherapy (50 Gy in 25 fractions) within 4–8 weeks after surgery (group 3). A preliminary analysis including 303 patients suggested that there was no difference in the acute toxicity between the two groups, although short-course radiotherapy with immediate surgery may be associated with a higher tendency for postoperative complications if surgery is delayed for 10 days from the start of radiotherapy (46%, 40%, and 32%; $P = .164$).[39]

An interim analysis of this study[40] examined the preoperative treatment outcomes of short-course radiotherapy followed by delayed surgery (4–8 weeks) vs. short-course radiotherapy and immediate surgery. Patients who underwent delayed surgery had higher downstaging, higher pCR rate (11.8% vs. 1.7%; $P = .001$), and higher tumor regression (10.1% vs. 1.7%; $P < .001$) than patients who underwent short-course radiotherapy and immediate surgery.

The Colorectal Cancer Group Trial[16] included 83 patients with resectable adenocarcinoma of the rectum (stage II and III) and compared downstaging between patients treated with short-course radiotherapy and radiochemotherapy, followed by surgery at 6 weeks in both groups. The radiochemotherapy arm had higher downstaging and downsizing than the short-course radiotherapy arm, as well as a higher rate of pCRs, with no differences between the two in terms of the percentage of R0 resections, sphincter-sparing surgery, and postoperative complications. In 2016 the final results of the study[41] with 150 patients were published; the radiochemotherapy arm had better rates of pCR, downstaging, 3-year survival, and overall survival; however, there were no differences found, except in the disease-free survival in favor of the radiochemotherapy arm (59% in the RT arm vs. 75.1% in the radiochemotherapy arm; $P = .022$). The perioperative complications were similar in both groups.

A recent meta-analysis[42] comparing between short-course and long-course neoadjuvant radiotherapies and including eight studies with a total of 1475 patients found no significant differences in recurrence outcomes, 5-year survival, and complication rates between the two treatment modalities. Subgroup analyses indicated that the outcome in terms of metastatic disease was significantly higher in the long course than in the short course (OR: 2.65, 95% CI: 1.05–6.68), suggesting that the long course of radiotherapy may increase the risk of distant metastases compared to the short course.

Watch and wait

Watch and wait to avoid a definitive stoma or the morbidity and mortality of standard treatment in locally advanced rectal cancer is an attractive alternative as long as equivalent oncological outcomes can be achieved.

This approach is based on the diagnosis of a complete clinical response after neoadjuvant treatment, where there is no evidence of detectable tumor on clinical endoscopic and radiological reassessment.[43] The role of molecular biology and markers is important in the decision-making for this case.

The 2020 guidelines from the American Society of Colon and Rectal Surgeons state that patients with an apparent complete clinical response to neoadjuvant therapy should generally be offered radical resection, although a watch and wait approach may be considered in highly selected patients in the context of a protocolized setting.[44]

One of the challenges in patients with a complete clinical response is the type of close follow-up that should be performed to identify possible recurrences. Some groups propose quarterly monitoring with digital and endoscopic examinations and tumor marker determination with associated biopsy of any suspicious areas. Most recurrences occur in the first 12 months. It seems that surgery at recurrence is feasible; however, whether this delay in the surgical procedure might work against patients who would have been amenable to radical tumor surgery with anastomosis remains unclear.

Randomized trials on this approach are lacking. Although retrospective and prospective series and systematic reviews have been conducted, the heterogeneity of these studies does not allow drawing of firm conclusions. In the recently published study by Dattani (2018)[45] conducted in 692 patients, the rate of recurrence was 21.6%; surgical salvage after recurrence, 88%; distant metastasis, 8.2%; and 3-year survival, 93.5%. Meanwhile, in the studies by Van der Valk (2018)[46–48] conducted in 1000 patients, the rate of recurrence was 25%; salvage after recurrence, 86%; distant metastasis, 8%; and 5-year survival, 85%.

At this stage, the decision must be individualized, and mutual agreement must be reached between the doctor and patient; further, the decision must be based on the context of clinical trials, given that the watch and wait approach is a nonstandardized treatment that may pose a potential risk to the patient's cure and involve extended or radical surgery in the event of regrowth of the lesion.

Total neoadjuvant treatment

With preQ RT, we have managed to minimize locoregional recurrence to 5%; the ideal setup would be to associate systemic QT before surgery to control potential micrometastatic disease using the same schemes that have been effective for colon cancer and thus achieve a distant relapse rate of <25%, which are the values currently being used in the medium and long terms.

Therefore administering systemic QT before surgery would allow early treatment of micrometastases and improve compliance with administration, as it is better tolerated, and delaying surgery after RT seems to favor downstaging and R0 resection.

The Dutch Rapid study,[49] which was presented last year, showed promising results with this regimen. In a multicenter, prospective, randomized trial involving 920 patients, patients with middle and distal rectal cancers: T4a/b +/− N2, treated with the short-course sequence of RT, systemic QT (FOLFOX or CAPEOX) at 4 months, and radical surgery with TME or with the classically used standard sequence of long course, TME and systemic QT with identical drugs, were compared.

This study showed that patients in the short-course arm had a higher rate of pCR (almost double, 28.4% vs. 14.3%) and lower percentage of metastases (20% vs. 26.6%); the 3-year survival (89%) and toxicity were similar in both groups; no differences in the postoperative complications or quality of life were found.

These results confirm a significant impact in reducing the risk of developing metastases and excellent disease control; the high rate of complete pathological responses opens the prospect of future development of further conservative treatment.

Neoadjuvant chemotherapy vs. neoadjuvant radiochemotherapy

The role of systemic chemotherapy as an alternative treatment to radiochemotherapy in the neoadjuvant treatment of locally advanced rectal cancer is a subject for research.[50] This therapeutic scheme would aim to reduce the risk of local recurrence and risk of metastasis and improve long-term survival, while avoiding the possible effects of pelvic irradiation. The more favorable profile of preoperative chemotherapy compared to postoperative chemotherapy seen in previous studies[50] and the lower risk of local recurrence in patients undergoing high-quality TME would allow a more selective use of radiotherapy.

The phase III FOWARC[51] study was a randomized study in which 495 patients with clinical stage II or III rectal cancer with tumors up to 12 cm from the anal margin were randomized to receive preoperative RT with concurrent FU (control), preoperative RT with modified FOLFOX6, or modified FOLFOX6 alone. The group receiving chemotherapy alone had a lower pCR rate than the other two groups (7% vs. 28% for FOLFOX + RT and 14% for FU + RT); however, the local recurrence rates were similar (8% vs. 7% and 8%, respectively), as were the 3-year disease-free survival (77% with FOLFOX vs. 74% with FOLFOX + RT and 73% with 5-FU + RT) and overall survival (91% with FOLFOX vs. 89% with FOLFOX + RT and 91% with FU + RT). The authors concluded that modified FOLFOX6 with or without radiotherapy did not significantly improve the outcomes compared to RT + FU.

These findings have led to the development of several phase II and III studies of neoadjuvant chemotherapy (with aggressive regimens and more selective use of radiotherapy).

The PROSPECT study (N1048 trial) was a phase II/III study comparing between neoadjuvant FOLFOX with selective use of radiochemotherapy and combination radiochemotherapy for the treatment of locally advanced rectal carcinoma in patients undergoing low anterior resection with TME. Its purpose is to compare the effects of standard chemotherapy and radiation treatment with chemotherapy using a combination regimen known as FOLFOX, 5-FU, oxaliplatin, and LV and selective use of standard treatment (RT/QT), depending on the response to FOLFOX. The study has completed its recruitment phase, and the results are pending publication.

This type of therapeutic management has been analyzed in the study by Cassidy et al[52] wherein 21,707 patients with clinical stage T2N1 (cT2N1), cT3N0, or cT3N1 rectal cancer who underwent neoadjuvant treatment with radiochemotherapy or neoadjuvant chemotherapy alone followed by surgical resection were included. Patients who did not receive neoadjuvant radiochemotherapy had worse overall survival (the 5-year overall survival was 75.0% for patients who received neoadjuvant radiochemotherapy vs. 67.2% for patients who received neoadjuvant chemotherapy; $P < .01$).

To date, this scheme remains under investigation and is not recommended for use outside of clinical trials or in exceptional cases not suitable for pelvic RT.

Meta-analyses and systematic reviews of supplementary treatment of rectal cancer

Several meta-analyses have evaluated the role of radiotherapy in stage II and III rectal cancers. The meta-analysis by Camma et al[53] included 14 clinical trials randomizing 6426 patients with resectable rectal cancer to receive surgery vs. radiotherapy followed by surgery. Their conclusion is that neoadjuvant radiotherapy decreased local recurrence ($P = .001$; OR: 0.49), cancer-related mortality ($P = .001$; OR: 0.71), and overall mortality ($P = .03$; OR: 0.84), resulting in a modest but significant increase in cancer-specific survival and overall survival. However, it did not change the rate of distant metastases ($P = .564$; OR: 0.93). The Colorectal Cancer Group meta-analysis[54] included 8507 randomized patients in 22 phase III trials of preoperative (14 studies; 6350 patients) and postoperative (8 studies; 2157 patients) radiotherapies. Using individual patient data (Camma et al. used only published data), this study showed a decreased risk of local recurrence regardless of the timing of radiotherapy, although the effect was superior to the neoadjuvant strategy (46% vs. 37% decrease; $P = .002$), with a minimal advantage in terms of the 5-year overall survival vs. surgery alone (45% vs. 42%). The benefit in cancer-specific mortality was diminished by a small but definite increase in mortality in the first year after surgery owing to cardiovascular and infectious causes. When evaluating these studies, it is important to note that they included a large number of older studies using poor patient selection, suboptimal radiation doses, and large treatment fields with a high risk of toxicity.

A synthesis of the literature on the role of radiotherapy in the treatment of rectal cancer is brilliantly summarized in the paper published by Glimelius et al,[55] based on data from more than 700 articles; the main conclusions are as follows:

- Strong evidence that preoperative radiotherapy decreases the risk of local failure by 50%–70%, while postoperative radiotherapy decreases it by 30%–40%.
- Strong evidence that preoperative radiotherapy is more effective than postoperative radiotherapy.
- Strong evidence that preoperative radiotherapy increases survival by 10%.
- Moderate evidence that preoperative radiotherapy decreases the risk of local failure in patients undergoing optimal surgery.
- No evidence that postoperative radiotherapy increases survival.
- Evidence that postoperative radiochemotherapy increases survival, with the addition of 5-FU being responsible for the increase in survival.

A synthesis of the literature in the paper published by Sajid et al[56] concluded that short-course preoperative radiotherapy is as effective as long-course preoperative radiochemotherapy for lower rectal tumors in terms of overall survival, recurrence, postoperative complications, and sphincter preservation, but with no difference in toxicity.

A Cochrane review including five randomized studies compared preoperative radiochemotherapy with preoperative radiotherapy alone in patients with resectable stage II and III rectal cancers. While chemotherapy resulted in increased pCR rates (OR: 2.12–5.84; $P < .00001$) and decreased local recurrence rates (OR: 0.39–0.72; $P < .001$), there was no impact found on sphincter-sparing surgery (OR: 0.92–1.30; $P = .32$), disease-free survival (OR: 0.92–1.30; $P = .32$), and overall survival (OR: 0.79–1.14; $P = .58$). A higher rate of grade 3–4 acute toxicity was also detected in the chemotherapy arm (OR: 1.68–10; $P = .002$).[57]

A Cochrane review including 19 randomized trials comparing exclusive surgery with preoperative radiotherapy in patients with resectable disease revealed superiority of neoadjuvant treatment in terms of local recurrence (HR: 0.71; $P < .0001$) and cause-specific mortality (HR: 0.87; $P < .02$), as well as a modest survival benefit in the order of 2% (HR: 0.93; $P = .04$).[58] An update of this review was published in 2018,[58] which included four studies and a total of 4663 patients; the studies focused on assessing the effect of short-course preoperative radiotherapy followed by surgery vs. surgery alone. Based on their findings, they concluded the following:

- Preoperative radiotherapy reduced local recurrence in patients with locally advanced rectal cancer compared to surgery alone.
- Preoperative radiotherapy reduced overall mortality; however, this effect was not confirmed in patients undergoing surgery with SMT.
- No differences were found in cause-specific mortality between the two therapeutic modalities nor in the occurrence of distant metastases.
- No differences were found in the rate of curative surgery and sphincter preservation between the groups.
- The risk of sepsis and postoperative complications may be higher with preoperative radiotherapy.

Another systematic review on preoperative radiochemotherapy for the treatment of rectal cancer was published by Bin Ma et al,[59] which included 106 articles with data from a total of 41,121 patients. In their conclusions, the major achievement of preoperative radiochemotherapy is the increase in local control of the disease and the improvement in regional locoregional failure-free survival; however, whether it increases overall survival has not yet clearly been demonstrated. In the future, a better understanding of the clinical and molecular factors that influence the response to preoperative treatment, as well as the development of better optimized and individualized treatments, will lead to a reduction in the adverse effects of these treatments and an improvement in metastasis-free survival, as well as overall survival.

New agents in combination with radiotherapy

The positive impact of the combination of 5-FU-based chemotherapy and preoperative radiotherapy in inducing higher pCR rates has led to several studies analyzing the potential impact of new agents, such as oral fluoropyrimidines, oxaliplatin, and irinotecan, and new biologic agents, such as bevacizumab and cetuximab, in neoadjuvant treatment. These studies are based on the premise that combining several agents with radiotherapy may have a synergistic effect, which could improve the results of conventional radiochemotherapy with 5-FU. Preoperative treatment with oral capecitabine (vs. 5-FU in ic) and concurrent radiotherapy, followed by total mesorectal surgery, is the current standard of care for locally advanced rectal cancer. When compared with postoperative 5-FU-based radiochemotherapy, this strategy is associated with lower rates of local recurrence and lower toxicity.[60]

Capecitabine

Capecitabine is an oral prodrug that is converted to 5-FU by intracellular thymidine phosphorylase. Substituting capecitabine for infusional 5-FU is an attractive option, as it yields a simpler administration schedule and possibly greater efficacy. Indeed, the incorporation of capecitabine into neoadjuvant combination therapy in locally advanced rectal cancer has been the subject of intense research over the past 10 years. A regimen of 825 mg/m^2 capecitabine twice daily for 7 days a week of continuous oral administration in combination with RT has been shown in phase I and II trials to be an active and well-tolerated regimen and is currently the most widely used regimen in the world.[61] Definitive demonstration that the efficacy of capecitabine in combination with radiotherapy is similar to 5-FU/radiotherapy has been provided by the German Phase III trial[60] and by NSABP-R-04[61] study.

The German Group study[62] directly compared preoperative radiotherapy (50.4 Gy) and capecitabine (825 mg/m^2 twice daily from days 1 to 38) with preoperative radiotherapy and 5-FU (1000 mg/m^2 continuous infusion on days 1 to 5 and 29 to 33) in 401 patients diagnosed with locally advanced rectal cancer. Patients treated with capecitabine were observed to have a higher rate of hand-foot syndrome but a lower rate of neutropenia. At a median follow-up of 52 months, no difference was observed in terms of local recurrence (6% vs. 7%; $P = .067$) between the capecitabine and 5-FU arms, although a lower rate of distant metastases was obtained in the capecitabine arm (19% vs. 28%; $P = .004$). The German Group study confirms that capecitabine is not inferior to 5-FU in terms of overall survival (75% vs. 67%; $P = .0004$).

The NSABP R-04 trial[50, 61, 63–67] included 1608 patients with rectal cancer (clinical stage II or III) undergoing preoperative radiotherapy (45 Gy in 25 fractions over 5 weeks +540 cGy boost in 3 fractions) who were randomly assigned to one of the following chemotherapy regimens: 5-FU continuous infusion (225 mg/m^2 5 days a week), with or without intravenous oxaliplatin (50 mg/m^2/week ×5); and oral capecitabine (825 mg/m^2 twice daily 5 days a week), with or without oxaliplatin (50 mg/m^2/week ×5). A preliminary analysis presented at ASCO 2011 showed that administration of capecitabine together with preoperative radiotherapy achieved similar results in terms of the pCR, downstaging, and sphincter preservation. The results presented at the 2014 Gastrointestinal Cancer Symposium[50, 61, 63–67] showed that there was no difference between capecitabine and continuous infusion 5-FU in terms of 3-year local control (11.2% vs. 11.8%; $P = .98$), disease-free survival, and overall survival, establishing capecitabine as the standard for neoadjuvant treatment in rectal cancer.

One approach to improving outcomes in rectal cancer has been to couple radiotherapy with a second drug with systemic activity. Therefore oxaliplatin and irinotecan would be good candidates. These drugs have been investigated in several trials with a role yet to be determined. Several randomized trials and two meta-analyses have studied the association of oxaliplatin with fluoropyrimidines in combination with radiotherapy. The vast majority have shown no improvement in oncological outcomes, reporting instead an increase in toxicities with this combination.[64]

Oxaliplatin

The benefit of integrating oxaliplatin into concurrent radiochemotherapy schedules is still controversial. The STAR-1[67] and ACCORD 12/0405[62, 68] studies have not demonstrated benefits of oxaliplatin by increasing toxicity without increasing the rate of pCRs. The final results of the PETTACC-66 study[69, 70] presented at ASCO 2020 are that the association of oxaliplatin to radiochemotherapy schedules decreases adherence to treatment and increases toxicity, without improving surgical outcomes, disease-free survival, or overall survival.

In the NSABP R-04 trial,[71] the addition of oxaliplatin to 5-FU or capecitabine did not improve the outcomes but significantly increased toxicity, mainly grade 3–4 diarrhea ($P < .001$).

However, Rodel et al[71] randomized 1265 patients with cT3-4/N0/+ rectal cancer into a control group of preoperative RT and concomitant 5-FU (1000 mg/m^2 for days 1 to 5 and 29 to 33), followed by surgery and four cycles of 5-FU, and an experimental group of preoperative RT, with 5-FU and oxaliplatin, followed by surgery and eight cycles of oxaliplatin. In their study, they reported a significant increase in the pCR rate and disease-free survival with this combination.

Jiao[72] randomized 206 patients to receive oxaliplatin in combination with capecitabine (experimental group) vs. capecitabine and radiotherapy (control group); the 3-year metastasis rate was significantly better in the experimental group (16.50% vs. 28.16%; $P = .045$).

Another trial that also reported a significant increase in the pCR rate in its preliminary results is the FOWARC multicenter study,[50] in which 495 patients with stage II/III rectal cancer were randomized into three treatment arms: one arm with neoadjuvant treatment with 5-FU/LV + radiotherapy, one arm with mFOLFOX6 + radiotherapy, and one arm with mFOLFOX6 without radiotherapy, followed by surgery with TME and postoperative chemotherapy. There was a higher rate of pCR in the mFOLFOX6 and radiotherapy arm than in the 5-FU/LV + radiotherapy arm (OR: 0.428; 95% CI: 0.237–0.776; $P = .005$). In the final results published in 2109,[73] no differences in the oncological outcomes were found, concluding that the addition of oxaliplatin to neoadjuvant radiochemotherapy and surgery with TME yields better pCR rates but does not reduce local recurrence or improve disease-free survival or overall survival at 3 years.

Zheng[74, 75] analyzed eight randomized studies with a total of 5597 patients and concluded that the addition of oxaliplatin in neoadjuvant treatment can increase the pCR rate and reduce the rate of distant metastases but that these benefits do not translate into increased overall survival, but rather into increased grade 3 and 4 toxicities.

Huntter[76] analyzed the results of 10 randomized studies with data from 5599 patients included; the results of these studies indicated that the addition of platinum derivatives increases the pCR and reduces the rate of distant metastases but does not improve overall survival or disease-free survival. However, grade 3–4 toxicities are increased.

In conclusion, the use of oxaliplatin in the neoadjuvant treatment of locally advanced rectal cancer is not recommended, as it yields greater toxicity and has an unproven efficacy.

Irinotecan

Nonrandomized trials showed a slight benefit with the use of irinotecan in radiochemotherapy regimens. This regimen showed encouraging results in phase I and II clinical trials[77]; however, these findings were not confirmed by the RTOG trial,[77] in which 106 patients with T3/4 rectal cancer were randomized to receive 5-FU (225 mg/m^2 continuous infusion) combined with hyperfractionated radiotherapy (55.2–60 Gy in 1.2-Gy fractions), with or without irinotecan (50 mg/m^2 weekly), or 5-FU (225 mg/m^2 continuous infusion) with irinotecan (50 mg/m^2 weekly for 4 weeks) and conventionally fractionated radiotherapy (50.4–54 Gy in 180-cGy fractions). The rate of pCRs was similar in both arms, as was for acute and chronic toxicities.

The results of the phase III study "ARISTOTLE"[78, 79] have been presented in abstract form at ASCO 2020. The study randomized 564 patients with locally advanced rectal cancer to two treatment arms: a standard arm with preoperative RT/QT with capecitabine and an experimental arm with the same schedule with the addition of weekly intravenous irinotecan. The study concluded that adding irinotecan to preoperative RT + capecitabine does not increase the rate of pCRs, decreases adherence to treatment, and increases the rate of adverse effects.

Biological agents

The incorporation of agents directed at new targets, such as antibodies against the epidermal growth factor receptor or antiangiogenic agents, into preoperative combination regimens is hampered by the results of early trials in which efficacy results with cetuximab were poor, while an excessive rate of surgical complications was observed with bevacizumab.[78] The lack of efficacy improvements with the addition of cetuximab or bevacizumab in adjuvant treatment of colon cancer has raised doubts on further development of these drugs for rectal cancer. The addition of other new antineoplastic agents, such as panitumumab or gefitinib,[80] has also been investigated, and although the trials showed a low toxicity profile, the results in terms of pCR were modest (pCR rate: 0%–27%, mean: 14%).

A recent review by Xi Zhong[80] published in 2018 analyzed 804 studies including a total of 1196 patients. The results in terms of pCR rates were encouraging; however, the data on efficacy and postoperative complications were contradictory. The study concluded that larger-scale, better-designed randomized trials are needed to clarify the role of new biologics in neoadjuvant treatment of locally advanced rectal cancer.

Intraoperative radiotherapy (IORT)

The main role of radiotherapy in cancer treatment is locoregional disease control, and most advances in the clinical application of radiotherapy are based on obtaining different dose distributions in tumor and normal tissue. For most malignancies, the ability of radiotherapy to control disease is considered to increase as its ability to deliver high doses of radiation to the tumor volume increases. However, the total dose of radiation that can be safely delivered is limited by the tolerance of adjacent healthy tissues.

From this need to achieve higher effective doses without a parallel increase in toxicity arises the philosophy of IORT. Direct visualization of the tumor or tumor bed allows both greater precision in defining areas of potential risk of recurrence and the displacement and protection of radiosensitive healthy tissues. These advantages are obvious; however, if conventional treatment methods (radiotherapy, chemotherapy, and surgery) achieve high rates of local control with minimal complications, the addition of IORT may be unnecessary. However, this situation is far from being achieved in several abdominal and pelvic tumors (unresectable locally advanced rectal cancer) where local relapse rates are very high and the presence of organs at risk prevents the use of high doses of radiation.

IORT refers to the administration of radiotherapy during surgery with the aim of increasing local tumor control. In a broader sense, it can be defined as a radiotherapy modality based on the administration of a single high dose of radiation to the surgical site or to an unresectable tumor, while moving the surrounding healthy tissues out of the radiation beam. In short, it is a technique of high precision and dosimetric quality (homogeneity in the deposition of the dose), which makes it possible to intensify treatment.[80]

The main biological advantage of IORT is the possibility to separate healthy tissues from the radiation field with a consequent decrease in toxicity. However, the administration of a single fraction also has a number of disadvantages compared to fractionated radiotherapy, such as the possible limitation of cellular processes of repair, repopulation, redistribution, and reoxygenation. These theoretical limitations can be minimized by combining IORT with external beam radiation therapy, which would allow both coverage of areas of possible subclinical disease and selective tumor inclusion. The advantages of this association can be summarized as follows:

- Increased locoregional disease control owing to decreased risk of marginal recurrence and the radiobiological advantage of fractionated irradiation.
- Decreased risk of damage to normal tissues owing to decreased overprint volume by direct visualization of the tumor and exclusion of radiosensitive structures by manipulation, shielding, or use of certain radiation energies.
- Intensification of treatment.

Therefore a good alternative in patients with locally advanced tumors is to use a preoperative radiochemotherapy schedule and to administer a radiation supplement during surgery. Conversely, the biological effectiveness of a dose of IORT is equivalent to two to three times the same dose with fractionated radiotherapy; thus, the effective dose of 45–50 Gy of RTE plus 10 Gy of IORT is 65–80 Gy, rising to 75–95 Gy with doses of 15 Gy of IORT and 85–110 Gy with doses of 20 Gy.[81–83]

Evaluating the effectiveness of IORT in the treatment of rectal cancer based on the available scientific evidence, both in terms of local control and survival, is complicated. The main limitation lies in the difficulty to perform multicenter randomized trials owing to the intrinsic peculiarities of the technique and complexity of the procedure; therefore, most of the available data come from single-institutional experiences with small samples of patients. In fact, with the exception of two small clinical trials, all available literature is reduced to studies of limited methodological quality. Moreover, comparison with other treatments is limited by factors, such as candidate selection and the retrospective nature of the studies. An additional problem is that IORT is part of the multidisciplinary management of patients with cancer, making it difficult to analyze its effectiveness separately from other therapeutic modalities.

In general, most of the results obtained with IORT combined with surgery and neoadjuvant radiochemotherapy in locally advanced (unresectable or recurrent) colorectal cancer show a general trend toward improvement in local control and, in some cases, in survival in patients who underwent optimal surgical resection.[80]

In the treatment of locally advanced tumors, randomized trials do not demonstrate a significant improvement in terms of efficacy, with 5-year local control rates of 90%–92% and overall survival of 65%–70%.[80] Other observational studies showed a very low incidence of recurrence in the area treated with IORT, with a local control rate of over 90% and overall survival ranging from 50% to 80%.[84] In recurrent disease, IORT combined with neoadjuvant radiochemotherapy and surgery is associated with 5-year local control rates ranging from 45% to 70%.[84]

A systematic review published in 2012, including 29 studies (prospective: 14; retrospective: 15) and 3003 patients with advanced or recurrent colorectal cancer, concluded that IORT decreases the incidence of local recurrences and may help improve local control (OR: 0.22; 95% CI: 0.05–0.86; $P = .03$), disease-free survival (HR: 0.51; 95% CI: 0.31–0.85; $P = .009$), and overall survival (HR: 0.33; 95% CI: 0.20–0.54; $P = .001$), without a significant increase in complications (OR: 1.13; 95% CI: 0.77–1.65; $P = .57$).[85]

Another systematic review[86] has recently been published, which included 3 randomized and 12 observational studies with 1460 patients comparing the use of IORT vs. no IORT in the treatment of locally advanced rectal cancer. Treatment with IORT was associated with better local control at 5 years (OR:3.07, 95% CI:1.66–5.66; $P = .000$). No significant differences were found between the treatment groups in overall survival and 5-year disease-free survival (HR: 0.80, 95% CI: 0.60–1.06; $P = .189$ and HR: 0.94, 95% CI: 0.73–1.22; $P = .650$, respectively). There was also no significant difference in the incidence of adverse effects (abscesses, fistulas, anastomotic leakage, surgical wound infection, or bladder dysfunction) between the treatment groups. The authors concluded that the addition of IORT to multidisciplinary rectal cancer treatment may increase local disease control but does not significantly increase overall survival or postoperative complications.

The adverse effects of IORT are difficult to differentiate from disease-related toxicity and are fundamentally influenced by dose, associated treatments, radiation volume, organs at risk included in the radiation field, and ability to shield them. In general, IORT is considered to yield an acceptable toxicity without a significant increase in the frequency and severity of complications typical of conventional treatment. In the short term, the most frequent complications are gastrointestinal in nature (surgical wound infections or intestinal pseudo-occlusion), while in the long term, the most relevant and frequent complications are peripheral neuropathy and ureter stenosis.[80]

IORT is currently included in the clinical guidelines of high scientific quality (National Cancer Compressive Network and European Society for Medical Oncology) as an integral part of rectal cancer treatment. Recently, the ESTRO and its advisory committee (ESTRO-ACROP) have published recommendations for its use[86] that aim to define the indications, patient selection criteria, and technical aspects of the technique, as well as to standardize its use in centers already using IORT and to guide and assist institutions planning to initiate the development of IORT programs, all within the multidisciplinary setting of the treatment of colorectal cancer in advanced rectal cancer.

References

1. Gómez Caamaño A, Candamio Folgar S, López López R. Tratamiento neoadyuvante y adyuvante en cáncer de recto. In: *Cáncer colorectal*. You & US; 2009.
2. Nagtegaal ID, Quirke P. What is the role for the circumferential margin in the modern treatment of rectal cancer? *J Clin Oncol* 2008;**26**(2): 303–12. https://doi.org/10.1200/jco.2007.12.7027.
3. Willet CG. *Adjuvant therapy for resected rectal cancer*; 2008 [Up to date, Published] www.uptodate.com.
4. van Gijn W, Marijnen CAM, Nagtegaal ID, et al. Preoperative radiotherapy combined with total mesorectal excision for resectable rectal cancer: 12-year follow-up of the multicentre, randomised controlled TME trial. *Lancet Oncol* 2011;**12**(6):575–82. https://doi.org/10.1016/S1470-2045(11)70097-3.
5. Tumor Study Group G. Prolongation of the disease-free interval in surgically treated rectal carcinoma. *N Engl J Med* 1985;**312**(23):1465–72. https://doi.org/10.1056/NEJM198506063122301.
6. Krook JE, Moertel CG, Gunderson LL, et al. Effective surgical adjuvant therapy for high-risk rectal carcinoma. *N Engl J Med* 1991;**324**(11): 709–15. https://doi.org/10.1056/NEJM199103143241101.
7. O'connell MJ, Martenson JA, Wieand HS, et al. Improving adjuvant therapy for rectal cancer by combining protracted-infusion fluorouracil with radiation therapy after curative surgery. *N Engl J Med* 1994;**331**(8):502–7. https://www.nejm.org/doi/full/10.1056/nejm199408253310803.
8. Wolmark N, Wieand HS, Hyams DM, et al. Randomized trial of postoperative adjuvant chemotherapy with or without radiotherapy for carcinoma of the rectum: National Surgical Adjuvant Breast and bowel project protocol R-02. *J Natl Cancer Inst* 2000;**92**(5):388–96. https://academic.oup.com/jnci/article-abstract/92/5/388/2606692.
9. Tepper JE, O'Connell M, Niedzwiecki D, et al. Adjuvant therapy in rectal cancer: analysis of stage, sex, and local control—final report of intergroup 0114. *J Clin Orthod* 2002;**20**(7):1744–50. https://doi.org/10.1200/JCO.2002.07.132.
10. Smalley SR, Benedetti JK, Williamson SK, et al. Phase III trial of fluorouracil-based chemotherapy regimens plus radiotherapy in postoperative adjuvant rectal cancer: GI INT 0144. *J Clin Oncol* 2006;**24**(22):3542–7. https://doi.org/10.1200/JCO.2005.04.9544.
11. Rectal Cancer Trial S. Improved survival with preoperative radiotherapy in resectable rectal cancer. *N Engl J Med* 1997;**336**(14):980–7. https://doi.org/10.1056/NEJM199704033361402.
12. Folkesson J, Birgisson H, Pahlman L, Cedermark B, Glimelius B, Gunnarsson U. Swedish Rectal Cancer Trial: long lasting benefits from radiotherapy on survival and local recurrence rate. *J Clin Oncol* 2005;**23**(24):5644–50. https://doi.org/10.1200/JCO.2005.08.144.
13. Kapiteijn E, Marijnen CA, Nagtegaal ID, et al. Preoperative radiotherapy combined with total mesorectal excision for resectable rectal cancer. *N Engl J Med* 2001;**345**(9):638–46. https://doi.org/10.1056/NEJMoa010580.
14. Bosset JF, Pavy JJ, Hamers HP, et al. Determination of the optimal dose of 5-fluorouracil when combined with low dose d,l-leucovorin and irradiation in rectal cancer: Results of three consecutive phase II studies. *Eur J Cancer* 1993;**29**(10):1406–10. https://doi.org/10.1016/0959-8049(93)90012-5.
15. Glynne-Jones R, Wyrwicz L, Tiret E, et al. Rectal cancer: ESMO clinical practice guidelines for diagnosis, treatment and follow-up. *Ann Oncol* 2017;**28**(suppl_4):iv22–40. https://doi.org/10.1093/annonc/mdx224.
16. Chen K, Xie G, Zhang Q, Shen Y, Zhou T. Comparison of short-course with long-course preoperative neoadjuvant therapy for rectal cancer: a meta-analysis. *J Cancer Res Ther* 2018;**14**(Supplement):S224–31. https://doi.org/10.4103/0973-1482.202231.
17. Francois Y, Nemoz CJ, Baulieux J, et al. Influence of the interval between preoperative radiation therapy and surgery on downstaging and on the rate of sphincter-sparing surgery for rectal cancer: the Lyon R90-01 randomized trial. *J Clin Oncol* 1999;**17**(8):2396. https://citeseerx.ist.psu.edu/viewdoc/download?doi=10.1.1.1010.4888&rep=rep1&type=pdf.
18. Cotte E, Passot G, Decullier E, et al. Pathologic response, when increased by longer interval, is a marker but not the cause of good prognosis in rectal cancer: 17-year follow-up of the lyon R90-01 randomized trial. *Int J Radiat Oncol Biol Phys* 2016;**94**(3):544–53. https://doi.org/10.1016/j.ijrobp.2015.10.061.
19. Erlandsson J, Holm T, Pettersson D, et al. Optimal fractionation of preoperative radiotherapy and timing to surgery for rectal cancer (Stockholm III): a multicentre, randomised, non-blinded, phase 3, non-inferiority trial. *Lancet Oncol* 2017;**18**(3):336–46. https://doi.org/10.1016/S1470-2045(17)30086-4.
20. Erlandsson J, Lörinc E, Ahlberg M, et al. Tumour regression after radiotherapy for rectal cancer—results from the randomised Stockholm III trial. *Radiother Oncol* 2019;**135**:178–86. https://doi.org/10.1016/j.radonc.2019.03.016.
21. Lefevre JH, Mineur L, Kotti S, et al. Effect of interval (7 or 11 weeks) between neoadjuvant radiochemotherapy and surgery on complete pathologic response in rectal cancer: a multicenter, randomized, controlled trial (GRECCAR-6). *J Clin Oncol* 2016;**34**(31):3773–80. https://ascopubs.org/doi/abs/10.1200/JCO.2016.67.6049.
22. Garcia-Aguilar J, Chow OS, Smith DD, et al. Effect of adding mFOLFOX6 after neoadjuvant chemoradiation in locally advanced rectal cancer: a multicentre, phase 2 trial. *Lancet Oncol* 2015;**16**(8):957–66. https://doi.org/10.1016/S1470-2045(15)00004-2.
23. Fokas E, Allgäuer M, Polat B, et al. Randomized phase II trial of chemoradiotherapy plus induction or consolidation chemotherapy as total neoadjuvant therapy for locally advanced rectal cancer: CAO/ARO/AIO-12. *J Clin Oncol* 2019;**37**(34):3212–22. https://ascopubs.org/doi/abs/10.1200/JCO.19.00308.
24. Kim MJ, Cho JS, Kim EM, Ko WA, Oh JH. Optimal time interval for surgery after neoadjuvant chemoradiotherapy in patients with locally advanced rectal cancer: analysis of health insurance review and assessment service data. *Ann Coloproctol* 2018;**34**(5):241–7. https://doi.org/10.3393/ac.2018.01.01.
25. Roh MS. Phase III randomized trial of preoperative versus postoperative multimodality therapy in patients with carcinoma of the rectum (NSABPR-03). *Proc ASCO* 2001;**20**:123a. https://ci.nii.ac.jp/naid/10023919958/. [Accessed 24 February 2021].

26. Sauer R, Becker H, Hohenberger W, et al. Preoperative versus postoperative chemoradiotherapy for rectal cancer. *N Engl J Med* 2004;**351**(17): 1731–40. https://doi.org/10.1056/NEJMoa040694.
27. Sauer R, Liersch T, Merkel S, et al. Preoperative versus postoperative chemoradiotherapy for locally advanced rectal cancer: results of the German CAO/ARO/AIO-94 randomized phase III trial after a median follow-up of 11 years. *J Clin Oncol* 2012;**30**(16):1926–33. http://mauriciolema.webhost4life.com/rolmm/downloads/files/ChemoRTPreVsPostOpRectalCa.pdf.
28. Rödel C, Martus P, Papadoupolos T, et al. Prognostic significance of tumor regression after preoperative chemoradiotherapy for rectal cancer. *J Clin Oncol* 2005;**23**(34):8688–96. https://doi.org/10.1200/JCO.2005.02.1329.
29. Sprenger T, Beißbarth T, Sauer R, et al. Long-term prognostic impact of surgical complications in the German Rectal Cancer Trial CAO/ARO/AIO-94. *Br J Surg* 2018;**105**(11):1510–8. https://www.ingentaconnect.com/content/jws/bjs/2018/00000105/00000011/art00020.
30. Bosset J-F, Collette L, Calais G, et al. Chemotherapy with preoperative radiotherapy in rectal cancer. *N Engl J Med* 2006;**355**(11):1114–23. https://doi.org/10.1056/NEJMoa060829.
31. Bosset J-F, Calais G, Mineur L, et al. Fluorouracil-based adjuvant chemotherapy after preoperative chemoradiotherapy in rectal cancer: long-term results of the EORTC 22921 randomised study. *Lancet Oncol* 2014;**15**(2):184–90. https://doi.org/10.1016/S1470-2045(13)70599-0.
32. Gérard J-P, Conroy T, Bonnetain F, et al. Preoperative radiotherapy with or without concurrent fluorouracil and leucovorin in T3-4 rectal cancers: results of FFCD 9203. *J Clin Oncol* 2006;**24**(28):4620–5. https://doi.org/10.1200/JCO.2006.06.7629.
33. Braendengen M, Tveit KM, Berglund A, et al. Randomized phase III study comparing preoperative radiotherapy with chemoradiotherapy in nonresectable rectal cancer. *J Clin Oncol* 2008;**26**(22):3687–94. https://doi.org/10.1200/JCO.2007.15.3858.
34. Bujko K, Nowacki MP, Nasierowska-Guttmejer A, Michalski W, Bebenek M, Kryj M. Long-term results of a randomized trial comparing preoperative short-course radiotherapy with preoperative conventionally fractionated chemoradiation for rectal cancer. *Br J Surg* 2006;**93**(10):1215–23. https://www.ingentaconnect.com/content/jws/bjs/2006/00000093/00000010/art00007.
35. Cisel B, Pietrzak L, Michalski W, et al. Long-course preoperative chemoradiation versus 5 × 5 Gy and consolidation chemotherapy for clinical T4 and fixed clinical T3 rectal cancer: long-term results of the randomized polish II study. *Ann Oncol* 2019;**30**(8):1298–303. https://www.sciencedirect.com/science/article/pii/S0923753419312967.
36. Ngan SY, Burmeister B, Fisher RJ, et al. Randomized trial of short-course radiotherapy versus long-course chemoradiation comparing rates of local recurrence in patients with T3 rectal cancer: trans-tasman radiation oncology group trial 01.04. *J Clin Oncol* 2012;**30**(31):3827–33. https://www.researchgate.net/profile/Stephen_Ackland/publication/231176339_Randomized_Trial_of_Short-Course_Radiotherapy_Versus_Long-Course_Chemoradiation_Comparing_Rates_of_Local_Recurrence_in_Patients_With_T3_Rectal_Cancer_Trans-Tasman_Radiation_Oncology_Group_Trial_0104/links/57fcdef008aed4ab46fe6730/Randomized-Trial-of-Short-Course-Radiotherapy-Versus-Long-Course-Chemoradiation-Comparing-Rates-of-Local-Recurrence-in-Patients-With-T3-Rectal-Cancer-Trans-Tasman-Radiation-Oncology-Group-Trial-0104.pdf.
37. Pettersson D, Cedermark B, Holm T, et al. Interim analysis of the Stockholm III trial of preoperative radiotherapy regimens for rectal cancer. *Br J Surg* 2010;**97**(4):580–7. https://doi.org/10.1002/bjs.6914.
38. Pettersson D, Lörinc E, Holm T, et al. Tumour regression in the randomized Stockholm III Trial of radiotherapy regimens for rectal cancer. *Br J Surg* 2015;**102**(8):972. https://www.ncbi.nlm.nih.gov/pmc/articles/pmc4744683/.
39. Latkauskas T, Pauzas H, Gineikiene I, et al. Initial results of a randomized controlled trial comparing clinical and pathological downstaging of rectal cancer after preoperative short-course radiotherapy or long-term chemoradiotherapy, both with delayed surgery. *Colorectal Dis* 2012;**14**(3):294–8. https://onlinelibrary.wiley.com/doi/abs/10.1111/j.1463-1318.2011.02815.x.
40. Latkauskas T, Pauzas H, Kairevice L, et al. Preoperative conventional chemoradiotherapy versus short-course radiotherapy with delayed surgery for rectal cancer: results of a randomized controlled trial. *BMC Cancer* 2016;**16**(1):927. https://doi.org/10.1186/s12885-016-2959-9.
41. Dalton RSJ, Velineni R, Osborne ME, et al. A single-Centre experience of chemoradiotherapy for rectal cancer: is there potential for nonoperative management? *Colorectal Dis* 2012;**14**(5):567–71. https://onlinelibrary.wiley.com/doi/abs/10.1111/j.1463-1318.2011.02752.x.
42. You YN, Hardiman KM, Bafford A, et al. The American Society of Colon and Rectal Surgeons clinical practice guidelines for the management of rectal cancer. *Dis Colon Rectum* 2020;**63**(9):1191–222. https://doi.org/10.1097/DCR.0000000000001762.
43. Dattani M, Heald RJ, Goussous G, et al. Oncological and survival outcomes in watch and wait patients with a clinical complete response after neoadjuvant chemoradiotherapy for rectal cancer: a systematic review and pooled analysis. *Ann Surg* 2018;**268**(6):955. https://doi.org/10.1097/SLA.0000000000002761.
44. van der Valk MJM, Hilling DE, Bastiaannet E, et al. Long-term outcomes of clinical complete responders after neoadjuvant treatment for rectal cancer in the International Watch & Wait Database (IWWD): an international multicentre registry study. *Lancet* 2018;**391**(10139):2537–45. https://doi.org/10.1016/S0140-6736(18)31078-X.
45. Bahadoer RR, Dijkstra EA, van Etten B, et al. Short-course radiotherapy followed by chemotherapy before total mesorectal excision (TME) versus preoperative chemoradiotherapy, TME, and optional adjuvant chemotherapy in locally advanced rectal cancer (RAPIDO): a randomised, open-label, phase 3 trial. *Lancet Oncol* 2021;**22**(1):29–42. https://www.sciencedirect.com/science/article/pii/S1470204520305556.
46. Schrag D, Weiser MR, Goodman KA, et al. Neoadjuvant chemotherapy without routine use of radiation therapy for patients with locally advanced rectal cancer: a pilot trial. *J Clin Oncol* 2014;**32**(6):513–8. https://doi.org/10.1200/jco.2013.51.7904.
47. Fernandez-Martos C, Brown G, Estevan R, et al. Preoperative chemotherapy in patients with intermediate-risk rectal adenocarcinoma selected by high-resolution magnetic resonance imaging: the GEMCAD 0801 phase II multicenter trial. *Oncologist* 2014;**19**(10):1042–3. https://doi.org/10.1634/theoncologist.2014-0233.
48. Bensignor T, Brouquet A, Dariane C, et al. Pathological response of locally advanced rectal cancer to preoperative chemotherapy without pelvic irradiation. *Colorectal Dis* 2015;**17**(6):491–8. https://doi.org/10.1111/codi.12879.

49. Fernández-Martos C, Pericay C, Aparicio J, et al. Phase II, randomized study of concomitant chemoradiotherapy followed by surgery and adjuvant capecitabine plus oxaliplatin (CAPOX) compared with induction CAPOX followed by concomitant chemoradiotherapy and surgery in magnetic resonance imaging–defined, locally advanced rectal cancer: grupo cáncer de recto 3 study. *J Clin Oncol* 2010;**28**(5):859–65. https://www.researchgate.net/profile/Bartomeu_Massuti/publication/40907761_Phase_II_Randomized_Study_of_Concomitant_Chemoradiotherapy_Followed_by_Surgery_and_Adjuvant_Capecitabine_Plus_Oxaliplatin_CAPOX_Compared_With_Induction_CAPOX_Followed_by_Concomitant_Chemoradiotherapy_/links/565649b408aefe619b1d3478/Phase-II-Randomized-Study-of-Concomitant-Chemoradiotherapy-Followed-by-Surgery-and-Adjuvant-Capecitabine-Plus-Oxaliplatin-CAPOX-Compared-With-Induction-CAPOX-Followed-by-Concomitant-Chemoradiotherapy.pdf.
50. Deng Y, Chi P, Lan P, et al. Neoadjuvant modified FOLFOX6 with or without radiation versus fluorouracil plus radiation for locally advanced rectal cancer: final results of the Chinese FOWARC trial. *J Clin Oncol* 2019;**37**(34):3223–33. https://doi.org/10.1200/JCO.18.02309.
51. Cammà C, Giunta M, Fiorica F, Pagliaro L, Craxì A, Cottone M. Preoperative radiotherapy for resectable rectal cancer: a meta-analysis. *JAMA* 2000;**284**(8):1008–15. https://doi.org/10.1001/jama.284.8.1008.
52. Group CCC. Adjuvant radiotherapy for rectal cancer: a systematic overview of 8507 patients from 22 randomised trials. *Lancet* 2001;**358**(9290):1291–304. https://doi.org/10.1016/S0140-6736(01)06409-1.
53. Glimelius B, Grönberg H, Järhult J, Wallgren A, Cavallin-Ståhl E. A systematic overview of radiation therapy effects in rectal cancer. *Acta Oncol* 2003;**42**(5–6):476–92. https://doi.org/10.1080/02841860310012301.
54. Sajid MS, Siddiqui MRS, Kianifard B, Baig MK. Short-course versus long-course neoadjuvant radiotherapy for lower rectal cancer: a systematic review. *Ir J Med Sci* 2010;**179**(2):165–71. https://doi.org/10.1007/s11845-009-0382-9.
55. De Caluwé L, Van Nieuwenhove Y, Ceelen WP. Preoperative chemoradiation versus radiation alone for stage II and III resectable rectal cancer. *Cochrane Database Syst Rev* 2013;**2**:CD006041. https://doi.org/10.1002/14651858.CD006041.pub3.
56. Wong RKS, Tandan V, De Silva S, Figueredo A. Pre-operative radiotherapy and curative surgery for the management of localized rectal carcinoma. *Cochrane Database Syst Rev* 2007;**2**. https://doi.org/10.1002/14651858.CD002102.pub2.
57. Abraha I, Aristei C, Palumbo I, et al. Preoperative radiotherapy and curative surgery for the management of localised rectal carcinoma. *Cochrane Database Syst Rev* 2018;**10**:CD002102. https://doi.org/10.1002/14651858.CD002102.pub3.
58. Ma B, Gao P, Wang H, et al. What has preoperative radio (chemo) therapy brought to localized rectal cancer patients in terms of perioperative and long-term outcomes over the past decades? A systematic review and meta-analysis based on 41,121 patients. *Int J Cancer* 2017;**141**(5):1052–65. https://onlinelibrary.wiley.com/doi/abs/10.1002/ijc.30805.
59. Dunst J, Reese T, Sutter T, et al. Phase I trial evaluating the concurrent combination of radiotherapy and capecitabine in rectal cancer. *J Clin Oncol* 2002;**20**(19):3983–91. https://doi.org/10.1200/JCO.2002.02.049.
60. Hofheinz R-D, Wenz F, Post S, et al. Chemoradiotherapy with capecitabine versus fluorouracil for locally advanced rectal cancer: a randomised, multicentre, non-inferiority, phase 3 trial. *Lancet Oncol* 2012;**13**(6):579–88. https://doi.org/10.1016/S1470-2045(12)70116-X.
61. Roh MS, Yothers GA, O'Connell MJ, et al. The impact of capecitabine and oxaliplatin in the preoperative multimodality treatment in patients with carcinoma of the rectum: NSABP R-04. *J Clin Oncol* 2011;**29**(15_suppl):3503. https://doi.org/10.1200/jco.2011.29.15_suppl.3503.
62. 2014 Gastrointestinal Cancers Symposium (GICS). Abstract 390. In: *2014 gastrointestinal cancers symposium (GICS)*; 2014.
63. Aschele C, Cionini L, Lonardi S, et al. Primary tumor response to preoperative chemoradiation with or without oxaliplatin in locally advanced rectal cancer: pathologic results of the STAR-01 randomized phase III trial. *J Clin Oncol* 2011;**29**(20):2773–80. https://www.academia.edu/download/42003820/Primary_Tumor_Response_to_Preoperative_C20160203-5548-u9gm62.pdf.
64. Gérard J-P, Azria D, Gourgou-Bourgade S, et al. Comparison of two neoadjuvant chemoradiotherapy regimens for locally advanced rectal cancer: results of the phase III trial ACCORD 12/0405-Prodige 2. *J Clin Oncol* 2010;**28**(10):1638–44. https://doi.org/10.1200/JCO.2009.25.8376.
65. Schmoll H-J, Haustermans K, Price TJ, et al. Preoperative chemoradiotherapy and postoperative chemotherapy with capecitabine and oxaliplatin versus capecitabine alone in locally advanced rectal cancer: First results of the PETACC-6 randomized phase III trial. *J Clin Orthod* 2013;**31**(15_suppl):3531. https://doi.org/10.1200/jco.2013.31.15_suppl.3531.
66. Aschele C, Lonardi S, Cionini L, et al. Final results of STAR-01: a randomized phase III trial comparing preoperative chemoradiation with or without oxaliplatin in locally advanced rectal cancer. *J Clin Orthod* 2016;**34**(15_suppl):3521. https://doi.org/10.1200/JCO.2016.34.15_suppl.3521.
67. Schmoll H-J, Stein A, Van Cutsem E, et al. Pre-and postoperative capecitabine without or with oxaliplatin in locally advanced rectal cancer: PETACC 6 Trial by EORTC GITCG and ROG, AIO, AGITG, BGDO, and FFCD. *J Clin Oncol* 2020;**39**(1):17–29. https://ascopubs.org/doi/abs/10.1200/JCO.20.01740.
68. Allegra CJ, Yothers G, O'Connell MJ, et al. Neoadjuvant 5-FU or capecitabine plus radiation with or without oxaliplatin in rectal cancer patients: a phase III randomized clinical trial. *J Natl Cancer Inst* 2015;**107**(11). https://doi.org/10.1093/jnci/djv248.
69. Rödel C, Liersch T, Becker H, et al. Preoperative chemoradiotherapy and postoperative chemotherapy with fluorouracil and oxaliplatin versus fluorouracil alone in locally advanced rectal cancer: initial results of the German CAO/ARO/AIO-04 randomised phase 3 trial. *Lancet Oncol* 2012;**13**(7):679–87. https://doi.org/10.1016/S1470-2045(12)70187-0.
70. Rödel C, Graeven U, Fietkau R, et al. Oxaliplatin added to fluorouracil-based preoperative chemoradiotherapy and postoperative chemotherapy of locally advanced rectal cancer (the German CAO/ARO/AIO-04 study): final results of the multicentre, open-label, randomised, phase 3 trial. *Lancet Oncol* 2015;**16**(8):979–89. https://www.sciencedirect.com/science/article/pii/S147020451500159X.
71. Jiao D, Zhang R, Gong Z, et al. Fluorouracil-based preoperative chemoradiotherapy with or without oxaliplatin for stage II/III rectal cancer: a 3-year follow-up study. *Chin J Cancer Res* 2015;**27**(6):588–96. https://doi.org/10.3978/j.issn.1000-9604.2015.12.05.
72. Deng Y, Chi P, Lan P, et al. Modified FOLFOX6 with or without radiation versus fluorouracil and leucovorin with radiation in neoadjuvant treatment of locally advanced rectal cancer: initial results of the Chinese FOWARC multicenter, open-label, randomized three-arm phase III trial. *J Clin Oncol*

2016;34(27):3300–7. https://www.researchgate.net/profile/Yue_Cai14/publication/305787126_Modified_FOLFOX6_With_or_Without_Radiation_Versus_Fluorouracil_and_Leucovorin_With_Radiation_in_Neoadjuvant_Treatment_of_Locally_Advanced_Rectal_Cancer_Initial_Results_of_the_Chinese_FOWARC_Multicente/links/5db8b663a6fdcc2128eb979a/Modified-FOLFOX6-With-or-Without-Radiation-Versus-Fluorouracil-and-Leucovorin-With-Radiation-in-Neoadjuvant-Treatment-of-Locally-Advanced-Rectal-Cancer-Initial-Results-of-the-Chinese-FOWARC-Multicente.pdf.

73. Hüttner FJ, Probst P, Kalkum E, et al. Addition of platinum derivatives to fluoropyrimidine-based neoadjuvant chemoradiotherapy for stage II/III rectal cancer: systematic review and meta-analysis. *J Natl Cancer Inst* 2019;**111**(9):887–902. https://doi.org/10.1093/jnci/djz081.
74. Gollins S, Myint AS, Haylock B, et al. Preoperative chemoradiotherapy using concurrent capecitabine and irinotecan in magnetic resonance imaging-defined locally advanced rectal cancer: impact on long-term clinical outcomes. *J Clin Oncol* 2011;**29**(8):1042–9. https://www.researchgate.net/profile/Arthur_Sun_Myint/publication/49781805_Preoperative_Chemoradiotherapy_Using_Concurrent_Capecitabine_and_Irinotecan_in_Magnetic_Resonance_Imaging-Defined_Locally_Advanced_Rectal_Cancer_Impact_on_Long-Term_Clinical_Outcomes/links/57c35ee708aeda1ec39195fb.pdf.
75. Willeke F, Horisberger K, Kraus-Tiefenbacher U, et al. A phase II study of capecitabine and irinotecan in combination with concurrent pelvic radiotherapy (CapIri-RT) as neoadjuvant treatment of locally advanced rectal cancer. *Br J Cancer* 2007;**96**(6):912–7. https://doi.org/10.1038/sj.bjc.6603645.
76. Navarro M, Dotor E, Rivera F, et al. A phase II study of preoperative radiotherapy and concomitant weekly irinotecan in combination with protracted venous infusion 5-fluorouracil, for resectable locally advanced rectal cancer. *Int J Radiat Oncol Biol Phys* 2006;**66**(1):201–5. https://www.sciencedirect.com/science/article/pii/S0360301606006511.
77. Sebag-Montefiore D, Adams R, Gollins S, et al. ARISTOTLE: a phase III trial comparing concurrent capecitabine with capecitabine and irinotecan (Ir) chemoradiation as preoperative treatment for MRI-defined locally advanced rectal cancer (LARC). *J Clin Orthod* 2020;**38**(15_suppl):4101. https://doi.org/10.1200/JCO.2020.38.15_suppl.4101.
78. Gollins S, West N, Sebag-Montefiore D, et al. Preoperative chemoradiation with capecitabine, irinotecan and cetuximab in rectal cancer: significance of pre-treatment and post-resection RAS mutations. *Br J Cancer* 2017;**117**(9):1286–94. https://doi.org/10.1038/bjc.2017.294.
79. Pinto C, Di Bisceglie M, Di Fabio F, et al. Phase II study of preoperative treatment with external radiotherapy plus Panitumumab in low-risk, locally advanced Rectal Cancer (RaP study/STAR-03). *Oncologist* 2018;**23**(8):912–8. https://doi.org/10.1634/theoncologist.2017-0484.
80. Gunderson LL, Willett CG, Calvo FA, Harrison LB. *Intraoperative irradiation: techniques and results*. Springer Science & Business Media; 2011. https://play.google.com/store/books/details?id=NKbr_my9EkgC.
81. Dubois J-B, Bussieres E, Richaud P, et al. Intra-operative radiotherapy of rectal cancer: results of the French multi-institutional randomized study. *Radiother Oncol* 2011;**98**(3):298–303. https://doi.org/10.1016/j.radonc.2011.01.017.
82. Masaki T, Takayama M, Matsuoka H, et al. Intraoperative radiotherapy for oncological and function-preserving surgery in patients with advanced lower rectal cancer. *Langenbecks Arch Surg* 2008;**393**(2):173–80. https://doi.org/10.1007/s00423-007-0260-8.
83. Masaki T, Matsuoka H, Kobayashi T, et al. Quality assurance of pelvic autonomic nerve-preserving surgery for advanced lower rectal cancer—preliminary results of a randomized controlled trial. *Langenbecks Arch Surg* 2010;**395**(6):607–13. https://doi.org/10.1007/s00423-010-0655-9.
84. Mirnezami R, Chang GJ, Das P, et al. Intraoperative radiotherapy in colorectal cancer: systematic review and meta-analysis of techniques, long-term outcomes, and complications. *Surg Oncol* 2012;**22**(1):22–35. https://doi.org/10.1016/j.suronc.2012.11.001.
85. Liu B, Ge L, Wang J, et al. Efficacy and safety of intraoperative radiotherapy in rectal cancer: a systematic review and meta-analysis. *World J Gastrointest Oncol* 2021;**13**(1):69–86. https://doi.org/10.4251/wjgo.v13.i1.69.
86. Calvo FA, Sole CV, Rutten HJ, et al. ESTRO/ACROP IORT recommendations for intraoperative radiation therapy in primary locally advanced rectal cancer. *Clin Transl Radiat Oncol* 2020;**25**:29–36. https://doi.org/10.1016/j.ctro.2020.09.001.

Further reading

Sebag-Montefiore D, Stephens RJ, Steele R, et al. Preoperative radiotherapy versus selective postoperative chemoradiotherapy in patients with rectal cancer (MRC CR07 and NCIC-CTG C016): a multicentre, randomised trial. *Lancet* 2009;**373**(9666):811–20. https://doi.org/10.1016/S0140-6736(09)60484-0.

Stephens RJ, Thompson LC, Quirke P, et al. Impact of short-course preoperative radiotherapy for rectal cancer on patients' quality of life: data from the Medical Research Council CR07/National Cancer Institute of Canada Clinical Trials Group C016 randomized clinical trial. *J Clin Oncol* 2010;**28**(27):4233–9. https://www.researchgate.net/profile/Richard_Stephens7/publication/44806793_Impact_of_Short-Course_Preoperative_Radiotherapy_for_Rectal_Cancer_on_Patients'_Quality_of_Life_Data_From_the_Medical_Research_Council_CR07National_Cancer_Institute_of_Canada_Clinical_Trials_Group_C01/links/55eef96208ae199d47bfea5a/Impact-of-Short-Course-Preoperative-Radiotherapy-for-Rectal-Cancer-on-Patients-Quality-of-Life-Data-From-the-Medical-Research-Council-CR07-National-Cancer-Institute-of-Canada-Clinical-Trials-Group-C01.pdf.

Du D, Su Z, Wang D, Liu W, Wei Z. Optimal interval to surgery after neoadjuvant chemoradiotherapy in rectal cancer: a systematic review and meta-analysis. *Clin Colorectal Cancer* 2018;**17**(1):13–24. https://doi.org/10.1016/j.clcc.2017.10.012.

Lefevre JH, Parc Y, Tiret E, French Research Group of Rectal Cancer Surgery (GRECCAR). Increasing the interval between neoadjuvant chemoradiotherapy and surgery in rectal cancer. *Ann Surg* 2015;**262**(6):e116. https://doi.org/10.1097/SLA.0000000000000771.

Ryan ÉJ, O'Sullivan DP, Kelly ME, et al. Meta-analysis of the effect of extending the interval after long-course chemoradiotherapy before surgery in locally advanced rectal cancer. *Br J Surg* 2019;**106**(10):1298–310. https://doi.org/10.1002/bjs.11220.

Cassidy RJ, Liu Y, Patel K, et al. Can we eliminate neoadjuvant chemoradiotherapy in favor of neoadjuvant multiagent chemotherapy for select stage II/III rectal adenocarcinomas: analysis of the national cancer data base. *Cancer* 2017;**123**(5):783–93. https://acsjournals.onlinelibrary.wiley.com/doi/abs/10.1002/cncr.30410.

Zheng J, Feng X, Hu W, Wang J, Li Y. Systematic review and meta-analysis of preoperative chemoradiotherapy with or without oxaliplatin in locally advanced rectal cancer. *Medicine* 2017;**96**(13):e6487. https://doi.org/10.1097/MD.0000000000006487.

Weiss C, Arnold D, Dellas K, et al. Preoperative radiotherapy of advanced rectal cancer with capecitabine and oxaliplatin with or without cetuximab: a pooled analysis of three prospective phase I-II trials. *Int J Radiat Oncol Biol Phys* 2010;**78**(2):472–8. https://doi.org/10.1016/j.ijrobp.2009.07.1718.

Crane CH, Eng C, Feig BW, et al. Phase II trial of neoadjuvant bevacizumab, capecitabine, and radiotherapy for locally advanced rectal cancer. *Int J Radiat Oncol Biol Phys* 2010;**76**(3):824–30. https://doi.org/10.1016/j.ijrobp.2009.02.037.

Chapter 38

Radiotherapy (stereotactic body radiotherapy) for oligometastatic disease

Paula Peleteiro Higuero, Patricia Calvo Crespo, and Ana María Carballo Castro
Department of Radiation Oncology, Complexo Hospitalario Universitario de Santiago de Compostela, A Coruña, Spain

The concept of oligometastatic disease (OMD), initially proposed by Hellman and Weichselbaum in 1995, involves solid tumors with limited metastatic disease.[1]

With respect to metastatic colorectal cancer (CRC), the definition of OMD is not unambiguous. Recently, the European Society for Medical Oncology defined oligometastatic CRC as a disease characterized by metastases affecting two to three sites and usually no more than five lesions and recommend considering local ablative treatment in the treatment strategy for these patients either initially or after systemic treatment.[2] The European consensus recommendations on OMD consider a lesion diameter of ≤3 cm and a maximum number of lesions per organ of 3.[3]

Patients in the oligometastatic state may have a better prognosis than patients with disseminated metastatic disease, and the goal of treatment may be directed toward curative intent. Systemic therapy is the main treatment for metastatic disease, while surgery is increasingly used in oligometastatic CRC.[4]

There is a high percentage of patients who do not undergo surgery because of either unresectable metastases or comorbidities, in which metastasis-directed treatments, such as radiofrequency ablation (RFA) and stereotactic body radiotherapy (SBRT),[5] play an important role.

A phase II study showed that aggressive management of metastases with RFA in a subgroup of patients with initially unresectable colorectal liver metastases improved the overall survival (OS) compared to systemic treatment alone.[6]

Data from surgical cohorts with CRC demonstrate an improved 5-year OS in patients with one to three resectable metastases compared to four to six or more than six metastases, suggesting the importance of disease burden.[7]

The paradigm of local SBRT of metastases in OMD is well established.[8] Retrospective studies suggest that patients with a small number of metastases treated with SBRT achieve high rates of local control with minimal toxicity, which may delay the need for systemic treatments and improve progression-free survival (PFS).[9]

Control of CRC oligometastases treated with SBRT appears to be a positive predictor for the PFS and OS, and early treatment of metastases with SBRT could improve local control.[10]

The SBRT-COMET study, a recently published phase II study of SBRT in patients with oligometastatic disease, enrolled 99 patients (18 with breast cancer, 18 with lung cancer, 18 with CRC, and 16 with prostate cancer and 29 of other primary tumors) between 2012 and 2016 and stratified them into a standard treatment arm and a standard treatment and SBRT arm. A total of 191 metastases were treated. With a median follow-up of 27 months, better OS (28 months vs. 41 months, $p=0.09$) and better EFS (6 months vs. 12 months, $p=0.001$) at the expense of higher Gr2 toxicity (9% vs. 30%), especially asthenia, dyspnea, and bone pain, were observed in favor of the SBRT arm. No differences in quality of life were found between both arms.[11] At a longer follow-up of 51 months, the OS benefit was maintained (50 months vs. 28 months; HR: 0.47, 95% CI: 0.27–0.81).[12]

The phase III SABR-COMET-3 study will investigate SBRT in patients with up to three metastatic lesions and the phase III SABR-COMET-10 study in patients with up to 10 lesions with lower doses of SBRT.

A meta-analysis of 21 prospective trials included 943 patients with OMD and different primary tumors treated with SBRT. The most frequently treated metastatic sites were the bones (44.8%), lungs (29.2%), liver (13.1%), and lymph nodes (12.2%). The most frequent primary tumor was prostate cancer (22.9%), followed by CRC (16.6.9%). Based on the data, 1-year OS ranged from 65.9% to 100%. The 1-year PFS ranged from 33.3% to 80%, and acute Gr3–5 toxicity was observed in less than 10% of the patients.[13]

The results of an observational study from the English Health System that evaluated the efficacy and safety of SBRT in OMD (one to three lesions) in 1422 patients with solid tumors have recently been published. A total of 1421 metastases were treated with SBRT. The most frequent primary tumor was prostate cancer (28.6%), followed by CRC (27.9%). Approximately 75.6% of patients had a single metastasis. The most frequently treated site was nodal (31.3%). The 1- and 2-year OS in patients with CRC was 92% and 80.3%, respectively.[14]

Visceral metastases are the most common CRC metastases treated with SBRT, and much of the literature to date has focused on liver or lung metastases.[15]

A previous systematic review has evaluated the efficacy of SBRT in terms of local control in oligometastatic CRC (lung and liver). Patients with liver metastases had local control rates ranging from 50% to 100% at 1 year and from 32% to 91% at 2 years. Meanwhile, patients with lung metastases had local control rates ranging from 62% to 92% at 1 year and from 53% to 92% after 2 years, with biological equivalent doses (BEDs) between 51.3 and 262.5 Gy.[16]

Role of SBRT in lung metastases from CRC

In cases of CRC, 20%–30% of patients are known to develop pulmonary metastases. Survival in patients undergoing metastasectomy with R0 margins at 5 years is 40% compared to 5% in those undergoing excision with affected margins.[17]

SBRT is considered a treatment option for oligometastatic lung metastases.[18]

The Okunieff study included 125 metastatic lung lesions (excluding lesions in other locations), and a maximum of five metastases were treated with curative intent. Whether the lesions were centrally or peripherally located was not mentioned. An excellent local control rate of 91% was reported at 3 years with 50–55 Gy in 5-Gy fractions (BED: 100–120 Gy). No local failures were reported after a follow-up period of 12.2 months, and the mean diameter of the longest tumor was 2.1 cm.[19]

Hof et al. reported local control rates of 63.1% at 3 years in a series of 71 metastatic lung lesions treated with a single dose of 20–30 Gy.[20]

Dose and fractionation depend on tumor size and proximity to risk areas, such as the chest wall, heart, and main bronchus. In one study, 60 patients were treated with doses of 48–60 Gy in three to four fractions, demonstrating 2-year local control rates of 80%, with a median time to lung progression of 7 months.[21]

Another study investigated 44 patients (69 lung metastases from CRC) who were treated with SBRT. The 2-year OS was 67.7%, and the 2-year PFS was 60.2%. No CTCAE toxicities of grade 3 or higher were reported.[22]

Another study evaluated the outcomes of patients with lung oligometastases after SBRT and compared them with those of lung metastasectomy. Despite the selection bias (SBRT was the second choice in this study when resection was not possible), the 2-year local control rates did not differ significantly between the two study groups: 94% for SBRT and 90% for resection.[23]

The prospective multicenter phase II TROG 13.01 SAFRON II study by Siva et al. is currently addressing the equivalence of single-fraction 28-Gy and 48-Gy SBRT in four fractions for noncentral lung oligometastatic tumors smaller than 5 cm.[24] CRC is the most common primary tumor (47%), followed by lung (11%) and kidney tumors (10%). The prespecified primary endpoint for safety was met. The preliminary results point toward equivalence between the two schemes in terms of local control, OS, and disease-free survival, although maturation is needed for these and other secondary endpoints, such as quality of life and cost-effectiveness.

Role of SBRT in lymph node metastases from CRC

There are few data on the efficacy of SBRT in lymph node metastases from CRC.

A retrospective study of 18 patients with different primary tumors (7 patients with CRC) with oligometastases treated with SBRT (dose: 36–51 Gy in three to five fractions) observed a 1-year local control rate of 94% and a 1-year OS of 89%. No CTCAE toxicities of grade 3 or higher were observed.[25]

One study analyzed a prospective multicenter database of 163 patients with oligometastatic CRC (one to three sites) treated with SBRT. The primary endpoints were local control, PFS, and OS. A total of 162 lesions were treated. They included lymph node (53%), liver (23%), lung (21%), and other (5%) metastases. With a median follow-up of 16 months, the 1-year local control rate was 83.8% (CI: 76.4%–91.9%), and the PFS was 55% (CI: 47%–64.7%). The 1-year local control rate was 90% (CI: 83%–99%) for lymph node metastases and 75% (63%–90%) for visceral metastases. In a multivariate analysis, lymph node metastases and ECOG 0 were found to be significant good prognostic factors. An exploratory analysis suggests that KRAS WT is also a good prognostic factor.[26]

ROLE of SBRT in liver metastases from CRC

SBRT has shown promising results in the treatment of liver metastases owing to the ability of this procedure to deliver a high dose of radiation to the tumor and a minimal dose to adjacent tissues.[27]

Classically, radiotherapy (RT) directed to the liver has been limited by the low tolerance of liver tissue to radiation. The liver is a radiobiological model of parallel architecture, i.e., it does not tolerate high average doses; however, small volumes can receive very high doses. Radio induced liver disease (RILD) usually occurs 2–16 weeks after the end of RT and depends on the average dose to healthy liver tissue; 700 mL of healthy liver should receive <15 Gy. This syndrome is characterized by anicteric ascites with elevated alkaline phosphatase and liver transaminase levels and can lead to liver failure and death.[28]

SBRT is a technological advance that allows a higher dose to be delivered with great precision to the tumor with ablative intent, significantly limiting the dose to the healthy liver and surrounding tissues and resulting in less liver toxicity (thus minimizing RILD). Very high doses can be administered in a single session or in a limited number of sessions.[29]

Retrospective and prospective studies have demonstrated high efficacy and safety with SBRT for liver metastases. The first experience was published in 1995 by Blomberg and included multiple sites, including the liver. In 2001 Herfarth presented his dose escalation study in liver metastases; since then, there has been an exponential increase in the number of publications.[30]

Table 38.1 lists the most relevant retrospective and prospective phase 1 and 2 studies.[31–41] Most studies included patients with good performance status and absent or stable extrahepatic disease and a limited number of treated liver metastases, between one and five lesions, with a maximum size of 6 cm. There was great heterogeneity observed in the treatment schedules, and the fraction regimens ranged from 1 to 6, with a median dose of 45 Gy.

High local control rates (67%–100% at 1 year and 55%–95% at 2 years) and promising OS (70%–100% at 1 year and 30%–83% at 2 years) have been achieved.[31–41]

Local control seems to be mainly influenced by size (3–5 cm), radiation dose (effective biological dose should be above 100 Gy), and dose escalation.[42] When only high-dose regimens such as 75 Gy in three fractions (Scorsetti) were analyzed, the 2-year local control rate was observed to have reached 91% and the 2-year OS, 70%.[31–41]

The toxicity profile is minimal with grades 3 or higher, at a toxicity rate of 1%–10% (RILD incidence: <1%) and with grades 1–2, at 0%–28%. Long-term results have also shown low toxicity (grade 3, <5%).[31–41]

SBRT has also been shown to maintain patients' quality of life (I. Thibault. Toronto Univ. ASTRO2014). A multidisciplinary assessment is recommended to identify patients with liver metastases who are candidates for SBRT. According to current literature, based on the number of metastases and their diameter, distance to organs at risk (rib, stomach, kidney, intestine, and heart), liver function (BT of <3 mg/dL, albumin level of ≥2.5 g/dL, and transaminase level of <3 times the normal value), and free liver volume, the selection criteria would be the following (Table 38.2).[43]

Age is not considered among these selection criteria, as even elderly and frail patients can safely undergo SBRT. Patients are also required to have a good general condition (ECOG 0–1) and adequate laboratory values; patients with potentially treatable extrahepatic disease are also allowed to undergo this intervention.

This technique requires the integration of imaging (computed tomography, magnetic resonance imaging, and positron emission tomography-computed tomography) to adequately define metastases, conformal dosimetry to further minimize the radiation dose to healthy tissues, and a precise immobilization control system to ensure intrafraction stability of liver movement to deliver dose to metastases with high precision and high dose gradient.[44]

The phase III trial NCT01233544 (radiofrequency vs. SBRT) was closed owing to lack of recruitment. Comparative randomized clinical trials on other local treatments and studies that would investigate the associations of this technique with chemotherapy and targeted therapy are needed.

TABLE 38.1 Most relevant prospective and retrospective studies on oligometastatic liver disease related to CRC treated with SBRT.

Author	Number of patients with CRC	Number of injuries	Number of metastases per patient	Metastasis size	Monitoring (months)	Total dose (Gy); number of fractions (fx)	Toxicity	LC	OS
Hoyer Phase II[31]	64	141	1–6	<6 cm	51	45 Gy, 3 fx	3 cases G3 (gastrointestinal, liver failure)	79% at 2 years (per tumor) 64% at 2 years per patient	67% at 1 year 38% at 2 years
Lee Phase I–I[32]	68	143		75.9 mL (median)	10.8	27.7–60 Gy, 6 fx	No RILD 10% G2	70% at 1 year	Median 18 months
Herfarth Phase I–II[33]	37	56	1–3	≤6 cm	18	14–26 Gy, 1 fx	Nonlate >G3	71% at 1 year 66% at 18 months	72% at 1 year 55% at 2 years
Méndez Romero Phase I–II[34]	25	34	1–3	1.1–322 mL	12.9	30–37.5 Gy, 3 fx	2 cases G3 (liver)	100% at 1 year 86% at 2 years	85% at 1 year 62% at 2 years
Rusthoven Phase I–II[35]	47	63	1–3	<6 cm	16	36–60 Gy, 3 fx	<2% G3 No RILD	95% at 1 year 90% at 2 years (100% <3 cm)	30% at 2 years
Scorsetti Phase II[36]	42	52	1–3	<6 cm	24	75 Gy, 3 fx	No RILD No toxicity ≥G3	91% at 1 year 95% at 2 years	70% at 1 year 80% at 2 years
Chang Phase I[37]	65	102	1–4	0.6–3088 mL	14.4	22–60 Gy, 1–6 fx	4 cases G3 (3 gastrointestinal, 2 liver)	67% at 1 year 55% at 2 years	72% at 1 year 38% at 2 years
Andratschke Retrospective[38]	74	91	1–4		15	15–62.5 Gy, 3–5 fx	None G3/4	74.7% at 1 year 48.3% at 2 years 48.3% at 3 years	77% at 1 year 30% at 2 years 27% at 3 years

TABLE 38.1 Most relevant prospective and retrospective studies on oligometastatic liver disease related to CRC treated with SBRT—cont'd

Author	Number of patients with CRC	Number of injuries	Number of metastases per patient	Metastasis size	Monitoring (months)	Total dose (Gy); number of fractions (fx)	Toxicity	LC	OS
Goodman Phase I[39]	81	106	1–3	≤6	33	54 Gy, 3–5 fx	4.9% G3	96% at 1 year 91% at 4 years	69% at 2 years 44% at 3 years 28% at 4 years
Van der Pool Retrospective[40]	20	31	1–4	0.7–6.2 cm	26	37.5–45 Gy, 3 fx	2 cases G3 (liver) 1 G2 rib fracture	100% at 1 year 74% at 2 years	100% at 1 year 83% at 2 years
Rubio Retrospective[41]	21	101	3–14	<8	23.2	36–60 Gy, 3–5 fx	No >G3	94.4% at 1 year 80.6% at 2 years 65% at 4 years	57.6% at 5 years OS: 62 months

CRC, colorectal cancer; SBRT, stereotactic body radiotherapy; LC, local control; OS, overall survival; RILD, radio induced liver disease.

TABLE 38.2 Selection criteria for stereotactic body radiotherapy for liver metastases.

Patients	Number of metastases	Diameter of metastases (cm)	Distance to organs at risk (mm)	Liver function	Healthy liver volume (mL)
Optimal	<3	1–3	>8	Child A	>1000
Intermediate	4	3–6	5–8	Child B	700–1000
Inadequate	>5	>6	< 5	Child C	<700

Conclusions

SBRT allows an opportunity for curative intent in CRC in patients with oligometastatic disease who are not susceptible to surgical intervention owing to unresectability, inoperability, or refusal of surgery. It is a safe and effective procedure with an impact on local control and OS and with low toxicity, which is why it is increasingly implemented in centers that have the infrastructure for this technique.

References

1. Hellman S, Weichselbaum RR. Oligometastases. *J Clin Oncol* 1995;**13**(1):8–10. https://doi.org/10.1200/JCO.1995.13.1.8.
2. Van Cutsem E, Cervantes A, Adam R, et al. ESMO consensus guidelines for the management of patients with metastatic colorectal cancer. *Ann Oncol* 2016;**27**(8):1386–422. https://doi.org/10.1093/annonc/mdw235.
3. Guckenberger M, Lievens Y, Bouma AB, et al. Characterisation and classification of oligometastatic disease: a European society for radiotherapy and oncology and European organisation for research and treatment of cancer consensus recommendation. *Lancet Oncol* 2020;**21**(1):e18–28. https://www.sciencedirect.com/science/article/pii/S1470204519307181.
4. Chua TC, Liauw W, Chu F, Morris DL. Viewing metastatic colorectal cancer as a curable chronic disease. *Am J Clin Oncol* 2012;**35**(1):77–80. https://doi.org/10.1097/COC.0b013e3181fe4444.
5. Lancia A, Zilli T, Achard V, et al. Oligometastatic prostate cancer: the game is afoot. *Cancer Treat Rev* 2019;**73**:84–90. https://doi.org/10.1016/j.ctrv.2019.01.005.
6. Ruers T, Van Coevorden F, Punt CJA, et al. Local treatment of unresectable colorectal liver metastases: results of a randomized phase II trial. *J Natl Cancer Inst* 2017;**109**(9). https://doi.org/10.1093/jnci/djx015.
7. Elias D, Liberale G, Vernerey D, et al. Hepatic and extrahepatic colorectal metastases: when resectable, their localization does not matter, but their total number has a prognostic effect. *Ann Surg Oncol* 2005;**12**(11):900–9. https://doi.org/10.1245/ASO.2005.01.010.
8. Corbin KS, Hellman S, Weichselbaum RR. Extracranial oligometastases: a subset of metastases curable with stereotactic radiotherapy. *J Clin Oncol* 2013;**31**(11):1384–90. https://doi.org/10.1200/JCO.2012.45.9651.
9. Tree AC, Khoo VS, Eeles RA, et al. Stereotactic body radiotherapy for oligometastases. *Lancet Oncol* 2013;**14**(1):e28–37. https://doi.org/10.1016/S1470-2045(12)70510-7.
10. Franzese C, Comito T, Toska E, et al. Predictive factors for survival of oligometastatic colorectal cancer treated with stereotactic body radiation therapy. *Radiother Oncol* 2019;**133**:220–6. https://doi.org/10.1016/j.radonc.2018.10.024.
11. Palma DA, Olson R, Harrow S, et al. Stereotactic ablative radiotherapy versus standard of care palliative treatment in patients with oligometastatic cancers (SABR-COMET): a randomised, phase 2, open-label trial. *Lancet* 2019;**393**(10185):2051–8. https://doi.org/10.1016/S0140-6736(18)32487-5.
12. Palma DA, Olson R, Harrow S, et al. Stereotactic ablative radiotherapy for the comprehensive treatment of oligometastatic cancers: long-term results of the SABR-COMET phase II randomized trial. *J Clin Oncol* 2020;**38**(25):2830–8. https://doi.org/10.1200/JCO.20.00818.
13. Lehrer EJ, Singh R, Wang M, et al. Safety and survival rates associated with ablative stereotactic radiotherapy for patients with oligometastatic cancer: a systematic review and meta-analysis. *JAMA Oncol* 2021;**7**(1):92–106. https://doi.org/10.1001/jamaoncol.2020.6146.
14. Chalkidou A, Macmillan T, Grzeda MT, et al. Stereotactic ablative body radiotherapy in patients with oligometastatic cancers: a prospective, registry-based, single-arm, observational, evaluation study. *Lancet Oncol* 2021;**22**(1):98–106. https://doi.org/10.1016/S1470-2045(20)30537-4.
15. Aranda E, Abad A, Carrato A, et al. Treatment recommendations for metastatic colorectal cancer. *Clin Transl Oncol* 2011;**13**(3):162–78. https://doi.org/10.1007/s12094-011-0636-7.
16. Kobiela J, Spychalski P, Marvaso G, et al. Ablative stereotactic radiotherapy for oligometastatic colorectal cancer: systematic review. *Crit Rev Oncol Hematol* 2018;**129**:91–101. https://doi.org/10.1016/j.critrevonc.2018.06.005.
17. van Halteren HK, van Geel AN, Hart AA, Zoetmulder FA. Pulmonary resection for metastases of colorectal origin. *Chest* 1995;**107**(6):1526–31. https://doi.org/10.1378/chest.107.6.1526.
18. Londero F, Grossi W, Morelli A, et al. Surgery versus stereotactic radiotherapy for treatment of pulmonary metastases. A systematic review of literature. *Future Sci OA* 2020;**6**(5):FSO471. https://doi.org/10.2144/fsoa-2019-0120.

19. Okunieff P, Petersen AL, Philip A, et al. Stereotactic body radiation therapy (SBRT) for lung metastases. *Acta Oncol* 2006;**45**(7):808–17. https://doi.org/10.1080/02841860600908954.
20. Hof H, Hoess A, Oetzel D, Debus J, Herfarth K. Stereotactic single-dose radiotherapy of lung metastases. *Strahlenther Onkol* 2007;**183**(12):673–8. https://doi.org/10.1007/s00066-007-1724-z.
21. Comito T, Cozzi L, Clerici E, et al. Stereotactic ablative radiotherapy (SABR) in inoperable oligometastatic disease from colorectal cancer: a safe and effective approach. *BMC Cancer* 2014;**14**:619. https://doi.org/10.1186/1471-2407-14-619.
22. Agolli L, Bracci S, Nicosia L, Valeriani M, De Sanctis V, Osti MF. Lung metastases treated with stereotactic ablative radiation therapy in oligometastatic colorectal cancer patients: outcomes and prognostic factors after long-term follow-up. *Clin Colorectal Cancer* 2017;**16**(1):58–64. https://doi.org/10.1016/j.clcc.2016.07.004.
23. Widder J, Klinkenberg TJ, Ubbels JF, Wiegman EM, Groen HJM, Langendijk JA. Pulmonary oligometastases: metastasectomy or stereotactic ablative radiotherapy? *Radiother Oncol* 2013;**107**(3):409–13. https://doi.org/10.1016/j.radonc.2013.05.024.
24. Siva S, Bressel M, Kron T, et al. Stereotactic ablative fractionated radiotherapy versus radiosurgery for oligometastatic neoplasia to the lung: a randomized phase II trial. *Int J Radiat Oncol Biol Phys* 2020;**108**(3):S3–4. https://doi.org/10.1016/j.ijrobp.2020.07.2072.
25. Yeung R, Hamm J, Liu M, Schellenberg D. Institutional analysis of stereotactic body radiotherapy (SBRT) for oligometastatic lymph node metastases. *Radiat Oncol* 2017;**12**(1):105. https://doi.org/10.1186/s13014-017-0820-1.
26. O'Cathail SM, Smith T, Owens R, et al. Superior outcomes of nodal metastases compared to visceral sites in oligometastatic colorectal cancer treated with stereotactic ablative radiotherapy. *Radiother Oncol* 2020;**151**:280–6. https://doi.org/10.1016/j.radonc.2020.08.012.
27. Vera R, González-Flores E, Rubio C, et al. Multidisciplinary management of liver metastases in patients with colorectal cancer: a consensus of SEOM, AEC, SEOR, SERVEI, and SEMNIM. *Clin Transl Oncol* 2019;**22**(5):647–62. https://doi.org/10.1007/s12094-019-02182-z.
28. Dawson LA, Normolle D, Balter JM, McGinn CJ, Lawrence TS, Ten Haken RK. Analysis of radiation-induced liver disease using the Lyman NTCP model. *Int J Radiat Oncol Biol Phys* 2002;**53**(4):810–21. https://doi.org/10.1016/s0360-3016(02)02846-8.
29. Elias D, Viganò L, Orsi F, et al. New perspectives in the treatment of colorectal metastases. *Liver Cancer* 2017;**6**(1):90–8. https://doi.org/10.1159/000449492.
30. Comito T, Clerici E, Tozzi A, D'Agostino G. Liver metastases and SBRT: a new paradigm? *Rep Pract Oncol Radiother* 2015;**20**(6):464–71. https://doi.org/10.1016/j.rpor.2014.10.002.
31. Hoyer M, Roed H, Traberg Hansen A, et al. Phase II study on stereotactic body radiotherapy of colorectal metastases. *Acta Oncol* 2006;**45**(7):823–30. https://doi.org/10.1080/02841860600904854.
32. Lee MT, Kim JJ, Dinniwell R, et al. Phase I study of individualized stereotactic body radiotherapy of liver metastases. *J Clin Oncol* 2009;**27**(10):1585–91. https://doi.org/10.1200/JCO.2008.20.0600.
33. Herfarth KK, Debus J, Wannenmacher M. Stereotactic radiation therapy of liver metastases: update of the initial phase-I/II trial. *Front Radiat Ther Oncol* 2004;**38**:100–5. https://doi.org/10.1159/000078271.
34. Mendez Romero A, Wunderink W, Hussain SM, et al. Stereotactic body radiation therapy for primary and metastatic liver tumors: a single institution phase i–ii study. *Acta Oncol* 2006;**45**(7):831–7. https://www.tandfonline.com/doi/abs/10.1080/02841860600897934.
35. Rusthoven KE, Kavanagh BD, Cardenes H, et al. Multi-institutional phase I/II trial of stereotactic body radiation therapy for liver metastases. *J Clin Oncol* 2009;**27**(10):1572–8. https://doi.org/10.1200/JCO.2008.19.6329.
36. Scorsetti M, Comito T, Tozzi A, et al. Final results of a phase II trial for stereotactic body radiation therapy for patients with inoperable liver metastases from colorectal cancer. *J Cancer Res Clin Oncol* 2015;**141**(3):543–53. https://doi.org/10.1007/s00432-014-1833-x.
37. Chang DT, Swaminath A, Kozak M, et al. Stereotactic body radiotherapy for colorectal liver metastases: a pooled analysis. *Cancer* 2011;**117**(17):4060–9. https://doi.org/10.1002/cncr.25997.
38. Andratschke NH, Nieder C, Heppt F, Molls M, Zimmermann F. Stereotactic radiation therapy for liver metastases: factors affecting local control and survival. *Radiat Oncol* 2015;**10**:69. https://doi.org/10.1186/s13014-015-0369-9.
39. Goodman BD, Mannina EM, Althouse SK, Maluccio MA, Cárdenes HR. Long-term safety and efficacy of stereotactic body radiation therapy for hepatic oligometastases. *Pract Radiat Oncol* 2016;**6**(2):86–95. https://doi.org/10.1016/j.prro.2015.10.011.
40. van der Pool AEM, M\endez Romero A, Wunderink W, et al. Stereotactic body radiation therapy for colorectal liver metastases. *Br J Surg* 2010;**97**(3):377–82. https://doi.org/10.1002/bjs.6895.
41. Rubio C, Hernando-Requejo O, Zucca Aparicio D, et al. Image guided SBRT for multiple liver metastases with ExacTrac® adaptive gating. *Rep Pract Oncol Radiother* 2017;**22**(2):150–7. https://doi.org/10.1016/j.rpor.2016.07.006.
42. Dewas S, Bibault J-E, Mirabel X, et al. Prognostic factors affecting local control of hepatic tumors treated by stereotactic body radiation therapy. *Radiat Oncol* 2012;**7**:166. https://doi.org/10.1186/1748-717X-7-166.
43. Scorsetti M, Clerici E, Comito T. Stereotactic body radiation therapy for liver metastases. *J Gastrointest Oncol* 2014;**5**(3):190–7. https://www.ncbi.nlm.nih.gov/pmc/articles/pmc4074953/.
44. Wild AT, Yamada Y. Treatment options in oligometastatic disease: stereotactic body radiation therapy—focus on colorectal cancer. *Visc Med* 2017;**33**(1):54–61. https://doi.org/10.1159/000454685.

Chapter 39

Palliative radiotherapy in CRC

Patricia Calvo-Crespo, Begoña Taboada-Valladares, and Antonio Gómez-Caamaño
Department of Radiation Oncology, Santiago de Compostela University Hospital Complex, Santiago de Compostela, Spain

Introduction

Over the last decades, advances in both diagnosis and treatment modalities have contributed to improved outcomes in rectal carcinoma, including increased survival in metastatic disease.[1]

In patients treated with curative intent, 15% develop pelvic recurrence.[2] Both patients with primary rectal tumors and pelvic recurrences may present with symptoms that cause significant pelvic morbidity, including pain, obstruction, urgency, bleeding, or incontinence. There is a group of patients who are not candidates for surgery or with unresectable pelvic recurrences in whom quality of life may be affected by these local symptoms.[1] Palliative radiotherapy is used to relieve symptoms in patients with locally advanced or recurrent inoperable rectal tumors.[3]

The administration of palliative radiotherapy in terms of indication, dose, timing, in palliative treatment is not well established,[1] although there are numerous publications that attempt to clarify this.

Retrospective studies

In 1998 Wong et al.[4] published a study on the role of radiotherapy in the management of pelvic recurrence, a retrospective study of 519 patients treated with palliative radiotherapy between 1975 and 1985 for recurrence of rectal carcinoma without prior radiotherapy or chemotherapy. The dose of radiotherapy varied between 4.4 and 65 Gy and the daily fractionation between 1.8 and 2.5 Gy. The median survival was 14 months and the time to local progression after radiotherapy was 5 months. The 5-year overall survival was 5% and pelvic disease-free progression was 7%. In multivariate analysis overall survival was significantly related to performance status, absence of extra-pelvic metastases, long interval from surgery to radiotherapy, total dose of radiotherapy, and absence of obstructive uropathy. Pelvic progression-free disease was associated with performance status and total radiotherapy dose. In analgesic treatment 48% responded after doses <20 Gy, 77% with doses 30–45 Gy, and 89% with doses >45 Gy.

They conclude that pelvic radiotherapy plays an important role in symptom relief but do not determine optimal doses or fractionations.

A study of old techniques by Wang et al.[5] analyzed 111 patients between 1940 and 1960 with inoperable or recurrent rectosigmoid carcinoma with symptoms of pain, mass effect, hemorrhage, or discharge. They found that symptoms at 6 months were better controlled with increasing dose (12% with 21–30 Gy dose, 31% with 31–40 Gy dose, 58% with 41–50 Gy dose).

Crane et al.[6] analyzed 80 patients with metastatic rectal carcinoma receiving hypofractionated radiotherapy with 3 different dose schedules (30 Gy/6 fr, 35 Gy/14 fr, and 45 Gy/25 fr). In multivariate analysis a DBE < 35 Gy demonstrated a higher risk of symptomatic pelvic progression, $P = .009$.

A study was published in 2011[7] of 80 patients with symptomatic pelvic masses due to metastatic rectal tumors, with BED greater or less than 40 Gy. Radiotherapy was effective with symptom control if BED >40 Gy, compared to doses <40 Gy. However, only 58 patients had rectal carcinoma and 23% were treated with radio-chemotherapy.

In 2016 Chia et al.[8] published a retrospective study of locally advanced rectal carcinoma or rectal recurrence including patients from the National Cancer Institute of Singapore Radiation Therapy Centers. Patients had at least one symptom: pain, bleeding, or obstruction. All patients received radiation therapy only, those receiving chemotherapy were excluded. The GTV (gross tumor volume) was the rectal tumor with/without regional nodes. The GTV was plus 1–2 cm margin for PTV (planning target volume). Response to bleeding was considered as such if patients maintained hemoglobin levels without the need for transfusion, response to pain if they had decreased pain or less need for analgesia, response to obstruction if they managed to improve constipation or decrease the use of laxatives. The study objectives were

TABLE 39.1 Summary of symptoms and response characteristics.

Symptoms	No. patients	Response to RT	Medium response time	% of relief	No benefit from RT
Bleeding	83	86.7%	5.4 months	91.4%	7 patients
Pain	29	79.3%	4.5 months	84.8%	4 patients
Obstruction	8	62.5%	4.2 months	80.3%	1 patients

symptomatic response (response rates, duration of response), median survival, and toxicity. Patients were compared according to the DBE achieved, with a median dose of 39 Gy, between those receiving BED > 39 Gy and BED < 39 Gy. Ninety-nine patients and 120 symptoms were included, with a follow-up duration of 6.9 months. Symptoms and response characteristics are summarized in Table 39.1.

The median survival of patients treated with palliative RT was 6.9 months (0–33). The 12-month survival was 30.6%. No statistically significant differences were observed in terms of dose received (BED < or > 39 Gy). No prognostic factors (age, RT dose, PS, gender) were associated with survival and symptom control. There was 3% Gr3 toxicity and no Gr4 toxicity.

They conclude that RT alone is effective and well tolerated in the palliative treatment of symptoms associated with rectal cancer, obtaining response rates of 90%, with a maintained response in most cases.

Mohiunddin et al.[9] reviewed 103 patients treated with reirradiation with concomitant QT, reaching 22% Gr3 and 6% Gr4 toxicity. They suggest that the dose of reirradiation should be set according to the time passed since previous RT, 35 Gy for 3–12 weeks, 40–45 Gy for 12–24 weeks, 45–50 Gy for 24–36 weeks, 50–55 Gy for more than 36 weeks. Table 39.2 summarizes the retrospective studies.

Prospective studies

Cameron et al.[10] conducted a prospective phase II study of palliative RT with doses of 30–39 Gy. The primary endpoint was the palliative effect of RT in symptomatic or recurrent primary rectal carcinoma at 12 weeks. Other study objectives

TABLE 39.2 Summary of retrospective studies.

Author/year	Design study	Number of patients	RT dose and fractionation	Response to RT palliative
Wang et al 1962[5]	Retrospective	111		Better with dose increase
Wong et al 1998[4]	Retrospective	519	4.4–66.5 Gy/1.8–2.5 Gy	89% responses with doses >45 Gy
Crane et al 2001[6]	Retrospective	80	30 Gy/6 fx (50%) 45 Gy/25 fx (16%) 35 Gy/14 fx (14%)	BED < 35 Gy more risk of symptomatic pelvic progression, $P = .009$
Bae et al 2011[7]	Retrospective	80	Median RT dose 36 Gy (8–60 Gy) Median dose per fraction 2.5 Gy (1.8–8 Gy).	RT was effective with symptom control if DBE >40 Gy
Mohiunddin et al 2002[9]	Retrospective	103	Reirradiation RT dose according to previous RT	35 Gy if 3–12 weeks prior RT 40–45 Gy for 12–24 weeks 45–50 Gy for 24–36 weeks 50–55 Gy for over 36 weeks
Chia et al 2016[8]	Retrospective	99	BED > 39 Gy and BED < 39 Gy	Survival at 12 m of 30.6% No differences for total dose or prognostic factors

included quality of life and toxicity at 6 and 12 weeks after RT. Eight centers in Norway recruited 51 patients between 2009 and 2015, with primary or recurrent rectal carcinoma with pelvic symptoms, with or without metastases, with life expectancy greater than 3 months. Doses ranged from 30 to 39 Gy in fractions of 3 Gy/fraction. The GTV included the pelvic tumor with or without nodes; the GTV was given 1–2 cm to create PTV. The median age of the patients was 79 years, 24% were pelvic recurrences, 80% had metastases outside the pelvis, 63% had not received previous chemotherapy, and 45% were ileostomy/colostomy carriers. Forty-one percent of patients had background opioids. The most frequent symptoms were 47% pain, 31% rectal dysfunction (obstruction, incontinence, diarrhea, mucus production), and 18% hemorrhage Thirty-three patients were evaluated at 12 weeks, 52% had complete responses, 33% improvement of symptoms, 12% no change, and 1% worsening of clinical symptoms. Sixty percent of evaluable patients had symptom response at the end of radiotherapy (about 3 days) and 85% at 6 and 12 weeks. No Gr4 toxicities were reported. The most frequent Gr1-2 toxicities were proctitis, diarrhea, nausea, dysuria, and increased urinary frequency. This is the first documented study of symptoms and toxicity of radiotherapy in primary tumor or recurrence using quality of life questionnaires. Radiotherapy at doses of 30–39 Gy contributes to the palliative treatment of pelvic symptoms, including rectal dysfunction, pain, and hematochezia with acceptable toxicity.

Picardi et al.[11] conducted another phase II trial evaluating the efficacy of short course radiotherapy in patients with obstructing rectal carcinoma. Patients who were not candidates for surgery either due to multiple synchronous metastases, age, or comorbidities were included. A dose of 25 Gy was administered in 5 fractions. Chemotherapy was suspended during radiotherapy. They recruited 18 patients between 2003 and 2012. Median follow-up was 11.5 months. At 4 weeks of treatment, a complete response was observed in 38.9% of patients and a partial response in 50%, while 11.1% did not respond to treatment. The rates of resolution or reduction of bleeding and pain were 100% and 87.5%, respectively. Colostomy-free survival at 1, 2, and 3 years was 85.2%, 53%, and 39.8%, respectively. No treatment was discontinued due to toxicity. They conclude that the short course of 5 × 5 Gy represents an effective and safe treatment in patients with rectal obstruction who are not candidates for curative treatment, avoiding colostomy in a significant proportion of patients.

Another similar phase II by Tyc-Szczepaniak et al.[12] studied in which situations of symptomatic E.IV. rectal carcinoma surgery could be avoided. They included 40 patients with symptomatic rectal carcinoma and distant metastases who received 25 Gy in five fractions followed by oxaliplatin-based chemotherapy. Before starting treatment 35% had lesions with almost complete obstructions. The palliative effect was assessed with patient questionnaires. Median follow-up was 26 months. Median OS was 11.5 months. Twenty percent of patients required surgery throughout the course of the disease. Thirty percent of patients had complete resolution of symptoms and 35% had significant improvement. The probability of requiring palliative surgery at 2 years was 17.5% and the probability of having good symptom control with radiotherapy was 67%. The authors conclude that short course radiotherapy followed by chemotherapy avoids surgery for most patients, including those with near complete obstructions. Table 39.3 summarizes the prospective studies.

Systematic reviews

The utility of radiotherapy has been well established in locally advanced nonmetastatic rectal carcinoma and its pre- and postoperative sequence.[13, 14]

TABLE 39.3 Summary of prospective studies.

Author/year	Design study	Number of patients	RT dosage and fractionation	Response to palliative RT
Tyc-Szczepaniak et al 2013[12]	Prospective	40	25 Gy/5 fx (97.5%) 30 Gy/6 fx (2.5%)	Median OS 11.5 months 30% of patients complete response 35% significant improvement
Picardi et al 2016[11]	Prospective	18	25 Gy/5 fx	At 4 weeks of treatment: 38.9% of patients CR and 50% RP
Cameron et al 2016[10]	Prospective	51	30–39 Gy at 3 Gy/fraction	52% obtained CR 33% PR 12% no change 1% worsening of symptoms

In the setting of metastatic rectal cancer disease, improvement in local control of the primary tumor using local therapies is associated with improved prognosis and survival.[15, 16]

Pelvic radiotherapy is effective in controlling symptoms of E.IV. tumors There are two reviews from different periods that confirm the efficacy of palliative radiotherapy in controlling symptoms of the primary tumour[1, 17] or recurrences.

In 2014 Cameron et al.[1] published a systematic review including studies of palliative radiotherapy (between 1949 and 1999) in rectal carcinoma assessing symptom response and quality of life. They included a total of 27 studies, 4 prospective and 23 retrospective. Three of the studies included only patients with primary rectal carcinoma, 14 with recurrence or residual disease and 10 studies with a combination of both. They included a total of 1759 patients, with a median age of 65 years. The most frequent symptoms were pain, rectorrhagia, mass effect, and rectal dysfunction. Radiotherapy definition, doses, schedules, and volumes were highly variable, ranging from 5 to 70 Gy in fractions of 1.5–5 Gy/fraction. Daily fractions of 2 Gy to a total dose of 30–60 Gy were most common. Response rates ranged from 56% to 100%. The authors conclude that high effectiveness is obtained with palliative radiotherapy on pelvic symptoms, such as pain, bleeding, mass effect, with acceptable toxicity. However, the heterogeneity in terms of doses and treatment volumes limits the extrapolation of results.

Symptomatic responses were found with low doses of radiotherapy (<20 Gy) during fractionated treatment[18] or after single fractions of 5–10 Gy.[19]

Pahlman et al.[20] reported that palliation was observed at 20–30 Gy, and that these patients were symptom free at 46 Gy, so did not benefit from escalation to 64 Gy.

Three retrospective studies[5, 21, 22] analyzed the duration of response to radiotherapy treatment, whereby the patients who obtained a longer duration of response were those who received higher doses of radiotherapy, although this was not statistically significant.

Buwenge et al.[17] conducted another review of more recent publications (2010–16) of palliative RT with modern techniques in symptomatic E.IV. rectal carcinoma. They included 9 studies (6 retrospective and 3 phase II) in which RT achieved response rates of 79%, 87%, and 78% for pain, bleeding, and obstruction, respectively. RT was well tolerated, and the most common side effect was diarrhea/proctitis.

Conclusions

RT is effective and safe in the treatment of local symptoms in inoperable primary tumors or pelvic recurrences, achieving significant rates of symptomatic control.

Prospective studies are needed to analyze the benefit of palliative RT in patients with rectal cancer, determining the optimal dose and fractionation, although it appears that both 5×5 Gy fractionation and other fractionations with DBE > 39 Gy are associated with better local control with acceptable toxicity.

References

1. Cameron MG, Kersten C, Vistad I, Fosså S, Guren MG. Palliative pelvic radiotherapy of symptomatic incurable rectal cancer—a systematic review. *Acta Oncol* 2014;**53**(2):164–73. https://www.tandfonline.com/doi/abs/10.3109/0284186X.2013.837582.
2. Pilipshen SJ, Heilweil M, Quan SH, Sternberg SS, Enker WE. Patterns of pelvic recurrence following definitive resections of rectal cancer. *Cancer* 1984;**53**(6):1354–62. https://doi.org/10.1002/1097-0142(19840315)53:6<1354.
3. Navrátilová P, Hynková L, Šlampa P. The role of palliative radiotherapy in bleeding from locally advanced gastrointestinal tumors. *Klin Onkol* 2017;**30**(6):433–6. https://doi.org/10.14735/amko2017433.
4. Wong CS, Cummings BJ, Brierley JD, et al. Treatment of locally recurrent rectal carcinoma—results and prognostic factors. *Int J Radiat Oncol Biol Phys* 1998;**40**(2):427–35. https://doi.org/10.1016/s0360-3016(97)00737-2.
5. Wang CC, Schulz MD. The role of radiation therapy in the management of carcinoma of the sigmoid, rectosigmoid, and rectum. *Radiology* 1962;**79**: 1–5. https://doi.org/10.1148/79.1.1.
6. Crane CH, Janjan NA, Abbruzzese JL, et al. Effective pelvic symptom control using initial chemoradiation without colostomy in metastatic rectal cancer. *Int J Radiat Oncol Biol Phys* 2001;**49**(1):107–16. https://doi.org/10.1016/s0360-3016(00)00777-x.
7. Bae SH, Park W, Choi DH, et al. Palliative radiotherapy in patients with a symptomatic pelvic mass of metastatic colorectal cancer. *Radiat Oncol* 2011;**6**:52. https://doi.org/10.1186/1748-717X-6-52.
8. Chia D, Lu J, Zheng H, et al. Efficacy of palliative radiation therapy for symptomatic rectal cancer. *Radiother Oncol* 2016;**121**(2):258–61. https://doi.org/10.1016/j.radonc.2016.06.023.
9. Mohiuddin M, Marks G, Marks J. Long-term results of reirradiation for patients with recurrent rectal carcinoma. *Cancer* 2002;**95**(5):1144–50. https://doi.org/10.1002/cncr.10799.

10. Cameron MG, Kersten C, Vistad I, et al. Palliative pelvic radiotherapy for symptomatic rectal cancer—a prospective multicenter study. *Acta Oncol* 2016;**55**(12):1400–7. https://www.tandfonline.com/doi/abs/10.1080/0284186X.2016.1191666.
11. Picardi V, Deodato F, Guido A, et al. Palliative short-course radiation therapy in rectal cancer: a phase 2 study. *Int J Radiat Oncol Biol Phys* 2016; **95**(4):1184–90. https://doi.org/10.1016/j.ijrobp.2016.03.010.
12. Tyc-Szczepaniak D, Wyrwicz L, Kepka L, et al. Palliative radiotherapy and chemotherapy instead of surgery in symptomatic rectal cancer with synchronous unresectable metastases: a phase II study. *Ann Oncol* 2013;**24**(11):2829–34. https://doi.org/10.1093/annonc/mdt363.
13. Sauer R, Becker H, Hohenberger W, et al. Preoperative versus postoperative chemoradiotherapy for rectal cancer. *N Engl J Med* 2004;**351**(17): 1731–40. https://doi.org/10.1056/NEJMoa040694.
14. Sauer R, Liersch T, Merkel S, et al. Preoperative versus postoperative chemoradiotherapy for locally advanced rectal cancer: results of the German CAO/ARO/AIO-94 randomized phase III trial after a median follow-up of 11 years. *J Clin Oncol* 2012;**30**(16):1926–33. http://mauriciolema.webhost4life.com/rolmm/downloads/files/ChemoRTPreVsPostOpRectalCa.pdf.
15. Fossum CC, Alabbad JY, Romak LB, et al. The role of neoadjuvant radiotherapy for locally-advanced rectal cancer with resectable synchronous metastasis. *J Gastrointest Oncol* 2017;**8**(4):650–8. https://doi.org/10.21037/jgo.2017.06.07.
16. Takada T, Tsutsumi S, Takahashi R, et al. Control of primary lesions using resection or radiotherapy can improve the prognosis of metastatic colorectal cancer patients. *J Surg Oncol* 2016;**114**(1):75–9. https://doi.org/10.1002/jso.24255.
17. Buwenge M, Giaccherini L, Guido A, et al. Radiotherapy for the primary tumor in patients with metastatic rectal cancer. *Curr Colorectal Cancer Rep* 2017;**13**(3):250–6. https://doi.org/10.1007/s11888-017-0371-8.
18. Stearns Jr MW, Whiteley Jr HW, Leaming RH, Deddish MR. Palliative radiation therapy in patients with localized cancer of the colon and rectum. *Dis Colon Rectum* 1970;**13**(2):112–5. https://doi.org/10.1007/BF02617638.
19. Allum WH, Mack P, Priestman TJ, Fielding JW. Radiotherapy for pain relief in locally recurrent colorectal cancer. *Ann R Coll Surg Engl* 1987;**69** (5):220–1. https://www.ncbi.nlm.nih.gov/pubmed/2445237.
20. Påhlman L, Glimelius B, Ginman C, Graffman S, Adalsteinsson B. Preoperative irradiation of primarily non-resectable adenocarcinoma of the rectum and rectosigmoid. *Acta Radiol Oncol* 1985;**24**(1):35–9. https://doi.org/10.3109/02841868509134362.
21. Gescher FM, Keijser AH. Palliative irradiation for recurrent rectal carcinoma. *Ned Tijdschr Geneeskd* 1987;**131**(13):533–5. Accessed 7 March 2021 https://inis.iaea.org/search/search.aspx?orig_q=RN:18090319.
22. Wise RE, Smedal MI. Palliative treatment of recurrent rectosigmoidal neoplasms with two million volt radiation. *Surg Clin North Am* 1959;**39**(3): 775–80. https://doi.org/10.1016/s0039-6109(16)35803-0.

Chapter 40

Immunology and immunotherapy in CRC

Oscar J. Cordero[a], Rubén Varela-Calviño[a], Begoña Graña-Suárez[b], and Alba García-López[c]

[a]*University of Santiago de Compostela, Department of Biochemistry and Molecular Biology. Santiago de Compostela, Spain,* [b]*Galician Health System (Sergas). Service of Oncology, Universitary Hospital Complex of A Coruña (CHUAC), A Coruña, Spain,* [c]*Galician Health System (Sergas). Service of Pharmacy, Universitary Hospital Complex of Santiago de Compostela (CHUS), Santiago de Compostela, Spain*

Immunity, inflammation, intestinal microbiota, and colorectal cancer

CRC is initially developed from genetic alterations that affect genes that encode intestinal homeostatic regulators or repair mismatch factors in DNA, primarily in intestinal stem cells that become as a result cancer stem cells (CSCs).[1,2]

When the transformed cells grow, they develop an organized and complex tissue formed by both transformed and normal stromal cells in a symbiotic relationship that supports the growth of the tumor and favors its dissemination. This is reminiscent of the relationship of an organism with infectious pathogens during a chronic infection, where surrounding tissues are often reorganized into specific anatomical structures such as granulomas. Although the main function of the immune system and inflammation is the elimination of pathogens or, at least, the control of their dissemination, simultaneously, these physiological responses induce repair mechanisms to recover the function and integrity of the tissue.

The similarity between cancer and inflammation was described many years ago: how they share some basic mechanisms of development (angiogenesis), tissue infiltration by different types of immune cells, and how tumors act as "wounds that don't heal,"[3,4] since growing tumors are not immunologically silent but there may be successful malignant cells that evolve together in that environment.[5,6] Consequently, Hanahan and Weinberg[7] in their 2011 update of the "Hallmarks of Cancer" introduced the immune attributes of cancer, focusing on immunosuppression on the one hand and on promoting inflammation in the tumor on the other.[8–10] Although the majority of human tumors originate in tissues with sterile chronic inflammation (only 15% of tumors can be attributed to a carcinogenic infection by pathogens such as human papillomavirus (HPV) or Epstein-Barr virus (EBV), which directly induce cell transformation), other pathogens can contribute to this situation of **chronic inflammation** of infected tissues,[11,12] for example hepatitis virus or bacteria such as *Helicobacter pylori*.

In this context, the mechanisms of inflammation involved in cell transformation and the onset of cancer are of two types: an intrinsic inflammation, when the transformation mechanisms are also responsible for the activation of a proinflammatory program in the tumor cell, by for example, overexpression of oncogenes (in particular Ras, Myc, Src, RET, and microRNAs such as mi-R155), mutation of other genes, DNA damage, or mitochondrial production of reactive oxygen species (ROS, *"reactive oxygen species"*).[8,10,13] On the contrary, extrinsic inflammation is activated by damage to the tissues surrounding the malignant cells. Again, several mechanisms such as activation of oncogenes, endoplasmic reticulum (ER) stress, oxidative damage, or cellular senescence will cause cellular stress, in turn inducing the synthesis of pro-inflammatory factors such as cytokines, chemokines, interferons (IFN), and enzymes that reorganize the tissue.

This inflammation is mediated primarily by infiltrating immune cells, although inflammation induced by mechanical, genetic, chemical, or radiation syndromes is also important.[14–16] Among the hematopoietic inflammatory cells recruited by these tissue-specific inflammatory responses, which we will discuss in more depth, are neutrophils, dendritic cells (DCs), macrophages, and lymphocytes. All these cells are capable of secreting molecules, such as cytokines and ROS, that contribute to the inflammatory environment and participate in the growth and dissemination of transformed cells.[17–20] For example, macrophages contribute to the elimination of transformed cells, but also produce potent angiogenic and growth factors such as VEGF-C and D that enhance neoplastic progression.[17,19]

In turn, this created inflammatory environment will favor the initial genetic mutation, functional modifications of the proteins, and/or epigenetic mechanisms that drive cell transformation and the onset of cancer. For example, ROS and reactive nitrogen species (RNS) can induce breaks and other mutations in DNA, as well as epigenetic modifications in proto-oncogenes and tumor suppressor genes, among others, primarily in CSCs.[7,8,21,22]

These immune cells express multiple specialized families of *pattern recognition receptors* (PRRs), including Toll-like receptors (TLRs), NOD (oligomerization binding domain of nucleotides)-like receptors (NLR, "*NOD-like receptors*"), and many others that may be found assembled to form signaling complexes including *inflammasomes*, *myddosomes*, and filament formation by MAVS (*mitochondrial antiviral signaling protein*) initiated by RIG-1.[20, 23] In addition, some PRRs (e.g., TLR 2 and 4 and NLRP3) have recently been identified in tumor cells themselves and also in metabolic tissues in what may represent feedback cycles.[24]

Innate antigen receptors are very promiscuous. Thus the same or very similar ligands can bind to different receptors in the same cell or in a different cell population. They can bind to *molecular patterns associated with cellular damage* (DAMPs) related to endogenous components released from cells damaged by molecular stress.[17, 20, 23] Recent results point to tumor-derived DNA, detected by the STING pathway in a certain population of intra-tumor DCs, as the main DAMP inducer of the production of interferon (IFN) type I and the host's antitumor immune responses (reviewed in Corrales, Matson & Flood, 2017[25]).

However, these PRRs were initially identified as receptors of molecular patterns associated with pathogens (PAMPs, *Pathogen-Associated Molecular Patterns*). The commensal microbiota and the effects of dysbiosis on local inflammation and carcinogenesis, at the interfaces between body tissues and the outside environment, can now be easily explained by the direct interaction of bacteria or their products with innate receptors in epithelial cells. This also applies to the upper gastrointestinal tract, the oral and bronchio-alveolar mucosa, and the skin. How the interaction between risk factors and commensal flora cooperates with colorectal carcinogenesis initiated by genetic predisposition or environmental causes has also been reviewed.[13, 17–20, 26, 27]

We emphasize that these interactions have additional profound systemic effects through the cells of the immune system and the alteration of the inflammatory environment. In addition, systemic metabolic disorders such as obesity, diabetes, and cachexia also cause inflammation, as we will comment. For example, dietary fats, such as saturated fatty acids (SFA), are well-characterized nutrients that promote metabolic inflammation through the NLRP3 inflammasome in macrophages. In contrast, some polyunsaturated fatty acids actively reduce inflammation through resolvins and protectins.[24, 28]

Inflammatory bowel disease and dysbiosis

In the case of the intestinal epithelial barrier we must remember that these circumstances are different, since the **commensal microbiota** of the intestine present in a healthy individual achieve homeostasis, regeneration, and healing of the bowel by establishing complex molecular and cellular interactions with epithelial, mesenchymal, and immune cells (reviewed in Kurashima and Kiyono, 2017[26]). The way in which exposure to developing commensal flora affects the maturation of both innate and adaptive mucosal immunity, or systemic inflammation/immunity after birth, has been extensively studied.[17, 25]

However, the discovery that patients with inflammatory bowel disease (IBD) have an increased susceptibility to develop tumors made that colorectal cancers in IBD patients (with Crohn's disease or with ulcerative colitis) were considered typical examples of tumors related to inflammation (cancers associated with colitis, CAC).[29, 30] Alterations in the percentages of the composition of the flora of the intestine and/or changes in the presence of specific bacterial species could modulate that environment through the catabolism of natural mutagens and carcinogens. In addition, modifications in the microbiota induce changes in the specific enzymatic activities present in the commensal flora with both pro- and antiinflammatory effects.[31]

Interestingly, patients with IBD also have a greater susceptibility to lymphomas/leukemia, hepatocarcinomas, and other tumors, suggesting that intestinal inflammation due to a pathological immune response against the commensal microbiota is responsible for tumor-promoting effects at a local but also at a systemic level.[32]

Many of the innate immune receptors share the signal transduction pathway mediated by the MyD88 adapter protein (associated with most TLR signaling pathways). Although MyD88 has a protective role in the development of colonic tumors in mouse models,[33, 34] this signaling pathway and specific microRNAs control the chemistry resistance of the CRC. Some mice deficient for individual TLRs also show increased susceptibility to colitis (reviewed in Saleh and Trinchieri, 2011[31]). Similarly, mice deficient in IL-18 and the IL-18 receptor (IL-18R), which require MyD88 for signaling, also show a greater susceptibility to CAC development.[17, 23–25, 28, 35–38]

The nuclear factor-κB (NF-κB) and the signal transducer and activator of transcription-3 (STAT-3) are among the best characterized transcription factors induced by inflammatory mediators.[1, 17–25, 28, 36] The activation of NF-κB in response to chronic inflammation is particularly relevant for the development of gastrointestinal cancer. In addition, the number of cells that show NF-κB activation correlates with the degree of mucosal inflammation.[37] In fact, the NF-κB signaling pathway has been identified as a target of nonsteroidal antiinflammatory drugs (NSAIDs), independent of the cyclooxygenase (COX) inhibition (see later).

Many factors released by the tumor and tumor stroma such as VEGF, interleukin-6 (IL-6), IL-10, IL-11, and IL-23, activate STAT-3 in tumor, stromal, and immune cells. In fact, several of these factors are regulated transcriptionally by STAT-3 creating a positive feedback circuit.[38] These factors also act as tumor promoters by establishing a tissue microenvironment that allows tumor progression and metastasis. In addition, some of these factors establish immunosuppressive mechanisms that prevent an effective immune response against the tumor.

Several experimental models of human CRC in animals used to study the role of inflammation in the development of CAC established: first, that there is no tumor development in germ-free animals; second, that immune cells, cytokines, and the STAT-3-NF-κB axis participate in the induction of inflammatory colitis that predisposes an individual to the development of tumors[13, 17–20, 23]; third, that at least two species of enterobacteria, *Klebsiella pneumoniae* and *Proteus mirabilis*, were responsible for colitis and cancer in this model,[39, 40] although the colonization of germ-free mice with only these two species was insufficient to induce colitis, which suggests that its role is the modification of the composition of the normal flora and intestinal physiology, or that they act in synergy with the normal components of the intestinal flora to activate the inflammation of the colon.[41, 42] Systemic effects of tumor promotion with *Helicobacter hepaticus* intestinal infection have been demonstrated, showing not only increases in colon and small intestine cancer, but also in breast adenocarcinoma in a mouse model of inherited intestinal cancer,[43] similar to that of humans infected with *Helicobacter pylori* in the development of gastric cancer.[44]

Functionally, the enterotoxigenic bacterium *Bacteroides fragilis* activated the Wnt and NF-κB signaling pathways in cancer cells promoting colon cancer in mice, and also inducing a Th17, "*T helper 17*" immune response.[45, 46] In addition, pathogenic strains of *Escherichia coli*, also associated with an increased risk of colorectal carcinoma, decrease the expression of mismatch repair genes in DNA in vitro.[47–49]

The expansion of the Bacteroidetes (*Prevotellaceae*) and TM7 (*Saccharibacteria*) in the fecal microbiota also correlates with susceptibility to colitis,[50] which suggests a role for the innate Nlrp6 receptor as a regulator of colonic microbial ecology.[17, 20, 23]

The other main known risk factors for CRC in addition to IBD, particularly those related to nutrition such as obesity, high-fat diets, lack of exercise, and alcohol and tobacco consumption[18, 51] also affect the commensal microbiota.[13, 17, 24, 25, 28, 36] It is very likely that these agents are responsible for the geographic variation in the incidence of CRC, since these wide geographic variations are lost in migrant populations in only one generation.[52, 53] This last fact does not support the alternative hypothesis that the same genetic alterations that affect inflammatory and immunological intestinal homeostasis also predispose an individual to carcinogenesis in other tissues.[29, 30]

Finally, we must emphasize that through all the routes reviewed before, inflammation also affects the response to therapies.[1, 25, 27]

Dysbiosis. Oral microbiota

Although IBDs became the main risk factors for CRC, the relative increase in CRC risk in these patients is less than threefold higher compared to healthy subjects. In addition, tumors usually appear after many years of intestinal pathology, with a cumulative lifetime risk of only 18%.[32]

The development of high-performance DNA sequencing technology also triggered the identification of the microbiota composition associated with the progression of the CRC, first in the experimental animal models discussed[54] and later in biopsies or human intestinal stool. These results showed a marked overrepresentation of sequences from *Fusobacterium nucleatum* in CRC tumors. Fusobacteria are rare components in the fecal microbiota, but they had been previously cultured from biopsies of inflamed intestinal mucosa[26, 55] and had previously been linked to periodontitis and appendicitis but not to cancer.

F. nucleatum is an invasive anaerobic and, interestingly, other anaerobic species of the same or similar genera, such as *Bacteroides*, or typically oral species such as some *Selenomonas*, *Prevotella*, *Parvimonas micra*, and *Peptostreptococcus stomatis*, are bacteria capable of forming biofilms. Although there is not a universally microbial characteristic in the CRC, the accumulation of so many species from the Human Oral Microbiome Database (http://www.homd.org/) supports a role of this microbiota in a significant percentage of CRC.[56–58]

It was hypothesized that a simple antibiotic treatment could stop CRC development as seen in the case of *Helicobacter pylori* and stomach cancer. In fact, the use of metronidazole reduced the burden of *Fusobacterium nucleatum* and overall tumor growth.[44] However, several questions arise regarding this particular study. First, the lack of a specific antibiotic for *F. nucleatum*. Second, not only *Fusobacterium* but other members of the genus are involved. Finally, a new study in humans showed that the abundance of *F. nucleatum* in feces was strongly associated with the presence of CRC, but no association was observed with the presence of advanced or nonadvanced adenomas, nor were associations observed with diets or life habits,[59] supporting the hypothesis that *Fusobacterium* is more a passenger that multiplies in favorable environmental conditions of the tumor, rather than a causal factor in the development of CRC.[59]

However, the presence of *Fusobacterium* has been related to the process of metastasis, at least to the liver[44] and to the lymph node.[55] Although the mechanism is unknown, these bacteria can interact with metastatic cells[44] and not only be present in the mucosal microbiota, where they can adhere to, or be found within epithelial tumor cells.[44,60] In fact, recent findings suggest that *F. nucleatum* organizes in the tumor cells a signaling network mediated by TLR4 and MyD88, as well as specific microRNAs that activate the autophagy pathway, controlling chemoresistance of the CRC.[60] This indicates the possibility that conventional chemotherapeutic regimens may be affected and/or modified by the microbiota present in the CRC.

This is why emerging intervention strategies on human microbiota are being developed.[61] For example, if the use of probiotics has a hypothetical long-term effect on metastasis or recurrence in patients with this microbiota-associated CRC.[60,62]

Other risk factors and their relationship with inflammation

The modern lifestyle and diet, although responsible for a general improvement in health and life expectancy, have also brought about a substantial worldwide increase in overweight (body mass index (BMI) of 25–30 kg/m^2) or obesity (BMI > 30 kg/m^2), affecting approximately half of the population in developed countries. Since the beginning of the millennium, the most recent epidemiological studies have confirmed that obesity is associated with a greater susceptibility to cancer development and patient survival,[63–66] despite the so-called obesity paradox: level I overweight or obesity (BMI: 30.0–34.9 kg/m^2) at diagnosis is associated with better survival compared to normal weight patients.[67]

Among the challenges in the study of the association of cancer with overweight and obesity, one is the quantification of adiposity and its potentially cumulative effect throughout life, since BMI is an imperfect measure of body fat, and it seems that waist circumference (abdominal obesity) is a more informative parameter. The response to the degree and timing of obesity, the presence of other risk factors for health, as well as the scale and profile of cancer in all countries of the world or the relationship with type 2 diabetes (DM2) are issues still to be resolved.[51–53,68,69] Despite this, there is an agreement that long-term prevention efforts should remain focused, together with a reduction in smoking, in promoting healthy lifestyles, balanced diets, and regular physical exercise.[65,68,69]

Metabolic syndrome is a controversial concept because it associates abdominal obesity with lipid imbalances in the circulation, inflammation, insulin resistance or diabetes, as well as with an increased risk of developing cardiovascular diseases. Abdominal obesity is the most frequent manifestation of metabolic syndrome and is a marker of "dysfunctional adipose tissue."[70,71] In parallel, the concept of **metaflammation**[72] was coined to define an inflammatory metabolic state where chronic low-grade inflammation is observed, not only in the adipocytes, but also in the stromal and inflammatory cells of the different metabolic tissues including the liver, muscles, pancreas, and brain, caused by an excess of nutrients and energy.[72] There are multiple reviews about the complex signaling networks that relate the immune response to metabolism and cellular and molecular events that take place in the context of overexposure to nutrients and energy.[13,24,25,28,36,73,74] Proper maintenance of this delicate balance is crucial for health and has important implications for the pathological conditions mentioned before.

In inflamed or "dysfunctional" adipose tissue, in addition to own resident cells such as adipocytes and fibroblasts, an infiltrate of immune cells such as lymphocytes and macrophages is observed. The inflammatory response determines the type and class of the inflammatory infiltrate and this can be resolved if nutrient levels are drastically altered.[75] This adipose tissue actively secretes adipokines and cytokines responsible for the induction of the proinflammatory and procoagulant environment, in addition to insulin resistance. All these factors can promote all stages of tumorigenesis, probably the main cause of the higher incidence of cancer associated with obesity. In addition, these mediators are not only limited to adipose tissue, but also affect systemically the inflammation and immune response, oxidative metabolism, and energy balance.[71]

The first identified proinflammatory cytokine produced by adipose tissue was the tumor necrosis factor (TNF), capable of suppressing insulin signaling and maintaining DM2.[72] Subsequently, other cytokines have been identified such as IL-6, IL-1β, IL-18, CCL2 (monocyte chemotactic protein-1, MCP-1) among others.[72] Markers such as IL-6 and C-reactive protein (CRP) are not only present in inflamed adipose tissue, but their levels also increase in blood. More restricted to adipose tissue are adipokines such as leptin, involved in appetite control, or adiponectin, an insulin sensitizer and an antiinflammatory mediator. Other molecules such as lipocalin 2 and resistin are secreted by both adipocytes and immune cells, both with proinflammatory and chemotactic functions promoting insulin resistance.[76] How obesity alters the adipokine secretion with effects on distant tissues has been recently reviewed.[77]

It is also known that an excess of nutrients induces an increased production of mitochondrial oxygen reactive species (ROS). ROS induce proinflammatory transcription factors such as NF-κB and activation of protein-1 (AP-1) by prior activation of IκBα and JNK kinases.[78, 79] All signaling transduction downstream of these transcription factors promote tumor development. NF-κB, in addition to regulating the synthesis of tumor-promoting cytokines, regulates the expression of genes such as BCL2 and BCLXL, and enhances the JAK/STAT signaling network[25, 36–38] with effects on the tumor transition from epithelial to mesenchymal cells.[80] All these events link the **inflammasome** with mitochondrial dysfunction,[24, 25, 36–38, 81, 82] and in turn with colon cancer.[83] As mentioned in the next section, it appears that the NF-κB network is one of the targets inhibited by NSAIDs. In addition, the mechanisms by which mitochondrial dysfunction may predispose to cancer appear to include a deterioration of oxidative phosphorylation, a metabolic pathway directly affected by adipose tissue secretions in obesity.[84]

In relation to the aforementioned, adipocytes, like tumor cells and macrophages, also express TLR 2 and 4. Not only the nonbacterial agonists of these receptors, such as saturated dietary fatty acids or lipopeptides, but also commensal flora or its bacterial products such as LPS and lipopeptides can reach fat cells and activate TLRs.[17, 26, 28, 85, 86] It has been suggested that the imbalance of the intestinal microbiota (dysbiosis or dysmicrobism) may be the link between IBD and CRC (already discussed) with DM2.[27, 34, 35, 50, 87]

Obesity is also associated with various changes in the composition of the commensal flora. In general, obese individuals exhibit a reduced bacterial diversity associated with changes at the Phylum level, such as an increase in *Firmicutes* and a corresponding decrease in *Bacteroidetes*, and an altered representation of bacterial genes and metabolic pathways that favor energy harvesting.[88] To note, with the arrival of the concept of **enterotype** this knowledge will be expanded soon.[89, 90] Although there is no scientific evidence that links the presence of *Fusobacterium nucleatum* at the intestinal level with obesity, a possible relationship between obesity and the subgingival microbiota in periodontal disease has recently been described.[91]

Another important link between inflammation and energy metabolism is the so-called waste syndrome or cachexia in cancer, which is observed in most patients with advanced cancers of certain types, such as in 80% of patients with pancreas or other upper gastrointestinal cancers, 60% of patients with lung cancer, and also in a proportion of patients with other chronic diseases, such as kidney diseases. However, it is less common in other types of cancer.[92] Weight loss and, in children, lack of growth are the main manifestations of cachexia. It has a profound effect on the morbidity and mortality of patients, with a decrease in the quality of life due to the decrease in physical activity, the ability to interact socially, and the perception of body image.[13, 51–53] Adipose tissue and skeletal muscle wasting, and reprogramming of liver metabolism, are hallmarks of metastatic cancer.[67–69] Both the use of carbohydrates and the incorporation of amino acids decrease in the muscles of patients with cancer cachexia. Cancer cells affect host metabolism in two ways: (a) through its own metabolism that generates other metabolites from different nutrients and (b) circulating factors secreted by themselves or induced in other host cells. Accelerated glycolysis, lactate production, and the resulting increase in the activity of the Cori cycle are the most studied metabolic effects.[93, 94]

In addition to favoring anorexia, insulin resistance, and muscle protein atrophy, observations on cancer cachexia are more similar to those associated with infections than with malnutrition. Various data have confirmed the association between systemic inflammation and cancer cachexia, both in patients and in experimental animal models. Thus circulating levels of CRP correlate with the degree of cachexia in patients as well as with the risk of death. Cytokine levels such as TNF, IL-1β, and IL-6 are elevated, either systemically or in the tumor microenvironment of cachectic patients,[92, 95] although none has proven to be a dominant factor.[93, 96] The metabolism of the liver in cachexia also leads to the suppression of antitumor immunity, mainly through IL-6.[97] And we must not forget that the interactive axis between metabolic pathways and the immune system can be altered by psychological stress, particularly through the signaling of the sympathetic nervous system.[98] The close coordination between the immune and nervous systems has been previously reviewed.[99]

The nervous system also influences the intestinal microbiota through various routes of gastrointestinal physiology, such as the functions of intestinal permeability or mucous production and antibacterial peptides. In the opposite direction, the intestinal microbiota that interacts with local and systemic inflammatory and immune responses activates the production of several metabolites including inflammatory cytokines, which end up affecting the brain and even behavior.[100–102] Obviously, this bidirectional relationship between the central nervous system and the intestinal microbiota can also be targeted by stress.

It is likely that these immune alterations alter innate responses to pathogens and enterotypes of the commensal flora, thereby modifying the inflammatory environment both in the tumor and at the systemic level and, with the subsequent feedback cycle, affecting energy metabolism.

NSAIDs, inflammation, and CRC

Although many studies had shown that the use of aspirin and other NSAIDs (nonsteroidal antiinflammatory drugs) reduce the risk of CRC, the toxicity of these molecules at the doses used prevented the U.S. Preventive Services Task Force (USPSTF) to recommend them in cancer prevention.[103]

But in 2015, the USPSTF reversed that position to include aspirin, at low doses, in its indications for routine CRC prevention.[104, 105] Aspirin is therefore the first pharmacological agent approved by the USPSTF for the chemoprevention of a cancer in a population not characterized as high risk.[106] Subsequently, some factors not previously considered, such as the additional impact of aspirin use in the context of CRC detection, including colonoscopy, which is already associated with a significantly lower risk of CRC,[106–109] were addressed in a 2016 study.[110] In this study, regular use of aspirin (for more than 10 years and between 0.5 and 1.5 tablets of standard aspirin per week) could prevent many thousands of gastrointestinal tumors, not just CRC, only in the US every year. Although the detection of CRC accounted for 50% of the overall decrease in the incidence of CRC in the USA, in the last two decades only 58% of the eligible population underwent an accepted screening option.[51, 111] In addition, regular aspirin use can prevent CRC not only among adults 50–75 years old who did not undergo CRC tests, but also among those who did.

It should be noted that the risk of developing distant metastases (for cancer in general) is also reduced in aspirin users, which suggests a potential benefit for patients with established disease, and the benefits of aspirin also include the syndrome of Lynch (the most common type of inherited colon cancer).[110, 112, 113]

We have recently reviewed the mechanisms of aspirin and other NSAIDs such as ibuprofen.[114] The main mechanism of action is the inhibition of COX-1 and COX-2 enzymes. The cyclooxygenase enzymes (COX) -1 and COX-2 catalyze the production of prostaglandins from fatty acids. The COX1 gene is expressed constitutively and induces basal levels of prostaglandins, thus contributing to homeostasis of the gastrointestinal mucosa. In contrast, COX2 gene expression is inducible, predominantly expressed in stromal cells such as fibroblasts and macrophages.

In carcinogenesis, the effects of the COX-2 enzyme and prostaglandins are complex and affect both the inflammatory microenvironment and the transformed epithelial cells.[115–117] Prostaglandin E2 (PGE2) directly affects innate and adaptive immune cells, mainly by facilitating tumor progression by developing Th17 type responses instead of the Th1 type responses that prevent the development of tumors. It also promotes inflammation, in part by dilating blood vessels and allowing immune cells to migrate to tissues, and also by regulating angiogenesis, improving the mobility of endothelial cells, and the accommodation of hematopoietic cells, i.e., leading progenitor cells to the damaged tissue where they differentiate into various immune cells. On the other hand, PGE2 promotes several carcinogenic pathways in epithelial cells.[115–117] Interestingly, COX-2 inhibitors not only prevent tumor formation, but also decrease the number of polyps already established in patients with familial adenomatous polyposis, a hereditary disorder characterized by the early onset of colon cancer.

It soon became clear that NSAIDs had also independent effects of COX enzyme inhibition[118, 119] that include the WNT, AMPK, and MTOR signaling pathways, the transcriptional activity of NF-κB and apoptosis. Exclusive effects on platelets have also been described.[120, 121]

In general, all these clinical trials of cancer prevention with NSAIDs support that inflammation is an underlying cause of cancer, even in some types of tumors that had traditionally been assumed not to originate in chronically inflamed tissues, such as the lung or prostate.

In that mentioned review[114] various approaches to limit the side effects of NSAIDs,[122–125] were also commented. Other antiinflammatory drugs directed against different inflammatory pathways, or that act in a different way, may therefore be important in preventing the onset and progression of gastrointestinal cancers and other solid organs. In fact, numerous in vitro and in vivo studies have shown that several phytochemicals have antioxidant, antiinflammatory, and anticancer effects by regulating the signaling pathways specified in previous sections and the molecular markers of the onset and development of CRC in particular, sulforaphane, curcumin, resveratrol, and its derivative 2,3,5,4′-tetrahydroxystilbene-2-O-β-glucoside (THSG).[126, 127] The latter, a polyphenol extracted from *Polygonum multiflorum*, has been proposed for possible complementary treatments or innovative therapeutic strategies for *Porphyromonas gingivalis*-induced periodontitis in human gingival fibroblasts.[128]

In addition, aspirin and other NSAIDs are efficient inhibitors of biofilm formation[129–135] and can control periodontitis.[136] In relation to the previous section, a role of inflammation inhibition related to gingival tissue microbiota may also be suspected in the antitumor effects of aspirin.[91] We also point out that platelets improve biofilm formation,[135] and again the functional relationship between platelets and tumors, and therefore the chemopreventive properties of aspirin may result, in part, from the direct modulation of platelet biology and biochemistry.[119, 120]

Suppression mechanisms of the immune system in the tumor microenvironment

We will first review the immune participants in the human CRC tissue, normally infiltrated in the margins of the transformed tissue. Both innate immune system cells such as neutrophils,[137] macrophages,[138] natural killer cells (NK),[139] or dendritic cells (CD),[140] as well as adaptive immune system cells such as CD4 helper (helper) T cells or CD8 cytotoxic[141] and also B lymphocytes can be found there. The set of these lymphocytes called TIL (tumor infiltrating lymphocytes, including B cells), as well as the total set of infiltrated immune cells, have an independent prognostic impact on CRC.[141–145]

This occurs because lymphocytic infiltration is antigen specific and therefore inhibits tumor growth.[141, 143–145] In other words, the host's immune system is activated by the colorectal tumor and works, what we call Immune Surveillance. Obviously, the role of the complex tumor microenvironment is not so clear because many times tumors develop. It must be taken into account in this context that in a fully functional immune system the immune cells release many molecules, including proangiogenic and prometastatic factors.[19] This is a double-edged sword so any advance in the knowledge of these mechanisms will serve to develop new immunotherapies.

To explain the changes of the immune system during the development of the tumor, the concept of Immunoedition has been developed.[146] The process has been divided into three phases: elimination, balance, and escape.[147] The phase of elimination is what has historically been designated as the Immune Surveillance of the tumor, during which the immune cells detect and eliminate the transformed cells. But if the removal is incomplete, some tumor cells remain inactive. Others may continue to evolve by accumulating more mutations that end up modulating the expression of tumor-associated antigens (TAA), or of different factors that increase the ability of these cells to suppress the immune response. During the equilibrium phase, the immune system continues to exert a selective pressure that eliminates some of these transformed clones, but if this elimination is again incomplete, the process ends with the selection of tumor cell variants that can resist, avoid, or even suppress the antitumor response, which leads to the escape phase.[147]

The lymphocytes that infiltrate, in most cases along the area of the invasive margin of the cancerous tissue, are CD4 and/or CD8 T cells.[148] During an immune response, CD4 T lymphocytes can be differentiated into several phenotypic subtypes: The main subtypes T helper 1 (Th1) and Th2 secrete different combinations of cytokines, thereby activating different types of immune responses. Th1 lymphocytes through the secretion of, among other cytokines, IFN-gamma and TNF-alpha produce the activation of CD8 cytotoxic T lymphocytes (CTL), NK cells, macrophages, and monocytes, all of which in an inflammatory context contribute to the cellular immune response that kills tumor cells.

Th2 lymphocytes, by secreting cytokines such as IL-4, IL-5, IL-10, or IL-13, divert the immune response to a humoral (antibody production), which seems less effective in eliminating tumor cells. In patients with CRC, the change toward a Th2 response (from Th1) is more significant the more the disease progresses.[149, 150] The mechanism by which CRC cells can change the immune response of T cells could be due in part to the secretion of cytokines by the tumor cells or by fibroblasts (discussed later), tumor-associated cells.[151]

Th17 cells, in addition to IL-17, release cytokines similar to those of Th1 and are also proinflammatory. They are closely related to the activation of neutrophil and macrophage migration to tissues, and their responses.[150] It seems that their normal function would be similar to that of ILC (innate lymphoid cells), another group of cells recently described in the gastrointestinal tract[152] which release IL-17 and IL-22.[153] There is no increase of these cells in CRC infiltrates,[154] but it does in the circulating blood of patients.[155] The higher production of IL-17 is related to a worse prognosis in patients with CRC since it helps to carcinogenesis,[150] at least in part because its favor neutrophil infiltration, which is related to an adverse prognosis.[137, 156, 157] From studies in the mouse, it has been deduced that the colonic microbiota of the CRC favors the presence of these Th17 cells.[150] However, the pathogenic role of Th17, and its association with metastases,[158–160] can be counteracted by the presence of Bregs regulatory cells, this subset associated with a better prognosis in the CRC.[158]

Another subset of TIL is composed of gamma delta T cells (γδTc) with cytotoxic capabilities similar to CTL. However, it has been suggested that in breast and colorectal cancer these cells have regulatory roles and functional plasticity depending on the context.[161] As the main function of γδTc is antimicrobial, its role in carcinogenesis can be caused by the presence of bacterial species mentioned in previous sections.

Dendritic cells are key APC cells because of their central role in inducing immune responses, including antitumor responses. The activated and mature DCs loaded with antigens are those that induce the specific antigen responses that lead to the proliferation and differentiation of CD4 and CD8 T cells in helper and cytotoxic lymphocytes. However, if the DCs are immature, they lead the T cells to tolerance.

In patients with CRC, DCs infiltrate the tumor or surrounding tissue forming groups with T lymphocytes and this infiltration seems to correlate with a better prognosis.[162] However, there is controversy about their role, since tumor cells can affect the function of these cells: Infiltrated DCs show an immature phenotype[163, 164] and correlate with the infiltration of Treg cells and with a lack of systemic response.[140] In fact, culture media explants of tumor tissue of CRC inhibit DC maturation (reduced levels of CD54, CD86, HLA-DR, and CD83), as well as induce secretion of IL-10 and inhibition of secretion of IL-12, thus inhibiting Th1 immune responses.[163]

The increase in DC infiltration in tumors shows a positive correlation with survival in CRC[162] probably because of the participation of these cells in a network with other subsets of regulatory, usually suppressor, MRCs (myeloid regulatory cells), such as Immature myeloid DCs, plasmacytoid DCs, and MDSC (myeloid-derived suppressor cells), which counter effective antitumor responses of at least the myeloid DCs.[165, 166]

Treg cells are critical for the prevention of autoimmunity and the regulation of immune responses to foreign and own antigens.[167] Treg are also often found in high frequencies in peripheral blood and in tumors of cancer patients, and for many of these human cancers the high densities of Tregs in the tumor correlate with a poor outcome of the disease,[168] as expected given its arsenal of immunosuppressive mechanisms.

However, this is not the case with CRC. There are several studies that indicate that the high density of T cell Foxp3+ infiltrating the tumor is associated with a favorable survival of patients with CRC.[168, 169] This apparent contradiction is explained by the fact that Foxp3+ T cells suppress inflammation and immune responses activated by commensal microflora through the secretion of IL-10 and TGF-beta and also by physical contacts.[170]

But there are other clones of Treg cells that are Foxp3-, which include the so-called Tr1[168, 171] and are now identified as adaptive or iTreg, different to nTreg (Foxp3+), which develop and function in response to pathological situations such as cancer.[172] It is the iTreg subset that should be monitored in cancer because iTreg plasticity is controlled and driven by the microenvironment, so they also play a double role in cancer and for now they should not be part of the available immunotherapy strategies.

Alternatively, plasticity may be related to the presence of other regulatory populations, since the network of regulatory cell types that dominate a given tumor and patient is not yet known.[173, 174] Apart from the aforementioned Treg or Breg subgroups, natural killer T cells (NKT) not only have cytotoxic antitumor activity, but even those of type I (there are types II and other subgroups) may have regulatory or suppressive activity in the CRC.[174–176]

In summary, there are three main mechanisms of escape, i.e. immune suppression induced by the tumor, each being fought by new targeted immunotherapy strategies: (1) alterations in the components of the antigen presentation machinery and the signaling pathways dependent on TCR, (2) secretion of pro-apoptotic or immunosuppressive factors, and (3) activation of inhibitory signaling pathways and recruitment of regulatory subsets.

(1) The classic mechanism of transformed cells to avoid the host's immune response is to negatively regulate key, soluble or membrane molecules involved in the immune response. The activation of CTL antitumor responses requires the recognition of immunogenic epitopes presented in various types of HLA class I molecules in the tumor. CRC tumors show high levels of class I HLA alterations due to multiple mechanisms,[163, 164, 177] discussed in the next section. Although the expression of HLA class I antigens is associated with a poor prognosis in many cancers, this is not the case with respect to CRC, probably because NK, NKT, or γδTc cells are effectors of the most important antitumor response in this cancer.[148, 174–176]

(2) Among the key soluble molecules that inhibit a protective response in CRC, TGF-beta plays a fundamental role in carcinogenesis[178, 179] because, although the effects of TGF-beta are inhibitory, many tumor cells derived from the epithelium, such as CRC, become resistant by mutations in SMAD proteins or TGF-beta receptors, and they are even stimulated to proliferate.[180] At the same time TGF-beta contributes to immunosuppression: it inhibits the proliferation and differentiation of T lymphocytes that prevent naïve T cells from acquiring effector functions[179, 180]; inhibits the ability of tumor infiltrating lymphocytes (TIL) to kill cancer cells, including cytotoxic T responses[180]; and also has powerful effects on APCs as macrophages.[179]

IL-10 is not secreted only by tumor cells, also by infiltrating monocytes, DCs of the lamina propria, and by Foxp3+ T cells (also secretors of TGF). IL-10 inhibits DCs (including the HLA class I function), regulates the Treg subsets previously discussed,[172] and together with TGF-beta participates in the change to a Th2 response.[181] In fact, elevated plasma IL-10 levels are associated with a poor prognosis and their levels return to normal in resected patients.[182]

Other molecules secreted by tumor cells with similar functions are VEGF (the angiogenic factor par excellence), PGE2, galectin-1, and gangliosides, while upregulation of IDO (indoleamine 2,3 dioxygenase) and the consequent deletion of essential amino acids such as tryptophan block the proliferation of T cells. IDO inhibitors are being evaluated for CRC,[183] although they are not mentioned in subsequent headings.

Without going deeper here, we note the probably important role of exosomes in immunosuppression.[184]

(3) Among the inhibitory molecules that actively inhibit a protective response, not only of the effector T cells but also of the NK, B, and pDC cells, we highlight the expression of the so-called immune control point molecules (ICM) since they

constitute the target of the most successful immunotherapeutics so far: (a) the "programmed death receptor ligand 1" (PD-L1) and the CD80 or CD86 proteins bind, respectively, to the "programmed death receptor-1" (PD-1) and the protein 4 associated with cytotoxic T lymphocytes (CTLA4)[185]; (b) HLA class II molecules, with the lymphocyte activation gene 3 protein (LAG3, CD223); and (c) galectin-9 (and other glycans), with the T cell immunoglobulin and mucin 3 domain (TIM3).[186]

Similarly, upregulation of TRAIL, RCAS1, and Fas in tumor cells eventually triggers apoptosis of effector immune cells when those ligands bind to their receptors.

Immunotherapy in CRC

Until recently, colorectal cancer had not benefited from immunotherapy, apart from the preventive use of antiinflammatory drugs (NSAIDs) mentioned before. In the last 15 years, the main treatment option still is cytotoxic chemotherapy based on fluorouracil in moderately toxic combinations, such as FOLFOX and FOLFIRI. Yet treatments focused on the alteration of the immune system through immune control point molecules (ICM) have been widely introduced in the practice of clinical oncology, although only the subgroup of patients with hypermutated colorectal cancers can benefit from these inhibitors. The impressive response rates experienced by patients with CRC dMMR/MSI-H treated with anti-PD-1 therapies contrast with the lack of response shown by those with stable CRC (competent pMMR).

Molecular subtypes consensus in CRC (meaning of MMR and MSI)

A classification with application to research and distinct from that routinely used in clinic has obtained a consensus on the differentiation into four molecular subtypes (CMS) of CRC: CMS 1 tumors (or immune MSI, 14%) are characterized by hypermutation, microsatellite instability (MSI) and strong immune activation. CMS 2 (canonical, 37%) are epithelial, with chromosomal instability (CIN) and prominent activation of WNT and MYC signaling. CMS 3 (metabolic, 13%) are epithelial, characterized by metabolic deregulation. Finally, CMS 4 (mesenchymal, 23%) possess a prominent activation of transforming growth factor β (TGF-β), stromal invasion, and angiogenesis. The remaining 13% have mixed characteristics.[187]

The microsatellite instability (MSI) is a marker of dysfunctional mismatch of repair proteins (MMR) within a tumor and was used clinically for the first time to identify patients who should undergo germline tests for Lynch syndrome.[188] The classical test for MSI is based on testing five specific microsatellites (BAT25, BAT26, D2S123, D5S346, and D17S250) through the polymerase chain reaction, and they are called MSI-H if there is instability (variation in length between tumor and normal) in more than 30% of the microsatellites tested. Only about 3% of all colorectal cancers have an MMR mutation in the germ line (Lynch syndrome). Therefore most MSI-H colorectal cancers are sporadic, due to somatic defects acquired by MMR gene dysfunction.[189]

The different immunotherapy protocols have not only been successful in the CMS 1 group with MSI-H, but in cancers with markedly high mutational rates in general, which occurs in those tumors with dysfunction of mismatch repair genes (MMR in English). MMR is one of many mechanisms that cells use to repair damaged DNA, particularly insertions, deletions, and erroneous incorporation of nucleotides during DNA replication. Deficiency in the MMR system (dMMR) versus the competent MMR system (pMMR) leads to the accumulation of mutations. The identification of dMMR can be achieved by immunohistochemical staining for the complete loss of one of the four most common MMR proteins: MLH1, MSH2, MSH6, and PMS2 or also TACSTD1/EPCAM. A tumor negative for one or two MMR proteins is classified as a mismatch repair deficiency (dMMR). This immunohistochemical assay has some potential advantages over the MSI assay.[190, 191]

But MSI-H/dMMR tumors constitute a minority of colorectal cancers, with a decreasing frequency in more advanced stage disease. The prevalence of MSI-H in stage II, III, and IV colorectal cancers is 22%, 12%, and 3%, respectively.

Key immunotherapeutic assays in CRC

As part of the initial clinical investigation of Nivolumab (monoclonal antibody (mAb) anti-PD-1), among the many patients with CRC who did not obtain any benefit, one patient achieved a complete response that was lasting without therapy for more than 3 years.[192, 193] The patient had a tumor of high microsatellite instability (MSI-H), with macrophages and infiltrating lymphocytes positive for the PD-1 ligand (PD-L1+). This finding supported the hypothesis that perhaps these inflamed tumors could be treated with PD-1 inhibitors. Immunotherapy has now been introduced in the treatment of metastatic CRC (mCRC) characterized as MSI-H or poor mismatch repair (dMMR).

In CheckMate 142, a phase II trial, Nivolumab (3 mg/kg q2 week) was combined with Ipilimumab (mAb anti-CTLA-4) (1 mg/kg q3 week × 4 doses), followed by monotherapy with Nivolumab to evaluate efficacy in terms of objective response rate (ORR) in patients with MSI-H and non-MSI-H with metastatic CRC.[194] Patients had progressed or were intolerant of ≥1 previous line of therapy. The ORR for patients with MSI-H was 31% and 15% for nivolumab and nivolumab + ipilimumab, respectively. Responses were observed regardless of PD-L1 expression or mutation status of BRAF or KRAS genes. Serious adverse events included acute kidney injury, increased ALT, colitis, and stomatitis.

In a phase II trial with Pembrolizumab (anti-PD-1 mAb),[192] it was administered (10 mg/kg every 2 weeks) to patients with dMMR (cohort A) and pMMR (cohort B) colorectal cancers, as well as patients with noncolorectal dMMR cancers (cohort C). The coprimary endpoints were ORR and the progression-free survival rate (PFS) at 20 weeks. The selected patients had progressive metastatic cancer refractory to treatment (97.5% of the patients had ≥2 prior therapy lines). The ORR for cohorts A and C was 40% and 71%, respectively, compared to 0% in cohort B. The PFS at 20 weeks for cohorts A and C was 78% and 67%, respectively, versus 11% in cohort B. Pembrolizumab caused rash/pruritus, thyroiditis (hypo- or hyperthyroidism), and asymptomatic pancreatitis. Based on these initial data, both Pembrolizumab and Nivolumab have obtained FDA approval for the treatment of patients with MSI-H/dMMR metastatic colorectal cancer as postfailure therapy in standard treatment lines. Patients who progress with any of these medications should not be offered the other.

However, not all patients with MSI-H/dMMR are positive for diagnosis with the PD-1 antibody although it is the biomarker for treatment with PD-1 antibodies, about 50%–60% of patients with MSI-H/dMMR are insensitive.[195] Therefore the tumor mutation load (TML) may be another biomarker of response to PD-1 therapy.[196,197]

We have already commented that COX (cyclooxygenase) inhibitors have been shown to reduce the risk of developing colorectal adenocarcinoma in many cases, including familial adenomatous polyposis (FAP)[198] and that COX-2 expression increases in up to 90% of colorectal carcinomas.[199] For this reason, an ongoing phase II trial will evaluate the efficacy and safety of the combination of the PD-1 antibody and the COX inhibitor in patients with MSI-H/dMMR or high TML colorectal cancer (NCT03638297). In this context, numerous additional combinations with radiotherapy, chemotherapy, or molecularly directed drugs are being studied. A summary of these combinations is presented in Table 40.1.

Other immunotherapeutic approaches tested in humans

The best genomic explanation underlying the different patterns of immune response against dMMR and pMMR tumors is the much higher frequency of neoantigens available in dMMR tumors, as a result of their high mutational load that correlates with the extent of infiltrating lymphocytes in the tumor. In addition, there is a higher percentage of immunoinhibitory cells such as regulatory T cells (Treg) within pMMR tumors that can explain their poor immune response.[200] Therefore the new strategies are based on the evaluation of combinations with chemotherapy, vaccines, or elimination of suppressor cells and Treg lymphocytes. In cancer in general, the different procedures to reactivate the suppressed immune response are aimed at counteracting: (a) the deterioration of antigen presentation by CPAs (DC, macrophages, B lymphocytes); (b) the activation of negative costimulatory signals; (c) the development of an immunosuppressive network; and (d) the recruitment of regulatory cell populations such as Tregs, NKT cells, or others.

Other monoclonal antibodies

Other mAbs have been designed to (i) inhibit the intrinsic signaling pathways of cancer cells or the immunosuppressive network of soluble factors, TGF, IL-10, VEGF, galectin-1, and other signaling molecules, (ii) bring toxins closer to cancer cells or (iii) interfere with tumor-stroma interaction.[173]

For the past 15 years, the FDA- and EMA-approved mAbs for the treatment of colorectal cancer are Cetuximab, Panitumumab, and Bevacizumab. Bevacizumab is a mAb that blocks VEGF and therefore angiogenesis, also interfering with the tumor-stroma interaction and indirectly inhibiting tumor growth. It is currently also used for breast, kidney, and lung cancer therapy[201] and has been tested in advanced CRC in combination with chemotherapy.[202,203]

Both Cetuximab and Panitumumab are mAbs that directly inhibit the autonomous pro-survival cascades of tumor cells, inhibiting the epidermal growth factor receptor, EGFR. Both have been approved for metastatic CRC in KRAS NRAS positive patients (nonmutated RAS gene) and can be used alone or in combination, depending on certain conditions (see the EMA, FDA, or company websites for more details). However, the effectiveness in other conditions is modest. Trastuzumab, which binds to a receptor of the same family (HER2), is being used for metastatic gastric cancer but not for CRC. Other antibodies that reached phase I/II have also been reviewed.[103]

TABLE 40.1 Registered clinical trials for the treatment of colorectal cancer.

ClinicalTrials.gov identifier	Drugs(s)	Phase	Type of patient	Primary result	Expected date
NCT03396926	Pembrolizumab + bevacizumab + capecitabine	II	pMMR[a] mCRC	ORR	April 2021
NCT03631407	Vicriviroc + Pembrolizumab	II	MSS mCRC	ORR	March 2025
NCT02563002	Pembrolizumab	III	MSI-H/dMMR mCRC	PFS, OS	March 2025
NCT02437071	Pembrolizumab + RT	II	pMMR mCRC	ORR	September 2019
NCT02227667	Durvalumab	II	mCRC MSI-H	BRR	August 2021
NCT02870920	Durvalumab + Tremelimumab	II	Refractory mCRC	OS	February 2019
NCT02997228	Atezolizumab +/− (Bevacizumab + mFOLFOX6)	III	dMMR mCRC	PFS	March 2022
NCT02873195	Atezolizumab + Capecitabine + Bevacizumab	II	Refractory mCRC	PFS	November 2022
NCT02291289	Atezolizumab	II	mCRC	PFS	April 2019
NCT03050814	Avelumab + vaccine Ad-CEA	II	mCRC	PFS	November 2020
NCT03642067	Nivolumab + Relatlimab	II	MSS mCRC	ORR	November 2021
NCT03638297	PD-1 antibody + Cox inhibitor	II	MSI-H/dMMR or High TMB	RR	January 2023

[a]mCRC, metastatic colorectal cancer; MSI, microsatellite instability; MSS, microsatellite stability; pMMR, competent in MMR; ORR, objective response rate; SLP, progression-free survival; OS, overall survival; RR, response rate; BRR, better RR. Details available at: www.clinicaltrials.gov.

Cell therapies

Today it is clear that CD8+ T lymphocytes that infiltrate the neoplastic epithelium (TILs) have antitumor activity and, therefore, are activated by tumor associated antigens (TAA) if immunosuppression does not block them,[204, 205] and at least 80% of patients with epithelial cancer have them.[206] Although adoptive cell therapy (ACT) with TILs is a promising modality for cancer treatment,[207] the generation of autologous TILs capable of inducing cancer regression for each individual patient is a logistic and economical challenge, since almost all the neoantigens (or TAA) of a patient are personal and the therapy must be developed for each individual, using massive genetic sequencing. With this approach only 15% of patients have responded so far, but a strategy to treat epithelial tumors, such as liver, colon, cervix, breast, which are the cause of 90% of all deaths from cancer, it is expected to be found. In fact, a recent article shows the case of a woman with disease-free metastatic colon cancer for almost 5 years after treatment.[206]

An alternative to the previous approach is therapy with CAR-T cells, a real revolution in the treatment of malignant hematological diseases due to its exceptional efficacy, and which is already being tested in solid tumors such as CRC.[208] Basically, it is about redirecting T cells through an antibody-based chimeric antigen receptor (CAR), generating "universal effector T cells," capable of recognizing targets regardless of restriction by histocompatibility molecules.[207]

Vaccinations

CPAs, particularly DCs, have been used in the development of therapeutic antitumor vaccines for a long time,[209] trying to generate an antitumor response by vaccinating the patient with the neoantigen, or incubating autologous DCs in vitro with said neoantigen followed by adoptive cell transfer. However, inadequate activation of DCs—the blockage of mature DC differentiation and the accumulation of immature DC and plasmacytoid DC is the dominant mechanism underlying the development of T cell tolerance—in addition to the immunosuppressive effects exerted by the tumor microenvironment fails to eradicate the tumor in most cases.[210] Thus the identification of these TAA or neoantigens is essential for the development of a cancer vaccine and we know now that they are very specific for each person.[206]

Importantly, many of the TAA in CRC were originally identified due to the presence of plasma autoantibodies in cancer patients. The humoral responses against some TAA correlate with the CD8+ lymphocyte responses in these patients[211] supporting the idea that the immune response that occurs in patients with CRC requires coordinated responses of CD4+ and CD8+ T cells, and B cells.[212]

For these reasons, up to three types of strategies for cancer vaccines based on DCs have been developed,[213] and in 2010 the FDA approved a therapeutic vaccine based on them (Sipuleucel-T, Provenge) for use in patients with asymptomatic or minimally symptomatic metastatic prostate cancer. Very recent results have shown promising results when mixing vaccination with anti-PD-1 therapies,[214–217] although these new approaches have not yet been tested in CRC. It is likely that advances in personalized medicine, an epigenomic firm that may identify the best patients for each approach, can help to develop this field.[218]

Another interesting field of research is the use of innate immune activators (molecular patterns associated with pathogens, PAMPs; molecular patterns associated with cell damage, DAMPs) that can be used as adjuvants, alone as monotherapy (e.g., as signals of maturation of DCs or in combination with chemotherapy[219, 220]). In colorectal cancer, several products reached the phase I/II clinical trial status (references NCT00773097, NCT00785122, NCT00780988, NCT00719199, NCT01208194, NCT00403052), in many cases with favorable overall survival results. TLR9 agonists such as IMO-2125 are being studied in therapeutic combinations with Ipilimumab or Pembrolizumab but not yet in CRC (NCT02644967).[221, 222]

References

1. Oskarsson T, Batlle E, Massagué J. Metastatic stem cells: sources, niches, and vital pathways. *Cell Stem Cell* 2014;**14**(3):306–21. https://doi.org/10.1016/j.stem.2014.02.002.
2. Melo FS, Kurtova AV, Harnoss JM, et al. A distinct role for Lgr5+ stem cells in primary and metastatic colon cancer. *Nature* 2017;**543**(7647):676–80. https://doi.org/10.1038/nature21713.
3. Reedy J. Galen on cancer and related diseases. *Clio Med* 1975;**10**(3):227–38. https://www.ncbi.nlm.nih.gov/pubmed/50913.
4. Dvorak HF. Tumors: wounds that do not heal. Similarities between tumor stroma generation and wound healing. *N Engl J Med* 1986;**315**(26):1650–9. https://doi.org/10.1056/nejm198612253152606.
5. Bindea G, Mlecnik B, Fridman W-H, Galon J. The prognostic impact of anti-cancer immune response: a novel classification of cancer patients. *Semin Immunopathol* 2011;**33**(4):335–40. https://doi.org/10.1007/s00281-011-0264-x.
6. Bissell MJ, Hines WC. Why don't we get more cancer? A proposed role of the microenvironment in restraining cancer progression. *Nat Med* 2011;**17**(3):320–9. https://doi.org/10.1038/nm.2328.
7. Hanahan D, Weinberg RA. Hallmarks of cancer: the next generation. *Cell* 2011;**144**(5):646–74. https://doi.org/10.1016/j.cell.2011.02.013.
8. Fouad YA, Aanei C. Revisiting the hallmarks of cancer. *Am J Cancer Res* 2017;**7**(5):1016–36. https://www.ncbi.nlm.nih.gov/pubmed/28560055.
9. Schreiber RD, Old LJ, Smyth MJ. Cancer immunoediting: integrating immunity's roles in cancer suppression and promotion. *Science* 2011;**331**(6024):1565–70. https://doi.org/10.1126/science.1203486.
10. Smyth MJ, Ngiow SF, Ribas A, Teng MWL. Combination cancer immunotherapies tailored to the tumour microenvironment. *Nat Rev Clin Oncol* 2016;**13**(3):143–58. https://doi.org/10.1038/nrclinonc.2015.209.
11. Plummer M, de Martel C, Vignat J, Ferlay J, Bray F, Franceschi S. Global burden of cancers attributable to infections in 2012: a synthetic analysis. *Lancet Glob Health* 2016;**4**(9):e609–16. https://doi.org/10.1016/S2214-109X(16)30143-7.
12. Shield KD, Marant Micallef C, de Martel C, et al. New cancer cases in France in 2015 attributable to infectious agents: a systematic review and meta-analysis. *Eur J Epidemiol* 2017;**33**(3):263–74. https://doi.org/10.1007/s10654-017-0334-z.
13. Trinchieri G. Cancer and inflammation: an old intuition with rapidly evolving new concepts. *Annu Rev Immunol* 2012;**30**:677–706. https://doi.org/10.1146/annurev-immunol-020711-075008.
14. Wall BM, Dmochowski RR, Malecha M, Mangold T, Bobal MA, Cooke CR. Inducible nitric oxide synthase in the bladder of spinal cord injured patients with a chronic indwelling urinary catheter. *J Urol* 2001;**165**(5):1457–61. https://www.ncbi.nlm.nih.gov/pubmed/11342896.
15. Ameille J, Brochard P, Letourneux M, Paris C, Pairon J-C. Asbestos-related cancer risk in patients with asbestosis or pleural plaques. *Rev Mal Respir* 2011;**28**(6):e11–7. https://doi.org/10.1016/j.rmr.2011.04.008.
16. Rebours V, Boutron-Ruault M-C, Jooste V, et al. Mortality rate and risk factors in patients with hereditary pancreatitis: uni- and multidimensional analyses. *Am J Gastroenterol* 2009;**104**(9):2312–7. https://doi.org/10.1038/ajg.2009.363.
17. Dzutsev A, Badger JH, Perez-Chanona E, et al. Microbes and Cancer. *Annu Rev Immunol* 2017;**35**:199–228. https://doi.org/10.1146/annurev-immunol-051116-052133.
18. Triantafillidis JK, Nasioulas G, Kosmidis PA. Colorectal cancer and inflammatory bowel disease: epidemiology, risk factors, mechanisms of carcinogenesis and prevention strategies. *Anticancer Res* 2009;**29**(7):2727–37. https://www.ncbi.nlm.nih.gov/pubmed/19596953.
19. Coussens LM, Werb Z. Inflammation and cancer. *Nature* 2002;**420**(6917):860–7. https://doi.org/10.1038/nature01322.

20. Brubaker SW, Bonham KS, Zanoni I, Kagan JC. Innate immune pattern recognition: a cell biological perspective. *Annu Rev Immunol* 2015;**33**(1):257–90. https://doi.org/10.1146/annurev-immunol-032414-112240.
21. Schetter AJ, Okayama H, Harris CC. The role of MicroRNAs in colorectal cancer. *Cancer J* 2012;**18**(3):244–52. https://doi.org/10.1097/ppo.0b013e318258b78f.
22. Cooks T, Harris CC, Oren M. Caught in the cross fire: p53 in inflammation. *Carcinogenesis* 2014;**35**(8):1680–90. https://doi.org/10.1093/carcin/bgu134.
23. Vajjhala PR, Ve T, Bentham A, Stacey KJ, Kobe B. The molecular mechanisms of signaling by cooperative assembly formation in innate immunity pathways. *Mol Immunol* 2017;**86**:23–37. https://doi.org/10.1016/j.molimm.2017.02.012.
24. Ralston JC, Lyons CL, Kennedy EB, Kirwan AM, Roche HM. Fatty acids and NLRP3 inflammasome-mediated inflammation in metabolic tissues. *Annu Rev Nutr* 2017;**37**(1):77–102. https://doi.org/10.1146/annurev-nutr-071816-064836.
25. Corrales L, Matson V, Flood B, Spranger S, Gajewski TF. Innate immune signaling and regulation in cancer immunotherapy. *Cell Res* 2017;**27**(1):96–108. https://doi.org/10.1038/cr.2016.149.
26. Kurashima Y, Kiyono H. Mucosal ecological network of epithelium and immune cells for gut homeostasis and tissue healing. *Annu Rev Immunol* 2017;**35**:119–47. https://doi.org/10.1146/annurev-immunol-051116-052424.
27. Roy S, Trinchieri G. Microbiota: a key orchestrator of cancer therapy. *Nat Rev Cancer* 2017;**17**(5):271–85. https://doi.org/10.1038/nrc.2017.13.
28. Kirwan AM, Lenighan YM, O'Reilly ME, McGillicuddy FC, Roche HM. Nutritional modulation of metabolic inflammation. *Biochem Soc Trans* 2017;**45**(4):979–85. https://doi.org/10.1042/BST20160465.
29. Bernstein CN, Blanchard JF, Rawsthorne P, Yu N. The prevalence of extraintestinal diseases in inflammatory bowel disease: a population-based study. *Am J Gastroenterol* 2001;**96**(4):1116–22. https://doi.org/10.1111/j.1572-0241.2001.03756.x.
30. Pedersen N, Duricova D, Elkjaer M, Gamborg M, Munkholm P, Jess T. Risk of extra-intestinal cancer in inflammatory bowel disease: meta-analysis of population-based cohort studies. *Am J Gastroenterol* 2010;**105**(7):1480–7. https://doi.org/10.1038/ajg.2009.760.
31. Saleh M, Trinchieri G. Innate immune mechanisms of colitis and colitis-associated colorectal cancer. *Nat Rev Immunol* 2011;**11**(1):9–20. https://doi.org/10.1038/nri2891.
32. Adami J, Gäbel H, Lindelöf B, et al. Cancer risk following organ transplantation: a nationwide cohort study in Sweden. *Br J Cancer* 2003;**89**(7):1221–7. https://doi.org/10.1038/sj.bjc.6601219.
33. Rakoff-Nahoum S, Medzhitov R. Regulation of spontaneous intestinal tumorigenesis through the adaptor protein MyD88. *Science* 2007;**317**(5834):124–7. https://doi.org/10.1126/science.1140488.
34. Salcedo R, Worschech A, Cardone M, et al. MyD88-mediated signaling prevents development of adenocarcinomas of the colon: role of interleukin 18. *J Exp Med* 2010;**207**(8):1625–36. https://doi.org/10.1084/jem.20100199.
35. Allen IC, TeKippe EM, Woodford R-MT, et al. The NLRP3 inflammasome functions as a negative regulator of tumorigenesis during colitis-associated cancer. *J Exp Med* 2010;**207**(5):1045–56. https://doi.org/10.1084/jem.20100050.
36. Marelli G, Sica A, Vannucci L, Allavena P. Inflammation as target in cancer therapy. *Curr Opin Pharmacol* 2017;**35**:57–65. https://doi.org/10.1016/j.coph.2017.05.007.
37. Rogler G, Brand K, Vogl D, et al. Nuclear factor kappaB is activated in macrophages and epithelial cells of inflamed intestinal mucosa. *Gastroenterology* 1998;**115**(2):357–69. https://doi.org/10.1016/s0016-5085(98)70202-1.
38. Yu H, Lee H, Herrmann A, Buettner R, Jove R. Revisiting STAT3 signalling in cancer: new and unexpected biological functions. *Nat Rev Cancer* 2014;**14**(11):736–46. https://doi.org/10.1038/nrc3818.
39. Garrett WS, Punit S, Gallini CA, et al. Colitis-associated colorectal cancer driven by T-bet deficiency in dendritic cells. *Cancer Cell* 2009;**16**(3):208–19. https://doi.org/10.1016/j.ccr.2009.07.015.
40. Garrett WS, Gallini CA, Yatsunenko T, et al. Enterobacteriaceae act in concert with the gut microbiota to induce spontaneous and maternally transmitted colitis. *Cell Host Microbe* 2010;**8**(3):292–300. https://doi.org/10.1016/j.chom.2010.08.004.
41. Uronis JM, Mühlbauer M, Herfarth HH, Rubinas TC, Jones GS, Jobin C. Modulation of the intestinal microbiota alters colitis-associated colorectal cancer susceptibility. *PLoS One* 2009;**4**(6):e6026. https://doi.org/10.1371/journal.pone.0006026.
42. Vannucci L, Stepankova R, Kozakova H, Fiserova A, Rossmann P, Tlaskalova-Hogenova H. Colorectal carcinogenesis in germ-free and conventionally reared rats: different intestinal environments affect the systemic immunity. *Int J Oncol* 2008;**32**(3):609–17. https://www.ncbi.nlm.nih.gov/pubmed/18292938.
43. Rao VP, Poutahidis T, Ge Z, et al. Innate immune inflammatory response against enteric bacteria *Helicobacter hepaticus* induces mammary adenocarcinoma in mice. *Cancer Res* 2006;**66**(15):7395–400. https://doi.org/10.1158/0008-5472.CAN-06-0558.
44. Bullman S, Pedamallu CS, Sicinska E, et al. Analysis of Fusobacterium persistence and antibiotic response in colorectal cancer. *Science* 2017;**358**(6369):1443–8. https://doi.org/10.1126/science.aal5240.
45. Wu S, Rhee K-J, Albesiano E, et al. A human colonic commensal promotes colon tumorigenesis via activation of T helper type 17 T cell responses. *Nat Med* 2009;**15**(9):1016–22. https://doi.org/10.1038/nm.2015.
46. Sears CL. Enterotoxigenic *Bacteroides fragilis*: a rogue among symbiotes. *Clin Microbiol Rev* 2009;**22**(2):349–69. https://doi.org/10.1128/CMR.00053-08.
47. Maddocks ODK, Short AJ, Donnenberg MS, Bader S, Harrison DJ. Attaching and effacing *Escherichia coli* downregulate DNA mismatch repair protein in vitro and are associated with colorectal adenocarcinomas in humans. *PLoS One* 2009;**4**(5):e5517. https://doi.org/10.1371/journal.pone.0005517.
48. Mármol I, Sánchez-de-Diego C, Pradilla Dieste A, Cerrada E, Rodriguez Yoldi MJ. Colorectal carcinoma: a general overview and future perspectives in colorectal cancer. *Int J Mol Sci* 2017;**18**(1). https://doi.org/10.3390/ijms18010197.

49. Gagnaire A, Nadel B, Raoult D, Neefjes J, Gorvel J-P. Collateral damage: insights into bacterial mechanisms that predispose host cells to cancer. *Nat Rev Microbiol* 2017;**15**(2):109–28. https://doi.org/10.1038/nrmicro.2016.171.
50. Elinav E, Strowig T, Kau AL, et al. NLRP6 inflammasome regulates colonic microbial ecology and risk for colitis. *Cell* 2011;**145**(5):745–57. https://doi.org/10.1016/j.cell.2011.04.022.
51. Edwards BK, Ward E, Kohler BA, et al. Annual report to the nation on the status of cancer, 1975-2006, featuring colorectal cancer trends and impact of interventions (risk factors, screening, and treatment) to reduce future rates. *Cancer* 2010;**116**(3):544–73. https://doi.org/10.1002/cncr.24760.
52. Kamangar F, Dores GM, Anderson WF. Patterns of cancer incidence, mortality, and prevalence across five continents: defining priorities to reduce cancer disparities in different geographic regions of the world. *J Clin Orthod* 2006;**24**(14):2137–50. https://doi.org/10.1200/JCO.2005.05.2308.
53. Siegel RL, Fedewa SA, Anderson WF, et al. Colorectal cancer incidence patterns in the United States, 1974–2013. *J Natl Cancer Inst* 2017;**109**(8). https://doi.org/10.1093/jnci/djw322.
54. Arthur JC, Perez-Chanona E, Mühlbauer M, et al. Intestinal inflammation targets cancer-inducing activity of the microbiota. *Science* 2012;**338**(6103):120–3. https://doi.org/10.1126/science.1224820.
55. Castellarin M, Warren RL, Freeman JD, et al. *Fusobacterium nucleatum* infection is prevalent in human colorectal carcinoma. *Genome Res* 2012;**22**(2):299–306. https://doi.org/10.1101/gr.126516.111.
56. Flemer B, Warren RD, Barrett MP, et al. The oral microbiota in colorectal cancer is distinctive and predictive. *Gut* 2017;**67**(8):1454–63. https://doi.org/10.1136/gutjnl-2017-314814.
57. Eklöf V, Löfgren-Burström A, Zingmark C, et al. Cancer-associated fecal microbial markers in colorectal cancer detection. *Int J Cancer* 2017;**141**(12):2528–36. https://doi.org/10.1002/ijc.31011.
58. Drewes JL, White JR, Dejea CM, et al. High-resolution bacterial 16S rRNA gene profile meta-analysis and biofilm status reveal common colorectal cancer consortia. *NPJ Biofilms Microbiomes* 2017;**3**:34. https://doi.org/10.1038/s41522-017-0040-3.
59. Amitay EL, Werner S, Vital M, et al. Fusobacterium and colorectal cancer: causal factor or passenger? Results from a large colorectal cancer screening study. *Carcinogenesis* 2017;**38**(8):781–8. https://doi.org/10.1093/carcin/bgx053.
60. Yu T, Guo F, Yu Y, et al. *Fusobacterium nucleatum* promotes chemoresistance to colorectal cancer by modulating autophagy. *Cell* 2017;**170**(3):548–63 [e16] https://doi.org/10.1016/j.cell.2017.07.008.
61. Kundu P, Blacher E, Elinav E, Pettersson S. Our gut microbiome: the evolving inner self. *Cell* 2017;**171**(7):1481–93. https://doi.org/10.1016/j.cell.2017.11.024.
62. Hibberd AA, Lyra A, Ouwehand AC, et al. Intestinal microbiota is altered in patients with colon cancer and modified by probiotic intervention. *BMJ Open Gastroenterol* 2017;**4**(1):e000145. https://doi.org/10.1136/bmjgast-2017-000145.
63. Renehan AG, Soerjomataram I, Tyson M, et al. Incident cancer burden attributable to excess body mass index in 30 European countries. *Int J Cancer* 2010;**126**(3):692–702. https://doi.org/10.1002/ijc.24803.
64. Calle EE, Rodriguez C, Walker-Thurmond K, Thun MJ. Overweight, obesity, and mortality from cancer in a prospectively studied cohort of U.S. adults. *N Engl J Med* 2003;**348**(17):1625–38. https://doi.org/10.1056/NEJMoa021423.
65. Campbell PT, Newton CC, Freedman ND, et al. Body mass index, waist circumference, diabetes, and risk of liver cancer for U.S. adults. *Cancer Res* 2016;**76**(20):6076–83. https://doi.org/10.1158/0008-5472.CAN-16-0787.
66. Arnold M, Karim-Kos HE, Coebergh JW, et al. Recent trends in incidence of five common cancers in 26 European countries since 1988: analysis of the European cancer observatory. *Eur J Cancer* 2015;**51**(9):1164–87. https://doi.org/10.1016/j.ejca.2013.09.002.
67. Lennon H, Sperrin M, Badrick E, Renehan AG. The obesity paradox in cancer: a review. *Curr Oncol Rep* 2016;**18**(9):56. https://doi.org/10.1007/s11912-016-0539-4.
68. Renehan AG, Zwahlen M, Egger M. Adiposity and cancer risk: new mechanistic insights from epidemiology. *Nat Rev Cancer* 2015;**15**(8):484–98. https://doi.org/10.1038/nrc3967.
69. Arnold M, Leitzmann M, Freisling H, et al. Obesity and cancer: an update of the global impact. *Cancer Epidemiol* 2016;**41**:8–15. https://doi.org/10.1016/j.canep.2016.01.003.
70. Després J-P, Lemieux I. Abdominal obesity and metabolic syndrome. *Nature* 2006;**444**(7121):881–7. https://doi.org/10.1038/nature05488.
71. Renehan AG, Roberts DL, Dive C. Obesity and cancer: pathophysiological and biological mechanisms. *Arch Physiol Biochem* 2008;**114**(1):71–83. https://doi.org/10.1080/13813450801954303.
72. Gregor MF, Hotamisligil GS. Inflammatory mechanisms in obesity. *Annu Rev Immunol* 2011;**29**:415–45. https://doi.org/10.1146/annurev-immunol-031210-101322.
73. Hotamisligil GS. Foundations of immunometabolism and implications for metabolic health and disease. *Immunity* 2017;**47**(3):406–20. https://doi.org/10.1016/j.immuni.2017.08.009.
74. Hotamisligil GS. Inflammation, metaflammation and immunometabolic disorders. *Nature* 2017;**542**(7640):177–85. https://doi.org/10.1038/nature21363.
75. Wellen KE, Thompson CB. Cellular metabolic stress: considering how cells respond to nutrient excess. *Mol Cell* 2010;**40**(2):323–32. https://doi.org/10.1016/j.molcel.2010.10.004.
76. Ouchi N, Parker JL, Lugus JJ, Walsh K. Adipokines in inflammation and metabolic disease. *Nat Rev Immunol* 2011;**11**(2):85–97. https://doi.org/10.1038/nri2921.
77. Fuster JJ, Ouchi N, Gokce N, Walsh K. Obesity-induced changes in adipose tissue microenvironment and their impact on cardiovascular disease. *Circ Res* 2016;**118**(11):1786–807. https://doi.org/10.1161/CIRCRESAHA.115.306885.
78. Kamp DW, Shacter E, Weitzman SA. Chronic inflammation and cancer: the role of the mitochondria. *Oncology* 2011;**25**(5):400–10. 413 https://www.ncbi.nlm.nih.gov/pubmed/21710835.

79. Naik E, Dixit VM. Mitochondrial reactive oxygen species drive proinflammatory cytokine production. *J Exp Med* 2011;**208**(3):417–20. https://doi.org/10.1084/jem.20110367.
80. Jurjus A, Eid A, Al Kattar S, et al. Inflammatory bowel disease, colorectal cancer and type 2 diabetes mellitus: the links. *BBA Clin* 2015;**5**:16–24. https://doi.org/10.1016/j.bbacli.2015.11.002.
81. Lamkanfi M, Dixit VM. Mechanisms and functions of inflammasomes. *Cell* 2014;**157**(5):1013–22. https://doi.org/10.1016/j.cell.2014.04.007.
82. Broz P, Dixit VM. Inflammasomes: mechanism of assembly, regulation and signalling. *Nat Rev Immunol* 2016;**16**(7):407–20. https://doi.org/10.1038/nri.2016.58.
83. Xue X, Bredell BX, Anderson ER, et al. Quantitative proteomics identifies STEAP4 as a critical regulator of mitochondrial dysfunction linking inflammation and colon cancer. *Proc Natl Acad Sci U S A* 2017;**114**(45):E9608–17. https://doi.org/10.1073/pnas.1712946114.
84. Schwartz B, Yehuda-Shnaidman E. Putative role of adipose tissue in growth and metabolism of colon cancer cells. *Front Oncol* 2014;**4**:164. https://doi.org/10.3389/fonc.2014.00164.
85. Engin A. The pathogenesis of obesity-associated adipose tissue inflammation. *Adv Exp Med Biol* 2017;**960**:221–45. https://doi.org/10.1007/978-3-319-48382-5_9.
86. Engin AB. Adipocyte-macrophage cross-talk in obesity. *Adv Exp Med Biol* 2017;**960**:327–43. https://doi.org/10.1007/978-3-319-48382-5_14.
87. Sinagra E, Morreale GC, Mohammadian G, et al. New therapeutic perspectives in irritable bowel syndrome: targeting low-grade inflammation, immuno-neuroendocrine axis, motility, secretion and beyond. *World J Gastroenterol* 2017;**23**(36):6593–627. https://doi.org/10.3748/wjg.v23.i36.6593.
88. Heiss CN, Olofsson LE. Gut microbiota-dependent modulation of energy metabolism. *J Innate Immun* 2018;**10**(3):163–71. https://doi.org/10.1159/000481519.
89. Costea PI, Hildebrand F, Arumugam M, et al. Enterotypes in the landscape of gut microbial community composition. *Nat Microbiol* 2018;**3**(1):8–16. https://doi.org/10.1038/s41564-017-0072-8.
90. Johnson EL, Heaver SL, Walters WA, Ley RE. Microbiome and metabolic disease: revisiting the bacterial phylum Bacteroidetes. *J Mol Med* 2017;**95**(1):1–8. https://doi.org/10.1007/s00109-016-1492-2.
91. Maciel SS, Feres M, Gonçalves TED, et al. Does obesity influence the subgingival microbiota composition in periodontal health and disease? *J Clin Periodontol* 2016;**43**(12):1003–12. https://doi.org/10.1111/jcpe.12634.
92. Donohoe CL, Ryan AM, Reynolds JV. Cancer cachexia: mechanisms and clinical implications. *Gastroenterol Res Pract* 2011;**2011**:601434. https://doi.org/10.1155/2011/601434.
93. Shyh-Chang N. Metabolic changes during cancer cachexia pathogenesis. *Adv Exp Med Biol* 2017;**1026**:233–49. https://doi.org/10.1007/978-981-10-6020-5_11.
94. Dong M, Lin J, Lim W, Jin W, Lee HJ. Role of brown adipose tissue in metabolic syndrome, aging, and cancer cachexia. *Front Med* 2018;**12**(2):130–8. https://doi.org/10.1007/s11684-017-0555-2.
95. Maddocks M, Jones LW, Wilcock A. Immunological and hormonal effects of exercise: implications for cancer cachexia. *Curr Opin Support Palliat Care* 2013;**7**(4):376–82. https://doi.org/10.1097/SPC.0000000000000010.
96. O'Sullivan Coyne G, Burotto M. MABp1 for the treatment of colorectal cancer. *Expert Opin Biol Ther* 2017;**17**(9):1155–61. https://doi.org/10.1080/14712598.2017.1347631.
97. Flint TR, Fearon DT, Janowitz T. Connecting the metabolic and immune responses to cancer. *Trends Mol Med* 2017;**23**(5):451–64. https://doi.org/10.1016/j.molmed.2017.03.001.
98. Repasky EA, Eng J, Hylander BL. Stress, metabolism and cancer: integrated pathways contributing to immune suppression. *Cancer J* 2015;**21**(2):97–103. https://doi.org/10.1097/PPO.0000000000000107.
99. Talbot S, Foster SL, Woolf CJ. Neuroimmunity: physiology and pathology. *Annu Rev Immunol* 2016;**34**:421–47. https://doi.org/10.1146/annurev-immunol-041015-055340.
100. Bercik P, Collins SM. The effects of inflammation, infection and antibiotics on the microbiota-gut-brain axis. *Adv Exp Med Biol* 2014;**817**:279–89. https://doi.org/10.1007/978-1-4939-0897-4_13.
101. De Palma G, Lynch MDJ, Lu J, et al. Transplantation of fecal microbiota from patients with irritable bowel syndrome alters gut function and behavior in recipient mice. *Sci Transl Med* 2017;**9**(379). https://doi.org/10.1126/scitranslmed.aaf6397.
102. Collins SM. The intestinal microbiota in the irritable bowel syndrome. *Int Rev Neurobiol* 2016;**131**:247–61. https://doi.org/10.1016/bs.irn.2016.08.003.
103. Varela-Calviño R, Cordero OJ. Immunology and immunotherapy of colorectal cancer. In: Rezaei N, editor. *Cancer immunology: cancer immunotherapy for organ-specific tumors*. Springer: Berlin Heidelberg; 2015. p. 217–36. https://doi.org/10.1007/978-3-662-46410-6_11.
104. Chubak J, Whitlock EP, Williams SB, et al. Aspirin for the prevention of cancer incidence and mortality: systematic evidence reviews for the U.S. preventive services task force. *Ann Intern Med* 2016;**164**(12):814. https://doi.org/10.7326/m15-2117.
105. Dehmer SP, Maciosek MV, Flottemesch TJ. *Aspirin use to prevent cardiovascular disease and colorectal cancer: a decision analysis: technical report*. Agency for Healthcare Research and Quality (US); 2015. https://www.ncbi.nlm.nih.gov/pubmed/26491755.
106. Chan AT, Ladabaum U. Where do we stand with aspirin for the prevention of colorectal cancer? The USPSTF recommendations. *Gastroenterology* 2015;**150**(1):14–8. https://doi.org/10.1053/j.gastro.2015.11.018.
107. Whitlock EP, Williams SB, Burda BU, Feightner A, Beil T. *Aspirin use in adults: cancer, all-cause mortality, and harms: a systematic evidence review for the u.s. preventive services task force*. Agency for Healthcare Research and Quality (US); 2015. https://www.ncbi.nlm.nih.gov/pubmed/26491756.

108. Sutcliffe P, Connock M, Gurung T, et al. Aspirin for prophylactic use in the primary prevention of cardiovascular disease and cancer: a systematic review and overview of reviews. *Health Technol Assess* 2013;**17**(43):1–253. https://doi.org/10.3310/hta17430.
109. Nishihara R, Wu K, Lochhead P, et al. Long-term colorectal-cancer incidence and mortality after lower endoscopy. *N Engl J Med* 2013;**369**(12):1095–105. https://doi.org/10.1056/NEJMoa1301969.
110. Cao Y, Nishihara R, Wu K, et al. Population-wide impact of long-term use of aspirin and the risk for cancer. *JAMA Oncol* 2016;**2**(6):762–9. https://doi.org/10.1001/jamaoncol.2015.6396.
111. National Center for Health Statistics (US). *Health, United States, 2014: with special feature on adults aged 55–64*. National Center for Health Statistics (US); 2015. https://www.ncbi.nlm.nih.gov/pubmed/26086064.
112. Cuzick J, Thorat MA, Bosetti C, et al. Estimates of benefits and harms of prophylactic use of aspirin in the general population. *Ann Oncol* 2015;**26**(1):47–57. https://doi.org/10.1093/annonc/mdu225.
113. Cook NR, Lee I-M, Zhang SM, Moorthy MV, Buring JE. Alternate-day, low-dose aspirin and cancer risk: long-term observational follow-up of a randomized trial. *Ann Intern Med* 2013;**159**(2):77–85. https://doi.org/10.7326/0003-4819-159-2-201307160-00002.
114. Cordero OJ, Varela-Calviño R. Oral hygiene might prevent cancer. *Heliyon* 2018;**4**(10):e00879. https://doi.org/10.1016/j.heliyon.2018.e00879.
115. Su C-W, Zhang Y, Zhu Y-T. Stromal COX-2 signaling are correlated with colorectal cancer: a review. *Crit Rev Oncol Hematol* 2016;**107**:33–8. https://doi.org/10.1016/j.critrevonc.2016.08.010.
116. Wang D, DuBois RN. PPARδ and PGE2 signaling pathways communicate and connect inflammation to colorectal cancer. *Inflamm Cell Signal* 2014;**1**(6). https://doi.org/10.14800/ics.338.
117. Wang D, DuBois RN. Role of prostanoids in gastrointestinal cancer. *J Clin Invest* 2018;**128**(7):2732–42. https://doi.org/10.1172/JCI97953.
118. Chen J, Stark LA. Aspirin prevention of colorectal cancer: focus on NF-κB signalling and the nucleolus. *Biomedicines* 2017;**5**(3). https://doi.org/10.3390/biomedicines5030043.
119. Drew DA, Cao Y, Chan AT. Aspirin and colorectal cancer: the promise of precision chemoprevention. *Nat Rev Cancer* 2016;**16**(3):173–86. https://doi.org/10.1038/nrc.2016.4.
120. Lasry A, Zinger A, Ben-Neriah Y. Inflammatory networks underlying colorectal cancer. *Nat Immunol* 2016;**17**(3):230–40. https://doi.org/10.1038/ni.3384.
121. Tsioulias GJ, Go MF, Rigas B. NSAIDs and colorectal cancer control: promise and challenges. *Curr Pharmacol Rep* 2015;**1**(5):295–301. https://doi.org/10.1007/s40495-015-0042-x.
122. Cuzick J. Preventive therapy for cancer. *Lancet Oncol* 2017;**18**(8):e472–82. https://doi.org/10.1016/S1470-2045(17)30536-3.
123. Penning TM. *Aldo-keto reductase (AKR) 1C3 inhibitors: a patent review. Expert opinion on therapeutic patents*; 2017. Published online https://www.tandfonline.com/doi/abs/10.1080/13543776.2017.1379503.
124. Rostom A, Dubé C, Lewin G, et al. Nonsteroidal anti-inflammatory drugs and cyclooxygenase-2 inhibitors for primary prevention of colorectal cancer: a systematic review prepared for the US preventive services task force. *Ann Intern Med* 2007;**146**(5):376–89. https://www.acpjournals.org/doi/abs/10.7326/0003-4819-146-5-200703060-00010.
125. Carnevali S, Buccellati C, Bolego C, Bertinaria M, Rovati GE, Sala A. Nonsteroidal anti-inflammatory drugs: exploiting bivalent COXIB/ TP antagonists for the control of cardiovascular risk. *Curr Med Chem* 2017;**24**(30):3218–30. https://doi.org/10.2174/0929867324666170602083428.
126. Yin T-F, Wang M, Qing Y, Lin Y-M, Wu D. Research progress on chemopreventive effects of phytochemicals on colorectal cancer and their mechanisms. *World J Gastroenterol* 2016;**22**(31):7058–68. https://doi.org/10.3748/wjg.v22.i31.7058.
127. Lin H-Y, Hsieh M-T, Cheng G-Y, et al. Mechanisms of action of nonpeptide hormones on resveratrol-induced antiproliferation of cancer cells: hormones block resveratrol-induced apoptosis. *Ann N Y Acad Sci* 2017;**1403**(1):92–100. https://doi.org/10.1111/nyas.13423.
128. Chin Y-T, Cheng G-Y, Shih Y-J, et al. Therapeutic applications of resveratrol and its derivatives on periodontitis. *Ann N Y Acad Sci* 2017;**1403**(1):101–8. https://doi.org/10.1111/nyas.13433.
129. Naqvi AZ, Mu L, Hasturk H, Van Dyke TE, Mukamal KJ, Goodson JM. Impact of docosahexaenoic acid therapy on subgingival plaque microbiota. *J Periodontol* 2017;**88**(9):887–95. https://doi.org/10.1902/jop.2017.160398.
130. Madariaga-Venegas F, Fernández-Soto R, Duarte LF, et al. Characterization of a novel antibiofilm effect of nitric oxide-releasing aspirin (NCX-4040) on *Candida albicans* isolates from denture stomatitis patients. *PLoS ONE* 2017;**12**(5):e0176755. https://doi.org/10.1371/journal.pone.0176755.
131. Marvasi M, Durie IA, Henríquez T, Satkute A, Matuszewska M, Prado RC. Dispersal of human and plant pathogens biofilms via nitric oxide donors at 4 °C. *AMB Express* 2016;**6**(1):49. https://doi.org/10.1186/s13568-016-0220-1.
132. Rosato A, Catalano A, Carocci A, et al. In vitro interactions between anidulafungin and nonsteroidal anti-inflammatory drugs on biofilms of *Candida* spp. *Bioorg Med Chem* 2016;**24**(5):1002–5. https://doi.org/10.1016/j.bmc.2016.01.026.
133. El-Mowafy SA, Abd El Galil KH, El-Messery SM, Shaaban MI. Aspirin is an efficient inhibitor of quorum sensing, virulence and toxins in *Pseudomonas aeruginosa*. *Microb Pathog* 2014;**74**:25–32. https://doi.org/10.1016/j.micpath.2014.07.008.
134. Abdelmegeed E, Shaaban MI. Cyclooxygenase inhibitors reduce biofilm formation and yeast-hypha conversion of fluconazole resistant *Candida albicans*. *J Microbiol* 2013;**51**(5):598–604. https://doi.org/10.1007/s12275-013-3052-6.
135. Jung C-J, Yeh C-Y, Shun C-T, et al. Platelets enhance biofilm formation and resistance of endocarditis-inducing streptococci on the injured heart valve. *J Infect Dis* 2012;**205**(7):1066–75. https://doi.org/10.1093/infdis/jis021.
136. Van Dyke TE. Control of inflammation and periodontitis. *Periodontol 2000* 2007;**45**:158–66. https://doi.org/10.1111/j.1600-0757.2007.00229.x.
137. Rao H-L, Chen J-W, Li M, et al. Increased intratumoral neutrophil in colorectal carcinomas correlates closely with malignant phenotype and predicts patients' adverse prognosis. *PLoS ONE* 2012;**7**(1):e30806. https://doi.org/10.1371/journal.pone.0030806.

138. Algars A, Irjala H, Vaittinen S, et al. Type and location of tumor-infiltrating macrophages and lymphatic vessels predict survival of colorectal cancer patients. *Int J Cancer* 2012;**131**(4):864–73. https://doi.org/10.1002/ijc.26457.
139. Papanikolaou IS, Lazaris AC, Apostolopoulos P, et al. Tissue detection of natural killer cells in colorectal adenocarcinoma. *BMC Gastroenterol* 2004;**4**:20. https://doi.org/10.1186/1471-230X-4-20.
140. Nagorsen D, Voigt S, Berg E, Stein H, Thiel E, Loddenkemper C. Tumor-infiltrating macrophages and dendritic cells in human colorectal cancer: relation to local regulatory T cells, systemic T-cell response against tumor-associated antigens and survival. *J Transl Med* 2007;**5**(1):62. https://doi.org/10.1186/1479-5876-5-62.
141. Koch M, Beckhove P, Op den Winkel J, et al. Tumor infiltrating T lymphocytes in colorectal cancer: tumor-selective activation and cytotoxic activity in situ. *Ann Surg* 2006;**244**(6):986–92 [discussion 992–3] https://doi.org/10.1097/01.sla.0000247058.43243.7b.
142. Ogino S, Nosho K, Irahara N, Meyerhardt JA. Lymphocytic reaction to colorectal cancer is associated with longer survival, independent of lymph node count, microsatellite instability, and CpG island methylator. *Clin Cancer Res* 2009;**15**:6412–20. Published online https://clincancerres.aacrjournals.org/content/15/20/6412.short.
143. Pagès F, Berger A, Camus M, et al. Effector memory T cells, early metastasis, and survival in colorectal cancer. *N Engl J Med* 2005;**353**(25): 2654–66. https://doi.org/10.1056/NEJMoa051424.
144. Diederichsen ACP, Hjelmborg JvB, Christensen PB, Zeuthen J, Fenger C. Prognostic value of the CD4+/CD8+ ratio of tumour infiltrating lymphocytes in colorectal cancer and HLA-DR expression on tumour cells. *Cancer Immunol Immunother* 2003;**52**(7):423–8. https://doi.org/10.1007/s00262-003-0388-5.
145. Banerjea A, Bustin SA, Dorudi S. The immunogenicity of colorectal cancers with high-degree microsatellite instability. *World J Surg Oncol* 2005;**3**:26. https://doi.org/10.1186/1477-7819-3-26.
146. Dunn GP, Bruce AT, Ikeda H, Old LJ, Schreiber RD. Cancer immunoediting: from immunosurveillance to tumor escape. *Nat Immunol* 2002; **3**(11):991–8. https://doi.org/10.1038/ni1102-991.
147. Dunn GP, Old LJ, Schreiber RD. The three Es of cancer immunoediting. *Annu Rev Immunol* 2004;**22**:329–60. https://doi.org/10.1146/annurev.immunol.22.012703.104803.
148. Menon AG, Janssen-van Rhijn CM, Morreau H, et al. Immune system and prognosis in colorectal cancer: a detailed immunohistochemical analysis. *Lab Investig* 2004;**84**(4):493–501. https://doi.org/10.1038/labinvest.3700055.
149. Tosolini M, Kirilovsky A, Mlecnik B, et al. Clinical impact of different classes of infiltrating T cytotoxic and helper cells (Th1, th2, treg, th17) in patients with colorectal cancer. *Cancer Res* 2011;**71**(4):1263–71. https://doi.org/10.1158/0008-5472.CAN-10-2907.
150. Hurtado CG, Wan F, Housseau F, Sears CL. Roles for interleukin 17 and adaptive immunity in pathogenesis of colorectal cancer. *Gastroenterology* 2018;**155**(6):1706–15. https://doi.org/10.1053/j.gastro.2018.08.056.
151. Hawinkels LJAC, Paauwe M, Verspaget HW, et al. Interaction with colon cancer cells hyperactivates TGF-β signaling in cancer-associated fibroblasts. *Oncogene* 2014;**33**(1):97–107. https://doi.org/10.1038/onc.2012.536.
152. Fung KY, Nguyen PM, Putoczki T. The expanding role of innate lymphoid cells and their T-cell counterparts in gastrointestinal cancers. *Mol Immunol* 2017;**110**:48–56. https://doi.org/10.1016/j.molimm.2017.11.013.
153. Kirchberger S, Royston DJ, Boulard O, et al. Innate lymphoid cells sustain colon cancer through production of interleukin-22 in a mouse model. *J Exp Med* 2013;**210**(5):917–31. https://doi.org/10.1084/jem.20122308.
154. Wu Y, Yuan L, Lu Q, Xu H, He X. Distinctive profiles of tumor-infiltrating immune cells and association with intensity of infiltration in colorectal cancer. *Oncol Lett* 2018;**15**(3):3876–82. https://doi.org/10.3892/ol.2018.7771.
155. Yan G, Liu T, Yin L, Kang Z, Wang L. Levels of peripheral Th17 cells and serum Th17-related cytokines in patients with colorectal cancer: a meta-analysis. *Cell Mol Biol* 2018;**64**(6):94–102. https://www.ncbi.nlm.nih.gov/pubmed/29808807.
156. Arelaki S, Arampatzioglou A, Kambas K, et al. Gradient infiltration of neutrophil extracellular traps in colon cancer and evidence for their involvement in tumour growth. *PLoS ONE* 2016;**11**(5):e0154484. https://doi.org/10.1371/journal.pone.0154484.
157. Galdiero MR, Bianchi P, Grizzi F, et al. Occurrence and significance of tumor-associated neutrophils in patients with colorectal cancer: significance of tumor-associated neutrophils in colorectal cancer. *Int J Cancer* 2016;**139**(2):446–56. https://doi.org/10.1002/ijc.30076.
158. Mao H, Pan F, Wu Z, et al. CD19loCD27hi plasmablasts suppress harmful Th17 inflammation through interleukin 10 pathway in colorectal cancer. *DNA Cell Biol* 2017;**36**(10):870–7. https://doi.org/10.1089/dna.2017.3814.
159. Lee JY, Seo E-H, Oh C-S, et al. Impact of circulating T helper 1 and 17 cells in the blood on regional lymph node invasion in colorectal cancer. *J Cancer* 2017;**8**(7):1249–54. https://doi.org/10.7150/jca.18230.
160. Sharp SP, Avram D, Stain SC, Lee EC. Local and systemic Th17 immune response associated with advanced stage colon cancer. *J Surg Res* 2017;**208**:180–6. https://doi.org/10.1016/j.jss.2016.09.038.
161. Wesch D, Peters C, Siegers GM. Human gamma delta T regulatory cells in cancer: fact or fiction? *Front Immunol* 2014;**5**:598. https://doi.org/10.3389/fimmu.2014.00598.
162. Palucka K, Ueno H, Roberts L, Fay J, Banchereau J. Dendritic cells: are they clinically relevant? *Cancer J* 2010;**16**(4):318–24. https://doi.org/10.1097/PPO.0b013e3181eaca83.
163. Michielsen AJ, Hogan AE, Marry J, et al. Tumour tissue microenvironment can inhibit dendritic cell maturation in colorectal cancer. *PLoS ONE* 2011;**6**(11):e27944. https://doi.org/10.1371/journal.pone.0027944.
164. Lindenbergh MFS, Stoorvogel W. Antigen presentation by extracellular vesicles from professional antigen-presenting cells. *Annu Rev Immunol* 2018;**36**:435–59. https://doi.org/10.1146/annurev-immunol-041015-055700.
165. Umansky V, Adema GJ, Baran J, et al. Interactions among myeloid regulatory cells in cancer. *Cancer Immunol Immunother* 2018;**68**(4): 645–60. https://doi.org/10.1007/s00262-018-2200-6.

166. Legitimo A, Consolini R, Failli A, Orsini G, Spisni R. Dendritic cell defects in the colorectal cancer. *Hum Vaccin Immunother* 2014;**10**(11): 3224–35. https://doi.org/10.4161/hv.29857.
167. Josefowicz SZ, Lu L-F, Rudensky AY. Regulatory T cells: mechanisms of differentiation and function. *Annu Rev Immunol* 2012;**30**:531–64. https://doi.org/10.1146/annurev.immunol.25.022106.141623.
168. deLeeuw RJ, Kost SE, Kakal JA, Nelson BH. The prognostic value of FoxP3+ tumor-infiltrating lymphocytes in cancer: a critical review of the literature. *Clin Cancer Res* 2012;**18**(11):3022–9. https://doi.org/10.1158/1078-0432.CCR-11-3216.
169. Ladoire S, Martin F, Ghiringhelli F. Prognostic role of FOXP3+ regulatory T cells infiltrating human carcinomas: the paradox of colorectal cancer. *Cancer Immunol Immunother* 2011;**60**(7):909–18. https://doi.org/10.1007/s00262-011-1046-y.
170. Erdman SE, Poutahidis T, Tomczak M, et al. CD4+ CD25+ regulatory T lymphocytes inhibit microbially induced colon cancer in Rag2-deficient mice. *Am J Pathol* 2003;**162**(2):691–702. https://doi.org/10.1016/S0002-9440(10)63863-1.
171. Ward-Hartstonge KA, Kemp RA. Regulatory T-cell heterogeneity and the cancer immune response. *Clin Transl Immunol* 2017;**6**(9):e154. https://asi.onlinelibrary.wiley.com/doi/abs/10.1038/cti.2017.43.
172. Whiteside TL. Regulatory T cell subsets in human cancer: are they regulating for or against tumor progression? *Cancer Immunol Immunother* 2014;**63**(1):67–72. https://doi.org/10.1007/s00262-013-1490-y.
173. Rabinovich GA, Gabrilovich D, Sotomayor EM. Immunosuppressive strategies that are mediated by tumor cells. *Annu Rev Immunol* 2007;**25**(1):267–96. https://doi.org/10.1146/annurev.immunol.25.022106.141609.
174. Izhak L, Ambrosino E, Kato S, et al. Delicate balance among three types of T cells in concurrent regulation of tumor immunity. *Cancer Res* 2013;**73**(5):1514–23. https://doi.org/10.1158/0008-5472.CAN-12-2567.
175. Wang Y, Cardell SL. The Yin and Yang of invariant natural killer T cells in tumor immunity—suppression of tumor immunity in the intestine. *Front Immunol* 2018;**8**:1945. https://doi.org/10.3389/fimmu.2017.01945.
176. Wang Y, Sedimbi S, Löfbom L, Singh AK, Porcelli SA, Cardell SL. Unique invariant natural killer T cells promote intestinal polyps by suppressing TH1 immunity and promoting regulatory T cells. *Mucosal Immunol* 2018;**11**(1):131–43. https://doi.org/10.1038/mi.2017.34.
177. Nagorsen D, Thiel E. HLA typing demands for peptide-based anti-cancer vaccine. *Cancer Immunol Immunother* 2008;**57**(12):1903–10. https://doi.org/10.1007/s00262-008-0493-6.
178. Li MO, Wan YY, Sanjabi S, Robertson A-KL, Flavell RA. Transforming growth factor-β regulation of immune responses. *Annu Rev Immunol* 2006;**24**(1):99–146. https://doi.org/10.1146/annurev.immunol.24.021605.090737.
179. Batlle E, Massagué J. Transforming growth factor-β signaling in immunity and cancer. *Immunity* 2019;**50**(4):924–40. https://doi.org/10.1016/j.immuni.2019.03.024.
180. Gorelik L, Flavell RA. Transforming growth factor-beta in T-cell biology. *Nat Rev Immunol* 2002;**2**(1):46–53. https://doi.org/10.1038/nri704.
181. Landskron G, De la Fuente M, Thuwajit P, Thuwajit C, Hermoso MA. Chronic inflammation and cytokines in the tumor microenvironment. *J Immunol Res* 2014;**2014**:149185. https://doi.org/10.1155/2014/149185.
182. Olsen RS, Nijm J, Andersson RE, Dimberg J, Wågsäter D. Circulating inflammatory factors associated with worse long-term prognosis in colorectal cancer. *World J Gastroenterol* 2017;**23**(34):6212–9. https://doi.org/10.3748/wjg.v23.i34.6212.
183. Kalyan A, Kircher S, Shah H, Mulcahy M, Benson A. Updates on immunotherapy for colorectal cancer. *J Gastrointest Oncol* 2018;**9**(1):160–9. https://doi.org/10.21037/jgo.2018.01.17.
184. Syn NL, Wang L, Chow EK-H, Lim CT, Goh B-C. Exosomes in cancer nanomedicine and immunotherapy: prospects and challenges. *Trends Biotechnol* 2017;**35**(7):665–76. https://doi.org/10.1016/j.tibtech.2017.03.004.
185. Gobbini E, Charles J, Toffart AC, Leccia MT, Moro-Sibilot D, Giaj LM. Current opinions in immune checkpoint inhibitors rechallenge in solid cancers. *Crit Rev Oncol Hematol* 2019;**144**:102816. https://doi.org/10.1016/j.critrevonc.2019.102816.
186. Huang Y-H, Zhu C, Kondo Y, et al. Corrigendum: CEACAM1 regulates TIM-3-mediated tolerance and exhaustion. *Nature* 2015;**517**:386–90. https://doi.org/10.1038/nature17421.
187. Guinney J, Dienstmann R, Wang X, et al. The consensus molecular subtypes of colorectal cancer. *Nat Med* 2015;**21**(11):1350–6. https://doi.org/10.1038/nm.3967.
188. Giardiello FM, Allen JI, Axilbund JE, et al. Guidelines on genetic evaluation and management of Lynch syndrome: a consensus statement by the US multi-society task force on colorectal cancer. *Am J Gastroenterol* 2014;**109**(8):1159–79. https://doi.org/10.1038/ajg.2014.186.
189. Yurgelun MB, Kulke MH, Fuchs CS, et al. Cancer susceptibility gene mutations in individuals with colorectal cancer. *J Clin Oncol* 2017;**35**(10):1086–95. https://doi.org/10.1200/jco.2016.71.0012.
190. Hampel H, Pearlman R, Beightol M, et al. Assessment of tumor sequencing as a replacement for lynch syndrome screening and current molecular tests for patients with colorectal cancer. *JAMA Oncol* 2018;**4**(6):806–13. https://doi.org/10.1001/jamaoncol.2018.0104.
191. Koopman M, Kortman GAM, Mekenkamp L, et al. Deficient mismatch repair system in patients with sporadic advanced colorectal cancer. *Br J Cancer* 2009;**100**(2):266–73. https://doi.org/10.1038/sj.bjc.6604867.
192. Le DT, Uram JN, Wang H, et al. PD-1 blockade in tumors with mismatch-repair deficiency. *N Engl J Med* 2015;**372**(26):2509–20. https://doi.org/10.1056/NEJMoa1500596.
193. Lipson EJ, Sharfman WH, Drake CG, et al. Durable cancer regression off-treatment and effective reinduction therapy with an anti-PD-1 antibody. *Clin Cancer Res* 2013;**19**(2):462–8. https://doi.org/10.1158/1078-0432.CCR-12-2625.
194. Overman MJ, Kopetz S, McDermott RS, et al. Nivolumab ± ipilimumab in treatment (tx) of patients (pts) with metastatic colorectal cancer (mCRC) with and without high microsatellite instability (MSI-H): CheckMate-142 interim results. *J Clin Orthod* 2016;**34**(15_suppl):3501. https://doi.org/10.1200/JCO.2016.34.15_suppl.3501.

195. Brahmer JR, Tykodi SS, Chow LQM, et al. Safety and activity of anti–PD-L1 antibody in patients with advanced cancer. *N Engl J Med* 2012;**366**(26):2455–65. https://doi.org/10.1056/NEJMoa1200694.
196. Bourdais R, Rousseau B, Pujals A, et al. Polymerase proofreading domain mutations: new opportunities for immunotherapy in hypermutated colorectal cancer beyond MMR deficiency. *Crit Rev Oncol Hematol* 2017;**113**:242–8. https://doi.org/10.1016/j.critrevonc.2017.03.027.
197. Gong J, Wang C, Lee PP, Chu P, Fakih M. Response to PD-1 blockade in microsatellite stable metastatic colorectal cancer harboring a POLE mutation. *J Natl Compr Cancer Netw* 2017;**15**(2):142–7. https://doi.org/10.6004/jnccn.2017.0016.
198. Steinbach G, Lynch PM, Phillips RK, et al. The effect of celecoxib, a cyclooxygenase-2 inhibitor, in familial adenomatous polyposis. *N Engl J Med* 2000;**342**(26):1946–52. https://doi.org/10.1056/NEJM200006293422603.
199. Gonzalez-Angulo AM, Fuloria J, Prakash O. Cyclooxygenase 2 inhibitors and colon cancer. *Ochsner J* 2002;**4**(3):176–9. https://www.ncbi.nlm.nih.gov/pubmed/22822342.
200. Liu S-S, Yang Y-Z, Jiang C, et al. Comparison of immunological characteristics between paired mismatch repair-proficient and -deficient colorectal cancer patients. *J Transl Med* 2018;**16**(1):195. https://doi.org/10.1186/s12967-018-1570-z.
201. Javan MR, Khosrojerdi A, Moazzeni SM. New insights into implementation of mesenchymal stem cells in cancer therapy: prospects for anti-angiogenesis treatment. *Front Oncol* 2019;**9**:840. https://doi.org/10.3389/fonc.2019.00840.
202. Romera A, Peredpaya S, Shparyk Y, et al. Bevacizumab biosimilar BEVZ92 versus reference bevacizumab in combination with FOLFOX or FOLFIRI as first-line treatment for metastatic colorectal cancer: a multicentre, open-label, randomised controlled trial. *Lancet Gastroenterol Hepatol* 2018;**3**(12):845–55. https://www.sciencedirect.com/science/article/pii/S2468125318302693.
203. Ushida Y, Shinozaki E, Chin K, et al. Two cases of long-term survival of advanced colorectal cancer with synchronous lung metastases treated with mFOLFOX6/XELOX + bevacizumab. *Case Rep Oncol* 2018;**11**(2):601–8. https://doi.org/10.1159/000492568.
204. Chiba T, Ohtani H, Mizoi T, et al. Intraepithelial CD8+ T-cell-count becomes a prognostic factor after a longer follow-up period in human colorectal carcinoma: possible association with suppression of micrometastasis. *Br J Cancer* 2004;**91**(9):1711–7. https://doi.org/10.1038/sj.bjc.6602201.
205. Deschoolmeester V, Baay M, Van Marck E, et al. Tumor infiltrating lymphocytes: an intriguing player in the survival of colorectal cancer patients. *BMC Immunol* 2010;**11**:19. https://doi.org/10.1186/1471-2172-11-19.
206. Zacharakis N, Chinnasamy H, Black M, et al. Immune recognition of somatic mutations leading to complete durable regression in metastatic breast cancer. *Nat Med* 2018;**24**(6):724–30. https://doi.org/10.1038/s41591-018-0040-8.
207. Marcus A, Waks T, Eshhar Z. Redirected tumor-specific allogeneic T cells for universal treatment of cancer. *Blood* 2011;**118**(4):975–83. https://doi.org/10.1182/blood-2011-02-334284.
208. Zhang C, Wang Z, Yang Z, et al. Phase I escalating-dose trial of CAR-T therapy targeting CEA+ metastatic colorectal cancers. *Mol Ther* 2017;**25**(5):1248–58. https://doi.org/10.1016/j.ymthe.2017.03.010.
209. Tacken PJ, de Vries IJM, Torensma R, Figdor CG. Dendritic-cell immunotherapy: from ex vivo loading to in vivo targeting. *Nat Rev Immunol* 2007;**7**(10):790–802. https://doi.org/10.1038/nri2173.
210. Turriziani M, Fantini M, Benvenuto M. *Carcinoembryonic antigen (CEA)-based cancer vaccines: recent patents and antitumor effects from experimental models to clinical trials*; 2012. Recent patents on. Published online https://www.ingentaconnect.com/content/ben/pra/2012/00000007/00000003/art00004.
211. Jäger E, Gnjatic S, Nagata Y, et al. Induction of primary NY-ESO-1 immunity: CD8+ T lymphocyte and antibody responses in peptide-vaccinated patients with NY-ESO-1+ cancers. *Proc Natl Acad Sci U S A* 2000;**97**(22):12198–203. https://doi.org/10.1073/pnas.220413497.
212. Nagorsen D, Keilholz U, Rivoltini L, et al. Natural T-cell response against MHC class I epitopes of epithelial cell adhesion molecule, her-2/neu, and carcinoembryonic antigen in patients with colorectal cancer. *Cancer Res* 2000;**60**(17):4850–4. https://www.ncbi.nlm.nih.gov/pubmed/10987297.
213. Galluzzi L, Senovilla L, Vacchelli E, et al. Trial watch: dendritic cell-based interventions for cancer therapy. *Oncoimmunology* 2012;**1**(7):1111–34. https://doi.org/10.4161/onci.21494.
214. Kranz LM, Diken M, Haas H, et al. Systemic RNA delivery to dendritic cells exploits antiviral defence for cancer immunotherapy. *Nature* 2016;**534**(7607):396–401. https://doi.org/10.1038/nature18300.
215. Hu Z, Ott PA, Wu CJ. Towards personalized, tumour-specific, therapeutic vaccines for cancer. *Nat Rev Immunol* 2018;**18**(3):168–82. https://doi.org/10.1038/nri.2017.131.
216. Sahin U, Türeci Ö. Personalized vaccines for cancer immunotherapy. *Science* 2018;**359**(6382):1355–60. https://doi.org/10.1126/science.aar7112.
217. Sahin U, Derhovanessian E, Miller M, et al. Personalized RNA mutanome vaccines mobilize poly-specific therapeutic immunity against cancer. *Nature* 2017;**547**(7662):222–6. https://doi.org/10.1038/nature23003.
218. Duruisseaux M, Martínez-Cardús A, Calleja-Cervantes ME, et al. Epigenetic prediction of response to anti-PD-1 treatment in non-small-cell lung cancer: a multicentre, retrospective analysis. *Lancet Respir Med* 2018;**6**(10):771–81. https://doi.org/10.1016/S2213-2600(18)30284-4.
219. Shekarian T, Valsesia-Wittmann S, Brody J, et al. Pattern recognition receptors: immune targets to enhance cancer immunotherapy. *Ann Oncol* 2017;**28**(8):1756–66. https://doi.org/10.1093/annonc/mdx179.
220. Vanacker H, Vetters J, Moudombi L, Caux C, Janssens S, Michallet M-C. Emerging role of the unfolded protein response in tumor immunosurveillance. *Trends Cancer Res* 2017;**3**(7):491–505. https://doi.org/10.1016/j.trecan.2017.05.005.
221. Thomas M, Ponce-Aix S, Navarro A, et al. Immunotherapeutic maintenance treatment with toll-like receptor 9 agonist lefitolimod in patients with extensive-stage small-cell lung cancer: results from the exploratory, controlled, randomized, international phase II IMPULSE study. *Ann Oncol* 2018;**29**(10):2076–84. https://www.sciencedirect.com/science/article/pii/S0923753419341997.
222. Wang D, Jiang W, Zhu F, Mao X, Agrawal S. Modulation of the tumor microenvironment by intratumoral administration of IMO-2125, a novel TLR9 agonist, for cancer immunotherapy. *Int J Oncol* 2018;**53**(3):1193–203. https://doi.org/10.3892/ijo.2018.4456.

Section D.IV

Anesthetic treatment and postoperative management

Chapter 41

Multimodal rehabilitation: Pre- and intraoperative optimization in CRC surgery

Manuel Núñez Deben, Miguel Pereira Loureiro, Vanesa Vilanova Vázquez, and Gerardo Baños Rodríguez

Department of Anesthesiology and Reanimation, Vigo University Hospital Complex, Vigo, Spain

Introduction

Colorectal surgery (CRC), understood as surgery involving colonic resection and anastomosis, is performed for the treatment of various pathologies such as ulcerative colitis, Crohn's disease, mechanical obstruction, recurrent diverticulitis, etc. However, it is colon cancer, which is the second most frequent neoplasm in both sexes, that is the main cause of surgical indication. CRC is a high-risk surgery that usually entails an average hospital stay of 12–14 days, which in most cases was not due to high morbidity but was in fact owing to a global perioperative approach responsible for this length of hospital stay.

Advances in surgical techniques in colorectal surgery and a different therapeutic approach to the process by healthcare staff have led to the emergence of techniques aimed at reducing hospitalization time, known as early or multimodal rehabilitation, which propose a different global strategy with the aim of reducing morbidity and mortality, shortening hospital stays, and improving the quality of life of these patients.

This concept (Enhanced recovery after surgery, or ERAS) also known as Fast-track was introduced by Wilmore and Kehlet[1] and endorsed by the favorable results obtained in subsequent randomized studies. It is based on evidence-based medicine and its implementation implies knowledge of the mechanisms involved in the pathophysiology of the surgical process and has forced surgeons and anesthesiologists to make changes in our actions in order to favor the early recovery of these patients.

Its fundamental objectives focus on minimizing bowel dysfunction, attenuating the response to surgical stress, careful management of fluid therapy, and effective pain control. The optimization begins at the time of diagnosis and aims to recognize the individual needs of each patient, adjusting treatment before, during, and after surgery. The close collaboration of all elements involved in the process is essential, including first and foremost the patient and family members.

In the preoperative period, care plans are aimed at informing the patient, optimizing vital conditions and comorbidities, and identifying risk factors that could predict perioperative complications.

During the intraoperative phase, the fundamental objectives, in addition to facilitating an adequate surgical field, are reducing the immunological response to stress, maintaining adequate colonic perfusion and oxygenation, meticulous fluid management, along with water and electrolyte therapy, providing adequate multimodal analgesia, preventing PONV and paralytic ileus.

The early or multimodal rehabilitation techniques we will describe for pre- and intraoperative rehabilitation are associated with reduced postoperative morbidity but there is no evidence that they decrease procedural mortality compared to conventional techniques.[2] The increasing use of minimally invasive surgical procedures and anesthetic techniques that are more protective of bowel function means that perioperative care chapters are in continuous development and therefore in need of frequent updating.

Preoperative

The classic preoperative period has evolved with measures that have been shown to favor postoperative recovery, trying to ensure that the patient arrives at the surgery in the best possible condition. Care plans are aimed at informing the patient (informed consent) and identifying risk factors that could predict perioperative complications, optimizing those that can be acted upon.

Premedication should be systematically avoided (moderate evidence, strong recommendation). Detailed information to the patient and their environment about their disease, the surgical procedure, and the phases they will undergo facilitates their identification with multimodal rehabilitation programs and reduces anxiety.[3] An adequate anamnesis is necessary to identify comorbidities that may condition recovery and even predict the appearance of complications during the perioperative period.

In order to clearly define the surgical risk in colorectal surgery, a series of scores have been defined to predict morbidity and mortality. The most commonly used by anesthesiologists is the ASA of the American Society of Anesthesiologists, which assesses the patient according to the underlying disease. Other surgical scores are also used, such as POSSUM (The Physiology and Operative Severity Score for Enumeration of Mortality and Morbidity) which calculates expected morbidity and mortality based on 12 physiological and 6 operative variables. Colorectal POSSUM (CR-POSSUM) more accurately predicts mortality compared to POSSUM and is based on age, cardiac function, blood pressure, pulse, blood urea nitrogen, hemoglobin, and surgical severity. Risk assessment is influenced by many factors, including those of the healthcare staff themselves, which make it difficult to choose the ideal score. The surgeon's instinct has been found to be a good predictor of risk, and when compared to POSSUM, a greater overestimation of risk was observed with POSSUM.[4]

The American College of Surgeons has developed an open-access surgical risk calculator (http://oldriskcalculator.facs.org) that receives an average of 1500 visits/day and calculates the likelihood of an unfavorable postoperative event. The risk is determined based on information obtained from the patient in the preoperative consultation and estimates are assessed using data from a large number of patients who underwent a similar surgical procedure. Hospitals that used this technique had encouraging results and, as reported by the researchers, between 2007 and 2016, infections decreased from 13.7% to 4.7% and the need for reoperation was 12% lower. The odds of earlier discharge were 12% higher. This marker can only be used in planned surgery because in emergency cases it underestimates serious complications and length of admission.

Factors that cannot be influenced preoperatively

- **Age**
  The mortality rate in geriatric patients is low and we know that neoplastic pathologies do not significantly increase complications in these patients. The risk will be determined by the type of surgery and the presence of associated comorbidity. All organic imbalances must be corrected; this is the time for preoperative optimization. There are situations that affect colonic perfusion and oxygenation such as smoking, atherosclerosis, heart failure, anemia, water and electrolyte imbalances, nutritional deficiencies (hypoalbuminemia), or weight loss that can lead to serious complications.
- **Previous abdominal surgery.**
- **Sex**
  Males present a higher number of perioperative complications in colorectal surgery with a higher rate of anastomotic leakage, both in open and laparoscopic surgery.

Factors that can be influenced preoperatively

- **Preoperative fasting**. Presurgery sugary oral fluids.
  Patients who are candidates for colorectal surgery often have poor preoperative nutritional status, especially hypoalbuminemia. This has been associated with a greater number of complications such as infections and worse clinical results postsurgery, associated with hypohydration, to which is added the preoperative fasting time, classically set at 6h to avoid the risk of aspiration. The American Society of Anesthesiologists and the European Society of Anesthesiology consider it safe to administer 200 mL of clear sweetened liquids 2h prior to surgery, that, without increasing the residual gastric content, decrease the response to surgical stress,[5] reduce intraoperative hypothermia, prevent peripheral insulin resistance, produce a decreased incidence of PONV, reduce weight loss and postoperative fatigue,[6] although no conclusive studies have confirmed a clear effect on postoperative morbidity and mortality.
  Preoperative administration of clear sugary fluids is considered routine in preparation for elective abdominal surgery. Low quality of evidence, strong recommendation.
  Contraindicated in diabetes and patients with impaired gastric emptying.
- **Preoperative anemia control**
  Using WHO criteria, preoperative anemia may be present in up to 50% of surgical patients, depending on the pathology to be treated, and is the most frequent blood disorder in neoplastic patients. In colorectal cancer, it is often

considered a consequence of the tumor pathology that will resolve itself after treatment, but in surgical patients, other causes can lead to anemia: blood loss, vitamin B12 or folic acid deficiency, anemia due to chronic disease with or without relation to the reason for surgery, or a combination of factors. Also, a relationship has been found between the type of preoperative anemia, tumor characteristics, systemic inflammation, and survival; proximal tumor location is predominantly associated with microcytic anemia with worse overall survival that does not improve with intraoperative blood administration and systemic inflammation is associated with normocytic anemia.[7]

In patients with preoperative colorectal cancer, there is a high prevalence of iron deficiency due to blood loss or chronic inflammatory process that should be treated by iron supplementation with the aim of maintaining a Hb between 10 and 13 g/dL depending on the presence of associated pathologies. Oral iron therapy is a slow process and in this type of patient may be poorly tolerated or not have a good response due to continuous bleeding and the presence of associated chronic disease, so intravenous iron therapy is of special interest in this group due to the low risk of adverse reactions and its greater effectiveness in correcting iron deficiency anemia as in the anemia of chronic disease. Erythropoietin administration is a controversial issue and there is no evidence on the advantages of its preoperative use.

One factor to consider is the association between perioperative transfusion and increased risk of tumor recurrence in patients undergoing curative neoplastic colorectal surgery.[8] These long-term effects justify a more restrictive transfusion policy.

- **Mechanical colon preparation**

    Colonic preparation frequently causes intra- and postoperative water and electrolyte disturbances and a higher incidence of dehiscence and infectious processes.[9] Its use is only recommended in patients with ileostomy and in lesions of the rectum, middle, and lower third. High quality of evidence, strong recommendation. If mechanical preparation of the colon is necessary, losses should be replaced with Ringer Lactate/Acetate type solutions.

- **Nutritional status**

    Preoperative malnutrition has been associated with increased postoperative morbidity and mortality as well as poor oncological outcomes in gastrointestinal cancer surgery with an increased prevalence of infectious complications and anastomotic leaks.[10]

    Various anthropometric measurements, biochemical markers, immunological tests, and body composition analyses have been used to assess nutritional status. However, there is no single test that can detect the degree of malnutrition. This is mainly because most of the parameters used to determine malnutrition can be influenced by other nonnutritional factors. For practical purposes, the Spanish Multimodal Rehabilitation Group (GERM) considers a patient to be malnourished when he/she presents analytical alterations according to the CONUT filter (decrease in cholesterol, lymphocytes, and albumin) and/or has lost more than 5% of his/her body weight.

    In elective surgery and in malnourished patients, it is important to administer hyperprotein supplements at least twice a day during the week before surgery. Immunomodulatory enteral nutrition, based on the provision of complete diets enriched with pharmaconutrients (L-arginine, RNA, glutamine, ω3 fatty acids, etc.) with the capacity to reduce the response to surgical aggression, is considered the treatment of choice in this type of patient.[11] In its latest 2017 guidelines, The European Society for Clinical Nutrition and Metabolism (ESPEN) recommends the administration of these immunonutrients in malnourished patients undergoing major surgery for colorectal cancer.

- **Obesity**

    Patients with a BMI greater than $25 kg/m^2$ have an increased risk of complications, such as incisional hernias and a higher rate of surgical infection. In elective colorectal surgery, preoperative weight loss is recommended in overweight patients to decrease complications.

- **Thromboembolism prophylaxis**

    Prophylaxis with low molecular weight heparin is indicated for at least 30 days, adjusting the dose according to the patient's risk factors and comorbidities. High evidence, strong recommendation.

- **Antibiotic prophylaxis**

    Adequate antibiotic prophylaxis with coverage for aerobic and anaerobic germs reduces operative wound infection by 75%. The choice of antibiotic will depend on the resistance of each center and according to the protocols established in each hospital. One option may be amoxiclavulanic acid, or a combination of metronidazole and cephalosporins, avoiding nephrotoxic drugs such as aminoglycosides. A single preoperative dose is recommended for all patients undergoing colorectal surgery, with a second dose administered if the surgery is prolonged beyond the half-life of the antibiotic or if there is an intraoperative complication (bleeding, etc.). High evidence, strong recommendation.

    The use of oral antibiotics associated with preoperative intravenous antibiotics has low evidence and its use is not recommended systematically. A recent metaanalysis[12] found a lower incidence of infections in patients who had received this dual therapy and prior bowel preparation; however, there are no studies on their use in patients without bowel preparation.

Preoperative skin decontamination with chlorhexidine has been shown to decrease the incidence of surgical site infection.[13] However, there is no evidence to support the practice of preoperative antiseptic showers.

- **Tobacco and alcohol**

  Smoking is known to be a risk factor for the development of perioperative respiratory complications. Several metaanalyses show that intensive nicotine replacement therapy is effective, but leave doubt as to the length of time of abstinence before surgery. It seems that a minimum time of no less than 4 weeks is necessary because it is not clear if, below this period, smoking cessation reduces the risk of postoperative respiratory complications.[14] A 4-week preoperative abstinence from alcohol intake is recommended and there is evidence of an increased infection rate, without affecting mortality, in heavy drinkers, but the impact in patients with lower alcohol consumption is unknown.[15]

Intraoperative

One of the key goals is to modify the immunological response to surgical stress, which results in the activation of the adrenocortical axis with the release of endogenous catecholamines that may lead to increased morbidity. In general, the magnitude and duration of the metabolic response is proportional to the surgical injury and the development of complications such as sepsis. An exaggerated stress response has been associated with postoperative bowel dysfunction, slow wound healing, infectious complications, increased incidence of anastomotic leakage and suture dehiscence, and increased cardiopulmonary complications due to fluid overload. In the long term, there is increased adhesion formation.[16]

Various strategies have been tried to attenuate the catabolic stress response. Glycemic control and shorter fasting periods, administration of preoperative dexamethasone and NSAIDs, the use of laparoscopic surgery and epidural analgesia have all been shown to be beneficial in reducing overall postoperative morbidity. Low-dose benzodiazepines improve the recovery of these patients, reducing anxiety and providing comfort and satisfaction, although their use in elderly adults is systematically contraindicated due to the possibility of increasing the occurrence of postoperative delirium. Adjuvants such as beta-blockers are available to attenuate the catecholaminergic response to surgical stress and prevent cardiovascular events in elderly patients undergoing noncardiac surgery. Evidence suggests that they are most effective in those patients with known coronary artery disease. They have anticatabolic properties, and in the critically ill patient, combined with parenteral nutrition, decrease protein catabolism. Alpha-2 agonists provide greater hemodynamic stability and reduce postoperative pain. Premedication with clonidine or dexmedetomidine has been associated with reduced opioid use, lower incidence of PONV, and less intraoperative bleeding. They facilitate glycemic control in patients with type 2 diabetes mellitus and reduce myocardial ischemia after surgery. Regardless of the strategy used, there is growing evidence that a multimodal approach to perioperative care significantly reduces surgical stress and improves recovery after major surgery.[17]

Perioperative management of nausea and vomiting

Their incidence ranges from 20% to 30% of patients undergoing surgery, up to 80% in high-risk patients. They increase the risk of hematoma, hemorrhage, suture dehiscence, and water and electrolyte disturbances, delay the introduction of diet and oral medication, thus prolonging hospital stay and recovery.

There are independent risk predictors related to the patient (female sex, nonsmoker, previous history of PONV, or motion sickness), anesthesia-related (volatile anesthetics, nitrous oxide, intra- and postoperative opioids) and surgery-related factors (duration of surgery: every 30 min increases the risk by 60%, laparoscopy, abdominal surgery). Other related risk factors, but not considered independent predictors, are fasting, anxiety, gastroparesis, peritoneal irritation, and neostigmine use.

In order to assess the risk of PONV, the Apfel scale has been developed, which assigns one point for each of the following circumstances: female sex, nonsmoker, history of PONV or motion sickness, postoperative opioids, and considers that prophylaxis should be done with:

- Apfel 0–1: Dexamethasone or Droperidol.
- Apfel 2–3: Dexamethasone + Droperidol and general risk reduction measures.
- Apfel 4: Dexamethasone + Droperidol + Ondansetron.

As a general rule, a multimodal approach to PONV prophylaxis and treatment should be considered (high quality of evidence, strong recommendation). Patients with 1–2 risk factors should receive 2-drug prophylaxis and if they have ≥2 risk factors they should receive 2–3. Dexamethasone administration is considered to be most effective at doses of 4–8 mg during anesthetic induction. Droperidol 1.25 mg and ondansetron 4 mg should be administered about 30 min before the end of surgery.

The absolute risk reduction achieved by the application of a given therapy depends on the a priori baseline risk of PONV.

Treatment will depend on whether or not prior prophylaxis has been done. In patients without primary prevention, ondansetron 4 mg is indicated. If previous prophylaxis has been performed, another family of drugs (corticosteroids, dopaminergic antagonists, serotonin receptor antagonists, neurokinergic antagonists, muscarinic antagonists, histaminergic antagonists) should be used and wait at least 6 h to repeat the dose of ondansetron if it has already been administered. Avoiding the onset of pain reduces the incidence of nausea[18] and the use of prophylactic paracetamol before surgery was shown to be effective, although without reducing the dose of postoperative opioids. Intravenous Propofol 20 mg has been shown to be effective for the treatment of PONV occurring during the first 6 h postsurgery. Studies by Greif et al.[19] have shown that an increase in inspiratory fraction >80% has been associated with a lower incidence of postoperative nausea and vomiting, as well as a decreased risk of infection.

Intraoperative **hyperglycemia** is an independent risk predictor for postoperative complications. Maintaining adequate blood glucose levels may attenuate the systemic inflammatory response and improve surgical outcomes. Blood glucose control at <180 mg/dL provides better outcomes and less associated risks than the tight blood glucose control advocated by van den Berghe et al. who recommended maintaining normoglycemia between control values of 81 and 108 mg/dL.

Intraoperative **normothermia** can have important benefits for the patient by decreasing morbidity, reducing intraoperative blood loss, cardiovascular events, and the incidence of perioperative infection. The use of warming blankets, hot air systems, and saline fluids can reduce postoperative pain, opioid consumption, and antiemetic therapy after laparoscopic surgery.

The fundamental pillars of intraoperative anesthesia are traditionally amnesia, analgesia, and muscle relaxation to achieve optimal conditions to facilitate the operation. The appearance of drugs with a better pharmacokinetic and pharmacodynamic profile has brought about a significant advance in early rehabilitation techniques.

In colorectal surgery, several anesthetic techniques have been proposed:

- General anesthesia ± local infiltration (at incision sites or continuous infusion of local anesthetic with a catheter over the surgical wound).
- General ± peripheral blocks
- Neuraxial techniques (epidural catheter or intradural techniques).

The choice will be conditioned by the surgical technique chosen, the most widespread being laparoscopic with the need for general anesthesia, with or without the use of neuraxial or peripheral blocks or surgical wound infiltration.

Drugs used to achieve rapid recovery

- Rapidly eliminated **halogenated gases** such as Sevoflurane (0.75%–1.5%) or Desflurane (3%–6%) may increase the incidence of postoperative nausea and vomiting. Propofol, used for both induction and maintenance, is a drug that is easy to handle and can be controlled by TCI pumps that allow more accurate dose adjustment. It has an antiemetic effect, although there is no evidence of a shorter awakening than with halogenated drugs.
- **Analgesics**: the appearance of opioid drugs with rapid effect and early elimination allow a more dynamic management adapted to the surgical moment. This reduces the incidence of the classic side effects of opioids such as PONV and paralytic ileus. The most commonly used is remifentanil. This ultra-short half-life necessitates multimodal analgesic rescue, use of adjuvant drugs (alpha-2 agonists) and less aggressive techniques such as infiltration of trocar entry ports, or continuous wound infusion catheters. The multimodal approach is performed with NSAIDs, COX2 inhibitors, alpha-2 agonists, glucocorticoids, ketamine, and local anesthetics and is effective in the control of acute perioperative pain. Beta-blockers (esmolol, labetalol, etc.) can be used to control the autonomic response during surgery.
- **Epidural techniques**: improve postoperative pain and comfort, decrease paralytic ileus, and increase blood flow by sympatholysis. Their position can vary from T6 to T10 depending on the area of resection. When thoracic epidurals are combined with multimodal analgesia techniques for colonic surgery, a clear reduction in paralytic ileus and hospital stay is observed,[20] although their use for minimally invasive surgery, such as laparoscopic colectomy, is highly questionable.
- **Muscle relaxants**: short half-life muscle relaxants of choice. Their indication in multimodal rehabilitation techniques is favored by the appearance of new reversers of neuromuscular relaxation (sugammadex), which avoid the

anticholinergic effects of the neostigmine-atropine combination, thus facilitating early extubation and reducing pulmonary complications related to the existence of residual muscle paralysis.

In conclusion, the combination of general anesthesia ± epidural catheter vs in-filtration of local anesthetic and multimodal analgesia, using ultra-short half-life anesthetics, to facilitate recovery and early extubation, will depend on the patient's characteristics and comorbidities, as well as on the surgical technique employed (laparoscopy vs. open surgery).

Infiltration of the trocar entry ports with local anesthetic or the use of wound catheters can facilitate early recovery in these patients, improving analgesia, reducing opioid consumption and the risk of chronic pain syndromes, and reducing the incidence of nausea and vomiting in the immediate postoperative period. It has the disadvantage of a higher incidence of postoperative wall bleeding episodes.

Opioid free analgesia

Opioid free analgesia (OFA) is a new approach to general anesthesia, which suppresses or minimizes as much as possible the use of opioids to avoid their adverse effects. The classical system provides hypnosis with amnesia and muscle relaxation and uses opioids to suppress sympathetic hyperactivity. OFA replaces these drugs with others with activity on the sympathetic system to maintain tissue perfusion while ensuring adequate hemodynamic stability. These drugs are not considered analgesics intrinsically, but they can cause analgesia by themselves or enhance the effects of opioid analgesics and therefore reduce their doses. The most commonly used are α-2 agonists (Clonidine and Dexmedetomidine), β-blockers (Esmolol), local anesthetics (Lidocaine), NMDA receptor blockers (MG sulfate and Ketamine), Gabapentinoids (Gabapentin and Pregabalin), NSAIDs, and Corticosteroids.

The available evidence supports the use of these drugs individually and a lower incidence of postoperative vomiting, ileus, delirium, and hyperalgesia has been observed but there are insufficient trials to determine the ideal combination for their single or multimodal administration and effective dosing to reduce morbidity and facilitate early discharge from hospital.

Its most accepted indications are obstructive sleep apnea, obesity, opioid addiction, and chronic pain syndromes. However, its use in anesthesia for colorectal surgery is fully justified because it suppresses or attenuates the adverse effects of opioids (nausea and vomiting, bowel obstruction, constipation, and tolerance) that negatively influence the goals sought in the optimization of colorectal surgery.

The advantage of such an approach is also important postoperatively to prevent immediate hyperalgesia secondary to massive opioid administration and to avoid the so-called opioid paradox: the more opioids are administered intraoperatively, the more opioids are needed postoperatively. Most notable effect with Remifentanil.[21]

Perioperative fluid therapy

The concept of perioperative fluid therapy is one of the variables that has undergone most changes in recent years and constitutes one of the basic pillars on which early or multimodal rehabilitation techniques are based. The effects of inadequate fluid management have been known for a long time, but it is only in recent years that the magnitude of these alterations has been quantified. The aim of perioperative fluid therapy is to restore and maintain the water-electrolyte balance altered in the perioperative period in order to achieve adequate perfusion and tissue oxygenation, replenishing or replacing losses and avoiding interstitial fluid overload due to hyperhydration This objective extends to the entire perioperative period, taking into account basal needs, preoperative fasting phase, intraoperative losses, and alterations secondary to surgical stress and postoperative intake.

Baseline fluid and electrolyte needs

There is a wide variation in salt and water intake under normal conditions. In the absence of associated pathology, the kidneys are able to maintain adequate homeostasis without allowing large water and electrolyte disturbances. Basal fluid requirements in healthy patients, excluding blood losses, are estimated at 40 mL/kg in young adults and 26–30 mL/kg for elderly patients due to their lower functional, organ, and cardiac reserve, which can be further aggravated by associated pathology. It should be taken into account that males have a higher fluid content (55%–65%) than females (45%–55%) due to the different proportion of body fat. A routine of 1.5 mL/kg/h is accepted for young adults, irrespective of gender, throughout the perioperative period, including also the hours of preoperative restraint.[22]

The basic daily electrolyte requirements in healthy adults are estimated to be 0.2–05 g Na^+, 1.8 g Cl^-, 2.1 g K^+, and 0.8–1.2 g Ca^{++}.

Changes in the surgical patient

All patients undergoing surgery will have their water and electrolyte status altered depending on the type and duration of the surgery, the anesthetic technique, the associated pathology, and the previous baseline status, which in the case of abdominal surgery acquires specific characteristics. Several factors influence perioperative fluid management.

- **Stress**. Surgical trauma produces alterations in the hemodynamic, metabolic, and immunological responses of patients in the perioperative period. There is an activation of the renin-angiotensin-aldosterone system which releases catecholamines and vasopressin with vasopressor and antidiuretic effects resulting in Na^+ and water retention and difficulty in their elimination by the kidney. The administration of saline solution in an attempt to promote diuresis can cause hyperchloremic acidosis with greater renal vasoconstriction and a decrease in glomerular filtration, further altering the excretion of water and sodium, causing an increase in interstitial edema and weight gain. This process is prolonged during the immediate postoperative period, generating an excess of free water in the organism that can give rise to the most frequent alteration in the perioperative period, dilutional hyponatremia.

  Dextrose 5% and saline are important sources of water and should be used with caution due to the risk of hyponatremia. Their use is restricted to situations of significant water depletion such as diabetes insipidus.

  At the systemic level, there is a prolonged increase in capillary permeability that favors the passage of albumin to the interstitial fluid by oncotic gradient, dragging intravascular fluid, and increasing interstitial edema. In colorectal surgery, this stress-induced systemic response may be enhanced locally by surgical manipulation, as disruption of endothelial integrity leads to increased fluid passage to the interstitial space, thus favoring increased edema formation and thus increased morbidity and mortality and length of hospital stay.[23] These alterations last approximately 24 h after surgery and are followed by a catabolic phase lasting up to 2 weeks and a final reparative phase characterized by anabolic metabolism.

- **Anesthetic techniques**. Depending on the technique used, intraoperative hemodynamic disturbances may occur. Sympathetic block during neuraxial anesthesia may cause hypotension as a consequence of increased venous capacitance and/or dilatation of resistance vessels. Inhalation anesthetics cause water retention due to decreased diuresis. Respiratory fluid loss associated with mechanical ventilation can be minimized by using humidified circuits and avoiding hyperventilation.

- **Glycocalyx**. The vascular endothelium consists of a fragile layer, the glycocalyx, which together with the endothelial cells, acts as a physiological barrier regulating the passage of ultrafiltration into the interstitial space. It is considered to be 1 μ thick and is composed of glycoproteins, proteoglycans, and other plasma proteins that generate oncotic pressure, which is the true limiting factor for transcapillary fluid loss, rather than the overall difference in hydrostatic and oncotic pressure between plasma and interstitium. Its existence explains why colloids remain longer in the intravascular space than crystalloids and provides the basis for the rationale for administering colloids for intravascular losses and crystalloids for extravascular losses. Alterations in the integrity of the glycocalyx and the composition of ions and products between this endothelial barrier and plasma will lead to pathological ultrafiltration into the interstitial space with a consequent increase in edema at this level.

  Fluid transport into the interstitium can take place in two ways:
  - Type I. Physiological displacement with an intact glycocalyx consisting of the passage of nonosmotic substances into the interstitium. The administration of large quantities of isotonic crystalloids causes rapid passage into the interstitial space, with the consequent formation of edema which is usually transient with good response to treatment.
  - Type II. Pathological displacement with altered glycocalyx. It is usually related to surgery and the administration of large amounts of preoperative fluids. During the perioperative period, some situations alter the integrity of the endothelium. Release of inflammatory mediators such as tumor necrosis factor α, low-density lipoproteins, and atrial natriuretic peptide (ANP) degrades the glycocalyx,[23] and situations such as hypervolemia, due to ANP release, ischemia/reperfusion, and hypoxia/reoxygenation states also lead to alteration of the glycocalyx.

  Prevention can be done in two ways: in the case of intact glycocalyx by using crystalloids in an amount similar to extracellular losses and isooncotic colloids to replace blood losses; in the case of an altered glycocalyx, try to reduce the intensity of surgical stress.

Calculation of perioperative fluid loss and replenishment

In addition to the basal and maintenance requirements described before, other losses in the surgical patient must also be calculated:

Expected surgical losses. These are calculated based on patient weight and surgical trauma[24]:
- Minor tissue trauma: 2 cc/kg/h.
- Moderate tissue trauma: 4–6 cc/kg/h.
- Major tissue trauma: 6–8 cc/kg/kg/h.

Unexpected losses. Basically blood. They are quantified by field aspirates and indirectly by the weight of the surgical drapes. Replacement aims to achieve an hourly diuresis of 0.5–1 cc/kg/h and the standard recommendation of administering 1 cc of crystalloid for each cc of blood lost is not currently indicated due to the volume overload it causes.

Concept of third space. Traditionally, the existence of a nonanatomical third space separated from the interstitial space and caused by surgical aggression was considered to exist, giving rise to fluid sequestration at this level. The different techniques used to quantify it have not demonstrated its existence, so it is not taken into account in the management of perioperative fluid therapy. The only perioperative fluid movement in the extravascular compartment is from the intravascular to the interstitial space, the magnitude of which is difficult to estimate and is directly related to a fluid supply.[25] Consequently, preoperative volume loading in normovolemic patients and routine replacement of high insensible and third space losses should be abolished in favor of demand-related fluid regimens.

Types of fluids

The controversy over the type of fluid to be administered has not yet been resolved and the literature has seen many papers recommending the use of colloids and/or crystalloids and even their use together.

Crystalloids. Crystalloids are freely distributed across the vascular barrier and only one-fifth of the infused amount remains in the intravascular space and the remaining 4/5 pass into the extracellular space. In the case of an intact vascular barrier, for each cc of colloid, 4 cc of crystalloid is necessary to achieve similar effects. If there is capillary leakage the ratio is balanced. They have no oncotic power and their distribution depends on the ionic concentration.[26]

Saline 0.9% is the simplest crystalloid and is often incorrectly referred to as physiological saline, because of its balance of the positive and negative charge. In plasma, because of its high chloride content, it causes increases in negative charge with the consequent formation of hydrogen ions, leading to hyperchloremic acidosis. Although the clinical consequences of this acidosis are unproven, perioperative replacement therapy with acetate/lactate Ringer's solution is associated with a lower incidence of metabolic disturbances, but no change in morbidity and renal function in healthy patients who do not receive large amounts of 0.9% saline,[27] with the exception of hypochloremia due to gastric drainage or vomiting which should be performed with saline. Low level of evidence, strong recommendation.

Saline depletion due to, for example, diuretic use, should be managed with Hartmann's type balanced electrolyte solutions. Ringer's solution is used with the addition of buffer in the form of lactate or acetate, of which the former is the most common. Lactate is metabolized in the liver and kidneys and may cause secondary metabolic alkalosis. Acetate is metabolized more rapidly and, in most tissues, consuming only half the amount of oxygen per mole of bicarbonate produced compared to lactate. Therefore lactate slightly increases oxygen consumption and may also raise plasma glucose.[28] Although the differences between lactate and acetate are generally negligible, acetate is considered to be a more balanced solution. It is the best buffer in the presence of compromised circulation and shock, and in general in all pathologies with increased lactic acidosis.

This type of fluid can be used to replace preoperative fluid losses due to diarrhea, and its administration must be controlled when used to reverse intraoperative hypotension. If the administration of a crystalloid bolus does not produce the desired effect, consideration should be given to varying the anesthetic technique or to instituting treatment with adrenergic drugs to avoid indiscriminate fluid administration.

Five percent glucose serum is free water without proteins and therefore has little oncotic power. Because of its ionic concentration, it is hypotonic concerning plasma. It is metabolized in the liver to water which is freely distributed to all compartments. Only 7%–10% of the administered dose remains in the intravascular space. Its perioperative use is only indicated in situations of significant water depletion such as diabetes insipidus.

Colloids. These are substances with macromolecules with oncotic and osmotic capacity that facilitate greater intravascular permanence. They have plasma-expanding properties depending on the volume and concentration of the solution administered and its infusion rate. They have a maximum dose of use since the administration of large volumes of colloids without sufficient free water (Dextrose 5%) can cause a hyperoncotic state.

- *Hydroxyethyl starch.* Plant origin: Available in different molecular weights and molar substitutions that determine its expansive effect. Duration of action of 4–6 h. In situations of hypovolemia, 60% remains in the intravascular space, while in normovolemia this percentage is close to 90%. Contraindicated in patients with oliguria not related to hypovolemia and in renal failure. They present a risk of anaphylaxis and dose-dependent coagulation alteration. They can be administered in saline or in a more balanced Rhine-ger solution.

A recent study[29] found that the use of synthetic colloids in critically ill patients increased the need for dialysis compared to isotonic saline. This increased renal toxicity and effects on hemostasis, coupled with their higher cost and lack of evidence demonstrating their clinical superiority, mean that their use is restricted in the critically ill patient.

- *Gelatins.* They have an expansion effect of 70%–80% with minimal effects on coagulation and renal function. Duration of action of 2–3 h. It is the colloid with the highest incidence of allergic reactions, which has limited its use.
- *Dextran.* Its use is limited to thromboembolic prophylaxis in vascular procedures due to its effects on coagulation. It has a volume effect of 100%–200%, depending on its molecular weight and concentration. It is rarely used because of its anaphylactic, renal, and hematological complications.
- *Albumin.* Albumin is the only natural colloid. Its use is highly controversial and it does not currently appear to have a role in emergency volume replacement. In addition to being more expensive, there is no evidence that albumin reduces mortality compared to cheaper options in patients with hypovolaemia.[30]

The use of "balanced solutions" is now advocated, meaning those formulated to have a neutral pH and concentrations of ions and electrolytes similar to those of human plasma. The choice of fluid type for elective surgery is based on the specific clinical situation and fluids are considered as drugs with their indications, contraindications, and side effects. It is generally accepted that crystalloids are indicated for replacement of the extracellular compartment (fasting, diuresis, and evaporative losses from the airway, skin, and surgical wound) and colloids for replacement of the intravascular compartment (bleeding and displacement losses).

In a normohydrated patient, fluid therapy is not indicated to treat intraoperative hypotension caused by general anesthetics nor to prevent hypotension secondary to neuraxial anesthesia due to the volume overload involved. In both cases, a vasopressor should be used.[31]

Fluid management schemes

Restrictive vs liberal therapy

In general terms, candidate patients for abdominal surgery were considered similar to dehydrated and hypovolemic patients and treated under a liberal approach to fluid administration. Consequently, patients undergoing major intra-abdominal procedures received up to 10–15 mL/(kg/h) of fluid. This resulted in a significant increase in body weight that was directly related to increased perioperative mortality.[23]

With the aim of reducing postoperative complications due to overhydration, a shift in the trend occurred and studies based on a more restrictive approach to fluid therapy began to appear. They presented data stating that this approach significantly reduced postoperative morbidity and mortality, with no acute deterioration of renal function despite decreased fluid intake and decreased urine output.[32] This restrictive regimen is aimed at maintaining unchanged body weight and advocates a fluid therapy of less than 4 mL/kg/h. The restrictive designation has been misleading and controversial, and the current trend is to refer to this regimen as zero-balance fluid therapy.

This different view is becoming a major topic of debate in surgical patients, and more particularly in patients undergoing colorectal surgery, because of conflicting results. In a recent literature review,[33] nine randomized clinical trials comparing the results of both regimens were studied. In five studies the restrictive regimen improved outcomes in colorectal surgery, two studies showed better results for the liberal regimen, and the remaining two showed no difference.[34] developed a mathematical model to quantify interstitial edema by bioelectrical impedance and observed that abdominal surgery lasting less than 3 h did not show large changes in interstitial space. On the other hand, in surgeries of longer duration, interstitial alterations with edema formation and its consequences for the patient appeared. Based on these data, and analyzing the comparative studies between the two methods of fluid therapy, a more liberal fluid therapy was accepted in minor and short-term procedures, and in more aggressive and longer-lasting surgeries the results would be favored by a restrictive therapy.

The debate moved to the strict application of this thesis and the development of guidelines and protocols. There is no consistent definition of liberal or restrictive, and it is known that a fixed fluid regimen does not work equally well for all patients. These concepts show the difficulty of developing evidence-based guidelines according to the earlier mentioned criteria and the conclusions drawn from the studies performed suggest nonspecifically that administration of balanced solutions slightly exceeding zero balance does not harm the patient. The RELIEF trial[35] on 3000 patients undergoing major abdominal surgery compared a restrictive regimen attempting to achieve zero total balance with a liberal regimen administering two times the amount of fluid as the previous regimen and the investigators found that a restrictive regimen was not associated with a higher survival rate and was associated with a higher rate of impaired renal function.

This controversy led to the emergence of goal-guided fluid therapy (GFD), which uses individualized regimens focused on achieving normovolemia to achieve hemodynamic parameters that maintain the most favorable ratio of O_2 intake to O_2 consumption.

The strategy of this technique, by supplementing routine monitoring, is based on the assumption that normovolemia in a supine patient is considered to be the preload necessary to achieve a maximal stroke volume or cardiac output. When small fluid loads fail to increase these parameters, normovolemia is considered to have been reached and this point is taken as a reference point for therapy. It has the advantage that it predicts the response to a fluid and, ideally, avoids unnecessary bolus administration. The most commonly used monitoring for its development is Transeophageal Doppler Echo, which has been associated with fewer postoperative complications and fewer readmissions to intensive care wards, as well as a faster return of bowel function in patients undergoing abdominal surgery and therefore a shorter hospital stay.[36]

The technique allows for continuous monitoring of cardiac output, preload, after load, contractility, and Systemic Vascular Resistance (SVR), by measuring blood flow velocity and the diameter of the descending thoracic aorta. This provides us with a "beat-by-beat" view of the patient's hemodynamic status allowing early detection of changes and appropriate treatment. We use an esophageal probe to obtain a continuous signal by emitting and capturing ultrasound, which is sent to a monitor in the form of a digitized image of the blood velocity profile over time. The aortic area is calculated by nomogram (weight, age, and height) without specific measurement.

There are multiple algorithms that can be applied to this technique, but one of the most widely used[37] consists of administering a bolus of 3–4 mL/kg after induction and observing the change in systolic volume on the Doppler monitor. If it increases by more than 10%, we consider the patient to be fluid responsive and continue to administer boluses every 10 min until the increase in Cardiac Output (CO) is less than 10%. At this point, we set the reference of normovolemia and aim for a mean arterial pressure (MAP) greater than 70 mmHg and a cardiac index greater than 2.5 L/min/m^2, using, if necessary, vasopressors and/or positive inotropes to achieve the target. At this point, we follow a maintenance therapy according to the type of surgery and regimen chosen, checking the monitor every 15 min and administering boluses according to the patient's fluid response as described before.

Although the correlation of this method with the standard thermodilution method is over 80%, this technique has some limitations. Turbulent aortic flow (severe valvulopathies, aortic root disturbances, balloon counterpulsation, etc.) and poor positioning can lead to erroneous readings. Likewise, anesthetic drugs and regional techniques may alter its use due to their vasodilator effect. This technique has been basically reduced to patients on mechanical ventilation who do not need frequent repositioning (such as left/ right tilt, Trendelenburg/reverse Trendelenburg, etc.). Esophageal Doppler is contraindicated in situations that restrict free access to the patient's head, esophageal pathologies, thoracic aortic aneurysms, surgeries in which laser is being used and in patients with carcinoma of the pharynx, larynx, or esophagus.

Another method of predicting response to fluid therapy has been described as a passive raising of the legs for at least 1 min from a horizontal position. A gravitational displacement of blood from the lower extremities into the intrathoracic cavity mimics a fluid load.[38] It is a reversible method by returning the legs to the horizontal position. If a greater than 10% increase in CO is observed when the maneuver is performed, it allows us to predict a greater than 15% increase in CO when an equivalent volume load is administered. Although not evaluated during the perioperative period, it has the potential to minimize the risk of fluid overload. It should be noted that intra-abdominal pressure greater than 16 mmHg impairs venous return and reduces the ability of this technique to detect fluid responsiveness.

Goal-guided fluid therapy clearly improves cardiac indices and tissue perfusion, but it remains unclear whether the clinical benefits are due to the overall improvement in perioperative care or to the technique itself. Its advantage over restrictive therapy is that it only replaces fluids lost during surgery, but it has the disadvantage of requiring much more invasive monitoring. There is a lack of conclusive data to say whether it is superior to other regimens and no major changes have been observed for the zero-balance fluid therapy strategy in patients undergoing elective abdominal surgery.[39] DGO has evolved from a standard intraoperative fluid therapy strategy in elective abdominal surgery to a strong recommendation only in high-risk patients undergoing surgery with large intravascular fluid losses.

Hemodynamic monitoring

Determining the intravascular volume of the patient undergoing surgery remains a challenge. This is due to the underlying condition that indicated the surgery and the changing intraoperative physiology (drugs, stress, and surgical losses), which means that alterations are often not immediately reflected by the usual screening measures (heart rate, noninvasive blood pressure, peripheral oxygen saturation, and diuresis). This makes it difficult to make an early and appropriate therapeutic decision.

It is recommended that volemia be monitored using measurements based on response to fluid therapy. When this is not possible, other data should be used, such as pulse, peripheral perfusion and capillary refill, blood pressure, acid-base and lactate balance, and intrathoracic pressures.

Diuresis of less than 0.5 cc/kg/h is commonly used as an indicator of hypovolemia, but it is not a good predictor of acute kidney injury,[40] because situations such as surgical aggression and some anesthetic drugs can reduce diuresis, even in situations of euvolemia and adequate tissue perfusion. Thus an extra volume supply would aggravate a preexisting hemodynamic disturbance. Creatinine is a late marker of renal dysfunction.

Diagnosis of hemodynamic disturbances is difficult when compensatory mechanisms (peripheral and splenic blood sequestration) increase morbidity. Tachycardia and hypotension may be absent during hypovolemic shock until volume loss reaches 20% or more.

If we consider that blood pressure is the product of flow times resistance, and that flow, in turn, is conditioned by preload and cardiac contractility, only one value is known from the equation. Moreover, we do not know which factor is responsible for changes in blood pressure. Therefore we cannot make an early and appropriate therapeutic decision (volume, vasopressors, both, etc.).

Measurements of global oxygenation, mixed venous saturation (SvO_2), and central venous saturation ($ScvO_2$) do not always reflect actual changes, because O_2 consumption varies during the preoperative period.[41]

Monitoring should start with an understanding of the patient's condition, medical history, physical and laboratory examination, and type of surgery that will form the basis of goal-directed therapy, in order to decide whether to add more advanced methods to the basic safety measures[42] and the usual monitoring described before. Invasive BP will be considered in patients with significant losses, and also facilitates frequent sampling. The use of more advanced monitoring will be based on patient risk and surgical aggressiveness.

Advanced monitoring

Static variables

- *Filling pressures*

    Central venous pressure (CVP) and pulmonary artery wedge pressure (PAOP) are considered guides to right and left ventricular preload, respectively, because of the theoretical relationship between ventricular diastolic pressure and ventricular diastolic volume. They are considered the gold standard and constitute the standard against which other techniques are compared. However, they are not good predictors of fluid loading behavior,[43] because the response to preload may be absent in heart failure and there is not such an exact correlation between filling pressure and diastolic volume. Also, filling pressure is highly dependent on lung compliance. Mechanical ventilation causes an increase in cardiac filling pressures which interferes with the interpretation of its results.

    Its routine use is currently interpreted within the clinical context of the patient.
- *Volume variables*

    Left end-diastolic volume obtained by echocardiography. A highly observer-dependent method that provides a quick overview of left ventricular function and dimensions. Theoretically, left end-diastolic volume would be a good variable to measure preload, but a poor predictor of fluid responsiveness.[44] It has the disadvantage of cost, low availability, and high staff training. Does not allow continuous monitoring.

    Global end-diastolic volume (GEDV). It can be obtained by a series of measurements that share in common the calculation of cardiac output from the analysis of the area under the curve of the arterial pulse wave.

Dynamic fluid response variables

These reflect changes in left ventricular volume induced by mechanical ventilation. The most commonly used are pulse pressure variation (PPV) and stroke volume variation (SVV). Monitors are used that measure PWV and SVV by analyzing the area under the pressure curve. The differences found during a mechanical breath at maximum and minimum values are shown in the form of waves on the monitor indicating the response to fluids.

Monitoring of these dynamic parameters based on functional hemodynamics is currently considered to be the best predictor of fluid responsiveness in mechanically ventilated patients without cardiac arrhythmias.[45]

The PICCO system is an invasive monitor that, by analyzing the arterial pulse curve and arterial thermodilution, allows us to continuously assess cardiac output, global end-diastolic volume GEDV (sum of the volumes of the four cardiac chambers), and extravascular lung water. It reflects preload and response to fluid administration and is not limited by the patient's spontaneous respiratory movements. The PICCO system measures volumes, unlike the Swan-Ganz system which measures pressures and gives equally concordant data and is less invasive. It requires permanent calibration. It may give altered values in case of arrhythmias, left-right shunt, and severe valvular insufficiency.

Other systems such as LiDCO and FLO TRAC VIGILEO are also used, but their use is more limited to intensive care units.

Other nonhemodynamic monitoring measures are anesthetic depth monitoring devices (BIS, Entropy), which allow dose adjustments in order to avoid intraoperative awakening, as well as overdosing or slow awakening. Monitoring of neuromuscular blockade (neurostimulator) and infusion pumps (TIVA) to maximally adjust the dosage of anesthetics according to the requirements at any given moment, thereby facilitating early postanesthetic recovery.

Emergency surgery

There are few published data for managing fluid therapy in emergency situations. Some studies of subgroups of patients admitted to hospital in critical condition seem to provide data indicating a decrease in mortality when goal-directed therapy, such as in elective surgery, is initiated early in the process. Until conclusive data are available in these situations, it is recommended to follow the guidelines outlined in the "Surviving Sepsis" campaign.

References

1. Wilmore DW, Kehlet H. Management of patients in fast track surgery. *BMJ* 2001;**322**(7284):473–6. https://doi.org/10.1136/bmj.322.7284.473.
2. Teeuwen PHE, Bleichrodt RP, Strik C, et al. Enhanced recovery after surgery (ERAS) versus conventional postoperative care in colorectal surgery. *J Gastrointest Surg* 2010;**14**(1):88–95. https://doi.org/10.1007/s11605-009-1037-x.
3. Monagle J, Waxman B, Abourizk S, Sparrow M, Shearer B. Preadmission processes may improve length of stay for colorectal surgery. *ANZ J Surg* 2003;**73**(4):210–2. https://doi.org/10.1046/j.1445-1433.2002.02583.x.
4. Markus PM, Martell J, Leister I, Horstmann O, Brinker J, Becker H. Predicting postoperative morbidity by clinical assessment. *Br J Surg* 2005;**92**(1):101–6. https://doi.org/10.1002/bjs.4608.
5. Zelić M, Stimac D, Mendrila D, et al. Influence of preoperative oral feeding on stress response after resection for colon cancer. *Hepatogastroenterology* 2012;**59**(117):1385–9. https://doi.org/10.5754/hge10556.
6. Henriksen MG, Hessov I, Dela F, Vind Hansen H, Haraldsted V, Rodt SÅ. Effects of preoperative oral carbohydrates and peptides on postoperative endocrine response, mobilization, nutrition and muscle function in abdominal surgery. *Acta Anaesthesiol Scand* 2003;**47**(2):191–9. https://doi.org/10.1034/j.1399-6576.2003.00047.x.
7. Väyrynen JP, Tuomisto A, Väyrynen SA, et al. Preoperative anemia in colorectal cancer: relationships with tumor characteristics, systemic inflammation, and survival. *Sci Rep* 2018;**8**(1):1126. https://doi.org/10.1038/s41598-018-19572-y.
8. Amato A, Pescatori M. Transfusiones de sangre perioperatorias para la recidiva del cáncer colorrectal (Revisión Cochrane traducida). In: *La biblioteca cochrane plus*; 2008. p. 2.
9. Wille-Jørgensen P, Guenaga KF, Matos D, Castro AA. Pre-operative mechanical bowel cleansing or not? An updated meta-analysis. *Colorectal Dis* 2005;**7**(4):304–10. https://onlinelibrary.wiley.com/doi/abs/10.1111/j.1463-1318.2005.00804.x.
10. Waitzberg DL, Saito H, Plank LD, et al. Postsurgical infections are reduced with specialized nutrition support. *World J Surg* 2006;**30**(8):1592–604. https://doi.org/10.1007/s00268-005-0657-x.
11. Moya P, Miranda E, Soriano-Irigaray L, et al. Perioperative immunonutrition in normo-nourished patients undergoing laparoscopic colorectal resection. *Surg Endosc* 2016;**30**(11):4946–53. https://doi.org/10.1007/s00464-016-4836-7.
12. Chen M, Song X, Chen L-Z, Lin Z-D, Zhang X-L. Comparing mechanical bowel preparation with both oral and systemic antibiotics versus mechanical bowel preparation and systemic antibiotics alone for the prevention of surgical site infection after elective colorectal surgery: a meta-analysis of randomized controlled clinical trials. *Dis Colon Rectum* 2016;**59**(1):70–8. https://doi.org/10.1097/DCR.0000000000000524.
13. Zhang D, Wang X-C, Yang Z-X, Gan J-X, Pan J-B, Yin L-N. RETRACTED: preoperative chlorhexidine versus povidone-iodine antisepsis for preventing surgical site infection: a meta-analysis and trial sequential analysis of randomized controlled trials. *Int J Surg* 2017;**44**:176–84. https://doi.org/10.1016/j.ijsu.2017.06.001.
14. Wong J, Lam DP, Abrishami A, Chan MTV, Chung F. Short-term preoperative smoking cessation and postoperative complications: a systematic review and meta-analysis. *Can J Anaesth* 2012;**59**(3):268–79. https://doi.org/10.1007/s12630-011-9652-x.
15. Shabanzadeh DM, Sørensen LT. Alcohol consumption increases post-operative infection but not mortality: a systematic review and meta-analysis. *Surg Infect (Larchmt)* 2015;**16**(6):657–68. https://doi.org/10.1089/sur.2015.009.
16. Cahill RA, Redmond HP. Cytokine orchestration in post-operative peritoneal adhesion formation. *World J Gastroenterol* 2008;**14**(31):4861–6. https://doi.org/10.3748/wjg.14.4861.

17. Kehlet H, Dahl JB. Anaesthesia, surgery, and challenges in postoperative recovery. *Lancet* 2003;**362**(9399):1921–8. https://doi.org/10.1016/S0140-6736(03)14966-5.
18. Apfel CC, Turan A, Souza K, Pergolizzi J, Hornuss C. Intravenous acetaminophen reduces postoperative nausea and vomiting: a systematic review and meta-analysis. *Pain* 2013;**154**(5):677–89. https://doi.org/10.1016/j.pain.2012.12.025.
19. Greif R, Laciny S, Rapf B, Hickle RS, Sessler DI. Supplemental oxygen reduces the incidence of postoperative nausea and vomiting. *Anesthesiology* 1999;**91**(5):1246. https://doi.org/10.1097/00000542-199911000-00014.
20. Liu SS, Carpenter RL, Mackey DC, et al. Effects of perioperative analgesic technique on rate of recovery after colon surgery. *Anesthesiology* 1995;**83**(4):757–65. https://doi.org/10.1097/00000542-199510000-00015.
21. Fletcher D, Martinez V. Opioid-induced hyperalgesia in patients after surgery: a systematic review and a meta-analysis. *Br J Anaesth* 2014;**112**(6):991–1004. https://doi.org/10.1093/bja/aeu137.
22. Hahn RG. *Clinical fluid therapy in the perioperative setting*. Cambridge University Press; 2016. https://play.google.com/store/books/details?id=nfa7DAAAQBAJ.
23. Lowell JA, Schifferdecker C, Driscoll DF, Benotti PN, Bistrian BR. Postoperative fluid overload: not a benign problem. *Crit Care Med* 1990;**18**(7):728–33. https://doi.org/10.1097/00003246-199007000-00010.
24. Stoelting RK, Miller RD, Garfield JM. Basics of anesthesia. *Anesthesiology* 1990;**73**(3):590. https://doi.org/10.1097/00000542-199009000-00063.
25. Chappell D, Jacob M, Hofmann-Kiefer K, Conzen P, Rehm M. A rational approach to perioperative fluid management. *Anesthesiology* 2008;**109**(4):723–40. https://doi.org/10.1097/ALN.0b013e3181863117.
26. Robarts WM. Nature of the disturbance in the body fluid compartments during and after surgical operations. *Br J Surg* 1979;**66**(10):691–5. https://doi.org/10.1002/bjs.1800661006.
27. Bampoe S, Odor PM, Dushianthan A, et al. Perioperative administration of buffered versus non-buffered crystalloid intravenous fluid to improve outcomes following adult surgical procedures. *Cochrane Database Syst Rev* 2017. https://doi.org/10.1002/14651858.cd004089.pub3.
28. Ahlborg G, Hagenfeldt L, Wahren J. Influence of lactate infusion on glucose and FFA metabolism in man. *Scand J Clin Lab Invest* 1976;**36**(2):193–201. https://doi.org/10.1080/00365517609055248.
29. Myburgh JA, Finfer S, Bellomo R, et al. Hydroxyethyl starch or saline for fluid resuscitation in intensive care. *N Engl J Med* 2012;**367**(20):1901–11. https://doi.org/10.1056/NEJMoa1209759.
30. The Albumin Reviewers, Alderson P, Bunn F, Li Wan Po A, Li L, Roberts I, et al. Human albumin solution for resuscitation and volume expansion in critically ill patients. *Cochrane Database Syst Rev* 2004. https://doi.org/10.1002/14651858.cd001208.pub2.
31. Brandstrup B. Fluid therapy for the surgical patient. *Best Pract Res Clin Anaesthesiol* 2006;**20**(2):265–83. https://doi.org/10.1016/j.bpa.2005.10.007.
32. Brandstrup B. Effects of intravenous fluid restriction on postoperative complications: comparison of two perioperative fluid regimens: a randomized assessor-blinded multicenter trial. *Ann Surg* 2004;**240**(2):386. https://www.ncbi.nlm.nih.gov/pmc/articles/PMC1356424/. [Accessed 16 February 2021].
33. Klein Y, Matot I. Intra-abdominal surgery. In: *Clinical fluid therapy in the perioperative*; 2011. https://books.google.com/books?hl=en&lr=&id=vNRb1sLM5DMC&oi=fnd&pg=PA29&dq=Yifat+K+Idit+M+Clinical+Fluid+Therapy+in+the+Perioperative+Setting&ots=CTxTV4COmy&sig=jIX-etE4K5XKrFgvFqEec4-NXyY.
34. Tatara T, Nagao Y, Tashiro C. The effect of duration of surgery on fluid balance during abdominal surgery: a mathematical model. *Anesth Analg* 2009;**109**(1):211–6. https://doi.org/10.1213/ane.0b013e3181a3d3dc.
35. Myles PS, Bellomo R, Corcoran T, et al. Restrictive versus liberal fluid therapy for major abdominal surgery. *N Engl J Med* 2018;**378**(24):2263–74. https://doi.org/10.1056/NEJMoa1801601.
36. Walsh SR, Tang T, Bass S, Gaunt ME. Doppler-guided intra-operative fluid management during major abdominal surgery: systematic review and meta-analysis: doppler-guided intra-operative fluid management. *Int J Clin Pract* 2008;**62**(3):466–70. https://doi.org/10.1111/j.1742-1241.2007.01516.x.
37. Feldheiser A, Conroy P, Bonomo T, Cox B, Garces TR, Spies C. Development and feasibility study of an algorithm for intraoperative goal-directed haemodynamic management in noncardiac surgery. *J Int Med Res* 2012;**40**(4):1227–41. https://doi.org/10.1177/147323001204000402.
38. Lafanechère A, Pène F, Goulenok C, et al. Changes in aortic blood flow induced by passive leg raising predict fluid responsiveness in critically ill patients. *Crit Care* 2006;**10**(5):R132. https://doi.org/10.1186/cc5044.
39. Brandstrup B, Svendsen PE, Rasmussen M, et al. Which goal for fluid therapy during colorectal surgery is followed by the best outcome: near-maximal stroke volume or zero fluid balance? *Br J Anaesth* 2012;**109**(2):191–9. https://doi.org/10.1093/bja/aes163.
40. Alpert RA, Roizen MF, Hamilton WK, et al. Intraoperative urinary output does not predict postoperative renal function in patients undergoing abdominal aortic revascularization. *Surv Anesthesiol* 1985;**29**(2):130–1. https://doi.org/10.1097/00132586-198504000-00041.
41. Renner J, Scholz J, Bein B. Monitoring fluid therapy. *Best Pract Res Clin Anaesthesiol* 2009;**23**(2):159–71. https://doi.org/10.1016/j.bpa.2008.12.001.
42. Merry AF, Cooper JB, Soyannwo O, Wilson IH, Eichhorn JH. An iterative process of global quality improvement: the International Standards for a Safe Practice of Anesthesia 2010. *Can J Anesth* 2010;**57**(11):1021–6. https://doi.org/10.1007/s12630-010-9380-7.
43. Osman D, Ridel C, Ray P, et al. Cardiac filling pressures are not appropriate to predict hemodynamic response to volume challenge. *Crit Care Med* 2007;**35**(1):64–8. https://doi.org/10.1097/01.ccm.0000249851.94101.4f.
44. Renner J, Gruenewald M, Brand P, et al. Global end-diastolic volume as a variable of fluid responsiveness during acute changing loading conditions. *J Cardiothorac Vasc Anesth* 2007;**21**(5):650–4. https://doi.org/10.1053/j.jvca.2007.05.006.
45. Yang X, Du B. Does pulse pressure variation predict fluid responsiveness in critically ill patients? A systematic review and meta-analysis. *Crit Care* 2014;**18**(6). https://doi.org/10.1186/s13054-014-0650-6.

Chapter 42

Postoperative control: Complications and management in critical care units

Susana López Piñeiro
Anaesthesia and Resuscitation Department, Hospital Complex of Pontevedra, Pontevedra, Spain

Introduction

Perioperative complications arising from general and digestive surgery are one of the main reasons for admission to postsurgical critical care units, with those related to colorectal cancer reaching 25.6%.[1] In addition, diagnostic and endoscopic therapeutic techniques, applied to colorectal cancer, can cause bleeding, or hollow viscera perforations, which when caused by a major hemorrhage or peritonitis lead to urgent admission to these intensive care units.

When facing the emergency admissions described in the previous paragraph, three-quarters of admissions to critical care units, related to colorectal cancer, are planned in advance. Patients with known comorbidities that, by themselves, justify a high likelihood of postoperative decompensation, and those subjected to highly complex surgical techniques that require close postoperative monitoring, as is the case with lung metastases surgery and liver surgery, are included here. Table 42.1 lists the reasons for admission in Critical Care Units.

Suture dehiscence with anastomotic leak

Surgical resections performed to treat colon and rectum cancer range from a total proctocolectomy in familial adenomatous polyposis to partial resections such as a left or right hemicolectomy, sigmoidectomy, and abdominoperineal amputation. In these partial resections, when primary anastomosis is performed without a definitive colostomy or temporary protection stoma, poor clinical outcomes with sepsis data should make the clinician strongly suspect suture line dehiscence. In the metaanalysis conducted by Snijders et al.[2] the average incidence of an anastomotic leak after a low anterior resection was 9%. Low levels of albumin in blood, use of steroids, and long operation times have been described as risk factors for men,[3] in addition to the use of diclofenac, performing rectal anastomosis versus colonic and blood derivative transfusions.[4]

Anastomotic suture failure in colorectal cancer is associated with significant mortality that ranges from 12% to 25%.[5] The clinical picture usually appears between 5 and 7 days although it may be sooner and more striking in cases of ischemia due to tension of the proximal end, poor perfusion of the distal end, or the inadvertent failure of the stapler. An anastomosis leak can present itself from low to high severity such as an intraabdominal localized abscess, an enterocutaneous fistula, or generalized peritonitis.

Intraabdominal abscess

An intraabdominal abscess is usually located in subphrenic, paracolic, or pelvic spaces. It should be suspected in patients with persistent fever, prolonged ileus, and a slower recovery than expected after surgery. The increased leukocyte count, plus data provided from the ultrasound and, preferably the CT scan, will lead us to the diagnosis. The treatment consists of controlling the focus by means of drainage, controlling the cause, and systemic antibiotic therapy. Percutaneous drainage or using endoscopic ultrasonography (EUS)[6] inside of the intestinal lumen can be chosen in single and well-located abscesses. Those located in the pelvis, apparent in a rectal or vaginal exam, can be drained by directly targeting that area.[7] Deep-rooted abscesses, associated with suture dehiscence, will be dealt with by means of a laparotomy. In this new surgery, the deposit will be drained and the continuity solution will be repaired by means of an ileostomy or colostomy, with or without a mucous fistula.

TABLE 42.1 Reasons for admission in critical care units.

Scheduled	Concomitant diseases	Advanced heart disease Advanced kidney failure Severe respiratory disease Diabetes
	Technical complexity	Lung metastases and liver surgery Peritonectomies
Emergency	Intraoperative complications	Anesthetic problems Unforeseen technical difficulties Bleeding
	Postoperative complications	Respiratory failure Perioperative myocardial infarction Anastomotic suture dehiscence Peritonitis Bleeding Intestinal obstruction Surgical wound infection Surgical wound dehiscence Abdominal compartment syndrome Injury other abdominal organs (ureter, spleen)
	After endoscopic techniques	Bleeding Perforated hollow viscera

Enterocutaneous fistula

We can see intestinal fluid flow outward through the surgical wound, through enterocutaneous fistulas. They generate, in addition to surgical wound infection, a systemic inflammatory response that can reach the same grades as sepsis—septic shock as demonstrated in suture dehiscence with peritonitis.

Generalized peritonitis

The emergence of florid sepsis is usually related to generalized peritonitis and requires an urgent laparotomy to control the focus of infection and administration, ideally in the first hour, of empirical antibiotic therapy. If the dehiscence exceeds 50% or there is necrosis of the edges, the anastomosis will be disassembled, and there will be a Hartmann intervention. In addition, when the inflammatory reaction is very significant and the closure of the abdominal wall would generate a lot of tension, it may be necessary to leave the abdomen open using a temporary laparostomy with a mesh, Bogota bag, or VAC system in order to avoid Abdominal Compartment Syndrome. In cases with a small dehiscence and with just a slight systemic and peritoneal affect, more conservative techniques can be used by performing discharge ostomies on the ileum or transverse colon and drainage of deposits.

Secondary and tertiary peritonitis are the forms of peritonitis that we diagnose in patients with colorectal cancer.

Secondary peritonitis is the most common form and includes the postoperative nosocomial kind due to suture dehiscence, postendoscopic due to perforation, and also community-based cases which produce a spontaneous perforation of the tumor mass in the peritoneal cavity.

Tertiary peritonitis is defined as a postinfection intraabdominal infection (IAI). It usually affects patients who have had repeated surgical procedures and have been admitted to intensive care units. Although the treatment will be similar, it should not be confused with persistent and/or recurrent peritonitis due to poor control of the focus or inadequacy of the initial empirical antibiotic treatment. The pathogens responsible are usually nosocomial. In patients with postoperative peritonitis who have had reoperations, and who have received prior antibiotic coverage, microorganisms with a very high resistance pattern of the type, *Escherichia coli* and *Klebsiella* spp. resistant to β-lactams, methicillin-resistant *Staphylococcus aureus* (MRSA), *Acinetobacter* spp., *Enterococcus faecium*, and *Candida* spp. have been identified in cultures.[8]

Sepsis and septic shock

The Third Consensus Definition for Sepsis and Septic Shock (Sepsis-3)[9] was published recently. Sepsis is now defined as a potentially fatal organ dysfunction caused by a deregulated response by the host to infection. The presence of sepsis is suspected when the SOFA (Sequential Organ Failure Assessment) score increases by 2 points or more. At the foot of the bed, and without delay as laboratory tests are not required, the qSOFA determination allows clinicians to select patients who may need admission to intensive care and to initiate care and treatment earlier. The criteria for a positive qSOFA are respiratory rate higher than 22 breaths per minute, along with altered mental state and systolic blood pressure below 100 mmHg.

With the intention of reducing the mortality associated with sepsis, the Society of Critical Care Medicine and the European Society of Intensive Care Medicine has been working in collaboration with the Surviving Sepsis Campaign since 2003 to educate the medical community on the right treatment of these cases. To do this, a review of previous recommendations from 2012 was recently published.[10]

- Initial Resuscitation
    (1) Start treatment and resuscitation immediately.
    (2) In the presence of hypoperfusion, administer at least 30 mL/kg of intravenous crystalloids within the first 3 h. More liquids will be administered according to the frequent reassessment of the hemodynamic status. Assess heart function if the clinical examination does not lead to a clear diagnosis.
    (3) Use dynamic variables instead of static ones to predict the response to the administration of liquids.
    (4) Search for a MAP target of 65 mmHg in those requiring vasopressors.
- Systemic Detection of Sepsis and Improved Performance
    (1) Recommend that hospitals have a performance improvement program for sepsis.
- Diagnosis
    (1) Obtaining adequate microbiological cultures before starting antibiotic treatment if this does not substantially delay the initiation of antibiotics.
- Antibiotic Treatment
    (1) Start intravenous antibiotic treatment as soon as possible and within a maximum period of 1 h.
    (2) Use wide-spectrum empirical treatment, with one or more antibiotics in order to cover all likely pathogens.
    (3) Reduce the antibiotic treatment, once the pathogen or sensitivities have been identified or clinical improvement is seen.
- Source Control
    (1) Identification and anatomical diagnosis of the source as soon as possible, and control of the source of infection as soon as it is medically and logistically possible.
    (2) Removal of vascular devices that may be a potential source of sepsis after another vascular access point has been established.
- Treatment With Liquids
    (1) Continue to administer intravenous fluid therapy in those cases in which its administration improves the hemodynamic parameters.
    (2) The recommended fluid is balanced crystalloids and albumin can be used for the initial resuscitation, and in those who require large volumes. Do not use hydroxyethyl starches.
- Vasoactive Drugs
    (1) The vasopressor chosen will be noradrenaline, as vasopressin or adrenaline can be added, with the intention of raising mean arterial blood pressure, and reducing the noradrenaline doses.
    (2) In selected patients with bradycardia and a low risk of tachycardias, dopamine may be an alternative vasopressor agent. Its use in low doses for renal protection is not recommended.
- Mechanical Ventilation
    (1) Low flow volumes will be used (6 mL/kg of body weight) and plateau pressures limited to 30 cm of H_2O.
    (2) Use higher PEEP and recruitment maneuvers in adult patients with Acute Respiratory Distress Syndrome (ARDS).
    (3) Lying face down is recommended if PaO_2/FiO_2 is less than 150.
    (4) Limit the use of neuromuscular blockers to less than 48 h.
    (5) Elevate the head of the bed between 30 and 45 degrees to prevent the development of ventilator-associated pneumonia.
    (6) The use of gradual disconnection from the ventilator protocols is recommended.

- Other recommendations from the Surviving Sepsis Campaign 2016 include:
  (1) Do not use hydrocortisone for septic shock if hemodynamic stability is restored with rehydration and vasopressors. If it is not achieved, a dose of 200 mg/day is suggested.
  (2) Administer a red blood cell transfusion, when hemoglobin drops below 7 g/dL in the absence of circumstances such as myocardial ischemia or severe hypoxemia.
  (3) Minimize the use of sedation.
  (4) Treatment of hyperglycemia with insulin when 180 mg/dL is exceeded in two measurements for a blood glucose target of $\leq$180 mg/dL.
  (5) Replacement of renal function if acute kidney failure develops with continuous or intermittent hemodialysis techniques.
  (6) The use of sodium bicarbonate in lactic acidosis by hypoperfusion if pH is below 7.15.
  (7) Venous thromboembolism prophylaxis with Low Molecular Weight Heparin (LMWH) in combination with mechanical prophylaxis.
  (8) Prophylaxis of acute gastroduodenal ulcers due to stress in patients with risk factors for gastrointestinal bleeding.
  (9) Administering parenteral nutrition in the first 7 days to patients with sepsis or septic shock, in whom enteral feeding is not possible, is not recommended. Administration of oral or enteral nutrition as soon as it is tolerated.

Following the recommendations of the Surviving Sepsis Campaign 2016 (SSC),[11] a package of measures to be performed at "time-1" was presented. This emphasizes the need to start resuscitation and management immediately. Perform the following 5 steps in this first hour:

- Determine lactate levels and continue with the determinations if the initial lactate is >2 mmol/L.
- Obtain blood cultures after administering antibiotics.
- Administer broad-spectrum antibiotic therapy.
- Quickly administer 30 mL/kg of crystalloid if there is hypotension or lactate is $\geq$4 mmol/L.
- Manage vasopressors if the patient is hypotensive during or after resuscitation with fluids to maintain mean arterial blood pressure at $\geq$65 mmHg.

Empirical antibiotic treatment for an IAI

Inadequate empirical antibiotic therapy will result in a significant increase in morbidity and mortality. In this regard, the recommendations of the different consensuses emphasize that antibiotic therapy should be administered as early as possible, in the proper dosage, taking into account resistance profiles evidenced in the local epidemiological studies and also by assessing the clinical impact on the patient.

Microbiology of an intraabdominal infection depending on its origin

Depending on the origin of the infection (community origin, or healthcare associated origin), a different profile is found in the pathogenic germs involved.

When origin is community based: Infections are often mixed, with a predominance of gram-negative bacilli, such as *Escherichia coli*, *Klebsiella*, and *Pseudomonas aeruginosa*. Anaerobic microorganisms from the Bacteroides group and gram-positive cocci are also frequent, usually *Streptococcus*, *Staphylococcus*, and *Enterococcus*.

In nosocomial or healthcare-associated origin: Cultures find *Escherichia coli* decreases and increases other enterobacteria of the *Enterobacter*, *Acinetobacter*, *Klebsiella pneumonia*, or *Stenotrophomonas maltophilia* type. The presence of Enterococcus is higher and the prevalence of *P. aeruginosa* is also higher than with a larger pattern of resistances.

Risk factors for poor outcomes

The initial severity of the condition, age, presence of comorbidities, poor control of focus, and the risk of infection by multiresistant germs are considered to be risk factors for poor outcomes.

IAI is classified as serious and has a worse prognosis when the patient initially presents two or more SRIS criteria, together with an organ failure parameter (hypotension, oliguria, or altered mental state), a venous lactate greater than 2 mmol/L, or an APACHE II score $\geq$15, and indicates the need to administer a broader spectrum antibiotic regimen. Being aged over 65 years is considered a risk factor, as well as the presence of comorbidities such as immunosuppression, malnutrition, diabetes, chronic kidney disease, malignant disease, COPD, and liver cirrhosis. The lack of control of the focus, either by delaying the initial surgery by more than 24 h, due to widespread peritonitis, or due to an adequate debridement being impossible, indicates a worse outcome. Finally, patients coming from a social healthcare center, those with recurrent urinary tract infections, or who have undergone invasive procedures such as an endoscopy or who have received antibiotic treatment in the last 3 months, have a higher risk of infection from multiresistant germs; germs that produce extended-spectrum beta-lactamases (ESBL), germs that produce carbapenemases (*Klebsiella pneumoniae*), multiresistant *Pseudomonas*, Vancomycin-resistant *Enterococcus*, or fluconazole-resistant *Candida*, which worsen the prognosis.

Empirical antibiotic treatment[12]

An adequate initial empirical antibiotic treatment is one of the keys to improving the prognosis of these patients. When facing the impossibility of knowing the germ responsible immediately, an approximation is performed, trying to cover the most common germs, taking into account the origin of the infection, the patient's clinical condition and their medical history. We will thus get four groups, assessing whether IAI has been acquired in the community or is a nosocomial infection; and if the patient has associated severity criteria, or not.

IAI acquired in the community in a noncritical patient:
- Amoxicillin-clavulanic acid, if there is a low residual inoculum and the focus is controlled.
  or
- Third generation cephalosporin plus metronidazole
  In patients with a high risk of infection by ESBL-producing enterobacteria
- Ertapenem
  or
- Tigecycline
  In patients allergic to beta-lactam antibiotics
- Ciprofloxacin plus metronidazole
  or
- Moxifloxacin

IAI acquired in the community in a critical patient
- Piperacillin-Tazobactam
  or
- Cefepime plus metronidazole plus ampicillin
  In patients with a high risk of infection by ESBL-producing enterobacteria
- Meropenem, Doripenem, or Imipenem/Cilastatin

IAI nosocomial or associated with healthcare in a noncritical patient
- Piperacillin-Tazobactam
  In patients with a high risk of infection by multiresistant germs
- Meropenem plus Ampicillin
  or
- Doripenem plus Ampicillin
  or
- Imipenem/Cilastatin
  As a guideline, to avoid carbapenem antibiotics, you could use
- Piperacillin-Tazobactam plus tigecycline
  In patients at risk of invasive candidiasis add
- Fluconazole

IAI nosocomial or associated with healthcare in a critical patient
- Meropenem plus Vancomycin or Teicoplanin
  or
- Doripenem plus Vancomycin or Teicoplanin
  or
- Imipenem/Cilastatin plus Vancomycin or Teicoplanin
  As a guideline, to avoid carbapenems, you could use
- Ceftolozane/Tazobactam plus Metronidazole plus Vancomycin or Teicoplanin
  or

- Ceftazidime/Avibactam plus Metronidazole plus Vancomycin or Teicoplanin

  In patients at risk of infection by Vancomycin-resistant enterococci, the glycopeptide should be replaced by Linezolid or Daptomycin.

  In patients at risk of invasive candidiasis, add either echinocandins (caspofungin, anidulafungin, or micafungin) or Liposomal amphotericin B.

  In patients with suspected or proven infection by *P. aeruginosa*, a producer of nonmetallo-beta-lactamase, consider the use of antibiotic combinations with Ceftolozane/Tazobactam.

  In patients with suspected or proven infection caused by *K. pneumoniae*, a producer of carbapenemase, consider the use of combinations of antibiotics with Ceftazidime/Avibactam.

  In patients with a documented allergy to beta-lactam antibiotics, consider combinations of antibiotics with Amikacin.

Empirical antifungal treatment

The isolation of candida in a peritoneal sample should be considered a serious candidiasis and is an indication of antifungal treatment. A high percentage of therapeutic failures have also been documented in relation to the gastroduodenal focus and the presence of peritoneal candida is not covered in the initial empirical treatment. On this basis, before the availability of low-toxicity drugs (azoles and echinocandins) and the possibility of suspending empirical antifungal treatment if the culture does not confirm a candida infection, early empirical antifungal treatment and its subsequent suspension is justified.

In the presence of yeasts in the gram stain, or when the candida score is greater than 3, antifungal treatment will begin with fluconazole, except in the cases of septic shock or prophylaxis/prior treatment with an azole, in which an echinocandin will be indicated initially.[8]

Duration of antibiotic treatment[13]

Three days of treatment will be sufficient for a mild-moderate intraabdominal infection without any risk factors of poor outcomes and with appropriate focus control.

Antibiotic treatment of the established infection should be limited to 4–7 days, except in cases in which adequate control of the focus has not been achieved. Longer treatment periods are not associated with better results.

In patients with a persistent or recurrent case of intraabdominal infection after 4–7 days of treatment, the IDSA guides recommend performing diagnostic tests looking for the underlying cause. Effective antibiotic treatment for the microorganisms initially identified should be continued.

Abdominal compartment syndrome

The abdominal cavity, together with the retroperitoneal space, can be considered a watertight compartment whose pressure is determined by the volume of the content (abdominal viscera and liquids) and the distensibility of the continent (abdominal wall). According to the recent consensus document,[14] intraabdominal pressure (IAP) must be measured intermittently, via the bladder and with an instillation of no more than 25 mL of sterile saline. The patient will be in the supine position and with the transducer calibrated to zero at the level of the mid-axillary line. Normal is considered to be a value between 5 and 7 mmHg in critically ill patients, and intraabdominal hypertension (IAH) is defined as maintained or repeated measurements of IAP $\geq$ 12 mmHg. This increased pressure at abdominal level has physiopathological implications in many organs and can end up in multiple organ failure.[15] As regards breathing, lung distensibility is reduced, peak pressures rise, and oxygenation is altered. On a cardiocirculatory level, cardiac output is decreased due to a reduction of the preload when the inferior vena cava is compressed and the postload of the left ventricular is increased. In the kidneys the glomerular filtration rate goes down, due to the reduced cardiac output and the direct compression of renal vessels. At the level of splanchnic circulation, liver hypoperfusion with ischemia, it will result in coagulation alterations and subsequent bleeding. Intestinal ischemia and edema of the wall of the loops will cause loss of the intestinal barrier with bacterial translocation which will contribute to septic processes (Table 42.2). Both intestinal ischemia and edema also hinder the postsurgery healing process.

The elevation of abdominal pressures to levels higher than 20 mmHg, associated with the failure or dysfunction of any organ, is defined as Abdominal Compartment Syndrome (ACS) with a mortality rate that reaches 40%. Of the recognized risk factors for this syndrome, many are present in patients with colorectal carcinoma. Of them, we emphasize abdominal surgery, gastroparesis, ileus, colon subocclusion, hemoperitoneum and poly-transfusion, intraabdominal infection and

TABLE 42.2 Pathophysiology of IAH.

Respiratory	Increased peak pressure Reduced oxygenation
Cranial	Increased intracranial pressure
Cardiocirculatory	Compression of the inferior vena cava Reduced preload Increased postload Decrease in cardiac output
Renal	Compression of renal vessels Reduced glomerular filtration rate
Splanchnic	Hepatic ischemia <ul><li>Coagulopathy</li><li>Acidosis</li><li>Intraabdominal bleeding</li></ul>Intestinal ischemia <ul><li>Wall edema</li><li>Bacterial translocation</li><li>Sepsis</li><li>Suture failure</li></ul>

abscesses, peritonitis, sepsis, laparoscopy with excessive insufflation pressures and also, as a complication, after endoscopic examinations due to unsuspected perforations.[16]

Given the growing concerns about this clinical picture, the World Society of the Abdominal Compartment Syndrome recommends measuring IAP every 4–6 h and in a protocolized way in any critical patient who presents any risk factor for IAH/ACS. So if the pressure value increases above 12 mmHg, medical treatment will be initiated to keep the pressure at values below 15 mmHg. These measures consist of:

(1) Improving the distensibility of the abdominal wall, with adequate sedation and analgesia, periods of neuromuscular blockade, and avoiding raising the head of the bed by more than 30 degrees.
(2) Evacuate intraluminal content, with a nasogastric and rectal decompression, with prokinetic agents such as neostigmine, minimizing enteral nutrition and administering enemas.
(3) Evacuate abdominal deposits of liquids with paracentesis or percutaneous drainage
(4) Correct positive balances, by avoiding excess fluid in resuscitation and using diuretics and ultrafiltration techniques, if necessary.

In cases in which IAH is refractory to medical treatment and causes the dysfunction of an organ not previously existing, we are faced with an ACS and must consider surgical abdominal decompression. Any surgical abdominal decompression requires temporary abdominal wall closure techniques, which must comply with the premises of protecting the bowel and abdominal viscera, to control losses to the third space and, at the same time, maintain a tension-free suture. There are different strategies which include the interposition of Dexon, polypropylene or polytetrafluoroethylene meshes between the edges of the fascia, a Bogota bag, Wittmann Patch temporary closure set, or the sandwich-vacuum pack technique. It has been seen that negative pressure prevents visceral adhesions to the abdominal wall, while maintaining medial traction of the fascia. In addition, it allows the elimination of proinflammatory cytokines of the peritoneum decreasing the systemic inflammatory response. Although there is concern that it might increase the incidence of enterocutaneous fistulas, the recent consensus of 2013 recommends the use of a technique that uses negative pressure for abdominal techniques that only insert bioprosthetic meshes.

It is important to remember that sudden decompression of the abdomen can produce a reperfusion syndrome with a high mortality. When the pressure in the abdomen is reduced, abdominal and pelvic veins expand again, which reduces cardiac preload. This, added to the massive release of potassium and hydrogen ions from ischemic areas, can lead to cardiac arrest due to asystole. For this reason, the decompression should be performed in the operating room, with good venous access, replenishment of fluids, and administration of bicarbonate and mannitol in order to buffer the excess of acids and force osmotic diuresis. Once decompression is performed, organic dysfunctions progressively improve if other pathophysiological processes such as sepsis are controlled. Usually, after several surgical revisions, when the pressures are normalized, the wall will be definitively closed.

In summary, and although there are many reasons why patients affected by colorectal carcinoma may be admitted to critical care units, the most common is that the postoperative period passes without incident in an admission ward. We have performed a more detailed review of the most complex conditions, such as suture dehiscence with peritonitis, sepsis, and Abdominal Compartment Syndrome, because they are the conditions most directly related to abdominal surgery, which lengthen hospital stays and generate the most morbidity and mortality.

References

1. Bos MMEM, Bakhshi-Raiez F, Dekker JWT, de Keizer NF, de Jonge E. Outcomes of intensive care unit admissions after elective cancer surgery. *Eur J Surg Oncol* 2013;**39**(6):584–92. https://doi.org/10.1016/j.ejso.2013.02.014.
2. Snijders HS, Wouters MWJM, van Leersum NJ, et al. Meta-analysis of the risk for anastomotic leakage, the postoperative mortality caused by leakage in relation to the overall postoperative mortality. *Eur J Surg Oncol* 2012;**38**(11):1013–9. https://doi.org/10.1016/j.ejso.2012.07.111.
3. Suding P, Jensen E, Abramson MA, Itani K, Wilson SE. Definitive risk factors for anastomotic leaks in elective open colorectal resection. *Arch Surg* 2008;**143**(9):907–11. discussion 911-2 https://doi.org/10.1001/archsurg.143.9.907.
4. Klein M, Gögenur I, Rosenberg J. Postoperative use of non-steroidal anti-inflammatory drugs in patients with anastomotic leakage requiring reoperation after colorectal resection: cohort study based on prospective data. *BMJ* 2012;**345**. https://doi.org/10.1136/bmj.e6166, e6166.
5. Guirao X. Infección intraabdominal postoperatoria. In: Guirao X, editor. *Infección intraabdominal complicada. protocolos diagnósticos y terapéuticos*. Arán Ediciones S.L; 2009. p. 115–26.
6. Fernandez-Urien I, Vila JJ, Jimenez FJ. Endoscopic ultrasound-guided drainage of pelvic collections and abscesses. *World J Gastrointest Endosc* 2010;**2**(6):223–7. https://doi.org/10.4253/wjge.v2.i6.223.
7. Arias J. Tipos de infección intraabdominal: peritonitis y abscesos intraabdominales. In: Guirao X, editor. *Infección intraabdominal complicada. Protocolos diagnósticos y terapéuticos*. Arán Ediciones S.L; 2009. p. 13–29.
8. Guirao X, Arias J, Badía JM, et al. Recomendaciones en el tratamiento antibiótico empírico de la infección intraabdominal. *Rev Esp Quimioter* 2009;**22**(3):151–72. https://seq.es/wp-content/uploads/2009/10/consensoguirao.pdf.
9. Singer M, Deutschman CS, Seymour CW, et al. The third international consensus definitions for sepsis and septic shock (sepsis-3). *JAMA* 2016;**315**(8):801–10. https://doi.org/10.1001/jama.2016.0287.
10. Rhodes A, Evans LE, Alhazzani W, et al. Surviving sepsis campaign: international guidelines for management of sepsis and septic shock: 2016. *Crit Care Med* 2017;**45**(3):486–552. https://doi.org/10.1097/CCM.0000000000002255.
11. Levy MM, Evans LE, Rhodes A. The surviving sepsis campaign bundle: 2018 update. *Intensive Care Med* 2018;**44**(6):925–8. https://doi.org/10.1007/s00134-018-5085-0.
12. Sartelli M, Chichom-Mefire A, Labricciosa FM, et al. The management of intra-abdominal infections from a global perspective: 2017 WSES guidelines for management of intra-abdominal infections. *World J Emerg Surg* 2017;**12**(1). https://doi.org/10.1186/s13017-017-0141-6.
13. Solomkin JS, Mazuski JE, Bradley JS, et al. Diagnosis and management of complicated intra-abdominal infection in adults and children: guidelines by the Surgical Infection Society and the Infectious Diseases Society of America. *Clin Infect Dis* 2010;**50**(2):133–64. https://doi.org/10.1086/649554.
14. Kirkpatrick AW, Roberts DJ, De Waele J, et al. Intra-abdominal hypertension and the abdominal compartment syndrome: updated consensus definitions and clinical practice guidelines from the World Society of the Abdominal Compartment Syndrome. *Intensive Care Med* 2013;**39**(7):1190–206. https://doi.org/10.1007/s00134-013-2906-z.
15. Castellanos G, Piñero A, Fernández JA. Intra-abdominal hypertension and abdominal compartment syndrome. What should surgeons know and how should they manage these entities? *Cir Esp* 2007;**81**(1):4–11. https://doi.org/10.1016/s0009-739x(07)71249-6.
16. Chiapponi C, Stocker U, Körner M, Ladurner R. Emergency percutaneous needle decompression for tension pneumoperitoneum. *BMC Gastroenterol* 2011;**11**(1):48. https://doi.org/10.1186/1471-230X-11-48.

Chapter 43

Pain units: Symptom control

Pilar Díaz Parada[a] and M. del Carmen Corujeira Rivera[b]
[a]Anestesiology Service, Pontevedra University Hospital Complex, Pontevedra, Spain, [b]Anestesiology Service, Vigo University Hospital Complex, Vigo, Spain

Acute postoperative pain

The term Acute Postoperative Pain (APP) encompasses unpleasant sensory, emotional, and mental perceptions associated with autonomic, psychological, and behavioral responses that have been triggered by a surgical operation. Its onset is caused by nociceptive stimulation resulting from the surgical aggression on the different organs and tissues. It has no biological function.

APP is the maximum expression of acute pain. It has a rapid onset after the surgical aggression and a limited duration, with maximum intensity in the first 24 h. It diminishes progressively as the patient recovers from the surgery,[1] although if it is not treated correctly, it can persist and lead to chronic, treatment-resistant pain.[2,3]

APP is caused by direct mechanisms (sectioned nerve endings in the different structures affected by surgical manipulation) and indirect mechanisms (release of algogenic substances capable of activating the receptors in charge of processing nociception). Depending on the level where these mechanisms take place, pain can be divided into the following three types: (1) somatic (superficial and deep), (2) visceral, and (3) neuropathic. APP is generally of the somatic nociceptive type, although visceral and neuropathic components are sometimes involved.

The direct nociceptive stimuli behind APP are not only caused by manipulation of the skin, deep structures, and viscera during surgical incision, but also by intraoperative tissue retraction and traction of the mesentery; orotracheal intubation; positioning during surgery; placement of probes, drains, and intravascular catheters; and anxiety and sleep disorders associated with the surgical procedure.[4]

Nociceptive impulses originate in these levels, and when they reach the CNS, trigger a cascade reaction that affects various organs and systems (respiratory, cardiocirculatory, digestive, and endocrine-metabolic) defining the body's reaction to the surgical aggression. The response is detrimental and considerably increases postoperative morbidity, prolonging hospitalization and raising economic costs. It is therefore necessary to treat APP, accelerating the patient's recovery, allowing them to breathe, cough, and move more easily, which in turn reduces the incidence of pulmonary, cardiovascular, and thromboembolic complications, and also leads to better postoperative results and earlier hospital discharges.[5]

APP occurs in all types of surgery, to a greater or lesser degree depending on the type of surgical aggression. Its treatment should be planned from the moment we visit the patient in the preanesthesia consultation[6] and should be personalized, assessing the risk factors that condition the onset of more intense acute pain and determining the most appropriate analgesic protocol and guideline for each case.[7]

APP does not only depend on the surgical intervention but it is also linked to several factors that could explain the varying degrees of pain experienced by each patient. The main factors are as follows[8]:

Patient-dependent factors: personality type, level of anxiety prior to surgery, or previous experiences. Ethnic and sociocultural factors: the higher the sociocultural level, the easier it is to apply sophisticated systems to treat APP.

Surgery-dependent factors: prolonged interventions, excessive traction with leaflets and retractors, or the onset of surgical complications all increase the intensity of pain. Laparoscopic interventions produce distension of the phrenic nerve and diaphragm when CO_2 is insufflated into the abdominal cavity, producing nausea and pain in the region of the scapula.

Factors dependent on the anesthetic technique: the infusion of drugs at the neuraxial level, performance of locoregional blocks, or infiltration of the trocar insertion points, and of the surgical wound can improve APP. It is important to maintain the patient's systemic stability and adequate anesthetic depth.

It is estimated that the incidence of moderate and severe APP after abdominal surgery can affect up to 25%–75% of patients.

Strategies for APP management

The main cause of inadequate control of APP is the lack of routine and systematic evaluation. The following subjective and unidimensional Acute Pain Measurement Scales[2] can be used for assessing APP:

- numerical scale
- visual analogue scale (VAS)
- Wong-Baker Faces scale, for children from the age of three and adults for whom communication is difficult (Fig. 43.1).

Nowadays, we cannot be satisfied only with the assessment of pain intensity. We should also evaluate functional capacity, the impact pain has on it, and its recovery with the different physical measures and therapies developed.[9] We should also monitor the effectiveness of the treatment used.

Although the treatment of choice for APP continues to be intravenous opioids as they are the fastest, most effective, and predictable option, there is currently scientific consensus on the clinical efficacy of multimodal analgesia. It consists of combining drugs with different mechanisms of action and analgesic techniques. The aim is to achieve a synergistic effect and minimize the frequency and intensity of adverse effects using different modes and routes of administration. However, the use of multimodal analgesia is not currently widespread.[3]

This concept of multimodal analgesia should be included in a broader concept: perioperative multimodal rehabilitation, which also includes physiotherapy, enteral nutrition and early mobilization, and whose objective is to accelerate the patient's return home in the best conditions, reducing hospitalization times and facilitating a rapid return to normal life.

The introduction of minimally invasive surgical techniques, such as laparoscopy, has modified the analgesic protocols in abdominal surgery. Nowadays, the guidelines for the treatment of APP tend to be drawn up according to the degree of surgical aggressiveness, which will determine the degree of pain.

The most outstanding advances in colorectal surgery are the consolidation of laparoscopic surgery (LPS)[10] and the early recovery or fast-track protocols, focused on shortening postoperative recovery and hospitalization. LPS has proven to be useful and safe in the treatment of oncologic disease, with results similar to those of open surgery.

Multimodal early recovery protocols have been developed aiming to optimize different aspects of perioperative care, such as personalizing prefasting, reducing the administration of sedatives and long-acting opioids, adjusting fluid therapy, avoiding gastric probing, promoting intestinal motility and early intake, as well as multimodal analgesia for intraoperative and postoperative pain management.

Despite the progress made in pharmacological, organizational, and treatment protocols, the data show that the prevalence of APP has hardly changed over the last 20 years. Most of the challenges of APP refer to pain at rest and there are little data on dynamic pain control, which is the one that will allow patients to walk and start rehabilitation or respiratory physiotherapy, thereby accelerating postoperative recovery.

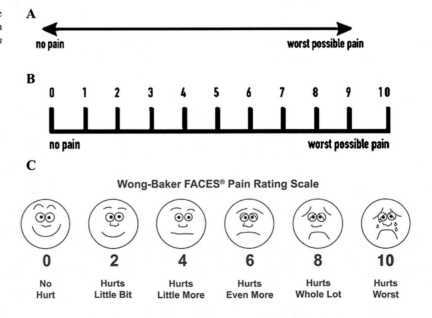

FIG. 43.1 (A) Visual analog scale (VAS), (B) The numerical rating scale (NRS), (C) Faces pain rating scale. *Reproduced with permission from Elsevier.*

Most frequent causes of inadequate analgesia

(1) Insufficient or inadequate medication.
(2) Inadequate training of health care professionals in pain management.
(3) Inadequate communication between health care professionals and patients.
(4) Lack of personalized treatments.
(5) Delay in starting analgesia.
(6) Inadequate methods of drug administration.

Specific analgesic protocols for each type of surgical intervention are a guarantee for personalizing treatments and responding adequately to the analgesic demands of each patient.

The update of the American Pain Society (APS) Postoperative Acute Pain Management Guidelines[11] has been published with 32 recommendations, including the use of multimodal analgesia, regional and epidural analgesia in specific procedures, minimum opioid doses, preference of the oral route over the intravenous route, and the PCA modality.

Analgesic techniques for CRC surgery

Fast-track surgery or early or intensified rehabilitation programs have changed the perioperative course of CRC patients. They are based on the application of 10–20 perioperative measures with the aim of accelerating postoperative recovery, reducing the response to surgical stress, complications, average length of hospitalization, and readmissions. The Feroci study[12] conducted on 606 patients who underwent CRC surgery recorded five key elements for achieving good results: (1) laparoscopic surgery, (2) early urinary catheter removal, (3) selective use of drains, (4) solid oral nutrition, and (5) early mobilization.

However, it is not clear which postoperative analgesia would be best in fast-track surgery programs.

In recent years, postoperative analgesic protocols have been modified due to the development of LPS, and robotic and minimally invasive surgery, which is associated with lower analgesic requirements than open surgery. Intensified recovery programs involving early patient mobility, and the progress of ultrasound-guided regional analgesia that improved the efficacy and safety of regional blocks have also played a role.

The following analgesic guidelines and techniques associated with fast-track surgery have been described:

(1) **Thoracic epidural analgesia**[13, 14]

A metaanalysis performed by Pöpping et al.[15] studied the impact of epidural analgesia on morbimortality after surgery, recording a decrease in mortality in the Epidural Analgesia (EA) group versus systemic analgesia, with a decrease in the incidence of arrhythmias, respiratory depression, atelectasis, pneumonia, ileus, postoperative nausea and vomiting, and acceleration of recovery of intestinal transit. Other beneficial effects of this technique include mesenteric vasodilatation (which favors adequate consolidation of the anastomoses) and reduction of the endocrine-metabolic response to surgical stress, reducing associated complications. It also seems to reduce thromboembolic complications, although this effect is clearer in its application in lower limb surgery than in abdominal surgery.

EA is more effective in reducing dynamic pain and postoperative ileus in major abdominal surgery.[16]

The impact of laparoscopy on APP is significant, with abdominal surgery showing a reduction in pain both at rest and during movement and a reduction in opioid consumption compared with laparotomy.

EA provides better analgesic quality than systemic EA in open surgery,[17] while in LPS surgery it delays recovery due to decreased patient mobility.[18]

EA is not currently recommended in laparoscopic approaches in enhanced recovery programs (ERAS).[19] Other alternatives with fewer side effects are proposed, such as multimodal analgesia, intravenous patient-controlled analgesia (PCA) with low doses of opioids and/or ultrasound-guided abdominal wall blocks (TAP), and/or surgical wound infiltration.

In open surgery, thoracic EA is recommended over lumbar EA, since the latter is associated with a high incidence of motor block, and patient-controlled infusion over continuous infusion because the latter also favors motor block.

Main factors increasing the efficacy and safety of postoperative thoracic epidural analgesia[20]
- Personalized indication according to comorbidities, risk, type, and approach of the surgical intervention and in the context of postoperative recovery guidelines.
- Standardized and pharmacy preparation of the epidural solution, with low doses of local anesthetic and opioids.
- Standardized PCA guidelines of 3–5 mL/h with 3–5 mL boluses and closure time of 10–15 min.

- Monitoring of analgesic effectiveness and side effects. Early removal of the catheter once ambulation has begun and pain is under control.
- Complementary multimodal analgesia guidelines and rescue opioids.
- Algorithms for side effects: motor block, lumbar pain, insufficient analgesia, etc.

Epidural opioids[13]
- Epidural administration of a lipid-soluble opioid (fentanyl) produces rapid, segmental analgesia by means of a correctly placed catheter in the corresponding dermatome, with a low tendency to cephalic migration and shorter duration of action.
- The use of a poorly lipid-soluble opioid (morphine) results in a slower onset of analgesia, which in addition has a poor segmental effect as it has a tendency to cephalic migration toward the supraspinal structures and a prolonged duration of action.

The administration of spinal opioids is an excellent option for separating desirable analgesic effects from expected, dose-limiting side effects and for improving postoperative analgesia. Study results support the use of mainly neuraxial morphine[21] which has a higher bioavailability than lipophilic opioids.

Complications associated with epidural analgesia[13]
- **Pain**: generally due to inadequate dosage. It is necessary to confirm the proper functioning of the system and its connections. If it is correct and the sensory level is low, the continuous perfusion is increased, and a rescue dose is administered. If severe pain appears and does not respond to treatment, after checking the system, catheter migration should be suspected, and the possibility of surgical complications should be considered.
- **Urinary retention**: manifested by the presence of suprapubic discomfort with bladder overdistension on abdominal palpation and anuria. The insertion of a bladder catheter for voiding is usually sufficient. If it persists, it may be necessary to maintain bladder catheterization until the device is removed.
- **Motor block**: first rule out the possibility that the catheter has migrated into the intradural space. Consider decreasing perfusion or even suspending it if the clinical condition does not resolve.
- **Pruritus**: improves with the administration of antihistamines and, if severe, may require naloxone perfusion.
- **Nausea and vomiting**: antiemetics are usually effective. When there is no response, naloxone may be considered.
- **Infection**: it is rare. Aseptic handling, monitoring of the catheter entry point, and temperature control reduce the risk of infectious complications.
- **Respiratory depression**: very infrequent if there is pain but very severe. It is treated with naloxone and ventilatory maintenance maneuvers.

Safety measures[13, 14]
- Do not administer other IV opioids or sedatives without the approval of an anesthesiologist. This will avoid serious complications, such as respiratory depression.
- Monitor respiratory rate regularly.
- Assess the level of sedation.
- Monitor sensory level, degree of motor block, BP, and HR.
- Assess urinary flow and bladder distension to rule out urinary retention.
- Inspect the catheter insertion site for signs of infection and check the connections.
- All patients who receive an epidural infusion should have a venous line available for the duration of the infusion and for at least 4 h after catheter removal.

Catheter removal in LMWH patients[13]

The catheter should be removed 12 h after the last dose of prophylactic heparin (24 h if the dose is anticoagulant) or 6 h before the next dose. In the case of fondaparinux, the catheter should be removed 36 h after the last dose, leaving a 12-h interval between removal and the next dose.

In patients treated with intravenous heparin sodium, the catheter should be withdrawn 4 h after the last dose, leaving a 60-min interval between removal and the next dose of heparin.

Perform neurological monitoring after catheter removal (lower limb mobility and sensitivity and onset of back pain).

Epidural infusion should be discontinued if
- Respiratory rate < 10
- Ramsay sedation scale of 5
- Sensory level above T6
- Mental disorientation/confusion

- Systolic blood pressure < 90 mmHg
- Accidental disconnection of epidural catheter
- Fever maintained for four consecutive hours.

(2) Intravenous analgesia

Opioids are the first line of treatment for moderate to severe APP. Continuous infusion of morphine is the most widely used and with proven efficacy. The use of patient-controlled infusion systems makes it possible to optimize treatment by allowing patients to administer additional doses by themselves, when needed. However, their adverse effects limit doses, delay postoperative recovery, and endanger the patient's life if not adequately controlled.[22]

Fast-track programs or early or intensified multimodal rehabilitation (ERAS) have promoted the concept of opioid-free analgesia to avoid their side effects[23] (nausea, vomiting, postoperative ileus, sedation, and delirium), which slow down recovery, delaying hospital discharge and the return to functional normality.

Current recommendations are based on minimizing the postoperative opioid dose, applying multimodal techniques, and early withdrawal of opioids when they can be replaced by other analgesics.

The American Pain Society (APS) Postoperative Acute Pain Management Guidelines[11] recommend the use of multimodal analgesia, minimal opioid doses, preference of oral over intravenous route, and the use of patient-controlled infusion.

Multimodal analgesia, based on a broad concept of combining analgesics, adjuvant drugs, and analgesic techniques, makes it possible to minimize doses and side effects, thereby increasing effectiveness. It is presented as a safe and effective alternative.[24]

In addition to regional anesthesia, many nonopioid drugs that can be administered perioperatively[21] reduce postoperative opioid requirements, including local infiltration and/or intravenous infusions of local anesthetics, paracetamol, NSAIDs, NMDA receptor antagonists (ketamine), magnesium sulfate, alpha agonists (clonidine and dexmedetomidine), anticonvulsants (gabapentin and pregabalin), glucocorticoids, and beta-blockers. Multimodal analgesia strategies are based on treating pain at a variety of receptors beyond the mu (opioid) receptors.

We should consider the operating room as an environment where we can safely control pain by using highly specific agents and thus decrease patient exposure to opioids, thereby decreasing tolerance to these drugs, which can develop in the early postoperative period. It has been shown, for example, that the combination of paracetamol and an NSAID decreases opioid consumption by 30%–50%.

New, noninvasive[25] or "needle-free" opioid delivery devices have been developed in order to eliminate the disadvantages of intravenous morphine PCA, namely:

- Sublingual Sufentanil Tablet System (SSTS). It has a high therapeutic index, high bioavailability, and crosses the blood-brain barrier quickly. It lacks active metabolites, and its metabolism is not affected by age, BMI, or hepatic or renal failure.
- Transdermal Fentanyl (Fentanyl Iontophoretic Transdermal System) (ITS). It has a rapid onset of action and a higher therapeutic index than morphine but lower than sufentanil.

These systems provide advantages by avoiding the intravenous route, increasing patient mobility and comfort, and cannot cause phlebitis or bacteremia. They are preprogrammed devices, thus avoiding human error in pump programming and drug preparation. They are opioids with rapid onset and prolonged action, without active metabolites, so theoretically they have an effective and safe pharmacological profile.

They also have disadvantages, namely: increased costs (cost-effectiveness studies are still pending), similar side effects to morphine, the preprogramming of the devices does not allow doses to be adjusted to individual demands, and the application is exclusively hospital based.

Studies have yet to be completed. They could play a role in transitional analgesia in the early withdrawal of epidural catheters.

Prescribing guidelines for analgesic drugs[26, 27]

- Prevent the onset of pain: administer analgesics as soon as they are needed. Prevention is more effective than treatment.
- Choose the simplest route of administration: intravenous is the route of choice and the patient should be switched to the oral route as soon possible.
- Prescribe the correct dose at the right intervals: remember that NSAIDs and minor opioids have a ceiling effect.
- Prescribe according to the intensity of the pain: major opioids from the beginning if the pain is severe.
- Periodically evaluate efficacy and detect possible adverse effects.
- Adjust according to response.

TABLE 43.1 Usual doses of the most frequently used intravenous nonopioid and weak opioid drugs.[a]

	Dose (mg)	Interval (h)	Maximum dose (mg/day)
Paracetamol	1000	6–8	4000 mg
Metamizole Mg	1000–2000	6–8	5000 mg
Dexketoprofen	50	8–12	150 mg Recommended maximum 2 days
Ibuprofen	400	6–8	Maximum recommended dose 1200 mg/day Maximum dose 1600 mg/day Recommended maximum 3 days
Tramadol	100 (lowest effective dose)	6–8	Maximum dose 400 mg/day (may be increased in certain situations)

[a]Source: Compiled from AEMPS

- Pay attention to contraindications: nephrotoxicity and gastrointestinal toxicity of NSAIDs.
- Use of adjuvant drugs if there is a neuropathic component.

In all of them, the lowest effective dose should be used for the shortest possible time. In null the most frequent doses are shown (Table 43.1.)

(3) **TAP block (transversus abdominis plane)**

TAP Block has been shown to be effective in LPS Colorectal Surgery.[28] It decreases morphine consumption and time to reinitiation of oral food intake.

(4) **Incisional analgesia**

The infiltration of local analgesia (LA) into the surgical wound is a technique widely used in recent years and it appears to significantly reduce pain and morphine requirements in open colorectal surgery. It can be administered as a single bolus or through the insertion of a multiperforated catheter at the end of surgery, both for continuous and intermittent postoperative administration.

However, a metaanalysis by Vnetham[29] concludes that there is insufficient evidence to support the analgesic effectiveness of continuous surgical wound infiltration. More studies per surgical procedure are needed.

Chronic pain in colorectal cancer

Together with fatigue, alterations in bowel habits, nausea, and psychological discomfort, pain is one of the symptoms that most frequently accompanies colorectal cancer, affecting up to 70% of patients to varying degrees.

One of the main factors involved in the high incidence of pain is that it is underreported by patients who perceive it as something inherent to the disease, bearing less importance for the clinician than other, more physical or tangible symptoms. The fear of drug dependence, the development of tolerance, that treatment will not be effective if the pain worsens, or that it will be a sign of progression of the disease, are some of the concerns that lead patients not to report the presence of pain.

The elderly are clearly the group most affected by undertreatment of pain. The frequent presence of organ dysfunction, comorbidities, and concomitant ingestion of other drugs usually leads to undertreatment.

Pain has a marked impact on the patient's functional and psychological state, resulting in a worsening of quality of life and personal relationships, as well as anxiety, depression, and sleep disorders. It is necessary to adopt adequate analgesic treatments and to develop coping measures to prevent patients from falling into catastrophism and to help them take an active role in pain management. It is, therefore, essential for the medical-surgical team and the nursing staff in charge of the patient to maintain a constant assessment of the patient's level of comfort and monitor for indirect signs of pain.[30]

The causes of pain include pain originating from the diseased organ, as well as disease progression with infiltration of adjacent or distant tissues, as in the case of metastases.[31] Oncological treatment measures, including chemotherapy, radiotherapy, and surgery, can also be a source of persistent pain, and prevention is essential whenever possible.

Pain can manifest in very different locations and forms depending on the mechanism involved. In CRC patients, pain can be classified as follows:

- Nociceptive pain: caused by tissue injury. It can be divided into:
  - Somatic: caused by bone, joint, or muscle lesions, frequently due to metastatic lesions. It is described as a deep, dull pain.
  - Visceral: caused by invasion of abdominal or pelvic organs. It manifests as diffuse pain, poorly localized and referred to dermatomes innervated by the corresponding segments of the spinal cord.
  - Neuropathic pain: due to peripheral or central nerve lesion caused by compression, section, hemorrhage, chemical mechanisms, etc. It causes burning pain, with a sensation of electricity, hypoesthesia, or dysesthesia. It may appear as a complication after surgery, radiotherapy, or chemotherapy, as well as due to involvement of the lumbosacral plexus.
- Psychogenic pain: pain mediated primarily by psychological factors. It is rare in cancer patients and should be a diagnosis of exclusion once other causes of pain have been ruled out.

Because of the implications for analgesic management, the possible temporal patterns of pain should also be assessed:

- Baseline pain: continuous pain present throughout the day, even though it may vary in intensity.
- Breakthrough pain: this is defined as an episode of intense pain of rapid onset and short duration (average 30 min) that is clearly differentiated within a relatively stable baseline pain. It occurs in 40%–80% of patients diagnosed with cancer. It may occur without an apparent cause (spontaneous pain) or be triggered by a specific event, such as movement, coughing, or defecation (incidental pain). These episodes usually occur between 1 and 4 times a day.
- End-of-dose pain: occurs near the end of the dosing interval, before the next dose, and is a sign of underdosing.

Frequent types of pain in colorectal cancer[32]

Abdominal pain: the form of presentation may vary greatly according to its origin.

- Intestinal obstruction: this is produced by the growth of the actual tumor or by the formation of adherences after surgery. It causes pain due to intestinal distension in the prestenotic segment, ischemia of the affected intestinal wall, or even traction of the mesentery. It can cause partial obstruction, developing subocclusive pictures that manifest as episodes of colicky pain of acute onset and self-limited in time. They are frequently accompanied by nausea, vomiting, and constipation. The management of these patients differs depending on whether they are eligible for surgery or not. In cases in which a surgical approach is ruled out, stent placement to bypass the obstruction should be considered. Conservative treatment includes fluid therapy and nasogastric catheterization, analgesia with opioids, glucocorticoids to reduce peritumoral edema, and anticholinergic drugs or octreotide to reduce intraluminal secretions and peristaltic movements.
- Liver metastases: they can cause pain due to traction of the capsule, triggering the so-called hepatic distension syndrome, with subcostal pain and pain in the right flank and right side of the back.
- Peritoneal carcinomatosis: causes peritoneal inflammation, formation of adhesions and ascites. It presents as focal or diffuse pain that worsens after ingestion of food and is accompanied by abdominal distension, nausea, or constipation.

Pelvic pain: pelvic pain can occur because of the onset of bone metastases, invasion of tissues at the presacral level, or in the form of neuropathic pain due to involvement of the lumbosacral plexus. The development of a colovesical or rectovesical fistulas may also present as suprapubic pain accompanied by pneumaturia, fecaluria, or bacteriuria.

Perineal pain: frequently accompanied by tenesmus or episodes of severe pain secondary to bladder spasm. Pain is exacerbated by prolonged sitting and/or standing.

Lumbosacral plexopathy: usually caused by direct extension of the tumor. It usually presents as an intense, lancinating pain, with or without radicular symptoms and with a variable distribution depending on the location of the tumor. The pain may be aggravated by supine decubitus, lifting weights, or straining to defecate.

Pain secondary to chemotherapy[33]

Chemotherapy-induced neuropathy (CIN) is the most frequent neurological complication of cancer treatment and affects approximately one-third of patients. It causes a deterioration in functionality, compromises quality of life, and frequently leads to the reduction or suppression of treatment, which affects the curative potential of the treatment and the patient's prognosis.

The main risk factors involved in the development of CIN are the dose and duration of treatment, although demographic, comorbidity, or genetic factors have also been involved, as well as the presence of previous neuropathy, even subclinical, with these patients being more vulnerable to the neurotoxic effect of chemotherapy. Changes in renal and hepatic functions related to the elimination of the toxic agent could also be risk factors.

Oxaliplatin, a third-generation platinum-derived compound frequently used in the treatment of colorectal cancer, causes sensory neurotoxicity that can become disabling. It can cause two forms of neuropathy:

— Acute and transient: this is the most frequent form, appearing in 95% of patients. It appears during the infusion of the drug or a few hours after the end of the infusion. It is usually triggered or aggravated by exposure to the cold. It is self-limited and patients recover rapidly in hours or days. Patients usually report paresthesia and dysesthesia in the distal portion of the limbs and in the perioral region and laryngopharyngeal dysesthesia in 1%–2% of cases. Transient exacerbation of sensory symptoms is frequent when undergoing surgery due to the release of intra-erythrocyte accumulations of the drug when hemolysis associated with surgery occurs. It is usually reversible.
— Chronic: Some cases present persistent symptoms of neurotoxicity. This occurs with cumulative drug doses >300 mg/m^2. Recovery is more prolonged and often incomplete and is considered irreversible if symptoms persist 9 months after completion of treatment.

Although many drugs are being studied as potential neuroprotectants, early recognition and subsequent dose reduction or discontinuation of the neurotoxic agent is the only way of minimizing the onset of this complication and preventing its progression to disabling neuropathy. The response to traditional antineuropathic drugs, such as amitriptyline, nortriptyline, lamotrigine, and gabapentin is poor.

Another painful process that is associated with chemotherapy treatment is mucositis.[34] It is usually limited to the oral mucosa but can affect the entire gastrointestinal tract. Its frequency and severity are dose dependent and may vary from mild irritation to severe erosive mucositis with intense pain and limitation of oral ingestion. It usually appears around 7 days after treatment and in general its course is self-limited, with resolution of the lesions within 10–14 days.

Although there is no consensus on their efficacy, formulations combining lidocaine with sodium bicarbonate, aluminum hydroxide, or diphenhydramine in the form of mouthwashes are frequently used. Other alternatives are dilutions of morphine sulfate or doxepin and mucosal protective gels.

Pain secondary to radiotherapy

Patients undergoing abdominal radiotherapy may develop enteritis or radicular proctitis, in which pain may be accompanied by digestive symptoms such as nausea, vomiting, diarrhea, or tenesmus. However, the onset of these complications in colorectal cancer is infrequent as radiotherapy of the colon is rare and there are very strict guidelines governing application to the rectum. In cases in which preoperative radiotherapy is administered, the development of proctitis is rare and the lesions potentially causing chronic pain are eliminated with the surgical specimen. In patients treated palliatively, the doses received are much lower than those administered in neoadjuvant treatment, so the risk of associated adverse effects is much lower.

Treatment of baseline pain

Despite its marked limitations, the WHO analgesic ladder continues to be the basic pillar in the treatment of pain in cancer patients.[35–37] Criticism of the delays in pain control caused by following certain pharmacological steps has led to the proposal of a modification of this strategy: the analgesic elevator. In this strategy, the step from the first step to the use of potent opioids in cases of moderate pain that do not respond to NSAIDs is considered.

Paracetamol and nonsteroidal antiinflammatory drugs are the analgesics of choice in cases of mild pain, as well as in moderate or severe pain in combination with opioids since their combined use enhances their effect. They have a ceiling effect, which means that their efficacy does not increase above their maximum dose, but the risk of side effects does. It is essential to know the possible adverse effects, including, but not limited to, cardiovascular, renal, and gastrointestinal effects, when prescribing them to patients with risk factors (known cardiovascular disease, hypertension, renal or hepatic failure, history of ulcer or gastrointestinal bleeding, etc.).

In case of moderate pain, the second step of the WHO Analgesic Ladder should be considered: combination with weak opioids, such as tramadol or codeine. Their analgesic effect is sometimes overshadowed by the onset of drowsiness or gastrointestinal disorders, such as nausea, vomiting, and constipation.

Treatment with potent opioids is reserved for moderate to severe pain, with morphine being the first choice due to its wide availability and favorable cost-effectiveness ratio. Despite this, there are still many obstacles that limit the prescription of morphine, ranging from the lack of knowledge about its pharmacological group to the bureaucratic procedures associated with its use. Fear of addiction, inappropriate use, the often-misunderstood tolerance, and adverse effects are other factors that negatively influence its use.

Due to its ease of administration and effectiveness, it is advisable to use the oral route whenever possible, reserving the subcutaneous and intravenous routes for when the oral route is not feasible or for treating exacerbations. In general, it is advisable to start with presentations of rapid oral morphine administered every 4 h and adjust the dose in 50% increments every 24 h until the effective dose is reached, taking into account that there is no ceiling dose. Once the pain has been controlled, an extended-release morphine formulation can be used every 12 or 24 h for greater patient comfort.

In recent years an increasing number of formulations have been developed with different routes of administration and bioavailability that try to adapt to the different needs of patients. Some of the presentations currently marketed are as follows:

- Oral route: immediate-release presentations (morphine and oxycodone) that can be administered every 4 h, and prolonged for 12 h (oxycodone, tapentadol, and hydromorphone) or 24 h (hydromorphone). A combination of oxycodone and naloxone is marketed and has been shown to reduce the incidence of constipation associated with its use.
- Transdermal route: fentanyl and buprenorphine patches are currently marketed. They are particularly useful in patients with swallowing difficulties, gastrointestinal disorders, or when adequate compliance cannot be guaranteed (as in the case of elderly patients). It is important to place the dressing on clean, dry skin to optimize adherence and to rotate the site of application to reduce the risk of local skin reactions. This treatment is not suitable for unstable pain or pain with a frequent breakthrough component because its slow onset of action (approximately 12 h) and its residual action of up to 17 h after withdrawal do not allow rapid dose adjustments. The main advantage of this formulation lies in the stability of the plasma levels reached, which allows its administration every 72 h, although in up to 25% of patients it may be necessary to apply it every 48 h.

Transdermal fentanyl and buprenorphine differ fundamentally in that while the former is a pure opioid agonist, the latter is an agonist-antagonist. This means that, unlike fentanyl, buprenorphine has an analgesic ceiling, with a maximum daily dose of 140 μg/h (two 70 μg/h patches). It also presents a risk of developing withdrawal symptoms when administered to patients treated with pure opioids.

Analgesic coadjuvants[38, 39]

- Corticosteroids: useful for their antiinflammatory action, the most potent being dexamethasone. Indicated mainly in colorectal cancer in case of pain due to bone metastases, nerve compression, intestinal obstruction, or hepatomegaly. It should be adjusted to the minimum dose necessary to reduce the risk of adverse effects.
- Anticonvulsants: considered as the first line of treatment in neuropathic pain and used as coadjuvants within a multimodal therapy. Gabapentin and pregabalin are the most widely used. They have been associated with improved sleep quality and pregabalin could be an option in patients with anxiety due to its anxiolytic effect. Their most frequent side effects include drowsiness, dizziness, nausea, weight gain, and peripheral edema. Other anticonvulsants, such as carbamazepine, oxcarbazepine, valproate, topiramate, or lacosamide are considered second-line drugs in oncology patients.
- Antidepressants: useful as coadjuvants and in the treatment of neuropathic pain. They are considered the drugs of choice in neuropathic pain in patients who also present depressed mood, although they have an analgesic effect independent of their antidepressant action. Serotonin and noradrenaline reuptake inhibitors (e.g., duloxetine) and secondary amine tricyclic antidepressants (mainly desipramine) are usually recommended for this group of patients due to their safety profile and their analgesic properties that have been widely studied in patients with chronic pain.
- Topical lidocaine: useful in oncologic neuropathic pain, especially in cases in which allodynia plays an important role. Its use has been proposed for painful surgical wound scars and in neuropathy secondary to chemotherapy. Except for skin irritation, which can sometimes limit its use, adverse effects are rare.
- Topical capsaicin: this drug has been shown to relieve nondiabetic peripheral neuropathic pain for up to 12 weeks. Recent publications suggest its possible indication in patients with neuropathic pain secondary to chemotherapy. One of its main advantages is the convenience of application when administered every 3 months, although prior application of topical local anesthetic is usually necessary to reduce the discomfort associated with it.

- Cannabinoids: currently, in Spain they have only been approved for use in the treatment of spasticity secondary to multiple sclerosis. Their antiemetic and appetite-stimulating effects have led to compassionate use in oncology patients, and in many countries their use has been approved as analgesic adjuvants in oncological pain refractory to opioids.
- Bisphosphonates: they relieve pain caused by bone metastases. Due to the risk of pathological fractures, osteonecrosis of the jaw, stomatitis, and ocular inflammatory disorders, the use of bisphosphonates is reserved for cases refractory to the usual analgesics (they are not recommended in patients who have had a recent fracture). Their use has also been associated with esophageal cancer and, although it has not been confirmed, prescription is not recommended in patients with Barrett's esophagus.[40]

Treatment of breakthrough pain

The treatment of breakthrough pain is based, as far as possible, on its prevention, either by avoiding triggering events or by administering analgesic drugs prior to their occurrence. This would be the case of pain caused by wound dressing or washing the patient. It is essential to adjust the basic treatment and its intervals, using rescue drugs in episodes of end-of-dose pain and in episodes that cannot be avoided.[41]

The search in recent years for formulations with a rapid, potent, and safe effect has led to a decrease in the use of rapid-release morphine and oxycodone in favor of fentanyl. It is a lipophilic opioid of rapid absorption through the oral and nasal mucosa with an onset of analgesic action in the first 5–10 min. The following presentations are available:

- Oral transmucosal fentanyl: available in tablet form with an integrated applicator that allows withdrawal of the drug when the pain subsides or in case of adverse effects. Its efficacy in the control of breakthrough pain compared with placebo and morphine has been confirmed in the literature.
- Fentanyl citrate: its effervescent formulation allows greater bioavailability (up to 65%) and speed of action due to the action of citric acid, with analgesic effect within 10 min of administration. The tablet should be placed between the cheek and the gum near a molar tooth and should not be swallowed or chewed as this would reduce its efficacy.
- Sublingual fentanyl: high bioavailability (70%) and effective pain relief after 10 min, with analgesic effect after 5 min in 70% of patients.
- Intranasal fentanyl: effective 5 min after application, with a bioavailability that exceeds that of oral presentations. It is especially useful in patients with poor salivary secretion or mucositis.

The main difficulty in the management of these drugs lies in the absence of a clear correlation between the patient's baseline opioid dose and the effective dose, which requires individual titration of the dose according to the response to treatment. Table 43.2 presents reference values that may be helpful for starting the dose and how it could be increased.

Interventional techniques

They should be evaluated early and associated with any analgesic step in order to optimize pain control, reduce consumption of analgesics and, consequently, improve the patient's quality of life. Their most relevant role is in cases in which analgesic control is not achieved despite adequate systemic treatment or when the side effects are intolerable and do not allow effective drug doses to be achieved.[42]

Intrathecal therapy

Intrathecal therapy consists of infiltrating drugs at intrathecal level using an infusion pump connected to a catheter. The most commonly used treatments are opioids, local anesthetics, α-2 agonists, and ziconotide. The blockade of nociceptive transmission at this level produces considerable pain relief in patients refractory to other treatments; its use is limited by the risk of major complications that restrict its being adopted by centers with experience and capacity for patient follow-up. The most feared complications, although infrequent, are still infection (usually limited to the surgical wound, but attention should be paid to the onset of meningeal signs) and neurological damage due to accidental nerve root injury.[43, 44]

The characteristics of the device and its more or less permanent nature must be assessed according to the patient's life expectancy and whether the device will be used only in the hospital or at home. The options vary from the placement of catheters that allow the administration of drugs from an external perfusion system to the use of fully implantable devices in which the infusion pump is surgically inserted in the abdominal wall or supragluteal region. The implantation of these devices has been shown to be effective in controlling pain and reducing the side effects of analgesic treatment and has also been associated with an increase in survival at 6 months compared with patients receiving medical treatment alone.

TABLE 43.2 Starting doses and usual management of the main drugs used in pain treatment.[a]

	Initial dose	Dose increase	Maximum dose
1st and 2nd step			
Paracetamol	1 g/6–8 h		4 g/24 h
Ibuprofen	400–600 mg/8 h		1200 mg/24 h
Tramadol	50 mg/8–12 h		400 mg/24 h
Codeine	15–30 mg/4 h		200 mg/24 h
Opioids			
Morphine per os	5–10 mg/4 h	↑ 25%–50% every 24 h	
Fentanyl TD	12–25 µg/h every 72 h	↑ 25 µg/h every 72 h	
Buprenorphine TD	35 µg/72 h	↑ to next concentration patch after 72 h	140 µg/72 h
Oxycodone	5–10 mg/12 h	↑ 25%–50% every 24–48 h	
Oxycodone-Naloxone	5–2.5 mg to 10–5 mg/12 h	↑ in increments of 5–2.5 mg/12 h or 10–5 mg/12 h every 24–48 h	80–40 mg/24 h
Tapentadol	25–50 mg/12 h	↑ 50 mg/12 h every 3 days	500 mg/24 h
Hydromorphone	4–8 mg/24 h	↑ 4–8 mg/24 h	
Analgesic coadjuvants			
Gabapentin	100–300 mg/24 h	↑100–300 mg/24 h (in batches every 8 h)	3600 mg/24 h
Pregabalin	25–75 mg/24 h	↑75–150 mg/3 days	600 mg/24 h
Amitriptyline	10–25 mg/24 h	↑25 mg/7 days	150 mg/24 h
Duloxetine	30 mg/24 h	↑30 mg/7 days	120 mg/24 h
Topical lidocaine			3 dressings for 12 h/day
Topical capsaicin			4 dressings/90 days
Dexamethasone	8–40 mg/24 h		Adjust dose to minimum necessary

[a]Source: Compiled from AEMPS

Neuroablative techniques

These attempt to modify nerve transmission of the painful stimulus at different levels by means of alcohol or phenol infiltration or radiofrequency thermal lesioning. It is necessary to use imaging techniques (CT, MRI, or fluoroscopy depending on the approach) to ensure the correct localization and avoid injury to adjacent structures. Although the evidence for these techniques in oncologic patients is limited due to the scarcity of studies, their use is supported by experience in acute and chronic nononcologic pain. These techniques include:

- Celiac plexus ablation: useful in abdominal pain of visceral origin located at the supraumbilical level. The celiac plexus is in charge of receiving visceral afferents from several organs in the upper abdomen, including part of the transverse colon. Infiltration of the plexus with a neurolytic drug (usually alcohol and local anesthetic) has traditionally been used for pain caused by pancreatic cancer, although it may be useful in patients with neoplastic processes in the upper abdomen, such as some patients with colorectal cancer or liver metastases. It provides good to excellent pain control in 89% of patients in the first 2 weeks after its performance, and its analgesic effect is maintained in 90% of patients who are still alive 3 months after the procedure. The most common complications are hypotension, diarrhea, or lumbago, and, to a lesser extent, neurological alterations, including paraplegia.

- Upper hypogastric plexus block: it prevents sensitivity in the urogenital organs, descending colon, and rectum, and is useful in case of pain associated with neoplastic disease in the pelvic region. It provides good analgesic control in 72% of patients and moderate control in 28%, maintaining its effects after 3 months. It is associated with a decrease in the consumption of opioids.
- Lumbar sympathectomy: indicated in visceral pain caused by infraumbilical tumors. The lumbar sympathetic chain ganglia are blocked with alcohol or phenol or by radiofrequency thermocoagulation. Hypotension may occur after the procedure, which usually responds to fluid therapy. Complications include renal injury, intravascular, intramuscular, or intradural injection, and neuritis.
- Ganglion impar block: it alleviates painful afferents from the sacrococcygeal and pelvic-perineal region, relieving pain caused by rectal cancer, as well as pain secondary to abdominoperineal amputation surgery and radicular proctitis.

Trigger point deactivation

The onset of musculoskeletal disease is common in oncology patients, with the appearance of areas of painful contraction and associated functional limitation (the so-called trigger points). Their deactivation by infiltration of local anesthetic or dry needling can provide adequate pain control, and the administration of botulinum toxin should be considered in cases in which a sustained effect is not achieved over time.

Radiotherapy in the treatment of pain from bone metastases

Several studies have demonstrated the usefulness of radiotherapy in the treatment of pain caused by bone metastases, offering pain relief in 50%–80% of patients and complete pain control in up to one-third of cases. Its effect is sustained over time, maintaining the analgesic effect for more than a year in most patients and even permanently in some cases. This clinical improvement is accompanied by a decrease in analgesic requirements. It has not been possible to clearly determine the superiority of some regimens over others, although it seems that those in which high doses of radiotherapy are applied could be associated with longer periods of relief compared with shorter cycles.[45,46]

Complementary treatments

A wide variety of alternative treatments can be taken into consideration for the treatment of cancer patients with pain. Physiotherapy can be useful in the management of these patients, especially in cases in which the progression of the disease or its complications lead to prolonged immobilization. Other options include techniques such as acupuncture, aromatherapy, hypnosis, massage, music therapy, reflexology, or relaxation techniques, although none of these have been proven to be effective.

Conclusions

Pain is frequently present in the different phases of colorectal cancer, markedly affecting the physical and psychological well-being of the patient, as well as their own perception of the course of the disease. It is therefore essential that the staff in charge of following-up patients be alert to the possible onset of pain and make a conscious effort to control it. There is a wide arsenal of drugs aimed at the treatment of both baseline pain and breakthrough pain peaks, depending on the characteristics and origin of the pain. The combination of different techniques within a multimodal therapy has been shown to improve the degree of analgesia obtained and to reduce analgesic needs and associated side effects.

References

1. Vallano A, Payrulet P, Malouf J, Baños JE, Grupo catalán de Investigación del Dolor Hospitalario. Estudio multicéntrico de la evaluación del dolor en el medio hospitalario. *Rev Esp Anestesiol Reanim* 2007;**54**(3):140–6.
2. Dura Navarro R, de Andrés JA. Bibliografía de evidencias clínicas sobre la prevención de síndromes de dolor crónico post-quirúrgicos. *Rev Esp Anestesiol Reanim* 2004;**51**:205–12.
3. Montes A, Aguilar JL, Benito C, Caba F, Margarit C. Management of postoperative pain in Spain: a nationwide survey of practice. *Acta Anaesthesiol Scand* 2017;**61**(5):480–91. https://doi.org/10.1111/aas.12876.
4. Ortega JL, Neira F. Etiopatogenia y efectos funcionales del dolor postoperatorio. In: En Torres LM, editor. *Tratamiento del dolor postoperatorio*. Madrid: Ergón; 2003. p. 3–29.

5. Aguilar JL, Montes A, Benito C, Caba Fy, Margarit C. Manejo farmacológico del dolor agudo postoperatorio en España. Datos de la encuesta nacional de la Sociedad Española del Dolor (SED). *Rev Soc Esp Dolor* 2018;**25**(2):70–85. https://doi.org/10.20986/resed.2017.3593/2017.
6. Vickers A, Bali S, Baxter A, Bruce G, England J, Heafield R, et al. Consensus statement on the anticipation and prevention of acute postoperative pain: multidisciplinary RADAR approach. *Curr Med Res Opin* 2009;**25**(10):2557–69.
7. García-Miguel FJ, Serrano-Aguilar PG, López-Bastida J. Preoperative assessment. *Lancet* 2003;**362**(9397):1749–57.
8. Aliaga L. Dolor agudo y postoperatorio. In: *Teoría y práctica*. Caduceo Multimedia SL; 2005.
9. Baker DW. Statement of pain management: understanding how joint commission standards address pain. *Jt Comm Perspect* 2016;**36**(6):10–2.
10. Buunen M, Veldkamp R, Hop WC, Kuhry E, Jeekel J, Haglind E, et al. Survival after laparoscopic surgery versus open surgery for colon cancer: long-term outcome of a randomised clinical trial. *Lancet Oncol* 2009;**10**(1):44–52.
11. Chou R, Gordon DB, de Leon-Casasola OA, Rosenberg JM, Bickler S, Brennan T, et al. Management of postoperative pain: a clinical practice guideline from the American Pain Society, the American Society of Regional Anesthesia and Pain Medicine, and the American Society of Anesthesiologists'. Committee on Regional Anesthesia, Executive Committee, and Administrative Council. *J Pain* 2016 Feb;**17**(2):131–57.
12. Feroci F, Lenzi E, Baraghini M, Garzi A, Vannucchi A, Cantafio S, et al. Fast-track colorectal surgery: protocol adherence influences postoperative outcomes. *Int J Colorectal Dis* 2013 Jan;**28**(1):103–9.
13. Catalá E, Ferrandiz M, Genové M, editors. *Manual de tratamiento del dolor*. 2ª edicion. Barcelona: publicaciones Permanyer; 2008.
14. Ramón JM My. *Guía de Dolor Agudo Postoperatorio*. Editorial de la SED; 2011.
15. Pöpping DM, Elia N, Van Aken HK, Marret E, Schug SA, Kranke P, et al. Impact of epidural analgesia on mortality and morbidity after surgery: systematic review and meta-analysis of randomized controlled trials. *Ann Surg* 2014;**259**(6):1056–67. https://doi.org/10.1097/SLA.0000000000000237.
16. Guay J, Nishimori M, Kopp SL. Epidural local anesthetics versus opioid-based analgesic regimens for postoperative gastrointestinal paralysis, vomiting, and pain after abdominal surgery: a cochrane review. *Anesth Analg* 2016;**123**(6):1591–602.
17. Roeb MM, Wolf A, Gräber SS, Meißner W, Volk T. Epidural against systemic analgesia: an international registry analysis on postoperative pain and related perceptions after abdominal surgery. *Clin J Pain* 2017;**33**(3):189–97. https://doi.org/10.1097/AJP.0000000000000393.
18. Hübner M, Blanc C, Demartines N. Reply to letter: "does thoracic epidural analgesia impede recovery after laparoscopic colorectal surgery?". *Ann Surg* 2016;**264**(2):e9–e10. https://doi.org/10.1097/SLA.0000000000001243.
19. Ljungqvist O, Scott M, Fearon K. Enhanced recovery after surgery: a review. *JAMA Surg* 2017;**152**(3):292–8. https://doi.org/10.1001/jamasurg.2016.4952.
20. Esteve Pérez N, Mora Fernández C. Analgesia epidural postoperatoria. *Rev Soc Esp Dolor* 2018;**25**(1):1–3. https://doi.org/10.20986/resed.2018.3654/2018.
21. Mugabure BB, González SS, Uría AA, Conejero MG, y González JN. Coadyuvantes farmacológicos con efecto ahorrador de opioides en el periodo perioperatorio. *Rev Soc Esp Dolor* 2018;**25**(4):278–90. https://doi.org/10.20986/resed.2018.3663/2018.
22. Rawal N. Current issues in postoperative pain management. *Eur J Anaesthesiol* 2016 Mar;**33**(3):160–71.
23. Tan M, Law LS, Gan TJ. Optimizing pain management to facilitate enhanced recovery after surgery pathways. *Can J Anaesth* 2015 Feb;**62**(2):203–18.
24. Manworren RC. Multimodal pain management and the future of a personalized medicine approach to pain. *AORN J* 2015;**101**(3):308–14.
25. Esteve Pérez N, Sansaloni Perelló C, Verd Rodríguez M, Leclerc R, Y Mora Fernández C. Nuevos enfoques en el tratamiento del dolor agudo postoperatorio. *Rev Soc Esp Dolor* 2017;**24**(3):132–9. https://doi.org/10.20986/resed.2017.3542/2016.
26. Perez-Guerrero AC, Aragón MC, y Torres LM. Dolor postoperatorio: ¿hacia dónde vamos? *Rev Soc Esp Dolor* 2017;**24**(1):1–3. https://doi.org/10.20986/resed.2017.3566/2017.
27. Rodríguez Manuel J, Antoni C, Ramón G-EJ, Ramón José María My, María R. *Valoración y manejo del dolor. Guías Clínicas de la Sociedad Española del Dolor*. Arán ediciones; 2006.
28. Pedrazzani C, Menestrina N, Moro M, Brazzo G, Mantovani G, Polati E, et al. Local wound infiltration plus transversus abdominis plane (TAP) block versus local wound infiltration in laparoscopic colorectal surgery and ERAS program. *Surg Endosc* 2016;**30**(11):5117–25. https://doi.org/10.1007/s00464-016-4862-5.
29. Ventham NT, Hughes M, O'Neill S, Johns N, Brady RR, Wigmore SJ. Systematic review and meta-analysis of continuous local anaesthetic wound infiltration versus epidural analgesia for postoperative pain following abdominal surgery. *Br J Surg* 2013 Sep;**100**(10):1280.
30. Börjeson S, Starkhammar H, Unosson M, Berterö C. Common symptoms and distress experienced among patients with colorectal cancer: a qualitative part of mixed method design. *Open Nurs J* 2012;**6**:100–7.
31. Portenoy RK. Cancer pain. Epidemiology and syndromes. *Cancer* 1989;**63**(11):2298–307.
32. Foley KM. *Acute and chronic cancer pain syndromes: Oxford textbook of palliative medicine*. 3rd ed. New York: Oxford University Press; 2004. p. 298.
33. Boland EG, Ahmedzai SH. Persistent pain in cancer survivors. *Curr Opin Support Palliat Care* 2017;**11**(3):181–90.
34. Peterson DE, Bensadoun RJ, Roila F. Management of oral and gastrointestinal mucositis: ESMO Clinical Practice Guidelines. *Oncologia* 2011;**22**(Suppl. 6):vi78–84.
35. Araujo AM, Gómez M, Pascual J. Tratamiento del dolor en el paciente oncológico. *An Sist Sanit Navar* 2004;**27**(3):63–75.
36. Portenoy RK. Treatment of cancer pain. *Lancet* 2011;**377**:2236–47.
37. Fallon M, Giusti R, Aielli F, Hoskin P, Rolke R, Sharma M, et al. Management of cancer pain in adult patients: ESMO Clinical Practice Guidelines. *Ann Oncol* 2018;**29**(Suppl. 4):iv166–91.

38. Portenoy RK. Adjuvant analgesics in pain management. In: Hanks G, Cherny NI, Christakis N, Fallon M, Kaasa S, Portenoy RK, editors. *Textbook of palliative medicine*. 4th ed. Oxford: Oxford University Press; 2010. p. 361.
39. Fallon MT. Neuropathic pain in cancer. *Br J Anaesth* 2013;**111**(1):105–11.
40. Vidal MA, Medina C, Torres LM. Seguridad de los bifosfonatos. *Rev Soc Esp Dolor* 2011;**1**:43–55.
41. Cánovas L, Rodríguez AB, Ml C. Tratamiento del dolor irruptivo. *Rev Soc Esp Dolor* 2012;**19**(6):318–24.
42. De Courcy JG. Interventional techniques for cancer pain management. *Clin Oncol (R Coll Radiol)* 2011;**23**(6):407–17.
43. Deer TR, Hayek SM, Pope JE, Lamer TJ, Hamza M, Grider JS, et al. The polyanalgesic consensus conference (PACC): recommendations for trialing of intrathecal drug delivery infusion therapy. *Neuromodulation* 2017;**20**(2):133–54.
44. Smith TJ, Staats PS, Deer T. Randomized clinical trial of an implantable drug delivery system compared with comprehensive medical management for refractory cancer pain: impact on pain, drug-related toxicity, and survival. *J Clin Oncol* 2002;**20**(19):4040–9.
45. Berk LS, Chang E, Chow E. Palliative radiotherapy for bone metastases: an ASTRO evidence-based guideline. *Int J Radiat Oncol Biol Phys* 2011;**79**(4):965–76.
46. Chow E, Zeng L, Salvo N, et al. Update on the systematic review of palliative radiotherapy trials for bone metastases. *Clin Oncol (R Coll Radiol)* 2012;**24**(2):112–24.

Section E

Microbiota, molecular and biological mechanisms of CRC

Chapter 44

The role of intestinal microbiota in the colorectal carcinogenesis

Alejandra Cardelle-Cobas[a], Beatriz I. Vázquez[a], José Luis Ulla Rocha[b], Carlos N. Franco[a], Margarita Poza[c], Nieves Martínez Lago[d], and Luis M. Antón Aparicio[d]

[a]Department of Analytical Chemistry, Nutrition and Bromatology, University of Santiago de Compostela, Lugo, Spain, [b]Digestive Disease Unit, Pontevedra University Hospital Complex, Pontevedra, Spain, [c]Microbiology Research Group, University Hospital Complex (CHUAC)—Institute of Biomedical Research (INIBIC), University of A Coruña (UDC), A Coruña, Spain, [d]Medical Oncology Service, A Coruña University Hospital (CHUAC), A Coruña, Spain

Introduction

Human gut microbiota

In general, when the scientific community refers to gut microbiota, it is referring to bacteria, as for the last few years, the focus of the field has largely been on the bacterial members of the microbiota. However, human intestinal microbiota consist of approximately 100 trillion microorganisms, including bacteria, archaea, fungi, and other unicellular eukaryotes and, in some cases, helminths as well as families of viruses (Fig. 44.1). All of them build a highly complex network of interactions between each other and the host, becoming a relevant aspect of human health.[1–3]

Microbial colonization runs in parallel with immune system maturation and plays a key role in intestinal physiology and regulation. The composition of the microbiota is rather stable along the length of the gut, but the absolute number of microorganisms varies considerably between the mouth and the rectum[4] (Fig. 44.2A).

The gut microbiota differs among individuals. In the last few years, the question about what is the first inoculum in the human being has been subject of controversy. Thus one of the theories contact suggests that the placenta could be the origin,[5,6] having this inoculation taking place in an intrauterine way, before birth. However, recent publications seem to indicate the contrary, namely, the placenta in healthy pregnancies, is not a microbial reservoir, being sterile.[7,8] These two theories continue to be discussed by the scientific community leaving the question about the first human inoculum without answer. The second theory defends that the first inoculum is produced at the time of birth from the maternal microbiota. These two theories continue to be discussed by the scientific community leaving the question about the first human inoculum without answer. As the first theory is controversial, the second one is the one that, for the moment, has more strength. Taking this theory as valid, the composition of the first inoculum, from the maternal microbiota, would depend of the mode of delivery (vaginal microbiota, skin, etc.), drugs administration, and later from the diet (breastfeeding, formula, etc.). Other perinatal factors such as genetics and intestinal mucin glycosylation all contribute to influence microbial colonization.[9] Microbiota development is the result of interactions between physiological process in the host and microorganisms that are introduced from the environment[10–12] (Fig. 44.2B).

The microbial diversity increases and converses toward an adult-like microbiota by the end of the first 3–5 years of life. At this point, the microbiota stabilizes and maintains a consistent composition, despite some fluctuations throughout adulthood in response to environmental, developmental, and pathological events.[9,13] In the elderly, the microbiota composition changes gradually but can maintain similar physiological functions; however, it has been reported to show greater interindividual variation than adults and certain studies have also shown a less diverse microbiota in this sector of the population.[14]

Some works talk about "a part" of the microbiota which adopt a more stable role within the individuals, known as the "core microbiota," whereas there is another part fluctuant in its composition, depending on such variables as age, geography, and other environmental factors.[15]

The early acquisition of a diverse and balanced microbiota is likely critical for the development and maturation of a healthy immune system. Shifts in this complex microbial system have been reported to increase the risk of disease.

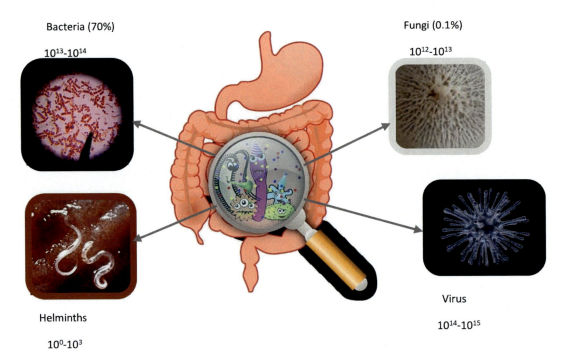

FIG. 44.1 Human intestinal organisms.

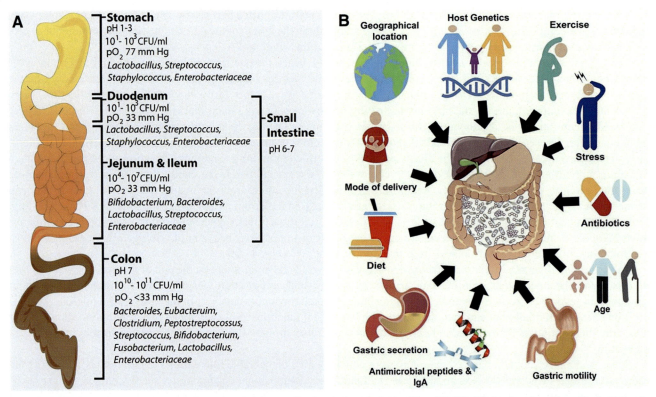

FIG. 44.2 Microbiota composition along the GIT (A) and factors affecting the intestinal microbiota (B). *(Modified and traduced from Clarke G, Sandhu KV, Griffin BT, Dinan TG, Cryan JF, Hyland NP. Gut reactions: breaking down xenobiotic–microbiome interactions. Pharmacol Rev 2019;71(2): 198–224. https://doi.org/10.1124/pr.118.015768.)*

Diet and human gut microbiota

Diet plays an important role in microbiota composition (Fig. 44.3), since it is the entry point for most of the exogenous factors accessing the digestive tract. The intestinal microbiota is the key for the nutrient processing and, in consequence, for the synthesis of important metabolites. Metabolites are necessary for the immune processes and the generation of important signals for the cellular function. Microbiota, mucosal epithelial cells, foodborne probiotic components, and small molecules including hormones, enzymes, mucus, and bile salts constitute a complex intestinal microecosystem, which plays an important role in the host immunity, and may influence numerous critical biological processes in carcinogenesis, including the balance between cell proliferation and death.

The research studies developed to date show that different patterns of dietary consumption are responsible for key changes in the microbiota; thus, if nutritional dietary interventions are introduced to individuals with a Western diet toward a consumption pattern rich in fruits and vegetables, the composition of the microbiota is changed.[16] The research studies developed to date show that different patterns of dietary consumption are responsible for key changes in the microbiota; thus, if nutritional dietary interventions are introduced to individuals with a Western diet toward a consumption pattern rich in fruits and vegetables, the composition of the microbiota change toward a more heathy and equilibrated profile.[16] Specially if a reduction in the consumption of red meat is produced since that reduction will decrease the sulfur-reducing bacteria which are promoters of cell proliferation.[17]

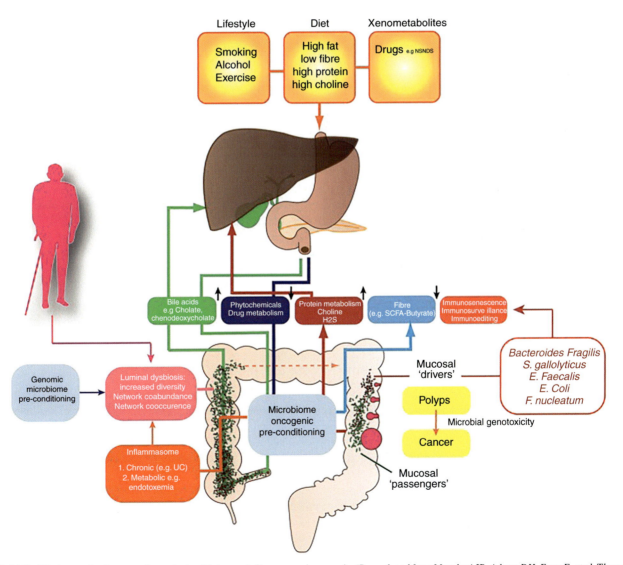

FIG. 44.3 The interaction between diet and microbiota may influence carcinogenesis. *(Reproduced from Marchesi JR, Adams DH, Fava F, et al. The gut microbiota and host health: a new clinical frontier. Gut 2016;65(2):330–339. https://doi.org/10.1136/gutjnl-2015-309990.)*

Conversely, an increase in fiber consumption seems to modify the microbiota composition in order to favor the proliferation of short-chain fatty acids (SCFAs) producing bacteria.[18] These SCFAs directly modulate host health through a range of tissue-specific mechanisms related to gut barrier function, glucose homeostasis, immunomodulation, appetite regulation, and obesity[19]: functions that exert a protecting effect against CRC. Additionally, a high consumption of omega-3 fatty acids present in fish seems to cause changes in the composition of our microbiota favoring the growth of these bacterial species producing SCFAs.[20]

Similarly, an increase in the consumption of fermented milk products is associated with the decrease of the bacteria *Bilophila wadsworthia*, which causes an increase in the production of SCFAs of proven beneficial effect.[21]

In addition, recent evidences seem to indicate that known risk factors for the appearance of CRC can act through changes generated in the microbiota. Thus obesity induces changes in the composition of the microbiota and through epigenetic changes, mainly through the mechanism of acetylation that would induce changes in DNA, could trigger CRC[22]; these changes are reversible when an effective weight loss is achieved.

Excessive consumption of alcohol is another factor that increases the risk of CRC. It seems to favor the growth of Proteobacteria and Actinobacteria which facilitate the permeability of the intestinal barrier,[23] but also the increase in conversion of ethanol to acetaldehyde, which may play a key role in possible carcinogenesis; this process is reversible when the increase in alcohol consumption is reversed.

A more unexplored but equally interesting part is that of diet-microbiome-host interactions in relation to its nonbacterial members: viruses, fungi, archaea, protozoa, and multicellular parasites, as well as the complex network of interdependencies between kingdoms within the intestinal microbiota. The communication between kingdoms could occur through the host through malabsorption, inflammation, or bleeding, or through syntrophism, so that waste products from one microorganism nourish another; for example, the bacterium *Bacteroides thetaiotaomicron* is able to use the yeast mannan.[24]

The human virome shows a high degree of interpersonal stability over time.[25] Small-scale studies in humans revealed that alterations in the normal development of the human gastrointestinal virome could be related to malnutrition in the neonatal life and that the human virome can change after alterations in the content of fat, sugar, and fiber in the diet.[26] Another study in mice suggested that these changes are more pronounced in the mucosal-associated virome than in the luminal one.[27] Nutritional deficiency can exert selective pressure on virome members to directly affect the host; for example, selenium deficiency triggered genomic evolution in a virulent strain of Coxsackievirus, which caused myocarditis in a study in mice.[28] In addition, dietary modulation of the viral repertoire can influence the host through the integration of bacteriophage chromosomes into bacterial genomes, thus altering the composition[26] and functionality[29,30] of the bacterial microbiota. Until now, it has been shown that this mechanism affects bacterial virulence factors; however, its ability to alter bacterial metabolism and its subsequent effects on the host deserve further investigation.

Smoking is another known risk factor, which seems to trigger an increase in the concentrations of Bacteroides-Prevotella, which affect changes in the mucin composition of the wall of the digestive tract[31] and it seems that these changes are reversible when the smoking ceases.

On the contrary, a protective mechanism against the appearance of CRC is physical exercise. It is considered that variations in the physical activity level of an individual play a role in the composition of their microbiota, increasing the butyrate-producing bacteria, a SCFA.[32]

Advanced technologies for the human gut microbiota analysis

Although microbiota studies often include an ecological component, it is noteworthy that most of the research performed to date has focused on bacteria and not all the micro- and macro-organisms present in the gut. This represents a logical approach because bacteria comprise most of the microbiome. However, even the microorganisms representing a small proportion of the microbiome may play important roles in the ecosystem.[33] Therefore researchers need to start shifting their approach to include eukaryotic, prokaryotic, and viral interactions in efforts to elucidate the roles of all components of the microbiota.[33,34]

Around a decade ago, most knowledge about the adult human gut microbiota stemmed from labor-intensive culture-based methods.[35] The advances produced through culture-independent approaches such as high-throughput and low-cost sequencing methods allowed an increase in the knowledge of the bacteria living in our intestinal tract. Targeting of the bacterial 16S ribosomal RNA (rRNA) gene is the most popular approach since this gene is present in all bacteria and archaea and contains nine highly variable regions (V1-V9) which allow species to be easily distinguished. Although the first studies developed were based on sequencing the entire 16S rRNA gene, recently, the focus of 16S rRNA sequencing has shifted to analyzing shorter subregions of the gene in greater depth[36]; however, the utilization of shorter read lengths can introduce errors.[37] More reliable estimates of microbiota composition and diversity may be provided by whole-genome shotgun metagenomics, due to the higher resolution and sensitivity of these techniques.[37]

Two larger projects, the MetaHit and the Human Microbiome Project, have provided the most comprehensive view of the human-associated microbial repertoire to date.[38, 39] Compiled data from these studies identified 2172 species isolated from human beings, classified into 12 different phyla, of which 93.5% belonged to *Proteobacteria*, *Firmicutes*, *Actinobacteria*, and *Bacteroidetes*. Three of the 12 identified phyla contained only one species isolated from humans, including an intestinal species, *Akkermansia muciniphila*, the only known representative of the Verrucomicrobia phyla. In humans, 386 of the identified species are strictly anaerobic and hence will generally be found in mucosal regions such as the oral cavity and the gastrointestinal (GI) tract.[38]

Indeed, the human microbiota also contains other more neglected components such as archaea, viruses, fungi, and other Eukarya (such as *Blastocystis* and *Amoebozoa*).[40, 41] Despite the fact that little is still known about commensal fungi, archaea, and protozoa,[40, 41] some emerging microbiological data on yeast composition and functions have clarified their subsequent clinical use in the modulation of gut microflora. In this sense, *Saccharomyces boulardii* is currently used with significant efficacy over placebo in the treatment of postinfectious and postantibiotic diarrhoea.[42] This yeast has shown to be effective in the prevention of infections produced by *Clostridium difficile*.[43]

Regarding viruses, knowledge on the composition of the gut virome has evolved from a niche of the gut microbiome populated only by pathogens (e.g., Norwalk, Rotavirus, Enterovirus, etc.) to an enlarged list of undetectable giant viruses (derived mainly from protozoa and parasites), and more recently to plant-derived viruses and bacteriophages, thanks to new metagenomic methods.[41, 44]

Recent approaches to determine the mycobiome focus on the analysis of the nuclear ribosomal Internal Transcribed Spacers (ITS), ITS1 and ITS2 regions employing Next Generation Sequencing (NGS) methodologies.[45–47]

The most important challenge for microbiome analysis is to homogenize and validate the methods used, starting from the extraction of nucleotides protocol, and to improve the Next-Generation Sequencing libraries.

All this advance in the knowledge of microbiome composition, made basically to the new sequencing methods, has allowed deepening in the role that this community has in the development of CRC and nowadays became in a new and interesting tool not only for the diagnosis but also for the prevention and treatment of this pathology.

Human gut microbiota and colorectal cancer

Geographic variability of the incidence of CRC highly suggests the involvement of certain environmental risk factors, such as high fat diets, obesity, or living in a Western country.[48] In fact, multiple epidemiological studies have suggested that excessive animal protein and fat intake, especially red meat and processed meat, could increase the risk of developing CRC, while fiber could protect against colorectal tumorigenesis.[49] Moreover, Knudson's two-hit hypothesis suggests that host factors play an important role in the predisposition to carcinogenesis. In this scenario, a second environmental hit can lead to uncontrolled cellular proliferation.[50]

In recent decades, growing attention has been given to the role of microbial infection in carcinogenesis and microbes are suspected to be involved in approximately a 20% of cancers, especially CRC. It is well known that in digestive cancers that some pathogens, such as *Helicobacter pylori*, have been directly and strongly linked to gastric cancer; however, the possible role of infection in CRC remains controversial.

The colon is colonized by approximately 10^3 different microbial species. Bacteria are present in 10^{14} ufc/g and represent the 70% of the host's microorganisms.[36, 37, 51, 52] Although viruses can be present at higher concentrations, to date, there are no studies indicating the percentage in the total of the microbiota. Only some studies indicate that the number of virus seems to be in a ratio 10:1 with regard to the number of bacterial cells.[53] The microbiota, as indicated before, can be divided according to location in the gut. Specifically, microbes in the lumen are referred to "luminal flora," whereas microbes that penetrate the mucosal layer overlying the intestinal epithelium are referred to "mucosa-associated flora."[52] Indeed, thick mucus layers protect enterocytes from excessive exposure to microorganisms and dietary antigens along the length of the intestine, particularly in the colon; thus, preventing hypersensitivity responses.[38]

When the microbiota equilibrium is altered, the intestinal epithelium and the mucus may be altered and be more exposed to pathogenic microorganisms and other aggressive factors that can damage the colon and be the beginning of the development of CRC. Once the tumor is initiated, its interaction with its local microenvironment and its systemic effects on the host are critical for the development and treatment of CRC. Emerging evidence has indicated that these microbes may induce inflammation, facilitate cell proliferation, and provide a microenvironment for host cells to alter stem cell dynamics and produce metabolites that affect glycolysis or the immune response. The molecular mechanism of the microbiota remains to be characterized.

Over the last 20 years, there have been a series of landmarks in the possible correlation of the microbiota and the CRC.

Thus, in 2001, alterations in the composition of the microbiota are discovered in patients with CRC. Genomic analyses identified a link between the *Fusobacterium* family and the presence of CRC, with this species being more frequently detected in carcinogenic tissues than in healthy tissues.[54]

In 2013, the knowledge about how the microbiota is able to modulate tumorigenesis was expanded. Researchers from the University of Michigan revealed that by extracting intestinal microbiota samples from mice with CRC and transplanting it into germ-free mice, the development of double the number of tumors was higher in these mice than in their homonymous mice free transplanted with fecal microbiota from healthy mice.[55]

A significant breakthrough appeared in 2014 when a European team was able to predict the presence of CRC only using stool samples through metagenomic sequencing of 22 bacterial species, obtaining the same precision than a blood test for early detection.[56] This finding is particularly promising given that its further development will allow early diagnosis of the presence of CRC through noninvasive samples such as a simple analysis of stool samples.

Particularly interesting are the latest advances on how the microbiota can influence the response in patients treated with immunotherapy, specifically with the PD1 checkpoint inhibitors. Those patients with greater bacterial diversity had a higher response rate to this type of immunotherapy.[57]

Due to all, these data indicating the strong relationship between the human gut microbiota and the CRC, the destruction of its microecological balance has become a hot spot in the study of the pathogenesis of CRC and may provide a new direction for its treatment, besides gene mutation and genetic factors.[58]

Bacteria

Bacterial taxa associated to colorectal cancer

As noted, CRC is a complex association of tumoral cells, nonneoplastic cells, and a large amount of microorganisms. The involvement of the microbiota in colorectal carcinogenesis is becoming increasingly clear and numerous studies have reported many changes in the bacterial composition of the gut microbiota in CRC, suggesting a major role of dysbiosis in colorectal carcinogenesis (as we can see in Fig. 44.4). In this sense, some bacterial species have been identified and

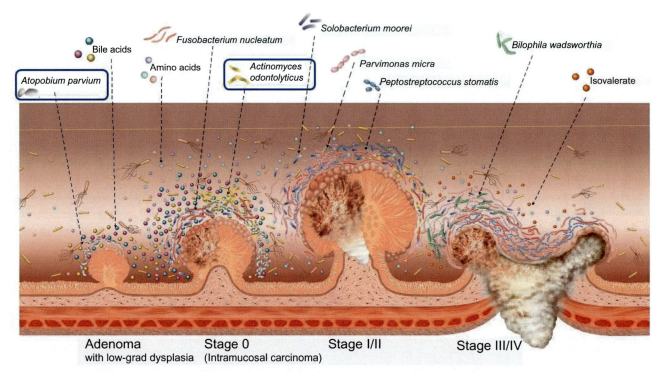

FIG. 44.4 Graphical representation of major microbial and metabolomic alterations during multistep CRC progression. *(Reproduced from Yachida S, Mizutani S, Shiroma H, et al. Metagenomic and metabolomic analyses reveal distinct stage-specific phenotypes of the gut microbiota in colorectal cancer. Nat Med 2019;25(6):968–976. https://doi.org/10.1038/s41591-019-0458-7: con permiso de Springer Nature.)*

suspected of playing a role in colorectal carcinogenesis, such as *Streptococcus bovis/gallolyticus, Helicobacter pylori, Bacteroides fragilis, Enterococcus faecalis, Clostridium septicum, Fusobacterium* spp., and *Escherichia coli*. The studies developed to date indicate that these bacteria could be used as biomarkers of a prognosis factor in CRC.

Fusobacterium nucleatum: This is an anaerobic Gram-negative pathogenic bacterium. It is a common member of the oral microbiota and possess a symbiotic activity with the host. However, when its presence increases it is associated with periodontitis. Under disease conditions, *F. nucleatum* is one of the most prevalent species found in extra-oral sites[59] and recently, many researchers have demonstrated that *F. nucleatum* is related to CRC development and pathogenicity.[54] *F. nucleatum* is a heterogeneous species with five proposed subspecies (subsp.), i.e., subsp. *animalis*, subsp. *fusiforme*, subsp. *nucleatum*, subsp. *polymorphum*, and subsp. *vincentii*, whose prevalence in disease vary. A recent work has shown that in individuals with CRC, the same subspecies were found both in saliva and in CRC samples.[60] Pathogenicity of *F. nucleatum* can be related to the release of enterotoxins, which stimulate the immune response causing alteration in intestinal epithelial cells.

Escherichia coli: Although *Escherichia coli* is a commensal bacteria of the human microbiota and represents the most common cultivable, Gram-negative, aero-anaerobic bacteria in the gut, various studies have demonstrated a clear link between mucosa-adherent *Escherichia coli* and CRC.[61, 62]

Some studies have reported higher levels of colonic colonization by mucosa-associated *Escherichia coli* in patients with CRC compared to healthy patients.[61] A study developed in 2004 by Martin et al. reported than more than 70% of mucosa samples from patients with CRC were colonized by bacteria, generally, *Escherichia coli*. A recent study about the intestinal bacteria detected in CRC and adjacent tissue shows that *Escherichia coli* was the bacteria more prevalent in stage III/IV of CRC compared with stage I and an association was identified between poor prognosis of CRC and mucosal colonization by *Escherichia coli*.[63] These results suggest that *Escherichia coli* could serve as a prognostic factor in CRC.[64]

Bacteroides fragilis: *B. fragilis* are common anaerobic bacteria that are detected in up to 80% of children and adults but which represent less than 1% of gut microbiota.[65] This commensal bacterium in the gut can be classified into two subtypes: nontoxigenic *B. fragilis* and enterotoxigenic *B. fragilis*.[65, 66] The latter is known to be responsible for diarrheal illnesses in humans, especially children. To date, different studies have reported increased colonic colonization by *B. fragilis* in patients presenting CRC. Most of the enterotoxigenic *B. fragilis* strains detected in colorectal mucosal samples harbored the *bft* gene, which encodes the metalloprotease toxin (*B. fragilis* toxin or BFT).

The involvement of this toxin in colorectal carcinogenesis is becoming evident as it affects pathways that lead to alterations in intestine epithelial cells,[67] increase cell proliferation, the epithelial release of proinflammatory effectors, and DNA damage in in vitro and colorectal-predisposed mouse models.[68] These preneoplastic CRC effects may occur in association with the action of other microorganisms such as *F. nucleatum*.

Clostridium septicum: *Clostridium septicum* is a rare cause of bacteremia. Its infections have been clinically linked to CRC.[69] Forty percent of patients presenting with *Clostridium septicum* infections present with some colorectal malignancies. The hypothetical mechanism is that the microenvironment of the tumor favors the germination of *Clostridium septicum* spores via ingestion of contaminated food. No direct involvement of *Clostridium septicum* in colorectal carcinogenesis has been well defined.

Helicobacter pylori: *H. pylori* has been classified as a carcinogen of the GI tract by the international Agency for Research of Cancer. The role of *H. pylori* in colorectal carcinogenesis remains controversial due to the differences with respect to the *H. pylori* strains and their specific virulence factors. Although the role of *H. pylori* in gastric cancer has been better studied and described than its involvement in colorectal carcinogenesis, some hypothesis can be extrapolated from the pathophysiology of bacterial-linked gastric cancer.

Some *H. pylori* strains can produce certain cytotoxins, which induce the activation of inflammation pathways and cellular proliferation in gastric cancer. Seropositivity to these strains (CAgA and VacA positive) in patients is related to an increased risk of CRC.[70]

Some studies also indicate that the direct production of pro-oxidative reactive oxygen and nitrogen species by some *H. pylori* strains could be involved in the development of CRC.

Streptococcus bovis/gallolyticus: This bacterium was the first bacteria associated with CRC and the studies carried out to date suggest the involvement of *Streptococcus bovis* at an early step of CRC development, which could be a good option as a biomarker in early detection of CRC.

Enterococcus faecalis: *Enterococcus faecalis* as much other bacteria is a facultative anaerobic commensal bacterium of the oral cavity and the GI tract. However, in 2004, in a study published by Pillar and Gillmore it emerged as a human pathogen. Different studies have indicated significant higher population of this bacterium in patients with CRC compared to healthy controls.[71, 72]

Metastases-associated bacteria

All the studies developed to date have shown that tumorous tissues possess a microbial environment different to the adjacent and normal tissues, and that this microenvironment may favor the proliferation of the tumor. In addition, the studies revealed that primary and metastatic microbial populations are most closely related in composition to one another. In this sense, a recent study developed at Harvard University[73] has found evidence that suggests a certain type of bacteria found in colon cancer tumors makes its way to tumors in other body parts by traveling with the metastasizing cells. Their studies show that patients containing *F. nucleatum* in their colon tumor cells also had the bacteria in the metastasized liver tumor cells. However, patients with no evidence of bacteria in their tumor cells did not show the presence of bacteria in the metastasized ones. *F. nucleatum* in the colon travels together with the tumor and aides in the colonization in the new site.

This study also revealed that the use of antibiotics could aid in the treatment of cancer since the bacteria associated with the cancer cells could be creating a microenvironment favorable to its proliferation.

Although more studies are necessary, these finds open the opportunity of developing new strategies for the treatment of cancer by targeting the tumor-associated bacteria.

Therapeutic strategies

In addition to its role in carcinogenesis, the gut microbiota have demonstrated an important role in the response to cancer therapy since both the microbiota and its environment may or may not contribute to therapeutic efficacy.

Published studies have implicated the gut microbiota in influencing response, as well toxicity, across a range of treatments, including chemotherapy, immune checkpoint blockade, and stem cell transplant, via a variety of proposed mechanisms, although a deeper understanding is clearly needed and is the focus of current studies.[74]

The studies developed to date have shown that patients submitted to immunotherapy with checkpoint blockade have different responses to treatment depending on their microbiota. Thus it has been shown that this "good response" could be transmitted to the patients with a bad response by manipulation of their microbiota. A good response was related with an enrichment in the phyla Actinobacteria and Firmicutes whereas a "none or bad response" was associated with Proteobacteria and Bacteroidetes.

This means that manipulation of gut microbiota with specific bacterial taxa could enhance therapeutic response. The mechanisms continue to be elucidated but data suggest that these could include the interaction of microbial components or products with antigens, innate effectors, etc. to cytokine produced by lymphocytes, among others.[74]

Based on these studies there is a tremendous interest in developing an optimal combination of bacteria to use in combination with checkpoint blockade to improve the efficacy of the treatment.

Using bacteria for colorectal cancer treatment

Bacteria as anticancer agents through released substances: Bacteria secrete substances such as bacteriocins which can inhibit the growth of tumors. Bacteriocins are cationic peptides synthesized ribosomally by almost all groups of bacteria. Their main advantages are that they are nonimmunogenic, biodegradable, and contain specific toxicities against cancer cells. Some of the studies developed to date have shown that bacteriocins have the potential to serve as synergistic agents to conventional cancer drugs.[75] On the other hand, cancer cell membranes predominantly carry a negative charge, thus bacteriocins preferentially bind to cancer cell membranes rather than normal cell membranes, which are neutral in charge and selective for binding of bacteria.[76] Some of these bacteriocins are colicins (Enterobacteriaceae, e.g.,: *Escherichia coli*, *Klebsiella pneumoniae*), Pediocins (*Pediococcus acidilactici* K2a2-3), Nisins (*Lactobacillus lactis*), Pyocin (90% *Pseudomonas aeruginosa* strains).

Bacteria as anticancer agents through biofilms: Notwithstanding the etiopathogenesis of biofilms and their protective role that allows bacteria to escape from the host defense system,[77] recent discoveries have revealed the potential ability and efficacy of biofilms in cancer therapy.

Anticancer drugs cause the induction of biofilm formation during cancer treatment, which results in metastasis distraction. Similarly, the formation of bacteria biofilm on cancer cells during the SOS response results in metastasis disruption. Thus bacteria biofilm shows the potential usefulness in cancer treatment.[77] Bacterial biofilm can affect colon cancer development and progression through modifying cancer metabolism to produce a regulator of cellular proliferation.[78] Also, the bacterial macromolecules necessary for biofilm formation, such as proteins and DNA, coat cancer cells to block metastasis.[79] For example, polysaccharides released by *Streptococcus agalactiae* inhibit adhesion of cancer cells to endothelial cells, an essential step in cancer metastasis.[80] Furthermore, another study carried out in 2016 demonstrated the potential application of iron oxide nanowires from a biofilm waste produced by bacteria (*Mariprofundus ferrooxydans*) as a new multifunctional drug carrier for cancer therapy and cancer hyperthmia.[81] While the earlier hypotheses avow for the potential of bacterial biofilm in cancer therapy, the evidence is relatively insufficient to build-up the case. However, the efficacy of bacteria biofilm for metastasis distraction calls for the further examination and investigation of the anticancer ability of bacteria biofilm[82] and it is necessary to carry out a critical evaluation of the tests carried out in this area before designing and carrying out experiments to test the hypothesis.[83]

Bacteria as a carrier for cancer therapeutic agents: Apart from their direct anticancer effect, tumor-targeting bacteria can also be used as carriers for cancer therapeutic agents in cancer treatments. Recent studies have revealed that bacteria are capable of targeting both primary tumors and metastasis.[84-86]

Bacteriobots: In all the previous cases, the bacteria can be vehiculated through bacteria-based microrobot (bacteriobot).[82] The bacteriobot technique is a new innovative theranostic (or more personalized medicine) methodology of bacteria-based fabrication for tumor therapy. The technique uses bacteria that, together with the use of microelectromechanical and nano- and biotechnology techniques, allow the bacteria to protect themselves from the immune system. The technique uses bacteria as microactuators and microsensors to deliver microstructures for targeting and treating solid tumors.[87] This also protects the patient from the harmful effects of the direct administration of the bacteria.

Bacteria are the double-edged sword in cancer therapy. Using bacteria for cancer therapy is feasible and its potential to treat solid tumors has been known for decades but they have always been the subject of controversy.[88,89] The clinical application of this therapy never became routine because of the adverse uncontrollable side effects.[17] In an effort to overcome these side effects, some attenuated species of bacteria capable of treating cancer have been recently identified and studied.[90] These species of bacteria are considered safe for cancer therapeutic application with little or no side effects. While bacteria alone may not demonstrate fully therapeutic potential, their modifications as antitumor agents, antioncogenes, or immunogenic antigens, and their combination with other therapeutic processes will improve their potential for cancer therapy. The arena of using bacteria as an anticancer agent is still new; further studies are imperative to scrutinize the clinical significance of bacteria-based cancer therapy.[17] Aspects related to the dose, the schedule, and the route of administration among others are aspects that must be elucidated in this new strategy, based on bacteria, of cancer therapy.[90]

Bacteria targeting treatment in colorectal cancer

Targeting the intestinal microbiota

The strategies to target the gut microbiota in the context of cancer therapy include the aspects shown in Fig. 44.5.

- **Targeted modulation**: This could be achieved by using antibiotic therapies or bacteriophages. The main advantages include a focused approach, fewer off-target effects whereas one the problems or disadvantages associated would be the lack of knowledge about the optimal targets.
- **Diet and lifestyle**: It is already known that changes in gut microbiota can be produced in a relatively short timeframe through intense dietary changes.[91] These changes not only involve composition changes but also transcriptomic and metabolomic profile changes. For that, diet, together with the administration of other microbiota modulators (pre- and/or probiotics) may be considered in the approach to cancer therapies.
- **Use of pre-, pro-, and postbiotics**: Beyond diet, pre-, pro-, and postbiotics could be used for the modulation of microbiota with the objective of improving cancer therapies. Although more studies are necessary, there is more and more evidence about the effectiveness of these agents in modeling the microbiota in CRC patients[92,93] and for example, augmenting the effect of multiple common chemotherapeutic agents as well as radiotherapy in murine models[94] and in the prevention of CRC. However, no optimal targets are yet defined.

  The impact of probiotic administration in patients with cancer has been studied in different clinical trials; however, these assays were mainly addressed as microbiota modification and not an improvement in the efficacy of cancer therapies. Some studies have been developed with the main objective of establishing if probiotic administration could reduce the toxicity and the side effects of treatments, some of them showing success. It is clear that through probiotics microbiota modification is produced, but more studies are necessary to find adequate consortium of bacteria that can, through their impact on the microbiota, modulate it with the objective of ameliorating therapies.

  Another of the most important aspects to address is also the safety of the use of probiotics in these patients. Although all of them are considered as safe (GRAS status or Generally Recognized as Safe, according to the Federal Food Drug and Cosmetic Act), as living microorganisms they are, the main concerns regarding their use are related to translocation, bacteremia, and sepsis, and in this sense, cases have already been described in critical and immunocompromised patients after the use of probiotics, as is the case of some strains of Lactobacillus[95] and others.[96]
- **Fecal microbiota transplantation**: Although it seems that the use of fecal microbiota transplantation is a modern methodology to treat some gastrointestinal diseases, the truth is that it was already in use 2000 years ago to treat severe

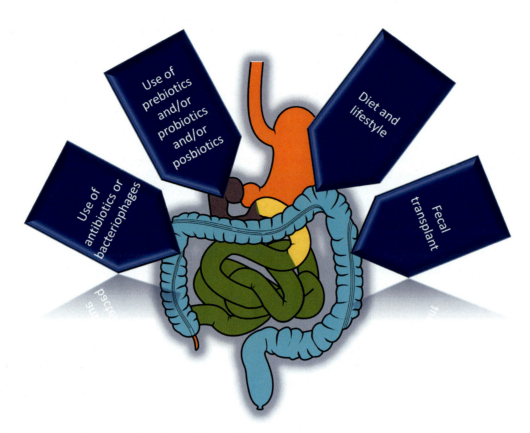

FIG. 44.5 Therapeutic strategies to target the gut microbiota in CRC.

diarrheas.[74] Currently, it is successfully used in patients with resistant *Clostridium difficile* infection and recently, there have been some preclinical trials involving patients with different kinds of cancer (melanoma, acute myeloid leukemia).[97] Although no report exists to date in CRC this could be a good alternative to modulate the microbiota. Added to the success in the treatment of *Clostridium difficile*, the efficacy showed in preclinical trials and that the modification is more or less stable with time, the only disadvantage is apparently, the nonscalability.

Targeting the tumor microbiota

All the later strategies indicated before have been addressed to modulate the intestinal microbiota with the main objective of preventing or influencing cancer therapy; however, recently, the scientific and medical efforts have addressed targeting the tumor microbiota with the main objective of not letting the cancer progress and enhancing responses to cancer therapy. Different studies have shown a deleterious effect of intratumoral bacteria on the therapeutic response in CRC.[98, 99]

The use of antibiotic in these studies with the objective of attenuating or eliminating this effect has shown an improvement in responses to chemotherapy as well as immune checkpoint blockade.

A study developed in 2017[73] showing the presence of *F. nucleatum* both in the primary tumor and in metastases also revealed that the use of antibiotics could ameliorate the treatment of cancer. The problem associated with the use of antibiotics in treating the tumor microbiota is that antibiotics would also affect the intestinal microbiota. It is for that that new alternatives are being studied with the objective of targeting the tumor microbiota. Some of them include the use of the predilection of some bacteria to home in on tumors, such as was explained before, in "Using bacteria for colorectal cancer treatment" section.

Fungi

The little-known human mycobiota

Fungi are also a part of the flora that colonizes humans, yet most scientific studies of human microbiota focus on evaluating the activity of bacterial communities alone. Specifically, if in a database as well known as Scopus a search is made for scientific articles related to the key words "microbiota" versus "mycobiota" and "microbiome" versus "mycobiome" it is found that those referring to fungi represent 2.7% and 1.4%, respectively, of the published works. In fact, the term "mycobiome" appears for the first time in an article published in 2010.[100]

Although the mechanisms that cause a fungus to move from mutualism to commensalism and from there to parasitism are not well known, it is generally considered that its presence is a risk factor leading to chronic colonization and safe damage to the host.[101] However, in a recent clinical trial by Jiang et al.[102] in mice that had been given a broad spectrum of antibiotics to treat dextran sulfate sodium (DSS)-induced colitis, it was observed that commensal fungi such as *Saccharomyces cerevisiae* and *Candida albicans* present in the intestine are able to replace the protective benefits exerted by bacteria that were eliminated and help overcome the disease. Therefore what is clearly demonstrated in recent studies is that fungi can influence both the individual's disease and health,[103] so they cannot be ignored.

Mycobiota is the part (less than 0.1%) of the microbiota made up of fungal species. Fungal communities can be found in the mouth,[104] nose,[105] ear canals,[106] lungs,[107] in the vaginal area[108] and of course, on the surface of the skin[109,110] but also within the GI tract.[111] Some fungi are already acquired at birth and colonize mucous membranes and the skin,[112] but most species come from the environment.[101]

Recent research reveals the importance of fungal activity in both the homeostatic balance and the host immune system.[113–115] However, the findings related to fungi-bacteria interactions in both their symbiotic and competitive aspects[116] are also very interesting. Furthermore, fungal species not only communicate with each other but also communicate with bacteria or viruses[111] and it seems that these signals could travel from communities of the intestine to, for example, those of the lungs.[117] In this sense, there are also researchers who are interested in the possible influence of mycobiome on neuronal diseases.[118]

Thanks to modern NGS sequencing techniques, it has been possible to show that the mycobiota profile is altered in individuals suffering from diseases such as inflammatory bowel disease,[119] Crohn's disease, autoimmune diseases,[101] and there is evidence that it is also involved in some way in CRC. Although it has not been possible to elucidate the meaning of these alterations in the mycobiome of those patients, it is suspected that a deep knowledge of it could, on the one hand, help to determine the truthful composition of mycobiome in health and, on the other hand, potentially lead to future strategies for the treatment of these diseases.

Mycobiota dysbiosis in colorectal cancer

In this emerging field of study on mycobiota, there are few publications yet in this regard but they show very interesting findings. It is found that in CRC there is also dysbiosis in the mycobiome of the gut, that is, there is a deviation of the balance of the fungal community with respect to that presented by a healthy individual.

The dysbiosis was observed by Luan et al.[120] when comparing the mycobiome found in 27 biopsy samples of colorectal adenomas with that found in biopsies of adjacent healthy tissues. The adenomas presented a greater presence of opportunistic pathogenic fungi at the level of phyla, genus, and species. Statistical comparison of relative abundance revealed the greatest difference in phylum. Glomeromycota ($P=25$), as well as *Phoma*, a known opportunistic pathogen genus (54% in adenomas versus 39% adjacent tissue) and Candida (7% in adenomas versus 1% adjacent tissue). The most abundant species found in the 54 samples was *Candida tropicalis* (3.6% in adenomas versus 2.1% adjacent tissue), a species also related to severe ulcerative colitis. Although at these levels the *t*-test did not justify that these differences were statistically significant, they were at the level of the Operational Taxonomic Unit (OTU). The analysis of main components revealed that even OTUs form clusters that separate advanced and nonadvanced adenoma states according to the American Society for Gastrointestinal Endoscopy (ASGE) guideline.

These findings are consistent with those reported by Gao et al.[121] highlighting the lowest fungal diversity already present in Chinese subjects with colon polyps ($n=29$) and the differences between the composition of mycobiome between patients with CRC at TNM stage II and III ($n=74$) compared to healthy individuals ($n=28$).

Showing a more recent example, Coker et al.[122] analyzed the mycobiome present in stool samples from 73 patients with CRC (47 male, 26 female, 65.90 average age, 24.07 average BMI) and 92 control subjects (51 male, 41 female, 58.51 average age, 23.87 average BMI). The results of the unsupervised multivariate statistical analysis revealed that different gut mycobiome discriminated healthy control subjects and CRC patients into two significantly distinct groups ($P=.001$ and $P=.0006$, respectively), including the supervised redundancy analysis which revealed that the fungal profile formed different clusters for control subjects ($P<.0001$) and early stage CRC (TNM stage I and II, $n=20$) and late stage CRC (TNM stage III and IV, $n=54$) ($P=.0048$).

When the mycobiome was assigned in a particular taxonomic group, differences were found at all levels between cases and controls. For example at the level of phyla, it was observed that in CRC patients the ratio *Basidiomycota*: *Ascomycota* was higher than in the controls ($P = .0042$), that also the presence of the class Malasseziomycetes increased as did that of the families Pisolithaceae, Marasmiaceae, Malasseziaceae, Erysiphaceae, Psedorotiaceae, and Chaetomicaceae and the genera Malassezia, Moniliophthora, Rhodotorula, Acremonium, Thielaviopsis, and Pisolithus.

However, the challenge for these investigators still remains to elucidate what effect a particular specie of fungus, present on the gut, has on the promotion and/or on the evolution of CRC.

Mycobiota as a possible CRC marker

Coker et al.[122] were able to selected 14 fungal species as possible CRC metagenomic markers. When comparing the presence of 38 fungal species between controls ($n = 92$), CRC patients ($n = 73$), and early stage CRC patients ($n = 20$), they realized that 14 of them showed an absolute log2 fold change greater than one. These markers classified controls and CRC with an Area Under the Curve (AUC) of 0.93 and also distinguished controls from early CRC with an AUC of 0.91. These species were *Aspergillus flavus, A. ochraceoroseus, A. rambellii, A. sydowii, Debaromyces fabryi, Kwoniella mangrovensis, Kwoniella heavenensis, Malassezia globose, Moniliophthora perniciosa, Nosemia apis, Pneumocystis murina, Pseudogymnoascus* sp. VKM F-4518 and VKM F-4520, *Talaromyces islandicus*.

These markers were subsequently validated in different ethnic cohorts. When used in an independent Asian cohort V1 (Chinese individuals, control subjects $n = 112$, CRC $n = 111$ and patients with early stages CRC $n = 61$) an AUC of 0.82 was obtained between controls and patients with CRC and a 0.81 AUC between controls and patients with adenoma. When used in an independent European cohort V2 (German and French individuals, controls $n = 66$, CRC $n = 90$ and CRC in early stages $n = 32$), these markers were also able to classify them separately with an AUC of 0.74 between controls/CRC and with an AUC 0.72 for CRC controls/early stage CRC. In contrast, only controls of adenoma patients could be distinguished with an AUC of 0.63 for the Chinese cohort V1 (adenoma patients $n = 197$) and with an AUC of 0.60 for the European cohort V2 (adenoma patients $n = 42$).

Against this background, it seems very interesting to continue carrying out these types of studies in order to be able to use the fecal fungal community as a possible complementary tool for CRC diagnosis.

Virus

Most studies about microbiota and CRC have focused, as it has been shown in this chapter, almost exclusively on bacteria and not on the viruses. However, recent studies seem that the initiator of the chain of events that eventually leads to the appearance of a CRC is a group of unidentified viruses.

According to previous evidence on pathogenic viruses, the human gut harbors plant-derived viruses, giant viruses and, only recently, abundant bacteriophages. New metagenomic methods have allowed the reconstitution of entire viral genomes from the genetic material spread in the human gut, opening new perspectives on the understanding of the gut virome composition, the importance of gut microbiome, and potential clinical applications.[123] This last subset of findings has been mostly unexpected because of the common representation of the gut virome as a source of pathogens. *Enteroviruses*, Norwalk, *Rotaviruses* are well known in daily clinical practice and are known to be responsible for common infectious gastroenteritis.[124] However, because gut viruses are not amenable to culture with common microbiological techniques, the development of nonculture-based metagenomic methods has allowed the reconstitution of viral particles from single genetic sequences from almost every environment. This has moved our idea of gut viruses from a mere source of pathogens to a physiological component of the healthy human microbiota.

Considering CRC, a study developed in 2018 showed as a chronic infection in humans caused by a pathogenic intestinal virus in nonvaccinated patients, such as in the case of persons developing poliomyelitis, is inversely associated with the development of CRC.[125] A possible explanation could be that this poliovirus may be occupying an ecological niche that prevents the development of the potential oncogenic virus responsible for CRC.

Gradually researchers have been trying to identify a possible relationship between known oncogenic viruses so far and their possible association with CCR. In 2014 a systematic review found no genome of the major oncogenic viruses known until then in tumor tissue.[126] Currently, only 1% of the human virome is characterized. In Table 44.1, a classification of the intestinal viruses is shown.

Some researchers have been able to exclude recently that the JC virus is responsible or has a direct relationship with the appearance of RCC.[127]

Also in recent years, some researchers have found that certain bacteriophage viruses are associated with the presence of CRC and possibly have an impact on tumor progression because they can alter the bacterial communities of the microbiota.[128] In this way, viruses could act well indirectly by controlling and modifying those harmful bacterial populations.

TABLE 44.1 Virus communities within the human gut.

Gut bacteriophages	
Mostly double-stranded and simple-stranded DNA phages:	
Myoviridae, Podoviridae, Siphoviridae, Inoviridae, and Microviridae	
DNA viruses:	
Double-stranded	Single-stranded
Adenoviridae	Anelloviridae
Herpesviridae	Circoviridae
Iridoviridae	
Marseilleviridae	
Mimiviridae	
Papillomaviridae	
Polyomaviridae	
Poxviridae	
RNA viruses	
Double-stranded	Single-stranded
Picobimaviridae	Caliciviridae
Reoviridae	Astroviridae
	Virgaviridae
	Picornaviridae
	Retroviridae
	Togaviridae
Definitive pathogenic eukaryotic viruses infecting the gut	
Rotavirus, norovirus, astrovirus, adenovirus (serotipos 40 and 41), enterovirus (only adenovirus is DNA virus, rest are all RNA viruses)	

Modified from Mukhopadhya I, Segal JP, Carding SR, Hart AL, Hold GL. The gut virome: the "missing link" between gut bacteria and host immunity? *Therap Adv Gastroenterol* 2019;*12*:1756284819836620. https://doi.org/10.1177/1756284819836620.

An added problem is that viruses exist in different genetic forms, so they differ in their structural composition of nucleic acids, they can be DNA or RNA, they differ in the number of chains, single or double strands, and they can also present a positive sense or negative in its composition in terms of spatial orientation. This complexity confers an important challenge when configuring a sequencing strategy. Viruses present in the intestinal tract constitute a very extensive virome, in such a way, that it is estimated that they exceed in a 10:1 ratio the number of intestinal bacterial cells, which gives an idea of how extremely complex their study is.[53]

It is estimated that 95% of the microbiota virome composition is currently unknown, which is referred to as dark matter.[129] Bacteria and phage viruses have evolved together over millions of years and their interaction with the individual is extremely complex.[130]

The study of the human intestinal virome needs to be standardized in terms of the samples to choose and the most appropriate controls to carry out the different studies, so far these limitations are difficult to ignore and we can consider that we are still starting at very early stages of the study of the virology of the human intestine.

Gradually, researchers have been trying to identify a possible relationship between the oncogenic viruses known so far and their possible association with CRC. In 2014 a systematic review found no genome of the major oncogenic viruses known until then in tumor tissue.[126]

Plasmids

Based on epidemiological studies carried out worldwide in the areas with the highest incidence of CRC, a higher risk of cancer has been found in populations with a high intake of beef or beef products obtained from cows. This has led to speculations about the possibility of transmissible infectious agents that may be present in these products.[131]

These transmitted viral infections through foods from bovine animals, specifically red meat; already in early life could end up triggering long-term CRC even decades after the first infection.[132]

In recent years, some researchers have identified unknown infectious agents, which have been called plasmids, which would have evolved to cause an independent infection.[133] These plasmids are suspected to be present in cattle, specifically in the most common species in the western world, *Bos taurus*, in this way with the ingestion of food products containing such agents, could lead to chronic inflammation in the own lamina of the colon next to the crypts of Lieberkühn. Through this process, reactive oxygen species and nitrogen radicals would be generated that would induce long-term mutations. This process is not always inexorable since it can affect a series of factors that exist in the affected individual, mainly the quality of the immune system.[132]

In cases where negative variables such as alcohol intake or smoking or the presence of severe obesity are present, the process of carcinogenicity can be accelerated causing the appearance of neoplasia at younger ages.

The search for possible infectious agents involved in the genesis of CRC is intense in laboratories throughout the world.

References

1. Savage DC. Microbial ecology of the gastrointestinal tract. *Annu Rev Microbiol* 1977;**31**(1):107–33. https://doi.org/10.1146/annurev.mi.31.100177.000543.
2. Suau A, Bonnet R, Sutren M, et al. Direct analysis of genes encoding 16S rRNA from complex communities reveals many novel molecular species within the human gut. *Appl Environ Microbiol* 1999;**65**(11):4799–807. https://doi.org/10.1128/AEM.65.11.4799-4807.1999.
3. Rowan-Nash AD, Korry BJ, Mylonakis E, Belenky P. Cross-domain and viral interactions in the microbiome. *Microbiol Mol Biol Rev* 2019;**83**(1). https://doi.org/10.1128/MMBR.00044-18.
4. Neish AS. Microbes in gastrointestinal health and disease. *Gastroenterology* 2009;**136**(1):65–80. https://doi.org/10.1053/j.gastro.2008.10.080.
5. Aagaard K, Ma J, Antony KM, Ganu R, Petrosino J, Versalovic J. The placenta harbors a unique microbiome. *Sci Transl Med* 2014;**6**(237). https://doi.org/10.1126/scitranslmed.3008599, 237ra65.
6. Collado MC, Rautava S, Aakko J, Isolauri E, Salminen S. Human gut colonisation may be initiated in utero by distinct microbial communities in the placenta and amniotic fluid. *Sci Rep* 2016;**6**:23129. https://doi.org/10.1038/srep23129.
7. Kuperman AA, Zimmerman A, Hamadia S, et al. Deep microbial analysis of multiple placentas shows no evidence for a placental microbiome. *BJOG* 2020;**127**(2):159–69. https://doi.org/10.1111/1471-0528.15896.
8. Segata N. No bacteria found in healthy placentas. *Nature* 2019;**572**(7769):317–8. https://doi.org/10.1038/d41586-019-02262-8.
9. Rodríguez JM, Murphy K, Stanton C, et al. The composition of the gut microbiota throughout life, with an emphasis on early life. *Microb Ecol Health Dis* 2015;**26**. https://doi.org/10.3402/mehd.v26.26050.
10. Goncharova GI, Dorofeĭchuk VG, Smolianskaia AZ, Sokolova KI. Microbial ecology of the intestines in health and in pathology. *Antibiot Khimioter* 1989;**34**(6):462–6. https://www.ncbi.nlm.nih.gov/pubmed/2802880.
11. Dominguez-Bello MG, Blaser MJ, Ley RE, Knight R. Development of the human gastrointestinal microbiota and insights from high-throughput sequencing. *Gastroenterology* 2011;**140**(6):1713–9. https://doi.org/10.1053/j.gastro.2011.02.011.
12. Mulder IE, Schmidt B, Lewis M, et al. Restricting microbial exposure in early life negates the immune benefits associated with gut colonization in environments of high microbial diversity. *PLoS One* 2011;**6**(12). https://doi.org/10.1371/journal.pone.0028279, e28279.
13. Stanghellini V, Barbara G, Cremon C, et al. RETRACTED ARTICLE: gut microbiota and related diseases: clinical features. *Intern Emerg Med* 2010;**5**(S1):57–63. https://doi.org/10.1007/s11739-010-0451-0.
14. Claesson MJ, Jeffery IB, Conde S, et al. Gut microbiota composition correlates with diet and health in the elderly. *Nature* 2012;**488**(7410):178–84. https://doi.org/10.1038/nature11319.
15. Sirisinha S. The potential impact of gut microbiota on your health: current status and future challenges. *Asian Pac J Allergy Immunol* 2016;**34**(4):249–64. https://doi.org/10.12932/AP0803.
16. O'Keefe SJD, Li JV, Lahti L, et al. Fat, fibre and cancer risk in African Americans and rural Africans. *Nat Commun* 2015;**6**:6342. https://doi.org/10.1038/ncomms7342.
17. Song M, Chan AT. Environmental factors, gut microbiota, and colorectal cancer prevention. *Clin Gastroenterol Hepatol* 2019;**17**(2):275–89. https://doi.org/10.1016/j.cgh.2018.07.012.
18. Donohoe DR, Holley D, Collins LB, et al. A gnotobiotic mouse model demonstrates that dietary fiber protects against colorectal tumorigenesis in a microbiota- and butyrate-dependent manner. *Cancer Discov* 2014;**4**(12):1387–97. https://doi.org/10.1158/2159-8290.CD-14-0501.
19. Chambers ES, Preston T, Frost G, Morrison DJ. Role of gut microbiota-generated short-chain fatty acids in metabolic and cardiovascular health. *Curr Nutr Rep* 2018;**7**(4):198–206. https://doi.org/10.1007/s13668-018-0248-8.

20. Watson H, Mitra S, Croden FC, et al. A randomised trial of the effect of omega-3 polyunsaturated fatty acid supplements on the human intestinal microbiota. *Gut* 2018;**67**(11):1974–83. https://doi.org/10.1136/gutjnl-2017-314968.
21. Veiga P, Pons N, Agrawal A, et al. Changes of the human gut microbiome induced by a fermented milk product. *Sci Rep* 2015;**4**(1). https://doi.org/10.1038/srep06328.
22. Qin Y, Roberts JD, Grimm SA, et al. An obesity-associated gut microbiome reprograms the intestinal epigenome and leads to altered colonic gene expression. *Genome Biol* 2018;**19**(1):7. https://doi.org/10.1186/s13059-018-1389-1.
23. Mutlu EA, Gillevet PM, Rangwala H, et al. Colonic microbiome is altered in alcoholism. *Am J Physiol Gastrointest Liver Physiol* 2012;**302**(9):G966–78. https://doi.org/10.1152/ajpgi.00380.2011.
24. Cuskin F, Lowe EC, Temple MJ, et al. Human gut Bacteroidetes can utilize yeast mannan through a selfish mechanism. *Nature* 2015;**517**(7533):165–9. https://doi.org/10.1038/nature13995.
25. Reyes A, Haynes M, Hanson N, et al. Viruses in the faecal microbiota of monozygotic twins and their mothers. *Nature* 2010;**466**(7304):334–8. https://doi.org/10.1038/nature09199.
26. Minot S, Sinha R, Chen J, et al. The human gut virome: inter-individual variation and dynamic response to diet. *Genome Res* 2011;**21**(10):1616–25. https://doi.org/10.1101/gr.122705.111.
27. Kim M-S, Bae J-W. Spatial disturbances in altered mucosal and luminal gut viromes of diet-induced obese mice. *Environ Microbiol* 2016;**18**(5):1498–510. https://doi.org/10.1111/1462-2920.13182.
28. Beck MA, Shi Q, Morris VC, Levander OA. Rapid genomic evolution of a non-virulent Coxsackievirus B3 in selenium-deficient mice results in selection of identical virulent isolates. *Nat Med* 1995;**1**(5):433–6. https://doi.org/10.1038/nm0595-433.
29. Willner D, Furlan M, Schmieder R, et al. Metagenomic detection of phage-encoded platelet-binding factors in the human oral cavity. *Proc Natl Acad Sci U S A* 2011;**108**(Suppl. 1):4547–53. https://doi.org/10.1073/pnas.1000089107.
30. Wang X, Kim Y, Ma Q, et al. Cryptic prophages help bacteria cope with adverse environments. *Nat Commun* 2010;**1**:147. https://doi.org/10.1038/ncomms1146.
31. Biedermann L, Brülisauer K, Zeitz J, et al. Smoking cessation alters intestinal microbiota: insights from quantitative investigations on human fecal samples using FISH. *Inflamm Bowel Dis* 2014;**20**(9):1496–501. https://doi.org/10.1097/MIB.0000000000000129.
32. Cronin O, Barton W, Skuse P, et al. A prospective metagenomic and metabolomic analysis of the impact of exercise and/or whey protein supplementation on the gut microbiome of sedentary adults. *mSystems* 2018;**3**(3). https://doi.org/10.1128/mSystems.00044-18.
33. Mills S, Shanahan F, Stanton C, Hill C, Coffey A, Ross RP. Movers and shakers: influence of bacteriophages in shaping the mammalian gut microbiota. *Gut Microbes* 2013;**4**(1):4–16. https://doi.org/10.4161/gmic.22371.
34. Cadwell K. The virome in host health and disease. *Immunity* 2015;**42**(5):805–13. https://doi.org/10.1016/j.immuni.2015.05.003.
35. Moore WEC, Holdeman LV. Human fecal flora: the normal flora of 20 Japanese-Hawaiians. *Appl Microbiol* 1974;**27**(5):961–79. https://doi.org/10.1128/aem.27.5.961-979.1974.
36. Hakansson A, Molin G. Gut microbiota and inflammation. *Nutrients* 2011;**3**(6):637–82. https://doi.org/10.3390/nu3060637.
37. Marchesi JR. Human distal gut microbiome. *Environ Microbiol* 2011;**13**(12):3088–102. https://doi.org/10.1111/j.1462-2920.2011.02574.x.
38. Kelly D, Mulder IE. Microbiome and immunological interactions. *Nutr Rev* 2012;**70**:S18–30. https://doi.org/10.1111/j.1753-4887.2012.00498.x.
39. Chen W, Liu F, Ling Z, Tong X, Xiang C. Human intestinal lumen and mucosa-associated microbiota in patients with colorectal cancer. *PLoS One* 2012;**7**(6). https://doi.org/10.1371/journal.pone.0039743, e39743.
40. Mai V, Draganov PV. Recent advances and remaining gaps in our knowledge of associations between gut microbiota and human health. *World J Gastroenterol* 2009;**15**(1):81–5. https://doi.org/10.3748/wjg.15.81.
41. Lozupone CA, Stombaugh JI, Gordon JI, Jansson JK, Knight R. Diversity, stability and resilience of the human gut microbiota. *Nature* 2012;**489**(7415):220–30. https://doi.org/10.1038/nature11550.
42. Dinleyici EC, Eren M, Ozen M, Yargic ZA, Vandenplas Y. Effectiveness and safety of Saccharomyces boulardii for acute infectious diarrhea. *Expert Opin Biol Ther* 2012;**12**(4):395–410. https://doi.org/10.1517/14712598.2012.664129.
43. Tung JM, Dolovich LR, Lee CH. Prevention of Clostridium difficile infection with Saccharomyces boulardii: a systematic review. *Can J Gastroenterol* 2009;**23**(12):817–21. https://doi.org/10.1155/2009/915847.
44. Reyes A, Semenkovich NP, Whiteson K, Rohwer F, Gordon JI. Going viral: next-generation sequencing applied to phage populations in the human gut. *Nat Rev Microbiol* 2012;**10**(9):607–17. https://doi.org/10.1038/nrmicro2853.
45. Blaalid R, Kumar S, Nilsson RH, Abarenkov K, Kirk PM, Kauserud H. ITS1 versus ITS2 as DNA metabarcodes for fungi. *Mol Ecol Resour* 2013;**13**(2):218–24. https://doi.org/10.1111/1755-0998.12065.
46. Huseyin CE, Rubio RC, O'Sullivan O, Cotter PD, Scanlan PD. The fungal frontier: a comparative analysis of methods used in the study of the human gut mycobiome. *Front Microbiol* 2017;**8**:1432. https://doi.org/10.3389/fmicb.2017.01432.
47. Hoggard M, Vesty A, Wong G, et al. Characterizing the human mycobiota: a comparison of small subunit rRNA, ITS1, ITS2, and large subunit rRNA genomic targets. *Front Microbiol* 2018;**9**:2208. https://doi.org/10.3389/fmicb.2018.02208.
48. Alexander DD, Cushing CA, Lowe KA, Sceurman B, Roberts MA. Meta-analysis of animal fat or animal protein intake and colorectal cancer. *Am J Clin Nutr* 2009;**89**(5):1402–9. https://doi.org/10.3945/ajcn.2008.26838.
49. Yang J, Yu J. The association of diet, gut microbiota and colorectal cancer: what we eat may imply what we get. *Protein Cell* 2018;**9**(5):474–87. https://doi.org/10.1007/s13238-018-0543-6.
50. Hutchinson E. Alfred Knudson and his two-hit hypothesis. *Lancet Oncol* 2001;**2**(10):642–5. https://doi.org/10.1016/s1470-2045(01)00524-1.
51. Claesson MJ, Cusack S, O'Sullivan O, et al. Composition, variability, and temporal stability of the intestinal microbiota of the elderly. *Proc Natl Acad Sci U S A* 2011;**108**(Suppl._1):4586–91. https://doi.org/10.1073/pnas.1000097107.

52. Sekirov I, Russell SL, Antunes LCM, Brett FB. Gut microbiota in health and disease. *Physiol Rev* 2010;**90**(3):859–904. https://doi.org/10.1152/physrev.00045.2009.
53. Mukhopadhya I, Segal JP, Carding SR, Hart AL, Hold GL. The gut virome: the "missing link" between gut bacteria and host immunity? *Therap Adv Gastroenterol* 2019;**12**. https://doi.org/10.1177/1756284819836620, 1756284819836620.
54. Kostic AD, Gevers D, Pedamallu CS, et al. Genomic analysis identifies association of Fusobacterium with colorectal carcinoma. *Genome Res* 2012;**22**(2):292–8. https://doi.org/10.1101/gr.126573.111.
55. Zackular JP, Baxter NT, Iverson KD, et al. The gut microbiome modulates colon tumorigenesis. *MBio* 2013;**4**(6). https://doi.org/10.1128/mBio.00692-13, e00692–13.
56. Raskov H, Burcharth J, Pommergaard H-C. Linking gut microbiota to colorectal cancer. *J Cancer* 2017;**8**(17):3378–95. https://doi.org/10.7150/jca.20497.
57. Vetizou M, Trinchieri G. Anti-PD1 in the wonder-gut-land. *Cell Res* 2018;**28**(3):263–4. https://doi.org/10.1038/cr.2018.12.
58. Lin C, Cai X, Zhang J, et al. Role of gut microbiota in the development and treatment of colorectal cancer. *Digestion* 2019;**100**(1):72–8. https://doi.org/10.1159/000494052.
59. Han YW. Fusobacterium nucleatum: a commensal-turned pathogen. *Curr Opin Microbiol* 2015;**23**:141–7. https://doi.org/10.1016/j.mib.2014.11.013.
60. Komiya Y, Shimomura Y, Higurashi T, et al. Patients with colorectal cancer have identical strains of Fusobacterium nucleatum in their colorectal cancer and oral cavity. *Gut* 2019;**68**(7):1335–7. https://doi.org/10.1136/gutjnl-2018-316661.
61. Martin HM, Campbell BJ, Hart CA, et al. Enhanced Escherichia coli adherence and invasion in Crohn's disease and colon cancer. *Gastroenterology* 2004;**127**(1):80–93. https://doi.org/10.1053/j.gastro.2004.03.054.
62. Arthur JC, Perez-Chanona E, Mühlbauer M, et al. Intestinal inflammation targets cancer-inducing activity of the microbiota. *Science* 2012;**338**(6103):120–3. https://doi.org/10.1126/science.1224820.
63. Bonnet M, Buc E, Sauvanet P, et al. Colonization of the human gut by E. coli and colorectal cancer risk. *Clin Cancer Res* 2014;**20**(4):859–67. https://doi.org/10.1158/1078-0432.ccr-13-1343.
64. Gagnière J. Gut microbiota imbalance and colorectal cancer. *World J Gastroenterol* 2016;**22**(2):501. https://doi.org/10.3748/wjg.v22.i2.501.
65. Huang JY, Melanie Lee S, Mazmanian SK. The human commensal Bacteroides fragilis binds intestinal mucin. *Anaerobe* 2011;**17**(4):137–41. https://doi.org/10.1016/j.anaerobe.2011.05.017.
66. Boleij A, Hechenbleikner EM, Goodwin AC, et al. The Bacteroides fragilis toxin gene is prevalent in the colon mucosa of colorectal cancer patients. *Clin Infect Dis* 2015;**60**(2):208–15. https://doi.org/10.1093/cid/ciu787.
67. Purcell RV, Pearson J, Aitchison A, Dixon L, Frizelle FA, Keenan JI. Colonization with enterotoxigenic Bacteroides fragilis is associated with early-stage colorectal neoplasia. *PLoS One* 2017;**12**(2). https://doi.org/10.1371/journal.pone.0171602, e0171602.
68. Housseau F, Sears CL. Enterotoxigenic Bacteroides fragilis(ETBF)-mediated colitis in Min (Apc /−) mice: a human commensal-based murine model of colon carcinogenesis. *Cell Cycle* 2010;**9**(1):3–5. https://doi.org/10.4161/cc.9.1.10352.
69. Mirza NN, McCloud JM, Cheetham MJ. Clostridium septicum sepsis and colorectal cancer—a reminder. *World J Surg Oncol* 2009;**7**(1). https://doi.org/10.1186/1477-7819-7-73.
70. Shmuely H, Passaro D, Figer A, et al. Relationship between Helicobacter pylori CagA status and colorectal cancer. *Am J Gastroenterol* 2001;**96**(12):3406–10. https://doi.org/10.1111/j.1572-0241.2001.05342.x.
71. Wang T, Cai G, Qiu Y, et al. Structural segregation of gut microbiota between colorectal cancer patients and healthy volunteers. *ISME J* 2012;**6**(2):320–9. https://doi.org/10.1038/ismej.2011.109.
72. Balamurugan R, Rajendiran E, George S, Samuel GV, Ramakrishna BS. Real-time polymerase chain reaction quantification of specific butyrate-producing bacteria, Desulfovibrio and Enterococcus faecalis in the feces of patients with colorectal cancer. *J Gastroenterol Hepatol* 2008;**23**(8 Pt. 1):1298–303. https://doi.org/10.1111/j.1440-1746.2008.05490.x.
73. Bullman S, Pedamallu CS, Sicinska E, et al. Analysis of Fusobacterium persistence and antibiotic response in colorectal cancer. *Science* 2017;**358**(6369):1443–8. https://doi.org/10.1126/science.aal5240.
74. Helmink BA, Khan MAW, Hermann A, Gopalakrishnan V, Wargo JA. The microbiome, cancer, and cancer therapy. *Nat Med* 2019;**25**(3):377–88. https://doi.org/10.1038/s41591-019-0377-7.
75. Kaur S, Kaur S. Bacteriocins as potential anticancer agents. *Front Pharmacol* 2015;**6**:272. https://doi.org/10.3389/fphar.2015.00272.
76. Chumchalová J, Smarda J. Human tumor cells are selectively inhibited by colicins. *Folia Microbiol* 2003;**48**(1):111–5. https://doi.org/10.1007/BF02931286.
77. Komor U, Bielecki P, Loessner H, et al. Biofilm formation by Pseudomonas aeruginosa in solid murine tumors—a novel model system. *Microbes Infect* 2012;**14**(11):951–8. https://doi.org/10.1016/j.micinf.2012.04.002.
78. Adnan M, Khan S, Al-Shammari E, Patel M, Saeed M, Hadi S. In pursuit of cancer metastasis therapy by bacteria and its biofilms: history or future. *Med Hypotheses* 2017;**100**:78–81. https://doi.org/10.1016/j.mehy.2017.01.018.
79. Weitao T. Bacteria form biofilms against cancer metastasis. *Med Hypotheses* 2009;**72**(4):477–8. https://doi.org/10.1016/j.mehy.2008.11.012.
80. Miyake K, Yamamoto S, Iijima S. Blocking adhesion of cancer cells to endothelial cell types by S. agalactiae type-specific polysaccharides. *Cytotechnology* 1996;**22**(1–3):205–9. https://doi.org/10.1007/BF00353940.
81. Kumeria T, Maher S, Wang Y, et al. Naturally derived Iron oxide nanowires from Bacteria for magnetically triggered drug release and cancer hyperthermia in 2D and 3D culture environments: bacteria biofilm to potent cancer therapeutic. *Biomacromolecules* 2016;**17**(8):2726–36. https://doi.org/10.1021/acs.biomac.6b00786.

82. Song S, Vuai MS, Zhong M. The role of bacteria in cancer therapy—enemies in the past, but allies at present. *Infect Agent Cancer* 2018;**13**:9. https://doi.org/10.1186/s13027-018-0180-y.
83. Huang Q. Comment on "Bacteria form biofilms against cancer metastasis". *Med Hypotheses* 2010;**74**(1):203. https://doi.org/10.1016/j.mehy.2009.06.045.
84. Weibel S, Stritzker J, Eck M, Goebel W, Szalay AA. Colonization of experimental murine breast tumours by Escherichia coli K-12 significantly alters the tumour microenvironment. *Cell Microbiol* 2008;**10**(6):1235–48. https://doi.org/10.1111/j.1462-5822.2008.01122.x.
85. Min J-J, Kim H-J, Park JH, et al. Noninvasive real-time imaging of tumors and metastases using tumor-targeting light-emitting Escherichia coli. *Mol Imaging Biol* 2008;**10**(1):54–61. https://doi.org/10.1007/s11307-007-0120-5.
86. Min J-J, Nguyen VH, Kim H-J, Hong Y, Choy HE. Quantitative bioluminescence imaging of tumor-targeting bacteria in living animals. *Nat Protoc* 2008;**3**(4):629–36. https://doi.org/10.1038/nprot.2008.32.
87. Park SJ, Park S-H, Cho S, et al. New paradigm for tumor theranostic methodology using bacteria-based microrobot. *Sci Rep* 2013;**3**:3394. https://doi.org/10.1038/srep03394.
88. Payette PJ, Davis HL. History of vaccines and positioning of current trends. *Curr Drug Targets Infect Disord* 2001;**1**(3):241–7. https://doi.org/10.2174/1568005014606017.
89. Cann SAH, Van Netten JP, Van Netten C. Dr William Coley and tumour regression: a place in history or in the future. *Postgrad Med J* 2003;**79**(938):672–80. https://pmj.bmj.com/content/79/938/672.short.
90. Kramer MG, Masner M, Ferreira FA, Hoffman RM. Bacterial therapy of cancer: promises, limitations, and insights for future directions. *Front Microbiol* 2018;**9**:16. https://doi.org/10.3389/fmicb.2018.00016.
91. David LA, Maurice CF, Carmody RN, et al. Diet rapidly and reproducibly alters the human gut microbiome. *Nature* 2014;**505**(7484):559–63. https://doi.org/10.1038/nature12820.
92. Hibberd AA, Lyra A, Ouwehand AC, et al. Intestinal microbiota is altered in patients with colon cancer and modified by probiotic intervention. *BMJ Open Gastroenterol* 2017;**4**(1). https://doi.org/10.1136/bmjgast-2017-000145, e000145.
93. Gianotti L, Morelli L, Galbiati F, et al. A randomized double-blind trial on perioperative administration of probiotics in colorectal cancer patients. *World J Gastroenterol* 2010;**16**(2):167–75. https://doi.org/10.3748/wjg.v16.i2.167.
94. Taper HS, Roberfroid MB. Possible adjuvant cancer therapy by two prebiotics—inulin or oligofructose. *In Vivo* 2005;**19**(1):201–4. https://www.ncbi.nlm.nih.gov/pubmed/15796175.
95. Land MH, Rouster-Stevens K, Woods CR, Cannon ML, Cnota J, Shetty AK. Lactobacillus sepsis associated with probiotic therapy. *Pediatrics* 2005;**115**(1):178–81. https://doi.org/10.1542/peds.2004-2137.
96. Besselink MGH, van Santvoort HC, Buskens E, et al. Probiotic prophylaxis in predicted severe acute pancreatitis: a randomised, double-blind, placebo-controlled trial. *Lancet* 2008;**371**(9613):651–9. https://doi.org/10.1016/s0140-6736(08)60207-x.
97. Mohty M, Malard F, D'Incan E, et al. Prevention of dysbiosis complications with autologous fecal microbiota transplantation (auto-FMT) in acute myeloid leukemia (AML) patients undergoing intensive treatment (ODYSSEE study): first results of a prospective multicenter trial. *Blood* 2017;**130**(Suppl. 1):2624. https://ashpublications.org/blood/article/130/Supplement%201/2624/80304.
98. Geller LT, Barzily-Rokni M, Danino T, et al. Potential role of intratumor bacteria in mediating tumor resistance to the chemotherapeutic drug gemcitabine. *Science* 2017;**357**(6356):1156–60. https://doi.org/10.1126/science.aah5043.
99. Yu J, Feng Q, Wong SH, et al. Metagenomic analysis of faecal microbiome as a tool towards targeted non-invasive biomarkers for colorectal cancer. *Gut* 2017;**66**(1):70–8. https://doi.org/10.1136/gutjnl-2015-309800.
100. Ghannoum MA, Jurevic RJ, Mukherjee PK, et al. Characterization of the Oral fungal microbiome (mycobiome) in healthy individuals. *PLoS Pathog* 2010;**6**(1). https://doi.org/10.1371/journal.ppat.1000713, e1000713.
101. Hall RA, Noverr MC. Fungal interactions with the human host: exploring the spectrum of symbiosis. *Curr Opin Microbiol* 2017;**40**:58–64. https://doi.org/10.1016/j.mib.2017.10.020.
102. Jiang TT, Shao T-Y, Gladys Ang WX, et al. Commensal fungi recapitulate the protective benefits of intestinal bacteria. *Cell Host Microbe* 2017;**22**(6):809–816.e4. https://doi.org/10.1016/j.chom.2017.10.013.
103. Klimesova K, Jiraskova Zakostelska Z, Tlaskalova-Hogenova H. Oral bacterial and fungal microbiome impacts colorectal carcinogenesis. *Front Microbiol* 2018;**9**:774. https://doi.org/10.3389/fmicb.2018.00774.
104. Chandra J, Retuerto M, Mukherjee PK, Ghannoum M. The fungal biome of the oral cavity. *Methods Mol Biol* 2016;**1356**:107–35. https://doi.org/10.1007/978-1-4939-3052-4_9.
105. Jung WH, Croll D, Cho JH, Kim YR, Lee YW. Analysis of the nasal vestibule mycobiome in patients with allergic rhinitis. *Mycoses* 2015;**58**(3):167–72. https://doi.org/10.1111/myc.12296.
106. Singh TD, Dinesh Singh T, Sudheer CP. Otomycosis: a clinical and mycological study. *Int J Otorhinolaryngol Head Neck Surg* 2018;**4**(4):1013. https://doi.org/10.18203/issn.2454-5929.ijohns20182704.
107. Nguyen LDN, Viscogliosi E, Delhaes L. The lung mycobiome: an emerging field of the human respiratory microbiome. *Front Microbiol* 2015;**6**:89. https://doi.org/10.3389/fmicb.2015.00089.
108. Bradford LL, Ravel J. The vaginal mycobiome: a contemporary perspective on fungi in women's health and diseases. *Virulence* 2017;**8**(3):342–51. https://doi.org/10.1080/21505594.2016.1237332.
109. Bożena D-K, Iwona D, Ilona K. The mycobiome—a friendly cross-talk between fungal colonizers and their host. *Ann Parasitol* 2016;**62**(3):175–84. https://doi.org/10.17420/ap6203.51.
110. Jo J-H, Kennedy EA, Kong HH. Topographical and physiological differences of the skin mycobiome in health and disease. *Virulence* 2017;**8**(3):324–33. https://doi.org/10.1080/21505594.2016.1249093.

111. Hager CL, Ghannoum MA. The mycobiome: role in health and disease, and as a potential probiotic target in gastrointestinal disease. *Dig Liver Dis* 2017;**49**(11):1171–6. https://doi.org/10.1016/j.dld.2017.08.025.
112. Ward TL, Dominguez-Bello MG, Heisel T, Al-Ghalith G, Knights D, Gale CA. Development of the human mycobiome over the first month of life and across body sites. *mSystems* 2018;**3**(3). https://doi.org/10.1128/mSystems.00140-17.
113. Cui L, Morris A, Ghedin E. The human mycobiome in health and disease. *Genome Med* 2013;**5**(7):63. https://doi.org/10.1186/gm467.
114. Ghannoum M. The mycobiome. *Sci Mag* 2016;**30**(2):32–7.
115. Iliev ID, Underhill DM. Striking a balance: fungal commensalism versus pathogenesis. *Curr Opin Microbiol* 2013;**16**(3):366–73. https://doi.org/10.1016/j.mib.2013.05.004.
116. Krüger W, Vielreicher S, Kapitan M, Jacobsen ID, Niemiec MJ. Fungal-bacterial interactions in health and disease. *Pathogens* 2019;**8**(2). https://doi.org/10.3390/pathogens8020070.
117. Shibuya A, Shibuya K. Exploring the gut fungi-lung allergy axis. *Cell Host Microbe* 2018;**24**(6):755–7. https://doi.org/10.1016/j.chom.2018.11.012.
118. Forbes JD, Bernstein CN, Tremlett H, Van Domselaar G, Knox NC. A fungal world: could the gut mycobiome be involved in neurological disease? *Front Microbiol* 2018;**9**:3249. https://doi.org/10.3389/fmicb.2018.03249.
119. Ott SJ, Kühbacher T, Musfeldt M, et al. Fungi and inflammatory bowel diseases: alterations of composition and diversity. *Scand J Gastroenterol* 2008;**43**(7):831–41. https://doi.org/10.1080/00365520801935434.
120. Luan C, Xie L, Yang X, et al. Dysbiosis of fungal microbiota in the intestinal mucosa of patients with colorectal adenomas. *Sci Rep* 2015;**5**:7980. https://doi.org/10.1038/srep07980.
121. Gao R, Kong C, Li H, et al. Dysbiosis signature of mycobiota in colon polyp and colorectal cancer. *Eur J Clin Microbiol Infect Dis* 2017;**36**(12):2457–68. https://doi.org/10.1007/s10096-017-3085-6.
122. Coker OO, Nakatsu G, Dai RZ, et al. Enteric fungal microbiota dysbiosis and ecological alterations in colorectal cancer. *Gut* 2019;**68**(4):654–62. https://doi.org/10.1136/gutjnl-2018-317178.
123. Scarpellini E, Ianiro G, Attili F, Bassanelli C, De Santis A, Gasbarrini A. The human gut microbiota and virome: potential therapeutic implications. *Dig Liver Dis* 2015;**47**(12):1007–12. https://doi.org/10.1016/j.dld.2015.07.008.
124. Khetsuriani N, LaMonte-Fowlkes A, Steven Oberste M, Pallansch MA. *Enterovirus surveillance—United States, 1970-2005. PsycEXTRA Dataset*. Published online; 2006. https://doi.org/10.1037/e540562006-001.
125. Lehrer S, Rheinstein PH. Inverse relationship between polio incidence in the US and colorectal cancer. *In Vivo* 2018;**32**(6):1485–9. https://doi.org/10.21873/invivo.11404.
126. Fiorina L, Ricotti M, Vanoli A, et al. Systematic analysis of human oncogenic viruses in colon cancer revealed EBV latency in lymphoid infiltrates. *Infect Agent Cancer* 2014;**9**:18. https://doi.org/10.1186/1750-9378-9-18.
127. Navand AH, Teimoori A, Makvandi M, et al. Study on JV virus in patients with colon cancer type adenocarcinoma. *Asian Pac J Cancer Prev* 2019;**20**(4):1147–51. https://doi.org/10.31557/apjcp.2019.20.4.1147.
128. Hannigan GD, Duhaime MB, Ruffin 4th MT, Koumpouras CC, Schloss PD. Diagnostic potential and interactive dynamics of the colorectal cancer virome. *MBio* 2018;**9**(6). https://doi.org/10.1128/mBio.02248-18.
129. Handley SA, Devkota S. Going viral: a novel role for bacteriophage in colorectal cancer. *MBio* 2019;**10**(1). https://doi.org/10.1128/mBio.02626-18.
130. Mirzaei MK, Maurice CF. Ménage à trois in the human gut: interactions between host, bacteria and phages. *Nat Rev Microbiol* 2017;**15**(7):397–408. https://doi.org/10.1038/nrmicro.2017.30.
131. Zur Hausen H, de Villiers E-M. Dairy cattle serum and milk factors contributing to the risk of colon and breast cancers. *Int J Cancer* 2015;**137**(4):959–67. https://doi.org/10.1002/ijc.29466.
132. Zur HH. Cancers in humans: a lifelong search for contributions of infectious agents, autobiographic notes. *Annu Rev Virol* 2019;**6**(1):1–28. https://doi.org/10.1146/annurev-virology-092818-015907.
133. de Villiers E-M, de Villiers E-M, Gunst K, et al. A specific class of infectious agents isolated from bovine serum and dairy products and peritumoral colon cancer tissue. *Emerg Microbes Infect* 2019;**8**(1):1205–18. https://doi.org/10.1080/22221751.2019.1651620.

Chapter 45

Genetic susceptibility to CRC

Ceres Fernández-Rozadilla[a,b], Anael López-Novo[a,b], Ángel Carracedo[a,b,c,d], and Clara Ruiz-Ponte[a,b,c,d]

[a]Genomic Medicine Group, Santiago de Compostela University (USC), Santiago de Compostela, Spain, [b]Health Research Institute of Santiago de Compostela (IDIS), Santiago de Compostela, Spain, [c]Galician Public Foundation of Genomic Medicine, Santiago de Compostela, Spain, [d]Center for Biomedical Research Network on Rare Diseases (CIBERER), Barcelona, Spain

Introduction

The lifetime risk of developing colorectal cancer (CRC) in the general population is about 5%. The etiology of CRC, like other common diseases, is that of a complex disease. This implies that the disease is caused by the interaction between various environmental (low-fiber diets, red meat consumption, obesity, alcohol consumption, or smoking), epigenetic, and genetic factors.[1] Regarding the relevance of the latter, twin studies have estimated heritability in CRC to be 40%.[2] This heritability is represented by multiple genetic variants ranging from rare variants with high penetrance to common variants in the population that confer a small effect on the risk of developing the disease.[3]

Mendelian syndromes with strong familial aggregation, such as Familial Adenomatous Polyposis and Lynch syndrome, are caused by variants of the first type in genes such as *APC*, *MUTYH*, and DNA repair genes (*MLH1*, *MSH2*, *MSH6*, *PMS2*). These Mendelian syndromes account for less than 5% of CRCs, and it is believed that most cases are instead due to the interaction between common variants of low/moderate penetrance and environmental factors. The common variants identified to date in GWAS (genome-wide association studies) could be responsible for 29% of the heritability in CRC.[4]

However, there is still a significant percentage of unexplained heritability. It has been postulated, following the "Rare Variant-Common Disease" hypothesis, that this genetic risk may be due to rare variants that cause "common" non-Mendelian CRC, but share some phenotypic characteristics with Mendelian syndromes, such as familial aggregation and early onset of CRC.[5]

Currently, there are two fundamental strategies for the identification of genetic variants related to the risk of developing CRC: on the one hand, case-control association studies represent the best strategy for the identification of common variants with low penetrance, while next-generation sequencing (NGS) studies are extensively used in the search for rare variants with a moderate effect.

In this chapter, we will study the heritability of CRC in detail, as well as the most common techniques for the identification and characterization of the genetic variants responsible.

High penetrance rare variants: Hereditary colorectal cancer predisposition syndromes

To date, rare, high penetrance germline variants in more than 10 known genes that cause Mendelian CRC predisposition syndromes have been identified in approximately 5% of CRCs. These syndromes are characterized by strong familial aggregation and/or early disease onset (Table 45.1). In addition to the already known Lynch syndrome, *APC*-associated adenomatous polyposis, and *MUTYH*-associated polyposis, and hamartomatous and mixed polyposis syndromes,[6–8] new Mendelian syndromes causing different types of hereditary polyposis have been reported in recent years. These include those caused by germline mutations in *POLE* and *POLD1* (PPAP, Polymerase Proofreading-Associated Polyposis),[9] *NTHL1* (NAP, *NTHL1*-associated polyposis),[10] or *MSH3*.[11] Other susceptibility genes have also been identified, such as *RPS20*, associated with hereditary nonpolyposis-type colorectal cancer cases,[12] and the *RNF43* gene identified in serrated polyposis.[13–17]

TABLE 45.1 Mendelian syndromes of hereditary predisposition to CRC.

Genes	Syndrome	Heredity	CRC ratio
MLH1 MSH2 MSH6 PMS2	Lynch syndrome (constitutional mismatch repair deficiency syndrome)	Dominant (recessive)	~3% (unknown)
APC	Familial adenomatous polyposis (attenuated)	Dominant	~1%
MUTYH	MUTYH-associated polyposis	Recessive	<1%
POLE POLD1	Adenomatous polyposis associated with polymerase test reading	Dominant	Unknown
NTHL1	NTHL1-associated polyposis	Recessive	Unknown
SMAD4 BMPR1A	Juvenile polyposis	Dominant	<1%
STK11	Peutz-Jeghers syndrome	Dominant	<1%
GREM1	Hereditary mixed polyposis syndrome	Dominant	<1%
MSH3	MSH3-associated polyposis	Recessive	Unknown
RNF43	Serrated polyposis syndrome	Dominant	Unknown
PTEN	PTEN hamartoma tumor syndrome	Dominant	<1%
TP53	Li-Fraumeni syndrome	Dominant	Unknown

Note: *AFAP*, attenuated familial adenomatous polyposis; *CMMRD*, constitutional mismatch repair deficiency; *FAP*, familial adenomatous polyposis; *HMPS*, hereditary mixed polyposis syndrome; *JPS*, juvenile polyposis; *LFS*, Li-Fraumeni syndrome; *LS*, Lynch syndrome; *MAP*, MUTYH-associated polyposis; *NAP*, NTHL1-associated polyposis; *PHTS*, PTEN tumor hamartoma syndrome; *PJS*, Peutz-Jeghers syndrome; *PPAP*, polymerase correction-associated polyposis; *SPS*, serrated polyposis syndrome.
Modified from Schubert SA, Morreau H, de Miranda NFCC, van Wezel T. The missing heritability of familial colorectal cancer. *Mutagenesis* 2019:gez027. https://doi.org/10.1093/mutage/gez027.

Other hereditary cancer syndromes

With the recent implementation of gene panels in the molecular diagnosis of hereditary CRC syndromes, pathogenic variants have been identified in hereditary cancer genes not traditionally associated with colorectal cancer, such as *BRCA1*, *BRCA2*, *CHEK2*, *ATM*, *PALB2* that could confer low/moderate CRC risks.[18,19] There could be a phenotypic overlap between traditionally different syndromes that would allow us to identify new genetic risk factors for CRC in the future.[7]

Candidate CRC susceptibility genes and NGS

Over the years, various methodologies have been used to identify the hereditary determinants of CRC development. These range from classical linkage studies in large families, which allowed the identification of the first genes responsible for hereditary CRC syndromes, to recent massive sequencing or NGS (Next Generation Sequencing) studies that have proposed a large number of candidate genes.

The best strategy for identifying rare variations of moderate penetrance is NGS sequencing. This technology has revolutionized genomic research by enabling the characterization of large regions of the genome in shorter periods of time and in a highly cost-effective manner.[20] The huge investments of research organizations in the extensive characterization of whole genomes in thousands of individuals (the 1000 and 10,000 Genomes projects, the UK Biobank, or the Haplotype Reference Consortium-HRC) have provided unprecedented genomic information that has facilitated the identification and characterization of these types of variants through the comprehensive large-scale characterization of genomic variability in diverse populations. This valuable information on the frequency of genetic variants identified in different populations is essential in the prioritization of variants for the identification of new disease susceptibility genes.

NGS has multiple applications, ranging from the analysis of specific regions of the genome to the sequencing of genomes and exomes (protein-coding regions corresponding to 1% of the genome).

Whole exome studies have been very successful in identifying new Mendelian CRC syndromes, such as PPAP or NAP, described before, and presumably caused by high penetrance variants. Although these NGS approaches have also allowed the identification of rare variants of moderate penetrance in *RPS20* in familial CRC without MMR pathway disruption,[12] many of the proposed candidate susceptibility genes, such as *FANCM, FAN1, TP53, BUB1, BUB3, LRP6,* and *PTPN12* have not been confirmed in external validation cohorts.[21,22] Basically, these whole exome studies have prioritized genetic variants that are very rare in the population (MAF < 0.1%) and have a high functional impact, and selected genes based on their function and involvement in disease-related pathways. The final corroboration of candidate variants was done using family cosegregation, recurrence and functional studies, somatic analyses, or assessments of the absence of the variant in the control population.[22] Further omics approaches integrating germline and tumor data will be necessary for the identification of new susceptibility genes.

Low-penetrance variants and association studies

Given our previous description that cases caused by rare variants of high penetrance represent a very low proportion of total CRC events, common variants of moderate/low penetrance are postulated as being particularly relevant to CRC heritability.

A piece of evidence supporting this hypothesis is the occurrence of CRC in familial aggregation groups, where the frequency of tumors is higher than expected by population incidence.

The importance of these common variants was established from the Common Disease-Common Variant (CDCV) hypothesis that suggests that the genetic component of complex diseases is mainly due to low/moderate effect variants that appear frequently in the population.[1]

Traditionally, these variants are identified using association studies, the most common of which is the case-control study, in which the frequency of the potential susceptibility variant is compared between two groups: one of affected individuals and one of healthy individuals. Differences in the prevalence of these variants could be indicative that the change is related to the onset of the disease.

Although the association strategy of case-control studies is technically simple, it inherently requires knowledge of the identification and location of variable positions in the genome. This only became possible after the completion of the Human Genome Project (HGP) and the discovery that the genome is highly variable in the form of single nucleotide polymorphisms (SNPs). SNPs are the most abundant genetic markers in the genome (more than 12 million) and constitute the greatest source of genetic and phenotypic variation between individuals. The construction of large SNP maps and databases after the termination of the HGP prompted the use of association studies in the discovery of new susceptibility variants for CRC and other complex diseases.

Candidate gene studies

Initially, the strategy used to search for new susceptibility variants consisted of a direct approach and observation of a candidate gene, which was based on the evaluation of specific, potentially relevant SNPs within certain genes considered important in the development of the disease. The main advantage of this strategy is the simple biological interpretation of the associations, as changes in these genes can be easily related to the neoplastic process. For CRC, several approaches screened common variants in relevant genes, such as *APC*, and examined the most known pathways related to carcinogenesis.[23]

Despite this, few variants could be irrefutably attributed to CRC risk. This setback is probably due to several reasons: firstly, candidate gene studies were restricted to potentially relevant loci based on theoretical estimates of biological mechanisms. Secondly, most of these early studies showed reduced statistical power for detecting variants with the expected risk effect under the CDCV hypothesis. In addition, methodological inconsistencies between different studies resulted in the inability to replicate most of these events.

Subsequently, the continuation of the HGP in the HapMap (Haplotype Map) project represented a milestone, as it showed us that most genome variants are inherited together in linkage disequilibrium (LD) blocks and not independently. This fact provided us with a new tool for conducting more informative association studies, since knowledge of the distribution of these blocks allows the identification of genetic variations in large regions by genotyping a relatively small set of informative markers or tag SNPs. The implementation of this indirect approach (compared with the direct approach of candidate gene studies, where it is assumed that the genotyped variant is the functional cause of disease susceptibility), together with the development of massive genotyping platforms and the consequent reduction in costs, made it possible for association studies to be performed on a genomic scale, giving rise to the so-called pangenomic association studies or GWAS (Genome-Wide Association Studies).

The knowledge of LD patterns obtained from the HapMap project also provided us with another essential tool for association studies, which is the ability to indirectly "impute" or infer the genotypes of different common variants by inspecting the SNPs that are inherited together, through the analysis of the most frequent haplotypic combinations observed in HapMap populations.

Genome-wide association studies (GWAS)

GWAS have represented a major advance over candidate gene studies, as they are a more comprehensive yet unbiased option for association strategies. This is because they do not make any prior assumptions about the location or functionality of variants that determine the risk of CRC. Furthermore, they represent an excellent method for the identification of common SNPs with low or even moderate effects on phenotype. Since their implementation at the end of 2007, they have resulted in the identification of multiple risk loci in several complex diseases and phenotypic traits, which have justified the creation of a proprietary database called the GWAS catalog (https://www.ebi.ac.uk/gwas/).

In the case of CRC, more than 100 CRC susceptibility SNPs have been described to date in European populations and an increasing number in East Asian populations.[4,24]

As expected from the complex disease model, all these associations show modest effects on disease risk, with effect magnitudes (Odds Ratios) generally below 1.5. Taken together, all these variants are estimated to be able to explain 29% of heritability, and the sample size needed to identify 80% of heritability in CRC would have to exceed 300,000 patients, far short of our current studies.[4]

Despite the considerable success obtained by GWAS studies in CRC, published studies have estimated that the total number of SNPs identified to date does not explain most of the heritability attributed to CRC.[4,24] This fact may be mainly due to two causes: (i) the CRC susceptibility gene variants present even smaller effects than those described, which would require further studies with larger numbers of samples or even in other populations—with different LD patterns—and the combination of data from the different individual GWAS in the form of metaanalyses, in the hope of identifying new variants that explain the remaining proportion[1,25] and (ii) the CDCV hypothesis does not capture all the heritability attributed to CRC, i.e., that there are other types of genomic variants contributing to the risk. Among the latter, it was initially postulated that copy number variants (or CNVs) could also significantly contribute to risk, since like SNPs, they presented a high level of genomic polymorphism. However, to date, studies have failed to reliably establish the link between these types of variants and CRC.[26,27]

One step further: strategies for causal variant identification and fine-mapping

Despite the fact that GWAS have generally been quite successful in discovering new risk variants, and that it is estimated that the progressive increase in the number of patients included in these studies will allow us to identify up to 73% of the genetic variants linked to the development of CRC, we must bear in mind that in most cases the variants identified are not the functional ones.

Moreover, given the complexity and large number of loci, in most cases the molecular mechanisms underlying the associations found have not yet been determined with complete certainty. In the case of CRC, it is however exceptional that several loci are close to genes belonging to pathways relevant to colorectal tumorigenesis, such as the transforming growth factor beta (TGF-β) signaling pathway,[28] the BMP pathway, linked to cell communication between the mesenchyme and the intestinal epithelium,[29] or the maintenance of telomere integrity, a fundamental mechanism of cellular aging and, by extension, cancer.[30]

With all this, it is still fundamental for prevention and population screening strategies to determine the variants through which the risk effect is exerted. The first approach in these cases entails extensive subsequent analyses of detailed studies of the associated regions, or fine-mapping strategies. In these, each locus is inspected independently (by sequencing or saturated genotyping), observing the common variation present and statistically evaluating the variable most associated with CRC, in order to identify the functional variant, although this is often complicated by complex LD patterns that generate a high correlation between variants in the same block.

In recent years, new analysis methodologies have also emerged, such as Mendelian randomization (MR), which allow us to establish causal relationships between the variants identified in GWAS and the phenotype (in our case, CRC).[31] In MR, genetic markers (mainly SNPs) are used as indirect markers of an exposure or risk factor, to determine the effect of the risk factor on the phenotype.[32]

Fine-mapping and MR studies are also often accompanied in a complementary manner by others of a more functional nature designed to directly observe the functional impact that the allelic change produces on cellular functions.[33]

Among these, the most frequent are chromatin immunoprecipitation tags (ChIP), which assess the interaction of specific genome sequences (in this case, CRC-associated variants) with specific proteins, and chromosome configuration capture techniques (3C technologies), which capture the interaction between genomic sequences to identify target sequences of potentially causal variants.[34]

Multiomic integration

On the other hand, the fact that the sample sizes needed for identifying the entire heritability of CRC in the near future are unattainable has encouraged the emergence of strategies complementary to association studies and NGS that allow us to identify new loci more easily without losing statistical rigor. Most of these strategies are linked to multiomics integration, i.e., the integration of information at different levels (transcriptomics, proteomics, methylomics, lipidomics, etc.), in order to reduce the genome to more likely regions, where some functional evidence of change mediated by gene variants is observed.

In their simplest form, these studies are based on the initial evaluation of markers (mainly SNPs) that cause changes in the expression of a given gene (eQTLs) or in methylation levels (meQTLs), to subsequently relate the presence of these events to that of risk markers, generally through colocalization techniques based on logistic regression or Bayesian methods.[35]

One of the most relevant characteristics of these studies is that in general, their effectiveness is limited by the type of tissue being studied. It has been observed that, despite there being multiple eQTLs that are shared in several tissues, the power to identify new susceptibility genes and loci is always greater when performed in the theoretically biologically relevant tissue (the intestinal epithelium in our case). There are now widely available resources that allow us to investigate the presence of eQTLs in most tissues in order to integrate this information with that of the risk markers obtained in GWAS. The most important of these resources is the GTEx (Genotype-Tissue Expression Project) portal.[36]

TWAS (Transcriptome-Wide Association Studies) represent improvement of this type of strategy. In these studies, information is obtained at the whole transcriptome (RNA-seq) and genome level (by complete sequencing or genotyping) from a series of reference samples that are used to establish the relationship between gene variants and expression levels, in order to then infer these in GWAS cohorts with a high number of patients, and thus identify susceptibility loci more easily.[37,38]

Both eQTLs and TWAS strategies have an advantage over GWAS, as they allow us not only to identify new susceptibility loci, but also provide us with candidate genes through which genetic variants are exerting their effect. In addition, the use of information obtained from expression levels allows us to improve the statistical power of the study and identify loci that would not be statistically significant if only genomic information were used.

Similarly, although still incipient, the strategies defined for expression can be applied mostly to the integration of other types of data, mainly tissue-specific methylation (MWAS).

Polygenic risk scores

Although the SNPs identified from GWAS and association studies in general are not individually informative given their low magnitude of effect, in recent years it has been observed that their discovery is not at all futile, since the information they provide as a whole can be very useful for identifying risk groups in the population. In fact, it has been calculated that if the total number of genetic variants of susceptibility to CRC were known, the top decile of the population would have a risk of developing CRC 7.7 times higher than the average. These predictive models may play a determinant role in improving population-based CRC screening strategies in the future.[30,39]

References

1. Houlston RS. Members of COGENT. COGENT (COlorectal cancer GENeTics) revisited. *Mutagenesis* 2012;**27**(2):143–51.
2. Graff RE, Möller S, Passarelli MN, et al. Familial risk and heritability of colorectal cancer in the Nordic twin study of cancer. *Clin Gastroenterol Hepatol* 2017;**15**(8):1256–64.
3. Peters U, Bien S, Zubair N, et al. Genetic architecture of colorectal cancer. *Gut* 2015;**64**(10):1623–36. https://doi.org/10.1136/gutjnl-2013-306705.
4. Law PJ, Timofeeva M, Fernandez-Rozadilla C, et al. Association analyses identify 31 new risk loci for colorectal cancer susceptibility. *Nat Commun* 2019;**10**(1):2154.
5. Hahn MM, de Voer RM, Hoogerbrugge N, Ligtenberg MJL, Kuiper RP, van Kessel AG. The genetic heterogeneity of colorectal cancer predisposition—guidelines for gene discovery. *Cell Oncol* 2016;**39**(6):491–510.
6. Lynch HT, de la Chapelle A. Hereditary colorectal cancer. *N Engl J Med* 2003;**348**(10):919–32.

7. Schubert SA, Morreau H, de Miranda NFCC, van Wezel T. The missing heritability of familial colorectal cancer. *Mutagenesis* 2020;**35**(3):221–31.
8. Valle L. Genetic predisposition to colorectal cancer: where we stand and future perspectives. *World J Gastroenterol* 2014;**20**(29):9828–49.
9. Palles C, Cazier J-B, Howarth KM, et al. Germline mutations affecting the proofreading domains of POLE and POLD1 predispose to colorectal adenomas and carcinomas. *Nat Genet* 2013;**45**(2):136–44.
10. Weren RDA, Ligtenberg MJL, Kets CM, et al. A germline homozygous mutation in the base-excision repair gene NTHL1 causes adenomatous polyposis and colorectal cancer. *Nat Genet* 2015;**47**(6):668–71.
11. Adam R, Spier I, Zhao B, et al. Exome sequencing identifies biallelic MSH3 germline mutations as a recessive subtype of colorectal adenomatous polyposis. *Am J Hum Genet* 2016;**99**(2):337–51.
12. Nieminen TT, O'Donohue M-F, Wu Y, et al. Germline mutation of RPS20, encoding a ribosomal protein, causes predisposition to hereditary nonpolyposis colorectal carcinoma without DNA mismatch repair deficiency. *Gastroenterology* 2014;**147**(3):595–598.e5.
13. Buchanan DD, Clendenning M, Zhuoer L, et al. Lack of evidence for germline mutations in patients with serrated polyposis syndrome from a large multinational study. *Gut* 2017;**66**(6):1170–2.
14. Gala MK, Mizukami Y, Le LP, et al. Germline mutations in oncogene-induced senescence pathways are associated with multiple sessile serrated adenomas. *Gastroenterology* 2014;**146**(2):520–9.
15. Quintana I, Mejías-Luque R, Terradas M, et al. Evidence suggests that germline mutations are a rare cause of serrated polyposis. *Gut* 2018;**67**(12):2230–2.
16. Taupin D, Lam W, Rangiah D, et al. A deleterious RNF43 germline mutation in a severely affected serrated polyposis kindred. *Hum Genome Var* 2015;**2**:15013.
17. Yan HHN, Lai JCW, Ho SL, et al. RNF43 germline and somatic mutation in serrated neoplasia pathway and its association with BRAF mutation. *Gut* 2017;**66**(9):1645–56.
18. Pearlman R, Frankel WL, Swanson B, et al. Prevalence and spectrum of germline cancer susceptibility gene mutations among patients with early-onset colorectal cancer. *JAMA Oncol* 2017;**3**(4):464–71.
19. Yurgelun MB, Kulke MH, Fuchs CS, et al. Cancer susceptibility gene mutations in individuals with colorectal cancer. *J Clin Oncol* 2017;**35**(10):1086–95.
20. Sahasrabudhe R, Lott P, Bohorquez M, et al. Germline mutations in PALB2, BRCA1, and RAD51C, which regulate DNA recombination repair, in patients with gastric cancer. *Gastroenterology* 2017;**152**(5):983–986.e6.
21. Broderick P, Dobbins SE, Chubb D, et al. Validation of recently proposed colorectal cancer susceptibility gene variants in an analysis of families and patients-a systematic review. *Gastroenterology* 2017;**152**(1):75–77.e4.
22. Valle L, de Voer RM, Goldberg Y, et al. Update on genetic predisposition to colorectal cancer and polyposis. *Mol Aspects Med* 2019;**69**:10–26.
23. van Wezel T, Middeldorp A, Wijnen JT, Morreau H. A review of the genetic background and tumour profiling in familial colorectal cancer. *Mutagenesis* 2012;**27**(2):239–45.
24. Huyghe JR, Bien SA, Harrison TA, et al. Discovery of common and rare genetic risk variants for colorectal cancer. *Nat Genet* 2019;**51**(1):76–87.
25. Stadler ZK, Thom P, Robson ME, et al. Genome-wide association studies of cancer. *J Clin Oncol* 2010;**28**(27):4255–67.
26. Fernandez-Rozadilla C, Cazier JB, Tomlinson I, et al. A genome-wide association study on copy-number variation identifies a 11q11 loss as a candidate susceptibility variant for colorectal cancer. *Hum Genet* 2014;**133**(5):525–34.
27. Wellcome Trust Case Control Consortium, Craddock N, Hurles ME, et al. Genome-wide association study of CNVs in 16,000 cases of eight common diseases and 3,000 shared controls. *Nature* 2010;**464**(7289):713–20.
28. Derynck R, Akhurst RJ, Balmain A. TGF-beta signaling in tumor suppression and cancer progression. *Nat Genet* 2001;**29**(2):117–29.
29. Tomlinson IPM, Carvajal-Carmona LG, Dobbins SE, et al. Multiple common susceptibility variants near BMP pathway loci GREM1, BMP4, and BMP2 explain part of the missing heritability of colorectal cancer. *PLoS Genet* 2011;**7**(6), e1002105.
30. Fernandez-Rozadilla C, Kartsonaki C, Woolley C, et al. Telomere length and genetics are independent colorectal tumour risk factors in an evaluation of biomarkers in normal bowel. *Br J Cancer* 2018;**118**(5):727–32.
31. Cornish AJ, Tomlinson IPM, Houlston RS. Mendelian randomisation: a powerful and inexpensive method for identifying and excluding non-genetic risk factors for colorectal cancer. *Mol Aspects Med* 2019;**69**:41–7.
32. van Rheenen W, Peyrot WJ, Schork AJ, Lee SH, Wray NR. Genetic correlations of polygenic disease traits: from theory to practice. *Nat Rev Genet* 2019;**20**(10):567–81.
33. Tuupanen S, Turunen M, Lehtonen R, et al. The common colorectal cancer predisposition SNP rs6983267 at chromosome 8q24 confers potential to enhanced Wnt signaling. *Nat Genet* 2009;**41**(8):885–90.
34. Orlando G, Kinnersley B, Houlston RS. Capture Hi-C library generation and analysis to detect chromatin interactions. *Curr Protoc Hum Genet* 2018; e63.
35. Zhu Z, Zhang F, Hu H, et al. Integration of summary data from GWAS and eQTL studies predicts complex trait gene targets. *Nat Genet* 2016;**48**(5):481–7.
36. GTEx Consortium. The genotype-tissue expression (GTEx) project. *Nat Genet* 2013;**45**(6):580–5.
37. Gamazon ER, Wheeler HE, Shah KP, et al. A gene-based association method for mapping traits using reference transcriptome data. *Nat Genet* 2015;**47**(9):1091–8.
38. Gusev A, Ko A, Shi H, et al. Integrative approaches for large-scale transcriptome-wide association studies. *Nat Genet* 2016;**48**(3):245–52.
39. Frampton MJE, Law P, Litchfield K, et al. Implications of polygenic risk for personalised colorectal cancer screening. *Ann Oncol* 2016;**27**(3):429–34.

Chapter 46

Signaling pathways in CRC

Víctor Sacristán Santos[a], Nieves Martínez Lago[a], Carla Pazos García[b], Alejandro Pazos García[c], and Luis M. Antón Aparicio[a]

[a]Medical Oncology Service, A Coruña University Hospital (CHUAC), A Coruña, Spain, [b]New Vision University (NVU), Tbilisi, Georgia, [c]Medical University of Bialystock (MUB), Białystok, Poland

Introduction

In common solid tumors, such as those affecting the colon, breast, brain, or pancreas, an average of 33 to 66 genes shows somatic mutations that could alter their protein products.

Tumors start off as benign lesions, which become malignant when they mutate over time. This process has been particularly well studied in colorectal tumors.[1,2] The first mutation, or "gatekeeping," provides a selective growth advantage to a normal epithelial cell, allowing it to overcome the surrounding cells and become a microscopic clone (Fig. 46.1).

Most gatekeeping mutations in the colon occur in the *APC* gene.[3] The small adenoma that results from this mutation grows slowly, but a second mutation in another gene, *KRAS*, triggers a second round of clonal growth that allows cells to proliferate. Cells with only the *APC* mutation may persist, but their number is small compared with cells that carry both gene mutations. This mutation process is followed by clonal expansion and continues with mutations in the *PIK3CA*, *SMAD4*, or *TP53* genes. This could generate a malignant tumor that can invade the underlying basement membrane and metastasize lymph nodes and distant organs, such as the liver.[4] Mutations that confer a selective growth advantage to the tumor cell are called "conductive" mutations. It has been estimated[5] that each conductive mutation provides only a small selective growth advantage to the cell, with a difference of up to 0.4% between the birth and death of the cell. However, this slight increase over the years, which is aggravated once or twice per week, can result in a large mass, containing billions of cells.

The number of mutations in self-renewal tissues of certain tumors directly correlates with age.[6] When this situation was evaluated by lineal regression, more than half of the somatic mutations identified in these tumors occurred during the paraneoplastic phase, meaning this occurs during the growth of normal cells that are continuously being replaced in gastrointestinal and genitourinary epithelium and other tissues. All these paraneoplastic mutations are "transient" mutations that have no effect on the neoplastic process. This explains why a colorectal tumor in a 90-year-old patient has almost twice as many mutations as a morphologically identical colorectal tumor in a 45-year-old patient. This finding also partly explains why some advanced brain tumors (glioblastomas) and pancreatic cancers (ductal pancreatic adenocarcinomas) have fewer mutations than colorectal tumors. Unlike the epithelial cells in colon crypts, brain glial cells and pancreas duct epithelial cells do not replicate. From a genetic perspective, it seems that mutations need to be present in order to turn a localized cancer into a metastatic one, just as there are mutations that transform a normal cell into a benign tumor, or a benign tumor into a malignant one (Fig. 46.1). However, despite the intense efforts that have been made to date, solid genetic mutations that distinguish cancers that will metastasize from cancers that will not, have not been identified yet.

The mutations depicted in Fig. 46.1 are clonal, meaning they are present in most tumor neoplastic cells. Additionally, there are other types of mutations (subclonal mutations) that are important for understanding the development of the tumor.

Four types of genetic heterogeneity relevant to tumorigenesis have been identified (Fig. 46.2):

(1) **Intratumoral:** heterogeneity between tumor cells. This type of heterogeneity has been recognized for decades. For example, it is rare to see a cytogenetic study of a solid tumor in which all tumor cells show the same karyotype.[7]
(2) **Intermetastatic:** heterogeneity between different metastatic lesions in the same patient.

Most cancer patients die because their tumors were located in surgically inaccessible sites, such as the liver, brain, lung, or bones, and were therefore not removed before metastasizing. Patients who relapse with a single metastatic lesion can often be cured with surgery or radiation therapy, but this is an exception rather than the rule. A typical patient in a clinical trial has a dozen or more metastatic lesions large enough to be visualized by imaging tests, and probably

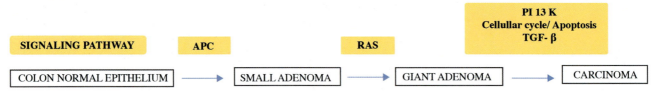

FIG. 46.1 Genetic mutations.

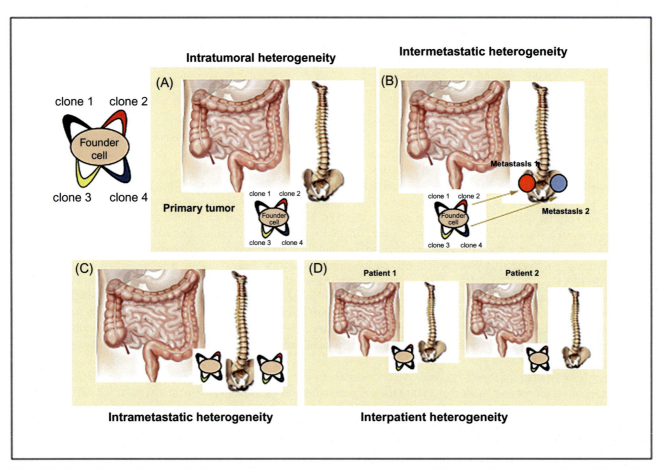

FIG. 46.2 Types of genetic heterogeneity in tumors, illustrated by a primary colonic tumor and its metastatic lesions in the liver. *(Adapted from Vogelstein B, Papadopoulos N, Velculescu VE, Zhou S, Diaz LA Jr, Kinzler KW. Cancer genome landscapes. Science 2013;339(6127):1546–1558. https://doi.org/10.1126/science.1235122.)*

also many smaller ones. If every metastatic lesion in a single patient were formed by a cell with a very different genetic constitution, then chemotherapy would not be the most appropriate way of eradicating a subset of metastatic lesions and would not increase a patient's long-term survival.[8]

How much heterogeneity is present in different metastatic lesions? In summary, a lot. It is not uncommon for a metastatic lesion to have 20 genetic clone alterations that are not shared by other metastases in the same patient.[9,10] Since they are clonal, these mutations have occurred in the founder cell of the metastasis, that is, the cell that escaped from the primary tumor and multiplied to form metastases. The founder cell of each metastasis is present in different geographical areas of primary tumors.[9]

This potentially disastrous situation is mitigated by the fact that heterogeneity appears to be largely limited to transient mutations.

(3) **Intrametastatic:** heterogeneity between the cells of a metastasis. Each metastasis is established by a single cell (or small group of cells) with a set of founder mutations. When they grow, metastases acquire new mutations with each cell division. Although the founder mutations can make the lesion sensitive to antitumor agents, the newer mutations

provide the roots of drug resistance. Unlike primary tumors, metastatic lesions cannot generally be removed by surgery and must be treated with systemic therapies. Patients with complete response to targeted therapies will undoubtedly relapse over time. Most of the initial lesions will usually relapse, and the period in which they do so is remarkably similar. This fact can be explained by the presence of intrametastatic resistance provided by mutations that have existed within each metastasis prior to the onset of targeted therapy.[11, 12]

Estimations show that any metastatic lesion that is large enough to be seen on an a radiologic image has thousands of cells (among billions present) that are resistant to any possible imaginable medication.[11, 13] Therefore recurrence is simply a matter of time and is totally predictable based on known mutation frequencies and tumor cell growth rates.

(4) **Between patients:** heterogeneity between different patients' tumors. This type of heterogeneity has been observed by all oncologists in routine clinical practice; no two cancer patients have the same clinical course, with or without treatment. Some of these differences could be related to host factors, such as germline variants, that can determine the drug's half-life or vascular permeability to medications or cells. On the other hand, some could be related to non-genetic factors.[14] However, this heterogeneity between patients is largely related to somatic mutations within tumors.

Signaling pathways

The immense complexity of the cancer genomes that can be deduced from the data described before is somewhat misleading. After all, even advanced tumors are not completely out of control, as shown by the dramatic responses to agents that target BRAF mutations in melanomas[15, 16] or mutant ALK in lung cancers.

CRC is an ideal model in the investigation of the molecular pathogenesis of cancer, because it is easy to biopsy and the development of invasive carcinoma is well understood, starting as normal epithelium, developing into a polyp, and then turning into a carcinoma. Currently, CRC is thought to be caused by genetic mutations. Mutations in the genetic material conform the base of hereditary syndromes, while sporadic cancers derive from somatic mutations. Many pathways are involved in colorectal carcinogenesis, including chromosomal instability pathways, microsatellite instability pathways, serrated pathways, epigenetic mechanisms of colon carcinogenesis, and different genes related to invasion and metastasis.

CRC is one of the most common causes of cancer-related deaths, representing about 10%, only surpassed by lung cancer. It is a disease that involves several stages, and several genes have to mutate for it to progress.

This chapter highlights some of the biological mechanisms that make CRC difficult to treat.

TGF-β and its role in colon cancer

TGF-βs are 25-kDa cytokines, which play a unique and fundamental role in homeostasis, wound healing, fibrosis, angiogenesis, carcinogenesis, and cell differentiation.[17] There are five isoforms of TGF-βs called TGF-β1, TGF-β2, TGF-β3, TGF-β4, and TGF-β5, which belong to a large superfamily which includes, but is not limited to, activins, inhibins, bone morphogenetic proteins (BMPs), or myostatin.[18]

TGF-β1 signaling through its receptor, a transmembrane serine-threonine kinase, plays an important but ambiguous role in carcinogenesis. TGF-β1 interacts with TGF-βRII (type II receptor), which in turn attracts and activates TGF-βRI (type I receptor). Smad 2 and Smad 3 are phosphorylated in the carboxyl terminal serines by the activated TGF-βRI receptor forming heteromeric complexes with Smad 4. Ultimately, the Smad 2/3/4 complex translocates to the nucleus and binds to specific regulatory elements of target genes.[19] Smad 4 can be translocated to the core only when it forms a complex with R-Smads, while Smad 2 and Smad3 can do so independently of Smad 4. This implies that Smad 4 has a regulatory role instead of a simple signal transmission from the cytoplasm to the nucleus (Fig. 46.3). Once in the nucleus, the Smad 2/3/4 complex activates p21, a cyclin-dependent kinase inhibitor (CDK), which stops growth.

TGF-β and its signaling effectors influence the biological behavior of cancer. The TGF-β signaling pathway has been considered both a tumor suppressor and a cancer promoter.[20] TGF-β1 goes from being a tumor cell growth inhibitor to a growth and invasion stimulator during colon cancer progression. It has been observed that metastatic CRC cells respond to proliferating TGF-β, while in moderate and well-differentiated primary colon tumors growth was inhibited by TGF-β. Intense staining for TGF-β1 correlates significantly with disease progression and is independent of the lymph node status and the degree of primary tumor differentiation. Patients with high protein levels of TGF-β in their primary CRC were more likely to experience recurrence of their disease than patients whose tumors exhibited lower levels of TGF-β.[21]

The overall incidence of mutations in TGF-βRII is close to 30% in CRC and is the most common mechanism identified so far, which eventually results in the alteration of TGF-β signaling.

Given the fact that both TGF-βRI and RII are equally essential for signal transduction by TGF-β, it can be deduced that mutations in any of the genes can produce equivalent functional results.

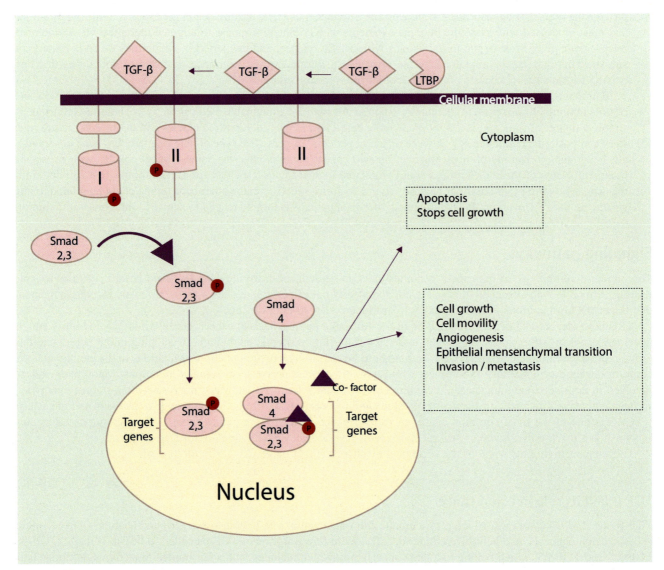

FIG. 46.3 TGF-β/Smad signaling pathway.

A predisposition to malignancy, namely to colon cancer, was observed in both homozygous TGF-βRI (6A) and heterozygous TGF-βRI (6A). Therefore TGF-βRI (6A) acts as a tumor susceptibility allele that could contribute to cancer development, especially colon cancer, reducing growth inhibition mediated by TGF-β.[22] A meta-analysis that included 13,113 people concluded that TGF-R1 * 6A is associated with an increased risk of cancer.[23]

Smads and its role in colon cancer

Deterioration of TGF-β signaling can also occur through mutations in Smad. Mutation in the Smad4 gene plays an important role in carcinogenesis of the colon. Loss of activation or expression of Smad occurs in approximately 10% of CRCs. This subset has a poor prognosis due to its association with advanced disease and the presence of nodal metastases at diagnosis.[24] Loss of Smad4 function occurs in later stages of malignancy, playing a role in the acquisition of advanced phenotypes. The frequency of Smad4 gene mutations increases with carcinogenesis progression and is 0% in adenomas, 10% in intramucosal carcinomas, 7% in invasive carcinomas without distant metastases, 35% in primary invasive carcinomas with distant metastases, and 31% in carcinomas that metastasize to the liver and distant or disseminated lymph nodes.[25] In addition, some other studies have reported frequent somatic mutations of DPC4 in human colorectal tumors, suggesting an important role of Smad4 in colorectal carcinogenesis.[26] In addition, transfection of DPC4 results in the constant expression of Smad4, suggesting that DPC4 inhibits cell growth, not only via the TGF-β signal transduction pathway, but also via other cascades.[27] Some researchers have revealed that overexpression of wild-type DPC4 could rebuild a new negative regulation pathway for cell proliferation apart from the TGF-β signal transduction cascade.[28]

The stem cell adapter protein ELF (β spectrin) is involved in the intestinal lineage. ELF activates and modulates Smad4 conferring cell polarity, maintaining architecture, and inhibiting epithelial-mesenchymal transition. It has been suggested that by modulating Smad4, ELF plays a key role in TGF-β signaling in nonmetastatic colon cancer suppression.[29]

Mutations in Smad2 are specifically associated with sporadic colorectal carcinoma and therefore suggest that Smad2 is a potential tumor suppressor.[30] Regardless of the mutational mechanism, Smad2-mutated protein provides an escape mechanism for the TGF-β pathway. By doing so, it detracts the antiproliferative effect of TGF-β, leading to cancer. Mutations in Smad4 occur in advanced-stage cancers (lymph node involvement and metastasis) while mutations in Smad2 occur in early stages. Smad3 is another powerful tumor suppressor of the colonic epithelium. Interestingly, to date, no mutations in Smad3 have been detected in human CRC. A logical explanation for this could be that tumor studies in humans are not sensitive enough to detect a broad spectrum of inactivating mutations.

EGFR and its role in colon cancer

In normal cells, the epidermal growth factor receptor (EGFR) signaling cascade begins with EGFR ligand activation (Fig. 46.4). Up to eleven ligands can bind to the ErbB family of receptors, including EGF and transforming growth factor alpha (TGF-α).[31] Ligand binding induces receptor dimerization with homodimer and heterodimer formation, leading to tyrosine kinase activation. Subsequently, the autophosphorylated intracellular tyrosine kinase residues induce the activation of multiple signal transduction pathways. The two main intracellular pathways activated by EGFR are mitogen-activated protein kinase (MAPK) and phosphatidylinositol 3-kinase (PI3K) protein kinase B (AKT) pathway. These pathways lead to the activation of various transcription factors, which will later affect cell responses, such as proliferation, migration, differentiation, and apoptosis.[32] Signaling through the EGFR pathway is a complex process that requires strict regulation.[32] The first level of complexity is at the receptor level, where several ligands are shared and collateral signaling occurs between members of the ErbB family. Then there are positive and negative feedback loops incorporated into the pathways and selective activation of transcription factors, depending on the type of cell. When this tightly regulated system goes wrong, it can contribute to malignant transformation and tumor progression through increased cell proliferation, prolonged survival, angiogenesis, antiapoptosis, invasion, and metastasis.[33]

The wide range of EGFR expression in CRC reported in the literature, as well as the uncertain meaning of EGFR expression as a prognostic indicator, may be related to the methodology used to detect the EGFR protein. Most studies use immunohistochemistry to detect the expression of EGFR in colorectal cancers. Based on the experience of HER2 expression in breast cancer, immunohistochemistry is highly dependent on the antibody clone used, the staining protocols, the selection of qualification methods, and the cutoff values. Until a standard EGFR staining method is adopted, the relevance of EGFR protein expression in CRC will remain controversial.

Mutations that affect the extracellular domain of EGFR, and that are often accompanied by gene amplification, are frequent in glioblastomas.[34] Moreover, mutations in the tyrosine kinase domain of EGFR are frequently associated with increased numbers of EGFR gene copies, and those are clinically relevant in lung adenocarcinoma.[35] Unlike lung cancer and other tumors, EGFR gene mutations are rare in colorectal cancers.[36]

The significance of EGFR gene amplification is more difficult to summarize. Some studies indicate that amplification of the EGFR gene (evaluated by in situ hybridization) is uncommon in CRC.[37] In contrast, in more recent studies on chemotherapy-resistant colon cancer, it appears that small increases in the number of copies (three to five times) are present in up to 50% of cases.[38] It seems, however, that increased EGFR protein expression does not always translate into an increase in the amount of the EGFR gene.[37, 39]

Similarly, the predictive meaning of EGFR gene amplification is also confusing and uncertain. A study of 47 patients with metastatic colorectal cancer treated with cetuximab showed that the increase in the number of EGFR gene copies, assessed by fluorescent in situ hybridization, had no correlation with the objective response rate, progression-free survival, or overall survival.[40] In contrast, another study of 173 patients with wild-type KRAS metastatic colorectal cancer treated with cetuximab showed that EGFR amplification, present in 17.7% of patients, was associated with better response to anti-EGFR therapy.[41] These contradictory results may be explained by the fact that there are no established guidelines for EGFR gene amplification. Since there are no specific indications, tests for amplification of the EGFR gene in colorectal cancer are not routinely carried out.

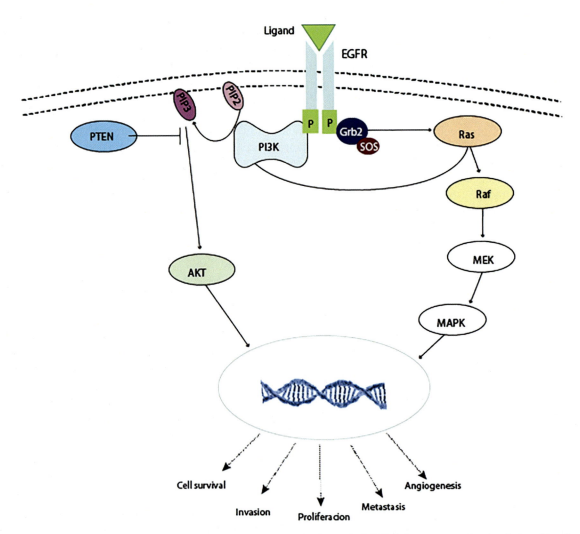

FIG. 46.4 EGFR signaling pathway. Ligand binding induces dimerization and activates the EGFR. Subsequent autophosphorylation of tyrosine residues activates downstream signaling. In the Ras-Raf-MEK-MAPK pathway, one axis of the EGFR signaling cascade, the process begins with phosphorylation of an adaptor protein complex composed of growth factor receptor-bound protein 2 adapter protein (Grb2), which harbors a tyrosine phosphate-docking site, and son of sevenless (SOS), a Ras GDP/GTP exchange factor, then activates the Ras GTPase. After activation, Ras (i.e., KRas) recruits and activates the serine protein Raf (i.e., B-Raf), and subsequent phosphorylation and activation of MEK and then MAPK occurs, resulting in activation of transcription factors in the cell nucleus. The Ras-Raf-MAPK signaling pathway is thought to control cell growth, differentiation, and survival (apoptosis). The other axis of the EGFR signaling cascade that is important in colorectal carcinogenesis is the PI3K-AKT pathway. Once the EFGR tyrosine residues are phosphorylated, PI3K is translocated to the cell membrane and binds to tyrosine phosphate (through its adaptor subunit p85) which triggers the PI3K catalytic subunit p110 to produce phosphatidylinositol-3,4,5-triphosphate (PIP3). PI3K then promotes AKT activation. Activated AKT (p-AKT), present within the cytoplasm, then activates various targets that result in cell growth, proliferation, and survival (parallel to the Ras-Raf-MEK-MAPK signaling pathway). Importantly, these two axes are closely related and have some overlap. For example, the p110 subunit of PI3K can also be activated via interaction with Ras. It should be noted that phosphatase with tensin homology (PTEN) is a phosphatase that converts PIP3 back to phosphatidylinositol (4, 5) bisphosphate (PIP2), thereby negatively regulating the PI3K-AKT pathway.

MAP kinase (MAPK) pathways and their role in colon cancer

Overactivation of the MAPK pathway is a very important phenomenon at the start of the CRC metabolic pathway. Normally, the activation of the signaling cascade only begins if the tyrosine kinase domain of any membrane effector receptor is phosphorylated. Next, the added phosphate groups can attract the SH2 domains of some KRAS activating proteins, such as GEPs (Guanine Exchanging Factors). Subsequently, the GEPs are phosphorylated, whereby their composition changes, making them attractive so that they can bind to the RAS family proteins. These proteins (KRAS, HRAS, and

NRAS) are GTPases, and their objective is to bind to and dephosphorylate guanosine-5′-triphosphate (GTP) molecules. During this process, they change their composition in order to interact with other proteins. Then, the RAS protein recovers its initial composition through GTP molecule dephosphorylation. In CRC, the most important RAS protein is KRAS and its main effector is BRAF. KRAS is constitutively active because of mutations in its primary structure in 33% of CRCs.[42] Up to 82% of the changes are due to mutations in codon 12, 17% occur in 13, and the rest in 61 or 146. The fact that KRAS is continuously activated produces constant cellular stimulation of several oncogenic functions, such as growth, progression of the cell cycle through the G1-S phase, cellular metabolism directed toward actions that are mainly anabolic and support the high demands derived from tumor behavior, stimulation of angiogenesis, cell immortalization or growth in motility—all of which could eventually confer metastatic potential. However, the prognostic role of KRAS mutations is unclear. Some studies do not attribute any prognostic implications, while others have observed a clearly detrimental role. These differences could be explained by the fact that there are different prognostic implications for specific KRAS mutations. It should be noted that the G12D mutation could be more aggressive than the others. In addition, KRAS mutations manifest a negative predictive role in response to anti-EGFR treatment, even though the absence of mutation does not demonstrate a positive predictive role.

Looking at the remaining RAS proteins, NRAS is mutated in between 1% and 3% of CRCs, especially in codon 61. This phenomenon is mutually exclusive with KRAS mutations and it is also a negative predictive factor for treatment with anti-EGFR antibodies.[42] HRAS does not play a significant role in the CRC.

BRAF, from the RAF family of proteins (A, B, C-RAF), is a serine-threonine kinase, the main effector of KRAS in the MAPK pathway. If KRAS activates BRAF, it interacts with CRAF and activates MEK, which activates ERK that translocates at the same time to the cell nucleus to achieve these effects. In CRC, BRAF is active because of mutations in its primary structure in between 5% and 15% of cases. This is incompatible with the presence of mutations in KRAS in the same cell. The most frequent activating alteration is V600E, which replaces the valine at position 600 with glutamic acid. Patients with CRC with a V600E mutation have a worse prognosis than those who do not. They are also refractory to anti-EGFR treatments.[43, 44] At the same time, BRAF mutations establish a biomarker to administer BRAF inhibitor therapies, which have very good results in other types of tumors (such as melanoma), but do not affect CRC to the same extent. This is probably due to differences between the two diseases with regard to functionality and levels of expression of the remaining members of the pathway.[45]

PI3K-AKT-mTOR and its role in colon cancer

The PI3K-AKT-mTOR pathway is a main cascade in cellular signaling for healthy tissues and for most tumor types. Membrane receptor stimulation (tyrosine kinase domain) triggers the signal. This domain, once phosphorylated, attracts the SH2 domain of the p85 part of PI3K (Phospho-Inositol 3 Kinase). PI3K is a lipid kinase formed by two subunits, the p85 regulatory part and p110 catalytic subunit. If p85 is linked to the transmembrane receptor phosphate group, a change in composition is generated on p110 to interact with the phospho-inositol bisphosphate (PIP2) molecules and thus phosphorylating them to form phospho-inositol triphosphate (PIP3). This fact can also occur by direct binding of KRAS to p110. PIP3 can be recycled to PIP2 by homologous tensinogen phosphatase (PTEN) that antagonizes the function of PI3K by acting as a pathway suppressor. PIP3 can phosphorylate AKT and PDK1, proteins located downstream in the pathway. The main function of AKT and PDK is to activate complexes 1 and 2 of mTOR and P70S6 kinase. They also suppress FOXO3a, activating the metabolism, resistance acquisition to apoptosis, increased cell motility, increased angiogenesis, and cell cycle progression.

In CRC, PI3K is altered by mutations in 10%–20% of cases[42] and is amplified less frequently. Eighty percent of oncogenic mutations occur in two hot spots found in the catalytic subunit: exon 9, where 65% of mutations occur (E545K and E542K), and exon 20 where the other 15% occur (H1047K). Different prognostic roles have been conferred to PI3K mutations in CRC. Liao, in 2012, determined that, by themselves, mutations in different exons of PI3K do not change the prognosis in CRC, but for tumors with coexistent mutations in exons 9 and 20 (about 4% of patients with PI3K alterations) the prognosis is shortened significantly.[46] Moreover, exon 20 mutations and KRAS mutations are mutually exclusive and cause resistance to treatment with cetuximab.[43]

PTEN, the main suppressor of the pathway, can be reduced by inactivating mutations or by decreasing its expression either by allelic losses or methylation of its promoter. Between 19% and 36% of CRCs show a decrease in PTEN expression,[47] which has been related in several clinical series to a worse prognosis as well as a negative predictive value of response to treatment with cetuximab.[48, 49]

PI3K-AKT-mTOR and its role in colon cancer

The APC tumor suppressor gene has a critical role in the development of sporadic CRC and is mutated in more than 70% of colorectal cancers. Although germline mutations are distributed throughout the gene, somatic mutations are located at the 5′ end of exon 15.[50] This region is involved in the interaction with β-catenin.[51]

Progress in understanding the function of APC has been made by studying its interaction with β-catenin and with a glycogen synthase kinase (GSK-3β) as a component of the Wnt signaling pathway. Normally, in the cytoplasm of a cell, APC together with GSK-3β bind to β-catenin, which undergoes phosphorylation that leads to its degradation in the proteasome. APC mutations prevent this gene from binding to β-catenin and thus it remains free in the cytoplasm. Free β-catenin translocates to the nucleus where it can form a complex with the T-cell factor (TCF) and activate the expression of the c-Myc and cyclin D1 oncogenes, as well as c-jun, fra-1, PPARδ, MMP-7, and uPA. Free β-catenin also forms a complex with α-catenin, which is simultaneously associated with the cytoplasmic region of E-cadherin, a transmembrane protein whose extracellular region can bind with a neighboring cell causing intercellular adhesion.

The main mutations found in CTNNB1 (gene encoding β-catenin) are located in residues susceptible to phosphorylation by GSK-3β. As it is believed that APC mutations promote colorectal carcinogenesis by preventing degradation of β-catenin,[52] it has been speculated that mutations in CTNNB1 are important in the small proportion of CRCs that lack these APC mutations.[53]

References

1. Nowell PC. The clonal evolution of tumor cell populations. *Science* 1976;**194**(4260):23–8. https://doi.org/10.1126/science.959840.
2. Fearon ER, Vogelstein B. A genetic model for colorectal tumorigenesis. *Cell* 1990;**61**(5):759–67. https://doi.org/10.1016/0092-8674(90)90186-i.
3. Kinzler KW, Vogelstein B. Gatekeepers and caretakers. *Nature* 1997;**386**(6627):761–3. https://doi.org/10.1038/386761a0.
4. Jones S, Chen W-D, Parmigiani G, et al. Comparative lesion sequencing provides insights into tumor evolution. *Proc Natl Acad Sci U S A* 2008;**105**(11):4283–8. https://doi.org/10.1073/pnas.0712345105.
5. Bozic I, Antal T, Ohtsuki H, et al. Accumulation of driver and passenger mutations during tumor progression. *Proc Natl Acad Sci U S A* 2010;**107**(43):18545–50. https://doi.org/10.1073/pnas.1010978107.
6. Tomasetti C, Vogelstein B, Parmigiani G. Half or more of the somatic mutations in cancers of self-renewing tissues originate prior to tumor initiation. *Proc Natl Acad Sci U S A* 2013;**110**(6):1999–2004. https://doi.org/10.1073/pnas.1221068110.
7. Höglund M, Gisselsson D, Säll T, Mitelman F. Coping with complexity. multivariate analysis of tumor karyotypes. *Cancer Genet Cytogenet* 2002;**135**(2):103–9. https://doi.org/10.1016/s0165-4608(01)00645-8.
8. Vogelstein B, Papadopoulos N, Velculescu VE, Zhou S, Diaz Jr LA, Kinzler KW. Cancer genome landscapes. *Science* 2013;**339**(6127):1546–58. https://doi.org/10.1126/science.1235122.
9. Yachida S, Jones S, Bozic I, et al. Distant metastasis occurs late during the genetic evolution of pancreatic cancer. *Nature* 2010;**467**(7319):1114–7. https://doi.org/10.1038/nature09515.
10. Campbell PJ, Yachida S, Mudie LJ, et al. The patterns and dynamics of genomic instability in metastatic pancreatic cancer. *Nature* 2010;**467**(7319):1109–13. https://doi.org/10.1038/nature09460.
11. Turke AB, Zejnullahu K, Wu Y-L, et al. Preexistence and clonal selection of MET amplification in EGFR mutant NSCLC. *Cancer Cell* 2010;**17**(1):77–88. https://doi.org/10.1016/j.ccr.2009.11.022.
12. Diaz Jr LA, Williams RT, Wu J, et al. The molecular evolution of acquired resistance to targeted EGFR blockade in colorectal cancers. *Nature* 2012;**486**(7404):537–40. https://doi.org/10.1038/nature11219.
13. Durrett R, Moseley S. Evolution of resistance and progression to disease during clonal expansion of cancer. *Theor Popul Biol* 2010;**77**(1):42–8. https://doi.org/10.1016/j.tpb.2009.10.008.
14. Kreso A, O'Brien CA, van Galen P, et al. Variable clonal repopulation dynamics influence chemotherapy response in colorectal cancer. *Science* 2013;**339**(6119):543–8. https://doi.org/10.1126/science.1227670.
15. Chapman PB, Hauschild A, Robert C, et al. Improved survival with vemurafenib in melanoma with BRAF V600E mutation. *N Engl J Med* 2011;**364**(26):2507–16. https://doi.org/10.1056/NEJMoa1103782.
16. Kwak EL, Bang Y-J, Camidge DR, et al. Anaplastic lymphoma kinase inhibition in non–small-cell lung cancer. *N Engl J Med* 2010;**363**(18):1693–703. https://doi.org/10.1056/NEJMoa1006448.
17. Derynck R, Feng XH. TGF-beta receptor signaling. *Biochim Biophys Acta* 1997;**1333**(2):F105–50. https://doi.org/10.1016/s0304-419x(97)00017-6.
18. Massagué J. The transforming growth factor-beta family. *Annu Rev Cell Biol* 1990;**6**:597–641. https://doi.org/10.1146/annurev.cb.06.110190.003121.
19. Miyazono K, ten Dijke P, Heldin CH. TGF-beta signaling by Smad proteins. *Adv Immunol* 2000;**75**:115–57. https://doi.org/10.1016/s0065-2776(00)75003-6.
20. Akhurst RJ, Derynck R. TGF-β signaling in cancer—a double-edged sword. *Trends Cell Biol* 2001;**11**:S44–51. https://doi.org/10.1016/s0962-8924(01)82259-5.
21. Friedman E, Gold LI, Klimstra D, Zeng ZS, Winawer S, Cohen A. High levels of transforming growth factor beta 1 correlate with disease progression in human colon cancer. *Cancer Epidemiol Biomarkers Prev* 1995;**4**(5):549–54. https://www.ncbi.nlm.nih.gov/pubmed/7549813.

22. Pasche B, Kolachana P, Nafa K, et al. TβR-I(6A) is a candidate tumor susceptibility allele. *Cancer Res* 1999;**59**(22):5678–82. Accessed March 22, 2021 https://cancerres.aacrjournals.org/content/59/22/5678.short.
23. Zhang H-T, Zhao J, Zheng S-Y, Chen X-F. Is TGFBR1*6A really associated with increased risk of cancer? *J Clin Oncol* 2005;**23**(30):7743–4. author reply 7744-6 https://doi.org/10.1200/JCO.2005.02.9108.
24. Xie W, Rimm DL, Lin Y, Shih WJ, Reiss M. Loss of Smad signaling in human colorectal cancer is associated with advanced disease and poor prognosis. *Cancer J* 2003;**9**(4):302–12. https://doi.org/10.1097/00130404-200307000-00013.
25. Miyaki M, Kuroki T. Role of Smad4 (DPC4) inactivation in human cancer. *Biochem Biophys Res Commun* 2003;**306**(4):799–804. https://doi.org/10.1016/s0006-291x(03)01066-0.
26. Miyaki M, Iijima T, Konishi M, et al. Higher frequency of Smad4 gene mutation in human colorectal cancer with distant metastasis. *Oncogene* 1999;**18**(20):3098–103. https://doi.org/10.1038/sj.onc.1202642.
27. Xiao D-S, Wen J-F, Li J-H, Hu Z-L, Zheng H, Fu C-Y. Effect of deleted pancreatic cancer locus 4 gene transfection on biological behaviors of human colorectal carcinoma cells. *World J Gastroenterol* 2005;**11**(3):348–52. https://doi.org/10.3748/wjg.v11.i3.348.
28. Yan Z, Kim G-Y, Deng X, Friedman E. Transforming growth factor beta 1 induces proliferation in colon carcinoma cells by Ras-dependent, smad-independent down-regulation of p21cip1. *J Biol Chem* 2002;**277**(12):9870–9. https://doi.org/10.1074/jbc.M107646200.
29. Tang Y, Katuri V, Srinivasan R, et al. Transforming growth factor-beta suppresses nonmetastatic colon cancer through Smad4 and adaptor protein ELF at an early stage of tumorigenesis. *Cancer Res* 2005;**65**(10):4228–37. https://doi.org/10.1158/0008-5472.CAN-04-4585.
30. Eppert K, Scherer SW, Ozcelik H, et al. MADR2 maps to 18q21 and encodes a TGFβ–regulated MAD–related protein that is functionally mutated in colorectal carcinoma. *Cell* 1996;**86**(4):543–52. https://doi.org/10.1016/S0092-8674(00)80128-2.
31. Hynes NE, Lane HA. ERBB receptors and cancer: the complexity of targeted inhibitors. *Nat Rev Cancer* 2005;**5**(5):341–54. https://doi.org/10.1038/nrc1609.
32. Citri A, Yarden Y. EGF–ERBB signalling: towards the systems level. *Nat Rev Mol Cell Biol* 2006;**7**(7):505–16. https://doi.org/10.1038/nrm1962.
33. Mitsudomi T, Yatabe Y. Epidermal growth factor receptor in relation to tumor development: EGFR gene and cancer. *FEBS J* 2010;**277**(2):301–8. https://doi.org/10.1111/j.1742-4658.2009.07448.x.
34. Frederick L, Wang XY, Eley G, James CD. Diversity and frequency of epidermal growth factor receptor mutations in human glioblastomas. *Cancer Res* 2000;**60**(5):1383–7. https://www.ncbi.nlm.nih.gov/pubmed/10728703.
35. Dacic S. EGFR assays in lung cancer. *Adv Anat Pathol* 2008;**15**(4):241–7. https://doi.org/10.1097/PAP.0b013e31817bf5a9.
36. Lee JW, Soung YH, Kim SY, et al. Absence of EGFR mutation in the kinase domain in common human cancers besides non-small cell lung cancer. *Int J Cancer* 2005;**113**(3):510–1. https://doi.org/10.1002/ijc.20591.
37. Shia J, Klimstra DS, Li AR, et al. Epidermal growth factor receptor expression and gene amplification in colorectal carcinoma: an immunohistochemical and chromogenic in situ hybridization study. *Mod Pathol* 2005;**18**(10):1350–6. https://www.nature.com/articles/3800417.
38. Spindler K-LG, Lindebjerg J, Nielsen JN, et al. Epidermal growth factor receptor analyses in colorectal cancer: a comparison of methods. *Int J Oncol* 2006;**29**(5):1159–65. https://www.ncbi.nlm.nih.gov/pubmed/17016647.
39. Cappuzzo F, Finocchiaro G, Rossi E, et al. EGFR FISH assay predicts for response to cetuximab in chemotherapy refractory colorectal cancer patients. *Ann Oncol* 2008;**19**(4):717–23. https://doi.org/10.1093/annonc/mdm492.
40. Italiano A, Follana P, Caroli F-X, et al. Cetuximab shows activity in colorectal cancer patients with tumors for which FISH analysis does not detect an increase in EGFR gene copy number. *Ann Surg Oncol* 2008;**15**(2):649–54. https://doi.org/10.1245/s10434-007-9667-2.
41. Laurent-Puig P, Cayre A, Manceau G, et al. Analysis of PTEN, BRAF, and EGFR status in determining benefit from cetuximab therapy in wild-type KRAS metastatic colon cancer. *J Clin Oncol* 2009;**27**(35):5924–30. https://doi.org/10.1200/JCO.2008.21.6796.
42. De Roock W, Claes B, Bernasconi D, et al. Effects of KRAS, BRAF, NRAS, and PIK3CA mutations on the efficacy of cetuximab plus chemotherapy in chemotherapy-refractory metastatic colorectal cancer: a retrospective consortium analysis. *Lancet Oncol* 2010;**11**(8):756–62. https://doi.org/10.1016/S1470-2045(10)70130-3.
43. Di Nicolantonio F, Martini M, Molinari F, et al. Wild-type BRAF is required for response to panitumumab or cetuximab in metastatic colorectal cancer. *J Clin Oncol* 2008;**26**(35):5705–12. https://doi.org/10.1200/JCO.2008.18.0786.
44. Cappuzzo F, Varella-Garcia M, Finocchiaro G, et al. Primary resistance to cetuximab therapy in EGFR FISH-positive colorectal cancer patients. *Br J Cancer* 2008;**29**(1):83–9. https://doi.org/10.1038/sj.bjc.6604439.
45. Prahallad A, Sun C, Huang S, et al. Unresponsiveness of colon cancer to BRAF(V600E) inhibition through feedback activation of EGFR. *Nature* 2012;**483**(7387):100–3. https://doi.org/10.1038/nature10868.
46. Liao X, Morikawa T, Lochhead P, et al. Prognostic role of PIK3CA mutation in colorectal cancer: cohort study and literature review. *Clin Cancer Res* 2012;**18**(8):2257–68. https://doi.org/10.1158/1078-0432.CCR-11-2410.
47. Goel A, Arnold CN, Niedzwiecki D, et al. Frequent inactivation of PTEN by promoter hypermethylation in microsatellite instability-high sporadic colorectal cancers. *Cancer Res* 2004;**64**(9):3014–21. https://doi.org/10.1158/0008-5472.can-2401-2.
48. Sartore-Bianchi A, Di Nicolantonio F, Nichelatti M, et al. Multi-determinants analysis of molecular alterations for predicting clinical benefit to EGFR-targeted monoclonal antibodies in colorectal cancer. *PLoS One* 2009;**4**(10). https://doi.org/10.1371/journal.pone.0007287, e7287.
49. Loupakis F, Pollina L, Stasi I, et al. PTEN expression and KRAS mutations on primary tumors and metastases in the prediction of benefit from cetuximab plus irinotecan for patients with metastatic colorectal cancer. *J Clin Oncol* 2009;**27**(16):2622–9. https://ascopubs.org/doi/abs/10.1200/jco.2008.20.2796.
50. Chung DC. The genetic basis of colorectal cancer: insights into critical pathways of tumorigenesis. *Gastroenterology* 2000;**119**(3):854–65. https://doi.org/10.1053/gast.2000.16507.

51. Su LK, Vogelstein B, Kinzler KW. Association of the APC tumor suppressor protein with catenins. *Science* 1993;**262**(5140):1734–7. https://doi.org/10.1126/science.8259519.
52. Kikuchi A. Regulation of beta-catenin signaling in the Wnt pathway. *Biochem Biophys Res Commun* 2000;**268**(2):243–8. https://doi.org/10.1006/bbrc.1999.1860.
53. Wong NACS, Pignatelli M. Beta-catenin—a linchpin in colorectal carcinogenesis? *Am J Pathol* 2002;**160**(2):389–401. https://doi.org/10.1016/s0002-9440(10)64856-0.

Chapter 47

CRC: A Darwinian model of cellular immunoselection

Mónica Bernal[a,b], Natalia Aptsiauri[b,c], María Otero[d,e], Ángel Concha[f,g], Federico Garrido[a,b,c], and Francisco Ruíz-Cabello[a,b,c]

[a]UGC of Laboratories, Hospital Universitario Virgen de las Nieves, Granada, Spain, [b]Department of Biochemistry and Molecular Biology III/Immunology, University of Granada, Granada, Spain, [c]Institute of Biosanitary Research of Granada (IBS), Granada, Spain [d]Pathological Anatomy Service Santiago de Compostela University Hospital Complex, Santiago de Compostela, Spain, [e]Institute of Biosanitary Research of Santiago de Compostela (IDIS), Santiago de Compostela, Spain, [f]Pathology Anatomy and Pathology Department, A Coruña University Hospital Complex, A Coruña, Spain, [g]Biobank A Coruña, Institute of Biosanitary Research of A Coruña (INIBIC), A Coruña, Spain

Cancer immunoedition: From immunosurveillance to tumor escape

Numerous reports describe that human tumors are usually infiltrated by immune cells.[1,2] These immune infiltrates can vary between tumors in size and composition, but their presence is considered an attempt by the host to detect and destroy emerging tumor cells in order to stop tumor progression.[3] This concept of the immune control over cancer cells was already proposed in 1909 by Paul Ehrlich,[4] and reformulated in the late 1950s by Burnet and Thomas as the theory of cancer immunosurveillance.[5,6] According to it, the immune system is able to specifically recognize and eliminate cancerous and precancerous cells that arise in the host before they can develop to detectable tumors and cause damage. However, the idea of cancer immunosurveillance was not actually accepted until the 1990s, when experimental animal models using knock-out mice validated the existence of immune control in both spontaneous and chemically induced tumors.[7,8]

In humans it is difficult to assess the existence of natural immunosurveillance, but there are convincing clinical data that indirectly demonstrate the importance of the immune system in cancer prevention. Individuals with acquired or inherited immunodeficiencies and patients treated with immunosuppressants following an organ transplant present a higher risk of developing cancer than healthy individuals.[9,10] On the other hand, cases of spontaneous regression of metastasis in melanoma and colorectal cancer have been reported,[11,12] as well as numerous papers have reported the presence of lymphocyte infiltrates in human tumors with a better prognosis and improved survival of patients.[13,14] More recent data obtained from the "microarray" analysis of immune markers suggested that the type, density, and location of immune cells in the tumor microenvironment of colorectal carcinomas have an important prognostic value.[14–16]

Immune cells within the tumor infiltrate include those that mediate adaptive immunity, such as T cells, dendritic cells (DCs), and occasional B lymphocytes, as well as innate immunity cells, such as macrophages, polymorphonuclear leukocytes, and NK cells.[17] Tumor-Infiltrating Lymphocytes ("TILs") are CD3+ T cells, including mainly CD8+ or cytotoxic T cells ("Cytotoxic T Lymphocytes," CTLs) and CD4+ "helper" T cells. CD4+T cells specifically recognize, via membrane receptors ("T Cell Receptors," TCRs), tumor-derived antigenic peptides that are presented by antigen-presenting cells ("Antigen Presenting Cells"), APCs (DCs and macrophages) through HLA Class II molecules. Consequently, a Th1 response is triggered that facilitates inflammation and tumor rejection by increasing cytotoxic activity of CD8+ T cells. These CTLs, in turn, recognize tumor-specific antigenic peptides presented by the cancer cell itself by HLA Class I molecules, resulting in lymphocyte activation, release of the contents of their cytolytic granules, and tumor cell destruction.

Both lymphocyte subpopulations are considered to be an essential part of the immune reaction responsible for inhibition of tumor growth and development. For this reason, many immunotherapy strategies are aimed at activating these lymphocytes to achieve tumor rejection, and at promoting long-term immune memory that prevents recurrence of primary tumor and the development of metastasis.

However, despite immunosurveillance, tumors develop in the presence of a fully functional immune system. On the other hand, there are publications demonstrating a lack of significant positive correlation between the intensity of lymphocyte infiltrate and prognosis. Moreover, they associate tumor infiltration by immune cells with a worse prognosis.[18,19] In fact, important observations indicate that both innate and adaptive immunity have a "dark" side that can promote tumor progression. On the one hand, tumor succeeds to benefit from infiltrating inflammatory cells, modifying its behavior to create a microenvironment that favors tumor progression. Thus it takes over a normal process of inflammation and reprograms functions, such as cell proliferation and differentiation, extracellular matrix remodeling, angiogenesis, and cell migration/recruitment, to benefit cancer growth and survival.[20] At the same time, it is known that the immune system, through its interaction with the tumor and as a result of intense immunosurveillance, edits cancer phenotype and promotes the immune selection of less immunogenic tumor variants that eventually facilitates tumor propagation.

This complex role of the immune system in tumor development has led to the new concept that refines and completes the theory of immunosurveillance: immunoediting of cancer (Fig. 47.1).[21] The immunoediting process occurs from immunosurveillance to immunoescape and has three phases: elimination, equilibrium, and escape.[21,22] During the elimination phase, as a result of immune surveillance, innate immunity cells, inflammatory mediators, and CD4+ and CD8+ effector T cells manage to suppress tumor growth in the earliest stages. In the equilibrium phase, the host immune system and tumor cells that have survived the elimination phase enter a temporary state of dynamic balance. The pressure exerted by immunosurveillance is sufficient to control tumor progression, but it does not completely eliminate malignant cells. During this period, tumor cells are thought to remain dormant or accumulate changes (DNA mutations or changes in gene expression), which alter their phenotype. Finally, this process results in the selection of tumor variants with alterations that allow them to resist, avoid, or suppress the antitumor immune response. This equilibrium phase is the longest and can

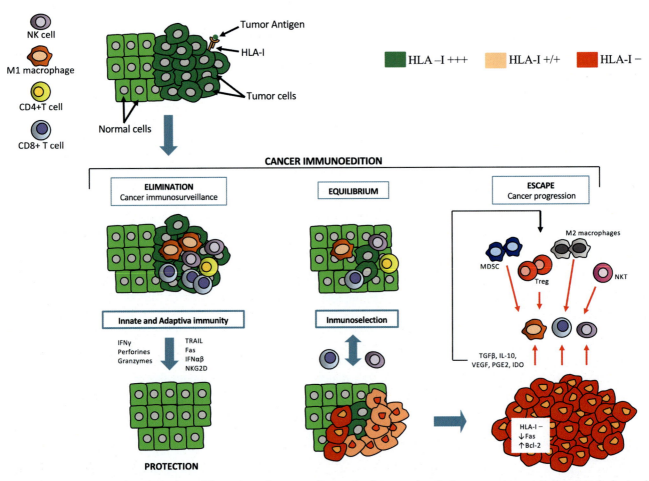

FIG. 47.1 Three phases of cancer immunoediting, a dynamic process of tumor development when the immune system can both protect the host and promote tumor growth. This process begins with the phase of immunosurveillance and cancer elimination, followed by equilibrium between tumor cells and immunity, which controls tumor expansion. During the last immune escape phase, less immunogenic tumor variants survive and proliferate, while the tumor microenvironment develops immunosuppressive mechanisms that attenuate the immune response and lead to cancer progression. These phases have been known as the "three Es of immuno-editing." Tregs—Regulatory T cells; NKTs—Natural Killer T cells; MDSCs—Myeloid Derived Suppressor Cells; IDO—Indolamine 2,3-dioxygenase; VEGF—Vascular endothelial growth factor; IL-10—Interleukin-10; PGE2—prostaglandin E2; TGF-β—Transforming growth factor-β; MICA/B—MHC class I chain-related antigens A and B.

last years.[23] During the escape phase, tumor variants that have survived begin to proliferate uncontrollably. At this point, the immune system is no longer able to contain the tumor that continues to grow progressively, clinically translating into the malignant development of the disease.

Tumor escape strategies

Numerous publications have focused on analyzing the complex interaction between the tumor and the immune system, and the diversity of tumor escape mechanisms trying to design new strategies that improve immunotherapy in cancer. Tumor escape mechanisms involve numerous elements of the tumor microenvironment (Table 47.1).

(i) Various soluble tumor-derived factors ("Tumor Derived Factors," TDFs) that can promote tumor escape, suppress innate and adaptive immunity, and facilitate tumor progression, including vascular endothelial growth factor (VEGF), interleukine-10 (IL-10), reactive oxygen species (ROS), indoamine-2,3-dioxygenase (IDO), prostaglandins (e.g., PGE2), transforming growth factor-β (TGF-β), and galectin-3.[24, 25]

TABLE 47.1 Principal mechanisms implicated in tumor escape from immunosurveillance.[24–36]

Principal mechanisms of tumor escape		
Elements of tumor microenvironment		Function
Tumor-derived soluble factors (TDFs)	TGF-β	Cytokine with pleiotropic effect, including: promotion of angiogenesis, suppression of the production of IFN by Th1 and T CD8+ cells, promotion of the differentiation of Treg and Th17 cells, inhibition of maturation, antigen presentation and costimulation of macrophages and dendritic cells (DCs), suppression of differentiation, proliferation and function of effector T cells, promotion of survival and resistance to apoptosis of tumor cells
	IL-10	Complex role in tumor development. It helps regulate the immune response by preventing inflammation, and, at the same time, by suppressing the activation of effector T cells and enhancing the survival of tumor cells, as a result favoring the development of cancer
	VEGF	Promotion of angiogenesis
	PGE2	Stimulation of angiogenesis and tumor progression
	Galectin-3	Carbohydrate binding lectin with antiapoptotic function, stimulates tumor progression by activation of cell migration and invasion, induces T cell apoptosis
Tumor-infiltrating cells	M2 macrophages, immunosuppressors	Production of IL-10, TGF-β, and Arginase (ARG). Promotion of Th2 differentiation, angiogenesis, and tissue remodeling
	MDSC (immunosuppressor cells of the myeloid origin)	Blocking CD maturation and polarization toward a Th2 response via IL-10 release and IL-12 inhibition. Production of inhibitory enzymes, such as Arginase 1 (ARG1), Nitric Oxide Synthase (NOS), and Indolamine 2,3 mediated T cell activation, as well as its survival. The action of IDO inhibits the tumor capacity of dioxygenase (IDO). The activity of ARG1 and NOS affects the signaling of TCR-activated T cells and enhances the suppressive activity of Treg cells
	Treg cells, CD4+CD25+ Foxp 3+	Suppression of the activation of effector antitumor T cells by producing IL-10 and TGF-β
	Other: immature CDs, CDs producing IDO, Th17 cells	Induced by the TDFs, which contribute to the creation of an immunosuppressive or peripheral tolerance network that compromises the effectiveness of the innate and adaptive immune response

Continued

TABLE 47.1 Principal mechanisms implicated in tumor escape from immunosurveillance—cont'd

Elements of tumor microenvironment	Principal mechanisms of tumor escape	Function
Cell surface molecules involved in the interaction between T cells and tumor cells.	Costimulatory molecules (B7 family)	Lack of expression of costimulating molecules necessary for the correct activation of T cells, which leads to T cell anergy
	CD3 ζ chain and tyrosine kinases	Decrease in the expression of the CD3 ζ of TCR and tyrosine kinases p56lck and p59fyn. These defects, influenced by the activity of ARG1, affect TCR-mediated signaling and T cell activation
	PD-L1 (Programmed cell death ligand)	Increased tumor PD-L1 expression, which negatively regulates the proliferation and production of cytokines by T cells
	Fas receptors in activated T cells	Interacts with the Fas Ligand (Fas-L) expressed or released by certain types of tumor, inducing the apoptosis of the effector T cells
	Death molecules and receptors in tumor cells.	Modulation of pro and antiapoptotic factors (Bcl-2, Bcl-x2), low expression or loss of Fas or TRAIL receptors, or deregulation of apoptosis signaling pathways—all lead to resistance to cell death in cancer cells
	Molecules involved in tumor antigen recognition	Loss or downregulation of HLA Class I molecules and antigen processing machinery (APM) molecules, which leads to the loss of the ability of tumor cells to directly present antigenic peptides derived from the tumor to CTLs

(ii) Cells of the innate and adaptive immunity that contribute to creating an immunosuppressive or peripheral tolerance network through its soluble products or membrane molecules. These cell populations include immunosuppressor M2 macrophages, regulatory T cells (Tregs), "Natural Killer" T cells (NKTs), immature dendritic cells (iDCs), IDO-producing plasmocytoid dendritic cells (pDC), and myeloid suppressor cells (MDSCs).[26, 27, 37]

(iii) Cell surface molecules that mediate the interaction of tumor cells with T lymphocytes, altered expression of which leads to the tolerance against tumor antigens or to a functional suppression of effector lymphocytes. Three key strategies are observed in these escape mechanisms that rely on intercellular physical contacts: (a) the aberrant expression of cell surface ligand receptors (downregulation of costimulating molecules, decreased expression of the CD3 chain of TCR, increased expression of the programmed death receptor ligand [PD-L1]), whose interactions affect the effector functions of lymphocytes[28, 29, 38]; (b) acquisition of genetic or epigenetic changes that increase the resistance to apoptosis mediated by death receptors in tumor cells (modulation of pro- and antiapoptotic factors, decrease or loss of expression of Fas or TRAIL receptors)[29]; (c) total loss or downregulation of the molecules involved in the recognition of tumor antigens (HLA Class I molecules, molecules of antigen processing machinery [APM]).[30, 39–41]

Defects in the presentation of tumor antigens: Alterations in HLA class I molecules

It has long been known that the malignant transformation of cells is associated with an alteration in the expression and/or function of HLA Class I molecules, and that these abnormalities provide mechanisms to escape recognition by the immune system. HLA molecules bind antigenic peptides derived from tumor endogenous proteins ("Tumor Associated Antigens," TAAs) and present them to CD8+ cytotoxic T lymphocytes (CTLs) to be recognized by their membrane receptors (TCRs).[42] They are also involved in the interaction of malignant cells with NK cells. Specifically, these "Natural Killer" lymphocytes detect tumor cells that do not express one or more HLA-I alleles.[43]

HLA Class I antigens are cell surface glycoproteins consisting of a 45-KDa polymorphic heavy chain and a 12-KDa nonpolymorphic light chain called beta2-microglobulin (B2m). The correct assembly of the "heavy chain/B2m/antigenic peptide" complex and its transport to the cell surface follows a process dependent on antigen processing machinery ("Antigen Processing Machinery," APM) which includes the subunits of the proteasome LMP2 and LMP7, the peptide transporters TAP1 and TAP2, and a number of chaperones of the endoplasmic reticulum, such as calnexin, calreticulin, and ERp57.[44] Molecular mechanisms responsible for the alterations in tumor HLA Class I expression may either directly affect heavy chain or the B2m, or the components of APM. The existence of mutations, chromosomal aberrations (loss of heterozygosity), or alterations in transcriptional regulation affecting heavy chain genes lead to the generation of altered HLA phenotypes consisting of allelic, locus, or HLA class I haplotype loss.[39, 45–49] HLA Class I downregulation is mostly due to low expression or loss of APM components.[41] The total loss of HLA Class I expression is mainly caused by the biallelic inactivation of the B2m gene resulting in the lack of expression of a functional B2m light chain of the HLA class I complex.[50, 51] Mutations in the B2m gene have been mainly described in melanoma and some cases of colorectal cancer.[52–57] In fact, the total HLA-I loss frequency in CRCs is approximately 27%. However, this mutation frequency increases significantly (up to 60%) in a subgroup of CRCs with a very characteristic phenotype from the molecular, clinical, and pathological point of view, which is called the phenotype with a Microsatellite Instability (MSI).[58–60] In CRCs with MSI, the accumulation of mutations in B2m gene is the best validation of tumor cell selection as a result of strong immune selective pressure in the tumor microenvironment.

An important structural mechanism that impairs antigen presentation in tumor cells is the loss of one HLA haplotype caused by the loss of heterozygosity at chromosome 6 (LOH-6) and inactivation of one B2m gene copy due to the LOH at chromosome 15 (LOH-15). LOH is a major mechanism for inactivation of tumor-suppressor genes results in a permanent loss of one allele in a polymorphic locus in somatic cells, resulting in homozygosity. The HLA heavy chain and light chain regions are located at chromosomes 6 and 15, respectively, and loss or alteration of this region provides tumor cells with a mechanism to escape from the immune system. LOH involving HLA genes is a frequent (up to 50%) and a common event described in a variety of tumors, including colorectal cancer, melanoma, breast cancer, bladder cancer, laryngeal cancer, and prostate cancer.[61–65] Under normal conditions six HLA class I alleles are expressed on the cell membrane. HLA-I is a highly polymorphic gene with each molecule capable of presenting a restricted set of antigenic peptides to T cells. It maximizes the variety of peptides that can be presented and improving responses to pathogens and tumors.[66] LOH at the 6p21.3 region leads to the simultaneous loss of three HLA class I genes and reduces the variety of antigens, including tumor-associated antigens and neoantigens, capable of stimulating T cell mediated cytotoxicity. Importantly, even the loss of one HLA locus can have an impact on the overall survival in cancer. Furthermore, as mentioned before, LOH in HLA increases its susceptibility to a second genomic alteration that completely inactivates the affected HLA-I locus, further decreasing the probability of tumor detection and destruction by the immune system. It has been reported that in lung cancer LOH in HLA was under selective pressure by the immune system and occurred in 40% of analyzed tumors.[67]

Colorectal carcinomas with microsatellite instability: Role of the B2M gene

CRC is a heterogeneous disease that can be divided in two major groups: 15% of CRC tumors are "hypermutated with microsatellite instability (MSI-H)" because of a defect in the DNA mismatch repair (MMR) system and ubiquitous somatic mutations in repeated DNA sequences that occur during colon carcinogenesis; and 85% of CRC cases belong to the "non-mutated, microsatellite stable (MSS) group."[39] These two groups of CRC also have different molecular mechanism responsible for their HLA-I alterations. Recently, based on a large-scale analysis of six previously published bulk-transcriptome data sets from CRC tumors (stage I–IV) four types CMSs have been proposed with distinct molecular and clinical features. To be precise, the MSI-H tumors were classified as CMS1, while MSS tumors were divided into CMS2, CMS3, and CMS4 tumors.[68]

Approximately 15% of sporadic CRCs develop from a molecular pathway of genomic destabilization known as microsatellite instability (MSI). This route of carcinogenesis is characterized by the accumulation of mutations (deletions, insertions, or substitutions) in short tandem repetition sequences (microsatellites) as a result of an altered mismatch repair system ("MMR"), unable to correct errors made by DNA-polymerase following DNA replication.[69, 70] As a result, there are hundreds of thousands of new somatic mutations in microsatellite sequences distributed throughout the genome. Many of these repeat sequences are located in gene-coding regions associated with cancer, mainly tumor-suppressing genes whose expressions and/or functions may be affected. CRC tumors with high degree of MSI that affects a very large number of genes involved in colorectal carcinogenesis are classified as tumors with high rate of microsatellite instability (MSI-H) with "mutator" phenotype and "hypermutable" areas.

In addition, this genetic defect that characterizes CRCs with MSI determines the high immunogenicity observed in these tumors: the accumulation of thousands of deletions/insertions in microsatellite sequences contained in the coding regions of many genes leads to a high frequency of frame shift mutations. These alterations promote the appearance of premature termination codons during translation, generating numerous truncated proteins that provide a source of new aberrant peptides ("Frame Shift Peptides," FSPs). Many of these new tumor-derived peptides are presented via HLA Class I molecules to CTLs that recognize them through their membrane receptors (TCRs). As a result, lymphocyte activation is triggered that

FIG. 47.2 Immunohistochemistry of tumor-infiltrating lymphocytes in colorectal carcinoma: (A) in CRC with microsatellite instability (MSI) and (B) in CRC with genomic stability (MSS). As shown in the image, CRCs with MSI show high infiltration by CD3+ T cells, which are mostly CD8+ T cells. This presence of CTLs in CRCs with MSI is much higher than that seen in MSS tumors, with few or even absent CD8+ T cells.

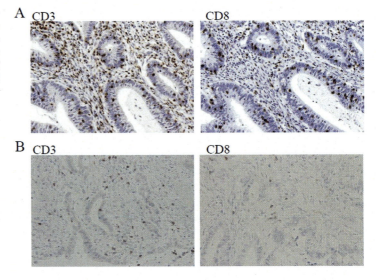

culminates in the release of the contents of the cytotoxic granules causing the death of neoplastic cells. This explanation reinforces the immunohistochemical and molecular findings that have been obtained in the last decade and confirms the existence of an activated immune reaction to the tumor. On the one hand, there are numerous studies describing that CRCs with MSI are widely infiltrated by TILs ("Tumor-Infiltrating Lymphocytes), identified as activated cytotoxic CD3+ CD8+ T lymphocytes, as compared to tumors showing microsatellite stability ("Micro Satellite Stability," MSS)[58, 71–73] (Fig. 47.2). In addition, it has been described that the significant CTL infiltration found in CRCs with MSI-H is independent of tumor HLA Class I expression, tumor stage, and certain molecular characteristics, such as mutations in BRAF or KRAS oncogenes.[58] In addition, recent genomic studies comparing the gene expression profiles of MSI-H CRCs and MSS CRCs using "microarrays" have found numerous relevant genes expressed differently in these two groups of tumors.[74–76] Among them, the genes related to immune response (genes involved in inflammatory response and cytokine production, macrophage marker coding genes, gene sequences involved in the activation, proliferation, and cytotoxic activity of T cells) have been reported to be upregulated in the microenvironment of MSI-H CRCs as compared to MSS tumors.[75] Consequently, it has been proposed that the "overexpression" of a considerable number of genes involved in the innate and antigen-specific immune response in MSI-H CRCs determines a gene expression profile that could be considered as a biomarker of cancers with a high degree of genomic instability.[75]

Therefore the strong infiltration with CTLs and high expression of immune response genes in CRCs with MSI-H support the concept that in the tumor microenvironment there is a strong antitumor response as a result of immunosurveillance that indicates the important role played by immunoediting in these carcinomas with MSI. The strong immune pressure in the tumor microenvironment of MSI-H CRCs as compared to MSS tumors reflects differences in tumor-host interaction between these tumor types, and explains the most frequently observed escape mechanism in genetically unstable carcinomas: the loss of expression of HLA Class I molecules caused by inactivating B2m mutations, which prevents the presentation of tumor-specific antigens to CTLs. Mutations affecting the B2m gene in MSI CRCs are mainly deletions and insertions that accumulate in the microsatellite sequences localized in exons 1 and 2 of the gene. These mutations are associated with the genomic instability of these tumors, i.e., they are the result of an ineffective DNA repair system. At the same time, they are a reflection of the pressure exerted by immunosurveillance that favors the selection of tumor variants with loss of immunogenicity due to the total loss of expression of HLA-I caused by mutations in the B2m gene.[39, 58, 77]

Mutations in the B2m gene: An immunoselection model in MSI-H carcinogenesis

A total loss of HLA-I has been seen in up to 70% of MSI-H CRC tumors due to the mutations in B2m gene, LOH-15 and LOH-6. It appears that inactivation of the B2m gene can play a crucial role in the development of MSI-H CRCs. In fact, there is accumulating evidence demonstrating the functional implications of B2m gene mutations, confirming its role as a "driver gene" (gene with mutations that provide a selective advantage for tumor growth) not only in MSI CRCs, but also in other human tumors.[78] First of all, different studies show a high frequency of B2m gene mutations in MSI-H CRCs (up to 60%), while in MSS CRCs B2m alterations are less common.[58–60, 79] In addition, the frequency of mutations in coding repetition sequences of the B2m gene in MSI-H CRCs is much higher than the incidence of random mutations, which is less than 1% in microsatellites with a length of 5 nucleotides.[80] Similarly, no mutations have been observed in nucleotide repeats longer than microsatellite sequences of the B2m gene, and which are present in both coding and noncoding regions of other genes in tumors with altered MMR system. For example, in MSI-H CRCs with B2m mutations no mutations have been detected in the regions of microsatellites located in exons of the TAP1 and TAP2 genes that, in principle, would be equally susceptible to accumulating alterations associated with MSI.[58] Biallelic inactivation of the B2m gene has also been observed to be a common event in MSI-H CRCs. However, it has not been documented in all target genes of MSI-H carcinogenesis,[59, 60] which reinforces the role of the immune system in selecting these mutations.

Another important fact is that mutations affecting the B2m gene have been detected in very early stages of MSI-H tumorigenesis. In colorectal adenomas the mutation frequency is 16%, but it continues to increase during tumor progression and reaches 40% in MSI-H CRCs with lymph nodes metastases.[60] These data support the idea that mutations in B2m play a role in tumor promotion, not only at early stages of cancer development, but also during local invasion and expansion, coinciding with what has been described for other relevant genes inactivated in MSI-H cancers, such as TGFBR2.

Objective evidence of the importance of the inactivating mutations in the HLA molecules in CRCs with MSI-H has been obtained by immunohistochemical tumor analysis with the use of specific monoclonal antibodies to the free cytoplasmic B2m and HLA-BC heavy chain, and the HLA-ABC-B2m complex on the cell surface. In these tumors, mostly a homogeneous pattern of total loss of the B2m protein is observed throughout tumor tissue, corresponding to a complete absence of HLA Class I antigen expression on the surface of neoplastic cells, even if the cytoplasmic expression of the HLA heavy chain is positive (Fig. 47.3).[58] This pattern of dominance of B2m mutations to all cancer cells suggests that, at some stage of MSI-H tumor development, mutations in the repeat sequences of B2m gene exons are positively selected because they confer an advantage in growth that eventually promotes clonal expansion of tumor cells carrying these B2m alterations.

Therefore MSI-H-specific mutations in B2m during the local development of the tumor and its expansion have a high functional relevance because they establish a fundamental molecular mechanism that promotes tumor escape from the action of the immune system in CRCs with MSI-H.

Clinical implication of B2m gene mutations in CRCs with MSI-H

Immune evasion mediated by the alterations associated with microsatellite instability in B2m has significant prognostic involvement in MSI-H CRCs and in other human neoplasms. Interestingly, an increase in the frequency of B2m mutations has been detected in the progression of colorectal cancer from adenoma to CRCs, and within CRCs with MSI-H, from Stage I (27%) to stage III (43%). However, no alterations in B2m have been found in stage IV MSI-H-positive colorectal carcinomas.[60] These data suggest that although they play a role in promoting local tumor growth and developing lymph nodes metastasis, B2m mutations may interfere with the formation of liver metastases in MSI-H CRCs. This would indicate that patients with MSI-H-positive CRCs and lack of HLA class I expression due to B2m mutations have low metastatic potential, and that the formation of distal metastasis requires the positive expression of HLA class I antigens by tumor cells.[81] These data are supported by the evidence demonstrating the control of the metastatic dissemination of HLA class I-deficient tumor cells via blood by NK lymphocytes.[82, 83] At the same time, cytotoxic activity of T lymphocytes would

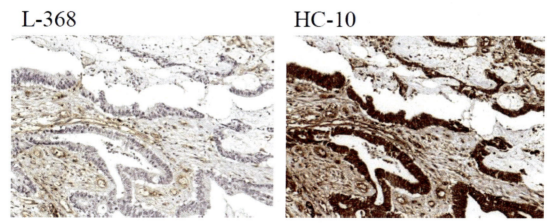

FIG. 47.3 Analysis of HLA class I expression in colorectal carcinoma (CRC) with microsatellite instability phenotype (MSI) using paraffin-embedded tissue sections and monoclonal antibodies against the B2m chain (L-368), and the HLA-BC free heavy chain (HC-10). The tumor shows positive expression of the intracytoplasmic HLA-BC heavy chain and loss of B2m expression, which causes a total loss of HLA class I molecules on the surface of tumor cells. Molecular studies revealed that this MSI tumor has a mutation in the B2m gene that is responsible for the loss of B2m expression.

promote the destruction of tumor cells positive for HLA class I expression in the primary tumor or metastases. In this context, the total absence of tumor HLA class I has been described to correlate with a favorable prognosis in patients with CRCs, unlike those with tumors showing an intermediate expression,[84, 85] which may help tumor cells to escape route from the cytotoxicity mediated by both NK and T cells.[30, 85] In conclusion, the fact that mutations in B2m are associated with an absence of metastasis and a lower risk of relapse in MSI-H cancers[86] suggests that tumor B2m alterations should be included to the list of prognostic factors in the clinical applications of cancer immunotherapy.

Recent progress in cancer immunotherapy has stimulated the interest to investigate the mechanisms of acquired tumor resistance to treatment. Novel immunogenomic methods of tumor examination allow for a more detailed molecular and cellular analysis of the tumor-immune microenvironment at the time of acquired resistance. There is accumulating evidence indicating that HLA class I alterations accumulate in tumor cells and provide an immune escape route, as well as resistance to immunotherapy.[39] Various research groups reported an increased frequency of altered antigen presentation associated with genomic alterations in HLA genes in recurrent tumors in patients treated with antibodies blocking immune checkpoint pathways. The efficiency of this therapy depends on cytotoxic CD8+ T cell (CTL) recognition of cancer-specific antigens presented on HLA class I complexes, which are composed of a heavy chain and beta-2-microglobulin (B2M), an important subunit required for to the assembly of all HLA class I complexes and for the stable presentation of antigens by the tumor cells.

Apart from being a predictive marker for the success of immune therapies, LOH in HLA- or B2M-carrying chromosomes can result in secondary immune evasion which has been described in several cancers, including bladder, melanoma, and lung cancer. These genomic aberrations may change the expression of HLA-ABC/b2m complex on the tumor cell membrane, but deletion of HLA genes by LOH may enable the clonal expansion of HLA negative tumor cells and this selective pressure could explain the increased frequency of LOH within the HLA genes after immunotherapy.

Studies by our laboratory showed that Bacillus Calmette-Guerin immunotherapy of bladder cancer induced the selection of human leukocyte antigen class I-deficient tumor cells with LOH in chromosome 6 and 15.[87] Similarly, melanoma patients, receiving anti-PD-1 therapy (pembrolizumab), developed immune escape lesions with LOH in Jak2 and a truncating mutation in B2M.[88] Sade-Feldman and colleagues analyzed escape lesions from five melanoma patients treated with Ipilimumab (anti-CTLA4 Abs) and found LOH in B2M in 29.4% of the cases.[89] Another example of acquire resistance to ICIs was described by Gettinger et al. who demonstrated a complete genomic loss of B2M (a copy number variation) in lung cancer lesions after anti-PD-L1 and anti-CTLA-4 treatment.[90] In colorectal cancer it was reported that LOH can occur at sites of disruptive mutations in β2m, resulting in biallelic β2m inactivation.[91] Importantly, the impact of such a lesion on the efficacy of ICI therapy was recently demonstrated in a case study where a frameshift mutation, followed by LOH, of the β2m locus rendered a MSI-H CRC tumor resistant to anti-PD-1 immune therapy.[92]

Moreover, it should be taken into account the importance of recovering a functional B2m gene by gene therapy in order to reexpress or enhance the expression of Class I HLA with the intention of achieving tumor rejection.[93]

The prognostic value of tumor lymphocyte infiltration: The importance of the "Immunoscore"

In recent years, several studies have confirmed the hypothesis that tumor development is influenced by the host's immune system.[14–16, 94, 95] To date, tumor staging (IUCCC-TNM classification) based mainly on histopathological characteristics has been used to estimate the prognosis of patients with different typed of cancer. However, it is known that this classification has some limitation in the evaluation of cancer prognosis, development, and clinical outcome, which can vary significantly between patients with the same tumor stage. Some advanced-stage cancer patients may remain stable for years, or even experience partial or total spontaneous regression of tumor metastases. In contrast, early stage cancer patients may experience rapid tumor progression and death, even after complete surgical removal of the visible tumor and without evidence of distal metastasis. Tumor progression has been assumed to be exclusively linked to the intrinsic characteristics of cancer cells, without taking into account the important role of host immune response.[94] Meanwhile, HLA genomic alterations responsible for tumor immune escape are produced under the pressure induced by T lymphocytes.

Histopathological tumor analysis demonstrates that tumors are usually infiltrated by inflammatory and immune cells at various degrees. In particular, more detailed studies of CRCs have revealed that immune cells are not randomly distributed but are organized into different infiltrates at the center of the tumor area (CT), the invasive margin (IM) of tumor nests, and lymphoid islets adjacent to the tumor.[15] As we have mentioned earlier, the interaction of the immune system with the tumor can influence tumor biology and determine the development and progression of cancer.[96, 97] Numerous data from a large number of human cancers showed that the number, type, and location of the immune infiltrate in primary tumors are essential prognostic factors of disease-free and overall survival.[98] Specifically, the presence of Th1 response markers, cytotoxic T cells and memory T cells, both in CT and IM of the primary tumor, have been strongly associated with a favorable prognosis in CRC patients.[15, 96] Similarly, this high lymphocyte density as well as the expression of antitumor cytotoxic response genes and Th1 differentiation genes correlates with the absence of histological markers of metastatic invasion (vascular embolism, lymphatic invasion, perineural invasion ("VELIPI")) and a lower risk of relapse in patients with CRC.[94] These results suggest that once the tumor becomes clinically detectable, adaptive immune response plays a role in preventing tumor recurrence and metastasis.[15, 16] Consequently, the analysis of the "immune contexture," i.e., the nature, density, and location of adaptive immune cells in different regions of the tumor, is essential for the estimation of the prognosis of patients with colorectal cancer.[14, 96, 99, 100] The impact of the cytotoxic and memory T cell infiltration has been initially demonstrated in CRC and later confirmed in other types of human tumors, and in all cases it has been associated with better survival.[96] Therefore it has been suggested to include an immune score as a component of tumor classification, quantifying the density and location of immune cells in the tumor.[97, 100, 101] The "immune score" (Im) determines the density of CD8+ T cells and memory T cells (CD45RO+) in CT and IM of surgically extracted tumors to provide a score ranging from Im0, when cell infiltration is low in both regions, to Im4, when the densities of these lymphocyte subtypes are high. In a work by Pagès et al., the analysis of 599 CRCs in early stages (stages I and II) revealed a highly significant correlation between disease-free and overall survival and the tumor "immune score".[95] Patients with low Im (Im0 and Im1) had very poor prognosis, while patients with a high Im (Im3 and Im4) experienced a very low incidence of disease recurrence. Detailed genomic studies have shown that this immune classification has a prognostic value that may be better than the histopathological methods normally used in tumor staging.[14–16] This immune score-based classification would help identify high-risk patients who could benefit from new therapeutic strategies, including immunotherapy.

In recent clinical follow-up studies it has been shown that Immunoscore has an important prognostic value in early and advanced stage CRC and the analysis of the Immunoscore in metastatic lesions provides valuable information on patient evolution.[102] In addition, Immunoscore has an important predictive value of the response to chemotherapy[103] and immunotherapy using monoclonal antibodies directed against the "immune checkpoint" molecules (CTLA-4, PD-1, PD-L1). Patients with high-scoring tumors may have a better response to these therapeutic strategies so tumors can be selected for such treatment.[104]

Consequently, an International Consensus has been reached for the validation of tumor Immunoscore as a diagnostic tool in clinical practice and an important player in the new precision medicine in CRC. The use of digital image analysis of tumor tissues and establishing accurate technique and evaluation criteria allows the Immunoscore to be applied in a standardized and reproducible way.[105–107]

In conclusion, the presented data underline the relevance of two new biomarkers with prognostic value as future tools to be used in clinical practice for the classification of CRCs: the immune score and the analysis of mutations that inactivate the B2m gene. Although they are considered to be prognostic and potentially predictive factors in colorectal cancer, they could be applied to most human tumors. Hence, a great effort has been made to standardize and validate various clinical trials aimed, above all, at determining the immune score before implementing it in the clinical routine. This would help to improve the future use of these biomarkers for better defining the prognosis of colorectal cancer patients, helping to identify patients at high risk of recurrence and facilitating decision-making regarding the care of the patient, helping to select those patients who would benefit from immunotherapy. It is also important to investigate how these two prognostic factors would provide new targets for immunotherapy, favoring the development of this field of treatment that has so far had little success in colorectal cancer.

References

1. Balkwill F, Mantovani A. Inflammation and cancer: back to Virchow? *Lancet* 2001;**357**(9255):539–45. https://doi.org/10.1016/S0140-6736(00)04046-0.
2. Whiteside TL. *Tumor-infiltrating lymphocytes in human malignancies*. RG Landes Company; 1993.
3. Zitvogel L, Tesniere A, Kroemer G. Cancer despite immunosurveillance: immunoselection and immunosubversion. *Nat Rev Immunol* 2006;**6**(10):715–27. https://doi.org/10.1038/nri1936.
4. Ehrlich P. Über den jetzigen Stand der Chemotherapie. *Ber Dtsch Chem Ges* 1909;**42**(1):17–47. https://doi.org/10.1002/cber.19090420105.
5. Burnet FM. Cancer-a biological approach. IV. Practical application. *BMJ* 1957;**1**:844–7.
6. Thomas L. Delayed hypersensitivity in health and disease. In: Lawrence HS, editor. *Cellular and humoral aspects of the hypersensitive states*. USA: Hoeber-Harper; 1959. p. 529–32. Published online.
7. Dighe AS, Richards E, Old LJ, Schreiber RD. Enhanced in vivo growth and resistance to rejection of tumor cells expressing dominant negative IFNγ receptors. *Immunity* 1994;**1**(6):447–56. https://doi.org/10.1016/1074-7613(94)90087-6.

8. Kaplan DH, Shankaran V, Dighe AS, et al. Demonstration of an interferon-dependent tumor surveillance system in immunocompetent mice. *Proc Natl Acad Sci* 1998;**95**(13):7556–61. https://doi.org/10.1073/pnas.95.13.7556.
9. Swann JB, Smyth MJ. Immune surveillance of tumors. *J Clin Invest* 2007;**117**(5):1137–46. https://doi.org/10.1172/JCI31405.
10. Reiman JM, Kmieciak M, Manjili MH, Knutson KL. Tumor immunoediting and immunosculpting pathways to cancer progression. *Semin Cancer Biol* 2007;**17**(4):275–87. https://doi.org/10.1016/j.semcancer.2007.06.009.
11. Kalialis LV, Drzewiecki KT, Klyver H. Spontaneous regression of metastases from melanoma: review of the literature. *Melanoma Res* 2009;**19**(5):275–82. https://doi.org/10.1097/CMR.0b013e32832eabd5.
12. Bir AS, Fora AA, Levea C, Fakih MG. Spontaneous regression of colorectal cancer metastatic to retroperitoneal lymph nodes. *Anticancer Res* 2009;**29**(2):465–8. https://www.ncbi.nlm.nih.gov/pubmed/19331187.
13. Baxevanis CN, Dedoussis GV, Papadopoulos NG, Missitzis I, Stathopoulos GP, Papamichail M. Tumor specific cytolysis by tumor infiltrating lymphocytes in breast cancer. *Cancer* 1994;**74**(4):1275–82. https://doi.org/10.1002/1097-0142(19940815)74:4<1275::aid-cncr2820740416>3.0.co;2-q.
14. Pagès F, Berger A, Camus M, et al. Effector memory T cells, early metastasis, and survival in colorectal cancer. *N Engl J Med* 2005;**353**(25):2654–66. https://doi.org/10.1056/NEJMoa051424.
15. Galon J, Costes A, Sanchez-Cabo F, et al. Type, density, and location of immune cells within human colorectal tumors predict clinical outcome. *Science* 2006;**313**(5795):1960–4. https://doi.org/10.1126/science.1129139.
16. Galon J, Fridman W-H, Pagès F. The adaptive immunologic microenvironment in colorectal cancer: a novel perspective. *Cancer Res* 2007;**67**(5):1883–6. https://doi.org/10.1158/0008-5472.CAN-06-4806.
17. Whiteside TL. The local tumor microenvironment. Published online, *General principles of tumor immunotherapy: basic and clinical applications of tumor immunology*. Springer; 2007. p. 145–56.
18. Nakano O, Sato M, Naito Y, Suzuki K, Orikasa S. Proliferative activity of intratumoral CD8+ T-lymphocytes as a prognostic factor in human renal cell carcinoma: clinicopathologic demonstration of antitumor immunity. *Cancer Res* 2001. Published online https://cancerres.aacrjournals.org/content/61/13/5132.short.
19. Sheu BC, Hsu SM, Ho HN, Lin RH, Torng PL, Huang SC. Reversed CD4/CD8 ratios of tumor-infiltrating lymphocytes are correlated with the progression of human cervical carcinoma. *Cancer* 1999;**86**(8):1537–43. 10.1002/(sici)1097-0142(19991015)86:8<1537::aid-cncr21>3.0.co;2-d.
20. Whiteside TL. The tumor microenvironment and its role in promoting tumor growth. *Oncogene* 2008;**27**(45):5904–12. https://doi.org/10.1038/onc.2008.271.
21. Dunn GP, Bruce AT, Ikeda H, Old LJ, Schreiber RD. Cancer immunoediting: from immunosurveillance to tumor escape. *Nat Immunol* 2002;**3**(11):991–8. https://doi.org/10.1038/ni1102-991.
22. Dunn GP, Old LJ, Schreiber RD. The three Es of cancer immunoediting. *Annu Rev Immunol* 2004;**22**:329–60. https://doi.org/10.1146/annurev.immunol.22.012703.104803.
23. Teng MWL, Swann JB, Koebel CM, Schreiber RD, Smyth MJ. Immune-mediated dormancy: an equilibrium with cancer. *J Leukoc Biol* 2008;**84**(4):988–93. https://doi.org/10.1189/jlb.1107774.
24. Califice S, Castronovo V, Van Den Brûle F. Galectin-3 and cancer. *Int J Oncol* 2004;**25**(4):983–1075. https://www.spandidos-publications.com/10.3892/ijo.25.4.983.
25. Stewart TJ, Abrams SI. How tumours escape mass destruction. *Oncogene* 2008;**27**(45):5894–903. https://doi.org/10.1038/onc.2008.268.
26. Ochoa AC, Zea AH, Hernandez C, Rodriguez PC. Arginase, prostaglandins, and myeloid-derived suppressor cells in renal cell carcinoma. *Clin Cancer Res* 2007;**13**(2 Pt. 2):721s–6s. https://doi.org/10.1158/1078-0432.CCR-06-2197.
27. Wei S, Kryczek I, Zou W. Regulatory T-cell compartmentalization and trafficking. *Blood* 2006;**108**(2):426–31. https://doi.org/10.1182/blood-2006-01-0177.
28. Freeman GJ, Long AJ, Iwai Y, et al. Engagement of the PD-1 immunoinhibitory receptor by a novel B7 family member leads to negative regulation of lymphocyte activation. *J Exp Med* 2000;**192**(7):1027–34. https://doi.org/10.1084/jem.192.7.1027.
29. Stewart TJ, Greeneltch KM, Christine Lutsiak ME, Abrams SI. Immunological responses can have both pro- and antitumour effects: implications for immunotherapy. *Expert Rev Mol Med* 2007;**9**(04). https://doi.org/10.1017/s1462399407000233.
30. Garrido F, Ruiz-Cabello F, Cabrera T, et al. Implications for immunosurveillance of altered HLA class I phenotypes in human tumours. *Immunol Today* 1997;**18**(2):89–95. https://doi.org/10.1016/s0167-5699(96)10075-x.
31. Andreola G, Rivoltini L, Castelli C, et al. Induction of lymphocyte apoptosis by tumor cell secretion of FasL-bearing microvesicles. *J Exp Med* 2002;**195**(10):1303–16. https://doi.org/10.1084/jem.20011624.
32. Kim R, Emi M, Tanabe K, Uchida Y, Toge T. The role of Fas ligand and transforming growth factor beta in tumor progression: molecular mechanisms of immune privilege via Fas-mediated apoptosis and potential targets for cancer therapy. *Cancer* 2004;**100**(11):2281–91. https://doi.org/10.1002/cncr.20270.
33. Mantovani A, Sozzani S, Locati M, Allavena P, Sica A. Macrophage polarization: tumor-associated macrophages as a paradigm for polarized M2 mononuclear phagocytes. *Trends Immunol* 2002;**23**(11):549–55. https://doi.org/10.1016/s1471-4906(02)02302-5.
34. Nagaraj S, Gabrilovich DI. Myeloid-derived suppressor cells. *Adv Exp Med Biol* 2007;213–23. https://doi.org/10.1007/978-0-387-72005-0_22. Published online.
35. Pollard JW. Tumour-educated macrophages promote tumour progression and metastasis. *Nat Rev Cancer* 2004;**4**(1):71–8. https://doi.org/10.1038/nrc1256.

36. Thomas DA, Massagué J. TGF-β directly targets cytotoxic T cell functions during tumor evasion of immune surveillance. *Cancer Cell* 2005;**8**(5):369–80. https://doi.org/10.1016/j.ccr.2005.10.012.
37. Bingle L, Brown NJ, Lewis CE. The role of tumour-associated macrophages in tumour progression: implications for new anticancer therapies. *J Pathol* 2002;**196**(3):254–65. https://onlinelibrary.wiley.com/doi/abs/10.1002/path.1027.
38. Whiteside T. Immune suppression in cancer: effects on immune cells, mechanisms and future therapeutic intervention. *Semin Cancer Biol* 2006;**16**(1):3–15. https://doi.org/10.1016/j.semcancer.2005.07.008.
39. Anderson P, Aptsiauri N, Ruiz-Cabello F, Garrido F. HLA class I loss in colorectal cancer: implications for immune escape and immunotherapy. *Cell Mol Immunol* 2021;**18**(3):556–65. https://doi.org/10.1038/s41423-021-00634-7.
40. Garrido F, Algarra I. MHC antigens and tumor escape from immune surveillance. *Adv Cancer Res* 2001;117–58. https://doi.org/10.1016/s0065-230x(01)83005-0. Published online.
41. Seliger B, Maeurer MJ, Ferrone S. Antigen-processing machinery breakdown and tumor growth. *Immunol Today* 2000;**21**(9):455–64. https://doi.org/10.1016/s0167-5699(00)01692-3.
42. Romero P, Valmori D, Pittet MJ, et al. Antigenicity and immunogenicity of Melan-a/MART-1 derived peptides as targets for tumor reactive CTL in human melanoma. *Immunol Rev* 2002;**188**:81–96. https://doi.org/10.1034/j.1600-065x.2002.18808.x.
43. Moretta L, Bottino C, Pende D, Mingari MC, Biassoni R, Moretta A. Human natural killer cells: their origin, receptors and function. *Eur J Immunol* 2002;**32**(5):1205–11. https://doi.org/10.1002/1521-4141(200205)32:5<1205::AID-IMMU1205>3.0.CO;2-Y.
44. Cresswell P, Bangia N, Dick T, Diedrich G. The nature of the MHC class I peptide loading complex. *Immunol Rev* 1999;**172**(1):21–8. https://doi.org/10.1111/j.1600-065x.1999.tb01353.x.
45. Griffioen M, Ouwerkerk IJM, Harten V, Schrier PI. HLA-B locus-specific downregulation in human melanoma requires enhancer A as well as a sequence element located downstream of the transcription initiation site. *Immunogenetics* 2000;**52**(1–2):121–8. https://doi.org/10.1007/s002510000262.
46. Imreh MP, Zhang Q-J, De Campos-Lima PO, et al. Mechanisms of allele-selective down-regulation of HLA class I in Burkitt's lymphoma. *Int J Cancer* 1995;**62**(1):90–6. https://doi.org/10.1002/ijc.2910620117.
47. Seliger B, Cabrera T, Garrido F, Ferrone S. HLA class I antigen abnormalities and immune escape by malignant cells. *Semin Cancer Biol* 2002;**12**(1):3–13. https://doi.org/10.1006/scbi.2001.0404.
48. Serrano A, Brady CS, Jimenez P, et al. A mutation determining the loss of HLA-A2 antigen expression in a cervical carcinoma reveals novel splicing of human MHC class I classical transcripts in both tumoral and normal cells. *Immunogenetics* 2000;**51**(12):1047–52. https://doi.org/10.1007/s002510000239.
49. Torres MJ, Ruiz-Cabello F, Skoudy A, et al. Loss of an HLA haplotype in pancreas cancer tissue and its corresponding tumor derived cell line. *Tissue Antigens* 1996;**47**(5):372–81. https://doi.org/10.1111/j.1399-0039.1996.tb02572.x.
50. Benitez R, Godelaine D, Lopez-Nevot MA, et al. Mutations of the β2-microglobulin gene result in a lack of HLA class I molecules on melanoma cells of two patients immunized with MAGE peptides. *Tissue Antigens* 1998;**52**(6):520–9. https://doi.org/10.1111/j.1399-0039.1998.tb03082.x.
51. Paschen A, Méndez RM, Jimenez P, et al. Complete loss of HLA class I antigen expression on melanoma cells: a result of successive mutational events. *Int J Cancer* 2003;**103**(6):759–67. https://doi.org/10.1002/ijc.10906.
52. Bernal M, Ruiz-Cabello F, Concha A, Paschen A, Garrido F. Implication of the β2-microglobulin gene in the generation of tumor escape phenotypes. *Cancer Immunol Immunother* 2012;**61**(9):1359–71. https://doi.org/10.1007/s00262-012-1321-6.
53. Bicknell DC, Rowan A, Bodmer WF. Beta 2-microglobulin gene mutations: a study of established colorectal cell lines and fresh tumors. *Proc Natl Acad Sci* 1994;**91**(11):4751–5. https://doi.org/10.1073/pnas.91.11.4751.
54. Bicknell DC, Kaklamanis L, Hampson R, Bodmer WF, Karran P. Selection for beta 2-microglobulin mutation in mismatch repair-defective colorectal carcinomas. *Curr Biol* 1996;**6**(12):1695–7. https://doi.org/10.1016/s0960-9822(02)70795-1.
55. Hicklin DJ, Dellaratta DV, Kishore R, Liang B, Kageshita T, Ferrone S. Beta2-microglobulin gene mutations in human melanoma cells: molecular characterization and implications for immune surveillance. *Melanoma Res* 1997;**7**(Suppl. 2):S67–74. https://www.ncbi.nlm.nih.gov/pubmed/9578419.
56. Pérez B, Benitez R, Fernández MA, et al. A new β2 microglobulin mutation found in a melanoma tumor cell line. *Tissue Antigens* 1999;**53**(6):569–72. https://doi.org/10.1034/j.1399-0039.1999.530607.x.
57. del Campo AB, Kyte JA, Carretero J, et al. Immune escape of cancer cells with beta2-microglobulin loss over the course of metastatic melanoma. *Int J Cancer* 2014;**134**(1):102–13. https://doi.org/10.1002/ijc.28338.
58. Bernal M, Concha A, Sáenz-López P, et al. Leukocyte infiltrate in gastrointestinal adenocarcinomas is strongly associated with tumor microsatellite instability but not with tumor immunogenicity. *Cancer Immunol Immunother* 2011;**60**(6):869–82. https://doi.org/10.1007/s00262-011-0999-1.
59. Cabrera CM, Jiménez P, Cabrera T, Esparza C, Ruiz-Cabello F, Garrido F. Total loss of MHC class I in colorectal tumors can be explained by two molecular pathways: β2-microglobulin inactivation in MSI-positive tumors and LMP7/TAP2 downregulation in MSI-negative tumors. *Tissue Antigens* 2003;**61**(3):211–9. https://doi.org/10.1034/j.1399-0039.2003.00020.x.
60. Kloor M, Becker C, Benner A, et al. Immunoselective pressure and human leukocyte antigen class I antigen machinery defects in microsatellite unstable colorectal cancers. *Cancer Res* 2005;**65**(14):6418–24. https://doi.org/10.1158/0008-5472.can-05-0044.
61. Maleno I, Cabrera C, Cabrera T, et al. Distribution of HLA class I altered phenotypes in colorectal carcinomas: high frequency of HLA haplotype loss associated with loss of heterozygosity in chromosome region 6p21. *Immunogenetics* 2004;**56**(4). https://doi.org/10.1007/s00251-004-0692-z.

62. Maleno I, Romero JM, Cabrera T, et al. LOH at 6p21.3 region and HLA class altered phenotypes in bladder carcinomas. *Immunogenetics* 2006;**58**(7):503–10. https://doi.org/10.1007/s00251-006-0111-8.
63. Maleno I, Aptsiauri N, Cabrera T, et al. Frequent loss of heterozygosity in the β2-microglobulin region of chromosome 15 in primary human tumors. *Immunogenetics* 2011;**63**(2):65–71. https://doi.org/10.1007/s00251-010-0494-4.
64. Carretero FJ, del Campo AB, Flores-Martín JF, et al. Frequent HLA class I alterations in human prostate cancer: molecular mechanisms and clinical relevance. *Cancer Immunol Immunother* 2016;**65**(1):47–59. https://doi.org/10.1007/s00262-015-1774-5.
65. Garrido MA, Rodriguez T, Zinchenko S, et al. HLA class I alterations in breast carcinoma are associated with a high frequency of the loss of heterozygosity at chromosomes 6 and 15. *Immunogenetics* 2018;**70**(10):647–59. https://doi.org/10.1007/s00251-018-1074-2.
66. Chowell D, Morris LGT, Grigg CM, et al. Patient HLA class I genotype influences cancer response to checkpoint blockade immunotherapy. *Science* 2018;**359**(6375):582–7. https://doi.org/10.1126/science.aao4572.
67. McGranahan N, Rosenthal R, Hiley CT, et al. Allele-specific HLA loss and immune escape in lung cancer evolution. *Cell* 2017;**171**(6):1259–1271.e11. https://doi.org/10.1016/j.cell.2017.10.001.
68. Guinney J, Dienstmann R, Wang X, et al. The consensus molecular subtypes of colorectal cancer. *Nat Med* 2015;**21**(11):1350–6. https://doi.org/10.1038/nm.3967.
69. Ionov Y, Peinado MA, Malkhosyan S, Shibata D, Perucho M. Ubiquitous somatic mutations in simple repeated sequences reveal a new mechanism for colonic carcinogenesis. *Nature* 1993;**363**(6429):558–61. https://doi.org/10.1038/363558a0.
70. Thibodeau S, Bren G, Schaid D. Microsatellite instability in cancer of the proximal colon. *Science* 1993;**260**(5109):816–9. https://doi.org/10.1126/science.8484122.
71. Buckowitz A, Knaebel H-P, Benner A, et al. Microsatellite instability in colorectal cancer is associated with local lymphocyte infiltration and low frequency of distant metastases. *Br J Cancer* 2005;**92**(9):1746–53. https://doi.org/10.1038/sj.bjc.6602534.
72. Dolcetti R, Viel A, Doglioni C, et al. High prevalence of activated intraepithelial cytotoxic T lymphocytes and increased neoplastic cell apoptosis in colorectal carcinomas with microsatellite instability. *Am J Pathol* 1999;**154**(6):1805–13. https://doi.org/10.1016/S0002-9440(10)65436-3.
73. Drescher KM, Sharma P, Watson P, Gatalica Z, Thibodeau SN, Lynch HT. Lymphocyte recruitment into the tumor site is altered in patients with MSI-H colon cancer. *Fam Cancer* 2009;**8**(3):231–9. https://doi.org/10.1007/s10689-009-9233-0.
74. Banerjea A, Ahmed S, Hands RE, et al. Colorectal cancers with microsatellite instability display mRNA expression signatures characteristic of increased immunogenicity. *Mol Cancer* 2004;**3**:21. https://doi.org/10.1186/1476-4598-3-21.
75. Bernal M, García-Alcalde F, Concha A, et al. Genome-wide differential genetic profiling characterizes colorectal cancers with genetic instability and specific routes to HLA class I loss and immune escape. *Cancer Immunol Immunother* 2012;**61**(6):803–16. https://doi.org/10.1007/s00262-011-1147-7.
76. Giacomini CP, Leung SY, Chen X, et al. A gene expression signature of genetic instability in colon cancer. *Cancer Res* 2005;**65**(20):9200–5. https://doi.org/10.1158/0008-5472.can-04-4163.
77. Kloor M, Michel S, von Knebel Doeberitz M. Immune evasion of microsatellite unstable colorectal cancers. *Int J Cancer* 2010;**127**(5):1001–10. https://doi.org/10.1002/ijc.25283.
78. Vogelstein B, Papadopoulos N, Velculescu VE, Zhou S, Diaz Jr LA, Kinzler KW. Cancer genome landscapes. *Science* 2013;**339**:1546–58.
79. Yamamoto H, Yamashita K, Perucho M. Somatic mutation of the beta2-microglobulin gene associates with unfavorable prognosis in gastrointestinal cancer of the microsatellite mutator phenotype. *Gastroenterology* 2001;**120**(6):1565–7. https://doi.org/10.1053/gast.2001.24497.
80. Woerner SM, Yuan YP, Benner A, Korff S, von Knebel Doeberitz M, Bork P. SelTarbase, a database of human mononucleotide-microsatellite mutations and their potential impact to tumorigenesis and immunology. *Nucleic Acids Res* 2010;**38**(Database issue):D682–9. https://doi.org/10.1093/nar/gkp839.
81. Kloor M, Michel S, Buckowitz B, et al. Beta2-microglobulin mutations in microsatellite unstable colorectal tumors. *Int J Cancer* 2007;**121**(2):454–8. https://doi.org/10.1002/ijc.22691.
82. Ericsson C, Seregard S, Bartolazzi A, et al. Association of HLA class I and class II antigen expression and mortality in uveal melanoma. *Invest Ophthalmol Vis Sci* 2001;**42**(10):2153–6. https://www.ncbi.nlm.nih.gov/pubmed/11527924.
83. Garrido ML, Pérez M, Delgado C, et al. Immunogenicity of h-2 positive and h-2 negative clones of a mouse tumour, GR9. *Eur J Immunogenet* 1986;**13**(2–3):159–67. https://doi.org/10.1111/j.1744-313x.1986.tb01097.x.
84. Menon AG, Rob AE, van de Velde CJH, et al. p53 and HLA class-I expression are not down-regulated in colorectal cancer liver metastases. *Clin Exp Metastasis* 2004;**21**(1):79–85. https://doi.org/10.1023/b:clin.0000017206.08931.42.
85. Watson NFS, Ramage JM, Madjd Z, et al. Immunosurveillance is active in colorectal cancer as downregulation but not complete loss of MHC class I expression correlates with a poor prognosis. *Int J Cancer* 2006;**118**(1):6–10. https://doi.org/10.1002/ijc.21303.
86. Tikidzhieva A, Benner A, Michel S, et al. Microsatellite instability and Beta2-microglobulin mutations as prognostic markers in colon cancer: results of the FOGT-4 trial. *Br J Cancer* 2012;**106**(6):1239–45. https://doi.org/10.1038/bjc.2012.53.
87. Carretero R, Romero JM, Ruiz-Cabello F, et al. Analysis of HLA class I expression in progressing and regressing metastatic melanoma lesions after immunotherapy. *Immunogenetics* 2008;**60**(8):439–47. https://doi.org/10.1007/s00251-008-0303-5.
88. Zaretsky JM, Garcia-Diaz A, Shin DS, et al. Mutations associated with acquired resistance to PD-1 blockade in melanoma. *N Engl J Med* 2016;**375**(9):819–29. https://doi.org/10.1056/NEJMoa1604958.
89. Sade-Feldman M, Jiao YJ, Chen JH, et al. Resistance to checkpoint blockade therapy through inactivation of antigen presentation. *Nat Commun* 2017;**8**(1):1136. https://doi.org/10.1038/s41467-017-01062-w.

90. Gettinger S, Choi J, Hastings K, et al. Impaired HLA class I antigen processing and presentation as a mechanism of acquired resistance to immune checkpoint inhibitors in lung Cancer. *Cancer Discov* 2017;**7**(12):1420–35. https://doi.org/10.1158/2159-8290.CD-17-0593.
91. Grasso CS, Giannakis M, Wells DK, et al. Genetic mechanisms of immune evasion in colorectal cancer. *Cancer Discov* 2018;**8**(6):730–49. https://doi.org/10.1158/2159-8290.CD-17-1327.
92. Gurjao C, Liu D, Hofree M, et al. Intrinsic resistance to immune checkpoint blockade in a mismatch repair–deficient colorectal cancer. *Cancer Immunol Res* 2019;**7**(8):1230–6. https://doi.org/10.1158/2326-6066.cir-18-0683.
93. del Campo AB, Carretero J, et al. Adenovirus expressing β2-microglobulin recovers HLA class I expression and antitumor immunity by increasing T-cell recognition. *Cancer Gene Ther* 2014;**21**(8):317–32. https://doi.org/10.1038/cgt.2014.32.
94. Mlecnik B, Bindea G, Pagès F, Galon J. Tumor immunosurveillance in human cancers. *Cancer Metastasis Rev* 2011;**30**(1):5–12. https://doi.org/10.1007/s10555-011-9270-7.
95. Pagès F, Kirilovsky A, Mlecnik B, et al. In situ cytotoxic and memory T cells predict outcome in patients with early-stage colorectal cancer. *J Clin Oncol* 2009;**27**(35):5944–51. https://doi.org/10.1200/JCO.2008.19.6147.
96. Galon J, Pagès F, Marincola FM, et al. The immune score as a new possible approach for the classification of cancer. *J Transl Med* 2012;**10**:1. https://doi.org/10.1186/1479-5876-10-1.
97. Galon J, Pagès F, Marincola FM, et al. Cancer classification using the immunoscore: a worldwide task force. *J Transl Med* 2012;**10**:205. https://doi.org/10.1186/1479-5876-10-205.
98. Pagès F, Galon J, Dieu-Nosjean M-C, Tartour E, Sautès-Fridman C, Fridman W-H. Immune infiltration in human tumors: a prognostic factor that should not be ignored. *Oncogene* 2010;**29**(8):1093–102. https://doi.org/10.1038/onc.2009.416.
99. Fridman WH, Pagès F, Sautès-Fridman C, Galon J. The immune contexture in human tumours: impact on clinical outcome. *Nat Rev Cancer* 2012;**12**(4):298–306. https://doi.org/10.1038/nrc3245.
100. Fridman W-H, Dieu-Nosjean M-C, Pagès F, et al. The immune microenvironment of human tumors: general significance and clinical impact. *Cancer Microenviron* 2013;**6**(2):117–22. https://doi.org/10.1007/s12307-012-0124-9.
101. Anitei M-G, Zeitoun G, Mlecnik B, et al. Prognostic and predictive values of the immunoscore in patients with rectal cancer. *Clin Cancer Res* 2014;**20**(7):1891–9. https://doi.org/10.1158/1078-0432.CCR-13-2830.
102. Van den Eynde M, Mlecnik B, Bindea G, et al. The link between the multiverse of immune microenvironments in metastases and the survival of colorectal cancer patients. *Cancer Cell* 2018;**34**(6):1012–1026.e3. https://doi.org/10.1016/j.ccell.2018.11.003.
103. Galon J, Lanzi A. Immunoscore and its introduction in clinical practice. *Q J Nucl Med Mol Imaging* 2020;**64**(2):152–61. https://doi.org/10.23736/S1824-4785.20.03249-5.
104. Galon J, Bruni D. Approaches to treat immune hot, altered and cold tumours with combination immunotherapies. *Nat Rev Drug Discov* 2019;**18**(3):197–218. https://doi.org/10.1038/s41573-018-0007-y.
105. Pagès F, Mlecnik B, Marliot F, et al. International validation of the consensus Immunoscore for the classification of colon cancer: a prognostic and accuracy study. *Lancet* 2018;**391**(10135):2128–39. https://doi.org/10.1016/S0140-6736(18)30789-X.
106. Ogino S, Giannakis M. Immunoscore for (colorectal) cancer precision medicine. *Lancet* 2018;**391**(10135):2084–6. https://doi.org/10.1016/s0140-6736(18)30953-x.
107. Angell HK, Bruni D, Barrett JC, Herbst R, Galon J. The Immunoscore: colon cancer and beyond. *Clin Cancer Res* 2020;**26**(2):332–9. https://doi.org/10.1158/1078-0432.CCR-18-1851.

Chapter 48

Epithelial-mesenchymal transition and CRC

Angélica Figueroa

Epithelial Plasticity and Metastasis Group, A Coruña Biomedical Research Institute, A Coruña University Hospital Complex, University of A Coruña, A Coruña, Spain

Introduction

Despite the progress made in recent decades, colorectal cancer (CRC) continues to be one of the most common causes of cancer death in developed countries. Less than 10% of CRC cases occur in individuals who have inherited a predisposition to this disease, developing specific gene mutations in the germline. This is the case of familial adenomatous polyposis (FAP), whose most frequent mutations are found in the *APC* gene, leading to the development of hundreds or thousands of adenomatous polyps (benign tumors) in the colon and rectum. Therefore these individuals will develop CRC at some point in their lives, given that there is a high probability of malignancy after the age of 30 years, leading to colon cancer. The *APC* gene is located on chromosome 5q21 and produces a multidomain protein that acts as a tumor suppressor. *APC* is involved in cell migration, apoptosis, cell adhesion, and chromosome segregation. Mutations in this gene are present in over 40% of patients with this disease, although there are other types of adenomatous polyposis that do not involve mutations in this gene but in *MUTYH* or other, as yet unidentified genes. Other germline mutations are found in the *MMR* (Mismatch Repair) genes characterized by changes in the error repair system during DNA replication resulting in Hereditary Nonpolyposis Colorectal Carcinoma (HNPCC) or Lynch syndrome. These patients develop very few polyps; however, the transition from polyp to cancer is very frequent and rapid tumor progression occurs. Another case of familial colon cancer occurs in 25% of CRCs, affecting families that show an incidence of disease too frequent to be considered sporadic cancer, yet do not meet the clinical criteria of the syndromes described.[1,2]

The vast majority of CRCs (approximately 70%) develop sporadically in the population. Indeed, in recent decades, much attention has been paid to the study of sporadic colon cancer, which continues to be a paradigm in the development of solid tumors. The understanding of the molecular mechanisms of tumor progression has been crucial for achieving characterization of the different stages of colorectal cancers that led[3] to the proposal of a multistage model of human tumorigenesis. This model, depicted in Fig. 48.1, describes the accumulation of frequent genetic events (sequential somatic mutations in oncogenes, such as *K-RAS*, or in tumor suppressors, such as *APC* or *P53*, which play a critical role in tumor progression). Each of these genetic events confers a selective growth advantage to the colon epithelial cells, which ultimately results in uncontrolled cell proliferation, clonal tumor development, and progression from adenoma to invasive carcinoma.

Epithelial plasticity

Most common human tumors are carcinomas, originating from epithelial cells. Epithelial cells are connected to each other by cell-cell contacts grouped into different types of junctions: tight junctions, adherens junctions, desmosomes, and gap junctions. In mammalian epithelia, tight junctions form a permeability barrier preventing the free flow of substances between cells; adherens junctions are in contact with the actin cytoskeleton; and desmosomes are in contact with the intermediate filaments. These intercellular contacts of epithelial cells are regulated not only during embryonic development but also during tumor progression.

Epithelial cells are highly polarized. They are characterized by a localized distribution of adhesion molecules (such as cadherins and integrins), by the polarized organization of the cytoskeleton, and by the presence of a basement membrane. They show apicobasal polarity; the apical side is in contact with the external surface of the body or with the lumen of the

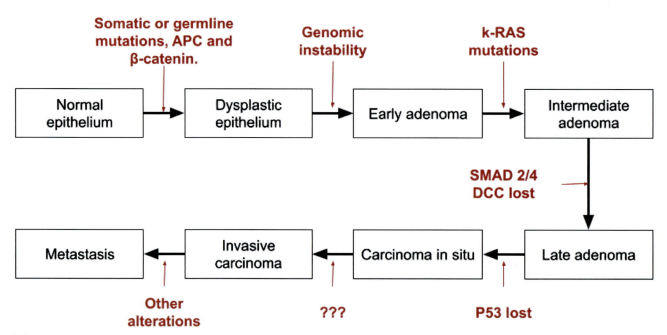

FIG. 48.1 Adaptation of the genetic model of colorectal carcinogenesis proposed by Fearon and Volgenstein in 1990. Somatic or germline mutations in the APC gene and mutations in β-catenin are necessary for tumor initiation. Tumor progression is accompanied by genomic instability and sequential mutations in different genes, as indicated in the figure above.

duct or cavity, and the basal side in contact with the basement membrane. Preserving this characteristic depends on the interactions between epithelial cells and between these cells and the extracellular matrix through the basement membrane. The cell-cell contacts that keep the cells together also contribute to determining cell polarity, participate in cell communication and differentiation, and in establishing and maintaining tissue homeostasis. Under normal conditions, epithelial cells present some motile capacity within the epithelium, but they do not separate from each other, nor do they have migratory capacity. They express epithelial markers, such as E-cadherin, desmoplakin, or cytokeratins.

In contrast, mesenchymal cells do not form organized layers, do not show apicobasal polarity, nor do they present an organization of the cytoskeleton similar to that of epithelial cells. Adhesions to adjacent mesenchymal cells are less frequent and are not associated with the basal lamina, which promotes their migratory capacity. Mesenchymal cells in culture have a fibroblast-like morphology and express characteristic mesenchymal markers, such as N-cadherin; intermediate filaments, such as vimentin; and extracellular matrix components, such as fibronectin.

Although it is necessary to maintain the epithelial phenotype for carrying out different cellular functions, it has been shown that differentiated epithelial cells can change their phenotype by activating a program called epithelial-to-mesenchymal transition (EMT), thus allowing the conversion of epithelial cells into mesenchymal cells (Fig. 48.2). The EMT process is an early step during carcinoma metastasis that resembles tissue remodeling during embryogenesis and organ morphogenesis. EMT is characterized by a set of cellular events that result in the loss of epithelial morphology and the acquisition of mesenchymal characteristics. These allow cells to become motile as a result of the loss of apicobasal polarity, loss of cell-cell contacts, and reorganization of the actin cytoskeleton, as well as the ability to invade the extracellular matrix as a single cell.[4–6] Many proteins are involved in the establishment of cell-cell contacts, but E-cadherin is the prototypical member of the cadherin family, the best characterized of the adherens junctions and most highly expressed in epithelial cells. E-cadherin forms adherens junctions, its extracellular domain establishes calcium-dependent homophilic interactions responsible for cell-cell contacts, and its intracellular domain interacts with proteins involved in cell signaling, such as catenins, α-catenin, β-catenin, and p120-catenin, and the actin cytoskeleton.[7] The epithelial-to-mesenchymal transition is characterized by the loss of E-cadherin, resulting in altered cell-cell contacts. E-cadherin is considered a tumor suppressor gene and its loss is indicative of poor prognosis.[8,9] In addition, other EMT markers have been described including markers of the epithelial phenotype, such as different cytokeratins (CK18, CK19, or CK20) and markers for the mesenchymal phenotype, such as Vimentin, Fibronectin, ZEB-1, Snail, Slug, or TWIST. In addition, molecular EMT markers are also associated with an increased capacity for cell migration and invasion, and by high resistance to anoikis/apoptosis. The transition of epithelial to mesenchymal cells is a reversible process, as cells also undergo a mesenchymal-epithelial (MET) process during both embryonic development and disease. Moreover, cells may partially

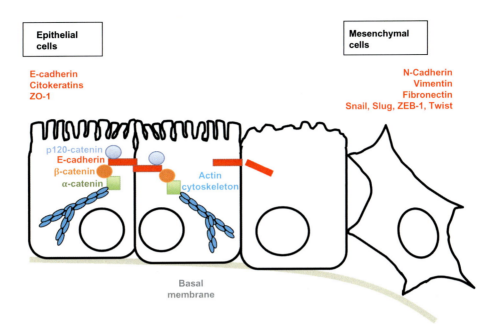

FIG. 48.2 Epithelial-to-mesenchymal transition. Epithelial cells present a localized distribution of adhesion molecules (such as E-cadherin linked to different catenins) and a polarized organization. Cells are connected to each other by cell-cell contacts. The apical side is in contact with the lumen of the duct or cavity, and the basal side is in contact with the basal membrane. Mesenchymal cells do not form organized layers, do not present apico-basolateral polarity, nor do they have the organization of the cytoskeleton. Adhesions between mesenchymal cells are less frequent and are not associated with the basal membrane, allowing the promotion of their migratory capacity.

lose their epithelial phenotype, by losing some of their epithelial traits or showing a combination of epithelial and mesenchymal features. Cells that undergo a partial MET have been referred to as 'metastable,' and display a high level of plasticity between epithelial and mesenchymal phenotypes. This ability of cells to reversibly change their phenotype and adopt intermediate epithelial and mesenchymal features has been recently termed epithelial-mesenchymal plasticity (EMP).[10–13]

Mechanisms responsible for E-cadherin inactivation during tumor progression

Given the importance of the loss of E-cadherin protein in EMT during tumor progression in most human epithelial cancers, the mechanisms involved in its functional inactivation have been studied in depth. Important mechanisms have been described including the genetic mechanisms (hereditary or somatic mutations in the *CDH1*/*E*-cadherin gene); epigenetic silencing of the promoter (such as hypermethylation of the E-cadherin promoter); transcriptional repression (mediated by transcriptional repressors, such as Snail, ZEB1/2, or TWIST and their respective micro-RNA regulators); and posttranslational modifications of the protein, such as phosphorylation, glycosylation, ubiquitination, or proteolysis.[11,14]

Mutations in the CDH1 gene

Mutations in the CDH1 gene that codes for the E-cadherin protein have been associated with various types of cancer. There are somatic inactivating mutations of the gene and germline mutations. The inactivating mutations of this gene are the most frequent genetic alterations in hereditary and sporadic diffuse gastric carcinoma. Somatic mutations cause loss of expression and function of E-cadherin. They are one of the important mechanisms of E-cadherin inactivation that occurs in adenocarcinoma and diffuse gastric carcinoma. This somatic mutation usually occurs in combination with loss of heterozygosity of the normal allele.

On the other hand, there are mutations of the *CDH1* gene in the germline, which predispose patients to suffering diffuse gastric carcinoma, being detected in families with this type of cancer. Germline mutations in the *CDH1* gene associated with the development of diffuse gastric carcinoma were first described in 1998. Hereditary diffuse gastric cancer is therefore an autosomal dominantly inherited neoplasm caused by germline mutations in the *CDH1* gene. This cancer accounts for about 3% of cases of gastric adenocarcinoma. Despite its low frequency, it is a major health problem due to its aggressiveness and the young age of the patients affected. The appearance of mutations in the early stages of tumorigenesis allows us to suggest that the *CDH1*/E-cadherin gene is a tumor suppressor gene in these two types of cancer. The *CDH1* gene is located on the long arm of chromosome 16 (16q22.1) and consists of 16 exons along which more than 50 mutations identified to date are located. Patients with germline *CDH1* gene alterations have a high risk of developing diffuse gastric cancer and, in addition, women are also at high risk of developing invasive lobular carcinoma. The estimated penetrance of *CDH1* mutations is 80% for diffuse gastric cancer and around 40%–52% for invasive lobular carcinoma. Unlike the sporadic forms of diffuse gastric cancer, the hereditary form can be predicted by genetic studies, which allows preventive actions to be implemented even before the onset of disease.

Different types of mutations have been identified in the *CDH1* gene. Diffuse invasive lobular carcinoma and diffuse gastric carcinoma frequently present germline mutations. Fifty-four percent are reading frame mutations or nonsense mutations, which result in a truncated protein (corresponding to the N-terminal region of the protein). On the other hand, somatic mutations frequently occur in diffuse gastric carcinoma: 28% of mutations occur at splicing sites, resulting in a loss of exons coding for calcium-binding motifs in the protein.[15]

Epigenetic silencing of the CDH1 gene by methylation of the promoter

Epigenetic silencing consists of the hypermethylation of CpG islands in the promoter region of the *CDH1* gene. This leads to the recruitment of methylated DNA binding proteins (MBD) and enzymes with histone deacetylase activity (HDAC), which together mediate chromatin compaction and gene silencing. Initially, this type of inactivation was described in prostate carcinoma and ductal breast carcinoma. It was observed in tumors that did not have *CDH1* gene mutations; therefore, the epigenetic inactivation was responsible for gene inactivation. Furthermore, it was observed that in these cases E-cadherin expression recovered upon treatment with demethylating agents, corroborating the importance of this epigenetic mechanism of inactivation. It is currently considered a general mechanism of *CDH1* gene inactivation in cancer.

Transcriptional repressors of the E-cadherin promoter

Several transcription factors have been described to regulate E-cadherin expression, including, but not limited to, Snail1, Snail2 (Slug), ZEB1 (δEF1), ZEB2 (Sip1), E12/E47, or TWIST.[16] The so-called E-boxes, which are constituted by short sequences of six nucleotides (-CACCTG-o-CAGGTG-), are in the proximal region of the E-cadherin promoter. The above-mentioned transcription factors can bind to the E-boxes of the E-cadherin promoter, acting as transcriptional repressors causing the silencing of mRNA expression and consequently of the E-cadherin protein. Usually, this repressor action is carried out through the recruitment of corepressors. However, this is not always the case since, for example, TWIST strongly silences E-cadherin by activating the expression of other E-cadherin repressors. The involvement of these transcriptional repressors by binding to the E-cadherin promoter causes epigenetic silencing of the gene by histone modifications. There are posttranslational modifications in histones that regulate gene expression. Histone modifications of nucleosomes include, but are not limited to: acetylations, methylations, phosphorylations, sumoylations, and ubiquitinations. Most of these modifications take place at the N-terminal end of these proteins. These modifications can alter the DNA-histone interaction and thus affect chromatin structure, and subsequently DNA hypermethylation. In a wide variety of human cancers, the E-cadherin gene is highly hypermethylated. However, the initial silencing of the gene promoter is converted into longer-term repression by DNA hypermethylation. The molecular basis for the reversibility or irreversibility of epigenetic silencing of E-cadherin in the context of EMT remains unclear.

Expression of these transcriptional repressors can be induced by a wide variety of stimuli, including activation of pathways mediated by TGF-β, HGF, EGF, or the Wnt pathway, as well as Notch signaling pathways. For example, TGF-β induces the EMT process during embryonic development and during tumor progression, inhibiting proteins such as E-cadherin, ZO-1, and certain keratins, and activating the expression of proteins such as fibronectin and vimentin. On the other hand, the alteration of cell-cell contacts during EMT has a major impact on extracellular matrix junctions through integrins. Thus, when TGF-β induces EMT, signaling via integrin β1 is required. Repression of E-cadherin results in activation of the Rap1 GTPase that activates cytoplasmic integrins important for activation of focal adhesions. The Wnt/β-catenin pathway also acts on the EMT process. When Wnt binds to its Frizzled and LRP5/6 receptors it results in inactivation of GSK3β, leading to stabilization of β-catenin and accumulation of β-catenin in the nucleus. The β-catenin in the nucleus transcriptionally activates proliferative genes, such as c-myc or cyclin D1. Another pathway that activates EMT is the Notch pathway, which after activation translocates to the nucleus and activates gene transcription in association with p300/CBP and MAM (Mastermind). Therefore there are extrinsic and intrinsic stimuli (such as mutations in signal transducer molecules) that are involved in the EMT process, which is a dynamic process.[17,18]

Posttranscriptional regulators of EMT

Although the transcriptional events that control the EMT process have been studied in depth, in recent years several critical posttranscriptional mechanisms in the regulation of EMT during tumor progression have been described. Posttranscriptional regulation is carried out by two main factors that directly influence EMT: microRNAs and RNA-binding proteins, called RBPs.[14,19]

MicroRNAs are small (~22-nucleotides), highly conserved, nonprotein-coding, single-stranded RNA molecules that modulate gene expression by binding to their target mRNAs, often to their 3′-noncoding region (3′-UTR). MicroRNAs have a major impact during development and during tumorigenesis. Recently, several microRNAs have been described that influence EMT. Thus miR-205 and the miR-200 family (miR-200a, miR-200b, miR-200c, miR-141, and miR-429, which share a consensus sequence) have become new epithelial markers and repressors of EMT. Members of the miR-200 family function by promoting mesenchyme-epithelial transition and inhibiting EMT induction through their direct action on the mRNAs of transcription factors ZEB1 and ZEB2. They therefore indirectly control E-cadherin expression. Furthermore, ZEB1 represses the transcription of miR-200 genes by directly binding to their promoter region, thereby forming a double negative feedback loop. In addition to the miR-200 family, additional miRNAs regulate EMT. For example, miR-9, a MYC/MYCN-induced micro-RNA, binds directly to the mRNA of the *CDH1* gene encoding for E-cadherin, leading to increased cell motility and invasiveness and induction of EMT.[19]

On the other hand, posttranscriptional regulation is also tightly controlled by RNA-binding proteins (RBPs). RBPs generally bind to specific untranslated regions of the RNA (the regulatory regions, called cis-elements). RBPs are present in all living cells and can bind to more than one messenger RNA with sequence specificity. Such interactions play an important role in regulating target RNA localization and expression and can coordinate the expression of messenger RNAs encoding functionally related proteins.

In recent years, it has been shown that the EMT process is strongly controlled by posttranscriptional regulatory processes mediated by RBPs. Different RBPs have been reported to modulate important target RNAs during EMT. Thus splicing regulatory proteins such as ESRP1, ZEPPO, SF2/ASF, and SAM68 have been described. These proteins exert their action on the splicing of their target RNAs (such as p120 for ZEPPO1, TCF4, or CD44 for ESRP1) and consequently have an impact on EMT. ESRP1 is also involved in the regulation of polyadenylation of a set of described ESRP targets, proposing that the regulation of polyadenylation by ESRP affects EMT. On the other hand, the influence of mRNA turnover during EMT has also been described. Thus the effect of HuR protein on EMT was described. HuR is a RBP that regulates the stability of its target mRNAs. It has been reported that HuR binds to the 3′UTR region of the Snail mRNA, inducing the stability of this transcriptional repressor of E-cadherin. In the presence of hydrogen peroxide, an oxidizing agent, HuR, binds to Snail mRNA, suggesting its involvement in the EMT process. Another level of posttranscriptional regulation has been described at the level of translation of mRNAs. Thus the hnRNPE1 protein exerts its action on the translation of its target mRNAs (Dab2 and ILEI) in response to TGFβ. TGFβ prevents hRNPE1 from binding to its target mRNAs, which in turn affects the translatability of mRNAs and the EMT process. Further understanding of posttranscriptional regulation during the EMT process will help to efficiently identify new targets that affect the EMT process and consequently invasion and metastasis.

Posttranslational regulators of E-cadherin

As we have seen before, E-cadherin-based cell-cell contacts are not static but dynamically regulated during various physiological and pathological processes, including mitosis, oncogenesis, and epithelial-mesenchymal transition, both during embryonic development and tumor progression. In all these processes, E-cadherin levels can also be regulated by endocytosis at the posttranslational level; however, the mechanism of endocytosis induction is still unclear. In in vitro culture models, two experimental stimuli induce endocytosis: the activation of tyrosine kinases and low Ca^{2+} concentrations. This mechanism has been shown to be important during the induction of cells from epithelial to mesenchymal phenotype in response to growth factors such as Hepatocyte Growth Factor (HGF).[20] Upon activation of tyrosine kinases, such as Src, E-cadherin is phosphorylated, then the E3 ubiquitin ligase Hakai interacts with the cytoplasmic domain of phosphorylated E-cadherin, which mediates its ubiquitination and degradation.[21] Thus Hakai was identified in 2002 as the first posttranslational regulator of E-cadherin stability,[15] and since then many articles have reported emerging biological functions for the Hakai protein, pointing out its influence on tumor progression and disease.[22-25] Hakai was identified as a novel E3 ubiquitin ligase, from the Cbl family of ubiquitin ligases, which act on the E-cadherin complex mediating its ubiquitination, endocytosis, and subsequent degradation via lysosomes, thus disrupting adherens junctions.[15] Excessive internalization or degradation of E-cadherin, as when Hakai is overexpressed, is related to the process of invasion and metastasis, suggesting its participation in carcinogenesis. The ubiquitination process involves the action of three different types of enzyme: ubiquitin-activating enzyme (E1), ubiquitin-conjugating enzyme (E2), and a variety of ubiquitin ligases (E3). The E3 ubiquitin ligase provides substrate specificity, as it recognizes the substrate through very specific protein-protein interactions. Hakai, also known as CBLL1, is the E3 ubiquitin ligase that maintains binding specificity on the substrate E-cadherin by binding to its cytoplasmic domain after being phosphorylated by the tyrosine kinase v-Src. The Src family plays a pivotal role in the regulation of several biological functions associated with changes in morphology, including malignant transformation, cell plasticity, and modulation of intercellular adhesion during EMT. Given the crucial role of Hakai during EMT and tumor progression, it was proposed as a novel drug target against cancer, and as a novel small molecule specifically designed to inhibit Hakai, Hakin-1.[26–32] Although ubiquitination is one of the most general mechanisms for driving cytosolic or nuclear proteins to degradation via proteasomes, it has been reported that many membrane proteins are degraded by lysosomes. The first published work on *Saccharomyces cerevisiae* demonstrated that Ste2p, a G protein-coupled cell surface receptor undergoing ligand-dependent ubiquitination, was internalized into vesicles to be ultimately degraded into lysosomes. Indeed, it has been shown that Src activation in Madin-Darby epithelial cells (MDCK) induces internalization of E-cadherin, which is transported to lysosomes for subsequent degradation, rather than following the normal nonubiquitinated E-cadherin pathway, in which E-cadherin is recycled back to the lateral plasma membrane to reform new cell-cell contacts. Indeed, modification of E-cadherin by ubiquitin is essential for its sorting into lysosomes, which occurs by a process mediated by the HGF receptor substrate and by activation of specific Rab GTPases (Rab5 and Rab7). Rab5-GTP can increase the transport rate of E-cadherin, and Rab7 activation serves to shift the balance of endocytic trafficking to lysosomes. Consequently, cell-cell contacts are not reestablished, and cells remain motile confirming that this is the first posttranslational mechanism to regulate E-cadherin depletion during EMT. Other E3 ubiquitin ligases have been reported to regulate EMT, but Hakai is a direct regulator of E-cadherin, which is a hallmark of the EMT.[33]

The cytoplasmic domain of E-cadherin contains two sequences, the CH2 and CH3 domains, preserved among the classical cadherins. The p120-catenin and β-catenin proteins can interact with these two domains, respectively.

Instead, Hakai interacts with the CH2 domain that contains three closed tyrosine residues (in mice: 756, 757, and 758), of which the first and second are specific for E-cadherin and the third is also preserved among other cadherins, including N-cadherin and OB-cadherin. Hakai cannot interact with these two cadherins, suggesting that its binding is specific to phosphorylated E-cadherin. By identifying the crystal structure of p120-catenin together with the cadherin fragment, it was suggested that p120-catenin could influence stability and function in cell-cell adhesion complexes. p120-catenin can interact with the juxtamembrane domain of E-cadherin (including tyrosine phosphorylation sites, where Hakai is also able to interact). It has been suggested that p120 associates with cadherin through dynamic and static interactions. The binding sites for Hakai and p120 are closely juxtaposed in the intracellular juxtamembrane domain of E-cadherin and, consequently, p120 is displaced by Hakai binding to E-cadherin prior to endocytosis. Moreover, p120 is also phosphorylated by Src kinase and receptor tyrosine kinases.

E-cadherin can also be controlled by posttranslational modifications, such as phosphorylation, glycosylation, and proteolytic processing. For example, casein kinase 1 (CK1) has been described, which is a serine/threonine kinase that has been evolutionarily preserved from yeast to mammals. At least seven CK1 isoforms (α, β, γ1, γ2, γ3, δ, and ε) are described, and ubiquitously expressed variants have been identified in mammals. All CK1 isoforms are homologous within their kinase domains but differ in the lengths and primary structures of their noncatalytic domains. CK1 phosphorylates many substrates involved in different cellular processes including cell differentiation, proliferation, membrane transport, and oncogenesis. For example, in *S. cerevisiae*, CK1 phosphorylates and enhances endocytosis of membrane proteins such as factor α receptor and uracil permease. In both invertebrates and vertebrates, several CK1 isoforms play a regulatory role in the Wnt signaling pathway. In 2007 it was demonstrated that the cytoplasmic domain of E-cadherin is a substrate of CK1, and that subsequent phosphorylation of E-cadherin negatively regulates cadherin-based cell-cell adhesions.[34] This study proposed a new mechanism that regulates E-cadherin-mediated cell-cell adhesion through phosphorylation of the cytoplasmic domain of E-cadherin by CK1. Upon phosphorylation, endocytosis of E-cadherin is increased, causing E-cadherin loss at cell-cell contact.

On the other hand, E-cadherin endocytosis can occur through clathrin- or caveolin-dependent mechanisms. A key player in clathrin-mediated E-cadherin endocytosis is Arf6. Arf6 is a small Ras-related GTPase. Here it promotes endocytosis through recruitment of nm23-H1, a nucleoside diphosphate kinase that in turn activates dynamin-dependent vesicle fission and destabilization of cortical actin through recruitment of the guanine nucleotide exchange factor (GEF) Tiam1, a Rac1 inhibitor. In addition, a GTPase-activating protein for Arf6, Smap1, has been identified as essential in E-cadherin endocytosis, thus revealing a novel posttranslational regulator of E-cadherin control.[35] Down-regulation of E-cadherin not only leads to mechanical cleavage of adherens junctions, but also releases proteins of the cell adhesion complex that exert ambivalent functions depending on their subcellular localization.

On the other hand, proteolytic cleavage of E-cadherin can also occur, producing fragments that exert functions in intracellular signaling. For example, γ-secretase-mediated cleavage of E-cadherin produces a C-terminal fragment, termed the cytoplasmic fragment (ctf2), which is transported to the nucleus in a p120-catenin-dependent manner. In the nucleus, the cleaved ctf2 fragment modulates the interaction between p120-catenin and Kaiso, which is a transcriptional repressor, affecting cell survival.[36] Activation of presenilin1/γ-secretase also induces dissociation of E-cadherin, α-catenin, and β-catenin from the cytoskeleton. This leads to a cytosolic increase in β-catenin, which is a key regulator of the Wnt signaling pathway.

Conclusions

Despite great advances in the understanding of the genetic and molecular mechanisms of colorectal cancer, the outcomes for many patients remain poor, especially for those who present advanced metastatic disease. Therefore it is of vital importance to clarify the biological mechanisms involved during tumor progression and metastasis. As demonstrated, the process of EMT represents a very important paradigm in colon cancer, and therefore it is crucial to investigate both the genetic and epigenetic aspects of late-stage tumor progression. Furthermore, since more than 80% of human cancers are carcinomas that arise from in epithelial cells, research on the EMT process will have an impact on many types of cancer. Understanding the molecular mechanisms involved in the EMT process that directly affect invasion and metastasis processes is very important. The development of new diagnostic and therapeutic strategies that focus on EMT process will contribute to the fight against colorectal cancer.

References

1. Markowitz SD, Dawson DM, Willis J, Willson JK. Focus on colon cancer. *Cancer Cell* 2002;**1**(3):233–6.
2. Vogelstein B, Fearon ER, Hamilton SR, Kern SE, Preisinger AC, Leppert M, et al. Genetic alterations during colorectal-tumor development. *N Engl J Med* 1988;**319**(9):525–32.
3. Fearon ER, Vogelstein B. A genetic model for colorectal tumorigenesis. *Cell* 1990;**61**(5):759–67.
4. Christofori G. New signals from the invasive front. *Nature* 2006;**441**(7092):444–50.
5. Yang J, Weinberg RA. Epithelial-mesenchymal transition: at the crossroads of development and tumor metastasis. *Dev Cell* 2008;**14**(6):818–29.
6. Brabletz T, Kalluri R, Nieto MA, Weinberg RA. EMT in cancer. *Nat Rev Cancer* 2018;**18**(2):128–34.
7. Nieto MA, Huang RYYJ, Jackson RAA, Thiery JPP. EMT: 2016. *Cell* 2016;**166**(1):21–45.
8. Christofori G, Semb H. The role of the cell-adhesion molecule E-cadherin as a tumour-suppressor gene. *Trends Biochem Sci* 1999;**24**(2):73–6.
9. Perl AK, Wilgenbus P, Dahl U, Semb H, Christofori G. A causal role for E-cadherin in the transition from adenoma to carcinoma. *Nature* 1998;**392**(6672):190–3.
10. Nieto MA. Epithelial plasticity: a common theme in embryonic and cancer cells. *Science* 2013;**342**(6159):1234850.
11. Aparicio LA, Blanco M, Castosa R, Concha Á, Valladares M, Calvo L, et al. Clinical implications of epithelial cell plasticity in cancer progression. *Cancer Lett* 2015;**366**(1):1–10. https://doi.org/10.1016/j.canlet.2015.06.007.
12. Pastushenko I, Brisebarre A, Sifrim A, Fioramonti M, Revenco T, Boumahdi S, et al. Identification of the tumour transition states occurring during EMT. *Nature* 2018;**556**(7702):463–8.
13. Yang J, Antin P, Berx G, Blanpain C, Brabletz T, Bronner M, et al. Guidelines and definitions for research on epithelial–mesenchymal transition. *Nat Rev Mol Cell Biol* 2020;**21**(6):341–52.
14. Aparicio LA, Abella V, Valladares M, Figueroa A. Posttranscriptional regulation by RNA-binding proteins during epithelial-to-mesenchymal transition. *Cell Mol Life Sci* 2013;**70**(23):4463–77. https://doi.org/10.1007/s00018-013-1379-0.
15. Berx G, Becker KF, Höfler H, van Roy F. Mutations of the human E-cadherin (CDH1) gene. *Hum Mutat* 1998;**12**(4):226–37.
16. García de Herreros A, Baulida J. Cooperation, amplification, and feed-back in epithelial-mesenchymal transition. *Biochim Biophys Acta* 2012;**1825**(2):223–8.
17. Dongre A, Weinberg RA. New insights into the mechanisms of epithelial–mesenchymal transition and implications for cancer. *Nat Rev Mol Cell Biol* 2019;**20**(2):69–84.
18. Derynck R, Muthusamy BP, Saeteurn KY. Signaling pathway cooperation in TGF-β-induced epithelial-mesenchymal transition. *Curr Opin Cell Biol* 2014;**31**:56–66.
19. Zhang J, Ma L. MicroRNA control of epithelial-mesenchymal transition and metastasis. *Cancer Metastasis Rev* 2012;**31**(3–4):653–62.
20. Palacios F, Tushir J, Fujita Y, D'Souza-Schorey C. Lysosomal targeting of E-cadherin: a unique mechanism for the down-regulation of cell-cell adhesion during epithelial to mesenchymal transitions. *Mol Cell Biol* 2005;**25**(1):389–402.
21. Fujita Y, Krause G, Scheffner M, Zechner D, Leddy H, Behrens J, et al. Hakai, a c-Cbl-like protein, ubiquitinates and induces endocytosis of the E-cadherin complex. *Nat Cell Biol* 2002;**4**(3):222–31.
22. Figueroa A, Kotani H, Toda Y, Mazan-Mamczarz K, Mueller E, Otto A, et al. Novel roles of hakai in cell proliferation and oncogenesis. *Mol Biol Cell* 2009;**20**(15):3533–42. https://doi.org/10.1091/mbc.e08-08-0845.

23. Figueroa A, Fujita Y, Gorospe M. Hacking RNA: Hakai promotes tumorigenesis by enhancing the RNA-binding function of PSF. *Cell Cycle* 2009;**8**(22):3648–51. https://doi.org/10.4161/cc.8.22.9909.
24. Rodríguez-Rigueiro T, Valladares-Ayerbes M, Haz-Conde M, Blanco M, Aparicio G, Fernández-Puente P, et al. A novel procedure for protein extraction from formalin-fixed paraffin-embedded tissues. *Proteomics* 2011;**11**:2555–9. https://doi.org/10.1002/pmic.201000809.
25. Rodríguez-Rigueiro T, Valladares-Ayerbes M, Haz-Conde M, Aparicio LA, Figueroa A. Hakai reduces cell-substratum adhesion and increases epithelial cell invasion. *BMC Cancer* 2011;**11**:474. https://doi.org/10.1186/1471-2407-11-474.
26. Aparicio LA, Valladares M, Blanco M, Alonso G, Figueroa A. Biological influence of Hakai in cancer: a 10-year review. *Cancer Metastasis Rev* 2012;**31**:375–86. https://doi.org/10.1007/s10555-012-9348-x.
27. Abella V, Valladares M, Rodriguez T, Haz M, Blanco M, Tarrío N, et al. miR-203 regulates cell proliferation through its influence on Hakai expression. *PLoS One* 2012;**7**(12):e52568. https://doi.org/10.1371/journal.pone.0052568.
28. Aparicio LA, Castosa R, Haz M, Rodriguez M, Blanco M, Valladares M, et al. Role of the microtubule-targeting drug vinflunine on cell-cell adhesions in bladder epithelial tumour cells. *BMC Cancer* 2014;**14**:507. https://doi.org/10.1186/1471-2407-14-507.
29. Castosa R, Martinez-Iglesias O, Roca-Lema D, Casas-Pais A, Diaz-Diaz A, Iglesias P, et al. Hakai overexpression effectively induces tumour progression and metastasis in vivo. *Sci Rep* 2018;**8**:3466. https://doi.org/10.1038/s41598-018-21808-w.
30. Roca-Lema D, Martinez-Iglesias O, Fernández de Ana-Portela C, Rodríguez-Blanco A, Valladares-Ayerbes M, Díaz-Díaz A, et al. In vitro anti-proliferative and anti-invasive effect of polysaccharide-rich extracts from trametes versicolor and grifola frondosa in colon cancer cells. *Int J Med Sci* 2019;**16**(2):231–40. https://doi.org/10.7150/ijms.28811.
31. Díaz-Díaz A, Roca-Lema D, Casas-Pais A, Romay G, Colombo G, Concha A, et al. Heat shock protein 90 chaperone regulates the E3 ubiquitin-ligase hakai protein stability. *Cancers (Basels)* 2020;**12**(01). https://doi.org/10.3390/cancers12010215.
32. Martinez-Iglesias O, Casas-Pais A, Castosa R, Díaz-Díaz A, Roca-Lema D, Concha A, et al. Hakin-1, a new specific small-molecule inhibitor for Hakai, inhibits carcinoma growth and progression. *Cancers (Basel)* 2020;**12**:E1340. https://doi.org/10.3390/cancers12051340. 32456234.
33. Rodríguez-Alonso A, Casas-Pais A, Roca-Lema D, Graña B, Romay G, Figueroa A. Regulation of epithelial-mesenchymal plasticity by the E3 ubiquitin-ligases in cancer. *Cancers (Basel)* 2020;**12**(11):3093. https://doi.org/10.3390/cancers12113093.
34. Dupre-Crochet S, Figueroa A, Hogan C, Ferber EC, Bialucha CU, Adams J, et al. Casein kinase 1 is a novel negative regulator of E-cadherin-based cell-cell contacts. *Mol Cell Biol* 2007;**27**(10):3804–16. https://doi.org/10.1128/MCB.01590-06.
35. Xu R, Zhang Y, Gu L, Zheng J, Cui J, Dong J, et al. Arf6 regulates EGF-induced internalization of E-cadherin in breast cancer cells. *Cancer Cell Int* 2015. 25678857.
36. Ferber C, Kajita M, Wadlow A, Tobiansky L, Niessen C, Ariga H, et al. A role for the cleaved cytoplasmic domain of E-cadherin in the nucleus. *J Biol Chem* 2008;**283**(19):12691–700. https://doi.org/10.1074/jbc.M708887200.

Chapter 49

Colorectal carcinoma: From molecular pathology to clinical practice

Catuxa Celeiro Muñoz[a], María Sánchez Ares[b], and José Ramón Antúnez López[c]

[a]Department of Pathological Anatomy, University Clinical Hospital of Santiago, Santiago de Compostela, Spain, [b]Health Research Institute of Santiago de Compostela Foundation, Santiago de Compostela, Spain, [c]Pathological Anatomy Department, University Clinical Hospital of Santiago de Compostela, Santiago de Compostela, Spain

Introduction

Despite improvements in early detection and treatment methods, colorectal carcinoma (CRC) remains the fourth leading cause of cancer death.[1] CRC is the second most common cancer diagnosed in women and the third most common in men.[2] Approximately 65% of CRCs are sporadic, with no apparent family history or genetic predisposition. The remaining cases are familial, stemming from a moderately pervasive inherited susceptibility, possibly interacting with environmental factors.[1]

Until recently, the diagnosis of these tumors was based solely on clinical and pathological anatomy. With advancements in the study of the molecular basis of cancer, several carcinogenic pathways have been described, revealing that it is a complex disease, with diverse phenotypic profiles that condition differences in histology, clinical assessment, prognosis, and therapeutic response. This opens up a new scenario in which the transition from a diagnosis based on clinical assessment, imaging, and histopathology to a clinical-pathological diagnosis based on imaging and molecular diagnosis is necessary. CRC, like many other solid tumors, is a heterogeneous disease in which different subtypes can be differentiated by their specific clinical and/or molecular features.[1]

Typically, CRCs are classified into sporadic (approximately 80% of cases) and those with a family history (around 20% of cases), indicating that genetic predisposition is an important risk factor in CRC. Between 5% and 7% of patients with colorectal carcinoma have a hereditary syndrome predisposing to CRC.[2] Hereditary CRC can be broadly divided into polyposis syndromes and nonpolyposis syndromes. Polyposis syndromes account for approximately 1% of cases of all cases of CRC. They include familial adenomatous polyposis (FAP) caused by mutations in the APC (adenomatous polyposis coli) gene, MUTYH-associated polyposis (MAP) caused by mutations in the *MUTYH* gene (mutY homologue), Peutz-Jeghers syndrome with mutations in *STK11* (serine/threonine kinase 11), juvenile polyposis with mutations in *SMAD4* (SMAD family member 4), and Cowden syndrome caused by mutations in *PTEN* (tensin phosphatase homologue). Nonpolyposis syndromes include Lynch syndrome (LS).[3]

Colorectal cancer: Molecular-genetic context

In 1990 Fearon and Vogelstein published the first model to attempt to describe colorectal carcinogenesis, which begins with a benign neoplasm (adenoma), which transforms into an invasive carcinoma (adenoma-carcinoma pathway). According to this model, an ordered combination of multiple genetic alterations, especially in oncogenes and tumor suppressor genes, is necessary for the eventual development of colorectal cancer (CRC), with the presence of a single genetic alteration being insufficient for malignant transformation.[4] Since this model was originally proposed, our understanding of CRC pathogenesis has advanced considerably, which has led to this prototype being modified. In addition to the adenoma-carcinoma pathway, which has been seen in 70%–90% of cases, the serrated pathway, responsible for 10%–20% of CRCs, has also been described.[2] To date, three pathways for the acquisition of genomic instability have been described: the suppressor pathway, also called the chromosomal instability pathway (CIN); the mutator or microsatellite instability pathway (MSI); and the methylator or CpG island methylation phenotype pathway (CIMP), also called the serrated pathway (Fig. 49.1).[1]

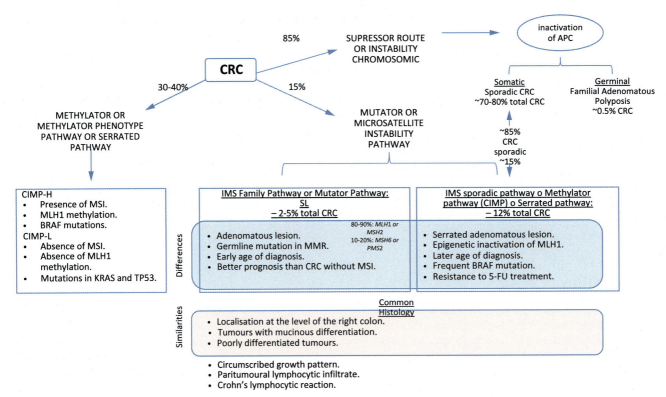

FIG. 49.1 Molecular pathways of CRC.

Chromosomal instability

The most frequent type of genomic instability is CIN, which is found in up to 85% of CRCs. Tumors with CIN are termed CIN positive (CIN+), while those lacking it are labeled CIN negative (CIN−). This instability is defined by the presence of numerical chromosomal changes or multiple structural aberrations of chromosomes.[5] These changes include gain or loss of chromosomal segments or whole chromosomes (aneuploidy or polyploidy), chromosomal rearrangements, and loss of heterozygosity (LOH), which causes copy number variations (CNV). These alterations affect the expression of tumor-associated genes, and/or genes that regulate cell proliferation or cell cycle checkpoints, which, in turn, may activate pathways essential for CRC initiation and progression. Also, CIN tumors are recognized by the accumulation of mutations in specific oncogenes, including the proto-oncogene GTPase KRAS (*KRAS*) and serine/threonine protein kinase B-Raf (proto-oncogene B-Raf, *BRAF*), and tumor suppressor genes, such as *APC* and p53 protein (*TP53*), thus contributing to CRC tumourigenesis.[1]

One of the first factors leading to the development of sporadic CRC is the inactivation of the *APC* gene that triggers the activation of the Wnt-β-catenin signaling pathway causing the activation of genes associated with cell growth and proliferation. Wnt pathway abnormalities characterize most sporadic CRCs (~70%–80%), but also tumors arising in patients with FAP (~0.5%) who have germline mutations in the *APC* gene.[6]

Microsatellite instability

CRC caused by microsatellite instability accounts for approximately 15% of CRCS. Only 3% of CRCs cases with MSI are associated with Lynch syndrome.[7]

Microsatellites (SSRs or STRs for simple sequence repeat and short tandem repeat) are repetitive DNA sequences distributed throughout the genome, which are located in both coding and noncoding regions. They consist of 1–6 base pairs, which are repeated in tandem a high number of times. Although they are highly polymorphic from one individual to another, they tend to have the same length in the patient's germline DNA as in the tumor's somatic DNA.[8]

These repetitive structures are particularly prone to errors during DNA replication in the case of deficiency in the genes of the DNA mismatch repair (MMR) system. The accumulation of errors in this microsatellite sequence is called MSI.

Cancers with these phenotypes are said to have MSI and, by extension, to be deficient in the MMR system (dMMR). Tumors lacking this feature are referred to as microsatellite stable (MSS) or MMR-competent (pMMR).[9]

The MMR system genes (*MLH1*, *MSH2*, *MSH6*, and *PMS2*) are involved in the correction of errors that occur during DNA replication, apoptosis, and cell cycle regulation. Deficiency in this system is one of the best studied causes of genetic instability in colorectal tumors.[7] MSI is generated as a consequence of the inactivation of mismatch repair genes through sporadic hypermethylation of the *MLH1* gene promoter and/or somatic/germline mutations in MMR genes. The MMR proteins function in pairs, MLH1 with PMS2 and MSH2 with MSH6. Loss of function of one of the four proteins leads to inactivation of the MMR system, resulting in a loss of replication fidelity and an accumulation of mutations.[9]

CpG island methylator phenotype (CIMP)

The CIMP mechanism is present in approximately 30%–40% of sporadic proximal CRCs and about 3%–12% of distal CRCs.[10] DNA methylation is a regulatory mechanism used by cells to silence gene expression. When a CpG site is methylated within the promoter region of a gene, its transcription is inhibited.[11] This phenotype is characterized by the presence of a high content of hypermethylated genes, causing transcriptional silencing within the promoter region, resulting in a loss of gene expression.[12] CpG islands are regions with a high concentration of phosphate-linked cytosine-guanine pairs found in many gene promoters. If genes are being expressed, these CpG sites are demethylated, and when methylated, can inactivate gene expression.[13] CIMP contributes to approximately 30%–35% of cases of colorectal adenomas, which occur at an early stage and act as a precursor to the serrated pathway of colorectal tumorogenesis.[12] CIMP can be graded as low (CIMP-L), high (CIMP-H), or negative (CIMP-0) depending on the degree of simultaneous hypermethylations occurring in CpG islands located near the promoter region of tumor suppressor genes. CRCs with CIMP-H are characterized by MSI, hypermethylation of *MLH1*, high *BRAF*, and low *TP53* mutation rates. CIMP-L CRCs are associated with the absence of MIS, the presence of *KRAS* and *TP53* mutations, as well as the absence of *MLH1* hypermethylation. CIMP-negative CRCs show an absence of *MLH1* methylation, microsatellite stability, and *TP53* mutations.[11]

Microsatellite instability and Lynch syndrome

Methods for determination of microsatellite instability

Molecular test

MSI analysis is performed on DNA extracted from fresh or paraffin-embedded tumor tissue.[9] Different panels have been used for the determination of MSI by multiplex PCR. In 1998 the National Cancer Institute established the so-called Bethesda panel, which is composed of 5 markers, two of mononucleotide type (BAT-25 and BAT-26) and 3 of dinucleotide type (D2S123, D5S346, and D17S250). When 2 or more of these markers are altered, tumors are classified as high MSI tumors (MSI-H), and cancers without instability in any of them are considered MSS.[9,14] The analysis is performed by comparing in each case the size of the tumor alleles with their corresponding nontumor DNA. An abnormality of only one of the markers is defined as low MSI (MSI-L), the clinical significance of which is under discussion. It is common to classify MSI-L cases in the same group as MSS cases.[10] Subsequently, a panel of 5 mononucleotide markers called pentaplex (BAT-25, BAT-26, NR-21, NR-24, and NR-27) has been proposed. Using this panel, a tumor is considered to have MSI if at least 3 of the 5 markers are unstable. In contrast to the Bethesda panel, a comparison of tumor tissue and normal tissue is not required.[15]

Determination of MSI status by next-generation sequencing (NGS) has been proposed as an alternative to the use of previous panels.[16] NGS allows the simultaneous sequencing of a large number of genes and can therefore also be used to identify other targeted alterations amenable to treatment.[8] However, NGS experiments are more time consuming and costly than standard techniques, in particular, due to the complex bioinformatics analysis required.[9]

Immunohistochemical study

Disruption of DNA repair proteins due to the presence of mutations or epigenetic silencing leads to alterations in the heterodimerization process of MMR system proteins and a loss of DNA error repair capacity. This results in small deletions/insertions causing MSI, i.e. a deficiency in the MMR system. Loss of function of this system correlates with loss of expression of its component proteins, MLH1, MSH2, MSH6, and PMS2.[17]

Each of these proteins can be detected in paraffin sections by immunohistochemical tests based on the use of validated staining panels composed of antibodies against the four MMR proteins. In CRCs with altered repair proteins, there is a loss

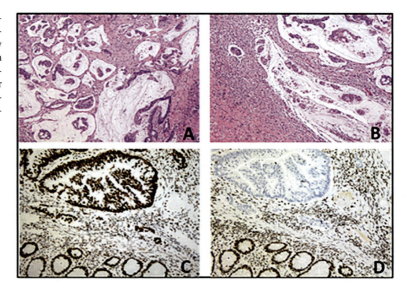

FIG. 49.2 Hematoxylin-Eosin images showing a low-grade mucinous adenocarcinoma (A), with a prominent peritumoral inflammatory infiltrate (B). Immunohistochemistry shows intense nuclear positivity in the tumor glands with MLH1, as well as in adjacent nonneoplastic glands and lymphocytes (C), while loss of staining is observed in the tumor cells for MSH6 (D). See how crypt epithelial cells and lymphocytes maintain their nuclear positivity, serving as a positive internal control.

of expression of these proteins in the tumor areas. In contrast, these proteins are normally expressed in normal colonic tissue, due to the expression of the wild-type allele. This feature makes nuclear expression in crypt epithelium and lymphocytes serve as a positive internal control to assess staining.[9]

Loss of nuclear staining for any of these proteins indicates the presence of MSI and correlates with the presence of germline mutation in the corresponding gene (Fig. 49.2). Therefore the identification of abnormal protein expression may be useful in identifying genes that should be sequenced.[18]

As mentioned before, the proteins of the MMR system work in heterodimers, MLH1 recruits its partner PMS2 to perform DNA repair, and the same is true for MSH2 and MSH6. Consequently, loss of expression of one of the MMR system proteins is often accompanied by the loss of its partner. Loss of normal MLH1 expression is associated with loss of PMS2 expression, just as the absence of MSH2 expression is associated with loss of MSH6. However, the reverse is not true as MLH1 and MSH2 proteins in their monomeric form can interact with other proteins in the MMR system and thus prevent their degradation. Loss of expression of MSH2 alone or in conjunction with MSH6 suggests the presence of a mutation in *MSH2*. Similarly, loss of MLH1 alone or with PMS2 indicates mutation or methylation of the *MLH1* gene.[9]

The immunohistochemical study is limited by the heterogeneity of staining in the tumor, which affects the sensitivity of the test. Also, it should be noted that a small group of LS-associated tumors may not show protein expression abnormalities of DNA repair genes by immunohistochemistry, even if they have a loss of function of the repair genes manifested with MSI by DNA testing.[7] Fig. 49.3 summarizes the strengths and weaknesses of this technique.

IMMUNOHISTOCHEMISTRY

STRENGTHS
- The result can guide molecular analysis by decreasing the chances of genes for germline mutation study.
- Ability to evaluate individual gene defects.
- Quick and easy to perform.
- Available in most pathology departments.

WEAKNESSES
- Heterogeneity of staining.
- Small percentage of false negative tumours.

FIG. 49.3 SWOT analysis of immunohistochemistry.

Solving a dilemma: Immunohistochemistry or molecular testing?

Both immunohistochemistry and the MSI molecular test show high sensitivity, but both have limitations and, in the case of questionable results, alternative tests are necessary.[19] The concordance between molecular test and IHC results varies according to the panel used for the MSI study but is generally very high (between 92% and 99%).[16] The molecular approach makes it possible to study the dysfunction of the MMR system and is not limited to protein expression alone.

Also, some point mutations allow normal protein expression but disrupt the MMR system causing MSI. However, MSI testing is more time consuming and costly than IHC assays and does not provide information as to which gene is deficient.[9]

Immunohistochemical analysis is simpler and cheaper, which is why it is usually the methodology used in most Pathology services. However, unusual expression patterns can sometimes be found due to technical or biological problems. Alterations in preanalytical variables such as fixation time or hypoxia can lead to false negatives. Also, aberrant staining patterns such as cytoplasmic, dot-like, or perinuclear staining may occur.[16] Weak, irregular, nucleolar staining, or even an absence of MSH6 expression has been observed in several cases after neoadjuvant treatment in the absence of MSI or mutation confirmed by molecular analysis. In the case of MLH1, although one-third of mutations generate nonfunctional proteins, their expression can be detected by immunohistochemistry.[9] Any loss of MMR system protein expression should be integrated with clinical information, family history, and additional tests such as *BRAF* mutation or *MLH1* methylation studies, as well as germline mutation studies.[17]

Characterization of Lynch syndrome

Lynch syndrome, previously known as hereditary nonpolyposis colorectal carcinoma (HNPCC), accounts for 2%–5% of colorectal carcinomas. It is an autosomal dominant familial syndrome caused by germline mutations that inactivate genes of the MMR system.[20]

This syndrome has unique features. Patients develop tumors at an early age and often multiple tumors. Although CRC is the most common cancer associated with this syndrome, cancers of the endometrium, small bowel, ureter and kidney, stomach, hepatobiliary tract, and ovary may also develop. Variants of LS include Muir-Torre syndrome (MTS) and Turcot syndrome (TS). MTS is characterized by the presence of multiple sebaceous adenomas and other skin tumors such as keratoacanthomas, but not all patients with this syndrome have LS. TS is characterized by the presence of multiple colorectal adenomas and glioblastomas.[21]

The clinical criteria for identifying patients at high risk for LS are known as the Amsterdam criteria, with very high specificity (98%) but low sensitivity (22%–42%), as more than 50% of families with SL do not meet these criteria. The revised, less stringent Bethesda guidelines have improved the identification of those cases with later onset and/or without a strong family history. These guidelines yielded better sensitivity results compared to the Amsterdam criteria, but lower specificity (82%–95% and 77%–93%, respectively). However, it has been shown that a significant proportion of patients with SL are not detected with these criteria and therefore improved screening strategies are needed.[16] The American Society of Clinical Oncology (ASCO), the American Society for Clinical Pathology (ASCP), the Association for Molecular Pathology (AMP), the College of American Pathologists (CAP), the National Comprehensive Cancer Network (NCCN), and the European Society for Medical Oncology (ESMO) recommend MSI analysis for all patients diagnosed with CRC.[21,22]

The main cause of LS is a pathogenic constitutional mutation affecting genes of the MMR system. In LS, mutations in the MMR system genes, when combined with the acquisition of a second pathogenic variant due to a somatic mutation in the nonmutated allele of the same MMR system gene, result in a complete loss of function of this repair system in the affected cells (Knudson hypothesis). This deficiency causes hypermutability, increasing the rate of mutations due to incorrect base pairing and MSI due to variation in the length of repetitive sequences.[23] Approximately 80%–90% of the mutations causing this syndrome involve *MLH1* or *MSH2*, while the other 10%–20% are caused by mutations in *MSH6* and *PMS2*, which is very infrequently mutated in LS.[24]

Up to 3% of cases of LS are due to variants involving the 3′ end of the *EPCAM* gene (immediately adjacent to *MSH2*). Another rare but important cause of LS is constitutional methylation of the *MLH1* gene promoter, which occurs in 1%–2% of cases. This methylation is usually sporadic in nature and therefore not heritable; however, in a small number of patients, hypermethylation may be secondary to a large deletion involving the *LRRFIP2* gene.[23]

Emerging concepts in Lynch syndrome

Familial colorectal cancer type X

Although the Amsterdam criteria are very restrictive, it is now known that a percentage of families meeting these criteria do not have germline mutations in any of the genes of the MMR system and also do not have MSI. Based on this, a new form of hereditary CRC has been defined in recent years, type X CRC (FCCTX), where "X" is used to describe the unknown nature of the etiology.[25,26] These patients are characterized by a late presentation of cancer, absence of extracolonic tumors, and lower relative risk of colorectal carcinoma compared to individuals with LS.[20] FCCTX is characterized by a lower rate of lymphocyte infiltration, peritumoral lymphocytes, synchronous and metachronous tumors, a higher adenoma/carcinoma ratio, and increased tumor cell differentiation compared to LS, indicating slower cancer progression in FCCTX. Also, these tumors have a more heterogeneous histology with a high frequency of tubular architecture, "dirty" necrosis (glands filled with necrotic debris), and a low frequency of mucin production. In contrast to LS tumors, which show expansive growth, these tumors show an infiltrative growth pattern. LS tumors are generally CIN-, whereas FCCTX tumors act like sporadic CRC, and up to 74% are CIN+.[27]

The absence of distinguishing features in FCCTX implies that family history is essential in order to identify these patients, as they will not be easily recognized based on histopathological features.[28]

Identification and differentiation of patients with FCCTX from those with sporadic CRC with MSS is challenging due to their similar clinical presentation and histopathological features. The age at diagnosis of patients with type X CRC is significantly lower than in sporadic CRC and they show higher recurrence. Despite this, only the identification of FCCTX-associated gene expression will help in the early diagnosis of FCCTX.[29]

Lynch syndrome without germline mutations or Lynch-like syndrome

There is a group of CRCs with MSI and loss of immunohistochemical staining of repair proteins, which do not have germline mutations in DNA repair genes, and do not show epigenetic inactivation of the *MLH1* gene or mutation in *BRAF*. It is estimated that this type of syndrome accounts for up to 60%–70% of cases in which there is clinical suspicion of LS.[30] Three hypotheses have been proposed to explain why germline mutations in MMR system genes are not detected in these tumors[1]: existence of unknown germline genetic mutations that may generate MSI,[2] germline mutations in MMR DNA genes that are not detectable by current tests (mutations in intronic or promoter regions),[3] existence of another mechanism that inactivates the DNA repair system, e.g., a somatic double mutation in MMR genes, a mechanism that is frequently detected in sporadic CRCs.[29] Studies by Mensenkamp et al. showed that somatic mutations in *MLH1* and *MSH2* are a frequent cause of inactivation of MMR DNA function and subsequent generation of MSI within Lynch-like syndrome cancers.[31]

Differential diagnosis of Lynch syndrome versus sporadic colorectal carcinoma with MSI

Sporadic MSI tumors and LS develop through different mechanisms of gene inactivation of the MMR system. As a consequence, sporadic MSI tumors show molecular features that are commonly used to distinguish them from LS-related tumors. In most sporadic tumors, MSI is due to the inactivation of the *MLH1* gene by methylation of its promoter. An immunohistochemical analysis of the MMR system proteins is therefore the first step in the discrimination of LS tumors and sporadic tumors. The combined loss of MSH2 and MSH6 or isolated loss of MSH6 or PMS2 will argue in favor of an LS-related tumor. In contrast, detection of MLH1 protein loss is nonspecific and requires complementary analysis to distinguish between an LS caused by a constitutional mutation in *MLH1* and a sporadic tumor caused by acquired somatic hypermethylation of the *MLH1* promoter. This methylation coincides with the CpG island methylator phenotype described before. Therefore analysis of the *MLH1* promoter in tumors showing loss of the MLH1 protein is very useful in distinguishing between the two types of tumors.

Another important difference is related to the presence of mutations in the *BRAF* gene. About 60% of tumors have a *BRAF* mutation, mainly p.V600E. In contrast, this mutation is found in a very low proportion in LS-related CRCs (1.4%). Therefore tumors with loss of MLH1 expression and detection of the V600E mutation are not usually associated with LS.

MLH1 methylation analysis is more expensive and technically more complex than *BRAF* mutational analysis. However, the clinical value of the methylation study is much greater for the identification of patients with CRC who are *MLH1* mutation carriers.

Different strategies have been proposed for the identification of patients with LS showing loss of MLH1 expression[1]: single analysis of the V600E mutation in *BRAF*,[2] single analysis of *MLH1* methylation, and[3] a *BRAF* mutation analysis

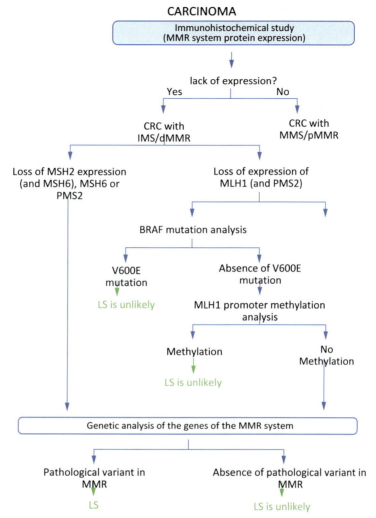

FIG. 49.4 Lynch syndrome (LS) diagnostic algorithm. *(Adapted from Leclerc, Vermaut, & Busine).*

followed by an *MLH1* methylation study in those cases without *BRAF* mutation. An *MLH1* hypermethylation analysis is more cost effective than *BRAF* mutational analysis. A one-time V600E mutation study would substantially increase the number of patients referred for a germline genetic study compared to an *MLH1* methylation study as a single analysis. The hybrid approach may facilitate the implementation of LS screening without generating a significant increase in cost.

Fig. 49.4 proposes a strategy to differentiate LS from sporadic CRC with MSI. However, it is important to note that no marker allows perfect discrimination of sporadic from SL-related cancers, even the detection of *MLH1* promoter hypermethylation or a *BRAF* mutation does not 100% exclude the possibility of SL, and it is important to consider the results of molecular testing in the context of clinical data, i.e., both the patient's personal and family history of cancer. Furthermore, although these markers are used to detect LS, they should not be considered as diagnostic markers for LS, as only the identification of a pathogenic germline variant (or epimutation) in one of the four genes of the MMR system or in *EPCAM* will confirm such a diagnosis.[16]

The proximity of the molecular changes of both tumors means that they share several morphological features: the tendency to occur in the right colon, medullary carcinoma phenotype, presence of mucinous or signet ring cell component, presence of peritumoral or intratumoral lymphocytes, Crohn's-type inflammatory response, and a circumscribed invasion front.[20,26,28] However, there are distinguishing features between them such as the age of onset which is usually lower in individuals with LS, a higher frequency of sporadic CRC in women, or the presence of significant familial clustering in LS. The precursor lesion of sporadic CRC with MSI is the serrated adenoma, with the tumor developing through the serrated adenoma pathway, whereas LS originates from an adenomatous polyp.[20,32] Furthermore, colorectal carcinomas that develop through this pathway generally have a better prognosis than tumors originating from the APC pathway and a poorer response to certain chemotherapies such as 5-fluorouracil.[33] Fig. 49.4 shows the different features of these pathways within the overall context of the molecular pathology of CRC.

Clinical application of molecular diagnostics: Biomarker analysis

Identification of microsatellite instability in colorectal carcinoma

Application in the diagnosis of Lynch syndrome

As previously described, MSI analysis is one of the reference studies for the diagnosis of Lynch syndrome. This instability is not unique to LS, with most tumors with MSI being of the sporadic type. When loss of MLH1 expression is observed, there are two different scenarios: a true case of LS with germline *MLH1* mutation or a sporadic cancer with methylation of the *MLH1* promoter. In this situation, before the germline study of variants in the genes of the MMR system, we must carry out a study of mutations in *BRAF* and methylation in the *MLH1* promoter. Eighty-five percent of sporadic tumors with IMS have mutations in this gene, while no mutation is detected in LS. The presence of the specific *BRAF* mutation p.V600E is a highly specific marker of sporadic origin. Patients with LS may benefit from different approaches to adjuvant therapy. This highlights the importance of differences between these two entities for appropriate management.[16]

Role as a prognostic marker

Published studies show that MSI is a strong prognostic marker in early stage CRC with a favorable impact on survival. CRC with MSI is most frequently diagnosed in early stage disease, while the incidence in stage II is 20%, in stage III it is 11%, and in metastatic disease it is 3.5%, suggesting that CRC with MSI has a reduced tendency to develop distant metastases.

The prognostic value of stage III MSI is less clear compared to stage II and there are conflicting data regarding this fact. It is envisaged that disease progression and the development of metastases mean that the immune evasion mechanisms developed by the tumor with MSI allow the tumor to evade immune surveillance and thus lose prognostic value. This aspect is observed in stage IV CRC in which MSI lacks prognostic value.[34]

Predictive value of response to chemotherapy

Current guidelines recommend that stage II patients with MSI should not receive adjuvant treatment with 5-fluorouracil.[22] This recommendation is based on clinical trials showing that this treatment does not improve survival in these patients. Other studies have found no evidence that MSI is predictive of a beneficial or harmful effect of chemotherapy and some show that adjuvant treatment in patients with MSI is beneficial. Several papers have attempted to clarify these associations, but controversy remains about the usefulness of MSI as a marker for predicting survival in the context of adjuvant chemotherapy.[35]

Adjuvant chemotherapy with a FOLFOX or CAPOX (capecitabine and oxaliplatin) regimen is the standard in stage III (node-positive) patients, regardless of whether or not MSI is present. However, several clinical trials have shown that patients with MSI have a longer survival.[36]

Predictive value of response to immunotherapy

Deficiency in the MMR system generates an MSI that leads to a high tumor mutational burden (TMB). Current scientific evidence suggests that tumors with high tumor burden carry a large number of tumor neoantigens that lead to increased tumor infiltrating lymphocyte density and T cell activation. Despite this increased immunogenicity, T cells are unable to eradicate these tumors, in part, due to overexpression of immune checkpoint proteins, such as PD-L1 (Programmed Death-ligand 1) or PD1 (programmed cell death receptor), in the tumour.[34]

Inhibitors of these checkpoints, such as pembrolizumab or nivolumab, anti-PD-L1 antibodies, have been evaluated in patients with metastatic CRC and MSI in whom previous treatment with cytotoxic agents has not been effective. Based on the good results obtained in these studies, the FDA approved the use of pembrolizumab and nivolumab alone or in combination with ipilimumab (anti-CTLA-4 antibody, Cytotoxic T-Lymphocyte Antigen 4) for the treatment of metastatic CRC with MSI, with pembrolizumab being the only drug approved as a first-line drug. It was also the first drug approved by the FDA for the treatment of advanced solid tumors with MMR pathway deficiency, regardless of tumor origin (tumor-agnostic indication).[36]

The fact that MSI has become a predictive biomarker of immunotherapy efficacy in metastatic CRC suggests that immunotherapy may be an effective treatment for earlier stage CRC as neoadjuvant or adjuvant therapy.[34]

Predicting response to anti-EGFR therapy in metastatic CRC

Analysis of RAS mutations

For several years, there has been scientific evidence that *RAS* and *BRAF* mutations have a predictive value for nonresponse to treatment with anti-EGFR monoclonal antibodies, such as cetuximab and panitumumab. About 50% of CRC patients have a clinically relevant mutation and should not be treated with anti-EGFR antibodies. Therefore international guidelines recommend analysis of the *KRAS* and *NRAS* genes at codons 12 and 13 of exon 2, 59 and 61 of exon 3, and 117 and 146 of exon 4.[17,37]

BRAF mutation analysis

BRAF mutation determination is used for both prognostic stratification and exclusion of Lynch syndrome, as previously described. Several studies have shown that patients with the p.V600E mutation do not benefit from anti-EGFR therapy. These mutations occur in 9%–20% of CRCs and are mutually exclusive with *RAS* mutations.[17]

Other predictive biomarkers

HER2

Approximately 1%–8% of CRCs have amplification of the *HER2* (human epidermal growth factor receptor 2) gene. Although the prognostic significance of *HER2* amplifications is controversial, several studies argue that it has a negative predictive value for anti-EGFR efficacy.[38] Preclinical studies show that *HER2* amplification is associated with primary and secondary resistance in response to anti-EGFR therapy.[39]

PIK3CA

Mutations in the *PIK3CA* gene, which encodes the catalytic subunit of PI3K p110α, have been identified in many solid tumors such as breast cancer, lung cancer, and CRC. Specifically, 15%–20% of CRCs have mutations in this gene. Several retrospective studies have investigated the role of these mutations in patients treated with anti-EGFR. Some of them suggest that these alterations may play a role in primary resistance to this treatment while others disagree with these results.[40]

Emerging biomarkers in colorectal carcinoma

Constitutive activation of different kinases as a consequence of gene translocations plays an essential role in the tumorigenesis of multiple neoplasms. Their pharmacological inhibition is one of the major therapeutic successes in the field of precision oncology. The most frequently identified fusions in CRC involve the *RET*, *NTRK*, *BRAF*, *ALK*, *ROS1*, and *FGFRs* genes. The presence of these fusions is associated with a worse prognosis due to a lower response to standard therapies.[40]

In 2018 treatment with the TRK inhibitor larotrectinib was approved for the treatment of metastatic solid tumors with fusion in the *NTRK1/2/3* genes. CRC-positive patients often lack mutations in *RAS* and *BRAF* and have a phenotype with MSI. These genetic alterations often occur in *BRAF* nonmutated tumors with methylation of the *MLH1* gene promoter.[38]

Role of liquid biopsy in CRC

Liquid biopsy is based on the analysis of circulating tumor cells (CTCs), free nucleic acids (DNA or RNA), and exosomes in a biological fluid, usually blood. In CRC patients, it has been used to determine mutations in *RAS* and *BRAF* genes at the time of diagnosis, showing a high correlation with tissue analysis; to monitor response to anti-EGFR treatment; and to detect the early onset of resistance.[41]

Recent studies have proposed its use for the early diagnosis of CRC, although it is still a preliminary option that requires future studies to determine its suitability as a population screening technique.[42]

Final considerations: the pathologist's point of view

The molecular basis and genetic alterations in CRC are often responsible for a specific phenotype that manifests itself in differences in histology, prognosis, or response to treatment. This means that nowadays it is essential for the pathologist, in addition to making a diagnosis of CRC with the best possible histopathological information, to accompany it with a molecular diagnosis based on the analysis of MSI, the presence of mutations in *RAS* and *BRAF*, or methylation of the

MLH1 promoter. This diagnosis will make it possible to discern between the different types of CRC, to select patients who are candidates for specific therapies, and to obtain prognostic and predictive information on response to treatment.

The MSI study in anatomical pathology departments is a simple technique that we believe should be performed in all CRC for three fundamental reasons:

a) It allows us to know if we are dealing with Lynch syndrome.
b) The existence of MSI in CRC determines the prognosis for treatment.
c) Tumors with MSI will be resistant to treatment with 5-FU, and in these patients, MSI will help us to decide which cases should not be treated with adjuvant treatment involving 5-FU.

References

1. Nguyen HT, Duong H-Q. The molecular characteristics of colorectal cancer: implications for diagnosis and therapy. *Oncol Lett* 2018;**16**(1):9–18. https://doi.org/10.3892/ol.2018.8679.
2. Dekker E, Tanis PJ, Vleugels JLA, Kasi PM, Wallace MB. Colorectal cancer. *Lancet* 2019;**394**(10207):1467–80. https://doi.org/10.1016/s0140-6736(19)32319-0.
3. Dominguez-Valentin M, Therkildsen C, Da Silva S, Nilbert M. Familial colorectal cancer type X: genetic profiles and phenotypic features. *Mod Pathol* 2015;**28**(1):30–6. https://doi.org/10.1038/modpathol.2014.49.
4. Fearon ER, Vogelstein B. A genetic model for colorectal tumorigenesis. *Cell* 1990;**61**(5):759–67. https://doi.org/10.1016/0092-8674(90)90186-i.
5. Grady WM, Markowitz SD. The molecular pathogenesis of colorectal cancer and its potential application to colorectal cancer screening. *Dig Dis Sci* 2015;**60**(3):762–72. https://doi.org/10.1007/s10620-014-3444-4.
6. Müller MF, Ibrahim AEK, Arends MJ. Molecular pathological classification of colorectal cancer. *Virchows Arch* 2016;**469**(2):125–34. https://doi.org/10.1007/s00428-016-1956-3.
7. Poulogiannis G, Frayling IM, Arends MJ. DNA mismatch repair deficiency in sporadic colorectal cancer and Lynch syndrome. *Histopathology* 2010;**56**(2):167–79. https://doi.org/10.1111/j.1365-2559.2009.03392.x.
8. Luchini C, Bibeau F, Ligtenberg MJL, et al. ESMO recommendations on microsatellite instability testing for immunotherapy in cancer, and its relationship with PD-1/PD-L1 expression and tumour mutational burden: a systematic review-based approach. *Ann Oncol* 2019;**30**(8):1232–43. https://www.sciencedirect.com/science/article/pii/S0923753419312694.
9. Evrard C, Tachon G, Randrian V, Karayan-Tapon L, Tougeron D. Microsatellite instability: diagnosis, heterogeneity, discordance, and clinical impact in colorectal cancer. *Cancers* 2019;**11**(10). https://doi.org/10.3390/cancers11101567.
10. Taghizadeh E, Hayat SMG. Molecular pathways, screening and follow-up of colorectal carcinogenesis: an overview. *Curr Cancer* 2020. https://www.ingentaconnect.com/content/ben/cctr/2020/00000016/00000002/art00003.
11. Palma FD, De Palma F, D'Argenio V, et al. The molecular hallmarks of the serrated pathway in colorectal cancer. *Cancers* 2019;**11**(7):1017. https://doi.org/10.3390/cancers11071017.
12. Ahmad R, Singh J, Wunnava A, Al-Obeed O, Abdulla M, Srivastava S. Emerging trends in colorectal cancer: dysregulated signaling pathways (Review). *Int J Mol Med* 2021;**47**(3). https://doi.org/10.3892/ijmm.2021.4847.
13. Baylln SB, Herman JG, Graff JR, Vertino PM, Issa J.-P. Alterations in DNA methylation: a fundamental aspect of neoplasia. In: Vande Woude GF, Klein G, Advances in cancer research. vol. 72. Academic Press; 1997:141-196. doi:https://doi.org/10.1016/S0065-230X(08)60702-2.
14. Boland CR, Thibodeau SN, Hamilton SR, Sidransky D. A National Cancer Institute Workshop on Microsatellite Instability for cancer detection and familial predisposition: development of international criteria for the .... *Cancer Res* 1998. https://cancerres.aacrjournals.org/content/58/22/5248.short.
15. Suraweera N, Duval A, Reperant M, et al. Evaluation of tumor microsatellite instability using five quasimonomorphic mononucleotide repeats and pentaplex PCR. *Gastroenterology* 2002;**123**(6):1804–11. https://doi.org/10.1053/gast.2002.37070.
16. Leclerc J, Vermaut C, Buisine MP. Diagnosis of lynch syndrome and strategies to distinguish lynch-related tumors from sporadic MSI/dMMR tumors. *Cancers* 2021;**13**:467. https://search.proquest.com/openview/d8b2ab3e8bfef849409939db951ad62b/1?pq-origsite=gscholar&cbl=2032421.
17. Sepulveda AR, Hamilton SR, Allegra CJ, et al. Molecular biomarkers for the evaluation of colorectal cancer: guideline from the American Society for Clinical Pathology, College of American Pathologists, Association for Molecular Pathology, and the American Society of Clinical Oncology. *J Clin Oncol* 2017;**35**(13):1453–86. https://doi.org/10.1200/JCO.2016.71.9807.
18. Lindor NM, Burgart LJ, Leontovich O, et al. Immunohistochemistry versus microsatellite instability testing in phenotyping colorectal tumors. *J Clin Oncol* 2002;**20**(4):1043–8. https://doi.org/10.1200/JCO.2002.20.4.1043.
19. Bedeir A, Krasinskas AM. Molecular diagnostics of colorectal cancer. *Arch Pathol Lab Med* 2011;**135**(5):578–87. https://doi.org/10.1043/2010-0613-RAIR.1.
20. Boland CR, Richard Boland C, Goel A. Microsatellite instability in colorectal cancer. *Gastroenterology* 2010;**138**(6). https://doi.org/10.1053/j.gastro.2009.12.064. 2073–2087.e3.
21. Sinicrope FA. Lynch syndrome-associated colorectal cancer. *N Engl J Med* 2018;**379**(8):764–73. https://doi.org/10.1056/NEJMcp1714533.
22. About the NCCN clinical practice guidelines in oncology (NCCN guidelines®). [Accessed 17 February 2021], https://www.nccn.org/professionals/default.aspx.
23. Cerretelli G, Ager A, Arends MJ, Frayling IM. Molecular pathology of Lynch syndrome. *J Pathol* 2020;**250**(5):518–31. https://doi.org/10.1002/path.5422.

24. Sobocińska J, Kolenda T, Teresiak A, et al. Diagnostics of mutations in MMR/EPCAM genes and their role in the treatment and care of patients with lynch syndrome. *Diagnostics (Basel)* 2020;**10**(10). https://doi.org/10.3390/diagnostics10100786.
25. Lindor NM. Familial colorectal cancer type X: the other half of hereditary nonpolyposis colon cancer syndrome. *Surg Oncol Clin N Am* 2009;**18**(4):637–45. https://doi.org/10.1016/j.soc.2009.07.003.
26. Lynch HT, de la Chapelle A. Hereditary colorectal cancer. *N Engl J Med* 2003;**348**(10):919–32. https://doi.org/10.1056/NEJMra012242.
27. Zetner DB, Bisgaard ML. Familial colorectal cancer type X. *Curr Genom* 2017;**18**(4). https://doi.org/10.2174/1389202918666170307161643.
28. Klarskov L, Holck S, Bernstein I, Nilbert M. Hereditary colorectal cancer diagnostics: morphological features of familial colorectal cancer type X versus Lynch syndrome. *J Clin Pathol* 2012;**65**(4):352–6. https://doi.org/10.1136/jclinpath-2011-200535.
29. Chen E, Xu X, Liu T. Hereditary nonpolyposis colorectal cancer and cancer syndromes: recent basic and clinical discoveries. *J Oncol* 2018;**2018**:3979135. https://doi.org/10.1155/2018/3979135.
30. Carethers JM, Stoffel EM. Lynch syndrome and Lynch syndrome mimics: the growing complex landscape of hereditary colon cancer. *World J Gastroenterol* 2015;**21**(31):9253–61. https://doi.org/10.3748/wjg.v21.i31.9253.
31. Mensenkamp AR, Vogelaar IP, van Zelst-Stams WAG, et al. Somatic mutations in MLH1 and MSH2 are a frequent cause of mismatch-repair deficiency in Lynch syndrome-like tumors. *Gastroenterology* 2014;**146**(3). https://doi.org/10.1053/j.gastro.2013.12.002. 643-646.e8.
32. Jass JR. Diagnosis of hereditary non-polyposis colorectal cancer. *Histopathology* 1998;**32**(6):491–7. https://doi.org/10.1046/j.1365-2559.1998.00442.x.
33. Zaanan A, Meunier K, Sangar F, Fléjou J-F, Praz F. Microsatellite instability in colorectal cancer: from molecular oncogenic mechanisms to clinical implications. *Cell Oncol* 2011;**34**(3):155–76. https://doi.org/10.1007/s13402-011-0024-x.
34. Jin Z, Sinicrope FA. Prognostic and predictive values of mismatch repair deficiency in non-metastatic colorectal cancer. *Cancers* 2021;**13**(2). https://doi.org/10.3390/cancers13020300.
35. Alwers E, Jansen L, Bläker H, et al. Microsatellite instability and survival after adjuvant chemotherapy among stage II and III colon cancer patients: results from a population-based study. *Mol Oncol* 2020;**14**(2):363–72. https://doi.org/10.1002/1878-0261.12611.
36. Zaanan A, Shi Q, Taieb J, et al. Role of deficient DNA mismatch repair status in patients with stage III colon cancer treated with FOLFOX adjuvant chemotherapy: a pooled analysis from 2 randomized clinical trials. *JAMA Oncol* 2018;**4**(3):379–83. https://doi.org/10.1001/jamaoncol.2017.2899.
37. Benson AB, Venook AP, Cederquist L, et al. Colon cancer, version 1.2017, NCCN Clinical Practice Guidelines in Oncology. *J Natl Compreh Cancer Netw* 2017;**15**(3):370–98. https://doi.org/10.6004/jnccn.2017.0036.
38. Cohen R, Pudlarz T, Delattre J-F, Colle R, André T. Molecular targets for the treatment of metastatic colorectal cancer. *Cancers* 2020;**12**(9). https://doi.org/10.3390/cancers12092350.
39. Sveen A, Kopetz S, Lothe RA. Biomarker-guided therapy for colorectal cancer: strength in complexity. *Nat Rev Clin Oncol* 2020;**17**(1):11–32. https://doi.org/10.1038/s41571-019-0241-1.
40. Rachiglio AM, Sacco A, Forgione L, Esposito C, Chicchinelli N, Normanno N. Colorectal cancer genomic biomarkers in the clinical management of patients with metastatic colorectal carcinoma. *Explor Target Anti-Tumor Ther* 2020;**1**(1):53–70. https://doi.org/10.37349/etat.2020.00004.
41. Kolenčík D, Shishido SN, Pitule P, Mason J, Hicks J, Kuhn P. Liquid biopsy in colorectal carcinoma: clinical applications and challenges. *Cancers* 2020;**12**(6):1376. https://doi.org/10.3390/cancers12061376.
42. Marcuello M, Vymetalkova V, Neves RPL, et al. Circulating biomarkers for early detection and clinical management of colorectal cancer. *Mol Aspects Med* 2019;**69**:107–22. https://doi.org/10.1016/j.mam.2019.06.002.

Section F

Biobanks

Chapter 50

The role of biobanks in the study of colorectal carcinoma

Vanesa Val Varela[a], Orlando Fernández Lago[b], Paula Vieiro Balo[b], Joaquín González-Carreró[a], Lydia Fraga Fontoira[b], and Máximo Fraga Rodríguez[b]

[a]Institute of Health Research in Southern Galicia Biobank, Galicia, Spain, [b]Santiago de Compostela University Hospital Complex Biobank, Santiago, Spain

Biobanks and biomedical research

Biomedical research is a key activity, both in terms of advances in basic knowledge of diseases and in the development of methods of diagnosis, prevention, and treatment to improve people's quality of life and life expectancy.

Due to the incalculable value of biological samples of human origin in this type of research, in recent years there has been a growing interest in developing structures and regulations to guarantee their adequate collection, storage, and use. For this reason, and in order to regulate and promote these and other practices appropriately, Spanish Law 14/2007, of July 3, 2007, regulating Biomedical Research (LIBM) was passed in 2007[1] and subsequently, in 2011, Royal Decree 1716/2011 was issued on November 18, 2011, establishing the basic requirements for the authorization and operation of biobanks for biomedical research purposes and the treatment of biological samples of human origin, and regulating the operation and organization of the National Register of Biobanks for biomedical research.[2] This legal framework reflects the important change that this type of research has undergone in recent times, marked notably by the rapid advances in knowledge and technology and by the arrival of new research tools.

Biobanks have been defined by the Biomedical Research Law (LIBM) as public or private, nonprofit establishments that house a collection of biological samples designed for diagnostic or biomedical research purposes and are organized into different units according to quality, order, and intended use.[1] They are therefore one of the most important support platforms for contributing to the building of bridges between basic and clinical research and health care practice, i.e., to the development of translational research. The main distinguishing feature of biobanks compared with the classic concept of sample collection is, therefore, their commitment to the transfer of these samples (and the information associated with them) to research groups in an open, transparent, and cooperative manner, for the benefit of science and, above all, for the benefit of biomedical research participants.

Collecting a significant number of human biological samples according to the quality and clinical information requires a major effort from the economic, planning, and coordination points of view. This requires time and funding, which are factors that can limit the development of certain research projects. For this reason, the promotion and implementation of biobanks, which provide researchers with access to samples and associated clinical information, is a very important aid: it shortens the time that elapses between the initial project and the application of its results, which substantially improves the effectiveness of the research.[2]

For this effort to achieve the desired impact, constant coordination and collaboration are required between the biobank, the local authorities, and the different professionals involved in the processing of samples and the information associated with them, so that no errors occur in the management or in any of the stages that form part of the process (identification of donors, obtaining samples, processing, storage, distribution, transfer, use of samples and associated information, etc.). Also, the growing ethical and regulatory requirements regarding the handling of human biological samples and the difficulty in obtaining the necessary quantity and quality of these samples are sometimes a major obstacle when conducting a research project, which is why biobanks often become facilitating tools that simplify the whole process. For all these reasons, biobanks constitute one of the infrastructures of the highest strategic value for the health care institutions of the National Health System (NHS), although this general conviction does not often translate into decisive support from the health care authorities.

Bearing this in mind, after the publication of the LIBM, the Carlos III Health Institute set up the National Biobanks Network, as a transversal structure within the Thematic Networks of Corporative Research, focusing mainly on NHS health care centers (State Secretariat for Research & ISCIII, 2009; ISCIII, 2009). This created the basis for the integration of the Spanish Biobank Network into international platforms, such as the BBMRI (Bio-Banking and BioMolecular Resources Research Infrastructure) project (BBMRI, 2013), financed by the 7th European Framework Program, with the aim of creating an infrastructure of biobanks and biomolecular research courses at a pan-European level, a project in which Spain is actively participating.

Currently, biomedical research of excellence revolves around the study of a large series of biological samples, whereby expansion to an increasingly larger geographical area is necessary. The samples must be subject to common and well-defined organizational criteria and be linked to the most relevant information, be it donor identification or clinical information. These requirements are particularly important in the following cases:

- Cancer, given that the most recent evidence demonstrates genetic heterogeneity between different patients.
- Infectious diseases, where both the disease causative agent and the genetic characteristics of the patient are important.
- Rare or uncommon diseases, where the participation of multiple institutions is essential to achieve sufficient caseloads for the development of biomedical research studies.
- Epidemiological studies, where it is necessary to have a representative series of the environmental and genetic differences of a personal/group nature involved in the development of the disease.

All the above justifies the growing interest in developing cooperative biobank networks, which seek to minimize the biases derived from heterogeneity in the quality of biological samples through the protocolization of procedures, the development of quality assurance policies, and the promotion of cooperative environments. In keeping with the above, an efficient effort to integrate the various information systems that manage the identification of patients, clinical data, and samples is required.

Based on the provisions of Law14/2007, 2007,[1] biobanks are called to ensure:

- The rights of patients and donors.
- The quality of biological samples and associated data made available to the scientific community.
- A process for managing these samples and data in a transparent, regulated, and efficient manner.
- Correct use of biological samples and associated data by this regulatory framework.

Organization and operation of the biobank

The LIBM establishes that the structure of the biobank will be composed of an Owner, who will be the natural or legal person, public or private, responsible for the operation of the biobank, and a Scientific Director who will have the following roles:

- To ensure compliance with current legislation.
- To maintain a record of the biobank's activities.
- To ensure the quality, safety, and traceability of the data and samples stored and of the procedures associated with the operation of the biobank.
- To prepare an annual activity report, including, inter alia, a reference to the agreements concluded for the collection and release of samples.
- To deal with any queries or complaints that may be addressed to the biobank.
- To direct the day-to-day management of the biobank.
- To draw up the biobank's good practice document.
- To manage the transfer of samples.

Moreover, the regulations establish the need for a Data Controller to deal with source subject requests to exercise the rights of access to personal data, rectification, cancellation, or opposition, in accordance with the provisions of the current personal data protection regulations.

The biobank will have two External Committees as advisory bodies to assist the Director in the fulfillment of their functions, one of which will be scientific and the other ethical. These committees, in accordance with their external nature, will be independent of the internal structure and organization of the biobank, and their members will not be linked to the biobank.

Another relevant organizational aspect is the definition of the biobank itself, which may be independent or associated with other biobanks. In the latter case, each biobank is integrated into a network, and its procedures will therefore be regulated by the procedures agreed by the network.

Biobank orientation

Determining the type of samples that are needed for research is a fundamental first step for streamlining the subsequent utilization process (Roche Institute). With this in mind, each biobank will first have to establish, based on its objectives, the criteria for selecting samples of interest for storage from those in which:

- There is leftover material from diagnostic studies.
- There is a specific IC from the patient.
- The time elapsed between sampling and storage complies with the quality standards for each type of sample.
- The criteria used to select a sample for storage in the biobank should be clearly disclosed to all those involved in its processing, from the collection of the sample to the end of the procedure.

Ultimately, the purpose of each biobank should be decided as early as possible, and this will determine the criteria for the selection of material:

- Indiscriminate sample collection.
- Collection restricted to certain types of samples: limited case series (clinical trials), samples from specific diseases (cardiovascular, neurodegenerative, oncological, etc.).

Where restricted storage of certain pathologies is intended for research purposes, those of known interest to the institution should be considered.

Integration of diagnostic anatomical pathology files in biobanks

The samples obtained during the care processes and kept in the archives of the Anatomical Pathology Service (APS) are of immense value for biomedical research. The LIBM and Royal Decree no. 1716/2011 that governs it regulate the conditions for the creation of collections of samples for research, regardless of whether they are included in a biobank system. It is advisable to promote and enable the use of this material in research, provided that the requirements of the applicable legislation are met, and it is preferable to organize it based on the biobank system. According to the SEAP (Spanish Anatomical Pathology Society) and the IAP (International Academy of Pathology), the organizational possibilities available for these cases are as follows:

1. The creation of collections of biological samples for research based on health care archives, comprising samples that are initially intended for diagnosis. These collections should:
   a. Correspond to remnants once the blocks with potential diagnostic utility have been selected.
   b. Have informed consent (IC) for a specific line of research.
   c. Have an investigator responsible for each collection.
   d. In this case, these collections must be registered individually under the Collections Section of the National Biobank Register.
2. The inclusion of samples from the health care archives into the hospital biobank, under the biobank system, given that these are samples of great value for research. These collections should:
   a. Correspond to remnants once the blocks necessary for diagnosis have been selected.
   b. Have the relevant informed consent (IC) for their inclusion into the biobank.
   c. The biobank must be authorized in accordance with applicable regulations.

It is important to emphasize that it is not compulsory for all APS samples to be placed in a biobank, as this would require an enormous effort without a significant scientific return. The inclusion should be done gradually, taking advantage of specific requests from researchers for specific samples, using the following strategy:

Prospective samples can be incorporated gradually from any diseases that represent a priority for the center, provided that they have the corresponding consents from the patients involved.

If the biobank requests retrospective samples from the diagnostic archive for a specific project, the request should be submitted to the biobank's committees for evaluation so that they can decide on the possibility of transfer and, consequently, the possibility of these samples becoming part of the biobank.

The latter has the advantage that only the samples from the historical archives of greatest interest, i.e., those most requested by researchers, would be included. Moreover, the effort involved in the process of including samples in the biobank (proving the IC, the need for reconsent, anonymization, etc.) would only be carried out the first time the samples were requested, since, once they have been integrated into the biobank system, subsequent transfers would be much simpler.

It is highly recommended to extend the use of IC to as many surgical samples as possible, regardless of whether the possible assignation of the samples to research is known in advance, as this will facilitate the future use of the health care archives for research purposes. This implies the need for a great deal of dissemination within the institution, as well as the involvement of the institution in organizational and financial aspects.

The role of biobanks in cancer research

The use of biological samples is essential for many research projects but, until recently, oncology researchers suffered from an almost dramatic lack of high-quality biological samples and their associated clinical information, which also met the high ethical and patient autonomy standards required by legislation. Biobanks are now helping to overcome this major obstacle.

In oncology research, there are many fields in which samples provided by biobanks, mainly from tumor banks, are of vital importance:

a) **Epidemiology**: although at first sight the epidemiology of cancer is far removed from the use of biological samples, large population-based DNA banks can provide very valuable information on the prevalence of germline mutations associated with the development of cancer (susceptibility biomarkers).
b) **Oncogenesis**: together with cell cultures and animal models, biological samples from biobanks, which provide first-rate genetic material, are an important tool in elucidating the molecular mechanisms that promote neoplasia, including epigenetic modifications, characteristics of microRNAs, or the interaction between genetic material and proteins.
c) **Taxonomy**: the current classification of neoplasms is based on the long-established identification of morphological, macro- and microscopic parameters. The use of samples from tumor banks, analyzed using high-throughput techniques (which will be reviewed in the following section), has led to the identification of new molecular types of neoplasms, with relevant prognostic and therapeutic implications.
d) **Tumor markers**: biobanks include numerous nontissue samples (whole blood, plasma, serum, serous effusions, etc.) from cancer patients. This material can be used to identify and test new tumor markers that facilitate early diagnosis and monitoring of the neoplasm.
e) **Personalized medicine**: biobanks also play an important role in this new therapeutic model defined by the design of drugs directed at a specific molecular target. Tumor samples are a first-rate source for identifying alterations in the signaling pathways inherent to the different neoplasms, which allows the first steps to be taken in obtaining compounds designed to act on the genetic-molecular alterations already known. On the other hand, although biobanks are nonprofit institutions, it is possible to find mutually beneficial channels of collaboration with the pharmaceutical industry, paying particular care to the wishes of past and future donors.

The International Cancer Genome Consortium (ICGC) is one of the most ambitious projects on a global scale, in which tumor samples are being used. It coordinates large-scale studies of 50 different types of cancer in order to identify the genetic changes involved in their development, through complete genome sequencing of tumor samples. All participants in the consortium, from the moment of their inclusion, undertake not to register patents or exercise intellectual property rights over the primary data obtained. Spain was assigned the study of one of the most frequent blood cancers, chronic lymphocytic leukemia (CLL), through a large team led by Dr. Elías Campo (Barcelona) and Dr. Carlos López Otín (Oviedo). The first results of this project have already been published, detecting recurrent mutations in four genes in CLL patients: NOTCH1, MYD88, XPO1, and KLHL6, a discovery that may have important therapeutic implications.

Tumor banking and high-throughput techniques

Formalin-fixed paraffin-embedded tissue (FFPE), a common component of the archives of anatomical pathology services, maintains high-quality proteins as evidenced by the widespread use of immunohistochemical (IHC) stains, aimed at identifying protein antigens in tissue sections or cytological samples. The DNA contained in FFPE tissue retains sufficient structural integrity for its study in health care practice and research, through widely known techniques such as PCR, in situ hybridization, etc. However, RNA is a very labile molecule that does not withstand the process of fixation and embedding in paraffin. Although great progress has been made recently in the extraction of useful RNA from FFPE tissue, the best source of this basic component for research is fresh tissue, with a short period of ischemia (time elapsed since sampling), frozen in a protocolized manner, and stored indefinitely in a deep freezer ($-80°C$) with the appropriate safety measures for guaranteeing its preservation.

One of the most consolidated and productive strategies in oncology research is the combination of two high-throughput techniques: expression microarray and tissue microarray (TMA). Expression microarray is the collection of thousands of short DNA sequences (probes) anchored to a solid surface. The mRNA from frozen tumor tissue and that from a paired sample of nontumor tissue from the same patient is extracted by the usual procedures and then transformed into DNA (cDNA) through a reverse transcriptase process.

The cDNA obtained from the tumor sample and from the nontumor sample, which is respectively labeled with red and green fluorochromes, is hybridized on the microarray. Thus, after hybridization, if the genes are overexpressed in the tumor, red fluorescence will be observed; if they are overexpressed in the nonneoplastic tissue, it will turn green; and if there are no differences in expression between the two samples, the color will be yellow. The results are analyzed using complex bioinformatics programs. In many cases, the overexpressed genes in the tumor will correspond to proteins of known or unknown function. By applying this procedure, several groups of researchers have succeeded in defining new types of cancers with a defined genetic signature, e.g., luminal A and B, HER-2, and basal types of breast carcinoma; subtypes of diffuse large B-cell lymphoma, etc.[3] However, access to expression microarrays is limited to the strongest research groups and cannot be expanded for several reasons, including high cost, complexity, technical demands, time, etc. Although not always possible, attempts should be made to transfer the identification of molecular subtypes of neoplasms to universal procedures applied in clinical care, such as immunohistochemistry (IHC). TMAs play an important role in this crucial step.

TMAs[4] are the result of constructing a paraffin block with hundreds of small tissue cylinders obtained from hundreds of archived tumors, together with the corresponding control cylinders. If a monoclonal or polyclonal antibody directed against the proteins identified by the expression microarrays is available, the presence of these markers in the hundreds of tumors contained in the TMA can be tested to see if the highly accessible IHC can reproduce the results of the expression analysis. Before the universal diffusion of this new high-throughput technique, immunostaining of whole tumor sections was necessary for the massive study of IHC expression of markers, resulting in a significant increase in costs, time invested by laboratory staff and researchers, and a high risk of losing valuable archive material. Ultimately, the use of TMAs is an inexpensive and easily accessible procedure that optimizes sample use. Theoretically, its most relevant drawback lies in the small size of the tissue cylinders included in the matrix, especially when studying heterogeneous neoplasms, but there are several strategies for minimizing possible sampling biases (different diameters and numbers of cylinders per tumor, studying very large series, etc.).

In many respects, expression microarray and TMA are complementary techniques. Expression microarray typically analyzes the expression of thousands of markers from a single donor tumor sample, whereas TMA is often used to determine the expression of a single marker in hundreds of tumor samples taken from hundreds of patients.

Naturally, the possibilities of high-throughput techniques are not exhausted in the strategy described. This is only one of the possible approaches that researchers working in the field of oncology can use, but the spectrum of possible uses of biological samples is very broad.

Colorectal cancer and biobanks

Colorectal cancer (CRC) is a neoplasm of very high incidence, it is the third most incident cancer worldwide, and 1.93 million new cases were diagnosed in 2020, with more than 935,000 associated deaths (WHO, 2020). It is, therefore, a central focus of biomedical research. In 2020 the PubMed search engine (http://www.ncbi.nlm.nih.gov/pubmed) had 15,710 entries under the heading "colorectal cancer."

The possible uses of biological samples in CRC research are almost infinite, covering epidemiological aspects (genetic epidemiology), basic research (carcinogenesis), translational research (identification of therapeutic targets), etc.

The biological samples that can be potentially used in CRC research vary greatly: primary tumor tissue, nontumor tissue, lymph nodes, regional lymph node metastases, distant metastases, different types of polyps, blood or serum from patients and family members, DNA extracted from these samples, etc. Ideally, tissue samples should be paired, i.e., with storage of both frozen tissue and formalin-fixed paraffin-embedded tissue. Samples can be obtained during ordinary health care practice, through screening programs, or from donors invited to participate in a specific project. The biobank guarantees that the samples have been collected and stored according to the strictest protocols, which translates into a high and uniform quality of the samples. Recently, the National Biobank Network, through the Tissue Quality Working Group, has

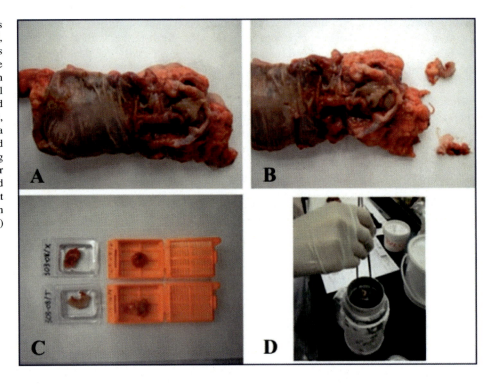

FIG. 50.1 The surgical specimen is received fresh, opened, and washed, checking whether there is any surplus tumor tissue that can be sent to the biobank (A). If so, samples are taken from the tumor and from the nontumoral intestinal wall (B). Part of these paired samples is selected for freezing, immersed in a special medium in a cryomold, and another portion is fixed in formalin for subsequent embedding in paraffin (C). The tissue intended for freezing is immersed in a precooled fluid, in this case, isopentane at −80°C, and subsequently stored in an ultra-low temperature freezer (ULT) until use (D).

standardized a series of preanalytical variables that allow for homogeneity in the process of collecting high-quality samples throughout the national system.[5]

Fig. 50.1 shows the standard procedure for collecting and preparing CRC tissue samples from a surgical specimen.

Researchers working in the field of CRC should be aware that the current legal framework for the use of biological samples is complex and strict, and the Research Ethics Committees (RECs) that assess projects are assigned the role of supervising compliance with the requirements set out in the applicable regulations.

Conclusions

The best way to access biological samples for research today is to request them from hospital biobanks or existing biobank networks. These institutions guarantee the provision of the necessary samples with clear advantages in several key aspects:

a) **Quality**: authorized biobanks use common, standardized protocols for obtaining and preserving samples, which offers researchers a quality guarantee.
b) **Associated information**: oncological research, in particular, requires abundant clinical data (type and histological grade of the neoplasm, staging, etc.) that can be linked to the information obtained from the samples. Biobanks collect a significant amount of clinical data in their information systems, which can be transferred to researchers, with some limitations based on ethical criteria and established in the regulations governing biomedical research.
c) **Informed consent (IC)**: donations of samples to biobanks allow the signature of a generic IC document by donors or patients, without having to specify the specific project or line of research in which they are going to be used. This makes their use by different groups of researchers much more versatile or flexible.
d) **Cooperative networks**: biobanks are integrated into networks at a regional, national, and international level; this makes it possible to quickly access samples from different biobanks, an aspect that is especially relevant when rare or difficult-to-collect samples are required.

References

1. Ley14/2007. *de Investigación biomédica de 4 de julio, Boletín Oficial del Estado (num 159)*. Madrid. https://www.boe.es/boe/dias/2007/07/04/pdfs/A28826-28848.pdf; 2007.
2. RealDecreto1716/2011. *de 18 de noviembre, Boletín Oficial del Estado de 2 de diciembre de 2011 (sección I num 290, pp 128434)*. Madrid. http://www.boe.es/boe/dias/2011/12/02/pdfs/BOE-A-2011-18919.pdf; 2011.

3. Kononen J, Bubendorf L, Kallioniemi A, et al. Tissue microarrays for high-throughput molecular profiling of tumor specimens. *Nat Med* 1998;**4**(7):844–7. https://doi.org/10.1038/nm0798-844.
4. Alizadeh AA, Eisen MB, Davis RE, et al. Distinct types of diffuse large B-cell lymphoma identified by gene expression profiling. *Nature* 2000;**403**(6769):503–11. https://doi.org/10.1038/35000501.
5. Esteva-Socias M, Artiga M-J, Bahamonde O, et al. In search of an evidence-based strategy for quality assessment of human tissue samples: report of the tissue Biospecimen Research Working Group of the Spanish Biobank Network. *J Transl Med* 2019;**17**(1). https://doi.org/10.1186/s12967-019-2124-8.

Section G

Approach to the sequelae and consequences of CRC

Chapter 51

Psychological approach and emotional management in CRC

Lucía Álvarez-Santullano, Lucía Barcia, Alba Burundarena, Marcos Calvo, Ainhoa Carrasco, Rosalía Fernández, Miriam Rojas, and Rosa Trillo

Spanish Association Against Cancer, Madrid, Spain

Keys and tools in the relationship between the health care team and the patient

Information will be key in helping patients adapt to the oncologic disease. However, high quality information from the teams will be as important as the information available through other informal channels, such as friends, family, or the internet. Handling of information will largely depend on the relationship of trust between the patient and their health care teams (Surgeons, Gastrointestinal Specialists, Medical and Radiation Oncology, Hospital Pharmacy, Psycho-oncology, Social Workers, etc.).

Tips for creating a trusting setting:

- Choose a quiet and private place.
- Allow sufficient time, do not rush.
- Make eye contact.
- Be receptive and have an open attitude.
- Observe the patient's verbal and nonverbal behavior.
- Avoid interruptions, subjective interpretations, advice, and/or phrases.
- Allow the patient to be accompanied by a support person.

Tips for facilitating emotional expression:

- Active empathic listening.
- Unconditional acceptance.
- Emotional exploration and validation.
- Collaborative and helpful attitude.
- Responding to feelings.

Tips for offering information:

- Use clear, simple, nontechnical language.
- Adapt to the patient's preferences and knowledge.
- Personalize the information given.
- Adapt the information to the specific time point of the oncologic process.
- Exposure of the professional's bias in communication.
- Encourage discussion, opinion, thoughts.

Tips on promoting understanding:

- Use supporting materials: videos, graphs, diagrams, etc.
- Use short sentences.
- Pay attention to emotions.
- Repeat, group, summarize, and emphasize information.
- Encourage the patient to ask questions.

Reinforce what has been expressed:

- Use verified technological resources (web pages, applications, videos, etc.).
- Use paper documents.

Provide access to appropriate psychosocial support services:

- Knowledge of the psychosocial network.

In any case, health care professionals must be aware of their own biases when presenting the information available, which are sometimes unintentional (e.g., surgeons tend to find surgical solutions the best). On the other hand, certain issues, such as sex-related ones, are given less attention than other problems the patient may experience, even though they may significantly affect the patient's quality of life and should be considered in clinical decision making, as pointed out by works such as those of Ubel et al. (2017).[1]

Searching for health information on the Internet

In a globalized world, with an excess of information at our fingertips, the ability to select relevant information is an exceptionally important factor in adapting to the disease and its treatments.

Some patient recommendations are as follows:

- Avoid entering the diagnosis in large search engines such as Google.
- Start from a prestigious page (such as that of the Spanish Association Against Cancer,[2] for example) and then, through the links, specify your search.
- Opt for patient- or family-oriented certified pages (e.g., HONcode).
- Choose an appropriate time when you are at ease. Avoid searching immediately after having received bad medical news.
- Immediately stop searching when you perceive that the information is harmful to you.
- Compare the information obtained with that provided by the health care staff, especially if the information you have found has caused pain, emotional suffering, or fear.

Psychological aspects in the different phases of the disease

Cancer generates significant psychological suffering that reaches levels of clinical emotional distress in fifty percent of those affected and is accompanied by psychopathological disorders in more than 30% of cases.[3] It is important to bear in mind that the prevalence of symptoms of anxiety, depression, and emotional distress is very high: between 2% and 25% of patients in different disease situations, time points, and treatment suffer from depressive symptoms.[3–6]

In Spain, colorectal cancer (CRC) is one of the main health problems, due to its high incidence in both men and women. The Council of the European Union recommends the fecal occult blood test (FOBT) in men and women aged 50–74 years, every two years. CRC is considered, in the majority of patients, as a chronic disease that requires effective and quality medical care and psychological intervention. It is, therefore, necessary to emphasize the importance of approaching the living situation of these patients from an integral and holistic perspective.

We must not lose sight of the fact that both patients who have been diagnosed with CRC and their families are forced to face different problems related to this type of diagnosis, such as eating disorders, loss of appetite, weight loss, gastrointestinal discomfort, diarrhea, constipation, alterations in body image, difficulties in sexual relations and other related events, events with a serious emotional impact on both the patient and their dear ones.

Emotional aspects in the prediagnostic phase

The prediagnostic phase sometimes requires hospital admission. The hospital is an institution that provides specialized medical services to patients; however, the hospitalization period stands out as one of the most stressful moments when there is a suspected diagnosis. Given the characteristics of CRC, the impact or relevance of early detection diagnostic tests at this stage is significant. The FOBT, a test associated with the early detection preventive program, is, together with colonoscopy, the most widespread test in the intermediate risk population (asymptomatic men and women aged between 50 and 74 years).

In general, the situations that generate the greatest concern for adult patients with respect to this initial phase revolve around the following points:

- The suspicion of a cancer diagnosis.
- The lack and delay of information ("not knowing the results or the reasons for the treatments, the analyses that are carried out," "not knowing when the tests are going to be done", etc.).
- The relationship with the health care staff in case of hospitalization ("seeing that the hospital staff is in a hurry," "that the doctors and nurses speak in a hurry or use words that I don't understand," etc.).

Continuity of care and information between different specialists in health care is key in this first moment of fear and uncertainty/lack of control.

Diagnostic phase

Facing a cancer diagnosis is a complex and very stressful experience. Undoubtedly, during this process many events will combine to generate tension, uncertainty, and anguish. For the patient, it is a major crisis, an unexpected and threatening situation that represents a complete change in their life. Many people diagnosed with cancer suffer an initial "emotional shock" when they receive the news. Some may feel very nervous, agitated, or irritable; others may experience a deep sadness, but we must remember that there are as many reactions as there are people. We could say that all these emotions, sensations, and alterations mean that the alarm has been set off, the patient perceives that there is a threat and reacts to it.

The vast majority of the above reactions are normal and predictable in these initial stages of the disease. It is advisable to inform patients that their reaction is normal, in order to avoid maladaptive behavior. In any case, the intensity of these reactions will be measured by the degree to which they interfere with daily life. Informing patients of psychosocial care networks is highly recommended.

The reported prevalence of emotional distress in cancer patients, at different times in the development of the disease, varies between 15% and 33% of cases, reaching 63% when patients who suffer from it at some point in the study are taken into account.[7]

Treatment phase

In addition to the impact of receiving a cancer diagnosis, cancer patients experience a series of emotional reactions of anxiety, fear, or helplessness, derived from the type of treatments that are usually used to control the disease.

In general, the changes produced by the different treatments are reflected in a significant modification to the daily life of the patient and their family, especially if hospitalization is necessary or if the patient lives outside the place where they receive treatment and has to frequently visit the hospital. Likewise, intrafamilial conflicts and the feeling of loss of control can be frequent. In addition, self-esteem problems derived from the consequences of the treatments (e.g., problems derived from changes in body image), as well as difficulties in sexual and couple relationships, may be frequently encountered.

In this block we will cover three of the most commonly used treatments to address CRC.

1. Surgery

Surgery remains the cornerstone of curative treatment for CRC. The surgical treatment of CRC has evolved greatly over the past 30 years with the aim of improving oncological and functional results, as well as quality of life by reducing surgical aggression, as has been pointed out by the Alianza para la Prevención del Cáncer de Colon[8] [Alliance for Colon Cancer Prevention].

From the psychological perspective, surgery is one of the situations most often described in the literature as a stressful life event. Some of the most common concerns that patients show in the face of surgery revolve around stress due to the hospitalization required for surgery, fear of anesthesia, fear of pain, fear of death during surgery, separation from their family, fear of losing control, fear of waking up during surgery, etc.

Feelings of hopelessness, helplessness, or hostility often accompany these fears. Surgical intervention in oncological diseases is generally aimed at the resection of the organ affected by the tumor, which can have serious psychological consequences. For example, mastectomy, genitourinary surgery, colostomy or head and neck surgery are some of the types of surgery that pose the most emotional problems for patients.

Having detailed information about the procedure before, during, and after according to the information needs of the patient plays a key role in good psychological adaptation and reduction of emotional discomfort.

2. Chemotherapy and new systemic treatments

Chemotherapy is another widely used therapy in the treatment of CRC. It is one of the most frightening treatments for patients and causes high levels of anxiety. In addition, the side effects of the treatment, such as alopecia, general malaise, nausea and vomiting, dizziness, weakness, fatigue, etc. are also strong stressors for both the patient and their family.

It is often a time when the person refers to feeling "really sick" due to many of the side effects that can limit their quality of life. Encouraging the patient to record the symptoms experienced and their intensity in a diary for later discussion with their care team will facilitate excellent symptom management and provide an increased sense of control associated with lower levels of psychological distress.

3. Radiotherapy

Although surgery is the main curative treatment for CRC, radiotherapy provides increased loco-regional control for certain types of tumor.

The most common reactions to this type of treatment are related to lack of knowledge. Symptoms of anxiety, depression, and hostility associated with the side effects of radiotherapy, such as tiredness, fatigue, dermatological changes, etc., are frequent.

According to several studies, patients undergoing radiotherapy may present fears, worries, apprehension, feelings of helplessness, nightmares and insomnia,[9–11] sexual dysfunction, and other problems[12,13] as well as short- and long-term neuropsychological sequelae that can affect 43% of patients depending on the type of tumor and radiotherapy.[3,14–17]

As for the course, in general, an increase in anxiety and depression is reported from the time preceding the start of radiotherapy until the end of treatment, but these decrease in the following months[18–21].

Physical sequelae of the treatments

Ostomy

Each treatment plan is tailored to the individual patient and can include curative-intent or palliative-intent surgery. In the former, the main objective is to remove the entire tumor, while in the latter, the aim is to control the disease or manage its symptoms when disease control is not possible.

Resection of the primary tumor includes the complete removal of the area of the colon where the tumor is located and the periregional nodes. The technique used will depend on the location of the tumor, so there are different types of surgery:

- Colectomy
- Metastasectomy
- Ileostomy or colostomy
- Lymphadenectomy
- Resection of neighboring structures or organs

As with any other major surgery, immediate postoperative complications such as fever, pain, infection of the abdominal or perineal wound, etc. may occur. Late-onset postoperative complications, such as changes in bowel habits with a tendency to diarrhea or constipation or discomfort and inconvenience in the pouch, are also possible.

Impact on quality of life

Ostomy is an important part of many patients with colon cancer.

Ostomies give patients quality of life, allowing them to live healthy lives, but due to their characteristics and implications, this process can cause physiological, psychological, and social problems for patients.

It is vitally important to take these factors into account since psychosocial aspects can negatively affect their quality of life. It is, therefore, necessary to encourage patients to develop coping strategies that allow a good adaptation to the ostomy and a correct management of their disease. Providing adequate information beforehand can help to adjust expectations and reduce the subsequent impact.

It seems that one of the problems that most concerns ostomates (76%) in their daily lives are leaks, odors, noises/flatulence, and pouch breakage or detachment as these are unpredictable. All this influences routines, daily life, and social interactions. This discomfort can be increased when they do not have a stable partner.

Emotional impact

In many people the ostomy experience provokes feelings of uncertainty, low self-esteem, frustration, and helplessness. Not knowing if it will be reversible or not, not knowing how to manage or care for it, and the feeling of lack of control are frequent issues that patients have to face in their daily lives. All this can frequently lead to symptoms of anxiety, depression, and difficulties in adapting and adjusting to the new life situation. As a result of the ostomy, there is an impact on the patient's own body image that can lead to problems of self-concept and self-esteem.

In his study, Ayaz-Alkaya[22] observed that 50% of stoma patients reported an increase in normal levels of anxiety, and 16% reported moderate levels of depression. Other articles also found a risk of suicide and death wishes. In relation to this, it was observed that 30% consumed drugs, especially anxiolytics and antidepressants, and this was more predominant in men than in women.

On the other hand, personal relationships and social life can sometimes be affected as the way in which the person relates to their environment changes, making them less independent and secure. Consequently, many people also experience changes in their work environment and difficulties in resuming their usual routines.

Due to the importance of adaptation to the ostomy in the patient's quality of life, comprehensive care by a multidisciplinary team is necessary, so that personal skills are acquired or maintained for coping with the psychosocial problems that arise after the intervention.

Therefore educational and psychoeducational interventions are useful and necessary for achieving the correct management of the disease, since they provide the patient with active coping strategies. In addition, the patient's reference health care team plays a fundamental role in helping patients to manage the emotional components of the process, by resolving doubts and providing adequate training on ostomy management and care. Autonomous behavior in ostomy care in the initial moments after surgery is an indicator of personal adjustment to the disease and it is advisable to promote it since it will act as a protective factor against psychological discomfort.

When symptoms of depression or anxiety persist and emotions are difficult to manage, specialized psychological care during the course of disease is usually beneficial and effective for the patient.

Finally, the social support of the close environment and partaking in pleasant activities are fundamental protective factors that significantly help to improve the quality of life of ostomized patients.

Impact on body image

Body image is understood as the idea one has about oneself based on perceptions, thoughts, and behaviors related to appearance, which allows the individual to feel they are in charge of their body and actions. This representation is not only shaped by the individual, but also by the society in which they live (relationships, religion, culture, and values). Body image is one of the components of self-esteem, which includes all the feelings and judgments, both positive and negative, that a person has of themselves.

A positive body image is a good indicator of physical and psychological well-being, as well as affecting personal development, self-esteem, self-worth, self-acceptance, and health-related behaviors and perspectives. Ostomized patients experience feelings of rejection and dislike toward their physique, which has been "mutilated" by the ostomy, therefore, their body image is negatively affected.

The most common feelings with respect to body image are usually mutilation, disfigurement, disgust, shame, feeling unattractive, etc. These people usually find it difficult or are unable to look in the mirror because of these negative psychological feelings toward the stoma, and due to esthetic reasons. The person may feel that their body has become the central theme of their existence. Discharge from hospital often causes discomfort related to the care of the stoma, which can lead to a prolongation of stoma care or to delegating this care to another person.

Beauty is given significant importance by our modern-day society and this creates a real stigma, since this continuous evaluation of one's physical appearance causes rejection and discrimination of all individuals who do not meet these beauty standards.

The body image that a person has of themselves therefore influences their daily life and social interactions and can even lead to social isolation. Within interpersonal relationships, it is worth mentioning the intimate relationship with the partner, who also experiences this new reality. The negative image that an ostomized patient has of their body is closely linked to sexuality, so that their sex life and relationship with their partner is also affected. It has been seen that younger people are more affected by the changes in body image and this affects their intimate, family, and social relationships more.

Impact on sexuality

Sexuality is more than a purely biological need; it is a higher need through which people express feelings and affection. It brings self-esteem, security, and acceptance and is closely related to quality of life.

Sexuality refers to how people express themselves sexually. It encompasses how they see, feel, and think of themselves as a sexual being, and the ways in which they show this through their actions, behaviors, and relationships. The experience is personal and therefore different for everyone.

- Stoma and sexuality

Regarding the impact of an ostomy on sex life, it has been seen that there may be some organic cause due to the surgery, but mostly there are psychological problems related to body image and self-esteem. In addition, anxiety and stress can be increased, as well as a more depressive mood that can lead to erectile dysfunction. There may also be a decrease in sexual activity for fear of damaging the ostomy (both the patient and their partner), or fear that something could happen to the pouch, such as leaks, flatulence, etc., during sexual intercourse.

To all the above we have to add the fear of rejection by their partners, even to the point of abstaining from sexual intercourse.

Sexual health is a multifactorial state, which has a strong impact on quality of life, both for the patient and for their partner. Dysfunction usually affects one in three ostomized men and one in two women.

Ten percent of patients with an ileostomy will experience mechanical difficulties during sexual intercourse, which can be improved by using belts or a girdle, as well as prior emptying of the pouch. In some cases, postural changes are recommended during intercourse.

According to the ASOE[23] [Spanish Association of Ostomates] the most frequent manifestations of sexual dysfunction are as follows:

1. Changes in desire or arousal.
2. Decrease or absence of orgasm.
3. In women:
   - vulvovaginal pain
   - vaginal dryness
   - vaginal atrophy.
4. In men:
   - erectile dysfunction
   - alterations in ejaculation.

- Causes of sexual dysfunction

Surgery, direct lesion of the nerves:

If the lesion occurs in the sympathetic plexus, it produces changes in orgasm and ejaculation. If the lesion occurs in the parasympathetic nerves, it produces changes in erection, blood flow to the clitoris and penis, and vaginal lubrication. If pudendal nerves are affected, this will cause sensory and motor alterations in the penis and clitoris.

Other factors to consider are as follows:

- Surgical technique and experience of the surgeon
- Radio-chemotherapy
- Psychological changes
- Patient's age and tumor stage.

From the patient's perspective there are beliefs and thoughts about sexuality that deserve our attention:

- Because the body is linked to sexuality, the fear of rejection may facilitate the onset of sexual dysfunction.
- Acceptance of the ostomy patient's condition will depend on the difficulties in social or sexual relations that they may have encountered before the ostomy.
- It is possible that we centralize this problem only on the stoma.
- All sexual difficulties have a psychological part, even if their origin is organic.

Factors to take into account for improving this situation

Providing enough information and establishing open communication with health care professionals is necessary because this will reduce anxiety. The health care professionals are the ones who will provide solutions and therefore they need to be informed of the difficulties faced by patients.

The partners of ostomates have to learn to share the new scenario in their sexual relations, this is a new stage in life, and new solutions need to be found.

It is essential to communicate, trust, and talk openly about the differences or changes that have occurred after the intervention. Looking for alternatives together and facing this uncertain future is a process, and as such it will involve different phases, very similar to those described by Elisabeth Kübler Ross in her classic book "On Death and the Dying" (1969).

On the other hand, the family is a facilitator, buffer, and repressor of anxiety. In this sense, the family should not be overprotective as this leads to a situation of dependence, which in turn will create feelings of uselessness. Table 51.1 summarizes recommendations to promote sexuality in ostomy patients.

Hygiene recommendations: general recommendations for ostomy patients are summarized on Table 51.2.

Sexuality and intimacy help people coping with cancer by helping them cope with feelings of distress and getting through treatment. However, the reality is that a person's sexual organs, sexual desire (sex drive or libido), sexual performance, well-being, and body image can be affected by cancer and its treatment.

Sexual problems often develop because of the physical and psychological side effects of cancer and cancer treatments. Some surgeries and treatments may have very little effect on a person's sexuality, sexual desire, and performance, but others may affect the functioning of certain parts of the body, change hormone levels, or damage nerve function, all of which can change a person's sexual performance. Certain types of treatments have side effects, such as tiredness, nausea, bowel or bladder problems, pain, skin problems, or other changes in appearance that may cause problems with sexuality. Some sexual problems improve or disappear with time; however, others are long lasting and may become chronic.

TABLE 51.1 Recommendations on sexuality for ostomy patients.

General recommendations on sexuality	Specific recommendations for women
- Know your body as best possible so you can show your partner what you like. Do not perform forced sexual acts as they tend to worsen the situation.	- Use lubricants for vaginal dryness. Consult your health care team about the most appropriate lubricants.
- Important: use your imagination and share it with your partner.	- Consider using sensual lingerie.
- Always practice safe sex.	- The woman should be on top (less uncomfortable).
- Concentrate on your feelings, not on the bag.	- Sex games and toys may be useful.
- Fear of stoma damage, pouch detachment, odors, noises, leaks, waste handling.	- Visualize the act as something normal.

TABLE 51.2 Recommendations on hygiene for ostomy patients.

Empty the bag, secure it in place, and use deodorant on it. Use a mini opaque bag, obturator, and stoma cover.
Wash properly, or even bathe/shower together. Wash hands, genital/anal area.
Urostomy: increase fluid intake/avoid odorous foods or carbonated liquids.
Empty bladder after intercourse.
Irrigation.
A girdle and other pouches may be used during intercourse.

Psychological aspects of the SARS-CoV-2 pandemic (COVID-19)

Diagnostic delays resulting from the patients' fear of going to health centers, interference in colon cancer screening programs, and the overburdening of the health system in both primary and specialized care have affected the diagnostic period in CRC during the COVID-19 pandemic.

Emotional burnout has been documented in several studies by the cancer observatory of the Spanish Association Against Cancer (2019). It has been especially related to the delay in diagnostic tests that has caused anxious symptomatology of varying intensity.

Another aspect that has been affected is related to the doctor-patient relationship, consultation times, and difficulties in verbal and nonverbal communication due to prevention measures (social distance and use of masks).

The repercussions of repeated periods of lockdown, mobility restrictions, and social and family limitations, as well as the reduction of physical and leisure activity in general, are to be studied in detail in the coming years. It is currently difficult to quantify their impact on the mental health of people who have been or are being treated for colon cancer and their families. It is, therefore, a challenge for psychological research in the field of oncology in the years to come.

References

1. Ubel PA, Scherr KA, Fagerlin A. Empowerment failure: how shortcomings in physician communication unwittingly undermine patient autonomy. *Am J Bioeth* 2017;**17**(11):31–9. https://doi.org/10.1080/15265161.2017.1378753.
2. AECC. *Asociación Española Contra el Cáncer*. https://www.aecc.es/es.
3. Hernández M, Antonio CJ. La atención psicológica a pacientes con cáncer: de la evaluación al tratamiento: 1er Premio de la XIX edición del Premio de Psicología Aplicada "fael Burgaleta" 2012. *Clín Salud* 2013;**24**(1):1–9. https://doi.org/10.5093/cl2013a1.
4. Carroll BT, Kathol RG, Noyes Jr. R, Wald TG, Clamon GH. Screening for depression and anxiety in cancer patients using the Hospital Anxiety and Depression Scale. *Gen Hosp Psychiatr* 1993;**15**(2):69–74. https://doi.org/10.1016/0163-8343(93)90099-a.
5. DeFlorio ML, Massie MJ. Review of depression in cancer: gender differences. *Depression* 1995;**3**(1-2):66–80. https://doi.org/10.1002/depr.3050030112.
6. Grassi L, Travado L, Moncayo FLG, Sabato S, Rossi E, SEPOS Group. Psychosocial morbidity and its correlates in cancer patients of the Mediterranean area: findings from the Southern European Psycho-Oncology Study. *J Affect Disord* 2004;**83**(2–3):243–8. https://doi.org/10.1016/j.jad.2004.07.004.
7. Cabrera Macías Y, López González E, López Cabrera E, Arredondo AB. La psicología y la oncología: en una unidad imprescindible. *Rev Finlay* 2017;**7**(2):115–27. http://scielo.sld.cu/scielo.php?pid=S2221-24342017000200007&script=sci_arttext&tlng=pt.
8. Morillas JD, Castells A, Oriol I, et al. Alianza para la Prevención del Cáncer de Colon en España: un compromiso cívico con la sociedad. *Gastroenterol Hepatol* 2012;**35**(3):109–28. https://doi.org/10.1016/j.gastrohep.2012.01.002.
9. Forester BM, Kornfeld DS, Fleiss J. Psychiatric aspects of radiotherapy. *Am J Psychiatry* 1978;**135**(8):960–3. https://doi.org/10.1176/ajp.135.8.960.
10. Munro AJ, Biruls R, Griffin AV, Thomas H, Vallis KA. Distress associated with radiotherapy for malignant disease: a quantitative analysis based on patients perceptions. *Br J Cancer* 1989;**60**(3):370–4. https://doi.org/10.1038/bjc.1989.287.
11. Peck A, Boland J. Emotional reactions to radiation treatment. *Cancer* 1977;**40**(1):180–4. https://doi.org/10.1002/1097-0142(197707)40:1<180::aid-cncr2820400129>3.0.co;2-5.
12. Helgason AR, Fredrikson M, Adolfsson J, Steineck G. Decreased sexual capacity after external radiation therapy for prostate cancer impairs quality of life. *Int J Radiat Oncol Biol Phys* 1995;**32**(1):33–9. https://doi.org/10.1016/0360-3016(95)00542-7.
13. Hervouet S, Savard J, Simard S, et al. Psychological functioning associated with prostate cancer: cross-sectional comparison of patients treated with radiotherapy, brachytherapy, or surgery. *J Pain Symptom Manage* 2005;**30**(5):474–84. https://doi.org/10.1016/j.jpainsymman.2005.05.011.
14. Ahles TA, Correa DD. Neuropsychological impact of cancer and cancer treatments. In: *Psycho-oncology*. Oxford University Press; 2010. p. 251–7. https://doi.org/10.1093/med/9780195367430.003.0034.
15. Crossen JR, Garwood D, Glatstein E, Neuwelt EA. Neurobehavioral sequelae of cranial irradiation in adults: a review of radiation-induced encephalopathy. *J Clin Oncol* 1994;**12**(3):627–42. https://doi.org/10.1200/JCO.1994.12.3.627.
16. Sanz A, Olivares ME, Barcia JA. Efectos cognitivos de la radioterapia en gliomas de bajo grado. *Psicooncologia (Pozuelo de Alarcon)* 2012;**8**(2–3):3. https://doi.org/10.5209/rev_psic.2011.v8.n2-3.37879.
17. Sheline GE, Wara WM, Smith V. Therapeutic irradiation and brain injury. *Int J Radiat Oncol Biol Phys* 1980;**6**(9):1215–28. https://doi.org/10.1016/0360-3016(80)90175-3.
18. Andersen BL, Karlsson JA, Anderson B, Tewfik HH. Anxiety and cancer treatment: response to stressful radiotherapy. *Health Psychol* 1984;**3**(6):535–51. https://doi.org/10.1037/0278-6133.3.6.535.
19. Bye A, Ose T, Kaasa S. Quality of life during pelvic radiotherapy. *Acta Obstet Gynecol Scand* 1995;**74**(2):147–52. https://doi.org/10.3109/00016349509008925.
20. Buick DL, Petrie KJ, Booth R, Probert J, Benjamin C, Harvey V. Emotional and functional impact of radiotherapy and chemotherapy on patients with primary breast cancer. *J Psychosoc Oncol* 2000;**18**(1):39–62. https://doi.org/10.1300/J077v18n01_03.

21. Janda M, Gerstner N, Obermair A, et al. Quality of life changes during conformal radiation therapy for prostate carcinoma. *Cancer* 2000;**89**(6):1322–8. https://doi.org/10.1002/1097-0142(20000915)89:6<1322::aid-cncr18>3.0.co;2-d.
22. Ayaz-Alkaya S. Overview of psychosocial problems in individuals with stoma: a review of literature. *Int Wound J* 2019;**16**(1):243–9. https://doi.org/10.1111/iwj.13018.
23. ASOE. Asociación Española de Ostomizados. https://www.asoeasociacion.org

Chapter 52

Nutritional status in patients with CRC: Assessment and recommendations

María Teresa García Rodríguez
Research in Nursing and Health Care, Institute of Biomedical Research of A Coruña (INIBIC), A Coruña University Hospital Complex (CHUAC), SERGAS, A Coruña, Spain

Introduction

Diet is an important part of both tumor formation and patient recovery from cancer. Different studies show that a high intake of red and processed meats is directly related to developing colorectal cancer, while foods such as fruits, vegetables, and fiber are considered to be protective.[1–3] Therefore it could be said that nutrition is a fundamental factor for the evolution and recovery of the patient since it influences the morbimortality of the tumor disease.

The presence of malnutrition is very high in cancer patients, although its prevalence varies depending on the studies.[4,5] Depending on the stage of the tumor, it is observed that malnutrition is present in 15%–20% of patients at the beginning of the disease, increasing to 80–90% in patients in the terminal phase.[6]

Nutritional deficiency gives rise to various problems and complications that will affect quality of life and the evolution of the disease. Due to insufficient food intake, the presence of malabsorption and/or digestive alterations, the alterations that appear are usually at both the caloric and protein level, leading to delayed healing recovery, the appearance of sarcopenia, anorexia, and weight loss or increased fragility.

For this reason, it is essential to carry out a nutritional assessment of cancer patients, even before undergoing treatment (surgery, radiotherapy, and/or chemotherapy).[7] In this way, adequate nutritional support could help to:

- Provide a good response to the treatment, since there would be a better overall tolerance, minimizing side effects, and thus being able to increase its effect.
- Avoid weight loss, to protect the immune status, through the preservation of muscle and lean mass.
- Maintain functional status to reduce complications, increase survival, and avoid prolonged periods of hospitalization.

Related complications

Different complications are brought about due to malnutrition and catabolism associated with the presence of cancer, among which are:

(a) *Tumor cachexia:* It is defined as "a multifactorial syndrome characterized by a loss of skeletal muscle mass that cannot be completely reversed with conventional nutritional support and leads to progressive functional deterioration. The pathophysiology is characterized by a negative protein and energy balance, due to a variable combination of reduced intake and altered metabolism."[8] It may or may not be accompanied by loss of fat mass and insulin resistance, anorexia, or increased muscle protein depletion may be found.

   It is associated with a lower tolerance and response to cancer treatment, with a worsening quality of life and lower patient survival.

It is classified into three stages:

- Precachexia where there is a weight loss that is less than 5% and in which there may be anorexia.
- Cachexia when the patient has a weight loss >5% in 6 months or a BMI < 20 kg/m^2 along with a weight loss > 2% or sarcopenia with current weight loss > 2%.
- Refractory cachexia where there is such active catabolism that guidelines or treatment to slow weight loss is ineffective.

(b) *Sarcopenia:* Sarcopenia is a syndrome in which there is a loss of muscle mass and consequently of strength. According to the European Working Group on Sarcopenia in Older People (EWGSOP),[9] the criteria for sarcopenia are as follows:
- Criterion 1: Low muscle mass
- Criterion 2: Low strength
- Criterion 3: Low functional performance

Therefore its diagnosis should be based on the presence of the first criterion together with one of the other two (loss of strength or low functional performance).

According to this group, there would be three stages:
- Presarcopenia: where there is a loss of muscle mass that does not affect either strength or functional performance.
- Sarcopenia where the loss of muscle mass will affect either lack of strength or functional performance.
- Severe sarcopenia in which the loss of body mass will affect both the patient's lack of strength and functional performance.

(c) *Frailty:* There is no consensus to define patient frailty. It is considered a state of vulnerability in which there is weight loss, loss of strength, reduced physical activity, exhaustion, or slow walking speed. According to Fried et al.[10] the diagnosis should be based on the presence of three or more of the identifying physical features.

Causes of malnutrition

According to various authors,[11,12] the causes of malnutrition in oncology patients are due to three reasons:

(a) *Tumor related:* tumors, especially those found in the digestive tract, in addition to causing intestinal occlusions, hemorrhages, fistulas, etc., can also cause problems of malabsorption of food and alterations in the metabolism of lipids, proteins, or glucose, which contributes to greater energy expenditure. In addition, the secretion of substances such as cytokines, interleukin, etc. causes anorexia, loss of weight, fat and muscle mass, and leads to cachexia.

(b) *Those related to the treatment*: both surgery and treatment with radiotherapy and chemotherapy will influence the patient's nutritional status.

Surgery is a procedure that is associated with an increase in metabolic stress, so both protein and energy requirements will increase due to the healing process and the fight against infections. Malnutrition caused by resection of the colon or rectum is due to the appearance of diarrhea, water and electrolyte disturbances, and malabsorption of vitamins and ions.

As for radiotherapy, its effects will depend, among others, on the site of radiation. At the colon and rectum level, the presence of enteritis and colitis can be observed, giving rise to diarrhea, malabsorption, enteropathy, or hydroelectrolytic alterations. Among the chronic alterations that can be found are the presence of fistulas, subocclusive pictures, stenosis, ulcers, etc. The effects of chemotherapy on the patient's nutritional status will depend on different factors such as the type of drug used, the dose and/or the duration of treatment, among others. The most frequent is the appearance of nausea and vomiting that give rise to water and electrolyte imbalance with weight loss and general weakness. Other consequences of chemotherapy are mucositis, changes in body composition, or alterations in taste.

(c) *Those related to the patient:* these are due to the unhealthy habits that the patient has acquired throughout his life. Thus the harmful effects of alcohol and tobacco, such as lack of appetite or decreased absorption of nutrients, will intensify if consumption continues. Poor dental hygiene will lead to oral alterations causing loss of teeth, bad taste of food, or gingivitis.

It should also be taken into account that psychological alterations may appear, motivated by maladaptive reactions to fear, anxiety, depression, and stress that will cause alterations in food intake.

Assessment of nutritional status in colorectal cancer patients

Since there is no specific method/parameter that can tell us categorically the nutritional status of patients, it is proposed in this section that a global assessment of nutritional status be performed. In other words, not only should a nutritional assessment be made, but a physical assessment should also be included due to their habitual deterioration.

Therefore, in this section, in addition to the usual methods of nutritional assessment, other methods that assess physical deterioration such as frailty, loss of muscle mass, or alteration of body composition will also be addressed.

(a) *Nutritional status screening methods:* With these, healthcare personnel can detect and recognize patients with alterations in nutritional status at an early stage and thus intervene in an attempt at recovery.

According to the guidelines on nutrition in cancer patients of the European Society for Clinical Nutrition and Metabolism (ESPEN),[13] the method of nutritional screening should be brief, inexpensive, and with high sensitivity and specificity. Among the most commonly used screening methods are the following:

a.1. *Nutritional Risk Screening (NRS-2002)*[14]: This method is recommended by ESPEN to determine the risk of malnutrition in hospitalized patients. It consists of an initial screening and a final screening. The first one takes into account: the BMI value <20.5, weight loss over time, and the patient's severity, moving on to the second screening if any of the aforementioned conditions are met.

The final screening takes into account the alteration of the nutritional status according to weight loss, BMI and intake in the previous week and the severity of the disease. After choosing and adding the scores for altered nutritional status and severity of disease, 1 point should be added if the patient's age is >70 years. Scores ≥3 points indicate that the patient is at nutritional risk. Available at: https://www.medicaa.hc.edu.uy/images/Curso_Soporte_Nutricional_2016/M%C3%B3dulo1/NRS_2002_herramienta.pdf

a.2. *Malnutrition Universal Screening Tool (MUST)*: Developed by the British Association of Parenteral and Enteral Nutrition (BAPEN), it consists of 5 steps. The first 3, like the previous method, take into account the BMI value, weight loss over a period of 3–6 months, and the effect of the disease on food intake in the last 5 days. The fourth adds the value obtained from the first three and the fifth assigns a nutritional risk along with guidelines for action. It categorizes the patient as being at low, medium, or high risk of malnutrition. It is available at: https://www.bapen.org.uk/images/pdfs/must/spanish/must-toolkit.pdf

a.3. *Brief Nutritional Assessment Questionnaire (SNAQ)*: Developed by Kruizenga et al.[15] for the detection of malnutrition in hospitalized patients, it consists of three questions where the answers are scored. If the patient obtains a score ≥2 it indicates that nutritional intervention is required. This method is available in different languages at: *www.fightmalnutrition.eu*

a.4. *Malnutrition Screening Tool (MST)*: This is a method that has been validated for both inpatients and outpatients undergoing radiotherapy or chemotherapy and is recommended for use by the Multidisciplinary Clinical Guidelines on Nutrition Management of the Cancer Patient. It consists of two items: unintentional weight loss and decreased dietary intake. Scores ≥2 indicate that the patient is at risk of malnutrition. Available at https://scielo.conicyt.cl/scielo.php?script=sci_arttext&pid=S0034-98872017000400005

(b) *Nutritional assessment methods:* With these methods, it will be possible to identify those patients who are malnourished or at risk of malnutrition. According to ESPEN,[13] the assessment should be repeated during treatment to determine and monitor nutritional requirements as needed at any given time.

b.1. *Subjective Patient-Generated Global Assessment (VGS-GP)*: It is the most widely used and recommended method of nutritional assessment.[13] This questionnaire is a modification of the original Detsky questionnaire by Ottery et al.[16] which includes common symptoms found in the oncology patient. It consists of two parts: the first is a personal assessment of the patient in which he/she is asked about the evolution of weight, changes in intake, type of food, food-related symptoms, and functional capacity; the second part is completed by the surveyor and refers to physical signs such as anthropometric parameters, metabolic demands, presence or absence of edema and ascites. The questionnaire has several answers and each of them has a score, globally classifying patients as well nourished, moderately malnourished or suspected malnutrition, and severely malnourished.

b.2. *Mini Nutritional Assessment (MNA)*[17]: The questionnaire contains two parts, one corresponding to the screening and consisting of seven items and the other for the evaluation consisting of 12 items. The latter section will only be carried out in the event that the patient obtains a score ≤11 in the screening part.

Among the variables assessed are anthropometric measurements, food consumption, lifestyle, mobility, or self-perception of health. Depending on the final score obtained, the patient will be classified as well nourished (scores between 24 and 30), at risk of malnutrition (17–23.5 points), or malnourished if the total score is <17 points. Available at: http://www.seom.org/seomcms/images/stories/recursos/infopublico/publicaciones/soporteNutricional/pdf/anexo_03.pdf

(c) *Methods for determining physical activity/fragility:* These methods show patients' risk of functional impairment. Among the most commonly used tests are:

c.1. *Short Physical Performance Battery (SPPB)*[18]: Also called the Guralnik test, it is based on different tests: a balance test, a gait speed test, and a sit-to-stand test. The order of execution cannot be changed and each test is scored independently. The sum of each of the parts will give us the total score, the scoring range being from 0 to 12 points. Scores below 10 are indicative of frailty, which means that patients will have a high probability of disability and falls. Available at https://www.fisterra.com/gestor/upload/guias/Anexo%203.pdf

c.2. *SHARE-FI*[19]: It is a frailty instrument validated for the Spanish population and can be applied in primary care. This instrument is available through the free access calculators available at https://sites.google.com/a/tcd.ie/share-frailty-instrument-calculators/.

There is a version for each sex and it is translated into other languages such as Italian, French, Polish, and German. The variables taken into account are appetite, lack of energy, muscular strength measured by grip strength, difficulties at the functional level (walking 100m and climbing a flight of stairs without resting), and the frequency with which physical activity is performed.

- **c.3.** *Cuestionario FRAIL*[20, 21]: It consists of 5 items where fatigue, endurance, aerobic performance, comorbidity, and weight loss are assessed. The scoring range is between 0 and 5 and the patient is classified as nonfragile if the score obtained is 0 points, probably fragile when a score between 1 and 2 points is obtained and fragile when scores are ≥3.
- **c.4.** *The Frailty Trait Scale (FTS)*[22, 23]: It was developed by Garcia-Garcia et al. in an elderly population in Toledo (Spain). Based on two frailty models, Fried's and the one proposed by Rockwood, it has 12 items assessing energy/nutrition, physical activity, vascular system, strength, endurance, and walking speed. The scoring range goes from 0 points (no frailty) to 100 points indicating maximum frailty.
- **c.5.** *Specific scales for oncogeriatric patients*[24]
  - Scale G8: Consists of 8 items that assess food intake, weight loss, mobility, presence of neuropsychological problems, BMI, amount of medication taken per day, condition, and age. Each item has several scored responses. The total sum is obtained from the sum of each item, with scores ≤14 points indicating frailty.
  - VES Scale 13: This is a self-administered questionnaire, consisting of 5 items, which assesses the patient's age, self-assessment of health, difficulty in performing activities such as walking 500m, lifting heavy objects, performing household chores, etc., and the need for help with activities of daily living. The scoring range is from 0 to 10 with the patient being considered disabled with scores ≥3 points.

(d) *Methods for determining muscle mass/function:* Among the various methods and techniques suggested by ESPEN are the following[13] for determining both muscle mass and grip strength:

- **d.1.** *Bioimpedanciometry (BIA)*[25]: It is an easy to perform, noninvasive and inexpensive method in which body composition (lean mass and fat mass volume) is estimated through total body water. With this method, the patient's weight is broken down into a value for each body component or element (water, proteins, minerals, and fat) determining its proportion in the organism. Special scales with metal plates or bioimpedance monitors containing surface electrodes can be used for this test.
- **d.2.** *Dynamometry*[26,27]: Muscle depletion secondary to malnutrition and disease contributes to weakening the patient at the functional level. One way to measure muscle function is through dynamometry or grip strength because it is also a simple, inexpensive, and easy-to-perform method. To determine the values of hand strength, a dynamometer is used, the Jamar dynamometer being the most commonly used in clinical practice. Measurements should be performed following the recommendations of the American Society of Hand Therapists, which suggest a standardized arm position.

It should be taken into account that the measurement will be influenced by different factors such as age, sex, patient's position, number of attempts, or the pause period between each determination.

(e) *Other determinations and methods to assess nutritional status:*

- **e.1.** *Analytical parameters*[13]: The determination of different analytical parameters such as protein or inflammatory markers (albumin, prealbumin, transferrin, total lymphocytes, retinol transporter protein, etc.) is common, with C-reactive protein and albumin being the parameters recommended by ESPEN to determine the degree of systemic inflammation caused by protein depletion.

  Even so, it should be taken into account that these values can be altered not only by malnutrition, but also by other factors such as infections, treatments, and so on.

  The Nutritional Control System (CONUT)[28] is a method that assesses nutritional status by means of the values of three analytical parameters (albumin, cholesterol, and total lymphocytes). Depending on the value of each parameter a score is given; the sum of all the scores gives a total score classifying patients as mildly malnourished (2 to 4 points), moderately malnourished (5 to 8 points), and severely malnourished (9 to 12 points).

  Another method that uses analytical parameters for the determination of nutritional status is the Onodera Prognostic Nutritional Index or Onodera Index (PNI-O).[29, 30] It is considered a prognostic factor for survival and an indicator of postoperative morbidity. The following mathematical formula is used for its calculation, which combines the values of albumin and total lymphocytes:

$$PNIO = 10 \times \text{Serum albumin (g/dL)} + 0.005 \times \text{Total lymphocytes (mm}^3\text{)}$$

Mathematical equation scores below 40 points are indicative of malnutrition.

- **e.2.** *Anthropometric parameters*[6]: One of the most commonly used anthropometric parameters is to determine the weight loss with respect to the usual weight in a specific period of time. The weight loss is established in %

TABLE 52.1 Relationship between weight loss and time.

Time	% Weight loss
1 month	>2%
3 months	>7'5%
6 months	>10%

and the time in which it has been lost in months (1, 3, or 6 months). The percentage of weight loss is calculated by the formula:

$$\% \text{ Weight loss} = (\text{Usual weight} - \text{Current weight}) \times 100/\text{Usual weights}$$

The patient is considered malnourished when there is the following relationship between weight loss and time of weight loss (Table 52.1).

Another anthropometric parameter that can be used is the arm circumference. It is measured with an inextensible and flexible anthropometric tape at the midpoint of the arm. The patient is considered to be malnourished if a value <20 cm is obtained or there is a decrease of >2 cm between two determinations.

As for the calculation of BMI, it is considered of little use due to two reasons: the high prevalence of obesity in the adult population and the redistribution of body fluids, since patients sometimes tend to retain fluids, which would affect its calculation.

e.3. *Methods using analytical and anthropometric parameters in the determination of nutritional status*[31]: The Spanish Society of Parenteral and Enteral Nutrition (SENPE) recommends that the methods used to establish the nutritional status of patients, in addition to being simple, should include both analytical and clinical parameters. Subsequently, the indicator based on SENPE recommendations is available and consists of the following criteria:
Criterion A: BMI < 18'5 kg/m^2 or significant weight loss (>5% in 1 month or > 10% in 6 months).
Criterion B: albumin <3.5 g/dL.
Criterion C: lymphocytes <1600 c/mm^3 or cholesterol <180 mg/dL.

It is established that the patient will be classified as malnourished if he/she meets two of the three criteria.

Another method is the Nutritional Risk Index (IRN). It is considered useful to determine the surgical risk of patients. Like the Onodera Nutritional Prognostic Index, it is based on a mathematical formula that relates the serum albumin value to the patient's weight as shown in the following equation:

$$\text{INR} = 1.519 \times \text{Albumin (g/L)} + 41.7 \,(\text{Current weight/Usual weight})$$

Current weight refers to the patient's weight at the time of assessment and usual weight refers to the patient's weight before the disease.

Results less than or equal to 100 points classify the patient as malnourished (mildly malnourished: 100–97.5 points, moderately malnourished: 97.4–83.5, and severely malnourished <83.5 points).

Dietary recommendations and nutritional support

According to the ESPEN clinical guidelines[13] on nutrition in cancer patients, the energy requirement of these patients would be 25–30 kcal/kg/day, similar to that of the healthy population. However, according to other authors, the intake should be between 30 and 40 kcal/kg/day depending on their weight.[32]

Like the rest of the population, the diet of a cancer patient should also be balanced, healthy, and varied. It must be adapted to their tastes and symptoms associated with the disease and/or treatment. Therefore, in order to achieve an adequate contribution to their nutritional needs, it is advisable to[32]:

- Eat 5–6 meals a day.
- Portions should not be too abundant as they are not as well tolerated. It is preferable to eat small portions where there is a high nutritional concentration to ensure the supply of all nutrients.
- Plan the schedules of the meals creating a food routine.

- The diet should be varied and adapted to the patient's taste, avoiding restrictive diets.
- Preferably consume fish and poultry, moderating the consumption of animal fat.

The nutritional support to be offered will depend on nutritional status, oral intake, type of treatment, symptoms due to the tumor or oncologic medication, etc. In this sense, there are different algorithms to determine the nutritional support or recommendations that the oncologic patient should receive.

For Gómez Candela et al[33] the nutritional recommendation will depend on the degree of malnutrition (determined by subjective global assessment) and the type of therapy (low risk, moderate risk, or high risk therapy), so that, if the patient is well nourished and depending on the type of therapy, he/she will be given nutritional education, general or specific dietary recommendations, symptom control, and/or high protein diet. If the patient is at risk of malnutrition or moderate malnutrition associated with low-risk or moderate-risk therapies, he/she will receive, in addition to the earlier mentioned recommendations, nutritional supplements. Meanwhile, patients with moderate malnutrition plus aggressive therapies, along with severely malnourished patients, will be started on enteral nutrition and if this is not possible, parenteral nutrition will be used.

Another algorithm is that of Colomer et al.[34] based on intake. Thus, patients with intakes above 75% of nutritional requirements can continue with their usual diet and will receive nutritional counseling. Patients with intakes between 75% and 50% will receive supplements or oral enteral nutrition, in addition to nutritional counseling. When the intake is less than 50%, enteral nutrition by tube will be added to nutritional advice and oral diet. Parenteral nutrition will only be initiated in case enteral nutrition is not sufficient or not possible.

Recommendations for symptoms that hinder food intake

It is frequent that oncological treatment and the tumor itself alter oral intake due to the appearance of symptoms such as anorexia, nausea and vomiting, taste alterations, diarrhea, and so on.[34, 35] Table 52.2 presents some recommendations for the most frequent symptoms that make food intake difficult.

TABLE 52.2 Recommendations according to symptoms associated with the treatment/tumor.

Symptom	Recommendations
Nausea and vomiting	Eat small meals on a frequent basis. Eat easily digestible foods (low fat, mild flavors, avoid acidic or spicy foods) at room temperature or cold. After eating, do not lie down and remain in a sitting posture. Drink outside of meals. Sometimes carbonated drinks, taken in small sips, can facilitate digestion. Eat dry food (cookies, toast, etc.). Eat slowly and in an odorless environment.
Anorexia	Ingest a higher energy and protein intake at the time of the day when you are hungriest. Eat foods with high energy and protein intake such as nuts, pasta, rice, hard-boiled eggs, cheese, etc. Increase the number of meals per day in small portions. Avoid low-calorie foods.
Taste/smell alterations	Rinse your mouth with tea, fruit juice, etc. both before and after meals to eliminate a metallic or bitter taste. Avoid high temperatures, red meat, and foods with strong odor (coffee, cauliflower, onion, etc.). Enhance the flavor with spices such as aromatic herbs, basil, oregano, lemon, etc.
Mucositis/Xerostomia	Avoid acidic, spicy, dry, fibrous or sticky foods (cookies, bread, nuts, etc.). Preferably eat soft foods. Maintain adequate oral hygiene and perform rinses with saline solution. Melt ice cubes in the mouth.
Diarrhea	Avoid fatty foods, insoluble fiber, extreme food temperatures, spices, irritants, and sugar. Drink plenty of liquids, lactose-free milk, skimmed yogurt, grated apples, ripe bananas, fruit in syrup, quince, etc.
Constipation	Increase the intake of foods rich in fiber (legumes, vegetables, fruits, etc.). Drink plenty of fluids. Avoid astringent foods such as bananas, tea, rice, etc. Drink unstrained orange juice, warm water, kiwi fruit, etc. on an empty stomach. As far as possible, increase physical activity.

References

1. Bouvard V, Loomis D, Guyton KZ, et al. Carcinogenicity of consumption of red and processed meat. *Lancet Oncol* 2015;**16**(16):1599–600. https://doi.org/10.1016/S1470-2045(15)00444-1.
2. Chan DSM, Lau R, Aune D, et al. Red and processed meat and colorectal cancer incidence: meta-analysis of prospective studies. *PLoS One* 2011;**6**(6). https://doi.org/10.1371/journal.pone.0020456, e20456.
3. Tollosa DN, Van Camp J, Huybrechts I, et al. Validity and reproducibility of a food frequency questionnaire for dietary factors related to colorectal cancer. *Nutrients* 2017;**9**(11). https://doi.org/10.3390/nu9111257.
4. Davies M. Nutritional screening and assessment in cancer-associated malnutrition. *Eur J Oncol Nurs* 2005;**9**(Suppl 2):S64–73. https://doi.org/10.1016/j.ejon.2005.09.005.
5. Hébuterne X, Lemarié E, Michallet M, de Montreuil CB, Schneider SM, Goldwasser F. Prevalence of malnutrition and current use of nutrition support in patients with cancer. *JPEN J Parenter Enteral Nutr* 2014;**38**(2):196–204. https://doi.org/10.1177/0148607113502674.
6. Virizuela JA, Camblor-Álvarez M, Luengo-Pérez LM, et al. Nutritional support and parenteral nutrition in cancer patients: an expert consensus report. *Clin Transl Oncol* 2018;**20**(5):619–29. https://doi.org/10.1007/s12094-017-1757-4.
7. Endo T, Momoki C, Yamaoka M, et al. Validation of skeletal muscle volume as a nutritional assessment in patients with gastric or colorectal cancer before radical surgery. *J Clin Med Res* 2017;**9**(10):844–59. https://doi.org/10.14740/jocmr3129w.
8. Fearon K, Strasser F, Anker SD, et al. Definition and classification of cancer cachexia: an international consensus. *Lancet Oncol* 2011;**12**(5):489–95. https://doi.org/10.1016/S1470-2045(10)70218-7.
9. Cruz-Jentoft AJ, Baeyens JP, Bauer JM, et al. Sarcopenia: European consensus on definition and diagnosis report of the European working group on sarcopenia in older people A. J. Cruz-Gentoft et al. *Age Ageing* 2010;**39**(4):412–23. https://doi.org/10.1093/ageing/afq034.
10. Fried LP, Tangen CM, Walston J, et al. Frailty in older adults: evidence for a phenotype. *J Gerontol A Biol Sci Med Sci* 2001;**56**(3):M146–56. https://doi.org/10.1093/gerona/56.3.m146.
11. Gómez Candela C, Rodríguez L, Luengo LM, et al. *Intervención nutricional en el paciente oncológico adulto*. Barcelona: Glosa; 2003. Published online.
12. García-Luna PP, Parejo Campos J, Pereira Cunill JL. Causes and impact of hyponutrition and cachexia in the oncologic patient. *Nutr Hosp* 2006;**21**(Suppl 3):10–6. https://www.ncbi.nlm.nih.gov/pubmed/16768026.
13. Arends J, Bachmann P, Baracos V, et al. ESPEN guidelines on nutrition in cancer patients. *Clin Nutr* 2017;**36**(1):11–48. https://doi.org/10.1016/j.clnu.2016.07.015.
14. Kondrup J, Rasmussen HH, Hamberg O, Stanga Z, Ad Hoc ESPEN Working Group. Nutritional risk screening (NRS 2002): a new method based on an analysis of controlled clinical trials. *Clin Nutr* 2003;**22**(3):321–36.
15. Kruizenga HM, Seidell JC, de Vet HCW, Wierdsma NJ, van der Schueren MAE van B. Development and validation of a hospital screening tool for malnutrition: the short nutritional assessment questionnaire (SNAQ©). *Clin Nutr* 2005;**24**(1):75–82. https://doi.org/10.1016/j.clnu.2004.07.015.
16. Ottery FD. Rethinking nutritional support of the cancer patient: the new field of nutritional oncology. *Semin Oncol* 1994;**21**(6):770–8. https://www.ncbi.nlm.nih.gov/pubmed/7992092.
17. Barker ME, Vellas BJ, Guigoz Y, Garry PJ, Albarede JL. Nutrition in the elderly: the mini-nutritional assessment, facts and research in gerontology, supplement 2, Serdi, Paris, 1994, 140 pp., ISSN 0990 2295. *Ageing Soc* 1995;**15**(3):452–4. https://doi.org/10.1017/s0144686x00002828.
18. Guralnik JM, Simonsick EM, Ferrucci L, et al. A short physical performance battery assessing lower extremity function: association with self-reported disability and prediction of mortality and nursing home admission. *J Gerontol* 1994;**49**(2):M85–94. https://doi.org/10.1093/geronj/49.2.m85.
19. Romero-Ortuno R, Walsh CD, Lawlor BA, Kenny RA. A frailty instrument for primary care: findings from the survey of health, ageing and retirement in Europe (SHARE). *BMC Geriatr* 2010;**10**:57. https://doi.org/10.1186/1471-2318-10-57.
20. Morley JE, Vellas B, Abellan van Kan G, et al. Frailty consensus: a call to action. *J Am Med Dir Assoc* 2013;**14**(6):392–7. https://doi.org/10.1016/j.jamda.2013.03.022.
21. Parra-Rodríguez L, Szlejf C, García-González AI, Malmstrom TK, Cruz-Arenas E, Rosas-Carrasco O. Cross-cultural adaptation and validation of the Spanish-language version of the SARC-F to assess sarcopenia in Mexican community-dwelling older adults. *J Am Med Dir Assoc* 2016;**17**(12):1142–6. https://doi.org/10.1016/j.jamda.2016.09.008.
22. García-García FJ, Carcaillon L, Fernandez-Tresguerres J, et al. A new operational definition of frailty: the Frailty Trait Scale. *J Am Med Dir Assoc* 2014;**15**(5):e371.e7–e371.e13. https://doi.org/10.1016/j.jamda.2014.01.004.
23. Pons Raventos M, Rebollo Rubio A, Jiménez Ternero JV. Fragilidad:¿'Cómo podemos detectarla? *Enfermería Nefrol* 2016;**19**(2):170–3. http://scielo.isciii.es/scielo.php?script=sci_arttext&pid=S2254-28842016000200010.
24. Soubeyran P, Bellera C, Goyard J, et al. Screening for vulnerability in older cancer patients: the ONCODAGE prospective multicenter cohort study. *PLoS One* 2014;**9**(12). https://doi.org/10.1371/journal.pone.0115060, e115060.
25. Alvero-Cruz JR, Gómez LC, Ronconi M, Vázquez RF, i Manzañido JP. La bioimpedancia eléctrica como método de estimación de la composición corporal: normas prácticas de utilización. *Rev Andal Med Deport* 2011;**4**(4):167–74. https://www.redalyc.org/pdf/3233/323327668006.pdf.
26. Norman K, Stobäus N, Gonzalez MC, Schulzke J-D, Pirlich M. Hand grip strength: outcome predictor and marker of nutritional status. *Clin Nutr* 2011;**30**(2):135–42. https://doi.org/10.1016/j.clnu.2010.09.010.
27. Fess EE. Grip strength. In: Casanova JS, editor. *Clinical assessment recommendations*. 2nd ed. Chicago: American Society of Hand Therapists; 1992. p. 41–5.
28. Daitoku N, Miyamoto Y, Tokunaga R, et al. Controlling nutritional status (CONUT) score is a prognostic marker in metastatic colorectal cancer patients receiving first-line chemotherapy. *Anticancer Res* 2018;**38**(8):4883–8. https://doi.org/10.21873/anticanres.12802.

29. Borda F, Borda A, Zozaya JM, Urman J, Jiménez J, Ibáñez B. Prognostic value of Onodera's index in colorectal cancer survival. *An Sist Sanit Navar* 2014;**37**(2):213–21. https://doi.org/10.4321/s1137-66272014000200004.
30. Seretis C, Kaisari P, Wanigasooriya K, Shariff U, Youssef H. Malnutrition is associated with adverse postoperative outcome in patients undergoing elective colorectal cancer resections. *J BUON* 2018;**23**(1):36–41. https://www.ncbi.nlm.nih.gov/pubmed/29552757.
31. García de Lorenzo A, Alvarez J, Calvo MV, et al. Conclusions of the II SENPE discussion forum on: hospital malnutrition. *Nutr Hosp* 2005;**20**(2):82–7. https://www.ncbi.nlm.nih.gov/pubmed/15813390.
32. Carmen GC, Samara PM, Coral CBS, Pilar RS, Robledo Sáenz PJ. *Alimentación, Nutrición Y Cáncer: Prevención Y Tratamiento*. Editorial UNED; 2016. https://play.google.com/store/books/details?id=DeehDAAAQBAJ.
33. Gómez Candela C, Luengo Pérez LM, Zamora Auñón P, et al. *Grupo de trabajo de Nutrición y Cáncer de la Sociedad Española de Nutrición Básica y Aplicada (SENBA). Algoritmos de evaluación y tratamiento nutricional en el paciente adulto con cáncer. Soporte nutricional en el paciente oncológico Segunda edición*. Madrid: BMS; 2004. Published online.
34. Colomer Bosch R. García de Lorenzo y Mateos A, Mañas Rueda A. Guía clínica multidisciplinar sobre el manejo de la nutrición en el paciente con cancer. *Nutr Hosp Supl* 2008;**1**(1):1–52. https://medes.com/publication/46703. [Accessed 15 February 2021].
35. García-Luna PP, Campos JP, Verdugo AA, Ibáñez JP, Aguayo PS, Cunill JLP. Nutrición y cáncer. *Nutr Hosp* 2012;**5**(1):17–32. https://www.redalyc.org/pdf/3092/309226797003.pdf.

Chapter 53

Fecal incontinence and CRC

José Luis Ulla Rocha[a], Pablo Parada Vázquez[a], Raquel Sardina Ferreiro[b], and Juan Turnes Vázquez[a]
[a]*Gastroenterology Department, Pontevedra University Hospital Complex, Pontevedra, Spain,* [b]*Department of Internal Medicine, Arquitecto Marcide Hospital, Ferrol, Spain*

Incontinence in relation to colorectal cancer

Fecal incontinence is defined as the lack of voluntary control over both solid and liquid feces.[1] The condition causes impairment of quality of life and great embarrassment to the patient. It is frequently underdiagnosed, owing to the fact that sufferers are reluctant to acknowledge their problem due to societal attitudes.[2] Although it is possible for incontinence be a direct result of CCR due to the tumor location itself, either invading or altering normal function of the anal sphincter complex, in this chapter, we will deal with incontinence resulting from treatments CCR patients have received, as outlined later.

The triggering mechanisms for incontinence following CCR treatment will depend on the type of surgery performed and the resulting long-term damage from radiotherapy.

Surgery type

There are three basic surgical procedures for rectal adenocarcinoma that are performed according to different circumstances:

- *Abdominoperineal resection.* This procedure is used in cases of tumor in the distal rectum or anal area. It involves creating a stoma and permanently closing the anus. It will not result in incontinence because the anal sphincter is removed and stools are evacuated through a stoma located in the abdominal wall. The use of the procedure has now declined due to generalized neoadjuvant preoperative radiotherapy and the performance of low anterior resection surgery.
- *Low anterior resection.* This procedure is the most frequent. Creation of a permanent stoma is avoided, but at the cost of reducing feces storage capacity in the neorectum and the development of the condition known as low anterior resection syndrome (LARS).[3] The symptoms include stool urgency, clustering (multiple bowel movements in a short period of time), and sometimes fecal incontinence. Not all patients will experience the same symptoms. Recovery is gradual, with the greatest improvement occurring during the first 6–9 months after the intervention. The level at which anastomosis is performed has great bearing on postoperative gastrointestinal function.[4] Thus the symptoms of incontinence—reduced ability to distinguish between gas and feces, and the need for dietary changes—occur in 40%–60% of patients who have undergone low anastomosis (below 5 cm from the anal margin).[5] Other factors are also linked to the development of incontinence, such as advanced age, neoadjuvant radiotherapy, and the use of the descending colon for anastomosis rather than the sigma.
- *Endoanal resection.* In essence the procedure involves resecting the lesion, gaining access through the anus, while trying not to cause damage to the anal sphincter complex.

Radiotherapy damage

Radiotherapy given as an adjuvant treatment can lead to long-term chronic tissue damage on two levels:

- *Postradiation or actinic proctitis.* This refers to the damage ionizing radiation used in the treatment of adenocarcinoma produces on the rectum wall over the long term. The risk of onset increases with smoking and the concomitant presence of diabetes mellitus.[6]

- *Anal sphincter affectation.* It has been observed that after radiotherapy a decrease in maximum resting pressure of the anal canal occurs, and this suggests a decrease in the strength of the internal sphincter that contributes to urgency in patients receiving radiotherapy.[7] In recent years there has been a trend to reduce long-term sphincter damage by minimizing high dosage radiation to the sphincters.[8]

Assessment

A history of CCR surgery and radiotherapy if applicable is likely to provide the etiological diagnosis of incontinence. In this context, the most important factor is patient follow-up to check for possible recurrence of neoplasia. Once an adequate period of time has elapsed since surgery—6–9 months for low anterior resection syndrome—and incontinence symptoms still persist, additional tests need to be carried out to rule out sphincter damage. The following methods are available for assessing incontinence.

Clinical history and physical examination

First of all, it is important to find out if the patient presents true incontinence and the severity, as well as to collect information on the duration, frequency, and severity of the symptoms, information on other variables relating to the kinds of rectal surgery and pelvic radiation received, and precipitating factors such as the presence of diabetes mellitus or neuromuscular disease. Perianal and digital rectal examination will give valuable information that may rule out any pathology out at that point and give an estimation of rectal tone.

Colonoscopy

Periodic colonoscopy is a firmly established practice in neoplasia follow-up. It can also provide information to rule out the presence of inflammation or postradiation (actinic) proctitis which has specific treatments.

Endoanal ultrasonography

This procedure can delimit damage to the anal sphincter complex, by establishing whether the internal and external sphincters and the puborectalis muscles are intact. It can also rule out the presence of other entities such as residual postsurgical fistulae or abscesses in the vicinity of the anastomosis.

Anorectal manometry

This process indirectly assesses sphincter function by determining the following parameters: maximum resting anal pressure (correlated with the internal anal sphincter), amplitude and duration of the maximum voluntary contraction (correlated with the external anal sphincter and the puborectalis muscles), the integrity of the rectoanal inhibitory reflex, the threshold for conscious rectal sensation, rectal distensibility and both anal and rectal pressure during defecatory effort. The technological innovation, known as high resolution anorectal manometry, provides greater anatomical detail. Determining the rectal sensory threshold is important because, in patients who have a low threshold, biofeedback treatment may not be of benefit.[9] Changes in the rectal neoreservoir capacity resulting from surgical resection lead to decreased distensibility and cause a reduction in rectal sensitivity volume. In addition, there are other physiological changes that take place following colorectal anastomosis that include lowering of resting anal pressure, loss of rectoanal inhibitory reflex, and a reduction in rectal capacity and distensibility.

High resolution and 3D manometry are new and promising tools for diagnosing fecal incontinence. Nonetheless, further research is needed to validate their clinical use in this context.[10]

Magnetic nuclear resonance

Magnetic nuclear resonance is a technique also capable of assessing sphincter integrity and ruling out the presence of other entities, such as fistulae and abscesses.

Defecography

This technique provides assessment of the anorectal angle, pelvic descent, and hidden reversible rectal prolapse.

Treatment and rehabilitation

At present, there are several different therapies available, although so far none have been established as an optimum treatment for restoring continence.[11] As a general rule, health professionals recommend that less invasive measures be tried first, such as dietary changes and exercise programs. More invasive measures, such as medication or surgery, would be reserved for use when other therapies have proved unsatisfactory. Details of the therapies are presented later—there are a number of simple measures that will significantly reduce incontinence, without being a cure:

Dietary regime and other general measures

Changes to gastrointestinal habits need to include the following elements:

- A bowel evacuation after each meal is advised to use the gastrocolic reflex mechanism, ensuring private facilities are available that allow sufficient time.
- Avoidance of foods that may worsen incontinence (Fig. 53.1).[12] It is important to advise patients to test different potentially harmful foods singly to be able to identify the key one causing the symptoms.
- A custom of frequent, small meals, spread throughout the day.
- Avoidance of medications that may worsen incontinence[9] (Fig. 53.1) and substituting them whenever possible for others free of side effects that compromise continence and gastrointestinal function.

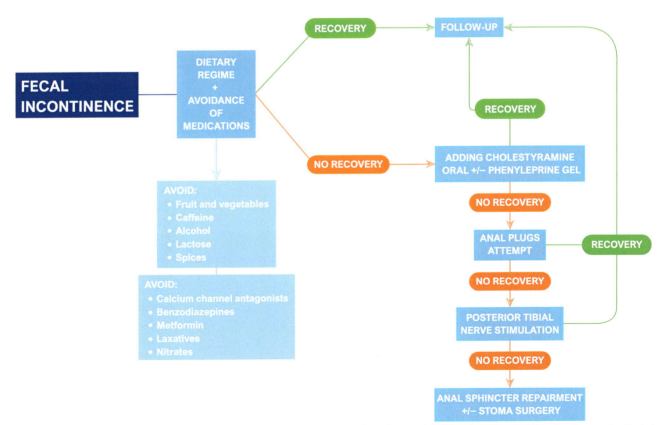

FIG. 53.1 Algorithm for treatment of post CRC surgery and fecal incontinence. *(From Pontevedra University Hospital Complex, Pontevedra, Spain.)*

Specific medications for incontinence

- The antidiarrheal medications, loperamide and diphenoxylate—loperamide proves more effective, reducing bowel urgency and having few central nervous system side effects; it also increases muscle tone of the internal anal sphincter and improves rectal distensibility.
- Bulking agents that create more solid stools and decrease liquid content, giving greater control—methylcellulose may be particularly useful in patients presenting radiation proctitis with reduced rectal distensibility or even postradiation stenosis. Other fiber supplements, such as psyllium husk, have demonstrated their capacity for lowering the number of fecal incontinence episodes.[13]
- Phenylephrine—a selective alpha-1 adrenergic receptor agonist that can be applied to the anal sphincter as a gel, aiding in increasing muscle tone of the internal sphincter.
- Ramosetron—a serotonin receptor antagonist that has demonstrated a reduction in symptoms in low anterior resection syndrome for a small group of patients, reducing incontinence as well as the degree of defecatory urgency.[14]

Specific treatments for radiation proctitis

In spite of the fact that the suffix of the name, proctitis, suggests inflammation, the treatment for proctitis considered the most suitable does not include corticoids or antiinflammatories. Better results have been produced with opioids, short 4 week courses of metronidazole, sucralfate enemas, a treatment course of Vitamin A, topical application of formalin and in the case of rectal bleeding, endoscopic treatment with argon gas or radiofrequencies.

Biofeedback

Biofeedback consists of retraining the pelvic floor to produce adequate defecation. The therapy is painless, free from risk and may be successful in patients with mild to midrange symptoms. There are various methods, such as manometric, electromyographic, or ultrasonographic monitoring. Effectiveness depends on the integrity of the anal sphincter complex[15] and has more chance of success if the incontinence is due to external anal sphincter or puborectalis muscle dysfunction.[16] A recent development is the use of a reduced protocol that seems to have similar results to the conventional.[17]

Pelvic floor rehabilitation

Pelvic floor rehabilitation comprises physical exercises carried out by progressive contraction of selective muscle groups with the object of reinforcing and strengthening the pelvic musculature to reduce incontinence. The disadvantage of this method is that without consistent, long-term practice, the symptoms may return.[18]

Transanal irrigation

Transanal irrigation consists of introducing 500–1000 mL of water through the anus by means of an irrigation system probe for the purpose of emptying the colon to avoid incontinence episodes for a certain period of time. It is a simple and effective option but further research is needed to assess its potential for rehabilitation or possible complications.[19]

Sacral nerve stimulation

Sacral neuromodulation occurs when an electrode is inserted into the S3 sacral foramen by means of minor surgical intervention.[20] Low voltage stimulation is sent from a stimulator; the mechanism of action is not clear, although some studies have shown improvement in internal and external anal sphincter pressure, as well as rectal distensibility. The procedure was approved by the US Food and Drug Administration (FDA) in 2011 and its use has expanded notably worldwide, with modest results at present.

Posterior tibial nerve stimulation

Neuromodulation of the posterior tibial consists of applying an electrical stimulus with a needle or patch to the posterior tibial nerve in the ankle. This stimulation modulates defecatory reflexes to correct them. The results published to date are encouraging,[21] but more research is needed to validate them.

Corrective surgery to the sphincters

If there is sphincter defect it is also possible to connect a local implant to a device that emits electrical stimulation. In the case of incomplete sphincters the technique of muscular transposition known as dynamic graciloplasty has been described.

The following are other less common treatment measures in summary.

Radiofrequencies to the anal canal

This procedure entails insertion of an electrode releasing radiofrequency energy into the anorectal junction; it creates thermal lesions in the muscle, maintaining it intact. The procedure takes less than an hour and is performed under local anesthesia.

Anal neosphincter

The use of devices that function locally to increase anal pressure on demand is experimental so far.[22] A clear improvement in symptoms has been described, but not without complications requiring its removal in many cases.

Biomaterial injection

This is an experimental treatment.[23] Various substances have been tried including silicone and hyaluronic acid, with diverse results.

Anal obturators

Anal obturators function using only a compression mechanism and are not generally well tolerated.

Permanent colostomy

If no other measure is effective, although the treatment that requires a stoma, with the corresponding effects described in the chapter on ostomies, permanent colostomy can on occasion be better tolerated than incontinence itself.

Algorithm for management of incontinence due to LARS (Fig. 53.1).

Conclusion

Over time, it has been possible not only to increase survival times for patients with rectal adenocarcinoma, but also to improve their long-term quality of life.[24] Taking conservative measures and prescribing appropriate medications in the treatment of fecal incontinence are fundamental to effectively alleviating symptoms.[25] The predicted incorporation of robotic surgery in the next few years is likely to further minimize the onset of symptomatology by bringing greater surgical precision that will mean less damage to the anal sphincter complex.[26]

References

1. Madoff RD, Parker SC, Varma MG, Lowry AC. Faecal incontinence in adults. *Lancet* 2004;**364**(9434):621–32. https://doi.org/10.1016/S0140-6736(04)16856-6.
2. Perry S. Prevalence of faecal incontinence in adults aged 40 years or more living in the community. *Gut* 2002;**50**(4):480–4. https://doi.org/10.1136/gut.50.4.480.
3. Emmertsen KJ, Laurberg S. Low anterior resection syndrome score: development and validation of a symptom-based scoring system for bowel dysfunction after low anterior resection for rectal cancer. *Ann Surg* 2012;**255**(5):922–8. https://doi.org/10.1097/SLA.0b013e31824f1c21.
4. Batignani G, Monaci I, Ficari F, Tonelli F. What affects continence after anterior resection of the rectum? *Dis Colon Rectum* 1991;**34**(4):329–35. https://doi.org/10.1007/BF02050593.
5. McDonald PJ, Heald RJ. A survey of postoperative function after rectal anastomosis with circular stapling devices. *Br J Surg* 1983;**70**(12):727–9. https://doi.org/10.1002/bjs.1800701211.
6. Fuccio L, Guido A, Andreyev HJN. Management of intestinal complications in patients with pelvic radiation disease. *Clin Gastroenterol Hepatol* 2012;**10**(12):1326–1334.e4. https://doi.org/10.1016/j.cgh.2012.07.017.
7. Putta S, Andreyev HJN. Faecal incontinence: a late side-effect of pelvic radiotherapy. *Clin Oncol* 2005;**17**(6):469–77. https://doi.org/10.1016/j.clon.2005.02.008.
8. van der Sande ME, Hupkens BJP, Berbée M, et al. Impact of radiotherapy on anorectal function in patients with rectal cancer following a watch and wait programme. *Radiother Oncol* 2019;**132**:79–84. https://doi.org/10.1016/j.radonc.2018.11.017.
9. Barnett JL, Hasler WL, Camilleri M. American Gastroenterological Association medical position statement on anorectal testing techniques. American Gastroenterological Association. *Gastroenterology* 1999;**116**(3):732–60. https://doi.org/10.1016/s0016-5085(99)70194-0.

10. Heinrich H, Misselwitz B. High-resolution anorectal manometry—new insights in the diagnostic assessment of functional anorectal disorders. *Visc Med* 2018;**34**(2):134–9. https://doi.org/10.1159/000488611.
11. Lal N, Simillis C, Slesser A, et al. A systematic review of the literature reporting on randomised controlled trials comparing treatments for faecal incontinence in adults. *Acta Chir Belg* 2019;**119**(1):1–15. https://doi.org/10.1080/00015458.2018.1549392.
12. Ahmad M, McCallum IJD, Mercer-Jones M. Management of faecal incontinence in adults. *BMJ* 2010;**340**. https://doi.org/10.1136/bmj.c2964, c2964.
13. Bliss DZ, Savik K, Jung H-JG, Whitebird R, Lowry A, Sheng X. Dietary fiber supplementation for fecal incontinence: a randomized clinical trial. *Res Nurs Health* 2014;**37**(5):367–78. https://doi.org/10.1002/nur.21616.
14. Koda K, Itagaki R, Yamazaki M, et al. Serotonin (5-HT3) receptor antagonists for the reduction of symptoms of low anterior resection syndrome. *Clin Exp Gastroenterol* 2014;47. https://doi.org/10.2147/ceg.s55410. Published online.
15. Norton C, Cody JD. Biofeedback and/or sphincter exercises for the treatment of faecal incontinence in adults. *Cochrane Database Syst Rev* 2012;(7). https://doi.org/10.1002/14651858.CD002111.pub3, CD002111.
16. Ho YH, Chiang JM, Tan M, Low JY. Biofeedback therapy for excessive stool frequency and incontinence following anterior resection or total colectomy. *Dis Colon Rectum* 1996;**39**(11):1289–92. https://doi.org/10.1007/BF02055125.
17. Mazor Y, Kellow JE, Prott GM, Jones MP, Malcolm A. Anorectal biofeedback: an effective therapy, but can we shorten the course to improve access to treatment? *Therap Adv Gastroenterol* 2019;**12**. https://doi.org/10.1177/1756284819836072, 1756284819836072.
18. Bocchini R, Chiarioni G, Corazziari E, et al. Pelvic floor rehabilitation for defecation disorders. *Tech Coloproctol* 2019;**23**(2):101–15. https://doi.org/10.1007/s10151-018-1921-z.
19. Martellucci J, Sturiale A, Bergamini C, et al. Role of transanal irrigation in the treatment of anterior resection syndrome. *Tech Coloproctol* 2018;**22**(7):519–27. https://doi.org/10.1007/s10151-018-1829-7.
20. Matzel KE, Stadelmaier U, Hohenfellner M, Gall FP. Electrical stimulation of sacral spinal nerves for treatment of faecal incontinence. *Lancet* 1995;**346**(8983):1124–7. https://doi.org/10.1016/s0140-6736(95)91799-3.
21. Vigorita V, Rausei S, Troncoso Pereira P, et al. A pilot study assessing the efficacy of posterior tibial nerve stimulation in the treatment of low anterior resection syndrome. *Tech Coloproctol* 2017;**21**(4):287–93. https://doi.org/10.1007/s10151-017-1608-x.
22. Faucheron J-L, Chodez M, Boillot B. Neuromodulation for fecal and urinary incontinence: functional results in 57 consecutive patients from a single institution. *Dis Colon Rectum* 2012;**55**(12):1278–83. https://doi.org/10.1097/DCR.0b013e31826c7789.
23. Altomare DF, Binda GA, Dodi G, et al. Disappointing long-term results of the artificial anal sphincter for faecal incontinence. *Br J Surg* 2004;**91**(10):1352–3. https://doi.org/10.1002/bjs.4600.
24. de la Portilla F, Fernández A, León E, et al. Evaluation of the use of PTQ implants for the treatment of incontinent patients due to internal anal sphincter dysfunction. *Colorectal Dis* 2008;**10**(1):89–94. https://doi.org/10.1111/j.1463-1318.2007.01276.x.
25. Tan JJY, Chan M, Tjandra JJ. Evolving therapy for fecal incontinence. *Dis Colon Rectum* 2007;**50**(11):1950–67. https://doi.org/10.1007/s10350-007-9009-2.
26. Luca F, Valvo M, Ghezzi TL, et al. Impact of robotic surgery on sexual and urinary functions after fully robotic nerve-sparing total mesorectal excision for rectal cancer. *Ann Surg* 2013;**257**(4):672–8. https://doi.org/10.1097/SLA.0b013e318269d03b.

Chapter 54

Ostomy care

Alba María Arceo Vilas[a], Antonio Jurjo Sieira[b], and Silvia Louzao Méndez[a]
[a]*University Hospital Complex of A Coruña, A Coruña, Spain,* [b]*Casa do Mar of A Coruña Health Center, A Coruña, Spain*

Introduction

Gastrointestinal ostomies are increasingly common surgical procedures that have a great impact on patients and their relatives.

The return to social, family, and work life should be seen as one of the primary objectives when treating a patient with an ostomy. To achieve this, the nursing staff should provide personalized, comprehensive, quality care during the perioperative period, from the moment the patient is informed of the treatment to follow until to when they are discharged home.

The lack of information, education, and communication prevents the individual from actively participating in their self-care and from detecting possible complications early on.

This type of surgery causes great changes that result in a loss of autonomy and self-esteem, denial of self-care, and depression. Self-help groups are very helpful for establishing coping strategies and achieving better adaptation.

Gastrointestinal ostomies

The term "stoma" means mouth or opening and refers to the exteriorization of an organ or viscera in an area other than the natural orifice of excretion.

An "ostomy" is the surgical procedure for bypassing said organ or viscus. Specifically, gastrointestinal ostomies are performed in the rectus abdominis muscle, where the patient can easily manipulate the collecting device, increasing their autonomy.

The most frequent causes of this type of ostomy are colorectal and bladder cancer, followed by inflammatory bowel diseases.[1]

They are classified according to different aspects, depending on the anatomical segment involved, the temporality, or the type of surgical technique used.

Ileostomy

An ileostomy is the opening of the distal ileum, usually located in the lower right quadrant of the abdomen, through the rectus abdominis muscle. It entails the loss of the ileocecal sphincter, which makes it difficult to regulate voiding, causes more rapid bowel transit, decreases absorption, and complicates evacuation control Fig. 54.1.

The daily ileal outflow is usually 500–800 mL, with a continuous flow, which increases slightly after meals or during episodes of gastroenteritis. It is very irritating for the skin, therefore, it is necessary for the stoma to protrude 2–3 cm above the cutaneous plane, avoiding bony prominences, skin folds, and scars, so that the device can be perfectly adjusted to the diameter of the ileum.

Temporary ileostomies

Certain gastrointestinal disorders can be treated by surgery, removing the affected part and putting part of the intestine at rest, where normal transit will be restored once the cause is solved,

To do this, a short-term ileostomy is created that will later be surgically removed and the intestine will begin to function as it did before. A temporary ileostomy is also performed in the first stage of construction of an ileoanal reservoir ("J" reservoir).

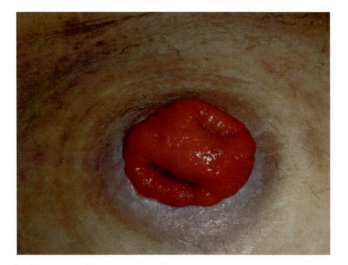

FIG. 54.1 Ileostomy.

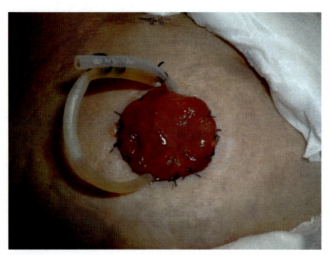

FIG. 54.2 Loop ileostomy with stomal bridge.

- **Loop ileostomy**: an intestinal loop is pulled out through an opening in the abdominal wall, which can be carried out with or without a stomal bridge Fig. 54.2.

Permanent ileostomies

The diseased part of the intestine and anus must be removed or irreversibly rendered useless. In this case, the ileostomy is considered permanent, and it is not expected to close in the future. The main causes of this type of ileostomy are ulcerative colitis, Crohn's disease, and familial adenomatous polyposis.

There are three types of permanent ileostomies:

- **Standard or Brooke ileostomy**: this is performed by pulling a portion of the distal ileum through an opening in the abdominal wall. It continuously evacuates liquid or pasty discharge with digestive enzymes, so it is important to protect the peristomal skin by permanently using an open system collection bag.
- **Continent or Kock pouch ileostomy**: a portion of the ileum is folded over itself, forming an intraabdominal reservoir or bag that constitutes a nipple-shaped valve, thereby avoiding the use an external bag. A catheter is placed to drain the waste out of the abdominal reservoir several times a day.
- **Ileoanal pouch**: this technique currently represents a great step forward in the surgical treatment of ulcerative colitis and familial polyposis, offering a better quality of life to patients who would otherwise have to carry a definitive

ileostomy. It is a bag made up of parts of the ileum and rectum located at the pelvic level. It is also known as a J reservoir, W reservoir, and S reservoir, depending on the surgical technique used. This pouch is connected to the anus, so waste is evacuated through it. The sphincter is preserved around the anal opening to prevent leakage.

Colostomies

A colostomy involves removing part of the large intestine through an opening in the abdominal wall. Due to its location, the nutrients in the food have already been absorbed by the small intestine.

Temporary colostomies are more common than permanent ones. They are indicated in emergency situations where the colon is involved or for complex anal problems.

As for permanent colostomies, they are mainly caused by rectal and anal neoplasms and complex perianal diseases.

Depending on the surgical procedure, the anatomy, or the external appearance of the stoma, one or two stomas may be observed at the abdominal level.

- **End colostomy**: this is the most frequent type. It consists of a single stoma through which feces and mucous secretions are evacuated.
- **Loop colostomy**: its appearance is that of a large stoma since an incomplete cut of the colon is made, so that two connected openings are visible. One is for evacuating feces and the other for mucous secretions.
- **Double-barrel colostomy**: the colon is completely divided, giving rise to two independent stomata whose function is the same as in loop colostomy, one for disposing of fecal matter and the other for mucous secretions.

Another classification of colostomies is based on the anatomical segment involved. They can be divided as follows:

- **Cecostomy**: located in the right lower quadrant, suturing the cecum to the abdominal wall. It works like an ileostomy, the content is not very thick and contains gastric juices, so a bag must be used continuously to protect the peristomal skin from the active enzymes. Cecostomies are not common.
- **Ascending colostomy**: located on the right side of the abdomen. These colostomies are not common. There is constant, unpredictable liquid outflow, which is highly irritating for the peristomal skin due to the presence of digestive enzymes.
- **Transverse colostomy**: located in the middle or right side of the body. The stools leave the colon before reaching the descending colon, rectum, and anus. They are usually temporary.
- **Descending or sigmoid colostomy**: located on the left lower side of the abdomen. The feces are solid and formed, more or less predictable and not highly irritating as they do not contain enzymes.

Main complications of gastrointestinal ostomies[2]

Complications can occur early, appearing in the immediate postoperative period, generally as a result of preoperative problems, technical errors, or as a direct consequence of the intervention. They can also occur later, appearing a while after the ostomy has been performed.

Immediate or early complications

- **Edema:** During the immediate postoperative period, it is common for the stomal mucosa to present a certain degree of physiological inflammation, produced by the manipulation and mobilization of the surgical act itself Fig. 54.3. It can last 1 or 2 weeks. Transparent bags will be used to visualize the stoma and follow its course. It is advisable to use adhesive plates that adjust to the stoma without compressing it (3 mm larger than the stoma), avoiding rubbing the stomal mucosa and its possible ulceration.

To reduce edema, cold hypertonic saline (never frozen) compresses can be placed on the stoma, or granulated sugar for osmotic therapy.

- **Hemorrhage**: this is a rare complication. It is usually due to bleeding from a blood vessel or a small ulcer in the mucosa of the stoma. Transparent devices allow us to observe its course and, above all, to rule out mucosal rubbing as a possible cause of bleeding.

  The amount of blood loss should be assessed, the origin (venous or arterial), and the hemodynamic status of the individual should be determined. If the bleeding is enterocutaneous, local hemostasis with silver nitrate or compression will suffice. If the bleeding is in the mucosa, it should be compressed, and local cold will be applied. If it persists, a dressing soaked in adrenaline or ferric chloride can be applied.

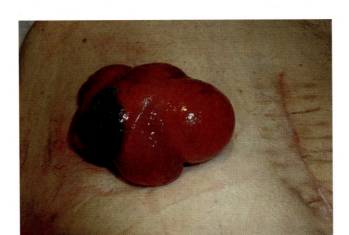

FIG. 54.3   Edema.

- **Ischemia and necrosis**: this can occur up to 24 h after surgery. It is considered the most serious complication since, depending on the extension, there is a high risk of perforation and peritonitis. It is caused by a sectioned artery or an unnoticed suture in a vessel, causing inadequate vascularization. The stoma becomes greenish-gray or blackish-gray without bleeding edges, the functionality, color, and appearance of which will be checked with a transparent device Fig. 54.4.
- **Infection and sepsis**: these usually manifest as pain in the peristomal area, inflammation, suppuration, and sometimes fever. If an abscess has formed, the affected area should be drained through an incision and washed with saline and antiseptic solution.
- **Mucocutaneous dehiscence**: this usually occurs in the first 7 postoperative days. There is a separation between the stoma and the peristomal skin that can be partial or affect the entire circumference of the stoma, with the consequent risk of fecal matter leakage and infection.

Late complications

- **Retraction of the stoma**: this refers to the stoma sinking below the level of the skin. It is usually due to excessive bowel tension, generally due to poor mobilization, although on many occasions the patient's weight gain can also favor it in the same way as abdominal hyperpressure, postoperative ileus, and peristomal septic complications. If the degree of retraction is significant, it may be accompanied by stenosis of the ostomy opening, resulting in difficulty in passing stools that justifies a surgical reconstruction.

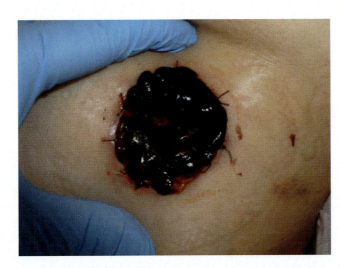

FIG. 54.4   Necrosis.

It is advisable to use flexible and sometimes convex devices, so that they adapt better to unevenness and skin folds. A protective paste and a girdle can also be used to prevent leaks and achieve greater safety.
- **Stenosis**: this is a narrowing of the stoma that makes it difficult or impossible to insert a finger through it and at simultaneously makes it difficult for stools to pass. It is more common in ileostomies than in colostomies Fig. 54.5.

  A diet in which the feces are pasty should be indicated, facilitating their evacuation.

  If the stenosis is partial, it is advisable to instruct the patient to perform periodic dilations with a gloved finger lubricated with petroleum jelly. If it is complete, the possibility of a surgical reintervention should be considered.
- **Hernia**: this is the most common late complication. When pulling out the intestinal portion, the muscles of the abdominal wall are affected, causing both the ostomy and the peristomal skin to protrude. The size can vary greatly, from small parastomal hernias to large hernias Fig. 54.6.

  One of the most important recommendations is to avoid efforts that involve the abdominal muscles. Sometimes adhesion to the skin is difficult, so it is advisable to use devices that are as flexible as possible to promote adaptability, evaluating the use of the girdle. The placement of the device in the supine position and the use of a soft compression girdle that does not affect the stoma are recommended.
- **Prolapse**: this is an excessive protrusion of the intestinal loops, which can reach 15–20 cm, over the skin plane of the abdomen. It generally occurs progressively, causing edema due to increased intraabdominal pressure. It usually occurs more frequently in colostomies, especially in the transverse ones Fig. 54.7.

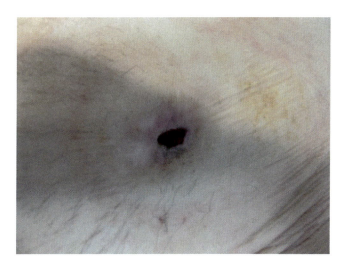

FIG. 54.5 Stenosis.

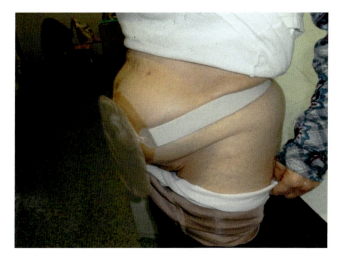

FIG. 54.6 Parastomal hernia.

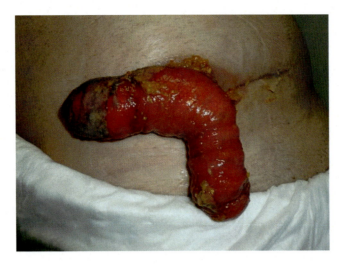

FIG. 54.7 Prolapse.

Physical efforts should be avoided. The local application of compresses with cold saline solution is recommended to reduce edema, as well as the use of devices that do not cause friction on the mucosa to avoid bleeding.

The patient should be taught the manual prolapse reduction procedure, which consists of lying relaxed in the supine position and gently massaging the prolapsed bowel in the direction of the stoma opening with gloved hands.

Skin complications

- **Peristomal dermatitis**: this can appear both in the immediate and late postoperative periods, and can be caused by irritative, mechanical, fungal, microbial, or allergic factors Fig. 54.8.

    Irritant contact dermatitis is caused by the direct action of irritants, such as poorly formed stools, on the skin. This chemical lesion should be treated mainly by avoiding such contact as much as possible, with an adequate adjustment of the device, leaving a minimum area of peristomal skin exposed.

    In irritant contact dermatitis, frequent changes of the adhesive should be avoided, keeping it glued to the skin as much as possible.

    Fungal or microbial dermatitis is caused by fungi or bacteria, so it will be necessary to consult a dermatologist for the appropriate treatment.

    In allergic dermatitis, the most advisable course of action is to change the device; this is sufficient to improve said complication as is the use of a barrier paste.

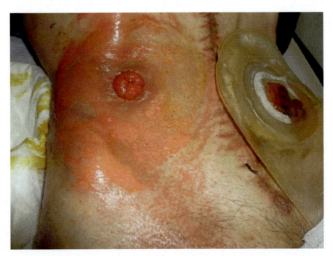

FIG. 54.8 Irritant contact dermatitis.

- **Ulcerations**: two of the main causes of this complication are bacterial infections and improper use of the devices, which should be checked to avoid rubbing.
- **Granulomas**: these are periosteal epidermal protrusions, not neoplastic. They are usually due to incomplete reabsorption of the suture material or caused by the adhesive edge of the device rubbing against the mucosa. They can be painful and bleed very easily and can be cauterized with silver nitrate if necessary.
- **Parastomal varicose veins**: these are caused by pathologies secondary to another disease, portal hypertension due to liver cirrhosis, or liver metastases. They can bleed massively, so abrupt detachment of the device should be avoided. Sometimes the local application of vasoconstrictor solutions will be necessary.

Preoperative consultation[3,4]

The preoperative ostomy consultation is essential for determining the baseline situation of the patient who is going to undergo stoma surgery. An initial interview is carried out to collect the necessary information for subsequently developing a Care Plan adapted to each individual.

Simple language should be used to address the patient, avoiding the use of technical language and being open to any questions the patient may have.

Initially, information should be collected about the patient and their environment, their habitual residence and family unit, identifying the caregiver (generally a family member or the person who accompanies the patient to the consultation); their degree of dependence on self-care (manual and visual dexterity); degree of knowledge and information about the surgery; doubts; etc.

Subsequently, the following issues should be clarified: type of stoma, prescribed period of use, hygiene and care of the peristomal area, the devices available, and the diet to be followed.

Stoma care and hygiene

The patient should be informed that slight bleeding is normal in the first days following the intervention.

As for the size of the stoma, it is usually reduced during the first months following the intervention, so it will be necessary to measure it frequently to adjust the diameter of the adhesive, ensuring that the device adapts perfectly to the stoma and that no skin is exposed to feces.

Train the patient and their family in basic skin care and placement of the drainage bag.

The removal of the device should be carried out as follows, depending on the type of ostomy:

- Ileostomy bags should be emptied whenever necessary. They are usually emptied every 4–6h or when they are 1/3 or half full.
- Colostomy bags should be removed when full. They are usually changed 2–3 times a day.

The bag should be gently removed from top to bottom, to avoid the discharge of fecal content, avoiding pulling. Remove the remains of fecal material with toilet paper. Clean the stoma skin with mild soap, a soft sponge, and warm water. Use circular movements from the outside inward.

Gently dry with a towel or tissue. If there is hair around the stoma, it should be cut and shaved. Slight bleeding may occur when handling the stoma.

After placing the bag, press for 30s to activate the adhesive.

When frequent bag changes are necessary, two-piece devices are recommended for reducing the risk of skin injury. You can also apply special pastes that act as a skin barrier.

Device types

The usefulness of the different devices will depend on the consistency of the stools and, above all, on the resistance of the ostomized patient's skin Fig. 54.9.

Depending on the voiding system of the bags, they can be either closed or open.

- Closed: these are heat-sealed bags; one is needed for each use. They are suitable for solid or pasty stools. In general, they usually have a filter for gases, which prevents bad smells.
- Open: the lower end has an opening that is closed with a clamp.

FIG. 54.9 Examples of collecting devices.

According to the fastening system:

- Single: adhesive, bag in one piece.
- 2-piece: adhesive plate (flat or convex) + bag.
- 3-piece: adhesive plate (flat or convex) + bag + safety clip that guarantees the union of the bag and the plate.

The bags may be opaque or transparent.

Skin protectors

Skin protectors help keep the stoma and peristomal skin in perfect condition and avoid complications that could affect the health and comfort of the ostomy patient Fig. 54.10.

- Barrier cream: maintains the pH balance of the skin.
- Protective film: forms a thin layer that protects the skin from waste products and prevents perspiration.
- Leveling paste: prevents leaks when there are skin folds and unevenness.
- Drying powders: treat excessive moisture in the peristomal skin.
- Zinc oxide and starch pastes: indicated for the prevention of irritation, not as a treatment.

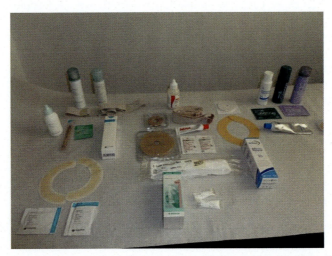

FIG. 54.10 Types of skin protectors.

Devices for controlling evacuations

Continent methods allow the patient to go without a collection bag for a time. These include:

- Obturators: obturators are similar to plugs. They are inserted into the stoma and allow gas to escape. Odors are eliminated through a filter. They are indicated for colostomized patients with solid stools, allowing continence for up to 12 h.
- Irrigation systems: these consist of an intestinal lavage for evacuating stools from the colon. Generally used in descending colostomies, obtaining continence for 48–72 h.

Stoma marking[5]

Stoma marking should be a personalized procedure performed prior to surgery, in order to confirm that the site selected for creating the stoma is visible and allows self-care, thus reducing the onset of complications. Folds, scars, bony prominences, and the natural waistline should be avoided, and physical characteristics and personal and cultural habits should be considered.

To locate the ideal point, draw a triangle in the left/right lower quadrant of the abdomen, using the navel, left/right iliac crest, and the midpoint of the pubis as vertices, then trace the bisectors. Their point of intersection indicates the preferred place for creating the stoma.

Once the point has been located on the abdomen, a dynamic test should be performed with an ostomy device, making the patient adopt different postures (supine, sitting, standing) and simulating activities of daily life (dressing, walking, bending over, etc.).

Postoperative follow-up[3, 6]

In the immediate postoperative period (the first 72 h), a global assessment of the patient should first be carried out upon arrival at the ward (vital signs, dressings, drains, probes, central line, etc.). The patient's device should be transparent for assessing the appearance of the stoma, without a filter so that start of functionality can be observed, and with an open bag.

Regarding the stoma, the mucosa should have a reddish-pink appearance, with some edema typical of the surgical intervention. The peristomal sutures and the condition of the skin should be monitored, always preventing and controlling possible complications.

The patient and their caregiver should be instructed in the hygiene and care of the stoma, generally in the immediate postoperative stage, including the 72-h postoperative period through to discharge from hospital. The patient should learn to manage their own stoma as soon as their general condition allows. To do this, the nursing staff will give them the necessary instructions regarding how to empty, remove, and subsequently replace the bag.

Depending on the needs of each patient, different ostomy accessories will be recommended (peristomal paste, drying powder, moldable ring, girdle, barrier spray, etc.)

It is very important to provide written information on care (ostomy kit, guidelines, dietary recommendations) as too much information will have been provided in a short period of time, during which the patient must come to terms with their new body image and self-care regime. Before discharge, check the patient's/caregiver's ostomy care skills and answer any doubts or questions they may have.

The ostomy appointment (1st visit) must be given with the stoma therapist nurse, contact telephone number, and email.

Ostomy consultation at discharge[3, 7]

The main objective of this consultation is to provide comprehensive care to the ostomy patient and their caregiver, to promote their independence and help them have a good quality of life, as a continuation of the care process, as well as to detect, prevent, and control possible late complications.

At the first visit after discharge, the stoma diameter and the peristomal skin should be assessed, and any peristomal stitches should be removed. The appropriateness of the device and accessories will be assessed.

By this stage, the patient should have gained greater autonomy in their self-care and the necessary information will be emphasized. Patients using continent methods should be taught the irrigation technique.

Clear and concise information will be given about their return to family, social, and work life, recovering their habits and daily activities, as well as assessing their adaptation to the new situation by identifying their concerns, fears, and/or rejection of body image, giving emotional support to the patient and caregiver throughout the process.

As for sports, it is possible to perform aerobic exercise, swimming, cycling, and walking using smaller, opaque, and discreet devices, with an abdominal girdle. Contact sports should be avoided.

When traveling, enough material should always be carried in a hand baggage, not checked baggage.

Sexual relations can be affected to a greater or lesser extent after surgery. It is important to talk to the patient about side effects, possible sexual disorders or alterations and, if necessary, refer them to a specialist.

References

1. Alterescu KB. Colostomy. *J Wound Ostomy Continence Nurs* 1982;**9**(2):39. https://doi.org/10.1097/00152192-198203000-00026.
2. Allen M (p). Selecting keywords: helping others find your article. *Nurse Author Ed* 1998;**8**(1):4–9. https://doi.org/10.1111/j.1750-4910.1998.tb00390.x.
3. Fingren J, Lindholm E, Petersen C, Hallen A-M, Carlsson E. A prospective, explorative study to assess adjustment 1 year after ostomy surgery among Swedish patients. *Ostomy Wound Manage* 2018;**64**(6):12–22. https://doi.org/10.25270/owm.2018.6.1222.
4. Almutairi D, LeBlanc K, Alavi A. Peristomal skin complications: what dermatologists need to know. *Int J Dermatol* 2018;**57**(3):257–64. https://doi.org/10.1111/ijd.13710.
5. Cronin E. Colostomies and the use of colostomy appliances. *Br J Nurs* 2008;**17**(Sup7):S12–9. https://doi.org/10.12968/bjon.2008.17.sup7.31117.
6. Burch J. Caring for peristomal skin: what every nurse should know. *Br J Nurs* 2010;**19**(3):166–72. https://doi.org/10.12968/bjon.2010.19.3.465387.
7. Gök AFK, Özgür I, Altunsoy M, et al. Complicated or not complicated: stoma site marking before emergency abdominal surgery. *Ulus Travma Acil Cerrahi Derg* 2019;**25**(1):60–5. https://doi.org/10.5505/tjtes.2019.48482.

Chapter 55

Sexual dysfunction among patients with ostomies

María Teresa García Rodríguez[a], Adriana Barreiro Trillo[b], and Sonia Pértega Díaz[c]

[a]Research in Nursing and Health Care, Institute of Biomedical Research of A Coruña (INIBIC), A Coruña University Hospital Complex (CHUAC), SERGAS, A Coruña, Spain, [b]University Hospital Complex of A Coruña (CHUAC), A Coruña, Spain, [c]Institute for Biomedical Research of A Coruña (INIBIC), University Hospital Complex of A Coruña (CHUAC), A Coruña, Spain

Introduction

Ostomy is "a term used to designate a surgical procedure whose objective is to create an artificial connection between two hollow organs or a hollow organ and the skin," the opening that is created is called a stoma; its role is to allow waste products to be excreted from the body.

Depending on the level at which the stoma is performed in the digestive system, one can speak of duodenostomy, jejunostomy, ileostomy, or colostomy.

The main reason for performing a digestive ostomy is colorectal cancer, although it may also be recommended in inflammatory diseases (Crohn's disease, ulcerative colitis), trauma, congenital diseases, malformations, and intestinal obstructions.[1]

It is classified according to its function, permanence, or the organ for which the stoma is created, as presented in Table 55.1.

In Spain there are more than 700,000 people who have an ostomy, and it is estimated that in Galicia there are around 4000 ostomy patients. Approximately 16,000 new cases are registered each year, with colostomies being more frequent (55.1%) followed by ileostomies (35.2%). This type of intervention affects all age groups and involves the alteration of the body image, self-esteem, and lifestyle of these patients, affecting their quality of life and sexual activity. Despite this, both the care and health education provided to these patients focus on the surgical procedure (recovery, self-care, previous pathology), and despite the fact that 70% of them report having an unsatisfactory sexual life, there is a low number of studies addressing the sexuality of the patient having undergone an ostomy surgery.[2]

Sexual impairment in patients with ostomies

According to Ang et al., the stressors after an ostomy surgery are different while hospitalized compared to when the patient is discharged. During admission, patients are more concerned about the self-care of the stoma, the disease that led them to it or the formation of the ostomy. However, once they are discharged, the main concerns are alterations in sexuality, body image, and daily activity.[3]

The resumption of sexual activity after the ostomy procedure can present difficulties to a large number of patients. This is due to different factors; in addition to the surgical procedure, cancer treatments (radiotherapy and chemotherapy), the psychological impact of having a stoma, or the nerve damage that may occur after resection and ostomy creation also influence the sexuality of the ostomy patients.

In this sense and based on different studies, sexual dysfunction is different according to gender. The main problem in men is erectile dysfunction; in women it is dyspareunia. In addition to these, other alterations were also found, such as lack of sexual desire, difficulty having an orgasm, impotence, less sexual satisfaction, vaginal dryness, etc.[4–10]

These alterations are caused by both physical and psychological factors (including self-esteem).

(a) Physical factors:
Organic lesion caused by surgery or the type of stoma is one of the main reasons influencing sexual performance. Different studies showed that patients with permanent ostomies, rectal cancer, low anterior resections, and abdominal

TABLE 55.1 Classification of ostomies.

Classification	Type of ostomy
According to its function	Nutritional ostomies
	Elimination ostomies
	Drainage ostomies
According to their permanence	Temporary ostomies
	Permanent ostomies
According to the organ	Urostomies, colostomies, ileostomies, gastrostomies, jejunostomies, etc.

perineal resections were at increased risk of suffering a greater deterioration in their sexual life than those who had temporary ostomies or had undergone a resection in the upper part of the colon.[5, 7, 8, 11] This is due to injuries encountered in the pelvic nerves (pudendal nerve, pelvic plexus, upper hypogastric plexus, etc.) leading to a lower blood supply to the penis or clitoris, thus preventing or hindering erection, vaginal lubrication, orgasm, or ejaculation.

Other physical reasons to take into account that will lead to sexual dysfunction are cancer treatments, the stage of the tumor, the patient's age, etc.[12]

(b) Psychological factors:

At a psychological level, the creation of the stoma also has an impact on sexuality, since the change in physical appearance implies a deterioration in body image. This combines with the lack of control over the body (there is no control over bowel movement and passing gas), the handling of the bag, worrying about the leaks, the uncertainty of whether they will be able to maintain a sexual relationship again and/or the acceptance of their partner, the fact that restoring sexual activity can be complicated, since self-esteem and self-confidence are reduced. All this will result in the isolation of the person, thus avoiding relationships with others.[1, 4, 13–17]

We should not forget that the impact of the ostomy can also be observed in the partners of the patients with ostomies. In spite of the fact that sexual relationships are generally reduced or stopped due to lack of sexual interest, it is important to have the acceptance, understanding, and support of the partners of the ostomy patients in order to resume sexual activity.[15–18]

Assessment of sexual dysfunction

For Albaugh et al., before moving on to discuss the assessment of sexual dysfunction, the most important thing is that the patient recognizes that he/she has problems when it comes to maintaining a sexual relationship. That is why, it is essential for them to know that, after the ostomy, it is common to experience sexual function alterations and that they are not going to be the only ones to mention this type of difficulty. As people are generally reluctant to talk about it, health workers should ask patients if they have any problems at this level. Once they admit it, they will be able to undergo an assessment of the situation.[12]

Although the majority of quality of life questionnaires for oncological patients include the sexual dimension, they do not deal with it in sufficient detail,[19] and there is no questionnaire specifically designed to assess sexual function in patients with ostomies, for which reason questionnaires previously validated for other populations were used. Different types of questionnaires and validated interviews are available to assess sexual dysfunction. This chapter refers to some of them, as presented in Table 55.2.

General questionnaires

Some of these questionnaires address different dimensions, including sexual function. There may be different versions for each gender.

TABLE 55.2 Questionnaires that assess sexuality.

General questionnaires		
Questionnaire	Acronyms	Assessment
Changes in sexual functioning questionnaire	CSFQ[20]	Changes in sexual function due to disease or side effects of treatment
Derogatis interview for sexual functioning	DISF/DISF-SR[21]	Measures the level and quality of sexual function
Golombok-Rust inventory of sexual satisfaction	GRISS[22]	Presence and severity of sexual problems
Sexual desire inventory	SDI[23]	Interest in sexual activity

Gender-specific questionnaires			
Gender	Questionnaire	Acronyms	Assessment
Women	Brief index of sexual functioning	BISF-W[24]	Sexual function and satisfaction in women
	Expanded sexual arousability inventory	SAI-E[25]	Sexual arousal and anxiety in women
	Female sexual function index	FSFI[26]	Women's sexual function over the past month
	Female sexual function	FSF[27]	Female sexual function
Men	Brief sexual function questionnaire	BSFQ[28]	Sexual functioning in men
	Brief sexual function inventory	BSFI[29]	Sexual functioning in men
	International index of erectile function	IIEF[30]	Erectile function

(a) Changes in sexual functioning questionnaire (CSFQ)[20]

It was developed to assess changes in the sexual function of sick and/or patients under treatment that result in sexual function alterations and identify specific or unusual side effects. It has been validated and translated into Spanish and although it is administered through an interview, it can also be self-administered.

There are two versions, one specific for men, consisting of 36 items, and another specific for women, made up 34 items, the first 21 items being common for both genders. It also has a version for the first interview (basal version) and another for follow-up, which is shorter.

The items, in addition to providing medical history information, also assess 5 dimensions of sexual functioning through 12 items: sexual desire/frequency, sexual desire/interest, sexual pleasure, arousal/emotion, and orgasm/climax.

The total score is obtained by adding the points of the five dimensions together with the score of two items, one related to arousal problems in women (in men it refers to painful erections) and the other to painful orgasms.

(b) Derogatis interview for sexual functioning (DISF/DISF-SR)[21]

This questionnaire can be developed as a personal interview (DISF) or as a self-administered questionnaire (DISF-SR) and it is recommended for assessing the level or quality of sexual function. Just like the previous questionnaire, it has been validated and translated into Spanish, and there is a version according to the patient's gender.

It consists of 26 items in which 5 dimensions are evaluated: sexual fantasy and cognition, arousal, sexual behaviors and experiences, orgasm, and sexual motivation/partner.

(c) Golombok-Rust inventory of sexual satisfaction (GRISS)[22]

It assesses the existence or absence of sexual dysfunction in each of the members of the couple, as well as their level of severity. It consists of 56 items, 28 items specific for men and 28 specific for women; it studies 12 dimensions, out of which 6 assess sexual function: premature ejaculation and impotence in men, anorgasmia and vaginismus in women, and avoidance in both men and women. An overall score is obtained for the quality of sexual functioning as a couple, in addition to the scores for each of the subscales. High scores indicate greater sexual dissatisfaction. It has not been validated or translated into Spanish.

(d) Sexual desire inventory (SDI)[23]

It was adapted to Spanish by Ortega, Zubeidat and Sierra in 2006. It is a self-administered questionnaire in which sexual desire is measured by distinguishing between sexual relations with a partner and masturbation. It consists of 13 items, in which the first 9 are related to sex with a partner and the rest to solo sexual satisfaction.

Specific questionnaires

Questionnaires specific for women

(a) Brief index of sexual functioning (BISF-W)[24]

The assessed dimensions are sexual desire/interest, sexual activity, sexual satisfaction, thoughts/desires, arousal, frequency of sexual activity, receptivity/initiation of sexual interaction, orgasm, satisfaction in sexual relations, and problems affecting sexual relations.

The questionnaire includes 22 items and assesses women's sexual functioning and satisfaction. The content of the items includes pain during the act, difficulties in achieving orgasm, the impact of health problems on sexual activity, or the frequency of certain sexual behaviors such as masturbation or sexual fantasies.

(b) Expanded sexual arousability inventory (SAI-E)[25]

Initially, it was designed to assess sexual arousal only. Chambless et al. added two scales, one referring to anxiety and the other to sexual satisfaction, resulting in a questionnaire that assesses sexual arousal and anxiety in women.

It consists of 28 items, each referring to sexual activities or sexually exciting situations. Each item will be assessed independently according to the degree of excitement, anxiety, and satisfaction it may cause. It was validated and translated into Spanish by Aluja, Torrubia, and Gallar in 1990.

(c) Female sexual function index (FSFI)[26]

The female sexual function index is made up of questions that include sociodemographic variables and 19 items that assess female sexual function over the past month. Therefore this questionnaire will not be useful for those patients who are not sexually active during the study period.

It is a simple and brief questionnaire that assesses desire, arousal, lubrication, orgasm, satisfaction, and pain. Each of the multiple choice questions will be scored with values ranging from 0 to 5. A score will be obtained for each domain which has to be multiplied by a factor, the total score being the sum of all domains. High scores will indicate better sexual quality of life. It has been translated and validated into Spanish.

(d) Female sexual function (FSF)[27]

It was designed and validated by Sánchez et al. in the field of primary care. It consists of 14 questions and assesses sexual desire, arousal, lubrication, orgasm, satisfaction of sexual activity, general sexual satisfaction, problems with vaginal penetration, anticipatory anxiety, sexual initiative, and the degree of sexual communication.

Questionnaires specific for men

(a) Brief sexual function questionnaire (BSFQ)[28]

It is a self-administered questionnaire that assesses male sexual function and takes into account sexual preference (homosexual vs heterosexual).

It consists of 21 questions and 4 dimensions are assessed: activity, satisfaction, interest, and physiological dysfunction. High scores indicate increased sexual interest, activity, and satisfaction.

It has a version for the couples, which attempts to assess whether the patient's perception of their own interest, activity, and sexual performance is the same as that referred to by their partner.

(b) Brief sexual function inventory (BSFI)[29]

It is a short, self-administered questionnaire consisting of 11 items. It has been validated and translated into Spanish and, the same as the previous one, it assesses male sexual function through the following dimensions: sexual motivation, erectile function, ejaculation, and sexual satisfaction.

An index of problems can also be obtained with this questionnaire. In order to do so, the scores obtained in the three dimensions should be added together: sexual motivation, erectile function, and ejaculation.

(c) International index of erectile function (IIEF)[30]

The international index of erectile function is a questionnaire that has been validated and translated into Spanish. It was developed to have a brief, reliable, self-administered and cross-cultural measure of erectile function.

It consists of 15 items that assess sexual desire, erectile function, orgasmic function, satisfaction with sexual intercourse, and global satisfaction. The answer to each item is through a Likert scale and refers to frequency, difficulty, intensity, or satisfaction.

The higher the score, the lower the sexual dysfunction, with the following cutoff points for the erectile dysfunction dimension: severe dysfunction ranging from 6 to 10 points, moderate dysfunction from 11 to 16 points, mild dysfunction from 17 to 25 points, and no erectile dysfunction with scores >26. There is also a reduced version of 5 items, the IIEF-5.

Recommendations to cope with the fear of sexual intimacy

After recovering from the surgery, the patient should be able to resume sexual activity. Facing this moment will not only depend on the organic limitations derived from the procedure, but also on the psychological factors mentioned in previous sections such as the deterioration of body image and self-esteem, the acceptance of their partner, etc.

At a general level, the recommendations for resuming sexual activity are based on increasing the security of control over the stoma, such as[7]:

- Emptying the bag before the act;
- Performing a colostomy irrigation for a gastric emptying in order to avoid leaks;
- Using opaque pouches or pouch covers;
- Using belts that support both the stoma disc and the colostomy bag.

Other recommendations that can be used are more specific to sexual dysfunction. In this sense, Albaugh et al.[12] make the following suggestions:

- Decreased libido:

    For this, the authors proposed a cognitive behavioral therapy or a sensory approach. The former implies that the patients must prepare themselves thinking about sex through reading material or the visualization of erotic clips. The sensory approach involves guidelines ranging from nongenital touching and pleasure to genital play and sexual intercourse. The patient will be gradually exposed with their partner to sensual and sexual situations in increasing order of difficulty and anxiety. It consists of six phases that go from touching the partner's body without including the erogenous zones to the normal accomplishment of coitus, without any restrictions.

    There is also hormone therapy with testosterone, which can be administered subcutaneously, transdermally, or by injection.

- Vaginal dryness and pain:

    The most commonly used is estrogen therapy, in the form of creams, vaginal rings, or oral medication. With this therapy, the thickness of the vaginal mucosa is improved, decreasing its atrophy.

    Pelvic floor therapy may also be helpful, as it improves sensitivity and decreases or prevents pain during intercourse, making it more pleasurable. This therapy consists of exercising the contraction of the pelvic floor muscles without contracting the muscles of the buttocks by performing exercises which tone the muscles of the area. These muscle exercises should be performed at different times of the day and especially during sexual intercourse. This therapy includes kegel exercises, specific exercises with biofeedback, exercises for pelvic and abdominal normalization, Baoding balls, etc.

    Other treatments for the pain during intercourse are local, such as topical lidocaine.

- Lack of sexual arousal

    This alteration is common among patients with dysfunction of the pelvic floor nerves, due to reduced blood flow in the genital area.

    There are different therapies for men such as oral medication, local treatment including vacuum devices or injections into the penis and surgery to implant prosthetic cylinders in the area of the corpus cavernosum, which is the most commonly used inflatable implant.

    In the case of women, a vacuum suction system, oral treatment, or topical treatment such as the Zestra stimulant gel can also be used.

Health education: Help to adapt to change

As discussed at the beginning of this chapter, the ostomy problem for patients is primarily addressed from a surgical perspective. If this is combined with the psychological and psychoemotional impact that it causes, along with the lack of information, it gives rise to a great insecurity that makes it difficult to adapt and face the change of the body.[1, 8, 14]

One of the main difficulties is that neither the possible sexual problems nor the treatment options that exist in the case of sexual dysfunction are usually addressed. Therefore it is essential that the health personnel start the conversation on this issue, so that the adequate resources could be offered for each problem. To promote dialog on the subject, models have been developed to determine not only the degree of existing commitment to sexual health, but also to establish the strategies to follow.

In this sense, Gomez et al.[1] recommended the PLISSIT model in order to carry out an effective intervention. This type of intervention is widely used in sexual care and is an acronym for Permission (P), Limited Information (LI), Specific Suggestions (SS), and Intensive Therapy (IT). The model presents four stages of intervention: in the first one, the patients are given permission (P) to talk about the sexual problem or the subject that worries them, they may be offered some limited information (LI) that can help to clarify the erroneous information or some specific suggestions (SS) related to a particular problem. However, sometimes it is necessary to resort to a specialized and personalized therapeutic intervention (TI) (psychological interventions, sexual therapies, etc.)

Another model is the 5-A, adapted for cancer patients by Park et al.[31] This model consists of five parts:

- Ask: the professional should mention the sexual complications that usually appear on a regular basis, then ask about their presence in the patient.
- Advise: make known the different resources to address the problems.
- Assess: using a specific checklist to have an overview of patient concerns.
- Assist: providing the patient with education, information and resources to acquire appropriate knowledge, and increase self-confidence. All this will be carried out by a multidisciplinary team.
- Arrange follow-up: the health personnel will follow up on the problems that have been identified.

Despite the importance of the topic, there is no common protocol for its approach. It would be necessary to draw up general guidelines establishing, for example, the ideal time to carry out educational and informative intervention about sexual health, how to assess more effectively the sexual dysfunction of ostomy patients, or how health personnel should ask questions about the existence of sexual function problems. This would diminish the negative effects on intimate relationships.[17]

References

1. del Río NG, Castro NM, Delgado CC, Rodríguez AMF, Clemente MJH, Fernández YG. Los cuidados de enfermería en el impacto psicológico del paciente ostomizado. *Revista Ene de Enfermería* 2013;**7**(3). http://www.ene-enfermeria.org/ojs/index.php/ENE/article/view/279. [Accessed 15 February 2021].
2. Vonk-Klaassen SM, de Vocht HM, den Ouden MEM, Eddes EH, Schuurmans MJ. Ostomy-related problems and their impact on quality of life of colorectal cancer ostomates: a systematic review. *Qual Life Res* 2016;**25**(1):125–33. https://doi.org/10.1007/s11136-015-1050-3.
3. Ang SGM, Chen H-C, Siah RJC, He H-G, Klainin-Yobas P. Stressors relating to patient psychological health following stoma surgery: an integrated literature review. *Oncol Nurs Forum* 2013;**40**(6):587–94. https://doi.org/10.1188/13.ONF.587-594.
4. Vural F, Harputlu D, Karayurt O, et al. The impact of an ostomy on the sexual lives of persons with stomas: a phenomenological study. *J Wound Ostomy Continence Nurs* 2016;**43**(4):381–4. https://doi.org/10.1097/WON.0000000000000236.
5. Yilmaz E, Çelebi D, Kaya Y, Baydur H. A descriptive, cross-sectional study to assess quality of life and sexuality in Turkish patients with a colostomy. *Ostomy Wound Manage* 2017;**63**(8):22–9. https://doi.org/10.25270/owm.2017.08.2229.
6. Alves RCP, Moreira KCR. A percepção do paciente portador de ostomia com relação a sua sexualidade. *Rev Interdiscip* 2013. Published online http://revistainterdisciplinar.uninovafapi.edu.br/index.php/revinter/article/view/90.
7. Melià ABR. La sexualidad en pacientes con ostomias digestivas y urinarias. *Enferm Integral: Revista científica del Colegio Oficial de Enfermería de Valencia* 2011;(95):42. https://dialnet.unirioja.es/servlet/articulo?codigo=3733528. [Accessed 15 February 2021].
8. Ozturk O, Yalcin BM, Unal M, Yildirim K, Ozlem N. Sexual dysfunction among patients having undergone colostomy and its relationship with self-esteem. *J Fam Med Community Health* 2015;**2**(1):1028. https://www.researchgate.net/profile/Onur_Ozturk2/publication/271833239_Sexual_Dysfunction_among_Patients_having_undergone_Colostomy_and_its_Relationship_with_Self-Esteem/links/54d32bd90cf25017918193ea.pdf.
9. Traa MJ, De Vries J, Roukema JA, Den Oudsten BL. Sexual (dys)function and the quality of sexual life in patients with colorectal cancer: a systematic review. *Ann Oncol* 2012;**23**(1):19–27. https://doi.org/10.1093/annonc/mdr133.
10. Anaraki F, Vafaie M, Behboo R, Maghsoodi N, Esmaeilpour S, Safaee A. Quality of life outcomes in patients living with stoma. *Indian J Palliat Care* 2012;**18**(3):176–80. https://doi.org/10.4103/0973-1075.105687.
11. Reese JB, Handorf E, Haythornthwaite JA. Sexual quality of life, body image distress, and psychosocial outcomes in colorectal cancer: a longitudinal study. *Support Care Cancer* 2018;**26**(10):3431–40. https://doi.org/10.1007/s00520-018-4204-3.
12. Albaugh JA, Tenfelde S, Hayden DM. Sexual dysfunction and intimacy for ostomates. *Clin Colon Rectal Surg* 2017;**30**(3):201–6. https://doi.org/10.1055/s-0037-1598161.
13. Kimura CA, Guilhem DB, Kamada I, de Abreu BS, Fortes RC. Oncology ostomized patients' perception regarding sexual relationship as an important dimension in quality of life. *J Coloproctol* 2017;**37**(3):199–204. https://doi.org/10.1016/j.jcol.2017.03.009.
14. Costa IKF, Liberato SMD, Freitas LS, Melo MDM, de Sena JF, de Medeiros LP. Distúrbio na imagem corporal: diagnóstico de enfermagem e características definidoras em pessoas ostomizadas. *Aquichan* 2017;**17**(3):270–83. https://doi.org/10.5294/aqui.2017.17.3.4.

15. de las Bonill Nieves C, Hueso Montoro C, Celdrán Mañas M, Rivas Marín C, Sánchez Crisol I, Morales Asencio JM. Viviendo con un estoma digestivo: la importancia del apoyo familiar. *Index Enferm* 2013;**22**(4):209–13. https://doi.org/10.4321/S1132-12962013000300004.
16. Calcagno Gomes G, Peres Bitencourt P, da Pizarro AR, Pereira Madruga A, Silva de Castro E, de Oliveira Gomes VL. Ser mujer con ostomía: la percepción de la sexualidad. *Enfermería Global* 2012;**11**(27):22–33. http://scielo.isciii.es/pdf/eg/v11n27/clinica2.pdf.
17. Silva AL, Monteiro PS, Sousa JB, Vianna AL, Oliveira PG. Partners of patients having a permanent colostomy should also receive attention from the healthcare team. *Colorectal Dis* 2014;**16**(12):O431–4. https://doi.org/10.1111/codi.12737.
18. Danielsen AK, Burcharth J, Rosenberg J. Spouses of patients with a stoma lack information and support and are restricted in their social and sexual life: a systematic review. *Int J Colorectal Dis* 2013;**28**(12):1603–12. https://doi.org/10.1007/s00384-013-1749-y.
19. Collado EJ, García P. Validación de un cuestionario específico de Calidad de Vida con una muestra de pacientes colostomizados o ileostomizados. *Eur J Health Res* 2015;**1**(3):107. https://doi.org/10.30552/ejhr.v1i3.9.
20. Bobes J, Gonzalez MP, Rico-Villandemoros F, Bascaran MT, Sarasa P, Clayton A. Validation of the Spanish version of the changes in sexual functioning questionnaire (CSFQ). *J Sex Marital Ther* 2000;**26**(2):119–31. https://doi.org/10.1080/009262300278524.
21. Derogatis LR. The derogatis interview for sexual functioning (DISF/DISF-SR): an introductory report. *J Sex Marital Ther* 1997;**23**(4):291–304. https://doi.org/10.1080/00926239708403933.
22. Ter Kuile MM, van Lankwd JJDM, Kalkhown P, van Egmond M. The golombok rust inventory of sexual satisfaction (GRISS): psychometric properties within a dutch population. *J Sex Marital Ther* 1999;**25**(1):59–71. https://doi.org/10.1080/00926239908403977.
23. Ortega V, Zubeidat I, Sierra JC. Further examination of measurement properties of Spanish version of the sexual desire inventory with undergraduates and adolescent students. *Psychol Rep* 2006;**99**(1):147–65. https://doi.org/10.2466/pr0.99.1.147-165.
24. Taylor JF, Rosen RC, Leiblum SR. Self-report assessment of female sexual function: psychometric evaluation of the brief index of sexual functioning for women. *Arch Sex Behav* 1994;**23**(6):627–43. https://doi.org/10.1007/BF01541816.
25. Aluja A, Torrubia R, Gallart S. Validación española del autoinforme de ansiedad y excitación sexual ampliado (SAI-E). *Rev Psiquiatr Fac Med Barc* 1990. Published online https://www.researchgate.net/profile/Anton_Aluja/publication/48907029_Validacion_espanola_del_autoinforme_de_ansiedad_y_excitacion_sexual_ampliado_SAI-E/links/0912f50c8b1b7c06d3000000/Validacion-espanola-del-autoinforme-de-ansiedad-y-excitacion-sexual-ampliado-SAI-E.pdf.
26. Blümel JE, Binfa L, Cataldo P, Carrasco A, Izaguirre H, Sarrá S. Índice de función sexual femenina: un test para evaluar la sexualidad de la mujer. *Rev Chil Obstet Ginecol* 2004;**69**(2):118–25. https://scielo.conicyt.cl/scielo.php?pid=S0717-75262004000200006&script=sci_arttext&tlng=p.
27. Sánchez F, Pérez Conchillo M, Borrás Valls JJ, Gómez Llorens O, Aznar Vicentee J. Caballero Martín de las Mulas A. Diseño y validación del cuestionario de Función Sexual de la Mujer (FSM). *Aten Primaria* 2004;**34**(6):286–94. https://doi.org/10.1016/S0212-6567(04)79497-4.
28. Reynolds III CF, Frank E, Thase ME, et al. Assessment of sexual function in depressed, impotent, and healthy men: factor analysis of a brief sexual function questionnaire for men. *Psychiatry Res* 1988;**24**(3):231–50. https://doi.org/10.1016/0165-1781(88)90106-0.
29. Vallejo-Medina P, Guillén-Riquelme A, Sierra JC. Análisis psicométrico de la versión española del Brief Sexual Function Inventory (BSFI) en una muestra de hombres con historia de abuso de drogas. *Adicciones* 2009;**21**(3):221. https://doi.org/10.20882/adicciones.232.
30. Zegarra L, Loza C, Pérez V. Validación psicométrica del instrumento índice internacional de función eréctil en pacientes con disfunción eréctil en Perú. *Rev Peru Med Exp Salud Publica* 2011;**28**(3):477–83. http://www.scielo.org.pe/scielo.php?script=sci_arttext&pid=S1726-46342011000300011.
31. Park ER, Norris RL, Bober SL. Sexual health communication during cancer care. *Cancer J* 2009;**15**(1):74–7. https://doi.org/10.1097/ppo.0b013e31819587dc.

Section H

Ethical and legal aspects in CRC

Chapter 56

Ethical and legal aspects in CRC: Research and clinical assistance

Natalia Cal Purriños[a,b,c], Isaac Martínez Bendayán[c,d,e], and Aliuska Duardo Sánchez[f]

[a]Novoa Santos Foundation, A Coruña, Spain [b]Institute for Biomedical Research A Coruña—INIBIC, A Coruña, Spain [c]Research Ethics Committee A Coruña, Ferrol, Spain [d]Cardiology Department, University Hospital of A Coruña, A Coruña and Cee Health Area, SERGAS, A Coruña, Spain [e]Structural and Congenital Heart Disease Research Group, Institute for Biomedical Research A Coruña—INIBIC, A Coruña, Spain [f]Department of Public Law, Faculty of Law UPV/EHU, G.I Chair in Law and the Human Genome, Leioa, Spain

Introduction: The relationship between ethics and law

The relationship between ethics and law is symbiotic; many legal principles were first conceived as ethical principles embodied in international declarations and codes of conduct. Thus, for example, article 7 of the International Covenant on Civil and Political Rights, which states: "… no one shall be subjected to medical or scientific experimentation without his free consent," clearly draws on the 1978 Belmont Commission Report and other documents of the same nature that preceded it. Although ethical reflection often precedes legal reflection, the law has also contributed to the former. Consider the evolution of the very concept of consent, now informed consent, which is rooted in legal principles such as the contractual party autonomy or the Freedom of decision.[1]

Specifically, bioethics as a systematic study of the ethical considerations of human actions on human, plant, and animal life, with the aim of favoring social consensus on what is good for humanity, future generations, and the ecosystem,[2] must constitute a legal premise with regard to biomedical research. At the same time, it contributes to its adaptation to new societal needs. However, the role of one and the other should not be confused.

When ethical principles are enshrined in a legal norm, they become law in a material sense, and thus cease to be voluntarily acceptable guidelines for behavior and become mandatory rules. Nonobservance of these rules can give rise to legal liability on the part of the scientific personnel involved: civil, administrative, even criminal. Think of the so-called genetic manipulation offenses under articles 159–162 of the Spanish Penal Code. Thus the fundamental difference between Ethics and Law, or Bioethics and Biolaw, to be more precise, lies in the coerciveness—the obligatory nature—of Law.

In any case, it is not a question of diminishing the role of ethics. In scientific research, and in particular biomedical research, it is necessary to bear in mind the evaluations provided by the ethical debate, as well as the contributions made by the theory of human rights, fundamental rights, and public freedoms. Taken together, these assumptions are essential to prevent the law from becoming an instrument of legitimization at the service of the needs of researchers and, where appropriate, of companies in the sector.

Ethical principles and research ethics committees

In 1964 the World Medical Association developed a code of research ethics called the Declaration of Helsinki, a document that became an extension of the Nuremberg Code and established ethical guidelines or principles to be followed in the conduct of research studies involving human subjects.[3]

The version approved at the 64th General Assembly held in Fortaleza, Brazil, in October 2013 is the most recent. Among other issues, it covers aspects relating to medical research on human subjects, especially human subjects who lack decision-making power, or with vulnerable subjects, and the need for research to be reviewed before it begins by an independent body known as a research ethics committee.

It is important to clarify that International declarations, such as the Universal Declaration on the Human Genome and Human Rights—UNESCO 1997—are more acts of goodwill than legal norms in the strict sense, since the signatory countries are not obliged to comply with their content. Hence, they are known as soft law. Nevertheless, these documents often contain ethical principles, guidelines, and important control mechanisms, which are later incorporated by individual States into their domestic legislation or binding international treaties. They constitute a further example of this relationship between ethics and law that we have called symbiotic.

Among the **ethical principles,** set out in the Declaration of Helsinki, which condition the conduct of medical research on human subjects are the following:

- Respect for all human subjects and protect their health and rights.
- Guarantee the right to self-determination.
- Protect the privacy of research subjects and the confidentiality of their personal information.
- To facilitate informed decision-making.
- Fulfilling the obligation to ensure the well-being of the individual over and above scientific or societal interests.
- Protect situations of special vulnerability and obtain the authorization of legal representatives to carry out studies on minors or the disabled.

On the other hand, it is also envisaged that research should be carried out by professionals with scientific expertise, and that a weighing of the risks and benefits of the research should be carried out, with the aim of minimizing the risks and maximizing the benefits, especially in those cases where it is not possible to guarantee zero risk.

Subsequent to the guidelines established in the ethical documents, different legislative texts were drafted that established the procedures for the constitution of research ethics committees and their operation. Initially, the **Clinical Research Ethics Committees (CREC) were** set up in the field of clinical trials with medicinal products. In 2007 Law 14/2007 of 3 July 2007 on Biomedical Research (**BRL**)[4] *included* **Research Ethics Committees (REC)** in its scope of regulation. And, in 2015, with the approval of the new Royal Decree 1090/2015, of 4 December, regulating clinical trials with medicinal products, the Research Ethics Committees with medicinal products and the Spanish Clinical Trials Register (**RD1090**),[5] which adapted the Spanish legal system to Regulation (EU) No. 536/2014 of the European Parliament and of the Council of 16 April 2014 on clinical trials on medicinal products for human use, and repealing Directive 2001/20/EC,[6] concerning clinical trials with medicinal products for human use conducted in Spain, established the **Medicines Research Ethics Committees (MREC)** and laid down the legal definitions of both the RECs and the MREC.

RECs are defined as independent, multidisciplinary bodies whose main purpose is to ensure the protection of the rights, safety, and well-being of subjects participating in a biomedical research project and to provide public assurance in this respect by providing an opinion on the relevant research project documentation, taking into account the views of lay persons, in particular patients, or patient organizations.

MREC are described as **REC** that are also accredited in accordance with the terms of RD1090 to issue an opinion on a clinical study involving medicinal products or a clinical investigation involving medical devices.

Currently, in Spain, the **CREC** have disappeared and **MREC** have been set up to review clinical trials with medicines or medical devices and **REC** to evaluate all health research studies that are not clinical trials or prospective observational studies with medicines or medical devices, i.e., studies regulated by the BRL or by the general data protection and health regulations.

Among the **functions** assigned to them are those of:

- Assess the qualifications of the principal investigator and the research team, as well as the feasibility of the project/trial.
- Review the methodological, ethical, and legal aspects of the research project.
- Weigh the balance of anticipated risks and benefits arising from the study.
- Ensure compliance with procedures to ensure the traceability of samples of human origin, without prejudice to the provisions of personal data protection legislation.
- To report, after evaluation of the project, on all biomedical research involving interventions on human beings or the use of biological samples of human origin, without prejudice to other reports to be issued.

Once the evaluation has been completed, a report shall be issued by the Committee, which shall be prior to and mandatory for the initiation of any investigation.

Regulations governing biomedical research

In Spain there is extensive legislation, both at national and regional level, which regulates the different types of research studies that can be carried out in the field of health.

In this chapter, without being exhaustive, we systematize the main requirements set out in the different regulatory texts in force that regulate bio-health research.

Firstly, **Law 14/2007, of 3 July, on Biomedical Research (BRL)**, regulates biomedical research, with full respect for human dignity and identity and the inherent rights of persons, establishing that the precepts contained therein shall be **applicable** to:

- Basic and clinical biomedical research, with the exception of clinical trials with medicines and medical devices, which shall be governed by their specific regulations.
- Human health-related research involving invasive procedures.
- The processing, storage, and movement of biological samples.
- Biobanks
- The donation and use of human oocytes, spermatozoa, preembryos and fetuses or their cells, tissues, and organs for biomedical research purposes and their possible clinical applications.
- The performance of genetic analyses and the processing of personal genetic data.

In the development of all these research activities, it is established that, at all times, the **principles and guarantees** recognized in the same law will be respected, which can be specified as follows:

- Protection of the **dignity and identity** of the human being. Respect for the **integrity, nondiscrimination** and all the fundamental rights and freedoms of persons participating in biomedical research. Guarantee **confidentiality** in the processing of human biological samples and data.
- **Protection** of their health, physical and psychological **integrity.** Principle of **precedence in favor of** the health, interest, and well-being of the human being over any interest of society or science.
- Ensure **freedom of research** and scientific production in the field of biomedical sciences.
- Need for **prior assessment**, with a mandatory favorable report from the ERC, in order to authorize and carry out any research project on human beings or their biological material.
- Obligation to evaluate the research carried out.
- Respect for the **precautionary principle** to prevent and avoid risks to life and health.

Respect for all these principles and guarantees is evident throughout the articles of the BRL, but in a more special way when it comes to biomedical research on those populations that are considered vulnerable. In these cases, more specific restrictions and requirements are applied.

Specifically, BRL establishes that, for the use of human biological samples and clinical data in biomedical research, the privacy of the participants must be protected at all times, guaranteeing the confidentiality of their information and samples. This guarantee is reinforced by requiring all professionals participating in the research to maintain **confidentiality** of all information concerning the persons subject to the study in question, even if the employment relationship has ceased. On the other hand, it includes the obligation not to use health data and biological samples for purposes other than those authorized by the subject who is the source of the data.

Individuals must **consent** to the collection, derivation, use, storage, and disclosure of their biological samples for the conduct of research studies; this consent must be given in writing, prior to the research being carried out, in an informed consent document specific to the particular research. The **information** that will need to be provided in writing to participants in research involving biological samples has been broken down by Article 59 of the BRL.

This **prior information,** for special cases such as Biobanks (Arts. 63 to 71 BRL), research in which invasive procedures on human beings are necessary (Arts. 13 to 27 BRL), research on human embryos, fetuses, their cells, tissues, or organs (Arts. 28 to 43 BRL), must comply with the specific requirements established by the BRL for these circumstances.

Royal Decree 1716/2011, of 18 November, which establishes the basic requirements for the authorization and operation of biobanks for biomedical research purposes and the treatment of biological samples of human origin, and regulates the operation and organization of the National Register of Biobanks for biomedical research (RD1716),[7] is the regulatory document established to develop the basic precepts set out in the BRL, with regard to the **use of biological samples for scientific and technical research purposes**, which includes innovation and development as the main or secondary purpose of obtaining, storing, or transferring them.

RD1716, in its introduction, establishes that the **subject's rights** must be respected whenever their biological material is used to obtain new scientific knowledge, confirm hypotheses, or carry out activities of technological adaptation, quality control, teaching, etc. And it distinguishes between a **general regime** for the processing of biological samples for biomedical research purposes and the **specific regime to be** applied when this processing is carried out in a biobank; in turn, it establishes a regime for obtaining and using samples from deceased persons in research.

Throughout the development of this RD1716, the principles and premises regulated by the BRL are maintained; and it is established as the object of the same:

- Develop the regime for the processing of biological samples of human origin for biomedical research purposes.
- Establish the basic requirements for the authorization and operation of biobanks for biomedical research purposes.
- **To** regulate the functioning and organization of the National Register of Biobanks for Biomedical Research.

- *Collection, storage, and use regimes for biological samples of human origin.*

  Biological samples of human origin intended for biomedical research may be stored in a biobank or kept for use in a specific research project or in a collection for biomedical research purposes.

  The **Biobank** is a public or private, nonprofit establishment that houses a collection of biological samples, conceived for diagnostic or biomedical research purposes and organized as a technical unit with criteria of quality, order, and destination. By definition, the samples from the biobank may be used in any biomedical research, provided that the legal regulations are complied with and that the source subject has given his or her consent under these terms.

  Biological specimen collections for biomedical research are an ordered set of biological samples collected for their special interest or value for biomedical research, which does not include biological samples of human origin that are kept exclusively for a specific research project and are not to be transferred to other researchers. Samples from a collection may only be used for the purpose stated in the specific consent, unless additional written consent is given by the source subject.

  With regard to the **storage and use of biological samples within the framework of a specific research project**, storage within the study may not extend beyond the end date of the study and samples may not be transferred to other researchers who are not involved in the specific research for which they were obtained. Their conservation for reuse in new projects will require their integration in the biobank or in a collection and the written consent of the source subject.

  With the approval and entry into force of RD1716, a different legal treatment has been established for the use in research of biological samples deposited in biobanks or sample collections.

  In relation to **biobanks**, it should be pointed out that they are set up as a public service aimed at providing biomedical research with human biological samples and associated clinical information of high quality, guaranteeing security and traceability. These samples are obtained with a broader consent than that established for other research purposes such as projects or collections of samples for the development of lines of research, where it is necessary to detail the specific purpose for which the samples are obtained. Biobanks may obtain and conserve human biological samples indefinitely, in order to make them available to the scientific community for use in future research, not foreseen at the time of collection and respecting in all cases the nature of the sample. The samples may be transferred for use in scientifically and ethically approved research projects, and such transfers will be carried out, as a general rule, anonymously or dissociated, with a necessary requirement for such a transfer being the establishment of a transfer agreement between the parties involved (the person responsible for the research project and the Biobank), in which the researcher undertakes not to use the samples for any use other than that for which they were requested, and it is assumed that the samples may not be transferred to third parties other than researchers.[8]

- *Informed Consent (IC) requirements for the collection, storage or conserve and use of biological samples of human origin and associated data.*

  RD1716 establishes that the collection of samples, their storage or conservation, and their subsequent use shall require the corresponding **prior consent** of the source subject, which shall indicate the purposes for which the samples are being collected, among other **information** specified in Article 23 of the same.

  If the **purposes** established for the samples are **several** (project, collection and/or biobank), these may appear in the same document, although the source subject's ability to grant consent for each purpose independently must be guaranteed in all cases.

  The document stating the source subject's consent to the collection and use of his/her biological samples for biomedical research purposes shall be issued in three copies. One of these shall be given to the source subject, another shall be kept at the center where the sample was obtained, and the third shall be kept by the biobank, or by the person responsible for the collection or research, as appropriate.

  Consent may **be withdrawn**, in whole or for certain purposes, at any time.

- *Referral of samples collected for healthcare purposes to research.*

  RD1716 also regulates the use for research purposes of human biological samples obtained in the exercise of a healthcare activity for diagnostic or therapeutic tests, as part of the medical care provided to the patient/donor. These samples may be used for research purposes provided that the following requirements are met:

  - Initial diagnostic or therapeutic purposes are not compromised.

- Whether it is the remnants of these samples, or the same samples after the end of the legal storage period for these samples.
- The prior IC of the donor and of the professional responsible for its use in healthcare is available, in order to refer these samples for research.

The donor of the sample or his family, when needed for health reasons, shall have priority in the use of the donated sample, provided that the sample is available and not anonymized.

Royal Decree 1090/2015, of 4 December, which regulates Clinical Trials with Medicines, the Ethics Committees for Research with Medicines and the Spanish Register of Clinical Studies, hereinafter—RD1090—is the Spanish regulation that has adapted our legal system to the new requirements regarding clinical trials established by the RDCT.

Among the **requirements to be** fulfilled when wishing to carry out a clinical trial with medicinal products or medical devices, it is established the need to have:

- **Authorization** by the Spanish Agency for Medicines and Health Products—**SAMHP**.
- The favorable opinion of an accredited MREC in Spain, which has been called **Single Opinion** per Member State. Regardless of the number of centers that will participate in Spain, only one MREC will assess the study and the suitability of the centers and research teams.
- A research **protocol** drawn up in accordance with the indications set out in Annex I of the RDCT.
- A specific **informed consent** for the clinical trial and with all the information sections established in this respect by both national and European regulations.
- A specific liability **insurance** for the trial. However, an exception has been made for **low-intervention clinical trials** not to require their own insurance, as long as they are covered by the liability policy of the healthcare institution where they will be conducted.
- A **contract** between the sponsor and the clinical trial development site.
- The **suitability report** of the research equipment and facilities.

For its part, **Royal Decree 957/2020, of 3 November, which regulates Observational Studies with Medicinal Products for Human Use (RD957)**[9] regulates the requirements necessary to carry out an observational drug study (**ODs**), i.e., any research that involves the collection of individual data relating to the health of persons, provided that it does not meet any of the conditions required to be considered a clinical trial established in article 2.1.i) of RD1090, and that it is conducted for one of the following **purposes**:

- Determine the beneficial effects of medicines, as well as their modifying factors, including the patient's perspective, and their relationship to the resources used to achieve them.
- Identify, characterize, or quantify adverse drug reactions and other patient safety risks related to the use of medicines, including potential risk factors or effect modifiers, and measure the effectiveness of risk management measures.
- Obtain information on patterns of medicine use in the population.

In order to be implemented, ODs must comply with a series of formalities and requirements, which are described as follows:

- A favorable **opinion** from a MREC accredited in Spain, which shall be **unique,** binding, and recognized throughout the national territory.
- Address feasible and relevant additional requirements that may have been stipulated by the competent health authorities.
- Prior **agreement** of the person in charge of the health **center**, which will be expressed by the signing of a **contract** with the sponsor. This contract will not be necessary in those cases in which the sponsor belongs to the health center, service, or establishment where the study is carried out, it being sufficient to obtain the express agreement of the person in charge of the same.
- Have a research **protocol** that justifies the need for the research and the work methodology to be developed. The protocol must describe in detail the sources of information from which the data to be used in the proposed research will be obtained and compliance with the requirements of both Regulation (EU) No. 679/2016 (2016) of the European Parliament and of the Council of 27 April 2016 on the protection of individuals with regard to the processing of personal data and on the free movement of such data and repealing Directive 95/46/EC—General Data Protection Regulation. (**GDPR**)[10] and by Organic Law 3/2018, of 5 December, on the Protection of Personal Data and Guarantee of Digital Rights (**OLPPGDR**),[11] specifically:

- ○ Have assessed and mitigated, through appropriate measures in each case, the impact that the conduct of the study may have on the protection of personal data.
- ○ To guarantee the confidentiality of the data of the participating subjects.
- ○ Detail in the protocol the conditions of access to personal data, including the conditions of their international transmission outside the European Economic Area, if foreseen.
- Where it is necessary to interview the subject participant, it will be necessary to obtain his or her **informed consent**. However, informed **consent** may be **waived** if the MREC considers that:
  - ○ observational research has an important social value
  - ○ its realization would not be feasible or practicable without such a waiver
  - ○ involves minimal risk to participants
- When the **consent** of the subject participant is **not** required because another legitimate basis for the processing of his/her personal data from among those referred to in articles 6.1 and 9.2 of the RGPD applies, the sponsor and the researchers must apply the criteria governing the processing of data in health research in accordance with the 17th Additional Provision of the OLPPGDR.
- In ODs where data are to be processed anonymized or have undergone pseudonymization processing, the **protocol** shall set out the **procedure** followed to achieve such **anonymization** or **pseudonymization**.

The new regulatory framework for personal data protection

No study on the regulation of biomedical research in the Spanish and EU context would be complete without devoting a section to the protection of the personal data of research subjects. In fact, the protection of personal data constitutes a fundamental right of European citizens, explicitly recognized in the EU Charter of Rights. The content of this right is developed in Regulation (EU) 2016/679 of the European Parliament and of the Council of 27 April 2016 on the protection of individuals with regard to the processing of personal data and on the free movement of such data and repealing Directive 95/46/EC (GDPR). According to this regulation, health data are included in the special category of "sensitive data," and this classification confers special protection on them. This is due to the fact that by their very nature "their processing could entail significant risks to fundamental rights and freedoms," GDPR Recital (51).

The GDPR, which came into force on 25 May 2018 in all EU Member States, signified a paradigm shift in the regulation of personal data processing. To begin with, it is a regulation and not a directive, which means that it does not need to be transposed into domestic law to be mandatory. In Spain, it implied the publication and entry into force of Organic Law 3/2018, of 5 December, on the protection of personal data and guarantees of digital rights (OLPPGDR) to adapt our regulatory framework to the new operating guidelines established by the GDPR.

Proactive responsibility, the most relevant principle that should govern the development of any activity involving the processing of personal data, requires that in all phases of the processing, compliance with data protection regulations is guaranteed and that all the necessary technical and organizational measures are established to guarantee the rights of the data subjects and to accredit such compliance when required.

The data **controller** is obliged to comply with the regulations and legally enforceable requirements and is responsible for making an appropriate selection of the service providers or collaborators who must process the data for which it is responsible on its behalf and who are known as data **processors**. Furthermore, it is established that it will be obliged to set up an **internal register in which** all personal data **processing activities** carried out will be included, and two new concepts are established, **privacy by design** and privacy **by default**, which implement the requirement to carry out a prior **assessment of the impact and risks** involved in the processing to be carried out, in order to determine and implement the **security measures** appropriate to the risks detected, at any stage of the processing of the data from its collection to the end of the purpose for which it is to be used.

Among other new features, the GDPR establishes the obligation to **notify and inform** the supervisory authority and data subjects of any security breach that affects or may affect the rights of data subjects, and also obliges public administrations and bodies, entities that process special categories of data on a large scale and all the entities described in article 34 of the OLPPGDR, to appoint a **Data Protection Officer** (DPO), a new professional figure who, among other functions, will be responsible for informing and advising data controllers or data processors on data protection matters, training the different professionals who have access to third-party information, supervising the proper functioning in accordance with the requirements demanded by current legislation or dealing with user queries.

With regard to the **principles, rights, and guarantees of users**, the GDPR maintains those already provided for in previous legislation, but introduces new concepts such as the principle of **data minimization, pseudonymization** by default for working with third-party data, the right to **be forgotten** or the right to **portability**, and specifically regulates the processing of data for scientific research purposes.

Pseudonymization becomes the technical measure to be established, by default, as a security measure for processing and is defined by the GDPR itself as the processing of personal data in such a way that they can no longer be attributed to a data subject without the use of additional information, provided that such additional information is separately identified and subject to technical and organizational measures designed to ensure that the personal data are not attributed to an identified or identifiable natural person, which could be described as "secure encryption."

In the development of data processing activities in the field of **scientific research**, the GDPR establishes particularities such as that the further processing of data for this purpose is considered compatible with the original purpose, a longer retention period is allowed than that established for other types of activities, or that the principle of data minimization must be guaranteed, for which it will be necessary to establish appropriate technical and organizational measures.

Biomedical research as scientific research carried out in the field of health requires the use of health data, genetic data, or information on life or sexual orientation, among others. These data are included in what Article 9 of the RPGD calls **special categories**, for which, as a general rule, their processing is prohibited, unless one of the **exceptions indicated in** section 2 of the aforementioned article applies. Among others, the processing operations that are established as lawful are those based on the explicit consent of the data subject (unless a law also prohibits it), on grounds of public interest in the field of public health, for the purpose of ensuring high standards of quality and safety of healthcare and of medicinal products or medical devices, or which are necessary for scientific research purposes. In the latter case, the processing must be proportionate to the objective pursued, essentially respect the right to data protection and establish adequate and specific measures to protect the interests and fundamental rights of the data subject, highlighting the obligation to work with pseudonymized data (encrypted) or in a way that no longer allows the identification of the data subjects (anonymized) whenever possible.

The current regulatory framework to be taken into consideration when carrying out biomedical research activities continues to be that of the sectoral regulations, those that specifically regulate the different research activities that can be carried out in the field of biomedicine and which we have described in detail in the previous sections. But, on the other hand, we must also comply with the regulations of the RGPD and the new provisions established by the OLPPGDR in its 17th Additional Provision, which specifically regulates the processing of data in health research.

Additional Provision 17th establishes a series of criteria that must govern the processing of data in health research and, among others, it is important to highlight the following:

— The possibility of granting broad consents for the use of data in research in general areas of medical speciality or research is established.
— In the field of public health surveillance, health authorities and public institutions with competences in this field may carry out scientific studies without consent in situations of exceptional public health relevance and gravity.
— It is considered lawful and compatible to reuse data in biomedical research on the basis of a consent previously granted for another initial study, which fits in with the purpose or research area of the new study, it being obligatory in this case to publish information on the processing on the corporate website of the center and to obtain a prior favorable report from the Research Ethics Committee (REC).
— The use of pseudonymized data in biomedical research without consent is considered lawful provided that the following requirements are met:
    ○ technical and functional separation between the pseudonymizer and the research group,
    ○ the research group signs an express commitment to confidentiality and nonre-identification,
    ○ the responsible entity puts in place security measures to prevent reidentification and possible unauthorized access
    ○ there is a prior favorable report from the IRB or the DPD of the responsible entity.
— It is established that the processing of health data for the purposes of biomedical research or public health requires
    ○ develop impact assessments that identify the risks arising from the processing,
    ○ comply with international quality and good clinical practice guidelines
    ○ adopt pseudonymization measures to prevent access to data subjects' identification data.

This new regulatory framework requires that the entities responsible for the processing of data carried out in the field of biomedical research adapt their operations and establish internal operating procedures that guarantee and allow evidence of compliance with the regulatory requirements that apply to the processing of personal data, especially with regard to the processing of health data. Furthermore, the implementation of all these operating guidelines should be accompanied by specific information and training activities in the field of data protection, aimed at the professionals who form part of the different entities responsible for research in the field of health.[12]

New technologies and research with health data

Today's society is experiencing a continuous and vertiginous advance in the field of new technologies in all areas of life. Technological developments are so rapid that what is considered new and innovative today can become obsolete in a few months. At the same time, alongside this technological explosion, what has been called the "data economy" is appearing, a new system in which people's information is quantified and priced, with prices varying according to the type of data that can be provided. This new situation requires us to bear in mind and apply the principles of prudence and caution in the introduction of new technologies in the field of data management, especially in everything related to health data. It is necessary to take into consideration that, in the 21st century, discrimination, social isolation, inequality, intolerance, etc., are realities that exist and are often induced by aspects related to the health of those who suffer from them. For this reason, when developing activities with personal data on a large scale, it must be a priority to guarantee the confidentiality and security of the people who hold the information.

As highlighted in the various reports published by the Observatory of Trends of the Roche Institute Foundation,[13] the healthcare sector uses a large amount of data from very heterogeneous sources to obtain information that can be of great relevance to researchers, healthcare professionals, and patients. The fact that data and the information derived from it are fundamental in the field of health has long been recognized.

Advances in the omics sciences[14] have made it possible to move away from generalized approaches to treatment or patient monitoring toward more personalized or precision medicine. Through the different technologies used by the aforementioned sciences, it is possible to analyze the large number of molecules present in a single sample, generating large amounts of highly relevant information in a very short time. This characteristic has meant an important change in the classical scientific method, as it has made it possible to first obtain the data on which theories based on the evidence generated can then be built. From a clinical point of view, these technologies have made it possible to learn about new physiological and pathological aspects from many different angles, and new tools have been developed for the diagnosis and monitoring of diseases based on genetic analysis, the design of programs for the prevention and early detection of pathologies based on blood biomarkers, personalized monitoring of treatment, etc. However, none of them alone can capture the biological complexity derived from the relationship that exists between the different molecular levels, in individual cells, tissues, and the organism as a whole. It is foreseeable that, in the future, the combination of knowledge derived from the different omic sciences will allow us to achieve a detailed vision of the individual from the molecular point of view; the omic sciences will allow us to approach an increasingly personalized medicine based on the individual characteristics of each patient.

The enormous accumulation of data in the field of medicine and biomedical research, with all the advances that have been made in obtaining molecular information, biomarkers, etc., has led to the need for technological tools that are sufficiently developed to allow an orderly storage and subsequent detailed analysis in an automated manner, given that it is not possible to deal with them in a traditional way. To this end, important advances and innovations have been developed in the field of big data and artificial intelligence (**AI**), which enable large-scale analyses of information to be carried out to extract data or conclusions that are not easy to obtain or deduce or would not even be feasible or manageable for human beings.

The AI-based tools being developed for application in medicine have been defined as systems for aiding diagnosis, monitoring patient progress, and choosing treatments, although other applications are also being developed for use in the field of biomedical research, health system management, and even developed for use by patients in areas such as self-care.[15]

The introduction in the field of health and biomedical research of all these new technologies that allow the storage and massive analysis of large volumes of data requires special protection of the privacy of the individuals who hold the information and compliance with all the demands and requirements established by the regulations currently in force on personal data protection, detailed in previous sections.

Specifically, the use of these tools may give rise to situations that significantly compromise individual rights and freedoms. Associated with them is the ability to infer information that has not been directly and voluntarily provided by the data subject/subject of the research, but which ultimately concerns him/her. This could affect their right to be informed and consequently their right to self-determination. Likewise, the use of the aforementioned technologies may lead to the reidentification of the data subject, in the event that the data have been anonymized or pseudoanonymized (art. 3 BRL) for processing, which may give rise to situations that jeopardize their right to privacy, including discriminatory situations proscribed by law.

Another risk inherent in the use of new technologies is the tendency toward automated decision-making and the danger of individuals being classified "into groups or subgroups according to their personal profiles."[16] These latter issues have been specifically addressed in the GDPR which identifies the risk of profiling, and specifically prohibits that decisions based solely on automated processing of personal data may have detrimental effects on the data subject (Art. 22). In this regard, the provisions of the GDPR, specifically provide for a so-called right to obtain human intervention, closely linked to the right not to be subject to a decision based solely on automated processing, and to the right to information about the making of such decisions and the logic that such processing entails.

Thus, in order for technological advances to reach their full potential in the field of biomedical research, researchers must bear in mind that it is essential and especially necessary to establish appropriate technical and organizational measures to guarantee the security and proper use of personal information. Preserving the confidentiality and privacy of the persons concerned and avoiding any kind of stigmatization is both an ethical and a legal requirement.

References

1. Romeo Casabona CM. *El Médico y El Derecho Penal. Tomo II Problemas Penales de La Biomedicina*. Rubinzal-Culzoni; 2011.
2. Romeo Casabona CM. Bioderecho y Bioética. In: *Enciclopedia de Bioderecho y Bioética*; 2019. Comares https://enciclopedia-bioderecho.com/voces/33.
3. World Medical Association. Declaration of Helsinki: ethical principles for medical research involving human subjects. *JAMA* 2000;**284**(23):3043–5. https://www.ncbi.nlm.nih.gov/pubmed/11122593.
4. Ley 14/2007. *de Investigación biomédica de 4 de julio, Boletín Oficial del Estado (num 159)*; 2007. Madrid http://www.boe.es/boe/dias/2007/07/04/pdfs/A28826-28848.pdf.
5. Real Decreto 1090/2015. *de 4 de diciembre, Boletín Oficial del Estado de 24 de diciembre de 2015*. sección I núm 307; 2015. p. 121923–64. Madrid https://www.boe.es/eli/es/rd/2015/12/04/1090/dof/spa/pdf.
6. *Regulation (EU) No 536/2014 of the European Parliament and of the Council of 16 April 2014 on clinical trials on medicinal products for human use, and repealing Directive 2001/20/EC Text with EEA relevance*. https://eur-lex.europa.eu/legal-content/EN/TXT/?uri=CELEX%3A32014R0536&qid=1615806194122.
7. Real Decreto 1716/2011. *de 18 de noviembre, Boletín Oficial del Estado de 2 de diciembre de 2011*. sección I núm 290; 2011. p. 128434. Madrid http://www.boe.es/boe/dias/2011/12/02/pdfs/BOE-A-2011-18919.pdf.
8. Arias-Diaz J, Martín-Arribas MC, García del Pozo J, Alonso C. Spanish regulatory approach for biobanking. *Eur J Hum Genet* 2013;**21**(7):708–12. https://doi.org/10.1038/ejhg.2012.249.
9. Real Decreto 957/2020. *de 3 de noviembre, Boletín Oficial del Estado de 26 de noviembre de 2020*. núm. 310; 2020. p. 104907. Madrid https://www.boe.es/boe/dias/2008/01/19/pdfs/A04103-04136.pdf.
10. *Regulation (EU) 2016/679 of the European Parliament and of the Council of 27 April 2016 on the protection of natural persons with regard to the processing of personal data and on the free movement of such data, and repealing Directive 95/46/EC (General Data Protection Regulation)*. https://eur-lex.europa.eu/legal-content/EN/ALL/?uri=CELEX%3A32016R0679.
11. Ley Orgánica 3/2018. *de Protección de Datos Personales y garantía de los derechos digitales, de 5 de diciembre, Boletín Oficial del Estado*. núm. 294; 2018. p. 119788. Madrid https://www.boe.es/buscar/pdf/2018/BOE-A-2018-16673-consolidado.pdf.
12. Purriños NC. El nuevo marco normativo de la protección de datos de carácter personal y la investigación biomédica. *Proyecto Lumbre* 2019;**18**:6–11. https://dialnet.unirioja.es/descarga/articulo/7012245.pdf.
13. Fundación Instituto Roche. *Informes Anticipando 1. Los Datos en la Era de la Medicina Personalizada de Precisión. Observatorio de Tendencias*. Fundación Instituto Roche; 2019. Published online https://www.institutoroche.es/observatorio/losdatos.
14. Fundación Instituto Roche. *Informes Anticipando 2. Ciencias Ómicas. Observatorio de Tendencias*. Fundación Instituto Roche; 2019. Published online https://www.institutoroche.es/observatorio/cienciasomicas.
15. Fundación Instituto Roche. *Informes Anticipando 3. Inteligencia Artificial en Salud: Retos Éticos y Legales. Observatorio de Tendencias*. Fundación Instituto Roche; 2020. Published online https://www.institutoroche.es/observatorio/retoseticosylegales.
16. Fundación Instituto Roche. *Retos éticos y necesidades normativas en la actividad asistencial en Medicina Personalizada de Precisión*. Fundación Instituto Roche; 2018. Published online https://www.institutoroche.es/recursos/publicaciones/180/retos_eticos_y_necesidades_normativas_en_la_actividad_asistencial_en_medicina_personalizada_de_precision.

Index

Note: Page numbers followed by *f* indicate figures and *t* indicate tables.

A

Abdominal compartment syndrome (ACS), 476–478
Abdominal pain, 485
Abdominoperineal resection, 593
Aberrant DNA methylation, 218–219
Acetylation, 27
Actinic proctitis, 115, 116*f*, 593
Acute and transient neuropathy, 486
Acute postoperative pain (APP), 479–484
  anesthetic technique, factors dependent on, 479
  inadequate analgesia, causes of, 481
  management strategies, 480
  patient-dependent factors, 479
  surgery-dependent factors, 479
Adenocarcinomas, 153
Adenoma-carcinoma pathway, 551
Adenoma detection rate (ADR), 165
Adenomas, 284
Adenomatous lesions, 162
Adenomatous polyposis syndromes, 195, 288
  Gardner syndrome, 199
  Turcot syndrome, 199
Adenomatous polyps, 61
Adenopathies, 144
Adjuvant chemotherapy, CRC, 381–386
  duration of, 385
  elderly patients, 386
  resected stage IV, 386
  stage II, 383
  stage III, 383–384
Adjuvant radiochemotherapy, rectal cancer
  vs. neoadjuvant radiochemotherapy, 406–407
  vs. neoadjuvant radiotherapy, 405
Advisory committee, 270–271, 270*t*
Alarm symptoms, 43
Albumin, 465
Alcohol consumption, 498
Allergic dermatitis, 604
Alternative Healthy Eating Index (AHEI), 35–36
American Cancer Society (ACS), 17
Amsterdam criteria, 187, 188*t*, 555
Anal canal, radiofrequencies to, 597
Analgesics, 461
Anal neosphincter, 597
Anal obturators, 597
Anal sphincter affectation, 594
Anatomical Pathology Service (APS), 567

Anatomopathological report, 220
Anesthetic techniques, 463
Angiotensin II inhibitors, 35
"Annotate" data, 233
Annular lesion, 100, 101*f*
Anorectal manometry, 594
Anterior peritoneal reflection, 133–134
Anticancer agents, bacteria as, 502
Anticonvulsants, 487
Antidepressants, 487
Anti-EGFR therapy, 559
Antigen processing machinery (APM), 532–533
Antiinflammatory diets, 35–36
Antioxidants, 34
Antispasmodics, 95–96
Antitumor drugs, 393–394
APP. *See* Acute postoperative pain (APP)
Arf6, 548
Argon plasma coagulation *(APC)* gene, 158, 216–217, 519, 526, 543, 552
Arm circumference, 589
Artificial intelligence (AI), 47, 626
ArtiSential, 308
Ascending colostomy, 601
Associated information, of biobanks, 570
Associating liver partition and portal vein ligation for staged hepatectomy (ALPPS), 336–337
Attenuated FAP (AFAP), 196
  diagnostics, 196
  follow-up of healthy individuals, 198
  treatment of, 199

B

Bacillus Calmette-Guerin immunotherapy, 536
Bacteria
  for colorectal cancer treatment, 500–503
  metastases-associated bacteria, 502
  targeting intestinal microbiota, 503–504
  targeting tumor microbiota, 504
  therapeutic strategies, 502
Bacteriobots, 503
Bacteroides fragilis, 501
Bannayan–Ruvalcaba–Riley syndrome, 201
Baseline fluid and electrolyte needs, 462
Baseline pain, 485
  treatment, 486–487
Benzodiazepines, 460

Beta2-microglobulin (B2m), 532–533
  role of, 533–536, 534*f*
  gene mutations in CRCs with MSI-H, 535–536
  immunoselection model in MSI-H carcinogenesis, 534–535, 535*f*
Bethesda criteria, 187, 188*t*
Bevacizumab, 367, 367*t*
B-group vitamins, 34
Bilophila wadsworthia, 498
Bio-Banking and BioMolecular Resources Research Infrastructure (BBMRI), 566
Biobanks, 622
  advantages, 570
  biological samples, collecting and preparing of, 569–570, 570*f*
  cancer research, 568
  diagnostic anatomical pathology files, 567–568
  integration, 566–568
  organization and operation, 566–567
  orientation, 567
Bioethics, 619
Biofeedback, 596
Bioimpedanciometry (BIA), 588
Bioinformatics tools, 231
  complex networks, 239
  databases, 233
    for oncogenomic research, 234–239
    ontological, 233–234
    for risk assessment, 240–242
Biological agents, rectal cancer, 413
Biological mechanisms, colorectal cancer (CRC)
  genetic mutations, 520*f*
  signaling pathways, 521–526
    epidermal growth factor receptor (EGFR), 523, 524*f*
    mitogen-activated protein kinase (MAPK) pathway, 524–525
    PI3K-AKT-mTOR pathway, 525–526
    smads, 522–523
    TGF-β signaling pathways, 521–522, 522*f*
Biological samples, 565–570, 620–622
Biomarkers, 49
  blood and stool, 211–215
  classical, 210
  colorectal cancer (CRC), 154 (*see also* Omics-based biomarkers, for CRC)
  concept of, 208

629

Biomarkers *(Continued)*
  determination of, 210–211
  nonclassical, 210
  pan-cancer, 209
Biomaterial injection, 597
Biomedical ontologies, 234
Biomedical research, 565–566
  regulations governing, 620–624
Biomedical Research Law (LIBM), 565–567
Bisphosphonates, 35, 488
Bladder cancer, 533
"Bleak nihilism,", 351
Blue Rubber Bleb Nevus Syndrome (BRBNS), 115
Bowel obstruction, 303
Bowel wall deformity
  aciform, 101, 102f
  "appel core,", 101, 102f
  trapezoid, 101, 102f
BRAF, 208, 215–216
  mutation analysis, 215–216, 525, 556, 559
Breakthrough pain treatment, 488
Breast cancer, 533
Breast Cancer Staging Ontology, 234
Bridge treatment, 295
Brief index of sexual functioning (BISF-W), 612
Brief Nutritional Assessment Questionnaire (SNAQ), 587
Brief sexual function inventory (BSFI), 612
Brief sexual function questionnaire (BSFQ), 612
Brooke ileostomy, 600
Budding tumor (BT), 220
Buprenorphine, 487

C

Cachexia, 585
Calcium supplementation, 33
Cancer-associated fibroblasts (CAFs), 222
Cancer Care Strategy, 269
Cancer-educated platelets (CEPs), 212–213
Cancer Genome Project, 235
Cancer immunoedition, 529–531, 530f
Cancer stem cells (CMCs), 435
Candidate colorectal cancer, 514–515
Candidate gene studies, 515–516
Candida tropicalis, 505
Cannabinoids, 488
Capecitabine, 411–412
Carcinoembryonic antigen (CEA), 49, 119, 122–123
Carcinogenesis, molecular pathways of, 42
Carpet lesion, 98
Case-control association, 513
Casein kinase 1 (CK1), 548
Catalan RDP program, 71–73
Catheter removal, 482
Causal variant identification strategies, 516–517
cBioPortal, 239
CDH1 gene
  epigenetic silencing, 546
  mutations in, 545–546
CEA. *See* Carcinoembryonic antigen (CEA)
Cecostomy, 601

Celiac plexus ablation, 489
Cell Lines Project, 235
Cetuximab, 207–208, 215
Changes in sexual functioning questionnaire (CSFQ), 611
Check-cap, 86
Checkpoint antigens, 210
Checkpoint inhibitors (CPI), 373, 373t
Chemoprevention
  of colorectal cancer (CRC), 34
  familial adenomatous polyposis (FAP), 199
  Lynch syndrome, 191
Chemotherapy, 558, 578
Chemotherapy-induced neuropathy (CIN), 485–486
Choline, 127
Chondroitin sulfate, 35
Chromatin immunoprecipitation tags (ChIP), 516–517
Chromatin structure remodeling, 27
Chromoendoscopy, 160–161, 163, 171–172
  conventional, 171–172
  indigocarmin, 171–172
  virtual, 172
Chromogenic in situ hybridization (CISH), 211
Chromosomal instability (CIN), 42, 552
Chronic diverticulitis, 103–104, 104f
Chronic inflammation, 160–161, 435
Chronic lymphocytic leukemia (CLL), 568
Chronic neuropathy, 486
Chronic pain, in colorectal cancer (CRC), 484–490
  abdominal pain, 485
  analgesic coadjuvants, 487–488
  baseline pain treatment, 486–487
  breakthrough pain, treatment of, 488
  chemotherapy-induced neuropathy (CIN), 485–486
  intrathecal therapy, 488
  lumbosacral plexopathy, 485
  neuroablative techniques, 489–490
  pelvic pain, 485
  perineal pain, 485
  physiotherapy, 490
  radiotherapy, 486, 490
  trigger point deactivation, 490
Circulating tumor cells (CTCs), 212, 250, 252
Circumferential mesorectum, 131–132
Circumferential resection margin (CRM), 91–92
Classic FAP
  clinical symptoms, 195
  diagnostics, 196
  follow-up of healthy individuals, 197–198
  genetics, 195
  treatment of, 198
Clathrin mediated E-cadherin endocytosis, 548
Cleveland Clinic's colon cancer risk assessment, 241
Clinical oncology committees, 269–271
Clinical Research Ethics Committees (CREC), 620
Clopidogrel, 158
Clostridium difficile, 499, 503
Clostridium septicum, 501

C-MET, 217
CNN4Polyps, 242–244, 243f
Coffee, 31
CO_2 insufflation, 95–96
Cold snare polypectomy, 275, 277f
Colloids, 464–465
Colon cancer, 303
  complication treatment, 304
  diagnostic tests, 303–304
  emergency department, 303
  factors, 303
  immunotherapy, 221–222
  tumor microenvironment, 222–223
  type of surgery, 304–305
  urgent surgery, 304
Colon cancer surgery
  contraindications, for laparoscopic surgery, 320
  laparoscopic approach to, 320
  nonmetastatic colon cancer, surgical treatment of
    general surgical technique, 317–318
  palliative procedures, 319–320
  treatment, surgical emergency, 318–319
    surgical options, for urgent CRC, 319
  tumors, 318
    surgical technique, specific locations and special situations, 318
Colon capsule endoscopy (CCE)
  vs. colonography, computerized axial tomography, 86
  vs. colonoscopy, 84–85
  current indications and contraindications, 86
  intestinal cleansing, 83–84
  perspectives, 86
  Pillcam COLON 2, 83
  technical characteristics, of second-generation colon capsule, 83, 85t
Colonic distention, 95–96, 96f
Colonic polyps, 32
Colonic prostheses
  contraindications, 296
  endoscopic technique, 296–297
  indications, 295
Colonography Reporting and Data System (C-RADS), 104, 105t
Colonoscopy, 61, 157–160, 283, 594
  in advanced colonic lesions, 177
  ancillary equipment and, 172
  classic FAP, 197
  clockwise torque maneuver and withdrawal of, 159, 159f
  colorectal polyps, classification of, 161–162
  in CRC diagnosis
    high-risk populations, 164
    medium-risk populations, 163–164
    serrated pathway, 162–163
  endoscopic description, 158–160
  high-definition, 171
  magnification techniques, 160–161
  MUTYH-associated polyposis (MAP), 198
  primary health care (PHC)
    performing technique, 61
    polyp detection, 61–62

quality indicators, 164–165
quality of, 171
ColonoTAC, 62
Colon tumors, 17
Colorectal Adenoma/Carcinoma Prevention Program (CAPP2), 191
Colorectal cancer (CRC), 157, 533
  adenocarcinomas, 153
  biobank (*see* Biobanks)
  biomarkers in, 154
  chronic pain in, 484–490
    abdominal pain, 485
    analgesic coadjuvants, 487–488
    baseline pain treatment, 486–487
    breakthrough pain, treatment of, 488
    chemotherapy-induced neuropathy (CIN), 485–486
    intrathecal therapy, 488
    lumbosacral plexopathy, 485
    neuroablative techniques, 489–490
    pelvic pain, 485
    perineal pain, 485
    physiotherapy, 490
    radiotherapy, 486, 490
    trigger point deactivation, 490
  classification, 551
  diagnosis, 551
  diagnostic delay, 20–21
  dysbiosis, 436–437
  emerging biomarkers, 559
  EMT (*see* Epithelial-mesenchymal transition (EMT))
  epidemiological analysis of, 15
  follow-up strategies, 18, 19*t*
  genetic biomarkers, 252
  genetic model, adaptation of, 543, 544*f*
  histopathological diagnosis, 149–151
  histopathological types, 153
  human gut microbiota and, 499–508
    bacteria (*see* Bacteria)
    fungi, 504–506
    plasmids, 508
    virus, 506–507
  immunity, 435–436
  immunotherapeutic approaches test, 444–446
  immunotherapy, 443–444
  incidence, 3, 4*t*, 5*f*
  inflammation, 435–436
  inflammatory bowel disease, 436–437
  intestinal microbiota, 435–436
  liquid biopsy, 559
  long-term quality of life, 19–20
  lymphomas, 153
  marker, mycobiota as, 506
  melanomas, 154
  mesenchymal tumors, 153–154
  metastatic tumors, 154
  molecular pathways, 551, 552*f*
  mortality, 3–15, 9–10*t*, 11*f*, 14*f*
  mycobiota dysbiosis in, 505–506
  neuroendocrine neoplasms, 153
  NSAIDs, 440
  oligometastatic disease, 421
  oral administration
    colonic drug delivery, 398–400
    systemic treatment, 393–398
    treatment of, 391–392
  pathological diagnosis, 151–153
  patients, survival and prognosis of, 17–18
  resection, surgical specimens of, 151–153
  surgery, 481–484
    analgesic drugs, 483–484
    catheter removal, 482
    epidural analgesia complications (*see* Epidural analgesia, complications of)
    epidural opioids, 482
    intravenous analgesia, 483
    safety measures, 482
    thoracic epidural analgesia, 481
  therapeutic delay, 20–21
Colorectal Cancer RISk Prediction tool (CRISP), 241
Colorectal cancer type X, 191
Colorectal carcinoma (CRC), 207. *See also* Colon cancer
  molecular staging, 220
Colorectal Risk Assessment Tool (CCRAT), 241
Colorectal tumors
  endoscopic polypectomy of
    clinical practice, 275–278
    endoscopic mucosal resection (EMR), 278–280
    endoscopic submucosal dissection (ESD), 280
    Paris classification, 276*t*
    preneoplastic lesions, 275–278
Colostomies, 601
Common Disease-Common Variant (CDCV) hypothesis, 515
Comprehensive action model, 270–271, 270*t*
Computed tomography (CT), 119–120, 123, 303–304
Computer-aided diagnosis (CAD), 97
Conductive mutations, 519
Condyloma acuminatum, 117
Confounding by indication bias, 21
Consensus molecular subtypes (CMS), 43*t*, 374–375
Continent ileostomy, 600
Cooperative networks, biobanks, 570
Coping strategies, 578–579, 581
Corticosteroids, 487
COSMIC tool, 235–236, 235*f*
COVID-19, psychological aspects of, 582
Cowden syndrome, 65, 196*t*, 200–201
CpG island methylator phenotype (CIMP), 42, 553
C-reactive protein (CRP), 438
Crizotinib, 208
Crystalloids, 464
CT. *See* Computed tomography (CT)
CT-colonography (CTC), 95, 109
  clinical findings
    annular lesion, 100, 101*f*
    chronic diverticulitis, 103–104, 104*f*
    C-RADS classification, 104, 105*t*
    endoluminal mass, 100, 100*f*
    polypoid or fungoid lesion, 100, 101*f*
    polyps, 98–100, 99*f*
    submucosal lesions, 102–103, 103*f*
  colonic distension, 109
  colonic distension, 95–96, 96*f*
  colonic preparation, 95, 109, 111*t*
  contraindications, 95, 96*t*, 109, 110*t*
  image acquisition, 96–97
  indications for, 98
  indications of, 109
  in inflammatory bowel diseases, 104
  postprocessing tools, 97–98, 97*f*
  rectal pathology
    artifacts, 116–117
    condyloma acuminatum, 117
    diverticula, 117
    hypertrophied anal papilla, 115, 116*f*
    inflammatory conditions, 115
    malignant neoplasms, 112, 113*f*
    polyps, 111, 111*t*, 112*f*
    postoperative changes, 115, 117*f*
    pseudolesions, 116–117
    submucosal lesions, 112–114, 113*t*, 114*f*
    vascular lesions, 114–115
    villous tumors, 111, 112*f*
  software and tools, 97
ctDNA, 213–215
Cuestionario FRAIL, 588
Cyclodextrins (CDs), 395
Cyclooxygenase enzymes, 440, 444
Cytoreductive surgery (CRS), 375

D

Dairy products, 30
Databases
  oncogenomic research, 234–239
  ontological, 233
Data economy, 626
Data Protection Officer (DPO), 624
Da Vinci Surgical System, 310–311
Deep learning (DL), 242–245, 242*f*, 245*f*
Defecography, 595
Deficient MMR (dMMR), 217
Denaturing gradient gel electrophoresis (DGGE), 185
Denaturing high performance liquid chromatography (DHPLC), 185
Dendritic cells (DCs), 435
Derogatis interview for sexual functioning (DISF/DISF-SR), 611
Descending colostomy, 601
Desmoid tumors, 198
  surgical excision, 199
Dextran, 465
Diagnostic anatomical pathology files, 567–568
Diet, 29, 585
  and human gut microbiota, 497–498
  recommendations, 589–590
Dietary fiber, 30
Dietary index, 35–36
Dietary inflammatory index (DII), 29, 35–36
Dietary patterns and colorectal cancer, 35
Differential in gel electrophoresis (DIGE), 257

Digital droplet PCR (ddPCR), 252
Dihydropyrimidine dehydrogenase (DPD), 368
Disease-free interval (DFI), 181, 349–350
Disease-free survival (DFS). See Disease-free interval (DFI)
Distant metastasis, 153
Distress, 576–578, 581
Ditrizoate, 95
Diverticulosis, 103–104, 104f
DNA-based microarrays, 252
DNA methylation, 27, 254–256, 254f
　biomarkers, 255–256
Docker Hub, 244
Doctor-patient communication, 575–576, 582
"Do not touch" technique, 317
Double-barrel colostomy, 601
"Down-to-up" approach, 326
Droplet digital PCR (ddPCR), 211
Dynamometry, 588
Dysbiosis, 219
　colorectal cancer, 436–437
　oral microbiota, 437–438

E

Early onset of colorectal cancer
　early ages, 41–44
　environmental factors, 41
　hereditary factors, 42
　microbiota alteration, 42
　molecular pathways, 42–43, 43t
　overview, 44
E-cadherin
　posttranslational regulators, 547–549
　proteolytic cleavage, 549
　transcriptional repressors, 546
　tumor progression, 545–549
Echoendoscopy, 89, 158
Edema, 601, 602f
Efflux, 394
Elimination phase, 530–531
Emergency surgery, 468
Emotional aspects. See Psychological aspects, in CRC
Empirical antibiotic treatment, 474
Empirical antifungal treatment, 476
EMR. See Endoscopic mucosal resection (EMR)
Encorafenib, 208
End colostomy, 601
Endoanal resection, 593
Endoanal ultrasonography, 594
EndoCuff, 172, 173f
End-of-dose pain, 485
Endoluminal mass, 100, 100f
Endorectal ultrasound, 131
EndoRing, 172, 173f
Endoscopic biopsy/polypectomy, 149
Endoscopic full thickness resection (EFTR), 312
Endoscopic iconography, 165–167
Endoscopic mucosal resection (EMR), 312
　for colorectal tumors, 278–280
　　materials, 279
　　variants, 279–280

Endoscopic polypectomy, 318
　of colorectal tumors
　　clinical practice, 275–278
　　endoscopic mucosal resection (EMR), 278–280
　　endoscopic submucosal dissection (ESD), 280
　　other techniques, 280
　　Paris classification, 276t
　　preneoplastic lesions, 275–278
　familial adenomatous polyposis (FAP), 199
　Juvenile polyposis, 200
　Peutz-Jeghers syndrome (PJS), 200
Endoscopic retrograde cholangiopancreatography (ERCP), 93
Endoscopic submucosal dissection (ESD), 280, 312
Endoscopic ultrasound (EUS), in CRC
　circumferential resection margin (CRM), 91–92
　EUS-FNA initial diagnosis and recurrence
　　differential diagnosis, 92
　　neoplastic affectation of adenopathies, 92
　　tumor recurrence diagnosis, 92–93
　extrahepatic bile duct drainage, 93
　interventional EUS, 92–93
　intestinal wall, degree of invasion, 89–91
　local adenopathies, affectation of, 91
　neoadjuvant therapy response, 92
　postsurgical fluid collections, drainage of, 93
　for rectal cancer staging, 89–92
Enhanced permeability and retention (EPR), 396
Enhanced recovery after surgery (ERAS), 457
Enterococcus faecalis, 501
Enterocutaneous fistula, 472
Enzymatic hydrolysis, 399
Epidemiology and biobanks, 568
Epidermal growth factor receptor (EGFR), 357–358, 391, 523, 524f
Epidural analgesia, complications of
　infection, 482
　motor block, 482
　nausea and vomiting, 482
　pain, 482
　pruritus, 482
　respiratory depression, 482
　urinary retention, 482
Epidural opioids, 482
Epidural techniques, 461
Epigenetics, 27
Epigenomics, 254–256
Epi ProColon, 219
Epithelial-mesenchymal transition (EMT), 544–547, 549
　posttranscriptional regulators, 547
Epithelial plasticity, 543–545
Epstein-Barr virus (EBV), 435
Equilibrium phase, 530–531
Escape phase, 530–531
Escherichia coli, 474, 501
ESD. See Endoscopic submucosal dissection (ESD)
Ethical principles, 619–620

Ethics vs. law, 619
Eubiosis, 219
EUROCARE-5 data, 17
European Center for Ontological Research (ECOR), 234
European Prospective Survey on Nutrition and Cancer (EPIC), 30
European Society for Clinical Nutrition and Metabolism (ESPEN), 586, 589
European Society for Gastrointestinal Endoscopy (ESGE), 275, 276f
European Society of Gastrointestinal and Abdominal Radiology (ESGAR), 95–96, 100
European Working Group on Sarcopenia in Older People (EWGSOP), 586
Europe, CRC in
　incidence rates, 3, 5–6t, 7f
　mortality rate, 6, 11–12t
Excessive alcohol consumption, 31
Excisional biopsy (polypectomy), 149–151
Exosomes, 212, 250
Expanded sexual arousability inventory (SAI-E), 612
Expected surgical losses, in surgical patient, 464
Experimental therapies, 597
Expression microarray technique, 569
Extramesorectal adenopathies, 144
Extramural tumor extension, 136, 136f, 144

F

Familial adenomatous polyposis (FAP), 42, 164, 196t, 513, 543
　chemoprevention, 199
　clinical symptoms, 195–196
　diagnostics, 196–197
　genetics, 195
　monitoring of healthy individuals, 197–198
　treatment of, 198–199
Familial colorectal cancer type X (FCCTX), 556
Family screenings, 289, 291
FAP. See Familial adenomatous polyposis (FAP)
Fastai, 244, 244f
Fast-track surgery, 271, 457, 481
18FDG PET/CT, 119, 121
　indications, 120
　recurrent disease detection, 122–123
　treatment response, assessment of, 124, 125f
Fecal immunochemical test (FIT), 43
Fecal incontinence
　clinical history, 594
　definition, 593
　dietary regime, 595
　physical examination, 594
　radiotherapy damage, 593–594
　rehabilitation, 595–597
　robotic surgery, 597
　specific medications, 596
　surgery type, 593
　treatment, 595–597
Fecal microbiota transplantation, 503
Fecal occult blood (FOB), 86
Fecal occult blood test (FOBT), 46, 48–49, 52, 576

Fecalomas, 117
Female sexual function (FSF), 612
Female sexual function index (FSFI), 612
Fentanyl, 487
Fentanyl citrate, 488
FICE, 160
Fine-mapping strategies, 516–517
Fish, 30
FIT, 48–49
5-A model, 614
Flat adenoma, 166f
Flat lesions, 162
FlexDex surgical instrument, 308
Flexible Single Incision Surgery (FSIS), 311
Flex Robotic System Technology, 311
Flow cytometry (FC), 211
Fluids
 management schemes, 465–466
 transport into interstitium, 463
 types of, 464–465
Fluorescence imaging, with indocyanine green (ICG), 309–310
Fluorescent in situ hybridization (FISH), 211
Fluoropyrimidine (FPO), 391
5-Fluorouracil (5-FU), 357–358, 381, 383–384, 391, 393
Folate, 34
FOLFOX scheme, 383–384
Folic acid, 34
Follow-up endoscopy, 287, 287f
Formalin-fixed paraffin-embedded tissue (FFPE), 568–569
Frailty, 586
Frailty Trait Scale (FTS), 588
Frame Shift Peptides (FSPs), 533–534
FreeHand surgical robotic, 311
Fungal dermatitis, 604
Fungi, 504–505
 as CRC marker, 506
 mycobiota dysbiosis, 505–506
Fungoid lesion, 100, 101f
5-FU plus oxaliplatin (FOLFOX), 357–358
Fusobacterium nucleatum, 219, 437, 501–502
Future Liver Remnant (FLR), 181

G

Galaxy, 241
Galaxy Community, 241–242
Galaxy Interactive Environments, 241
Galaxy ToolShed, 241
Galician Health Service (SERGAS), 71
Gamma delta T cells, 441
Ganglion impar block, 490
Gardner syndrome, 164, 199
Garlic, 31
5G-assisted telementored surgery, 310
Gastroduodenoscopy
 classic FAP, 196
 MUTYH-associated polyposis (MAP), 198
 Peutz-Jeghers syndrome (PJS), 200
Gastrointestinal ostomies, 599–601
 complications, 601–605
 consultation at discharge, 607–608

 nursing care, 599, 607
 perioperative care, 599
 postoperative follow-up, 607
 preoperative consultation, 605–607
 skin complications, 604–605
Gefitinib, 207–208
Gelatins, 465
Gene Ontology (GO), 234
General Data Protection Regulation (GDPR), 623–625
Generalized peritonitis, 472
Genetic heterogeneity, types of, 519–521, 520f
Genetics
 adenomatous polyposis syndromes, 195
 hereditary nonpolyposis colorectal cancer (HNPCC), 183–184
Genome-Wide Association Studies (GWAS), 515–516
Genomic Data Commons Data Portal, 236, 237f
Genomics, 250–252
 single cell, 234–235
Genotype-Tissue Expression Project (GTEx), 517
Genotyping, 232
GitHub, 245
Glandular crypts, 162, 162f
Glioma polyposis, 199
GLOBOCAN 2018 project, 3–6, 157
 in Europe, 6
 in Spain, 3, 10
Glucosamine, 35
Glucose transporters, 119
Glycocalyx, 463
Golombok-Rust inventory of sexual satisfaction (GRISS), 611
Google Glass, 310
Granulomas, 605
Graph theory, 239
Guaiac test, 60

H

Haggitt classification, 291–292
Hakai protein, 547–548
"Hallmarks of Cancer,", 435
Halogenated gases, 461
Hamartomatous polyposis syndrome, 200
Haplotype Map (HapMap) project, 515–516
Hartmann procedure, 319
HCC-Pred, 239, 240f
Health care team-patient relationship, 575–576
Health data, 619, 624–627
Health-related quality of life (HRQoL), 358
Health Services Interval (HSI), 20
Healthy Eating Index (HEI), 35–36
Healthy practices, 32
Helicobacter pylori, 501
Hematogenous metastases, 346
Hemorrhage, 601
Hepatectomy, 334–335, 334f
Hepatic flexure, 160
Hepatic metastasectomy, 181
HER-2. *See* Human epidermal growth factor receptor 2 (HER2) gene

HERACLES phase 2 trial, 359–360
Hereditary cancer
 chemoprevention, 67
 clinical suspicion of, 66–67
 family doctor approach role in, 67
 periodic medical surveillance, 67
 prophylactic surgery, 67
Hereditary colorectal cancer predisposition syndromes, 513, 514t
Hereditary colorectal cancer risk syndrome, 42
Hereditary hamartomatous polyposis syndromes, 201–202
Hereditary nonpolyposis colorectal cancer (HNPCC), 66, 164, 183. *See also* Lynch syndrome
 Amsterdam criteria, 187, 188t
 Bethesda criteria, 187, 188t
 characteristics of, 185, 186t
 chemoprevention, 191
 clinical controls and surveillance, 189–191
 clinical features, 185–187
 computer predictive models, 189
 diagnosis, 187–188
 follow-up
  of healthy individuals, 189–190
  in patients, 190
 general recommendations, 190
 genetics of, 183–184
 molecular identification, 184
  germline mutations, detection of, 185
  immunohistochemistry (IHC) techniques, 185
  microsatellite instability analysis, 184–185
  MLH1 methylation, 185
 molecular study, 188–189
 risk of, 185–186, 186t
 systemic treatment, 191
Hereditary polyposis CRC (HPCC), 195
Hereditary polyposis syndromes, 288
Heritability, of colorectal cancer (CRC), 513
 candidate, 514–515
 hereditary colorectal cancer predisposition syndromes, 513, 514t
 high penetrance rare variants, 513
 low-penetrance variants and association studies, 515–517
  candidate gene studies, 515–516
  causal variant identification strategies, 516–517
  fine-mapping strategies, 516–517
  pangenomic association studies, 516
 multiomic integration, 517
 new generation sequencing (NGS) susceptibility genes, 514–515
 polygenic risk scores, 517
Hernia, 603, 603f
High-grade dysplasia (HGD), 157
High microsatellite instability (MSI-H), 184, 392
High penetrance rare variants, 513
High performance liquid chromatography-ultraviolet (HPLC-UV), 255
High resolution anorectal manometry, 594
High-resolution consultations, 271

High-throughput genotyping (HTG), 232
Histone protein modification, 27
HLA class I molecules, 532–533
Hormone replacement therapy, 34
Hot avulsion, 279
Hot snare polypectomy, 277, 278f
Human Cell Atlas (HCA), 237, 238f
Human Colon Cancer Prediction, 239
Human epidermal growth factor receptor 2 (HER2) gene, 209, 216, 392, 559
Human Genome Project (HGP), 515
Human gut microbiota, 495–496
　advanced technologies, 498–499
　and colorectal cancer, 499–508
　　bacteria (see Bacteria)
　　fungi, 504–506
　　plasmids, 508
　　virus, 506–507
　diet and, 497–498
Human papillomavirus (HPV), 435
Hydroxyethyl starch, 464–465
Hydroxylation, 27
Hygiene, 581, 581t
Hyperglycemia, intraoperative, 461
Hypermethylation, 218
Hypermutated with microsatellite instability (MSI-H)
　B2m gene mutations in CRCs with, 535–536
　carcinogenesis, immunoselection model in, 534–535, 535f
Hyperplastic polyps (pH), 62, 283, 284f, 289
Hyperthermic intraperitoneal chemotherapy (HIPEC), 375
Hypertrophied anal papilla, 115, 116f
Hypomethylation, 218

I

IAH. See Intraabdominal hypertension (IAH)
Iinnate lymphoid cells (ILC), 441
IL-10, 442
Ileoanal pouch, 600
Ileostomy, 599–601, 600f
Immune cells, 436, 529
Immune contexture, 536–537
Immune escape, 536
Immune score (IM), 536–537
Immunity, of colorectal cancer, 435–436
Immunoediting process, 530–531
Immunofluorescence (IF), 211
Immunohistochemical study, 553–555
Immunohistochemistry (IHC) techniques, 151, 154, 185
Immunological test, 60
Immunoscore, 536–537
Immunosurveillance, 529–531, 534
Immunotherapeutic test
　cell therapies, 445
　monoclonal antibodies, 444
　vaccinations, 445–446
Immunotherapy, 558
　in colon cancer, 221–222
　in CRC
　　key immunotherapeutic assays, 443–444
　　molecular subtypes consensus, 443

Indigo carmine, 278
Indocyanine green (ICG), 309–310
Infection, 475–476, 482, 602
Inferior mesenteric artery (IMA), 324
Inflammation
　colorectal cancer, 435–436
　NSAIDs, 440
　risk factors, 438–439
Inflammatory bowel disease (IBD), 164, 436–437
Informed consent (IC), of biobanks, 570
Innovation and new technologies, in CRC
　definition, 307
　flexible single incision surgery, 311–313
　highly skilled laparoscopic instruments, 308
　in imaging systems, 309–310
　minimally invasive surgery (MIS), 307–308
　robotic surgery, 310–311
　"surgical value,", 307
InSiGHT. See International Society for Gastrointestinal Hereditary Tumors (InSiGHT)
Intermetastatic heterogeneity, 519
Internal hemorrhoids, 114–115
International Academy of Pathology (IAP), 567
International Agency for Research on Cancer (IARC), 15, 29–31
International Cancer Genome Consortium (ICGC), 237, 238f, 568
International Collaborative Group on Hereditary Non-Polyposis Colorectal Cancer (ICG-HNPCC), 236
International Duration Evaluation of Adjuvant Chemotherapy (IDEA), 385
International index of erectile function (IIEF), 612
International Registry of Lung Metastases (IRLM), 345
International Society for Gastrointestinal Hereditary Tumors (InSiGHT), 236, 236f
Internet, health information searching, 576
Intestinal microbiota
　bacteria targeting treatment, 503–504
　colorectal cancer, 435–436
Intestinal obstruction, 295, 485
IntOGen, 238
Intraabdominal abscess, 471
Intraabdominal hypertension (IAH), 476, 477t
Intraabdominal infection (IAI), 475–476
　antibiotic treatment duration, 476
　empirical antibiotic treatment, 474
　empirical antifungal treatment, 476
　microbiology, 474
　risk factors, 474–475
Intraabdominal pressure (IAP), 476
Intrametastatic heterogeneity, 520
Intramucosal adenocarcinoma, 149
Intranasal fentanyl, 488
Intraoperative radiotherapy (IORT), 413–414
Intraoperative ultrasound, 334–335
Intrathecal therapy, 488
Intratumoral heterogeneity, 519
Intravenous analgesia, 483
Intravenous chemotherapy, 392

Invendoscopy E200 system, 311
Ipilimumab, 210
Irinotecan, 413
Irrigation systems, 607
Irritant contact dermatitis, 604, 604f
Ischemia, 602

J

J-colon reservoir, 325
Juvenile polyposis syndrome, 65, 196t, 200

K

Kikuchi classification, 291–292
Kock pouch ileostomy, 600
KRAS, 208, 215, 361, 524–525
Kudo's classification, 161, 175, 176f

L

Laparoendoscopic fusion surgery, 311
Laparoendoscopic Single Site Surgery (LESS), 307
Laparoscopic surgery (LPS), 309, 480
Lapatinib, 209
Larotrectinib, 559
Laryngeal cancer, 533
Laterally spreading tumors (LST), 162, 174
Law 14/2007, of 3 July, on Biomedical Research (BRL), 621
Lead time bias, 17
Leeds Castle Polyposis Group (LCPG), 236
Lesions
　characterization of, 173–177
　glandular pattern of
　　Kudo's classification, 175, 176f
　　NICE classification, 175–176, 176f
　identification, 171–172
　morphological evaluation, 173–174
Lesions with granular surface (LST-G), 162
Lesions with nongranular surface (LST-NG), 162
Leucovorin (LV), 381, 383–384, 391
Liquid biopsy (LB), 182, 211, 250, 252, 559
　advantages, 221
　ctDNA, 213–215
　exosomes, 212
　messenger ribonucleic acid (mRNA), 213
　proteins, 213
　small noncoding ribonucleic acids, 213
Liquid nanochromatography coupled to tandem mass spectrometry (nLC-MS/MS), 257
Liver metastases, CRC, 331, 485
　chemotherapy after resection, 341
　follow-up after resection, 341
　local treatment, for metastatic disease, 337
　recurrence, treatment of, 340
　regional chemotherapy via hepatic artery, 337
　surgical technique, 334–340
　treatment, of metastatic disease
　　abdominal computed tomography (CT), 333
　　anatomical factors, 333
　　clinical guidelines, 331–332
　　colonoscopy, 334

magnetic resonance imaging (MRI), 333
patient-related factors, 332
positron emission tomography, 333
three-dimensional (3D) planning, 334
tumor-related factors, 332–333
tumor ablation
liver transplantation, 339–340
two-stage hepatectomy, 338–339
Local analgesia (LA), 484, 484t
Locally advanced rectal cancer, 406, 409
Locoregional recurrence, of rectal cancer, 145–146
Long-course neoadjuvant radiochemotherapy vs. short-course neoadjuvant radiotherapy, 407–409
Loop colostomy, 601
Loop ileostomy, 600, 600f
Loperamide, 596
Loss of heterozygosity (LOH), 533
Low anterior resection, 593
Low anterior resection syndrome (LARS), 593
Low microsatellite instability (MSI-L), 184
Low-penetrance variants and association studies, 515–517
Lumbar sympathectomy, 490
Lumbosacral plexopathy, 485
Lymphatic involvement, 292
Lymph node malignancy, 91, 91t
Lymphocytes, 435
Lymphomas, 114, 114f, 153
Lymphovascular infiltration, 292
Lynch syndrome, 42, 66, 191, 513, 543, 553–558. *See also* Hereditary nonpolyposis colorectal cancer (HNPCC)
characterization, 555
diagnosis, 558
differential diagnosis, 556–558
without germline mutations, 556

M

Macrophages, 435
Magnesium citrate, 158
Magnesium supplementation, 33
Magnetic nuclear resonance, 594
Magnetic resonance imaging (MRI), in rectal cancer (RC), 131
locoregional recurrence, 145–146
N-staging, 137–138
objective of, 131
regression, 145, 145t
restaging after neoadjuvant treatment
N-restaging, 142
T-restaging, 139–142
role, 131
structured radiology report, 143–145
adenopathies, 144
extramesorectal adenopathies, 144
extramural extension, 144
location, 143
mesorectal fascia, distance to, 144
mesorectal tumor deposits, 144
morphology, 143
restaging after neoadjuvant treatment, 144–145
T-staging, 144
T1 and T2 tumors, 135
T-staging, 134–135
T3 tumors, 136–137
T4 tumors, 137
Malignant neoplasms, CT rectal pathology, 112, 113f
Malignant polyps, 165, 291
Malnutrition, 585
causes, 586
complications, 585–586
dietary recommendations
food intake hindering symptoms, 590
nutritional support, 589–590
nutritional status, assessment of, 586–589
weight loss vs. time, 589, 589t
Malnutrition Screening Tool (MST), 587
Malnutrition Universal Screening Tool (MUST), 587
Mass sequencing, 220
Matrix-assisted laser desorption/ionization mass spectrometry imaging (MALDI-MSI), 258–259
Medicines Research Ethics Committees (MREC), 620
Medullary carcinoma, 153
Melanomas, 154, 533
Mendelian randomization (MR), 516
Mendelian syndromes, 513
Mesenchymal cells, 544
Mesenchymal-epithelial (MET) process, 544–545
Mesenchymal tumors, 153–154
Mesorectal fascia (MRF), 131–134, 144
Mesorectal tumor deposits, 144
Mesorectum, 131
Messenger ribonucleic acid (mRNA), 213
Metabolic syndrome, 438
Metabolite profile (metabolomics), 258
Metaflammation, 438
Metagenomics, 219
Metastases, SBRT
liver, 423–425
lung, 422
lymph node, 422–423
Metastasis, 544–545, 547–549
Metastasis-directed therapy (MDT), 181
Metastatic colorectal cancer (mCRC), 357
Metastatic CRC, 365
bevacizumab/anti-EGFR, chemotherapy, 368–369
BRAF V600E -variant metastatic CRC, 372
chemotherapy and metastases surgery, 374
duration and intensity, of first-line treatment, 370–371
first-line chemotherapy, 366–368
5-fluorouracilo, oxaliplatin and irinotecan, triplet combinations, 370
immunotherapy, 373–374
management, colorectal cancer and peritoneal metastases, 374–375
second-line treatment strategies, 371
treatment, 365–366
treatment-refractory, 371–372
Metastatic tumors, 154
Metformin, 35
Methylation
biomarkers, 255
DNA biomarkers, 255–256
Methylation-specific microarrays, 255
Methylation-specific quantitative PCR (MS-qPCR), 255
Methylator phenotype, 42
Methylcellulose, 596
Microbial dermatitis, 604
Microbial markers, 219–220
Microbiota, 219
alteration, 42
Micropapillary adenocarcinoma, 153
MicroRNAs, 27, 218, 253–254, 547
Microsatellite instability (MSI), 42, 184–185, 217–218, 383, 443, 532–533, 552–553, 556, 560
colorectal carcinomas with, 533–536, 534f
determination methods, 553–555
identification, 558
immunohistochemical study, 553–555
Lynch syndrome, 184–185
molecular testing, 553, 555
Microsatellite instability-high (MSI-H), 357
Microsatellites, 217
Microsatellite stability (MSS), 184
Migration, 298
Minimally invasive surgery (MIS), 307–308, 457
Mini Nutritional Assessment (MNA), 587
Mismatch repair (MMR) proteins, 183, 217, 533
Mitochondrial antiviral signaling protein (MAVS), 436
Mitogen-activated protein kinase (MAPK) pathway, 357, 524–525
MLH1 methylation analysis, 556–557
MMRPredict, 189
MMRPro, 189
Mmyeloid-derived suppressor cells (MDSC), 442
Molecularly targeted therapy, in metastatic CRC
anti- EGFR antibodies, 357–358
BRAF tyrosine kinase inhibitors, 359
HER -2 blockade, 359–360
inmmune checkpoint inhibitors, 358
KRAS inhibitors, 361
tyrosine kinase inhibitors, 360–361
Molecular staging, 220
Monoclonal antibodies, 391–392
mRNAs, 253
MSI-H. *See* High microsatellite instability (MSI-H)
MSI-L. *See* Low microsatellite instability (MSI-L)
Mucinous adenocarcinoma, 153
Mucocutaneous dehiscence, 602
Muir-Torre syndrome (MTS), 187, 555
Multidisciplinary committee
clinical oncology committees, 269–271, 270t
effect of, 272

Multidisciplinary committee (Continued)
  factors influencing the performance, 271–272
Multidrug resistance protein 1, 394
Multimodal rehabilitation technique
  advanced monitoring
    dynamic fluid response variables, 467–468
    static variables, 467
  baseline fluid and electrolyte needs, 462
  changes in surgical patient, 463
  drugs used for recovery, 461–462
  emergency surgery, 468
  fluid management schemes, 465–466
  fluids types, 464–465
  hemodynamic monitoring, 466–467
  intraoperative, 460
  perioperative fluid loss and replenishment calculation, 463–464
  perioperative fluid therapy, 462
  perioperative management of nausea and vomiting, 460–461
  preoperative, 457–458
    anemia control, 458
    antibiotic prophylaxis, 459
    fasting, 458
    mechanical colon preparation, 459
    nutritional status, 459
    obesity, 459
    thromboembolism prophylaxis, 459
    tobacco and alcohol, 460
Multiomic integration, 517
Muscle relaxants, 461
Mutator pathway, 42. See also Microsatellite instability (MSI)
Mutual adaptation, 270–271, 270t
MUTYH-associated polyposis (MAP), 196t
  follow-up of healthy individuals, 198
  genetics, 195
My CancerIQ, 241
Mycobiota dysbiosis, 505–506
MyPathway trial, 359–360

N

Nanocarriers, 397–398
Nanocrystals, 397
Nanomedicines, 397–398
Nanotransporters, 397–398
Narrow-band imaging (NBI), 160
National Biobanks Network, 566, 569–570
National Center for Biomedical Ontology (NCBO), 234
National Comprehensive Cancer Network (NCCN), 305
National Register of Biobanks for biomedical research, 565
Natural killer (NK) cells, 441
Natural killer T (NKT) cells, 442
Natural Orifice Surgery Consortium for Assessment and Research (NOSCAR), 307–308
Natural Orifice Transluminal Endoscopic Surgery (NOTES), 307–308
Nausea and vomiting, 460–461, 482

NCI Thesaurus (NCIt), 234
Necrosis, 602, 602f
Neoadjuvant chemotherapy, 335–336
  vs. neoadjuvant radiochemotherapy, 409–410
Neoadjuvant radiochemotherapy
  rectal cancer, 405–406
  vs. adjuvant radiochemotherapy, 406–407
  vs. neoadjuvant chemotherapy, 409–410
  vs. neoadjuvant radiotherapy, 407
Neoadjuvant radiotherapy
  rectal cancer, 404–405
  vs. adjuvant radiochemotherapy, 405
  vs. neoadjuvant radiochemotherapy, 407
NeoGuide Endoscopy System, 311
Neoplastic lesions
  high-definition colonoscopy, 171
  morphological evaluation, 173
  quality of colonoscopy, 171
Network, 239
Neuroablative techniques, 489–490
Neuroendocrine neoplasms, 153
Next-generation sequencing (NGS), 233, 251, 513–515, 553
NICE classification, 162, 163f, 175–176, 176f
Nivolumab, 210
Nociceptive pain, 485
Nodal staging system, 101
NOD-like receptors (NLR), 436
Nonanatomical third space concept, 464
Nonclassical biomarkers, 210
Noncoded ribonucleic acid (RNA), 27
Nonenzymatic Antioxidant Activity Capacity (NEAC), 35–36
Nonhemodynamic monitoring, 468
Nonhereditary attenuated polyposis
  definition, 288
  family screening, 289
  management, 288
  surveillance
    after colorectal lesions resection, 288
    for extracolonic lesions, 288
Nonhereditary polyposis, 288
  nonhereditary attenuated polyposis, 288–289
  serrated polyposis syndrome (SPS), 289–291
Noninvasive screening test, 46
Noninvasive screening tests, 49
Nonneoplastic lesions, 162
Nonpolypoid lesions, 162
Nonpolyposis syndromes. See Lynch syndrome
Nonsteroidal antiinflammatory drugs (NSAIDs), 33, 398, 440, 486
Normothermia, intraoperative, 461
Nosocomial infection, 475–476
Novel immunogenomic methods, 536
N-RAS, 215
N-restaging, rectal cancer, 142
N-staging, rectal cancer, 137–138
NTRK rearrangements, 221
Nutritional assessment methods, 587
Nutritional Control System (CONUT), 588
Nutritional deficiency. See Malnutrition
Nutritional Risk Index (IRN), 589
Nutritional Risk Screening (NRS-2002), 587

O

Obesity, 32
Objective response rate (ORR), 358
Observatory of Trends of the Roche Institute Foundation, 626
Obturators, 607
OC. See Optical colonoscopy (OC)
Oligometastatic disease (OMD), 181–182, 421, 424–425t
Omics, 515
Omics-based biomarkers, for CRC
  biological samples, 249–250
  clinical application, 249–250
  development phases, 259–261
    clinical translation, 261
    clinical utility validation, 260
    exploratory analysis, 259
    technical/analytical validation, 259
  epigenomics approaches, 254–256
    De novo identification, 255
    DNA methylation, 255–256
    methylation assessment, 255
    whole genome methylation profiling, 255
  genomics approaches, 250–252
    conventional polymerase chain reaction, 251
    De novo identification, 251
    digital droplet PCR (ddPCR), 252
    genetic biomarkers, 252
    polymerase chain reaction (PCR), 251
    real-time/quantitative PCR (qPCR), 251–252
  matrix-assisted laser desorption/ionization mass spectrometry imaging (MALDI-MSI), 258–259
  metabolite profile, 258
  proteomics approaches, 256–258
    differential in gel electrophoresis (DIGE), 257
    nLC-MS/MS, 257
    protein biomarkers, 257–258
    protein microarrays, 257
    SELDI-TOF-MS, 257
    two-dimensional electrophoresis (2-DE), 256–257
  transcriptomics approaches, 252–254
    identification, 252–253
    miRNAs, 253–254
    mRNAs, 253
Omic sciences, 626
Oncogenesis and biobanks, 568
Oncogenes, mutations in, 209
Oncogenomic research databases, 234–239
Oncology committees, 269–271, 270t
Onodera Prognostic Nutritional Index (PNI-O), 588
Ontological databases, 233
Ontology, 233
Open-access surgical risk calculator, 458
Opioid free analgesia (OFA), 462
Optical colonoscopy (OC), 95, 100
  contraindications, 109, 110t
Oral administration, CRC
  colonic drug delivery

enzyme-activated binding, 399–400
pH-dependent release, 399
time-dependent binding, 399
systemic treatment, 393–398
absorption, 394–395
antitumor drugs, 393–394
distribution and elimination, 396
nanostructures, 397–398
treatment of, 391–392
Oral fluoropyrimidine (FPO), 381
Oral microbiota, dysbiosis, 437–438
Oral transmucosal fentanyl, 488
Organic Law 3/2018, of 5 December, on the Protection of Personal Data and Guarantee of Digital Rights (OLPPGDR), 623–625
OSNA technology, 310
Ostomy, 578. *See also* Gastrointestinal ostomies
body image, 579
classification, 609, 610*t*
definition, 609
emotional impact, 579
health education, 613–614
hygiene recommendations, 581, 581*t*
quality of life, 578
sexual dysfunction, assessment of, 610–613
sexual impairment, 609–610
sexuality recommendations, 581, 581*t*
Oxaliplatin, 381, 412–413, 486

P

p53, 216
mutations, 216
Pain, 482
Palliative radiotherapy
prospective studies, 430–431
quality of life, 430–431
retrospective studies, 429–430
systematic review, 431–432
Palliative treatment, 295
Pan-cancer biomarkers, 209
Panchromoendoscopy, 172
Pangenomic association studies, 515–516
Panitumumab, 215
Paracetamol, 486
Parastomal hernia, 603, 603*f*
Parastomal varicose veins, 605
Paris classification, 161–162, 161*f*, 173–174, 174*f*
Pathogen-Associated Molecular Patterns (PAMPs), 436
Pathological complete response (pCR), 328
Patient interval (PI), 20
Pattern recognition receptors (PRRs), 436
PEGylation, 396
Pelvic floor rehabilitation, 596
Pelvic pain, 485
Pembrolizumab, 210
Perforation, 296, 298
Perineal pain, 485
Periodic colonoscopy, 594
Perioperative fluid loss and replenishment calculation, 463–464
Perioperative fluid therapy, 462
Peristomal dermatitis, 604, 604*f*
Peritoneal carcinomatosis, 485
Permanent colostomies, 597, 601
Permanent ileostomies, 600–601
Permission, Limited Information, Specific Suggestions and Intensive Therapy (PLISSIT) model, 614
Personal data protection, new regulatory framework for, 624–625
Personalized medicine and biobanks, 568
PET/CT colonography, 120–121
for detection of liver metastases, 123
treatment response, assessment of, 124
Peutz–Jeghers syndrome (PJS), 65, 196*t*, 200
PGY1, 394
Phenylephrine, 596
Phosphatase with tensin homology (PTEN), 525
Phosphatidyl-inositol-3 kinase (PI3K), 357
Physical exercise, 32
PICCO system, 468
PI3K-AKT-mTOR pathway, 525–526
PIK3CA gene, 559
Pillcam COLON 2, 83, 84*f*
Placental growth factor (PlGF) 1 and 2, 357
Planning target volume (PTV), 429–430
Plasmids, 508
Point mutations, 209
POLE, 222
Poly(ethylene oxide) (PEO), 395
Polyethylene glycol (PEG), 395
Polygenic risk scores, 517
Polymerase chain reaction (PCR), 211, 251
Polypectomy, 111, 158, 165
Polypoid adenoma, 162–163, 165*f*
Polypoid carcinoma, 167*f*
Polypoid lesions, 100, 101*f*, 161, 174
Polyposis syndromes, 288, 551
adenomatous, 195
hamartomatous, 200
juvenile, 200
serrated, 201–202
Polyps, 98–100, 283–286
CT colonography (CTC), 98–100, 99*f*
CT rectal pathology, 111, 111*t*, 112*f*
flat, 98, 99*f*
pediculate, 98, 99*f*
sessile, 98, 99*f*
size, 100
Positive predictive value (PPV), 51
Positron emission tomography (PET-CT), 347–349
Positron emission tomography–magnetic resonance (PET/MR), 120, 127
POSSUM, 458
Posterior tibial nerve stimulation, 596
Postmenopausal hormone therapy, 34
Postoperative colorectal cancer
abdominal compartment syndrome, 476–478
enterocutaneous fistula, 472
generalized peritonitis, 472
intraabdominal abscess, 471
intraabdominal infection (IAI)
empirical antibiotic treatment (*see* Intraabdominal infection (IAI))
sepsis and septic shock (Sepsis-3), 473–474
suture dehiscence, 471
Postradiation, 593
Precachexia, 585
Precision medicine, 207
Preneoplastic colonic lesions surveillance, 283–288
basal polyps characteristics, 283–286
initial colonoscopy, 283
Preneoplastic lesions, endoscopic polypectomy of, 275–278
Preoperative portal embolization, 336–337
Presarcopenia, 586
Primary care (PC), 70
Primary health care (PHC), CRC, 57
early diagnosis, 58–62
clinical practice guidelines, 59–60
colonoscopy, 60–61
evaluation, of results, 61–62
indications, 60
preparation, for tests, 61
sigmoidoscopy, 60
stool occult blood test, 60
symptomatology in, 59
tests for, 59–60
family/hereditary characteristics, 65–67
health promotion, measures of, 58–59
patients referral criteria, specialized care, 65
precursor injuries (polyps), 63–65
prevention, 58–59
surveillance recommendations, 63–64
Primary prevention of colorectal cancer (CRC), 27
coffee, 31
dairy products, 30
diet, 29
dietary fiber, 30
dietary index and, 35–36
dietary patterns, 35
dietary recommendations, 36, 36*t*
environmental factors, 28, 29*f*
epidemiology, 27–28, 28*t*
fish, 30
garlic, 31
healthy practices, 32
physical exercise, 32
lifestyle behaviors, 31–32
processed and red meat, 29–30
risk factors, 27
therapeutic strategies, 32–35
angiotensin II inhibitors, 35
antioxidants, 34
B-group vitamins, 34
bisphosphonates, 35
calcium and magnesium supplementation, 33
chondroitin sulfate, 35
folate, 34
folic acid, 34
glucosamine, 35
hormone replacement therapy, 34
metformin, 35

Primary prevention of colorectal cancer (CRC) (Continued)
  nonsteroidal antiinflammatory drugs (NSAIDs), 33
  salicylic acetyl acid (SAA), 32–33
  statins, 34
  vitamin D supplementation, 33–34
  unhealthy practices, 31–32
    excessive alcohol consumption, 31
    obesity, 32
    smoking, 31–32
    white and lean meat, 31
    whole grain cereals, 30
Primary tumor, 152
PRIME trial, 358
Proactive responsibility, 624
Processed meat, 29–30
Process mining, 75
PRODIGE 7 trial, 375
Programmed death ligand-1 (PD-L1), 392
Progression-free survival (PFS), 383, 421
Prolapse, 603, 604f
Prostaglandin E2 (PGE2), 440
Prostate cancer, 533
Prostate Cancer Ontology, 234
Prosthesis insertion technique, 93
Protein biomarkers, for colorectal cancer, 258
Protein markers, 49
Protein microarrays, 257
Proteins and liquid biopsy, 213
Protein truncation test (PTT), 185
Proteomics, 256–258
Protruded lesions, 161
Pruritus, 482
Pseudolesions, 116–117
Pseudonymization, 624–625
Psychogenic pain, 485
Psychological aspects, in CRC
  diagnostic phase, 577
  health care team-patient relationship, 575–576
  prediagnostic phase, 576–577
  treatment phase, 577–578
pT1 neoplastic lesions, after endoscopic treatment, 291–292
  follow-up, 292
  histology, 291
  lymphovascular infiltration, 292
  resection margin, 291
  submucosal infiltration, 291–292
  tumor budding, 292
Puborectalis muscle, 132, 132f, 134
Pulmonary metastasectomy (PM), CRC
  lung metastases, 345
  metachronous metastasectomy, 344
  metastatic patterns, 343
  pathophysiology, 346
  patient selection, 345–346
  perspectives, 351
  preoperative evaluation, 347–349
  prognostic factors, 349–351
  randomized controlled trial, 346
  results of, 343–344
  selection criteria, 344–345
  surgical indication, 345
  surgical resection, 343
  surgical technique, 351
  symptoms, 346
Pulse pressure variation (PPV), 467
Pyrophosphates (PPi), 255
Pyrosequencing, 255

Q
Quality, of biobanks, 570
QUASAR trial, 383

R
Radiation proctitis, 596
Radioinduced liver disease (RILD), 423
Radiology report, of rectal cancer, 143–145
Radiotherapy, 182, 337, 486, 578
  in pain treatment, 490
Radius Surgical System (Tuebingen Scientific), 308
Ramosetron, 596
Rapid diagnostic pathways (RDP), CRC
  clinical pathways, 69
  delayed diagnosis, 70
  effectiveness of, 74–75
  implementation, 70–71
  preventive and improved care plans, in oncology programs, 70
  published research
    cancer detection rates, 71–73
    care timing, effect on, 73–74
    diagnosed cancer cases, percentages of, 73
    meeting referral criteria, 71
    sensitivity and specificity of, 73
    tumor stage and survival, effect on, 74
  rapid referral criteria, 70, 70t
  risk factor, 70
Rare Variant-Common Disease hypothesis, 513
RAS mutations, 215, 559
  undruggable, 220
Reactive oxygen species (ROS), 435
Real-time/quantitative PCR (qPCR), 251–252
Rectal cancer
  adjuvant treatment, 403–404
  intraoperative radiotherapy, 413–414
  meta-analyses and systematic reviews, 410–411
  neoadjuvant chemotherapy vs. neoadjuvant radiochemotherapy, 409–410
  neoadjuvant radiochemotherapy, 405–406
    vs. adjuvant radiochemotherapy, 406–407
    vs. neoadjuvant radiotherapy, 407
  neoadjuvant radiotherapy, 404–405
    vs. adjuvant radiochemotherapy, 405
  nonadjuvant treatment, 404
  radiotherapy
    biological agents, 413
    capecitabine, 411–412
    irinotecan, 413
    oxaliplatin, 412–413
  short-course neoadjuvant radiotherapy vs. long-course neoadjuvant radiochemotherapy, 407–409
  standard treatment, 409
  surgery efficacy, 403
  total neoadjuvant treatment, 409
Rectal cancer surgery
  oncological and technical principles, 324–325
  preoperative assessment and staging, 323–324
  surgical techniques
    abdomino-perineal amputation, 327
    anterior resection, 325–326
    dysfunctionalizing stoma, 326–327
    Hartmann-type resection, 327
    intersphincteric resection, 326
    laparoscopic surgery for, 328
    local resection, 327–328
    nonsurgical treatment, 328
    robotic surgery, 328
    transanal mesorectal resection, 326
  surgical treatment, 323
Rectal catheter, 116, 117f
Rectal pathology, 109
  artifacts, 116–117
  condyloma acuminatum, 117
  diverticula, 117
  hypertrophied anal papilla, 115, 116f
  inflammatory conditions, 115
  malignant neoplasms, 112, 113f
  polyps, 111, 111t, 112f
  postoperative changes, 115, 117f
  pseudolesions, 116–117
  submucosal lesions, 112–114, 113t, 114f
  vascular lesions, 114–115
  villous tumors, 111, 112f
Rectal polyps, 111, 112f
Rectal varices, 115, 115f
Recto-colonic anastomosis, 115
Rectorrhagia, 43
Rectosigmoid carcinoma, 168f
Rectum, 132–134
  anorectal angulation, 132, 132f
  anterior peritoneal reflection, 133–134
  lower rectum, 132f, 134
  lymph nodes, 134
  mesorectal fascia, 133–134
  middle rectum, 132f, 134
  puborectalis muscle, 132, 132f
  upper rectum, 132f, 134
Red meat, 29–30
Reduced representation bisulfite sequencing (RRBS), 255
Refractory cachexia, 585
Regional lymph nodes, 153
Regulations governing biomedical research, 620–624
Reobstruction, 298
Research Ethics Committees (RECs), 570, 619–620
Respiratory depression, 482
Restrictive vs. liberal therapy, 465–466
Revo-i robotic surgical system, 311
RNA-binding proteins (RBPs), 547
RNA-sequencing, 252–253
Robotic surgery, 597
ROLARR study, 310
Royal Decree 957/2020, 623
Royal Decree 1090/2015, 623
Royal Decree 1716/2011, 565, 567, 621

S

Sacral nerve stimulation, 596
Salicylic acetyl acid (SAA), 32–33
Saliva, 250
Sanger sequencing, 233
Sarcopenia, 586
SARS-CoV-2 pandemic, psychological aspects of, 582
Scale G8, 588
"Scarless surgery,", 307
SCOT study, 385
Screening
  problems
    lack of awareness, 51
    participation, 51
  screening program implementation
    evaluation of, 50–51
    requirements for, 49–51
  secondary prevention and screening concept
    colorectal cancer screening, 45–47
    population screening conditions, WHO general principles, 45
  in Spain, CRC
    current situation, 52–53
    Galicia, CRC implementation in, 53
    implementation, 52
    improvement, in Spanish screening programs, 53
  tests, used in colorectal cancer screening
    colonoscopy, 47
    computed tomographic colonography (CTC), 48
    fecal occult blood test (FOBT), 48–49
    flexible sigmoidoscopy, 47–48
    noninvasive screening tests, 49
"Seed and soil theory,", 346
Self-emulsifying solid systems (SMEDDSs), 395
Senhance Surgical System, 311
Sepsis, 602
Sepsis and septic shock (Sepsis-3), 473–474
Septin-9 (SEPT9), 49
Sequential hepatectomy, 336–337
Serrated adenocarcinoma, 153
Serrated adenoma, 166f
Serrated pathway. See CpG island methylator phenotype (CIMP)
Serrated polyposis syndrome (SPS), 196t, 284–286, 286f
  characteristics, 289–290
  clinical features, 201–202
  definition, 289
  diagnostic criteria, 201
  endoscopic monitoring, 290–291
  epidemiology, 290
  family screenings, 291
  risk factors, 290
  types, 201, 289–290
Serrated polyps, 61
Serum markers, 49
Sessile serrated adenomas (SSA), 62, 289
Severe sarcopenia, 586
Sexual desire inventory (SDI), 612
Sexual dysfunction
  assessment of, 610–613
    questionnaires, 610–613, 611t
    physical factors, 609
    psychological factors, 610
Sexual intimacy, 613
Sexuality, 580
SHARE-FI, 587
Short-chain fatty acids (SCFAs), 498
Short-course neoadjuvant radiotherapy vs. long-course neoadjuvant radiochemotherapy, 407–409
Short Physical Performance Battery (SPPB), 587
Sigmoid colon, 158–160
Sigmoid colostomy, 601
Sigmoid neoplasm, 177, 178f
Signet ring cell carcinoma, 153
Silver in situ hybridization (SISH), 211
Single cell genomics, 234–235
Single nucleotide polymorphisms (SNPs), 515
Single-Port Instrument Delivery Extended Research (SPIDER) system, 311
Single strand conformation polymorphism (SSCP), 185
Skin protectors, 606, 606f
Smad4 gene, 522–523
Smads, 522–523
  signaling pathway, 521, 522f
Smoking, 31–32, 498
Sodium picosulfate, 158
SoloAssist II, 311
Sotorasib, 209, 361
Spain
  incidence rates of CRC in, 3, 8t, 9f
  mortality rate of CRC in, 10, 12–13t
Spanish Anatomical Pathology Society (SEAP), 567
Spanish Association Against Cancer, 582
Spanish Association of Ostomates (ASOE), 580
Spanish Health System, 269–270
Spanish Society of Parenteral and Enteral Nutrition (SENPE), 589
Specialist care (SC) process, 70
Sphincters, corrective surgery to, 597
Spigelman scores, 197t
Spigelman stage, 197t
SPIRITT study, 371
Sporadic colorectal carcinoma, 552, 556–558
SPS. See Serrated polyposis syndrome (SPS)
16S ribosomal RNA (rRNA) gene, 498–499
S2SNet, 239
Stained feces, 116
Standard ileostomy, 600
Statins, 34
Stem tumor cells, 222
Stenosing neoplasm, 177, 178f
Stenosis, 96, 103–104, 105f, 603, 603f
Stent, 304–305
Stereotactic Ablative Radiotherapy (SABR), 182
Stereotactic body radiotherapy (SBRT)
  liver metastases, 423–425
  lung metastases, 422
  lymph node metastases, 422–423
  metastasis-directed treatments, 421
  oligometastatic disease, 421, 424–425t
Stereotactic (ablative) body radiotherapy (SABR/SBRT), 351
Stereotactic radiotherapy, 337
Stoma
  care and hygiene, 605
  definition, 599
  marking, 607
  retraction, 602
  sexuality, 580
Stool-based miRNAs, 253
Stool test, noninvasive screening tests, 49
Streptococcus agalactiae, 502–503
Streptococcus bovis/gallolyticus, 501
Stroke volume variation (SVV), 467
Subjective Patient-Generated Global Assessment (VGS-GP), 587
Sublingual fentanyl, 488
Sublingual sufentanil tablet system (SSTS), 483
Submucosal dissection (SMD), 277
Submucosal infiltration, 291–292
Submucosal lesions
  CT colonography (CTC), 102–103, 103f
  CT rectal pathology, 112–114, 113t, 114f
  extramural, 102–103, 114, 114f
  intramural, 102–103, 114
Superficial neoplasia, 173
Suppressor pathway, 42. See also Chromosomal instability (CIN)
Surface-enhanced laser desorption/ionization time-of-flight mass spectrometry (SELDI-TOF-MS), 257
Surgery, 17
  for classic FAP, 198
Surgical planning in colorectal surgery, 3D reconstruction for, 310
Suture dehiscence, 471
Symptom Diagnosis Interval (SDI), 20
Symptom Treatment Interval (STI), 20

T

Taxonomy, of biobanks, 568
Temporary colostomies, 601
Temporary ileostomies, 599–600
Tenesmus, 298
2,3,5,4′-Tetrahydroxystilbene-2-O-b-glucoside (THSG), 440
TGF-β signaling pathways, 521–522, 522f
Th17 cells, 441
The Cancer Genome Atlas (TCGA), 236
T helper 1 (Th1), 441
T helper 17, 437
Thoracic epidural analgesia, 481–482
3D-HD display technology (4th generation), 309
Three-dimensional (3D) printing, 310
3D manometry, 594
Thymidine phosphorylase, 393
Tissue microarray (TMA) technique, 569
Toll-like receptors (TLRs), 436
Topical capsaicin, 487
Topical lidocaine, 487
Total mesorectal excision (TME), 131–132, 310, 324–326
Total mutational burden (TMB), 210, 222

Traditional/Sanger sequencing, 233
Traditional serrated adenomas (TSA), 62, 289
Transanal irrigation, 596
Transanal Minimally Invasive Surgery (TAMIS), 308, 312
Transanal Total Mesorectal Excision (TaTME), 308, 312
Transcriptome-Wide Association Studies (TWAS), 517
Transcriptomics, 252–254
Transdermal fentanyl, 483
Transverse colon, 160
Transverse colostomy, 601
Trastuzumab, 207
Treatment interval (TI), 20
T-restaging, of rectal cancer, 139–142
Trigger point deactivation, 490
Tripartite motif (TRIM), 221
Tropomyosin receptor kinase (TRK) inhibitor, 360
T-staging system, 100–101, 134–135, 144
Tubular adenoma polyp, 284, 285f
Tubulovillous polyp, 284, 285f
Tumor-associated antigens (TAA), 441
Tumor banking, 568–569
Tumor budding, 292
Tumor cachexia, 585
Tumor derived factors (TDFs), 531
Tumor escape strategies, 531–533, 531–532t
Tumor infiltrating lymphocytes (TIL), 441–442
Tumor lymphocyte infiltration, 536–537
Tumor markers, of biobanks, 568
Tumor microbiota, 504
Tumor microenvironment, in colon carcinoma, 222–223
Turcot syndrome (TS), 187, 555
Turnbull-Cutait technique, 326
Two-dimensional electrophoresis (2-DE), 256–257
2017 World Cancer Research Fund, 30
Type 2 diabetes (DM2), 438

U

Ulcerated proximal colon neoplasia, 168f
Ulcerations, 605
Ulcerative carcinoma, 167f
Ultra high-definition (4K) imaging systems, 309
Ultra-sequencing, 233
Undruggable RAS, 220
Unexpected losses, in surgical patient, 464
Unhealthy practices, 31–32
Unified Medical Language System (UMLS), 234
UNI-VEC device, 311–313, 312–313f
Unstained feces, 117
Upper hypogastric plexus block, 490
Urinary retention, 482
US Food and Drug Administration (FDA), 596
U.S. Preventive Services Task Force (USPSTF), 33, 440

V

Vascular endothelial growth factor (VEGF), 391
Vascular involvement, 292
Vascular lesions, CT rectal pathology
  internal hemorrhoids, 114–115
  rectal varices, 115, 115f
  venous malformations, 115
Vemurafenib, 207–208
Venous malformations, 115
Versius Surgical Robotic System, 311
VES Scale 13, 588
Video-assisted thoracoscopic surgery (VATS), 344, 351
Villous tumors, CT rectal pathology, 111, 112f
Virtual biopsy, 97, 97f
Virtual chromoendoscopy, 172
Virtual colonoscopy, 62. See also CT-colonography (CTC)
Virtual dissection, 97, 97f
Virus, 506–507
Vitamin D receptors, 33–34
Vitamin D supplementation, 33–34

W

Weight loss vs. time, 589, 589t
Whole exome sequencing (WES), 251
Whole genome bisulfite sequencing (WGBS), 255
Whole genome methylation profiling, 255
Whole genome sequencing (WGS), 251
Whole grain cereals, 30
Wnt signaling pathway, 526
Workgroup Serrated Polyps and Polyposis (WASP) classification, 177
World Health Organization (WHO), 14–15, 29–30, 33–34, 45, 46t

Printed in the United States
by Baker & Taylor Publisher Services